These safety symbols are used in laboratory and field investigations in this book to indicate possible hazards. Learn the meaning of each symbol and refer to this page often. *Remember to wash your hands thoroughly after completing lab procedures.*

PROTECTIVE EQUIPMENT Do not begin any lab without the proper

GOGGLES 	Proper eye protection must be worn when performing or observing science activities which involve items or conditions as listed below.	**APRON** 	Wear an approved apron when using substances that could stain, wet, or destroy cloth.	**SOAP** 			...oves when ... with biological ..., chemicals, ..., or materials that can stain or irritate hands.

LABORATORY HAZARDS

Symbols	Potential Hazards	Precaution	Response
DISPOSAL	contamination of classroom or environment due to improper disposal of materials such as chemicals and live specimens	• DO NOT dispose of hazardous materials in the sink or trash can. • Dispose of wastes as directed by your teacher.	• If hazardous materials are disposed of improperly, notify your teacher immediately.
EXTREME TEMPERATURE	skin burns due to extremely hot or cold materials such as hot glass, liquids, or metals; liquid nitrogen; dry ice	• Use proper protective equipment, such as hot mitts and/or tongs, when handling objects with extreme temperatures.	• If injury occurs, notify your teacher immediately.
SHARP OBJECTS	punctures or cuts from sharp objects such as razor blades, pins, scalpels, and broken glass	• Handle glassware carefully to avoid breakage. • Walk with sharp objects pointed downward, away from you and others.	• If broken glass or injury occurs, notify your teacher immediately.
ELECTRICAL	electric shock or skin burn due to improper grounding, short circuits, liquid spills, or exposed wires	• Check condition of wires and apparatus for fraying or uninsulated wires, and broken or cracked equipment. • Use only GFCI-protected outlets	• DO NOT attempt to fix electrical problems. Notify your teacher immediately.
CHEMICAL	skin irritation or burns, breathing difficulty, and/or poisoning due to touching, swallowing, or inhalation of chemicals such as acids, bases, bleach, metal compounds, iodine, poinsettias, pollen, ammonia, acetone, nail polish remover, heated chemicals, mothballs, and any other chemicals labeled or known to be dangerous	• Wear proper protective equipment such as goggles, apron, and gloves when using chemicals. • Ensure proper room ventilation or use a fume hood when using materials that produce fumes. • NEVER smell fumes directly. • NEVER taste or eat any material in the laboratory.	• If contact occurs, immediately flush affected area with water and notify your teacher. • If a spill occurs, leave the area immediately and notify your teacher.
FLAMMABLE	unexpected fire due to liquids or gases that ignite easily such as rubbing alcohol	• Avoid open flames, sparks, or heat when flammable liquids are present.	• If a fire occurs, leave the area immediately and notify your teacher.
OPEN FLAME	burns or fire due to open flame from matches, Bunsen burners, or burning materials	• Tie back loose hair and clothing. • Keep flame away from all materials. • Follow teacher instructions when lighting and extinguishing flames. • Use proper protection, such as hot mitts or tongs, when handling hot objects.	• If a fire occurs, leave the area immediately and notify your teacher.
ANIMAL SAFETY	injury to or from laboratory animals	• Wear proper protective equipment such as gloves, apron, and goggles when working with animals. • Wash hands after handling animals.	• If injury occurs, notify your teacher immediately.
BIOLOGICAL	infection or adverse reaction due to contact with organisms such as bacteria, fungi, and biological materials such as blood, animal or plant materials	• Wear proper protective equipment such as gloves, goggles, and apron when working with biological materials. • Avoid skin contact with an organism or any part of the organism. • Wash hands after handling organisms.	• If contact occurs, wash the affected area and notify your teacher immediately.
FUME	breathing difficulties from inhalation of fumes from substances such as ammonia, acetone, nail polish remover, heated chemicals, and mothballs	• Wear goggles, apron, and gloves. • Ensure proper room ventilation or use a fume hood when using substances that produce fumes. • NEVER smell fumes directly.	• If a spill occurs, leave area and notify your teacher immediately.
IRRITANT	irritation of skin, mucous membranes, or respiratory tract due to materials such as acids, bases, bleach, pollen, mothballs, steel wool, and potassium permanganate	• Wear goggles, apron, and gloves. • Wear a dust mask to protect against fine particles.	• If skin contact occurs, immediately flush the affected area with water and notify your teacher.
RADIOACTIVE	excessive exposure from alpha, beta, and gamma particles	• Remove gloves and wash hands with soap and water before removing remainder of protective equipment.	• If cracks or holes are found in the container, notify your teacher immediately.

Get Connected to ConnectED

connectED.mcgraw-hill.com

Your online portal to everything you need!

- **One-Stop Shop, One Personalized Password**
- **Easy Intuitive Navigation**
- **Resources, Resources, Resources**

For Students

Leave your books at school. Now you can go online and interact with your StudentWorks™ Plus digital Student Edition from any place, any time!

For Teachers

ConnectED is your one-stop online center for everything you need to teach, including: digital eTeacherEdition, lesson planning and scheduling tools, pacing, and assessment.

For Parents

Get homework help, help your student prepare for testing, and review science topics.

Log on today and get ConnectED!

Your online portal to everything you need

connectED.mcgraw-hill.com

Look for these icons to access exciting digital resources

Inquiry

WebQuest

Assessment

Concepts in Motion

McGraw Hill Education

PHYSICAL

Glencoe

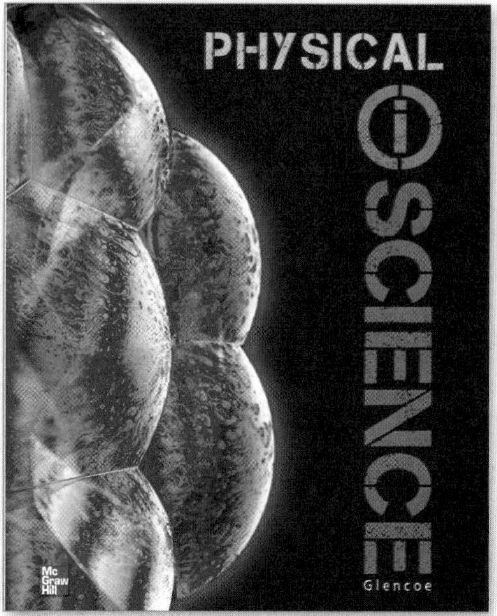

Bubbles

The iridescent colors of these soap bubbles result from a property called interference. Light waves reflect off both outside and inside surfaces of bubbles. When this happens, the waves interfere with each other and you see different colors. The thickness of the soap film that forms a bubble also affects interference.

The McGraw-Hill Companies

Education

Send all inquiries to:
McGraw-Hill Education
8787 Orion Place
Columbus, OH 43240-4027

ISBN: 978-0-07-888004-9
MHID: 0-07-888004-1

Printed in the United States of America.

1 2 3 4 5 6 7 8 9 10 DOW 15 14 13 12 11 10

Contents in Brief

Scientific Problem Solving .... NOS 2

Unit 1

Motion and Forces **2**
Chapter 1 Describing Motion 6
Chapter 2 The Laws of Motion 42
Chapter 3 Work and Simple Machines 84
Chapter 4 Forces and Fluids 120

Unit 2

Energy and Matter **154**
Chapter 5 Energy and Energy Resources 158
Chapter 6 Thermal Energy 194
Chapter 7 Foundations of Chemistry 228
Chapter 8 States of Matter 270

Unit 3

Atoms and Elements **306**
Chapter 9 Understanding the Atom 310
Chapter 10 The Periodic Table 342
Chapter 11 Elements and Chemical Bonds 378

Unit 4

Interactions of Matter **412**
Chapter 12 Chemical Reactions and Equations 416
Chapter 13 Mixtures, Solubility, and Acid/Base Solutions 450
Chapter 14 Carbon Chemistry 486

Unit 5

Waves, Electricity, and Magnetism **522**
Chapter 15 Waves 526
Chapter 16 Sound 562
Chapter 17 Electromagnetic Waves 598
Chapter 18 Light 632
Chapter 19 Electricity 676
Chapter 20 Magnetism 714

Authors and Contributors

Authors

American Museum of Natural History
New York, NY

Michelle Anderson, MS
Lecturer
The Ohio State University
Columbus, OH

Juli Berwald, PhD
Science Writer
Austin, TX

John F. Bolzan, PhD
Science Writer
Columbus, OH

Rachel Clark, MS
Science Writer
Moscow, ID

Patricia Craig, MS
Science Writer
Bozeman, MT

Randall Frost, PhD
Science Writer
Pleasanton, CA

Lisa S. Gardiner, PhD
Science Writer
Denver, CO

Jennifer Gonya, PhD
The Ohio State University
Columbus, OH

Mary Ann Grobbel, MD
Science Writer
Grand Rapids, MI

Whitney Crispen Hagins, MA, MAT
Biology Teacher
Lexington High School
Lexington, MA

Carole Holmberg, BS
Planetarium Director
Calusa Nature Center and Planetarium, Inc.
Fort Myers, FL

Tina C. Hopper
Science Writer
Rockwall, TX

Jonathan D. W. Kahl, PhD
Professor of Atmospheric Science
University of Wisconsin-Milwaukee
Milwaukee, WI

Nanette Kalis
Science Writer
Athens, OH

S. Page Keeley, MEd
Maine Mathematics and Science Alliance
Augusta, ME

Cindy Klevickis, PhD
Professor of Integrated Science and Technology
James Madison University
Harrisonburg, VA

Kimberly Fekany Lee, PhD
Science Writer
La Grange, IL

Michael Manga, PhD
Professor
University of California, Berkeley
Berkeley, CA

Devi Ried Mathieu
Science Writer
Sebastopol, CA

Elizabeth A. Nagy-Shadman, PhD
Geology Professor
Pasadena City College
Pasadena, CA

William D. Rogers, DA
Professor of Biology
Ball State University
Muncie, IN

Donna L. Ross, PhD
Associate Professor
San Diego State University
San Diego, CA

Marion B. Sewer, PhD
Assistant Professor
School of Biology
Georgia Institute of Technology
Atlanta, GA

Julia Meyer Sheets, PhD
Lecturer
School of Earth Sciences
The Ohio State University
Columbus, OH

Michael J. Singer, PhD
Professor of Soil Science
Department of Land, Air and Water Resources
University of California
Davis, CA

Karen S. Sottosanti, MA
Science Writer
Pickerington, Ohio

Paul K. Strode, PhD
I.B. Biology Teacher
Fairview High School
Boulder, CO

Jan M. Vermilye, PhD
Research Geologist
Seismo-Tectonic Reservoir Monitoring (STRM)
Boulder, CO

Judith A. Yero, MA
Director
Teacher's Mind Resources
Hamilton, MT

Dinah Zike, MEd
Author, Consultant, Inventor of Foldables
Dinah Zike Academy; Dinah-Might Adventures, LP
San Antonio, TX

Margaret Zorn, MS
Science Writer
Yorktown, VA

Consulting Authors

Alton L. Biggs
Biggs Educational Consulting
Commerce, TX

Ralph M. Feather, Jr., PhD
Assistant Professor
Department of Educational Studies and Secondary Education
Bloomsburg University
Bloomsburg, PA

Douglas Fisher, PhD
Professor of Teacher Education
San Diego State University
San Diego, CA

Edward P. Ortleb
Science/Safety Consultant
St. Louis, MO

Series Consultants

Science

Solomon Bililign, PhD
Professor
Department of Physics
North Carolina Agricultural and Technical State University
Greensboro, NC

John Choinski
Professor
Department of Biology
University of Central Arkansas
Conway, AR

Anastasia Chopelas, PhD
Research Professor
Department of Earth and Space Sciences
UCLA
Los Angeles, CA

David T. Crowther, PhD
Professor of Science Education
University of Nevada, Reno
Reno, NV

A. John Gatz
Professor of Zoology
Ohio Wesleyan University
Delaware, OH

Sarah Gille, PhD
Professor
University of California San Diego
La Jolla, CA

David G. Haase, PhD
Professor of Physics
North Carolina State University
Raleigh, NC

Janet S. Herman, PhD
Professor
Department of Environmental Sciences
University of Virginia
Charlottesville, VA

David T. Ho, PhD
Associate Professor
Department of Oceanography
University of Hawaii
Honolulu, HI

Ruth Howes, PhD
Professor of Physics
Marquette University
Milwaukee, WI

Jose Miguel Hurtado, Jr., PhD
Associate Professor
Department of Geological Sciences
University of Texas at El Paso
El Paso, TX

Monika Kress, PhD
Assistant Professor
San Jose State University
San Jose, CA

Mark E. Lee, PhD
Associate Chair & Assistant Professor
Department of Biology
Spelman College
Atlanta, GA

Linda Lundgren
Science writer
Lakewood, CO

Keith O. Mann, PhD
Ohio Wesleyan University
Delaware, OH

Charles W. McLaughlin, PhD
Adjunct Professor of Chemistry
Montana State University
Bozeman, MT

Katharina Pahnke, PhD
Research Professor
Department of Geology and Geophysics
University of Hawaii
Honolulu, HI

Jesús Pando, PhD
Associate Professor
DePaul University
Chicago, IL

Hay-Oak Park, PhD
Associate Professor
Department of Molecular Genetics
Ohio State University
Columbus, OH

David A. Rubin, PhD
Associate Professor of Physiology
School of Biological Sciences
Illinois State University
Normal, IL

Toni D. Sauncy
Assistant Professor of Physics
Department of Physics
Angelo State University
San Angelo, TX

Series Consultants, continued

Malathi Srivatsan, PhD
Associate Professor of Neurobiology
College of Sciences and Mathematics
Arkansas State University
Jonesboro, AR

Cheryl Wistrom, PhD
Associate Professor of Chemistry
Saint Joseph's College
Rensselaer, IN

Reading

ReLeah Cossett Lent
Author/Educational Consultant
Blue Ridge, GA

Math

Vik Hovsepian
Professor of Mathematics
Rio Hondo College
Whittier, CA

Series Reviewers

Thad Boggs
Mandarin High School
Jacksonville, FL

Catherine Butcher
Webster Junior High School
Minden, LA

Erin Darichuk
West Frederick Middle School
Frederick, MD

Joanne Hedrick Davis
Murphy High School
Murphy, NC

Anthony J. DiSipio, Jr.
Octorara Middle School
Atglen, PA

Adrienne Elder
Tulsa Public Schools
Tulsa, OK

Carolyn Elliott
Iredell-Statesville Schools
Statesville, NC

Christine M. Jacobs
Ranger Middle School
Murphy, NC

Jason O. L. Johnson
Thurmont Middle School
Thurmont, MD

Felecia Joiner
Stony Point Ninth Grade Center
Round Rock, TX

Joseph L. Kowalski, MS
Lamar Academy
McAllen, TX

Brian McClain
Amos P. Godby High School
Tallahassee, FL

Von W. Mosser
Thurmont Middle School
Thurmont, MD

Ashlea Peterson
Heritage Intermediate Grade Center
Coweta, OK

Nicole Lenihan Rhoades
Walkersville Middle School
Walkersvillle, MD

Maria A. Rozenberg
Indian Ridge Middle School
Davie, FL

Barb Seymour
Westridge Middle School
Overland Park, KS

Ginger Shirley
Our Lady of Providence Junior-Senior High School
Clarksville, IN

Curtis Smith
Elmwood Middle School
Rogers, AR

Sheila Smith
Jackson Public School
Jackson, MS

Sabra Soileau
Moss Bluff Middle School
Lake Charles, LA

Tony Spoores
Switzerland County Middle School
Vevay, IN

Nancy A. Stearns
Switzerland County Middle School
Vevay, IN

Kari Vogel
Princeton Middle School
Princeton, MN

Alison Welch
Wm. D. Slider Middle School
El Paso, TX

Linda Workman
Parkway Northeast Middle School
Creve Coeur, MO

Teacher Advisory Board

The Teacher Advisory Board gave the authors, editorial staff, and design team feedback on the content and design of the Student Edition. They provided valuable input in the development of *Glencoe Physical iScience.*

Frances J. Baldridge
Department Chair
Ferguson Middle School
Beavercreek, OH

Jane E. M. Buckingham
Teacher
Crispus Attucks Medical Magnet High School
Indianapolis, IN

Elizabeth Falls
Teacher
Blalack Middle School
Carrollton, TX

Nelson Farrier
Teacher
Hamlin Middle School
Springfield, OR

Michelle R. Foster
Department Chair
Wayland Union Middle School
Wayland, MI

Rebecca Goodell
Teacher
Reedy Creek Middle School
Cary, NC

Mary Gromko
Science Supervisor K–12
Colorado Springs District 11
Colorado Springs, CO

Randy Mousley
Department Chair
Dean Ray Stucky Middle School
Wichita, KS

David Rodriguez
Teacher
Swift Creek Middle School
Tallahassee, FL

Derek Shook
Teacher
Floyd Middle Magnet School
Montgomery, AL

Karen Stratton
Science Coordinator
Lexington School District One
Lexington, SC

Stephanie Wood
Science Curriculum Specialist, K–12
Granite School District
Salt Lake City, UT

Online Guide

connectED.mcgraw-hill.com

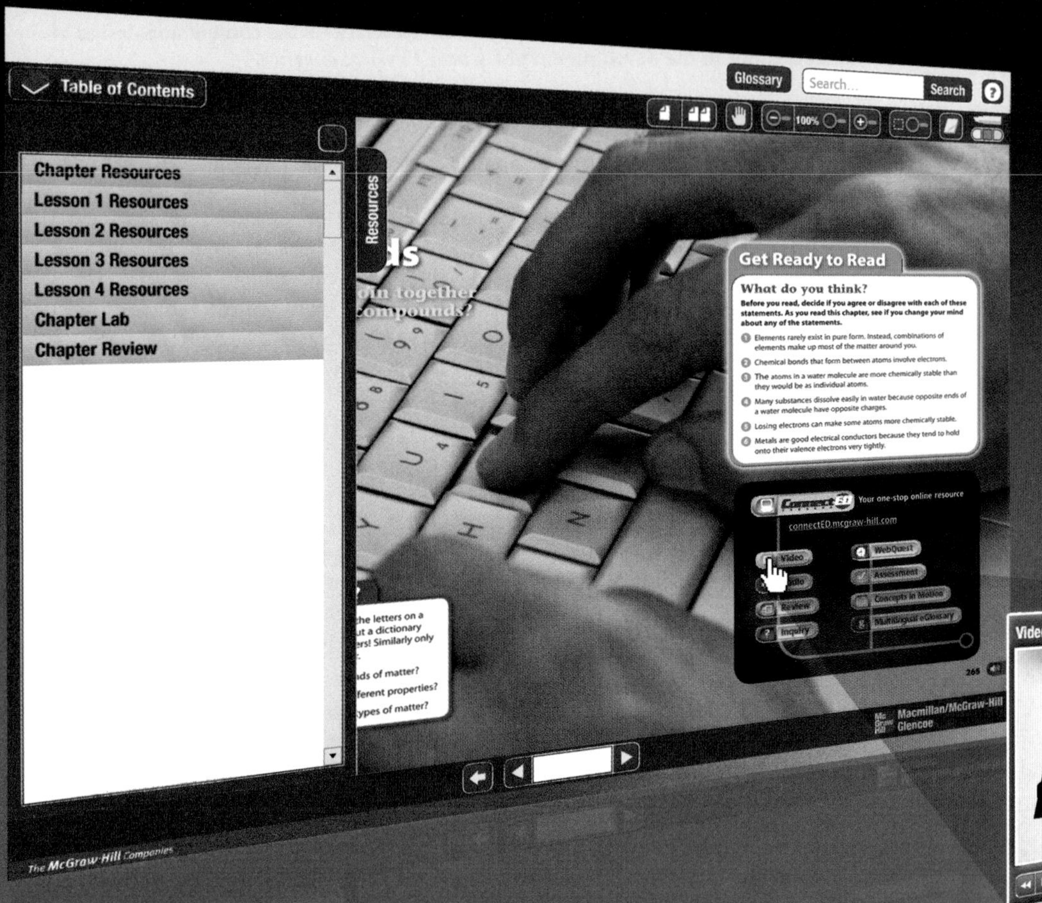

See the science in real life through these exciting videos.

Click the link and you can listen to the text while you follow along.

Try these interactive tools to help you review the lesson concepts.

Explore concepts through hands-on and virtual labs.

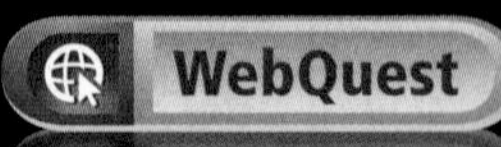

These web-based challenges relate the concepts you're learning about to the latest news and research.

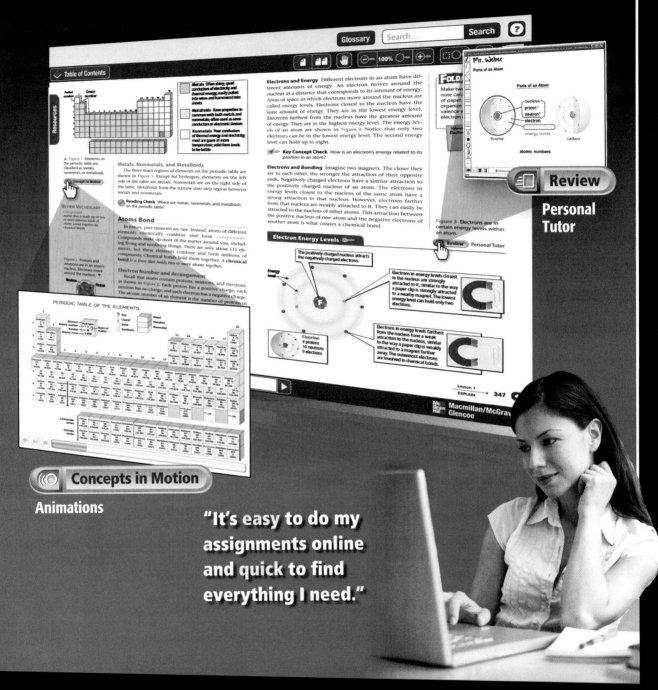

The icons in your online student edition link you to interactive learning opportunities. Browse your online student book to find more.

Review

Personal Tutor

Concepts in Motion

Animations

"It's easy to do my assignments online and quick to find everything I need."

Check how well you understand the concepts with online quizzes and practice questions.

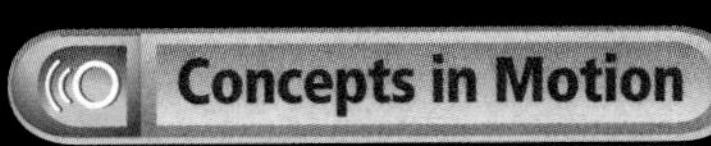

The textbook comes alive with animated explanations of important concepts.

Read key vocabulary in 13 languages.

Treasure Hunt

Your science book has many features that will aid you in your learning. Some of these features are listed below. You can use the activity at the right to help you find these and other special features in the book.

- THE BIG IDEA can be found at the start of each chapter.
- The Reading Guide at the start of each lesson lists **Key Concepts,** vocabulary terms, and online supplements to the content.
- ConnectED icons direct you to online resources such as animations, personal tutors, math practices, and quizzes.
- Inquiry Labs and Skill Practices are in each chapter.
- Your FOLDABLES help organize your notes.

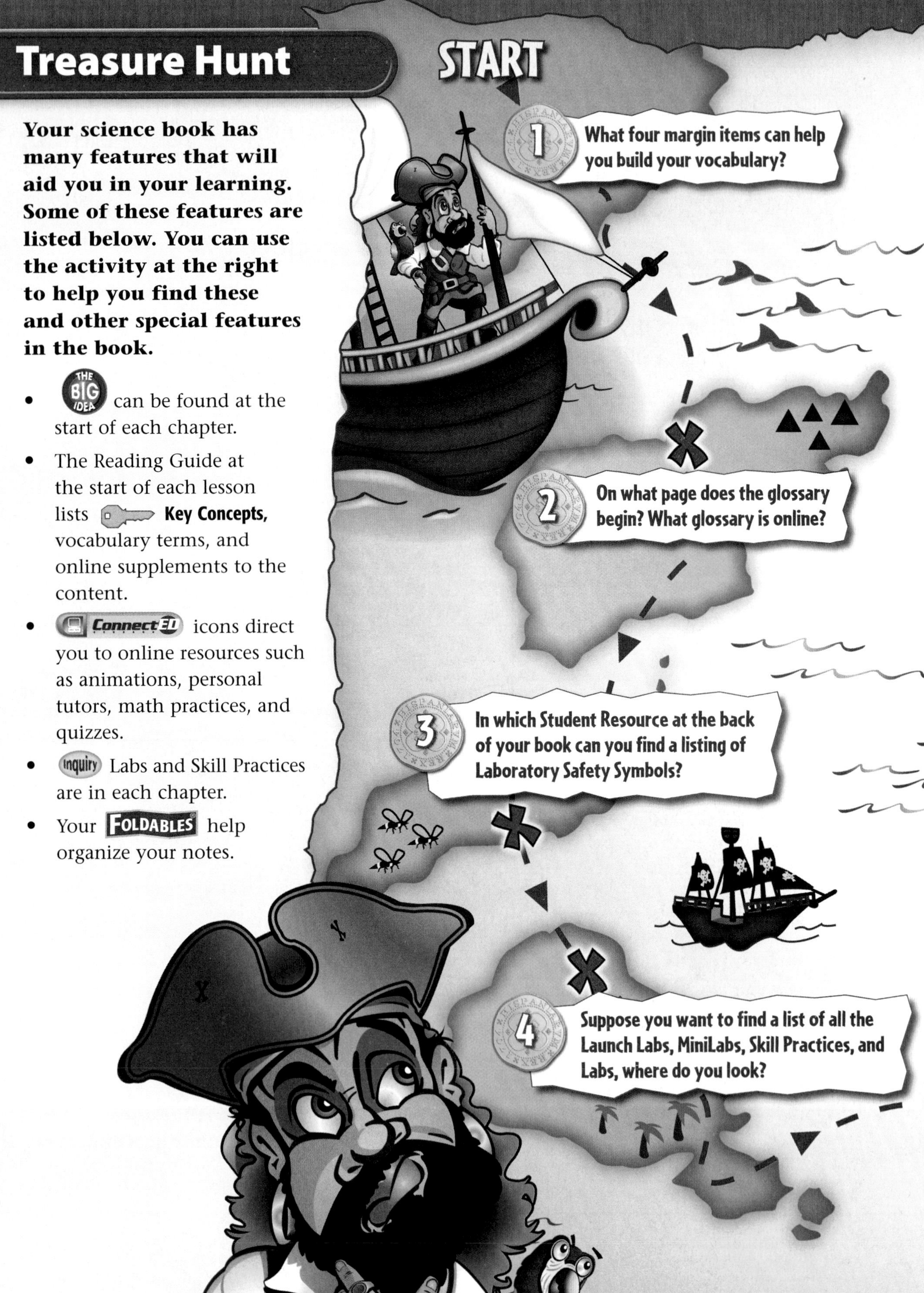

START
1 What four margin items can help you build your vocabulary?
2 On what page does the glossary begin? What glossary is online?
3 In which Student Resource at the back of your book can you find a listing of Laboratory Safety Symbols?
4 Suppose you want to find a list of all the Launch Labs, MiniLabs, Skill Practices, and Labs, where do you look?

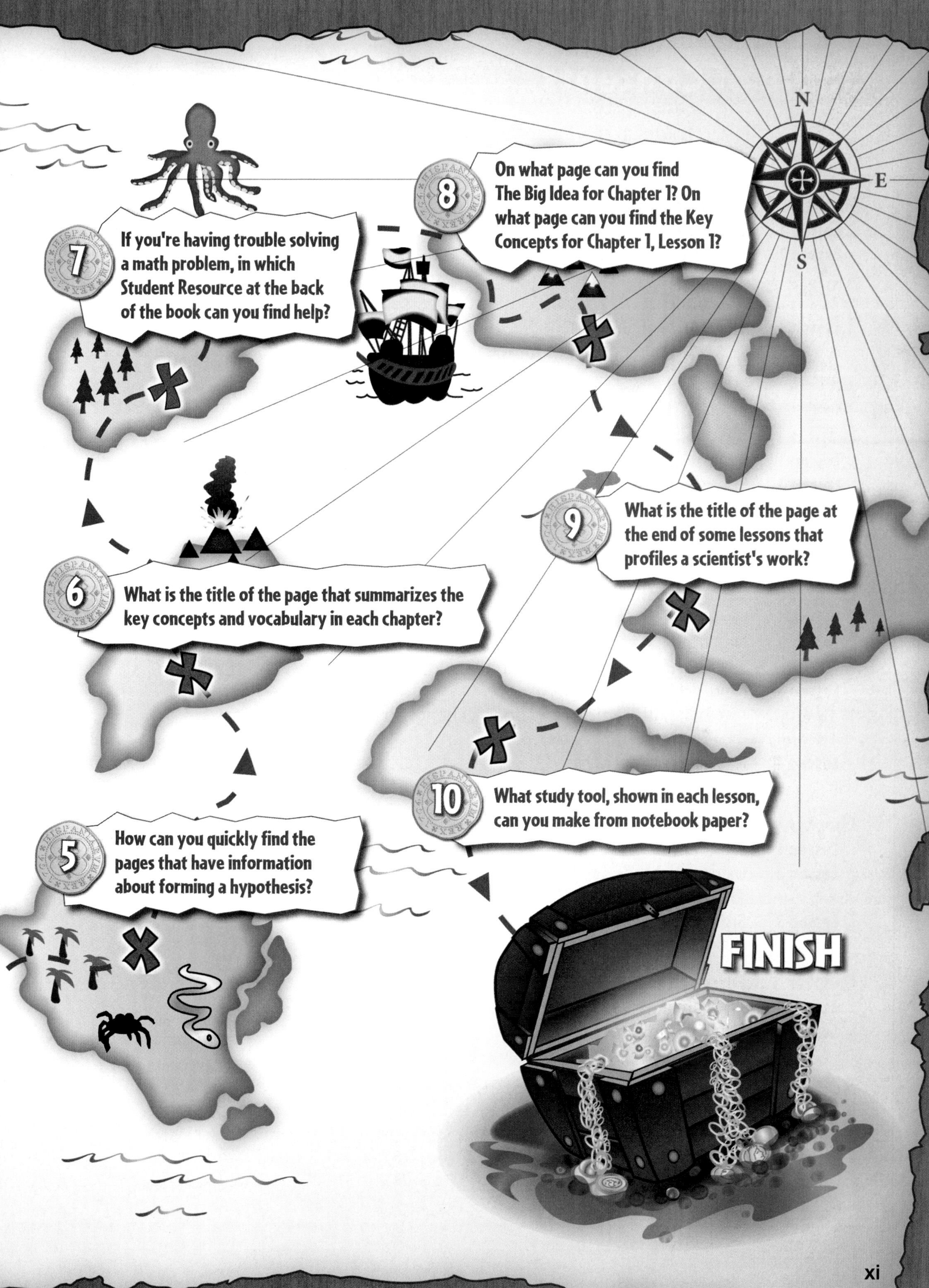
N
E
S
8
On what page can you find The Big Idea for Chapter 1? On what page can you find the Key Concepts for Chapter 1, Lesson 1?
7
If you're having trouble solving a math problem, in which Student Resource at the back of the book can you find help?
9
What is the title of the page at the end of some lessons that profiles a scientist's work?
6
What is the title of the page that summarizes the key concepts and vocabulary in each chapter?
10
What study tool, shown in each lesson, can you make from notebook paper?
5
How can you quickly find the pages that have information about forming a hypothesis?
FINISH

Table of Contents

Scientific Problem Solving .... **NOS 2**

- **Lesson 1** Scientific Problem Solving .... NOS 4
- **Lesson 2** Measurement and Scientific Tools .... NOS 12
  - Inquiry **Lab: Skill Practice** How does the strength of geometric shapes differ? .... NOS 19
- **Lesson 3** Case Study: The Minneapolis Bridge Failure .... NOS 20
  - Inquiry **Lab** Build and Test a Bridge .... NOS 28

Unit 1 **Motion and Forces** **2**

Chapter 1 **Describing Motion** **6**

- **Lesson 1** Position and Motion 8
- **Lesson 2** Speed and Velocity 16
  - Inquiry **Lab: Skill Practice** What do you measure to calculate speed? 25
- **Lesson 3** Acceleration 26
  - Inquiry **Lab** Calculate Average Speed from a Graph 34

Chapter 2 **Laws of Motion** **42**

- **Lesson 1** Gravity and Friction 44
- **Lesson 2** Newton's First Law 53
  - Inquiry **Lab: Skill Practice** How can you model Newton's first law of motion? 60
- **Lesson 3** Newton's Second Law 61
  - Inquiry **Lab: Skill Practice** How does a change in mass or force affect acceleration? 68
- **Lesson 4** Newton's Third Law 69
  - **Lab** Modeling Newton's Laws of Motion 76

Chapter 3 **Work and Simple Machines** **84**

- **Lesson 1** Work and Power 86
- **Lesson 2** Using Machines 94
  - Inquiry **Lab: Skill Practice** How does mechanical advantage affect a machine? 101
- **Lesson 3** Simple Machines 102
  - Inquiry **Lab** Comparing Two Simple Machines 112

Chapter 4 **Forces and Fluids** **120**

- **Lesson 1** Pressure and Density of Fluids 122
- **Lesson 2** The Buoyant Force 131
  - Inquiry **Lab: Skill Practice** Do heavy objects always sink and light objects always float? 138
- **Lesson 3** Other Forces from Fluids 139
  - Inquiry **Lab** Design a Cargo Ship 146

Unit 2 Energy and Matter 154

Chapter 5 Energy and Energy Resources 158
Lesson 1 Forms of Energy 160
Lesson 2 Energy Transformations 168
Inquiry **Lab: Skill Practice** Can you identify energy transformations? 175
Lesson 3 Energy Resources 176
Inquiry **Lab** Pinwheel Power 186

Chapter 6 Thermal Energy 194
Lesson 1 Thermal Energy, Temperature, and Heat 196
Inquiry **Lab: Skill Practice** How do different materials affect thermal energy transfer? 203
Lesson 2 Thermal Energy Transfers 204
Lesson 3 Using Thermal Energy 214
Inquiry **Lab** Design an Insulated Container 220

Chapter 7 Foundations of Chemistry 228
Lesson 1 Classifying Matter 230
Lesson 2 Physical Properties 239
Inquiry **Lab: Skill Practice** How can following a procedure help you solve a crime? 247
Lesson 3 Physical Changes 248
Inquiry **Lab: Skill Practice** How can known substances help you identify unknown substances? 254
Lesson 4 Chemical Properties and Changes 255
Inquiry **Lab** Design an Experiment to Solve a Crime 262

Chapter 8 States of Matter 270
Lesson 1 Solids, Liquids, and Gases 272
Lesson 2 Changes in State 281
Inquiry **Lab: Skill Practice** How does dissolving substances in water change its freezing point? 290
Lesson 3 The Behavior of Gases 291
Inquiry **Lab** Design an Experiment to Collect Data 298

Unit 3 Atoms and Elements 306

Chapter 9 Understanding the Atom 310
Lesson 1 Discovering Parts of the Atom 312
Lesson 2 Protons, Neutrons, and Electrons—How Atoms Differ 325
Inquiry **Lab** Communicate Your Knowledge about the Atom 334

Table of Contents

Chapter 10 The Periodic Table 342
Lesson 1 Using the Periodic Table 344
Inquiry **Lab: Skill Practice** How is the periodic table arranged? 353
Lesson 2 Metals 354
Lesson 3 Nonmetals and Metalloids 362
Inquiry **Lab** Alien Insect Periodic Table 370

Chapter 11 Elements and Chemical Bonds 378
Lesson 1 Electrons and Energy Levels 380
Lesson 2 Compounds, Chemical Formulas, and Covalent Bonds 389
Inquiry **Lab: Skill Practice** How can you model compounds? 396
Lesson 3 Ionic and Metallic Bonds 397
Inquiry **Lab** Ions in Solution 404

Unit 4 Interactions of Matter 412

Chapter 12 Chemical Reactions and Equations 416
Lesson 1 Understanding Chemical Reactions 418
Inquiry **Lab: Skill Practice** What can you learn from an experiment? 428
Lesson 2 Types of Chemical Reactions 429
Lesson 3 Energy Changes and Chemical Reactions 435
Inquiry **Lab** Design an Experiment to Test Advertising Claims 442

Chapter 13 Mixtures, Solubility, and Acid/Base Solutions 450
Lesson 1 Substances and Mixtures 452
Lesson 2 Properties of Solutions 460
Inquiry **Lab: Skill Practice** How does a solute affect the conductivity of a solution? 469
Lesson 3 Acid and Base Solutions 470
Inquiry **Lab** Can the pH of a solution be changed? 478

Chapter 14 Carbon Chemistry 486
Lesson 1 Elemental Carbon and Simple Organic Compounds 488
Lesson 2 Other Organic Compounds 498
Inquiry **Lab: Skill Practice** How do you test for vitamin C? 506
Lesson 3 Compounds of Life 507
Inquiry **Lab** Testing for Carbon Compounds 514

Unit 5 **Waves, Electricity, and Magnetism** **522**

Chapter 15 **Waves** **526**
Lesson 1 What are waves? 528
Lesson 2 Wave Properties 538
Inquiry **Lab: Skill Practice** How are the properties of waves related? 545
Lesson 3 Wave Interactions 546
Inquiry **Lab** Measuring Wave Speed 554

Chapter 16 **Sound** **562**
Lesson 1 Producing and Detecting Sound 564
Lesson 2 Properties of Sound Waves 572
Inquiry **Lab: Skill Practice** How can you use a wind instrument to play music? 581
Lesson 3 Using Sound Waves 582
Inquiry **Lab** Make Your Own Musical Instrument 590

Chapter 17 **Electromagnetic Waves** **598**
Lesson 1 Electromagnetic Radiation 600
Lesson 2 The Electromagnetic Spectrum 608
Inquiry **Lab: Skill Practice** What's at the edge of a rainbow? 613
Lesson 3 Using Electromagnetic Waves 614
Inquiry **Lab** Design an Exhibit for a Science Museum 624

Chapter 18 **Light** **632**
Lesson 1 Light, Matter, and Color 634
Lesson 2 Reflection and Mirrors 642
Inquiry **Lab: Skill Practice** How can you demonstrate the law of reflection? 648
Lesson 3 Refraction and Lenses 649
Inquiry **Lab: Skill Practice** How does a lens affect light? 659
Lesson 4 Optical Technology 660
Inquiry **Lab** Design Your Own Optical Illusion 668

Chapter 19 **Electricity** **676**
Lesson 1 Electric Charge and Electric Forces 678
Lesson 2 Electric Current 689
Inquiry **Lab: Skill Practice** What effect does voltage have on a circuit? 697
Lesson 3 Electric Circuits 698
Inquiry **Lab** Design an Elevator 706

Chapter 20 **Magnetism** **714**
Lesson 1 Magnets and Magnetic Fields 716
Lesson 2 Making Magnets Using Electric Current 726
Inquiry **Lab: Skill Practice** How can you measure current? 733
Lesson 3 Making Electric Current with Magnets 734
Inquiry **Lab** Build a Wind-Powered Generator 742

Table of Contents

Student Resources

Science Skill Handbook **SR-2**
- Scientific Methods SR-2
- Safety Symbols SR-11
- Safety in the Science Laboratory SR-12

Math Skill Handbook **SR-14**
- Math Review SR-14
- Science Application SR-24

Foldables Handbook **SR-29**

Reference Handbook **SR-40**
- Periodic Table of the Elements SR-40

Glossary **G-2**

Index **I-2**

Credits **C-2**

Inquiry Labs

Inquiry Launch Labs

1-1 How do you get there from here? 9
1-2 How can motion change? 17
1-3 In what ways can velocity change? 27
2-1 Can you make a ball move without touching it? 45
2-2 Can you balance magnetic forces? 54
2-3 What forces affect motion along a curved path? 62
2-4 How do opposite forces compare? 70
3-1 How do you know when work is done? 87
3-2 How do machines work? 95
3-3 How does a lever work? 103
4-1 What changes? What doesn't? 123
4-2 How does shape affect an object's ability to float? 132
4-3 How is force transferred through a fluid? 140
5-1 Can you make a change in matter? 161
5-2 Is energy lost when it changes form? 169
5-3 How are energy resources different? 177
6-1 How can you describe temperature? 197
6-2 How hot is it? 205
6-3 How can you transform energy? 215
7-1 How do you classify matter? 231
7-2 Can you follow the clues? 240
7-3 Where did it go? 249
7-4 What can colors tell you? 256
8-1 How can you see particles in matter? 273
8-2 Do liquid particles move? 282
8-3 Are volume and pressure of a gas related? 292
9-1 What's in there? 313
9-2 How many different things can you make? 326
10-1 How can objects be organized? 345
10-2 What properties make metals useful? 355
10-3 What are some properties of nonmetals? 363
11-1 How is the periodic table organized? 381
11-2 How is a compound different from its elements? 390
11-3 How can atoms form compounds by gaining and losing electrons? 398
12-1 Where did it come from? 419
12-2 What combines with what? 430
12-3 Where's the heat? 436
13-1 What makes black ink black? 453
13-2 How are they different? 461
13-3 What color is it? 471
14-1 Why is carbon a unique element? 489
14-2 How do functional groups affect compounds? 499
14-3 What does a carbon compound look like when its structure changes? 508
15-1 How can you make waves? 529
15-2 Which sounds have more energy? 539
15-3 What happens in wave collisions? 547
16-1 What causes sound? 565
16-2 How can sound blow out a candle? 573
16-3 Why didn't you hear the phone ringing? 583
17-1 How can you detect invisible waves? 601
17-2 How do electromagnetic waves differ? 609
17-3 How do X-rays show teeth? 615
18-1 How can you make a rainbow? 635
18-2 How can you read a sign behind you? 643
18-3 What happens to light that passes between transparent substances? 650
18-4 How do long distance phone lines work? 661
19-1 How can you bend water? 679
19-2 What is the path to turn on a light? 690
19-3 How would you wire a house? 699
20-1 What does magnetic mean? 717
20-2 When is a wire a magnet? 727
20-3 Why does the pendulum slow down? 735

Inquiry Labs

Inquiry MiniLabs

1-1 Why is a reference point useful? 11
1-2 How can you graph motion? 20
1-3 How is a change in speed related to acceleration? 30
2-1 How does friction affect motion? 50
2-2 How do forces affect motion? 57
2-3 How are force and mass related? 64
2-4 Is momentum conserved during a collision? 74
3-1 What affects work? 89
3-2 Does a ramp make it easier to lift a load? 96
3-3 Can you increase mechanical advantage? 109
4-1 How are force and pressure different? 125
4-2 Can you overcome the buoyant force? 134
4-3 How does air speed affect air pressure? 142
5-1 Can a moving object do work? 164
5-2 How does energy change form? 173
5-3 What energy resources provide our electric energy? 183
6-1 How do temperature scales compare? 201
6-2 How does adding thermal energy affect a wire? 209
6-3 Can thermal energy be used to do work? 217
7-1 How can you model an atom? 232
7-2 Can the weight of an object change? 242
7-3 Can you make ice without a freezer? 251
7-4 Can you spot the clues for chemical change? 258
8-1 How can you make bubble films? 277
8-2 How can you make a water thermometer? 288
8-3 How does temperature affect the volume? 295
9-1 How can you gather information about what you can't see? 320
9-2 How many penny isotopes do you have? 329
10-1 How does atom size change across a period? 351
10-2 How well do materials conduct thermal energy? 359
10-3 Which insulates better? 368
11-1 How does an electron's energy relate to its position in an atom? 386
11-2 How do compounds form? 394
11-3 How many ionic compounds can you make? 401
12-1 How does an equation represent a reaction? 423
12-3 Can you speed up a reaction? 440
13-1 Which one is the mixture? 456
13-2 How much is dissolved? 464
13-3 Is it an acid or a base? 476
14-1 How do carbon atoms bond with carbon and hydrogen atoms? 494
14-2 How can you make a polymer? 503
14-3 How many carbohydrates do you consume? 510
15-1 How do waves travel through matter? 531
15-2 How are wavelength and frequency related? 541
15-3 How can reflection be used? 549
16-1 How do you know a sound's direction? 569
16-2 How can you hear beats? 577
16-3 How fast is sound? 585
17-1 How are electric fields and magnetic fields related? 603
17-3 How does infrared imaging work? 620
18-1 What color is that? 638
18-2 Where is the image in a plane mirror? 645
18-3 How can water bend light? 652
18-4 How does a zoom lens work? 663
19-1 How can a balloon push or pull? 682
19-2 When is one more than two? 692
19-3 What else can a circuit do? 704
20-1 Where is magnetic north? 721
20-2 What is an electromagnet? 730
20-3 How many paper clips can you lift? 737

Inquiry Skill Practice

NOS 2 How does the strength of geometric shapes differ? ....NOS 19
1-2 What do you measure to calculate speed? 25
2-2 How can you model Newton's first law of motion? 60
2-3 How does a change in mass or force affect acceleration? 68
3-2 How does mechanical advantage affect a machine? 101
4-2 Do heavy objects always sink and light objects always float? 138
5-2 Can you identify energy transformations? 175
6-1 How do different materials affect thermal energy transfer? 203
7-2 How can following a procedure help you solve a crime? 247
7-3 How can known substances help you identify unknown substances? 254
8-2 How does dissolving substances in water change its freezing point? 290
10-1 How is the periodic table arranged? 353
11-2 How can you model compounds? 396
12-1 What can you learn from an experiment? 428
13-2 How does a solute affect the conductivity of a solution? 469
14-2 How do you test for vitamin C? 506
15-2 How are the properties of waves related? 545
16-2 How can you use a wind instrument to play music? 581
17-2 What's at the edge of a rainbow? 613
18-2 How can you demonstrate the law of reflection? 648
18-3 How does a lens affect light? 659
19-2 What effect does voltage have on a circuit? 697
20-2 How can you measure current? 733

Inquiry Labs

NOS 3 Build and Test a Bridge .... NOS 28
1-3 Calculate Average Speed from a Graph.... 34
2-4 Modeling Newton's Laws of Motion 76
3-3 Comparing Two Simple Machines....... 112
4-3 Design a Cargo Ship 146
5-3 Pinwheel Power 186
6-3 Design an Insulated Container 220
7-4 Design an Experiment to Solve a Crime .. 262
8-3 Design an Experiment to Collect Data... 298
9-2 Communicate Your Knowledge about the Atom............................... 334
10-3 Alien Insect Periodic Table 370
11-3 Ions in Solution.......................... 404
12-3 Design an Experiment to Test Advertising Claims 442
13-3 Can the pH of a solution be changed?... 478
14-3 Testing for Carbon Compounds......... 514
15-3 Measuring Wave Speed................. 554
16-3 Make Your Own Musical Instrument..... 590
17-3 Design an Exhibit for a Science Museum 624
18-4 Design Your Own Optical Illusion 668
19-3 Design an Elevator 706
20-3 Build a Wind-Powered Generator........ 742

Features

GREEN SCIENCE

5-1 Fossil Fuels & Rising CO_2 Levels 167
11-1 New Green Airships 388

HOW IT WORKS

1-1 GPS to the Rescue! 15
4-1 Submersibles 130
7-1 U.S. Mint 238
8-1 Freeze-Drying Foods 280
12-2 How does a light stick work? 434
16-1 Cochlear Implants.................... 571
17-1 Solar Sails............................ 607
19-1 Van de Graaff Generator.............. 688

SCIENCE & SOCIETY

2-1 Avoiding an Asteroid Collision..........52
3-1 What is horsepower?...................93
6-2 Insulating the Home 213
9-1 Subatomic Particles 324
10-2 Fireworks 361
13-1 Sports Drinks and Your Body 459
14-1 Carbon 497
18-1 Color!............................... 641
20-1 Roller Coasters 725

CAREERS in SCIENCE

15-1 Making a Computer Tsunami 537

Nature of Science

Scientific Problem Solving

What is scientific inquiry?

Sci-Fi Movie Scene?

This might look like a weird spaceship docking in a science-fiction movie. However, it is actually the back of an airplane engine being tested in a huge wind tunnel. An experiment is an important part of scientific investigations.

- Why is an experiment important?
- Does experimentation occur in all branches of science?
- What is scientific inquiry?

Nature of SCIENCE

This chapter begins your study of the nature of science, but there is even more information about the nature of science in this book. Each unit begins by exploring an important topic that is fundamental to scientific study. As you read these topics, you will learn even more about the nature of science.

Models	Unit 1
Technology	Unit 2
Patterns	Unit 3
Health	Unit 4
Graphs	Unit 5

Lesson 1 Scientific Inquiry

Reading Guide

Key Concepts
ESSENTIAL QUESTIONS

- What are some steps used during scientific inquiry?
- What are the results of scientific inquiry?
- What is critical thinking?

Vocabulary

science p. NOS 4
observation p. NOS 6
inference p. NOS 6
hypothesis p. NOS 6
prediction p. NOS 6
scientific theory p. NOS 8
scientific law p. NOS 8
technology p. NOS 9
critical thinking p. NOS 10

Video
- BrainPOP®
- Science Video

Understanding Science

A clear night sky is one of the most beautiful sights on Earth. The stars seem to shine like a handful of diamonds scattered on black velvet. Why do stars seem to shine more brightly some nights than others?

Did you know that when you ask questions, such as the one above, you are practicing science? **Science** *is the investigation and exploration of natural events and of the new information that results from those investigations.* Like a scientist, you can help shape the future by accumulating knowledge, developing new technologies, and sharing ideas with others.

Throughout history, people of many different backgrounds, interests, and talents have made scientific contributions. Sometimes they overcame a limited educational background and excelled in science. One example is Marie Curie, shown in **Figure 1.** She was a scientist who won two Nobel prizes in the early 1900s for her work with radioactivity. As a young student, Marie was not allowed to study at the University of Warsaw in Poland because she was a woman. Despite this obstacle, she made significant contributions to science.

Figure 1 Modern medical procedures such as X-rays, radioactive cancer treatments, and nuclear-power generation are some of the technologies made possible because of the pioneering work of Marie Curie and her associates.

Branches of Science

Scientific study is organized into several branches, or parts. The three branches that you will study in middle school are physical science, Earth science, and life science. Each branch focuses on a different part of the natural world.

Word Origin

science
from Latin *scientia,* means "knowledge" or "to know"

Physical Science

Physical science, or physics and chemistry, is the study of matter and energy. The physicist shown here is adjusting an instrument that measures radiation from the Big Bang. Physical scientists ask questions such as

- What happens to energy during chemical reactions?
- How does gravity affect roller coasters?
- What makes up protons, neutrons, and electrons?

Earth Science

Earth scientists study the many processes that occur on Earth and deep within Earth. This scientist is collecting a water sample in southern Mexico.

Earth scientists ask questions such as

- What are the properties of minerals?
- How is energy transferred on Earth?
- How do volcanoes form?

Life Science

Life scientists study all organisms and the many processes that occur in them. The life scientist shown is studying the avian flu virus.

Life scientists ask questions such as

- How do plant cells and animal cells differ?
- How do animals survive in the desert?
- How do organisms in a community interact?

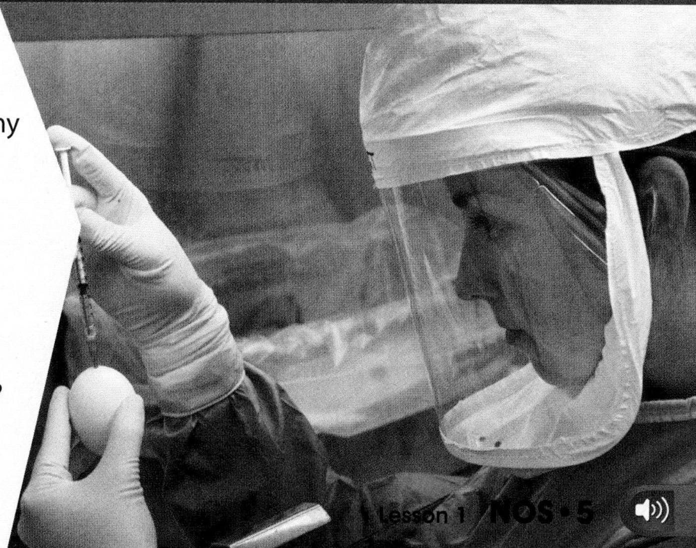

What is Scientific Inquiry?

When scientists conduct investigations, they often want to answer questions about the natural world. To do this, they use scientific inquiry—a series of skills used to answer questions. You might have heard these steps called "the scientific method." However, there is no one scientific method. In fact, the skills that scientists use to conduct an investigation can be used in any order. One possible sequence is shown in **Figure 2.** Like a scientist, you perform scientific investigations every day, and you will do investigations throughout this course.

 **Reading Check** What is scientific inquiry?

Ask Questions

Imagine warming yourself near a fireplace or a campfire? As you throw twigs and logs onto the fire, you see that fire releases smoke and light. You also feel the warmth of the thermal energy being released. These are **observations**—*the results of using one or more of your senses to gather information and taking note of what occurs.* Observations often lead to questions. You ask yourself, "When logs burn, what happens to the wood? Do the logs disappear? Do they change in some way?"

When observing the fire, you might recall from an earlier science course that matter can change form, but it cannot be created or destroyed. Therefore, you infer that the logs do not just disappear. They must undergo some type of change. An **inference** *is a logical explanation of an observation that is drawn from prior knowledge or experience.*

Hypothesize and Predict

After making observations and inferences, you decide to investigate further. Like a scientist, you might develop a **hypothesis**—*a possible explanation for an observation that can be tested by scientific investigations.* Your hypothesis about what happens to the logs might be: When logs burn, new substances form because matter cannot be destroyed.

When scientists state a hypothesis, they often use it to make predictions to help test their hypothesis. *A* **prediction** *is a statement of what will happen next in a sequence of events.* Scientists make predictions based on what information they think they will find when testing their hypothesis. For instance, based on your hypothesis, you might predict that if logs burn, then the substances that make up the logs change into other substances.

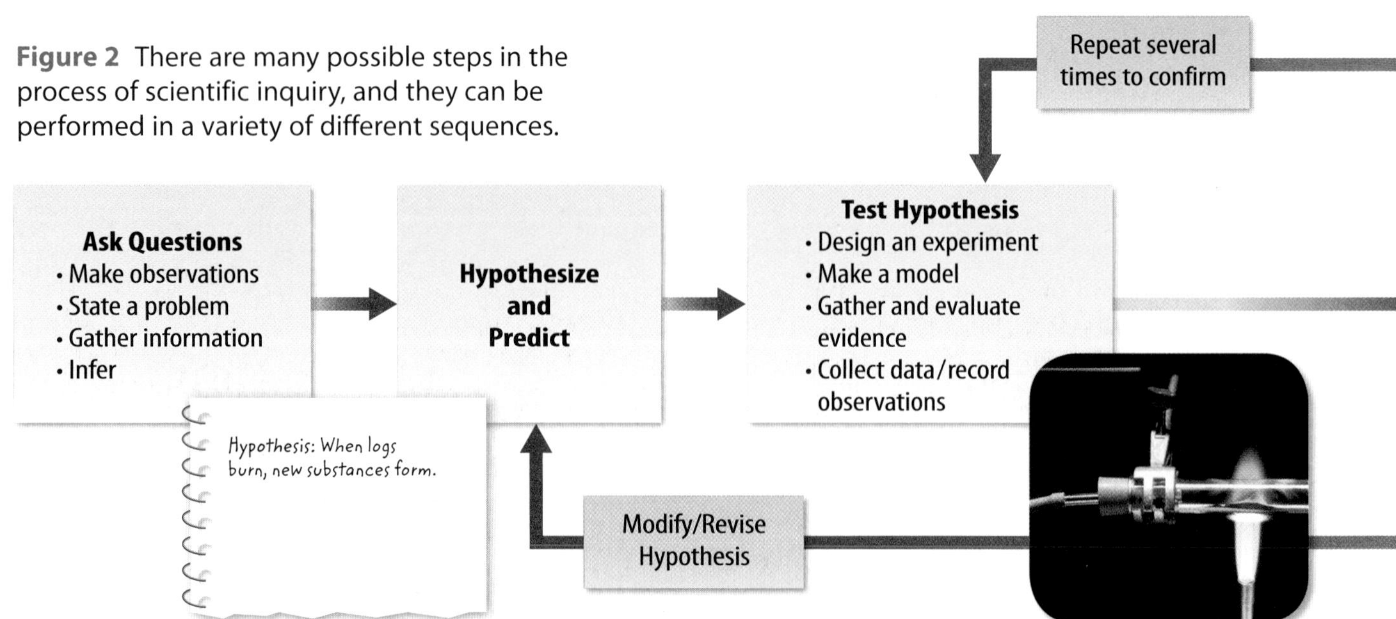

Figure 2 There are many possible steps in the process of scientific inquiry, and they can be performed in a variety of different sequences.

Test Hypothesis and Analyze Results

How could you test your hypothesis? When you test a hypothesis, you often test your predictions. If a prediction is confirmed, then it supports your hypothesis. If your prediction is not confirmed, you might modify your hypothesis and retest it.

To test your predictions and hypothesis, you could design an experiment to find out what substances make up wood. Then you could determine what makes up the ash, the smoke, and other products that formed after the burning process. You also could research this topic and possibly find answers on reliable science Web sites or in science books.

After doing an experiment or research, you need to analyze your results and findings. You might make additional inferences after reviewing your data. If you find that new substances actually do form when wood burns, your hypothesis is supported. If new products do not form, your hypothesis is not supported. Some methods of testing a hypothesis and analyzing results are shown in **Figure 2.**

Draw Conclusions

After analyzing your results, you can begin to draw conclusions about your investigation. A conclusion is a summary of the information gained from testing a hypothesis. Like a scientist does, you should test and retest your hypothesis several times to make sure the results are consistent.

Communicate Results

Sharing the results of a scientific inquiry is an important part of science. By exchanging information, scientists can evaluate and test others' work and make faster progress in their own research. Exchanging information is one way of making scientific advances as quickly as possible and keeping scientific information accurate. During your investigation, if you do research on the Internet or in science books, you use information that someone else communicated. Scientists exchange information in many ways, as shown below in **Figure 2.**

Key Concept Check What are some steps used during scientific inquiry?

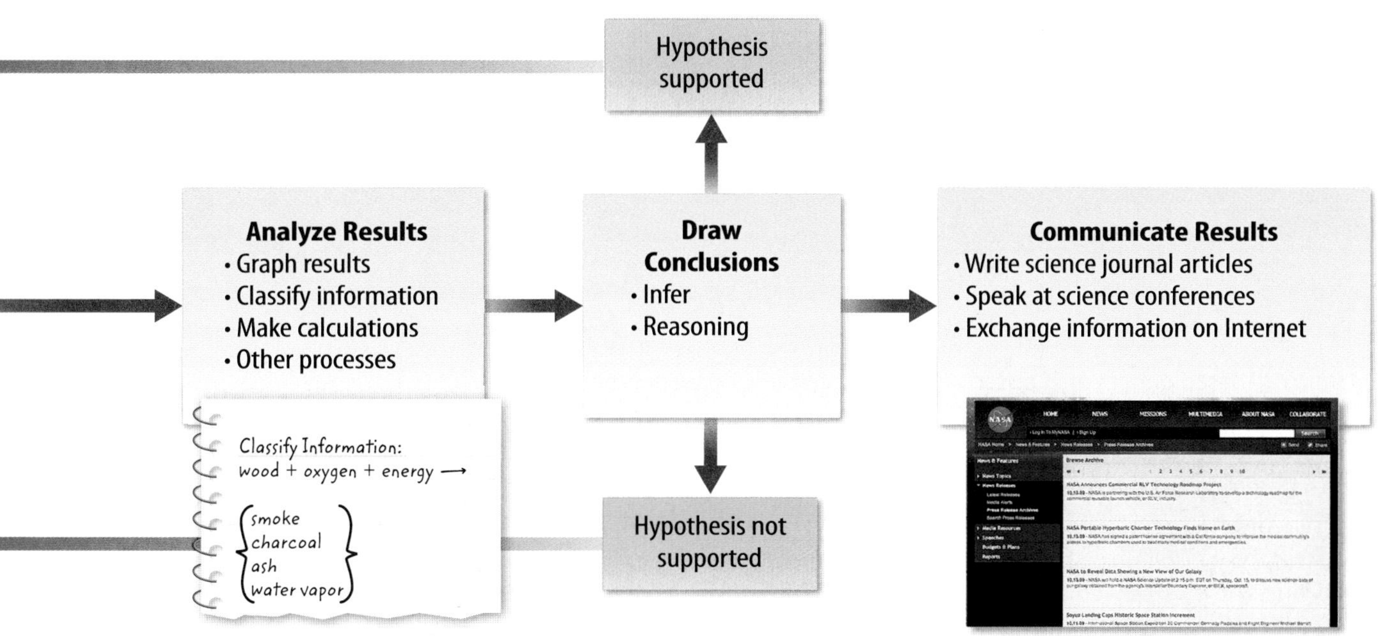

Unsupported or Supported Hypotheses

What happens if a hypothesis is not supported by an investigation? Was the scientific investigation a failure and a waste of time? Absolutely not! Even when a hypothesis is not supported, you gain valuable information. You can revise your hypothesis and test it again. Each time you test a hypothesis, you learn more about the topic you are studying.

Scientific Theory

When a hypothesis (or a group of closely related hypotheses) is supported through many tests over many years, a scientific theory can develop. A **scientific theory** *is an explanation of observations or events that is based on knowledge gained from many observations and investigations.*

A scientific theory does not develop from just one hypothesis, but from many hypotheses that are connected by a common idea. The kinetic molecular theory described below explains the behavior and energy of particles that make up a gas.

Scientific Law

A scientific law is different from a societal law, which is an agreement on a set of behaviors. A **scientific law** *is a rule that describes a repeatable pattern in nature.* A scientific law does not explain why or how the pattern happens, it only states that it will happen. For example, if you drop a ball, it will fall towards the ground every time. This is a repeated pattern that relates to the law of universal gravitation. The law of conservation of energy, described below, is also a scientific law.

Kinetic Molecular Theory

The kinetic molecular theory explains how particles that make up a gas move in constant, random motions. A particle moves in a straight line until it collides with another particle or with the wall of its container.

The kinetic molecular theory also assumes that the collisions of particles in a gas are elastic collisions. An elastic collision is a collision in which no kinetic energy is lost. Therefore, kinetic energy among gas particles is conserved.

Law of Conservation of Energy

The law of conservation of energy states that in any chemical reaction or physical change, energy is neither created nor destroyed. The total energy of particles before and after collisions is the same.

However, this scientific law, like all scientific laws, does not explain *why* energy is conserved. It simply states that energy *is* conserved.

Scientific Law v. Scientific Theory

Both are based on repeated observations and can be rejected or modified.

A scientific law states that an event *will* occur. For example, energy will be conserved when particles collide. It does not explain why an event will occur or how it will occur. Scientific laws work under specific conditions in nature. A law stands true until an observation is made that does not follow the law.

A scientific theory is an explanation of *why* or *how* an event occurred. For example, collisions of particles of a gas are elastic collisions. Therefore, no kinetic energy is lost. A theory can be rejected or modified if someone observes an event that disproves the theory. A theory will never become a law.

Results of Scientific Inquiry

Why do you and others ask questions and investigate the natural world? Just as scientific questions vary, so do the results of science. Most often, the purpose of a scientific investigation is to develop new materials and technology, discover new objects, or find answers to questions, as shown below.

Key Concept Check What are the results of scientific inquiry?

New Materials and Technology

Every year, corporations and governments spend millions of dollars on research and design of new materials and technologies. **Technology** *is the practical use of scientific knowledge, especially for industrial or commercial use.* For example, scientists hypothesize and predict how new materials will make bicycles and cycling gear lighter, more durable, and more aerodynamic. Using wind tunnels, scientists test these new materials to see whether they improve the cyclist's performance. If the cyclist's performance improves, their hypotheses are supported. If the performance does not improve or it doesn't improve enough, scientists will revise their hypotheses and conduct more tests.

New Objects or Events

Scientific investigations also lead to newly discovered objects or events. For example, NASA's *Hubble Space Telescope* captured this image of two colliding galaxies. They have been nicknamed the mice, because of their long tails. The tails are composed of gases and young, hot blue stars. If computer models are correct, these galaxies will combine in the future and form one large galaxy.

Answers to Questions

Often scientific investigations are launched to answer *who, what, when, where,* or *how* questions. For example, research chemists investigate new substances, such as substances found in mushrooms and bacteria, as shown on the right. New drug treatments for cancer, HIV, and other diseases might be found using new substances. Other scientists look for clues about what causes diseases, whether they can be passed from person to person, and when the disease first appeared.

FOLDABLES®

Create a two-tab book and label it as shown. Use it to discuss the importance of evaluating scientific information.

Why is it important to...

...be scientifically literate? | ...use critical thinking?

Evaluating Scientific Information

Do you ever you read advertisements, articles, or books that claim to contain scientifically proven information? Are you able to determine if the information is actually true and scientific instead of pseudoscientific (information incorrectly represented as scientific)? Whether you are reading printed media or watching commercials on TV, it is important that you are skeptical, identify facts and opinions, and think critically about the information. **Critical thinking** *is comparing what you already know with the information you are given in order to decide whether you agree with it.*

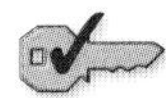

Key Concept Check What is critical thinking?

Be A Rock Star!

Do you dream of being a rock star?

Sing, dance, and play guitar like a rock star with the new Rocker-rific Spotlight. A new scientific process developed by Rising Star Laboratories allows you to overcome your lack of musical talent and enables you to perform like a real rock star.

This amazing new light actually changes your voice quality and enhances your brain chemistry so that you can sing, dance, and play a guitar like a professional rock star. Now, there is no need to practice or pay for expensive lessons. The Rocker-rific Spotlight does the work for you.

Dr. Sammy Truelove says, "Never before has lack of talent stopped someone from achieving his or her dreams of being a rock star. This scientific breakthrough transforms people with absolutely no talent into amazing rock stars in just minutes. Of the many patients that I have tested with this product, no one has failed to achieve his or her dreams."

Disclaimer: This product was tested on laboratory rats and might not work for everyone.

Skepticism
Have you heard the saying, if it sounds too good to be true, it probably is? To be skeptical is to doubt the truthfulness of something. A scientifically literate person can read information and know that it misrepresents the facts. Science often is self-correcting because someone usually challenges inaccurate information and tests scientific results for accuracy.

Identifying Facts and Misleading Information
Misleading information often is worded to sound like scientific facts. A scientifically literate person can recognize fake claims and quickly determine when information is false.

Critical Thinking
Use critical thinking skills to compare what you know with the new information given to you. If the information does not sound reliable, either research and find more information about the topic or dismiss the information as unreliable.

Identify Opinions
An opinion is a personal view, feeling, or claim about a topic. Opinions cannot be proven true or false. And, an opinion might contain inaccurate information.

Science cannot answer all questions.

It might seem that scientific inquiry is the best way to answer all questions. But there are some questions that science cannot answer. Questions that deal with beliefs, values, personal opinions, and feelings cannot be answered scientifically. This is because it is impossible to collect scientific data on these topics.

Science cannot answer questions such as

- Which video game is the most fun to play?
- Are people kind to others most of the time?
- Is there such a thing as good luck?

Figure 3 Always follow safety procedures when doing scientific investigations. If you have questions, ask your teacher.

Safety in Science

Scientists know that using safe procedures is important in any scientific investigation. When you begin a scientific investigation, you should always wear protective equipment, as shown in **Figure 3**. You also should learn the meaning of safety symbols, listen to your teacher's instructions, and learn to recognize potential hazards. For more information, consult the Science Skills Handbook at the back of this book.

Lesson 1 Review

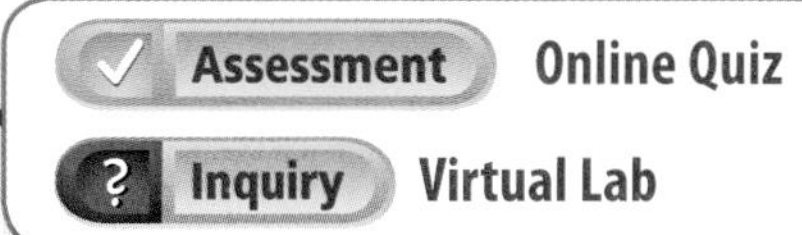

Use Vocabulary

1. **Define** *technology* in your own words.
2. **Use the term** *observation* in a sentence.

Understand Key Concepts

3. Which action is NOT a way to test a hypothesis?

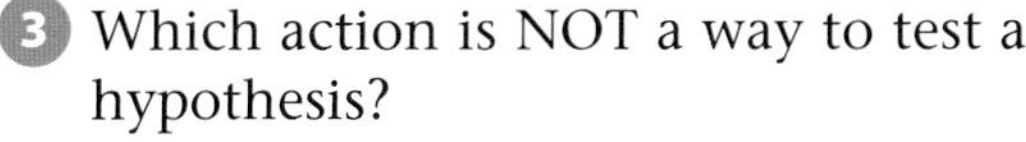

   A. analyze results
   B. design an experiment
   C. make a model
   D. gather and evaluate evidence

4. **Describe** three examples of the results of scientific inquiry.
5. **Give an example** of a time when you practiced critical thinking.

Interpret Graphics

6. **Compare** Copy and fill in the graphic organizer below. List some examples of how to communicate the results of scientific inquiry.

Critical Thinking

7. **Summarize** Your classmate writes the following as a hypothesis:

   Red is a beautiful color.

   Write a brief explanation to your classmate explaining why this is not a hypothesis.

Lesson 2

Measurement and Scientific Tools

Reading Guide

Key Concepts
ESSENTIAL QUESTIONS

- Why did scientists create the International System of Units (SI)?
- Why is scientific notation a useful tool for scientists?
- How can tools, such as graduated cylinders and triple-beam  balances, assist physical scientists?

Vocabulary

description p. NOS 12
explanation p. NOS 12
International System of Units (SI) p. NOS 13
scientific notation p. NOS 15
percent error p. NOS 15

g Multilingual eGlossary

Description and Explanation

Suppose you work for a company that tests cars to see how they perform during accidents, as shown in **Figure 4.** You might use various scientific tools to measure the acceleration of cars as they crash into other objects.

A **description** *is a spoken or written summary of observations.* The measurements you record are descriptions of the results of the crash tests. Later, your supervisor asks you to write a report that interprets the measurements you took during the crash tests. An **explanation** *is an interpretation of observations.* As you write your explanation, you make inferences about why the crashes damaged the vehicles in specific ways.

Notice that there is a difference between a description and an explanation. When you describe something, you report your observations. When you explain something, you interpret your observations.

The International System of Units

Different parts of the world use different systems of measurements. This can cause confusion when people who use different systems communicate their measurements. This confusion was eliminated in 1960 when a new system of measurement was adopted. *The internationally accepted system of measurement is the* **International System of Units (SI).**

Key Concept Check Why did scientists create the International System of Units?

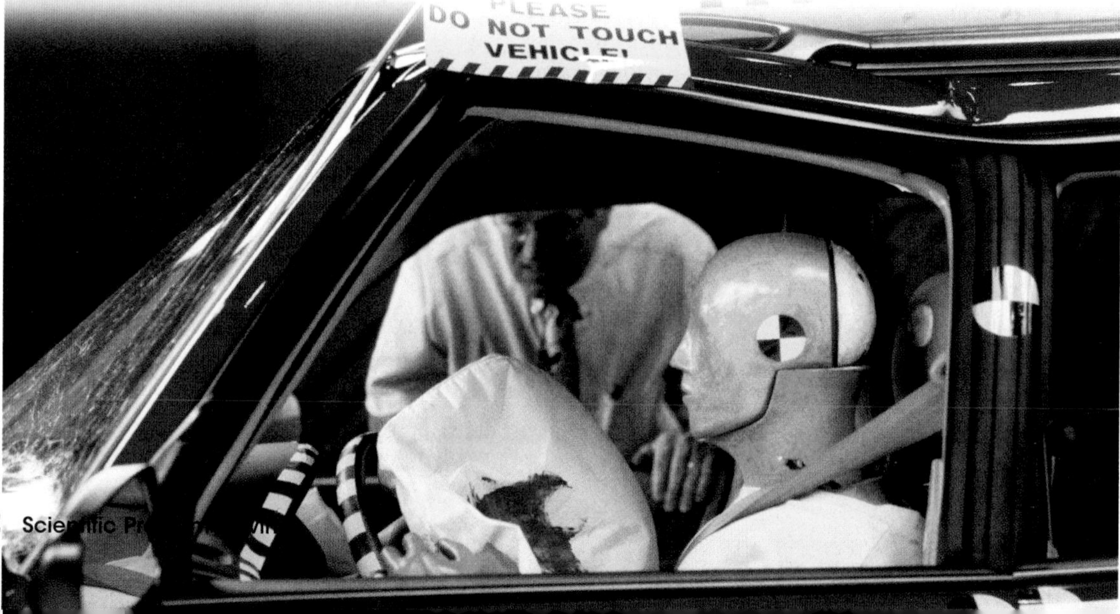

Figure 4 A description of an event details what you observed. An explanation explains why or how the event occurred.

SI Base Units

When you take measurements during scientific investigations and labs in this course, you will use the SI system. The SI system uses standards of measurement, called base units, as shown in **Table 1.** Other units used in the SI system that are not base units are derived from the base units. For example, the liter, used to measure volume, was derived from the base unit for length.

SI Unit Prefixes

In older systems of measurement, there usually was no common factor that related one unit to another. The SI system eliminated this problem.

The SI system is based on multiples of ten. Any SI unit can be converted to another by multiplying by a power of ten. Factors of ten are represented by prefixes, as shown in **Table 2.** For example, the prefix *milli-* means 0.001 or 10^{-3}. So, a milliliter is 0.001 L, or 1/1,000 L. Another way to say this is: 1 L is 1,000 times greater than 1 mL.

Converting Among SI Units

It is easy to convert from one SI unit to another. You either multiply or divide by a factor of ten. You also can use proportion calculations to make conversions. An example of how to convert between SI units is shown in **Figure 5.**

Interactive Table

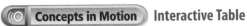

Table 1 SI Base Units

Quantity Measured	Unit (symbol)
Length	meter (m)
Mass	kilogram (kg)
Time	second (s)
Electric current	ampere (A)
Temperature	kelvin (K)
Substance amount	mole (mol)
Light intensity	candela (cd)

Table 2 Prefixes

Prefix	Meaning
Mega- (M)	1,000,000 or (10^6)
Kilo- (k)	1,000 or (10^3)
Hecto- (h)	100 or (10^2)
Deka- (da)	10 or (10^1)
Deci- (d)	0.1 or $\left(\frac{1}{10}\right)$ or (10^{-1})
Centi- (c)	0.01 or $\left(\frac{1}{100}\right)$ or (10^{-2})
Milli- (m)	0.001 or $\left(\frac{1}{1,000}\right)$ or (10^{-3})
Micro- (μ)	0.000001 or $\left(\frac{1}{1,000,000}\right)$ or (10^{-6})

Figure 5 The rock in the photograph has a mass of 17.5 grams. Convert that measurement to kilograms. ▼

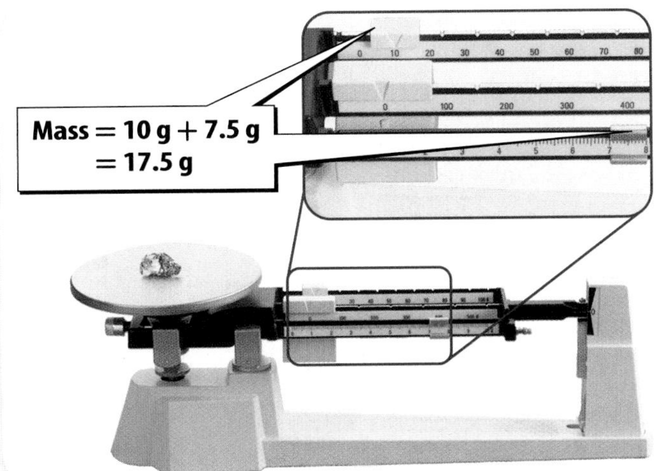

1. Determine the correct relationship between grams and kilograms. There are 1,000 g in 1 kg.

$$\frac{1\text{ kg}}{1{,}000\text{ g}}$$

$$\frac{x}{17.5\text{ g}} = \frac{1\text{ kg}}{1{,}000\text{ g}}$$

$$x = \frac{(17.5\text{ g})(1\text{ kg})}{1{,}000\text{ g}};\ x = 0.0175\text{ kg}$$

2. Check your units. The unit *grams* divides in the equation, so the answer is 0.0175 kg.

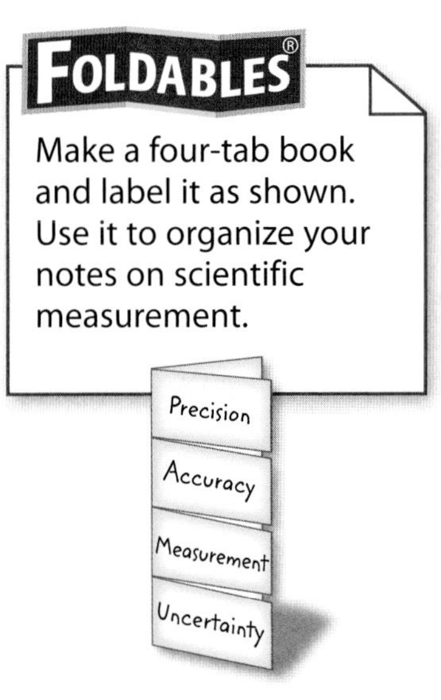

FOLDABLES
Make a four-tab book and label it as shown. Use it to organize your notes on scientific measurement.
Precision
Accuracy
Measurement
Uncertainty

Word Origin

notation
from Latin *notationem,* means "a marking or explanation"

Table 3 Student Density and Error Data (Accepted value: Density of sodium chloride, 21.7 g/cm^3)			
	Student A	**Student B**	**Student C**
	Density	**Density**	**Density**
Trial 1	23.4 g/cm^3	18.9 g/cm^3	21.9 g/cm^3
Trial 2	23.5 g/cm^3	27.2 g/cm^3	21.4 g/cm^3
Trial 3	23.4 g/cm^3	29.1 g/cm^3	21.3 g/cm^3
Mean	23.4 g/cm^3	25.1 g/cm^3	21.5 g/cm^3

Measurement and Uncertainty

You might be familiar with the terms *precision* and *accuracy*. In science, these terms have different meanings. Precision is a description of how similar or close repeated measurements are to each other. Accuracy is a description of how close a measurement is to an accepted value.

The difference between precision and accuracy is illustrated in **Table 3.** Students were asked to find the density of sodium chloride (NaCl). In three trials, each student measured the volume and the mass of NaCl. Then, they calculated the density for each trial and calculated the mean, or average. Student A's measurements are the most precise because they are closest to each other. Student C's measurements are the most accurate because they are closest to the scientifically accepted value. Student B's measurements are neither precise nor accurate. They are not close to each other or to the accepted value.

Tools and Accuracy

No measuring tool provides a perfect measurement. All measurements have some degree of uncertainty. Some tools or instruments produce more accurate measurements, as shown in **Figure 6.**

Figure 6 The graduated cylinder is graduated in 1-mL increments. The beaker is graduated in, or divided into, 50-mL increments. Liquid measurements taken with the graduated cylinder have greater accuracy.

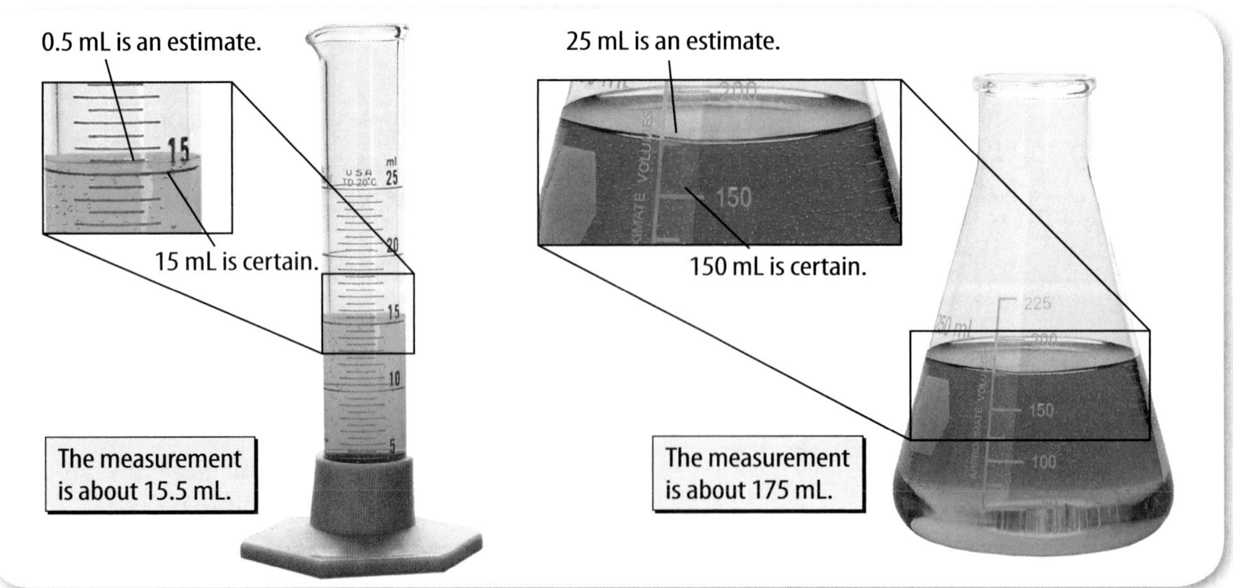

0.5 mL is an estimate.
15 mL is certain.

25 mL is an estimate.
150 mL is certain.

Scientific Notation

Suppose you are writing a report that includes Earth's distance from the Sun—149,600,000 km—and the density of the Sun's lower atmosphere—0.000000028 g/cm^3. These numbers take up too much space in your report, so you use **scientific notation**—*a method of writing or displaying very small or very large numbers in a short form.* To write the numbers above in scientific notation, use the steps in the beige box on the right.

Key Concept Check Why is scientific notation a useful tool for scientists?

Percent Error

The densities recorded in **Table 3** are experimental values because they were calculated during an experiment. Each of these values has some error because the accepted value for table salt density is 21.65 g/cm^3. Percent error can help you determine the size of your experimental error. **Percent error** *is the expression of error as a percentage of the accepted value.*

How to Write in Scientific Notation

1. Write the original number.
   - A. **149,600,000**
   - B. **0.000000028**
2. Move the decimal point to the right or the left to make the number between 1 and 10. Count the number of decimal places moved and note the direction.
   - A. **1.49600000** = 8 places to the left
   - B. **00000002.8** = 8 places to the right
3. Rewrite the number deleting all extra zeros to the right or to the left of the decimal point.
   - A. **1.496**
   - B. **2.8**
4. Write a multiplication symbol and the number *10* with an exponent. The exponent should equal the number of places that you moved the decimal point in step 2. If you moved the decimal point to the left, the exponent is positive. If you moved the decimal point to the right, the exponent is negative.
   - A. **1.496×10^8**
   - B. **2.8×10^{-8}**

Math Skills — Percent Error

Solve for Percent Error A student in the laboratory measures the boiling point of water at 97.5°C. If the accepted value for the boiling point of water is 100.0°C, what is the percent error?

1. This is what you know: **experimental value = 97.5°C**; **accepted value = 100.0°C**
2. This is what you need to find: percent error
3. Use this formula:

$$\text{percent error} = \frac{|\text{experimental value} - \text{accepted value}|}{\text{accepted value}} \times 100\%$$

4. Substitute the known values into the equation and perform the calculations

$$\text{percent error} = \frac{|97.5° - 100.0°|}{100.0°} \times 100\% = 2.50\%$$

Practice

Calculate the percent error if the experimental value of the density of gold is 18.7 g/cm^3 and the accepted value is 19.3 g/cm^3.

Review
- Math Practice
- Personal Tutor

Scientific Tools

As you conduct scientific investigations, you will use tools to make measurements. The tools listed here are some of the tools commonly used in science. For more information about the correct use and safety procedures for these tools, see the Science Skills Handbook at the back of this book.

◀ Science Journal

Use a science journal to record observations, write questions and hypotheses, collect data, and analyze the results of scientific inquiry. All scientists record the information they learn while conducting investigations. Your journal can be a spiral-bound notebook, a loose-leaf binder, or even just a pad of paper.

Balances ▶

A balance is used to measure the mass of an object. Units often used for mass are kilograms (kg), grams (g), and milligrams (mg). Two common types of balances are the electronic balance and the triple-beam balance. In order to get the most accurate measurements when using a balance, it is important to calibrate the balance often.

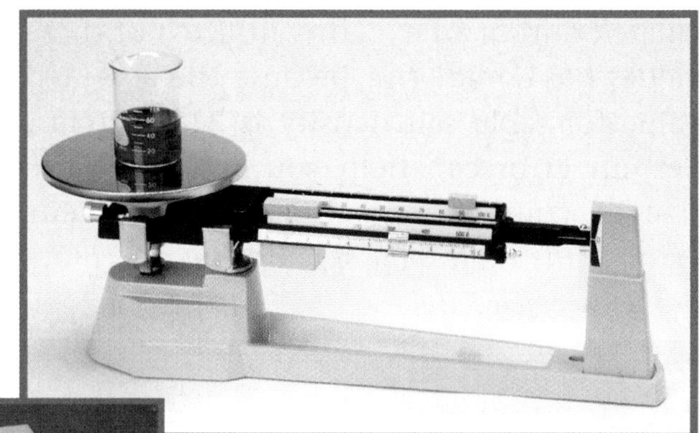

◀ Glassware

Laboratory glassware is used to hold or measure the volume of liquids. Flasks, beakers, test tubes, and graduated cylinders are just some of the different types of glassware available. Volume usually is measured in liters (L) and milliliters (mL).

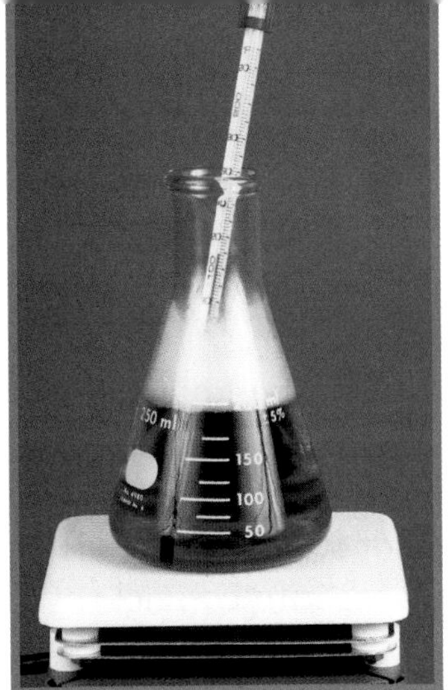

Thermometers ▶

A thermometer is used to measure the temperature of substances. Although Kelvin is the SI unit of measurement for temperature, in the science classroom, you often measure temperature in degrees Celsius (°C). Never stir a substance with a thermometer because it might break. If a thermometer does break, tell your teacher immediately. Do not touch the broken glass or the liquid inside the thermometer.

◀ Calculators

A hand-held calculator is a scientific tool that you might use in math class. But you also can use it in the lab and in the field (real situation outside the lab) to make quick calculations using your data.

Computers ▼

For today's students, it is difficult to think of a time when scientists—or anyone—did not use computers in their work. Scientists can collect, compile, and analyze data more quickly using a computer. Scientists use computers to prepare research reports and to share their data and ideas with investigators worldwide.

Hardware refers to the physical components of a computer, such as the monitor and the mouse. Computer software refers to the programs that are run on computers, such as word processing, spreadsheet, and presentation programs.

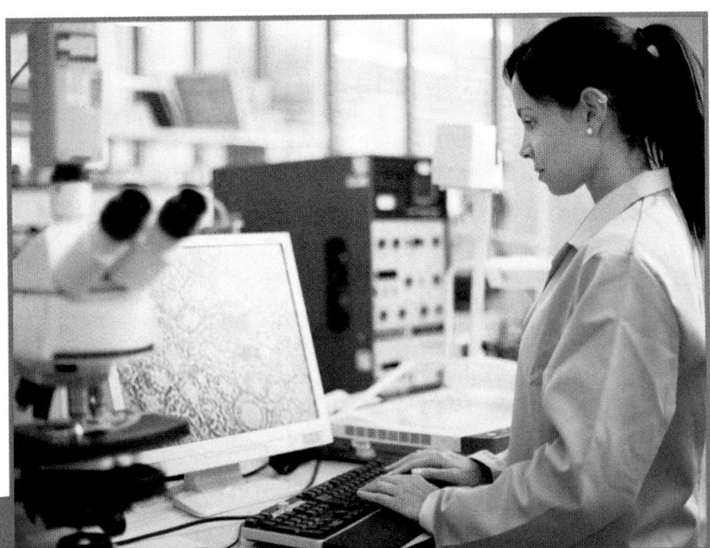

Electronic probes can be attached to computers and handheld calculators to record measurements. There are probes for collecting different kinds of information, such as temperature and the speed of objects.

Key Concept Check How can scientific tools, such as graduated cylinders and triple-beam balances, assist scientists?

Additional Tools Used by Physical Scientists

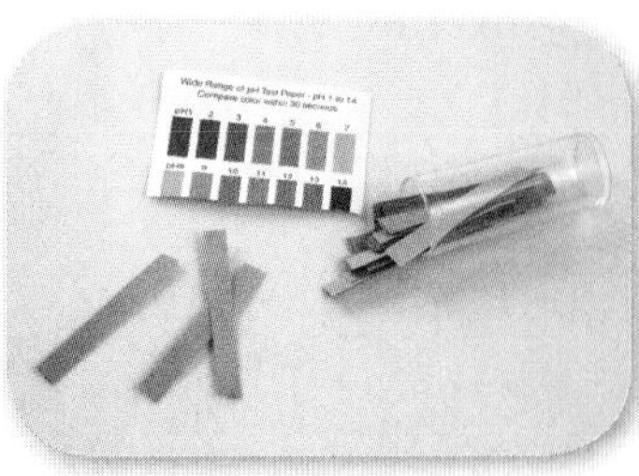

You can use pH paper to quickly estimate the acidity of a liquid substance. The paper changes color when it comes into contact with an acid or a base.

Scientists use stopwatches to measure the time it takes for an event to occur. The SI unit for time is seconds (s). However, for longer events, the units *minutes (min)* and *hours (h)* can be used.

A hot plate is a small heating device that can be placed on a table or desk. Hot plates are used to heat substances in the laboratory.

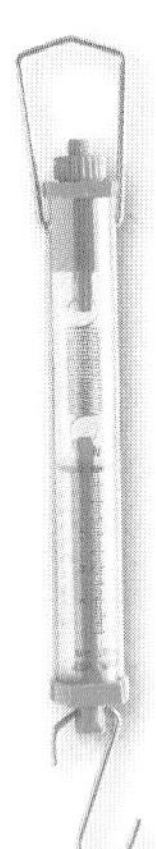

You use a spring scale to measure the weight or the amount of force applied to an object. The SI unit for weight is the newton (N).

Lesson 2 Review

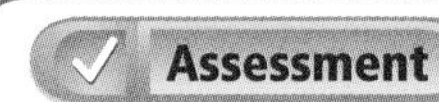

Online Quiz

Use Vocabulary

1. A spoken or written summary of observations is a(n) ________, while a(n) ________ is an interpretation of observations.

Understand Key Concepts 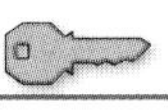

2. Which type of glassware would you use to measure the volume of a liquid?
   A. beaker
   B. flask
   C. graduated cylinder
   D. test tube

3. **Summarize** why a scientist measuring the diameter of an atom or the distance to the Moon would use scientific notation.

4. **Explain** why scientists use the International System of Units (SI).

Interpret Graphics

5. **Identify** Copy and fill in the graphic organizer below listing some scientific tools that you could use to collect data.

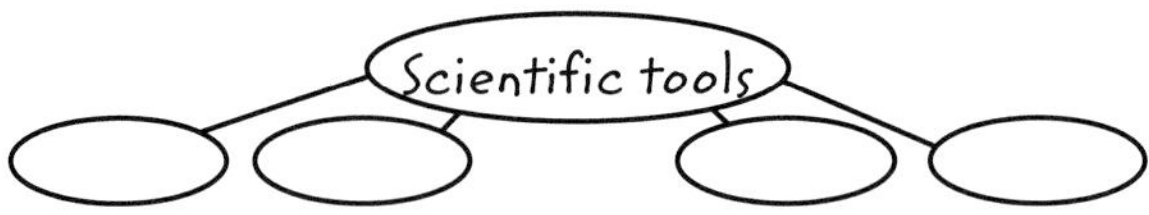

Critical Thinking

6. **Explain** why precision and accuracy should be reported in a scientific investigation.

Math Skills

Math Practice

7. Calculate the percent error if the experimental value for the density of zinc is 9.95 g/cm^3. The accepted value is 7.13 g/cm^3.

Skill Practice **Predict**

45 minutes

How do geometric shapes differ in strength?

Materials

plastic straws

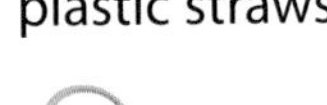

scissors

ruler

string

Safety

If you look at a bridge, a building crane, or the framework of a tall building, you will notice that various geometric shapes make up the structure. In this activity, you will observe the strength of several geometric shapes in terms of their rigidity, or resistance to changing their shape.

Learn It

When scientists make a hypothesis, they often then **predict** that an event will occur based on their hypothesis.

Try It

1. Read and complete a lab safety form.
2. You are going to construct a triangle and a square using straws. Predict which shape will be more resistant to changing shape and write your prediction in your Science Journal.
3. Measure and cut the straws into seven segments, each 6 cm long.
4. Measure and cut one 20-cm and one 30-cm length of string.
5. Thread the 30-cm length of string through four straw segments. Bend the corners to form a square. Tie the ends of the string together in a double knot to complete the square.
6. Thread the 20-cm string through three of the straw segments. Bend to form a triangle. Tie the ends of the string together to complete the triangle.
7. Test the strength of the square by gently trying to change its shape. Repeat with the triangle. Record your observations.
8. Propose several ways to make the weaker shape stronger. Draw diagrams showing how to modify the shape to make it more rigid.
9. Test your hypothesis. If necessary, refine your hypothesis and retest it. Repeat this step until you make the shape stronger.

Apply It

10. Look at the photograph at to the left. Which of your tested shapes is used the most? Based on your observations, why is this shape used?
11. What modifications made your shape stronger? Why?
12. **Key Concept** How might a scientist use a model to test a hypothesis?

Lesson 3 Case Study

Reading Guide

Key Concepts
ESSENTIAL QUESTIONS

- Why are evaluation and testing important in the design process?
- How is scientific inquiry used in a real-life scientific investigation?

Vocabulary

variable p. NOS 21
constant p. NOS 21
independent variable p. NOS 21
dependent variable p. NOS 21
experimental group p. NOS 21
control group p. NOS 21
qualitative data p. NOS 24
quantitative data p. NOS 24

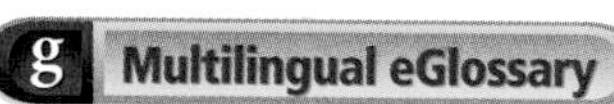

Science Video

The Minneapolis Bridge Failure

On August 1, 2007, the center section of the Interstate-35W (I-35W) bridge in Minneapolis, Minnesota, suddenly collapsed. A major portion of the bridge fell more than 30 m into the Mississippi River, as shown in **Figure 7.** There were more than 100 cars and trucks on the bridge at the time, including a school bus carrying over 50 students.

The failure of this 8-lane, 581-m long interstate bridge came as a surprise to almost everyone. Drivers do not expect a bridge to drop out from underneath them. The design and engineering processes that bridges undergo are supposed to ensure that bridge failures do not happen.

Controlled Experiments

After the 2007 bridge collapse, investigators had to determine why the bridge failed. To do this, they needed to use scientific inquiry, which you read about in Lesson 1. The investigators designed controlled experiments to help them answer questions and test their hypotheses. A controlled experiment is a scientific investigation that tests how one factor affects another. You might conduct controlled experiments to help discover answers to questions, to test a hypotheses, or to collect data.

Figure 7 A portion of the Interstate-35W bridge in Minneapolis, Minnesota, collapsed in August 2007. Several people were killed, and many more were injured.

Identifying Variables and Constants

When conducting an experiment, you must identify factors that can affect the experiment's outcome. A **variable** *is any factor that can have more than one value.* In controlled experiments, there are two kinds of variables. The **independent variable** *is the factor that you want to test. It is changed by the investigator to observe how it affects a dependent variable.* The **dependent variable** *is the factor you observe or measure during an experiment.* **Constants** *are the factors in an experiment that do not change.*

You can change the independent variable to observe how it affects the dependent variable. Without constants, two independent variables could change at the same time, and you would not know which variable affected the dependent variable.

Experimental Groups

A controlled experiment usually has at least two groups. The **experimental group** *is used to study how a change in the independent variable changes the dependent variable.* The **control group** *contains the same factors as the experimental group, but the independent variable is not changed.* Without a control, it is impossible to know whether your experimental observations result from the variable you are testing or some other factor.

This case study will explore how the investigators used scientific inquiry to determine why the bridge collapsed. Notebooks in the margin identify what a scientist might write in a science journal. The blue boxes contain additional helpful information that you might use.

Simple Beam Bridges

Before you read about the bridge-collapse investigation, think about the structure of bridges. The simplest type of bridge is a beam bridge, as shown in **Figure 8.** This type of bridge has one horizontal beam across two supports. A beam bridge often is constructed across small creeks. A disadvantage of beam bridges is that they tend to sag in the middle if they are too long.

Figure 8 Simple beam bridges span short distances, such as small creeks.

Figure 9 Truss bridges can span long distances and are strengthened by a series of interconnecting triangles called trusses. ▶

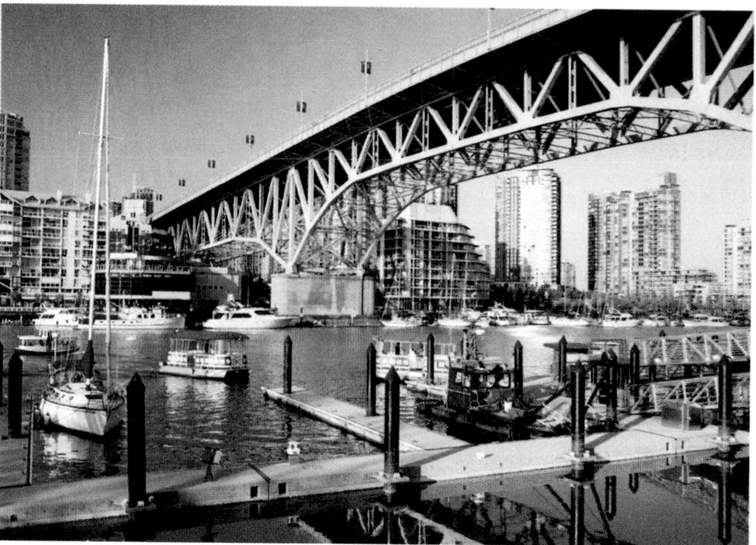

Truss Bridges

A truss bridge, shown in **Figure 9,** often spans long distances. This type of bridge is supported only at its two ends, but an assembly of interconnected triangles, or trusses, strengthens it. The I-35W bridge, shown in **Figure 10,** was a truss bridge designed in the early 1960s. The I-35W bridge was designed with straight beams connected to triangular and vertical supports. These supports held up the deck of the bridge, or the roadway. The beams in the bridge's deck and the supports came together at structures known as gusset plates, shown below on the right. These steel plates joined the triangular and vertical truss elements to the overhead roadway beams. These beams ran along the deck of the bridge. This area, where the truss structure connects to the roadway portion of the bridge at a gusset plate, also is called a node.

 **Reading Check** What are the gusset plates of a bridge?

Figure 10 Trusses were a major structural element of the I-35W bridge. The gusset plates at each node in the bridge, shown on the right, are critical pieces that hold the bridge together. ▼

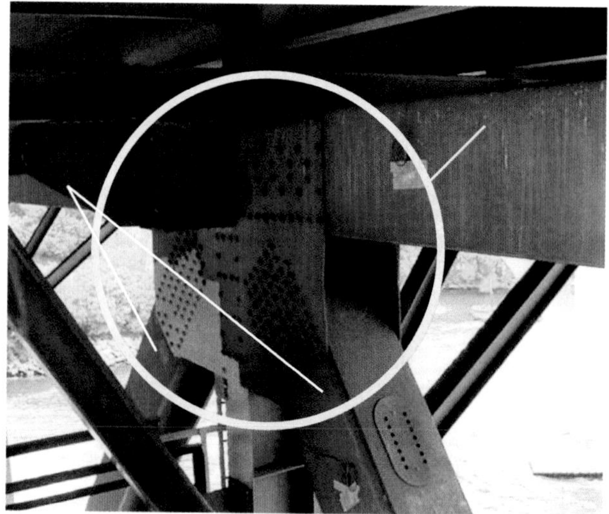

Figure 11 The collapsed bridge was further damaged by rescue workers trying to recover victims of the accident.

Scientists often observe and gather information about an object or an event before proposing a hypothesis. This information is recorded or filed for the investigation.

Observations:
- Recovered parts of the collapsed bridge
- A video showing the sequence of events as the bridge fails and falls into the river

Bridge Failure Observations

After the I-35W bridge collapsed, shown in **Figure 11**, the local sheriff's department handled the initial recovery of the collapsed bridge structure. Finding, freeing, and identifying victims was a high priority, and unintentional damage to the collapsed bridge occurred in the process. However, investigators eventually recovered the entire structure.

The investigators labeled each part with the location where it was found. They also noted the date when they removed each piece. Investigators then moved the pieces to a nearby park. There, they placed the pieces in their relative original positions. Examining the reassembled structure, investigators found physical evidence they needed to determine where the breaks in each section occurred.

The investigators found more clues in a video. A motion-activated security camera recorded the bridge collapse. The video showed about 10 seconds of the collapse, which revealed the sequence of events that destroyed the bridge. Investigators used this video to help pinpoint where the collapse began.

Asking Questions

One or more factors could have caused the bridge to fail. Was the original bridge design faulty? Were bridge maintenance and repair poor or lacking? Was there too much weight on the bridge at the time of the collapse? Each of these questions was studied to determine why the bridge collapsed. Did one or a combination of these factors cause the bridge to fail?

Asking questions and seeking answers to those questions is a way that scientists formulate hypotheses.

When gathering information or collecting data, scientists might perform an experiment, create a model, gather and evaluate evidence, or make calculations.

Qualitative data:
A thicker layer of concrete was added to the bridge to reinforce rods.

Quantitative data:
- The concrete increased the load on the bridge by 13.4 percent.
- The modifications in 1998 increased the load on the bridge by 6.1 percent.
- At the time of the collapse in 2007, the load on the bridge increased by another 20 percent.

Gathering Information and Data

Investigators reviewed the modifications made to the bridge since it opened in 1967. In 1977, engineers noticed that salt used to deice the bridge during winter weather was causing the reinforcement rods in the roadway to weaken. To protect the rods, engineers applied a thicker layer of concrete to the surface of the bridge roadway. Analysis after the collapse revealed that this extra concrete increased the dead load on the bridge by about 13.4 percent. A load can be a force applied to the structure from the structure itself (dead load) or from temporary loads such as traffic, wind gusts, or earthquakes (live load). Investigators recorded this qualitative and quantitative data. **Qualitative data** *uses words to describe what is observed.* **Quantitative data** *uses numbers to describe what is observed.*

In 1998, additional modifications were made to the bridge. The bridge that was built in the 1960s did not meet current safety standards. Analysis showed that the changes made to the bridge during this renovation further increased the dead load on the bridge by about 6.1 percent.

An Early Hypothesis

At the time of the collapse in 2007, the bridge was undergoing additional renovations. Four piles of sand, four piles of gravel, a water tanker filled with over 11,000 L of water, a cement tanker, a concrete mixer, and other equipment, supplies, and workers were assembled on the bridge. This caused the load on the bridge to increase by about 20 percent. In addition to these renovation materials, normal vehicle traffic was on the bridge. Did these renovations, materials, and traffic overload the bridge, causing the center section to collapse as shown in **Figure 12**? Only a thorough analysis could answer this question.

A hypothesis is a possible explanation for an observation that can be tested by scientific investigations.

Hypothesis:
The bridge failed because it was overloaded.

Figure 12 The center section of the bridge broke away and fell into the river.

Figure 13 Engineers used computer models to study the structure and loads on the bridge.

Computer Modeling

The analysis of the bridge was conducted using computer-modeling software, as shown in **Figure 13.** Using computer software, investigators entered data from the Minnesota bridge into a computer. The computer performed numerous mathematical calculations. After thorough modeling and analysis, it was determined that the bridge was not overloaded.

Revising the Hypothesis

Evaluations conducted in 1999 and 2003 provided additional clues as to why the bridge might have failed. As part of the study, investigators took numerous pictures of the bridge structure. The photos revealed bowing of the gusset plates at the eleventh node from the south end of the bridge. Investigators labeled this node *U10.* Gusset plates are designed to be stronger than the structural parts they connect. It is possible that the bowing of the plates indicated a problem with the gusset plate design. Previous inspectors and engineers missed this warning sign.

The accident investigators found that some recovered gusset plates were fractured, while others were not damaged. If the bridge had been properly constructed, none of the plates should have failed. But inspection showed that some of the plates failed very early in the collapse.

After evaluating the evidence, the accident investigators formulated the hypothesis that the gusset plates failed, which lead to the bridge collapse. Now investigators had to test this hypothesis.

Hypothesis:
~~1. The bridge failed because it was overloaded.~~
2. The bridge collapsed because the gusset plates failed.
Prediction:
If a gusset plate is not properly designed, then an overload on a bridge will cause a gusset plate to fail.

Test the Hypothesis:
- Compare the load on the bridge when it collapsed with the load limits of the bridge at each of the main gusset plates.
- Determine the demand-to-capacity ratios for the main gusset plates.
- Calculate the appropriate thicknesses of the U10 gusset plates.

Independent Variables: actual load on bridge and load bridge was designed to handle

Dependent Variable: demand-to-capacity ratio

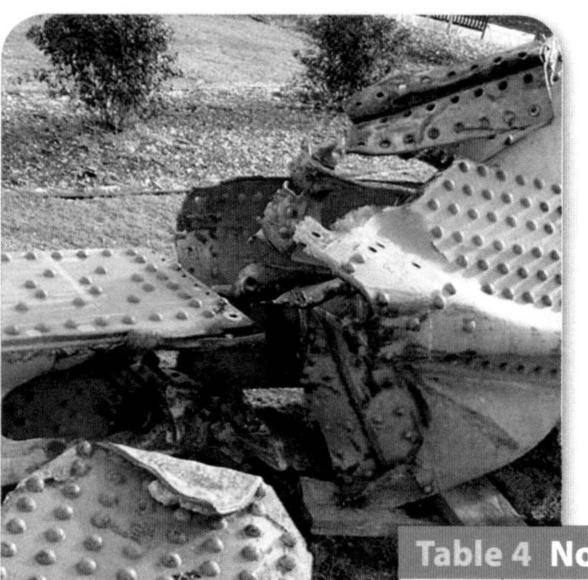

Figure 14 The steel plates, or gusset plates, at the U10 node were too thin for the loads the bridge carried.

Testing the Hypothesis

The investigators knew the load limits of the bridge. To calculate the load on the bridge when it collapsed, they estimated the combined weight of the bridge and the traffic on the bridge. The investigators divided the load on the bridge when it collapsed by the load limits of the bridge to find the demand-to-capacity ratio. The demand-to-capacity ratio provides a measure of a structure's safety.

Analyzing Results

As investigators calculated the demand-to-capacity ratios for each of the main gusset plates, they found that the ratios were particularly high for the U10 node. The U10 plate, shown in **Figure 14,** failed earliest in the bridge collapse. **Table 4** shows the demand-to-capacity ratios for a few of the gusset plates at some nodes. A value greater than 1 means the structure is unsafe. Notice how high the ratios are for the U10 gusset plate compared to the other plates.

Further calculations showed that the U10 plates were not thick enough to support the loads they were supposed to handle. They were about half the thickness they should have been.

Key Concept Check Why are evaluation and testing important in the design process?

Table 4 Node-Gusset Plate Analysis

Gusset Plate	Thickness (cm)	Demand-to-Capacity Ratios for the Upper-Node Gusset Plates					
		Horizontal loads			Vertical loads		
U8	3.5	0.05	0.03	0.07	0.31	0.46	0.20
U10	*1.3*	*1.81*	*1.54*	*1.83*	*1.70*	*1.46*	*1.69*
U12	2.5	0.11	0.11	0.10	0.71	0.37	1.15

Drawing Conclusions

Over the years, modifications to the I-35W bridge added more load to the bridge. On the day of the accident, traffic and the concentration of construction vehicles and materials added still more load. Investigators concluded that if the U10 gusset plates were properly designed, they would have supported the added load. When the investigators examined the original records for the bridge, they were unable to find any detailed gusset plate specifications. They could not determine whether undersized plates were used because of a mistaken calculation or some other error in the design process. The only thing that they could conclude with certainty was that undersized gusset plates could not reliably hold up the bridge.

The Federal Highway Administration and the National Transportation Safety Board published the results of their investigations. These published reports now provide scientists and engineers with valuable information they can use in future bridge designs. These reports are good examples of why it is important for scientists and engineers to publish their results and to share information.

Analyzing Results: The U10 gusset plates should have been twice as thick as they were to support the bridge.

Conclusions: The bridge failed because the gusset plates were not properly designed and they could not carry the load that they were supposed to carry.

Key Concept Check Give three examples of the scientific inquiry process that was used in this investigation.

Lesson 3 Review

Assessment Online Quiz

Use Vocabulary

1. **Distinguish** between qualitative data and quantitative data.
2. **Contrast** *variable, independent variable,* and *dependent variable.*

Understand Key Concepts 

3. Constants are necessary in a controlled experiment because, without constants, you would not know which variable affected the
   A. control group.
   B. experimental group.
   C. dependent variable.
   D. independent variable.
4. **Give an example** of a situation in your life in which you depend on adequate testing and evaluation in a product design to keep you safe.

Interpret Graphics

5. **Summarize** Copy and fill in the flow chart below and summarize the sequence of scientific inquiry steps that was used in one part of the case study.

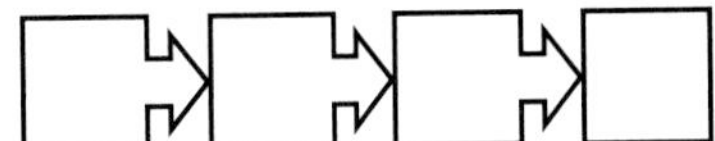

Critical Thinking

6. **Analyze** how the scientific inquiry process differs when engineers design a product, such as a bridge, and when they investigate design failure.
7. **Evaluate** why the gusset plates were such a critical piece in the bridge design.
8. **Recommend** ways that bridge designers and inspectors can prevent future bridge collapses.

1–2 class periods

Build and Test a Bridge

Materials

plastic straws

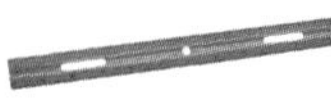

ruler

scissors

cotton string

cardboard

Also needed: notebook paper, books or other masses, balance (with a capacity of at least 2 kg)

Safety

In the Skill lab, you observed the relative strengths of two different geometric shapes. In the case study about the bridge collapse, you learned how scientists used scientific inquiry to determine the cause of the bridge collapse. In this investigation, you will combine geometric shapes to build model bridge supports. Then you will use scientific inquiry to determine the maximum load that your bridge will hold.

Ask a Question

What placement of supports produces the strongest bridge?

Make Observations

1. Read and complete a lab safety form.
2. Cut the straws into 24 6-cm segments.
3. Thread three straw segments onto a 1-m piece of string. Slide the segments toward one end of the string. Double knot the string to form a triangle. There should be very little string showing between the segments.
4. Thread the long end of the remaining string through two more straw segments. Double knot the string to one unattached corner to form another triangle. Cut off the remaining string, leaving at least a 1 cm after the knot. Use the string and one more straw segment to form a tetrahedron, as shown below.
5. Use the remaining string and straw segments to build three more tetrahedrons.
6. Set the four tetrahedrons on a piece of paper. They will serve as supports for your bridge deck, a 20-cm x 30-cm piece of cardboard.
7. With your teammates, decide where you will place the tetrahedrons on the paper to best support a load placed on the bridge deck.

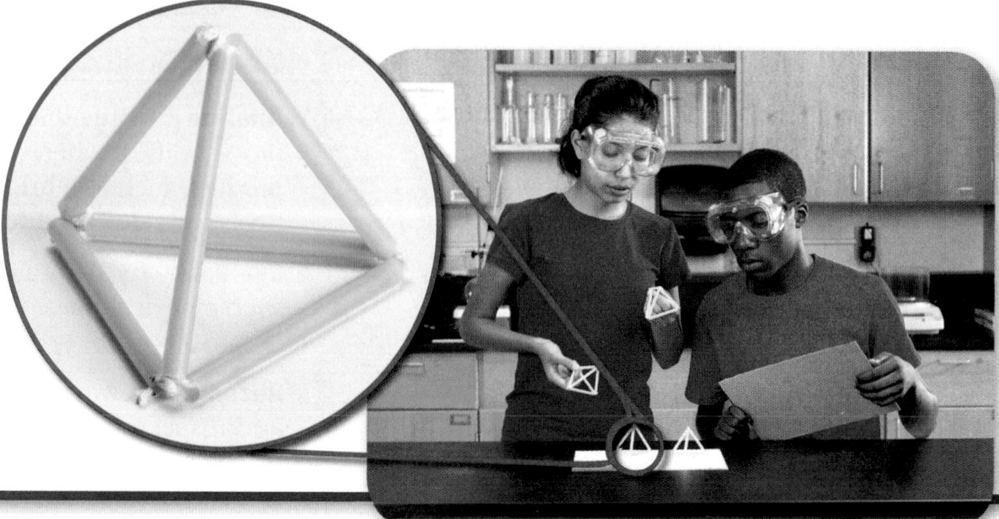

Form a Hypothesis

8. Form a hypothesis about where you will place your tetrahedrons and why that placement will support the most weight. Recall that a hypothesis is an explanation of an observation.

Test Your Hypothesis

9. Test your hypothesis by placing the tetrahedrons in your chosen locations on the paper. Lay the cardboard "bridge deck" over the top.
10. Use a balance to find the mass of a textbook. Record the mass in your Science Journal.
11. Gently place the textbook on the bridge deck. Continue to add massed objects until your bridge collapses. Record the total mass that collapsed the bridge.
12. Examine the deck and supports. Look for possible causes of bridge failure.

Analyze and Conclude

13. **Analyze** Was your hypothesis supported? How do you know?
14. **Compare and Contrast** Study the pictures of bridges in Lesson 3. How does the failure of your bridge compare to the failure of the I-35W bridge?
15. **The Big Idea** What steps of scientific inquiry did you use in this activity? What would you do next to figure out how to make a stronger bridge?

Communicate Your Results

Compare your results with those of several other teams. Discuss the placement of your supports and any other factors that may cause your bridge to fail.

Try building your supports with straw segments that are shorter (4 cm long) and longer (8 cm long). Test your bridges in the same way with each size of support.

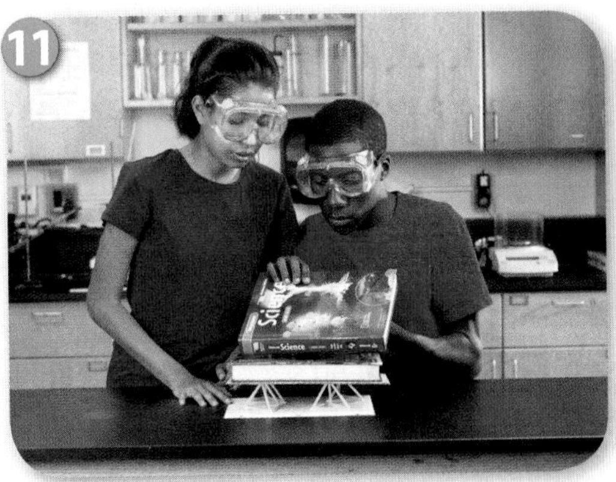

Lab Tips

- When building your tetrahedrons, make sure to double knot all connections and pull them tight. When you are finished, test each tetrahedron by pressing lightly on the top point.
- When adding the books to the bridge deck, place the books gently on top of the pile. Do not drop them.

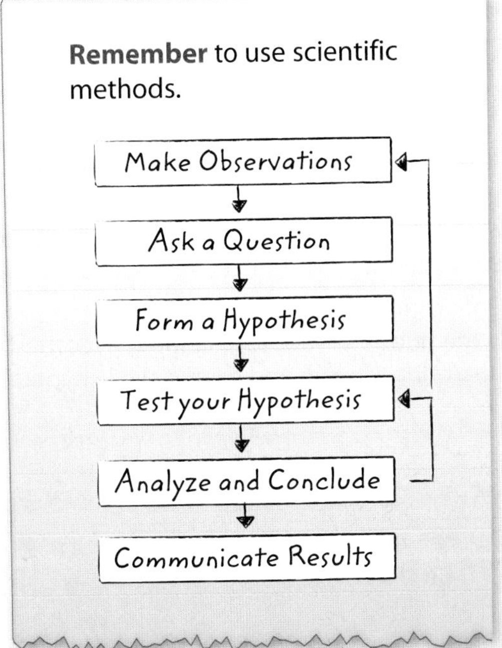

Study Guide and Review

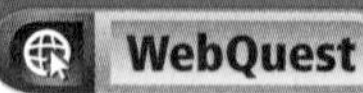

Scientific inquiry is a collection of methods that scientists use in different combinations to perform scientific investigations.

Key Concepts Summary	Vocabulary
Lesson 1: Scientific Inquiry • Some steps used during scientific inquiry are making **observations** and **inferences,** developing a **hypothesis,** analyzing results, and drawing conclusions. These steps, among others, can be performed in any order. • There are many results of scientific inquiry, and a few possible outcomes are the development of new materials and new technology, the discovery of new objects and events, and answers to basic questions. • **Critical thinking** is comparing what you already know about something to new information and deciding whether or not you agree with the new information.	**science** p. NOS 4 **observation** p. NOS 6 **inference** p. NOS 6 **hypothesis** p. NOS 6 **prediction** p. NOS 6 **scientific theory** p. NOS 8 **scientific law** p. NOS 8 **technology** p. NOS 9 **critical thinking** p. NOS 10
Lesson 2: Measurement and Scientific Tools • Scientists developed one universal system of units, the **International System of Units (SI),** to improve communication among scientists. • **Scientific notation** is a useful tool for writing large and small numbers in a shorter form. • Tools such as graduated cylinders and triple-beam balances make scientific investigation easier, more accurate, and repeatable.	**description** p. NOS 12 **explanation** p. NOS 12 **International System of Units (SI)** p. NOS 12 **scientific notation** p. NOS 15 **percent error** p. NOS 15
Lesson 3: Case Study—The Minneapolis Bridge Failure • Evaluation and testing are important in the design process for the safety of the consumer and to keep costs of building or manufacturing the product at a reasonable level. • Scientific inquiry was used throughout the process of determining why the bridge collapsed, including hypothesizing potential reasons for the bridge failure and testing those hypotheses.	**variable** p. NOS 21 **independent variable** p. NOS 21 **dependent variable** p. NOS 21 **constants** p. NOS 21 **qualitative data** p. NOS 21 **quantitative data** p. NOS 21 **experimental group** p. NOS 24 **control group** p. NOS 24

Use Vocabulary

1. The ________ contains the same factors as the experimental group, but the independent variable is not changed.
2. The expression of error as a percentage of the accepted value is ________.
3. The process of studying nature at all levels and the collection of information that is accumulated is ________.
4. The ________ are the factors in the experiment that stay the same.

Assessment

Online Test Practice

Understand Key Concepts

5 Which is NOT an SI base unit?

A. kilogram

B. liter

C. meter

D. second

6 While analyzing results from an investigation, a scientist calculates a very small number that he or she wants to make easier to use. Which does the scientist use to record the number?

A. explanation

B. inference

C. scientific notation

D. scientific theory

7 Which is NOT true of a scientific law?

A. It can be modified or rejected.

B. It states that an event will occur.

C. It explains why an event will occur.

D. It is based on repeated observations.

8 Which tool would a scientist use to find the mass of a small steel rod?

A. balance

B. computer

C. hot plate

D. thermometer

Critical Thinking

9 **Write** a brief description of the activity shown in the photo.

Writing in Science

10 **Write** a five-sentence paragraph that gives examples of how critical thinking, skepticism, and identifying facts and opinions can help you in your everyday life. Be sure to include a topic sentence and concluding sentence in your paragraph.

REVIEW THE BIG IDEA

11 What is scientific inquiry? Explain why it is a constantly changing process.

12 Which part of scientific inquiry does this photo demonstrate?

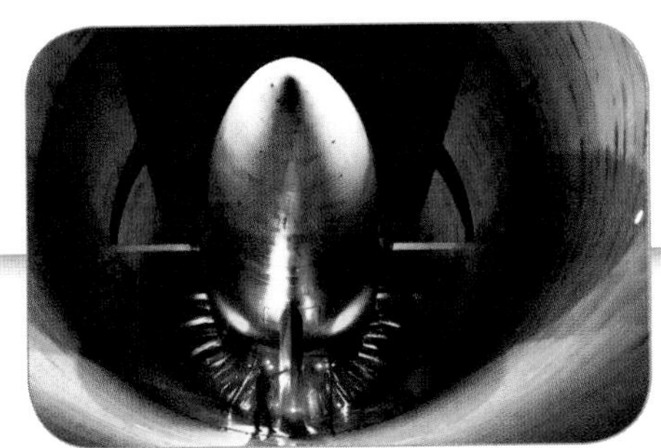

Math Skills

Review Math Practice

13 The accepted scientific value for the density of sucrose is 1.59 g/cm^3. You perform three trials to measure the density of sucrose, and your data is shown in the table below. Calculate the percent error for each trial.

Trial	Density	Percent Error
Trial 1	1.55 g/cm^3	
Trial 2	1.60 g/cm^3	
Trial 3	1.58 g/cm^3	

3500 B.C.
The oldest wheeled vehicle is depicted in Mesopotamia, near the Black Sea.

400 B.C.
The Greeks invent the stone-hurling catapult.

1698
English military engineer Thomas Savery invents the first crude steam engine while trying to solve the problem of pumping water out of coal mines.

1760–1850
The Industrial Revolution results in massive advances in technology and social structure in England.

1769
The first vehicle to move under its own power is designed by Nicholas Joseph Cugnot and constructed by M. Breszin. A second replica is built that weighs 3,629 kg and has a top speed of 3.2 km per hour.

1794
Eli Whitney receives a patent for the mechanical cotton gin.

1800

1900

1817
Baron von Drais invents a machine to help him quickly wander the grounds of his estate. The machine is made of two wheels on a frame with a seat and a pair of pedals. This machine is the beginning design of the modern bicycle.

1903
Wilbur and Orville Wright build their airplane, called the Flyer, and take the first successful, powered, piloted flight.

1976
The first computer for home use is invented by college dropouts Steve Wozniak and Steve Jobs, who go on to found Apple Computer, Inc.

? Inquiry

Visit ConnectED for this unit's STEM activity.

Unit 1

Nature of SCIENCE

Models

Have you ridden on an amusement park roller coaster such as the one in **Figure 1**? As you were going down the steepest hill or hanging upside down in a loop, did you think to yourself, "I hope I don't fly off this thing"? Before construction begins on a roller coaster, engineers build different models of the thrill ride to ensure proper construction and safety. A **model** is a representation of an object, an idea, or a system that is similar to the physical object or idea being studied.

Figure 1 Engineers use various models to design roller coasters.

Using Models in Physical Science

Models are used to study things that are too big or too small, happen too quickly or too slowly, or are too dangerous or too expensive to study directly. Different types of models serve different purposes. Roller-coaster engineers might build a physical model of their idea for a new, daring coaster. Using mathematical and computer models, the engineers can calculate the measurements of hills, angles, and loops to ensure a safe ride. Finally, the engineers might create another model called a blueprint, or drawing, that details the construction of the ride. Studying the various models allows engineers to predict how the actual roller coaster will behave when it travels through a loop or down a giant hill.

Types of Models

Physical Model

A physical model is a model that you can see and touch. It shows how parts relate to one another, how something is built, or how complex objects work. Physical models often are built to scale. A limitation of a physical model is that it might not reflect the physical behavior of the full-size object. For example, this model will not accurately show how wind will affect the ride.

Mathematical Model

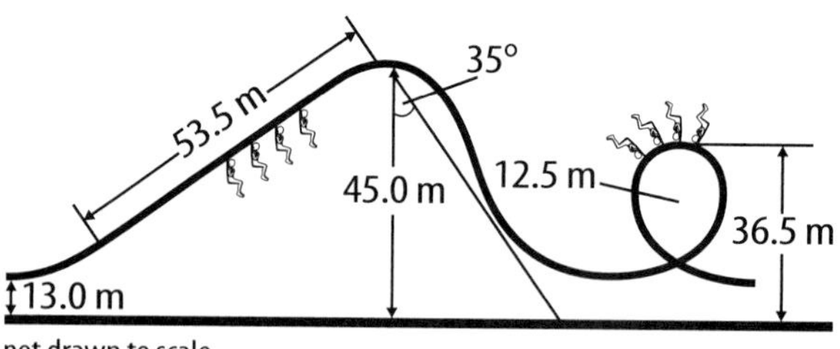

A mathematical model uses numerical data and equations to model an event or idea. Mathematical models often include input data, constants, and output data. When designing a thrill ride, engineers use mathematical models to calculate the heights, the angles of loops and turns, and the forces that affect the ride. One limitation of a mathematical model is that you cannot use it to model how different parts are assembled.

Making Models

An important factor in making a model is determining its purpose. You might need a model that physically represents an object. Or, you might need a model that includes only important elements of an object or a process. When you build a model, first determine the function of the model. What variables need to change? What materials should you use? What do you need to communicate to others? **Figure 2** shows two models of a glucose molecule, each with a different purpose.

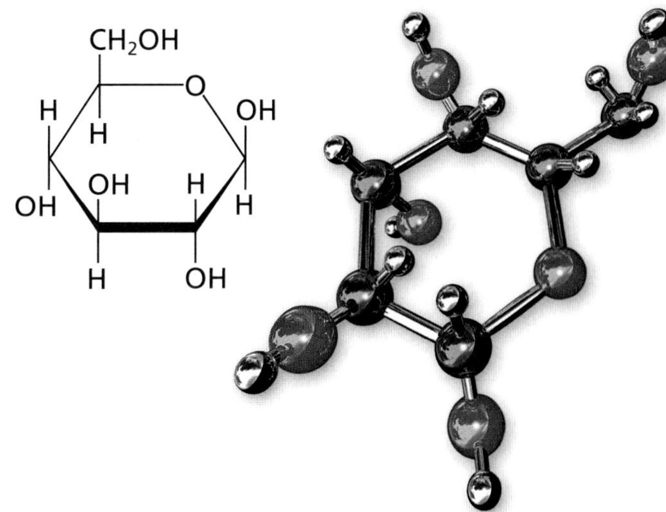

Figure 2 The model on the left is used to represent how the atoms in a glucose molecule bond together. The model on the right is a 3-D representation of the molecule, which shows how atoms might interact.

Limitations of Models

It is impossible to include all the details about an object or an idea into one model. All models have limitations. When using models to design a structure, an engineer must be aware of the information each model does and does not provide. For example, a blueprint of a roller coaster does not show the maximum weight that a car can support. However, a mathematical model would include this information. Scientists and engineers consider the purpose and the limitations of the model they use to ensure they draw accurate conclusions from models.

Computer Simulation

A computer simulation is a model that combines large amounts of data and mathematical models with computer graphic and animation programs. Simulations can contain thousands of complex mathematical models. When roller coaster engineers change variables in mathematical models, they use computer simulation to view the effects of the change.

Inquiry MiniLab — 30 minutes

Can you model a roller coaster?

You are an engineer with an awesome idea for a new roller coaster—the car on your roller coaster makes a jump and then lands back on the track. You model your idea to show it to managers at a theme park in hopes that you can build it.

1. Read and complete a lab safety form.
2. Create a blueprint of your roller coaster. Include a scale and measurements.
3. Follow your blueprint to build a scaled physical model of your roller coaster. Use **foam hose insulation, tape,** and other **craft supplies.**
4. Use a **marble** as a model for a roller-coaster car. Test your model. Record your observations in your Science Journal.

Analyze and Conclude

1. **Compare** your blueprint and physical model.
2. **Evaluate** After you test your physical model, list the design changes would make to your blueprint.
3. **Identify** What are the limitations of each of your models?

Chapter 1

Describing Motion

What are some ways to describe motion?

How is their motion changing?

Have you ever seen a group of planes zoom through the sky at an air show? When one plane speeds up, all the planes speed up. When one plane turns, all the other planes turn in the same direction.

- What might happen if all the planes did not move in the same way?
- How could you describe the positions of the planes in the photo?
- What are some ways you could describe the motion of the planes?

Get Ready to Read

What do you think?

Before you read, decide if you agree or disagree with each of these statements. As you read this chapter, see if you change your mind about any of the statements.

1. Displacement is the distance an object moves along a path.
2. The description of an object's position depends on the reference point.
3. Constant speed is the same thing as average speed.
4. Velocity is another name for speed.
5. You can calculate average acceleration by dividing the change in velocity by the change in distance.
6. An object accelerates when either its speed or its direction changes.

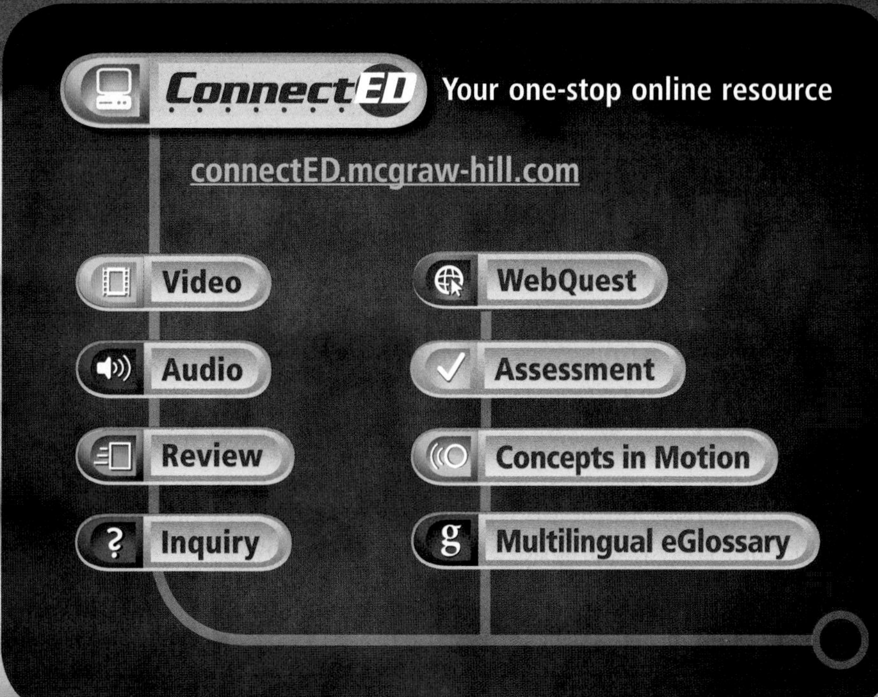

Lesson 1

Reading Guide

Key Concepts

ESSENTIAL QUESTIONS

- How does the description of an object's position depend on a reference point?
- How can you describe the position of an object in two dimensions?
- What is the difference between distance and displacement?

Vocabulary

reference point p. 9

position p. 9

motion p. 13

displacement p. 13

g Multilingual eGlossary

Position and Motion

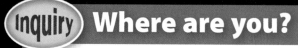

Inquiry Where are you?

A short time ago, people on this ship probably saw only open ocean. They knew where they were only by looking at the instruments on the ship. But the situation has changed. How can the lighthouse help the ship's crew guide the ship safely to shore?

Launch Lab

10 minutes

How do you get there from here?

How would you give instructions to a friend who was trying to walk from one place to another in your classroom?

1. Read and complete a lab safety form.
2. Place a sheet of **paper** labeled *North, East, South,* and *West* on the floor.
3. Walk from the paper to one of the three locations your teacher has labeled in the classroom. Have a partner record the number of steps and the directions of movement in his or her Science Journal.
4. Using these measurements, write instructions other students could follow to move from the paper to the location.
5. Repeat steps 3 and 4 for the other locations.

Think About This

1. How did your instructions to each location compare to those written by other groups?
2. **Key Concept** How did the description of your movement depend on the point at which you started?

Describing Position

How would you describe where you are right now? You might say you are sitting one meter to the left of your friend. Perhaps you would explain that you are at home, which is two houses north of your school. You might instead say that your house is ten blocks east of the center of town, or even 150 million kilometers from the Sun.

What do all these descriptions have in common? Each description states your location relative to a certain point. *A* **reference point** *is the starting point you choose to describe the location, or position, of an object.* The reference points in the first paragraph are your friend, your school, the center of town, and the Sun.

Science Use v. Common Use
relative
Science Use compared (to)
Common Use a member of your family

Each description of your location also includes your distance and direction from the reference point. Describing your location in this way defines your position. *A* **position** *is an object's distance and direction from a reference point.* A complete description of your position includes a distance, a direction, and a reference point.

Reading Check What are two ways you could describe your position right now?

Describing Position

Figure 1 The arrows indicate the distances and directions from different reference points.

Visual Check How do you know which reference point is farther from the table?

Using a Reference Point to Describe Position

Why do you need a reference point to describe position? Suppose you are planning a family picnic. You want your cousin to arrive at the park early to save your favorite picnic table. The park is shown in **Figure 1.** How would you describe the position of your favorite table to your cousin? First, choose a reference point that a person can easily find. In this park, the statue is a good choice. Next, describe the direction that the table is from the reference point—toward the slide. Finally, say how far the table is from the statue—about 10 m. You would tell your cousin that the position of the table is about 10 m from the statue, toward the slide.

FOLDABLES®

Fold a sheet of paper to make a half book. Use it to organize your notes about how position and motion are related.

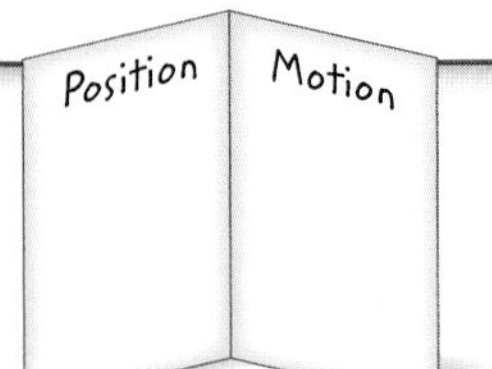

 **Reading Check** How could you describe the position of a different table using the statue as a reference point?

Changing the Reference Point

The description of an object's position depends on the reference point. Suppose you choose the drinking fountain in **Figure 1** as the reference point instead of the statue. You could say that the direction of the table is toward the dead tree. Now the distance is measured from the drinking fountain to the table. You could tell your cousin that the table is about 12 m from the drinking fountain, toward the dead tree. The description of the table's position changed because the reference point is different. Its actual position did not change at all.

 **Key Concept Check** How does the description of an object's position depend on a reference point?

Figure 2 If east is the reference direction, then the museum is in the positive direction from the bus stop. The library is in the negative direction.

The Reference Direction

When you describe an object's position, you compare its location to a reference direction. In **Figure 1,** the reference direction for the first reference point is toward the slide. Sometimes the words *positive* and *negative* are used to describe direction. The reference direction is the positive (+) direction. The opposite direction is the negative (–) direction. Suppose you specify east as the reference direction in **Figure 2.** You could say the museum's entrance is +80 m from the bus stop. The library's entrance is –40 m from the bus stop. To a friend, you would probably just say the museum is two buildings east of the bus stop and the library is the building west of the bus stop. Sometimes, however, using the words *positive* and *negative* to describe direction is useful for explaining changes in an object's position.

ACADEMIC VOCABULARY

specify
(verb) to indicate or identify

Inquiry MiniLab

10 minutes

Why is a reference point useful?

To find an object's position, you need to know its distance and direction from a reference point.

1. Read and complete a lab safety form.
2. Put a **sticky note** at the 50-cm mark of a **meterstick.** This is your reference point.
3. Place a **small object** at the 40-cm mark. It is 10 cm in the negative direction from the reference point.
4. Copy the table in your Science Journal. Continue moving the object and recording its distance, its reference direction, and its position to complete the table.

Position of Object

Distance (cm)	Reference Direction	Position (cm)
10 cm	negative	40 cm
40 cm	positive	
15 cm	positive	
	positive	75 cm
		30 cm

Analyze and Conclude

1. **Recognize Cause and Effect** How would the data in the table change if the positions were the same but the reference point was at the 40-cm mark?
2. **Key Concept** Why is a reference point useful in describing positions of an object?

Describing Position in Two Dimensions

Figure 3 You need two reference directions to describe the position of a building in the city.

Visual Check If the library is the reference point, how would you describe the position of your house in two dimensions?

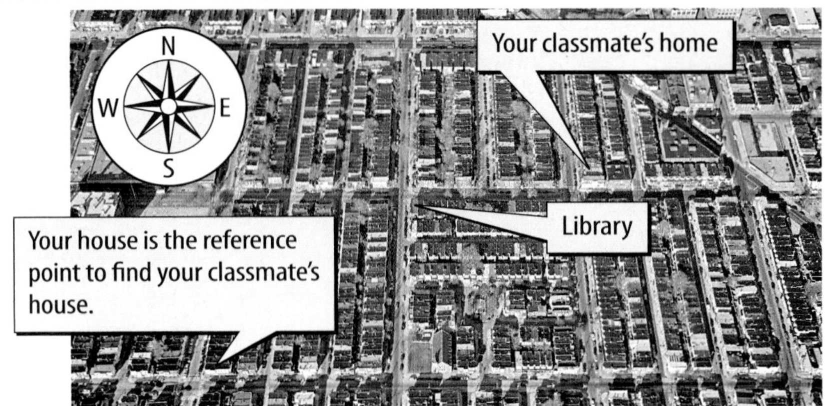

REVIEW VOCABULARY

dimension
distance or length measured in one direction

Describing Position in Two Dimensions

You were able to describe the position of the picnic table in the park in one **dimension.** Your cousin had to walk in only one direction to reach the table. But sometimes you need to describe an object's position using more than one reference direction. The city shown in **Figure 3** is an example. To describe the position of a house in a city might require two reference directions. When you describe a position using two directions, you are using two dimensions.

Reference Directions in Two Dimensions

To describe a position on the map in **Figure 3,** you might choose north and east or south and west as reference directions. Sometimes north, south, east, and west are not the most useful reference directions. If you are playing checkers and want to describe the position of a certain checker, you might use "right" and "forward" as reference directions. If you are describing the position of a certain window on a skyscraper, you might choose "left" and "up" as reference directions.

Reading Check What are two other reference directions you might use to describe the position of a building in a city?

Locating a Position in Two Dimensions

Finding a position in two dimensions is similar to finding a position in one dimension. First, choose a reference point. To locate your classmate's home on the map in **Figure 3,** you could use your home as a reference point. Next, specify reference directions—north and east. Then, determine the distance along each reference direction. In the figure, your classmate's house is one block north and two blocks east of your house.

Key Concept Check How can you describe the position of an object in two dimensions?

Describing Changes in Position

Sometimes you need to describe how an object's position changes. You can tell that the boat in **Figure 4** moved because its position changed relative, or compared, to the buoy. **Motion** *is the process of changing position.*

Word Origin

motion
from Latin *motere;* means "to move"

Motion Relative to a Reference Point

Is the man in the boat in **Figure 4** in motion? Suppose the fishing pole is the reference point. Because the positions of the man and the pole do not change relative to each other, the man does not move relative to the pole. Now suppose the buoy is the reference point. Because the man's distance from the buoy changes, he is in motion relative to the buoy.

◀ **Figure 4** The man in the boat is not in motion compared to his fishing pole. He is in motion compared to the buoy.

Distance and Displacement

Suppose a baseball player runs the bases, as shown in **Figure 5.** Distance is the length of the path the player runs, as shown by the red arrows. **Displacement** *is the difference between the initial (first) position and the final position of an object.* It is shown in the figure by the blue arrows. Notice that distance and displacement are equal only if the motion is in one direction.

 **Key Concept Check** What is the difference between distance and displacement?

Figure 5 Distance depends on the path taken. Displacement depends only on the initial and final positions. ▼

Distance and Displacement

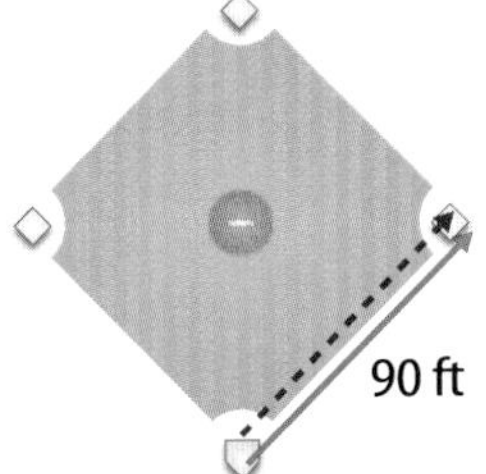

When a player runs to first base, the distance is 90 ft and the displacement is 90 ft toward first base.

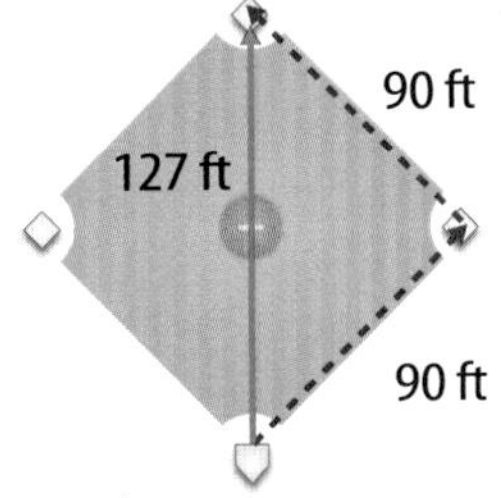

When a player runs to second base, the distance is 180 ft, but the displacement is 127 ft toward second base.

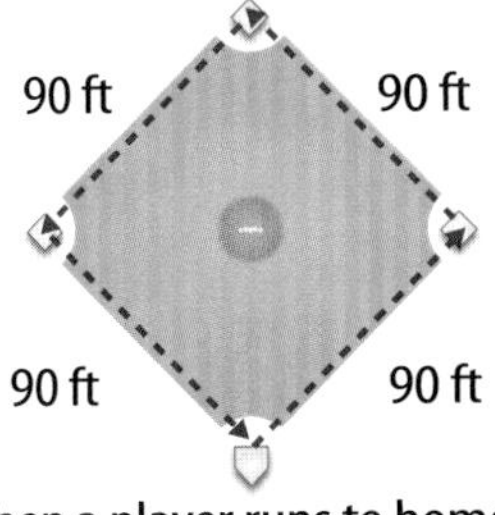

When a player runs to home base, the distance is 360 ft, but the displacement is 0 ft.

Lesson 1 Review

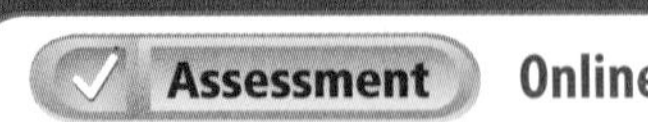

Visual Summary

A reference point, a reference direction, and distance are needed to describe the position of an object.

An object is in motion if its position changes relative to a reference point.

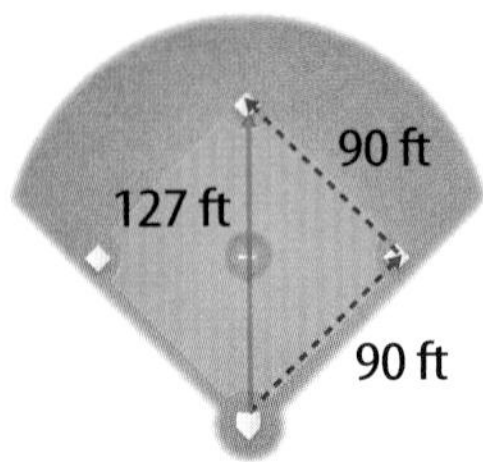

The distance an object moves and the object's displacement are not always the same.

FOLDABLES®

Use your lesson Foldable to review the lesson. Save your Foldable for the project at the end of the chapter.

What do you think NOW?

You first read the statements below at the beginning of the chapter.

1. Displacement is the distance an object moves along a path.
2. The description of an object's position depends on the reference point.

Did you change your mind about whether you agree or disagree with the statements? Rewrite any false statements to make them true.

Use Vocabulary

1. **Define** *motion* in your own words.
2. The difference between the initial position and the final position of an object is its ________.

Understand Key Concepts 

3. **Explain** why a description of position depends on a reference point.
4. To describe a position in more than one dimension, you must use more than one
   A. displacement.
   B. reference direction.
   C. reference point.
   D. type of motion.
5. **Apply** If you walk 2 km from your house to a store and then back home, what is your displacement?

Interpret Graphics

6. **Interpret** Using 12 as the reference point, how you can tell that the hands of the clock on the right have moved from their previous position, shown on the left?

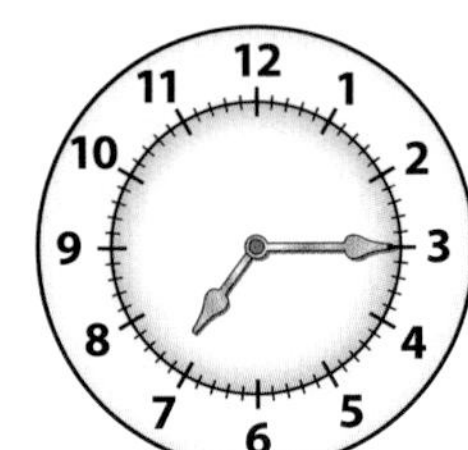

7. **Summarize** Copy and fill in the graphic organizer below to identify the three things that must be included in the description of position.

Critical Thinking

8. **Compare** Relative to some reference points, your nose is in motion when you run. Relative to others, it is not in motion. Give one example of each.

GPS to the Rescue!

How Technology Helps Bring Home Family Pets

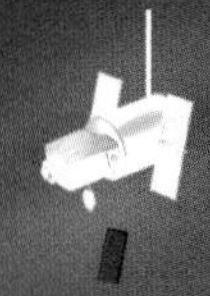

You've seen the signs tacked to streetlights and telephone poles: *LOST! Golden retriever. Reward. Please Call!* Losing a pet can be heartbreaking. Fortunately, there's an alternative to posting fliers—a pet collar with a Global Positioning System (GPS) chip that helps locate the pet. Here is how GPS can help you track or locate your pet:

Cell phone tracking display ▶

Cell phone tower

GPS Collar

1 GPS is a network of at least 24 satellites in orbit around Earth. Each satellite circles Earth twice a day and sends information to ground receivers.

2 GPS satellites act as reference points. Ground-based GPS receivers compare the time a signal was transmitted by a satellite to the time it is received on Earth. The difference indicates the satellite's distance. Signals from as many as four satellites are used to pinpoint a user's exact position.

3 GPS uses computer technology to calculate location, speed, direction, and time. The same GPS technology used to locate or guide airplanes, cars, and campers can help find a lost pet anywhere on Earth!

4 A GPS pet collar works much the same as any other GPS receiver. Once it is activated, the collar can transmit a message to a Web site or to the owner's cell phone.

It's Your Turn

DESIGN GPS technology has revolutionized the way people track and locate almost everything. Can you think of a new application for GPS technology? Write an advertisement or a TV commercial for a new idea that puts GPS technology to work!

Lesson 2

Reading Guide

Key Concepts
ESSENTIAL QUESTIONS

- What is speed?
- How can you use a distance-time graph to calculate average speed?
- What are ways velocity can change?

Vocabulary

speed p. 17
constant speed p. 18
instantaneous speed p. 18
average speed p. 19
velocity p. 23

Multilingual eGlossary

Speed and Velocity

Inquiry How Fast?

When you hear the word *cheetah,* you might think of how fast a cheetah can run. As the fastest land animal on Earth, it can reach a speed of 30 m/s for a short period of time. Other than how fast it runs, how might you describe the motion of a cheetah?

Launch Lab — 10 minutes

How can motion change?

Have you ever used a tube slide at a playground or at a water park? You can build a marble tube slide from foam tubes. You can then use the slide to observe how the motion of a marble changes as it rolls down the slide.

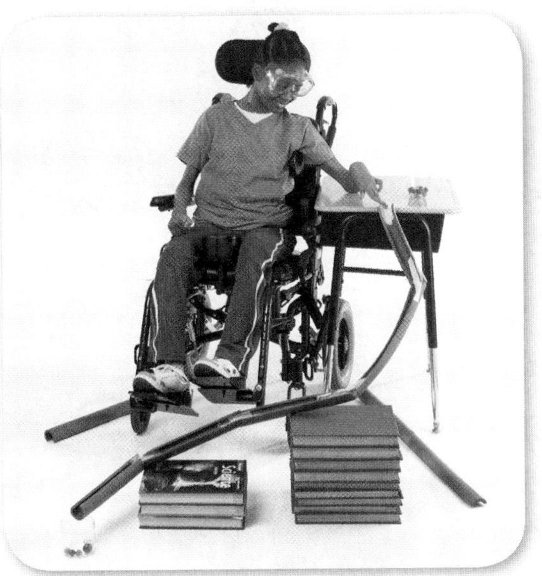

1. Read and complete a lab safety form.
2. Join **foam tubes** into a tube slide, using **masking tape** to connect the tubes. Use a desk and other objects to support the slide so that it changes direction and slopes toward the floor.
3. Drop a **marble** into the top of the slide. Observe the changes in its motion as it rolls through the tubes. Have the marble roll into a **container** at the bottom of the slide.

Think About This

1. In what ways did the motion of the marble change as it moved down the slide?
2. **Key Concept** At what parts of the tube slide did the marble move fastest? At what parts did it move slowest?

What is speed?

How fast do you walk when you are hungry and there is good food on the table? How fast do you move when you have a chore to do? Sometimes you move quickly, and sometimes you move slowly. One way you can describe how fast you move is to determine your speed. **Speed** *is a measure of the distance an object travels per unit of time.*

Key Concept Check What is speed?

Units of Speed

You can calculate speed by dividing the distance traveled by the time it takes to go that distance. The units of speed are units of distance divided by units of time. The SI unit for speed is meters per second (m/s). Other units are shown in **Table 1.** What units of distance and time are used in each example?

Table 1 Typical Speeds

Airplane 245 m/s 882 km/h 548 mph	
Car on a Highway 27 m/s 97 km/h 60 mph	
Person Walking 1.3 m/s 4.7 km/h 2.9 mph	

Table 1 Different units of distance and time can be used to determine units of speed.

Constant and Changing Speed

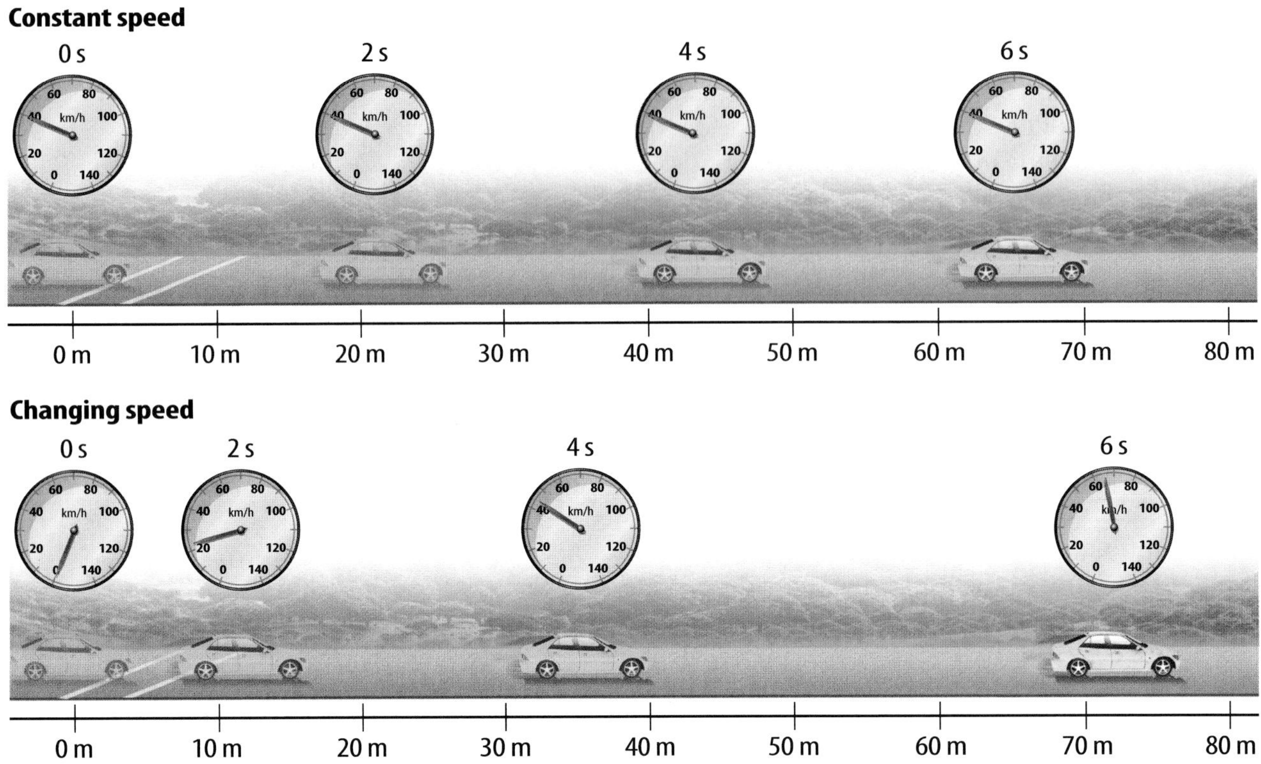

Figure 6 When the car moves with constant speed, it moves the same distance each period of time. When the car's speed changes, it moves a different distance each period of time.

Visual Check What is the bottom car's instantaneous speed at 6 s?

Constant Speed

What happens to your speed when you ride in a car? Sometimes you ride at a steady speed. If you move away from a stop sign, your speed increases. You slow down when you pull into a parking space.

Think about a time that a car's speed does not change. As the car at the top of **Figure 6** travels along the road, the speedometer above each position shows that the car is moving at the same speed at each location and time. Each second, the car moves 11 m. Because it moves the same distance each second, its speed is not changing. **Constant speed** *is the rate of change of position in which the same distance is traveled each second.* The car is moving at a constant speed of 11 m/s.

Changing Speed

How is the motion of the car at the bottom of **Figure 6** different from the motion of the car at the top? Between 0 s and 2 s, the car at the bottom travels about 10 m. Between 4 s and 6 s, however, the car travels more than 20 m. Because the car travels a different distance each second, its speed is changing.

If the speed of an object is not constant, you might want to know its speed at a certain moment. **Instantaneous speed** *is speed at a specific instant in time.* You can see a car's instantaneous speed on its speedometer.

Reading Check How would the distance the car travels each second change if it were slowing down?

Average Speed

Describing an object's speed is easy if the speed is constant. But how can you describe the speed of an object when it is speeding up or slowing down? One way is to calculate its average speed. **Average speed** *is the total distance traveled divided by the total time taken to travel that distance.* You can calculate average speed using the equation below.

Average Speed Equation

$$\textbf{average speed} \text{ (in m/s)} = \frac{\textbf{total distance} \text{ (in m)}}{\textbf{total time} \text{ (in s)}}$$

$$\bar{v} = \frac{d}{t}$$

The symbol $\bar{v}$ represents the term "average velocity." You will read more about velocity, and how it relates to speed, later in this lesson. However, at this point, $\bar{v}$ is simply used as the symbol for "average speed." The SI unit for speed, meters per second (m/s), is used in the above equation. You could instead use other units of distance and time in the average speed equation, such as kilometers and hours.

Math Skills — Average Speed Equation

Solve for Average Speed Melissa shot a model rocket 360 m into the air. It took the rocket 4 s to fly that far. What was the average speed of the rocket?

1. **This is what you know:** distance: $d = 360$ m; time: $t = 4$ s
2. **This is what you need to find:** average speed: $\bar{v}$
3. **Use this formula:** $\bar{v} = \frac{d}{t}$
4. **Substitute:** the values for *d* and *t* into the formula and divide. $\bar{v} = \frac{360 \text{ m}}{4 \text{ s}} = \textbf{90 m/s}$

Answer: The average speed was **90 m/s**.

Review
- Math Practice
- Personal Tutor

Practice

1. It takes Ahmed 50 s on his bicycle to reach his friend's house 250 m away. What is his average speed?
2. A truck driver makes a trip that covers 2,380 km in 28 hours. What is the driver's average speed?

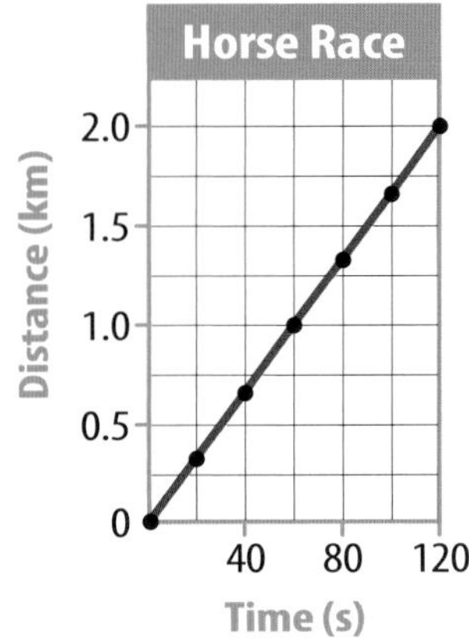

Figure 7 According to this distance-time graph, it took 2 min for the horse to run 2 km.

 **Visual Check** How would this distance-time graph be different if the horse's speed changed over time?

Distance-Time Graphs

The Kentucky Derby is often described as the most exciting two minutes in sports. The thoroughbred horses in this race run for a distance of 2 km. The speed of the horses changes many times during a race, but **Figure 7** describes what a horse's motion might be if its speed did not change. The graph shows the distance a horse might travel when distance measurements are made every 20 seconds. Follow the height of the line from the left side of the graph to the right side. You can see how the distance the horse ran changed over time.

Graphs like the one in **Figure 7** can show how one measurement compares to another. When you study motion, two measurements frequently compared to each other are distance and time. The graphs that show these comparisons are called distance-time graphs. Notice that the change in the distance the horse ran around the track is the same each second on the graph. This means the horse was moving with a constant speed. Constant speed is shown as a straight line on a distance-time graph.

Reading Check How is constant speed shown on a distance-time graph?

Inquiry MiniLab

15 minutes

How can you graph motion?

You can represent motion with a distance-time graph.

1. Read and complete a lab safety form.
2. Use **masking tape** to mark a starting point on the floor.
3. As you cross the starting point, start a **stopwatch.** Stop walking after 2 s. Measure the distance with a **meterstick.** Record the time and distance in your Science Journal.
4. Repeat step 3 by walking at about the same speed for 4 s and then for 6 s.
5. Use the graph in **Figure 7** as an example to create a distance-time graph of your data. The line on the graph should be as close to the points as possible.

Analyze and Conclude

1. **Predict** Based on the graph, how far would you probably walk at the same speed in 8 s?
2. **Key Concept** Look back at the average speed equation. Explain how you could use your graph to find your average walking speed.

Comparing Speeds on a Distance-Time Graph

You can use distance-time graphs to compare the motion of two different objects. **Figure 8** is a distance-time graph that compares the motion of two horses that ran the Kentucky Derby. The motion of horse A is shown by the blue line. The motion of horse B is shown by the orange line. Look at the far right side of the graph. When horse A reached the finish line, horse B was only 1.5 km from the starting point of the race.

Recall that average speed is distance traveled divided by time. Horse A traveled a greater distance than horse B in the same amount of time. Horse A had greater average speed. Compare how steep the lines are on the graph. The measure of steepness is the slope. The steeper the line, the greater the slope. The blue line is steeper than the orange line. Steeper lines on distance-time graphs indicate faster speeds.

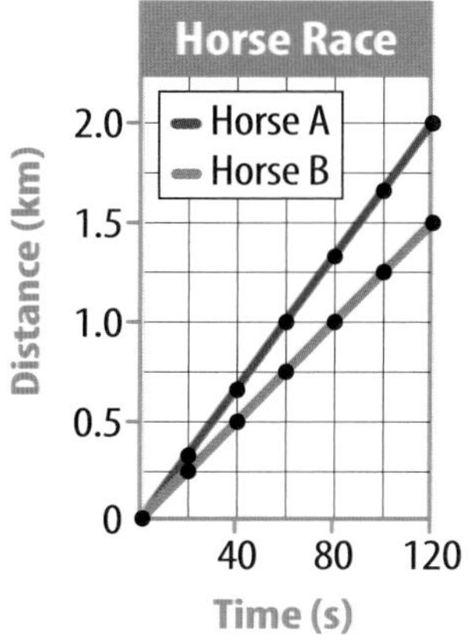

Figure 8 You can tell that horse A ran a faster race than horse B because the blue line is steeper than the orange line.

Using a Distance-Time Graph to Calculate Speed

You can use distance-time graphs to calculate the average speed of an object. The motion of a trail horse traveling at a constant speed is shown on the graph in **Figure 9.** The steps needed to calculate the average speed from a distance-time graph also are shown in the figure.

Key Concept Check How can you use a distance-time graph to calculate average speed?

Average Speed

Figure 9 The average speed of the horse from 60 s to 120 s can be calculated from this distance-time graph.

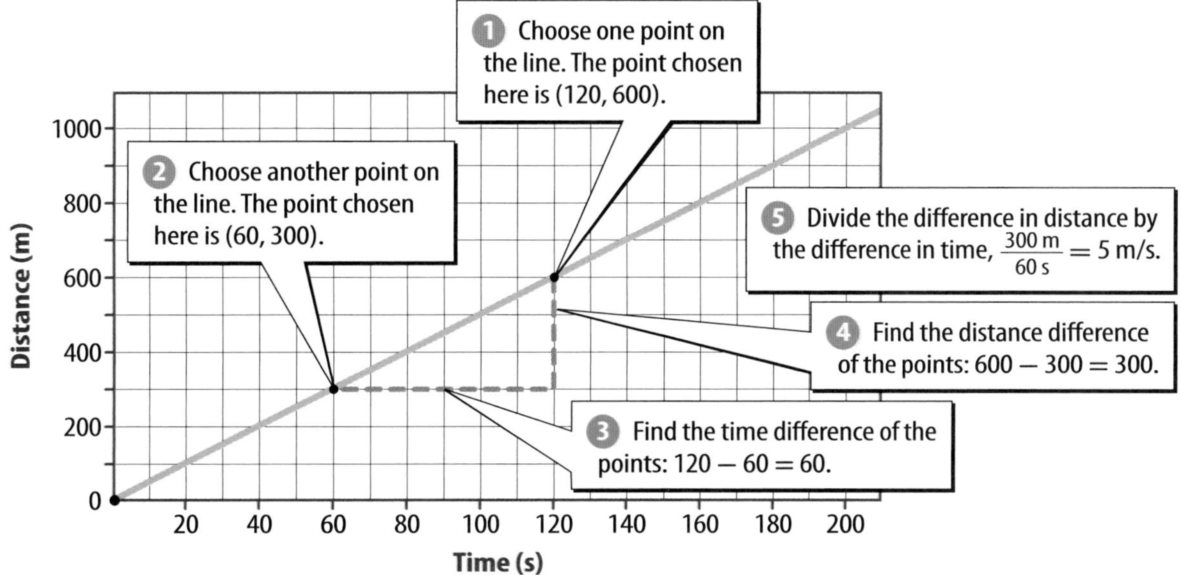

Visual Check How does the average speed of the horse from 60 s to 120 s compare to its average speed from 120 s to 180 s?

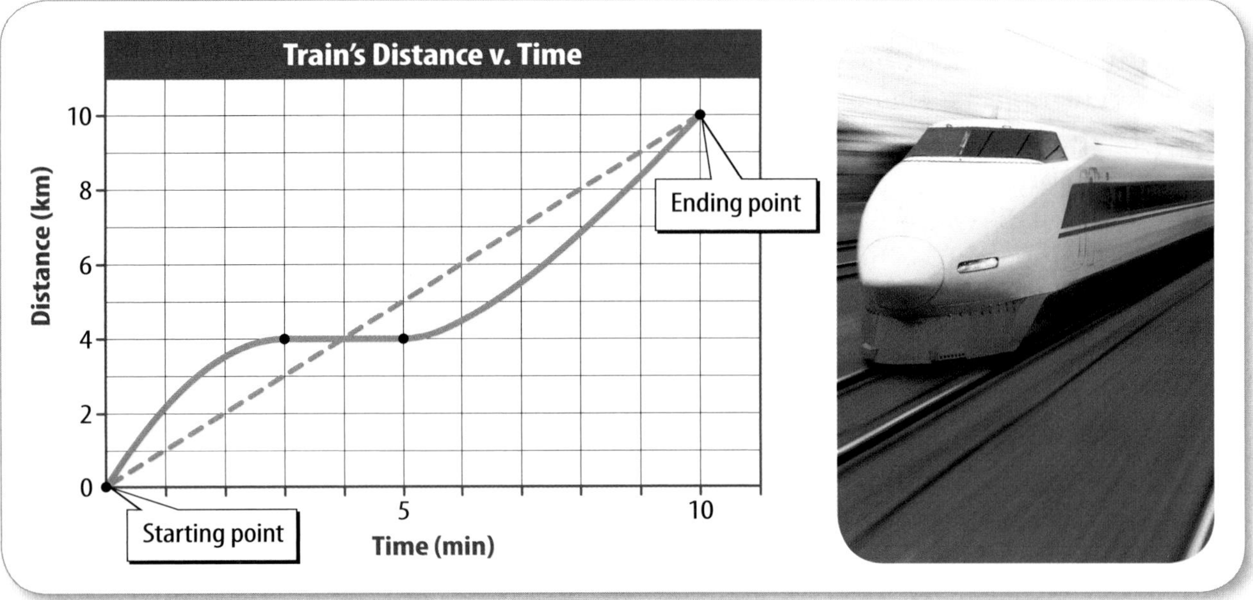

Figure 10 Even though the train's speed is not constant, you can calculate its average speed from a distance-time graph.

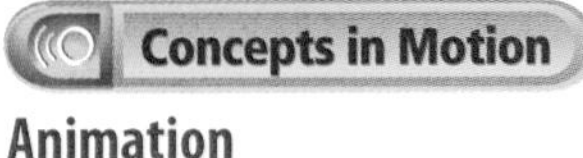

Animation

Distance-Time Graph and Changing Speed

So far, the distance-time graphs in this lesson have included straight lines. Distance-time graphs have straight lines only for objects that move at a constant speed. The graph in **Figure 10** for the motion of a train is different. Because the speed of the train changes instead of being constant, its motion on a distance-time graph is a curved line.

Slowing Down Notice how the shape of the line in **Figure 10** changes. Between 0 min and 3 min, its slope decreases. The downward curve indicates that the train slowed down.

Stopping What happened between 3 min and 5 min? The line during these times is horizontal. The train's distance from the starting point remains 4 km. A horizontal line on a distance-time graph indicates that there is no motion.

Speeding Up Between 5 min and 10 min, the slope of the line on the graph increases. The upward curve indicates that the train was speeding up.

Average Speed Even when the speed of an object changes, you can calculate its average speed from a distance-time graph. First, choose a starting point and an ending point. Next, determine the change in distance and the change in time between these two points. Finally, substitute these values into the average speed equation. The slope of the dashed line in **Figure 10** represents the train's average speed between 0 minutes and 10 minutes.

Reading Check What is the average speed of the train for the trip shown in **Figure 10?**

Make a horizontal three-tab book and label it as shown. Use it to summarize how changes in speed are represented on a distance-time graph.

Velocity

Often, describing just the speed of a moving object does not completely describe its motion. If you describe the motion of a bouncing ball, for example, you would also describe the direction of the ball's movement. Both speed and direction are part of motion. **Velocity** *is the speed and the direction of a moving object.*

Representing Velocity

In Lesson 1, an arrow represented the displacement of an object from a reference point. The velocity of an object also can be represented by an arrow, as shown in **Figure 11.** The length of the arrow indicates the speed. A greater speed is shown by a longer arrow. The arrow points in the direction of the object's motion.

In **Figure 11,** both students are walking at 1.5 m/s. Because the speeds are equal, both arrows are the same length. But the girl is walking to the left and the boy is walking to the right. The arrows point in different directions. The students have different velocities because each student has a different direction of motion.

1.5 m/s to the left

1.5 m/s to the right

▲ **Figure 11** The students are walking with the same speed but different velocities.

WORD ORIGIN

velocity
from Latin *velocitas;* means "swiftness, speed"

Changes in Velocity

Look at the bouncing ball in **Figure 12.** Notice how from one position to the next, the arrows showing the velocity of the ball change direction and length. The changes in the arrows mean that the velocity is constantly changing. Velocity changes when the speed of an object changes, when the direction that the object moves changes, or when both the speed and the direction change. You will read about changes in velocity in Lesson 3.

 **Key Concept Check** How can velocity change?

Figure 12 The velocity of the ball changes continually because both the speed and the direction of the ball change as the ball bounces. ▼

Changing Speed and Direction

Visual Check Are there two positions of the bouncing ball in which the velocity is the same? Explain.

Lesson 2 Review

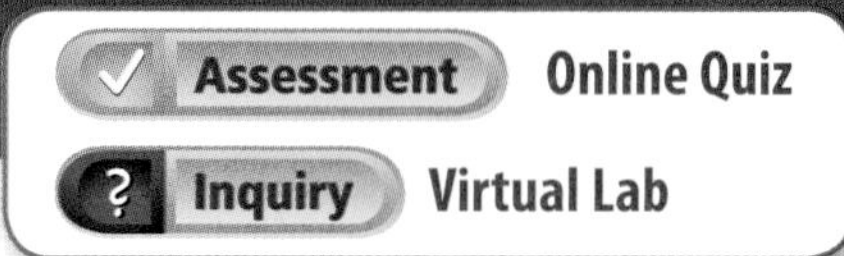

Visual Summary

Speed is a measure of the distance an object travels per unit of time. You can describe an object's constant speed, instantaneous speed, or average speed.

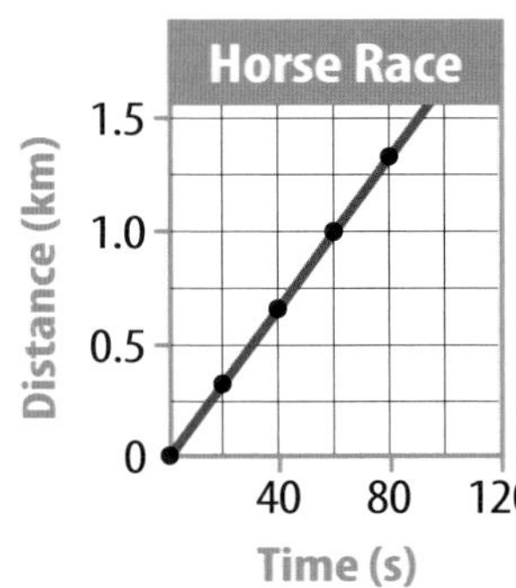

A distance-time graph shows the speed of an object.

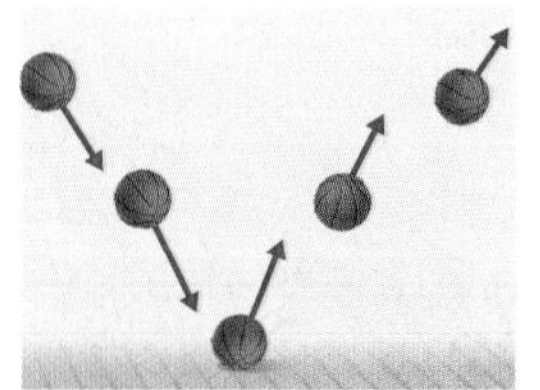

Velocity includes both the speed and the direction of motion.

FOLDABLES®

Use your lesson Foldable to review the lesson. Save your Foldable for the project at the end of the chapter.

What do you think 

You first read the statements below at the beginning of the chapter.

3. Constant speed is the same thing as average speed.

4. Velocity is another name for speed.

Did you change your mind about whether you agree or disagree with the statements? Rewrite any false statements to make them true.

Use Vocabulary

1. **Distinguish** between speed and velocity.
2. **Define** *constant speed* in your own words.

Understand Key Concepts 

3. **Recall** How can you calculate average speed from a distance-time graph?
4. **Analyze** Describe three ways a bicyclist can change velocity.
5. Which choice is a unit of speed?
   A. h/mi
   B. km/h
   C. m^2/s
   D. $N{\cdot}m^2$

Interpret Graphics

6. **Organize Information** Copy and fill in the graphic organizer below to show possible steps for making a distance-time graph.

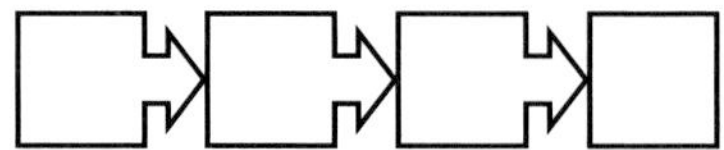

7. **Interpret** What does the shape of each line indicate about the object's speed?

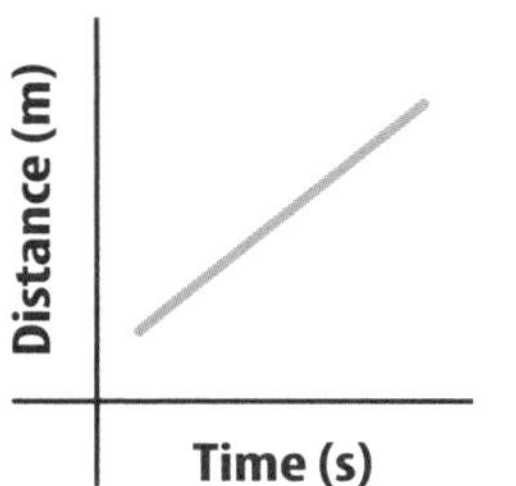

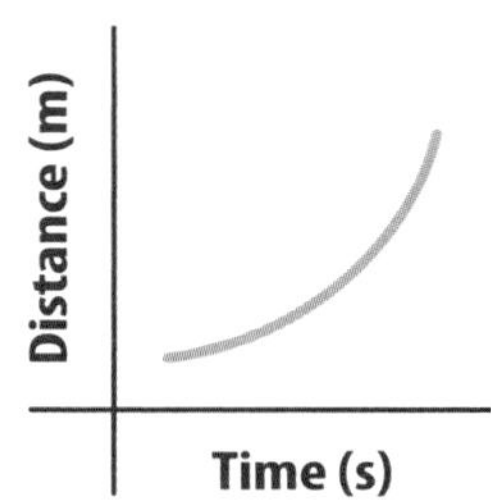

Critical Thinking

8. **Decide** Aaron leaves one city at noon. He has to be at another city 186 km away at 3:00 P.M. The speed limit the entire way is 65 km/h. Can he arrive at the second city on time? Explain.

Math Skills

Math Practice

9. A train traveled 350 km in 2.5 h. What was the average speed of the train?

What do you measure to calculate speed?

Materials

meterstick

stopwatch

wind-up toys (4)

calculator

graph paper

Safety

You turn on the television and see a news report. It shows trees that are bent almost to the ground because of a strong wind. Is it a hurricane or a tropical storm? The type of storm depends on the speed of the wind. A meteorologist must measure both distance and time before calculating the wind's speed.

Learn It

When you **measure,** you use a tool to find a quantity. To find the average speed of an object, you measure the distance it travels and the time it travels. You can then calculate speed using the average speed equation. In this lab, you use distance and time measurements to calculate speeds of moving toys.

Try It

1. Read and complete a lab safety form.
2. Copy the data table on this page into your Science Journal. Add more lines as you need them.
3. Choose appropriate starting and ending points on the floor. Use a meterstick to measure the distance between these points. Record this distance to the nearest centimeter.
4. Wind one toy. Measure in tenths of a second the time the toy takes to travel from start to finish. Record the time in the data table.
5. Repeat steps 3 and 4 for three more toys. Vary the distance from start to finish for each toy.

Apply It

6. **Calculate** the average speed of each toy. Record the speeds in your data table.
7. **Create** a bar graph of your data. Place the name of each toy on the *x*-axis and the average speed on the *y*-axis.
8. **Key Concept** Use the definition of *speed* to explain why the average speeds of the toys can be compared, even though the toys traveled different distances.

Toy Speeds

Toy	Distance (m)	Time (s)	Average Speed (m/s)

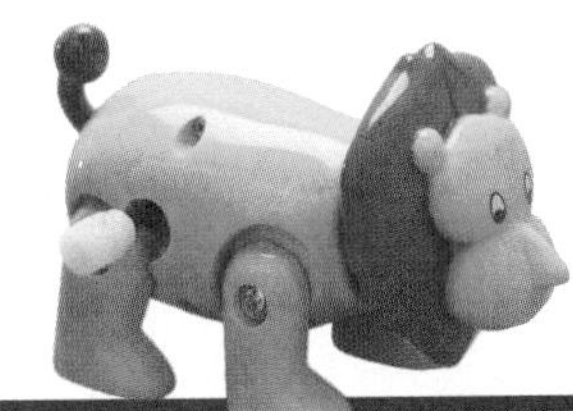

Lesson 3

Reading Guide

Key Concepts
ESSENTIAL QUESTIONS

- What are three ways an object can accelerate?
- What does a speed-time graph indicate about an object's motion?

Vocabulary

acceleration p. 27

Multilingual eGlossary

Video BrainPOP®

Acceleration

Inquiry Is velocity changing?

How does the velocity of this motorcycle racer change as he speeds along the track? As he enters a curve, he slows down, leans to the side, and changes direction. On a straightaway, he speeds up and moves in a straight line. How can the velocity of a moving object change?

Launch Lab

10 minutes

In what ways can velocity change?

As you walk, your motion changes in many ways. You probably slow down when the ground is uneven. You might speed up when you realize that you are late for dinner. You change direction many times. What would these changes in velocity look like on a distance-time graph?

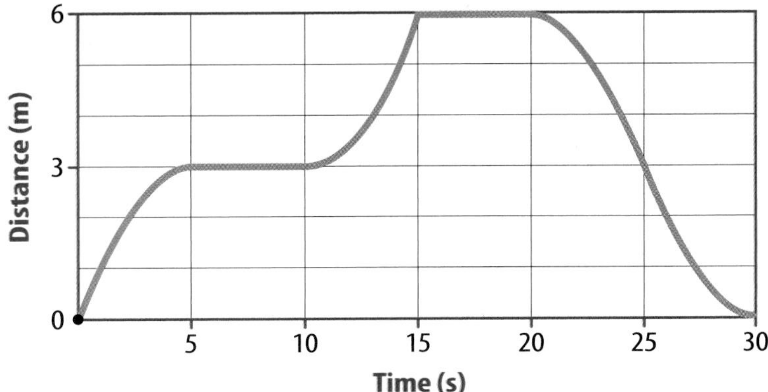

1. Read and complete a lab safety form.
2. Use a **meterstick** to measure a 6-m straight path along the floor. Place a mark with **masking tape** at 0 m, 3 m, and 6 m.
3. Look at the graph above. Decide what type of motion occurs during each 5-second period.
4. Try to walk along your path according to the motion shown on the graph. Have your partner time your walk with a **stopwatch.** Switch roles, and repeat this step.

Think About This

1. What does a horizontal line segment on a distance-time graph indicate?
2. **Key Concept** According to the graph, at what times do the following motions take place? **a.** You change direction. **b.** Your speed increases. **c.** Your speed decreases.

Acceleration—Changes in Velocity

Imagine riding in a car. The driver steps on the gas pedal, and the car moves faster. Moving faster means the car's velocity increases. The driver then takes her foot off the pedal, and the car's velocity decreases. Next the driver turns the steering wheel. The car's velocity changes because its direction changes. The car's velocity changes if either the speed or the direction of the car changes.

When a car's velocity changes, the car is accelerating. **Acceleration** *is a measure of the change in velocity during a period of time.* An object accelerates when its velocity changes as a result of increasing speed, decreasing speed, or changing direction.

You might have experienced a large acceleration if you have ever ridden a roller coaster. Think about all the changes in speed and direction you experience on a roller coaster ride. When you drop down a hill of a roller coaster, you reach a faster speed quickly. The roller coaster is accelerating because its speed is increasing. The roller coaster also accelerates any time it changes direction. It accelerates again when it slows down and stops at the end of the ride. Each time the velocity of the roller coaster changes, it accelerates.

 **Reading Check** What is acceleration?

Ways an Object Can Accelerate

Figure 13 Acceleration occurs when an object speeds up, slows down, or changes its direction of motion.

Speeding Up

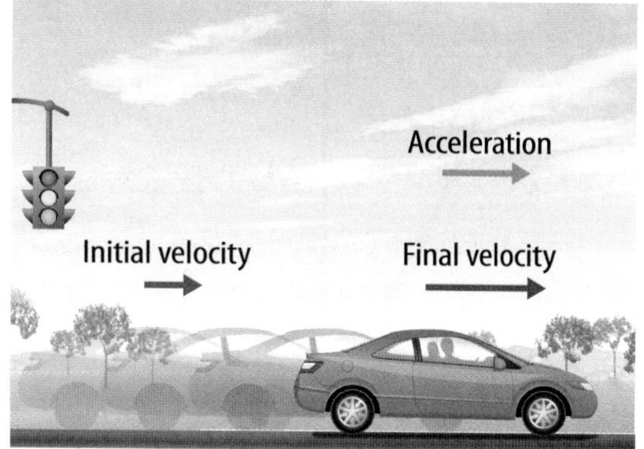

Slowing Down

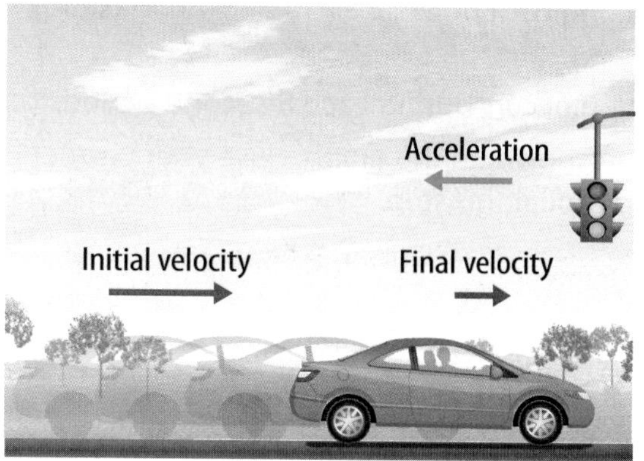

Changing Direction

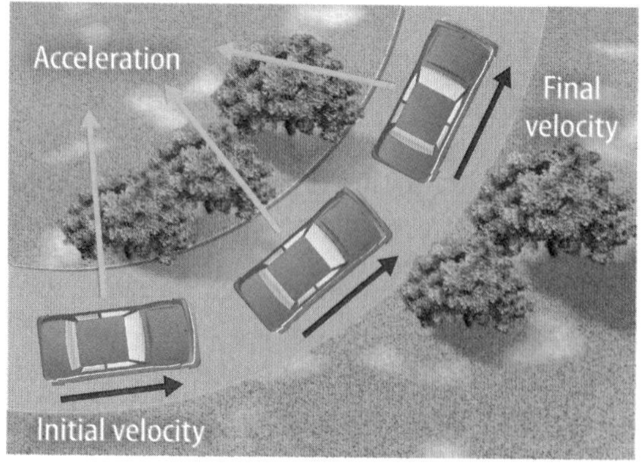

Visual Check If the car in the top picture moved faster, how would the acceleration arrow change?

Representing Acceleration

Like velocity, acceleration has a direction and can be represented by an arrow. Ways an object can accelerate are shown in **Figure 13.** The length of each blue acceleration arrow indicates the amount of acceleration. An acceleration arrow's direction depends on whether velocity increases or decreases.

Changing Speed

The car in the top picture of **Figure 13** is speeding up. At first it is moving slowly, so the arrow that represents its initial velocity is short. The car's speed increases, so the final velocity arrow is longer. As velocity increases, the car accelerates. Notice that the acceleration arrow points in the same direction as the velocity arrows.

The car in the middle picture of **Figure 13** is slowing down. At first it moves fast, so the arrow showing its velocity is long. After the car slows down, the arrow showing its final velocity is shorter. When velocity decreases, acceleration and velocity are in opposite directions. The arrow that represents acceleration is pointing in the direction opposite to the direction the car is moving.

Reading Check In what direction is acceleration if an object is slowing down?

Changing Direction

The car in the bottom picture of **Figure 13** has a constant speed, so the velocity arrow is the same length at each point in the turn. But the car's velocity changes because its direction changes. Because velocity changes, the car is accelerating. Notice the direction of the blue acceleration arrows. It might surprise you that the car is accelerating toward the inside of the curve. Recall that acceleration is the change in velocity. If you compare one velocity arrow with the next, you can see that the change is always toward the inside of the curve.

Key Concept Check What are three ways an object can accelerate?

Calculating Acceleration

Acceleration is a change in velocity divided by the time interval during which the velocity changes. Recall that "velocity" is the speed of an object in a given direction. However, if an object moves along a straight line, you can calculate its acceleration without considering the object's direction. In this lesson, "velocity" refers to only an object's speed. Positive acceleration can be thought of as speeding up in the forward direction. Negative acceleration is slowing down in the forward direction as well as speeding up in the reverse direction.

Acceleration Equation

$$\text{acceleration (in m/s}^2\text{)} = \frac{\text{final speed (in m/s)} - \text{initial speed (in m/s)}}{\text{total time (in s)}}$$

$$a = \frac{v_f - v_i}{t}$$

Acceleration has SI units of meters per second per second (m/s/s). This can also be written as meters per second squared (m/s^2).

FOLDABLES®

Make a horizontal two-tab Foldable and label it as shown. Use it to summarize information about the changes in velocity that can occur when an object is accelerating.

Changing speed | Changing direction

Acceleration

Math Skills: Acceleration Equation

Solve for Acceleration A bicyclist started from rest along a straight path. After 2.0 s, his speed was 2.0 m/s. After 5.0 s, his speed was 8.0 m/s. What was his acceleration during the time 2.0 s to 5.0 s?

1. **This is what you know:**
   - initial speed: $v_i = 2.0$ m/s
   - final speed: $v_f = 8.0$ m/s
   - total time: $t = 5.0 \text{ s} - 2.0 \text{ s} = 3.0 \text{ s}$
2. **This is what you need to find:** acceleration: a
3. **Use this formula:** $a = \frac{v_f - v_i}{t}$
4. **Substitute:** the values for v_i, v_f, and t into the formula; subtract; then divide.

   $a = \frac{8.0 \text{ m/s} - 2.0 \text{ m/s}}{3.0 \text{ s}} = \frac{6.0 \text{ m/s}}{3.0 \text{ s}} = 2.0 \text{ m/s}^2$

Answer: The acceleration of the bicyclist was 2.0 m/s^2.

Practice

Aidan drops a rock from a cliff. After 4.0 s, the rock is moving at 39.2 m/s. What is the acceleration of the rock?

- Math Practice
- Personal Tutor

Inquiry **MiniLab** 10 minutes

How is a change in speed related to acceleration?

What happens if the distance you walk each second increases? Follow these steps to demonstrate acceleration.

1. Read and complete a lab safety form.
2. Use **masking tape** to mark a course on the floor. Mark start, and place marks along a straight path at 10 cm, 40 cm, 90 cm, 160 cm, and 250 cm from the start.
3. Clap a steady beat. On the first beat, the person walking the course is at start. On the second beat, the walker should be at the 10-cm mark, and so on.

Analyze and Conclude

1. **Explain** what happened to your speed as you moved along the course.
2. **Key Concept** Suppose your speed at the final mark was 0.95 m/s. Calculate your average acceleration from start through the final segment of the course.

WORD ORIGIN

horizontal
from Greek *horizein,* means "limit, divide, separate"

vertical
from Latin *verticalis,* means "overhead"

Speed-Time Graphs

Recall that you can show an object's speed using a distance-time graph. You also can use a speed-time graph to show how speed changes over time. Just like a distance-time graph, a speed-time graph has time on the **horizontal** axis—the *x*-axis. But speed is on the **vertical** axis—the *y*-axis. The figures on the next few pages compare distance-time graphs and speed-time graphs for different types of motion.

Object at Rest

An object at rest is not moving, so its speed is always zero. As a result, the speed-time graph for an object at rest is a horizontal line at $y = 0$, as shown in **Figure 14.**

Figure 14 Both the distance-time graph and the speed-time graph are horizontal lines for an object at rest.

Distance / Time

The object's distance from the reference point does not change.

Speed / Time

The speed is zero and does not change.

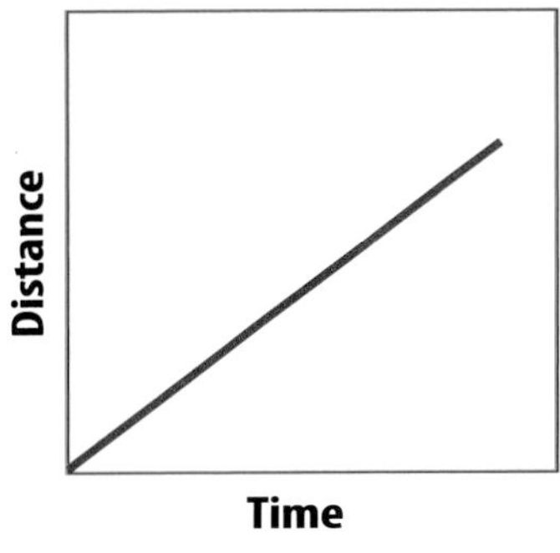

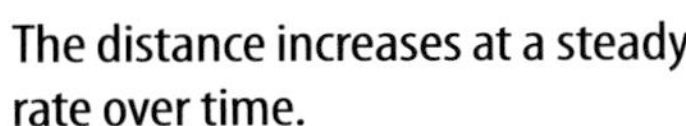

The distance increases at a steady rate over time.

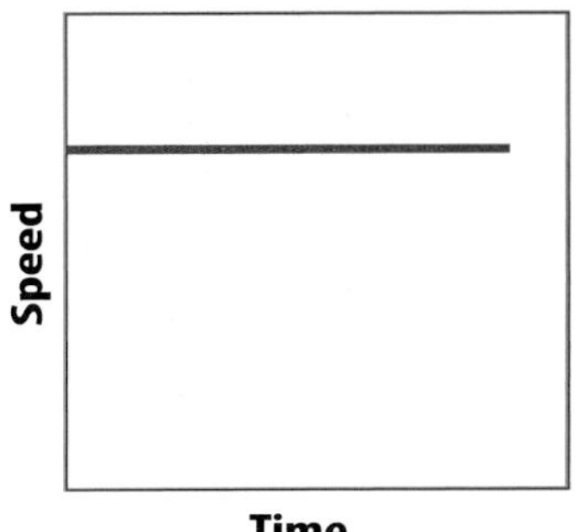

The object's speed does not change.

▲ **Figure 15** For an object moving at constant speed, the speed-time graph is a horizontal line.

Constant Speed

Think about a farm machine moving through a field at a constant speed. At every point in time, its speed is the same. If you plot its speed on a speed-time graph, the plotted line is horizontal, as shown in **Figure 15.** The speed of the object is represented by the distance the horizontal line is from the *x*-axis. If the line is farther from the *x*-axis, the object is moving at a faster speed.

Speeding Up

A plane speeds up as it moves down a runway and takes off. Suppose the speed of the plane increases at a steady rate. If you plot the speed of the plane on a speed-time graph, the line might look like the one in **Figure 16.** The line on the speed-time graph is closer to the *x*-axis at the beginning of the time period when the plane has a lower speed. It slants upward toward the right side of the graph as the speed increases.

 **Reading Check** Why does the speed-time graph of an object that is speeding up slope upward from left to right?

Figure 16 The line on the speed-time graph for an object that is speeding up has an upward slope. ▼

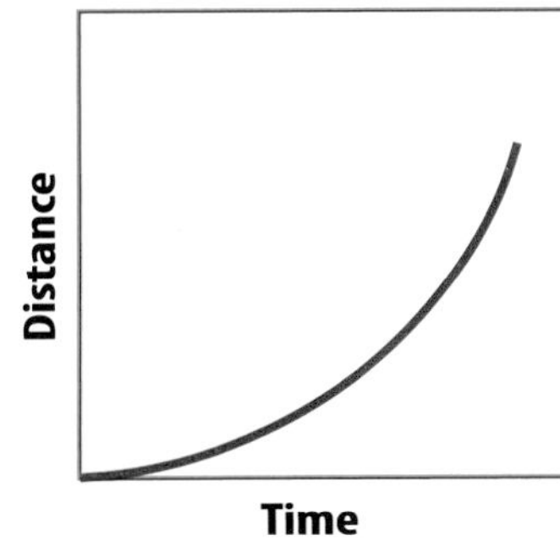

As the distance increases, the rate of increase gets larger over time.

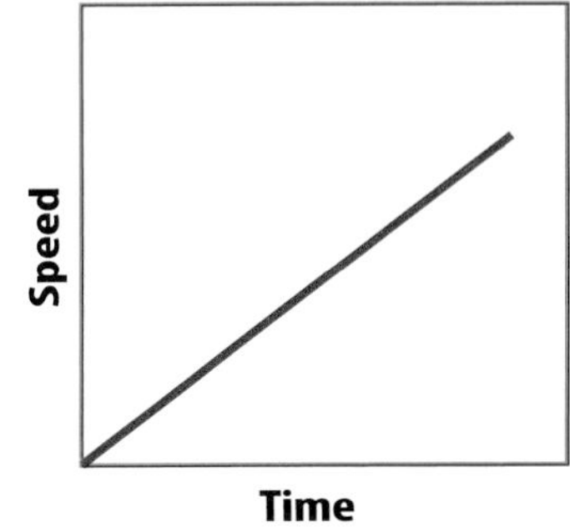

The speed of the object increases at a steady rate over time.

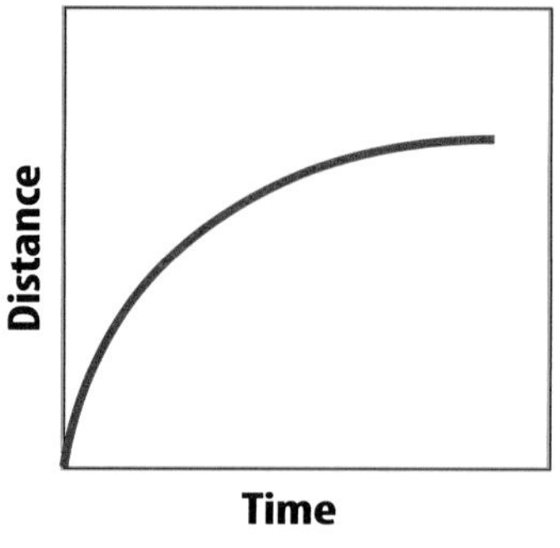

As the distance increases, the rate of increase gets smaller over time.

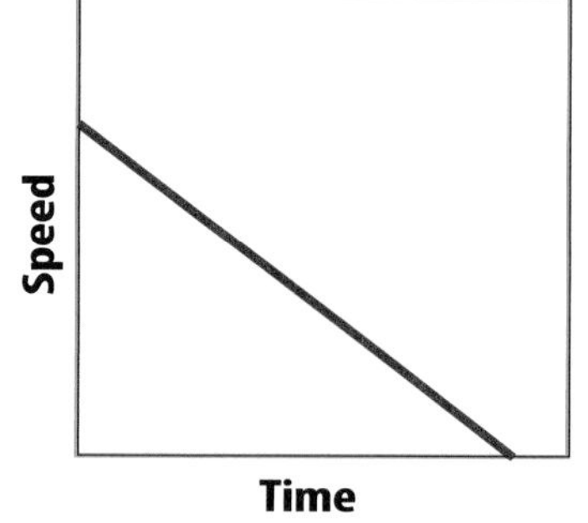

The speed of the object decreases at a steady rate over time.

▲ **Figure 17** 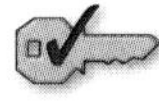 The line on the speed-time graph for an object that is slowing down has a downward slope.

Slowing Down

The speed-time graph in **Figure 17** shows the motion of a space shuttle just after it lands. It slows down at a steady rate and then stops. Initially, the shuttle is moving at a high speed. The point representing this speed is far from the *x*-axis. As the shuttle's speed decreases, the points representing its speed are closer to the *x*-axis. The line on the speed-time graph slopes downward to the right. When the line touches the *x*-axis, the speed is zero and the shuttle is stopped.

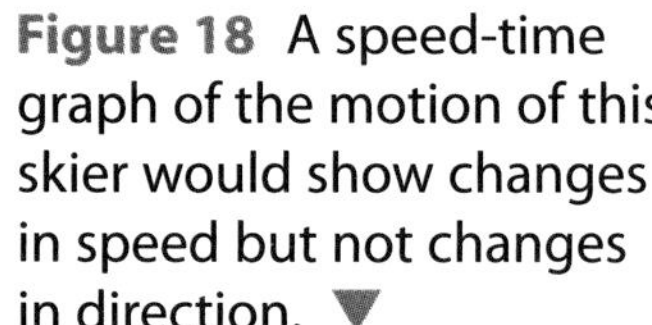

Key Concept Check What does a speed-time graph show about the motion of an object?

Limits of Speed-Time Graphs

You have read that distance-time graphs show the speed of an object. However, they do not describe the direction in which an object is moving. In the same way, speed-time graphs show only the relationship between speed and time. A speed-time graph of the skier in **Figure 18** would show changes in his speed. It would not show what happens when the skier's velocity changes as the result of a change in his direction.

Figure 18 A speed-time graph of the motion of this skier would show changes in speed but not changes in direction. ▼

Summarizing Motion

Now that you know about motion, how might you describe a walk down the hallway at school? You can describe your position by your direction and distance from a reference point. You can compare your distance and your displacement and find your average speed. You know that you have an instantaneous speed and can tell when you walk at a constant speed. You can describe your velocity by your speed and your direction. You know you are accelerating if your velocity is changing.

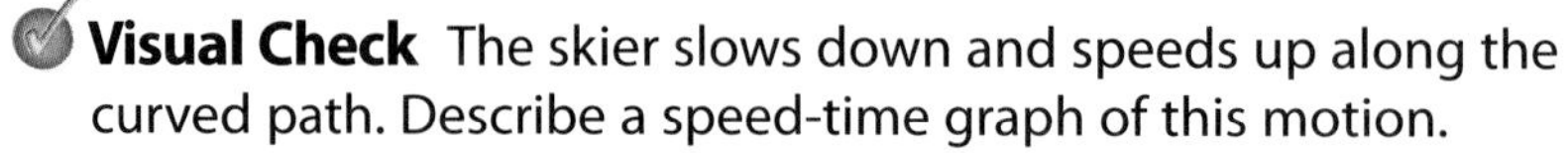

Visual Check The skier slows down and speeds up along the curved path. Describe a speed-time graph of this motion.

Lesson 3 Review

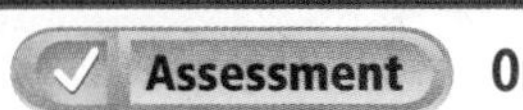

Visual Summary

An object accelerates if it speeds up, slows down, or changes direction.

Acceleration in a straight line can be calculated by dividing the change in speed by the change in time.

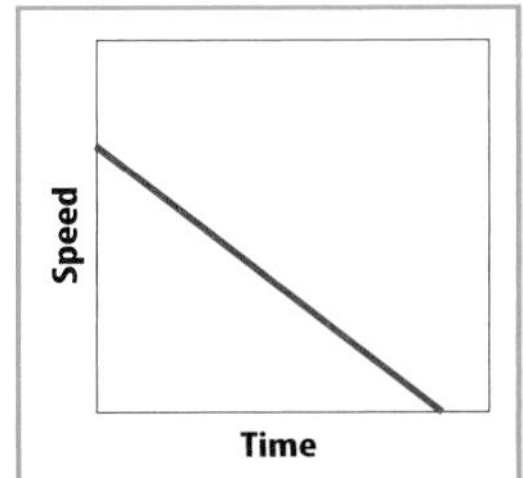

A speed-time graph shows how an object's speed changes over time.

FOLDABLES®

Use your lesson Foldable to review the lesson. Save your Foldable for the project at the end of the chapter.

What do you think NOW?

You first read the statements below at the beginning of the chapter.

5. You can calculate acceleration by dividing the change in velocity by the change in distance.

6. An object accelerates when either its speed or its direction changes.

Did you change your mind about whether you agree or disagree with the statements? Rewrite any false statements to make them true.

Use Vocabulary

1. **Define** *acceleration* in your own words.
2. **Use the term** *acceleration* in a complete sentence.

Understand Key Concepts 

3. **Recall** how a roller coaster can accelerate, even when it is moving at a constant speed.
4. A speed-time graph is a horizontal line with a *y*-value of 4. Which describes the object's motion?
   - **A.** at rest
   - **B.** constant speed
   - **C.** slowing down
   - **D.** speeding up

Interpret Graphics

5. **Organize Information** Copy and fill in the graphic organizer below for the four types of speed-time graphs. For each, describe the motion of the object.

Critical Thinking

6. **Evaluate** A race car accelerates on a straight track from 0 to 100 km/h in 6 s. Another race car accelerates from 0 to 100 km/h in 5 s. Compare the velocities and accelerations of the cars during their races.

Math Skills

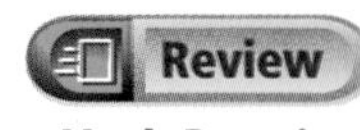

Math Practice

7. After 2.0 s, Isabela was riding her bicycle at 3.0 m/s on a straight path. After 5.0 s, she was moving at 5.4 m/s. What was her acceleration?
8. After 3.0 s, Mohammed was running at 1.2 m/s on a straight path. After 7.0 s, he was running at 2.0 m/s. What was his acceleration?

Calculate Average Speed from a Graph

Materials

meterstickS (6)

stopwatches (6)

masking tape

tennis ball

Safety

You probably do not walk the same speed uphill and downhill, or when you are just starting out and when you are tired. If you are walking and you measure and record the distance you walk every minute, the distances will vary. How might you use these measurements to calculate the average speed you walked? One way is to organize the data on a distance-time graph. In this activity, you will use such a graph to compare average speeds of a ball on a track using different heights of a ramp.

Ask a Question

How does the height of a ramp affect the speed of a ball along a track?

Make Observations

1. Read and complete a lab safety form.
2. Make a 3-m track. Place three metersticks end-to-end. Place three other metersticks end-to-end about 6 cm from the first set of metersticks. Use tape to hold the metersticks in place. Mark each half-meter with tape. Use books to make a ramp leading to the track.
3. A student should be at each half-meter mark with a stopwatch. Another student should be by the ramp to roll a ball along the track.
4. When the ball passes start, all group members should start their stopwatches. Each student should stop his or her stopwatch when the ball crosses the mark where the student is stationed.
5. Practice several times to get consistent rolls and times.

Form a Hypothesis

6. Create a hypothesis about how the number of books used as a ramp affects the speed of the ball rolling along the track.

Test Your Hypothesis

7. Write a plan for varying the number of books and making distance and time measurements.
8. Create a data table in your Science Journal that matches your plan. A sample is shown to the right.
9. Use your plan to make the measurements. Record them in the data table.
10. Plot the data for each height of the ramp on a graph that shows the distance the ball traveled on the *x*-axis and time on the *y*-axis. For each ramp height, draw a straight line that goes through the most points.
11. Choose two points on each line. Calculate the average speed between these points by dividing the difference in the distances for the two points by the difference in the times.

Analyze and Conclude

12. **Compare** the average speeds for each ramp height. Use this comparison to decide whether your results support your hypothesis.
13. **The Big Idea** How was the distance-time graph useful for describing the motion of the ball?

Communicate Your Results

Prepare a poster that shows your graph and describes how it can be used to calculate average speed.

Extension

Design and conduct an experiment comparing the average speed of different types of balls along the track.

8

Distance (m)	Time(s)		
	2 books	3 books	4 books
0.50			
1.00			
1.50			
2.00			
2.50			
3.00			

Lab Tips

- ☑ If the ball doesn't roll far enough, reduce the track length to 2 m.
- ☑ Practice using the stopwatches several times to gain experience in making accurate readings.

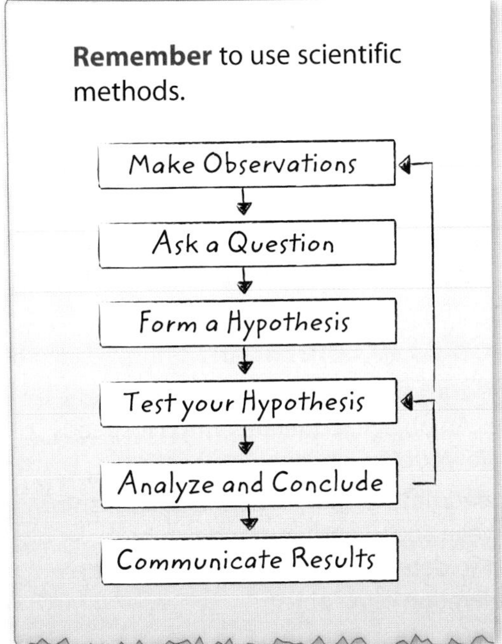

Chapter 1 Study Guide

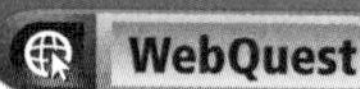

The motion of an object can be described by the object's position, velocity, and acceleration.

Key Concepts Summary

Lesson 1: Position and Motion

- An object's **position** is its distance and direction from a **reference point.**
- The position of an object in two dimensions can be described by choosing a reference point and two reference directions, and then stating the distance along each reference direction.
- The distance an object moves is the actual length of its path. Its **displacement** is the difference between its initial position and its final position.

Vocabulary

reference point p. 9
position p. 9
motion p. 13
displacement p. 13

Lesson 2: Speed and Velocity

- **Speed** is the distance an object moves per unit of time.
- An object moving the same distance each second is moving at a **constant speed.** The speed of an object at a certain moment is its **instantaneous speed.**
- You can calculate an object's **average speed** from a distance-time graph by dividing the distance the object travels by the total time it takes to travel that distance.
- **Velocity** changes when speed, direction, or both speed and direction change.

Vocabulary

speed p. 17
constant speed p. 18
instantaneous speed p. 18
average speed p. 19
velocity p. 23

Lesson 3: Acceleration

- **Acceleration** is a change in velocity over time. An object accelerates when it speeds up, slows down, or changes direction.
- A speed-time graph shows the relationship between speed and time and can be used to determine information about the acceleration of an object.

Vocabulary

acceleration p. 27
average acceleration p. 29

- Personal Tutor
- Vocabulary eGames
- Vocabulary eFlashcards

FOLDABLES® Chapter Project

Assemble your lesson Foldables as shown to make a Chapter Project. Use the project to review what you have learned in this chapter.

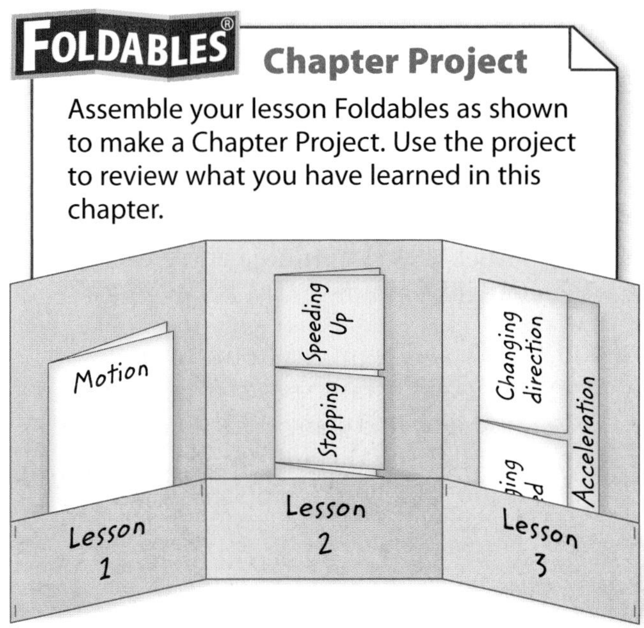

Use Vocabulary

1. A pencil's _______ might be described as 3 cm to the left of the stapler.
2. An object that changes position is in _______.
3. If an object is traveling at a _______, it does not speed up or slow down.
4. An object's _______ includes both its speed and the direction it moves.
5. An object's change in velocity during a time interval, divided by the time interval during which the velocity changed, is its _______.
6. A truck driver stepped on the brakes to make a quick stop. The truck's _______ is in the opposite direction as its velocity.

Link Vocabulary and Key Concepts

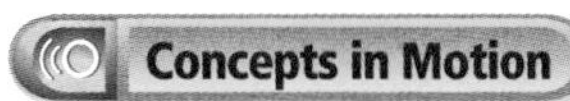

Interactive Concept Map

Copy this concept map, and then use vocabulary terms from the previous page to complete the concept map.

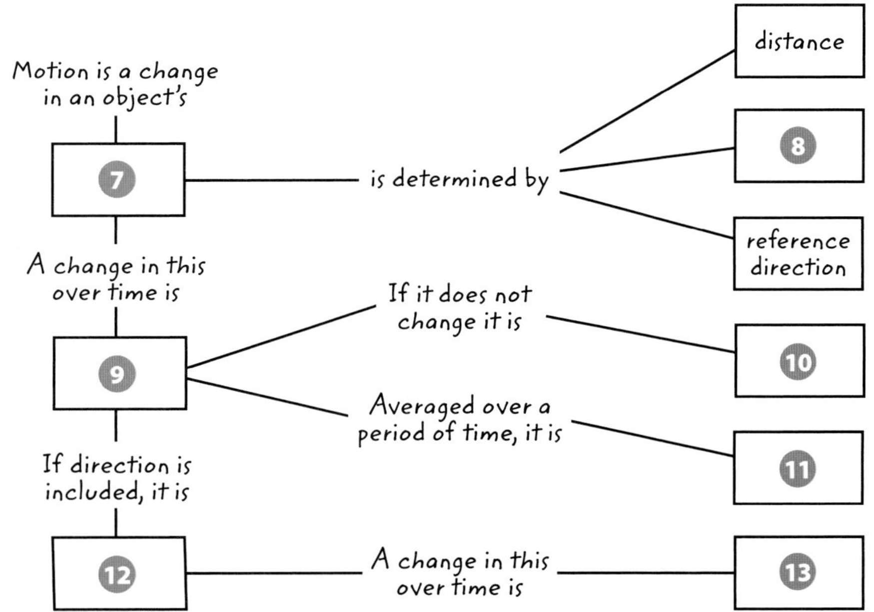

Chapter 1 Review

Understand Key Concepts

1. An airplane rolls down the runway. Compared to which reference point is the airplane in motion?
   - A. the cargo the plane carries
   - B. the control tower
   - C. the pilot flying the plane
   - D. the plane's wing

2. Which describes motion in two dimensions?
   - A. a car driving through a city
   - B. a rock dropping off a cliff
   - C. a sprinter on a 100-m track
   - D. a train on a straight track

3. Which line represents the greatest average speed during the 30-s time period?

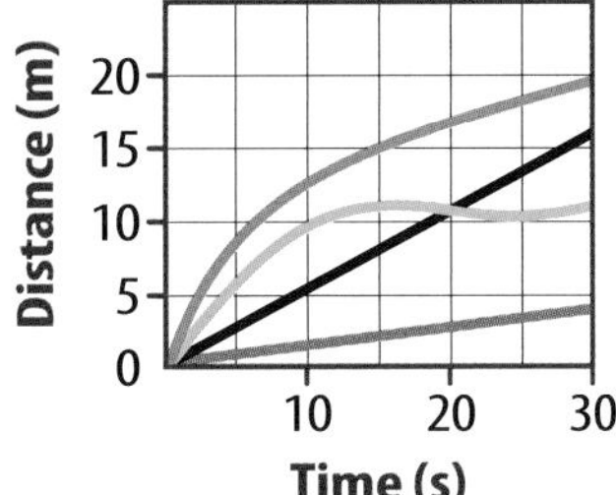

   - A. the blue line
   - B. the black line
   - C. the green line
   - D. the orange line

4. Which describes the greatest displacement?
   - A. walking 3 m east, then 3 m north, then 3 m west
   - B. walking 3 m east, then 3 m south, then 3 m east
   - C. walking 3 m north, then 3 m south, then 3 m north
   - D. walking 3 m north, then 3 m west, then 3 m south

5. Which has the greatest average speed?
   - A. a boat sailing 80 km in 2 hours
   - B. a car driving 90 km in 3 hours
   - C. a train traveling 120 km in 3 hours
   - D. a truck moving 50 km in 1 hour

6. Which describes motion in which the person or object is accelerating?
   - A. A bird flies straight from a tree to the ground without changing speed.
   - B. A dog walks at a constant speed along a straight sidewalk.
   - C. A girl runs along a straight path the same distance each second.
   - D. A truck moves around a curve without changing speed.

7. Richard walks from his home to his school at a constant speed. It takes him 4 min to travel 100 m. Which of the lines in the following distance-time graph could show Richard's motion on the way to school?

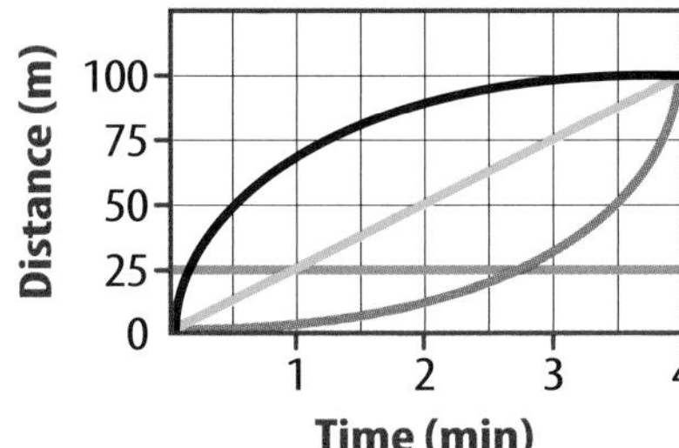

   - A. the black line
   - B. the blue line
   - C. the green line
   - D. the orange line

8. Which is a unit of acceleration?
   - A. kg/m
   - B. kg•m/s^2
   - C. m/s
   - D. m/s^2

9. Which have the same velocity?
   - A. a boy walking east at 2 km/h and a man walking east at 4 km/h
   - B. a car standing still and a truck driving in a circle at 4 km/h
   - C. a dog walking west at 3 km/h and a cat walking west at 3 km/h
   - D. a girl walking west at 3 km/h and a boy walking south at 3 km/h

Assessment
Online Test Practice

Critical Thinking

10 **Describe** A ruler is on the table with the higher numbers to the right. An ant crawls along the ruler from 6 cm to 2 cm in 2 seconds. Describe the ant's distance, displacement, speed, and velocity.

11 **Describe** a theme-park ride that has constant speed but changing velocity.

12 **Construct** a distance-time graph that shows the following motion: A person leaves a starting point at a constant speed of 4 m/s and walks for 4 s. The person then stops for 2 s. The person then continues walking at a constant speed of 2 m/s for 4 s.

13 **Calculate** A truck driver travels 55 km in 1 hour. He then drives a speed of 35 km/h for 2 hours. Next, he drives 175 km in 3 hours. What was his average speed?

14 **Interpret** Keisha measured the distance her friend Morgan ran on a straight track every 2 s. Her measurements are recorded in the table below. What was Morgan's average speed? What was her acceleration?

Time (s)	Distance (m)
0	0
2	2
4	6
6	8
8	14
10	20

Writing in Science

15 **Write** A friend tells you he is 30 m from the fountain in the middle of the city. Write a short paragraph explaining why you cannot identify your friend's position from this description.

REVIEW THE BIG IDEA

16 Nora rides a bicycle for 5 min on a curvy road at a constant speed of 10 m/s. Describe Nora's ride in terms of position, velocity, and acceleration. Compare the distance she rides and her displacement.

17 What are some ways to describe the motion of the jets in the photograph below?

Math Skills

Math Practice

Solve One-Step Equations

18 A model train moves 18.3 m in 122 s. What is the train's average speed?

19 A car travels 45 km in an hour. In each of the next two hours, it travels 78 km. What is the average speed of the car?

20 The speed of a car traveling on a straight road increases from 63 m/s to 75 m/s in 4.2 s. What is the car's acceleration?

21 A girl starts from rest and reaches a walking speed of 1.4 m/s in 3.0 s. She walks at this speed for 6.0 s. The girl then slows down and comes to a stop during a 10.0-s period. What was the girl's acceleration during each of the three time periods? What was her acceleration for the entire trip?

Standardized Test Practice

Record your answers on the answer sheet provided by your teacher or on a sheet of paper.

Multiple Choice

1 Radar tells an air traffic controller that a jet is slowing as it nears the airport. Which might represent the jet's speed?

A 700 h

B 700 h/km

C 700 km

D 700 km/h

Use the diagram below to answer question 2.

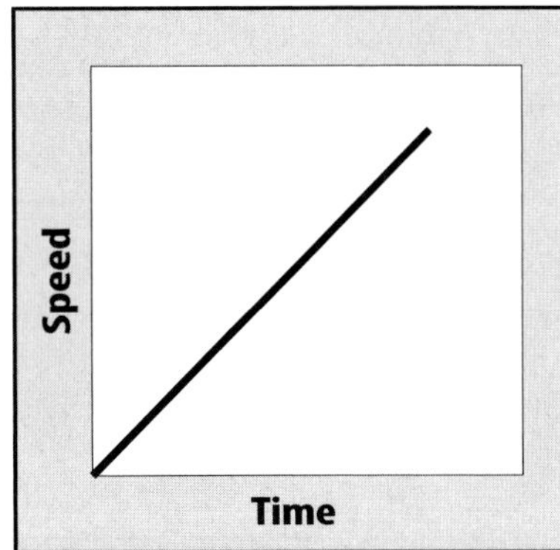

2 What does the graph above illustrate?

A average speed

B constant speed

C decreasing speed

D increasing speed

3 Why is a car accelerating when it is circling at a constant speed?

A It is changing its destination.

B It is changing its direction.

C It is changing its distance.

D It is changing its total mass.

4 Which is defined as the process of changing position?

A displacement

B distance

C motion

D relativity

5 Each diagram below shows two sliding boxes. Which boxes have the same velocity?

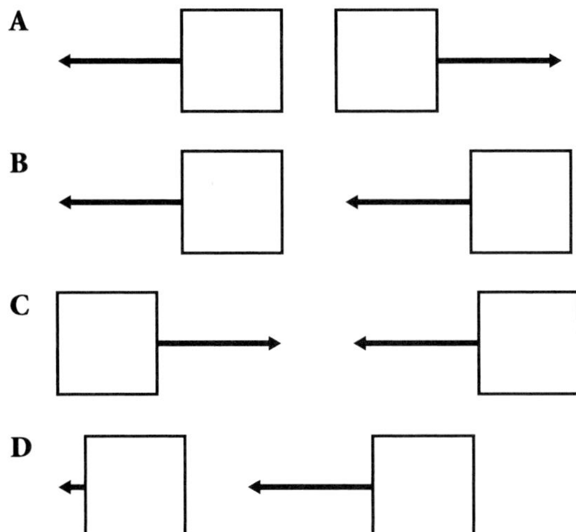

6 In the phrase "two miles southeast of the mall," what is the mall?

A a dimension

B a final destination

C a position

D a reference point

7 The initial speed of a dropped ball is 0 m/s. After 2 seconds, the ball travels at a speed of 20 m/s. What is the acceleration of the ball?

A 5 m/s^2

B 10 m/s^2

C 20 m/s^2

D 40 m/s^2

Assessment

Online Standardized Test Practice

8 Which could be described by the expression "100 m/s northwest"?

A acceleration

B distance

C speed

D velocity

Use the diagram below to answer question 9.

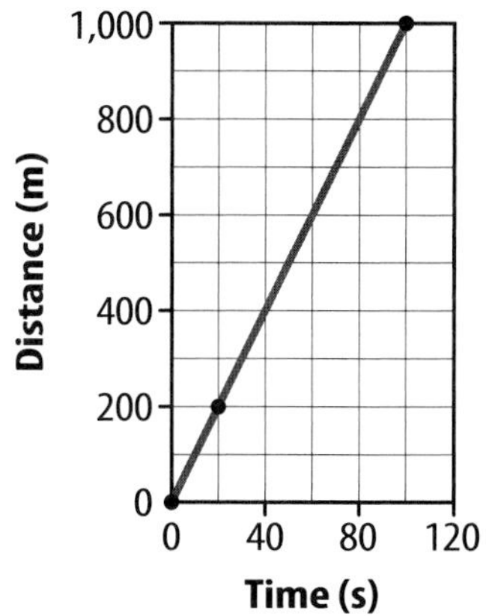

9 In the above graph, what is the average speed of the moving object between 20 and 60 seconds?

A 5 m/s

B 10 m/s

C 20 m/s

D 40 m/s

10 A car travels 250 km and stops twice along the way. The entire trip takes 5 hours. What is the average speed of the car?

A 25 km/h

B 40 km/h

C 50 km/h

D 250 km/h

Constructed Response

Use the diagram below to answer questions 11 and 12.

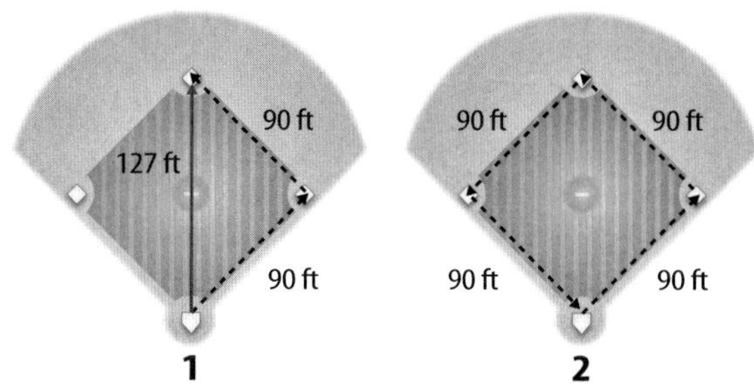

11 The dashed lines show the paths two players run on baseball diamonds. What distance does the player travel on diamond 1? How does it compare to the distance the player runs on diamond 2?

12 Calculate the displacement of the runners on diamonds 1 and 2. Explain your answers.

Use the diagram below to answer question 13.

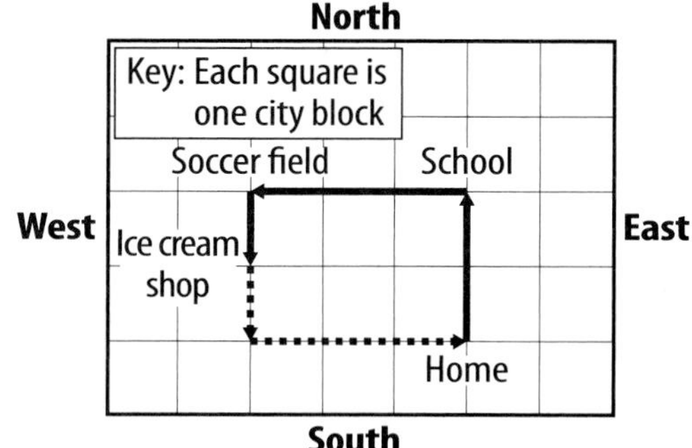

13 A student walks from home to school, on to a soccer field, then to an ice cream shop, and finally home. Use grid distances and directions to describe each leg of his trip. What is the distance between the student's home and the ice cream shop?

NEED EXTRA HELP?													
If You Missed Question...	1	2	3	4	5	6	7	8	9	10	11	12	13
Go to Lesson...	2	3	3	1	2	1	3	2	2	2	1	1	1

Chapter 2

The Laws of Motion

THE BIG IDEA

How do forces change the motion of objects?

Inquiry Why move around?

Imagine the sensations these riders experience as they swing around. The force of gravity pulls the riders downward. Instead of falling, however, they move around in circles.

- What causes the riders to move around?
- What prevents the riders from falling?
- How do forces change the motion of the riders?

Get Ready to Read

What do you think?

Before you read, decide if you agree or disagree with each of these statements. As you read this chapter, see if you change your mind about any of the statements.

1. You pull on objects around you with the force of gravity.
2. Friction can act between two unmoving, touching surfaces.
3. Forces acting on an object cannot be added.
4. A moving object will stop if no forces act on it.
5. When an object's speed increases, the object accelerates.
6. If an object's mass increases, its acceleration also increases if the net force acting on the object stays the same.
7. If objects collide, the object with more mass applies more force.
8. Momentum is a measure of how hard it is to stop a moving object.

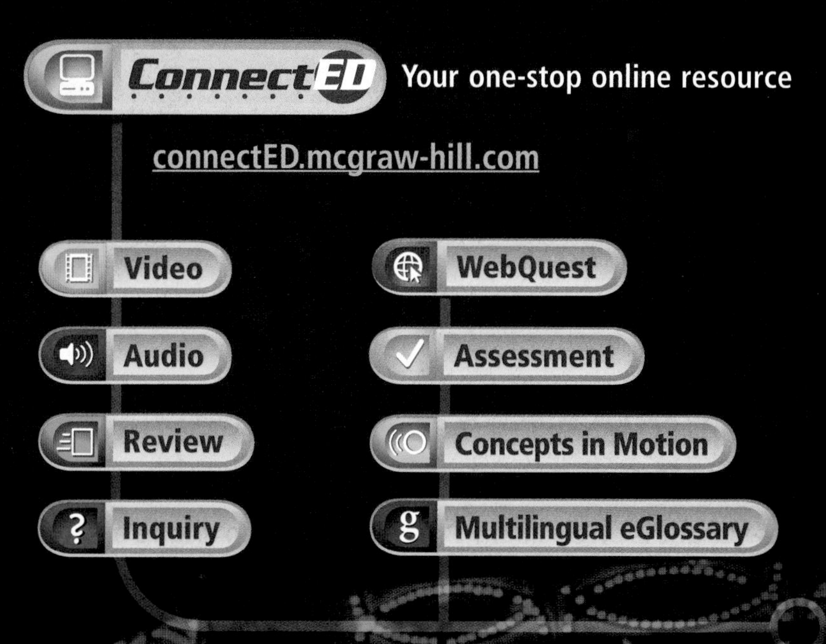

ConnectED
Your one-stop online resource
connectED.mcgraw-hill.com
Video
WebQuest
Audio
Assessment
Review
Concepts in Motion
Inquiry
Multilingual eGlossary

Lesson 1

Gravity and Friction

Reading Guide

Key Concepts
ESSENTIAL QUESTIONS

- What are some contact forces and some noncontact forces?
- What is the law of universal gravitation?
- How does friction affect the motion of two objects sliding past each other?

Vocabulary

force p. 45
contact force p. 45
noncontact force p. 46
gravity p. 47
mass p. 47
weight p. 48
friction p. 49

Multilingual eGlossary

Video

What's Science Got to do With It?

Inquiry
Why doesn't he fall?

This astronaut is on an aircraft that flies at steep angles and provides a sense of weightlessness. Why doesn't he fall? He does! Earth's gravity pulls the astronaut down, but the aircraft moves downward at the same speed.

Inquiry Launch Lab

5 minutes

Can you make a ball move without touching it?

You can make a ball move by kicking it or throwing it. Is it possible to make the ball move even when nothing is touching the ball?

1. Read and complete a lab safety form.
2. Roll a **tennis ball** across the floor. Think about what makes the ball move.
3. Toss the ball into the air. Watch as it moves up and then falls back to your hand.
4. Drop the ball onto the floor. Let it bounce once, and then catch it.

Think About This

1. What made the ball move when you rolled, tossed, and dropped it? What made it stop?
2. **Key Concept** Did something that was touching the ball or not touching the ball cause it to move in each case?

Types of Forces

Think about all the things you pushed or pulled today. You might have pushed toothpaste out of a tube. Maybe you pulled out a chair to sit down. *A push or a pull on an object is called a* **force.** An object or a person can apply a force to another object or person. Some forces are applied only when objects touch. Other forces are applied even when objects do not touch.

WORD ORIGIN

force
from Latin *fortis*, means "strong"

Contact Forces

The hand of the karate expert in **Figure 1** applied a force to the stack of boards and broke them. You have probably also seen a musician strike the keys of a piano and an athlete hit a ball with a bat. In each case, a person or an object applied a force to an object that it touched. *A* **contact force** *is a push or a pull on one object by another that is touching it.*

Contact forces can be weak, like when you press the keys on a computer keyboard. They also can be strong, such as when large sections of underground rock suddenly move, resulting in an earthquake. The large sections of Earth's crust called plates also apply strong contact forces against each other. Over long periods of time, these forces can create mountain ranges if one plate pushes another plate upward.

Figure 1 The man's hand applies a contact force to the boards.

FOLDABLES®

Make a vertical two-tab book from a sheet of paper. Label it as shown. Use it to organize your notes on gravity and friction.

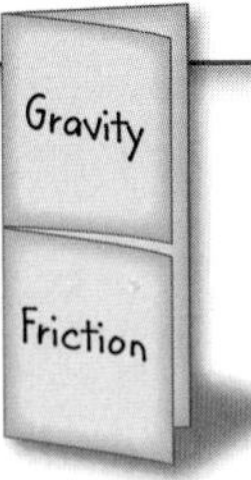

▲ **Figure 2** A noncontact force causes the girl's hair to stand on end.

Noncontact Forces

Lift a pencil and then release it. What happens? The pencil falls toward the floor. A parachutist falls toward Earth even though nothing is touching him. *A force that one object can apply to another object without touching it is a* **noncontact force.** Gravity, which pulled on your pencil and the parachutist, is a noncontact force. The magnetic force, which attracts certain metals to magnets, is also a noncontact force. In **Figure 2,** another noncontact force, called the electric force, causes the girl's hair to stand on end.

Key Concept Check What are some contact forces and some noncontact forces?

Strength and Direction of Forces

Forces have both strength and direction. If you push your textbook away from you, it probably slides across your desk. What happens if you push down on your book? It probably does not move. You can use the same strength of force in both cases. Different things happen because the direction of the applied force is different.

As shown in **Figure 3,** arrows can be used to show forces. The length of an arrow shows the strength of the force. Notice in the figure that the force applied by the tennis racquet is stronger than the force applied by the table-tennis paddle. As a result, the arrow showing the force of the tennis racquet is longer. The direction that an arrow points shows the direction in which force was applied.

The SI unit for force is the newton (N). You apply a force of about 1 N when lifting a stick of butter. You use a force of about 20 N when lifting a 2-L bottle of water. If you use arrows to show these forces, the water's arrow would be 20 times longer.

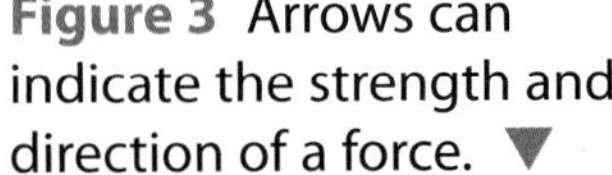

Figure 3 Arrows can indicate the strength and direction of a force. ▼

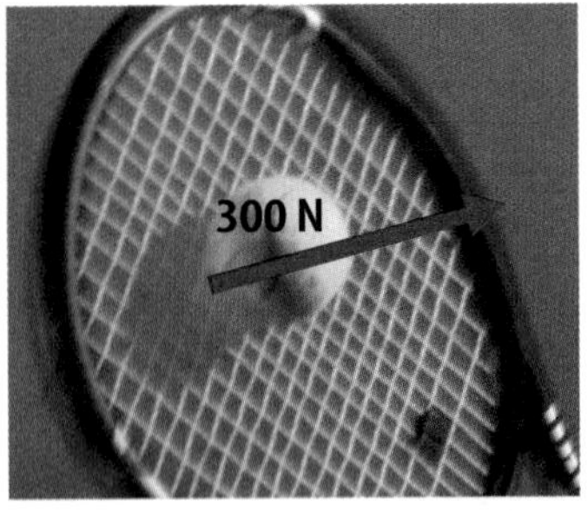

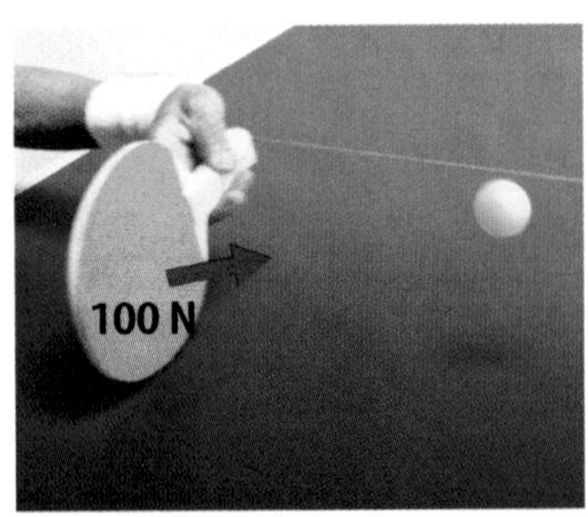

Visual Check How are the lengths of the arrows related to the different forces on the two balls?

Change in Mass

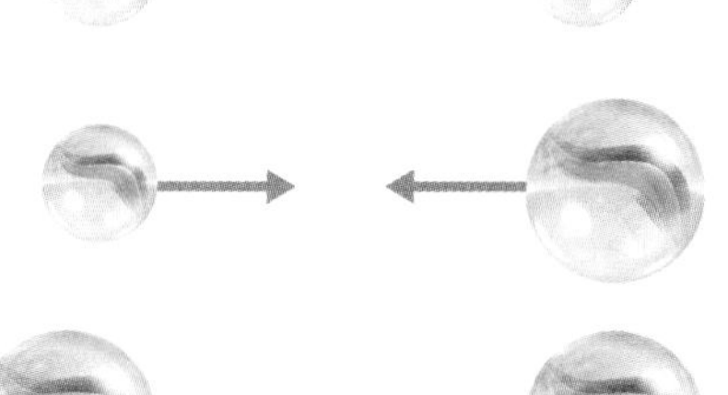

Gravitational force increases if the mass of at least one of the objects increases.

Change in Distance

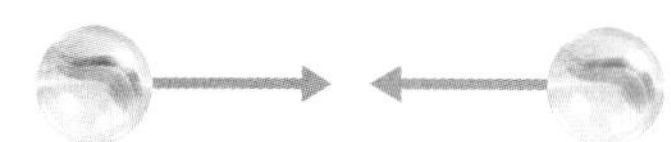

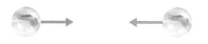

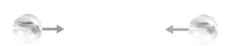

The gravitational force between objects decreases as the objects move apart.

▲ **Figure 4** The gravitational force between objects depends on the mass of the objects and the distance between them.

What is gravity?

Objects fall to the ground because Earth exerts an attractive force on them. Did you know that you also exert an attractive force on objects? **Gravity** *is an attractive force that exists between all objects that have mass.* **Mass** *is the amount of matter in an object.* Mass is often measured in kilograms (kg).

The Law of Universal Gravitation

In the late 1600s, an English scientist and mathematician, Sir Isaac Newton, developed the law of universal gravitation. This law states that all objects are attracted to each other by a gravitational force. The strength of the force depends on the mass of each object and the distance between them.

 **Key Concept Check** What is the law of universal gravitation?

Gravitational Force and Mass The way in which the mass of objects affects gravity is shown in **Figure 4.** When the mass of one or both objects increases, the gravitational force between them also increases. Notice that the force arrows for each pair of marbles are the same size even when one object has less mass. Each object exerts the same attraction on the other object.

Gravitational Force and Distance The effect that distance has on gravity is also shown in **Figure 4.** The attraction between objects decreases as the distance between the objects increases. For example, if your mass is 45 kg, the gravitational force between you and Earth is about 440 N. On the Moon, about 384,000 km away, the gravitational force between you and Earth would only be about 0.12 N. The relationship between gravitational force and distance is shown in the graph in **Figure 5.**

 **Reading Check** What effect does distance have on gravity?

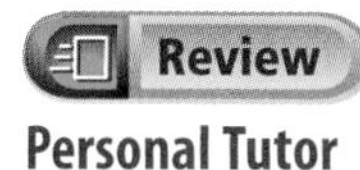

Personal Tutor

Effect of Distance on Gravity

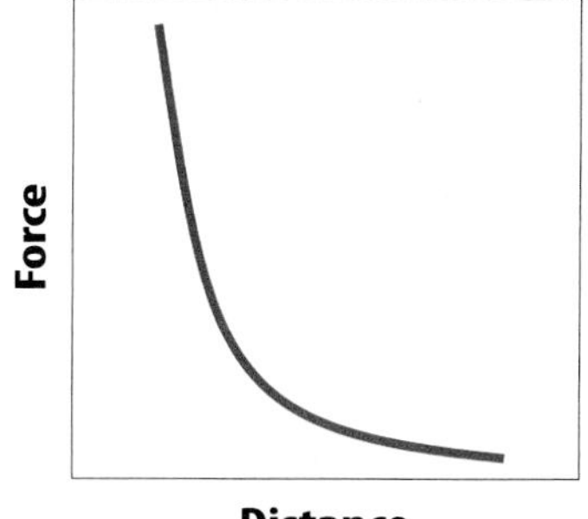

▲ **Figure 5** The gravitational force between objects decreases as the distance between the objects increases.

Weight—A Gravitational Force

Earth has more mass than any object near you. As a result, the gravitational force Earth exerts on you is greater than the force exerted by any other object. **Weight** *is the gravitational force exerted on an object.* Near Earth's surface, an object's weight is the gravitational force exerted on the object by Earth. Because weight is a force, it is measured in newtons.

The Relationship Between Weight and Mass An object's weight is proportional to its mass. For example, if one object has twice the mass of another object, it also has twice the weight. Near Earth's surface, the weight of an object in newtons is about ten times its mass in kilograms.

 **Reading Check** What is the relationship between mass and weight?

ACADEMIC VOCABULARY
significant
(adjective) important, momentous

Weight and Mass High Above Earth You might think that astronauts in orbit around Earth are weightless. Their weight is about 90 percent of what it is on Earth. The mass of the astronaut in **Figure 6** is about 55 kg. Her weight is about 540 N on Earth and about 500 N on the space station 350 km above Earth's surface. Why is there no **significant** change in weight when the distance increases so much? Earth is so large that an astronaut must be much farther away for the gravitational force to change much. The distance between the astronaut and Earth is small compared to the size of Earth.

 **Reading Check** Why is the gravitational force that a friend exerts on you less than the gravitational force exerted on you by Earth?

Figure 6 As she travels from Earth to the space station, the astronaut's weight changes, but her mass remains the same.

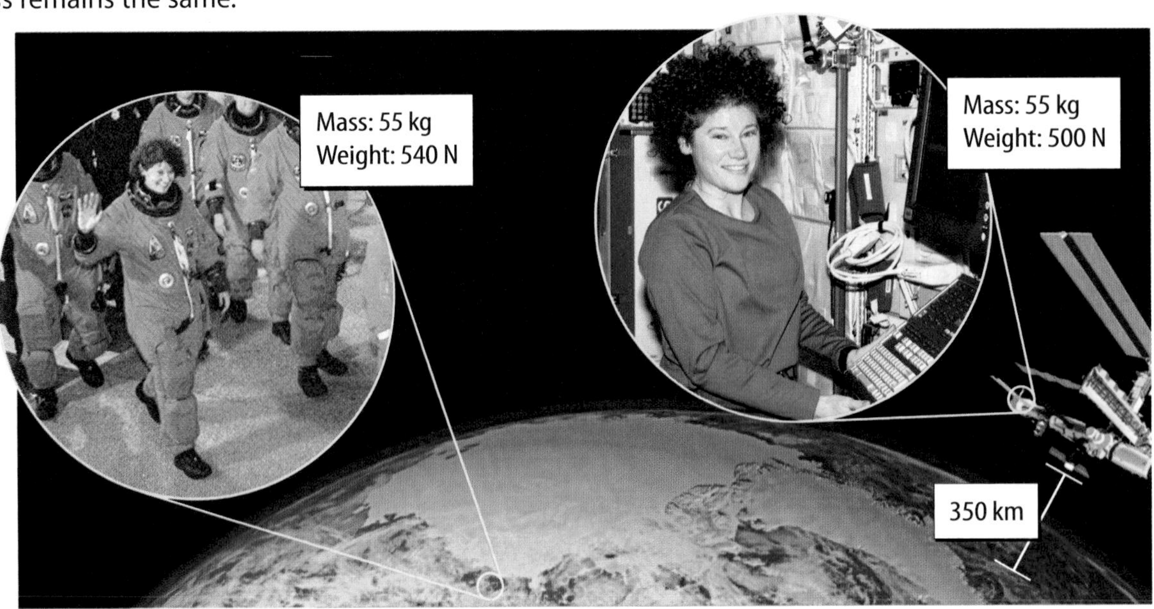

Visual Check What would be the weight of a 110-kg object on Earth? On the space station?

Static and Sliding Friction

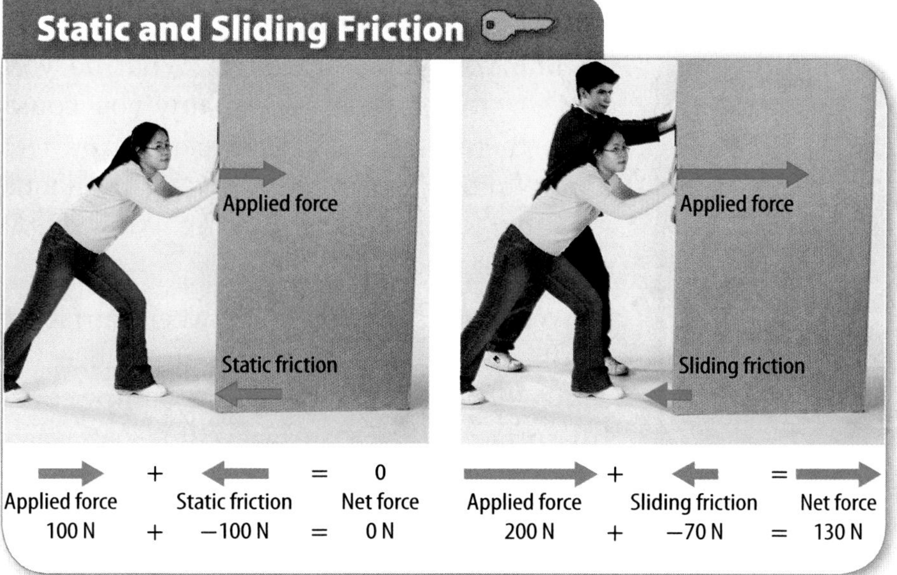

◀ **Figure 7** Static friction prevents the box on the left from moving. Sliding friction slows the motion of the box on the right.

Friction

If you slide across a smooth floor in your socks, you move quickly at first and then stop. The force that slows you is friction. **Friction** *is a force that resists the motion of two surface that are touching.* There are several types of friction.

Static Friction

The box on the left in **Figure 7** does not move because the girl's applied force is balanced by static friction. Static friction prevents surfaces from sliding past each other. Up to a limit, the strength of static friction changes to match the applied force. If the girl increases the applied force, the box still will not move because the static friction also increases.

Science Use v. Common Use
static
Science Use at rest or having no motion
Common Use noise produced in a radio or a television

Sliding Friction

When static friction reaches its limit between surfaces, the box will move. As shown in **Figure 7,** the force of two students pushing is greater than the static friction between the box and the floor. Sliding friction opposes the motion of surfaces sliding past each other. As long as the box is sliding, the sliding friction does not change. Increasing the applied force makes the box slide faster. If the students stop pushing, the box will slow and stop because of sliding friction.

Fluid Friction

Friction between a surface and a fluid—any material, such as water or air, that flows—is fluid friction. Fluid friction between a surface and air is air resistance. Suppose an object is moving through a fluid. Decreasing the surface area toward the oncoming fluid decreases the air resistance against the object. The crumpled paper in **Figure 8** falls faster than the flat paper because it has less surface area and less air resistance.

Figure 8 Air resistance is greater on the flat paper. ▼

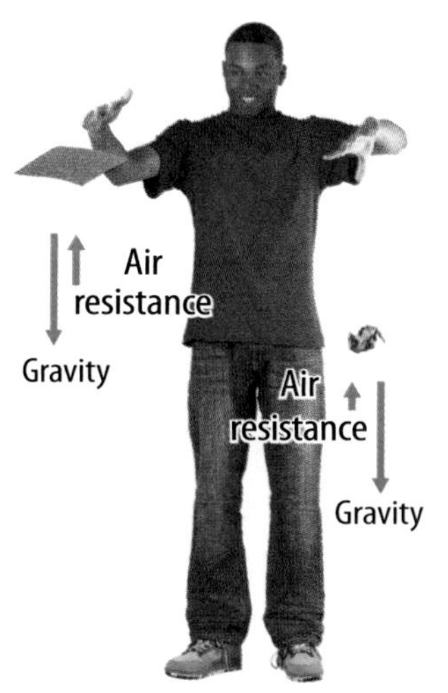

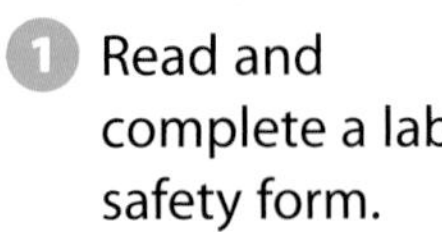

MiniLab

10 minutes

How does friction affect motion?

Friction affects the motion of an object sliding across a surface.

1. Read and complete a lab safety form.
2. Use **tape** to fasten **sandpaper** to a table. Attach a **spring scale** to a **wooden block** with an **eyehook** in it.
3. Record in your Science Journal the force required to gently pull the block at a constant speed on the sandpaper and then on the table.

Analyze and Conclude

1. **Compare** the forces required to pull the block across the two surfaces.
2. **Key Concept** How did reducing friction affect the motion of the block?

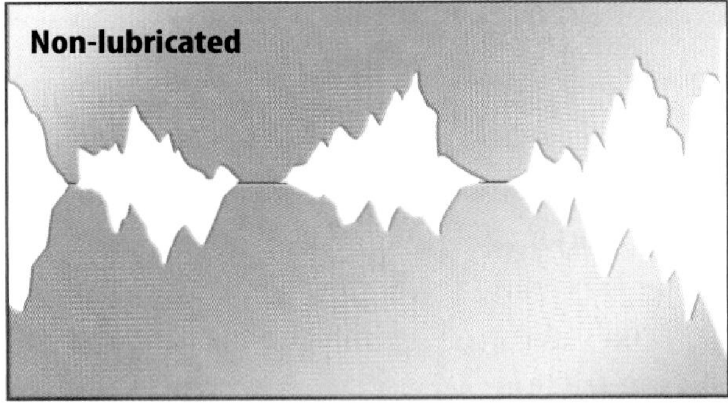

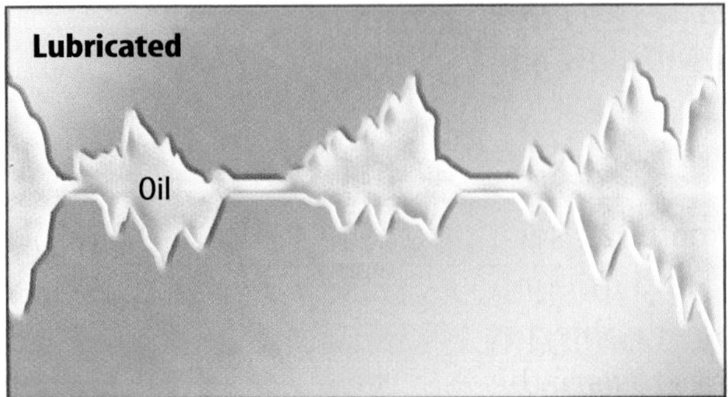

Figure 9 Lubricants such as oil decrease friction caused by microscopic roughness.

What causes friction?

Rub your hands together. What do you feel? If your hands were soapy, you could slide them past each other easily. You feel more friction when you rub your dry hands together than when you rub your soapy hands together.

What causes friction between surfaces? Look at the close-up view of surfaces in **Figure 9.** Microscopic dips and bumps cover all surfaces. When surfaces slide past each other, the dips and bumps on one surface catch on the dips and bumps on the other surface. This microscopic roughness slows sliding and is a source of friction.

Key Concept Check How does friction affect the motion of two objects sliding past each other?

In addition, small particles—atoms and molecules—make up all surfaces. These particles contain weak electrical charges. When a positive charge on one surface slides by a negative charge on the other surface, an attraction occurs between the particles. This attraction slows sliding and is another source of friction between the surfaces.

Reading Check What are two causes of friction?

Reducing Friction

When you rub soapy hands together, the soapy water slightly separates the surfaces of your hands. There is less contact between the microscopic dips and bumps and between the electrical charges of your hands. Soap acts as a lubricant and decreases friction. With less friction, it is easier for surfaces to slide past each other, as shown in **Figure 9.** Motor oil is a lubricant that reduces friction between moving parts of a car's engine.

Look again at the effect of air resistance on the falling paper in **Figure 8.** Reducing the paper's surface area reduces the fluid friction between it and the air.

Lesson 1 Review

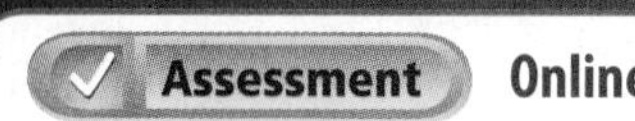

Visual Summary

Forces can be either contact, such as a karate chop, or noncontact, such as gravity. Each type is described by its strength and direction.

Gravity is an attractive force that acts between any two objects that have mass. The attraction is stronger for objects with greater mass.

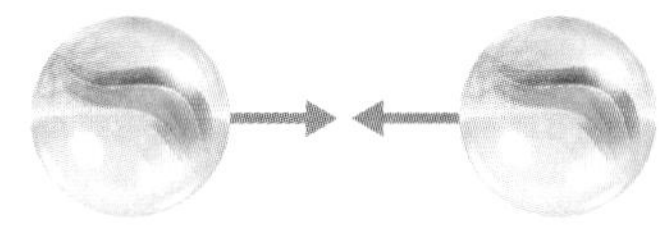

Friction can reduce the speed of objects sliding past each other. Air resistance is a type of fluid friction that slows the speed of a falling object.

FOLDABLES®

Use your lesson Foldable to review the lesson. Save your Foldable for the project at the end of the chapter.

What do you think NOW?

You first read the statements below at the beginning of the chapter.

1. You pull on objects around you with the force of gravity.
2. Friction can act between two unmoving, touching surfaces.

Did you change your mind about whether you agree or disagree with the statements? Rewrite any false statements to make them true.

Use Vocabulary

1. **Define** *friction* in your own words.
2. **Distinguish** between weight and mass.

Understand Key Concepts 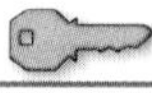

3. **Explain** the difference between a contact force and a noncontact force.
4. You push a book sitting on a desk with a force of 5 N, but the book does not move. What is the static friction?
   A. 0 N
   B. 5 N
   C. between 0 N and 5 N
   D. greater than 5 N
5. **Apply** According to the law of universal gravitation, is there a stronger gravitational force between you and Earth or an elephant and Earth? Why?

Interpret Graphics

6. **Interpret** Look at the forces on the feather.

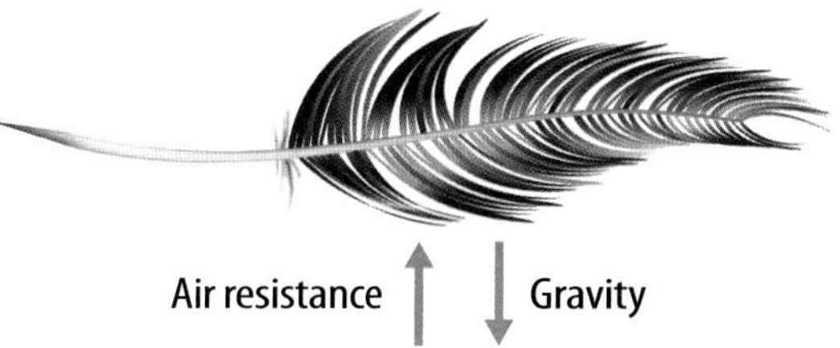

Air resistance ↑ ↓ Gravity

In terms of these forces, explain why the feather falls slowly rather than fast.

7. **Organize Information** Copy and fill in the table below to describe forces mentioned in this lesson. Add as many rows as you need.

Force	Description

Critical Thinking

8. **Decide** Is it possible for the gravitational force between two 50-kg objects to be less than the gravitational force between a 50-kg object and a 5-kg object? Explain.

SCIENCE & SOCIETY

AMERICAN MUSEUM of NATURAL HISTORY

Avoiding an Asteroid Collision

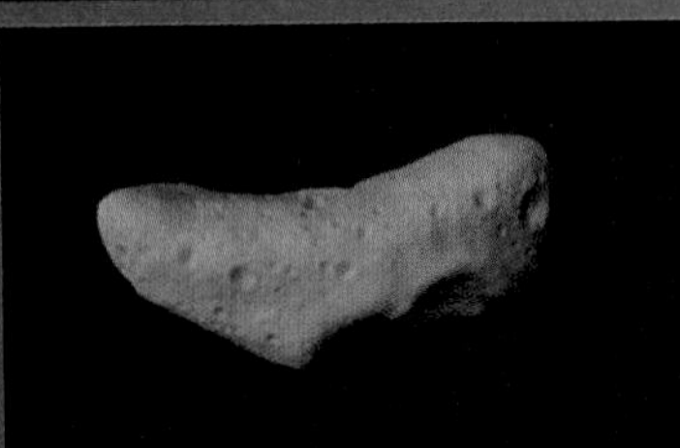

The force of gravity can change the path of an asteroid moving through the solar system.

Gravity to the rescue!

Everything in the universe—from asteroids to planets to stars—exerts gravity on every other object. This force keeps the Moon in orbit around Earth and Earth in orbit around the Sun. Gravity can also send objects on a collision course—a problem when those objects are Earth and an asteroid. Asteroids are rocky bodies found mostly in the asteroid belt between Mars and Jupiter. Jupiter's strong gravity can change the orbits of asteroids over time, occasionally sending them dangerously close to Earth.

Astronomers use powerful telescopes to track asteroids near Earth. More than a thousand asteroids are large enough to cause serious damage if they collide with Earth. If an asteroid were heading our way, how could we prevent the collision? One idea is to launch a spacecraft into the asteroid. The impact could slow it down enough to cause it to miss Earth. But if the asteroid broke apart, the pieces could rain down onto Earth!

Scientists have another idea. They propose launching a massive spacecraft into an orbit close to the asteroid. The spacecraft's gravity would exert a small tug on the asteroid. Over time, the asteroid's path would be altered enough to pass by Earth. Astronomers track objects now that are many years away from crossing paths with Earth. This gives them enough time to set a plan in motion if one of the objects appears to be on a collision course with Earth.

The Spacewatch telescope in Arizona scans the sky for near-Earth asteroids. Other U.S. telescopes with this mission are in Hawaii, California, and New Mexico.

Meteor Crater in Arizona was created when an asteroid about 50 m wide collided with Earth about 50,000 years ago.

PROBLEM SOLVING With a group, come up with a plan for avoiding an asteroid's collision with Earth. Present your plan to the class. Include diagrams and details that explain exactly how your plan will work.

Lesson 2

Reading Guide

Key Concepts
ESSENTIAL QUESTIONS

- What is Newton's first law of motion?
- How is motion related to balanced and unbalanced forces?
- What effect does inertia have on the motion of an object?

Vocabulary

net force p. 55
balanced forces p. 56
unbalanced forces p. 56
Newton's first law of motion p. 57
inertia p. 58

BrainPOP®

Newton's First Law

Inquiry How does it balance?

You probably would be uneasy standing under Balanced Rock near Buhl, Idaho. Yet this unusual rock stays in place year after year. The rock has forces acting on it. Why doesn't it fall over? The forces acting on the rock combine, and the rock does not move.

Inquiry Launch Lab

10 minutes

Can you balance magnetic forces?

Magnets exert forces on each other. Depending on how you hold them, magnets either attract or repel each other. Can you balance these magnetic forces?

1. Read and complete a lab safety form.
2. Have your lab partner hold a **ring magnet** horizontally on a **pencil,** as shown in the picture.
3. Place **another magnet** on the pencil, and use it to push the first magnet along the pencil.
4. Place a **third magnet** on the same pencil so that the outer magnets push against the middle one. Does the middle magnet still move along the pencil?

Think About This

1. Describe the forces that the other magnets exert on the first magnet in steps 3 and 4.
2. **Key Concept** Describe how the motion of the first magnet seemed to depend on whether each force on the magnet was balanced by another force.

Identifying Forces

Ospreys are birds of prey that live near bodies of water. Perhaps several minutes ago, the mother osprey in **Figure 10** was in the air in a high-speed dive. It might have plunged toward a nearby lake after sighting a fish in the water. As it neared the water, it moved its legs forward to grab the fish with its talons. It then stretched out its wings and used them to climb high into the air. Before the osprey comes to rest on its nest, it will slow its speed and land softly on the nest's edge, near the young birds waiting for food.

Figure 10 Forces change the motion of this osprey.

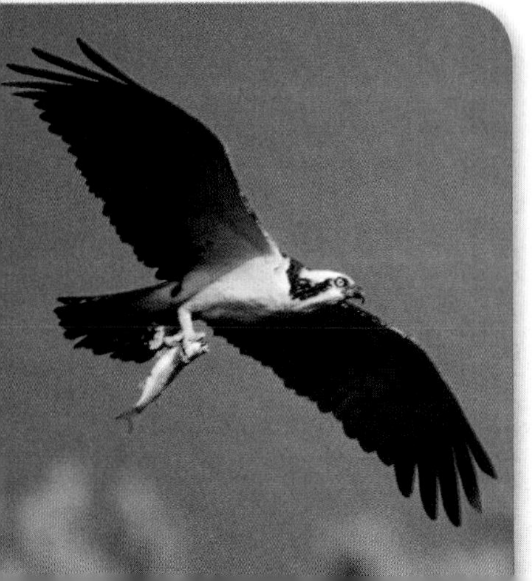

Forces helped the mother osprey change the speed and direction of its motion. Recall that a force is a push or a pull. Some of the forces were contact forces, such as air resistance. When soaring, the osprey spread its wings, increasing air resistance. In a dive, it held its wings close to its body, changing its shape, decreasing its surface area and air resistance. Gravity also pulled the osprey toward the ground.

To understand the motion of an object, you need to identify the forces acting on it. In this lesson you will read how forces change the motion of objects.

Combining Forces—The Net Force

Suppose you try to move a piece of heavy furniture, such as the dresser in **Figure 11.** If you push on the dresser by yourself, you have to push hard on the dresser to overcome the static friction and move it. If you ask a friend to push with you, you do not have to push as hard. When two or more forces act on an object, the forces combine. *The combination of all the forces acting on an object is the* **net force.** The way in which forces combine depends on the directions of the forces applied to an object.

Combining Forces in the Same Direction

When the forces applied to an object act in the same direction, the net force is the sum of the individual forces. In this case, the direction of the net force is the same as the direction of the individual forces.

Because forces have direction, you have to specify a **reference direction** when you combine forces. In **Figure 11,** for example, you would probably choose "to the right" as the positive reference direction. Both forces then would be positive. The net force on the dresser is the sum of the two forces pushing in the same direction. One person pushes on the dresser with a force of 200 N to the right. The other person pushes with a force of 100 N to the right. The net force on the dresser is 200 N + 100 N = 300 N to the right. The force applied to the dresser is the same as if one person pushed on the dresser with a force of 300 N to the right.

Reading Check How do you calculate the net force on an object if two forces are acting on it in the same direction?

REVIEW VOCABULARY

reference direction
a direction that you choose from a starting point to describe an object's position

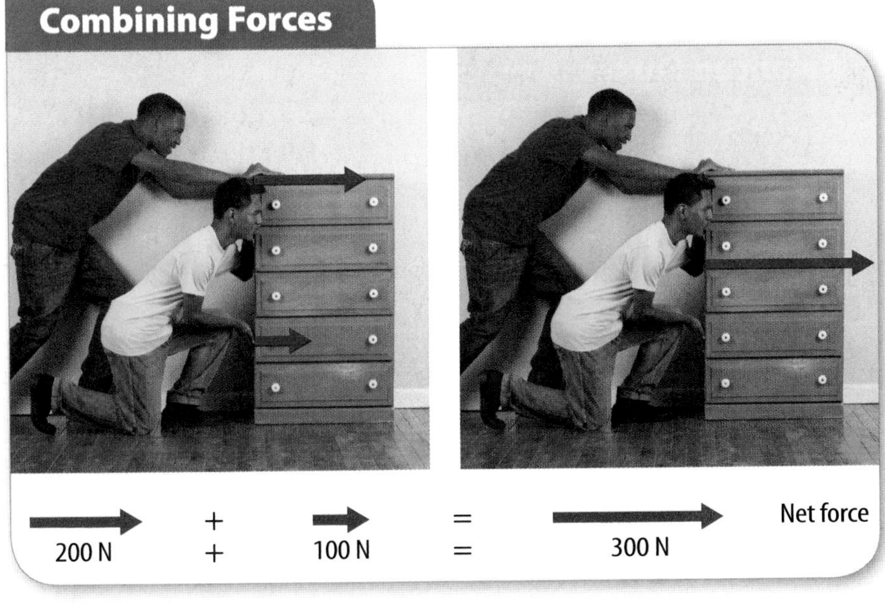

Figure 11 When forces in the same direction combine, the net force is also in the same direction. The strength of the net force is the sum of the forces.

Visual Check What would the net force be if one boy pushed with 250 N and the other boy pushed in the same direction with 180 N?

Personal Tutor

Figure 12 When two forces acting on an object in opposite directions combine, the net force is in the same direction as the larger force. The strength of the net force is the sum of the positive and negative forces. ▶

Combining Forces in Opposite Directions

When forces act in opposite directions on an object, the net force is still the sum of the forces. Suppose you choose "to the right" again as the reference direction in **Figure 12.** A force in that direction is positive, and a force in the opposite direction is negative. The net force is the sum of the positive and negative forces. The net force on the dresser is 100 N to the right.

Balanced and Unbalanced Forces

When equal forces act on an object in opposite directions, as in **Figure 13,** the net force on the object is zero. The effect is the same as if there were no forces acting on the object. *Forces acting on an object that combine and form a net force of zero are* **balanced forces.** Balanced forces do not change the motion of an object. However, the net force on the dresser in **Figure 12** is not zero. There is a net force to the right. *Forces acting on an object that combine and form a net force that is not zero are* **unbalanced forces.**

Figure 13 When two forces acting on an object in opposite directions are the same strength, the forces are balanced. ▶

Visual Check How are the force arrows for the balanced forces in the figure alike? How are they different?

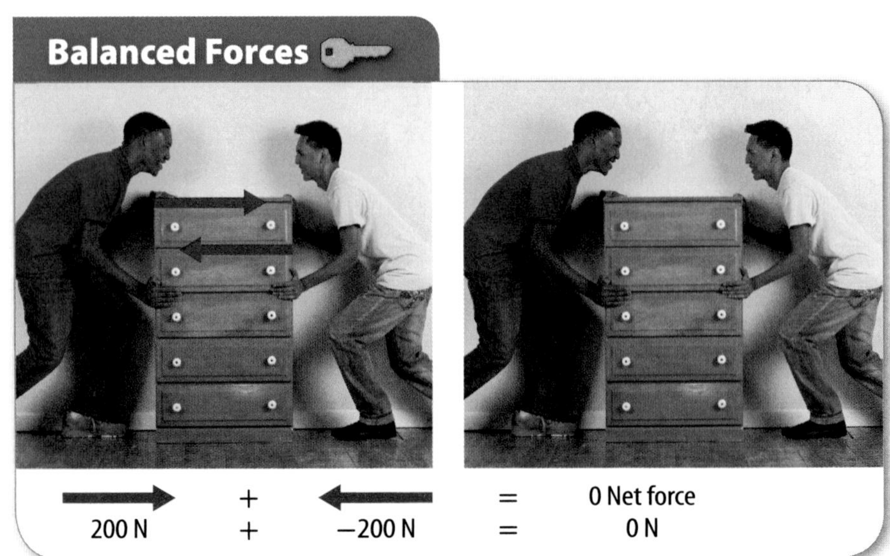

Newton's First Law of Motion

Sir Isaac Newton studied how forces affect the motion of objects. He developed three rules known as Newton's laws of motion. *According to* **Newton's first law of motion,** *if the net force on an object is zero, the motion of the object does not change.* As a result, balanced forces and unbalanced forces have different results when they act on an object.

 **Key Concept Check** What is Newton's first law of motion?

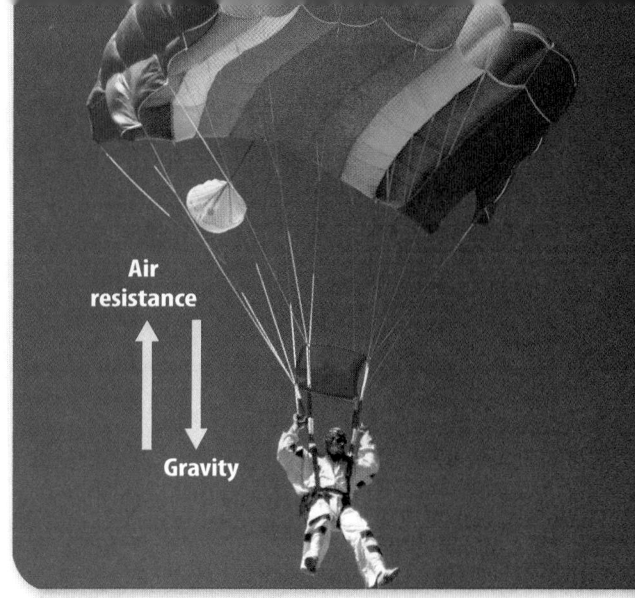

Figure 14 Balanced forces acting on an object do not change the object's speed and direction.

Balanced Forces and Motion

According to Newton's first law of motion, balanced forces cause no change an object's velocity (speed in a certain direction). This is true when an object is at rest or in motion. Look again at **Figure 13.** The dresser is at rest before the boys push on it. It remains at rest when they apply balanced forces. Similarly, because the forces in **Figure 14**—air resistance and gravity—are balanced, the parachutist moves downward at his terminal velocity. Terminal velocity is the constant velocity reached when air resistance equals the force of gravity acting on a falling object.

 **Reading Check** What happens to the velocity of a moving car if the forces on it are balanced?

Unbalanced Forces and Motion

Newton's first law of motion only applies to balanced forces acting on an object. When unbalanced forces act on an object at rest, the object starts moving. When unbalanced forces act on an already moving object, the object's speed, direction of motion, or both change. You will read more about how unbalanced forces affect an object's motion in the next lesson.

 **Key Concept Check** How is motion related to balanced and unbalanced forces?

Inquiry MiniLab — 15 minutes

How do forces affect motion?

The motion of an object depends on whether balanced or unbalanced forces act on it.

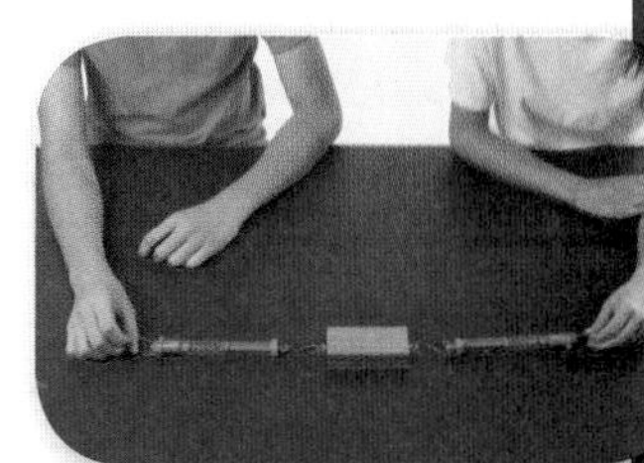

1. Read and complete a lab safety form.
2. Attach **spring scales** to opposite sides of a **wooden block** with **eyehooks.**
3. With a partner, gently pull the scales so that the block moves toward one of you. Sketch the setup in your Science Journal, including the force readings on each scale.
4. Repeat step 3, pulling on the block so that it does not move.

Analyze and Conclude

1. **Explain** Use Newton's first law of motion to explain what occurred in steps 3 and 4.
2. **Key Concept** How was the block's motion related to balanced and unbalanced forces?

Figure 15 Inertia causes the crash-test dummy to keep moving forward after the car stops.

Visual Check What effect would a shoulder belt and a lap belt have on the inertia of the crash-test dummy?

Inertia

According to Newton's first law, the motion of an object will not change if balanced forces act on it. *The tendency of an object to resist a change in its motion is called* **inertia** (ihn UR shuh). Inertia explains the motion of the crash-test dummy in **Figure 15.** Before the crash, the car and dummy moved with constant velocity. If no other force had acted on them, the car and dummy would have continued moving with constant velocity because of inertia. The impact with the barrier results in an unbalanced force on the car, and the car stops. The dummy continues moving forward because of its inertia.

WORD ORIGIN
inertia
from Latin *iners,* means "without skill, inactive"

 **Key Concept Check** What effect does inertia have on the motion of an object?

FOLDABLES®

Make a chart with six columns and six rows. Use your chart to define and show how this lesson's vocabulary terms are related. Afterward, fold your chart in half and label the outside *Newton's First Law.*

	Net Force	Balanced Forces	Unbalanced Forces	Newton's First Law	Inertia
Net Force					
Balanced Forces					
Unbalanced Forces					
Newton's First Law					
Inertia					

Why do objects stop moving?

Think about how friction and inertia together affect an object's movement. A book sitting on a table, for example, stays in place because of inertia. When you push the book, the force you apply to the book is greater than static friction between the book and the table. The book moves in the direction of the greater force. If you stop pushing, friction stops the book.

What would happen if there were no friction between the book and the table? Inertia would keep the book moving. According to Newton's first law, the book would continue to move at the same speed in the same direction as your push.

On Earth, friction can be reduced but not totally removed. For an object to start moving, a force greater than static friction must be applied to it. To keep the object in motion, a force at least as strong as friction must be applied continuously. Objects stop moving because friction or another force acts on them.

Lesson 2 Review

Visual Summary

Unbalanced forces cause an object to move.

According to Newton's first law of motion, if the net force on an object is zero, the object's motion does not change.

Inertia is a property that resists a change in the motion of an object.

FOLDABLES®

Use your lesson Foldable to review the lesson. Save your Foldable for the project at the end of the chapter.

What do you think NOW?

You first read the statements below at the beginning of the chapter.

3. Forces acting on an object cannot be added.

4. A moving object will stop if no forces act on it.

Did you change your mind about whether you agree or disagree with the statements? Rewrite any false statements to make them true.

Use Vocabulary

1. **Define** *net force* in your own words.
2. **Distinguish** between balanced forces and unbalanced forces.

Understand Key Concepts 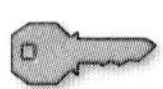

3. Which causes an object in motion to remain in motion?
   - **A.** friction
   - **B.** gravity
   - **C.** inertia
   - **D.** velocity
4. **Apply** You push a coin across a table. The coin stops. How does this motion relate to balanced and unbalanced forces?
5. **Explain** Use Newton's first law to explain why a book on a desk does not move.

Interpret Graphics

6. **Analyze** What is the missing force?

7. **Organize Information** Copy and fill in the graphic organizer below to explain Newton's first law of motion in each case.

Object at rest	
Object in motion	

Critical Thinking

8. **Extend** Three people push a piano on wheels with forces of 130 N to the right, 150 N to the left, and 165 N to the right. What are the strength and direction of the net force on the piano?
9. **Assess** A child pushes down on a box lid with a force of 25 N. At the same time, her friend pushes down on the lid with a force of 30 N. The spring on the box lid pushes upward with a force of 60 N. Can the children close the box? Why or why not?

Inquiry **Skill Practice** **Model** 20 minutes

How can you model Newton's first law of motion?

Materials

markers (red, blue, black, green)

According to Newton's first law of motion, balanced forces do not change an object's motion. Unbalanced forces change the motion of objects at rest or in motion. You can model different forces and their effects on the motion of an object.

Learn It

When you **model** a concept in science, you act it out, or imitate it. You can model the effect of balanced and unbalanced forces on motion by using movements on a line.

Try It

1. Draw a line across a sheet of lined notebook paper lengthwise. Place an X at the center. Each space to the right of the X will model a force of 1 N east, and each space to the left will model 1 N west.
2. Suppose a force of 3 N east and a force of 11 N west act on a moving object. Model these forces by starting at X and drawing a red arrow three spaces to the right. Then, start at that point and draw a blue arrow 11 spaces to the left. The net force is modeled by how far this point is from X, 8 N west.
3. Are the forces you modeled balanced or unbalanced? Will the forces change the object's motion?

Apply It

4. Suppose a force of 8 N east, a force of 12 N west, and a force of 4 N east act on a moving object. Use different colors of markers to model the forces on the object.
5. What is the net force on the object? Are the forces you modeled balanced or unbalanced? Will the forces change the object's motion?
6. **Model** other examples of balanced and unbalanced forces acting on an object. In each case, decide which forces will act on the object.
7. **Key Concept** For each of the forces you modeled, determine the net force, and decide if the forces are balanced or unbalanced. Then, decide if the forces will change the object's motion.

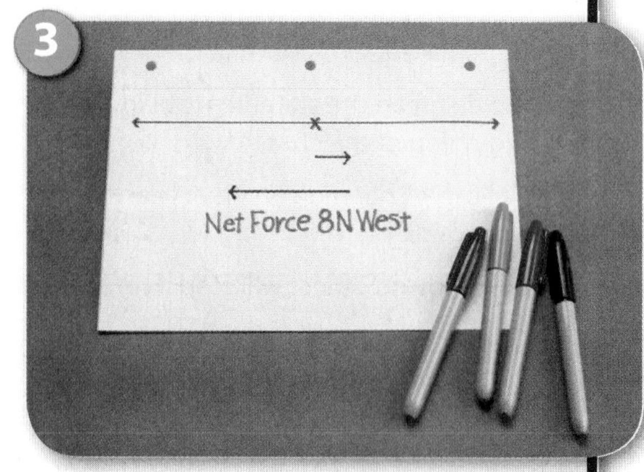

Lesson 3

Newton's Second Law

Reading Guide

Key Concepts
ESSENTIAL QUESTIONS

- What is Newton's second law of motion?
- How does centripetal force affect circular motion?

Vocabulary

Newton's second law of motion p. 65
circular motion p. 66
centripetal force p. 66

g Multilingual eGlossary

Inquiry What makes it go?

The archer pulls back the string and takes aim. When she releases the string, the arrow soars through the air. To reach the target, the arrow must quickly reach a high speed. How is it able to move so fast? The force from the string determines the arrow's speed.

Launch Lab

10 minutes

What forces affect motion along a curved path?

When traveling in a car or riding on a roller coaster, you can feel different forces acting on you as you move along a curved path. What are these forces? How do they affect your motion?

1. Read and complete a lab safety form.
2. Attach a piece of **string** about 1 m long to a rolled-up **sock.**

   WARNING: Find a spot away from your classmates for the next steps.

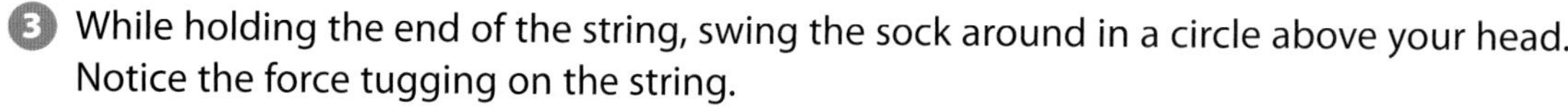

3. While holding the end of the string, swing the sock around in a circle above your head. Notice the force tugging on the string.

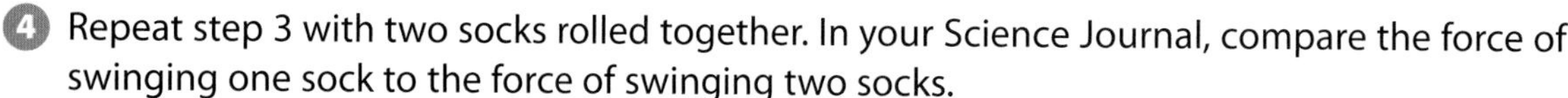

4. Repeat step 3 with two socks rolled together. In your Science Journal, compare the force of swinging one sock to the force of swinging two socks.

Think About This

1. Describe the forces acting along the string while you were swinging it. Classify each force as balanced or unbalanced.
2. **Key Concept** How does the force from the string seem to affect the sock's motion?

How do forces change motion?

Think about different ways that forces can change an object's motion. For example, how do forces change the motion of someone riding a bicycle? The forces of the person's feet on the pedals cause the wheels of the bicycle to turn faster and the bicycle's speed to increase. The speed of a skater slowly sliding across ice gradually decreases because of friction between the skates and the ice. Suppose you are pushing a wheelbarrow across a yard. You can change its speed by pushing with more or less force. You can change its direction by pushing it in the direction you want to move. Forces change an object's motion by changing its speed, its direction, or both its speed and its direction.

Unbalanced Forces and Velocity

Velocity is speed in a certain direction. Only unbalanced forces change an object's velocity. A bicycle's speed will not increase unless the forces of the person's feet on the pedals is greater than friction that slows the wheels. A skater's speed will not decrease if the skater pushes back against the ice with a force greater than the friction against the skates. If someone pushes the wheelbarrow with the same force but in the opposite direction that you are pushing, the wheelbarrow's direction will not change.

In the previous lesson, you read about Newton's first law of motion—balanced forces do not change an object's velocity. In this lesson you will read about how unbalanced forces affect the velocity of an object.

Unbalanced Forces on an Object at Rest

An example of how unbalanced forces affect an object at rest is shown in **Figure 16.** At first the ball is not moving. The hand holds the ball up against the downward pull of gravity. Because the forces on the ball are balanced, the ball remains at rest. When the hand moves out of the way, the ball falls downward. You know that the forces on the ball are now unbalanced because the ball's motion changed. The ball moves in the direction of the net force. When unbalanced forces act on an object at rest, the object begins moving in the direction of the net force.

Balanced Forces

Force exerted by hand

Force due to gravity

Unbalanced Forces

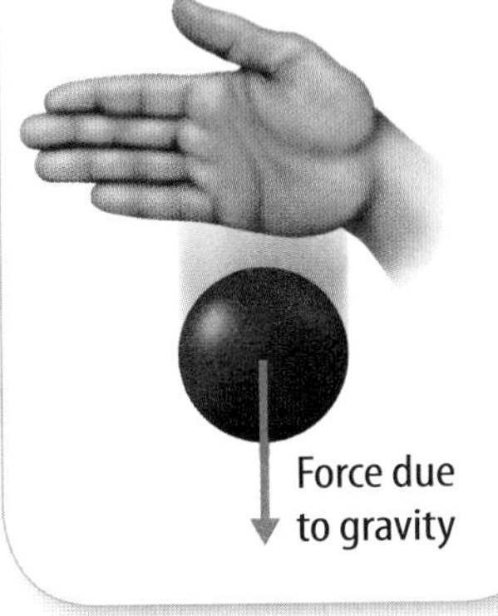

▲ **Figure 16** When unbalanced forces act on a ball at rest, it moves in the direction of the net force.

Unbalanced Forces on an Object in Motion

Unbalanced forces change the velocity of a moving object. Recall that one way to change an object's velocity is to change its speed.

Speeding Up If the net force acting on a moving object is in the direction that the object is moving, the object will speed up. For example, a net force acts on the sled in **Figure 17.** Because the net force is in the direction of motion, the sled's speed increases.

Slowing Down Think about what happens if the direction of the net force on an object is opposite to the direction the object moves. The object slows down. When the boy sliding on the sled in **Figure 17** pushes his foot against the snow, friction acts in the direction opposite to his motion. Because the net force is in the direction opposite to the sled's motion, the sled's speed decreases.

Reading Check What happens to the speed of a wagon rolling to the right if a net force to the right acts on it?

Slowing down

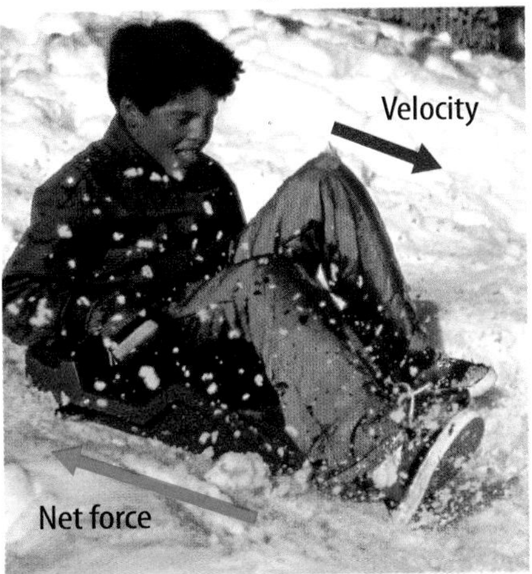

◀ **Figure 17** Unbalanced forces can cause an object to speed up or slow down.

Visual Check How would the net force and velocity arrows in the left photo change if the girl pushed harder?

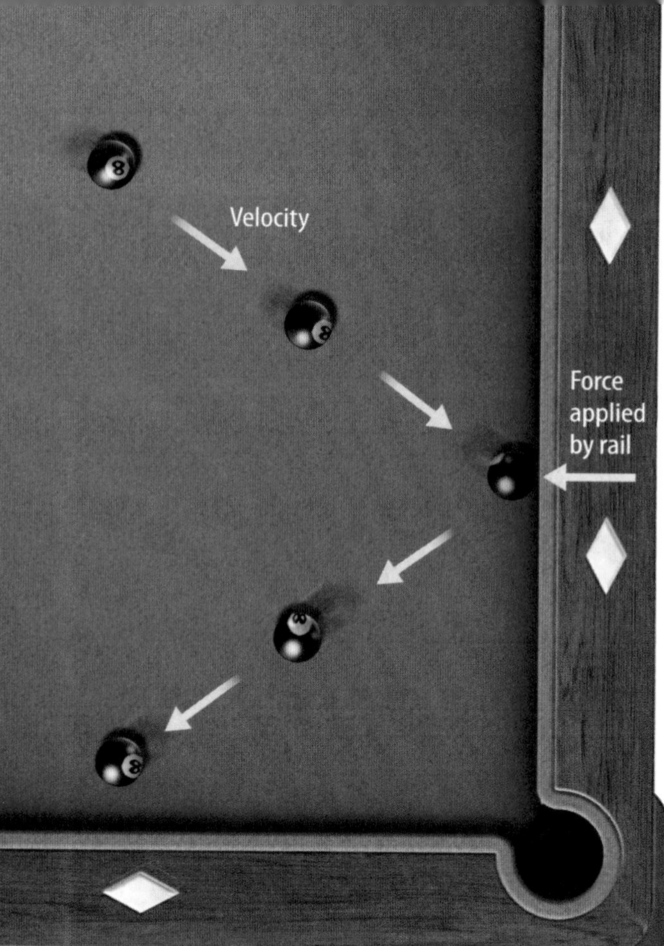

Figure 18 Unbalanced forces act on the billiard ball, causing its direction to change.

Changes in Direction of Motion

Another way that unbalanced forces can change an object's velocity is to change its direction. The ball in **Figure 18** moved at a constant velocity until it hit the rail of the billiard table. The force applied by the rail changed the ball's direction. Likewise, unbalanced forces change the direction of Earth's crust. Recall that the crust is broken into moving pieces called plates. The direction of one plate changes when another plate pushes against it with an unbalanced force.

Unbalanced Forces and Acceleration

You have read how unbalanced forces can change an object's velocity by changing its speed, its direction, or both its speed and its direction. Another name for a change in velocity over time is acceleration. When the girl in **Figure 17** pushed the sled, the sled accelerated because its speed changed. When the billiard ball in **Figure 18** hit the side of the table, the ball accelerated because its direction changed. Unbalanced forces can make an object accelerate by changing its speed, its direction, or both.

Reading Check How do unbalanced forces affect an object at rest or in motion?

Inquiry MiniLab

10 minutes

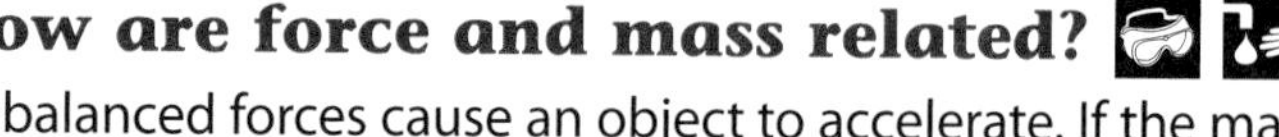

How are force and mass related?

Unbalanced forces cause an object to accelerate. If the mass of the object increases, how does the force required to accelerate the object change?

1. Read and complete a lab safety form.
2. Tie a **string** to a **small box.** Pull the box about 2 m across the floor. Notice the force required to cause the box to accelerate.
3. Put **clay** in the box to increase its mass. Pull the box so that its acceleration is about the same as before. Notice the force required.

Analyze and Conclude

1. **Compare** the strength of the force needed to accelerate the box each time.
2. **Key Concept** How did the mass affect the force needed to accelerate the box?

Newton's Second Law of Motion

Isaac Newton also described the relationship between an object's acceleration (change in velocity over time) and the net force that acts on an object. *According to* **Newton's second law of motion,** *the acceleration of an object is equal to the net force acting on the object divided by the object's mass.* The direction of acceleration is the same as the direction of the net force.

 **Key Concept Check** What is Newton's second law of motion?

FOLDABLES®

Make a half-book from a sheet of notebook paper. Use it to organize your notes on Newton's second law.

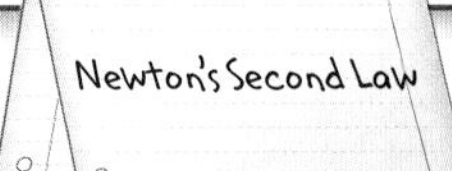

Newton's Second Law Equation

$$\text{acceleration (in m/s}^2) = \frac{\text{net force (in N)}}{\text{mass (in kg)}}$$

$$a = \frac{F}{m}$$

Notice that the equation for Newton's second law has SI units. Acceleration is expressed in meters per second squared (m/s^2), mass in kilograms (kg), and force in newtons (N). From this equation, it follows that a newton is the same as $kg \cdot m/s^2$.

Math Skills — Newton's Second Law Equation

Solve for Acceleration You throw a 0.5-kg basketball with a force of 10 N. What is the acceleration of the ball?

1. **This is what you know:** mass: $m = 0.5$ kg; force: $F = 10$ N or $10\ kg \cdot m/s^2$
2. **This is what you need to find:** acceleration: a
3. **Use this formula:** $a = \frac{F}{m}$
4. **Substitute:** the values for F and m into the formula and divide. $a = \frac{10\ \text{N}}{0.5\ \text{kg}} = 20\ \frac{kg \cdot m/s^2}{kg} = 20\ m/s^2$

Answer: The acceleration of the ball is 20 m/s^2.

Review
- Math Practice
- Personal Tutor

Practice

1. A 24-N net force acts on an 8-kg rock. What is the acceleration of the rock?
2. A 30-N net force on a skater produces an acceleration of 0.6 m/s^2. What is the mass of the skater?
3. What net force acting on a 14-kg wagon produces an acceleration of 1.5 m/s^2?

Circular Motion

Newton's second law of motion describes the relationship between an object's change in velocity over time, or acceleration, and unbalanced forces acting on the object. You already read how this relationship applies to motion along a line. It also applies to circular motion. **Circular motion** *is any motion in which an object is moving along a curved path.*

Centripetal Force

The ball in **Figure 19** is in circular motion. The velocity arrows show that the ball has a tendency to move along a straight path. Inertia—not a force—causes this motion. The ball's path is curved because the string pulls the ball inward. *In circular motion, a force that acts perpendicular to the direction of motion, toward the center of the curve, is* **centripetal** (sen TRIH puh tuhl) **force.** The figure also shows that the ball accelerates in the direction of the centripetal force.

WORD ORIGIN
centripetal
from Latin *centripetus,* means "toward the center"

 **Key Concept Check** How does centripetal force affect circular motion?

The Motion of Satellites and Planets

Another object that experiences centripetal force is a satellite. A satellite is any object in space that orbits a larger object. Like the ball in **Figure 19,** a satellite tends to move along a straight path because of inertia. But just as the string pulls the ball inward, gravity pulls a satellite inward. Gravity is the centripetal force that keeps a satellite in orbit by changing its direction. The Moon is a satellite of Earth. As shown in **Figure 19,** Earth's gravity changes the Moon's direction. Similarly, the Sun's gravity changes the direction of its satellites, including Eartch.

Figure 19 Inertia of the moving object and the centripetal force acting on the object produce the circular motion of the ball and the Moon.

Visual Check How does the direction of the velocity of a satellite differ from the direction of its acceleration?

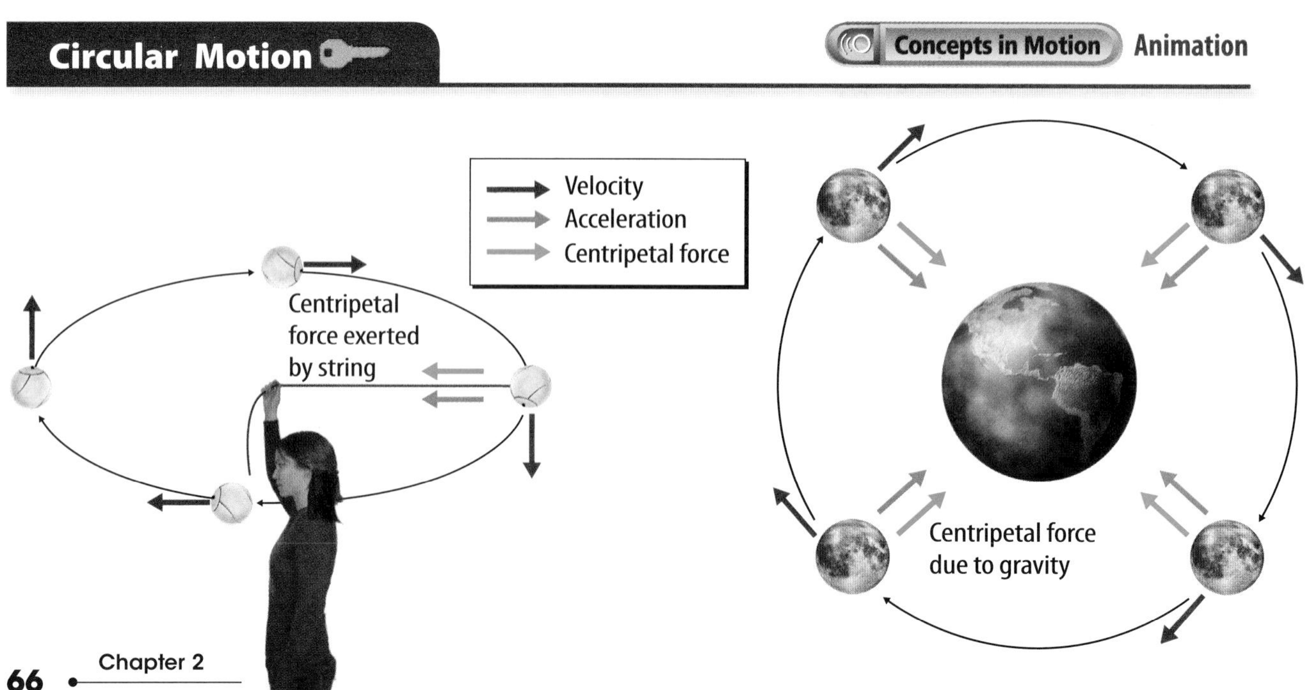

Lesson 3 Review

Visual Summary

Unbalanced forces cause an object to speed up, slow down, or change direction.

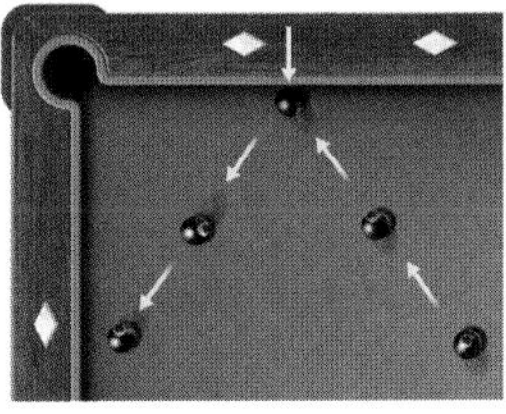

Newton's second law of motion relates an object's acceleration to its mass and the net force on the object.

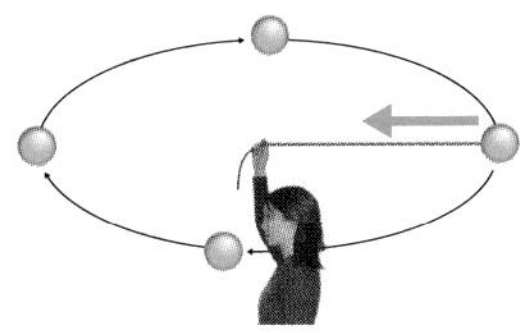

Any motion in which an object is moving along a curved path is circular motion.

FOLDABLES®

Use your lesson Foldable to review the lesson. Save your Foldable for the project at the end of the chapter.

What do you think NOW?

You first read the statements below at the beginning of the chapter.

5. When an object's speed increases, the object accelerates.

6. If an object's mass increases, its acceleration also increases if the net force acting on the object stays the same.

Did you change your mind about whether you agree or disagree with the statements? Rewrite any false statements to make them true.

Use Vocabulary

1. **Explain** Newton's second law of motion in your own words.
2. **Use the term** *circular motion* in a sentence.

Understand Key Concepts 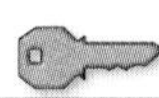

3. A cat pushes a 0.25-kg toy with a net force of 8 N. According to Newton's second law what is the acceleration of the ball?

   A. 2 m/s^2 **C.** 16 m/s^2
   B. 4 m/s^2 **D.** 32 m/s^2

4. **Describe** how centripetal force affects circular motion.

Interpret Graphics

5. **Apply** Copy and fill in the graphic organizer below. In each oval, give an example of unbalanced forces on an object causing the object to accelerate.

6. **Complete** each equation according to Newton's second law.

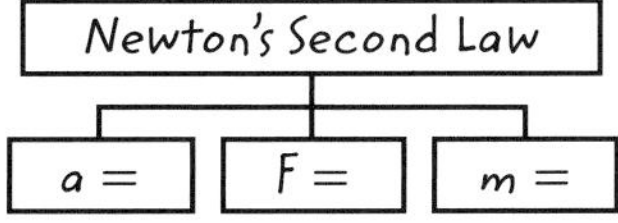

Critical Thinking

7. **Design** You need to lift an object that weighs 45 N straight up. Draw an illustration that explains the strength and direction of the force you must apply to lift the object.

Math Skills

Math Practice

8. The force of Earth's gravity is about 10 N downward. What is the acceleration of a 15-kg backpack if you lift it with a force of 15 N?

Inquiry **Skill Practice** **Use Variables** 25 minutes

How does a change in mass or force affect acceleration?

Materials

baseball

foam ball

meterstick

Safety

Force, mass, and acceleration are all related variables. In this activity, you will use these variables to study Newton's second law of motion.

Learn It

Vary means "to change." A **variable** is a quantity that can be changed. For example, the variables related to Newton's second law of motion are force, mass, and acceleration. You can find the relationship between any two of these variables by changing one of them and keeping the third variable the same.

Try It

1. Read and complete a lab safety form.
2. Hold a baseball in one hand and a foam ball in your other hand. Compare the masses of the two balls.
3. Lay both balls on a flat surface. Push a meterstick against the balls at the same time with the same force. Compare the accelerations of the ball.
4. Using only the baseball and the meterstick, lightly push the ball and observe its acceleration. Again observe the acceleration as you push the baseball with a stronger push. Compare the accelerations of the ball when you used a weak force and when you used a strong force.

Apply It

5. Answer the following questions for both step 3 and step 4. What variable did you change? What variable changed as a result? What variable did you keep the same?
6. Using your results, state the relationship between acceleration and mass if the net force on an object does not change. Then, state the relationship between acceleration and force if mass does not change.
7. **Key Concept** How do your results support Newton's second law of motion?

Lesson 4

Reading Guide

Key Concepts
ESSENTIAL QUESTIONS

- What is Newton's third law of motion?
- Why don't the forces in a force pair cancel each other?
- What is the law of conservation of momentum?

Vocabulary

Newton's third law of motion p. 71
force pair p. 71
momentum p. 73
law of conservation of momentum p. 74

Multilingual eGlossary

Newton's Third Law

Inquiry Why move up?

To reach the height she needs for her dive, this diver must move up into the air. Does she just jump up? No, she doesn't. She pushes down on the diving board and the diving board propels her into the air. How does pushing down cause the diver to move up?

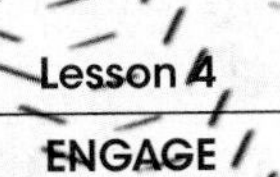

Inquiry Launch Lab

10 minutes

How do opposite forces compare? 

If you think about forces you encounter every day, you might notice forces that occur in pairs. For example, if you drop a rubber ball, the falling ball pushes against the floor. The ball bounces because the floor pushes with an opposite force against the ball. How do these opposite forces compare?

1. Read and complete a lab safety form.
2. Stand so that you face your lab partner, about half a meter away. Each of you should hold a **spring scale.**
3. Hook the two scales together, and gently pull them away from each other. Notice the force reading on each scale.
4. Pull harder on the scales, and again notice the force readings on the scales.
5. Continue to pull on both scales, but let the scales slowly move toward your lab partner and then toward you at a constant speed.

Think About This

1. Identify the directions of the forces on each scale. Record this information in your Science Journal.
2. **Key Concept** Describe the relationship you noticed between the force readings on the two scales.

Opposite Forces

Figure 20 When the skater pushes against the wall, the wall applies a force to the skater that pushes him away from the wall.

Have you ever been on in-line skates and pushed against a wall? When you pushed against the wall, like the boy is doing in **Figure 20,** you started moving away from it. What force caused you to move?

You might think the force of the muscles in your hands moved you away from the wall. But think about the direction of your push. You pushed against the wall in the opposite direction from your movement. It might be hard to imagine, but when you pushed against the wall, the wall pushed back in the opposite direction. The push of the wall caused you to accelerate away from the wall. When an object applies a force on another object, the second object applies a force of the same strength on the first object, but the force is in the opposite direction.

 **Reading Check** When you are standing, you push on the floor, and the floor pushes on you. How do the directions and strengths of these forces compare?

Newton's Third Law of Motion

Newton's first two laws of motion describe the effects of balanced and unbalanced forces on one object. Newton's third law relates forces between two objects. *According to* **Newton's third law of motion,** *when one object exerts a force on a second object, the second object exerts an equal force in the opposite direction on the first object.* An example of forces described by Newton's third law is shown in **Figure 21.** When the gymnast pushes against the vault, the vault pushes back against the gymnast. Notice that the lengths of the force arrows are the same, but the directions are opposite.

 **Key Concept Check** What is Newton's third law of motion?

Force Pairs

The forces described by Newton's third law depend on each other. *A* **force pair** *is the forces two objects apply to each other.* Recall that you can add forces to calculate the net force. If the forces of a force pair always act in opposite directions and are always the same strength, why don't they cancel each other? The answer is that each force acts on a different object. In **Figure 22,** the girl's feet act on the boat. The force of the boat acts on the girl's feet. The forces do not result in a net force of zero because they act on different objects. Adding forces can only result in a net force of zero if the forces act on the same object.

 **Key Concept Check** Why don't the forces in a force pair cancel each other?

Action and Reaction

In a force pair, one force is called the action force. The other force is called the reaction force. The girl in **Figure 22** applies an action force against the boat. The reaction force is the force that the boat applies to the girl. For every action force, there is a reaction force that is equal in strength but opposite in direction.

Newton's Third Law

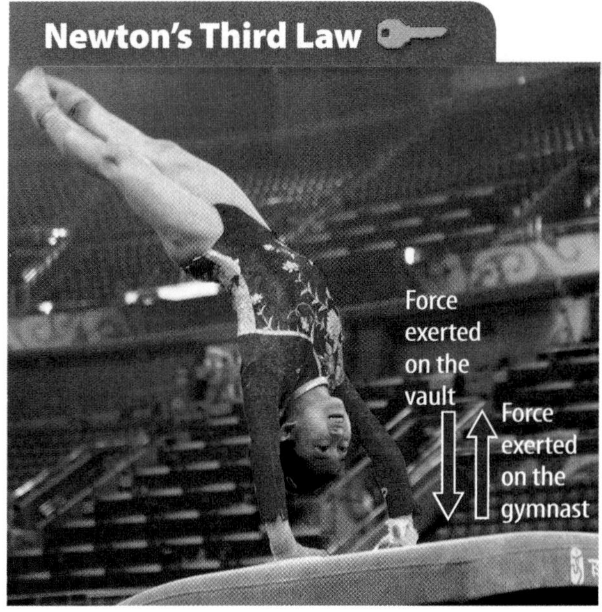

▲ **Figure 21** The force of the vault propels the gymnast upward.

Force Pairs

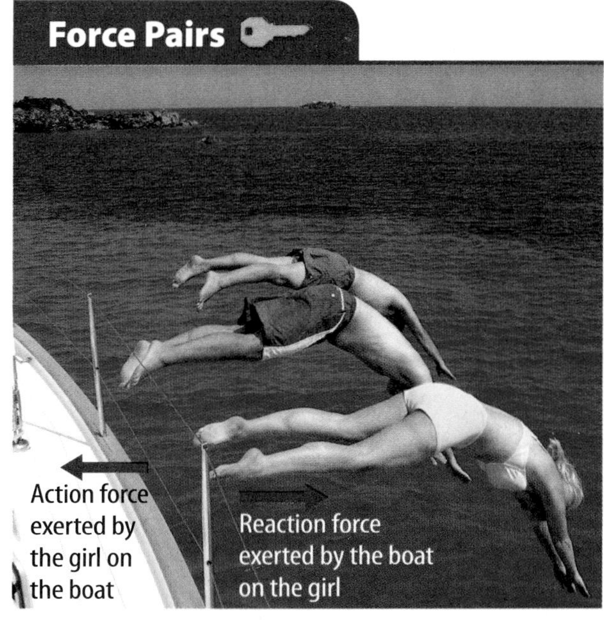

▲ **Figure 22** The force pair is the force the girl applies to the boat and the force that the boat applies to the girl.

Visual Check How can you tell that the forces don't cancel each other?

FOLDABLES®

Make a half-book from a sheet of notebook paper. Use it to summarize how Newton's third law explains the motion of a variety of common activities.

Using Newton's Third Law of Motion

When you push against an object, the force you apply is called the action force. The object then pushes back against you. The force applied by the object is called the reaction force. According to Newton's second law of motion, when the reaction force results in an unbalanced force, there is a net force, and the object accelerates. As shown in **Figure 23,** Newton's third law explains how you can swim and jump. It also explains how rockets can be launched into space.

Reading Check How does Newton's third law apply to the motion of a bouncing ball?

Action and Reaction Forces

Figure 23 Every action force has a reaction force in the opposite direction.

◀ **Swimming** When you swim, you push your arms against the water in the pool. The water in the pool pushes back on you in the opposite (forward) direction. If your arms push the water back with enough force, the reaction force of the water on your body is greater than the force of fluid friction. The net force is forward. You accelerate in the direction of the net force and swim forward through the water.

▶ **Jumping** When you jump, you push down on the ground, and the ground pushes up on you. The upward force of the ground combines with the downward force of gravity to form the net force acting on you. If you push down hard enough, the upward reaction force is greater than the downward force of gravity. The net force is upward. According to Newton's second law, your acceleration is in the same direction as the net force, so you accelerate upward.

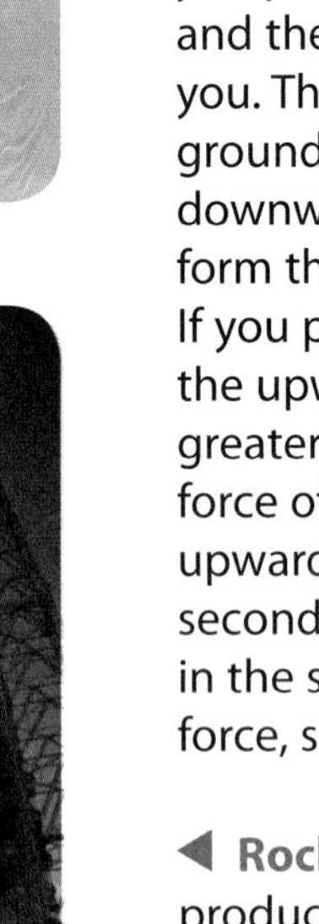

◀ **Rocket Motion** The burning fuel in a rocket engine produces a hot gas. The engine pushes the hot gas out in a downward direction. The gas pushes upward on the engine. When the upward force of the gas pushing on the engine becomes greater than the downward force of gravity on the rocket, the net force is in the upward direction. The rocket then accelerates upward.

Visual Check On what part of the swimmer's body does the water's reaction force push?

Momentum

Because action and reaction forces do not cancel each other, they can change the motion of objects. A useful way to describe changes in velocity is by describing momentum. **Momentum** *is a measure of how hard it is to stop a moving object.* It is the product of an object's mass and velocity. An object's momentum is in the same direction as its velocity.

WORD ORIGIN

momentum
from Latin *momentum,* means "movement, impulse"

Momentum Equation

momentum (in kg•m/s) = **mass** (in kg) × **velocity** (in m/s)

$$p = m \times v$$

If a large truck and a car move at the same speed, the truck is harder to stop. Because it has more mass, it has more momentum. If cars of equal mass move at different speeds, the faster car has more momentum and is harder to stop.

Newton's first two laws relate to momentum. According to Newton's first law, if the net force on an object is zero, its velocity does not change. This means its momentum does not change. Newton's second law states that the net force on an object is the product of its mass and its change in velocity. Because momentum is the product of mass and velocity, the force on an object equals its change in momentum.

Math Skills — Finding Momentum

Solve for Momentum What is the momentum of a 12-kg bicycle moving at 5.5 m/s?

1. **This is what you know:** mass: $m = 12$ kg; velocity: $v = 5.5$ m/s
2. **This is what you need to find:** momentum: p
3. **Use this formula:** $p = m \times v$
4. **Substitute:** the values for m and v into the formula and multiply. $p = \mathbf{12\ kg} \times \mathbf{5.5\ m/s} = 66$ kg•m/s

Answer: The momentum of the bicycle is 66 kg•m/s in the direction of the velocity.

Review
- Math Practice
- Personal Tutor

Practice

1. What is the momentum of a 1.5-kg ball rolling at 3.0 m/s?
2. A 55-kg woman has a momentum of 220 kg•m/s. What is her velocity?

Conservation of Momentum

You might have noticed that if a moving ball hits another ball that is not moving, the motion of both balls changes. The cue ball in **Figure 24** has momentum because it has mass and is moving. When it hits the other balls, the cue ball's velocity and momentum decrease. Now the other balls start moving. Because these balls have mass and velocity, they also have momentum.

The Law of Conservation of Momentum

In any collision, one object transfers momentum to another object. The billiard balls in **Figure 24** gain the momentum lost by the cue ball. The total momentum, however, does not change. *According to the* **law of conservation of momentum,** *the total momentum of a group of objects stays the same unless outside forces act on the objects.* Outside forces include friction. Friction between the balls and the billiard table decreases their velocities, and they lose momentum.

Figure 24 The total momentum of all the balls is the same before and after the collision.

 **Key Concept Check** What is the law of conservation of momentum?

Types of Collisions

Objects collide with each other in different ways. When colliding objects bounce off each other, it is an elastic collision. If objects collide and stick together, such as when one football player tackles another, the collision is inelastic. No matter the type of collision, the total momentum will be the same before and after the collision.

Inquiry MiniLab **15 minutes**

Is momentum conserved during a collision? 

1. Read and complete a lab safety form.
2. Make a track by using **masking tape** to secure two **metersticks** side by side on a table, about 4 cm apart.
3. Place two **tennis balls** on the track. Roll one ball against the other. Then, roll the balls at about the same speed toward each other.
4. Place the balls so that they touch. Observe the collision as you gently roll **another ball** against them.

Analyze and Conclude

1. **Explain** how you know that momentum was transferred from one ball to another.
2. **Key Concept** What could you measure to show that momentum is conserved?

Lesson 4 Review

Assessment Online Quiz

Inquiry Virtual Lab

Visual Summary

Newton's third law of motion describes the force pair between two objects.

For every action force, there is a reaction force that is equal in strength but opposite in direction.

In any collision, momentum is transferred from one object to another.

FOLDABLES

Use your lesson Foldable to review the lesson. Save your Foldable for the project at the end of the chapter.

What do you think NOW?

You first read the statements below at the beginning of the chapter.

7. If objects collide, the object with more mass applies more force.

8. Momentum is a measure of how hard it is to stop a moving object.

Did you change your mind about whether you agree or disagree with the statements? Rewrite any false statements to make them true.

Use Vocabulary

1. **Define** *momentum* in your own words.
2. The force of a bat on a ball and the force of a ball on a bat are a(n) ________.

Understand Key Concepts

3. **State** Newton's third law of motion.
4. A ball with momentum 16 kg•m/s strikes a ball at rest. What is the total momentum of both balls after the collision?

   A. –16 kg•m/s **C.** 8 kg•m/s
   B. –8 kg•m/s **D.** 16 kg•m/s

5. **Identify** A child jumps on a trampoline. The trampoline bounces her up. Why don't the forces cancel each other?

Interpret Graphics

6. **Predict** what will happen to the velocity and momentum of each ball when the small ball hits the heavier large ball?

7. **Organize** Copy and fill in the table.

Event	Action Force	Reaction Force
A girl kicks a soccer ball.		
A book sits on a table.		

Critical Thinking

8. **Decide** How is it possible for a bicycle to have more momentum than a truck?

Math Skills

Review

Math Practice

9. A 2.0-kg ball rolls to the right at 3.0 m/s. A 4.0-kg ball rolls to the left at 2.0 m/s. What is the momentum of the system after a head-on collision of the two balls?

Lab

40 minutes

Modeling Newton's Laws of Motion

Materials

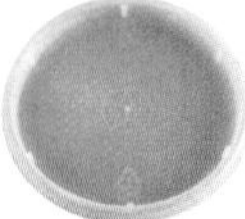

plastic lid

golf ball

modeling clay

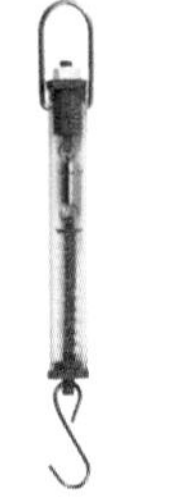

2.5-N spring scales (2)

Safety

Newton's first and second laws of motion describe the relationship between unbalanced forces and motion. These laws relate to forces acting on one object. Newton's third law describes the strength and direction of force pairs. This law relates to forces on two different objects. You can learn about all three of Newton's laws of motion by modeling them.

Question

How can you model Newton's laws of motion?

Procedure

1. Read and complete a lab safety form.
2. Attach a spring scale to a plastic lid. Add mass to the lid by placing a ball of modeling clay on it.
3. Slowly pull the lid along a table with the spring scale. Record the force reading on the scale in your Science Journal.
4. Try to use the spring scale to pull the lid with constant force and constant speed.
5. Try pulling the lid with increasing force and constant speed.
6. Pull the lid so that it accelerates quickly.
7. Increase the mass of the lid by adding more modeling clay. Repeat steps 3–6.
8. Replace the modeling clay with a golf ball. Try pulling the lid slowly. Then, try pulling it from a standstill quickly. What happens to the ball in each case? Record your results.

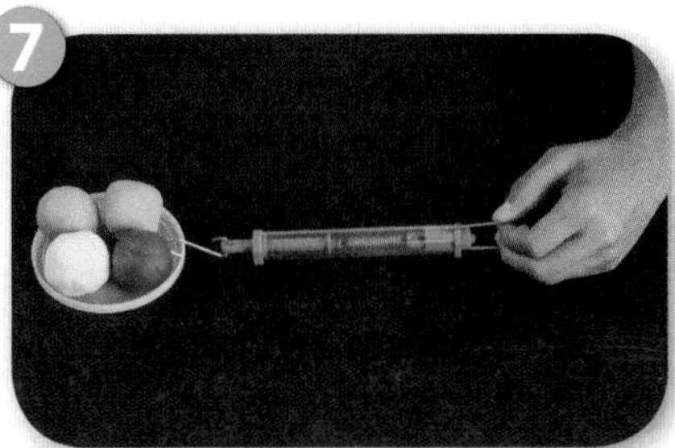

9 To model Newton's first law of motion, design an activity using the lid and the spring scales that shows that a net force of zero does not change the motion of an object.

10 To model Newton's second law of motion, design an activity that shows that if the net force acting on an object is not zero, the object accelerates.

11 To model Newton's third law of motion, plan an activity that shows action and reaction forces on an object.

12 After your teacher approves your plan, perform the activities.

Analyze and Conclude

13 **Identify** the variables in each of your models. Which variables changed and which remained constant?

14 **The Big Idea** For each law of motion that you modeled, how did the force applied to the lid relate to the motion of the lid?

Lab Tips

- ☑ Use a smooth surface so that the lid moves easily.
- ☑ You might want to make a data table in which you can record your observations and the force readings.

Communicate Your Results

Choose one of the laws of motion, and model it for the class. Compare your model with the method of modeling used by other lab groups.

Inquiry Extension

Describe another way you could model Newton's three laws of motion using materials other than those used in this lab. For example, for Newton's first law of motion, you could pedal a bicycle at a constant speed.

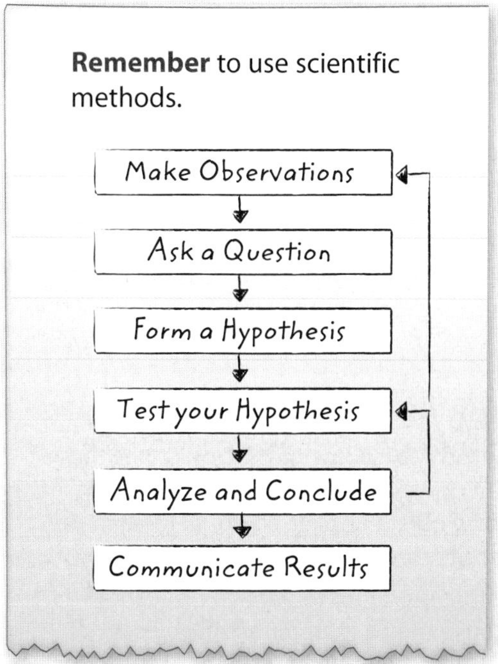

Chapter 2 Study Guide

An object's motion changes if a net force acts on the object.

Key Concepts Summary

Lesson 1: Gravity and Friction

- Friction is a **contact force.** Magnetism is a **noncontact force.**
- The law of universal gravitation states that all objects are attracted to each other by **gravity.**
- **Friction** can stop or slow down objects sliding past each other.

Vocabulary

force p. 45
contact force p. 45
noncontact force p. 46
gravity p. 47
mass p. 47
weight p. 48
friction p. 49

Lesson 2: Newton's First Law

- An object's motion can only be changed by **unbalanced forces.**
- According to **Newton's first law of motion,** the motion of an object is not changed by **balanced forces** acting on it.
- **Inertia** is the tendency of an object to resist a change in its motion.

Vocabulary

net force p. 55
balanced forces p. 56
unbalanced forces p. 56
Newton's first law of motion p. 57
inertia p. 58

Lesson 3: Newton's Second Law

- According to **Newton's second law of motion,** an object's acceleration is the net force on the object divided by its mass.
- In **circular motion,** a **centripetal force** pulls an object toward the center of the curve.

Vocabulary

Newton's second law of motion p. 65
circular motion p. 66
centripetal force p. 66

Lesson 4: Newton's Third Law

- **Newton's third law of motion** states that when one object applies a force on another, the second object applies an equal force in the opposite direction on the first object.
- The forces of a **force pair** do not cancel because they act on different objects.
- According to the **law of conservation of momentum,** momentum is conserved during a collision unless an outside force acts on the colliding objects.

Vocabulary

Newton's third law of motion p. 71
force pair p. 71
momentum p. 73
law of conservation of momentum p. 74

Study Guide

- Personal Tutor
- Vocabulary eGames
- Vocabulary eFlashcards

FOLDABLES® Chapter Project

Ass emble your lesson Foldables as shown to make a Chapter Project. Use the project to review what you have learned in this chapter.

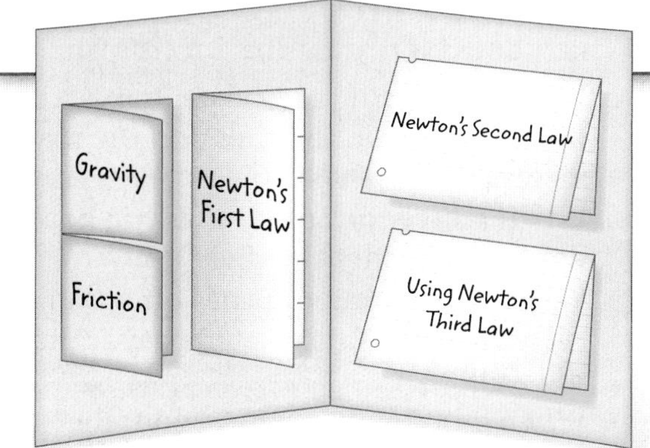

Use Vocabulary

1. Kilograms is the SI unit for measuring ________.
2. The force of gravity on an object is its ________.
3. The sum of all the forces on an object is the ________.
4. An object that has ________ acting on it acts as if there were no forces acting on it at all.
5. A car races around a circular track. Friction on the tires is the ________ that acts toward the center of the circle and keeps the car on the circular path.
6. A heavy train requires nearly a mile to come to a complete stop because it has a lot of ________.

Link Vocabulary and Key Concepts

Concepts in Motion Interactive Concept Map

Copy this concept map, and then use vocabulary terms from the previous page to complete the concept map.

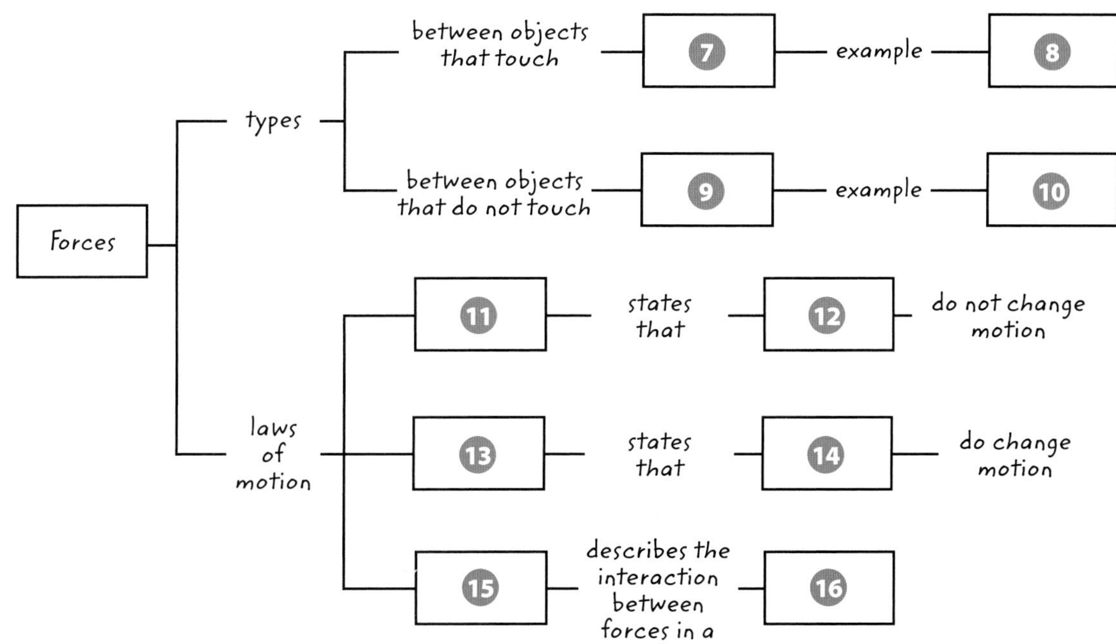

Chapter 2 Review

Understand Key Concepts

1 The arrows in the figure represent the gravitational force between marbles that have equal mass.

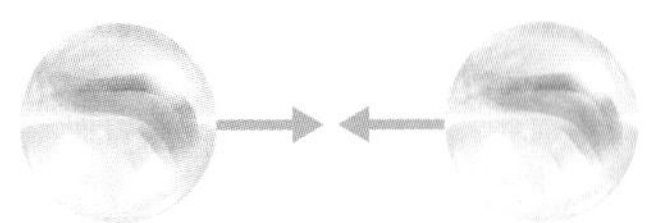

How should the force arrows look if a marble that has greater mass replaces one of these marbles?

A. Both arrows should be drawn longer.
B. Both arrows should stay the same length.
C. The arrow from the marble with less mass should be longer than the other arrow.
D. The arrow from the marble with less mass should be shorter than the other arrow.

2 A person pushes a box across a flat surface with a force less than the weight of the box. Which force is weakest?

A. the force of gravity on the box
B. the force of the table on the box
C. the applied force against the box
D. the sliding friction against the box

3 A train moves at a constant speed on a straight track. Which statement is true?

A. No horizontal forces act on the train as it moves.
B. The train moves only because of its inertia.
C. The forces of the train's engine balances friction.
D. An unbalanced force keeps the train moving.

4 The Moon orbits Earth in a nearly circular orbit. What is the centripetal force?

A. the pull of the Moon on Earth
B. the outward force on the Moon
C. the Moon's inertia as it orbits Earth
D. Earth's gravitational pull on the Moon

5 A 30-kg television sits on a table. The acceleration due to gravity is 10 m/s^2. What force does the table exert on the television?

A. 0.3 N
B. 3 N
C. 300 N
D. 600 N

6 Which does NOT describe a force pair?

A. When you push on a bike's brakes, the friction between the tires and the road increases.
B. When a diver jumps off a diving board, the board pushes the diver up.
C. When an ice skater pushes off a wall, the wall pushes the skater away from the wall.
D. When a boy pulls a toy wagon, the wagon pulls back on the boy.

7 A box on a table has these forces acting on it.

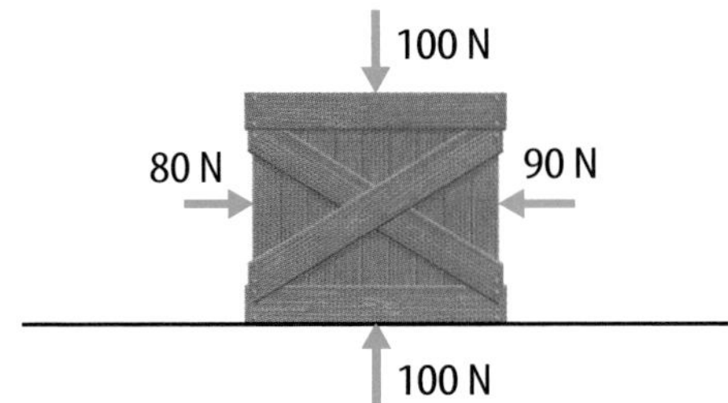

What is the static friction between the box and the table?

A. 0 N
B. 10 N
C. greater than 10 N
D. between 0 and 10 N

8 A 4-kg goose swims with a velocity of 1 m/s. What is its momentum?

A. 4 N
B. 4 kg•m/s^2
C. 4 kg•m/s
D. 4 m/s^2

Chapter Review

Assessment

Online Test Practice

Critical Thinking

9 **Predict** If an astronaut moved away from Earth in the direction of the Moon, how would the gravitational force between Earth and the astronaut change? How would the gravitational force between the Moon and the astronaut change?

10 **Analyze** A box is on a table. Two people push on the box from opposite sides. Which of the labeled forces make up a force pair? Explain your answer.

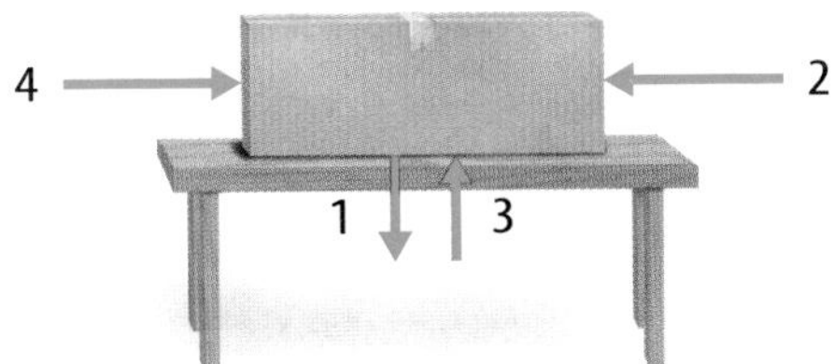

11 **Conclude** A refrigerator has a maximum static friction force of 250 N. Sam can push the refrigerator with a force of 130 N. Amir and André can each push with a force of 65 N. How could they all move the refrigerator? Will the refrigerator move with constant velocity? Why or why not?

12 **Give an example** of unbalanced forces acting on an object.

13 **Infer** Two skaters stand on ice. One weighs 250 N, and the other weighs 500 N. They push against each other and move in opposite directions. Which one will travel farther before stopping? Explain your answer.

Writing in Science

14 Imagine that you are an auto designer. Your job is to design brakes for different automobiles. Write a four-sentence plan that explains what you need to consider about momentum when designing brakes for a heavy truck, a light truck, a small car, and a van.

REVIEW THE BIG IDEA

15 Explain how balanced and unbalanced forces affect objects that are not moving and those that are moving.

16 The photo below shows people on a carnival swing ride. How do forces change the motion of the riders?

Math Skills

Math Practice

Solve One-Step Equations

17 A net force of 17 N is applied to an object, giving it an acceleration of 2.5 m/s^2. What is the mass of the object?

18 A tennis ball's mass is about 0.60 kg. Its velocity is 2.5 m/s. What is the momentum of the ball?

19 A box with a mass of 0.82 kg has these forces acting on it.

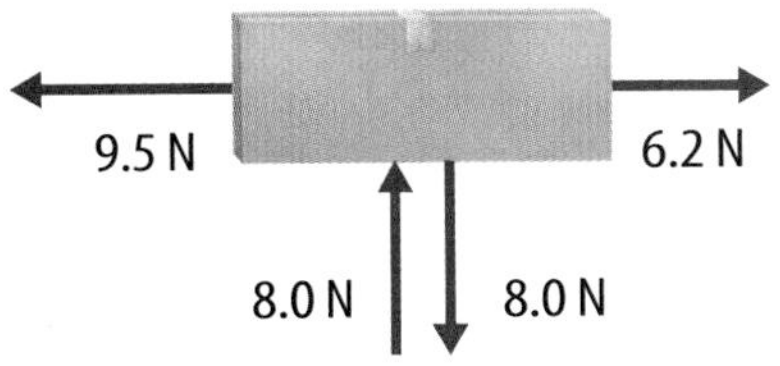

What is the strength and direction of the acceleration of the box?

Standardized Test Practice

Record your answers on the answer sheet provided by your teacher or on a sheet of paper.

Multiple Choice

1 A baseball has an approximate mass of 0.15 kg. If a bat strikes the baseball with a force of 6 N, what is the acceleration of the ball?

A 4 m/s^2

B 6 m/s^2

C 40 m/s^2

D 60 m/s^2

Use the diagram below to answer question 2.

2 The person in the diagram above is unable to move the crate from its position. Which is the opposing force?

A gravity

B normal force

C sliding friction

D static friction

3 The mass of a person on Earth is 72 kg. What is the mass of the same person on the Moon where gravity is one-sixth that of Earth?

A 12 kg

B 60 kg

C 72 kg

D 432 kg

4 A swimmer pushing off from the wall of a pool exerts a force of 1 N on the wall. What is the reaction force of the wall on the swimmer?

A 0 N

B 1 N

C 2 N

D 10 N

Use the diagram below to answer questions 5 and 6.

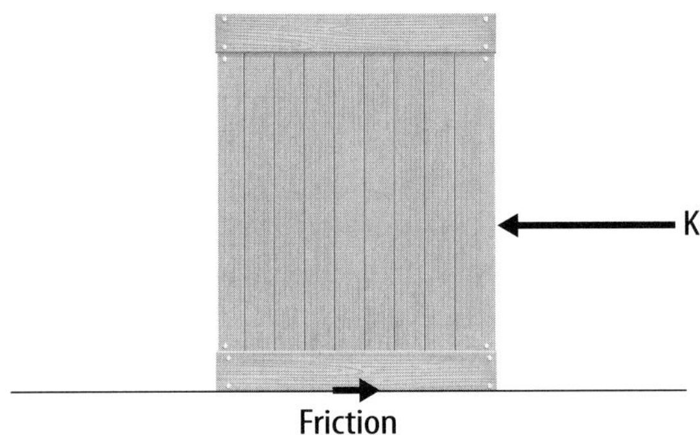

5 Which term applies to the forces in the diagram above?

A negative

B positive

C reference

D unbalanced

6 In the diagram above, what happens when force *K* is applied to the crate at rest?

A The crate remains at rest.

B The crate moves back and forth.

C The crate moves to the left.

D The crate moves to the right.

7 What is another term for change in velocity?

A acceleration

B inertia

C centripetal force

D maximum speed

Use the diagram below to answer question 8.

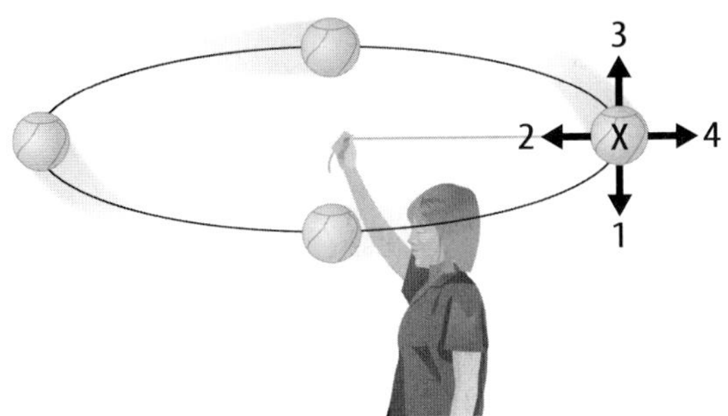

8 The person in the diagram is spinning a ball on a string. When the ball is in position *X*, what is the direction of the centripetal force?

A 1

B 2

C 3

D 4

9 Which is ALWAYS a contact force?

A electric

B friction

C gravity

D magnetic

10 When two billiard balls collide, which is ALWAYS conserved?

A acceleration

B direction

C force

D momentum

Constructed Response

Use the table below to answer question 11.

Newton's Laws of Motion	Explanation
First	
Second	
Third	

11 Explain each of Newton's laws of motion. What is one practical application of each law?

Use the diagram below to answer questions 12 and 13.

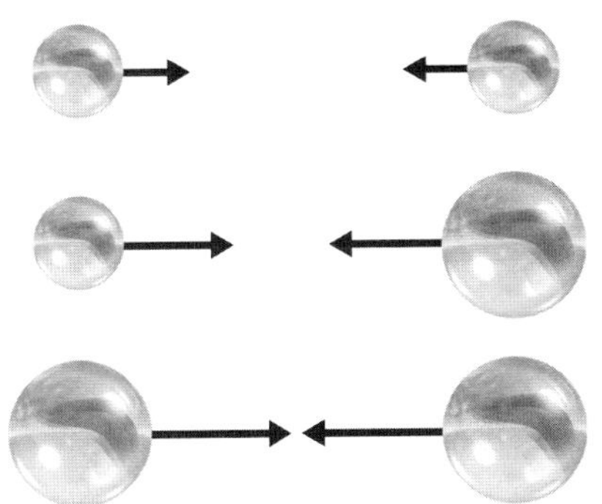

12 The arrows in the diagram above represent forces. What scientific law does the diagram illustrate? What does the law state?

13 Using the diagram, explain how marble mass affects gravitational attraction.

NEED EXTRA HELP?													
If You Missed Question...	1	2	3	4	5	6	7	8	9	10	11	12	13
Go to Lesson...	3	1	1	4	2	2	3	3	1	4	2–4	1	1

Chapter 3

Work and Simple Machines

THE BIG IDEA **How do machines make doing work easier?**

Inquiry Hard Work or Not?

Digging this hole with a hand shovel would be hard work. Using an earthmover makes the task easier to do. You probably think writing an essay for English class is hard work, but is it?

- What do you think work is?
- How do you think work and energy are related?
- How do you think machines make doing work easier?

Get Ready to Read

What do you think?

Before you read, decide if you agree or disagree with each of these statements. As you read this chapter, see if you change your mind about any of the statements.

1. You do work when you push a book across a table.
2. Doing work faster requires more power.
3. Machines always decrease the force needed to do a job.
4. A well-oiled, low-friction machine can be 100 percent efficient.
5. A doorknob is a simple machine.
6. A loading ramp makes it easier to lift a load.

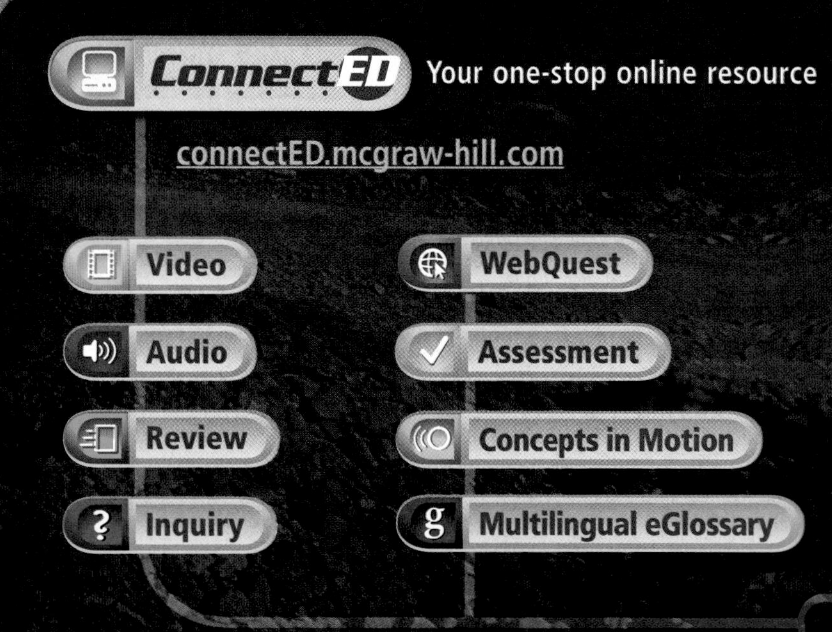

Lesson 1

Work and Power

Reading Guide

Key Concepts
ESSENTIAL QUESTIONS

- What must happen for work to be done?
- How does doing work on an object change its energy?
- How are work and power related?

Vocabulary

work p. 87

power p. 91

Multilingual eGlossary

Inquiry Enough Power?

A powerful tugboat can pull this large cargo ship. Could a much smaller boat also get this work done? As you will soon read, work and power are closely related.

Inquiry Launch Lab

10 minutes

How do you know when work is done?

Many things that seem like hard work to you are not work at all in the scientific sense. This activity explores the scientific meaning of work.

1. Read and complete a lab safety form.
2. Tie a **string** around a **book.** Center the string on the book, and place the book on a desk.
3. Attach a **spring scale** to the string. Have a partner read the scale as you pull on the scale as shown in the photo. Pull very gently at first and then slowly increase how hard you pull.
4. Record in your Science Journal the scale reading when the book begins to move.

Think About This

1. What amount of force was needed to make the book move? How do you know?
2. **Key Concept** How do you think force and motion might be related to doing work? Explain your reasoning.

What is work?

Students might say studying and doing homework is hard work. To a scientist, these things take no work at all. Why? Because the scientific meaning of work includes forces and motion, not concentration. In science, **work** *is the transfer of energy to an object by a force that makes an object move in the direction of the force. Work is only being done while the force is applied to the object.*

Word Origin
work
from Old English *weorc,* means "activity"

Imagine pushing your bicycle into a bike rack. Your push (a force) makes your bicycle (an object) move. Therefore, work is done. You do no work if you push on the bicycle and it does not move. A force that does not make the object move does no work. The weight lifter shown in **Figure 1** does work in one example but not in the other.

Figure 1 The weight lifter does work only when he moves the weights.

The weight lifter does work on the weights when he exerts a force that makes the weights move.

Although the weight lifter exerts force, no work is done on the weights because the weights are not moving.

FOLDABLES®

Use two sheets of paper to make a layered book. Label it as shown. Use it to organize your notes on work.

Calculating Work

The amount of work done is easy to calculate. You must know two things to calculate work—force and distance, as shown in the following equation.

Work Equation

Work (in joules) = **force** (in newtons) × **distance** (in meters)

$$W = Fd$$

The force must be in newtons (N) and distance must be in meters (m). When you multiply force and distance together, the result has units of newton-meter (N•m). The newton-meter is also known as the joule (J). Like other types of energy, work is measured in joules. The joule is the SI unit of work and energy.

The distance in the work equation is the distance the object moves while the force is acting on it. Suppose you push on a book over a distance of 0.25 m and then the book slides 3.0 m. Which distance do you use? You calculate the work done using 0.25 m because the force was applied along that distance.

 **Key Concept Check** How is work done?

Math Skills Work Equation

Solve for Work A student pushes a desk 2.0 m across the floor using a constant force of 25.0 N. How much work does the student do on the desk?

1. **This is what you know:** force: $F = 25.0$ N; distance: $d = 2.0$ m
2. **This is what you need to find:** work: W
3. **Use this formula:** $W = Fd$
4. **Substitute:** values for F and d into the formula and multiply: $W = (\mathbf{25.0\ N}) \times (\mathbf{2.0\ m}) = 50.0$ N•m
5. **Convert units:** (N) × (m) = N•m = J

Answer: The amount of work done is 50.0 J.

Practice

A child pushes a toy truck 2.5 m across a floor with a constant force of 22 N. How much work does the child do on the toy truck?

- Math Practice
- Personal Tutor

Forces and Work

Figure 2 A force that acts in the direction of the motion does work.

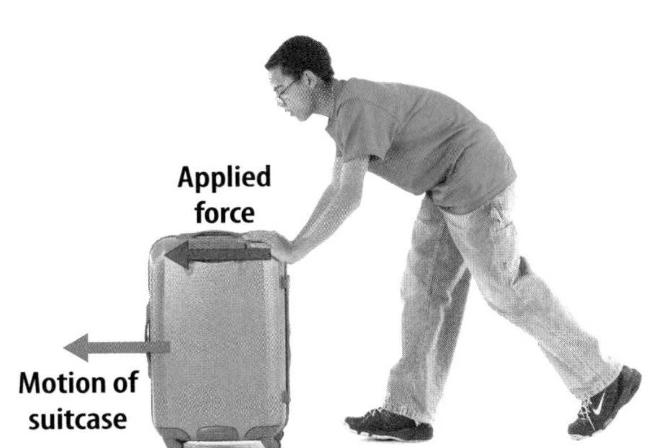

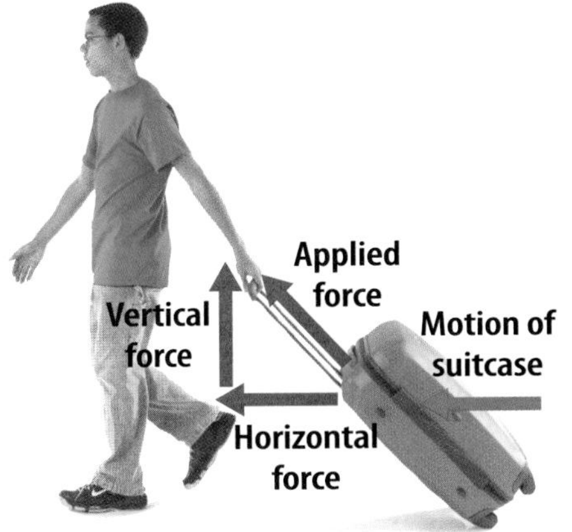

Visual Check How are the applied forces different in each example?

Factors That Affect Work

The work done on an object depends on the direction of the force applied and the direction of motion, as shown in Figure 2. Sometimes the force and the motion are in the same direction, such as when you push a suitcase along the floor. To calculate the work in this case, simply multiply the force and the distance.

Force at an Angle Now imagine pulling a wheeled suitcase, as shown in Figure 2. Note that the suitcase moves along the floor, but the force acts at an angle to the direction of the suitcase's motion. In other words, the force and the motion are not in the same direction. How do you calculate work?

Only the part of the force that acts in the direction of motion does work. Notice in Figure 2 that the force has a horizontal part and a vertical part. Only the horizontal part of the force moves the suitcase across the floor. Therefore, only the horizontal part of the force is used in the work equation. The vertical part of the force does no work on the suitcase.

Reading Check What part of a force does work when the direction of the force and the direction of motion are not the same?

Inquiry MiniLab — 20 minutes

What affects work?

Explore how different surfaces affect the work needed to pull an object across it.

1. Read and complete a lab safety form.
2. Tie a **string** around the center of a **book.** Attach a **spring scale** to the string.
3. Place the book on a desk next to a **meterstick.** Have a partner read the scale as you pull horizontally on it. Determine the least amount of force needed to slide the book 0.2 m at a constant speed. Record your data in your Science Journal in a table like the one below.
4. Repeat step 3 using the floor and other surfaces as instructed by your teacher.

Sliding Book Data

Surface	Force (N)	Work (J)
Desk		
Floor		

Analyze and Conclude

1. **Determine** the work done in each trial.
2. **Compare** the surface requiring the least force to the one requiring the most force.
3. **Key Concept** What caused the change in work done for each surface?

Upward force
Weight

▲ **Figure 3** The amount of force needed to lift the backpack is equal to the weight of the backpack.

Lifting Objects Lifting your backpack requires you to do work. The backpack has weight because of the downward force of gravity acting on it. To lift your backpack, you must pull upward with a force equal to or greater than the backpack's weight, as shown in **Figure 3.** The work done to lift any object is equal to the weight of the object multiplied by the distance it is lifted.

Reading Check How do you calculate the work done when lifting an object?

Work and Energy

Doing work on an object transfers energy to the object. This is important to scientists because it helps them predict how an object will act when forces are applied to it. Recall that moving objects have kinetic energy. The boy in **Figure 4** does work on the tray when he applies a force that makes the tray move. This work transfers energy to the tray. The added energy is the kinetic energy of the moving tray.

Work done when you lift an object also increases the object's energy. Recall that the gravitational potential energy of an object increases as its height above the ground increases. When the girl in **Figure 4** lifts the tray the gravitational potential energy of the tray increases. The tray also gains kinetic energy as the tray moves upward. The girl did work on the tray as she lifted it. She transferred energy to the tray, increasing the tray's potential and kinetic energy.

Key Concept Check How does doing work on an object change its energy?

Figure 4 Doing work on a tray transfers energy to it. The energy change can be in the form of kinetic energy or potential energy.

Visual Check How does the tray's energy change when it is lifted?

The girl does work on the tray as she lifts it upward. The moving tray has kinetic energy. As the tray is raised, it gains potential energy. The energy comes from the work done on the tray.
Motion of tray
Force
Force
Motion of tray
Force
Force
The boy does work on the tray as he pushes it along the counter. The kinetic energy of the moving tray comes from the work done on the tray.

What is power?

What does it mean to have power? If two weight lifters lift identical weights to the same height, they both do the same amount of work. How quickly a weight lifter lifts a weight does not change the amount of work done. Doing work more quickly, does affect **power**—*the rate at which work is done.* A weight lifter lifting a weight more quickly exerts more power.

Knowing the power required for a task enables engineers to properly size engines and motors. You can calculate power by dividing the work done by the time needed to do the work. In the power equation, work done is in joules (J) and time is in seconds (s). The SI unit of power is the watt (W). Note that 1 J/s = 1 W.

Science Use v. Common Use

power
Science Use the rate at which work is done

Common Use the ability to accomplish something or to command or control other people

Power Equation

$$\text{power (in watts)} = \frac{\text{work (in joules)}}{\text{time (in seconds)}}$$

$$P = \frac{W}{t}$$

You know that power is the rate at which work is done. You also know that work **transfers** energy. Thus, you can think of power as the rate at which energy is transferred to an object.

Academic Vocabulary

transfer
(verb) to move from one place to another

Key Concept Check How are work and power related?

Math Skills — Power Equation

Solve for Power A boy does 18 J of work in 2.0 s on his backpack as he lifts it from a table. How much power did the boy use on the backpack?

1. **This is what you know:** work: $W = 18$ J time: $t = 2.0$ s
2. **This is what you need to find:** power: P
3. **Use this formula:** $P = \frac{W}{t}$
4. **Substitute:** values for W and t into the formula and divide $P = \frac{18\text{ J}}{2.0\text{ s}} = 9.0$ J/s
5. **Convert units:** J/s = W $\frac{18\text{ J}}{2.0\text{ s}}$

The symbol for work, *W*, is usually italicized. However, the abbreviation for watt, W, is not italicized.

Answer: The amount of power used was 9.0 W.

Practice
A child pulls a wagon, doing 360 J of work in 8.0 s. How much power is exerted?

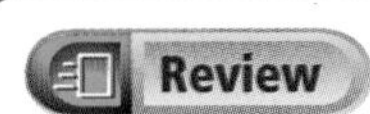
- Math Practice
- Personal Tutor

Lesson 1 Review

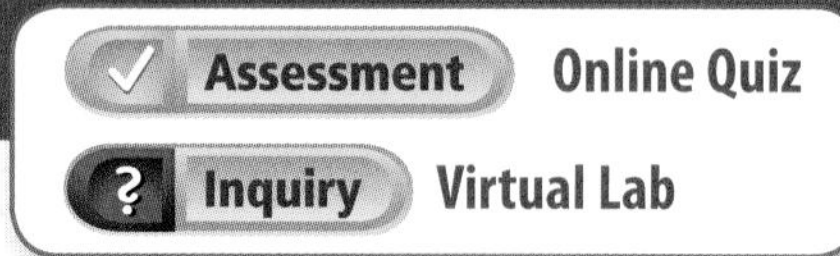

Visual Summary

Work is done on an object when the object moves in the direction of the applied force.

When work is done on an object, energy is transferred to the object.

To increase power, work must be done in less time.

FOLDABLES®

Use your lesson Foldable to review the lesson. Save your Foldable for the project at the end of the chapter.

What do you think NOW?

You first read the statements below at the beginning of the chapter.

1. Work is done when you push a book across a table.

2. Doing work faster requires more power.

Did you change your mind about whether you agree or disagree with the statements? Rewrite any false statements to make them true.

Use Vocabulary

1. **Use the term** *work* in a sentence.
2. **Define** *power* in your own words.

Understand Key Concepts 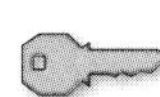

3. **Explain** how work and power are related.
4. Lifting a stone block 146 m to the top of the Great Pyramid required 146,000 J of work. How much work was done to lift the block halfway to the top?

   A. 36,500 J **C.** 146,000 J
   B. 73,000 J **D.** 292,000 J

5. **Give** an example of a situation in which doing work on an object changed its energy. Explain how the energy changed.

Interpret Graphics

6. **Explain** why this motionless weight lifter is not doing work.

7. **Determine Cause and Effect** Copy and complete the graphic organizer below to list two ways power can be increased.

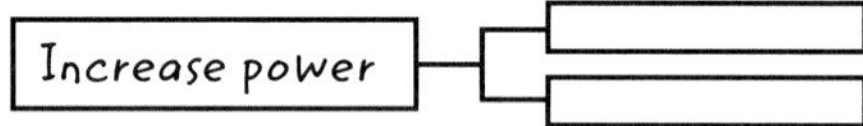

Critical Thinking

8. **Explain** When you type on a computer keyboard, do you do work on the computer keys? Explain your answer.

Math Skills

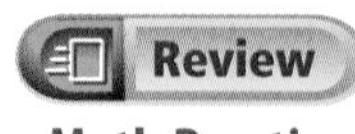

Math Practice

9. A motor applies a 5000-N force and raises an elevator 10 stories. If each story is 4 m tall, how much work does the motor do?
10. Calculate the power, in watts, needed to mow a lawn in 50 minutes if the work required is 500,000 J.

SCIENCE & SOCIETY

What is horsepower?

You might be surprised to learn that there is a connection between a horse and a steam engine.

In the early 1700s, horses did work on farms, powered factories, and moved vehicles. When people spoke of power, they literally were referring to horsepower. It was natural, then, for James Watt to think in terms of horsepower when he set out to improve the steam engine.

Watt did not invent the steam engine, but he realized its potential. He also realized that he needed a way to measure the power produced by steam engines. Watt knew that fabric mills used horses to power machinery. A worker attached a horse to a power wheel. The horse turned the wheel by walking in a 24-ft diameter circle at a rate of two revolutions per minute. From this information, Watt calculated the power supplied by a horse to be about 33,000 foot-pounds per minute. This amount of power became known as 1 horsepower (1 hp).

▼ A draft horse powering a mill

▲ An early steam engine developed by James Watt

Watt succeeded in making better steam engines. Eventually, some steam engines produced more than 200 hp. Something unexpected happened as a result of all this power—life changed. More work was done and done faster. The mills expanded. People moved to the cities to work in the mills. Populations of cities in industrialized countries increased. The world changed because steam engines easily could produce more horsepower than horses.

▼ With cheap power came factories, jobs, and pollution.

It's Your Turn

RESEARCH Steam engines are seldom used today. Research steam engines to determine why the power source that was key to the Industrial Revolution is no longer used.

Lesson 2

Reading Guide

Key Concepts 

ESSENTIAL QUESTIONS

- What are three ways a machine can make doing work easier?
- What is mechanical advantage?
- Why can't the work done by a machine be greater than the work done on the machine?

Vocabulary

mechanical advantage p. 98

efficiency p. 99

g Multilingual eGlossary

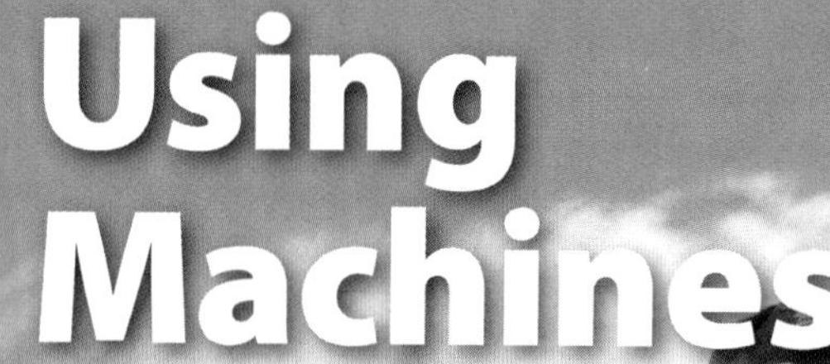

Using Machines

Inquiry Could it be worse?

Have you ever shoveled snow? You probably thought it was hard work. But imagine moving all of the snow only using your hands instead of a shovel. A shovel is a simple machine that makes moving snow easier.

Inquiry Launch Lab — 15 minutes

How do machines work?

Bicycles, pencil sharpeners, staplers, and doorknobs are machines that make specific tasks easier to do. How does the way a machine operates make work easier to do?

1. Read and complete a lab safety form.
2. Examine a **can opener** and answer these questions.
   - How many different ways does the can opener move?
   - Where do you apply force to the can opener?
   - Where does the can opener apply force to the can?
   - Is the amount of force you apply to the can opener different from the force the can opener applies to the can?
   - Does the can opener change the direction of the forces applied to it?
3. Use the can opener to open a **can.** Record the process step by step in your Science Journal.

Think About This

1. Why is a can opener considered a machine? What task does it make easier to do?
2. **Key Concept** Review your observations. Identify several ways the can opener makes opening a can easier.

What is a machine?

If you ride a bicycle to school, you have firsthand experience with a machine. A machine is any device that makes doing something easier. The snow shovel on the previous page is a machine that is used to move snow. The scissors and the watch shown in **Figure 5** are also machines. As these example show, some machines are simple and other machines are more complex.

Like a pair of scissors, a leaf rake, broom, screwdriver, baseball bat, shovel, and doorknob are all machines. Other machines, such as a watch, an automobile, a snowblower, and a lawn mower, are more complex. All machines make tasks easier, but they do not decrease the amount of work required. Instead, a machine changes the way in which the work is done.

Reading Check Do machines decrease the amount of work needed for a task? Explain.

Figure 5 A machine makes work easier to do regardless of whether it is simple or complex.

Inquiry MiniLab — 20 minutes

Does a ramp make it easier to lift a load?

In this lab, you will use a ramp and determine why it is useful for lifting heavy loads.

1. Read and complete a lab safety form.
2. Attach a **toy car** to the hook of a **spring scale.** Slowly lift the car off the table. Record in your Science Journal the force shown on the scale.
3. Lean a 30–40-cm **wood board** against a stack of three **textbooks** as shown. Use the scale to slowly pull the car up the ramp at a constant speed. Record the force.

Analyze and Conclude

1. **Compare** How did the force needed to lift the car off the table compare with the force needed to pull the car up the ramp?
2. **Key Concept** How does a ramp make work easier? Does the ramp decrease the amount of work done? Explain.

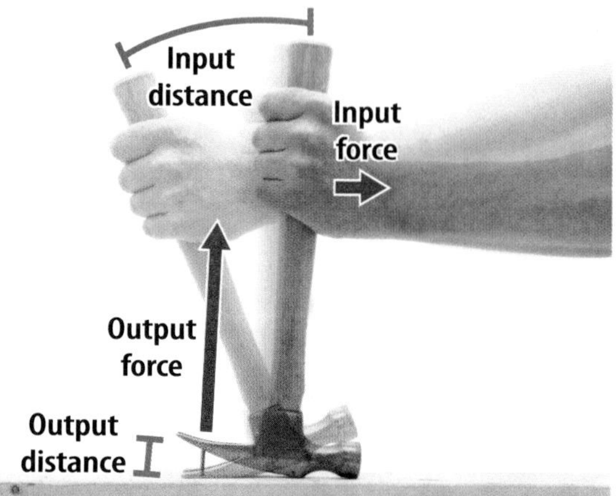

Figure 6 You exert an input force on the hammer. The hammer exerts an output force on the nail.

Input Force to Output Force

To use a machine, such as a hammer, you must apply a force to it. This force is the input force. The machine changes the input force into an output force, as shown in **Figure 6.** You apply an input force when you pull on the hammer's handle. The hammer changes the input force to an output force that pulls the nail out of the board.

Input Work to Output Work

Squeezing the handles of a pair of scissors makes the blades move. You apply an input force that moves part of the machine—the scissors—and does work. The work is called input work, W_{in}. It is the product of the input force and the distance the machine moves in the direction of the input force.

Machines convert, or change, input work to output work. They do this by applying an output force on something and making it move. The output work, W_{out}, is the product of the output force and the distance part of the machine moves in the direction of the output force. The examples in **Figure 7** on the next page show these relationships.

How do machines make work easier to do?

A machine can make work easier in three ways. It can change

- the size of a force;
- the distance the force acts;
- the direction of a force.

How hard it would be to pull a nail out of a board using your fingers? A hammer makes it easier in three ways. It changes the sizes of the forces and the distances the forces act. Notice that the person applies a smaller force over a longer distance, and the hammer exerts a greater force over a shorter distance. The hammer changes the direction of the input force. You pull back on the handle, and the hammer pulls up on the nail.

❶ Change the Size of a Force

Could you use only your hands to pull a nailed-down board from a deck? Probably not. However, a crowbar makes this task fairly easy, as shown in **Figure 7**. You would start by placing the tip of a crowbar under the edge of the board. Then you would press down on the opposite end of the crowbar. The tip of the crowbar would lift the board away from the supporting board below. Repeating this process, you could remove the length of the board from the deck.

A crowbar is a machine. It makes work easier by changing a input force into a larger output force. Note that although the output force is greater than the input force, it acts over a shorter distance.

Reading Check How does a crowbar make work easier?

❷ Change the Distance a Force Acts

Using a rake to gather leaves is an example of a machine that increases the distance over which a force acts. As shown in **Figure 7**, a person's hands move one end of the rake a short distance. The other end of the rake, however, sweeps through a greater distance making it easier to rake leaves. Note that the force applied by the rake (the output force) decreases as the distance over which the force acts (the output distance) increases. This relationship is true for all machines.

❸ Change the Direction of a Force

Machines also can make work easier by changing the direction of the input force. A machine is used in **Figure 7** to lift a load. A rope tied to an object passes through a pulley. As the free end of the rope is pulled down, the object tied to the other end of the rope is lifted up. The machine changes the direction of the applied force.

Key Concept Check In what three ways do machines make doing work easier?

Figure 7 Machines make work easier in three ways.

❶ When the output force is greater than the input force, the output force acts over a shorter distance.

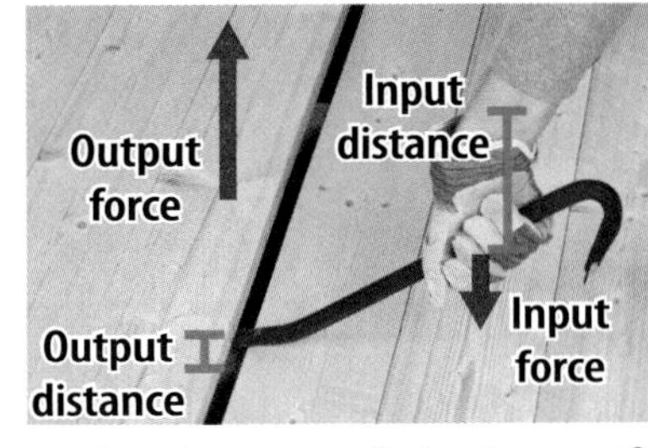

Input force × Input distance = Output force × Output distance

Input work = Output work

❷ When the output force acts over a longer distance than the input force, the output force is less than the input force.

Input force × Input distance = Output force × Output distance

Input work = Output work

❸ Equal output and input forces act over equal distances.

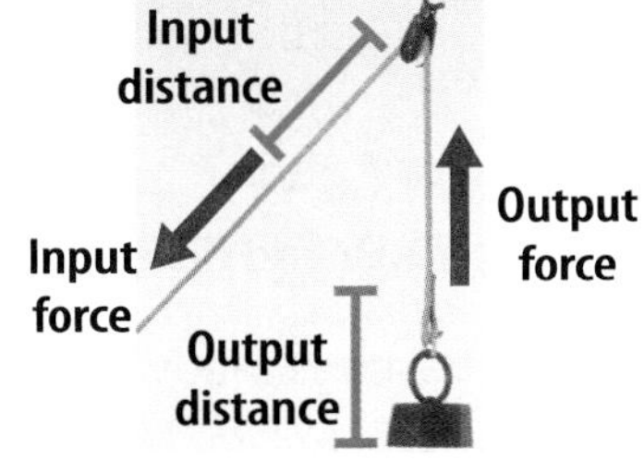

Input force × Input distance = Output force × Output distance

Input work = Output work

What is mechanical advantage?

WORD ORIGIN

mechanical
from Greek *mechanikos,* means "machine"

Most machines change the size of the force applied to them. *A machine's* **mechanical advantage** *is the ratio of a machine's output force produced to the input force applied.* The mechanical advantage, or *MA,* tells you how many times larger or smaller the output force is than the input force.

Mechanical Advantage Equation

$$\text{mechanical advantage (no units)} = \frac{\text{output force (in newtons)}}{\text{input force (in newtons)}}$$

$$MA = \frac{F_{out}}{F_{in}}$$

FOLDABLES®

Make a three-tab book from a sheet of horizontal paper. Label it as shown. Use it to summarize the three ways machines make doing work easier.

How Machines Make Work Easier
Changing Size of Force | Changing Distance | Changing Direction

Mechanical advantage can be less than 1, equal to 1, or greater than 1. A mechanical advantage greater than 1 means the output force is greater than the input force. A crowbar, for example, has a mechanical advantage much greater than 1.

The ideal mechanical advantage, or *IMA,* is the mechanical advantage if no friction existed. Machines cannot operate at ideal mechanical advantage because friction always exists.

Key Concept Check What is mechanical advantage?

Math Skills — Mechanical Advantage Equation

Solve for Mechanical Advantage A carpenter applies 525 N to the end of a crowbar. The force exerted on the board is 1,575 N. What is the mechanical advantage of the crowbar?

1. **This is what you know:** input force: $F_{in} = 525$ N; output force: $F_{out} = 1{,}575$ N
2. **This is what you need to find:** mechanical advantage: *MA*
3. **Use this formula:** $MA = \frac{F_{out}}{F_{in}}$
4. **Substitute:** the values for F_{in} and F_{out} into the formula and divide: $MA = \frac{1{,}575 \text{ N}}{525 \text{ N}} = 3$

Answer: The mechanical advantage is 3.

Practice

While raking leaves, a woman applies an input force of 32 N to a rake. The rake has an output force of 16 N. What is the mechanical advantage of the rake?

Review
- Math Practice
- Personal Tutor

What is efficiency?

The output work done by a machine never exceeds the input work of the machine. The reason for this is friction. Friction converts some of the input work to thermal energy. The converted energy cannot be used to do work.

The **efficiency** *of a machine is the ratio of the output work to the input work.* Efficiency is calculated using the equation below. Because output work is always less than input work, a machine's efficiency is always less than 100 percent. As shown in **Figure 8,** lubricating a machine's moving parts increases efficiency.

Efficiency Equation

$$\text{efficiency (in \%)} = \frac{\text{output work (in joules)}}{\text{input work (in joules)}} \times 100\%$$

$$\text{efficiency} = \frac{W_{out}}{W_{in}} \times 100\%$$

Key Concept Check Why can't the work done by a machine be greater than the work done on the machine?

Figure 8 A lubricant reduces friction by forming a thin film between moving surfaces.

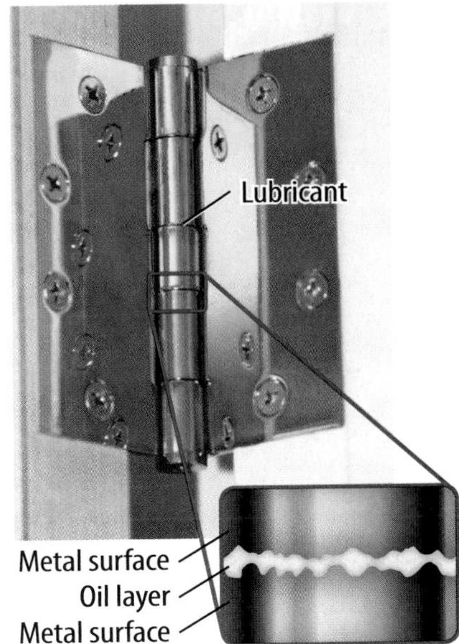

Math Skills — Efficiency Equation

Solve for Efficiency A mechanic does 78.0 J of work pulling the rope on a pulley to lift a motor. The output work of the pulley is 64.0 J. What is the efficiency of the pulley?

1. **This is what you know:** input work: $W_{in} = 78.0$ J; output work: $W_{out} = 64.0$ J
2. **This is what you need to find:** efficiency
3. **Use this formula:** $\text{efficiency} = \frac{W_{out}}{W_{in}} \times 100\%$
4. **Substitute:** the values for W_{in} and W_{out} into the formula and divide: $\text{efficiency} = \frac{64.0\text{ J}}{78.0\text{ J}} \times 100\% = 82.1\%$

Answer: The efficiency is 82.1%

Practice

A carpenter turns a handle to adjust a saw blade. The input work is 55 J and the output work is 51 J. What is the efficiency of the blade adjuster?

- Math Practice
- Personal Tutor

Lesson 2 Review

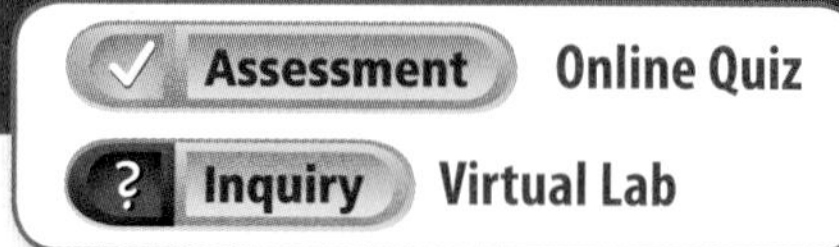

Visual Summary

A machine makes a task easier and it can be simple or complex.

The mechanical advantage of a machine indicates how it changes an input force.

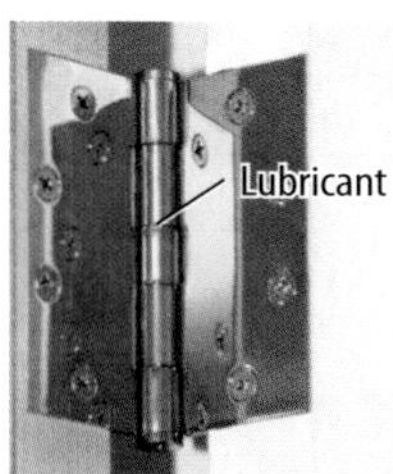

The efficiency of a machine is increased when a lubricant coats moving parts.

FOLDABLES®

Use your lesson Foldable to review the lesson. Save your Foldable for the project at the end of the chapter.

What do you think NOW?

You first read the statements below at the beginning of the chapter.

3. Machines always decrease the force needed to do a job.

4. A well-oiled, low-friction machine can be 100 percent efficient.

Did you change your mind about whether you agree or disagree with the statements? Rewrite any false statements to make them true.

Use Vocabulary

1. **Define** *efficiency* in your own words.
2. **Distinguish** between efficiency and mechanical advantage.

Understand Key Concepts 

3. **Explain** how friction reduces the efficiency of machines.
4. Which machine efficiency is impossible?
   - **A.** 1 percent
   - **B.** 80 percent
   - **C.** 99 percent
   - **D.** 100 percent
5. **Compare and contrast** input and output forces and input and output distances for a hammer pulling a nail out of a board.

Interpret Graphics

6. **Analyze** How does a rake make gathering leaves easier? Explain in terms of distances and forces.
7. **Organize** Copy and fill in the graphic organizer below to describe three ways machines can make work easier.

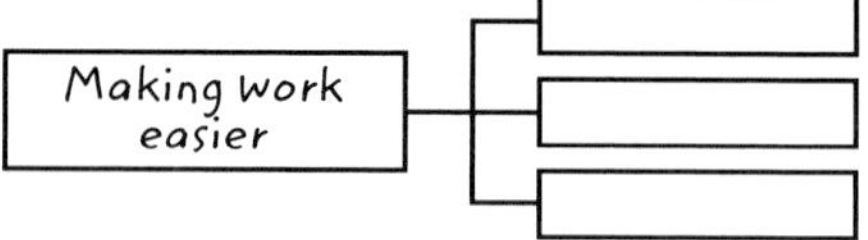

Critical Thinking

8. **Modify** the efficiency equation by writing it in terms of input force and output force.

Math Skills

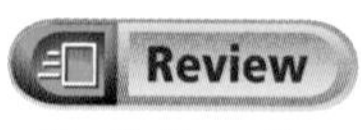

Math Practice

9. A pulley system uses a 250-N force to lift a 2,750-N crate. What is the mechanical advantage of the system?
10. An assembly line machine needs 150 J of input work to do 90 J of output work. What is the efficiency of the machine?

How does mechanical advantage affect a machine?

Materials

centimeter ruler

clear plastic tape

modeling clay

spring scale

Safety

Machines change forces and make work easier. In this lab, you build and use a simple machine to lift a load, and record force and distance data. You then interpret data and determine how the mechanical advantage of the machine affected the results.

Learn It

Measurements and observations are data. When you **interpret data,** you look for patterns and relationships in the data.

Try It

1. Read and complete a lab safety form.
2. Lay a metric ruler on a table so one end extends over an edge, as shown below. Tape the other end of the ruler to the table. Make sure the free end of the ruler moves up and down easily. This is your simple machine.
3. Use a spring scale to measure the mass of a handful-sized lump of clay. Use this weight as the output force (F_{out}) for your tests.
4. Push the clay onto the 15-cm mark of the ruler so that it sticks.
5. Attach a spring scale to the ruler's free end. Pull upward on the spring scale so the ruler lifts off of the table. Record the force reading as the input force (F_{in}).

Output force
Input force
Output distance
Input distance
Tape
Table
Centimeter ruler

6. Make a data table in your Science Journal. Record your measurements, including the distance from the taped end to the clay (d_{out}) and from the taped end to the spring scale (d_{in}).
7. Calculate the mechanical advantage (MA) of the machine from your data. $MA = \frac{F_{out}}{F_{in}}$.

Apply It

8. Move the clay to the 27-cm mark. Repeat steps 5 through 7.
9. Move the clay to two other locations on the ruler, repeating steps 5 through 7 each time.
10. How is the machine's mechanical advantage related to d_{out}, the distance from the taped end of the ruler to the clay?
11. When the spring scale lifts the end of the ruler up, is the clay lifted the same amount? Does the amount the clay is lifted depend on its location along the ruler? Explain.
12. **Key Concept** What is the relationship between the machine's mechanical advantage and how easy it is to lift the clay?

Lesson 3

Reading Guide

Key Concepts
ESSENTIAL QUESTIONS

- What is a simple machine?
- How is the ideal mechanical advantage of simple machines calculated?
- How are simple machines and compound machines different?

Vocabulary

simple machine p. 103
lever p. 104
fulcrum p. 104
wheel and axle p. 106
inclined plane p. 107
wedge p. 108
screw p. 108
pulley p. 109

Multilingual eGlossary

Video

- BrainPOP®
- Science Video

Simple Machines

Inquiry Hit or Miss?

A trebuchet (TRE bu shet) is a device from the Middle Ages designed to hurl large rocks. The trebuchet was effective at hurling rocks at castle walls, though it had limited accuracy. A trebuchet is a large simple machine.

Inquiry Launch Lab

15 minutes

How does a lever work?

Humans have used levers for thousands of years to make work easier. How does a lever change the force applied to it?

1. Read and complete a lab safety form.
2. Place a **ruler** on top of an **eraser** as shown. Place a **book** on one end of the ruler.
3. Lift the book by pushing down on the other end of the ruler. Record in your Science Journal how easy or difficult it is to lift the book.
4. Repeat steps 2 and 3 several times using different locations of the eraser along the ruler.

Think About This

1. Explain how changing the position of the eraser affected the force exerted by the lever.
2. **Key Concept** How would you describe the motion of the lever? Is the motion of the lever simple or complicated? Explain.

What is a simple machine?

Some machines, such as a trebuchet, have only a few parts. Six types of **simple machines,** shown in **Figure 9,** *do work using only one movement.* They are lever, wheel and axle, inclined plane, wedge, screw, and pulley.

Key Concept Check Describe a simple machine.

Figure 9 The six types of simple machines are shown below. Each simple machine does work with one motion.

Lever

Wheel and axle

Inclined plane

Wedge

Screw

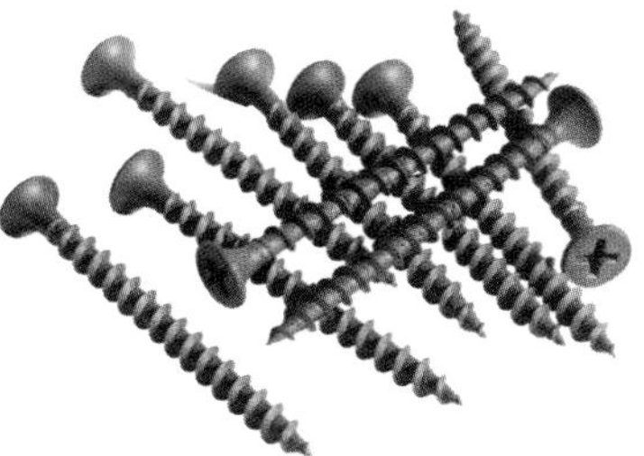

Pulley

The Three Classes of Levers

Figure 10 The location of the input force, the output force, and the fulcrum determine the class of lever.

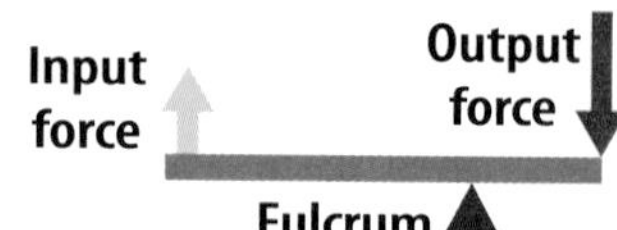

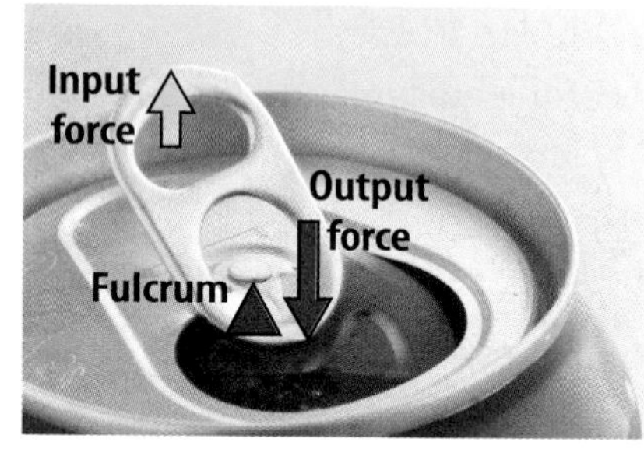

2 Second-Class Lever

Input force
Output force
Fulcrum

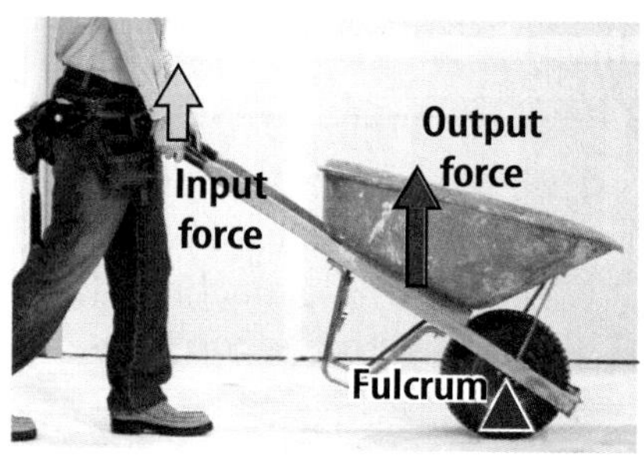

3 Third-Class Lever

Input force
Output force
Fulcrum

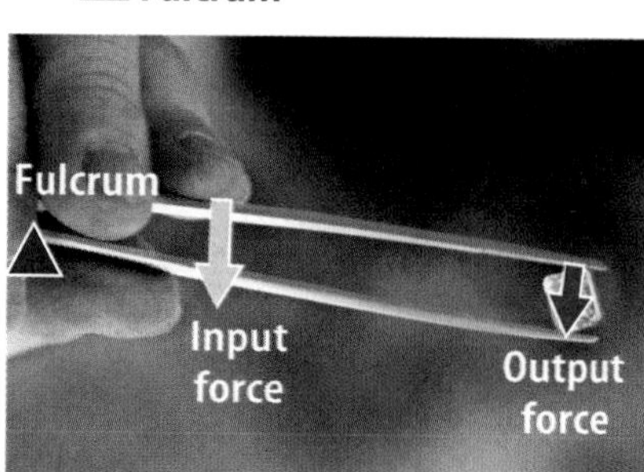

Levers

The next time you open an aluminum beverage can, watch how the finger tab works. The finger tab is a **lever**—*a simple machine made up of a bar that pivots, or rotates, about a fixed point. The point about which a lever pivots is called a* **fulcrum.** The fulcrum on the aluminum beverage can in **Figure 10** is where the finger tab attaches to the can.

Notice that the input force and the output force act on opposite ends of the finger tab. The distance from the fulcrum to the input force on the tab is the *input arm.* The distance from the fulcrum to the end of the tab that pushes down on the can is the *output arm.*

There are three types of levers, also shown in **Figure 10.** First-class, second-class, and third-class levers differ in where the input force and output force are relative to the fulcrum.

1 In a first-class lever, the fulcrum is between the input force and the output force. The direction of the input force is always different than the direction of the output force. A hammer is a first-class lever when it is used to pull a nail out of wood. A finger tab on a beverage can is also a first-class lever.

2 A second-class lever has the output force between the input force and the fulcrum. The output force and the input force act in the same direction. A second-class lever makes the output force greater than the input force. A wheelbarrow, a nut cracker, and your foot are second-class levers.

3 The input force is between the output force and the fulcrum in a third-class lever. The output force is less than the input force, though both forces act in the same direction. Tweezers, a rake, and a broom are examples of third-class levers.

Reading Check What is a lever?

Mechanical Advantage of Levers

Mechanical Advantage of a First-Class Lever

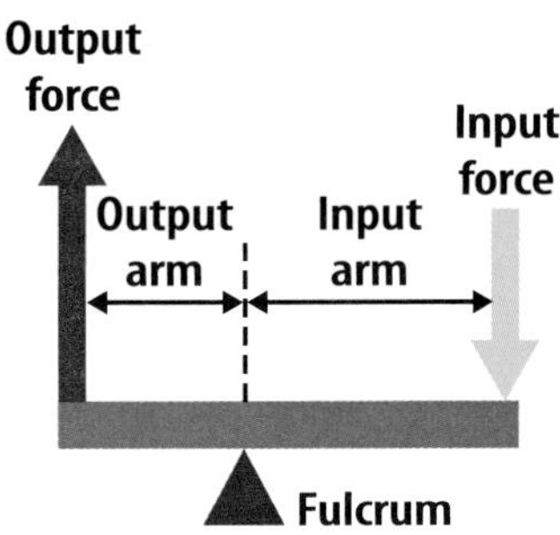

- Fulcrum location determines mechanical advantage.
- Mechanical advantage can be less than 1, equal to 1, or greater than 1.

Mechanical Advantage of a Second-Class Lever

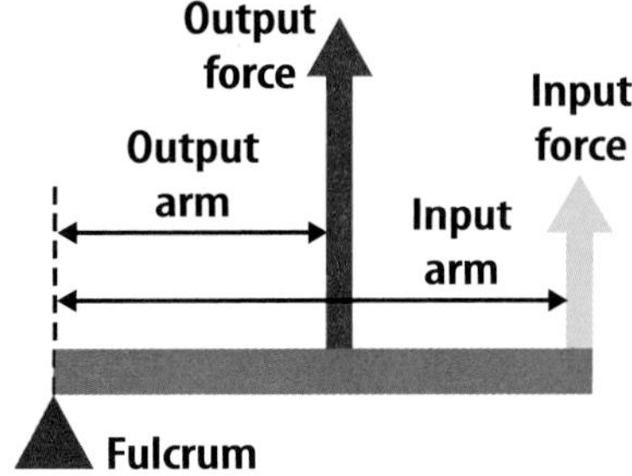

- Input arm is longer than output arm.
- Output force is greater than input force.
- Mechanical advantage is greater than 1.

Mechanical Advantage of a Third-Class Lever

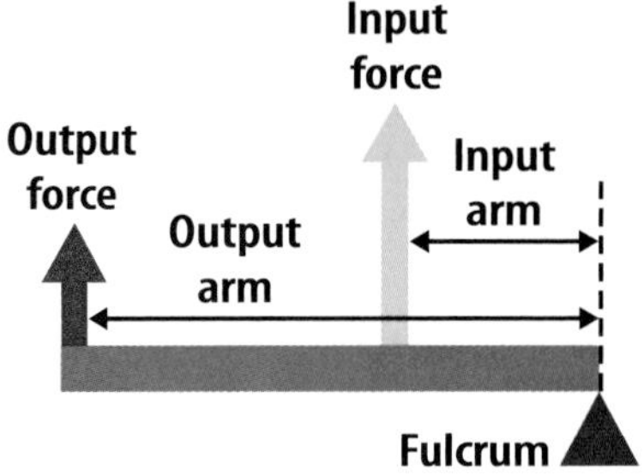

- Input arm is shorter than output arm.
- Output force is less than input force.
- Mechanical advantage is less than 1.

Figure 11 The mechanical advantage of a lever varies depending on the location of the fulcrum.

Visual Check How does the input arm compare to the output arm in a second-class lever? In a third-class lever?

Personal Tutor

Mechanical Advantage of Levers

The ideal mechanical advantage of a lever equals the length of the input arm divided by the length of the output arm.

Ideal Mechanical Advantage of a Lever

$$\text{ideal mechanical advantage} = \frac{\text{length of input arm (in meters)}}{\text{length of output arm (in meters)}}$$

$$IMA = \frac{L_{in}}{L_{out}}$$

First-Class Levers The location of the fulcrum determines the mechanical advantage in first-class levers. In **Figure 11,** the mechanical advantage of the lever is greater than 1 because the input arm is longer than the output arm. This makes the output force greater than the input force. When the mechanical advantage is equal to 1, the input and output arms and input and output forces are equal. If the mechanical advantage is less than 1, the input arm is shorter than the output arm. The output force then is less than the input force.

Second-Class Levers The input arm is longer than the output arm for all second-class levers. The mechanical advantage of a second-class lever is always greater than 1.

Third-Class Levers For third-class levers, the input arm is always shorter than the output arm. Thus, the mechanical advantage of a third-class lever is always less than 1.

Key Concept Check How is the ideal mechanical advantage of a lever calculated?

Levers in the Body

First-Class Lever

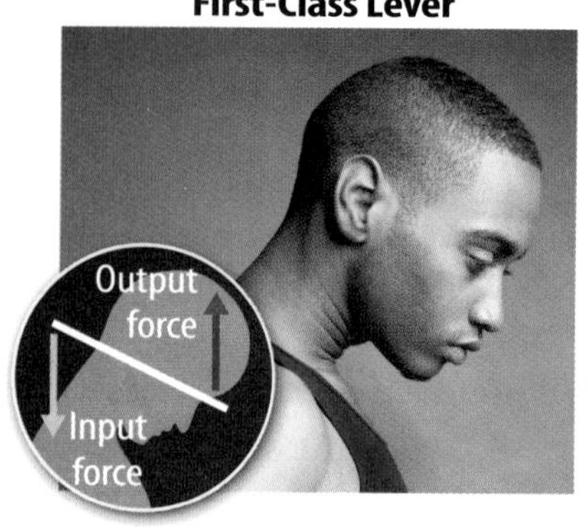

Second-Class Lever

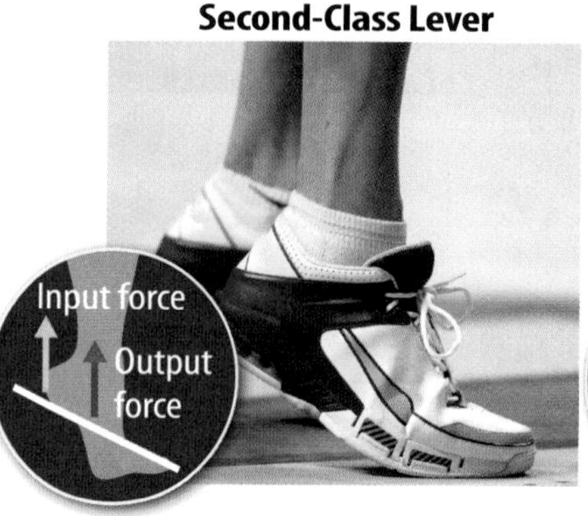

Third-Class Lever

▲ **Figure 12** The neck, foot, and arm are examples of first-, second-, and third-class levers in the human body.

Levers in the Human Body

The human body uses all three classes of levers to move. Muscles provide the input force for the levers. Examples of levers in the neck, foot, and arm are shown in **Figure 12.**

The Neck Your neck acts like a first-class lever. The fulcrum is the joint connecting the skull to the spine. The neck muscles provide the input force. The output force is applied to the head and helps support your head's weight.

The Foot When standing on your toes, the foot acts like a second-class lever. The ball of the foot is the fulcrum. The input force comes from muscles on the back of the lower leg.

The Arm Your forearm works like a third-class lever. The elbow is the fulcrum and the input force comes from muscles located near the elbow.

Figure 13 A screwdriver is a wheel and axle. The handle is the wheel and the shaft is the axle. ▼

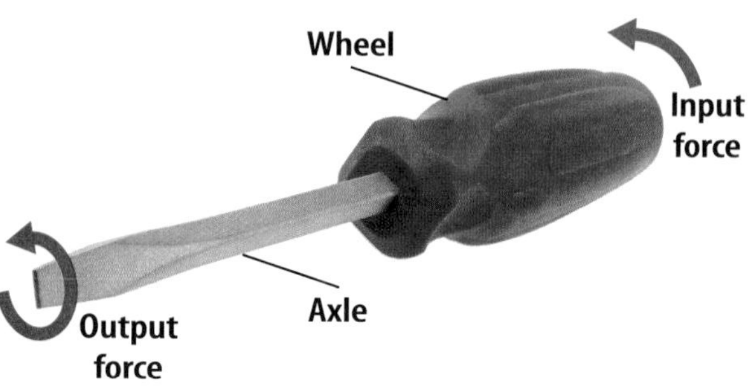

Wheel and Axle

A **wheel and axle** *is an axle attached to the center of a wheel and both rotate together.* Note that *axle* is another word for a shaft. The screwdriver shown in **Figure 13** is a wheel and axle. The handle is the wheel because it has the larger diameter. The axle is the shaft attached to the handle. Both the handle and the shaft rotate when the handle turns.

Mechanical Advantage of a Wheel and Axle

For a wheel and axle, the length of the input arm is the radius of the wheel. Likewise, the length of the output arm is the radius of the axle. These lengths give the ideal mechanical advantage as shown in the equation below.

Ideal Mechanical Advantage of Wheel and Axle

$$\text{ideal mechanical advantage} = \frac{\text{radius of wheel (in meters)}}{\text{radius of axle (in meters)}}$$

$$IMA = \frac{r_{wheel}}{r_{axle}}$$

FOLDABLES®

Make a six-tab book and label it as shown. Use your book to summarize information about six simple machines.

Using a Wheel and Axle

You know a screwdriver makes it easier to turn a screw. But how does a screwdriver work? Examining the mechanical advantage equation offers an explanation. When you turn a screwdriver, you apply an input force to the handle (the wheel). The output force is applied to the screw by the screwdriver's shaft (the axle). Because the wheel is larger than the axle, the mechanical advantage is greater than 1. This makes the screw easier to turn.

Inclined Planes

The ancient Egyptians built pyramids using huge stone blocks. Moving the blocks up the pyramid must have been difficult. To make the task easier, ramps were often used. *A ramp, or* **inclined plane,** *is a flat, sloped surface.* It takes less force to move an object upward along an inclined plane than it does to lift the object straight up. As shown in **Figure 14,** ramps are still useful for moving heavy loads.

Figure 14 Moving a sofa is easier using a ramp. As shown, using a ramp only requires a 100-N force to move the 500-N sofa. Because of friction, no ramp operates at its ideal mechanical advantage.

Figure 15 A wedge that splits wood is a simple machine. ▶

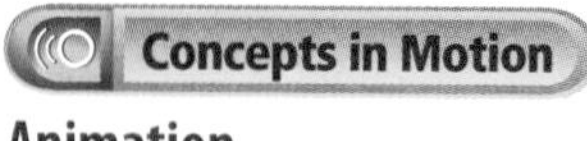

Animation

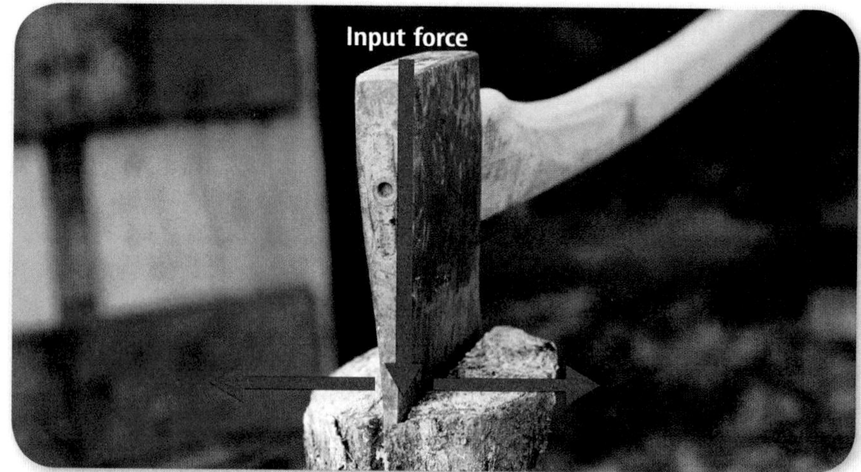

Mechanical Advantage of Inclined Planes

The ideal mechanical advantage of an inclined plane equals its length divided by its height. See the equation below.

Ideal Mechanical Advantage of an Inclined Plane

$$\text{ideal mechanical advantage} = \frac{\text{length of inclined plane (in meters)}}{\text{height of inclined plane (in meters)}}$$

$$IMA = \frac{\ell}{h}$$

Note that increasing the length and decreasing the height of the inclined plane increases its ideal mechanical advantage. The longer or less-sloped an inclined plane is, the less force is needed to move an object along its surface.

Wedges

A sloped surface that moves is called a **wedge.** A wedge is really a type of inclined plane with one or two sloping sides. A doorstop is a wedge with one sloped side. The wedge shown in **Figure 15** is a wedge with two sloped sides. Notice how the shape of the wedge gives the output forces a different direction than the input force.

Your front teeth also are wedges. As you push your front teeth into food, the downward force is changed by your teeth into a sideways force that pushes the food apart.

Reading Check How are a wedge and a ramp different?

Screws

As shown in **Figure 16,** a **screw** *is an inclined plane wrapped around a cylinder.* When you turn a screw, the screw threads change the input force to an output force. The output force pulls the screw into the material.

Figure 16 The groove, or thread, that wraps around a screw is an inclined plane. ▼

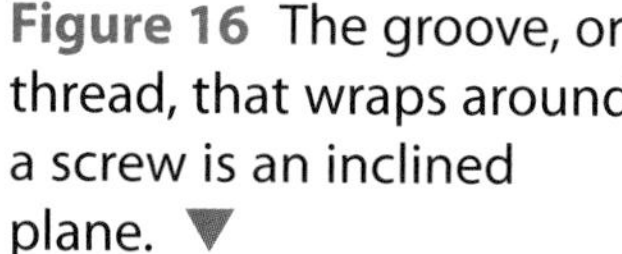

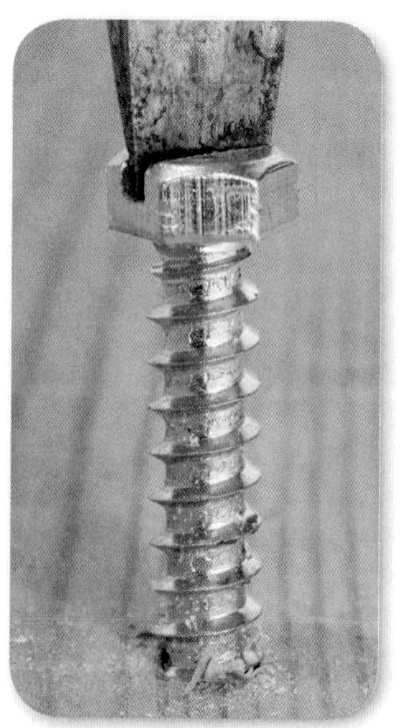

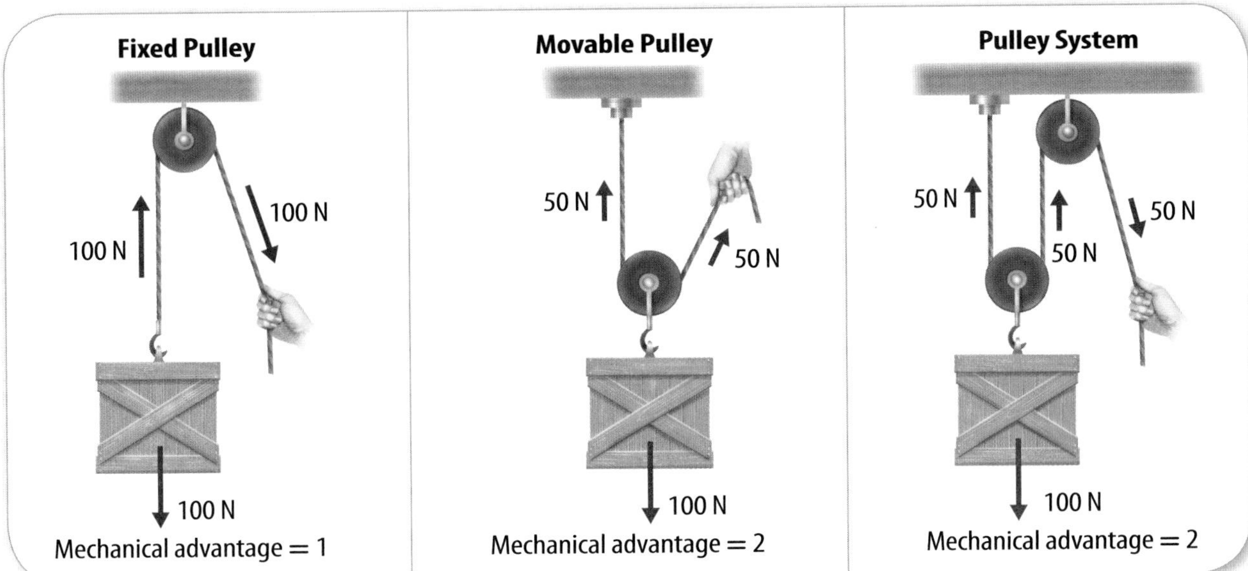

Figure 17 Pulleys can change force and direction.

Pulleys

You might have seen large cranes lifting heavy loads at construction sites. The crane uses a **pulley**—*a simple machine that is a grooved wheel with a rope or a cable wrapped around it.*

Fixed Pulleys Have you ever pulled down on a cord to raise a window blind? The cord passes through a fixed pulley mounted to the top of the window frame. A fixed pulley only changes the direction of the force, as shown **Figure 17.**

Movable Pulleys and Pulley Systems A pulley can also be attached to the object being lifted. This type of pulley, called a movable pulley, is shown in **Figure 17.** Movable pulleys decrease the force needed to lift an object. The distance over which the force acts increases.

A pulley system is a combination of fixed and movable pulleys that work together. An example of a pulley system is shown in **Figure 17.**

Mechanical Advantage of Pulleys The ideal mechanical advantage of a pulley or a pulley system is equal to the number of sections of rope pulling up on the object.

Inquiry MiniLab — 20 minutes

Can you increase mechanical advantage?

Pulley systems change the size and the direction of an applied force. How can you change the mechanical advantage of a pulley system?

1. Read and complete a lab safety form.
2. Use two **broomsticks** and an 8-m length of **rope** to make a pulley system as shown. Hold each broomstick waist-high. Tie the rope to one of the broomsticks and loop it around the other.
3. Have a student pull the free end of the rope as shown. Record in your Science Journal the forces acting on the broomsticks.
4. Take the excess rope and make two more loops around the broomsticks. Repeat step 3.

Analyze and Conclude

1. **Describe** how the forces applied on the broomsticks changed when the number of loops in the pulley system increased.
2. **Key Concept** Relate the number of rope segments in each pulley system to the mechanical advantage of each system.

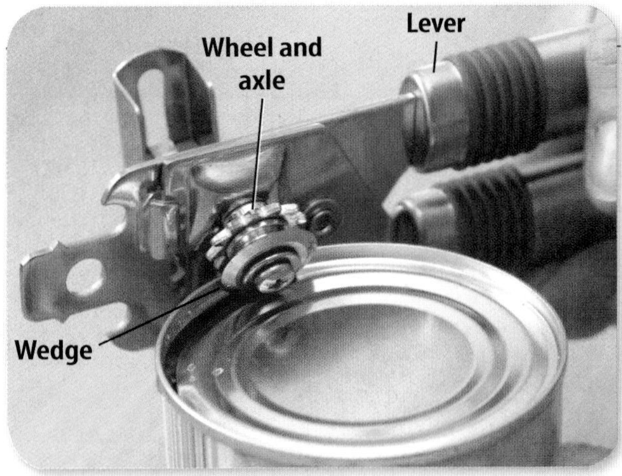

▲ **Figure 18** A can opener is a compound machine. It uses a second-class lever to move the handle, a wheel and axle to turn the blade, and a wedge to puncture the lid.

Visual Check How many simple machines are part of the can opener?

Figure 19 A system of gears is a compound machine. Notice how the direction of rotation changes from one gear to the next. ▼

What is a compound machine?

Two or more simple machines that operate together form a compound machine. The can opener in **Figure 18** is a compound machine.

 **Key Concept Check** How are simple machines and compound machines different?

Gears

A gear is a wheel and axle with teeth around the wheel. Two or more gears working together form a compound machine. When the teeth of two gears interlock, turning one gear causes the other to turn. The direction of motion, as shown in **Figure 19,** changes from one gear to the next.

Gears of different sizes turn at different speeds. Smaller gears rotate faster than larger gears. The amount of force transmitted through a set of gears is also affected by the size of the gears. The input force applied to a large gear is reduced when it is applied to a smaller gear.

Efficiency of Compound Machines

How can you determine the efficiency of a compound machine? The efficiency of a compound machine is calculated by multiplying the efficiencies of each simple machine together.

Consider the can opener in **Figure 18.** It is made up of three simple machines. Suppose the efficiencies are as follows:

- efficiency of lever = 95%
- efficiency of wheel and axle = 90%
- efficiency of wedge = 80%

The efficiency of the can opener is equal to the product of these efficiencies.

$$\text{efficiency of the can opener} = 95\% \times 90\% \times 80\% = 68\%$$

Each simple machine decreases the overall efficiency of the compound machine.

Lesson 3 Review

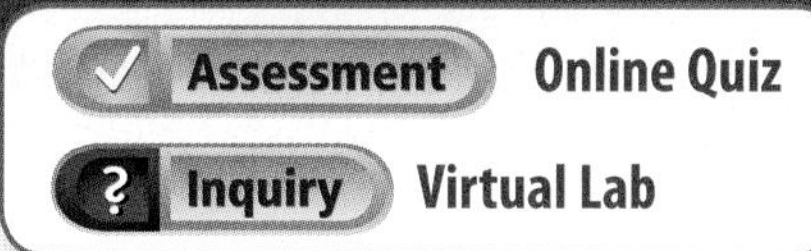

Visual Summary

Six simple machines are the lever, wheel and axle, inclined plane, wedge, screw, and pulley.

All levers rotate, or pivot, about the fulcrum.

The kind of wedge used to split logs is a simple machine.

FOLDABLES®

Use your lesson Foldable to review the lesson. Save your Foldable for the project at the end of the chapter.

What do you think NOW?

You first read the statements below at the beginning of the chapter.

5. A doorknob is a simple machine.

6. A loading ramp makes it easier to lift a load.

Did you change your mind about whether you agree or disagree with the statements? Rewrite any false statements to make them true.

Use Vocabulary

1. **Define** *wheel and axle* in your own words.

Understand Key Concepts 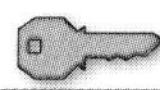

2. Identify the simple machine in which the fulcrum is between the input force and the output force.
   A. wheel and axle
   B. lever
   C. inclined plane
   D. pulley
3. **Determine** how the ideal mechanical advantage of a ramp changes if the ramp is made longer.
4. **Calculate** the ideal mechanical advantage of a screwdriver with a 24-mm radius handle and an 8-mm radius shaft.
5. **Explain** how simple and compound machines are different.

Interpret Graphics

6. **Observe** the scissors shown. They have a pin about which the blades rotate. Each blade has a sloped cutting surface. Identify the simple machines in a pair of scissors.

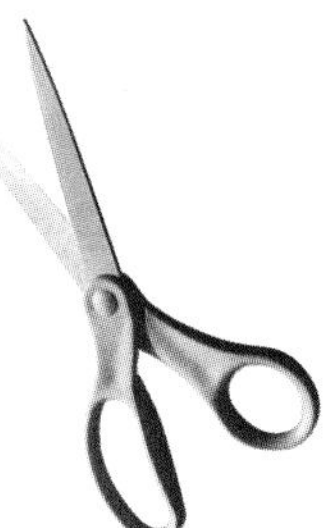

7. **Organize Information**
   Copy and fill in the graphic organizer below with the equations used to calculate the ideal mechanical advantage of each these simple machines.

Type of Simple Machine	Ideal Mechanical Advantage
Lever	
Wheel and axle	
Inclined plane	

Critical Thinking

8. **Suggest** a way in which a pulley could be considered a type of lever.

Inquiry **Lab**

40 minutes

Comparing Two Simple Machines

Materials

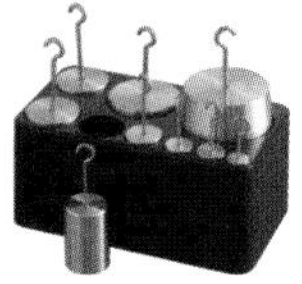

250-g hanging mass

5-N spring scale

small pulley

ring stand

meterstick

Also needed: 50-cm × 15-cm board, books, 2-m length of heavy string

Safety

You will use a pulley and an inclined plane to lift a 250-g mass to a height of 20 cm. Which simple machine makes the work of lifting the load the easiest?

Question

Will the pulley or the inclined plane have a greater mechanical advantage?

Procedure

1. Read and complete a lab safety form.
2. Examine the equipment and diagrams. With your group, discuss why each simple machine has a mechanical advantage when lifting the mass.
3. Predict whether the pulley or the inclined plane will provide the greatest mechanical advantage in lifting the mass. Record your prediction in your Science Journal.
4. Also in your Science Journal, make a data table with columns for data you plan to collect to calculate mechanical advantage for each simple machine.

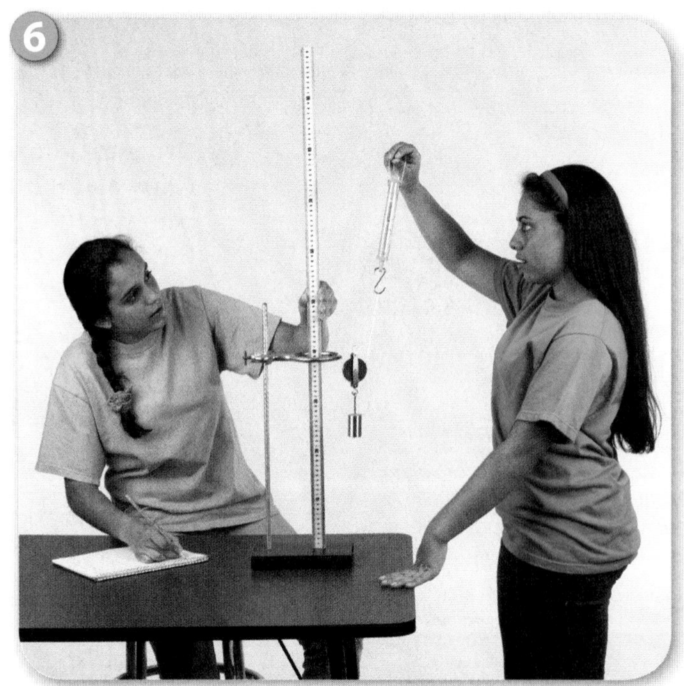

5 Use a spring scale to measure the weight of the 250-g mass. Record the weight in newtons.

6 Set up the pulley and the ring stand as shown on the previous page. Adjust the spring scale and the pulley so that the weight just clears the table. Use the spring scale as shown to slowly lift the weight to a height of 20 cm. Record the force in newtons. Repeat.

7 Set up the inclined plane as shown to the right so that the top of the ramp is 20 cm above the table. Attach the spring scale to the weight. Beginning at the bottom of the ramp, use the spring scale to pull the weight slowly up the ramp. Record the force in newtons. Repeat.

Analyze and Conclude

8 **Interpret Data** What is the mechanical advantage of the pulley and the inclined plane? Was your prediction correct? Explain.

9 **Explain** What other measurements would you need in order to calculate the efficiency of each machine?

10 **The Big Idea** Did the amount of work done on the weight to lift it to a height of 20 cm change with each machine? Explain. Did the machines make doing the work easier?

Communicate Your Results

Make a large data table on which all groups can display their data. Discuss similarities and differences between group data.

How can you increase the mechanical advantage of your inclined plane? To investigate your question, design a controlled experiment.

Lab Tips

- A movable pulley supports part of the weight, allowing less force to be used to lift the object.
- The ideal mechanical advantage of an inclined plane is the length of the ramp divided by the height of the ramp.

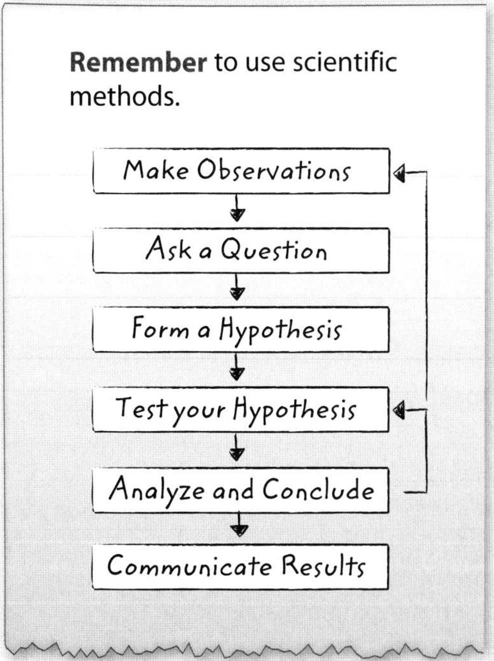

Chapter 3 Study Guide

A machine makes work easier by changing the size of the applied force, changing the distance over which the applied force acts, or changing the direction of the applied force.

Key Concepts Summary	Vocabulary
Lesson 1: Work and Power • For **work** to be done on an object, an applied force must move the object in the direction of the force. • When work is done on an object, the energy of the object increases. • **Power** is the rate at which work is done. 	**work** p. 87 **power** p. 91
Lesson 2: Using Machines • A machine can make work easier in three ways: changing the size of a force, changing the distance the force acts, or changing the direction of a force. • The **mechanical advantage** of a machine is the ratio of the output force to the input force. • Because of friction, the output work done by a machine is always less than the input work to the machine.  • Friction between moving parts converts some of the input work into thermal energy and decreases the **efficiency** of the machine.	**mechanical advantage** p. 98 **efficiency** p. 99
Lesson 3: Simple Machines • A **simple machine** does work using only one movement. • The ideal mechanical advantage of simple machines is calculated using simple formulas. • A compound machine is made up of two or more simple machines that operate together. 	**simple machine** p. 103 **lever** p. 104 **fulcrum** p. 104 **wheel and axle** p. 106 **inclined plane** p. 107 **wedge** p. 108 **screw** p. 108 **pulley** p. 109

- Personal Tutor
- Vocabulary eGames
- Vocabulary eFlashcards

FOLDABLES® Chapter Project

Assemble your lesson Foldables as shown to make a Chapter Project. Use the project to review what you have learned in this chapter.

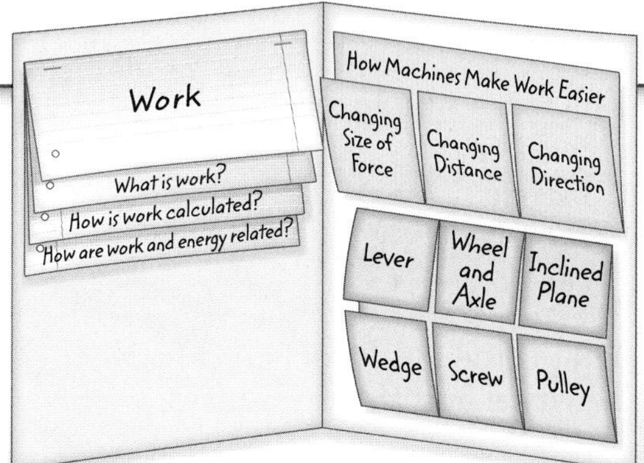

Use Vocabulary

Match each phrase with the correct vocabulary term from the Study Guide.

1. A wheel and axle is an example of a(n) ________.
2. As work is done more quickly, the ________ required increases.
3. A third-class lever has a(n) ________ that is always less than 1.
4. The ________ of a machine is always less than 100 percent.
5. A sloped road is an example of a(n) ________.
6. Your front teeth act like a(n) ________.
7. A(n) ________ is an inclined plane wrapped around a cylinder.

Concepts in Motion Interactive Concept Map

Link Vocabulary and Key Concepts

Copy this concept map, and then use vocabulary terms from the previous page and other terms in this chapter to complete the concept map.

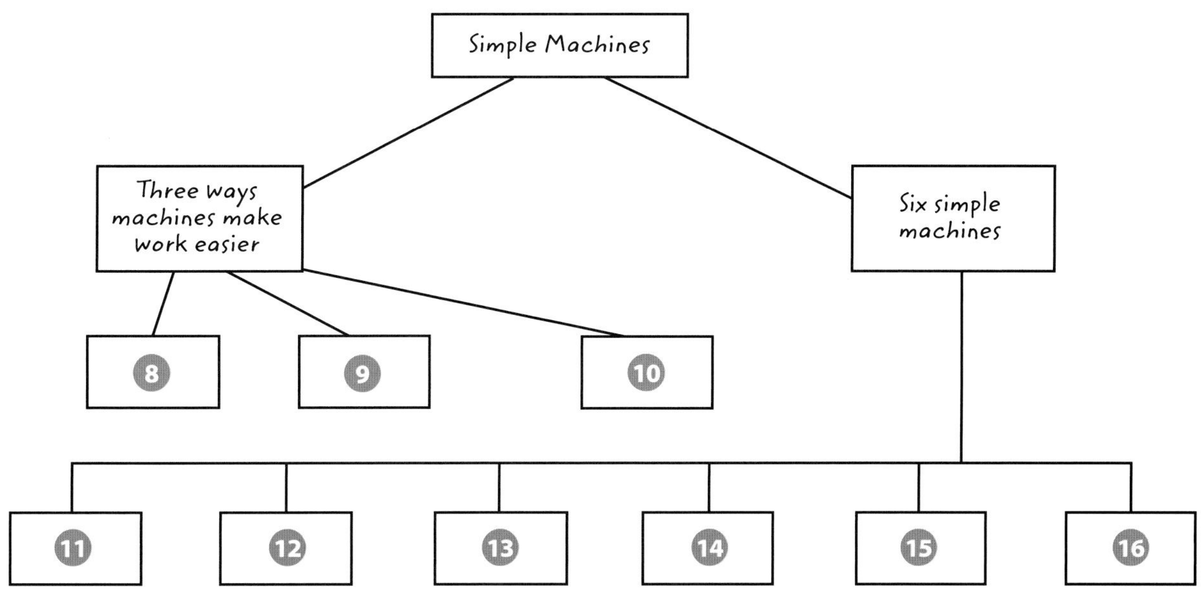

Chapter 3 Review

Understand Key Concepts 

1. Which must be true when work is done?
   A. No force acts on the object.
   B. An object slows down.
   C. An object moves quickly.
   D. An object moves in the same direction as a force exerted on it.

2. Which must be true when work is done on a resting object and the object moves?
   A. The energy of the object changes.
   B. The energy of the object is constant.
   C. The force on the object is constant.
   D. The force on the object is zero.

3. Which increases the power used in lifting the backpack?
   A. lifting at a constant speed
   B. lifting at an angle
   C. lifting more quickly
   D. lifting a lighter backpack

4. Which does more work on an object?
   A. pushing a 50-N object a distance of 2 m
   B. pushing a 120-N object a distance of 7 m
   C. pushing a 150-N object a distance of 5 m
   D. pushing a 300-N object a distance of 1 m

5. Which must be true if a machine's mechanical advantage is equal to 2?
   A. Direction of force is changed.
   B. Output force equals input force.
   C. Output force is greater than input force.
   D. Output work is greater than input work.

6. Which is ideal mechanical advantage of the pulley system?

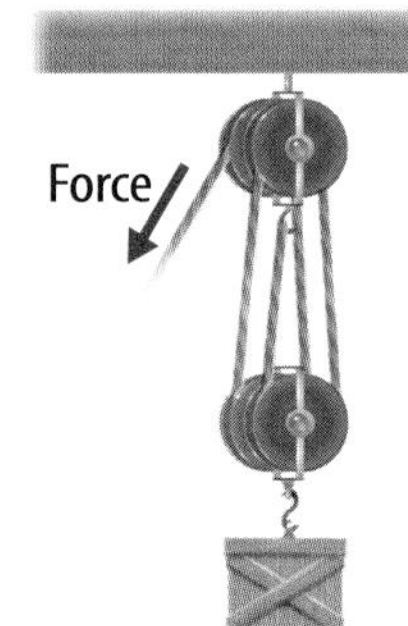

   A. 1
   B. 2
   C. 3
   D. 4

7. How does friction affect a machine?
   A. It converts input work into thermal energy.
   B. It converts output work into thermal energy.
   C. It converts thermal energy into input work.
   D. It converts thermal energy into output work.

8. Which is a simple machine?
   A. tweezers
   B. bicycle
   C. can opener
   D. car

9. What is the ideal mechanical advantage of the wheel and axle shown below?

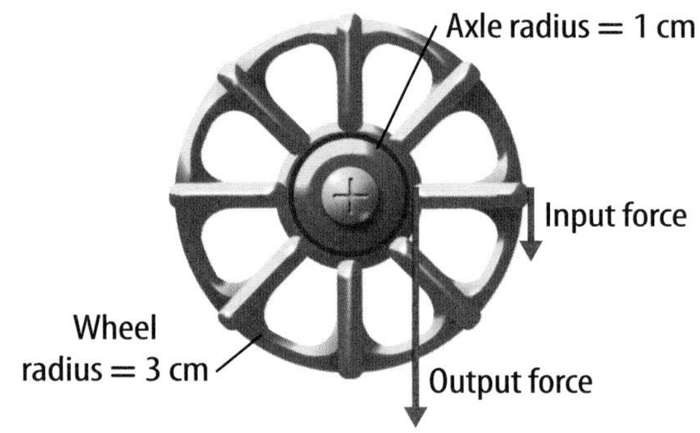

   A. 1
   B. 2
   C. 3
   D. 4

10. Which is most easily used as either a wheel and axle or as a lever?
    A. ax
    B. scissors
    C. screwdriver
    D. screw

11. Which quantity increases when a frictionless ramp is used to lift an object?
    A. input force
    B. input distance
    C. input work
    D. output work

Assessment
Online Test Practice

Critical Thinking

12 **Contrast** the work done when you lift a box with the work done when you carry the box across a room.

13 **Analyze** Does a weight lifter transfer more energy, less energy, or the same amount of energy when lifting the weight faster? Explain.

14 **Decide** First you lift an object from the floor onto a shelf. Then, you move the object from the shelf back to the floor. Do you perform the same amount of work each time? Explain.

15 **Explain** why the output work done by a machine is never greater than the input work to the machine.

16 **Suggest** a reason the efficiency of a machine used in a factory might decrease over time. What could be done to increase the efficiency?

17 **Explain** Imagine using a screwdriver to drive a screw into a piece of wood. Explain why turning the handle of the screwdriver is easier than turning its shaft.

18 **Explain** Can you determine if work is being done on the backpack simply by looking at this photo? Explain why or why not.

Writing in Science

19 **Write** a paragraph about using an inclined plane on the Moon. Objects weigh less on the Moon because the force of gravity is less than on Earth. Explain whether this affects the mechanical advantage of the inclined plane or the way it is used.

REVIEW THE BIG IDEA

20 How do machines make doing work easier? Describe several examples of how machines change forces and make work easier to do.

21 Explain why the earthmover below uses more power than a person moving the same amount of dirt with a shovel.

Math Skills

Review
Math Practice

Use Numbers

22 How much work is done when a force of 30 N moves an object a distance of 3 m?

23 How much power is used when 600 J of work is done in 10 s?

24 How much force would be needed to push a box weighing 30 N up a ramp that has a ideal mechanical advantage of 3? Assume there is no friction.

25 Calculate the efficiency of a machine that requires 20 J of input work to do 10 J of output work.

26 Calculate the mechanical advantage of a machine that changes an input force of 50 N into an output force of 150 N.

27 A compound machine is made up of four simple machines. If the efficiencies of the simple machines are 98%, 93%, 87%, and 92%, respectively, what is the overall efficiency of the compound machine?

Standardized Test Practice

Record your answers on the answer sheet provided by your teacher or on a sheet of paper.

Multiple Choice

1 Which is the SI unit of work?

A ampere

B joule

C newton

D watt

2 Which transfers both kinetic energy and potential energy to an object?

A lifting it

B lowering it

C pushing it

D rolling it

Use the chart below to answer questions 3 and 4.

Moving a Chair	
Force	20 newtons
Distance	5 meters
Time	2 seconds

3 How much work was involved in moving the chair?

A 4 J

B 10 J

C 40 J

D 100 J

4 How much power was used to move the chair?

A 10 W

B 20 W

C 50 W

D 200 W

5 Which is a third-class lever?

A a broom

B a hammer

C a nutcracker

D a wheelbarrow

Use the chart below to answer question 6.

Mechanical Advantage Equation
mechanical advantage $= \frac{\text{output force}}{?}$

6 Which correctly completes the mechanical advantage equation?

A distance

B time

C input force

D output work

7 A simple machine is NOT able to

A change the size of a force.

B decrease the amount of work required.

C exert an output force on an object.

D make work easier to perform.

8 Which MUST happen in order for work to be done?

A A force must move an object.

B A machine must transfer force.

C Force must be applied to an object.

D Output force must exceed input force.

9 Which increases the efficiency of a complex machine?

A adding more simple machines

B increasing input force

C putting more work in

D reducing friction by lubricating

Assessment
Online Standardized Test Practice

Use the diagram below to answer question 10.

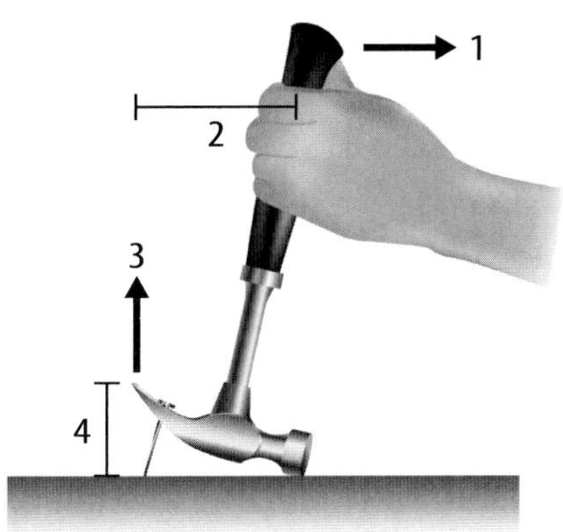

10 Which number represents the output distance in the diagram above?

A 1

B 2

C 3

D 4

11 What information is needed to calculate the ideal mechanical advantage of an inclined plane?

A its height and length

B its length and thickness

C its thickness and weight

D its weight and width

12 Which is a characteristic feature of a wedge?

A cable

B fulcrum

C slope

D thread

Constructed Response

Use the table below to answer questions 13 and 14.

Simple Machine	Example	Task
Lever		
Inclined plane		
Wheel and axle		
Pulley		
Wedge		
Screw		

13 In the table, the six types of simple machines are listed. Provide an example of each machine. Then, list an everyday task you have performed with the help of the machine.

14 How do each of the simple machines in the table above make work easier?

15 What is the difference between simple and compound machines?

16 Describe a real-life situation in which simple machines would make work easier. Identify two machines that would be helpful in the situation and explain how they would be used.

17 What two factors affect the amount of work done on an object? When does a force acting on an object NOT do work? Give an example.

NEED EXTRA HELP?																	
If You Missed Question...	1	2	3	4	5	6	7	8	9	10	11	12	13	14	15	16	17
Go to Lesson...	1	1	1	1	3	2	2	1	2	2	3	3	3	3	3	2,3	1

Forces and Fluids

THE BIG IDEA

In what ways do people use forces in fluids?

Inquiry What makes it float?

Did you know that the helium in parade balloons is considered a fluid? Forces acting on the helium in this balloon keep it in the air. What are those forces?

- What is a fluid?
- How do forces affect fluids and the objects within them?
- In what ways do people use forces in fluids?

Get Ready to Read

What do you think?

Before you read, decide if you agree or disagree with each of these statements. As you read this chapter, see if you change your mind about any of the statements.

1. Air is a fluid.
2. Pressure is a force acting on a fluid.
3. You can lift a rock easily under water because there is a buoyant force on the rock.
4. The buoyant force on an object depends on the object's weight.
5. If you squeeze an unopened plastic ketchup bottle, the pressure on the ketchup changes everywhere in the bottle.
6. Running with an open parachute decreases the drag force on you.

connectED.mcgraw-hill.com

WebQuest

Assessment

Concepts in Motion

Inquiry

Multilingual eGlossary

Lesson 1

Reading Guide

Key Concepts

ESSENTIAL QUESTIONS

- How do force and area affect pressure?
- How does pressure change with depth in the atmosphere and under water?
- What factors affect the density of a fluid?

Vocabulary

fluid p. 123

pressure p. 124

atmospheric pressure p. 126

Multilingual eGlossary

Pressure and Density of Fluids

Inquiry Why Not Straight?

Have you ever noticed that dams are thicker at the bottom than they are at the top? Why do you think people build dams this way?

15 minutes

What changes? What doesn't?

What happens to the volume of a liquid as you move the liquid from one container to another?

1. Read and complete a lab safety form.
2. As you do each of the following steps, use the markings on the containers to measure the liquid's volume. Record the measurements in your Science Journal.
3. Measure 100 mL of water in a **graduated cylinder.**
4. Pour the water into a **200-mL beaker.**
5. Pour the water into a **square plastic container.** Then pour it back into the graduated cylinder.

Think About This

1. What happened to the shape of the water as you moved it from one container to another? What happened to the volume of the water?
2. **Key Concept** How do you think the results would be different if you had used another liquid, such as honey? What would happen if you did this experiment with a gas?

What is a fluid?

When people tell you to drink plenty of fluids when you play a sport, they usually are referring to water or other liquids. It may surprise you to know that gases, such as the oxygen and the nitrogen in the air, are fluids, too. *A* **fluid** *is any substance that can flow and take the shape of the container that holds it.*

Liquids

You probably have poured milk from a carton into a glass or a bowl. Milk, like other liquids, flows and takes the shape of its container. The volume of a liquid remains the same in any container. When you pour all the milk from a carton into a glass, the milk in the glass occupies the same volume it occupied in the carton. It just has a different shape.

Gases

What happens when helium gas fills a balloon? If the balloon is long and thin, then the helium atoms flow into and fill the long, thin shape. If the balloon is spherical, then the helium atoms flow into and fill the spherical shape. Like liquids, helium is a fluid. It takes the shape of its container. However, unlike liquids, helium and other gases do not occupy the same volume in different containers. A gas fills its entire container no matter the size of the container.

Reading Check How are gases and liquids different?

WORD ORIGIN

fluid
from Latin *fluidus,* means "flowing"

FOLDABLES®

Make a vertical two-column chart with labels as shown. Use the chart to organize your notes about pressure and density of fluids.

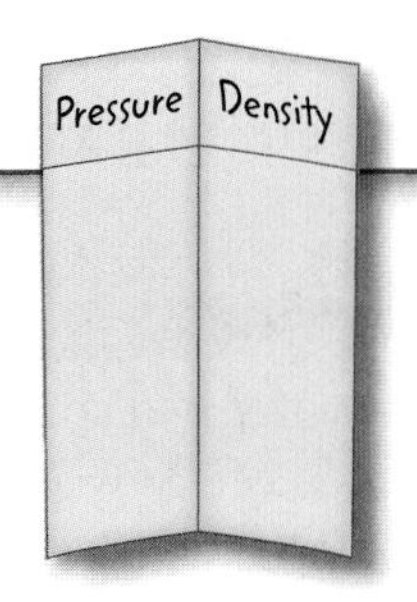

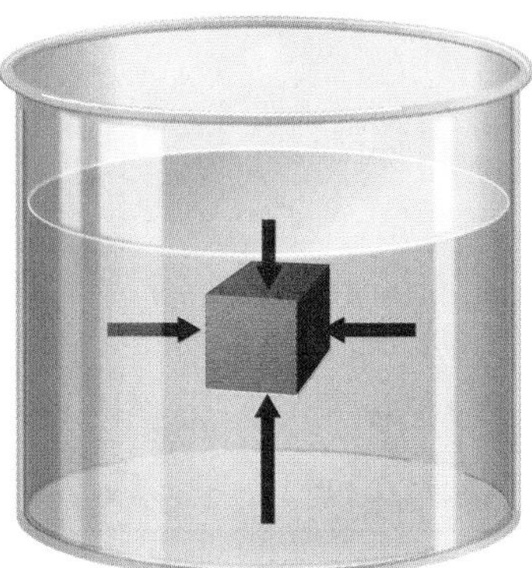

Figure 1 The pressure applied by a fluid is perpendicular to the surface of any object in contact with the fluid.

Pressure of Fluids

Have you ever heard someone talk about the air pressure in a car tire or the water pressure on a deep-sea diver? What is pressure? **Pressure** *is the amount of force per unit area applied to an object's surface.* All fluids, both liquids and gases, apply pressure. The air around you is applying pressure on you right now. Pressure applied on an object by a fluid is related to the weight of the fluid. Like all forces, weight is measured in newtons (N). Pressure can be calculated using the equation below. In the equation, *P* is pressure, *F* is the force applied to a surface, and *A* is the surface area over which the force is applied.

pressure = force ÷ area

$$P = \frac{F}{A}$$

The unit measurement for pressure is the pascal (Pa). A dollar bill lying on your hand would apply a pressure of about 1 Pa. A small carton of milk would apply about 1,000 Pa.

The Direction of Pressure

If the pressure applied by a fluid is related to the weight of the fluid, is the pressure only in the downward direction? No. A fluid applies pressure perpendicular to all sides of an object in contact with the fluid, as shown in **Figure 1.**

Math Skills — Solve a One-Step Equation

Solve for Pressure A surfer on his surfboard applies a force of 645 N on the water. The area of the surfboard is 1.5 m². What is the pressure applied on the water?

1. **This is what you know:** force: $F = 645$ N; area: $A = 1.5\ m^2$
2. **This is what you need to find:** pressure: P
3. **Use this formula:** $P = \frac{F}{A}$
4. **Substitute:** the values for *F* and *A* into the formula and divide. $P = \frac{645\ N}{1.5\ m^2} = 430$ Pa

Answer: The pressure is 430 Pa.

- Math Practice
- Personal Tutor

Practice

The water in a small plastic swimming pool applies a force of 30,000 N over an area of 75 m². What is the pressure on the pool?

Inquiry MiniLab

10 minutes

How are force and pressure different?

Sharp pins can pop a balloon. You might think ten pins could pop a balloon more easily than just one pin could. Is this true?

1. Read and complete a lab safety form.
2. Push a **straight pin** through a small piece of **foam board** so the sharp end of the pin extends 1–2 cm.
3. With a partner holding an inflated balloon, gently place the board on the **balloon** so the pin is touching the balloon.
4. Gently apply pressure on the board until the balloon pops.
5. Add nine **more pins** to the foam board. Space them about 0.5 cm apart.
6. While your partner steadies another inflated balloon, hold the board on top of the balloon and gently increase pressure on the board until the balloon pops. Estimate in your Science Journal the difference in pressure you used to pop the balloons.

Analyze and Conclude

1. **Analyze** What differences did you notice when trying to pop the balloons?
2. **Key Concept** Explain your results by contrasting the force and the pressure per pin on each balloon.

Pressure and Area

You read on the previous page that the amount of pressure on an object depends on the area over which a force is applied. If you tried to push the box of books shown in Figure 2 with one finger, your finger would probably bend. It might even hurt. Now suppose you use your entire hand to push the box with the same force. Why is it easier?

When you push a box with one finger, the force you apply spreads over a small area—the surface area of your fingertip. The pressure is great, and your finger hurts. If you push the box with your whole hand, you apply the same force, but the force is spread over a larger area. The pressure is less, and it is easier to push the box.

Pressure decreases when the surface area over which a force is applied increases. Pressure increases when the surface area over which a force is applied decreases.

Key Concept Check How does pressure change as surface area changes?

Figure 2 When a force is applied over a small area, the pressure is greater than when the same force is applied over a larger area.

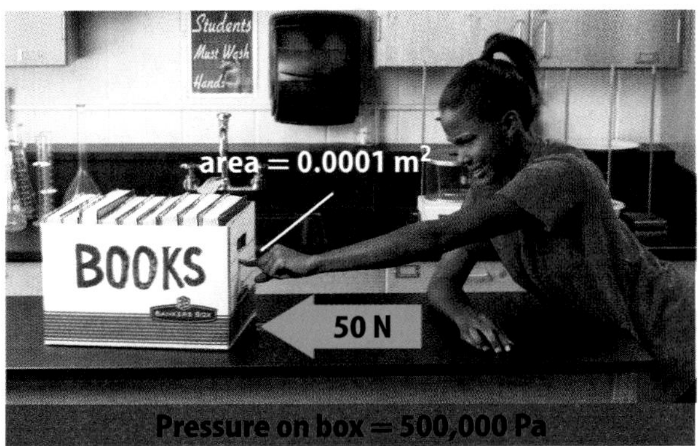

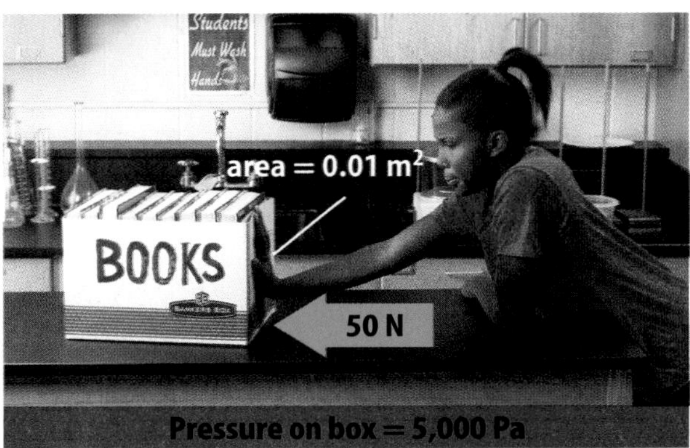

Figure 3 The weight of a column of air applies pressure downward. The greater the height above sea level, the lower the pressure.

REVIEW VOCABULARY
elevation
height above sea level

ACADEMIC VOCABULARY
sum
(noun) the whole amount

Pressure and Depth

Have you ever dived under water and felt pressure in your ears? The deeper you dive, the more pressure you feel because pressure applied by a fluid increases with depth.

Atmospheric Pressure Think of the air above you as a column that extends high into the atmosphere, as shown in **Figure 3.** The weight of the air in this column applies pressure downward. *The ratio of the weight of all the air above you to your surface area is* **atmospheric pressure.** At sea level, the atmospheric pressure on you and everything around you is about 100,000 Pa, or 100 kPa. Now imagine you are on the top of Mount Everest, nearly 9 km above sea level, as shown in **Figure 4.** Atmospheric pressure is only about 33 kPa. The column of air above you is not as tall as it was at sea level. Therefore, the weight of the air applies less pressure downward. Atmospheric pressure increases as you hike down a mountain, toward lower **elevation.** It decreases as you hike up, to higher elevation.

 **Key Concept Check** How does elevation affect atmospheric pressure?

Underwater Pressure How does the pressure on the divers in **Figure 4** change as the divers swim deeper? Underwater pressure depends on the **sum** of the weight of the column of air above an object and the weight of the column of water above the object. As a diver dives farther under water, the air column above him or her stays the same. But the water column increases in height, and it weighs more. Underwater pressure increases with depth. At the deepest part of the ocean, the Mariana Trench, the pressure is 108,600 kPa—more than 1,000 times greater than pressure at the ocean's surface. Increased pressure with depth explains why the wall of a dam is thicker at the bottom than at the top. The water behind the dam applies more pressure to the bottom of the wall.

Atmospheric and Underwater Pressure

Figure 4 On land, the atmospheric pressure depends on your elevation. Under water, the water pressure depends on your depth below the water's surface.

Visual Check What would be the pressure on an airplane flying at an altitude of 5,500 m?

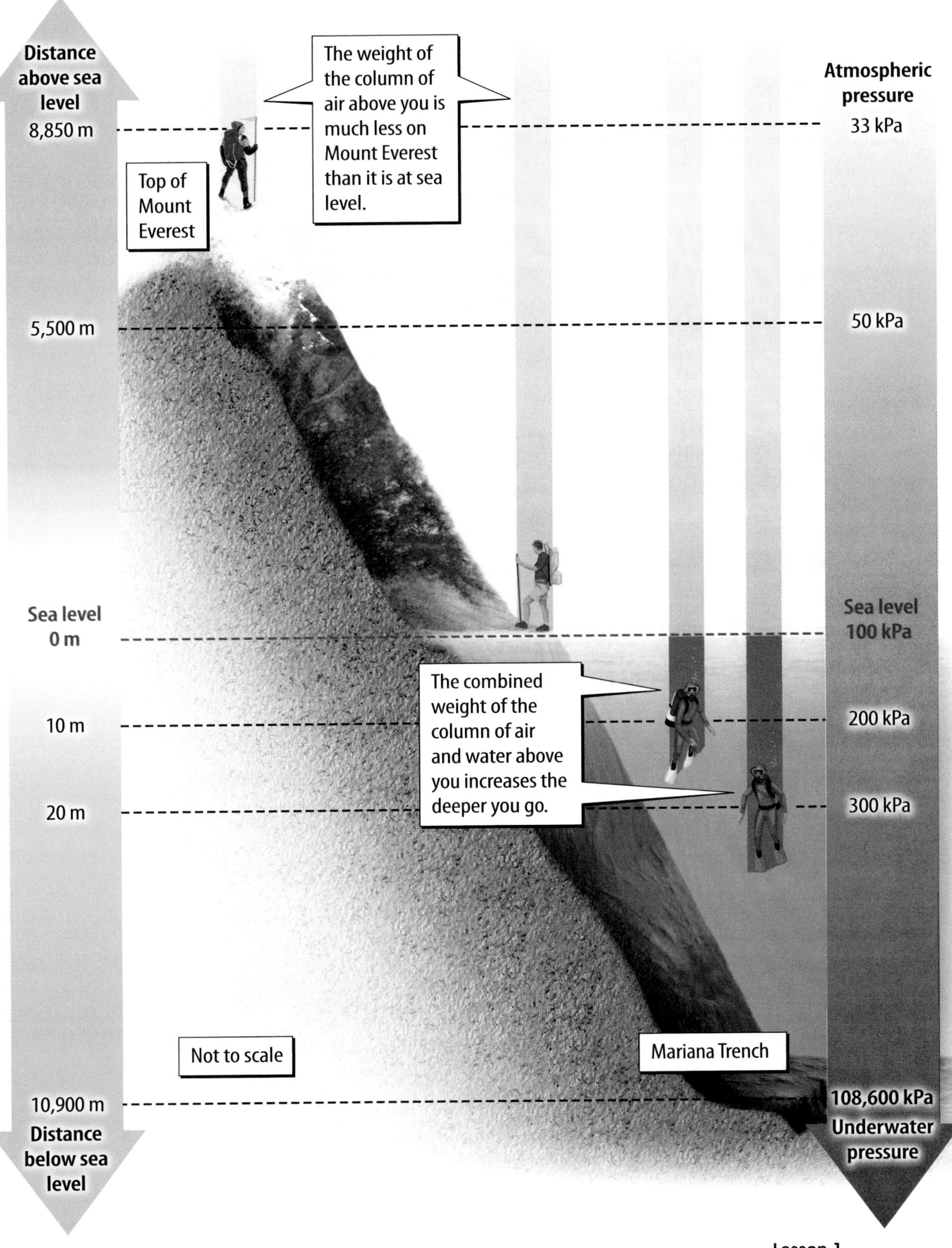

Density of Fluids

If you walked 10 m down a hill, you would not feel a change in pressure. But if you dived 10 m under water, you would feel a big pressure change. Why is there more pressure under water? The reason is density. If the volume of two fluids is the same, the fluid that weighs more is **denser.** A 10-m column of water weighs about 1,000 times more than a 10-m column of air. Therefore, the water column is denser.

SCIENCE USE V. COMMON USE
dense
Science Use describes the measure of the ratio of mass to volume

Common Use slow to learn or understand

Calculating Density

You can calculate density with the equation below, where D is density, m is mass, and V is volume.

Density = mass ÷ volume

$$D = \frac{m}{V}$$

Density is often measured in grams per cubic centimeter (g/cm^3). The density of water is 1.0 g/cm^3. The density of air at 100 kPa is about 0.001 g/cm^3.

Densities of Different Materials

Have you ever noticed that vinegar and oil form layers in a bottle of salad dressing? Each layer, such as those shown in **Figure 5,** has a different density. The density of a material is determined by the masses of the atoms or the molecules that make up the material and the distances between them.

A penny is made mostly of zinc. One zinc atom has greater mass than an entire water molecule. Also, zinc atoms are closer together than the atoms are in a water molecule. Therefore, a penny is denser than water.

Molecules in most solids are closer together than molecules in liquids or gases. Therefore, solids are usually denser than liquids or gases.

Review Personal Tutor

Figure 5 Materials have different densities because of differences in the masses of their molecules and in the distances between them.

Key Concept Check What factors determine the density of fluids?

Lesson 1 Review

Online Quiz

Visual Summary

Pressure is high when a force is applied over a small area.

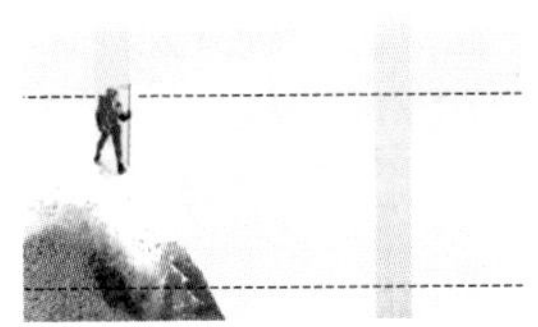

Atmospheric pressure decreases with elevation.

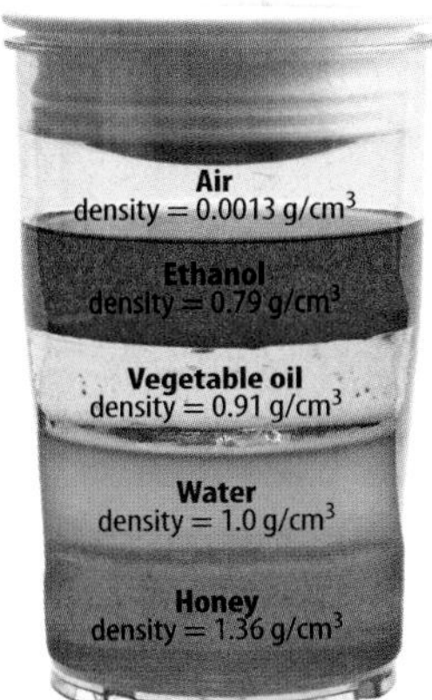

Fluids form layers depending on their densities.

FOLDABLES®

Use your lesson Foldable to review the lesson. Save your Foldable for the project at the end of the chapter.

What do you think NOW?

You first read the statements below at the beginning of the chapter.

1. Air is a fluid.

2. Pressure is a force acting on a fluid.

Did you change your mind about whether you agree or disagree with the statements? Rewrite any false statements to make them true.

Use Vocabulary

1. **Compare** pressure and atmospheric pressure.
2. Gases and liquids are both ________.

Understand Key Concepts

3. Ice floats in water. Which must be true?
   - **A.** Ice molecules have more mass than water molecules.
   - **B.** Ice molecules have less mass than water molecules.
   - **C.** A volume of ice has less mass than the same volume of water.
   - **D.** One gram of ice has less volume than 1 g of water.
4. **Compare** One person walks 5 m down a hill. Another dives 5 m in a pool. Which feels a greater pressure increase? Explain.
5. **Draw** a beaker filled with two liquids. One liquid is blue with a density of 0.93 g/cm^2. The other liquid is red with a density of 1.03 g/cm^2.

Interpret Graphics

6. **Organize Information** Copy and fill in the following graphic organizer to list two factors that affect density.

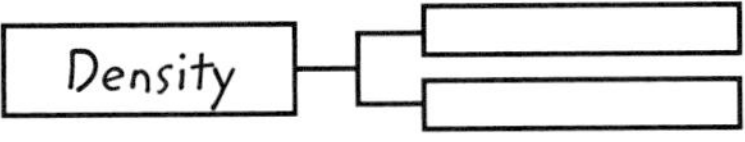

Critical Thinking

7. **Propose** As you walk across a field covered with deep snow, you keep sinking. What could you do to prevent this? Explain your reasoning.

Math Skills

Math Practice

8. What pressure does 50,000 N of water apply to a 0.25-m^2 surface area of coral?
9. How much pressure is on the bottom of a pot that holds 20 N of soup? The surface area of the pot is 0.05 m^2.

ͦOW IT WORKS

Submersibles

Deep-sea craft carry scientists to the depths of the ocean.

Exploring the deep ocean is exciting work. However, the pressure that ocean water applies increases rapidly as depth increases. This ›ssure limits how far humans can dive without special equipment. One y that humans can explore the deep ocean is to use a submersible.

A submersible is a special watercraft designed to function in deep water. me submersibles are remote-controlled and do not carry people. Others, :h as the *Johnson-Sea-Link* submersible shown below, can carry people. ey are designed to protect passengers from pressure in the deep ocean.

A submersible is carried to the dive location on a ship. A lift is used to lower the submersible into the water. At the conclusion of the dive, the submersible is reattached to this lift and hoisted back onto the ship.

HARBOR BRANCH

A steel sphere with an acrylic dome protects the scientists inside from the extreme deep-ocean pressure.

Lights attached to the *JSL* submersible enable scientists to see in the darkness of the deep ocean.

Scientists manipulate robotic arms in order to lift and grab various objects.

his submersible, one of two *Johnson-Sea-Link* nersibles, can transport humans safely to a depth of ly 1,000 m under water. It carries the oxygen that ›ilot and the passengers will need throughout the time. The *JSL* submersibles were key to exploring vreckage of the Civil War ironclad USS *Monitor* off oast of North Carolina as well as the space shuttle *enger* wreckage off the coast of Florida.

It's Your Turn

REPORT Recall that atmospheric pressure decreases with increased elevation. How do airplanes protect people from the decreased air pressure when flying at high elevation? Find out and report what you learn.

Lesson 2

Reading Guide

Key Concepts
ESSENTIAL QUESTIONS

- How are pressure and the buoyant force related?
- How does Archimedes' principle describe the buoyant force?
- What makes an object sink or float in a fluid?

Vocabulary

buoyant force p. 132

Archimedes' principle p. 134

g Multilingual eGlossary

The Buoyant Force

Inquiry Why doesn't it sink?

This boat is overloaded. Yet it still floats. What forces are on a boat sitting in water? What forces affect whether it floats or sinks?

Inquiry Launch Lab

10 minutes

How can objects denser than water float on water?

Many ships are made of aluminum. Aluminum is denser than water. Why does it float?

1. Read and complete a lab safety form.
2. Use **scissors** to cut three 10- × 10-cm squares of **aluminum foil.**
3. Form a boat shape from one square of foil. Squeeze another square into a tight ball. Fold the third square several times into a 2- × 2-cm square. Flatten it completely.
4. Predict whether each object will sink or float. Then, gently place each in a **tub of water.** Record your observations in your Science Journal.

Think About This

 **Key Concept** What do you think caused each object to float or sink?

What is a buoyant force?

Have you ever floated on a raft, like the person in **Figure 6?** What forces are acting on this person and his raft? The gravitational force pulls them down. Recall that the gravitational force on an object is the object's weight. The weight of the person and his raft includes the weight of the column of air above them. Why don't they sink? The total weight of the raft, the person, and the air is balanced by another, upward force. *A* **buoyant** (BOY unt) **force** *is an upward force applied by a fluid on an object in the fluid.* The buoyant force on the raft is equal to the total weight of the raft, so the raft floats.

Word Origin
buoyant
from Spanish *boyar,* means "to float"

Buoyant Forces in Liquids

You might have noticed that you can lift someone in water who you could not lift when you are both out of water. This is because of a buoyant force. A buoyant force acts on any object in a liquid. This includes floating objects, such as rafts, ships, and bath toys. It also includes submerged objects, such as rocks at the bottom of a lake.

Figure 6 A buoyant force balances the weight of this person and his raft and keeps them afloat.

Buoyant Forces in Air

Objects in a gas also experience a buoyant force. For example, the buoyant force from air keeps a helium balloon up even though a gravitational force pulls downward. An object does not need to be floating in air for a buoyant force to act on it. In fact, a buoyant force from air is acting on you right now. However, the buoyant force acting on you is less than the gravitational force acting on you. Therefore, you do not float.

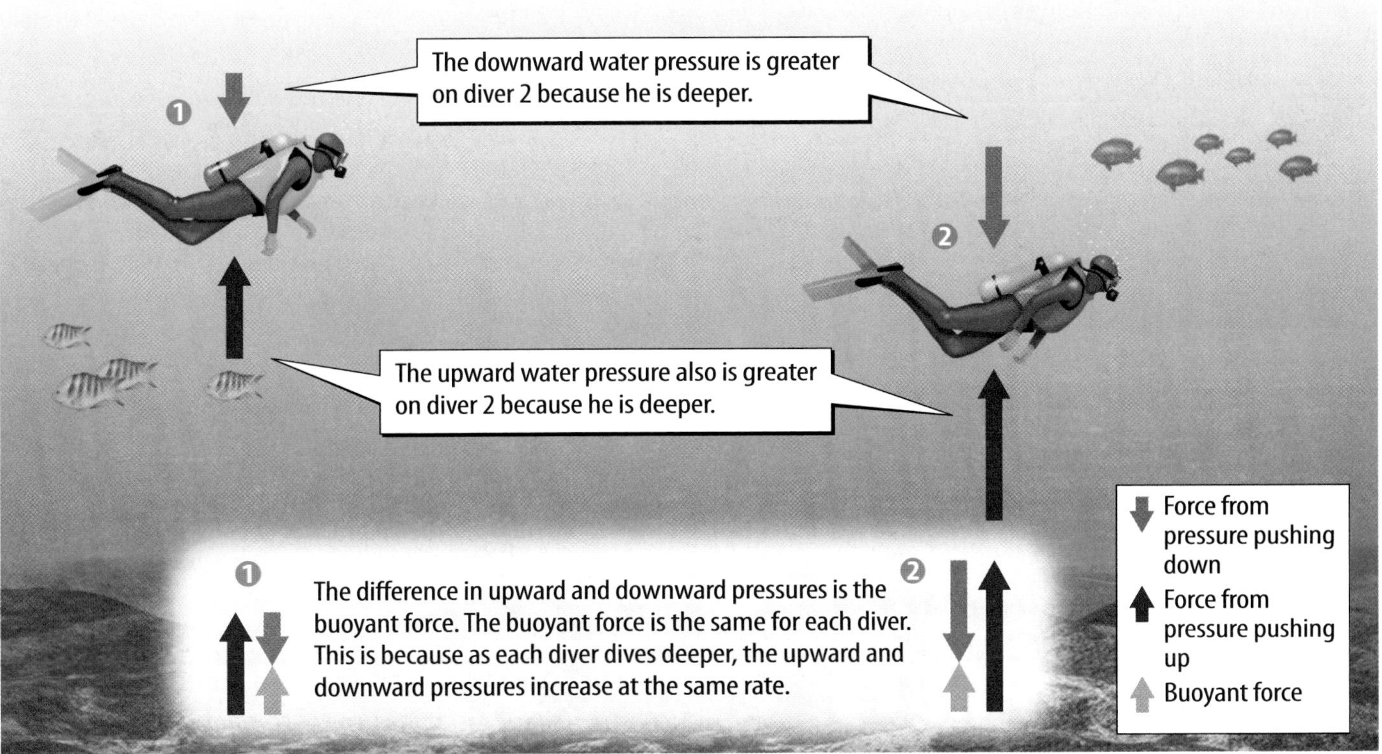

The downward water pressure is greater on diver 2 because he is deeper.
1
2
The upward water pressure also is greater on diver 2 because he is deeper.
1
2
The difference in upward and downward pressures is the buoyant force. The buoyant force is the same for each diver. This is because as each diver dives deeper, the upward and downward pressures increase at the same rate.
Force from pressure pushing down
Force from pressure pushing up
Buoyant force

Figure 7 The buoyant force on a diver is the difference between the force from pressure above and below the diver.

Buoyant Force and Pressure

How does a fluid, such as water, exert an upward force on any object in the fluid? You read in Lesson 1 that a fluid applies pressure perpendicular to all sides of an object within it. Therefore, the forces from water pressure on the divers in **Figure 7** are in the horizontal and vertical directions. The horizontal forces on the sides of each diver are equal. But, the vertical forces on the top and the bottom of each diver are not equal.

Recall that pressure increases with depth. The pressure at the bottom of each diver is greater than the pressure at the top of each diver. This is illustrated in **Figure 7** by the differences in length of the purple arrows. The dark purple arrow represents the pressure pushing up on each diver. The light purple arrow represents the pressure pushing down on each diver. Notice that the arrows are longer on the diver on the right. That is because this diver is deeper than the other diver. The pressure on both the top and the bottom of this diver is greater.

The difference between the upward force from pressure and the downward force from pressure on each diver is the buoyant force on each diver. The buoyant force is represented by the orange arrows in **Figure 7.** The buoyant force on an object is always in an upward direction because the pressure is always greater at an object's bottom.

 **Key Concept Check** How is pressure related to the buoyant force?

Buoyant Force and Depth

Does the buoyant force on a diver change as she dives deeper? Except with extreme changes of depth, the buoyant force does not significantly change. This is because the pressure of the water above the diver and the pressure below the diver increase by about the same amount as the diver goes deeper. The depth of an object submerged in a fluid has little effect on the buoyant force.

Figure 8 The buoyant force is greater on the balloon than on the tennis ball or the billiard ball because the balloon displaces more water.

Inquiry MiniLab 10 minutes

Can you overcome the buoyant force?

An air-filled balloon floats in water because the buoyant force pushing up on it balances the weight pushing down. Can you overcome the buoyant force and make the balloon sink?

1. Read and complete a lab safety form.
2. Inflate a **balloon** until it is 2 cm in diameter.
3. Place the balloon in a **tub of water.** Attach a **large paper clip** to the balloon.
4. Continue adding **paper clips,** one at a time. Record your observations in your Science Journal.

Analyze and Conclude

1. **Analyze** What happened to the balloon as you added paper clips? What happened to the buoyant force on the balloon?
2. **Key Concept** Were you able to make the balloon sink below the surface? Why or why not?

Archimedes' Principle

There is another way to think about the buoyant force. What happens to each object in **Figure 8** as it is pushed underwater? As each object is submerged in a large beaker, some water spills out, or is displaced, into a smaller beaker. Notice that the balloon displaces more water than the tennis ball or the billiard ball does. The volume of water displaced by the balloon is equal to the volume of the balloon. Similarly, the volume of water displaced by the tennis and the billiard ball is equal to the volume of each ball.

Which small beaker do you think weighs the most? The beaker on the left does because it has the most water in it. According to **Archimedes'** (ar kuh MEE deez) **principle,** *the weight of the fluid that an object displaces is equal to the buoyant force acting on the object.* Because the balloon displaces more water than the billiard ball or the tennis ball does, a greater buoyant force acts on it.

Key Concept Check What is the buoyant force on you if you displace 400 N of water as you dive under water?

Recall the image of the divers on the previous page. The buoyant force on each diver does not change as the divers go deeper because the volume of each diver does not change. Each displaces the same amount of water at any depth.

Buoyant Force and Weight

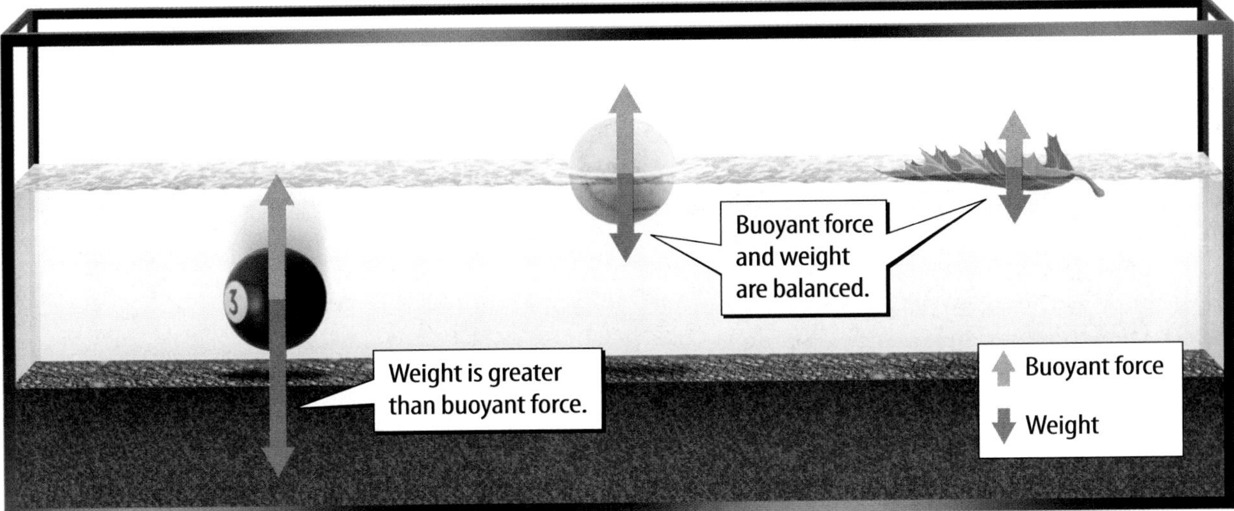

Buoyant force and weight are balanced.
Weight is greater than buoyant force.
Buoyant force
Weight

Figure 9 If the weight of an object is greater than the buoyant force acting on it, the object sinks.

Visual Check How do the buoyant forces on the billiard ball, the tennis ball, and the leaf compare?

Sinking and Floating

Now look at the tennis ball and the billiard ball in **Figure 9.** Both have the same volume. But, if you set each on the surface of water and let go, one floats and one sinks.

Buoyant Force and Weight

If you set a tennis ball on the surface of water, it displaces water. With only a little bit of the tennis ball submerged, the tennis ball stops displacing water and floats. Why is this? When the weight of the water displaced by an object equals the weight of the object, the object floats. So, the weight of the water displaced by the tennis ball, or the buoyant force, equals the weight of the tennis ball. The same is true for the leaf in **Figure 9.**

However, when you set a billiard ball on the surface of water, it sinks. Even after the billiard ball is completely submerged and has displaced the maximum volume of water that it can, it still weighs more than the water it displaces. When an object's weight is greater than the buoyant force, the object sinks.

The buoyant force on the billiard ball is greater than the buoyant force on the tennis ball because the billiard ball is completely submerged. The water displaced by the sunken billiard ball weighs more than the water displaced by the floating tennis ball.

Buoyant Force and Density

Because the billiard ball sinks, you know it weighs more than the water it displaces. Therefore, the billiard ball has more mass per volume, or a greater density, than water. If an object is more dense than the fluid in which it is placed, then the buoyant force on that object will be less than the object's weight, and the object will sink.

FOLDABLES®

Make a vertical two-column chart. Label it as shown. Use the chart to compare and contrast sinking and floating.

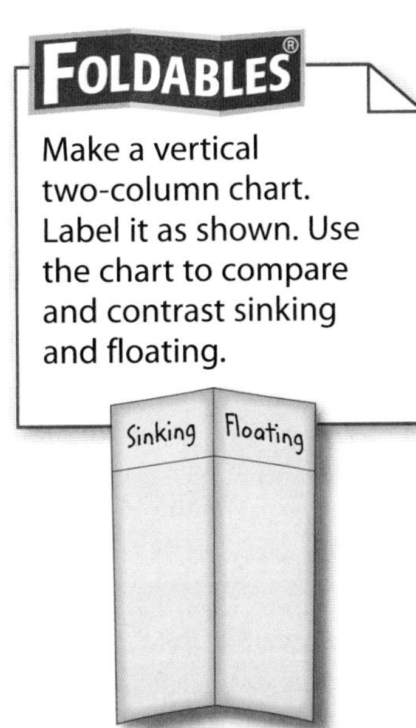

Sinking
Floating

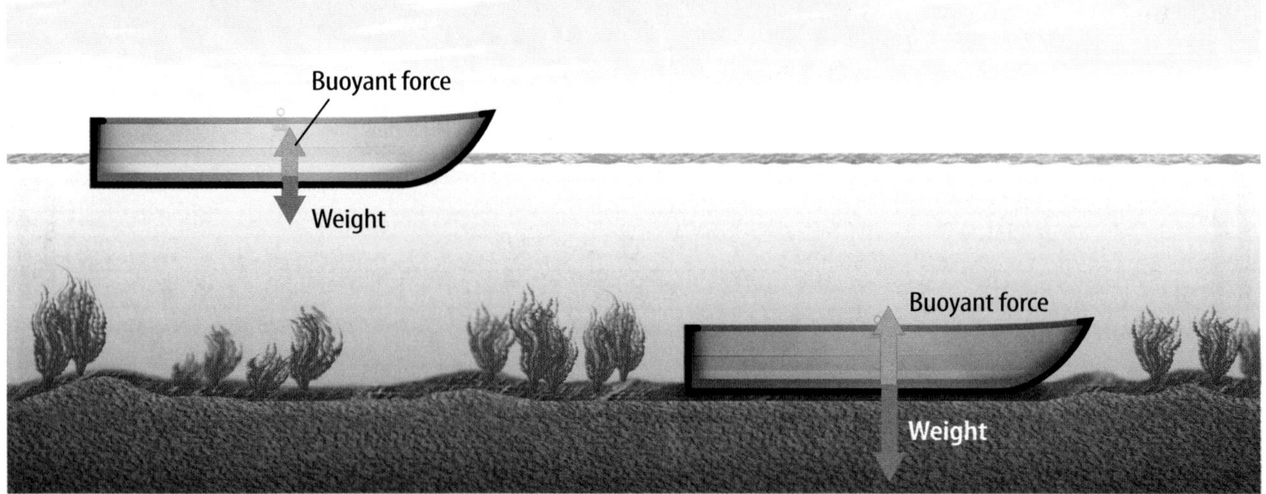

▲ **Figure 10**  The boat on the left floats because it is filled with air instead of water.

Metal Boats

Aluminum has more than twice the density of water. So, how does the aluminum boat in the left part of **Figure 10** stay afloat?

The boats in **Figure 10** are the same size and made from the same material. The boat on the left is filled with air. The weight of the water displaced equals the weight of the boat plus the air inside. Therefore, the boat floats.

The boat on the right is filled with water. In order for this boat to float, the buoyant force would need to equal the weight of the boat plus the water inside. However, the total weight of the boat and the water inside is greater than the water displaced by the boat, or the buoyant force. So, the boat sinks.

You can also think of this is in terms of density. The density of the boat and air on the left is less than the density of water, so the boat floats. However, the density of the boat and water on the right is greater than the density of water. It sinks. The buoyant force is still greater on the boat on the right because it displaces more water than the boat on the left.

Key Concept Check If an object weighing 14 N experiences a 12-N buoyant force, will it sink or float?

Balloons

Why does a helium balloon rise? Helium is less dense than either oxygen or nitrogen in the air. Therefore, the buoyant force acting on a balloon filled with helium is greater than the weight of the balloon, as illustrated in the top of **Figure 11.** The buoyant force pushes the balloon upward. After a day or two, a helium balloon begins to shrink. Helium atoms are so small that they pass between particles that make up the balloon. As the volume of the balloon decreases, the density of the balloon increases. Eventually, the balloon's density is greater than the air's density. The buoyant force on the balloon is less than the balloon's weight, and the balloon falls to the ground.

Figure 11 As a balloon loses helium, its density increases and its buoyant force decreases.

Visual Check How would the forces be different on the balloon if it were neither rising nor falling but was floating in the air? ▼

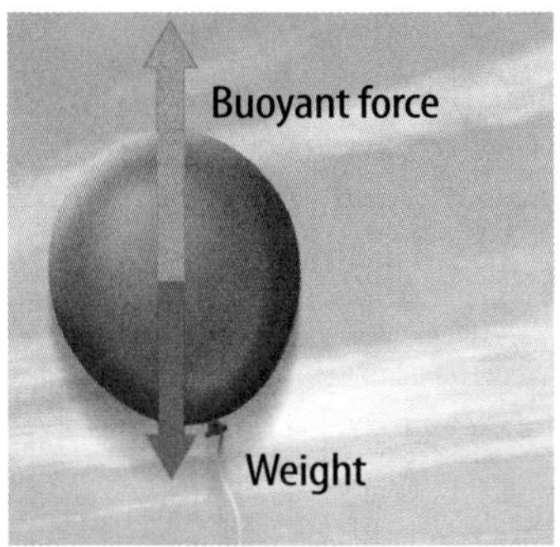

Lesson 2 Review

Assessment Online Quiz
Inquiry Virtual Lab

Visual Summary

A buoyant force results from the difference in pressure between the top and the bottom of an object.

Objects that have the same volume in a fluid experience the same buoyant force.

When the density of a balloon becomes greater than the density of air, the balloon sinks.

FOLDABLES

Use your lesson Foldable to review the lesson. Save your Foldable for the project at the end of the chapter.

What do you think NOW?

You first read the statements below at the beginning of the chapter.

3. You can lift a rock easily under water because there is a buoyant force on the rock.

4. The buoyant force on an object depends on the object's weight.

Did you change your mind about whether you agree or disagree with the statements? Rewrite any false statements to make them true.

Use Vocabulary

1. **State** Archimedes' principle in your own words.
2. **Relate** buoyant force to pressure.

Understand Key Concepts 

3. **Construct** a diagram of the forces on a jellyfish suspended in water. Include forces from pressure on the top, the sides, and the bottom of the jellyfish. Also include the buoyant force and the force from the weight of the jellyfish.
4. The density of water is 1.0 g/cm^3. Which of these floats in water?
   - **A.** a ball with a density of 1.7 g/cm^3
   - **B.** a ruler with a density of 1.1 g/cm^3
   - **C.** a cup that weighs 2 N and displaces 3 N of water when submerged
   - **D.** a toy that weighs 4 N and displaces 3 N of water when submerged

Interpret Graphics

5. **Identify** If the person and the raft shown below together weigh 550 N, then how large is the buoyant force on them?

6. **Organize Information** Copy and complete the graphic organizer stating the relative sizes of forces acting on a sinking object.

Critical Thinking

7. **Use** Archimedes' principle to explain why it is easier to lift a rock in water than it is to lift the same rock in air.

Predict

30 minutes

Do heavy objects always sink and light objects always float?

Materials

triple-beam balance

craft stick

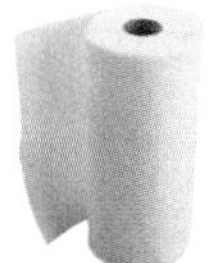

paper towels

tub

Also needed: Objects: sink or float?

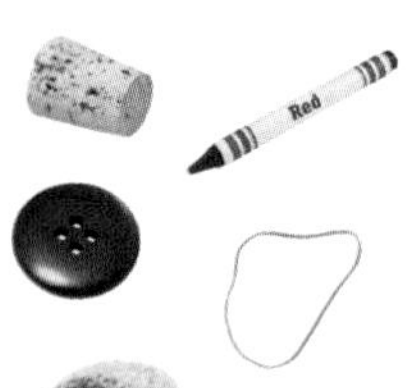

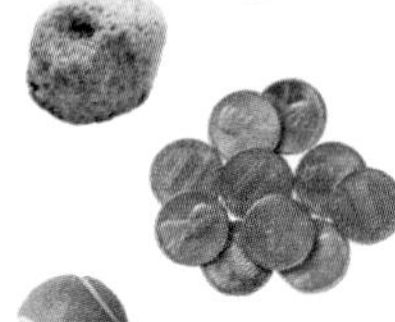

Safety

You have seen many objects that sink in water and many that float. Is it possible to predict whether an object will sink or float if you know its mass? In this lab, you will measure the mass of various objects and then predict whether the objects sink or float.

Learn It

When you **predict** the results of a scientific investigation, you tell what you think will happen. You should base your prediction on what you already know and on things you observe.

Try It

1. Read and complete a lab safety form.
2. Copy the data table below into your Science Journal. Add more rows as needed.
3. Measure the mass of a craft stick. Record it in your data table.
4. Predict whether the craft stick will sink or float. Record your prediction.
5. Place the craft stick in the tub of water. Does it sink, or does it float? Record your results.

Apply It

6. Measure and record the mass of each of the other objects.
7. Predict whether each object will sink or float. Record your predictions.
8. Place each object in the water, and observe whether it sinks or floats. Record your results.
9. **Key Concept** Do heavy objects always sink and light objects always float? Explain your reasoning.

Object	Mass (g)	Predict Sink or Float?	Observe Sink or Float?

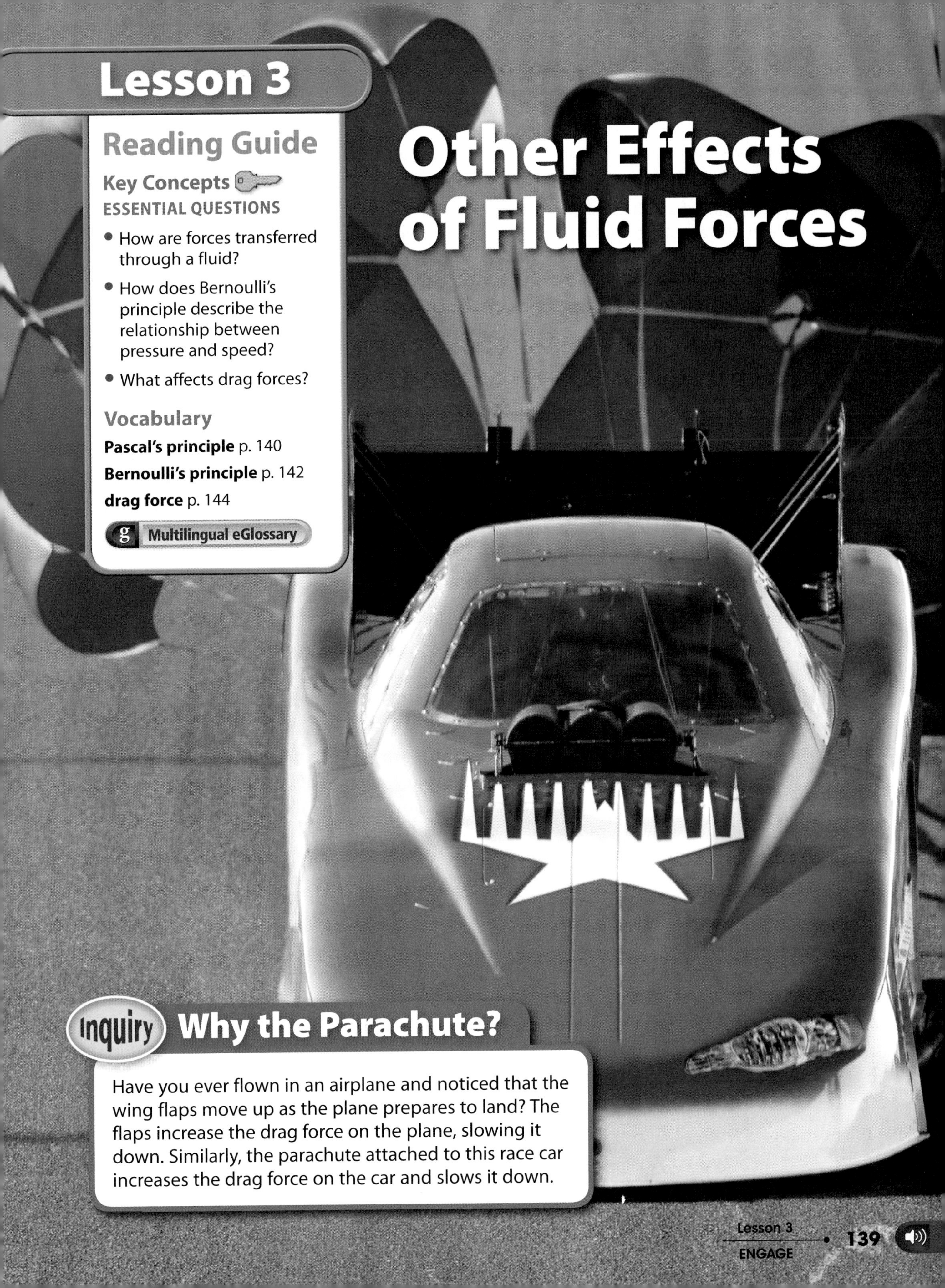

Lesson 3

Other Effects of Fluid Forces

Reading Guide

Key Concepts

ESSENTIAL QUESTIONS

- How are forces transferred through a fluid?
- How does Bernoulli's principle describe the relationship between pressure and speed?
- What affects drag forces?

Vocabulary

Pascal's principle p. 140

Bernoulli's principle p. 142

drag force p. 144

g Multilingual eGlossary

Inquiry Why the Parachute?

Have you ever flown in an airplane and noticed that the wing flaps move up as the plane prepares to land? The flaps increase the drag force on the plane, slowing it down. Similarly, the parachute attached to this race car increases the drag force on the car and slows it down.

Inquiry Launch Lab

15 minutes

How is force transferred through a fluid?

Fluids apply forces. How are forces transferred through fluids?

1. Read and complete a lab safety form.
2. With a partner, use a **nail** to carefully poke three small, identical holes about 1 cm apart along one side of a **drinking straw.** Use **tape** to connect one end of this straw to one end of a **bendable straw.**
3. Tape a **funnel** to the open end of the bendable straw.
4. Keeping the straws and funnel connected, place them in a **tub.** The straw with the holes should be horizontal, with the holes facing upward. Have your partner hold a finger over the open end of the horizontal straw.
5. Gently pour water into the funnel. Record your observations in your Science Journal.

Think About This

1. How did the streams of water pouring out of the holes differ?
2. **Key Concept** How do you think force from pressure is transferred through a fluid?

Fluid Forces—Benefits and Challenges

You use forces in fluids to accomplish everyday tasks. You produce a force when you water a garden with a hose or squeeze a plastic ketchup bottle. You make use of a buoyant force when you float on a raft or lift a rock under water.

Forces in fluids also can make tasks difficult or even dangerous. When a car moves at high speeds, air pushes against the car. This increases the amount of fuel the car uses. Fluid forces from floods, tornadoes, and hurricanes can cause damage that costs lives and billions of dollars. How are forces in fluids both useful and dangerous?

FOLDABLES®

Make a horizontal three-column chart book. Label it as shown. Use it to organize your notes on Pascal's principle, Bernoulli's principle, and drag forces.

Pascal's Principle	Bernoulli's Principle	Drag Forces

Pascal's Principle

Blaise Pascal (pas KAL), a 17th century French physicist, studied the pressures of fluids in closed containers. A fluid cannot flow into or out of a closed container. **Pascal's principle** *states that when pressure is applied to a fluid in a closed container, the pressure increases by the same amount everywhere in the container.* If you push on a closed ketchup bottle, the pressure on the ketchup increases not only under your fingers, it increases equally throughout the bottle.

Key Concept Check How does pressure change when force is applied to a fluid in a closed container?

Pushing on a Fluid

Pascal's principle applies to fluid power systems. A fluid power system like the one in **Figure 12** uses a pressurized fluid to transfer motion. The figure shows the piston on the left moving down as the piston on the right moves up. In this example, the surface area of the input piston is half that of the output piston. Any input force applied to the input piston results in an output force that is doubled by the output piston.

Why is the output force greater than the input force? Pascal's principle can be written as $F = P \times a$. If pressure (P) is always equal throughout the system, increasing the area of the output piston (a) results in a larger force (F). Does the output piston do more work than the input piston?

No. Recall that *work* = *force* × *distance*. In this example, the output force is twice that of the input force. And, if you were to build the fluid power system shown in **Figure 12,** you would find that the input piston is pushed down twice the distance the output piston moves up. If you use the work equation to solve for the input work and then for the output work of the system, you would find that the two are equal.

Hydraulic Lifts

Automobile mechanics use Pascal's principle to lift a car with a hydraulic lift, as shown in **Figure 13.** A hydraulic lift is an example of a fluid power system that uses a liquid for the fluid. In an oil-filled hydraulic lift, a narrow tube connects to a wider tube under a car. Pushing down on a piston in the narrow tube generates an upward force on a larger piston great enough to lift a car.

Like the fluid power system in **Figure 12,** a hydraulic lift does not reduce the amount of work needed to lift a car. To lift a car, the piston in the narrow tube travels a longer distance than the piston in the wider tube.

Reading Check On what principle do hydraulic lifts rely?

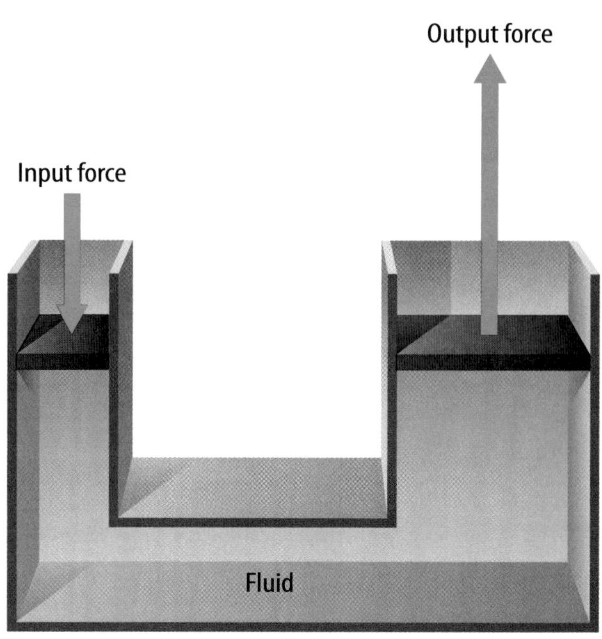

▲ **Figure 12** The left side of the piston has a surface area that is half that of the right side. A force pushing down on the left side generates a force twice as large on the right side.

Visual Check If you push down with a force of 100 N on the left side of the piston, what is the force on the right side?

Animation

Figure 13 People rely on Pascal's principle when they use hydraulic lifts. A hydraulic lift increases the force lifting up on a car. ▼

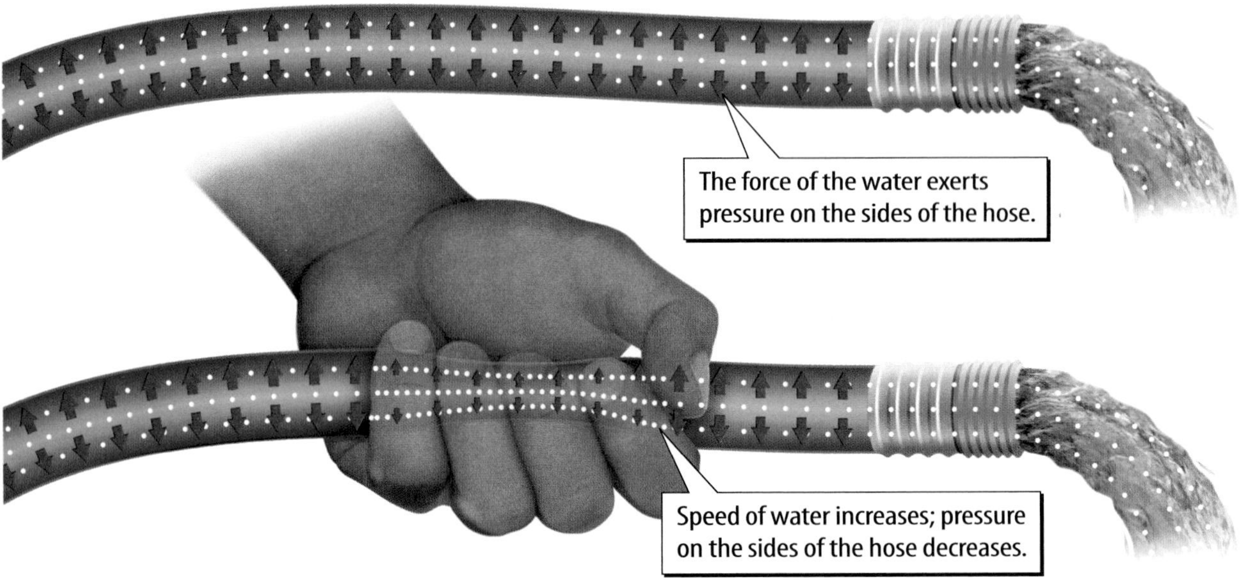
The force of the water exerts pressure on the sides of the hose.
Speed of water increases; pressure on the sides of the hose decreases.

Figure 14 According to Bernoulli's principle, when water moves faster in the pinched part of a hose, it applies less pressure on the sides of the hose.

Concepts in Motion
Animation

Bernoulli's Principle

Have you ever sprayed a friend with water from a hose? Garden hoses illustrate a principle that describes the relationship between speed and pressure in fluids. **Bernoulli's** (ber NEW leez) **principle** *states that the pressure of a fluid decreases when the speed of that fluid increases.*

Consider the hose shown at the top of **Figure 14.** As the water flows through the hose, it applies a constant pressure on the sides of the hose, and it comes out of the hose at a constant speed.

Suppose you pinch the hose slightly, as shown in the bottom of **Figure 14.** The water speeds up in the pinched part. After it has traveled through the pinched part, it returns to its regular speed. According to Bernoulli's principle, when speed increases, pressure decreases. This means the pressure of the water on the sides of the hose in the pinched part is less than it is elsewhere in the hose.

Key Concept Check What is the relationship between speed and pressure in a fluid?

Inquiry MiniLab — 10 minutes

How does air speed affect air pressure?

What happens when moving air changes the pressure on an object?

1. Read and complete a lab safety form.
2. Use **tape** to attach a 25-cm **string** to each of two empty **aluminum cans.** Hold the cans 1–2 cm apart in front of your face.
3. Gently blow between the cans. Record your observations in your Science Journal.

Analyze and Conclude

1. **Describe** how the cans moved.
2. **Key Concept** Explain how this activity demonstrates Bernoulli's principle.

Damage from High Winds

Have you ever seen a photograph of a house that has had its roof blown off during a windstorm? When strong winds blow over a house, the air outside the house has a high speed. But the air inside the house has almost no speed. According to Bernoulli's principle, increased speed means lowered pressure. Therefore, the pressure outside the house is lower than the pressure inside the house. This is illustrated in the left side of **Figure 15.** The force of the air pressure pushing down on the roof of this house is less than the force of the air pressure pushing up. When the upward force inside the house becomes greater than the combined downward force outside the house—including the force of gravity—the roof begins to rise.

 **Key Concept Check** How does Bernoulli's principle explain how wind can take the roof off a house?

Now think about what happens immediately after the roof lifts off the house, as shown in the right side of **Figure 15.** The wind continues to blow. Its speed is now the same below and above the roof. The air pressure is also the same below and above the roof. The force from the upward air pressure on the roof is balanced by the force from the downward air pressure on the roof. But, the force of gravity adds to the force from the downward air pressure. When the combined downward force is greater than the upward force, the roof crashes back down.

Figure 15 When the upward pressure on a roof is greater than the downward pressure, the roof moves up.

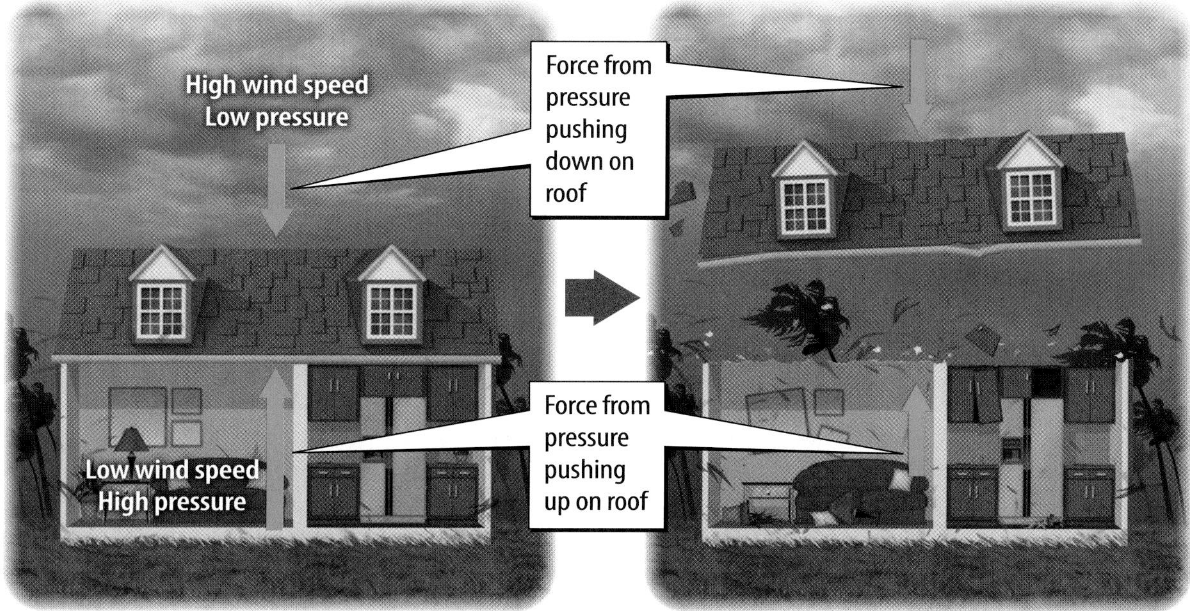

If wind is moving very quickly outside a house, such as during a tornado, the pressure outside will be less than the pressure inside. If the pressure inside the house is greater than the pressure outside, the roof can lift off.

If a roof lifts off a house, the pressure outside the house pushing down on the roof and the pressure inside the house pushing up on the roof are equal.

▲ **Figure 16** The higher pressure on the left side of the soccer ball causes the ball to curve right.

Visual Check Which way would the ball curve if it were spinning in the opposite direction?

Soccer Kicks

You might have seen a soccer player kick a soccer ball so that it curves around an opposing player. A soccer player who curves the ball this way is making use of Bernoulli's principle.

The soccer player puts a spin on the ball as he or she kicks it, as shown in **Figure 16.** This makes the speed of the air on one side of the ball greater than the speed of the air on the other side. According to Bernoulli's principle, the side with the lower speed has greater air pressure acting on it. Because air moves from areas of high to low pressure, the spinning ball curves toward the side with lower pressure.

WORD ORIGIN

drag force
from Old Norse *draga,* means "to draw"; and Latin *fortis,* means "force"

Drag Forces

Have you ever seen a runner using a parachute, like in **Figure 17?** Runners use parachutes to increase the drag force on them and build strength. The **drag force** *is a force that opposes the motion of an object through a fluid.* As the speed of an object in a fluid increases, the drag force on that object also increases. The faster the runner, the greater the drag force on him or her, and the harder he or she needs to work to resist it.

The drag force on an object also depends on the size and shape of that object. If two objects move in the same direction, the object with the greater surface area toward the direction of motion has a greater drag force on it. If the parachute were larger, the drag force on the runner would be even greater.

The drag force increases when the density of a fluid increases. You probably have felt the pull of water against your legs when you wade in shallow water. When you walk through air, which has a lower density, you barely notice the drag force. Whether you notice them or not, the drag force and other forces in fluids are all around you. They help you work and play.

Figure 17 The parachute increases the drag force on this runner and makes him work harder. ▼

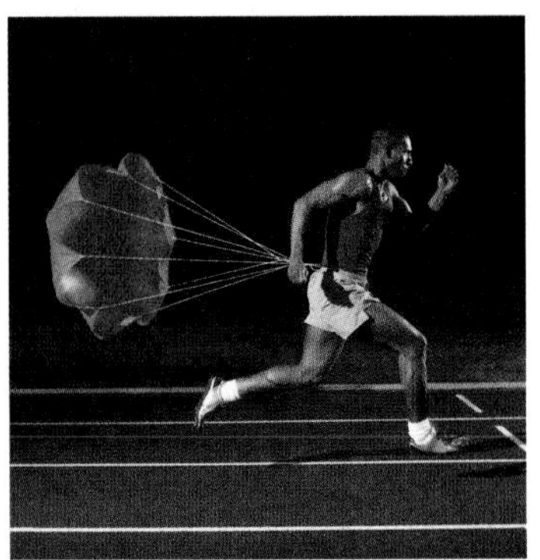

Key Concept Check What affects the drag force on an object?

Lesson 3 Review

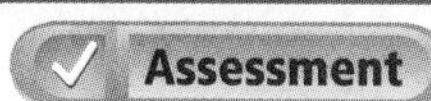

Visual Summary

People rely on Pascal's principle when they use hydraulic lifts.

The imbalance of pressures in fluids can cause a roof to lift off a house in a severe windstorm.

A soccer player who kicks a curved ball makes use of Bernoulli's principle.

Use your lesson Foldable to review the lesson. Save your Foldable for the project at the end of the chapter.

What do you think NOW?

You first read the statements below at the beginning of the chapter.

5. If you squeeze an unopened plastic ketchup bottle, the pressure on the ketchup changes everywhere in the bottle.

6. Running with an open parachute decreases the drag force on you.

Did you change your mind about whether you agree or disagree with the statements? Rewrite any false statements to make them true.

Use Vocabulary

1. **State** Bernoulli's principle in your own words.
2. The transfer of forces in fluids in closed containers is explained by ________.
3. **Use the term** *drag force* in a sentence.

Understand Key Concepts 

4. **Identify** Water flows through one pipe section at 4 m/s and then through a second section at 8 m/s. In which section is the pressure greater? Explain why.
5. **Determine** A closed container of liquid soap is 10 cm tall. You squeeze on the container's top with a pressure of 1,000 Pa. What is the pressure at the bottom?
6. **Explain** how drag force differs on two cars traveling at different speeds. Why does the car traveling faster use more fuel?

Interpret Graphics

7. **Analyze** In the diagram below, if you push the left side down 3 m, how far will the right side move up?

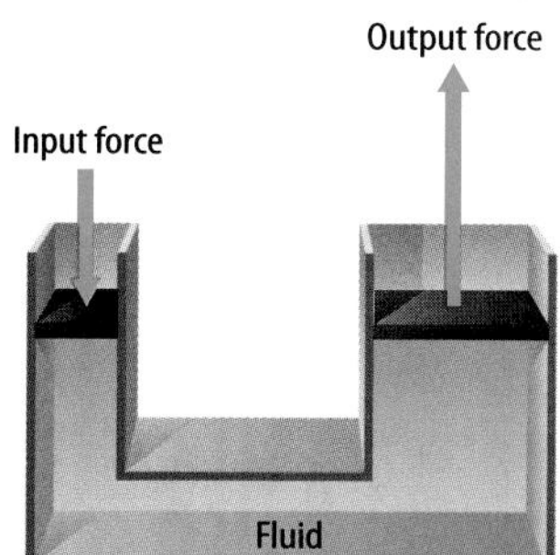

8. **Organize Information** Copy and fill in the graphic organizer below to list four things that affect drag forces.

Critical Thinking

9. **Design** a hydraulic lift that could lift 100 N 1 m with a 25-N force.

50 minutes

Design a Cargo Ship

Materials

craft supplies

creative building materials

office supplies

rocks

Safety

Modern cargo ships can carry thousands of freight containers. But cargo ships will sink if they are loaded with too much cargo. In this lab, you will build a model cargo ship. You then will test the ship's buoyancy to determine how much weight it can support.

Question

How is buoyant force related to weight in a model cargo ship?

Procedure

1. Read and complete a lab safety form.
2. Look at the available supplies. Discuss with your partner how you can build a cargo ship using the supplies.
3. Consider what you have learned about the buoyant force and about Archimedes' principle. What features should your boat have to enable it to carry heavy cargo?
4. Work with your partner to build your cargo ship.
5. Measure the weight of your ship. Record it in your Science Journal in newtons. Use this formula to convert grams to newtons:

   $N = \text{mass (in kg)} \times 9.8$
6. Place the ship in a large tub of water. How high on the water does the ship float? How much of the ship is below water?

7. Add rocks to your ship to model freight containers. Before you add a rock, measure it and record its weight in your Science Journal. Note how the position of the ship on the water changes with each rock.
8. Continue adding rocks until your ship sinks. Then, calculate the total weight of the ship and the rocks it was able to support before it sank. Show your calculations in your Science Journal.

Analyze and Conclude

9. **Measure** What is the weight of your ship?
10. **Measure** What was the total weight of the ship and the rocks before it sank?
11. **Assess** How did the buoyant force on your ship change each time you added a rock?
12. THE BIG IDEA **The Big Idea** Describe how the buoyant force kept your ship afloat.
13. **Relate** How did the buoyant force on your ship relate to its volume?
14. **Interpret** How can you explain your results in terms of density?

Communicate Your Results

With your partner, create a brochure advertising your ship. Include how much weight it can carry. Compare brochures with those of your classmates. How did the ability of your ship to support weight compare with the ships of other groups? What was different about the ships that could support the most weight?

Inquiry Extension

Place the small tub of water in a larger tub. Fill the small tub completely with water. Place two rocks on your ship, and set the ship on the water. Measure the weight of the water it displaces. Fill the small tub again. Add to the ship all the rocks your ship can support. Again measure the weight of the water it displaces. How did the buoyant force change? In your Science Journal, draw your ship with no rocks, with two rocks, and with all the rocks it could support. Label all forces on the ship in each drawing.

Lab Tips

- ☑ As you design your ship, think about how you can maximize the amount of water displaced by your ship while keeping the weight of the ship to a minimum.
- ☑ When calculating the total weight supported by your ship, be sure not to include the weight of the final rock that sank the ship.

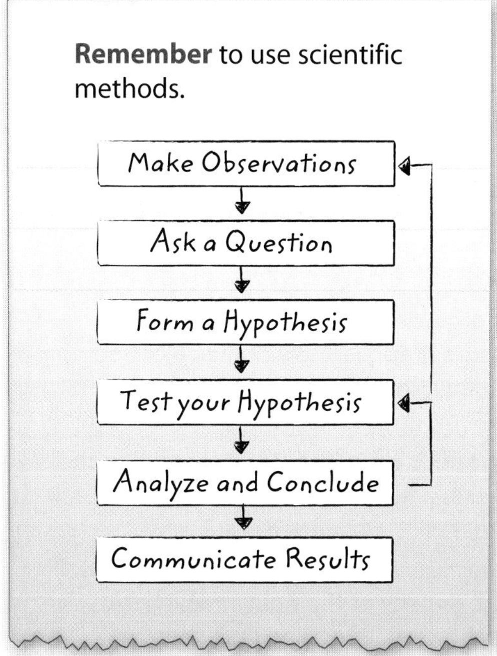

Chapter 4 Study Guide

People use forces in fluids to float objects on water and in air, to lift objects, and to affect the motions of objects.

Key Concepts Summary

Lesson 1: Pressure and Density of Fluids

- **Pressure** is the ratio of force to area.
- **Atmospheric pressure** decreases with elevation. Pressure under water increases with depth.
- The density of a **fluid** depends on the mass of the fluid and its volume.

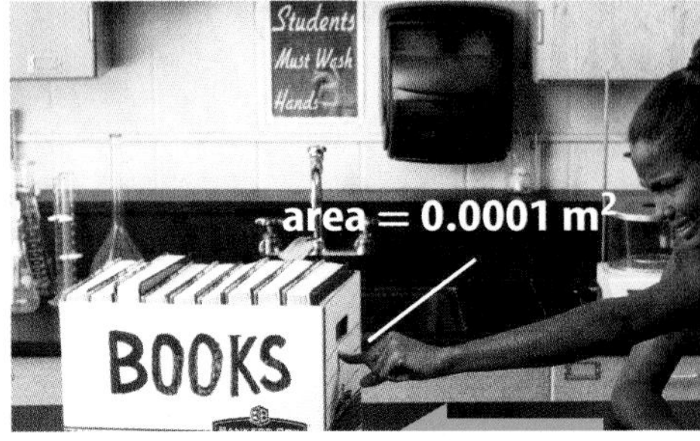

Vocabulary

fluid p. 123
pressure p. 124
atmospheric pressure p. 126

Lesson 2: The Buoyant Force

- The change in pressure between the top and the bottom of an object results in an upward force called the **buoyant force.**
- **Archimedes' principle** states that the weight of the fluid displaced by an object is equal to the buoyant force on that object.
- An object sinks if its weight is greater than the buoyant force on it. An object does not sink if the buoyant force on it is equal to its weight.

Vocabulary

buoyant force p. 132
Archimedes' principle p. 134

Lesson 3: Other Effects of Fluid Forces

- **Pascal's principle** states that when pressure is applied to a fluid in a closed container, the pressure increases by the same amount everywhere in the container.
- **Bernoulli's principle** states that when the speed in a fluid increases, the pressure decreases.
- Speed, size, and shape of an object, as well as the density of the fluid in which the object moves, affect the **drag force** on that object.

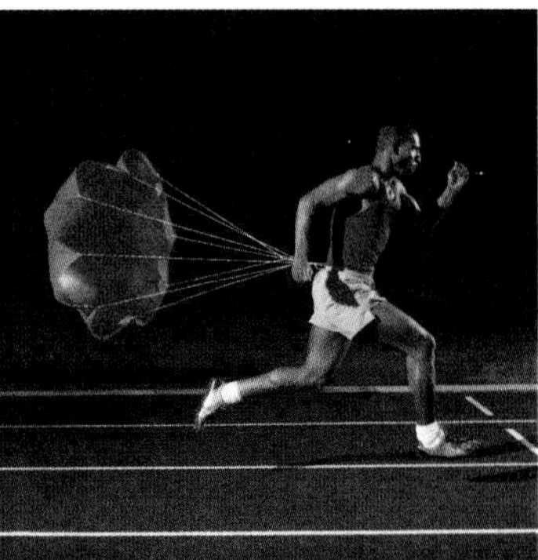

Vocabulary

Pascal's principle p. 140
Bernoulli's principle p. 142
drag force p. 144

- Personal Tutor
- Vocabulary eGames
- Vocabulary eFlashcards

FOLDABLES® Chapter Project

Assemble your lesson Foldables as shown to make a Chapter Project. Use the project to review what you have learned in this chapter.

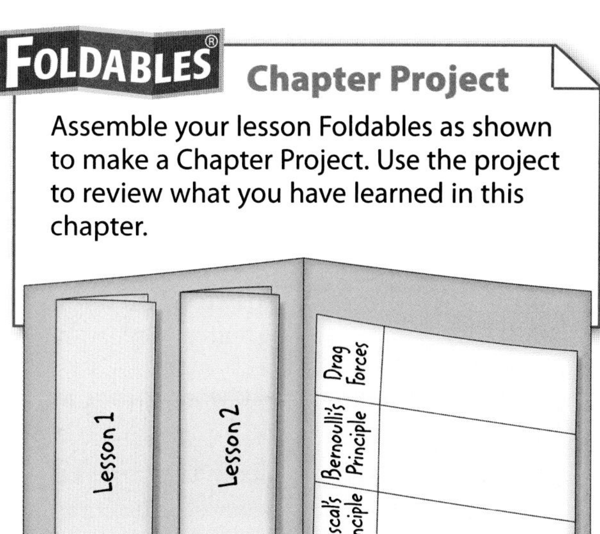

Use Vocabulary

1. Air and water are both ________.
2. A ship does not sink because a(n) ________ acts on it.
3. A column of air exerts ________ on you.
4. According to ________, two objects of equal volume in a fluid experience the same buoyant force.
5. A parachute with a 5-m^2 surface area experiences a much larger ________ than a parachute with a 3-m^2 surface area.
6. Fluid power systems work according to ________.

Link Vocabulary and Key Concepts

Concepts in Motion Interactive Concept Map

Copy this concept map, and then use vocabulary terms from the previous page to complete the concept map.

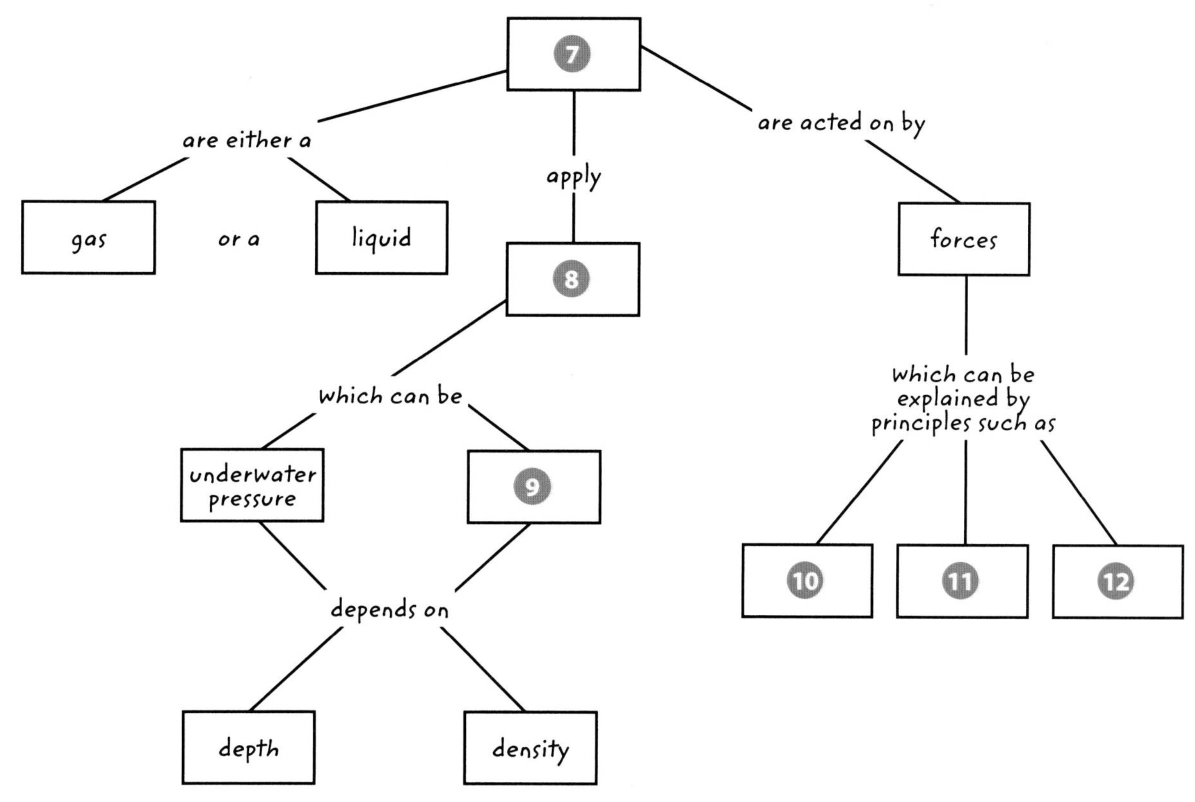

Chapter 4 Review

Understand Key Concepts 

1. Which is NOT a fluid?
   A. helium
   B. ice
   C. milk
   D. water

2. If you poured the following fluids into a container, which would float on top?
   A. maple syrup, with a density of 1.33 g/cm^3
   B. olive oil, with a density of 0.9 g/m^3
   C. seawater, with a density of 1.03 g/cm^3
   D. water, with a density of 1.0 g/cm^3

3. What pressure does Adam apply to a ball of dough when he pushes on it with a 25-N force? The area of his hand is 0.01 m^2.
   A. 0.0004 Pa
   B. 0.5 Pa
   C. 25 Pa
   D. 2,500 Pa

4. Which of these has the greatest pressure applied to it from the surrounding fluid?
   A. a fish swimming 20 m below the surface
   B. a hawk flying 300 m above sea level
   C. a mountain climber at an altitude of 4,400 m
   D. a person fishing off the coast of California

5. In the diagram of the beach ball floating on water below, what force does the blue arrow represent?

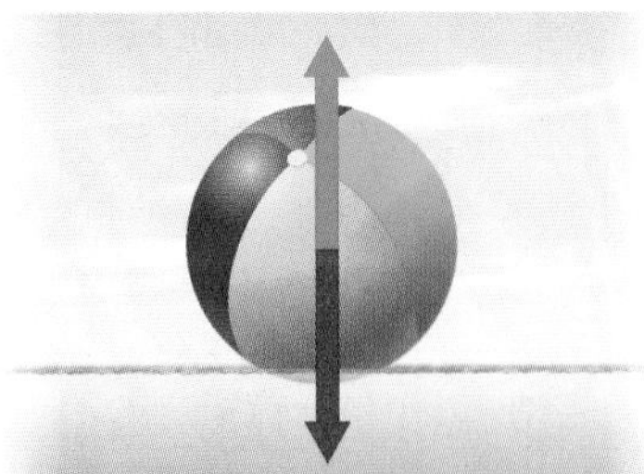

   A. pressure
   B. weight
   C. buoyant force
   D. drag force

6. Joseph weighs 290 N and displaces 300 N of water as he swims under water in a pool. What is the buoyant force on Joseph?
   A. 10 N upward
   B. 300 N upward
   C. 290 N downward
   D. 590 N downward

7. Which statement about boats is correct?
   A. A boat cannot be made from metal because metal has a greater density than water.
   B. A boat floats if its overall density is less than that of water.
   C. A boat floats only if its overall mass per volume is more than water's mass per volume.
   D. A boat floats only if the weight of water it displaces is less than the boat's weight.

8. In the diagram below, how large a force is applied by the piston on the right?

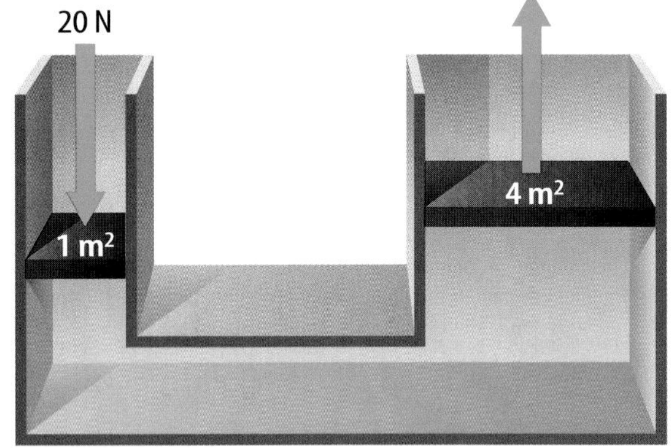

   A. 10 N
   B. 20 N
   C. 40 N
   D. 80 N

9. A leaf enters a drainpipe, and the pressure in the water increases from 10 Pa to 30 Pa. What happens to the leaf's speed?
   A. It decreases.
   B. It increases.
   C. It becomes exactly 30 m/s.
   D. It does not change.

Assessment

Online Test Practice

Critical Thinking

10 **Design an Experiment** You are given a material and asked to determine whether it is a fluid. What experiment could you do?

11 **Propose** The legs of your chair sink into the sand on a beach. What could you do to prevent this? Explain your reasoning.

12 **Compare** You drop a solid cube into water. The cube is 40 cm on each side. Then you drop another solid object into water that is 160 cm tall, 20 cm deep, and 20 cm wide and made from the same material as the cube. Which object experiences the greater buoyant force? Explain.

13 **Interpret** A sailor drops an anchor over the side of a ship. When the anchor is 10 m below the ocean's surface, the buoyant force on the anchor is 80 N. What is the buoyant force on the anchor when it sinks to 100 m below the surface?

14 **Assess** How does opening a parachute change the drag force on a skydiver?

15 **Explain** Woodchucks live in underground tunnels such as the one shown below. One opening has a dirt mound around it, and air flows across it quickly. The other opening is even with the ground. The air moves across it with less speed. How does this design help ventilate the tunnel?

Writing in Science

16 **Write** a paragraph explaining why the *Titanic* sank. Use what you have learned about forces and fluids.

REVIEW THE BIG IDEA

17 The braking systems in most automobiles rely on a hydraulic fluid power system. Use Pascal's principle to explain how the pressure of a foot on a car's brake can stop the car.

18 The photo below shows a helium balloon in a parade. Imagine you are holding a rope attached to this balloon. What forces do you encounter? Explain at least two ways in which you might encounter or use forces in fluids in your everyday life.

Math Skills

Review — Math Practice

Solve a One-Step Equation

19 The buoyant force on an inflatable pool raft with a surface area of 2 m^2 is 36 N. How large is the upward pressure on the raft?

20 A ballerina stands on her toes in a pointe shoe. The pressure on her toes is 454,000 Pa. When she stands on flat feet, the pressure is 22,700 Pa. If the ballerina's weight is 454 N, what is the surface area she stands on when she's on her toes, and what is the surface area she stands on when she's flat-footed?

Standardized Test Practice

Record your answers on the answer sheet provided by your teacher or on a sheet of paper.

Multiple Choice

1 The same force is applied over two areas that differ in size. Which is true of the pressure over these areas?

A The pressure is equal to the force multiplied by the area.

B The pressure on both areas is the same.

C The pressure on the larger area is greater.

D The pressure on the smaller area is greater.

Use the figure to answer question 2.

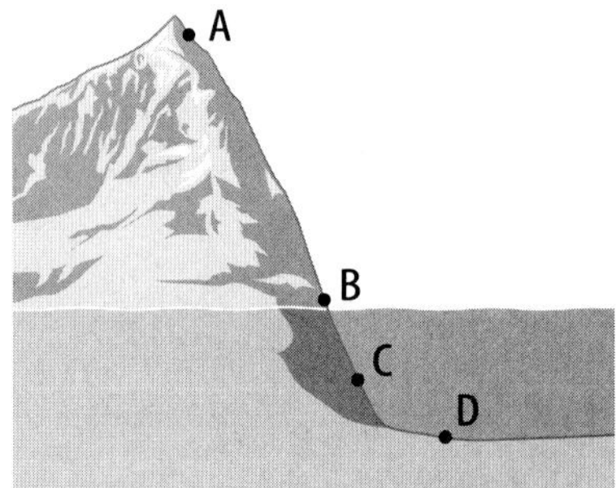

2 The figure shows a side view of a mountain that extends below the water's surface. At which point would the pressure be 100 kPa?

A A

B B

C C

D D

3 Which characteristics determine the density of a fluid?

A mass and energy of the particles

B mass of particles and distance between them

C number of particles and distance between them

D shape and energy of the particles

4 Which changes as a solid object moves downward within a fluid?

A the buoyant force acting on the object

B the mass of the object

C the pressure acting on the object

D the volume of the object

5 Which explains how the buoyant force on an object changes with the weight of the fluid it displaces?

A Archimedes' principle

B definition of density

C definition of pressure

D Pascal's principle

Use the figure to answer question 6.

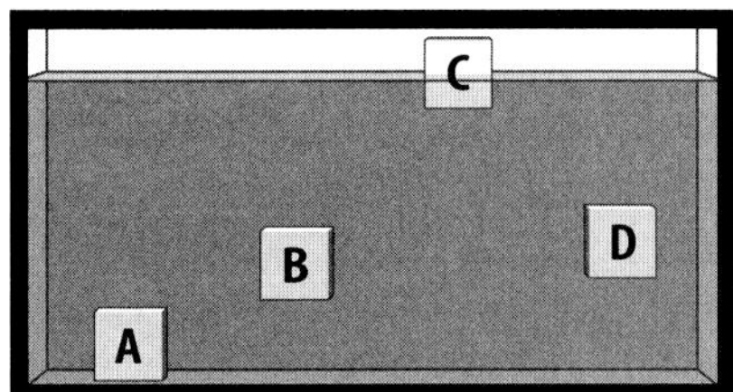

6 The figure shows an aquarium filled with water. Four cubes made out of different materials have been placed in the aquarium. For which object is the buoyant force acting on it equal to the object's weight?

A A

B B

C C

D D

7 Why are drag forces greater in water than they are in air?

A Air is denser than water.

B Air is a fluid, but water is not.

C Water is denser than air.

D Water is a fluid, but air is not.

Use the figure to answer questions 8 and 9.

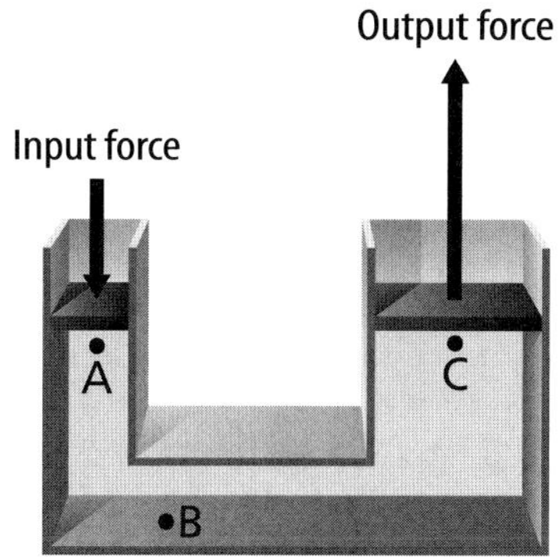

8 When an input force is applied above point A, which is true of the change in fluid pressure at points A, B, and C?

A The fluid pressure increases the most at point A.

B The fluid pressure increases the most at point B.

C The fluid pressure increases the most at point C.

D The fluid pressure increases by the same amount at all three points.

9 Which describes how the input force affects the fluid pressure at different points in the hydraulic lift?

A Bernoulli's principle

B definition of buoyant force

C equation for density

D Pascal's principle

Constructed Response

10 Explain how the buoyant force on an underwater diver is exactly the same at two different depths.

11 Use the terms *buoyant force* and *weight* to describe how a helium balloon floats in air.

Use the figure to answer question 12.

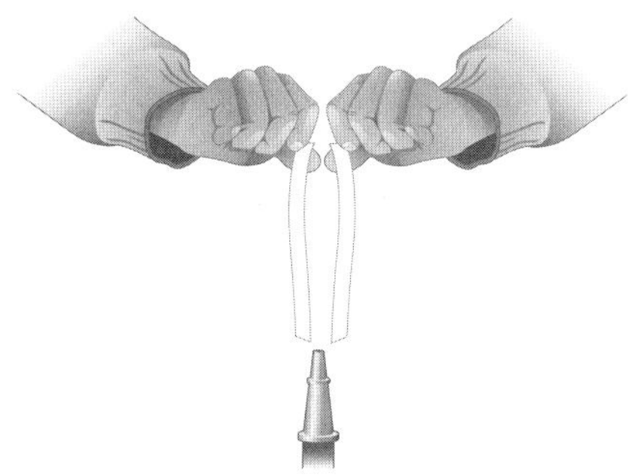

12 The figure shows two strips of paper held above an air source. When the air jet is turned on, the air between the pieces of paper will move quickly. The air on either side of each piece of paper will hardly move at all. When the air jet is turned on, how will the pressure between the pieces of paper compare to the pressure on the outer sides of each piece? Predict what will happen to the pieces of paper when the air jet is turned on.

13 Mercury is a fluid that has a density of 13.53 g/cm^3. Gold is a solid that has a density of 19.32 g/cm^3. Aluminum is a solid that has a density of 2.7 g/cm^3. Predict what will happen if 5-g samples of gold and aluminum are dropped into a flask of mercury.

NEED EXTRA HELP?													
If You Missed Question...	1	2	3	4	5	6	7	8	9	10	11	12	13
Go to Lesson...	1	1	1	2	2	2	3	3	3	2	2	3	1

1875 1900 1925

1895
The first X-ray photograph is taken by Wilhelm Konrad Roentgen of his wife's hand. It is now possible to look inside the human body without surgical intervention.

1898
Chemist Marie Curie and her husband Pierre discover radioactivity. They are later awarded the Nobel Prize in Physics for their discovery.

1917
Ernest Rutherford, the "father of nuclear physics," is the first to split atoms.

1934
Nuclear fission is first achieved experimentally in Rome by Enrico Fermi when his team bombards uranium with neutrons.

1939
The Manhattan Project, a code name for a research program to develop the first atomic bomb, begins. The project is directed by American physicist J. Robert Oppenheimer.

1950 1975 2000

1945
American-led atomic bomb attacks on the Japanese cities of Hiroshima and Nagasaki bring about the end of World War II.

1954
Obninsk Nuclear Power Plant, located in the former USSR, begins operating as the world's first nuclear power plant to generate electricity for a power grid. It produces around 5 megawatts of electric power.

2007
Fourteen percent of the world's electricity now comes from nuclear power.

? Inquiry
Visit ConnectED for this unit's STEM activity.

Technology

Figure 1 Nature cannot easily recycle many human-made materials. Much of our trash remains in landfills for years. Scientists are developing materials that degrade quickly. This will help decrease the amount of pollution..

Scientists use technology to develop materials with desirable properties. **Technology** is the practical use of scientific knowledge, especially for industrial or commercial use. In the late 1800s, scientists developed the first plastic material, called celluloid, from cotton. Celluloid quickly gained popularity for use as photographic film. In the 20th century, scientists developed other plastic materials, such as polystyrene, rayon, and nylon. These new materials were inexpensive, durable, lightweight, and could be molded into any shape.

New technologies can come with problems. For example, many plastics are made from petroleum and contain harmful chemicals. The high pressures and temperatures needed to produce plastics require large amounts of energy. Bacteria and fungi that easily break down natural materials do not easily decompose plastics. Often, plastics accumulate in landfills where they can remain for hundreds, or even thousands, of years, as shown in **Figure 1.**

Types of Materials

Figure 2 Some organisms produce materials with properties that are useful to people. Scientists are trying to replicate these materials for new technologies.

Most human-made adhesives attach to some surfaces, but not others. Mussels, which are similar to clams, produce a "superglue" that is stronger than anything people can make. It also works on any surface, wet or dry. Chemists are trying to develop a technology that will replicate the mussel glue. This glue would provide solutions to difficult problems: Ships could be repaired under water. The glue also would work on teeth and could be used to set broken bones.

Abalone and other mollusks construct a protective shell from proteins and seawater. The material is four times stronger than human-made metal alloys and ceramics. Using technology, scientists are working to duplicate this material. They hope to use the new product in many ways, including hip and elbow replacements. Automakers could use these strong, lightweight materials for automobile body panels.

Consider the Possibilities!

Chemists are looking to nature for ideas for new materials. For example, some sea sponges have skeletons that beam light deep inside the animal, similar to the way fiber-optic cables work. A bacterium from a snail-like nudibranch contains compounds that stop other sea creatures from growing on the nudibranch's back. These compounds could be used in paints to stop creatures from forming a harmful crust on submerged parts of boats and docks. Chrysanthemum flowers produce a product that keeps ticks and mosquitoes away. **Figure 2** includes other organisms that produce materials with remarkable properties.

Chemists and biologists are teaming up to understand, and hopefully replicate, the processes that organisms use to survive. Hopefully, these process can lead to technologies and materials with unique properties that are helpful to people.

MiniLab
15 minutes

How would you use it?

How would you use an adhesive that could stick to any surface? Invent a new purpose for mussel "superglue"!

1. Work with a partner to develop three tasks that could be accomplished using products produced by an organism.
2. Select one of your ideas, and develop it into an invention. Draw pictures of your new invention and explain how it works.
3. Write an advertisement for your invention including a description of the role of the material from nature used in your product.

Analyze and Conclude

1. **Explain** What task or problem did your invention solve?
2. **Infer** How did a material from an organism help you develop your invention?

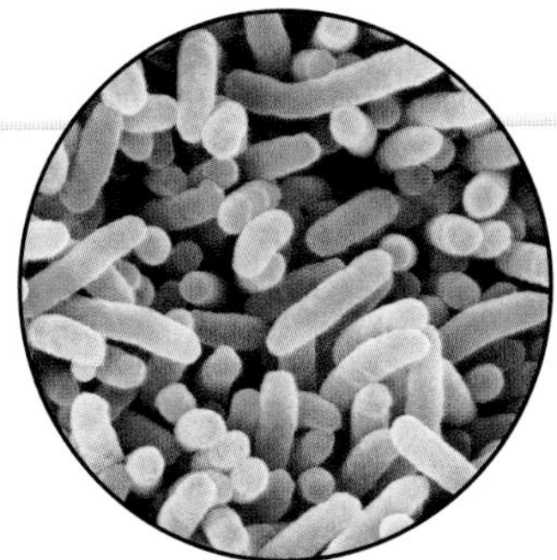

◀ A British company has developed bacteria that produce large amounts of hydrogen gas when fed a diet of sugar. Chemists are working to produce tanks of these microorganisms that produce enough hydrogen to replace other fuels used to heat homes. Bacteria may become the power plants of the future.

Under a microscope, the horn of a rhinoceros looks much like the material used to make the wings of a Stealth aircraft. However, the rhino horn is self-healing. Picture a car with technologically advanced fenders similar to the horn of a rhinoceros; such a car could repair itself if it were in a fender-bender! ▶

◀ Spider silk begins as a liquid inside the spider's body. When ejected through openings, called spinnerets, it becomes similar to a plastic thread. However, its properties include strength five times greater than steel, stretchability greater than nylon, and toughness better than the material in bulletproof vests! Chemists are using technology to make a synthetic spider silk. They hope to someday use the material for cables strong enough to support a bridge or as reinforcing fibers in aircraft bodies some day.

Chapter 5

Energy and Energy Resources

What is energy and what are energy resources?

Which objects have energy?

If your answer is "everything in the photo," you are right. All objects contain energy. Some objects contain more energy than other objects. The Sun contains so much energy it is considered an energy resource.

- Where do you think the energy comes from that powers the cars?
- Do you think the energy in the Sun and the energy in the green plants are related?
- What do the terms energy and energy resources mean to you?

Get Ready to Read

What do you think?

Before you read, decide if you agree or disagree with each of these statements. As you read this chapter, see if you change your mind about any of the statements.

1. A fast-moving baseball has more kinetic energy than a slow-moving baseball.
2. A book sitting on a shelf has no energy.
3. Energy can change from one form to another.
4. If you toss a baton straight up, total energy decreases as the baton rises.
5. Nuclear power plants release many dangerous pollutants into the air as they transform nuclear energy into electric energy.
6. Thermal energy from within Earth can be transformed into electric energy at a power plant.

ConnectED Your one-stop online resource

connectED.mcgraw-hill.com

Lesson 1

Reading Guide

Key Concepts
ESSENTIAL QUESTIONS

- What is energy?
- What are potential and kinetic energy?
- How is energy related to work?
- What are different forms of energy?

Vocabulary

energy p. 161
kinetic energy p. 162
potential energy p. 162
work p. 164
mechanical energy p. 165
sound energy p. 165
thermal energy p. 165
electric energy p. 165
radiant energy p. 165
nuclear energy p. 165

Multilingual eGlossary

Video BrainPOP®

Forms of Energy

Inquiry Why is this cat glowing?

A camera that detects temperature made this image. Dark colors represent cooler temperatures and light colors represent warmer temperatures. Temperatures are cooler where the cat's body emits less radiant energy and warmer where the cat's body emits more radiant energy.

20 minutes

Can you make a change in matter?

You observe many things changing. Birds change their positions when they fly. Bubbles form in boiling water. The filament in a lightbulb glows when you turn on a light. How can you cause a change in matter?

1. Read and complete the lab safety form.
2. Half-fill a **foam cup** with **sand.** Place the bulb of a **thermometer** about halfway into the sand. *Do not stir*. Record the temperature in your Science Journal.
3. Remove the thermometer and place a **lid** on the cup. Hold down the lid and shake the cup vigorously for 10 minutes.
4. Remove the lid. Measure and record the temperature of the sand.

Think About This

1. What change did you observe in the sand?
2. **Predict** how you could change your results.
3. **Key Concept** What do you think caused the change?

What is energy?

It might be exciting to watch a fireworks display like the one shown in **Figure 1.** Over and over, you hear the crack of explosions and see bursts of colors in the night sky. Fireworks release energy when they explode. **Energy** *is the ability to cause change.* The energy in the fireworks causes the changes you see as bursting flashes of light and hear as loud booms.

Energy also causes other changes. The plant in **Figure 1** uses the energy from the Sun and makes food that it uses for growth and other processes. Energy can cause changes in the motions and positions of objects, such as the nail in **Figure 1.** Can you think of other ways energy might cause changes?

Word Origin
energy
from Greek *energeia*, means "activity"

 **Key Concept Check** What is energy?

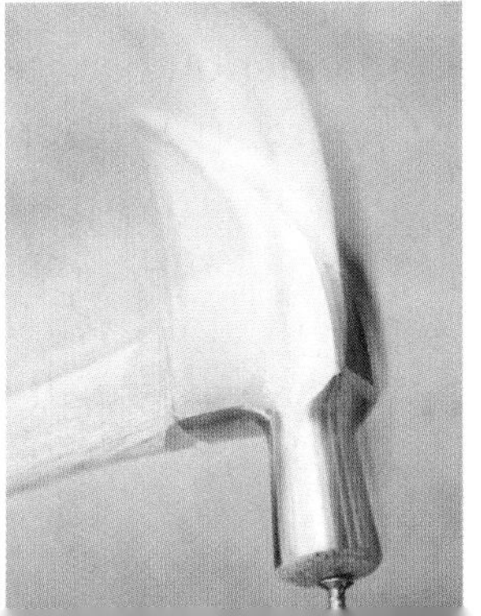

Figure 1 The explosion of the fireworks, the growth of the flower, and the motion of the hammer all involve energy.

Figure 2 The kinetic energy (KE) of an object depends on its speed and its mass. The vertical bars show the kinetic energy of each vehicle.

Kinetic Energy—Energy of Motion

Have you ever been to a bowling alley? When you rolled the ball and it hit the pins, a change occurred—the pins fell over. This change occurred because the ball had a form of energy called kinetic (kuh NEH tik) energy. **Kinetic energy** *is energy due to motion.* All moving objects have kinetic energy.

Kinetic Energy and Speed

An object's kinetic energy depends on its speed. The faster an object moves, the more kinetic energy it has. For example, the blue car has more kinetic energy than the green car in **Figure 2** because the blue car is moving faster.

Kinetic Energy and Mass

A moving object's kinetic energy also depends on its mass. If two objects move at the same speed, the object with more mass has more kinetic energy. For example, the truck and the green car in **Figure 2** are moving at the same speed, but the truck has more kinetic energy because it has more mass.

 **Key Concept Check** What is kinetic energy?

FOLDABLES®

Make a 3-in. fold along the long edge of a sheet of paper and make a two-pocket book. Label it as shown. Organize information about the forms of energy on quarter sheets and put them in the pockets.

Potential Energy—Stored Energy

Energy can be present even if objects are not moving. If you hold a ball in your hand and then let it go, the gravitational interaction between the ball and Earth causes a change to occur. Before you dropped the ball, it had a form of energy called potential (puh TEN chul) energy. **Potential energy** *is stored energy due to the interactions between objects or particles.* Gravitational potential energy, elastic potential energy, and chemical potential energy are all forms of potential energy.

Gravitational Potential Energy

Even when you are just holding a book, energy is stored between the book and Earth. This type of energy is called gravitational potential energy. The girl in **Figure 3** increases the gravitational potential energy between her backpack and Earth by lifting the backpack.

The gravitational potential energy stored between an object and Earth depends on the object's weight and height. Dropping a bowling ball from a height of 1 m causes a greater change than dropping a tennis ball from 1 m. Similarly, dropping a bowling ball from 3 m causes a greater change than dropping the same bowling ball from only 1 m.

Reading Check What factors determine the gravitational potential energy stored between an object and Earth?

Elastic Potential Energy

When you stretch a rubber band, like the one in **Figure 3,** you are storing another form of potential energy called elastic (ih LAS tik) potential energy. Elastic potential energy is energy stored in objects that are compressed or stretched, such as springs and rubber bands. When you release the end of a stretched rubber band, the stored elastic potential energy is transformed into kinetic energy.

Chemical Potential Energy

Food, gasoline, and other substances are made of atoms joined together by chemical bonds. Chemical potential energy is energy stored in the chemical bonds between atoms, as shown in **Figure 3.** Chemical potential energy is released when chemical reactions occur. Your body uses the chemical potential energy in foods for all its activities. People also use the chemical potential energy in gasoline to drive cars and buses.

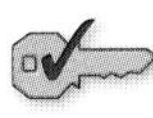

Key Concept Check In what way are all forms of potential energy the same?

Potential Energy

Figure 3 There are different forms of potential energy.

Gravitational Potential Energy
Gravitational potential energy increases when the girl lifts her backpack.

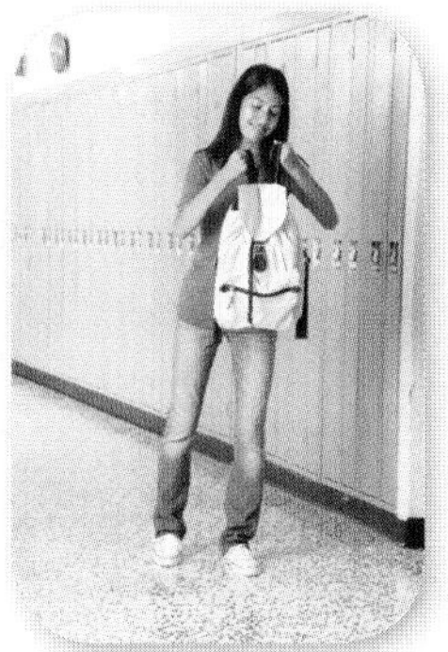

Elastic Potential Energy
The rubber band's elastic potential energy increases when you stretch the rubber band.

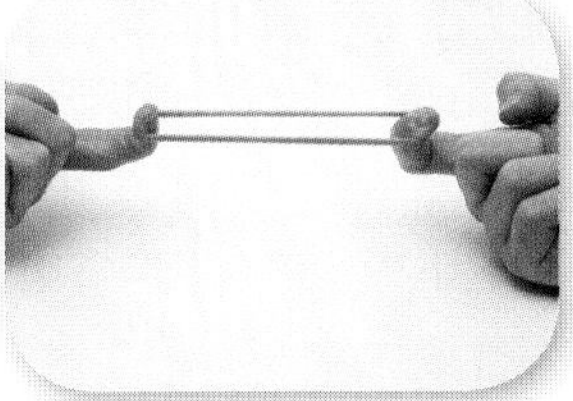

Chemical Potential Energy
Foods and other substances, including glucose, have chemical potential energy stored in the bonds between atoms.

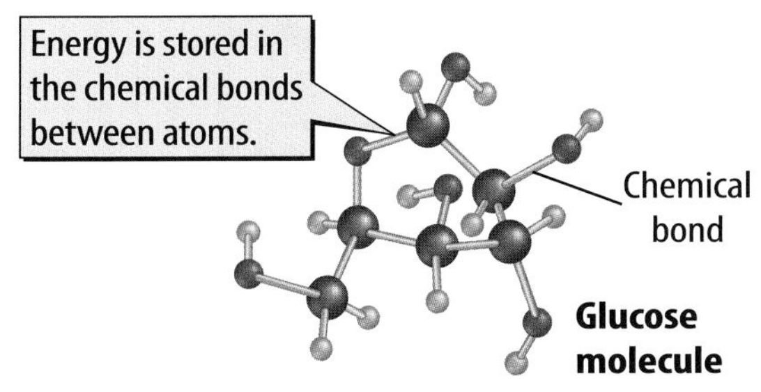
Energy is stored in the chemical bonds between atoms.
Chemical bond
Glucose molecule

Figure 4 The girl does work on the box as she lifts it. The work she does transfers energy to the box. The colored bars show the work that the girl does (W) and the box's potential energy (PE).

Energy and Work

You can transfer energy by doing work. **Work** *is the transfer of energy that occurs when a force is applied over a distance.* For example, the girl does work on the box in **Figure 4.** As the girl lifts the box onto the shelf, she transfers energy from herself to the gravitational interaction between the box and Earth.

Work depends on both force and distance. You only do work on an object if that object moves. Imagine that the girl in **Figure 4** tries to lift the box but cannot actually lift it off the floor. Then she does no work on the box and transfers no energy.

 **Key Concept Check** How is energy related to work?

An object that has energy also can do work. For example, when a bowling ball collides with a bowling pin, the bowling ball does work on the pin. Some of the ball's kinetic energy is transferred to the bowling pin. Because of this connection between energy and work, energy is sometimes described as the ability to do work.

Other Forms of Energy

Some other forms of energy are shown in **Table 1.** All energy can be measured in joules (J). A softball dropped from a height of about 0.5 m has about 1 J of kinetic energy just before it hits the floor.

Inquiry MiniLab

20 minutes

Can a moving object do work? 

Is work done when a moving object hits another object?

1. Read and complete a lab safety form.
2. **Tape** one end of a **30-cm grooved ruler** to the edge of a stack of **books** about 8 cm high. Put the lower end of the ruler in a **paper cup** laying on its side.
3. Release a **marble** in the groove at the top end of the ruler.
4. Record your observations in your Science Journal.

Analyze and Conclude

1. **Compare** the kinetic energy of the marble just before and after it hit the cup.
2. **Key Concept** Is work being done on the cup? Explain your answer.

Table 1 Forms of Energy

Mechanical Energy

The total energy of an object or group of objects due to large-scale motions and interactions is called **Mechanical energy**. For example, the mechanical energy of a basketball increases when a player shoots the basketball. However, the mechanical energy of a pot of water does not increase when you heat the water.

Sound Energy

When you pluck a guitar string, the string vibrates and produces sound. *The energy that sound carries is* **sound energy.** Vibrating objects emit sound energy. However, sound energy cannot travel through a vacuum such as the space between Earth and the Sun.

Thermal Energy

All objects and materials are made of particles that are always moving. Because these particles move, they have energy. **Thermal energy** *is energy due to the motion of particles that make up an object.* Thermal energy moves from warmer objects to colder objects. When you heat objects, you transfer thermal energy to those objects from their surroundings.

Electric Energy

An electric fan uses another form of energy—electric energy. When you turn on a fan, there is an electric current through the fan's motor. **Electric energy** *is the energy that an electric current carries.* Electric appliances, such as fans and dishwashers, change electric energy into other forms of energy.

Radiant Energy—Light Energy

The Sun gives off energy that travels to Earth as electromagnetic waves. Unlike sound waves, electromagnetic waves can travel through a vacuum. Light waves, microwaves, and radio waves are all electromagnetic waves. *The energy that electromagnetic waves carry is* **radiant energy.** Sometimes radiant energy is called light energy.

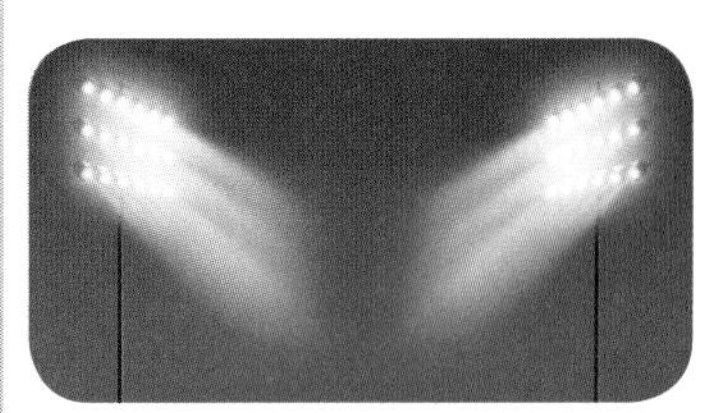

Nuclear Energy

At the center of every atom is a nucleus. **Nuclear energy** *is energy that is stored in the nucleus of an atom.* In the Sun, nuclear energy is released when nuclei join together. In a nuclear power plant, nuclear energy is released when the nuclei of uranium atoms are split apart.

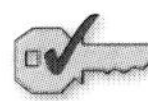

Key Concept Check Describe three forms of energy.

Lesson 1 Review

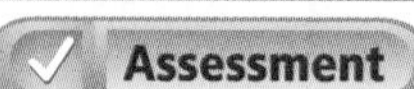

Online Quiz

Visual Summary

Energy is the ability to cause change.

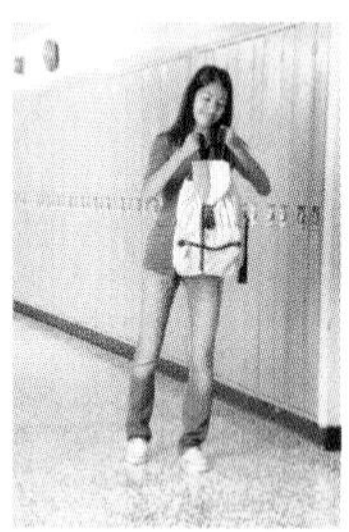

The gravitational potential energy between an object and Earth increases when you lift the object.

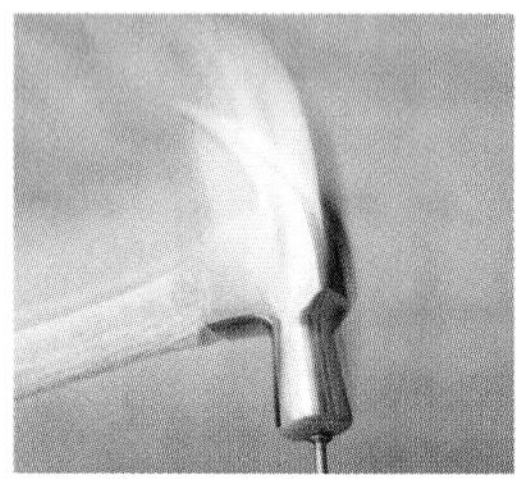

You do work on an object when you apply a force to that object over a distance.

FOLDABLES®

Use your lesson Foldable to review the lesson. Save your Foldable for the project at the end of the chapter.

What do you think NOW?

You first read the statements below at the beginning of the chapter.

1. A fast-moving baseball has more kinetic energy than a slow-moving baseball.
2. A book sitting on a shelf has no energy.

Did you change your mind about whether you agree or disagree with the statements? Rewrite any false statements to make them true.

Use Vocabulary

1. **Distinguish** between kinetic energy and potential energy.
2. **Write** a definition of work.

Understand Key Concepts 

3. Which type of energy increases when you compress a spring?
   A. elastic potential energy
   B. kinetic energy
   C. radiant energy
   D. sound energy
4. **Infer** How could you increase the gravitational potential energy between yourself and Earth?
5. **Infer** how a bicycle's kinetic energy changes when that bicycle slows down.
6. **Compare and contrast** radiant energy and sound energy.

Interpret Graphics

7. **Identify** Copy and fill in the graphic organizer below to identify three types of potential energy.

8. **Describe** where chemical potential energy is stored in the molecule shown below.

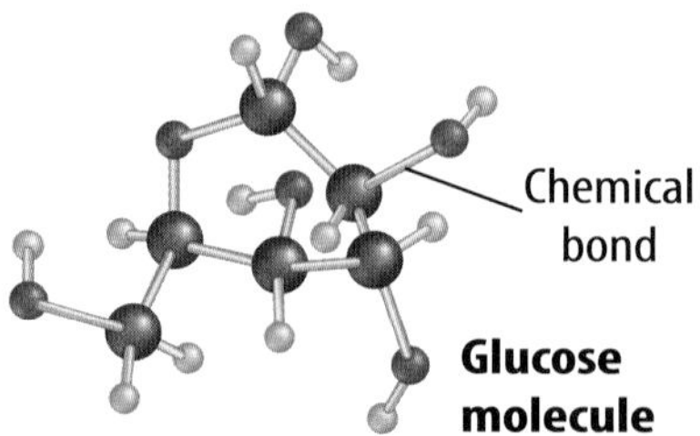

Critical Thinking

9. **Analyze** Will pushing on a car always change the car's mechanical energy? What must happen for the car's kinetic energy to increase?

Fossil Fuels and Rising CO_2 Levels

Investigate the link between energy use and carbon dioxide in the atmosphere.

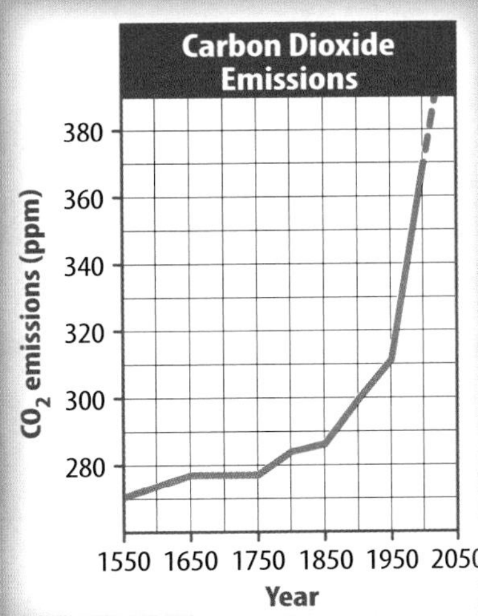

You use energy every day—when you ride in a car or on a bus, turn on a television, and even when you send an e-mail.

Much of the energy that produces electric current, heats and cools buildings, and powers engines, comes from burning fossil fuels—coal, oil, and natural gas. When fossil fuels burn, the carbon in them combines with oxygen in the atmosphere and forms carbon dioxide gas (CO_2). Carbon dioxide is one of the greenhouse gases. In the atmosphere, greenhouse gases absorb energy. This causes the atmosphere and Earth's surface to become warmer. Greenhouse gases make Earth warm enough to support life. Without greenhouse gases, Earth's surface would be frozen.

However, over the past 150 years, the amount of CO_2 in the atmosphere has increased faster than at any time in the past 800,000 years. Most of this increase is the result of burning fossil fuels. This additional carbon dioxide will cause average global temperatures to increase. As temperatures increase, weather patterns worldwide could change. More storms and heavier rainfall could occur in some areas, while other regions could become drier. Increased temperatures also will cause more of the polar ice sheets to melt and raise sea levels. Higher sea levels will cause more flooding in coastal areas.

Developing other energy sources, such as geothermal, solar, nuclear, wind, and hydroelectric power, would reduce the use of fossil fuels and slow the increase in atmospheric CO_2.

It's Your Turn

MAKE A LIST How can CO_2 emissions be reduced? Work with a partner. List five ways people in your home, school, or community could reduce their energy consumption. Combine your list with your classmates' lists to make a master list.

GREEN SCIENCE

300 Years OF CARBON DIOXIDE

1712

A new invention, the steam engine, is powered by burning coal that heats water to produce steam.

Early 1800s

Coal-fired steam engines, able to pull heavy trains and power steamboats, transform transportation.

1882

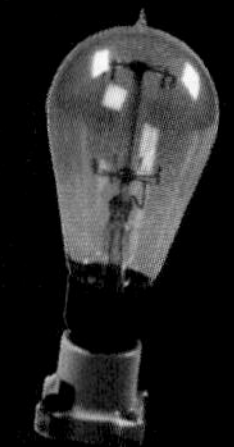

Companies make and sell electric energy from coal for everyday use. Electricity used to power the first lightbulbs, which give off 20 times the light of a candle.

1908

The first mass-produced automobiles are made available. By 1915, Ford was selling 500,000 cars a year. Oil becomes the fuel of choice for car engines.

Late 1900s

Electric appliances transform the way we live, work, and communicate. Most electricity is generated by coal-burning power plants.

2007

There are more than 800 million cars and light trucks on the world's roads.

Lesson 2

Reading Guide

Key Concepts
ESSENTIAL QUESTIONS

- What is the law of conservation of energy?
- How does friction affect energy transformations?
- How are different types of energy used?

Vocabulary

law of conservation of energy p. 170

friction p. 171

Multilingual eGlossary

Energy Transformations

Inquiry What's that sound?

Blocks of ice breaking off the front of this glacier can be bigger than a car. Imagine the loud rumble they make as they crash into the sea. But after the ice falls into the sea, it will gradually melt. All of these processes involve energy transformations—energy changing from one form to another.

Launch Lab

15 minutes

Is energy lost when it changes form?

Energy can have different forms. What happens when energy changes from one form to another?

1. Read and complete the lab safety form.
2. Three students should sit in a circle. One student has 30 **buttons,** one has 30 **pennies,** and one has 30 **paper clips.**
3. Each student should exchange 10 items with the student to the right and 10 items with the student to the left.
4. Repeat step 3.

Think About This

1. If the buttons, pennies, and paper clips represented different forms of energy, what represented changes from one form of energy to another?
2. **Key Concept** If each button, penny, and paper clip represented one unit of energy, did the total amount of energy increase, decrease, or stay the same? Explain your answer.

Changes Between Forms of Energy

It is the weekend and you are ready to make some popcorn in the microwave and watch a movie. Energy changes form when you make popcorn and watch TV. As shown in **Figure 5,** a microwave changes electric energy into **radiant** energy. Radiant energy changes into thermal energy in the popcorn kernels.

The changes from electric energy to radiant energy to thermal energy are called energy transformations. As you watch the movie, energy transformations also occur in the television. A television transforms electric energy into sound energy and radiant energy.

SCIENCE USE V. COMMON USE
radiant
Science Use transmitted by electromagnetic waves
Common Use bright and shining; glowing

Figure 5 Energy changes from one form to another when you use a microwave oven to make popcorn.

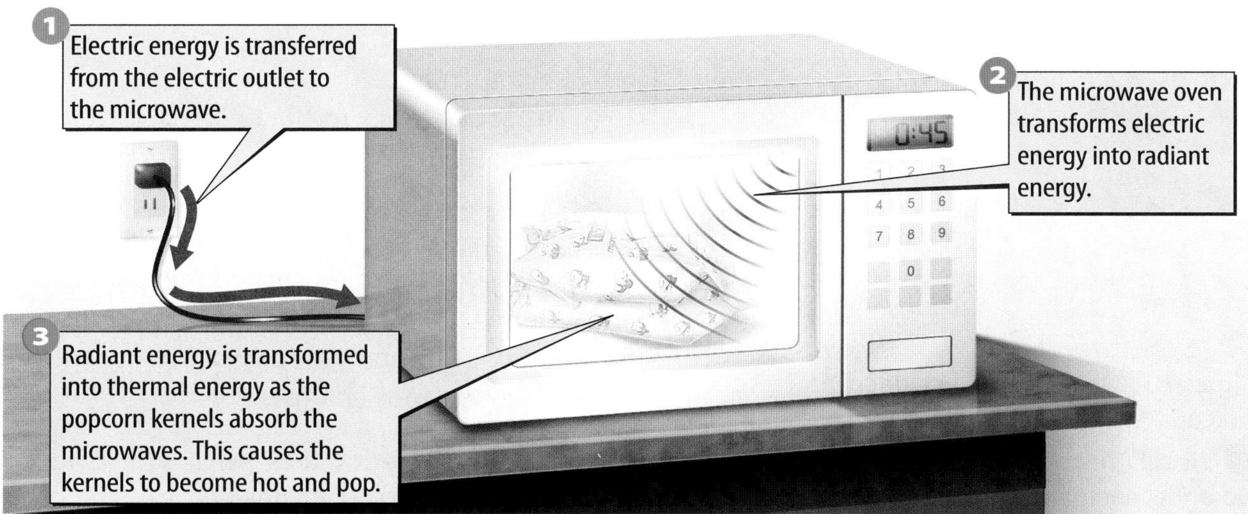

1 Electric energy is transferred from the electric outlet to the microwave.
2 The microwave oven transforms electric energy into radiant energy.
0:45
3 Radiant energy is transformed into thermal energy as the popcorn kernels absorb the microwaves. This causes the kernels to become hot and pop.

Conservation of Energy

Concepts in Motion Animation

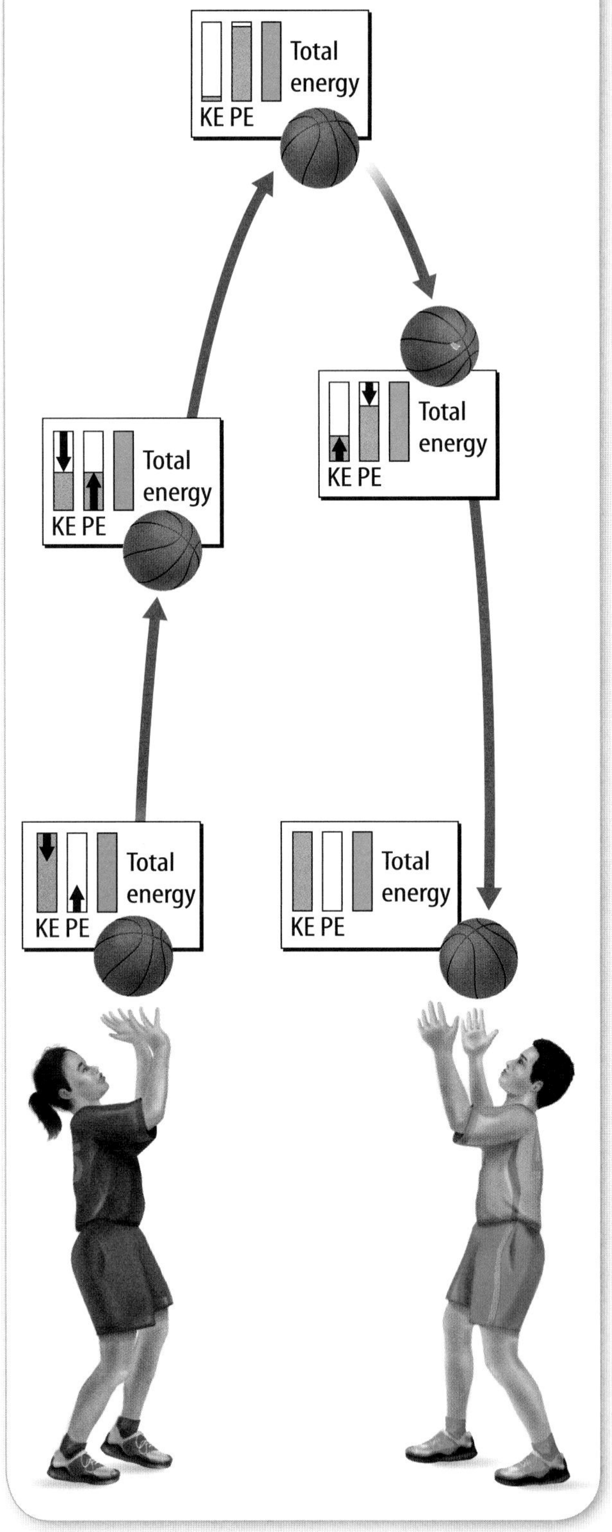

Figure 6 The ball's kinetic energy (KE) and potential energy (PE) changes as it moves.

Visual Check When is the gravitational potential energy the greatest?

Changes Between Kinetic and Potential Energy

Energy transformations also occur when you toss a ball upward, as shown in **Figure 6.** The ball slows down as it moves upward and then speeds up as it moves downward. The ball's speed and height change as energy changes from one form to another.

Kinetic Energy to Potential Energy

The ball is moving fastest and has the most kinetic energy as it leaves your hand, as shown in **Figure 6.** As the ball moves upward, its speed and kinetic energy decrease. However, the potential energy is increasing because the ball's height is increasing. Kinetic energy is changing into potential energy. At the ball's highest point, the gravitational potential energy is greatest, and the ball's kinetic energy is the least.

Potential Energy to Kinetic Energy

As the ball moves downward, potential energy decreases. At the same time, the ball's speed increases. Therefore, the ball's kinetic energy increases. Potential energy is transformed into kinetic energy. When the ball reaches the player's hand again, its kinetic energy is at the maximum value again.

Reading Check Why does the potential energy decrease as the ball falls?

The Law of Conservation of Energy

The total energy in the universe is the sum of all the different forms of energy everywhere. According to the **law of conservation of energy,** *energy can be transformed from one form into another or transferred from one region to another, but energy cannot be created or destroyed.* The total amount of energy in the universe does not change.

Key Concept Check What is the law of conservation of energy?

Friction and the Law of Conservation of Energy

Sometimes it may seem as if the law of conservation of energy is not accurate. Imagine riding a bicycle, as in **Figure 7.** The moving bicycle has mechanical energy. What happens to this mechanical energy when you apply the brakes and the bicycle stops?

When you apply the brakes, the bicycle's mechanical energy is not destroyed. Instead the bicycle's mechanical energy is transformed into thermal energy, as shown in **Figure 7.** The total amount of energy never changes. The additional thermal energy causes the brakes, the wheels, and the air around the bicycle to become slightly warmer.

Friction *is a force that resists the sliding of two surfaces that are touching.* Friction between the bicycle's brake pads and the moving wheels transforms mechanical energy into thermal energy.

Word Origin
friction
from Latin *fricare,* means "to rub"

Key Concept Check Friction causes what energy transformation?

There is always some friction between any two surfaces that are rubbing against each other. As a result, some mechanical energy is always transformed into thermal energy when two surfaces rub against each other.

It is easier to pedal a bicycle if there is less friction between the bicycle's parts. With less friction, less of the bicycle's mechanical energy gets transformed into thermal energy. One way to reduce friction is to apply a lubricant, such as oil, grease, or graphite, to surfaces that rub against each other.

Friction and Thermal Energy

Review Personal Tutor

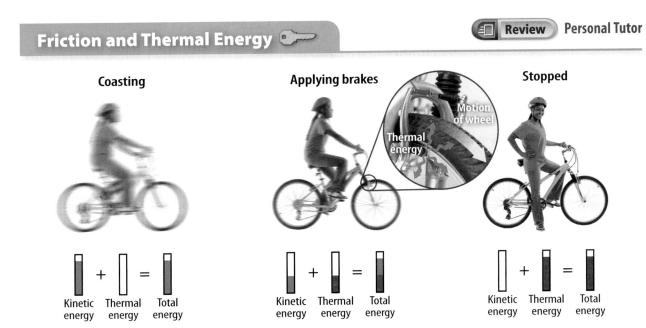

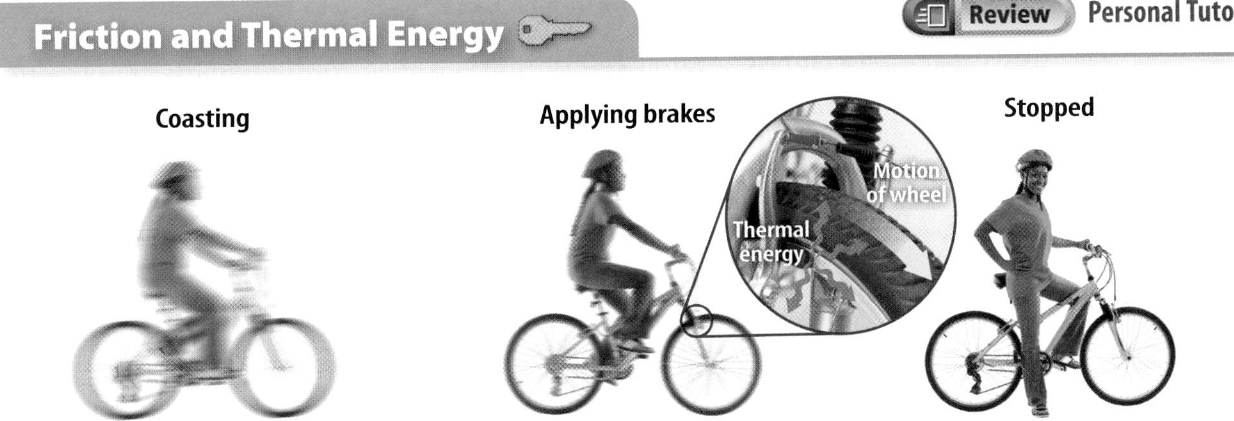

Figure 7 When the girl applies the brakes, friction between the bicycle's brake pads and its wheels transforms mechanical energy into thermal energy. As mechanical energy changes into thermal energy, the bicycle slows down. The total amount of energy does not change.

Using Energy

Every day you use different forms of energy to do different things. You might use the radiant energy from a lamp to light a room, or you might use the chemical energy stored in your body to run a race. When you use energy, you usually change it from one form into another. For example, the lamp changes electric energy into radiant energy and thermal energy.

Using Thermal Energy

All forms of energy can be transformed into thermal energy. People often use thermal energy to cook food or provide warmth. A gas stove transforms the chemical energy stored in natural gas into the thermal energy that cooks food. An electric space heater transforms the electric energy from a power plant into the thermal energy that warms a room. In a jet engine, burning fuel releases thermal energy that the engine transforms into mechanical energy.

Cut three sheets of paper in half. Use the six half sheets to make a side-tab book with five tabs and a cover. Use your book to organize your notes on energy transformations.

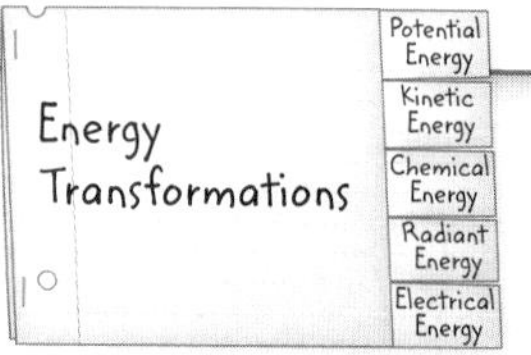

Using Chemical Energy

During photosynthesis, a plant transforms the Sun's radiant energy into chemical energy that it stores in chemical compounds. Some of these compounds become food for other living things. Your body transforms the chemical energy from your food into the kinetic energy necessary for movement. Your body also transforms chemical energy into the thermal energy necessary to keep you warm.

Using Radiant Energy

The cell phone in **Figure 8** sends and receives radiant energy using microwaves. When you are listening to someone on a cell phone, that cell phone is transforming radiant energy into electrical energy and then into sound energy. When you are speaking into a cell phone, it is transforming sound energy into electric energy and then into radiant energy.

Figure 8 A cell phone changes sound energy into radiant energy when you speak.

Sound waves carry energy into the cell phone.

The cell phone converts the energy carried by sound waves into radiant energy that is carried away by microwaves.

Using Electric Energy

Many of the devices you might use every day, such as handheld video games, mp3 players, and hair dryers, use electric energy. Some devices, such as hair dryers, use electric energy from electrical power plants. Other appliances, such as handheld video games, transform the chemical energy stored in batteries into electric energy.

Key Concept Check What happens to energy when it is used?

Waste Energy

When energy changes form, some thermal energy is always released. For example, a lightbulb converts some electric energy into radiant energy. However, the lightbulb also transforms some electric energy into thermal energy. This is what makes the lightbulb hot. Some of this thermal energy moves into the air and cannot be used.

Scientists often refer to thermal energy that cannot be used as waste energy. Whenever energy is used, some energy is transformed into useful energy and some is transformed into waste energy. For example, we use the chemical energy in gasoline to make the cars in **Figure 9** move. However, most of that chemical energy ends up as waste energy—thermal energy that moves into the air.

Reading Check What is waste energy?

Inquiry MiniLab — 20 minutes

How does energy change form?

When an object falls, energy changes form. How can you compare energies for falling objects?

1. Read and complete a lab safety form.
2. Place a piece of **clay** about 10 cm wide and 3 cm thick on a **small paper plate.**
3. Drop a **marble** into the clay from a height of about 20 cm and measure the depth of the depression caused by the marble. Record the measurement in your Science Journal.
4. Repeat step 3 with a heavier marble.

Analyze and Conclude

1. **Infer** Which marble had more kinetic energy just before it hit the clay? Explain your answer.
2. **Key Concept** For which marble was the potential energy greater just before the marble fell? Explain your answer using the law of conservation of energy.

Figure 9 Cars transform most of the chemical energy in gasoline into waste energy.

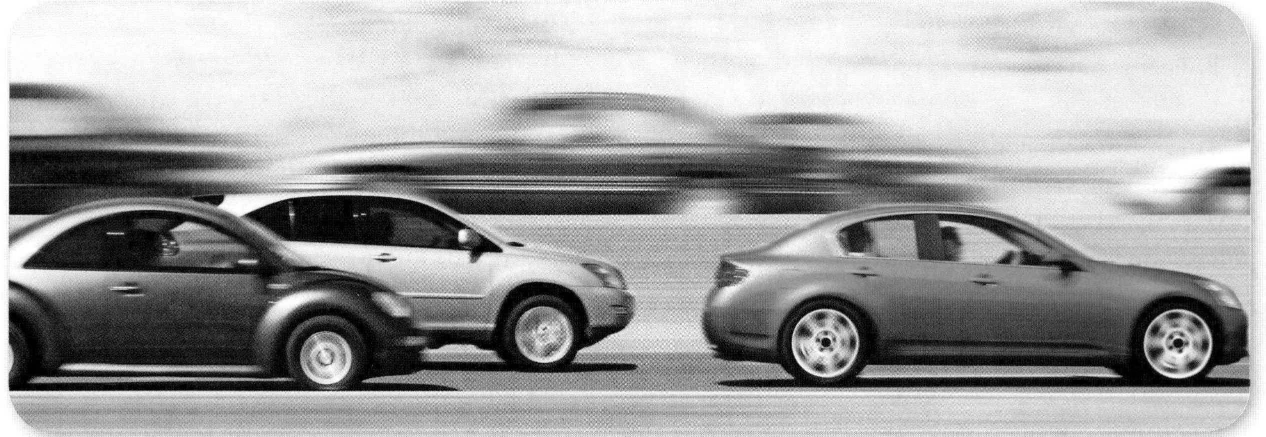

Lesson 2 Review

Assessment Online Quiz

Inquiry Virtual Lab

Visual Summary

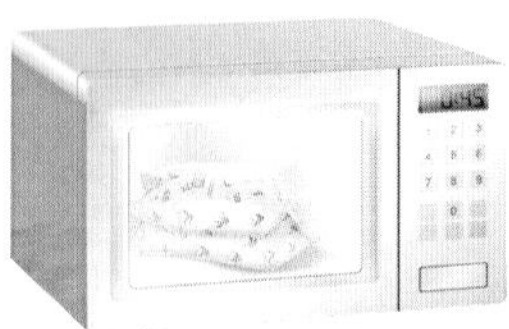

Energy can change form, but according to the law of conservation of energy, energy can never be created or destroyed.

Friction transforms mechanical energy into thermal energy.

Different forms of energy, such as sound and radiant energy, are used when someone talks on a cell phone.

FOLDABLES®

Use your lesson Foldable to review the lesson. Save your Foldable for the project at the end of the chapter.

What do you think NOW?

You first read the statements below at the beginning of the chapter.

3. Energy can change from one form to another.

4. If you toss a baton straight up, total energy decreases as the baton rises.

Did you change your mind about whether you agree or disagree with the statements? Rewrite any false statements to make them true.

Use Vocabulary

1. **Use the term** *friction* in a complete sentence.
2. **Explain** the law of conservation of energy in your own words.

Understand Key Concepts 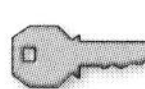

3. **Describe** the energy transformations that occur when a piece of wood burns.
4. **Identify** the energy transformation that takes place when you apply the brakes on a bicycle.
5. Which energy transformation occurs in a toaster?
   A. chemical to electrical
   B. electrical to thermal
   C. kinetic to chemical
   D. thermal to potential

Interpret Graphics

6. **Organize Information** Copy and fill in the graphic organizer below to show how kinetic and potential energy change when a ball is thrown straight up and then falls down.

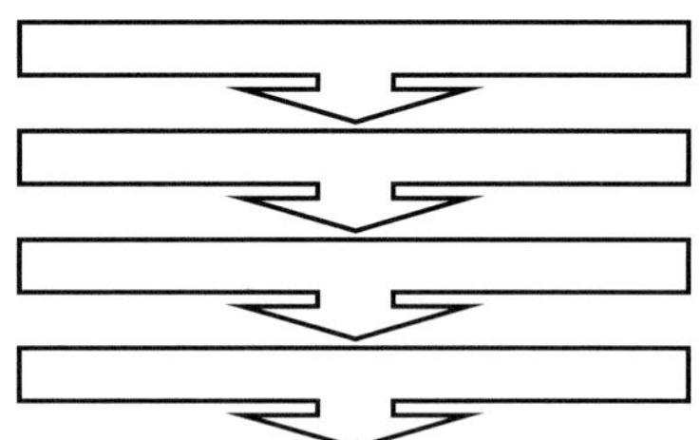

Critical Thinking

7. **Identify** Choose three electric appliances. For each appliance, identify the forms of energy that electric energy is transformed into.
8. **Judge** An advertisement states that a machine with moving parts will continue moving forever without having to add any energy. Can this advertisement be correct? Explain.

Can you identify energy transformations?

Materials

string

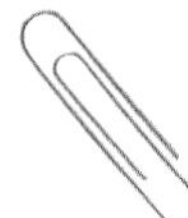
paper clip

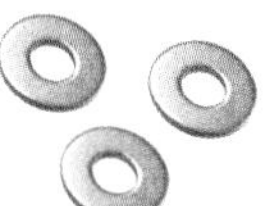
three large washers

meterstick

tape

small box

ruler

Safety

Energy cannot be created or destroyed. However, energy can be transformed from one type into another type, and energy can be transferred from one object to another object. In this lab, you will transform gravitational potential energy into kinetic energy as well as transfer energy from one object to another object.

Learn It

Before you can draw valid conclusions from any scientific experiment, you must **analyze** the results of that experiment. This means you must look for patterns in the results.

Try It

1. Read and complete a lab safety form.
2. Use the photo below to make a pendulum. Hang one washer on the paper clip. Place the box so it will block the swinging pendulum. Mark the position of the box with tape.
3. Pull the pendulum back until the bottom of the washer is 15 cm from the floor. Release the pendulum. Measure and record the distance the box moves in your Science Journal. Repeat two more times.
4. Repeat step 3 using pendulum heights of 30 cm and 45 cm.
5. Repeat steps 3 and 4 with two washers, then with three washers.

Apply It

6. How does the gravitational potential energy depend on the pendulum's weight and height?
7. How does the distance the box travels depend on the initial gravitational potential energy?
8. Does the pendulum do work on the box? Explain your answer.
9. What energy transformation occurred that caused the box to stop moving?
10. **Key Concept** Describe the energy transformations and transfers that took place as the pendulum fell, as the pendulum hit the box, and as the box slid along the floor.

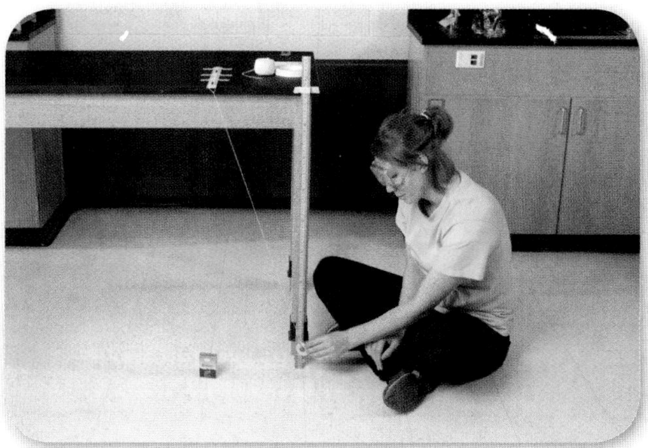

Lesson 3

Reading Guide

Key Concepts

ESSENTIAL QUESTIONS

- What are nonrenewable energy resources?
- What are renewable energy resources?
- Why is it important to conserve energy?

Vocabulary

nonrenewable energy resource p. 178

fossil fuel p. 178

renewable energy resource p. 180

inexhaustible energy resource p. 181

Multilingual eGlossary

Energy Resources

Inquiry Extracting Energy?

Where does the electric energy come from when you turn on the lights in your home? The answer to that question depends on where you live. Different energy resources are used in different parts of the United States.

Launch Lab

15 minutes

How are energy resources different?

Is there an infinite supply of usable energy, or could we someday run out of energy resources? In this activity, the red beans represent an energy resource that is available in limited amounts. The white beans represent an energy resource that is available in unlimited amounts.

1. Read and complete a lab safety form.
2. Place **40 red beans** and **40 white beans** in a **paper bag.** Mix the contents of the bag.
3. Each team should remove 20 beans from the bag without looking at the beans. Record the numbers of red and white beans in your Science Journal.
4. Put the red beans aside. They are "used up." Return all the white beans to the bag. Mix the beans in the bag. Repeat steps 3 and 4 three more times.

Think About This

1. What happened to the number of red beans drawn during each round?
2. What would eventually happen to the red beans in the bag?
3. **Key Concept** How would changing the number of beans drawn in each round make the red beans last longer? Explain your answer.

Sources of Energy

Every day, you use many forms of energy in many ways. According to the law of conservation of energy, energy cannot be created or destroyed. Energy can only change form. Where does all the energy that you use come from?

Almost all the energy you use can be traced back to the Sun, as shown in **Figure 10.** For example, the chemical energy in the food you eat originally came from the Sun. The energy in fuels, such as gasoline, coal, and wood, also came from the Sun. In addition, a small amount of energy that reaches Earth's surface comes from inside Earth. However, the amount of energy that comes from the Sun each day is about 5,000 times greater than the amount of energy that comes from inside Earth.

Figure 10 The energy contained in food, gasoline, and wood originally came from the Sun.

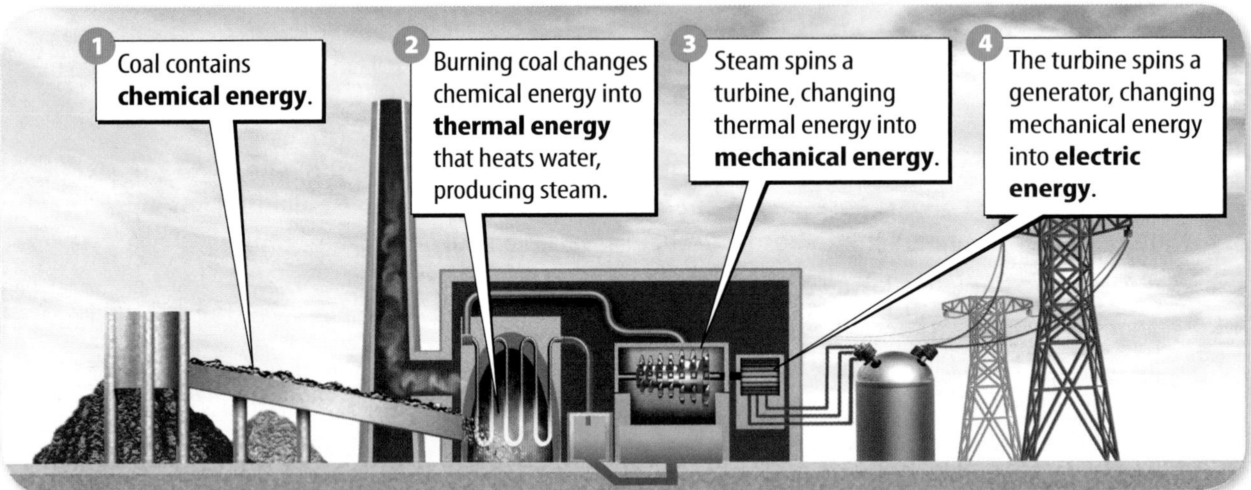

1
Coal contains **chemical energy**.
2
Burning coal changes chemical energy into **thermal energy** that heats water, producing steam.
3
Steam spins a turbine, changing thermal energy into **mechanical energy**.
4
The turbine spins a generator, changing mechanical energy into **electric energy**.

Figure 11 This coal-burning electric power plant transforms chemical energy stored in a fossil fuel into electric energy.

Visual Check What form of energy changes water into steam?

Electric Power Plants

Most of the energy you use every day does not come directly from the Sun. Instead, much of the energy you use every day is in the form of electric energy. Most of the electric energy you use comes from electric power plants.

An electric power plant transforms the energy in an energy source into electric energy. There are three main energy sources used in power plants. One source of energy comes from burning fuels, such as coal. The power plant shown in **Figure 11** uses coal as an energy source. Nuclear power plants use the nuclear energy contained in uranium. Hydroelectric power plants convert the kinetic energy in falling water into electric energy.

FOLDABLES®

Make a shutter fold from a sheet of paper. Label it as shown. Use it to organize your notes on nonrenewable and renewable energy resources.

Nonrenewable Energy Resources

Renewable Energy Resources

Nonrenewable Energy Resources

The coal that a power plant burns is an example of a nonrenewable energy resource. A **nonrenewable energy resource** *is an energy resource that is available in limited amounts or that is used faster than it is replaced in nature.*

Fossil Fuels

The most commonly used nonrenewable energy resources are fossil fuels. **Fossil fuels** *are the remains of ancient organisms that can be burned as an energy source.* Fossil fuels take millions of years to form. They are being used up much faster than they form. Three types of fossil fuels are coal, natural gas, and petroleum.

WORD ORIGIN

fossil
from Latin *fossilis,* means "dug up"

Key Concept Check Why are fossil fuels considered a nonrenewable energy resource?

Formation of Petroleum

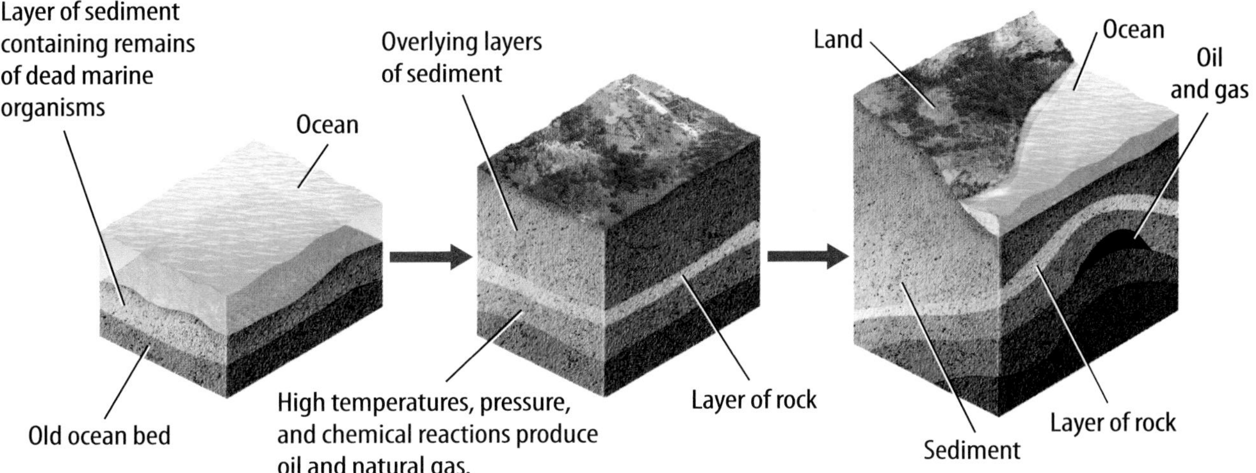

Figure 12 Petroleum and natural gas formed over millions of years from dead microscopic ocean organisms. Geologic processes buried these dead organisms under layers of sediment and rock. There, high temperature and pressure changed them into oil and natural gas.

The Formation of Fossil Fuels

The processes that formed fossil fuels began at Earth's surface. Petroleum and natural gas formed from microscopic ocean organisms that died and sank to the ocean floor, as shown in **Figure 12.** These organisms were gradually buried under layers of sediment—sand and mud—and rock.

Over many millions of years, increasing temperature and pressure from the weight of the sediment and rock layers changed the dead organisms into petroleum and natural gas. Coal formed on land from plants that died millions of years ago and were buried under thick layers of sediment and rock.

Using Fossil Fuels

Fossil fuels formed from organisms that changed radiant energy from the Sun to chemical potential energy. The chemical potential energy stored in fossil fuels changes to thermal energy when fossil fuels burn.

Using Petroleum Gasoline, fuel oil, diesel, and kerosene are made from petroleum. These fuels are burned mainly to power cars, trucks, and planes and to heat buildings. Petroleum is also used as a raw material in making plastics and other materials.

Using Coal Electric power plants burn about 90 percent of the coal used in the United States. Coal is also used to directly heat buildings and to produce steel and concrete. Burning coal produces more pollutants than burning other fossil fuels. In some places, these pollutants react with water vapor in the air and create acid rain. Acid rain can harm organisms, such as trees and fish.

Reading Check What percentage of the coal used in the United States is burned in electric power plants?

Using Natural Gas About half of all homes in the United States use natural gas for heating. Electric power plants burn about 30 percent of the natural gas used in the United States. Burning natural gas produces less pollution than burning other fossil fuels.

Fossil Fuels and Global Warming

Burning fossil fuels releases carbon dioxide gas into Earth's atmosphere. Carbon dioxide is one of the gases that helps keep Earth's surface warm. However, over the past 100 years, Earth's surface has warmed by about 0.7°C. Some of this warming is due to the increasing amount of carbon dioxide produced by burning fossil fuels.

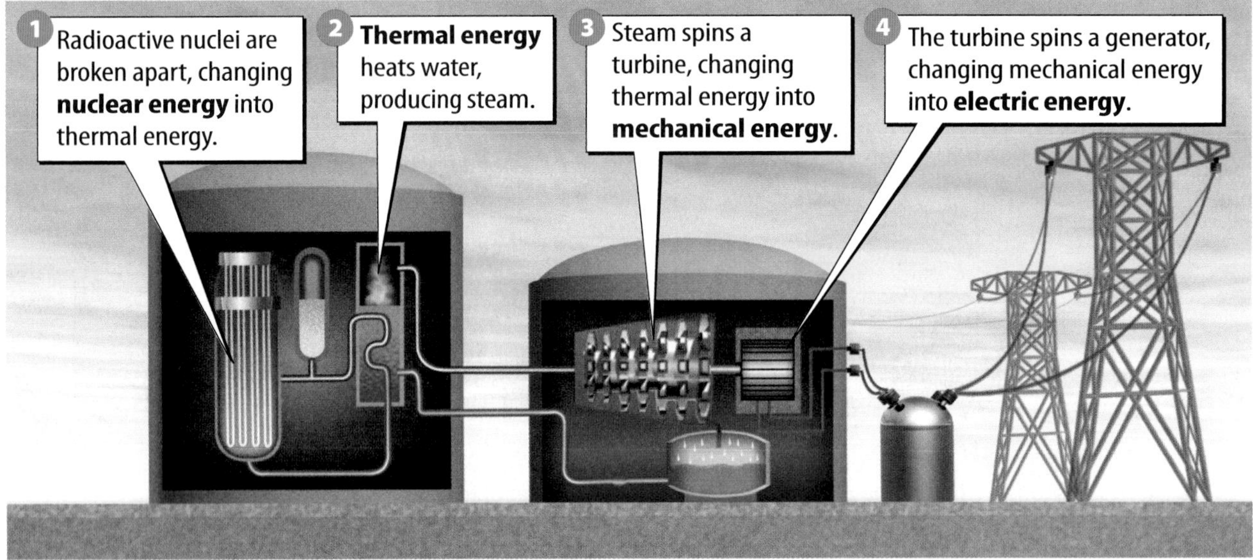
1 Radioactive nuclei are broken apart, changing nuclear energy into thermal energy.
2 Thermal energy heats water, producing steam.
3 Steam spins a turbine, changing thermal energy into mechanical energy.
4 The turbine spins a generator, changing mechanical energy into electric energy.

Figure 13 A nuclear power plant transforms nuclear energy into electric energy.

REVIEW VOCABULARY

nuclei
plural form of nucleus; the positively charged center of an atom that contains protons and neutrons.

Nuclear Energy

Humans can also transform the nuclear energy from uranium **nuclei** into thermal energy. Uranium is found in certain minerals, but significant amounts of uranium are no longer being formed inside Earth. As a result, nuclear energy from uranium is a nonrenewable energy resource.

Reading Check Why is nuclear energy released from uranium nuclei considered a nonrenewable energy resource?

Nuclear Power Plants In a nuclear power plant, breaking apart uranium nuclei transforms nuclear energy into thermal energy. As shown in **Figure 13,** this thermal energy changes water into the steam that spins the turbine. Unlike fossil fuel power plants, a nuclear power plant does not release pollutants into the air. However, a nuclear power plant does produce harmful nuclear waste.

Storing Nuclear Waste Nuclear waste contains radioactive materials that can damage living things. Some of these materials remain radioactive for thousands of years. Almost all nuclear waste in the United States is currently stored at the nuclear power plants where this waste is produced.

Math Skills

Solve a One-Step Equation

Electric energy is often measured in units called *kilowatt-hours* (kWh). To calculate the electric energy used by an appliance in kWh, use this equation:

$$\text{kWh} = (\text{watts}/1{,}000) \times \text{hours}$$

Appliances typically have a power rating measured in *watts.*

Practice

A hair dryer is rated at 1,200 watts. If you use the dryer for 0.25 hours, how much electric energy do you use?

- **Math Practice**
- **Personal Tutor**

Renewable Energy Resources

Fossil fuels and uranium are being used up faster than they are being replaced. However, there are other energy resources that are not being used up. A **renewable energy resource** *is an energy resource that is replaced as fast as, or faster than, it is used.*

Key Concept Check Contrast renewable and nonrenewable energy resources.

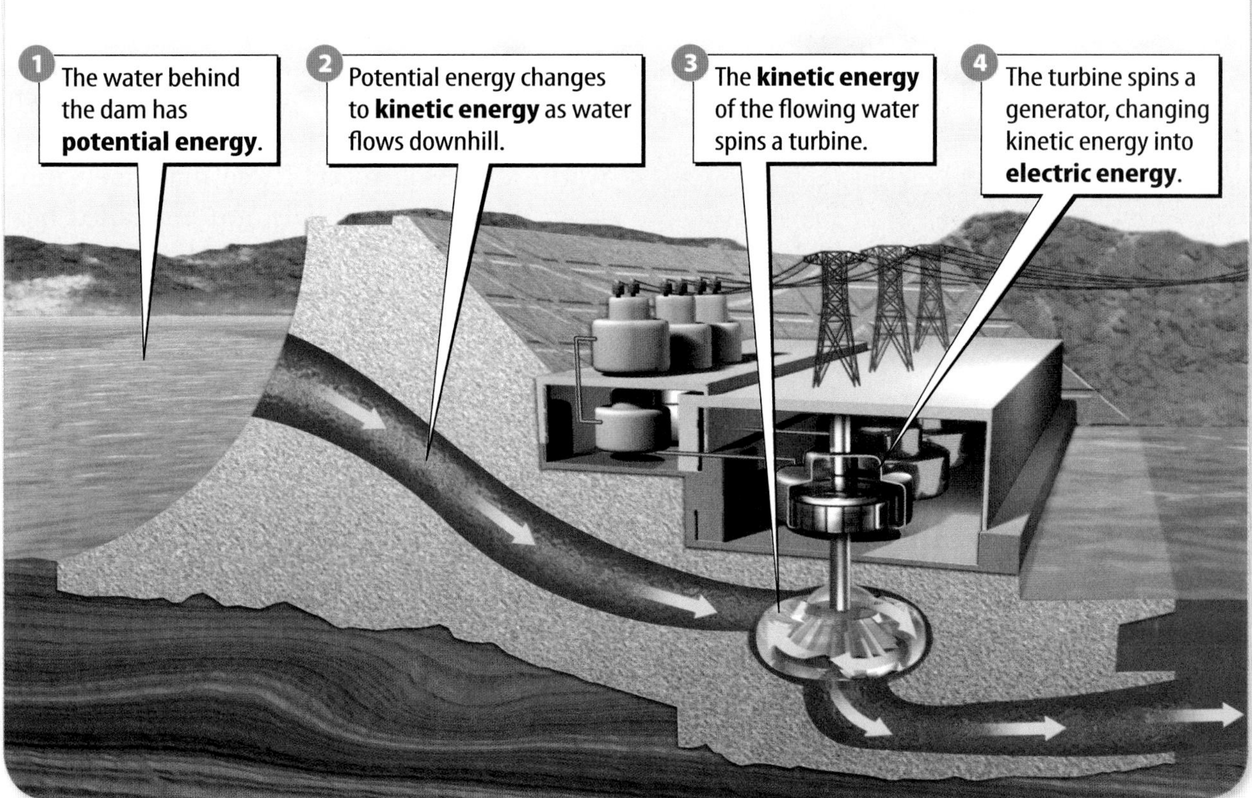
1 The water behind the dam has **potential energy**.
2 Potential energy changes to **kinetic energy** as water flows downhill.
3 The **kinetic energy** of the flowing water spins a turbine.
4 The turbine spins a generator, changing kinetic energy into **electric energy**.

▲ **Figure 14** A hydroelectric power plant converts the potential energy of the water stored behind the dam to electric energy.

Visual Check Infer what would happen to the turbine if the river stopped flowing.

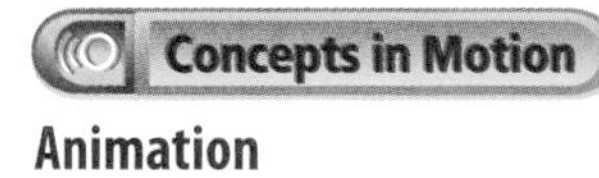
Concepts in Motion

Animation

Hydroelectric Power Plants

The most widely used renewable energy resource is falling water. To generate electric energy from falling water, a dam is built across a river, forming a reservoir (REH zuh vwor). As water falls through tunnels in the dam, the water's potential energy transforms into kinetic energy. **Figure 14** shows how a hydroelectric power plant transforms the kinetic energy in falling water into electric energy. Hydroelectric power plants do not emit pollutants. However, in some places, dams can disturb the life cycle of some wildlife, such as fish.

Solar Energy

Another renewable energy source is radiant energy from the Sun—solar energy. Because the Sun will produce energy for billions of years, solar energy is also an inexhaustible energy resource. An **inexhaustible energy resource** *is an energy resource that cannot be used up.* Less than about 0.1 percent of the energy used in the United States comes directly from the Sun.

Solar energy is converted directly into electric energy by solar cells. Solar cells contain materials that transform radiant energy into electric energy when sunlight strikes the solar cell. Solar cells can be placed on the roof of a building to provide electric energy, as shown in **Figure 15.**

Figure 15 The dark blue panels on this roof are made of materials that convert solar energy into electric energy. ▼

▲ **Figure 16** Wind turbines convert the kinetic energy in wind into electric energy. Some wind turbines are over 60 m high.

Concepts in Motion Animation

Wind Energy

Wind energy is another inexhaustible energy resource. Modern wind turbines, such as the ones in **Figure 16**, convert the kinetic energy in wind into electric energy. Wind spins a propeller that is connected to an electric generator.

Wind turbines produce no pollution. However, wind turbines are practical only in regions where the average wind speed is more than about 5 m/s. Also, many wind turbines covering a large area are needed to obtain as much electric energy as one fossil fuel–burning power plant.

Biomass

People have often burned materials such as wood, dried peat moss, and manure to stay warm and cook food. These materials come from plants and animals and are called biomass. Because plant and animal materials can be replaced as fast as they are used, biomass is a renewable energy resource.

Some biomass is converted into fuels that can be burned in the engines of cars and other vehicles. Fuels made from biomass are often called biofuels. Using biofuels in vehicles can reduce the use of gasoline and make the supply of petroleum last longer.

Geothermal Energy

Thermal energy from inside Earth is called geothermal energy. This energy comes from the decay of radioactive nuclei deep inside Earth. In some places, geothermal energy produces underground pockets of hot water and steam. These pockets are called geothermal reservoirs.

In a few places, wells can be drilled to reach geothermal reservoirs. **Figure 17** shows a geothermal power plant. The hot water and steam in the geothermal reservoir are piped to the surface where they spin a turbine attached to an electric generator.

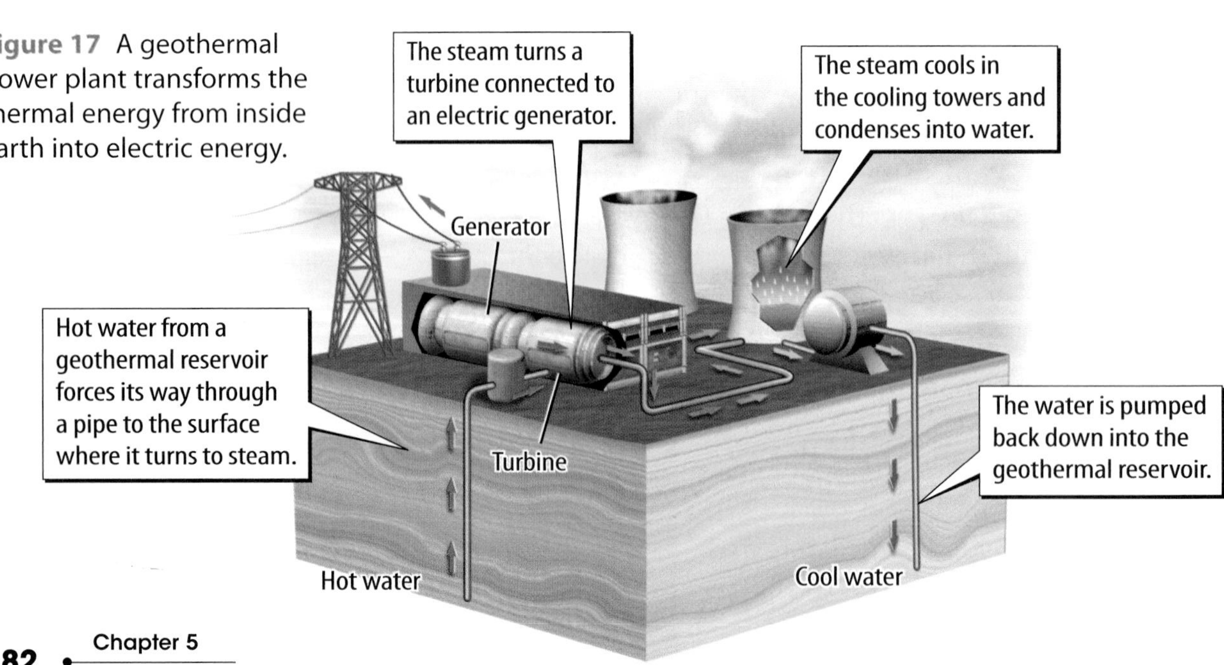

The steam turns a turbine connected to an electric generator.
The steam cools in the cooling towers and condenses into water.
Generator
Hot water from a geothermal reservoir forces its way through a pipe to the surface where it turns to steam.
Turbine
The water is pumped back down into the geothermal reservoir.
Hot water
Cool water

Figure 17 A geothermal power plant transforms the thermal energy from inside Earth into electric energy.

Sources of Energy Used in the U.S. in 2006

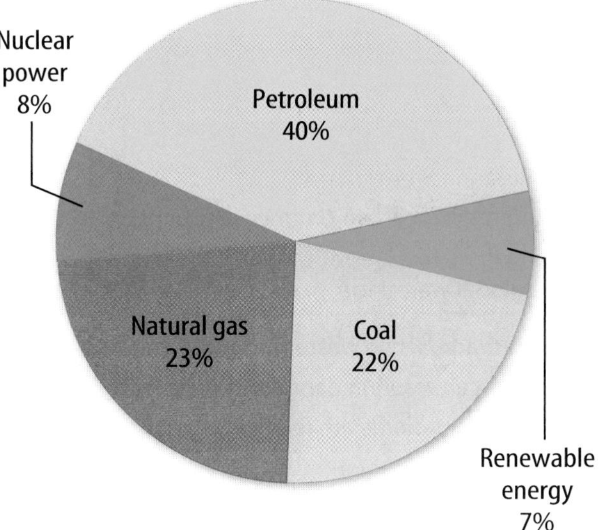

Figure 18 About 93 percent of the energy used in the United States comes from nonrenewable energy resources—fossil fuels and nuclear energy.

Conserving Energy Resources

The graph in **Figure 18** shows that fossil fuels provide about 85 percent of the energy used in the United States. Because fossil fuels are a nonrenewable energy resource, the supply decreases as they are used.

Because the supply of fossil fuels is decreasing, there could be shortages of fossil fuels in the future. Conserving energy is one way to reduce the rate at which all energy resources are used. Conserving energy means to avoid wasting energy. For example, turning off the lights when no one is in a room is a way to conserve energy.

 **Key Concept Check** How does conserving energy affect the rate at which energy resources are used?

In the future, energy resources besides fossil fuels might become more widely used. However, as **Table 2** on the next page shows, all energy resources have advantages and disadvantages. Comparing advantages and disadvantages will help determine which energy resources are used in the future.

Inquiry MiniLab 20 minutes

What energy resources provide our electric energy?

The data table below shows how much electric energy comes from different energy resources in the United States.

Data Table	
Energy Resource	Percentage Of Electric Energy Provided
Petroleum	2
Natural gas	22
Coal	49
Nuclear	19
Hydroelectric	6
Wind, geothermal, and biomass	2

On a sheet of **paper,** make a circle graph of the data in the data table. Label the pie chart *Sources of Electric Energy Used in the United States.*

Analyze and Conclude

1. **Identify** from your circle graph which energy resource provides about half the electric energy used in the United States.
2. **Interpret** Why is your circle graph different from the circle graph in **Figure 18?**
3. **Key Concept** What percentage of electric energy in the United States do renewable resources provide?

Table 2 Advantages and Disadvantages of Energy Resources

Energy Resource	Advantages	Disadvantages
Nonrenewable Energy Resources		
Fossil Fuels	• Easy to transport • Widely available • Relatively inexpensive • Fossil fuel power plants are relatively inexpensive to operate.	• Drilling and surface mining may damage land and wildlife habitats. • Oil spills and leaks can harm wildlife. • Burning fossil fuels can produce air pollution. • Burning fossil fuels produces carbon dioxide that can cause global warming.
Nuclear Energy	• Nuclear power plants are relatively inexpensive to operate. • Does not produce air pollution	• Produces radioactive waste that is difficult to store • Accidents can result in dangerous leaks of radiation. • Nuclear power plants are relatively expensive to build.
Renewable Energy Resources		
Hydroelectric	• Does not pollute the air or water • Hydroelectric power plants are relatively inexpensive to operate.	• Dams damage wildlife habitats. • Dams can affect water quality and reduce the flow of water downstream • Droughts can affect hydroelectric power plants.
Solar	• Does not pollute the air or water • Theoretically inexhaustible supply	• The amount of solar energy that reaches Earth's surface varies, depending on the location, time of day, season, and weather conditions. • A large area is needed to collect enough solar energy for a solar power plant to be viable.
Wind	• Does not pollute the air or water • Can be used in isolated areas where electricity is unavailable	• Wind turbines can be noisy. • Can be disruptive to wildlife • Only generates electricity when the wind is blowing
Geothermal	• Does not pollute the air • Geothermal power plants are relatively inexpensive to operate.	• Geothermal reservoirs are located primarily in the western United States, Alaska, and Hawaii. • Some geothermal plants produce solid wastes that require careful disposal.
Biomass	• Biofuels could replace petroleum fuels in most vehicles.	• Energy from fossil fuels is used to grow biomass. • Some farm land is used to grow biomass instead of food crops. • Burning some biomass produces pollutants, such as smoke.

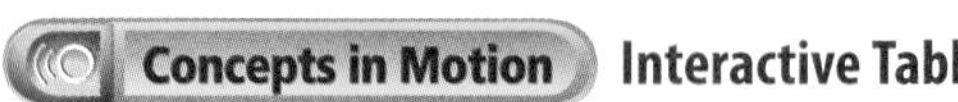

Interactive Table

Lesson 3 Review

Assessment **Online Quiz**

Visual Summary

Nonrenewable energy resources, such as fossil fuels, are used faster than they are replaced in nature.

Renewable energy resources, such as wind energy, are replaced in nature as fast as they are used.

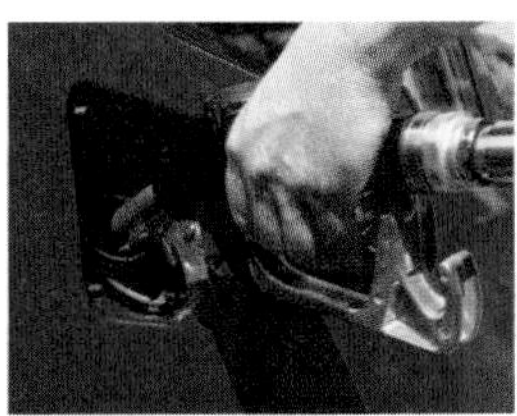

Conserving energy, such as driving fuel-efficient cars, is one way to reduce the rate at which energy resources are used.

FOLDABLES®

Use your lesson Foldable to review the lesson. Save your Foldable for the project at the end of the chapter.

What do you think NOW?

You first read the statements below at the beginning of the chapter.

5. Nuclear power plants release many dangerous pollutants into the air as they transform nuclear energy into electric energy.

6. Thermal energy from within Earth can be transformed into electric energy at a power plant.

Did you change your mind about whether you agree or disagree with the statements? Rewrite any false statements to make them true.

Use Vocabulary

1 **Define** *fossil fuel* in your own words.

Understand Key Concepts 

2 **Compare and contrast** fossil fuels and biofuels.

3 **Explain** Conserving energy makes which type of energy resources last longer?

4 Which of these energy resources is effectively inexhaustible?

A. biomass
B. fossil fuels
C. nuclear energy
D. solar energy

Interpret Graphics

5 **Compare and Contrast** Copy and fill in the graphic organizer below.

Energy Resource	Similarities	Differences
Coal		
Wind		

6 **Sequence** Copy and fill in the graphic organizer below to sequence the energy transformations that occur in a coal-burning electric power plant.

Critical Thinking

7 **Recommend** Fossil fuels form over millions of years. Use this information to explain why fossil fules are a nonrenewable energy resource.

Math Skills

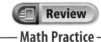

Review **Math Practice**

8 A 22-W compact fluorescent lightbulb (CFL) produces as much light as a 100-W regular lightbulb. How much electric energy in kWh does each bulb use in 10 hours?

Lab 45 minutes

Pinwheel Power

Materials

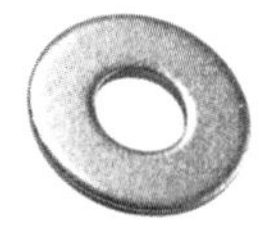

round pencil with unused eraser

metal washers

cardboard container

sand or small rocks

three-speed hair dryer

stopwatch

Also needed: manila folder, metric ruler, scissors, single-hole punch, thread

Safety

Moving air, or wind, is a renewable energy resource. In some places, wind turbines transform the kinetic energy of wind into electric energy. This electric energy can be used to do work by making an object move. In this lab, you will construct a pinwheel turbine and observe how changes in wind speed affect the rate at which your wind turbine does work.

Ask a Question

How does the wind speed affect the rate at which a wind turbine does work?

Make Observations

1. Read and complete a lab safety form.
2. Construct a pinwheel from a manila folder using the diagram below.
3. Use the plastic push pin to carefully attach the pinwheel to the eraser of the pencil.
4. Use the hole punch to make holes on opposite sides of the top of the container. Use your ruler to make sure the holes are exactly opposite one another. Weigh down the container with sand or small rocks.
5. Put the pencil through the holes and make sure it spins freely. Blow against the blades of the pinwheel with varying amounts of force to observe how the pinwheel moves. Record your observations in your Science Journal.
6. Measure and cut 100 cm of thread. Tie the washers to one end of the thread. Tape the other end of the thread to the pencil. Your wind turbine should resemble the one shown on the next page.

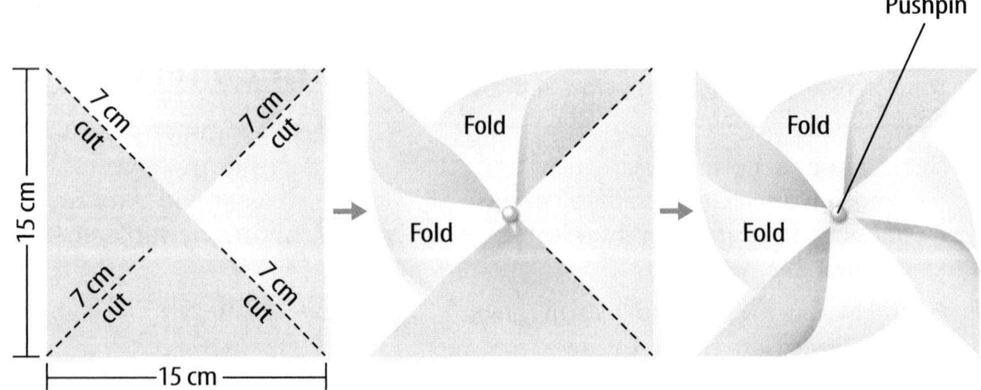

Form a Hypothesis

7. Use your observations from step 5 to form a hypothesis about the relationship between wind speed and the rate at which the wind turbine does work.

Test Your Hypothesis

8. Work with two other students to test your hypothesis. One person will use the hair dryer to model a "slow" wind speed. Another person will stop the pencil's movement after 5 seconds on the stopwatch. The third person will measure the length of thread remaining between the pencil and the top of the washers. Then, someone will unwind the thread and the group will repeat this procedure four more times with the dryer on low. Record all data in your Science Journal.
9. Repeat step 8 with the dryer on medium.
10. Repeat step 8 with the dryer on high.

Analyze and Conclude

11. **Interpret Data** Did your hypothesis agree with your data and observations? Explain.
12. **Sequence** Describe how energy was transformed from one form into another in this lab.
13. **Draw Conclusions** What factors might have affected the rate at which your pinwheel turbine did work?
14. **The Big Idea** Explain how wind is used as an energy resource.

Communicate Your Results

Use your data and observations to write a paragraph explaining how wind speed affects the rate at which a wind turbine can do work.

Extension

Research designs of real wind generators and then create a model of one. Write a short explanation of its advantages and disadvantages compared to other real wind generators.

Lab Tips

☑ You measure the rate at which the wind turbine does work by measuring how fast the turbine lifts the metal washers.

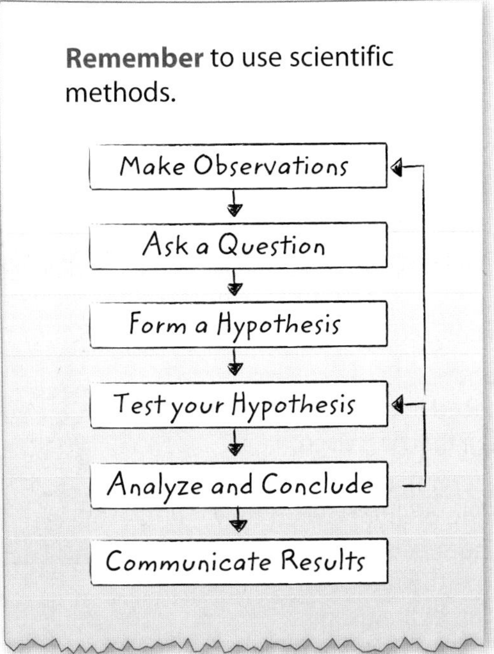

Chapter 5 Study Guide

Energy is the ability to cause change. Energy resources contain energy that can be transformed into other, more useful forms of energy.

Key Concepts Summary

Lesson 1: Forms of Energy

- **Energy** is the ability to cause change.
- **Kinetic energy** is the energy a body has because it is moving. **Potential energy** is stored energy.
- Different forms of energy include **thermal energy** and **radiant energy.**

Vocabulary

energy p. 161
kinetic energy p. 162
potential energy p. 162
work p. 164
mechanical energy p. 165
sound energy p. 165
thermal energy p. 165
electric energy p. 165
radiant energy p. 165
nuclear energy p. 165

Lesson 2: Energy Transformations

- Any form of energy can be transformed into other forms of energy.
- According to the **law of conservation of energy,** energy can be transformed from one form into another or transferred from one region to another, but energy cannot be created or destroyed.
- **Friction** transforms mechanical energy into thermal energy.

Vocabulary

law of conservation of energy p. 170
friction p. 171

Lesson 3: Energy Resources

- A **nonrenewable energy resource** is an energy resource that is available in a limited amount and can be used up.
- A **renewable energy resource** is replaced in nature as fast as, or faster, than it is used.
- Conserving energy, such as turning off lights when they are not needed, is one way to reduce the rate at which energy resources are used.

Vocabulary

nonrenewable energy resource p. 178
fossil fuel p. 178
renewable energy resource p. 180
inexhaustible energy resource p. 181

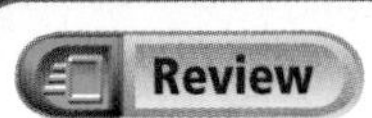

- Personal Tutor
- Vocabulary eGames
- Vocabulary eFlashcards

FOLDABLES® Chapter Project

Assemble your Lesson Foldables as shown to make a Chapter Project. Use the project to review what you have learned in this chapter.

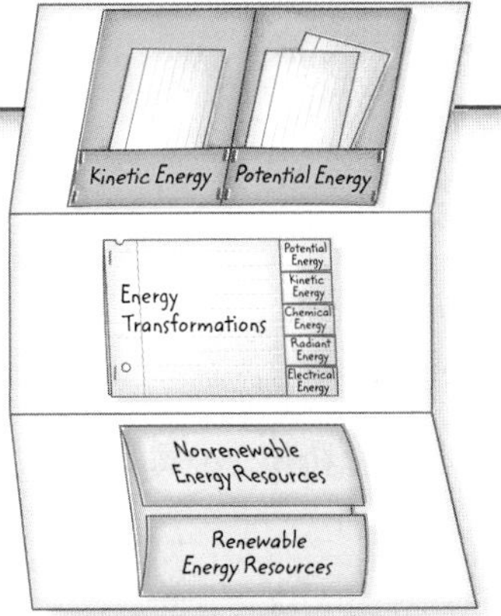

Use Vocabulary

Each of the following sentences is false. Make the sentence true by replacing the italicized word with a vocabulary term.

1. *Thermal energy* is the form of energy carried by an electric current.
2. The *chemical potential energy* of an object depends on its mass and speed.
3. *Friction* is the transfer of energy that occurs when a force is applied over a distance.
4. Natural gas is considered *an inexhaustible energy resource.*
5. *Radiant energy* is energy that is stored in the nucleus of an atom.

Link Vocabulary and Key Concepts

Concepts in Motion **Interactive Concept Map**

Copy this concept map, and then use vocabulary terms from the previous page to complete the concept map.

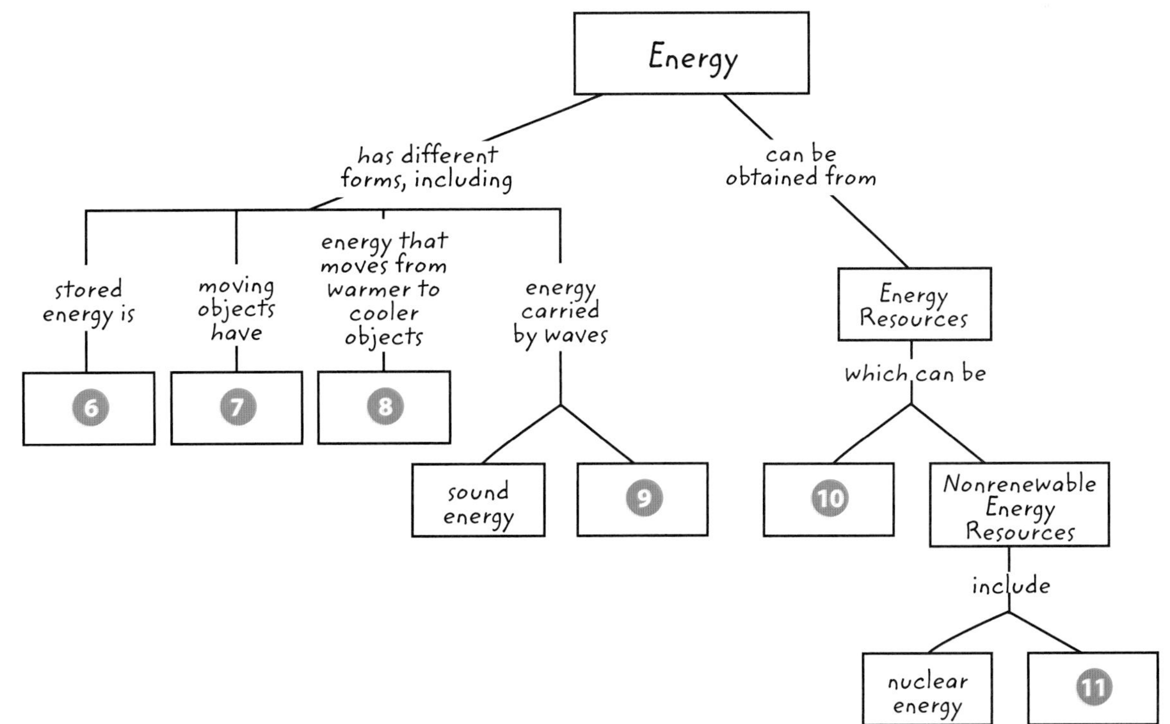

Chapter 5 Review

Understand Key Concepts 

1. What factors determine an object's kinetic energy?
   A. its height and mass
   B. its mass and speed
   C. its size and weight
   D. its speed and height

2. The gravitational potential energy stored between an object and Earth depends on
   A. the object's height and weight.
   B. the object's mass and speed.
   C. the object's size and weight.
   D. the object's speed and height.

3. When a ball is thrown upward, where does it have the least kinetic energy?
   A. at its highest point
   B. at its lowest point when it is moving downward
   C. at its lowest point when it is moving upward
   D. midway between its highest point and its lowest point

4. What energy transformation occurs in the panels on this roof?

   A. chemical energy to thermal energy
   B. nuclear energy to electric energy
   C. radiant energy to electric energy
   D. sound energy to thermal energy

5. According to the law of conservation of energy, which is always true?
   A. Almost all energy used on Earth comes from fossil fuels.
   B. Energy can never be created or destroyed.
   C. Nuclear energy is a renewable energy resource.
   D. Work is done when a force is exerted on an object.

6. Which energy resource is being formed in the picture below?

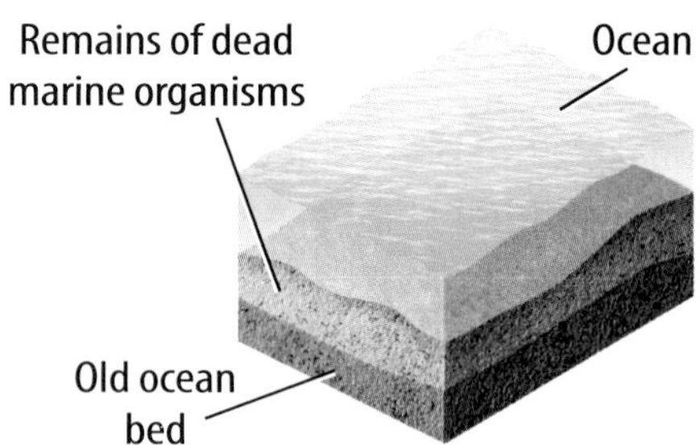

   A. biomass
   B. geothermal reservoirs
   C. petroleum
   D. uranium

7. In which situation would the gravitational potential energy between you and Earth be greatest?
   A. You are running down a hill.
   B. You are running up a hill.
   C. You stand at the bottom of a hill.
   D. You stand at the top of a hill.

8. Which produces the most air pollutants?
   A. burning coal
   B. burning natural gas
   C. hydroelectric power plants
   D. nuclear power plants

9. Which is an example of a renewable energy resource?
   A. geothermal energy
   B. natural gas
   C. nuclear energy
   D. petroleum

10. Which best describes coal?
    A. It burns without polluting the air.
    B. It formed from the remains of plants.
    C. It is a renewable energy resource.
    D. It is the energy source used at nuclear power plants.

Assessment
Online Test Practice

Critical Thinking

11 **Determine** if work is done on the nail if a person pulls the handle to the left and the handle moves. Explain your reasoning.

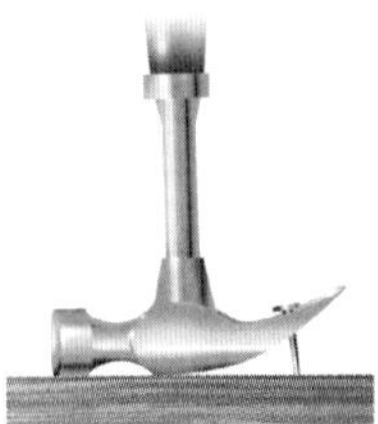

12 **Contrast** the energy transformations that occur in a electric toaster oven and in an electric fan.

13 **Infer** A red box and a blue box are on the same shelf. There is more gravitational potential energy between the red box and Earth than between the blue box and Earth. Which box weighs more? Explain your answer.

14 **Infer** Juanita moves a round box and a square box from a lower shelf to a higher shelf. The gravitational potential energy for the round box increases by 50 J. The gravitational potential energy for the square box increases by 100 J. On which box did Juanita do more work? Explain your reasoning.

15 **Explain** why a skateboard coasting on a flat surface slows down and comes to a stop.

16 **Describe** the difference between the law of conservation of energy and what is meant by conserving energy.

17 **Decide** Harold stretches a rubber band and lets it go. The rubber band flies across the room. He says this demonstrates the transformation of kinetic energy to elastic potential energy. Is Harold correct? Explain.

Writing in Science

18 **Write** a short essay explaining which energy resources you think will be most important in the future.

REVIEW THE BIG IDEA

19 Write an explanation of energy and energy resources for a fourth grader who has never heard of these terms.

20 Identify five energy transformations in the photo below.

Math Skills

Math Practice

21 An electric water heater is rated at 5,500 watts and operates for 106 hours per month. How much electric energy in kWh does the water heater use each month?

22 A family uses 1,303 kWh of electric energy in a month. If the power company charges $0.08 cents per kilowatt hour, what is the total electric energy bill for the month?

Standardized Test Practice

Record your answers on the answer sheet provided by your teacher or on a sheet of paper.

Multiple Choice

1 Which is the transfer of energy that occurs when a player throws a basketball toward a hoop?

A displacement

B force

C velocity

D work

Use the diagram below to answer questions 2 and 3.

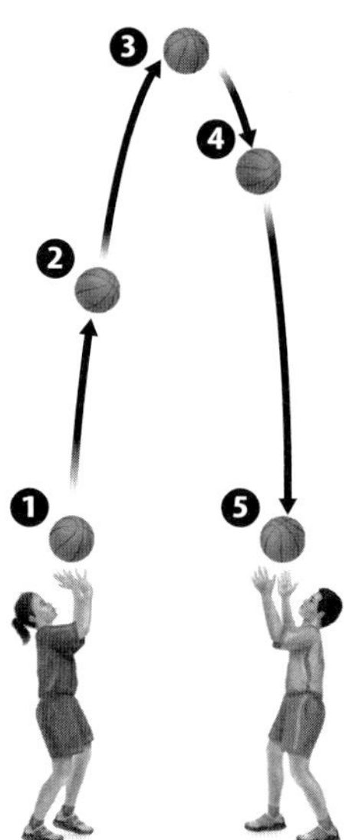

2 At which points is the kinetic energy of the basketball greatest?

A 1 and 5

B 2 and 3

C 2 and 4

D 3 and 4

3 At which point is the gravitational potential energy at a maximum?

A 1

B 2

C 3

D 4

Use the table below to answer question 4.

Vehicle	Mass	Speed
Car 1	1,200 kg	20 m/s
Car 2	1,500 kg	20 m/s
Truck 1	4,800 kg	20 m/s
Truck 2	6,000 kg	20 m/s

4 Which vehicle has the most kinetic energy?

A car 1

B car 2

C truck 1

D truck 2

5 Which form of energy is most abundant on Earth?

A geothermal

B fossil fuel

C solar

D wind

6 Which energy source has been linked to global warming?

A coal

B geothermal

C hydroelectric

D wind

7 A bicyclist uses brakes to slow from 3 m/s to a stop. What stops the bike?

A cohesion

B acceleration

C friction

D gravity

Online Standardized Test Practice

Use the diagram below to answer question 8.

8 The work being done in the diagram above transfers energy to

A the box.

B the floor.

C the girl.

D the shelf.

9 Which is true of energy?

A It cannot be created or destroyed.

B It cannot change form.

C Most forms cannot be conserved.

D Most forms cannot be traced to a source.

10 From which does biomass originate?

A animal and plant material

B nuclear fission

C manufacturing or industrial waste

D power plants

Constructed Response

Use the table below to answer questions 11 and 12.

Forms of Energy	Definition

11 List six forms of energy in the table. Briefly define each form.

12 Provide real-life examples of each of the listed forms of energy.

Use the diagram below to answer question 13.

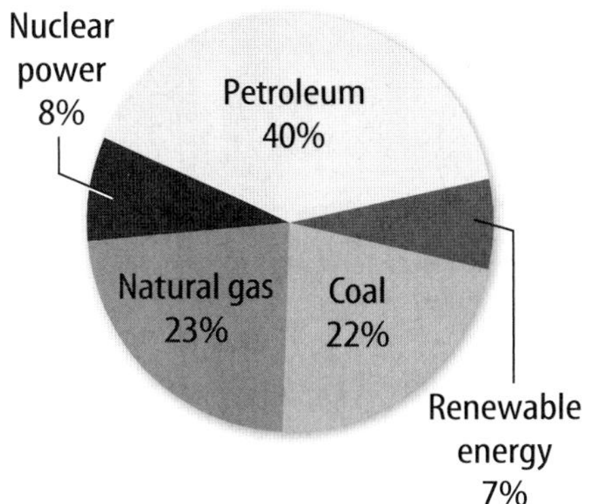

13 What percentage of U.S. energy usage comes from renewable resources? Why is energy conservation important?

NEED EXTRA HELP?													
If You Missed Question...	1	2	3	4	5	6	7	8	9	10	11	12	13
Go to Lesson...	1	2	2	1	3	3	2	1	2	3	1	2	3

Chapter 6

Thermal Energy

How can thermal energy be used?

Inquiry What are these colors?

This image shows the thermal energy of cars moving in traffic. The white indicates regions of high thermal energy, and the dark blue indicates regions of low thermal energy.

- What is thermal energy?
- How does thermal energy relate to temperature and heat?
- How can thermal energy be used?

Get Ready to Read

What do you think?

Before you read, decide if you agree or disagree with each of these statements. As you read this chapter, see if you change your mind about any of the statements.

1. Temperature is the same as thermal energy.
2. Heat is the movement of thermal energy from a hotter object to a cooler object.
3. It takes a large amount of energy to significantly change the temperature of an object with a low specific heat.
4. The thermal energy of an object can never be increased or decreased.
5. Car engines create energy.
6. Refrigerators cool food by moving thermal energy from inside the refrigerator to the outside.

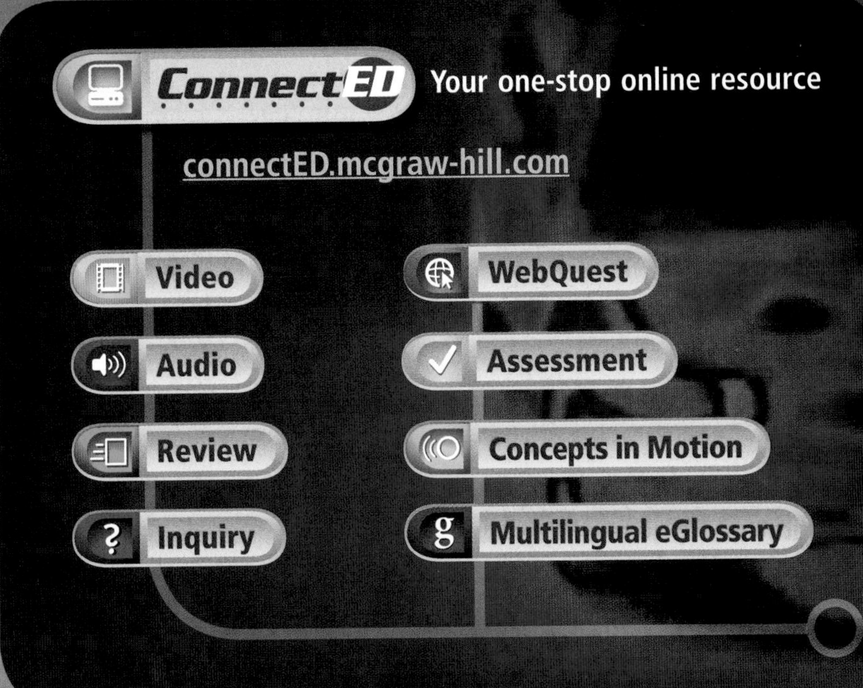

Lesson 1

Reading Guide

Key Concepts
ESSENTIAL QUESTIONS

- How are temperature and kinetic energy related?
- How do heat and thermal energy differ?

Vocabulary

thermal energy p. 198
temperature p. 199
heat p. 201

g Multilingual eGlossary

Thermal Energy, Temperature, and Heat

Inquiry How hot is it?

Forty gallons of sugar-maple sap must be heated to a very high temperature for several days to produce 1 gallon of maple syrup. What kind of energy is needed to achieve this very high temperature? Is there a difference between heat, temperature, and thermal energy?

Launch Lab

15 minutes

How can you describe temperature?

Have you ever used Fahrenheit or Celsius to describe the temperature? Why can't you just make up your own temperature scale?

1. Read and complete a lab safety form.
2. Use a **ruler** and a **permanent marker** to divide a **clear plastic straw** into equal parts. Number the lines. Give your scale a name.
3. Add a room-temperature **colored alcohol-water mixture** to an **empty plastic water bottle** until it is about $\frac{1}{4}$ full.
4. Place one end of the straw into the bottle with the tip just below the surface of the liquid. Seal the straw onto the bottle top with **clay.**
5. Place the bottle in a **hot water bath**, and observe the liquid in your straw.

Think About This

1. Why is it important for scientists to use the same scale to measure temperature?
2. Key Concept What are some ways to make the liquid in your thermometer rise or fall?

Kinetic and Potential Energy

What do a soaring soccer ball and the particles that make up hot maple syrup have in common? They all have energy, or the ability to cause change. What type of energy does a moving soccer ball have? Recall that any moving object has kinetic energy. When the athlete in Figure 1 kicks the ball and puts it in motion, the ball has kinetic energy.

In addition to kinetic energy, when the soccer ball is in the air, it has potential energy. Potential energy is stored energy due to the interaction between two objects. For example, think of Earth as one object and the ball as another. When the ball is in the air, it is attracted to Earth due to gravity. This attraction is called gravitational potential energy. In other words, since the ball has the potential to change, it has potential energy. And, the higher the ball is in the air, the greater the potential energy of the ball.

You also might recall that the potential energy plus the kinetic energy of an object is the mechanical energy of the object. When a soccer ball is flying through the air, you could describe the mechanical energy of the ball by describing both its kinetic and potential energy. On the next page, you will read about how the particles that make up maple syrup have energy, just like a soaring soccer ball.

Reading Check How could you describe the energy of a moving object?

REVIEW VOCABULARY

kinetic energy
the energy an object or a particle has because it is moving

potential energy
stored energy

Figure 1 This soccer ball has both kinetic energy and potential energy.

What is thermal energy?

Every solid, liquid, and gas is made up of trillions of tiny particles that are constantly moving. Moving particles make up the books you read, the air you breathe, and the maple syrup you pour on your pancakes. For example, the particles that make up a book, or any solid, vibrate in place. The particles that make up the air around you, or any gas, are spread out and move freely and quickly. Because particles are in motion, they have kinetic energy, like the soaring soccer ball in **Figure 2.** The faster particles move, the more kinetic energy they have.

The particles that make up matter also have potential energy. Like the interaction between a soccer ball and Earth, particles that make up matter interact with and are attracted to one another. The particles that make up solids usually are held very close together by attractive forces. The particles that make up a liquid are slightly farther apart than those that make up a solid. And, the particles that make up a gas are much more spread out than those that make up either a solid or a liquid. The greater the average distance between particles, the greater the potential energy of the particles.

Recall that a flying soccer ball has mechanical energy, which is the sum of its potential energy and its kinetic energy. The particles that make up the soccer ball, or any material, have a similar kind of energy called thermal energy. **Thermal energy** *is the sum of the kinetic energy and the potential energy of the particles that make up a material.* Thermal energy describes the energy of the particles that make up a solid, a liquid, or a gas.

Reading Check How are thermal energy and mechanical energy similar? How are they different?

Figure 2 The potential energy of the soccer ball depends on the distance between the ball and Earth. The potential energy of the particles of matter depends on their distance from one another.

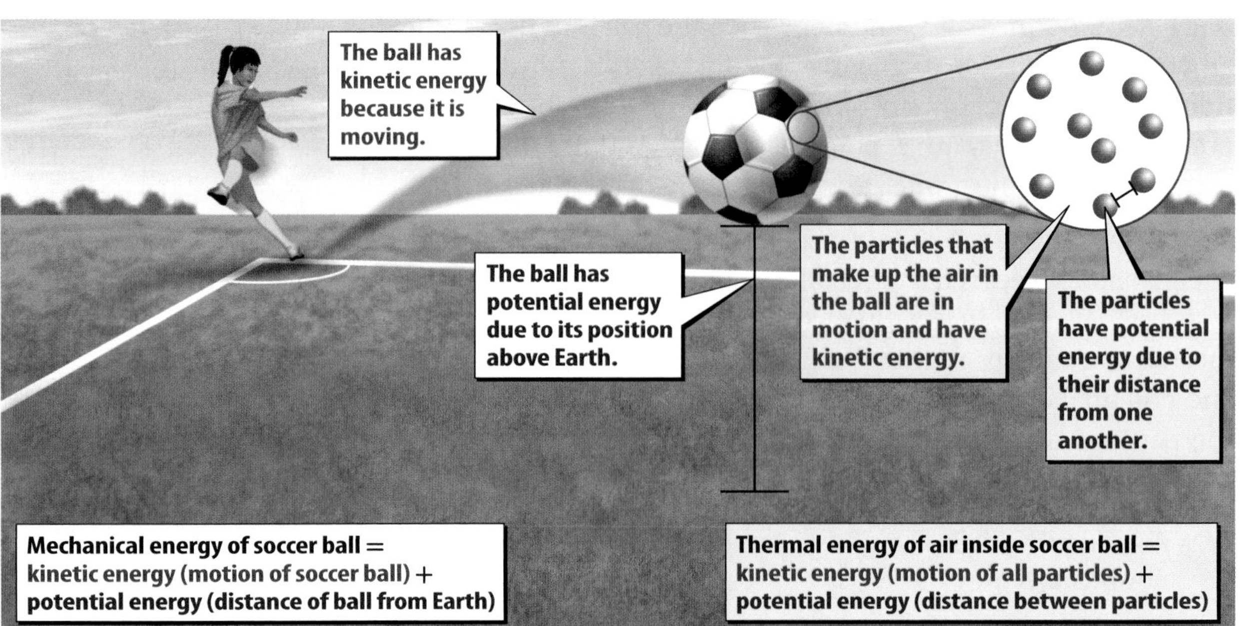

The ball has kinetic energy because it is moving.
The ball has potential energy due to its position above Earth.
The particles that make up the air in the ball are in motion and have kinetic energy.
The particles have potential energy due to their distance from one another.
Mechanical energy of soccer ball = kinetic energy (motion of soccer ball) + potential energy (distance of ball from Earth)
Thermal energy of air inside soccer ball = kinetic energy (motion of all particles) + potential energy (distance between particles)

Figure 3 The air's temperature depends on how fast the particles in the air move.

Visual Check What happens to the motion of the particles in the air as temperature increases?

Concepts in Motion **Animation**

What is temperature?

When you think of temperature, you probably think of it as a measurement of how warm or cold something is. However, scientists define temperature in terms of kinetic energy.

Average Kinetic Energy and Temperature

The particles that make up the air inside and outside the house in **Figure 3** are moving. However, they are not moving at the same speed. The particles in the air in the warm house move faster and have more kinetic energy than those outside on a cold winter evening. **Temperature** *represents the average kinetic energy of the particles that make up a material.*

The greater the average kinetic energy of particles, the greater the temperature. The temperature of the air inside the house is higher than the temperature of the air outside the house. This is because the particles that make up the air inside the house have greater average kinetic energy than those outside. In other words, the particles of air inside the house are moving at a greater average speed than those outside.

Key Concept Check How are temperature and kinetic energy related?

Thermal Energy and Temperature

Temperature and thermal energy are related, but they are not the same. For example, as a frozen pond melts, both ice and water are present and they have the same temperature. Therefore, the particles that make up the ice and the water have the same average kinetic energy, or speed. However, the particles do not have the same thermal energy. This is because the average distance of the particles that make up liquid water and ice are different. The particles that make up the liquid and the solid water have different potential energies and, therefore, different thermal energies.

WORD ORIGIN

temperature
from Latin *temperatura,* means "moderating, tempering"

FOLDABLES®

Make a vertical three-column chart book. Label it as shown. Use it to organize your notes on the properties of heat, temperature, and thermal energy.

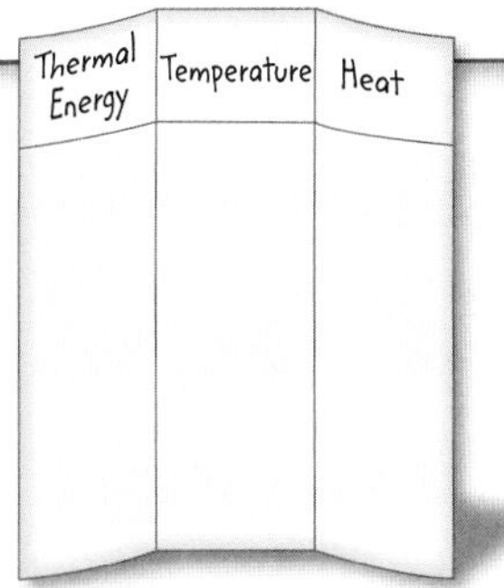

Math Skills 

Convert Between Temperature Scales

To convert Fahrenheit to Celsius, use the following equation:

$$°C = \frac{(°F - 32)}{1.8}$$

For example, to convert **176°F** to Celsius:

1. Always perform the operation in parentheses first.

   $176 - 32 = 144$

2. Divide the answer from Step 1 by 1.8.

   $\frac{144}{1.8} = 80°C$

To convert Celsius to Fahrenheit, follow the same steps using the following equation:

$$°F = (°C \times 1.8) + 32$$

Practice

1. Convert 86°F to Celsius.
2. Convert 37°C to Fahrenheit.

Review

- **Math Practice**
- **Personal Tutor**

Measuring Temperature

How can you measure temperature? It would be impossible to measure the kinetic energy of individual particles and then calculate their average kinetic energy to determine the temperature. Instead, you can use thermometers, such as the ones in **Figure 4,** to measure temperature.

A common type of thermometer is a bulb thermometer. A bulb thermometer is a glass tube connected to a bulb that contains a liquid such as alcohol. When the temperature of the alcohol increases, the alcohol expands and rises in the glass tube. When the temperature of the alcohol decreases, the alcohol contracts back into the bulb. The height of the alcohol in the tube indicates the temperature.

There are other types of thermometers too, such as an electronic thermometer. This thermometer measures changes in the resistance of an electric circuit and converts this measurement to a temperature.

Temperature Scales

You might have seen the temperature in a weather report given in degrees Fahrenheit and degrees Celsius. On the Fahrenheit scale, water freezes at 32° and boils at 212°. On the Celsius scale, water freezes at 0° and boils at 100°. The Celsius scale is used by scientists worldwide.

Scientists also use the Kelvin scale. On the Kelvin scale, water freezes at 273 K and boils at 373 K. The lowest possible temperature for any material is 0 K. This is known as absolute zero. If a material were at 0 K, the particles in that material would not be moving and would no longer have kinetic energy. Scientists have not been able to cool any material to 0 K.

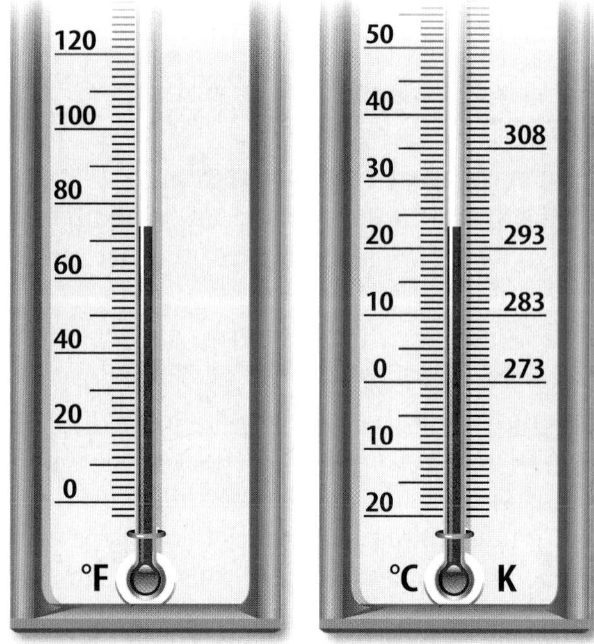

Figure 4 Thermometers measure temperature. Common temperature scales are Celsius, Kelvin, and Fahrenheit.

The hot cocoa has a high temperature. Thermal energy is transferred from the mug to its surroundings.
The heat from the hot cocoa to the air is greater than the heat from the hot cocoa to the girl's hands. This is because the temperature difference is greater from the hot cocoa to the air.

Figure 5 The hot cocoa heats the air and the girl's hands.

What is heat?

Have you ever held a cup of hot cocoa on a cold day like the girl in **Figure 5**? When you do, thermal energy moves from the warm cup to your hands. *The movement of thermal energy from a warmer object to a cooler object is called* **heat.** Another way to say this is that thermal energy from the cup heats your hands, or the cup is heating your hands.

Just as temperature and thermal energy are not the same thing, neither are heat and thermal energy. All objects have thermal energy. However, you heat something when thermal energy transfers from one object to another. The girl in **Figure 5** heats her hands because thermal energy transfers from the hot cocoa to her hands.

Key Concept Check How do heat and thermal energy differ?

The rate at which heating occurs depends on the difference in temperatures between the two objects. The difference in temperatures between the hot cocoa and the air is greater than the difference in temperatures between the hot cocoa and the cup. The hot cocoa heats the air more than it heats the cup. Heating continues until all objects in contact are the same temperature.

Inquiry MiniLab — 10 minutes

How do temperature scales compare?

If someone told you it was 2°C or 300 K outside, would you know whether it was warm or cold?

	Celsius (°C)	Fahrenheit (°F)	Kelvin (K)
Room temperature			
Light jacket weather			
Hot summer day			

1. Copy the table into your Science Journal.
2. Lay a **ruler** across **Figure 4** so that it lines up with the temperatures at which water freezes. Record the temperatures
3. Repeat step 2 for the three values in the table.

Analyze and Conclude

1. **Estimate** Imagine that it is snowing outside. What might the temperature be in degrees Celsius? In kelvin?
2. **Key Concept** Why doesn't the Kelvin scale include negative numbers?

Lesson 1 Review

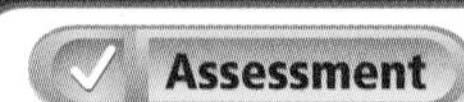

Online Quiz

Visual Summary

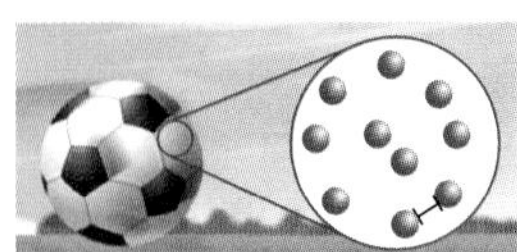

The greater distance between two particles or two objects, the greater the potential energy.

Heat is the movement of thermal energy from a warmer object to a cooler object.

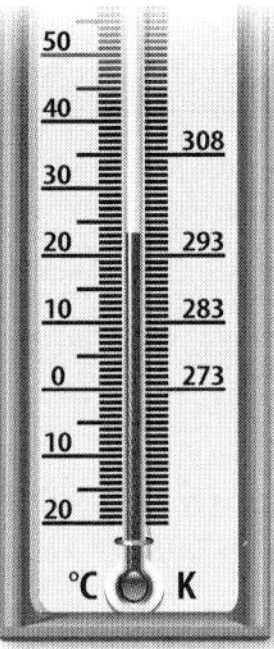

When thermal energy moves between a material and its environment, the material's temperature changes.

FOLDABLES®

Use your lesson Foldable to review the lesson. Save your Foldable for the project at the end of the chapter.

What do you think NOW?

You first read the statements below at the beginning of the chapter.

1. Temperature is the same as thermal energy.
2. Heat is the movement of thermal energy from a hotter object to a cooler object.

Did you change your mind about whether you agree or disagree with the statements? Rewrite any false statements to make them true.

Use Vocabulary

1. The sum of kinetic energy and potential energy of the particles in a material is ________.
2. **Relate** temperature to the average kinetic energy in a material.

Understand Key Concepts

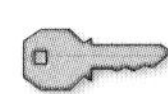

3. **Differentiate** between thermal energy and heat.
4. Which increases the kinetic energy of the particles that make up a bowl of soup?
   - **A.** dividing the soup in half
   - **B.** putting the soup in a refrigerator
   - **C.** heating the soup for 1 min on a stove
   - **D.** decreasing the distance between the particles that make up the soup
5. **Infer** Suppose a friend tells you he has a temperature of 38°C. Your temperature 37°C. Do the particles that make up your body or your friend's body have a greater average kinetic energy? Explain.

Interpret Graphics

6. **Identify** Copy and fill in the following graphic organizer to show the forms of energy that make up thermal energy.

Critical Thinking

7. **Explain** How could you increase the kinetic thermal energy of a liquid?

Math Skills

Math Practice

8. Maple sap boils at 104°C. At what Fahrenheit temperature does the sap boil?

How do different materials affect thermal energy transfer?

Materials

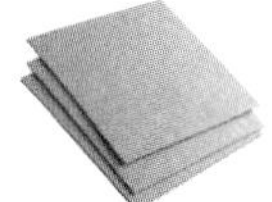

cardboard

100-mL graduated cylinder

2 thermometers

Also Needed
1-L square plastic container, test containers (metal, polystyrene, ceramic, glass, plastic), large rubber band, hot water

Safety

You might have noticed that thermal energy moves more easily through some substances than others. For example, juice stays colder in a foam cup than in a can. How does the container's material affect how quickly thermal energy moves through it?

Learn It

To **form a hypothesis** is to propose an explanation for an observation. The explanation should be testable. One way to **test a hypothesis** is by gathering data that shows whether the hypothesis is correct.

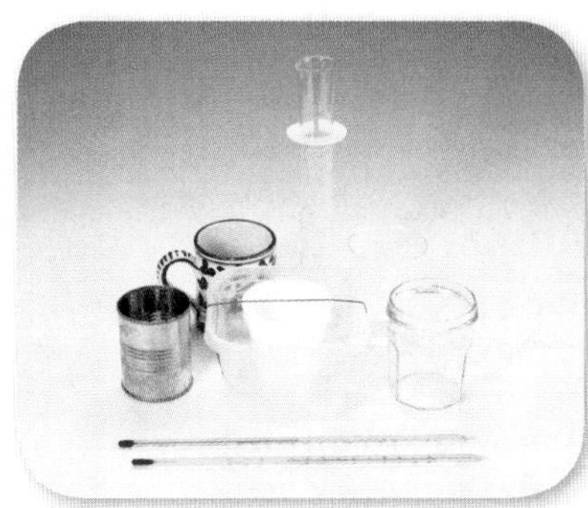

Try It

1. Read and complete a lab safety form.
2. Observe the test containers. Write a hypothesis in your Science Journal that explains why you think a certain material will slow the transfer of thermal energy more than others.
3. Copy the table below.
4. Each lab group will test one container. Stand your test container in the center of a 1-L plastic container.
5. Add 125 mL of hot water to the test container. Measure and record the water's temperature.
6. Add room-temperature water to the plastic container until the level in both containers is equal. Measure and record the room-temperature water's temperature.
7. Place a cardboard square over the test container. Use two thermometers to take the temperature of the water in both containers every 2 min for 20 min. Record your data in your table.
8. Compare your data with the data gathered by the other teams. Rank the test containers from slowest to fastest thermal energy transfer in your Science Journal.

Apply It

9. **Analyze Data** Did your data support your hypothesis? Why or why not?
10. **Key Concept** What happened to the thermal energy of the water in the test container? Why did this happen?

°C	0 min	2 min	4 min	6 min	8 min	10 min	12 min	14 min	16 min	18 min	20 min
Temp in test container											
Temp in outer container											

Lesson 2

Reading Guide

Key Concepts
ESSENTIAL QUESTIONS

- What is the effect of having a small specific heat?
- What happens to a material when it is heated?
- In what ways can thermal energy be transferred?

Vocabulary

radiation p. 205
conduction p. 206
thermal conductor p. 206
thermal insulator p. 206
specific heat p. 207
thermal contraction p. 208
thermal expansion p. 208
convection p. 210
convection current p. 211

Multilingual eGlossary

Video Science Video

Thermal Energy Transfers

Inquiry Keeping Warm?

Imagine camping in the mountains on a cold winter night. Your survival could depend on keeping warm. There are many things you could do to get warm and stay warm. In this picture, how is thermal energy transferred from the fire to the camper? Why does his coat keep him from losing thermal energy?

15 minutes

How hot is it?

When you touch an ice cube, you sense that it is cold. When you get inside a car on a warm day, you sense that it is hot. How accurate is your sense of touch in predicting temperature?

1. Read and complete a lab safety form.
2. Place the palm of one hand flat against a piece of **metal** and the other hand against a piece of **wood.** Observe which material feels colder, and record it in your Science Journal.
3. Repeat step 2 with other materials, including **cardboard, glass, plastic,** and **foam.**
4. Rank the materials from coldest to warmest in your Science Journal.
5. Place a **liquid crystal thermometer** on each material. Record the temperature of each material in your Science Journal.

Think About This

1. Were you able to accurately rank the materials by temperature only by touching them?
2. **Key Concept** Why might some of the materials in this experiment feel cooler than others even though they are in the same room?

How is thermal energy transferred?

Have you ever gotten into a car, such as the one in **Figure 6**, on a hot summer day? You can guess that the inside of the car is hot even before you touch the door handle. You open the door and hot air seems to pour out of the car. When you touch the metal safety-belt buckle, it is hot. How is thermal energy transferred between objects? Thermal energy is transferred in three ways—by radiation, conduction, and convection.

Radiation

The transfer of thermal energy from one material to another by electromagnetic waves is called **radiation.** All matter, including the Sun, fire, you, and even ice, transfers thermal energy by radiation. Warm objects emit more radiation than cold objects do. For example, when you place your hands near a fire, you can more easily feel the transfer of thermal energy by radiation than when you place your hands near a block of ice.

Thermal energy from the Sun heats the inside of the car in **Figure 6** by radiation. In fact, radiation is the only way thermal energy can travel from the Sun to Earth. This is because space is a **vacuum.** However, radiation also transfers thermal energy through solids, liquids, and gases.

 **Reading Check** How does the Sun heat the inside of a car?

Science Use v. Common Use

vacuum

Science Use a space that contains little or no matter

Common Use a device for cleaning carpets and rugs that uses suction

Figure 6 The Sun heats this car by radiation.

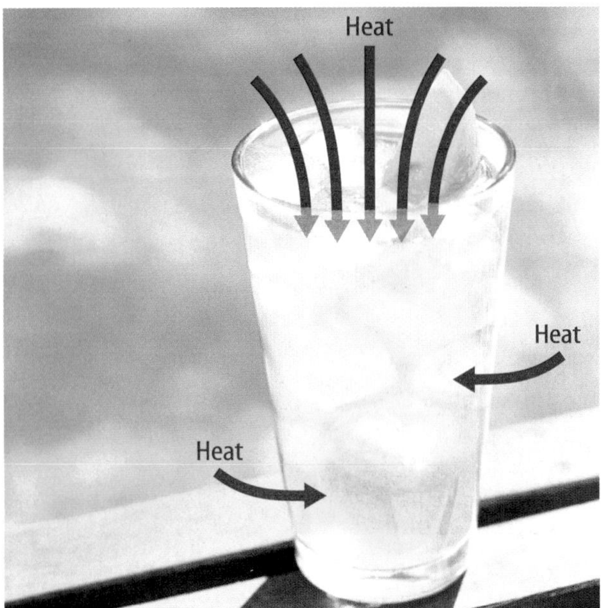

Figure 7 The hot air transfers thermal energy to, or heats, the cool lemonade by conduction. Eventually the kinetic thermal energy and temperature of the air and the lemonade will be equal.

Animation

FOLDABLES®

Make a vertical three-column chart book. Label it as shown. Use it to describe the ways thermal energy is transferred.

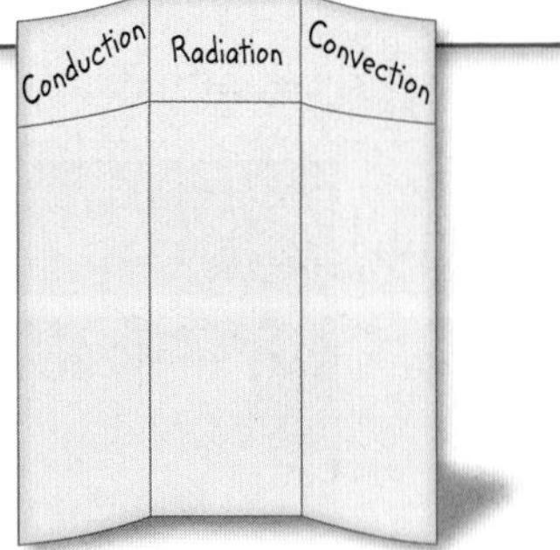

Conduction

Suppose it's a hot day and you have a cold glass of lemonade, such as the one in **Figure 7.** The lemonade has a lower temperature than the surrounding air. Therefore, the particles that make up the lemonade have less kinetic energy than the particles that make up the air. When particles with different kinetic energies collide, the particles with higher kinetic energy transfer energy to particles with lower kinetic energy.

In **Figure 7,** the particles that make up the air collide with and transfer kinetic energy to the particles that make up the lemonade. As a result, the average kinetic energy, or temperature, of the particles that make up the lemonade increases. Since kinetic energy is being transferred, thermal energy is being transferred. *The transfer of thermal energy between materials by the collisions of particles is called* **conduction.** Conduction continues until the thermal energy of all particles in contact is equal.

Thermal Conductors and Insulators

On a hot day, why does a metal safety-belt buckle in a car feel hotter than the safety belt. Both the buckle and safety belt receive the same amount of thermal energy from the Sun. The metal that makes up the buckle is a good thermal conductor. *A* **thermal conductor** *is a material through which thermal energy flows easily.* Atoms in good thermal conductors have electrons that move easily. These electrons transfer kinetic energy when they collide with other electrons and atoms. Metals are better thermal conductors than nonmetals. The cloth that makes up a safety belt is a good thermal insulator. *A* **thermal insulator** *is a material through which thermal energy does not flow easily.* The electrons in the atoms of a good thermal insulator do not move easily. These materials do not transfer thermal energy easily because fewer collisions occur between electrons and atoms.

Specific Heat

The amount of thermal energy required to increase the temperature of 1 kg of a material by 1°C is called **specific heat.** Every material has a specific heat. It is easy to change the temperature of a material with a low specific heat but hard to change the temperature of a material with high specific heat.

Academic Vocabulary

specific
(adjective) precise and detailed; belonging to a distinct category

Thermal conductors, such as the cloth of the safety-belt buckle in **Figure 8,** have a lower specific heat than thermal insulators, such as the cloth safety belt. This means it takes less thermal energy to increase the buckle's temperature than it takes to increase the temperature of the cloth safety belt by the same amount.

The specific heat of water is particularly high. It takes a large amount of energy to increase or decrease the temperature of water. The high specific heat of water has many beneficial effects. For example, much of your body is water. Water's high specific heat helps prevent your body from overheating. The high specific heat of water is one of the reasons why pools, lakes, and oceans stay cool in summer. Water's high specific heat also makes it ideal for cooling machinery, such as car engines and rock cutting saws.

Key Concept Check What does it mean if a material has a low specific heat?

Specific Heat

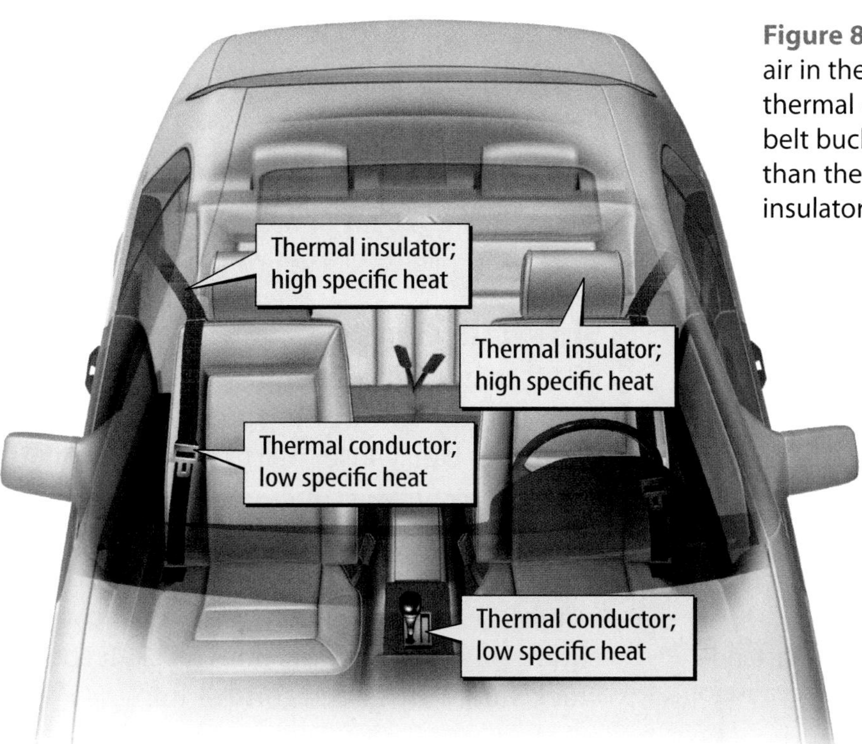

Figure 8 On a hot summer day, the air in the car is hot. The temperature of thermal conductors, such as the safety-belt buckles, increases more quickly than the temperature of thermal insulators, such as the seat material.

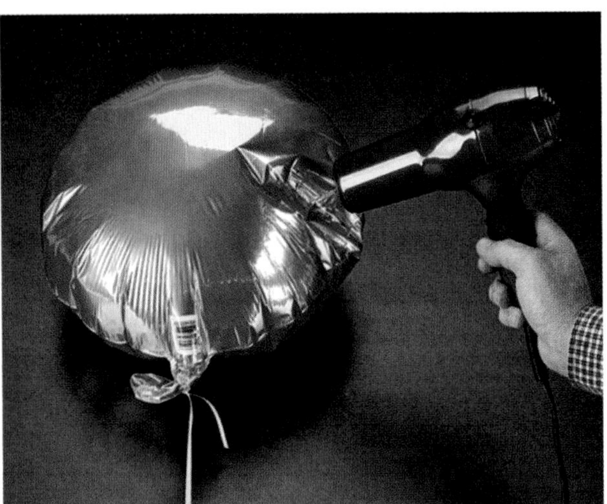

▲ **Figure 9** Air inside the balloon increases in volume when the temperature increases.

Thermal Expansion and Contraction

What happens if you take an inflated balloon outside on a cold day? Thermal energy transfers from the particles that make up the air inside the balloon to the particles that make up the balloon material and then to the cold outside air. As the particles that make up the air in the balloon lose thermal energy, which includes kinetic energy, they slow down and get closer together. This causes the volume of the balloon to decrease. **Thermal contraction** *is a decrease in a material's volume when its temperature decreases.*

How could you reinflate the balloon? You could heat the air inside the balloon with a hair dryer, like in **Figure 9.** The particles that make up the hot air coming out of the hair dryer transfer thermal energy, which includes kinetic energy, to the particles that make up the air inside the balloon. As the average kinetic energy of the particles increases, the air temperature increases. Also, as the average kinetic energy of the particles increases, they speed up and spread out, increasing the volume of air inside the balloon. **Thermal expansion** *is an increase in a material's volume when its temperature increases.*

Thermal expansion and contraction are most noticeable in gases, less noticeable in liquids, and the least noticeable in solids.

Key Concept Check What happens to the volume of a gas when it is heated?

Sidewalk Gaps

In many places, outdoor temperatures become very hot in the summer. High temperatures can cause thermal expansion in structures, such as concrete sidewalks. If the concrete expands too much or expands unevenly, it could crack. Therefore, control joints are cut into sidewalks, as shown in **Figure 10.** If the sidewalk does crack, it should crack smoothly at the control joint.

▲ **Figure 10** Sidewalks can withstand thermal expansion and contraction because of control joints.

Hot-Air Balloons

How do hot-air balloons work? As shown in **Figure 11,** a burner heats the air in the balloon, causing thermal expansion. The particles that make up the air inside the balloon move faster and faster. As the particles collide with one another, some are forced outside the balloon through the opening at the bottom. Now, there are fewer particles in the balloon than in the same volume of air outside the balloon. The balloon is less dense, and it begins to rise through denser outside air.

To land a hot-air balloon, the balloonist allows the air inside the balloon to gradually cool. The air undergoes thermal contraction. However, the balloon itself does not contract. Instead, denser air from outside the balloon fills the space inside. As the density of the balloon increases, it slowly descends.

Figure 11 Hot-air balloonists control their balloons using thermal expansion and contraction.

Ovenproof Glass

If you put an ordinary drinking glass into a hot oven, the glass might break or shatter. However, an ovenproof glass dish would not be damaged in a hot oven. Why is this so?

Different parts of ordinary glass expand at different rates when heated. This causes it to crack or shatter. Ovenproof glass is designed to expand less than ordinary glass when heated, which means that it usually does not crack in the oven.

Inquiry MiniLab

20 minutes

How does adding thermal energy affect a wire?

How could thermal energy help you remove a metal lid from a glass jar?

1. Read and complete a lab safety form.
2. Set up **two ring stands** so that the rings are 1–2 m apart. Tie the ends of a **2-m length of wire** to the rings so that the wire is straight and tight.
3. Use **thread** to tie a **weight** to the middle of the wire.
4. Use a **ruler** to measure the distance from the bottom of the weight to the table. Record your data in your Science Journal.
5. Using **matches** light two **candles.** Move the candle flames back and forth under the wire. Repeat step 4 every minute for 5 min. Blow out the candles.
6. Repeat step 4 again every minute for 5 min as the wire cools.

Analyze and Conclude

1. **Predict** What would happen if you continued to heat the wire? Explain.
2. **Apply** How could you use this idea to help you remove a metal lid from a glass jar?
3. **Key Concept** What happens to the particles that make up the wire when the wire is heated? How do you know?

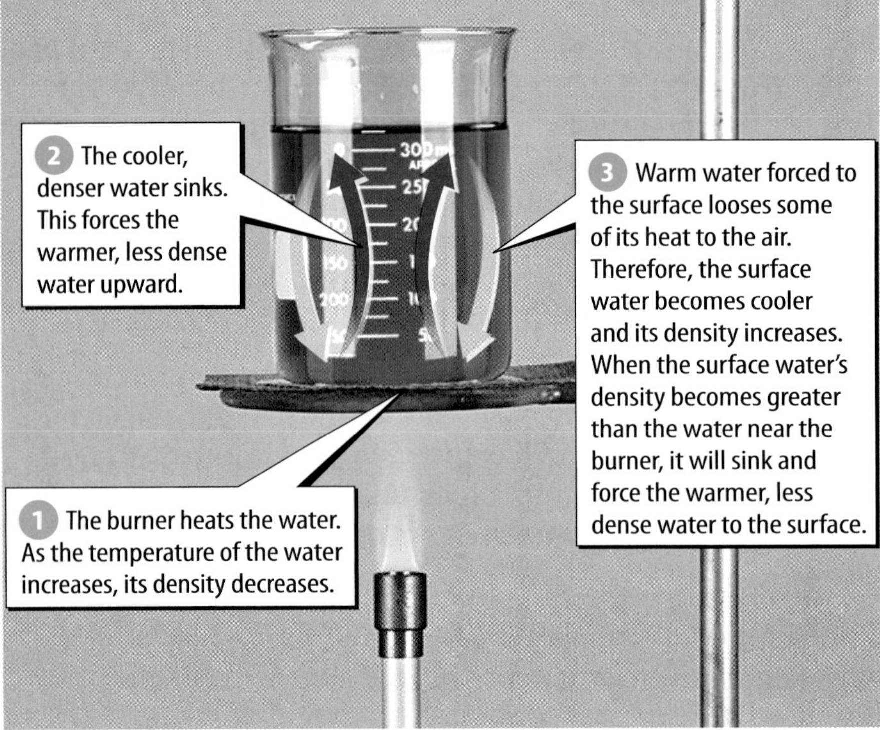

2 The cooler, denser water sinks. This forces the warmer, less dense water upward.
3 Warm water forced to the surface looses some of its heat to the air. Therefore, the surface water becomes cooler and its density increases. When the surface water's density becomes greater than the water near the burner, it will sink and force the warmer, less dense water to the surface.
1 The burner heats the water. As the temperature of the water increases, its density decreases.

Figure 12 This cycle of cooler water sinking and forcing warmer water upward is an example of convection.

Personal Tutor

Convection

When you heat a pan of water on the stove, the burner heats the pan by conduction. This process, shown in **Figure 12,** involves the movement of thermal energy within a fluid. The particles that make up liquids and gases move around easily. As they move, they transfer thermal energy from one location to another. **Convection** *is the transfer of thermal energy by the movement of particles from one part of a material to another.* Convection only occurs in fluids, such as water, air, magma, and maple syrup.

Word Origin

convection
from Latin *convectionem,* means "the act of carrying"

Key Concept Check What are the three processes that transfer thermal energy?

Density, Thermal Expansion, and Thermal Contraction

In **Figure 12,** the burner transfers thermal energy to the beaker, which transfers thermal energy to the water. Thermal expansion occurs in water nearest the bottom of the beaker. Heating increases the water's volume making it less dense.

At the same time, water molecules at the water's surface transfer thermal energy to the air. This causes cooling and thermal contraction of the water on the surface. The denser water at the surface sinks to the bottom, forcing the less dense water upward. This cycle continues until all the water in the beaker is at the same temperature.

Convection Currents in Earth's Atmosphere

The movement of fluids in a cycle because of convection is a **convection current.** Convection currents circulate the water in Earth's oceans and other bodies of water. They also circulate the air in a room, and the materials in Earth's interior. Convection currents also move matter and thermal energy from inside the Sun to its surface.

On Earth, convection currents move air between the equator and latitudes near 30°N and 30°S. This plays an important role in Earth's climates, as shown in **Figure 13.**

Convection Currents in Earth's Atmosphere

Figure 13 Convection currents in the atmosphere influence the locations of rain forests and deserts.

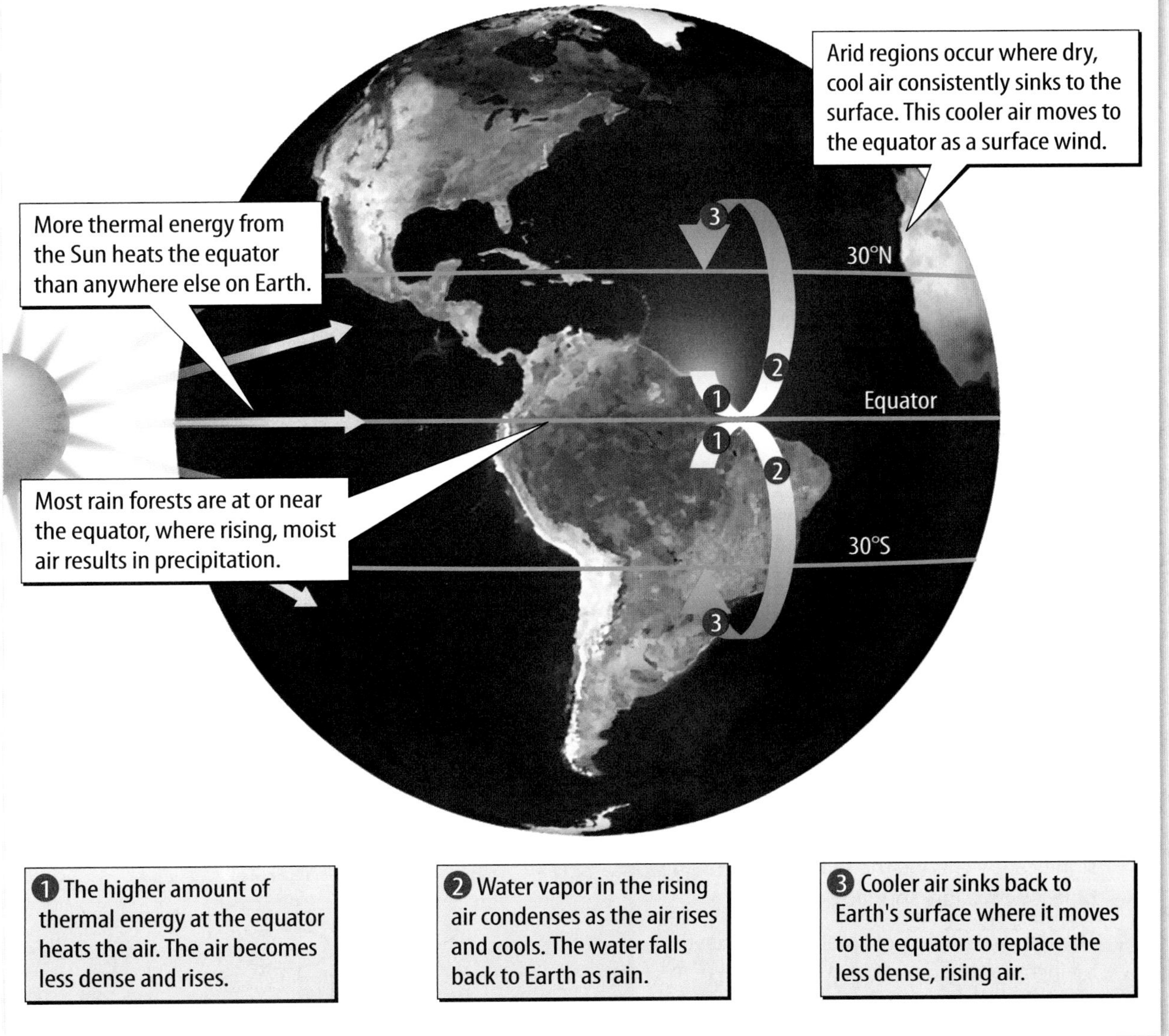

Arid regions occur where dry, cool air consistently sinks to the surface. This cooler air moves to the equator as a surface wind.
More thermal energy from the Sun heats the equator than anywhere else on Earth.
Most rain forests are at or near the equator, where rising, moist air results in precipitation.
30°N
Equator
30°S
1 The higher amount of thermal energy at the equator heats the air. The air becomes less dense and rises.
2 Water vapor in the rising air condenses as the air rises and cools. The water falls back to Earth as rain.
3 Cooler air sinks back to Earth's surface where it moves to the equator to replace the less dense, rising air.

Lesson 2 Review

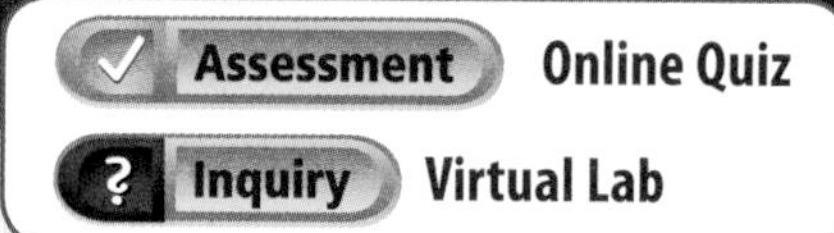

Visual Summary

When a material has a low specific heat, transferring a small amount of energy to the material increases its temperature significantly.

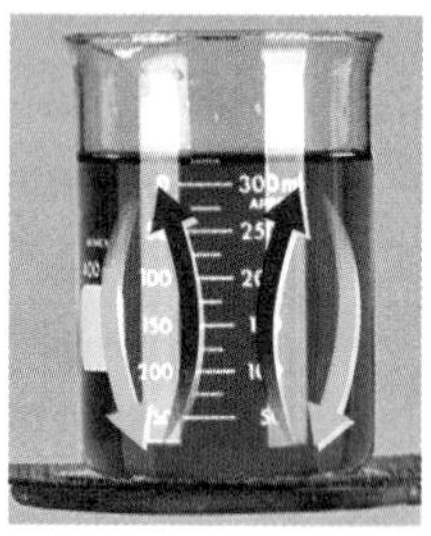

Thermal energy can be transferred through radiation, conduction, or convection.

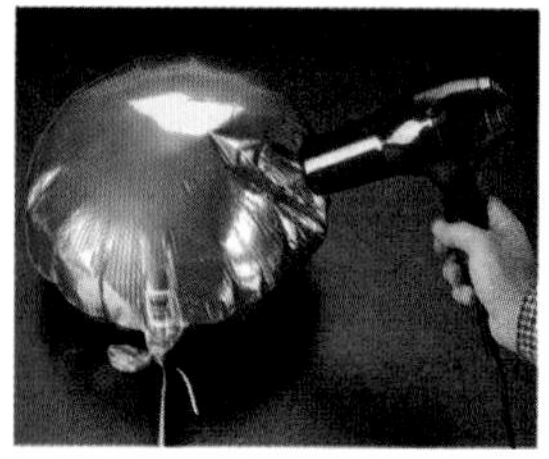

When a material is heated, the thermal energy of the material increases and the material expands.

FOLDABLES®

Use your lesson Foldable to review the lesson. Save your Foldable for the project at the end of the chapter.

What do you think NOW?

You first read the statements below at the beginning of the chapter.

3. It takes a large amount of energy to significantly change the temperature of an object with a low specific heat.

4. The thermal energy of an object can never be increased or decreased.

Did you change your mind about whether you agree or disagree with the statements? Rewrite any false statements to make them true.

Use Vocabulary

1. The transfer of thermal energy by electromagnetic waves is ________.
2. **Define** *convection* in your own words.

Understand Key Concepts 

3. **Contrast** radiation with conduction.
4. Why do hot-air balloons rise?
   - **A.** thermal conduction
   - **B.** thermal convection
   - **C.** thermal expansion
   - **D.** thermal radiation
5. **Infer** why the sauce on a hot pizza burns your mouth but the crust of the pizza does not burn your mouth.

Interpret Graphics

6. **Analyze** Two cubes with the same mass and volume are heated in the same pan of water. The graph below shows the change in temperature with time. Which cube has the higher specific heat?

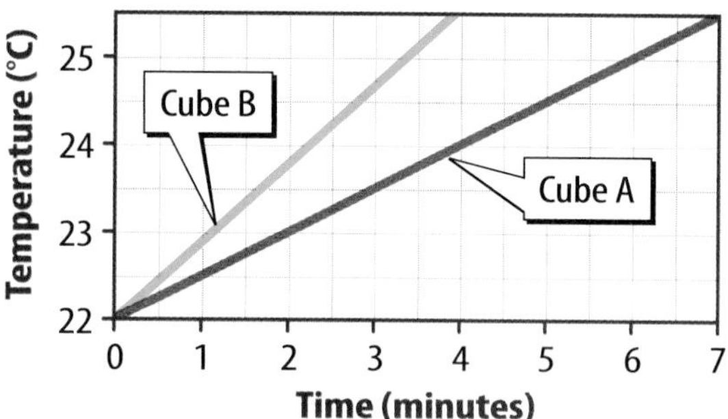

7. **Organize** Copy and fill in the graphic organizer to show how thermal energy is transferred.

Critical Thinking

8. **Explain** Why do you use a pot holder when taking hot food out of the oven?

SCIENCE & SOCIETY

Insulating the Home

It's what's between the walls that matters.

The first requirement of a shelter is to protect you from the weather. If the weather where you live is mild year-round, almost any kind of shelter will do. However, basic shelters, such as huts and tents, are not comfortable during cold winters or hot summers. Over many centuries, societies have experimented with thermal insulators that keep the inside of a shelter warm during winter and cool during summer.

One of the first thermal insulators used in shelters was air. Because air is a poor conductor of thermal energy, using cavity walls became a common form of insulating homes in the United States.

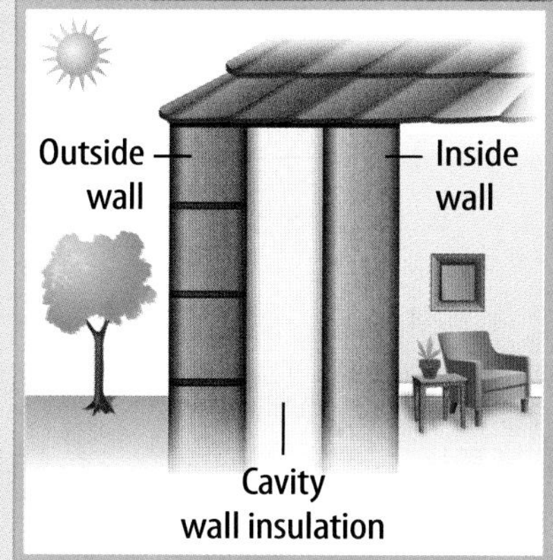

An air gap was not the perfect solution, however. Convection currents in the cavity carried some thermal energy across the gap. At first, no one seemed to mind. But, in the 1970s, the cost of heating and cooling homes suddenly increased. People began looking for a better way to reduce the transfer of thermal energy between the outside and the inside.

To meet the growing demand for better insulation, scientists began researching to find better insulation materials. If you could stop the convection currents, they reasoned, you could stop the transfer of thermal energy. One way was to use materials such as polymer foam or fiberglass. Each of these trapped air between the walls and held it there.

But how do you install insulation if your house is already built? You poke holes in the walls and blow it in! This process has little effect on your home's structure, and it decreases the cost of heating and cooling the house.

It's Your Turn

MAKE A POSTER A material's insulating ability is rated with an R-value. Find out what an R-value is. Then make a poster showing the ratings of common materials.

Lesson 3

Using Thermal Energy

Reading Guide

Key Concepts
ESSENTIAL QUESTIONS

- How does a thermostat work?
- How does a refrigerator keep food cold?
- What are the energy transformations in a car engine?

Vocabulary

heating appliance p. 215
thermostat p. 216
refrigerator p. 216
heat engine p. 218

g Multilingual eGlossary

Inquiry Concentrating Energy?

This power plant uses mirrors to focus light toward a tower. The tower then transforms some of the light into thermal energy. In what ways do we use thermal energy?

15 minutes

How can you transform energy?

If you rub your hands together very quickly, do they become warm? Where does the thermal energy that raises the temperature of your hands come from?

1. Read and complete a lab safety form.
2. Copy the table into your Science Journal.
3. Place a **thermometer strip** on the surface of a **block of wood.** Record the temperature after the thermometer stops changing color.
4. Rub the wood vigorously with **sandpaper** for 30 seconds. Quickly replace the thermometer, and record the temperature.
5. Repeat steps 3 and 4 on another part of the wood. This time, sand the wood for 60 seconds.

	Starting temp (°C)	Ending temp (°C)
30 s		
60 s		

Think About This

1. Did the temperature of the wood change? Why or why not?
2. When did the wood have the highest temperature? Explain this result.
3. **Key Concept** What energy transformations take place in this activity?

Thermal Energy Transformations

You can convert other forms of energy into thermal energy. Repeatedly stretching a rubber band makes it hot. Burning wood heats the air. A toaster gets hot when you turn it on.

You also can convert thermal energy into other forms of energy. Burning coal can generate electricity. Thermostats transform thermal energy into mechanical energy that switch heaters on and off. When you convert energy from one form to another, you can use the energy to perform useful tasks.

Remember that energy cannot be created or destroyed. Even though many devices transform energy from one form to another or transfer energy from one place to another, the total amount of energy does not change.

Heating Appliances

A device that converts electric energy into thermal energy is a **heating appliance.** Curling irons, coffeemakers, and clothes irons are some examples of heating appliances.

Other devices, such as computers and cell phones, also become warm when you use them. This is because some electric energy always is converted to thermal energy in an electronic device. However, the thermal energy that most electronic devices generate is not used for any purpose.

FOLDABLES®

Make a vertical four-tab book. Label it as shown. Use it to explain the energy transformation that occurs in each device.

Figure 14 The coil in a thermostat contains two different metals that expand at two different rates.

Thermostats

You might have heard the furnace in your house or in your classroom turn on in the winter. After the room warms, the furnace turns off. *A* **thermostat** *is a device that regulates the temperature of a system.* Kitchen refrigerators, toasters, and ovens are all equipped with thermostats.

WORD ORIGIN

thermostat
from Greek *therme*, meaning "heat"; and *statos*, meaning "a standing"

Most thermostats used in home heating systems contain a bimetallic coil. A bimetallic coil is made of two types of metal joined together and bent into a coil, as shown in **Figure 14.** The metal on the inside of the coil expands and contracts more than the metal on the outside of the coil. After the room warms, the thermal energy in the air causes the bimetallic coil to uncurl slightly. This moves a switch that turns off the furnace. As the room cools, the metal on the inside of the coil contracts more than the metal on the outside, curling the coil tighter. This moves the switch in the other direction, turning on the furnace.

 **Key Concept Check** How does the bimetallic coil in a thermostat respond to heating and cooling?

Refrigerators

A device that uses electric energy to transfer thermal energy from a cooler location to a warmer location is called a **refrigerator.** Recall that thermal energy naturally flows from a warmer area to a cooler area. The opposite might seem impossible. But, that is exactly how your refrigerator works. So, how does a refrigerator move thermal energy from its cold inside to the warm air outside? Pipes that surround the refrigerator are filled with a fluid, called a coolant, that flows through the pipes. Thermal energy from inside the refrigerator transfers to the coolant, keeping the inside of the refrigerator cold.

Vaporizing the Coolant

A coolant is a substance that evaporates at a low temperature. In a refrigerator, a coolant is pumped through pipes on the inside and the outside of the refrigerator. The coolant, which begins as a liquid, passes through an expansion valve and cools. As the cold gas flows through pipes inside the refrigerator, it absorbs thermal energy from the refrigerator compartment and vaporizes. The coolant gas becomes warmer, and the inside of the refrigerator becomes cooler.

Condensing the Coolant

The coolant flows to an electric compressor at the bottom of the refrigerator. Here, the coolant is compressed, or forced into a smaller space, which increases its thermal energy. Then, the gas is pumped through condenser coils. In the coils, the thermal energy of the gas is greater than that of the surrounding air. This causes thermal energy to flow from the coolant gas to the air behind the refrigerator. As thermal energy is removed from the gas, it condenses, or becomes liquid. Then, the liquid coolant is pumped up through the expansion valve. The cycle repeats.

 **Key Concept Check** How does a refrigerator keep food cold?

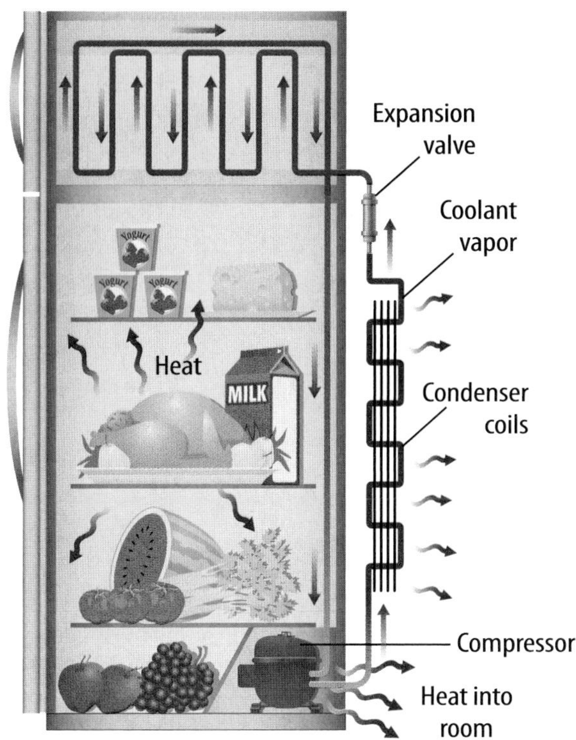

Figure 15 Coolant in a refrigerator moves thermal energy from inside to outside the refrigerator.

Inquiry MiniLab

10 minutes

Can thermal energy be used to do work?

You know you can raise the thermal energy of a substance by doing work on it. Is the opposite true? Can thermal energy cause something to move?

1. Read and complete a lab safety form.
2. Add 10 mL of water to a **100-mL beaker.**
3. Place a **small square of aluminum foil** over the top of the beaker.
4. Place the beaker on a **hot plate,** and turn it on. Observe the results and record them in your Science Journal.

Analyze and Conclude

1. **Infer** Is thermal energy used to do work in this lab? Explain your answer.
2. **Key Concept** Is thermal energy transformed into another form of energy in this experiment? If so, what is the other form of energy?

Internal Combustion Engine

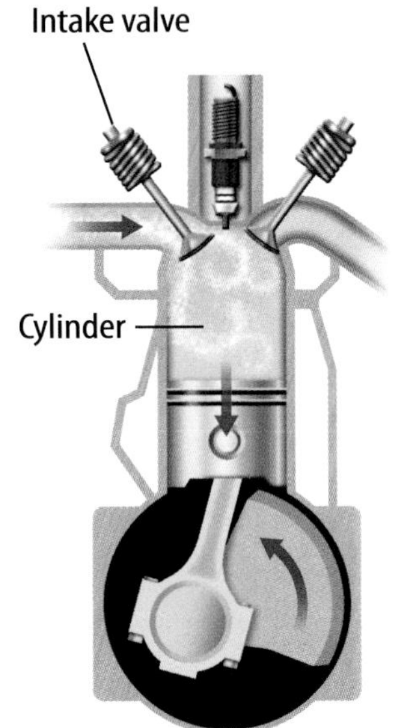

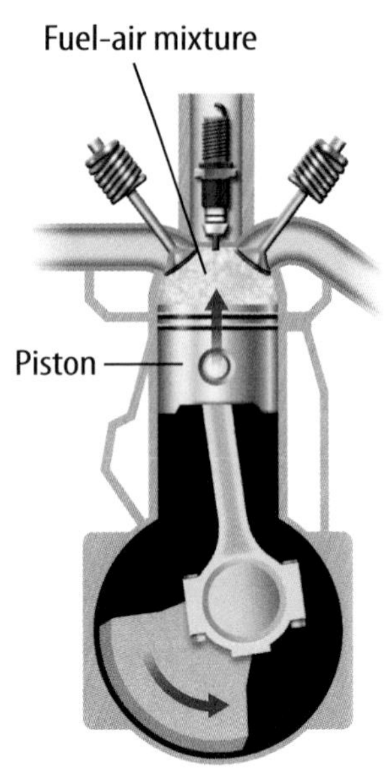

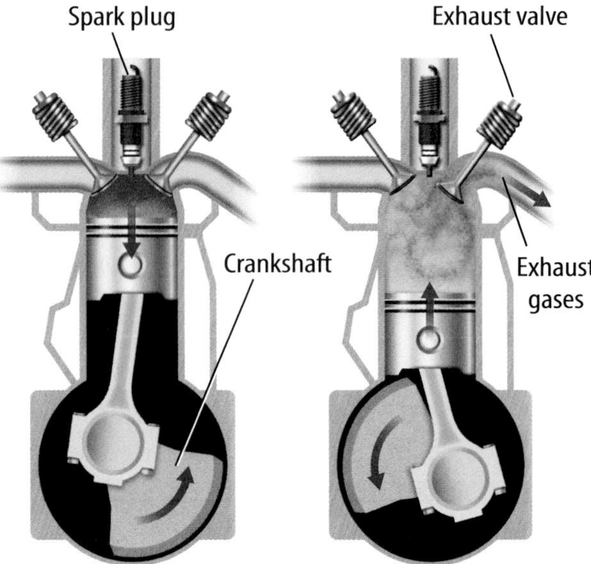

1. The intake valve opens as the piston moves downward, drawing a mixture of gasoline and air into the cylinder.
2. The intake valve closes as the piston moves upward, compressing the fuel-air mixture.
3. A spark plug ignites the fuel-air mixture. As the mixture burns, hot gases expand, pushing the piston down.
4. As the piston moves up, the exhaust valve opens, and the hot gases are pushed out of the cylinder.

Figure 16  Internal combustion engines transform the chemical energy from fuel to thermal energy, which then produces mechanical energy.

Heat Engines

A typical automobile engine is a heat engine. *A* **heat engine** *is a machine that converts thermal energy into mechanical energy.* When a heat engine converts thermal energy into mechanical energy, the mechanical energy moves the vehicle. Most cars, buses, boats, trucks, and lawn mowers use a type of heat engine called an internal combustion engine. **Figure 16** shows how one type of internal combustion engine converts thermal energy into mechanical energy.

Perhaps you have heard someone refer to a car as having a six-cylinder engine. A cylinder is a tube with a piston that moves up and down. At one end of the cylinder a spark ignites a fuel-air mixture. The ignited fuel-air mixture expands and pushes the piston down. This action occurs because the fuel's chemical energy converts to thermal energy. Some of the thermal energy immediately converts to mechanical energy.

A heat engine is not efficient. Most automobile engines only convert about 20 percent of the chemical energy in gasoline into mechanical energy. The remaining energy from the gasoline is lost to the environment.

Key Concept Check What is one form of energy that is output from a heat engine?

Lesson 3 Review

Assessment | Online Quiz

Visual Summary

A bimetallic coil inside a thermostat controls a switch that turns a heating or cooling device on or off.

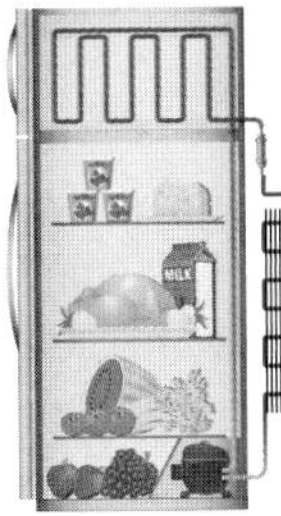

A refrigerator keeps food cold by moving thermal energy from the inside of the refrigerator out to the refrigerator's surroundings.

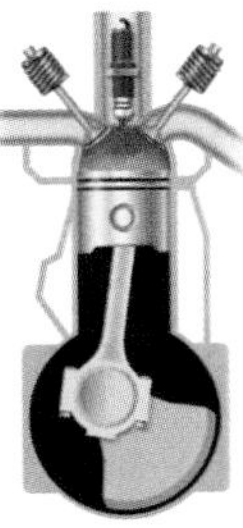

In a car engine, chemical energy in fuel is transformed into thermal energy. Some of this thermal energy is then transformed into mechanical energy.

FOLDABLES®

Use your lesson Foldable to review the lesson. Save your Foldable for the project at the end of the chapter.

What do you think NOW?

You first read the statements below at the beginning of the chapter.

5. Car engines create energy.

6. Refrigerators cool food by moving thermal energy from inside the refrigerator to the outside.

Did you change your mind about whether you agree or disagree with the statements? Rewrite any false statements to make them true.

Use Vocabulary

1. A ________ is a device that converts electric energy into thermal energy.
2. **Explain** how an internal combustion engine works.

Understand Key Concepts 

3. **Describe** the path of thermal energy in a refrigerator.
4. Which sequence describes the energy transformation in an automobile engine?
   - **A.** chemical→thermal→mechanical
   - **B.** thermal→kinetic→potential
   - **C.** thermal→mechanical→potential
   - **D.** thermal→chemical→mechanical
5. **Explain** how a thermostat uses electric energy, mechanical energy, and thermal energy.

Interpret Graphics

6. **Predict** Suppose you pointed a hair dryer at the device pictured below and turned on the hair dryer. What would happen?

7. **Sequence** Copy the graphic organizer below. Use it to show the steps involved in one cycle of an internal combustion engine.

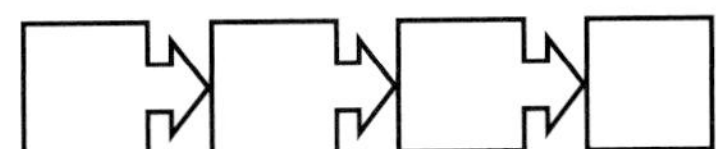

Critical Thinking

8. **Explain** how two of the devices you read about in this chapter could be used in one appliance.

Design an Insulated Container

Materials

aluminum foil

self-sealing plastic bag

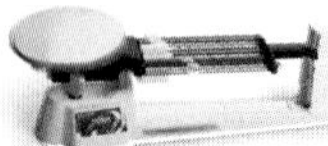

triple-beam balance

creative building materials

office supplies

Also Needed
frozen fruit pop, foam packing peanuts, rubber bands

Safety

Many refrigerated or frozen food products must be kept cold as they are transported long distances. Meat or fresh fruits might travel from South America to grocery stores in the United States. Imagine that you have been hired to design a container that will keep a frozen fruit pop from melting for as long as possible.

Ask a Question

How can you construct a container that will prevent a frozen fruit pop inside a plastic bag from melting? Think about thermal energy transfer by conduction, convection, and radiation. You will begin with a shoe box, but you can modify it in any way. Consider the materials you have available. Ask yourself what material you can bring from home that might slow the melting of a frozen fruit pop.

Make Observations

1. Read and complete a lab safety form.
2. In your Science Journal, write your ideas about
   - how you can you reduce the amount of thermal energy moving by conduction, convection, and radiation;
   - what materials you will use inside and outside your box;
   - what materials you will need to bring from home.
3. Outline the steps in preparing for your box. Have your teacher check your procedures. Decide who will obtain which materials before the next lab period. Design a logo for your container.
4. As a class, decide how many hours you will wait before checking the condition of your frozen fruit pop.

Form a Hypothesis

5. Formulate a hypothesis explaining why the materials you use inside your bag will be effective in insulating the frozen fruit pop. Remember, your hypothesis should be a testable explanation based on observations.

Test Your Hypothesis

6. On the second lab day, follow the steps you have outlined and prepare your container. Check it over one more time to be sure you have accounted for all ways that thermal energy could enter or leave the box.
7. Obtain a frozen fruit pop. Place it inside a self-sealing plastic bag. Seal the bag. Quickly measure and record its mass. Attach your logo and return the pop to the freezer.
8. On the third lab day, remove your frozen fruit pop from the freezer. Do not open the plastic bag. Place your frozen fruit pop in your container and seal it. Place your container in a location assigned by your teacher.
9. After the set amount of time, remove the fruit pop from the container. Open the plastic bag, and pour off any melted juice. Reseal the bag. Measure and record the mass.

Analyze and Conclude

10. **Calculate** What percentage of your fruit pop remained frozen? How long do you think it would take for the fruit pop to completely melt in your container? Justify your answer.
11. **Analyze** What are some possible ways thermal energy entered your bag? How could you improve the package on another try?
12. **The Big Idea** How would you modify your design to keep something hot inside the bag? Explain your answer.

Communicate Your Results

Make a class graph showing the percentages of the different frozen fruit pops remaining. Discuss why some packages were more or less effective.

Explore designs for portable coolers. What are the most effective portable packages that keep things hot or cold without external cooling or heating?

Lab Tips

- Keep in mind that you are trying to keep thermal energy out of the package.
- The length of the test time you decide on should be long enough to allow some of the fruit pop to melt.

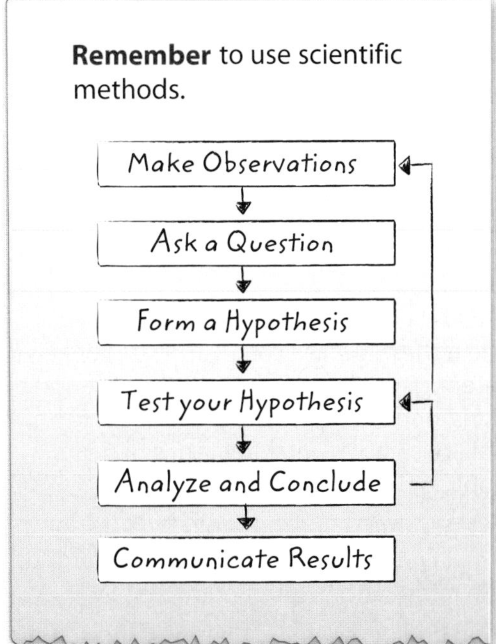

Chapter 6 Study Guide

Thermal energy can be transferred by conduction, radiation, and convection. Thermal energy also can be transformed into other forms of energy and used in devices such as thermostats, refrigerators, and automobile engines.

Key Concepts Summary

Lesson 1: Thermal Energy, Temperature, and Heat

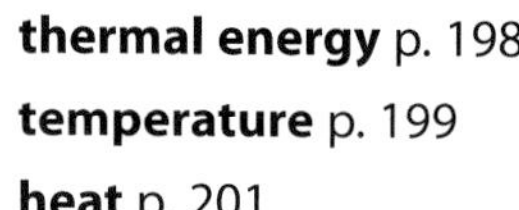

- The **temperature** of a material is the average kinetic energy of the particles that make up the material.
- **Heat** is the movement of **thermal energy** from a material or area with a higher temperature to a material or area with a lower temperature.
- When a material is heated, the material's temperature changes.

Vocabulary

thermal energy p. 198
temperature p. 199
heat p. 201

Lesson 2: Thermal Energy Transfers

- When a material has a low **specific heat,** transferring a small amount of energy to the material increases its temperature significantly.
- When a material is heated, the thermal energy of the material increases and the material expands.
- Thermal energy can be transferred by **conduction, radiation,** or **convection.**

Vocabulary

radiation p. 205
conduction p. 206
thermal conductor p. 206
thermal insulator p. 206
specific heat p. 207
thermal contraction p. 208
thermal expansion p. 208
convection p. 210
convection current p. 211

Lesson 3: Using Thermal Energy

- The two different metals in a bimetallic coil inside a **thermostat** expand and contract at different rates. The bimetallic coil curls and uncurls, depending on the thermal energy of the air, pushing a switch that turns a heating or cooling device on or off.
- A **refrigerator** keeps food cold by moving thermal energy from inside the refrigerator out to the refrigerator's surroundings.
- In a car engine, chemical energy in fuel is transformed into thermal energy. Some of this thermal energy is then transformed into mechanical energy.

Vocabulary

heating appliance p. 215
thermostat p. 216
refrigerator p. 216
heat engine p. 218

- Personal Tutor
- Vocabulary eGames
- Vocabulary eFlashcards

FOLDABLES® Chapter Project

Assemble your lesson Foldables as shown to make a Chapter Project. Use the project to review what you have learned in this chapter.

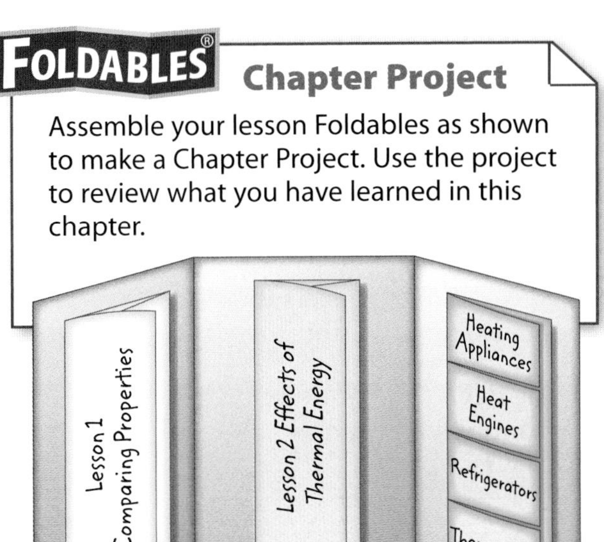

Use Vocabulary

1. When you increase the ________ of a cup of hot cocoa, you increase the average kinetic energy of the particles that make up the hot cocoa.
2. The increase in volume of a material when heated is ________.
3. A(n) ________ is used to control the temperature in a room.
4. Thermal energy is transferred by ________ between two objects that are touching.
5. A fluid moving in a circular pattern because of convection is a ________.
6. Define *heating appliance* in your own words.

Link Vocabulary and Key Concepts

Interactive Concept Map

Copy this concept map, and then use vocabulary terms from the previous page to complete the concept map.

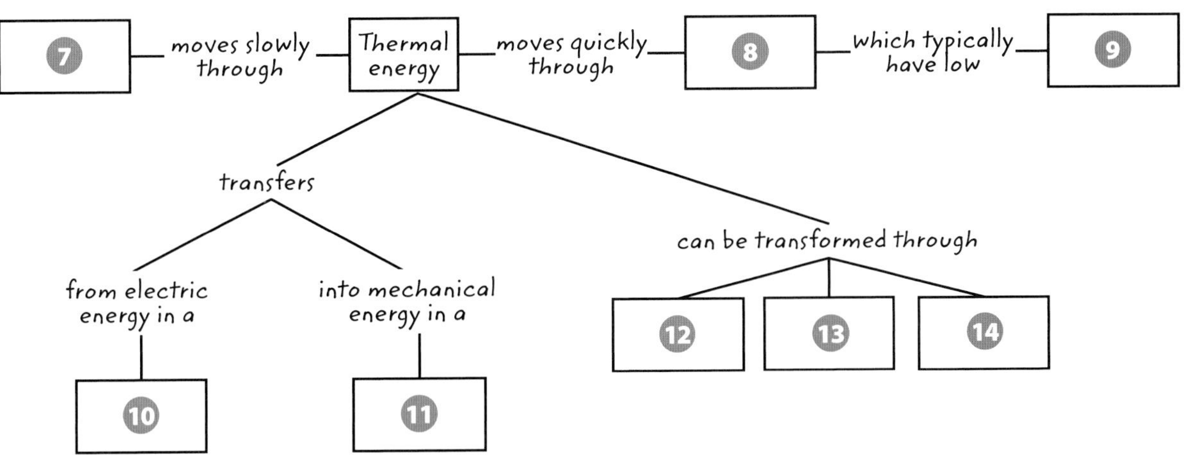

Chapter 6 Review

Understand Key Concepts 

1 Which would decrease a material's thermal energy?

A. heating the material
B. increasing the kinetic energy of the particles that make up the material
C. increasing the temperature of the material
D. moving the material to a location where the temperature is lower

2 You put a metal spoon in a bowl of hot soup. Why does the spoon feel hotter than the outside of the bowl?

A. The bowl is a better conductor than the spoon.
B. The bowl has a lower specific heat than the spoon.
C. The spoon is a good thermal insulator.
D. The spoon transfers thermal energy better than the bowl does.

3 In the picture to the right, thermal energy moves from the

A. glass to the air.
B. lemonade to the air.
C. ice to the lemonade.
D. air to the lemonade.

4 Which has the lowest specific heat?

A. an object that is made out of metal
B. an object that does not transfer thermal energy easily
C. an object with electrons that do not move easily
D. an object that requires a lot of energy to change its temperature

5 Which does NOT occur in an internal combustion engine?

A. Most of the thermal energy is wasted.
B. Thermal energy forces the piston downward.
C. Thermal energy is converted into chemical energy.
D. Thermal energy is converted into mechanical energy.

6 Which statement about radiation is correct?

A. In solids, radiation transfers electromagnetic energy, but not thermal energy.
B. Cooler objects radiate the same amount of thermal energy as warmer objects.
C. Radiation occurs in fluids such as gas and water, but not solids such as metals.
D. Radiation transfers thermal energy from the Sun to Earth.

7 The device below detects an increase in room temperature as

A. an increase in thermal energy causes a bimetallic coil to curl.
B. an increase in thermal energy causes a bimetallic coil to uncurl.
C. a switch causes a bimetallic coil to curl.
D. a switch causes a bimetallic coil to uncurl

8 Which is the lowest temperature?

A. 0°C
B. 0°F
C. 32°F
D. 273 K

9 Which energy conversion typically occurs in a heating appliance?

A. chemical energy to thermal energy
B. electric energy to thermal energy
C. thermal energy to chemical energy
D. thermal energy to mechanical energy

Online Test Practice

Critical Thinking

10 **Compare** A swimming pool with a temperature of 30°C has more thermal energy than a cup of soup with a temperature of 60°C. Explain why this is so.

11 **Contrast** A spoon made of aluminum and a spoon made of steel have the same mass. The aluminum spoon has a higher specific heat than the steel spoon. Which spoon becomes hotter when placed in a pan of boiling water?

12 **Describe** How do convection currents influence Earth's climate?

13 **Diagram** A room has a heater on one side and an open window letting in cool air on the opposite side. Diagram the convection current in the room. Label the warm air and the cool air.

14 **Evaluate** When engineers build bridges, they separate sections of the roadway with expansion joints such as the one below that allow movement between the sections. Why are expansion joints necessary?

15 **Explain** Why is conduction slower in a gas than in a liquid or a solid?

Writing in Science

16 **Research** various types of heat engines that have been developed throughout history. Write 3–5 paragraphs explaining the energy transformations in one of these engines.

REVIEW THE BIG IDEA

17 **Describe** each of the three ways thermal energy can be transferred. Give an example of each.

18 What do the different colors in this photograph indicate?

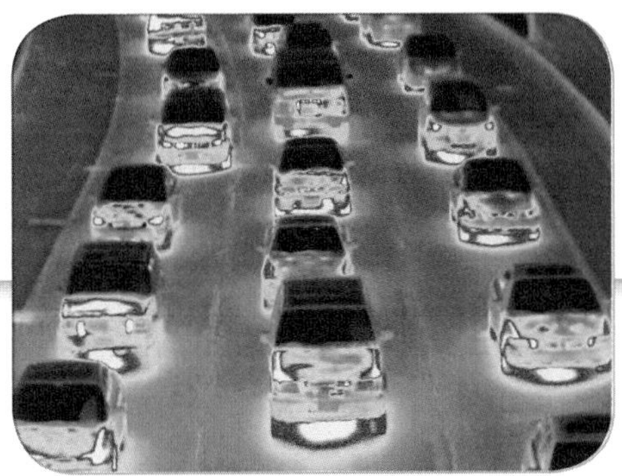

Math Skills

Math Practice

Convert Between Temperature Scales

19 If water in a bath is at 104°F, then what is the temperature of the water in degrees Celsius?

20 Convert −40°C to degrees Fahrenheit.

Standardized Test Practice

Record your answers on the answer sheet provided by your teacher or on a sheet of paper.

Multiple Choice

1 Which statement describes the thermal energy of an object?

A kinetic energy of particles + potential energy of particles

B kinetic energy of particles ÷ number of particles

C potential energy of particles ÷ number of particles

D kinetic energy of particles ÷ (kinetic energy of particles + potential energy of particles)

2 Which term describes a transfer of thermal energy?

A heat

B specific heat

C temperature

D thermal energy

Use the figures below to answer question 3.

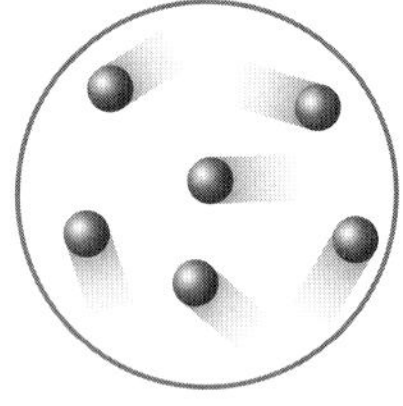

Sample X

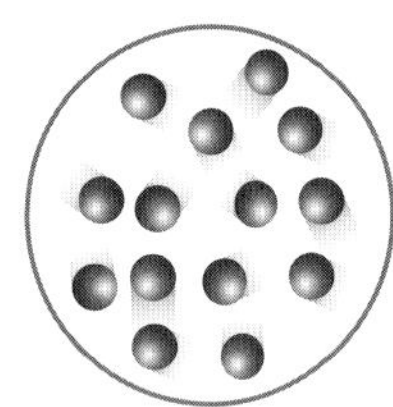

Sample Y

3 The figures show two different samples of air. In what way do they differ?

A Sample X is at a higher temperature than sample Y.

B Sample X has a higher specific heat than sample Y.

C Particles of sample Y have a higher average kinetic energy than those of sample X.

D Particles of sample Y have a higher average thermal energy than those of sample X.

Use the table below to answer question 4.

Material	Specific Heat (in J/g•K)
Air	1.0
Copper	0.4
Water	4.2
Wax	2.5

4 The table shows the specific heat of four materials. Which statement can be concluded from the information in the table?

A Copper is a thermal insulator.

B Wax is a thermal conductor.

C Air takes the most thermal energy to change its temperature.

D Water takes the most thermal energy to change its temperature.

5 Which term describes what happens to a cold balloon when placed in a hot car?

A thermal conduction

B thermal contraction

C thermal expansion

D thermal insulation

6 A girl stirs soup with a metal spoon. Which process causes her hand to get warmer?

A conduction

B convection

C insulation

D radiation

7 In a thermostat's coil, what causes the two metals in the strip to curl and uncurl?

A They contract at the same rate when cooled.

B They expand at different rates when heated.

C They have the same specific heat.

D They melt at different temperatures.

Use the figure to below to answer questions 8–10.

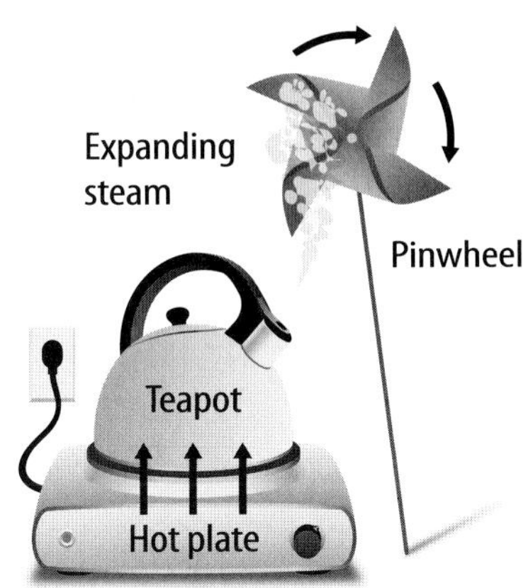

8 Which term describes the transfer of thermal energy between the hot plate and the teapot?

A conduction

B convection

C insulation

D radiation

9 Which energy transformations are taking place in this system?

A electrical → thermal → chemical

B electrical → thermal → mechanical

C thermal → electrical → chemical

D thermal → electrical → mechanical

10 What kind of machine is represented by the hot plate, the teapot, the steam, and the pinwheel working together?

A bimetallic coil

B heat engine

C refrigerator

D thermostat

Constructed Response

Use the figure to answer questions 11 and 12.

11 The foam cooler and the metal pan both contain ice. Describe the energy transfers that cause the ice to melt in each container.

12 After 1 hour, the ice in the metal pan had melted more than the ice in the foam cooler. What is it about the containers that could explain the difference in the melting rates?

13 What causes the air around a refrigerator to become warmer as the refrigerator is cooling the air inside it?

14 How does a car's internal combustion engine convert thermal energy to mechanical energy?

NEED EXTRA HELP?														
If You Missed Question...	1	2	3	4	5	6	7	8	9	10	11	12	13	14
Go to Lesson...	1	1	1	2	2	2	3	2	3	3	2	2	3	3

Chapter 7

Foundations of Chemistry

THE BIG IDEA **What is matter, and how does it change?**

Inquiry Why does it glow?

This siphonophore (si FAW nuh fawr) lives in the Arctic Ocean. Its tentacles have a very powerful sting. However, the most obvious characteristic of this organism is the way it glows.

- What might cause the siphonophore to glow?
- How do you think its glow helps the siphonophore survive?
- What changes happen in the matter that makes up the organism?

Get Ready to Read

What do you think?

Before you read, decide if you agree or disagree with each of these statements. As you read this chapter, see if you change your mind about any of the statements.

1. The atoms in all objects are the same.
2. You cannot always tell by an object's appearance whether it is made of more than one type of atom.
3. The weight of a material never changes, regardless of where it is.
4. Boiling is one method used to separate parts of a mixture.
5. Heating a material decreases the energy of its particles.
6. When you stir sugar into water, the sugar and water evenly mix.
7. When wood burns, new materials form.
8. Temperature can affect the rate at which chemical changes occur.

Your one-stop online resource

connectED.mcgraw-hill.com

WebQuest

Assessment

Concepts in Motion

Multilingual eGlossary

Lesson 1

Reading Guide

Key Concepts
ESSENTIAL QUESTIONS

- What is a substance?
- How do atoms of different elements differ?
- How do mixtures differ from substances?
- How can you classify matter?

Vocabulary

matter p. 231
atom p. 231
substance p. 233
element p. 233
compound p. 234
mixture p. 235
heterogeneous mixture p. 235
homogeneous mixture p. 235
dissolve p. 235

Multilingual eGlossary

Video

- BrainPOP®
- Science Video
- What's Science Got to do With It?

Classifying Matter

Inquiry Making Green?

You probably have mixed paints together. Maybe you wanted green paint and had only yellow paint and blue paint. Perhaps you watched an artist mixing several tints get the color he or she needed. In all these instances, the final color came from mixing colors together and not from changing the color of a paint.

15 minutes

How do you classify matter?

An object made of paper bound together might be classified as a book. Pointed metal objects might be classified as nails or needles. How can you classify an item based on its description?

1. Read and complete a lab safety form.
2. Place the **objects** on a table. Discuss how you might separate the objects into groups with these characteristics:
   a. Every object is the same and has only one part.
   b. Every object is the same but is made of more than one part.
   c. Individual objects are different. Some have one part, and others have more than one part.
3. Identify the objects that meet the requirements for group *a,* and record them in your Science Journal. Repeat with groups *b* and *c.* Any object can be in more than one group.

Think About This

1. Does any object from the bag belong in all three of the groups (*a, b,* and *c*)? Explain.
2. What objects in your classroom would fit into group *b?*
3. Key Concept What descriptions would you use to classify items around you?

Understanding Matter

Have you ever seen a rock like the one in **Figure 1**? Why are different parts of the rock different in color? Why might some parts of the rock feel harder than other parts? The parts of the rock look and feel different because they are made of different types of matter. **Matter** *is anything that has mass and takes up space.* If you look around, you will see many types of matter. If you are in a classroom, you might see things made of metal, wood, or plastic. If you go to a park, you might see trees, soil, or water in a pond. If you look up at the sky, you might see clouds and the Sun. All of these things are made of matter.

Everything you can see is matter. However, some things you cannot see also are matter. Air, for example, is matter because it has mass and takes up space. Sound and light are not matter. Forces and energy also are not matter. To decide whether something is matter, ask yourself if it has mass and takes up space.

An **atom** *is a small particle that is a building block of matter.* In this lesson, you will explore the parts of an atom and read how atoms can differ. You will also read how different arrangements of atoms make up the many types of matter.

WORD ORIGIN
matter
from Latin *materia,* meaning "material, stuff"

Figure 1 You can see different types of matter in this rock.

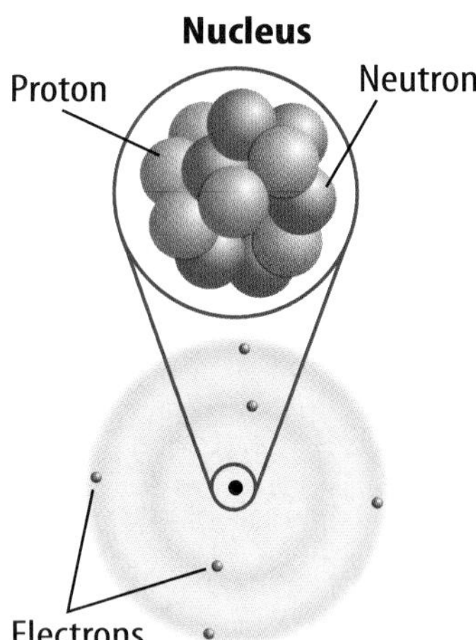

Figure 2 An atom has electrons moving in an area outside a nucleus. Protons and neutrons make up the nucleus.

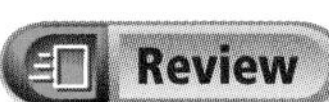 Review **Personal Tutor**

Atoms

To understand why there are so many types of matter, it helps if you first learn about the parts of an atom. Look at the diagram of an atom in **Figure 2.** At the center of an atom is a nucleus. Protons, which have a positive charge, and neutrons, which have a neutral charge, make up the nucleus. Negatively charged particles, or electrons, move quickly throughout an area around the nucleus called the electron cloud.

 **Reading Check** What are the parts of an atom?

Not all atoms have the same number of protons, neutrons, and electrons. Atoms that have different numbers of protons differ in their properties. You will read more about the differences in atoms on the next page.

An atom is almost too small to imagine. Think about how thin a human hair is. The diameter of a human hair is about a million times greater than the diameter of an atom. In addition, an atom is about 10,000 times wider than its nucleus! Even though atoms are so tiny, they determine the properties of the matter they compose.

Inquiry MiniLab

20 minutes

How can you model an atom? 

How can you model an atom out of its three basic parts?

1. Read and complete a lab safety form.
2. Twist the ends of a piece of **florist wire** together to form a ring. Attach two **wires** across the ring to form an *X*.
3. Use **double-sided tape** to join the **large pom-poms** (protons and neutrons), forming a nucleus. Hang the nucleus from the center of the *X* with **fishing line.**
4. Use fishing line to suspend each **small pom-pom** (electron) from the ring so they surround the nucleus.
5. Suspend your model as instructed by your teacher.

Analyze and Conclude

1. **Infer** Based on your model, what can you infer about the relative sizes of protons, neutrons, and electrons?
2. **Model** Why is it difficult to model the location of electrons?
3. **Key Concept** Compare your atom with those of other groups. How do they differ?

Substances

You can see that atoms make up most of the matter on Earth. Atoms can combine and arrange in millions of different ways. In fact, these different combinations and arrangements of atoms are what makes up the various types of matter. There are two main classifications of matter—substances and mixtures.

A **substance** *is matter with a composition that is always the same.* This means that a given substance is always make up of the same combination(s) of atoms. Aluminum, oxygen, water, and sugar are examples of substances. Any sample of aluminum is always made up of the same type of atoms, just as samples of oxygen, sugar, and water each are always made of the same combinations of atoms. To gain a better understanding of what makes up substances, let's take a look at the two types of substances—elements and compounds.

Key Concept Check What is a substance?

Elements

Look at the periodic table of elements on the inside back cover of this book. The substances oxygen and aluminum are on the table. They are both elements. *An* **element** *is a substance that consists of just one type of atom.* Because there are about 115 known elements, there are about 115 different types of atoms. Each type of atom contains a different number of protons in its nucleus. For example, each aluminum atom has 13 protons in its nucleus. The number of protons in an atom is the atomic number of the element. Therefore, the atomic number of aluminum is 13, as shown in **Figure 3.**

The atoms of most elements exist as individual atoms. For example, a roll of pure aluminum foil consists of trillions of individual aluminum atoms. However, the atoms of some elements usually exist in groups. For example, the oxygen atoms in air exist in pairs. Whether the atoms of an element exist individually or in groups, each element contains only one type of atom. Therefore, its composition is always the same.

Key Concept Check How do atoms of different elements differ?

Figure 3 Each element on the periodic table consists of just one type of atom.

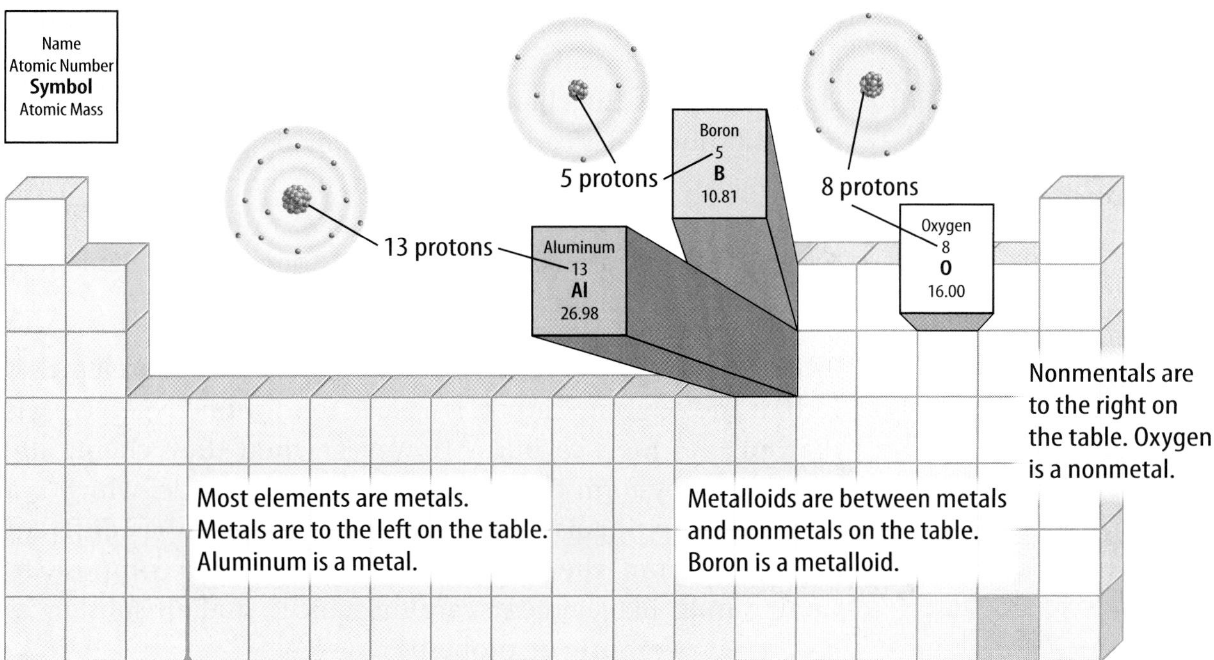

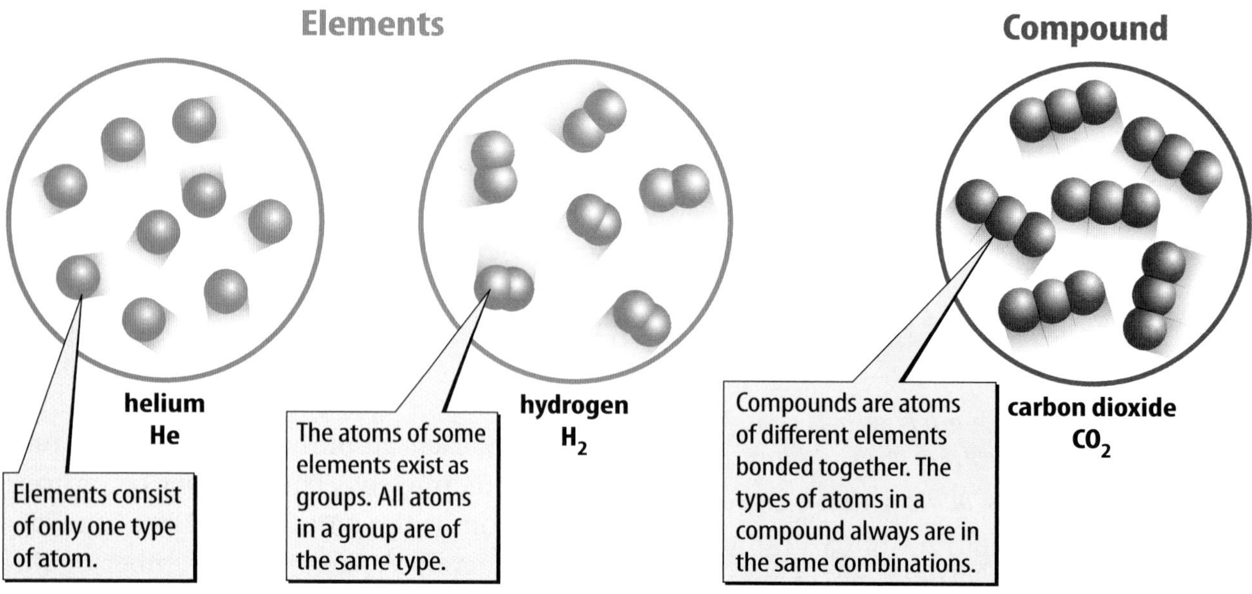
Elements
Compound
helium
He
Elements consist of only one type of atom.
The atoms of some elements exist as groups. All atoms in a group are of the same type.
hydrogen
H_2
Compounds are atoms of different elements bonded together. The types of atoms in a compound always are in the same combinations.
carbon dioxide
CO_2

▲ **Figure 4** If a substance contains only one type of atom, it is an element. If it contains more than one type of atom, it is a compound.

Figure 5 Carbon dioxide is a compound composed of carbon and oxygen atoms. ▼

Review
Personal Tutor

CO_2
This subscript means there are two oxygen atoms bonded to one carbon atom.

Compounds

Water is a substance, but it is not an element. It is a compound. *A* **compound** *is a type of substance containing atoms of two or more different elements chemically bonded together.* As shown in **Figure 4,** carbon dioxide (CO_2) is also a compound. It consists of atoms of two different elements, carbon (C) and oxygen (O), bonded together. Carbon dioxide is a substance because the C and the O atoms are always combined in the same way.

Chemical Formulas The combination of symbols and numbers that represents a compound is called a chemical formula. Chemical formulas show the different atoms that make up a compound, using their element symbols. Chemical formulas also help explain how the atoms combine. As illustrated in **Figure 5,** CO_2 is the chemical formula for carbon dioxide. The formula shows that carbon dioxide is made of C and O atoms. The small *2* is called a subscript. It means that two oxygen atoms and one carbon atom form carbon dioxide. If no subscript is written after a symbol, one atom of that element is present in the chemical formula.

Properties of Compounds Think again about the elements carbon and oxygen. Carbon is a black solid, and oxygen is a gas that enables fuels to burn. However, when they chemically combine, they form the compound carbon dioxide, which is a gas used to extinguish fires. A compound often has different properties from the individual elements that compose it. Compounds, like elements, are substances, and all substances have their own unique properties.

ACADEMIC VOCABULARY
unique
(adjective) having nothing else like it

Mixtures

Another classification of matter is mixtures. *A* **mixture** *is matter that can vary in composition.* Mixtures are combinations of two or more substances that are physically blended together. The amounts of the substances can vary in different parts of a mixture and from mixture to mixture. Think about sand mixed with water at the beach. The sand and the water do not bond together. Instead, they form a mixture. The substances in a mixture do not combine chemically. Therefore, they can be separated by physical methods, such as filtering.

Heterogeneous Mixtures

Mixtures can differ depending on how well the substances that make them up are mixed. Sand and water at the beach form a mixture, but the sand is not evenly mixed throughout the water. Therefore, sand and water form a heterogeneous mixture. *A* **heterogeneous mixture** *is a type of mixture in which the individual substances are not evenly mixed.* Because the substances in a heterogeneous mixture are not evenly mixed, two samples of the same mixture can have different amounts of the substances, as shown in **Figure 6.** For example, if you fill two buckets with sand and water at the beach, one bucket might have more sand in it than the other.

Homogeneous Mixtures

Unlike a mixture of water and sand, the substances in mixtures such as apple juice, air, or salt water are evenly mixed. *A* **homogeneous mixture** *is a type of mixture in which the individual substances are evenly mixed.* In a homogeneous mixture, the particles of individual substances are so small and well-mixed that they are not visible, even with most high-powered microscopes.

A homogeneous mixture also is known as a solution. In a solution, the substance present in the largest amount is called the solvent. All other substances in a solution are called solutes. The solutes dissolve in the solvent. *To* **dissolve** *means to form a solution by mixing evenly.* Because the substances in a solution, or homogeneous mixture, are evenly mixed, two samples from a solution will have the same amounts of each substance. For example, imagine pouring two glasses of apple juice from the same container. Each glass will contain the same substances (water, sugar, and other substances) in the same amounts. However, because apple juice is a mixture, the amounts of the substances from one container of apple juice to another might vary.

Key Concept Check How do mixtures differ from substances?

Figure 6 Types of mixtures differ in how evenly their substances are mixed.

Heterogeneous Mixture	Homogeneous Mixture
• The individual substances are not evenly mixed. • Different samples of a given heterogeneous mixture can have different combinations of the same substances.	• The individual substances are evenly mixed. • Different samples of a given homogeneous mixture will have the same combinations of the same substances.

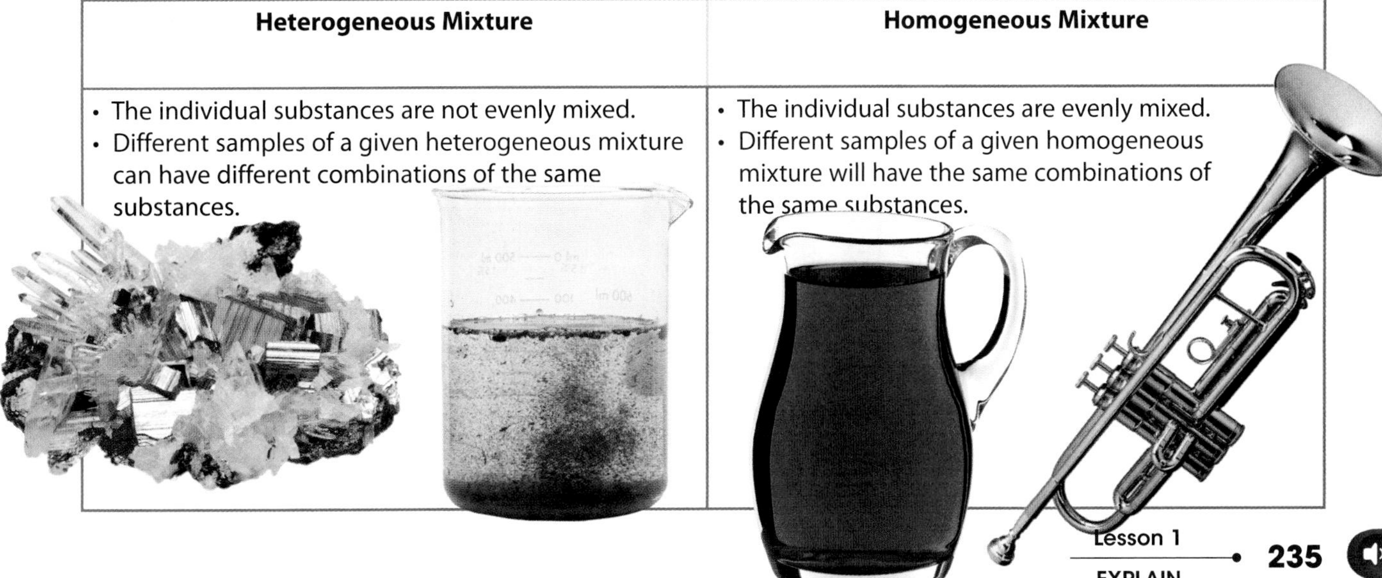

Use three sheets of copy paper to make a layered Foldable. Cut and label the tabs as illustrated. Use this Foldable to summarize the lesson.

Compounds v. Solutions

If you have a glass of pure water and a glass of salt water, can you tell which is which just by looking at them? You cannot. Both the compound (water) and the solution (salt water) appear identical. How do compounds and solutions differ?

Because water is a compound, its composition does not vary. Pure water is always made up of the same atoms in the same combinations. Therefore, a chemical formula can be used to describe the atoms that make up water (H_2O). Salt water is a homogeneous mixture, or solution. The solute (NaCl) and the solvent (H_2O) are evenly mixed but are not bonded together. Adding more salt or more water only changes the relative amounts of the substances. In other words, the composition varies. Because composition can vary in a mixture, a chemical formula cannot be used to describe mixtures.

Summarizing Matter

You have read in this lesson about classifying matter by the arrangement of its atoms. **Figure 7** is a summary of this classification system.

Key Concept Check How can you classify matter?

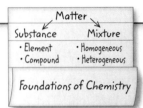

Figure 7 Scientists classify matter according to the arrangement of the atoms that make up the matter.

Classifying Matter

Matter
- Anything that has mass and takes up space
- Most matter on Earth is made up of atoms.
- Two classifications of matter: substances and mixtures

Substances
- Matter with a composition that is always the same
- Two types of substances: elements and compounds

Element
- Consists of just one type of atom
- Organized on the periodic table
- Each element has a chemical symbol.

Compound
- Two or more types of atoms bonded together
- Properties are different from the properties of the elements that make it up
- Each compound has a chemical formula.

Substances physically combine to form mixtures.

Mixtures can be separated into substances by physical methods.

Mixtures
- Matter that can vary in composition
- Substances are not bonded together.
- Two types of mixtures: heterogeneous and homogeneous

Heterogeneous Mixture
- Two or more substances unevenly mixed
- Different substances are visible by an unaided eye or a microscope.

Homogeneous Mixture—Solution
- Two or more substances evenly mixed
- Different substances cannot be seen even by a microscope.

Lesson 1 Review

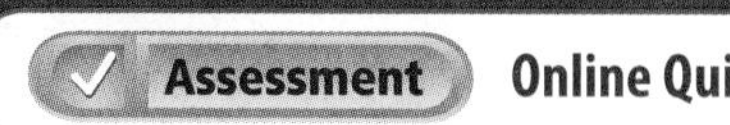

Visual Summary

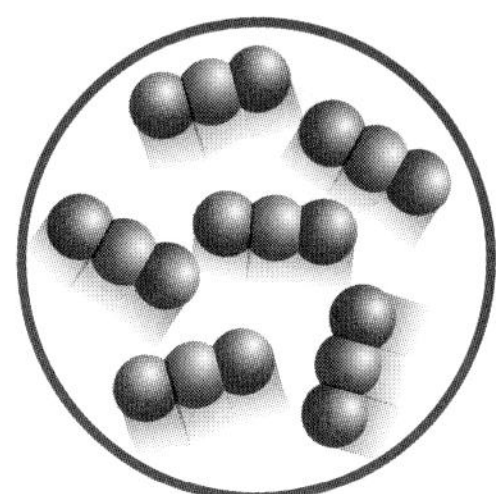

A substance has the same composition throughout. A substance is either an element or a compound.

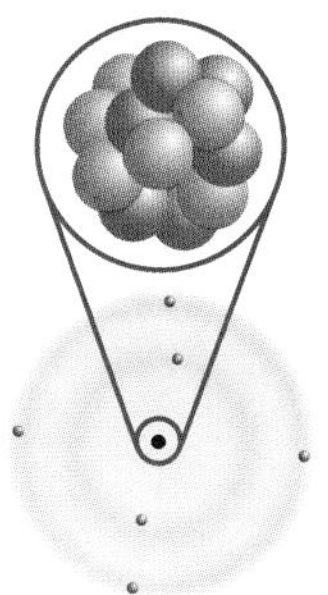

An atom is the smallest part of an element that has its properties. Atoms contain protons, neutrons, and electrons.

The substances in a mixture are not chemically combined. Mixtures can be either heterogeneous or homogeneous.

FOLDABLES®

Use your lesson Foldable to review the lesson. Save your Foldable for the project at the end of the chapter.

What do you think NOW?

You first read the statements below at the beginning of the chapter.

1. The atoms in all objects are the same.
2. You cannot always tell by an object's appearance whether it is made of more than one type of atom.

Did you change your mind about whether you agree or disagree with the statements? Rewrite any false statements to make them true.

Use Vocabulary

1. Substances and mixtures are two types of ________.
2. **Use the term** *atom* in a complete sentence.
3. **Define** *dissolve* in your own words.

Understand Key Concepts

4. **Explain** why aluminum is a substance.
5. The number of ________ always differs in atoms of different elements.
   - **A.** electrons
   - **B.** protons
   - **C.** neutrons
   - **D.** nuclei
6. **Distinguish** between a heterogeneous mixture and a homogeneous mixture.
7. **Classify** Which term describes matter that is a substance made of different kinds of atoms bonded together?

Interpret Graphics

8. **Describe** what each letter and number means in the chemical formula below.

$$C_6H_{12}O_6$$

9. **Organize Information** Copy and fill in the graphic organizer below to classify matter by the arrangement of its atoms.

Type of Matter	Description

Critical Thinking

10. **Reorder** the elements aluminum, oxygen, fluorine, calcium, and hydrogen from the least to the greatest number of protons. Use the periodic table if needed.
11. **Evaluate** this statement: Substances are made of two or more types of elements.

HOW IT WORKS

U.S. Mint

How Coins are Made

In 1793, the U.S. Mint produced more than 11,000 copper pennies and put them into circulation. Soon after, gold and silver coins were introduced as well. Early pennies were made of 95 percent copper and 5 percent zinc. Today's penny contains much more zinc than copper and is much less expensive to produce. Quarters, dimes, and nickels, once made of silver, are now made of copper-nickel alloy.

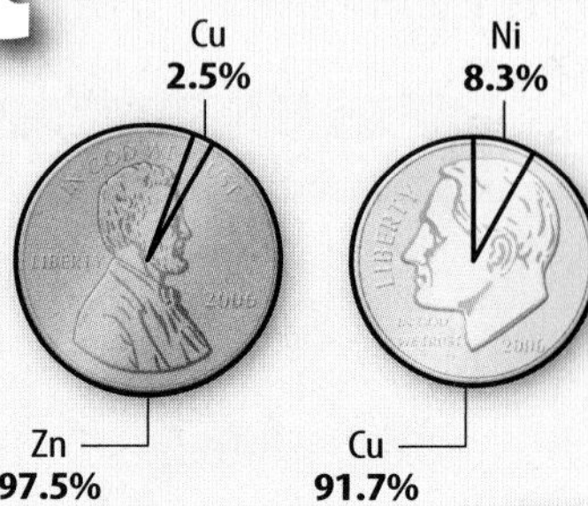

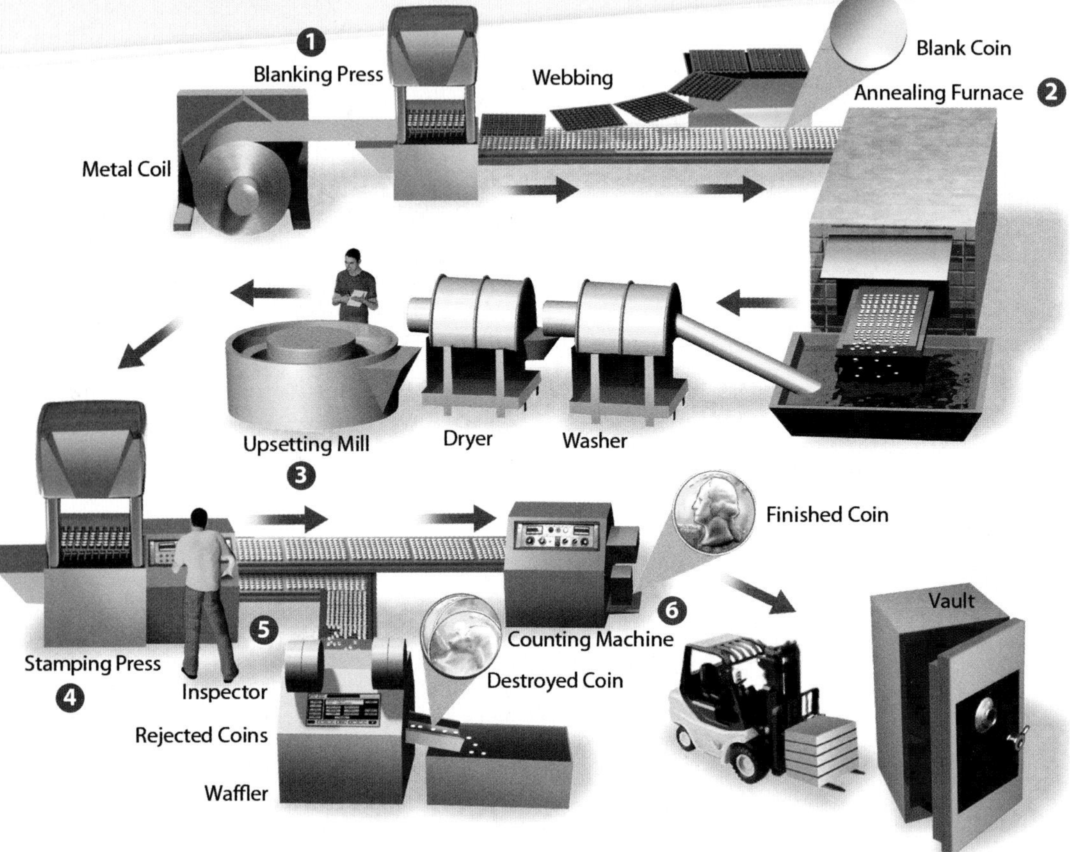

How Coins Are Made

1. **Blanking** For nickels, dimes, quarters, half-dollars, and coin dollars, a strip of 13-inch-wide metal is fed through a blanking press, which punches out round discs called blanks. The leftover webbing strip is saved for recycling. Ready-made blanks are purchased for making the penny.
2. **Annealing, Washing, Drying** Blanks are softened in an annealing furnace, which makes the metal less brittle. The blanks are then run through a washer and a dryer.
3. **Upsetting** Usable blanks are put through an upsetting mill, which creates a rim around the edges of each blank.
4. **Striking** The blanks then go to the stamping press, where they are imprinted with designs and inscriptions.
5. **Inspection** Once blanks leave the stamping press, inspectors check a few coins from each batch. Coins that are defective go to the waffler in preparation for recycling.
6. **Counting and Bagging** A machine counts the finished coins then drops them into large bags that are sealed shut. The coins are then taken to storage before being shipped to Federal Reserve Banks and then to your local bank.

It's Your Turn

COMPARE Collect a variety of coins that includes both older and current coins. Observe and compare their properties. Using the dates of the coins' production, utilize library or Internet sources to research the composition of metals used.

Lesson 2

Reading Guide

Key Concepts
ESSENTIAL QUESTIONS

- What are some physical properties of matter?
- How are physical properties used to separate mixtures?

Vocabulary

physical property p. 240
mass p. 242
density p. 243
solubility p. 244

Multilingual eGlossary

Video Science Video

Physical Properties

Inquiry Panning by Properties?

The man lowers his pan into the waters of an Alaskan river and scoops up a mixture of water, sediment, and hopefully gold. As he moves the pan in a circle, water sloshes out of it. If he is careful, gold will remain in the pan after the water and sediment are gone. What properties of water, sediment, and gold enable this man to separate this mixture?

Launch Lab

15 minutes

Can you follow the clues?

Clues are bits of information that help you solve a mystery. In this activity, you will use clues to help identify an object in the classroom.

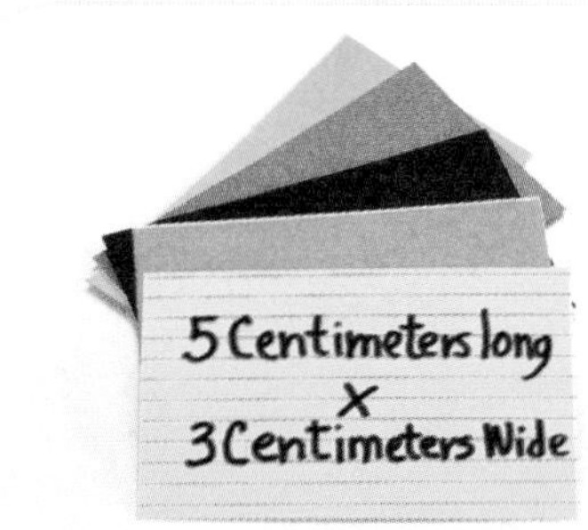

1. Read and complete a lab safety form.
2. Select one **object** in the room. Write a different clue about the object on each of five **index cards.** Clues might include one or two words that describe the object's color, size, texture, shape, or any property you can observe with your senses.
3. Stack your cards face down. Have your partner turn over one card and try to identify the object. Respond either "yes" or "no."
4. Continue turning over cards until your partner identifies your object or runs out of cards. Repeat for your partner's object.

Think About This

1. What kind of clues are the most helpful in identifying an object?
2. How would your clues change if you were describing a substance, such as iron or water, rather than an object?
3. **Key Concept** How do you think you use similar clues in your daily life?

Physical Properties

Review Vocabulary
property
a characteristic used to describe something

As you read in Lesson 1, the arrangement of atoms determines whether matter is a substance or a mixture. The arrangement of atoms also determines the properties of different types of matter. Each element and compound has a unique set of properties. When substances mix together and form mixtures, the properties of the substances that make up the mixture are still present.

You can observe some properties of matter, and other properties can be measured. For example, you can see that gold is shiny, and you can find the mass of a sample of iron. Think about how you might describe the different substances and mixtures in the photo on the previous page. Could you describe some of the matter in the photo as a solid or a liquid? Why do the water and the rocks leave the pan before the gold does? Could you describe the mass of the various items in the photo? Each of these questions asks about the physical properties of matter. *A* **physical property** *is a characteristic of matter that you can observe or measure without changing the identity of the matter.* There are many types of physical properties, and you will read about some of them in this lesson.

States of Matter

How do aluminum, water, and air differ? Recall that aluminum is an element, water is a compound, and air is a mixture. How else do these three types of matter differ? At room temperature, aluminum is a solid, water is a liquid, and air is a gas. Solids, liquids, and gases are called states of matter. The state of matter is a physical property of matter. Substances and mixtures can be solids, liquids, or gases. For example, water in the ocean is a liquid, but water in an iceberg is a solid. In addition, water vapor in the air above the ocean is a gas.

Did you know that the particles, or atoms and groups of atoms, that make up all matter are constantly moving and are attracted to each other? Look at your pencil. It is made up of trillions of moving particles. Every solid, liquid, and gas around you is made up of moving particles that attract one another. What makes some matter a solid and other matter a liquid or a gas? It depends on how close the particles in the matter are to one another and how fast they move, as shown in **Figure 8.**

Reading Check How do solids, liquids, and gases differ?

FOLDABLES®

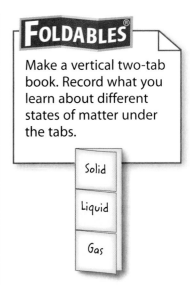

Make a vertical two-tab book. Record what you learn about different states of matter under the tabs.

Solid | Liquid | Gas

Figure 8 The three common states of matter on Earth are solid, liquid, and gas.

Concepts in Motion Animation

Solids, Liquids, and Gases

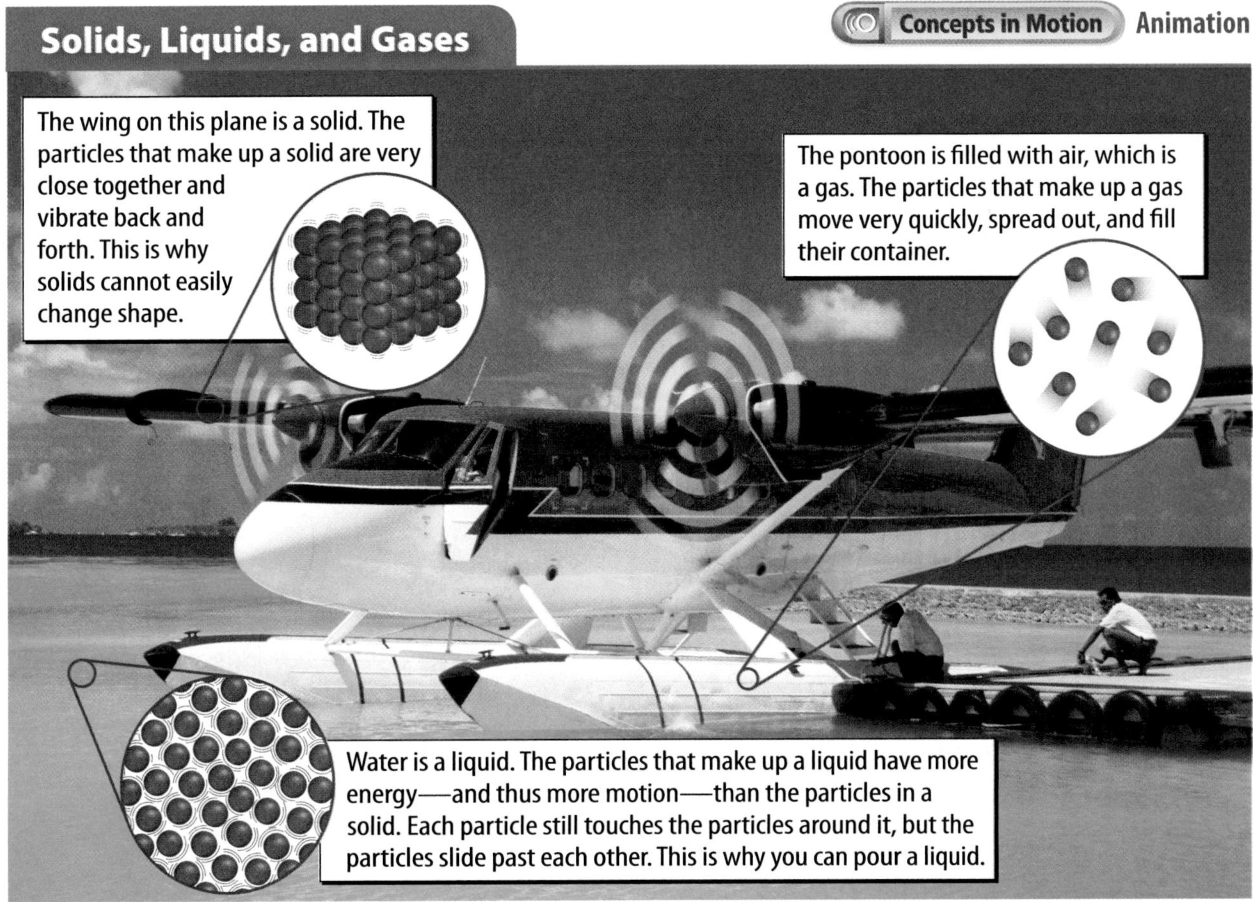

The wing on this plane is a solid. The particles that make up a solid are very close together and vibrate back and forth. This is why solids cannot easily change shape.
The pontoon is filled with air, which is a gas. The particles that make up a gas move very quickly, spread out, and fill their container.
Water is a liquid. The particles that make up a liquid have more energy—and thus more motion—than the particles in a solid. Each particle still touches the particles around it, but the particles slide past each other. This is why you can pour a liquid.

Visual Check Which state of matter flows, keeps the same volume, and takes the shape of its container?

Figure 9 The larger dumbbells have greater mass than the smaller dumbbells because they contain more matter.

Size-Dependent Properties

State is only one of many physical properties that you can use to describe matter. Some physical properties, such as mass and volume, depend on the size or amount of matter. Measurements of these properties vary depending on how much matter is in a sample.

Mass Imagine holding a small dumbbell in one hand and a larger one in your other hand. What do you notice? The larger dumbbell seems heavier. The larger dumbbell has more mass than the smaller one. **Mass** *is the amount of matter in an object.* Both small dumbbells shown in **Figure 9** have the same mass because they both contain the same amount of matter. Mass is a size-dependent property of a given substance because its value depends on the size of a sample.

Mass sometimes is confused with weight, but they are not the same. Mass is an amount of matter in something. Weight is the pull of gravity on that matter. Weight changes with location, but mass does not. Suppose one of the dumbbells in the figure was on the Moon. The dumbbell would have the same mass on the Moon that it has on Earth. However, the Moon's gravity is much less than Earth's gravity, so the weight of the dumbbell would be less on the Moon.

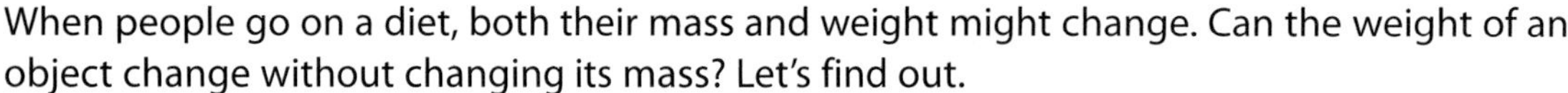

Can the weight of an object change?

When people go on a diet, both their mass and weight might change. Can the weight of an object change without changing its mass? Let's find out.

1. Read and complete a lab safety form.
2. Use a **balance** to find the mass of five **metal washers.** Record the mass in grams in your Science Journal.
3. Hang the washers from the hook on a **spring scale.** Record the weight in newtons.
4. Lower just the washers into a **500-mL beaker** containing approximately 300 mL water. Record the weight in newtons.

Analyze and Conclude

1. **Draw Conclusions** Did the weight of the washers change during the experiment? How do you know?
2. **Predict** In what other ways might you change the weight of the washers?
3. **Key Concept** What factors affect the weight of an object, but not its mass?

Volume Another physical property that depends on the size or the amount of a substance is volume. A unit often used to measure volume is the milliliter (mL). Volume is the amount of space something takes up. Suppose a full bottle of water contains 400 mL of water. If you pour exactly half of the water out, the bottle contains half of the original volume, or 200 mL, of water.

 **Reading Check** What is a common unit for volume?

Size-Independent Properties

Unlike mass, weight, and volume, some physical properties of a substance do not depend on the amount of matter present. These properties are the same for both small samples and large samples. They are called size-independent properties. Examples of size-independent properties are melting point, boiling point, density, electrical conductivity, and solubility.

Melting Point and Boiling Point The temperature at which a substance changes from a solid to a liquid is its melting point. The temperature at which a substance changes from a liquid to a gas is its boiling point. Different substances have different boiling points and melting points. The boiling point for water is 100°C at sea level. Notice in **Figure 10** that this temperature does not depend on how much water is in the container.

Density Imagine holding a bowling ball in one hand and a foam ball of the same size in the other. The bowling ball seems heavier because the density of the material that makes up the bowling ball is greater than the density of foam. **Density** *is the mass per unit volume of a substance.* Like melting point and boiling point, density is a size-independent property.

Math Skills

Use Ratios
When you compare two numbers by division, you are using a ratio. Density can be written as a ratio of mass and volume. What is the density of a substance if a 5-mL sample has a mass of 25 g?

1. Set up a ratio.

$$\frac{25\text{ g}}{5\text{ mL}}$$

2. **Divide the numerator by the denominator to get the mass (in g) of 1 mL.**

$$\frac{25\text{ g}}{5\text{ mL}} = \frac{5\text{ g}}{1\text{ mL}}$$

3. **The density is 5 g/mL.**

Practice
A sample of wood has a mass of 12 g and a volume of 16 mL. What is the density of the wood?

- **Math Practice**
- **Personal Tutor**

WORD ORIGIN

density
from Latin *densus,* means "compact"; and Greek *dasys,* means "thick"

Figure 10 The boiling point of water is 100°C at sea level. The boiling point does not change for different volumes of water.

 **Personal Tutor**

Conductivity Another property that is independent of the sample size is conductivity. Electrical conductivity is the ability of matter to conduct, or carry along, an electric current. Copper often is used for electrical wiring because it has high electrical conductivity. Thermal conductivity is the ability of a material to conduct thermal energy. Metals tend to have high electrical and thermal conductivity. Stainless steel, for example, often is used to make cooking pots because of its high thermal conductivity. However, the handles on the pan probably are made out of wood, plastic, or some other substance that has low thermal conductivity.

Reading Check What are two types of conductivity?

Solubility Have you ever made lemonade by stirring a powdered drink mix into water? As you stir, the powder mixes evenly in the water. In other words, the powder dissolves in the water.

What do you think would happen if you tried to dissolve sand in water? No matter how much you stir, the sand does not dissolve. **Solubility** *is the ability of one substance to dissolve in another.* The powdered drink mix is soluble in water, but sand is not. **Table 1** explains how physical properties such as conductivity and solubility can be used to identify objects and separate mixtures.

Key Concept Check What are five different physical properties of matter?

Table 1 This table contains the descriptions of several physical properties. It also shows examples of how physical properties can be used to separate mixtures.

Visual Check How might you separate a mixture of iron filings and salt?

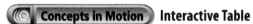

Concepts in Motion Interactive Table

Table 1 Physical Properties of Matter

	Property		
	Mass	**Conductivity**	**Volume**
Description of property	The amount of matter in an object	The ability of matter to conduct, or carry along, electricity or heat	The amount of space something occupies
Size-dependent or size independent	Size-dependent	Size-independent	Size-dependent
How the property is used to separate a mixture (example)	Mass typically is not used to separate a mixture.	Conductivity typically is not used to separate a mixture.	Volume could be used to separate mixtures whose parts can be separated by filtration.

Separating Mixtures

In Lesson 1, you read about different types of mixtures. Recall that the substances that make up mixtures are not held together by chemical **bonds**. When substances form a mixture, the properties of the individual substances do not change. One way that a mixture and a compound differ is that the parts of a mixture often can be separated by physical properties. For example, when salt and water form a solution, the salt and the water do not lose any of their individual properties. Therefore, you can separate the salt from the water by using differences in their physical properties. Water has a lower boiling point than salt. If you boil salt water, the water will boil away, and the salt will be left behind. Other physical properties that can be used to separate different mixtures are described in **Table 1.**

Physical properties cannot be used to separate a compound into the elements it contains. The atoms that make up a compound are bonded together and cannot be separated by physical means. For example, you cannot separate the hydrogen atoms from the oxygen atoms in water by boiling water.

Key Concept Check How are physical properties used to separate mixtures?

SCIENCE USE V. COMMON USE

bond

Science Use a force between atoms or groups of atoms

Common Use a monetary certificate issued by a government or a business that earns interest

Property				
Boiling/Melting Points	**State of matter**	**Density**	**Solubility**	**Magnetism**
The temperature at which a material changes state	Whether something is a solid, a liquid, or a gas	The amount of mass per unit of volume	The ability of one substance to dissolve in another	Attractive force for some metals, especially iron
Size-independent	Size-independent	Size-independent	Size-independent	Size-independent
Each part of a mixture will boil or melt at a different temperature.	A liquid can be poured off a solid.	Objects with greater density sink in objects with less density.	Dissolve a soluble material to separate it from a material with less solubility.	Attract iron from a mixture of materials.

Lesson 2 Review

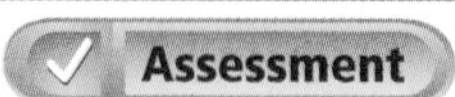

Online Quiz

Visual Summary

A physical property is a characteristic of matter that can be observed or measured without changing the identity of the matter.

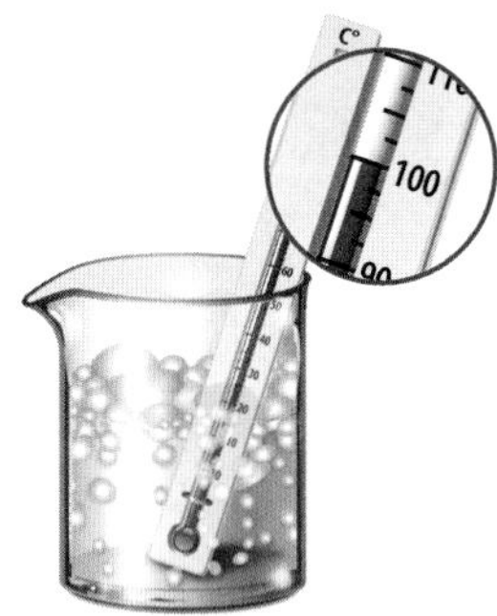

Examples of physical properties include mass, density, volume, melting point, boiling point, state of matter, and solubility.

Many physical properties can be used to separate the components of a mixture.

FOLDABLES®

Use your lesson Foldable to review the lesson. Save your Foldable for the project at the end of the chapter.

What do you think NOW?

You first read the statements below at the beginning of the chapter.

3. The weight of a material never changes, regardless of where it is.

4. Boiling is one method used to separate parts of a mixture.

Did you change your mind about whether you agree or disagree with the statements? Rewrite any false statements to make them true.

Use Vocabulary

1. **Distinguish** between mass and weight.
2. **Use the term** *solubility* in a sentence.
3. An object's ________ is the amount of mass per a certain unit of volume.

Understand Key Concepts 

4. **Explain** how to separate a mixture of sand and pebbles.
5. Which physical property is NOT commonly used to separate mixtures?
   - **A.** magnetism
   - **B.** conductivity
   - **C.** density
   - **D.** solubility
6. **Analyze** Name two size-dependent properties and two size-independent properties of an iron nail.

Interpret Graphics

7. **Sequence** Draw a graphic organizer like the one below to show the steps in separating a mixture of sand, iron filings, and salt.

Critical Thinking

8. **Examine** the diagram below.

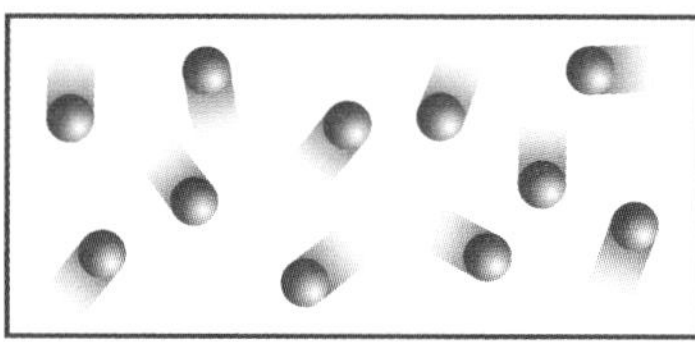

How can you identify the state of matter represented by the diagram?

Math Skills

Math Practice

9. A piece of copper has a volume of 10.0 cm^3. If the mass of the copper is 890 g, what is the density of copper?

Inquiry **Skill Practice** **Follow a Procedure** **30 minutes**

How can following a procedure help you solve a crime?

Materials

Plastic sealable bag

triple-beam balance

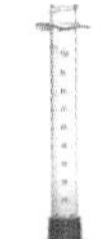

50-mL graduated cylinder

paper towels

Also needed:

Crime Scene Objects

Safety

Imagine that you are investigating a crime scene. You find several pieces of metal and broken pieces of plastic that look as if they came from a car's tail light. You also have similar objects collected from the suspect. How can you figure out if they are parts of the same objects?

Learn It

To be sure you do the same tests on each object, it is helpful to **follow a procedure.** A procedure tells you how to use the materials and what steps to take.

Try It

1. Read and complete a lab safety form.
2. Copy the table below into your Science Journal.
3. Use the balance to find the mass of an object from the crime scene. Record the mass in your table.
4. Place about 25 mL of water in a graduated cylinder. Read and record the exact volume. Call this volume V_1.
5. Carefully tilt the cylinder, and allow one of the objects to slide into the water. Read and record the volume. Call this volume V_2.
6. Repeat steps 3–5 for each of the other objects.

Apply It

7. Complete the table by calculating the volume and the density of each object.
8. What conclusions can you draw about the objects collected from the crime scene and those collected from the suspect?
9. **Key Concept** How could you use this procedure to help identify and compare various objects?

Object	Mass (M) (g)	V_1 (mL)	V_2 (mL)	Volume of Object (V) ($V_2 - V_1$) (mL)	Density of Object M/V (g/mL)
1					
2					
3					
4					
5					
6					

Lesson 3

Reading Guide

Key Concepts
ESSENTIAL QUESTIONS

- How can a change in energy affect the state of matter?
- What happens when something dissolves?
- What is meant by conservation of mass?

Vocabulary

physical change p. 249

g Multilingual eGlossary

Physical Changes

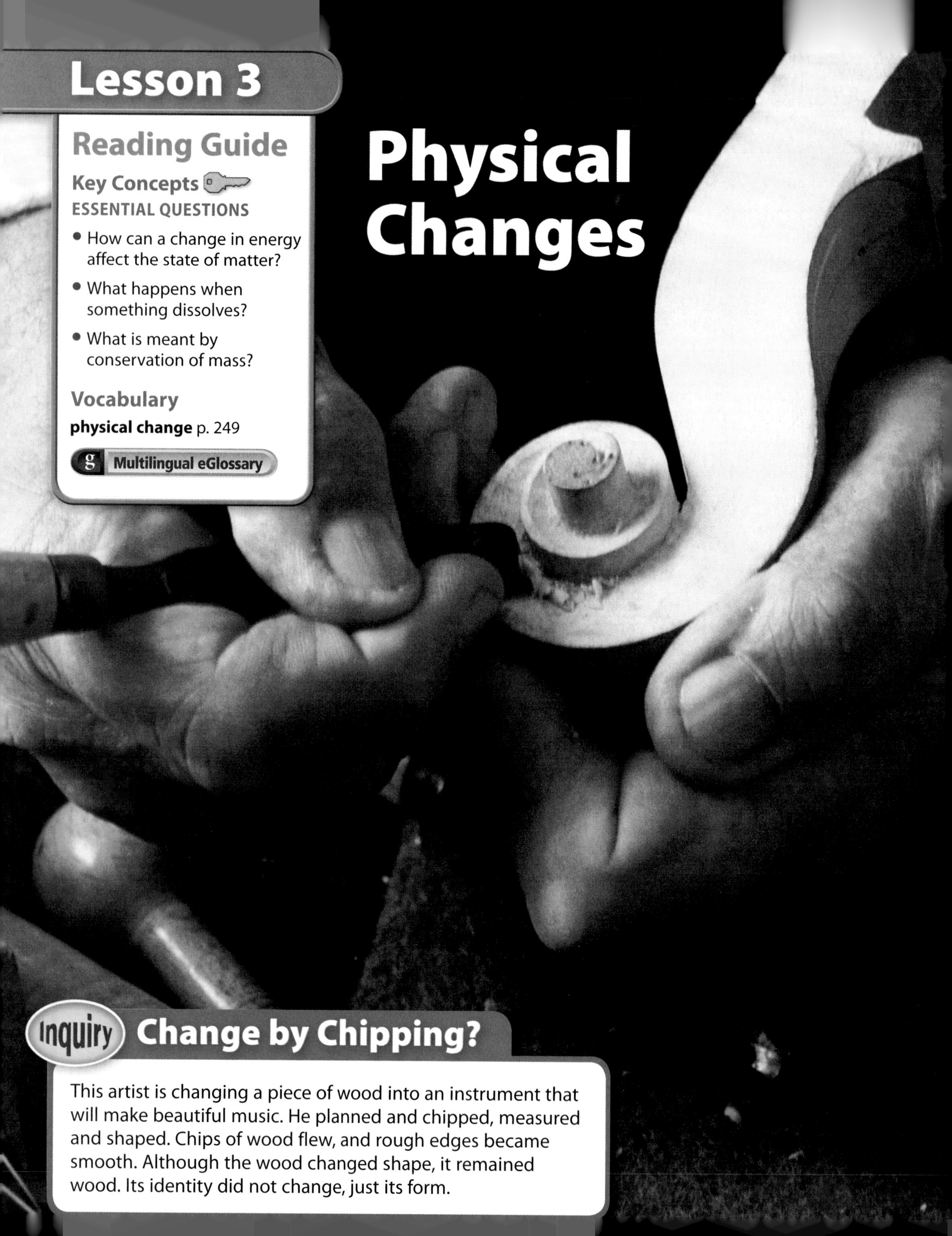

Inquiry Change by Chipping?

This artist is changing a piece of wood into an instrument that will make beautiful music. He planned and chipped, measured and shaped. Chips of wood flew, and rough edges became smooth. Although the wood changed shape, it remained wood. Its identity did not change, just its form.

 Launch Lab 20 minutes

Where did it go?

When you dissolve sugar in water, where does the sugar go? One way to find out is to measure the mass of the water and the sugar before and after mixing.

1. Read and complete a lab safety form.
2. Add **sugar** to a **small paper cup** until the cup is approximately half full. Bend the cup's opening, and pour the sugar into a **balloon.**
3. With the balloon hanging over the side, stretch the neck of the balloon over a **flask** half full of **water.**
4. Use a **balance** to find the mass of the flask-and-balloon assembly. Record the mass in your Science Journal.
5. Lift the end of the balloon, and empty the sugar into the flask. Swirl until the sugar dissolves. Measure and record the mass of the flask-and-balloon assembly again.

Think About This

1. Is the sugar still present after it dissolves? How do you know?
2.  **Key Concept** Based on your observations, what do you think happens to the mass of objects when they dissolve? Explain.

Physical Changes

How would you describe water? If you think about water in a stream, you might say that it is a cool liquid. If you think about water as ice, you might describe it as a cold solid. How would you describe the change from ice to water? As ice melts, some of its properties change, such as the state of matter, the shape, and the temperature, but it is still water. In Lesson 2, you read that substances and mixtures can be solids, liquids, or gases. In addition, substances and mixtures can change from one state to another. *A* **physical change** *is a change in size, shape, form, or state of matter in which the matter's identity stays the same.* During a physical change, the matter does not become something different even though physical properties change.

Change in Shape and Size

Think about changes in the shapes and the sizes of substances and mixtures you experience each day. When you chew food, you are breaking it into smaller pieces. This change in size helps make food easier to digest. When you pour juice from a bottle into a glass, you are changing the shape of the juice. If you fold clothes to fit them into a drawer, you are changing their shapes. Changes in shape and size are physical changes. The identity of the matter has not changed.

Word Origin

physical
from Greek *physika*, means "natural things"

change
from Latin *cambire*, means "to exchange"

FOLDABLES®

Make a vertical two-tab book. Label the tabs as illustrated. Record specific examples illustrating how adding or releasing thermal energy results in physical change.

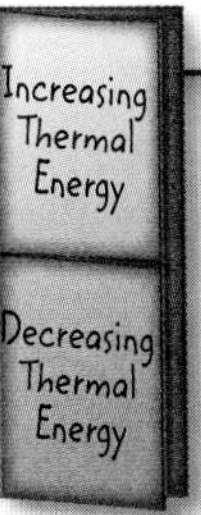

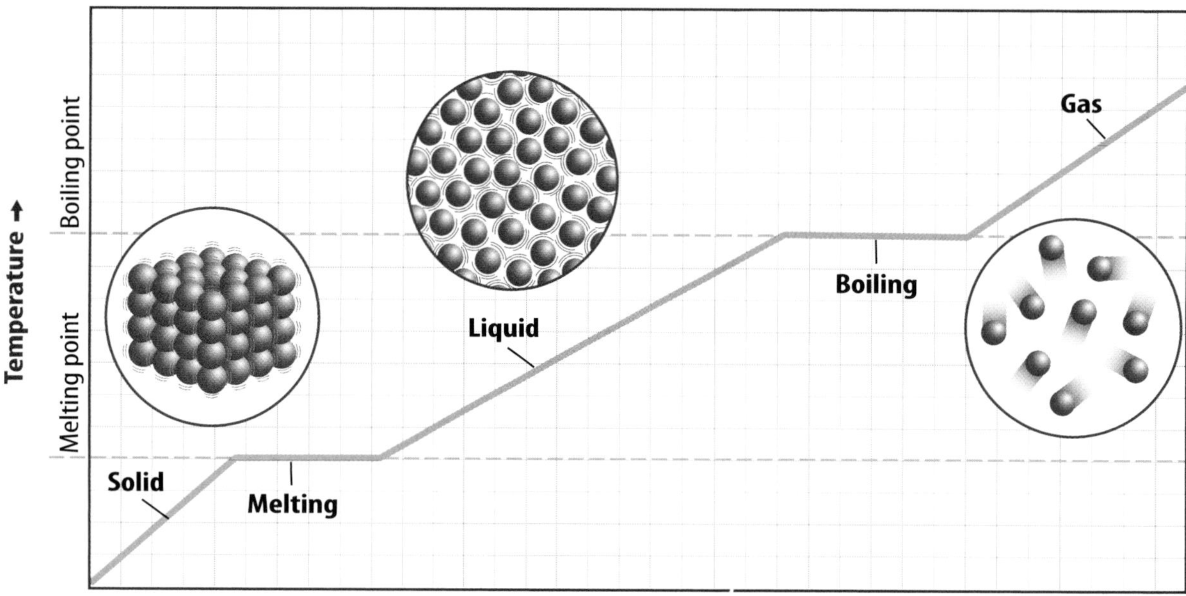

▲ **Figure 11** As thermal energy is added to a material, temperature increases when the state of the material is not changing. Temperature stays the same during a change of state.

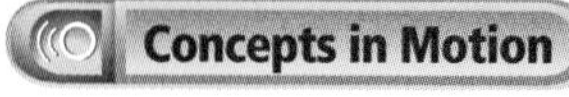

Animation

Figure 12 Solid iodine undergoes sublimation. It changes from a solid to a gas without becoming a liquid. ▼

Change in State of Matter

Why does ice melt in your hand? Or, why does water turn to ice in the freezer? Matter, such as water, can change state. Recall from Lesson 2 how the particles in a solid, a liquid, and a gas behave. To change the state of matter, the movement of the particles has to change. In order to change the movement of particles, thermal energy must be either added or removed.

Adding Thermal Energy When thermal energy is added to a solid, the particles in the solid move faster and faster, and the temperature increases. As the particles move faster, they are more likely to overcome the attractive forces that hold them tightly together. When the particles are moving too fast for attractive forces to hold them tightly together, the solid reaches its melting point. The melting point is the temperature at which a solid changes to a liquid.

After all the solid has melted, adding more thermal energy causes the particles to move even faster. The temperature of the liquid increases. When the particles are moving so fast that attractive forces cannot hold them close together, the liquid is at its boiling point. The boiling point is the temperature at which a liquid changes into a gas and the particles spread out. **Figure 11** shows how temperature and change of state relate to each other when thermal energy is added to a material.

Some solids change directly to a gas without first becoming a liquid. This process is called sublimation. An example of sublimation is shown in **Figure 12.** You saw another example of sublimation in **Figure 5** in Lesson 1.

Removing Thermal Energy When thermal energy is removed from a gas, such as water vapor, particles in the gas move more slowly and the temperature decreases. Condensation occurs when the particles are moving slowly enough for attractive forces to pull the particles close together. Recall that condensation is the process that occurs when a gas becomes a liquid.

After the gas has completely changed to a liquid, removing more thermal energy from the liquid causes particles move even more slowly. As the motion between the particles slows, the temperature decreases. Freezing occurs when the particles are moving so slowly that attractive forces between the particles hold them tightly together. Now the particles only can vibrate in place. Recall that freezing is the process that occurs when a liquid becomes a solid.

Freezing and melting are reverse processes, and they occur at the same temperature. The same is true of boiling and condensation. Another change of state is deposition. Deposition is the change from a gas directly to a solid, as shown in **Figure 13.** It is the process that is the opposite of sublimation.

Key Concept Check How can removing thermal energy affect the state of matter?

30 minutes

Can you make ice without a freezer?

What happens when you keep removing energy from a substance?

1. Read and complete a lab safety form.
2. Draw the data table below in your Science Journal. Half fill a **large test tube** with **distilled water.** Use a **thermometer** to measure the temperature of the water, then record it.
3. Place the test tube into a **large foam cup** containing **ice** and **salt.** Use a **stirring rod** to slowly stir the water in the tube.
4. Record the temperature of the water every minute until the water freezes. Continue to record the temperature each minute until it drops to several degrees below 0°C.

Time (min)	0	1	2	3	4	5	6	7	8
Temperature (C)									

Analyze and Conclude

1. **Organize Data** Graph the data in your table. Label time on the *x*-axis and temperature on the *y*-axis.
2. **Interpret Data** According to your data, what is the freezing point of water?
3. **Key Concept** What caused the water to freeze?

Figure 13 When enough thermal energy is removed, one of several processes occurs.

Freezing

Condensation

Deposition

▲ **Figure 14**  Salt dissolves when it is added to the water in this aquarium.

Dissolving

Have you ever owned a saltwater aquarium, such as the one shown in **Figure 14**? If you have, you probably had to add certain salts to the water before you added the fish. Can you see the salt in the water? As you added the salt to the water, it gradually disappeared. It was still there, but it dissolved, or mixed evenly, in the water. Because the identities of the substances—water and salt—are not changed, dissolving is a physical change.

Like many physical changes, dissolving is usually easy to reverse. If you boil the salt water, the liquid water will change to water vapor, leaving the salt behind. You once again can see the salt because the particles that make up the substances do not change identity during a physical change.

Key Concept Check What happens when something dissolves?

Conservation of Mass

During a physical change, the physical properties of matter change. The particles in matter that are present before a physical change are the same as those present after the physical change. Because the particles are the same both before and after a physical change, the total mass before and after the change is also the same, as shown in **Figure 15**. This is known as the conservation of mass. You will read in Lesson 4 that mass also is conserved during another type of change—a chemical change.

Key Concept Check What is meant by conservation of mass?

Figure 15 Mass is conserved during a physical change. ▼

Conservation of Mass 

Visual Check If a sample of water has a mass of 200 g and the final solution has a mass of 230 g, how much solute dissolved in the water?

Lesson 3 Review

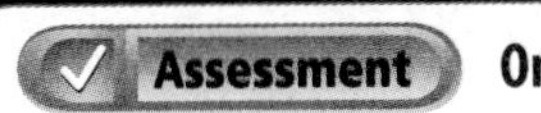

Visual Summary

During a physical change, matter can change form, shape, size, or state, but the identity of the matter does not change.

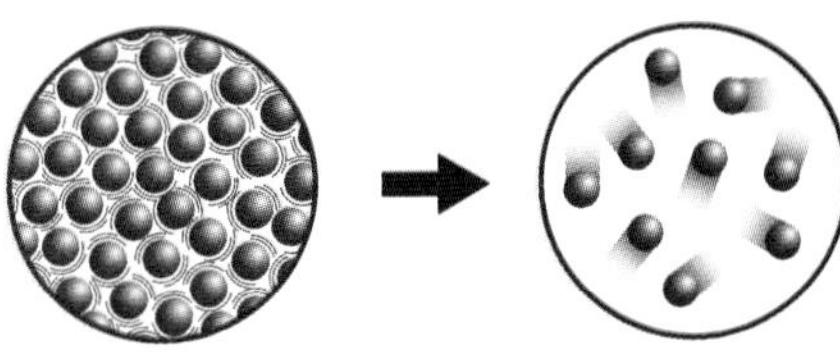

Matter either changes temperature or changes state when enough thermal energy is added or removed.

Mass is conserved during physical changes, which means that mass is the same before and after the changes occur.

FOLDABLES®

Use your lesson Foldable to review the lesson. Save your Foldable for the project at the end of the chapter.

What do you think NOW?

You first read the statements below at the beginning of the chapter.

5. Heating a material decreases the energy of its particles.

6. When you stir sugar into water, the sugar and water evenly mix.

Did you change your mind about whether you agree or disagree with the statements? Rewrite any false statements to make them true.

Use Vocabulary

1 **Use the term** *physical change* in a sentence.

Understand Key Concepts 

2 **Describe** how a change in energy can change ice into liquid water.

3 Which never changes during a physical change?

A. state of matter
B. temperature
C. total mass
D. volume

4 **Relate** What happens when something dissolves?

Interpret Graphics

5 **Examine** the graph below of temperature over time as a substance changes from solid to liquid to gas. Explain why the graph has horizontal lines.

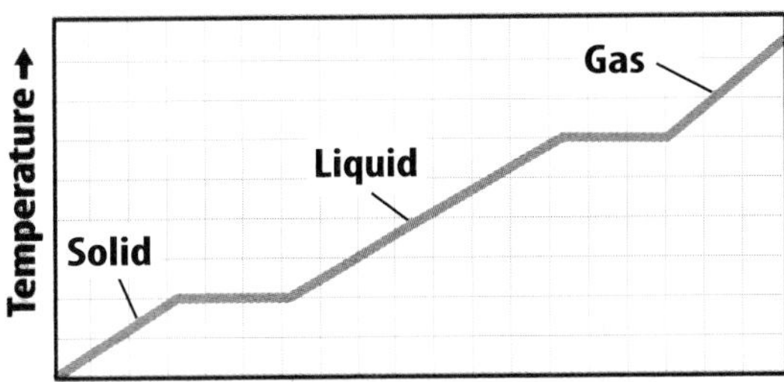

6 **Take Notes** Copy the graphic organizer below. For each heading, summarize the main idea described in the lesson.

Heading	Main Idea
Physical Changes	
Change in State of Matter	
Conservation of Mass	

Critical Thinking

7 **Design** a demonstration that shows that temperature remains unchanged during a change of state.

Inquiry **Skill Practice** Identify Variables and Controls 30 minutes

How can known substances help you identify unknown substances?

Materials

plastic spoons

magnifying lens

stirring rod

Also needed: known substances (baking soda, ascorbic acid, sugar, cornstarch) test tubes, test tube rack, watch glass, dropper bottles containing water, iodine, vinegar, and red cabbage indicator

Safety

While investigating a crime scene, you find several packets of white powder. Are they illegal drugs or just harmless packets of candy? Here's one way to find out.

Learn It

A **control** is something that stays the same. If you determine how a known substance reacts with other substances, you can use it as a control. Unknown substances are **variables.** They might or might not react in the same way.

Try It

1. Read and complete a lab safety form.
2. Copy the data table below into your Science Journal.
3. Use a magnifying lens to observe the appearance of each known substance.
4. Test small samples of each known substance for their reaction with a drop or two of water, vinegar, and iodine solution.
5. Feel the texture of each substance.
6. Mix each substance with water, and add the red cabbage indicator.
7. After you complete your observations, ask your teacher for a mystery powder. Repeat steps 3–6 using the mystery powder. Use the data you collect to identify the powder.

Apply It

8. What test suggests that a substance might be cornstarch?
9. Why should you test the reactions of the substances with many different things?
10. **Key Concept** How did you use the properties of the controls to identify your variable?

Substance	Appearance	Texture	Reaction to Water	Reaction to Iodine	Reaction to Vinegar	Red Cabbage Indicator
Baking soda						
Sugar						
Ascorbic acid						
Cornstarch						
Mystery powder						

Lesson 4

Reading Guide

Key Concepts
ESSENTIAL QUESTIONS

- What is a chemical property?
- What are some signs of chemical change?
- Why are chemical equations useful?
- What are some factors that affect the rate of chemical reactions?

Vocabulary

chemical property p. 256
chemical change p. 257
concentration p. 260

Multilingual eGlossary

Video

- BrainPOP®
- What's Science Got to do With It?

Chemical Properties and Changes

Inquiry A Burning Issue?

As this car burns, some materials change to ashes and gases. The metal might change form or state if the fire is hot enough, but it probably won't burn. Why do fabric, leather, and paint burn? Why do many metals not burn? The properties of matter determine how matter behaves when it undergoes a change.

Inquiry Launch Lab

15 minutes

What can colors tell you?

You mix red and blue paint to get purple paint. Iron changes color when it rusts. Are color changes physical changes?

1. Read and complete a lab safety form.
2. Divide a **paper towel** into thirds. Label one section *RCJ,* the second section *A,* and the third section *B.*
3. Dip one end of three **cotton swabs** into **red cabbage juice** (RCJ). Observe the color, and set the swabs on the paper towel, one in each of the three sections.
4. Add one drop of **substance A** to the swab in the *A* section. Observe any changes, and record observations in your Science Journal.
5. Repeat step 4 with **substance B** and the swab in the *B* section.
6. Observe **substances C** and **D** in their **test tubes.** Then pour C into D. Rock the tube gently to mix. Record your observations.

Think About This

1. What happened to the color of the red cabbage juice when substances A and B were added?
2. **Key Concept** Which of the changes you observed do you think was a physical change? Explain your reasoning.

Chemical Properties

Recall that a physical property is a characteristic of matter that you can observe or measure without changing the identity of the matter. However, matter has other properties that can be observed only when the matter changes from one substance to another. *A* **chemical property** *is a characteristic of matter that can be observed as it changes to a different type of matter.* For example, what are some chemical properties of a piece of paper? Can you tell by just looking at it that it will burn easily? The only way to know that paper burns is to bring a flame near the paper and watch it burn. When paper burns, it changes into different types of matter. The ability of a substance to burn is a chemical property. The ability to rust is another chemical property.

Comparing Properties

You now have read about physical properties and chemical properties. All matter can be described using both types of properties. For example, a wood log is solid, rounded, heavy, and rough. These are physical properties that you can observe with your senses. The log also has mass, volume, and density, which are physical properties that can be measured. The ability of wood to burn is a chemical property. This property is obvious only when you burn the wood. It also will rot, another chemical property you can observe when the log decomposes, becoming other substances. When you describe matter, you consider both its physical and its chemical properties.

Key Concept Check What are some chemical properties of matter?

Chemical Changes

Recall that during a physical change, the identity of matter does not change. However, *a* **chemical change** *is a change in matter in which the substances that make up the matter change into other substances with new physical and chemical properties.* For example, when iron undergoes a chemical change with oxygen, rust forms. The substances that undergo a change no longer have the same properties because they no longer have the same identity.

 **Reading Check** What is the difference between a physical change and a chemical change?

FOLDABLES®

Use a sheet of paper to make a chart with four columns. Use the chart throughout this lesson to explain how the identity of matter changes during a chemical change.

Action/ Matter	Signs of Chemical Change	Explain the Chemical Reaction	What affects the reaction rate?

Signs of Chemical Change

How do you know when a chemical change occurs? What signs show you that new types of matter form? As shown in **Figure 16,** signs of chemical changes include the formation of bubbles or a change in odor, color, or energy.

It is important to remember that these signs do not always mean a chemical change occurred. Think about what happens when you heat water on a stove. Bubbles form as the water boils. In this case, bubbles show that the water is changing state, which is a physical change. The evidence of chemical change shown in **Figure 16** means that a chemical change might have occurred. However, the only proof of chemical change is the formation of a new substance.

 **Key Concept Check** What are signs of a chemical change?

WORD ORIGIN

chemical
from Greek *chemeia,* means "cast together"

Some Signs of Chemical Change

Figure 16 Sometimes you can observe clues that a chemical change has occurred.

Bubbles

Energy change

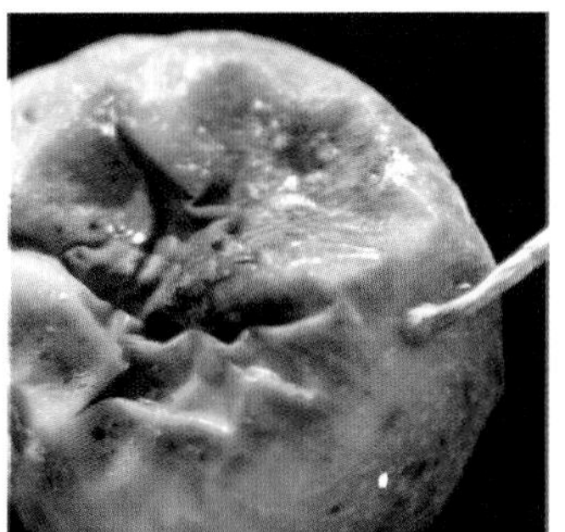

Odor change

Color change

Visual Check What signs show that a chemical change takes place when fireworks explode?

Inquiry MiniLab — 20 minutes

Can you spot the clues for chemical change?

What are some clues that let you know a chemical change might have taken place?

1. Read and complete a lab safety form.
2. Add about 25 mL of room-temperature water to a **self-sealing plastic bag.** Add two **dropperfuls** of **red cabbage juice.**
3. Add one **measuring scoop** of **calcium chloride** to the bag. Seal the bag. Tilt the bag to mix the contents until the solid disappears. Feel the bottom of the bag. Record your observations in your Science Journal.
4. Open the bag, and add one measuring scoop of **baking soda.** Quickly press the air from the bag and reseal it. Tilt the bag to mix the contents. Observe for several minutes. Record your observations.

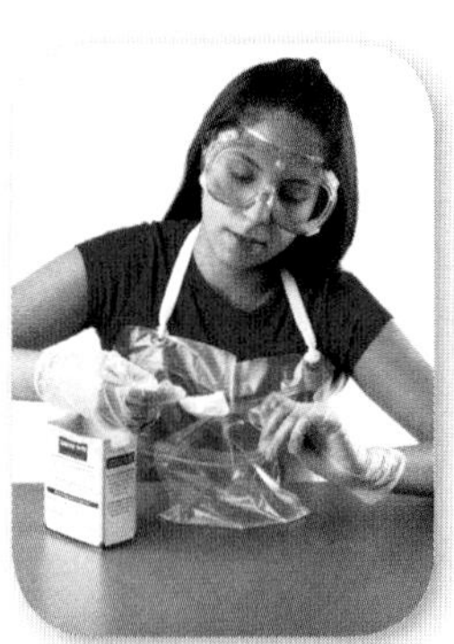

Analyze and Conclude

1. **Observe** What changes did you observe?
2. **Infer** Which of the changes suggested that a new substance formed? Explain.
3. **Key Concept** Are changes in energy always a sign of a chemical change? Explain.

Explaining Chemical Reactions

You might wonder why chemical changes produce new substances. Recall that particles in matter are in constant motion. As particles move, they collide with each other. If the particles collide with enough force, the bonded atoms that make up the particles can break apart. These atoms then rearrange and bond with other atoms. When atoms bond together in new combinations, new substances form. This process is called a reaction. Chemical changes often are called chemical reactions.

Reading Check What does it mean to say that atoms rearrange during a chemical change?

Using Chemical Formulas

A useful way to understand what happens during a chemical reaction is to write a chemical equation. A chemical equation shows the chemical formula of each substance in the reaction. The formulas to the left of the arrow represent the reactants. Reactants are the substances present before the reaction takes place. The formulas to the right of the arrow represent the products. Products are the new substances present after the reaction. The arrow indicates that a reaction has taken place.

Key Concept Check Why are chemical equations useful?

Figure 17 Chemical formulas and other symbols are parts of a chemical equation.

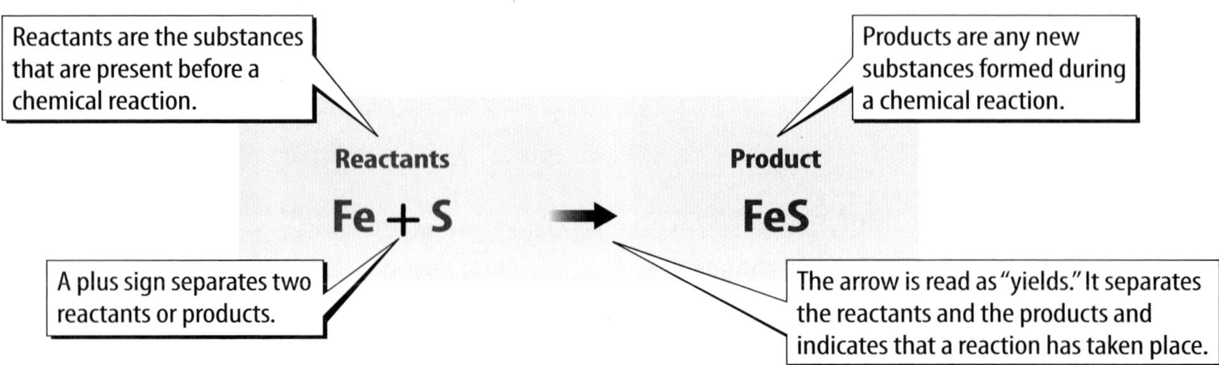
Reactants are the substances that are present before a chemical reaction.
Products are any new substances formed during a chemical reaction.
Reactants
Product
Fe + S → FeS
A plus sign separates two reactants or products.
The arrow is read as "yields." It separates the reactants and the products and indicates that a reaction has taken place.

Balancing Chemical Equations

Look at the equation in **Figure 17.** Notice that there is one iron (Fe) atom on the reactants side and one iron atom on the product side. This is also true for the sulfur (S) atoms. Recall that during both physical and chemical changes, mass is conserved. This means that the total mass before and after a change must be equal. Therefore, in a chemical equation, the number of atoms of each element before a reaction must equal the number of atoms of each element after the reaction. This is called a balanced chemical equation, and it illustrates the conservation of mass. **Figure 18** explains how to write and balance a chemical equation.

When balancing an equation, you cannot change the chemical formula of any reactants or products. Changing a formula changes the identity of the substance. Instead, you can place coefficients, or multipliers, in front of formulas. Coefficients change the amount of the reactants and products present. For example, an H_2O molecule has two H atoms and one O atom. Placing the coefficient *2* before H_2O ($2H_2O$) means that you double the number of H atoms and O atoms present:

$$2 \times 2 \text{ H atoms} = 4 \text{ H atoms}$$
$$2 \times 1 \text{ O atom} = 2 \text{ O atoms}$$

Note that $2H_2O$ is still water. However, it describes two water particles instead of one.

Figure 18 Equations must be balanced because mass is conserved during a chemical reaction.

Balancing Chemical Equations Example

When methane (CH_4)—a gas burned in furnaces—reacts with oxygen (O_2) in the air, the reaction produces carbon dioxide (CO_2) and water (H_2O). Write and balance a chemical equation for this reaction.

1 Write the equation, and check to see if it is balanced.

a. Write the chemical formulas with the reactants on the left side of the arrow and the products on the right side.

b. Count the atoms of each element in the reactants and in the products.
- Note which elements have a balanced number of atoms on each side of the equation.
- If all elements are balanced, the overall equation is balanced. If not, go to step 2.

a. $CH_4 + O_2 \rightarrow CO_2 + H_2O$ **not balanced**

b. reactants $\rightarrow$ products

reactants	products	
C=1	C=1	**balanced**
H=4	H=2	**not balanced**
O=2	O=3	**not balanced**

2 Add coefficients to the chemical formulas to balance the equation.

a. Pick an element in the equation whose atoms are not balanced, such as hydrogen. Write a coefficient in front of a reactant or a product that will balance the atoms of the chosen element in the equation.

b. Recount the atoms of each element in the reactants and the products, and note which are balanced on each side of the equation.

c. Repeat steps 2a and 2b until all atoms of each element in the reactants equal those in the products.

a. $CH_4 + O_2 \rightarrow CO_2 + 2H_2O$ **not balanced**

b.

C=1	C=1	**balanced**
H=4	H=4	**balanced**
O=2	O=4	**not balanced**

c. $CH_4 + 2O_2 \rightarrow CO_2 + 2H_2O$ **balanced**

C=1	C=1	**balanced**
H=4	H=4	**balanced**
O=4	O=4	**balanced**

3 Write the balanced equation that includes the coefficients: $CH_4 + 2O_2 \rightarrow CO_2 + 2H_2O$

Factors that Affect the Rate of Chemical Reactions

Figure 19 The rate of most chemical reactions increases with an increase in temperature, concentration, or surface area.

1 Temperature

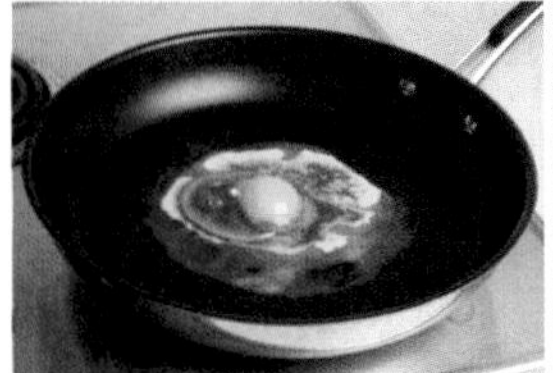

Chemical reactions that occur during cooking happen at a faster rate when temperature increases.

2 Concentration

Acid rain contains a higher concentration of acid than normal rain does. As a result, a statue exposed to acid rain is damaged more quickly than a statue exposed to normal rain.

3 Surface Area

When an antacid tablet is broken into pieces, the pieces have more total surface area than the whole tablet does. The pieces react more rapidly with water because more of the broken tablet is in contact with the water.

The Rate of Chemical Reactions

Recall that the particles that make up matter are constantly moving and colliding with one another. Different factors can make these particles move faster and collide harder and more frequently. These factors increase the rate of a chemical reaction, as shown in **Figure 19.**

1 A higher **temperature** usually increases the rate of reaction. When the temperature is higher, the particles move faster. Therefore, the particles collide with greater force and more frequently.

2 **Concentration** *is the amount of substance in a certain volume.* A reaction occurs faster if the concentration of at least one reactant increases. When concentration increases, there are more particles available to bump into each other and react.

3 **Surface area** also affects reaction rate if at least one reactant is a solid. If you drop a whole effervescent antacid tablet into water, the tablet reacts with the water. However, if you break the tablet into several pieces and then add them to the water, the reaction occurs more quickly. Smaller pieces have more total surface area, so more space is available for reactants to collide.

Key Concept Check List three factors that affect the rate of a chemical reaction.

Chemistry

To understand chemistry, you need to understand matter. You need to know how the arrangement of atoms results in different types of matter. You also need to be able to distinguish physical properties from chemical properties and describe ways these properties can change. In later chemistry chapters and courses, you will examine each of these topics closely to gain a better understanding of matter.

Lesson 4 Review

Assessment Online Quiz
Inquiry Virtual Lab

Visual Summary

A chemical property is observed only as a material undergoes chemical change and changes identity.

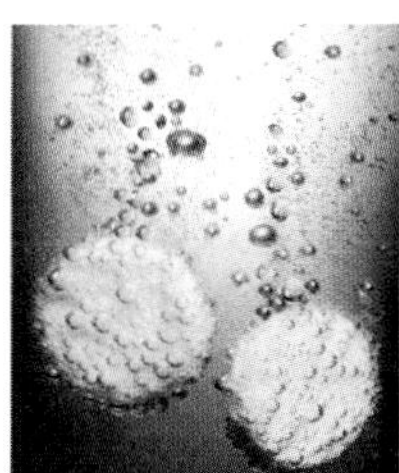

Signs of possible chemical change include bubbles, energy change, and change in odor or color.

Chemical equations show the reactants and products of a chemical reaction and that mass is conserved.

FOLDABLES®

Use your lesson Foldable to review the lesson. Save your Foldable for the project at the end of the chapter.

What do you think NOW?

You first read the statements below at the beginning of the chapter.

7. When wood burns, new materials form.

8. Temperature can affect the rate at which chemical changes occur.

Did you change your mind about whether you agree or disagree with the statements? Rewrite any false statements to make them true.

Use Vocabulary

1. The amount of substance in a certain volume is its ________.
2. **Use the term** *chemical change* in a complete sentence.

Understand Key Concepts 

3. **List** some signs of chemical change.
4. Which property of matter changes during a chemical change but does NOT change during a physical change?
   A. energy
   B. identity
   C. mass
   D. volume
5. **State** why chemical equations are useful.
6. **Analyze** What affects the rate at which acid rain reacts with a statue?

Interpret Graphics

7. **Examine** Explain how the diagram below shows conservation of mass.

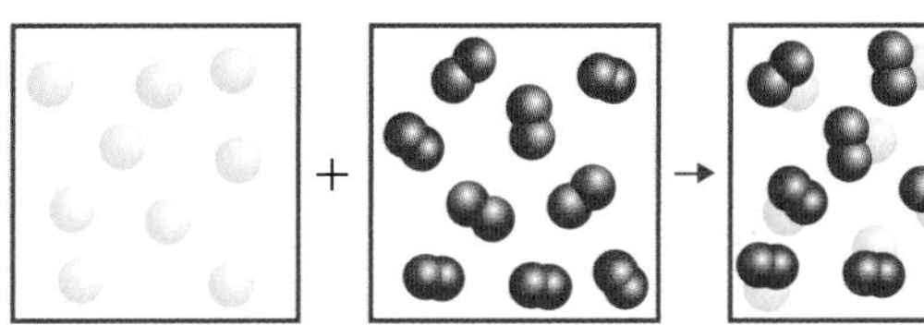

8. **Compare and Contrast** Copy and fill in the graphic organizer to compare and contrast physical and chemical changes.

Physical and Chemical Changes	
Alike	
Different	

Critical Thinking

9. **Compile** a list of three physical changes and three chemical changes you have observed recently.
10. **Recommend** How could you increase the rate at which the chemical reaction between vinegar and baking soda occurs?

1–2 class periods

Materials

triple-beam balance

50-mL graduated cylinder

magnifying lens

bar magnet

Also needed: crime scene evidence, unknown substances, dropper bottles containing water, iodine, cornstarch, and red cabbage indicator, test tubes, test tube rack, stirring rod

Safety

Design an Experiment to Solve a Crime

Recall how you can use properties to identify and compare substances. You now will apply those ideas to solving a crime. You will be given evidence collected from the crime scene and from the suspect's house. As the investigator, decide whether evidence from the crime scene matches evidence from the suspect. What tests will you use? What does the evidence tell you?

Question

Determine which factors about the evidence you would like to investigate further. Consider how you can describe and compare the properties of each piece of evidence. Evaluate the properties you will observe and measure, and decide whether it would be an advantage to classify them as physical properties or chemical properties. Will the changes that the evidence will undergo be helpful to you? Think about controls, variables, and the equipment you have available. Is there any way to match samples exactly?

Procedure

1. Read and complete a lab safety form.
2. In your Science Journal, write the procedures you will use to answer your question. Include the materials and steps you will use to test each piece of evidence. By the appropriate step in the procedure, list any safety procedures you should observe while performing the investigation. Organize your steps by putting them in a graphic organizer, such as the one below. Have your teacher approve your procedures.

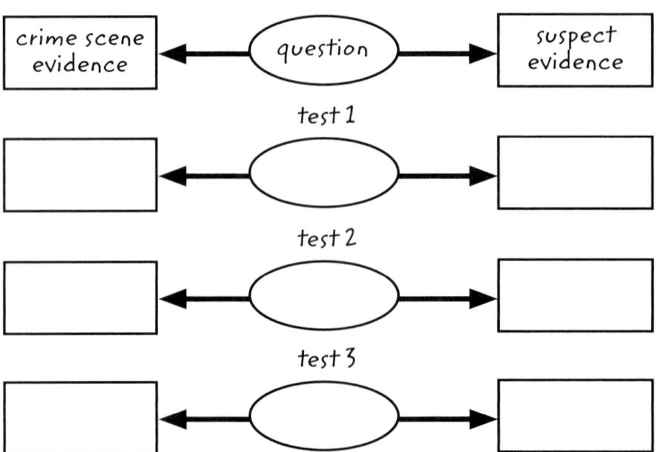

3 Begin by observing and recording your observations on each piece of evidence. What can you learn by comparing physical properties? Are any of the samples made of several parts?

4 Use the available materials to test the evidence. Accurately record all observations and data for each piece of evidence.

5 Add any additional tests you think you need to answer your questions.

Analyze and Conclude

6 Examine the data you have collected. What does the evidence tell you about whether the crime scene and the suspect are related?

7 Write your conclusions in your Science Journal. Be thorough because these are the notes you would use if you had to testify in court about the case.

8 **Analyze** Which data suggest that evidence from the crime scene was or wasn't connected to the suspect?

9 **Draw Conclusions** If you were to testify in court, what conclusions would you be able to state confidently based on your findings?

10 THE BIG IDEA **The Big Idea** How does understanding physical and chemical properties of matter help you to solve problems?

Communicate Your Results

Compare your results with those of other teams. Discuss the kinds of evidence that might be strong enough to convict a suspect.

Extension

Research the difference between individual and class evidence used in forensics. Decide which class of evidence your tests provided.

Lab TIPS

- ☑ Don't overlook simple ideas such as matching the edges of pieces.
- ☑ Can you separate any of the samples into other parts?
- ☑ Always get your teacher's approval before trying any new test.

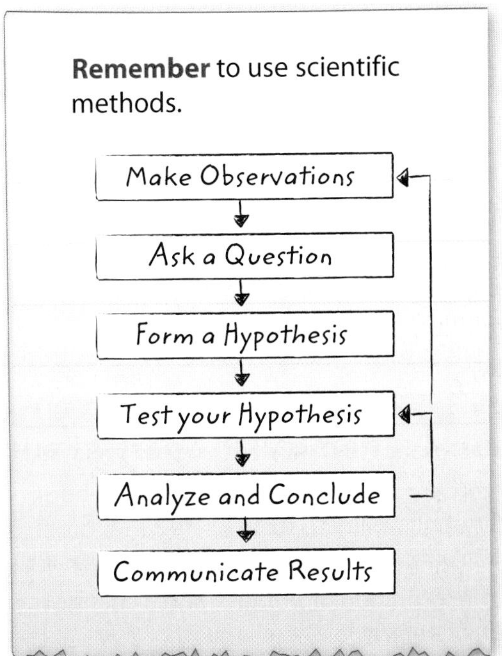

Chapter 7 Study Guide

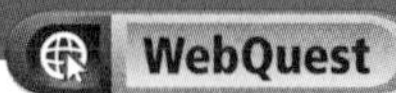

Matter is anything that has mass and takes up space. Its physical properties and its chemical properties can change.

Key Concepts Summary

Lesson 1: Classifying Matter

- A **substance** is a type of **matter** that always is made of atoms in the same combinations.
- **Atoms** of different elements have different numbers of protons.
- The composition of a substance cannot vary. The composition of a **mixture** can vary.
- Matter can be classified as either a substance or a mixture.

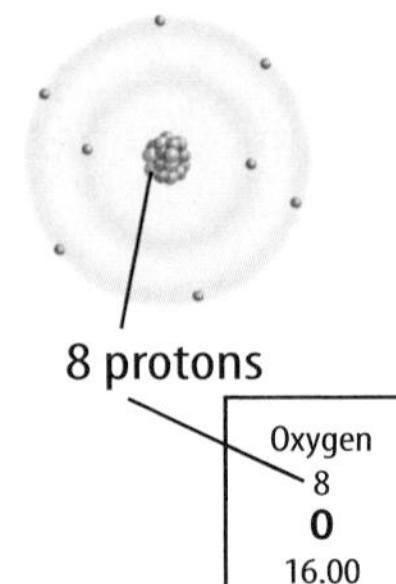

Vocabulary

matter p. 231
atom p. 231
substance p. 233
element p. 233
compound p. 234
mixture p. 235
heterogeneous mixture p. 235
homogeneous mixture p. 235
dissolve p. 235

Lesson 2: Physical Properties

- **Physical properties** of matter include size, shape, texture, and state.
- Physical properties such as **density,** melting point, boiling point, and size can be used to separate mixtures.

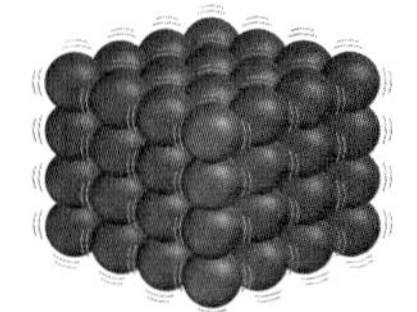

Vocabulary

physical property p. 240
mass p. 242
density p. 243
solubility p. 244

Lesson 3: Physical Changes

- A change in energy can change the state of matter.
- When something dissolves, it mixes evenly in a substance.
- The masses before and after a change in matter are equal.

Vocabulary

physical change p. 249

Lesson 4: Chemical Properties and Changes

- **Chemical properties** include ability to burn, acidity, and ability to rust.
- Some signs that might indicate **chemical changes** are the formation of bubbles and a change in odor, color, or energy.
- Chemical equations are useful because they show what happens during a chemical reaction.
- Some factors that affect the rate of chemical reactions are temperature, **concentration,** and surface area.

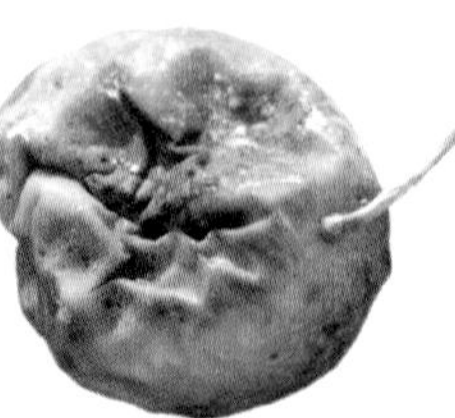

Vocabulary

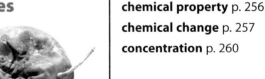

chemical property p. 256
chemical change p. 257
concentration p. 260

- Personal Tutor
- Vocabulary eGames
- Vocabulary eFlashcards

FOLDABLES® Chapter Project

Assemble your lesson Foldables as shown to make a Chapter Project. Use the project to review what you have learned in this chapter. Fasten the Foldable from Lesson 4 on the back of the board.

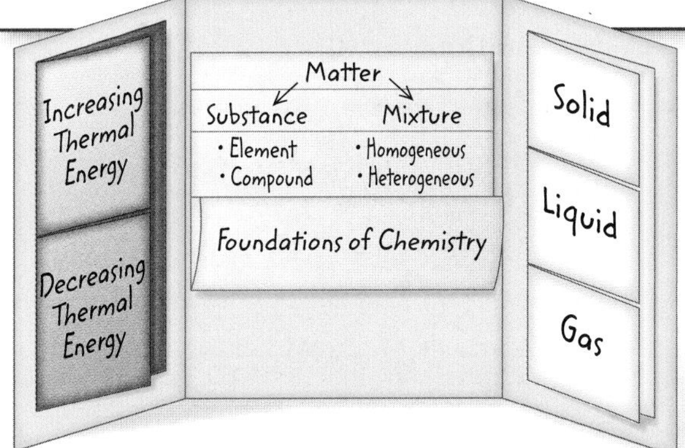

Use Vocabulary

Give two examples of each of the following.

1. element
2. compound
3. homogeneous mixture
4. heterogeneous mixture
5. physical property
6. chemical property
7. physical change
8. chemical change

Link Vocabulary and Key Concepts

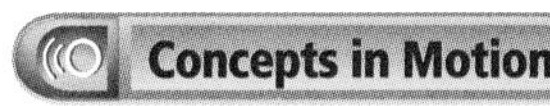

Interactive Concept Map

Copy this concept map, and then use vocabulary terms from the previous page to complete the concept map.

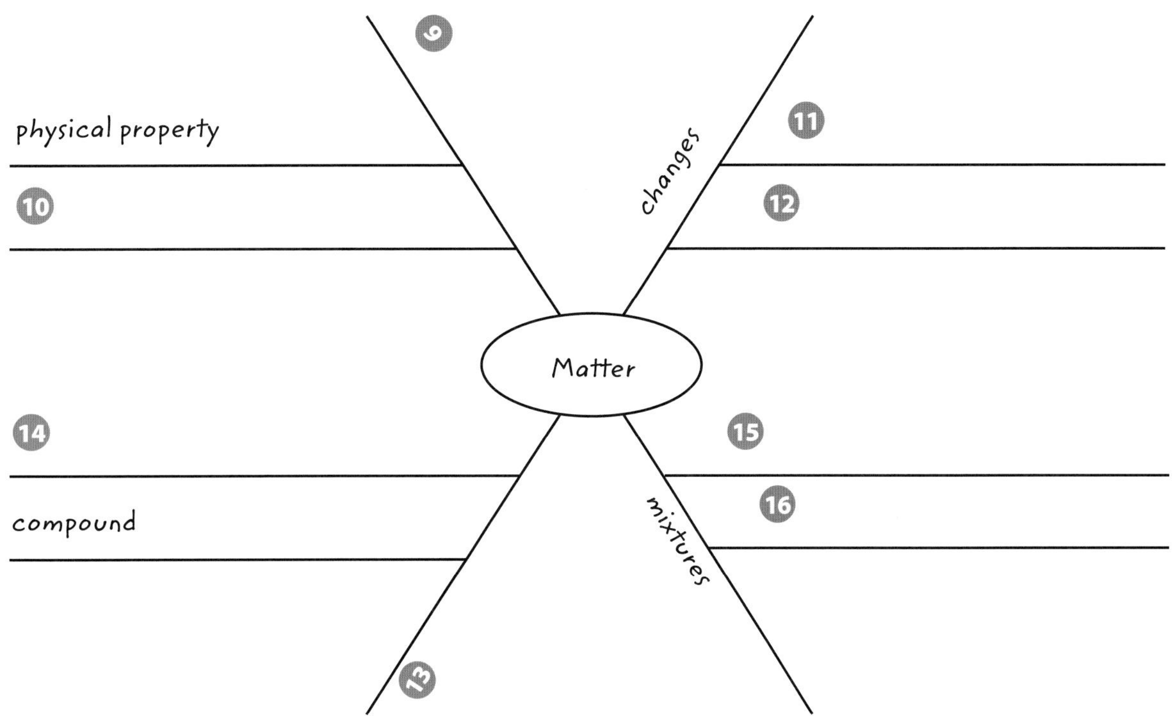

Chapter 7 Review

Understand Key Concepts 

1 The formula $AgNO_3$ represents a compound made of which atoms?

A. 1 Ag, 1 N, 1 O
B. 1 Ag, 1 N, 3 O
C. 1 Ag, 3 N, 3 O
D. 3 Ag, 3 N, 3 O

2 Which is an example of an element?

A. air
B. water
C. sodium
D. sugar

3 Which property explains why copper often is used in electrical wiring?

A. conductivity
B. density
C. magnetism
D. solubility

4 The table below shows densities for different substances.

Substance	Density (g/cm^3)
1	1.58
2	0.32
3	1.52
4	1.62

For which substance would a 4.90-g sample have a volume of 3.10 cm^3?

A. substance 1
B. substance 2
C. substance 3
D. substance 4

5 Which would decrease the rate of a chemical reaction?

A. increase in concentration
B. increase in temperature
C. decrease in surface area
D. increase in both surface area and concentration

6 Which physical change is represented by the diagram below?

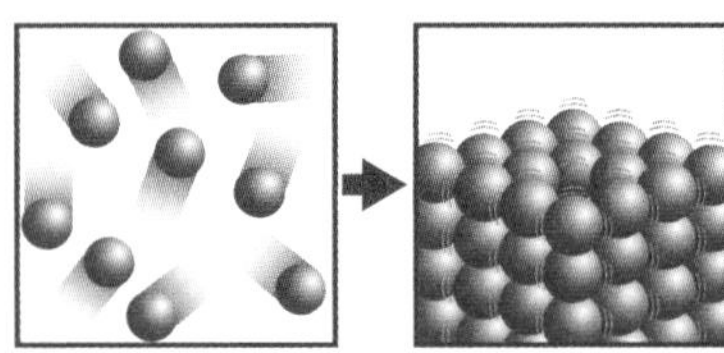

A. condensation
B. deposition
C. evaporation
D. sublimation

7 Which chemical equation is unbalanced?

A. $2KClO_3 \rightarrow 2KCl + 3O_2$
B. $CH_4 + 2O_2 \rightarrow CO_2 + 2H_2O$
C. $Fe_2O_3 + CO \rightarrow 2Fe + 2CO_2$
D. $H_2CO_3 \rightarrow H_2O + CO_2$

8 Which is a size-dependent property?

A. boiling point
B. conductivity
C. density
D. mass

9 Why is the following chemical equation said to be balanced?

$$O_2 + 2PCl_3 \rightarrow 2POCl_3$$

A. There are more reactants than products.
B. There are more products than reactants.
C. The atoms are the same on both sides of the equation.
D. The coefficients are the same on both sides of the equation.

10 The elements sodium (Na) and chlorine (Cl) react and form the compound sodium chloride (NaCl). Which is true about the properties of these substances?

A. Na and Cl have the same properties.
B. NaCl has the properties of Na and Cl.
C. All the substances have the same properties.
D. The properties of NaCl are different from the properties of Na and Cl.

Assessment
Online Test Practice

Critical Thinking

11 **Compile** a list of ten materials in your home. Classify each material as an element, a compound, or a mixture.

12 **Evaluate** Would a periodic table based on the number of electrons in an atom be as effective as the one shown in the back of this book? Why or why not?

13 **Develop** a demonstration to show how weight is not the same thing as mass.

14 **Construct** an explanation for how the temperature and energy of a material changes during the physical changes represented by the diagram below.

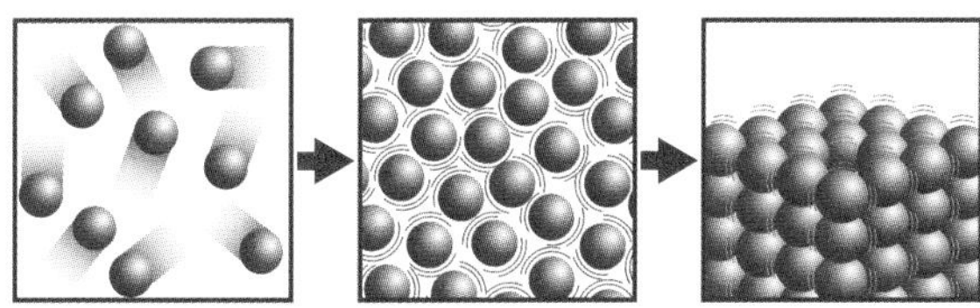

15 **Revise** the definition of physical change given in this chapter so it mentions the type and arrangement of atoms.

16 **Find an example** of a physical change in your home or school. Describe the changes in physical properties that occur during the change. Then explain how you know the change is not a chemical change.

17 **Develop** a list of five chemical reactions you observe each day. For each, describe one way that you could either increase or decrease the rate of the reaction.

Writing in Science

18 **Write** a poem at least five lines long to describe the organization of matter by the arrangement of its atoms. Be sure to include both the names of the different types of matter as well as their meanings.

REVIEW THE BIG IDEA

19 Explain how you are made of matter that undergoes changes. Provide specific examples in your explanation.

20 How does the photo below show an example of a physical change, a chemical change, a physical property, and a chemical property?

Math Skills

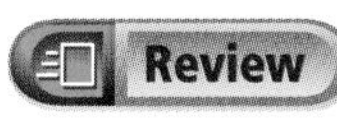

Math Practice

Use Ratios

21 A sample of ice at 0°C has a mass of 23 g and a volume of 25 cm^3. Why does ice float on water? (The density of water is 1.00 g/cm^3.)

22 The table below shows the masses and the volumes for samples of two different elements.

Element	Mass (g)	Volume (cm^3)
Gold	386	20
Lead	22.7	2.0

Which element sample in the table has greater density?

Standardized Test Practice

Record your answers on the answer sheet provided by your teacher or on a sheet of paper.

Multiple Choice

1 Which describes how mixtures differ from substances?

A Mixtures are homogeneous.

B Mixtures are liquids.

C Mixtures can be separated physically.

D Mixtures contain only one kind of atom.

Use the figure below to answer question 2.

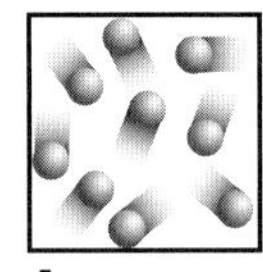
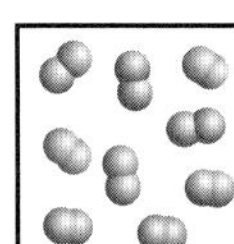
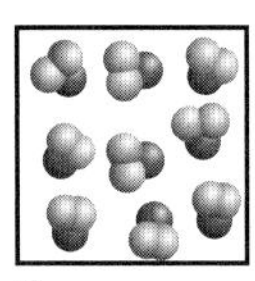

A B C D

2 Which image in the figure above is a model for a compound?

A A

B B

C C

D D

3 Which is a chemical property?

A the ability to be compressed

B the ability to be stretched into thin wire

C the ability to melt at low temperature

D the ability to react with oxygen

4 You drop a sugar cube into a cup of hot tea. What causes the sugar to disappear in the tea?

A It breaks into elements.

B It evaporates.

C It melts.

D It mixes evenly.

5 Which is an example of a substance?

A air

B lemonade

C soil

D water

Use the figure below to answer question 6.

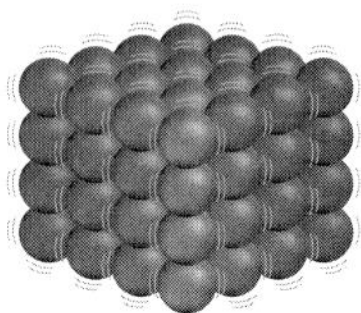

6 The figure above is a model of atoms in a sample at room temperature. Which physical property does this sample have?

A It can be poured.

B It can expand to fill its container.

C It has a high density.

D It has a low boiling point.

7 Which observation is a sign of a chemical change?

A bubbles escaping from a carbonated drink

B iron filings sticking to a magnet

C lights flashing from fireworks

D water turning to ice in a freezer

8 Zinc, a solid metal, reacts with a hydrochloric acid solution. Which will increase the reaction rate?

A cutting the zinc into smaller pieces

B decreasing the concentration of the acid

C lowering the temperature of the zinc

D pouring the acid into a larger container

Use the figure below to answer question 9.

9 In the figure above, what will be the mass of the final solution if the solid dissolves in the water?

A 5 g

B 145 g

C 150 g

D 155 g

10 Which is NOT represented in a chemical equation?

A chemical formula

B product

C conservation of mass

D reaction rate

Constructed Response

Use the graph below to answer questions 11 and 12.

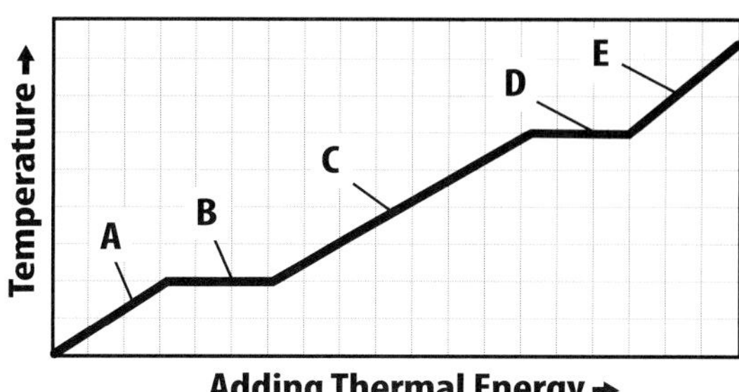

11 Use the graph above to explain why ice will keep water cold on a hot day.

12 Use two sections of the graph to explain what happens when you put a pot of cold water on a stove to boil. Specify which two sections you used.

13 Describe how you would separate a mixture of sugar, sand, and water.

14 The reaction of zinc metal with hydrochloric acid produces zinc chloride and hydrogen gas. A student writes the following to represent the reaction.

$$Zn + HCl \rightarrow ZnCl_2 + H_2$$

Is the equation correct? Use conservation of mass to support your answer.

NEED EXTRA HELP?														
If You Missed Question...	1	2	3	4	5	6	7	8	9	10	11	12	13	14
Go to Lesson...	1	1	4	3	1	2	4	4	3	4	3	3	2	4

Chapter 8

States of Matter

THE BIG IDEA What physical changes and energy changes occur as matter goes from one state to another?

Liquid Glass?

When you look at this blob of molten glass, can you envision it as a beautiful vase? The solid glass was heated in a furnace until it formed a molten liquid. Air is blown through a pipe to make the glass hollow and give it form.

- Can you identify a solid, a liquid, and a gas in the photo?
- What physical changes and energy changes do you think occurred when the glass changed state?

Get Ready to Read

What do you think?

Before you read, decide if you agree or disagree with each of these statements. As you read this chapter, see if you change your mind about any of the statements.

1. Particles moving at the same speed make up all matter.
2. The particles in a solid do not move.
3. Particles of matter have both potential energy and kinetic energy.
4. When a solid melts, thermal energy is removed from the solid.
5. Changes in temperature and pressure affect gas behavior.
6. If the pressure on a gas increases, the volume of the gas also increases.

Your one-stop online resource

connectED.mcgraw-hill.com

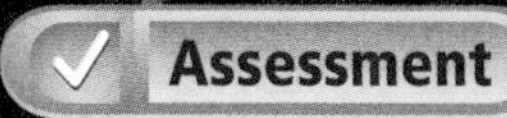

Concepts in Motion

Multilingual eGlossary

Lesson 1

Reading Guide

Key Concepts
ESSENTIAL QUESTIONS

- How do particles move in solids, liquids, and gases?
- How are the forces between particles different in solids, liquids, and gases?

Vocabulary

solid p. 275
liquid p. 276
viscosity p. 276
surface tension p. 277
gas p. 278
vapor p. 278

Multilingual eGlossary

Solids, Liquids, and Gases

Inquiry Giant Bubbles?

Giant bubbles can be made from a solution of water, soap, and a syrupy liquid called glycerine. These liquids change the properties of water. Soap changes water's surface tension. Glycerine changes the evaporation rate. How do surface tension and evaporation work?

Launch Lab

10 minutes

How can you see particles in matter?

It's sometimes difficult to picture how tiny objects, such as the particles that make up matter, move. However, you can use other objects to model the movement of these particles.

1. Read and complete a lab safety form.
2. Place about 50 **copper pellets** into a **plastic petri dish.** Place the cover on the dish, and secure it with **tape.**
3. Hold the dish by the edges. Gently vibrate the dish from side to side no more than 1–2 mm. Observe the pellets. Record your observations in your Science Journal.
4. Repeat step 3, vibrating the dish less than 1 cm from side to side.
5. Repeat step 3, vibrating the dish 3–4 cm from side to side.

Think About This

1. If the pellets represent particles in matter, what do you think the shaking represents?
2. In which part of the experiment do you think the pellets were like a liquid? Explain.
3.  **Key Concept** If the pellets represent molecules of water, what do you think are the main differences among molecules of ice, water, and vapor?

Describing Matter

Take a closer look at the photo on the previous page. Do you see **matter?** The three most common forms, or states, of matter on Earth are solids, liquids, and gases. The giant bubble contains air, which is a mixture of gases. The ocean water and the soap mixture used to make the bubble are liquids. The sand, sign, and walkway are a few of the solids in the photo.

Review Vocabulary

matter
anything that takes up space and has mass

There is a fourth state of matter, plasma, that is not shown in this photo. Plasma is high-energy matter consisting of positively and negatively charged particles. Plasma is the most common state of matter in space. It also is in lightning flashes, fluorescent lighting, and stars, such as the Sun.

There are many ways to describe matter. You can describe the state, the color, the texture, and the odor of matter using your senses. You also can describe matter using measurements, such as mass, volume, and density. Mass is the amount of matter in an object. The units for mass are often grams (g) or kilograms (kg). Volume is the amount of space that a sample of matter occupies. The units for liquid volume are usually liters (L) or milliliters (mL). The units for solid volume are usually cubic centimeters (cm^3) or cubic meters (m^3). Density is the mass per unit volume of a substance. The units are usually g/cm^3 or g/mL. Density of a given substance remains constant, regardless of the size of the sample.

Particles in Motion

Have you ever wondered what makes something a solid, a liquid, or a gas? Two main factors that determine the state of matter are particle motion and particle forces.

Particles, such as atoms, ions, or molecules, moving in different ways make up all matter. The particles that make up some matter are close together and vibrate back and forth. In other types of matter, the particles are farther apart, move freely, and can spread out. Regardless of how close particles are to each other, they all move random motion—movement in all directions and at different speeds. However, particles will move in straight lines until they collide with something. Collisions usually change the speed and direction of the particles' movements.

FOLDABLES®

Use a sheet of notebook paper to make a three-tab Foldable as shown. Record information about each state of matter under the tabs.

Forces Between Particles

Recall that atoms that make up matter contain positively charged protons and negatively charged electrons. There is a force of attractions between these oppositely charged particles, as shown in **Figure 1.**

You just read that the particles that make up matter move at all speeds and in all directions. If the motion of particles slows, the particles move closer together. This is because the attraction between them pulls them toward each other. Strong attractive forces hold particles close together. As the motion of particles increases, particles move farther apart. The attractive forces between particles get weaker. The spaces between them increase and the particles can slip past one another. As the motion of particles continues to increase, they move even farther apart. Eventually, the distance between particles is so great that there is little or no attractive forces between the particles. The particles move randomly and spread out. As you continue to read, you will learn how particle motion and particle forces determine whether matter is a solid, a liquid, or a gas.

Figure 1 The forces between particles of matter and the movement of particles determine the physical state of matter.

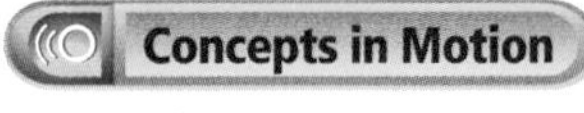

Animation

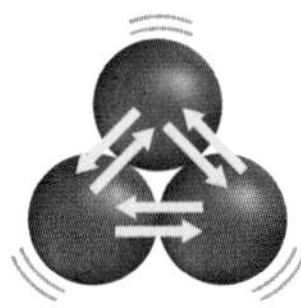

Particles move slowly and can only vibrate in place. Therefore, the attractive forces between particles are strong.

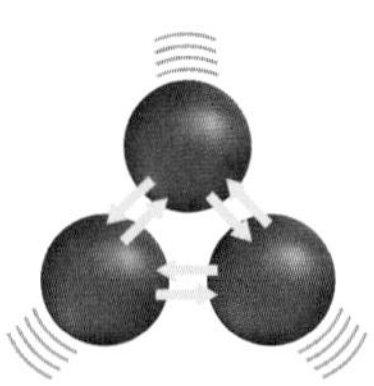

Particles move faster and slip past each other. The distance between particles increases. Therefore, the attractive forces between particles are weaker.

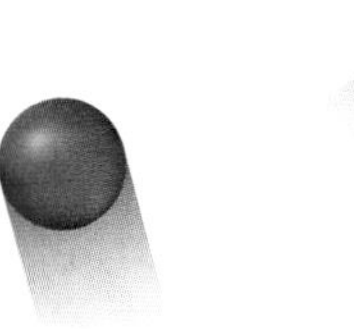

Particles move fast. The distance between the particles is great, and therefore, the attractive forces between particles are very weak.

Solids

If you had to describe a solid, what would you say? You might say, a **solid** *is matter that has a definite shape and a definite volume.* For example, if the skateboard in **Figure 2** moves from one location to another, the shape and volume of it do not change.

Particles in a Solid

Why doesn't a solid change its shape and volume? Notice in **Figure 2** how the particles in a solid are close together. The particles are very close to their neighboring particles. That's because the attractive forces between the particles are strong and hold them close together. The strong attractive forces and slow motion of the particles keep them tightly held in their positions. The particles simply vibrate back and forth in place. This arrangement gives solids a definite shape and volume.

Key Concept Check Describe the movement of particles in a solid and the forces between them.

Types of Solids

All solids are not the same. For example, a diamond and a piece of charcoal don't look alike. However, they are both solids made of only carbon atoms. A diamond and a lump of charcoal both contain particles that strongly attract each other and vibrate in place. What makes them different is the arrangement of their particles. Notice in **Figure 3** that the arrangement of particles in a diamond is different from that in charcoal. A diamond is a crystalline solid. It has particles arranged in a specific, repeating order. Charcoal is an amorphous solid. It has particles arranged randomly. Different particle arrangements give these materials different properties. For example, a diamond is a hard material, and charcoal is a brittle material.

Reading Check What is the difference between crystalline and amorphous solids?

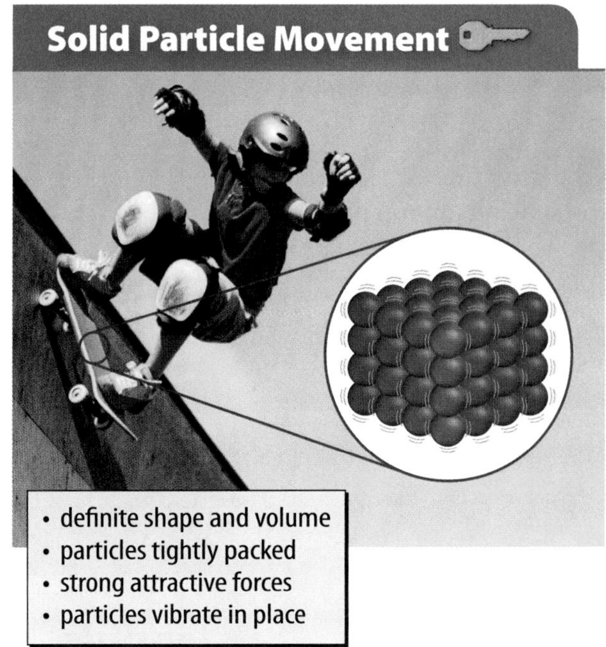

Solid Particle Movement
• definite shape and volume
• particles tightly packed
• strong attractive forces
• particles vibrate in place

▲ **Figure 2** The particles in a solid have strong attractive forces and vibrate in place.

Figure 3 Carbon is a solid that can have different particle arrangements. ▼

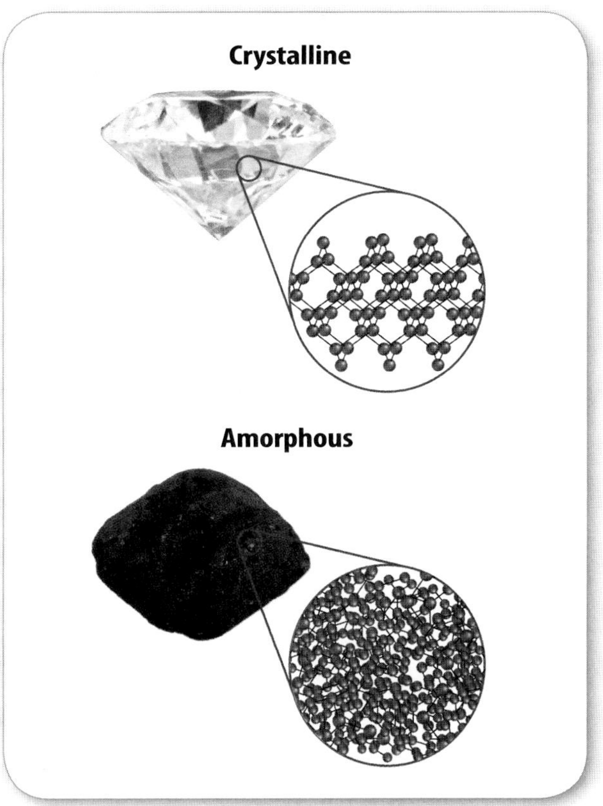

Figure 4 The motion of particles in a liquid causes the particles to move slightly farther apart. ▶

Visual Check How does the spacing among these particles compare to the particle spacing in **Figure 2?**

Liquid Particle Movement
• no definite shape, has definite volume
• particles free to move past other particles
• attractive forces weaker than those in solids

Liquids

You have probably seen a waterfall, such as the one in **Figure 4.** Water is a liquid. *A* **liquid** *is matter with a definite volume but no definite shape.* Liquids flow and can take the shape of their containers. The container for this water is the riverbed.

Particles in a Liquid

How can liquids change their shape? The particle motion in the liquid state of a material is faster than the particle motion in the solid state. This increased particle motion causes the particles to move slightly farther apart. As the particles move farther apart, the attractive forces between the particles decrease. The weaker attractive forces allow particles to slip past one another. The weather forces also enable liquids to flow and take the shape of their containers.

WORD ORIGIN

viscosity
from Latin *viscum,* means "sticky"

Viscosity

If you have ever poured or dipped honey, as shown in **Figure 5,** you have experienced a liquid with a high viscosity. **Viscosity** (vihs KAW sih tee) *is a measurement of a liquid's resistance to flow.* Honey has high viscosity, while water has low viscosity. Viscosity is due to particle mass, particle shape, and the strength of the attraction between the particles of a liquid. In general, the stronger the attractive forces between particles, the higher the viscosity. For many liquids, viscosity decreases as the liquid becomes warmer. As a liquid becomes warmer, particles begin to move faster and the attractive forces between them get weaker. This allows particles to more easily slip past one another. The mass and shape of particles that make up a liquid also affect viscosity. Large particles or particles with complex shapes tend to move more slowly and have difficulty slipping past one another.

Figure 5 Honey has a high viscosity. ▼

Figure 6 The surface tension of water enables this spider to walk on the surface of a lake.

Surface Tension

How can the nursery web spider in **Figure 6** walk on water? Believe it or not, it is because of the interactions between molecules.

The blowout in **Figure 6** shows the attractive forces between water molecules. Water molecules below the surface are surrounded on all sides by other water molecules. Therefore, they have attractive forces, or pulls, in all directions. The attraction between similar molecules, such as water molecules, is called cohesion.

Water molecules at the surface of a liquid do not have liquid water molecules above them. As a result, they experience a greater downward pull, and the surface particles become tightly stretched like the head of a drum. Molecules at the surface of a liquid have **surface tension**, *the uneven forces acting on the particles on the surface of a liquid.* Surface tension allows a spider to walk on water. In general, the stronger the attractive forces between particles, the greater the surface tension of the liquid.

Recall the giant bubbles at the beginning of the chapter. The thin water-soap film surrounding the bubbles forms because of surface tension between the particles.

 **Key Concept Check** Describe the movement of particles in a liquid and the forces between them.

Inquiry MiniLab 20 minutes

How can you make bubble films?

Have you ever observed surface tension? Which liquids have greater surface tension?

1. Read and complete a lab safety form.
2. Place about 100 mL of cool water in a **small bowl.** Lower a **wire bubble frame** into the bowl, and gently lift it. Use a **magnifying lens** to observe the edges of the frame. Write your observations in your Science Journal.
3. Add a full **dropper** of **liquid dishwashing soap** to the water. Stir with a **toothpick** until mixed. Lower the frame into the mixture and lift it out. Record your observations.
4. Use a toothpick to break the bubble film on one side of the thread. Observe.

Analyze and Conclude

1. **Recognize Cause and Effect** Explain what caused the thread to form an arc when half the bubble film broke.
2. **Key Concept** Explain why pure water doesn't form bubbles. What happens to the forces between water molecules when you add soap?

Gas Particle Movement

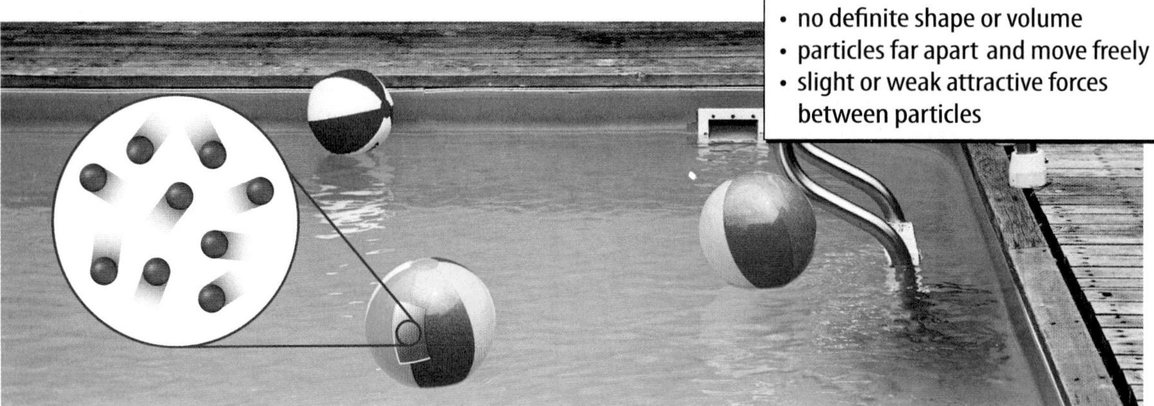

• no definite shape or volume
• particles far apart and move freely
• slight or weak attractive forces between particles

Figure 7 The particles in a gas are far apart, and there is little or no attractive forces are between particles.

Visual Check What are gas particles likely to hit as they move?

Gases

Look at the photograph in **Figure 7.** Where is the gas? *A* **gas** *is matter that has no definite volume and no definite shape.* It is not easy to identify the gas because you cannot see it. However, gas particles are inside and outside the inflatable balls. Air is a mixture of gases, including nitrogen, oxygen, argon, and carbon dioxide.

Reading Check What is a gas, and what is another object that contains a gas?

Particles in a Gas

Why don't gases have definite volumes or definite shapes like solids and liquids? Compare the particles in **Figures 2, 4,** and **7.** Notice how the distance between particles differs. As the particles move faster, such as when matter goes from the solid state to the liquid state, the particles move farther apart. When the particles in matter move even faster, such as when matter goes from the liquid state to the gas state, the particles move even farther apart. When the distances between particles changes, the attractive forces between the particles also changes.

Forces Between Particles

As a type of matter goes from the solid state to the liquid state, the distance between the particles increases and the attractive forces between the particles decreases. When the same matter goes from the liquid state to the gas state, the particles are even farther apart and the attractive forces between the particles are weak or absent. As a result, the particles spread out to fill their container. Because gas particles lack attractive forces between particles, they have no definite shape or definite volume.

Vapor

Have you ever heard the term *vapor? The gas state of a substance that is normally a solid or a liquid at room temperature is called* **vapor.** For example, water is normally a liquid at room temperature. When it is a gas state, such as in air, it is called water vapor. Other substances that can form a vapor are rubbing alcohol, iodine, mercury, and gasoline.

Key Concept Check How do particles move and interact in a gas?

Lesson 1 Review

Assessment Online Quiz

Visual Summary

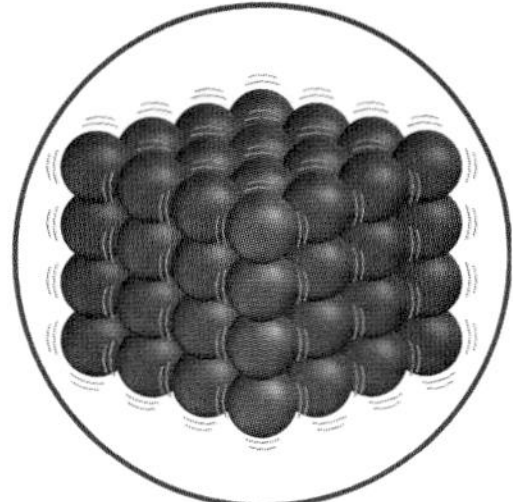

The particles that make up a solid can only vibrate in place. The particles are close together, and there are strong forces among them.

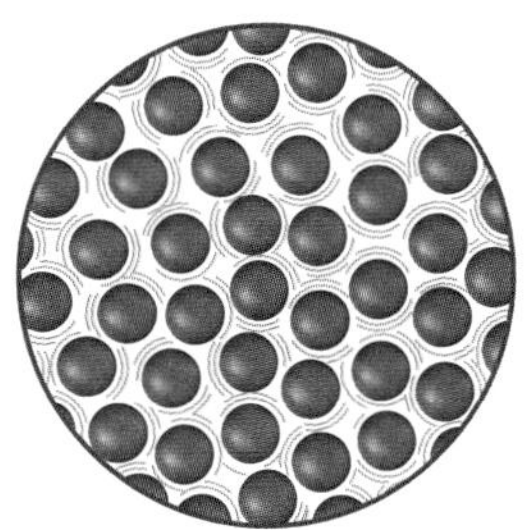

The particles that make up a liquid are far enough apart that particles can flow past other particles. The forces among these particles are weaker than those in a solid.

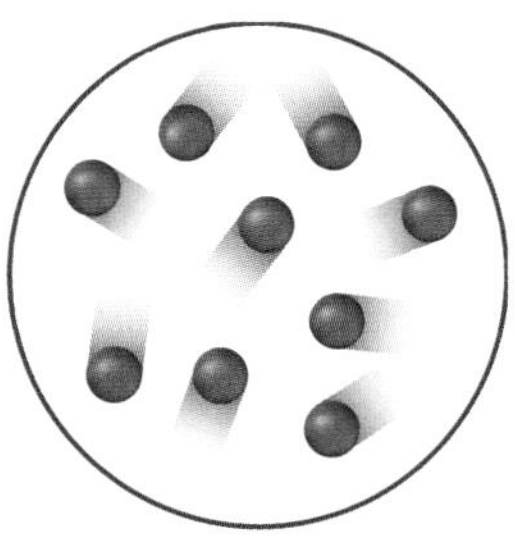

The particles that make up a gases are far apart. There is little or no attraction among the particles.

FOLDABLES®

Use your lesson Foldable to review the lesson. Save your Foldable for the project at the end of the chapter.

What do you think NOW?

You first read the statements below at the beginning of the chapter.

1. Particles moving at the same speed make up all matter.
2. The particles in a solid do not move.

Did you change your mind about whether you agree or disagree with the statements? Rewrite any false statements to make them true.

Use Vocabulary

1. A measurement of how strongly particles attract one another at the surface of a liquid is __________.
2. **Define** *solid, liquid,* and *gas* in your own words.
3. A measurement of a liquid's resistance to flow is known as __________.

Understand Key Concepts 

4. Which state of matter rarely is found on Earth?
   A. gas
   B. liquid
   C. plasma
   D. solid
5. **Compare** particle movement in solids, liquids, and gases.
6. **Compare** the forces between particles in a liquid and in a gas.

Interpret Graphics

7. **Explain** why the particles at the surface in the image below have surface tension while the particles below the surface do not.

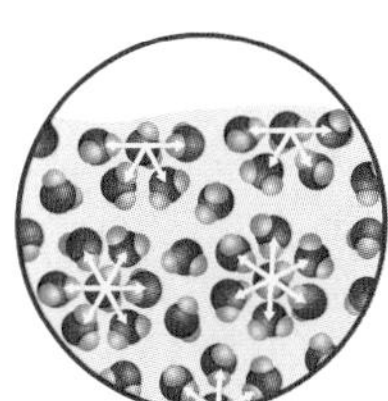

8. **Summarize** Copy and fill in the graphic organizer to compare two types of solids.

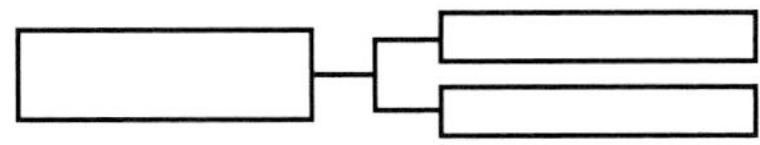

Critical Thinking

9. **Hypothesize** how you could change the viscosity of a cold liquid, and explain why your idea would work.
10. **Summarize** the relationship between the motion of particles and attractive forces between particles.

HOW IT WORKS

Freeze-Drying Foods

Have you noticed that the berries you find in some breakfast cereals are lightweight and dry—much different from the berries you get from the market or the garden?

Fresh fruit would spoil quickly if it were packaged in breakfast cereal, so fruits in cereals are often freeze-dried. When liquid is returned to the freeze-dried fruit, its physical properties more closely resemble fresh fruit. Freeze-drying, or lyophilization (lie ah fuh luh ZAY shun), is the process in which a solvent (usually water) is removed from a solid. During this process, a frozen solvent changes to a gas without going through the liquid state. Freeze-dried foods are lightweight and long-lasting. Astronauts have been using freeze-dried food during space travel since the 1960s.

How Freeze-Drying Works

1 Machines called freeze-dryers are used to freeze-dry foods and other products. Fresh or cooked food is flash-frozen, changing moisture in the food to a solid.

2 The frozen food is placed in a large vacuum chamber, where moisture is removed. Heat is applied to accelerate moisture removal. Condenser plates remove vaporized solvent from the chamber and convert the frozen food to a freeze-dried solid.

3 Freeze-dried food is sealed in oxygen- and moisture-proof packages to ensure stability and freshness. When the food is rehydrated, it returns to its near-normal state of weight, color, and texture.

ASTRONAUT
Ice Cream
NET WT .7 OZ (19g)

It's Your Turn

PREDICT/DISCOVER What kinds of products besides food are freeze-dried? Use library or internet resources to learn about other products that undergo the freeze-drying process. Discuss the benefits or drawbacks of freeze-drying.

Lesson 2

Changes in State

Reading Guide

Key Concepts
ESSENTIAL QUESTIONS

- How is temperature related to particle motion?
- How are temperature and thermal energy different?
- What happens to thermal energy when matter changes from one state to another?

Vocabulary

kinetic energy p. 282
temperature p. 282
thermal energy p. 283
vaporization p. 285
evaporation p. 286
condensation p. 286
sublimation p. 286
deposition p. 286

 Multilingual eGlossary

 Video

- BrainPOP®
- What's Science Got to do With It?

Inquiry Spring Thaw?

When you look at a snowman, you probably don't think about states of matter. However, water is one of the few substances that you frequently observe in three states of matter at Earth's temperatures. What energy changes are involved when matter changes state?

10 minutes

Do liquid particles move?

If you look at a glass of milk sitting on a table, it appears to have no motion. But appearances can be deceiving!

1. Read and complete a lab safety form.
2. Use a **dropper,** and place one drop of **2 percent milk** on a **glass slide.** Add a **cover slip.**
3. Place the slide on a **microscope** stage, and focus on low power. Focus on a single globule of fat in the milk. Observe the motion of the globule for several minutes. Record your observations in your Science Journal.

Think About This

1. Describe the motion of the fat globule.
2. What do you think caused the motion of the globule?
3. **Key Concept** What do you think would happen to the motion of the fat globule if you warmed the milk? Explain.

Kinetic and Potential Energy

When snow begins to melt after a snowstorm, all three states of water are present. The snow is a solid, the melted snow is a liquid, and the air above the snow and ice contains water vapor, a gas. What causes particles to change state?

Kinetic Energy

Recall that the particles that make up matter are in constant motion. These particles have **kinetic energy,** *the energy an object has due to its motion.* The faster particles move, the more kinetic energy they have. Within a given substance, such as water, particles in the solid state have the least amount of kinetic energy. This is because they only vibrate in place. Particles in the liquid state move faster than particles in the solid state. Therefore, they have more kinetic energy. Particles in the gaseous state move very quickly and have the most kinetic energy of particles of a given substance.

Temperature *is a measure of the average kinetic energy of all the particles in an object.* Within a given substance, a temperature increase means that the particles, on average, are moving at greater speeds, or have a greater average kinetic energy. For example, water molecules at 25°C are generally moving faster and have more kinetic energy than water molecules at 10°C.

Key Concept Check How is temperature related to particle motion?

Potential Energy

In addition to kinetic energy, particles have potential energy. Potential energy is stored energy due to the interactions between particles or objects. For example, when you pick up a ball and then let it go, the gravitational force between the ball and Earth causes the ball to fall toward Earth. Before you let the ball go, it has potential energy.

Potential energy typically increases when objects get farther apart and decreases when they get closer together. The basketball in the top part of **Figure 8** is farther off the ground than it is in the bottom part of the figure. The farther an object is from Earth's surface, the greater the gravitational potential energy. As the ball gets closer to the ground, the potential energy decreases.

You can think of the potential energy of particles in a similar way. The chemical potential energy is due to the position of the particles relative to other particles. The chemical potential energy of particles increases and decreases as the distances between particles increases or decreases. The particles in the top part of **Figure 8** are farther apart than the particles in the bottom part. The particles that are farther apart have greater chemical potential energy.

Thermal Energy

Thermal energy *is the total potential and kinetic energies of an object.* You can change an object's state of matter by adding or removing thermal energy. When you add thermal energy to an object, the particles either move faster (increased kinetic energy) or get farther apart (increased potential energy) or both. The opposite is true when you remove thermal energy from an object. If enough thermal energy is added or removed, a change of state can occur.

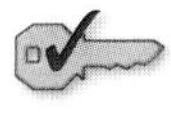

Key Concept Check How do thermal energy and temperature differ?

Figure 8 The potential energy of the ball depends on the distance between the ball and Earth. The potential energy of particles in matter depends on the distances between the particles.

Greater Potential Energy

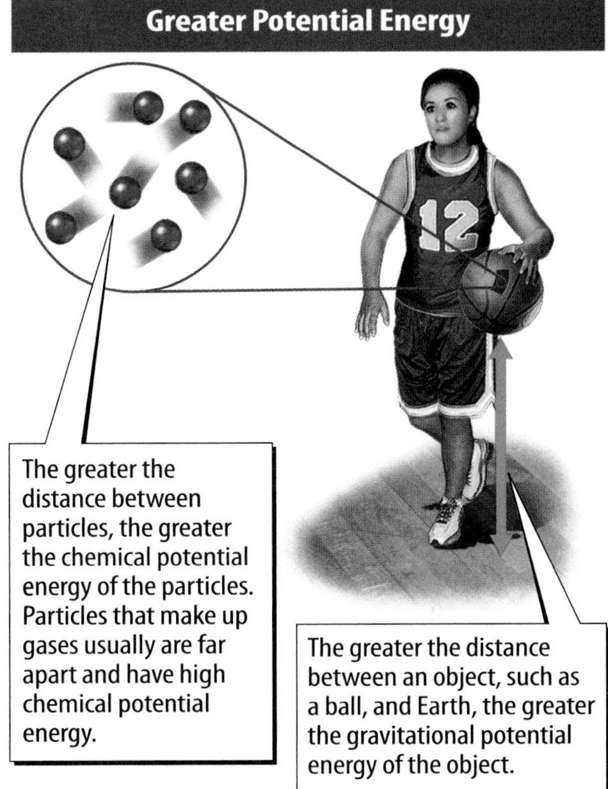

Less Potential Energy

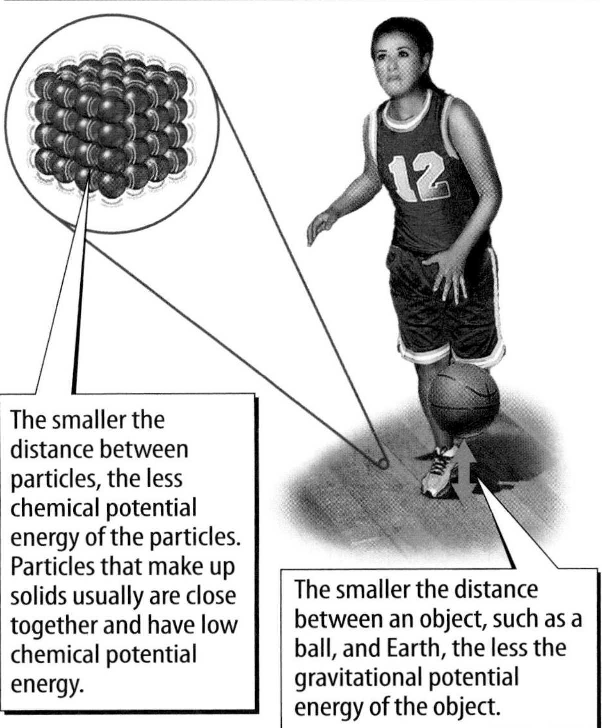

12
The smaller the distance between particles, the less chemical potential energy of the particles. Particles that make up solids usually are close together and have low chemical potential energy.
The smaller the distance between an object, such as a ball, and Earth, the less the gravitational potential energy of the object.

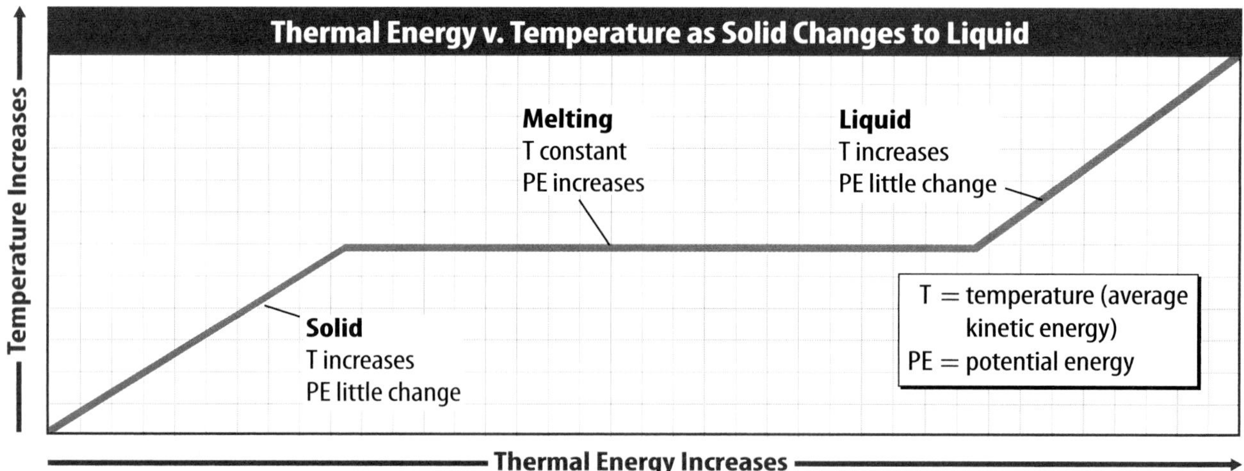

Figure 9 Adding thermal energy to matter causes the particles that make up the matter to increase in kinetic energy, potential energy, or both.

Visual Check During melting, which factor remains constant?

Solid to Liquid or Liquid to Solid

When you drink a beverage from an aluminum can, do you recycle the can? Aluminum recycling is one example of a process that involves changing matter from one state to another by adding or removing thermal energy.

Melting

The first part of the recycling process involves melting aluminum cans. To change matter from a solid to a liquid, thermal energy must be added. The graph in **Figure 9** shows the relationship between increasing temperature and increasing thermal energy (potential energy + kinetic energy).

At first, both the thermal energy and the temperature increase. The temperature stops increasing when it reaches the melting point of the matter, the temperature at which the solid state changes to the liquid state. As aluminum changes from solid to liquid, the temperature does not change. However, energy changes still occur.

Reading Check What is added to matter to change it from a solid to a liquid?

Energy Changes

What happens when a solid reaches its melting point? Notice the line on the graph is horizontal. This means that the temperature, or average kinetic energy, stops increasing. However, the amount of thermal energy continues to increase. How is this possible?

Once a solid reaches its melting point, the average speed of particles does not change, but the distance between the particles does change. The particles move farther apart and potential energy increases. Once a solid completely melts, the addition of thermal energy will cause the kinetic energy of the particles to increase again, as shown by a temperature increase.

Freezing

After the aluminum melts, it is poured into molds to cool. As the aluminum cools, thermal energy leaves it. Freezing is a process that is the reverse of melting. The temperature at which matter changes from the liquid state to the solid state is its freezing point. To observe the temperature and thermal energy changes that occur to hot aluminum blocks, move from right to left on the graph in **Figure 9.**

During evaporation, a liquid vaporizes only at its surface.
During boiling, a liquid vaporizes at its surface and within the liquid.
Bubbles, or vaporized particles, rise to the top of the liquid and escape from the container.

▲ **Figure 10** Boiling and evaporation are two kinds of vaporization.

Visual Check Why doesn't the evaporation flask have bubbles below the surface?

Personal Tutor

Liquid to Gas or Gas to Liquid

When you heat water, do you ever notice how bubbles begin to form at the bottom and rise to the surface? The bubbles contain water vapor, a gas. *The change in state of a liquid into a gas is* **vaporization.** **Figure 10** shows two types of vaporization—evaporation and boiling.

Boiling

Vaporization that occurs within a liquid is called boiling. The temperature at which boiling occurs in a liquid is called its boiling point. In **Figure 11,** notice the energy changes that occur during this process. The kinetic energy of particles increases until the liquid reaches its boiling point.

At the boiling point, the potential energy of particles begins increasing. The particles move farther apart until the attractive forces no longer hold them together. At this point, the liquid changes to a gas. When boiling ends, if thermal energy continues to be added, the kinetic energy of the gas particles begins to increase again. Therefore, the temperature begins to increase again as shown on the graph.

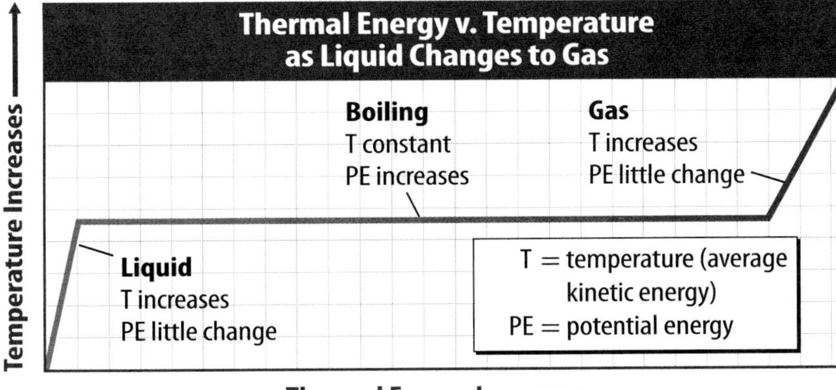

◀ **Figure 11** When thermal energy is added to a liquid, kinetic energy and potential energy changes occur.

WORD ORIGIN

evaporation
from Latin *evaporare,* means "disperse in steam or vapor"

Evaporation

Unlike boiling, **evaporation** *is vaporization that occurs only at the surface of a liquid.* Liquid in an open container will vaporize, or change to a gas, over time due to evaporation.

Condensation

Boiling and evaporation are processes that change a liquid to a gas. A reverse process also occurs. When a gas loses enough thermal energy, the gas changes to a liquid, or condenses. *The change of state from a gas to a liquid is called* **condensation.** Overnight, water vapor often condenses on blades of grass, forming dew.

Solid to Gas or Gas to Solid

Is it possible for a solid to become a gas without turning to a liquid first? Yes, in fact, dry ice does. Dry ice, as shown in **Figure 12,** is solid carbon dioxide. It turns immediately into a gas when thermal energy is added to it. The process is called sublimation. **Sublimation** *is the change of state from a solid to a gas without going through the liquid state.* As dry ice sublimes, it cools and condenses the water vapor in the surrounding air, creating a thick fog.

SCIENCE USE V. COMMON USE

deposition
Science Use the change of state of a gas to a solid without going through the liquid state

Common Use giving a legal testimony under oath

The opposite of sublimation is deposition. **Deposition** *is the change of state of a gas to a solid without going through the liquid state.* For deposition to occur, thermal energy has to be removed from the gas. You might see deposition in autumn when you wake up and there is frost on the grass. As water vapor loses thermal energy, it changes into a solid known as frost.

Reading Check Why are sublimation and deposition unusual changes of state?

Figure 12 Dry ice sublimes—goes directly from the solid state to the gas state—when thermal energy is added. Frost is an example of the opposite process—deposition.

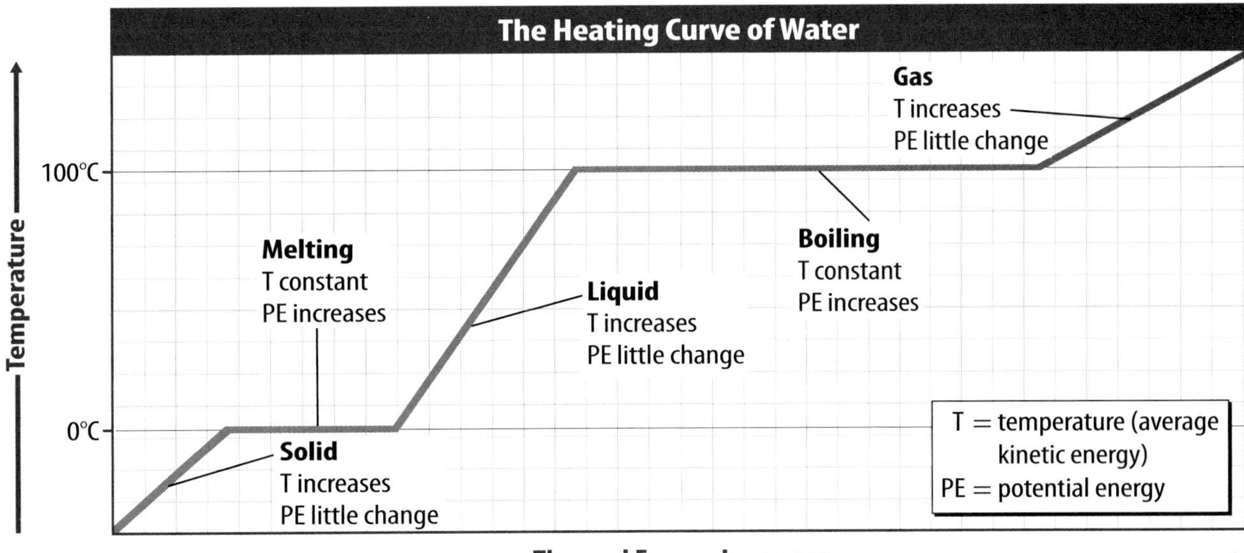

Figure 13 Water undergoes energy changes and state changes as thermal energy is added and removed.

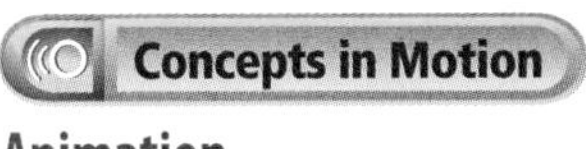

Animation

States of Water

Water is the only substance that exists naturally as a solid, a liquid, and a gas within Earth's temperature range. To better understand the energy changes during a change in state, it is helpful to study the heating curve of water, as shown in **Figure 13.**

Adding Thermal Energy

Suppose you place a beaker of ice on a hot plate. The hot plate transfers thermal energy to the beaker and then to the ice. The temperature of the ice increases. Recall that this means the average kinetic energy of the water molecules increases.

At 0°C, the melting point of water, the water molecules vibrate so rapidly that they begin to move out of their places. At this point, added thermal energy only increases the distance between particles and decreases attractive forces—melting occurs. Once melting is complete, the average kinetic energy of the particles (temperature) begins to increase again as more thermal energy is added.

When water reaches 100°C, the boiling point, liquid water begins to change to water vapor. Again, kinetic energy is constant as vaporization occurs. When the change of state is complete, the kinetic energy of molecules increases once more, and so does the temperature.

Key Concept Check Describe the changes in thermal energy as water goes from a solid to a liquid.

Removing Thermal Energy

The removal of thermal energy is the reverse of the process shown in **Figure 13.** Cooling water vapor changes the gas to a liquid. Cooling the water further changes it to ice.

FOLDABLES®

Fold a sheet of notebook paper to make a four-tab Foldable as shown. Label the tabs, define the terms, and record what you learn about each term under the tabs.

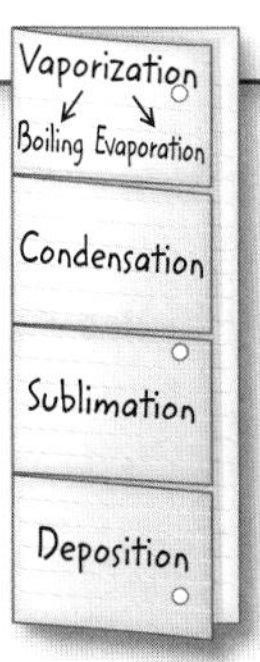

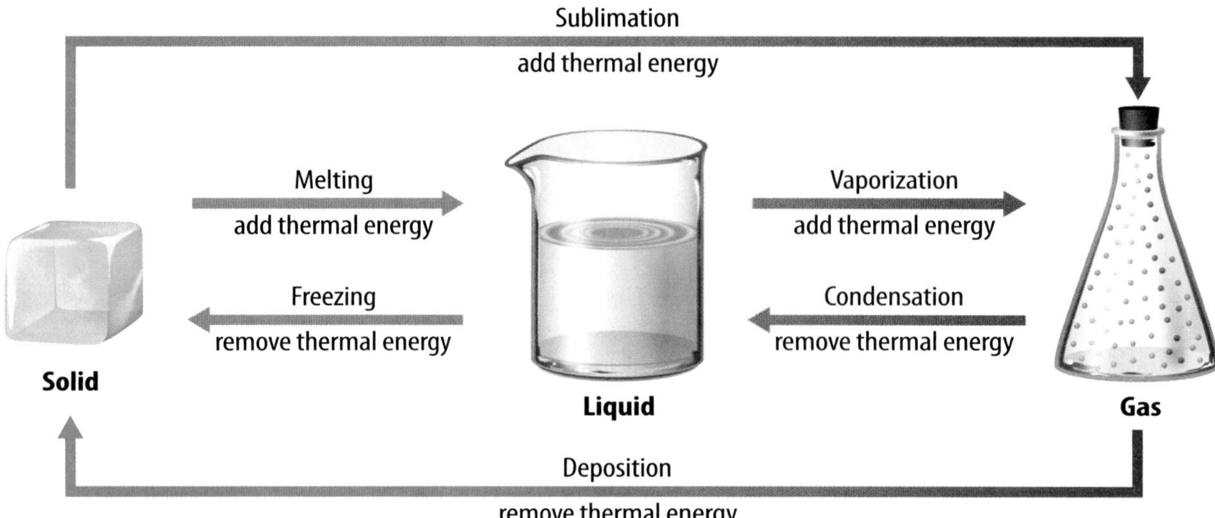

Figure 14 For a change of state to occur, thermal energy must move into or out of matter.

Animation

Conservation of Mass and Energy

The diagram in **Figure 14** illustrates the energy changes that occur as thermal energy is added or removed from matter. Notice that opposite processes, melting and freezing and vaporization and condensation, are shown. When matter changes state, matter and energy are always conserved.

When water vaporizes, it appears to disappear. If the invisible gas is captured and its mass added to the remaining mass of the liquid, you would see that matter is conserved. This is also true for energy. Surrounding matter, such as air, often absorbs thermal energy. If you measured all the thermal energy, you would find that energy is conserved.

Inquiry MiniLab

20 minutes

How can you make a water thermometer?

What causes liquid in a thermometer to rise and fall?

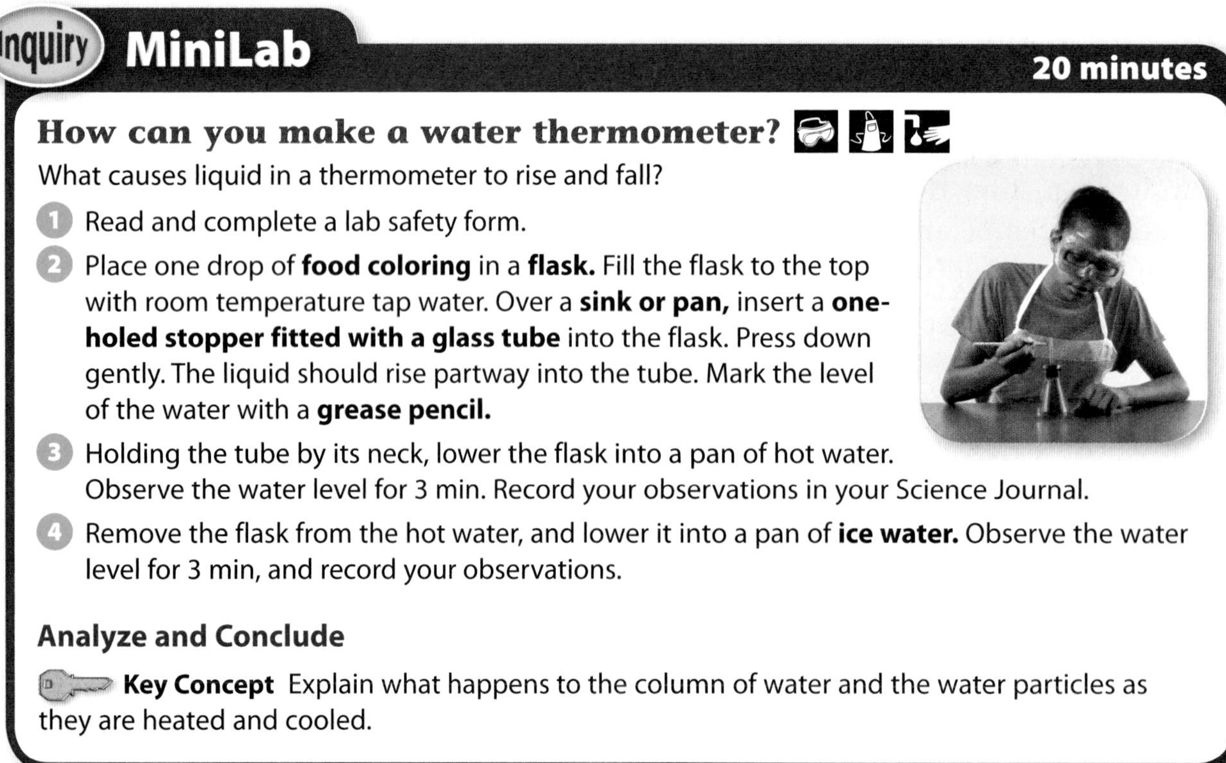

1. Read and complete a lab safety form.
2. Place one drop of **food coloring** in a **flask.** Fill the flask to the top with room temperature tap water. Over a **sink or pan,** insert a **one-holed stopper fitted with a glass tube** into the flask. Press down gently. The liquid should rise partway into the tube. Mark the level of the water with a **grease pencil.**
3. Holding the tube by its neck, lower the flask into a pan of hot water. Observe the water level for 3 min. Record your observations in your Science Journal.
4. Remove the flask from the hot water, and lower it into a pan of **ice water.** Observe the water level for 3 min, and record your observations.

Analyze and Conclude

Key Concept Explain what happens to the column of water and the water particles as they are heated and cooled.

Lesson 2 Review

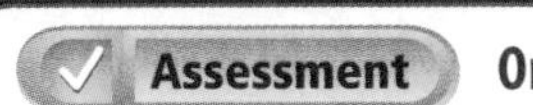

Visual Summary

All matter has thermal energy. Thermal energy is the sum of potential and kinetic energy.

When thermal energy is added to a liquid, vaporization can occur.

When enough thermal energy is removed from matter, a change of state can occur.

FOLDABLES®

Use your lesson Foldable to review the lesson. Save your Foldable for the project at the end of the chapter.

What do you think NOW?

You first read the statements below at the beginning of the chapter.

3. Particles of matter have both potential energy and kinetic energy.

4. When a solid melts, thermal energy is removed from the solid.

Did you change your mind about whether you agree or disagree with the statements? Rewrite any false statements to make them true.

Use Vocabulary

1. The measure of average kinetic energy of the particles in a material is ___________.
2. **Define** *kinetic energy* and *thermal energy* in your own words.
3. The change of a liquid into a gas is known as ___________.

Understand Key Concepts 

4. The process that is opposite of condensation is known as
   - **A.** deposition.
   - **B.** freezing.
   - **C.** melting.
   - **D.** vaporization.
5. **Explain** how temperature and particle motion are related.
6. **Describe** the relationship between temperature and thermal energy.
7. **Generalize** the changes in thermal energy when matter increases in temperature and then changes state.

Interpret Graphics

8. **Describe** what is occurring below.

9. **Summarize** Copy and fill in the graphic organizer below to identify the two types of vaporization that can occur in matter.

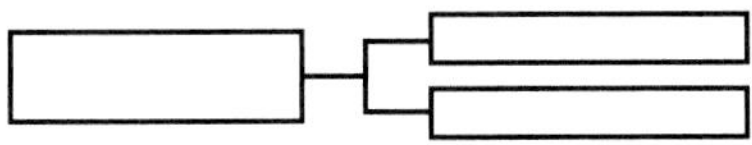

Critical Thinking

10. **Summarize** the energy and state changes that occur when freezing rain falls and solidifies on a wire fence.
11. **Compare** the amount of thermal energy needed to melt a solid and the amount of thermal energy needed to freeze the same liquid.

Inquiry Skill Practice Form and Test a Hypothesis 40 minutes

How does dissolving substances in water change its freezing point?

Materials

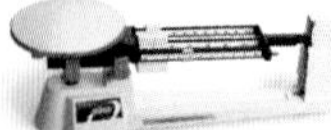
triple-beam balance

beaker

foam cup

50-mL graduated cylinder

distilled water

Also needed: ice-salt slush, test tubes, thermometers

Safety

You know that when thermal energy is removed from a liquid, the particles move more slowly. At the freezing point, the particles move so slowly that the attractive forces pull them together to form a solid. What happens if the water contains particles of another substance, such as salt? You will form a hypothesis and test the hypothesis to find out.

Learn It

To **form a hypothesis** is to propose a possible explanation for an observation that is testable by a scientific investigation. You **test the hypothesis** by conducting a scientific investigation to see whether the hypothesis is supported.

Try It

1. Read and complete a lab safety form.
2. Form a hypothesis that answers the question in the title of the lab. Write your hypothesis in your Science Journal.
3. Copy the data table in your Science Journal.
4. Use a triple-beam balance to measure 5 g of table salt (NaCl). Dissolve the 5 g of table salt in 50 mL of distilled water.
5. Place 40 mL of distilled water in one large test tube. Place 40 mL of the salt-water mixture in a second large test tube.
6. Measure and record the temperature of the liquids in each test tube.
7. Place both test tubes into a large foam cup filled with crushed ice-salt slush. Gently rotate the thermometers in the test tubes. Record the temperature in each test tube every minute until the temperature remains the same for several minutes.

Apply It

8. How does the data tell you when the freezing point of the liquid has been reached?
9. Was your hypothesis supported? Why or why not?
10. **Key Concept** Explain your observations in terms of how temperature affects particle motion and how a liquid changes to a solid.

Water	Time (min)	0	1	2	3	4	5	6	7	8
	Temperature (°C)									
Salt water	Time (min)	0	1	2	3	4	5	6	7	8
	Temperature (°C)									

Lesson 3

The Behavior of Gases

Reading Guide

Key Concepts

ESSENTIAL QUESTIONS

- How does the kinetic molecular theory describe the behavior of a gas?
- How are temperature, pressure, and volume related in Boyle's law?
- How is Boyle's law different from Charles's law?

Vocabulary

kinetic molecular theory p. 292

pressure p. 293

Boyle's law p. 294

Charles's law p. 295

Multilingual eGlossary

Video

What's Science Got to do With It?

Inquiry Survival Gear?

Why do some pilots wear oxygen masks? Planes fly at high altitudes where the atmosphere has a lower pressure and gas molecules are less concentrated. If the pressure is not adjusted inside the airplane, a pilot must wear an oxygen mask to inhale enough oxygen to keep the body functioning.

Launch Lab

15 minutes

Are volume and pressure of a gas related?

Pressure affects gases differently than it affects solids and liquids. How do pressure changes affect the volume of a gas?

1. Read and complete a lab safety form.
2. Stretch and blow up a **small balloon** several times.
3. Finally, blow up the balloon to a diameter of about 5 cm. Twist the neck, and stretch the mouth of the balloon over the opening of a **plastic bottle. Tape** the neck of the balloon to the bottle.
4. Squeeze and release the bottle several times while observing the balloon. Record your observations in your Science Journal.

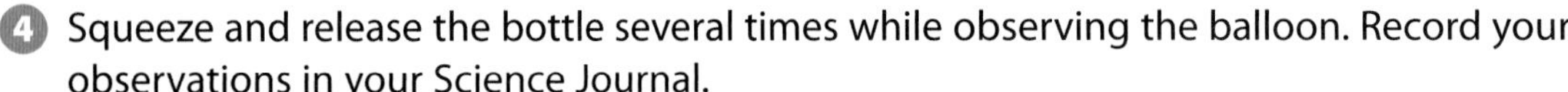

Think About This

1. Why doesn't the balloon deflate when you attach it to the bottle?
2. What caused the balloon to inflate when you squeezed the bottle?
3. **Key Concept** Using this lab as a reference, do you think pressure and volume of a gas are related? Explain.

Understanding Gas Behavior

Pilots do not worry as much about solids and liquids at high altitudes as they do gases. That is because gases behave differently than solids and liquids. Changes in temperature, pressure, and volume affect the behavior of gases more than they affect solids and liquids.

The explanation of particle behavior in solids, liquids, and gases is based on the kinetic molecular theory. The **kinetic molecular theory** *is an explanation of how particles in matter behave.* Some basic ideas in this theory are

- small particles make up all matter;
- these particles are in constant, random motion;
- the particles collide with other particles, other objects, and the walls of their container;
- when particles collide, no energy is lost.

ACADEMIC VOCABULARY
theory
(noun) an explanation of things or events that is based on knowledge gained from many observations and investigations

You have read about most of these, but the last two statements are very important in explaining how gases behave.

Key Concept Check How does the kinetic molecular theory describe the behavior of a gas?

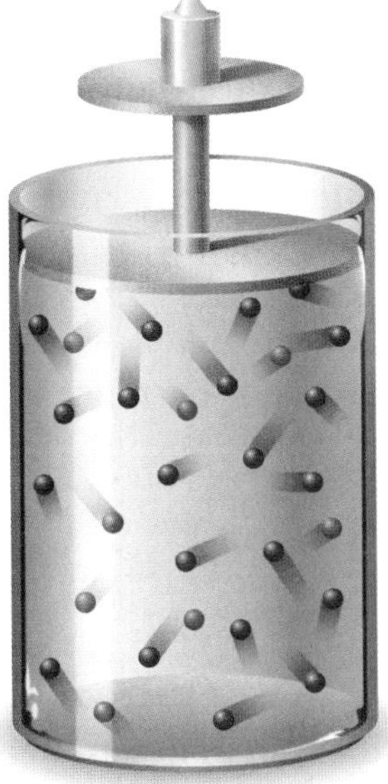

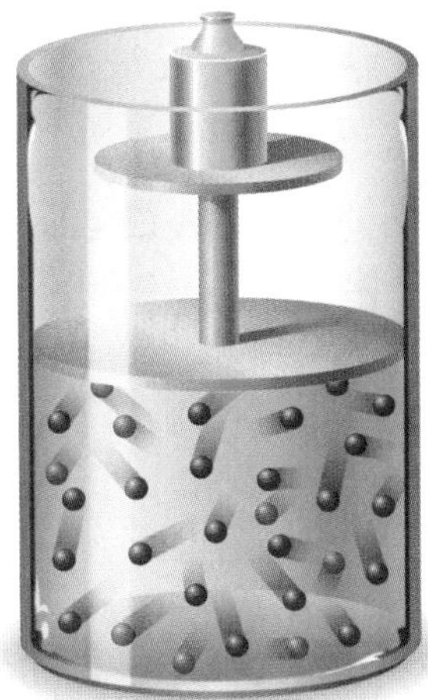

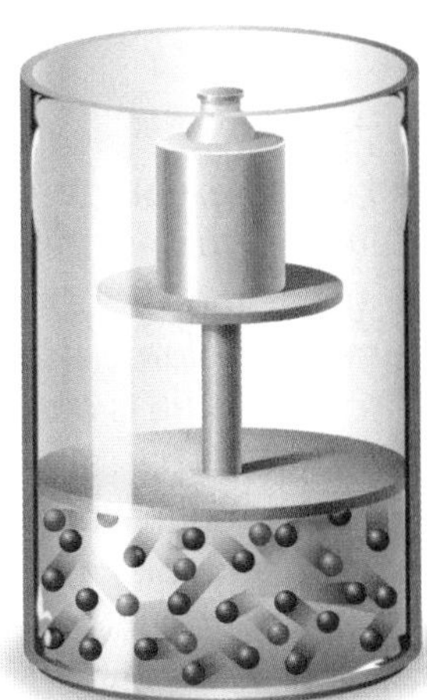

Figure 15 As pressure increases, the volume of the gas decreases.

What is pressure?

Particles in gases move constantly. As a result of this movement, gas particles constantly collide with other particles and their container. When particles collide with their container, pressure results. **Pressure** *is the amount of force applied per unit of area.* For example, gas in a cylinder, as shown in **Figure 15,** might contain trillions of gas particles. These particles exert forces on the cylinder each time they strike it. When a weight is added to the plunger, the plunger moves down, compressing the gas in the cylinder. With less space to move around, the particles that make up the gas collide with each other more frequently, causing an increase in pressure. The more the particles are compressed, the more often they collide, increasing the pressure.

WORD ORIGIN

pressure
from Latin *pressura,* means "to press"

Pressure and Volume

Figure 15 also shows the relationship between pressure and volume of gas at a constant temperature. What happens to pressure if the volume of a container changes? Notice that when the volume is greater, the particles have more room to move. This additional space results in fewer collisions within the cylinder, and pressure is less. The gas particles in the middle cylinder have even less volume and more pressure. In the cylinder on the right, the pressure is greater because the volume is less. The particles collide with the container more frequently. Because of the greater number of collisions within the container, pressure is greater.

FOLDABLES®

Fold a sheet of notebook paper to make a three-tab Foldable and label as shown. Use your Foldable to compare two important gas laws.

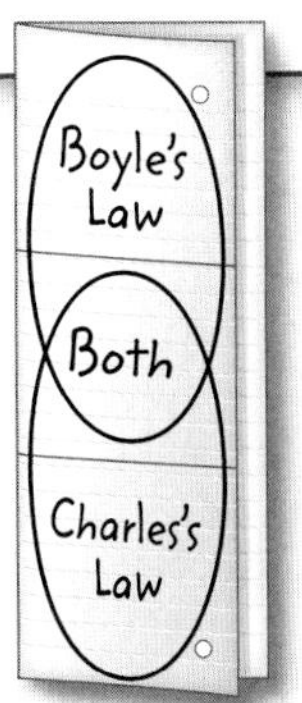

Math Skills 

Solve Equations

Boyle's law can be stated by the equation

$$V_2 = \frac{P_1V_1}{P_2}$$

P_1 and V_1 represent the pressure and volume before a change. P_2 and V_2 are the pressure and volume after a change. Pressure is often measured in kilopascals (kPa). For example, what is the final volume of a gas with an initial volume of **50.0 mL** if the pressure increases from **600.0 kPa** to **900.0 kPa**?

1. Replace the terms in the equation with the actual values.

$$V_2 = \frac{(600.0\ \text{kPa})(50.0\ \text{mL})}{(900.0\ \text{kPa})}$$

2. Cancel units, multiply, and then divide.

$$V_2 = \frac{(600.0\ \cancel{\text{kPa}})(50.0\ \text{mL})}{(900.0\ \cancel{\text{kPa}})}$$

$$V_2 = 33.3\ \text{mL}$$

Practice

What is the final volume of a gas with an initial volume of 100.0 mL if the pressure decreases from 500.0 kPa to 250.0 kPa?

- **Math Practice**
- **Personal Tutor**

Boyle's Law

You read that the pressure and volume of a gas are related. Robert Boyle (1627–1691), a British scientist, was the first to describe this property of gases. **Boyle's law** *states that pressure of a gas increases if the volume decreases and pressure of a gas decreases if the volume increases, when temperature is constant.* This law can be expressed mathematically as shown to the left.

Key Concept Check What is the relationship between pressure and volume of a gas if temperature is constant?

Boyle's Law in Action

You have probably felt Boyle's law in action if you have ever traveled in an airplane. While on the ground, the air pressure inside your middle ear and the pressure of the air surrounding you are equal. As the airplane takes off and begins to increase in altitude, the air pressure of the surrounding air decreases. However, the air pressure inside your middle ear does not decrease. The trapped air in your middle ear increases in volume, which can cause pain. These pressure changes also occur when the plane is landing. You can equalize this pressure difference by yawning or chewing gum.

Graphing Boyle's Law

This relationship is shown in the graph in **Figure 16.** Pressure is on the *x*-axis, and volume is on the *y*-axis. Notice that the line decreases in value from left to right. This shows that as the pressure of a gas increases, the volume of the gas decreases.

Figure 16 The graph shows that as pressure increases, volume decreases. This is true only if the temperature of the gas is constant.

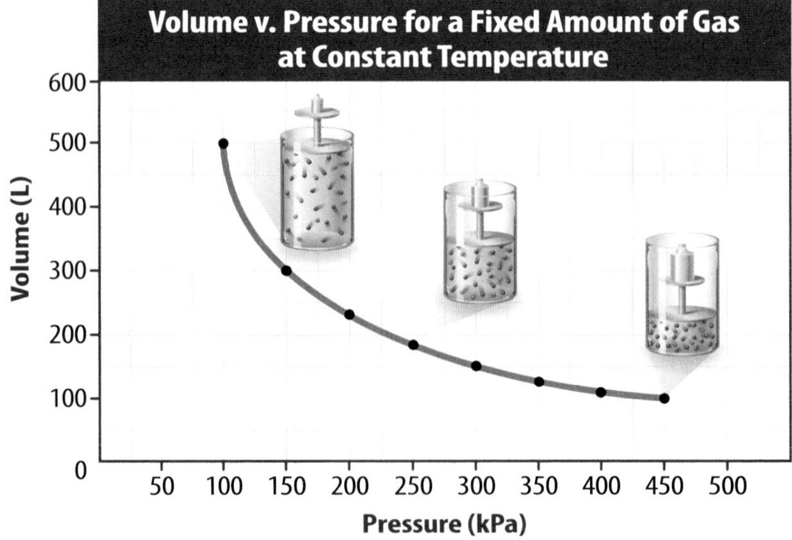

Concepts in Motion
Animation

Figure 17 As the temperature of a gas increases, the kinetic energy of the particles increases. The particles move farther apart, and volume increases.

Temperature and Volume

Pressure and volume changes are not the only factors that affect gas behavior. Changing the temperature of a gas also affects its behavior, as shown in **Figure 17.** The gas in the cylinder on the left has a low temperature. The average kinetic energy of the particles is low, and they move closer together. The volume of the gas is less. When thermal energy is added to the cylinder, the gas particles move faster and spread farther apart. This increases the pressure from gas particles, which push up the plunger. This increases the volume of the container.

Charles's Law

Jacque Charles (1746–1823) was a French scientist who described the relationship between temperature and volume of a gas. **Charles's law** *states that the volume of a gas increases with increasing temperature, if the pressure is constant.* Charles's practical experience with gases was most likely the result of his interest in balloons. Charles and his colleague were the first to pilot and fly a hydrogen-filled balloon in 1783.

 **Key Concept Check** How is Boyle's law different from Charles's law?

Inquiry MiniLab — 20 minutes

How does temperature affect the volume?

You can observe Charles's law in action using a few lab supplies.

1. Read and complete a lab safety form.
2. Stretch and blow up a **small balloon** several times.
3. Finally, blow up the balloon to a diameter of about 5 cm. Twist the neck and stretch the mouth of the balloon over the opening of an **ovenproof flask.**
4. Place the flask on a cold **hot plate.** Turn on the hot plate to low, and gradually heat the flask. Record your observations in your Science Journal.
5. ⚠ Use **tongs** to remove the flask from the hot plate. Allow the flask to cool for 5 min. Record your observations.
6. Place the flask in a **bowl of ice water.** Record your observations.

Analyze and Conclude

Key Concept What is the effect of temperature changes on the volume of a gas?

Charles's Law in Action

You have probably seen Charles's law in action if you have ever taken a balloon outside on a cold winter day. Why does a balloon appear slightly deflated when you take it from a warm place to a cold place? When the balloon is in cold air, the temperature of the gas inside the balloon decreases. Recall that a decrease in temperature is a decrease in the average kinetic energy of particles. As a result, the gas particles slow down and begin to get closer together. Fewer particles hit the inside of the balloon. The balloon appears partially deflated. If the balloon is returned to a warm place, the kinetic energy of the particles increases. More particles hit the inside of the balloon and push it out. The volume increases.

 **Reading Check** What happens when you warm a balloon?

Graphing Charles's Law

The relationship described in Charles's law is shown in the graph of several gases in **Figure 18.** Temperature is on the *x*-axis and volume is on the *y*-axis. Notice that the lines are straight and represent increasing values. Each line in the graph is extrapolated to −273°C. *Extrapolated* means the graph is extended beyond the observed data points. This temperature also is referred to as 0 K (kelvin), or absolute zero. This temperature is theoretically the lowest possible temperature of matter. At absolute zero, all particles are at the lowest possible energy state and do not move. The particles contain a minimal amount of thermal energy (potential energy + kinetic energy).

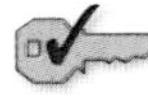 **Key Concept Check** Which factors must be constant in Boyle's law and in Charles's law?

Figure 18 The volume of a gas increases when the temperature increases at constant pressure.

Visual Check What do the dashed lines mean?

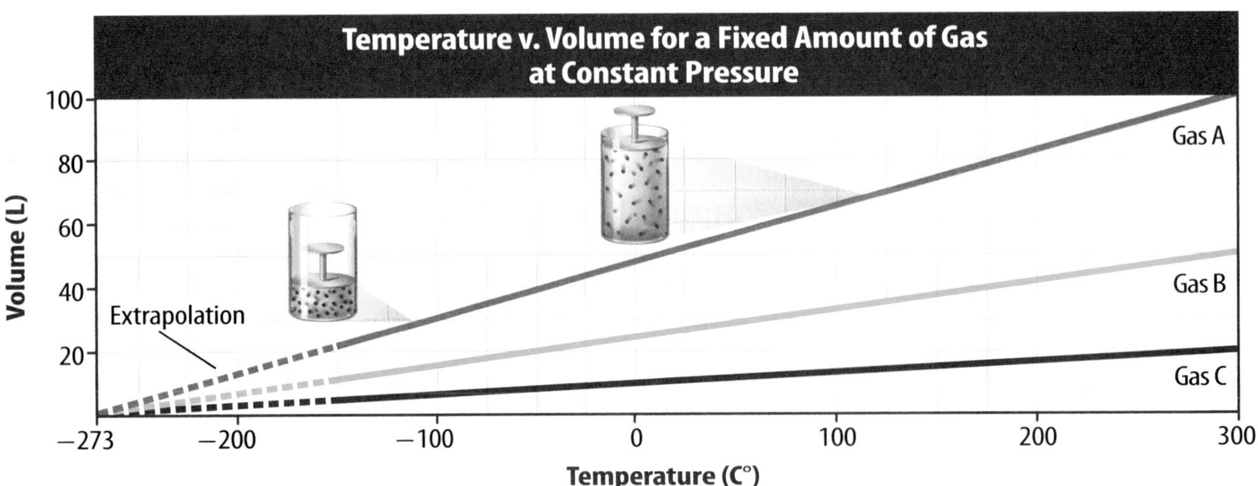

Lesson 3 Review

Assessment Online Quiz
? Inquiry Virtual Lab

Visual Summary

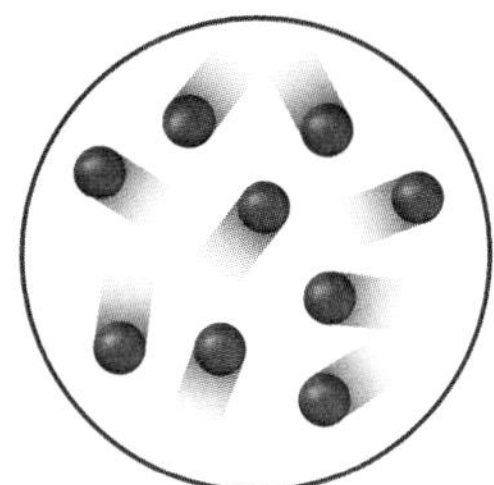

The explanation of particle behavior in solids, liquids, and gases is based on the kinetic molecular theory.

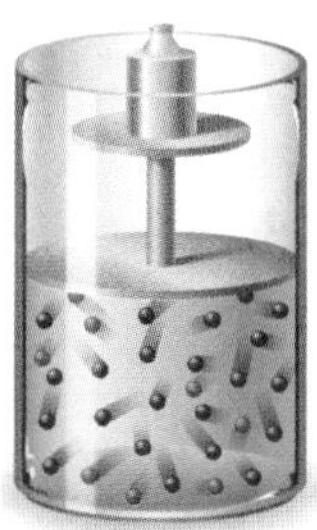

As volume of a gas decreases, the pressure increases when at constant temperature.

At constant pressure, as the temperature of a gas increases, the volume also increases.

FOLDABLES®

Use your lesson Foldable to review the lesson. Save your Foldable for the project at the end of the chapter.

What do you think NOW?

You first read the statements below at the beginning of the chapter.

5. Changes in temperature and pressure affect gas behavior.

6. If the pressure on a gas increases, the volume of the gas also increases.

Did you change your mind about whether you agree or disagree with the statements? Rewrite any false statements to make them true.

Use Vocabulary

1. **List** the basic ideas of the kinetic molecular theory.

2. ________ is force applied per unit area.

Understand Key Concepts 

3. Which is held constant when a gas obeys Boyle's law?
   - **A.** motion
   - **B.** pressure
   - **C.** temperature
   - **D.** volume

4. **Describe** how the kinetic molecular theory explains the behavior of a gas.

5. **Contrast** Charles's law with Boyle's law.

6. **Explain** how temperature, pressure, and volume are related in Boyle's law.

Interpret Graphics

7. **Explain** what happens to the particles to the right when more weights are added?

8. **Identify** Copy and fill in the graphic organizer below to list three factors that affect gas behavior.

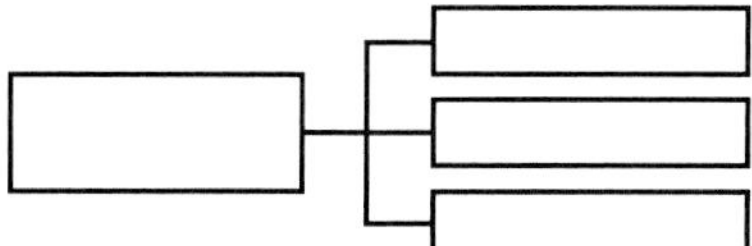

Critical Thinking

9. **Describe** what would happen to the pressure of a gas if the volume of the gas doubles while at a constant temperature.

Math Skills

Math Practice

10. **Calculate** The pressure on 400 mL of a gas is raised from 20.5 kPa to 80.5 kPa. What is the final volume of the gas?

Design an Experiment to Collect Data

Materials

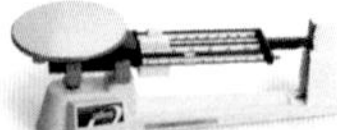

triple-beam balance

50-mL graduated cylinders

beakers

test tubes

thermometers

distilled water

Also needed: ice, salt

Safety

In this chapter, you have learned about the relationship between the motion of particles in matter and change of state. How might you use your knowledge of particles in real life? Suppose that you work for a state highway department in a cold climate. Your job is to test three products. You must determine which is the most effective in melting existing ice, the best at keeping melted ice from refreezing, and the best product to buy.

Question

How can you compare the products? What might make one product better than another? Consider how you can describe and compare the effect of each product on both existing ice and the freezing point of water. Think about controls, variables, and the equipment you have available.

Procedure

1. Read and complete a lab safety form.
2. In your Science Journal, write a set of procedures you will use to answer your questions. Include the materials and steps you will use to test the effect of each product on existing ice and on the freezing point of water. How will you record your data? Draw any data tables, such as the example below, that you might need. Have your teacher approve your procedures.

Distilled Water	Time (min)	0	1	2	3	4	5	6	7	8
	Temperature (°C)									
Product A	Time (min)	0	1	2	3	4	5	6	7	8
	Temperature (°C)									
Product B	Time (min)	0	1	2	3	4	5	6	7	8
	Temperature (°C)									
Product C	Time (min)	0	1	2	3	4	5	6	7	8
	Temperature (°C)									

3. Begin by observing and recording your observations on how each product affects ice. Does it make ice melt or melt faster?
4. Test the effect of each product on the freezing point of water. Think about how you will ensure that each product is tested in the same way.
5. Add any additional tests you think you might need to make your recommendation.

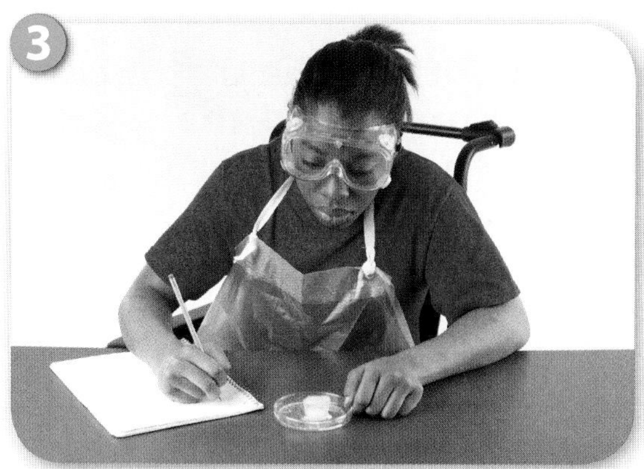

Analyze and Conclude

6. **Analyze the data** you have collected. Which product was most effective in melting existing ice? How do you know?
7. **Determine** which product was most effective in lowering the freezing point of water.
8. **Draw or make a model** to show the effect of dissolved solids on water molecules.
9. **Recognize Cause and Effect** In terms of particles, what causes dissolved solids to lower the freezing point of water?
10. **Draw Conclusions** In terms of particles, why are some substances more effective than others in lowering the freezing point of water?
11. **The Big Idea** Why is the kinetic molecular theory important in understanding how and why matter changes state?

Communicate Your Results

You are to present your recommendations to the road commissioners. Create a graphic presentation that clearly displays your results and justifies your recommendations about which product to buy.

Extension

In some states, road crews spray liquid deicer on the roads. If your teacher approves, you may enjoy testing liquids, such as alcohol, corn syrup, or salad oil.

Lab Tips

- ☑ To ensure fair testing, add the same mass of each product to the ice cubes at the same time.
- ☑ Be sure to add the same mass of each solid to the same volume of water. About 1 g of solid in 10 mL of water is a good ratio.
- ☑ Keep adding crushed ice/salt slush to the cup so that the liquid in the test tubes remains below the surface.

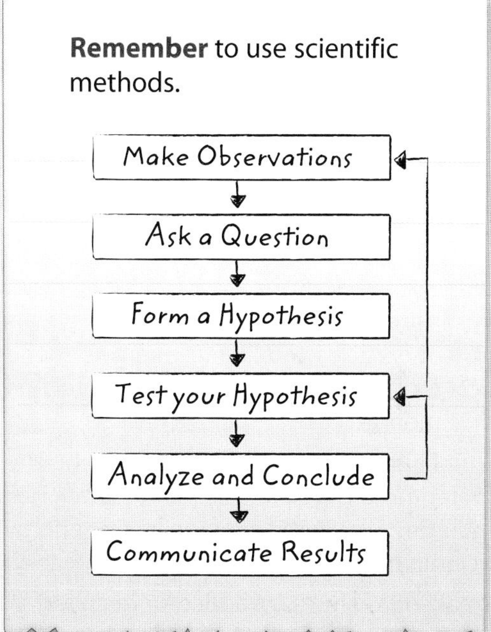

Chapter 8 Study Guide

As matter changes from one state to another, the distances and the forces between the particles change, and the amount of thermal energy in the matter changes.

Key Concepts Summary

Lesson 1: Solids, Liquids, and Gases

- Particles vibrate in **solids.** They move faster in **liquids** and even faster in **gases.**
- The force of attraction among particles decreases as matter goes from a solid, to a liquid, and finally to a gas.

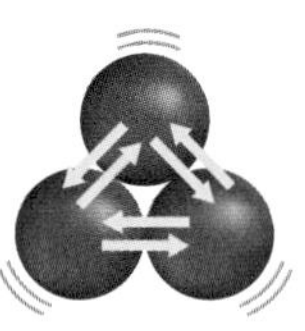
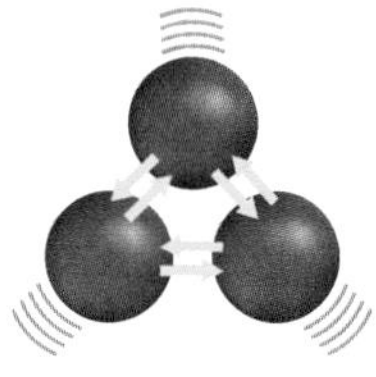

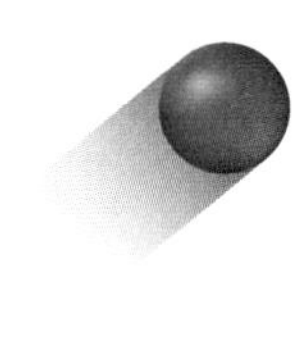

Solid Liquid Gas

Vocabulary

solid p. 275
liquid p. 276
viscosity p. 276
surface tension p. 277
gas p. 278
vapor p. 278

Lesson 2: Changes in State

- Because **temperature** is defined as the average **kinetic energy** of particles and kinetic energy depends on particle motion, temperature is directly related to particle motion.
- **Thermal energy** includes both the kinetic energy and the potential energy of particles in matter. However, temperature is only the average kinetic energy of particles in matter.
- Thermal energy must be added or removed from matter for a change of state to occur.

Vocabulary

kinetic energy p. 282
temperature p. 282
thermal energy p. 283
vaporization p. 285
evaporation p. 286
condensation p. 286
sublimation p. 286
deposition p. 286

Lesson 3: The Behavior of Gases

- The **kinetic molecular theory** states basic assumptions that are used to describe particles and their interactions in gases and other states of matter.
- **Pressure** of a gas increases if the volume decreases, and pressure of a gas decreases if the volume increases, when temperature is constant.
- **Boyle's law** describes the behavior of a gas when pressure and volume change at constant temperature. **Charles's law** describes the behavior of a gas when temperature and volume change, and pressure is constant.

Vocabulary

kinetic molecular theory p. 292
pressure p. 293
Boyle's law p. 294
Charles's law p. 295

- Personal Tutor
- Vocabulary eGames
- Vocabulary eFlashcards

FOLDABLES® Chapter Project

Assemble your lesson Foldables as shown to make a Chapter Project. Use the project to review what you have learned in this chapter.

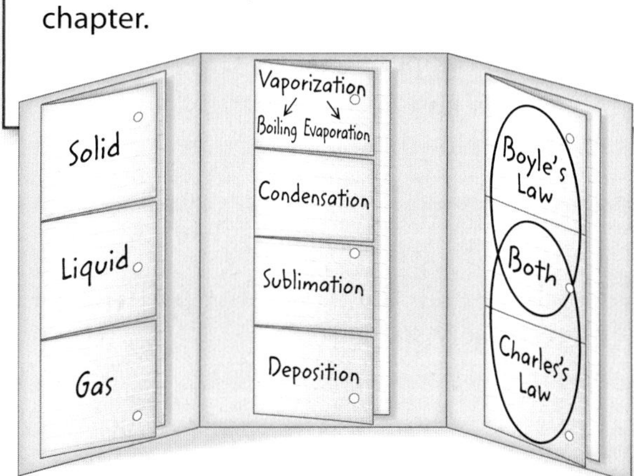

Use Vocabulary

Replace the underlined word with the correct term.

1. Matter with a definite shape and a definite volume is known as a <u>gas</u>.
2. <u>Surface tension</u> is a measure of a liquid's resistance to flow.
3. The gas state of a substance that is normally a solid or a liquid at room temperature is a <u>pressure</u>.
4. <u>Boiling</u> is vaporization that occurs at the surface of a liquid.
5. <u>Boyle's law</u> is an explanation of how particles in matter behave.
6. When graphing a gas obeying <u>Boyle's law</u>, the line will be a straight line with a positive slope.

Link Vocabulary and Key Concepts

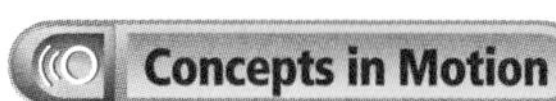

Interactive Concept Map

Copy this concept map, and then use vocabulary terms from the previous page to complete the concept map.

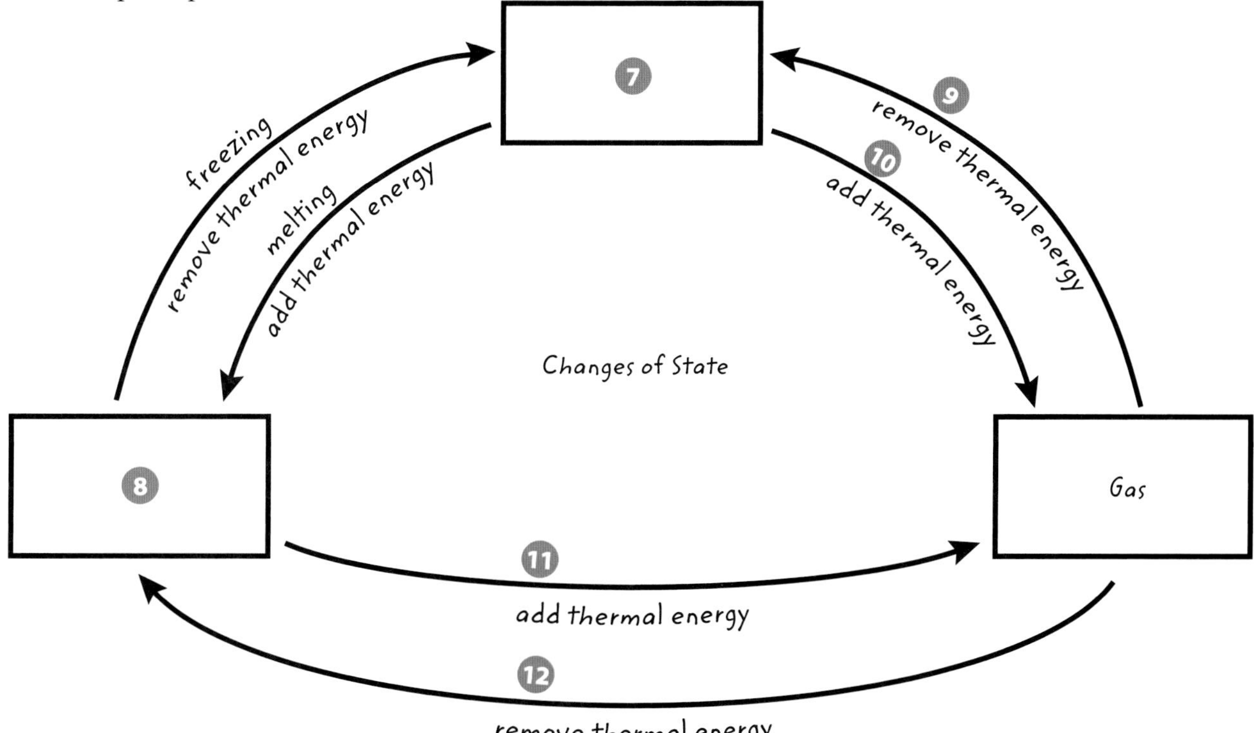

Chapter 8 Review

Understand Key Concepts 

1 What would happen if you tried to squeeze a gas into a smaller container?

A. The attractive forces between the particles would increase.
B. The force of the particles would prevent you from doing it.
C. The particles would have fewer collisions with the container.
D. The repulsive forces of the particles would pull on the container.

2 Which type of motion in the figure below best represents the movement of gas particles?

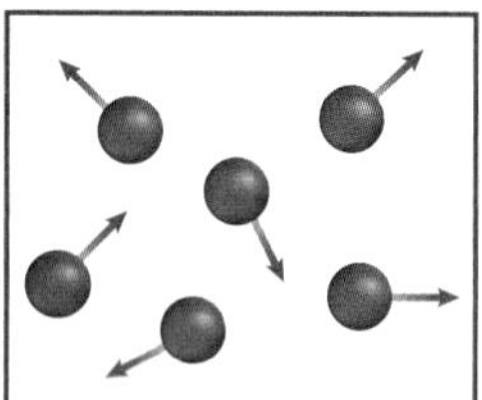

Motion 1

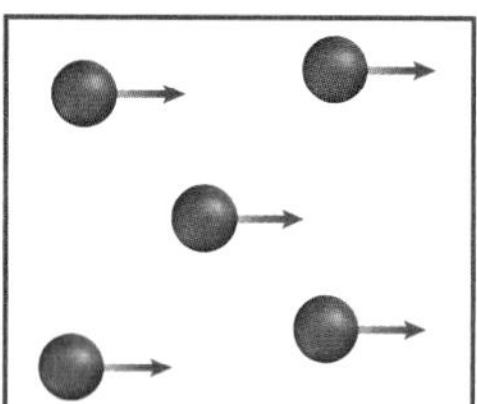

Motion 2

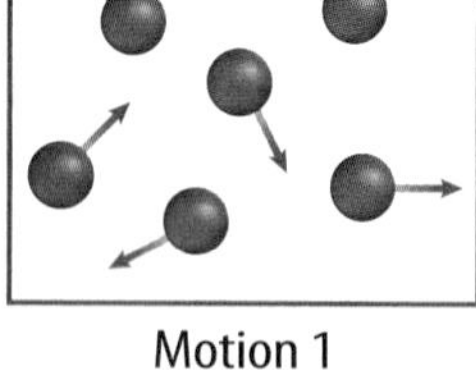

Motion 3

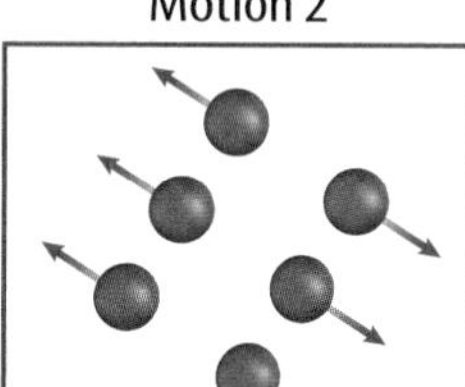

Motion 4

A. motion 1
B. motion 2
C. motion 3
D. motion 4

3 A pile of snow slowly disappears into the air, even though the temperature remains below freezing. Which process explains this?

A. condensation
B. deposition
C. evaporation
D. sublimation

4 Which unit is a density unit?

A. cm^3
B. cm^3/g
C. g
D. g/cm^3

5 Which is a form of vaporization?

A. condensation
B. evaporation
C. freezing
D. melting

6 When a needle is placed on the surface of water, it floats. Which idea best explains why this happens?

A. Boyle's law
B. molecular theory
C. surface tension
D. viscosity theory

7 In which material would the particles be most closely spaced?

A. air
B. brick
C. syrup
D. water

Use the graph below to answer questions 8 and 9.

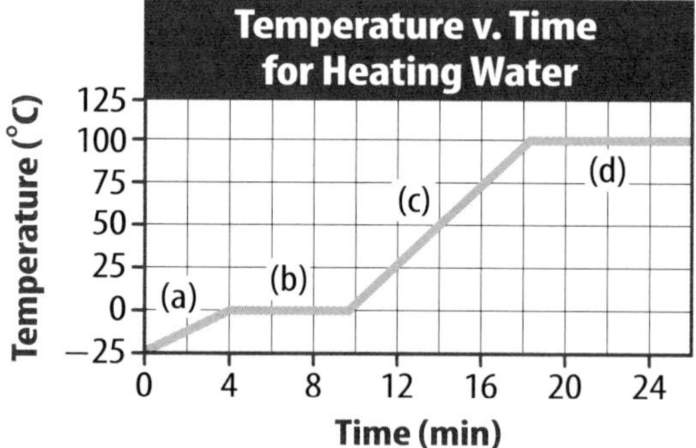

8 Which area of the graph above shows melting of a solid?

A. a
B. b
C. c
D. d

9 Which area or areas of the graph above shows a change in the potential energy of the particles?

A. a
B. a and c
C. b and d
D. c

Critical Thinking

10 **Explain** how the distances between particles in a solid, a liquid, and a gas help determine the densities of each.

11 **Describe** what would happen to the volume of a balloon if it were submerged in hot water.

12 **Assess** The particles of an unknown liquid have very weak attractions for other particles in the liquid. Would you expect the liquid to have a high or low viscosity? Explain your answer.

13 **Rank** these liquids from highest to lowest viscosity: honey, rubbing alcohol, and ketchup.

14 **Evaluate** Each beaker below contains the same amount of water. The thermometers show the temperature in each beaker. Explain the kinetic energy differences in each beaker.

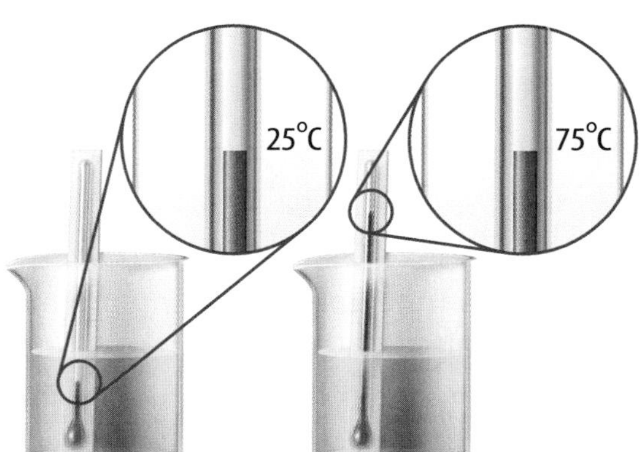

15 **Summarize** A glass with a few milliliters of water is placed on a counter. No one touches the glass. Explain what happens to the water after a few days.

Writing in Science

16 **Write** a paragraph that describes how you could determine the melting point of a substance from its heating or cooling curve.

REVIEW THE BIG IDEA

17 During springtime in Alaska, frozen rivers thaw and boats can navigate the rivers again. What physical changes and energy changes occur to the ice molecules when ice changes to water? Explain the process in which water in the river changes to water vapor.

18 In the photo below, explain how the average kinetic energy of the particles change as the molten glass cools. What instrument could you use to verify the change in the average kinetic energy of the particles?

Math Skills

Math Practice

Solve Equations

19 The pressure on 1 L of a gas at a pressure of 600 kPa is lowered to 200 kPa. What is the final volume of the gas?

20 A gas has a volume of 30 mL at a pressure of 5000 kPa. What is the volume of the gas if the pressure is lowered to 1,250 kPa?

Standardized Test Practice

Record your answers on the answer sheet provided by your teacher or on a separate sheet of paper.

Multiple Choice

1 Which property applies to matter that consists of particles vibrating in place?

A has a definite shape

B takes the shape of the container

C flows easily at room temperature

D Particles are far apart.

Use the figure below to answer questions 2 and 3.

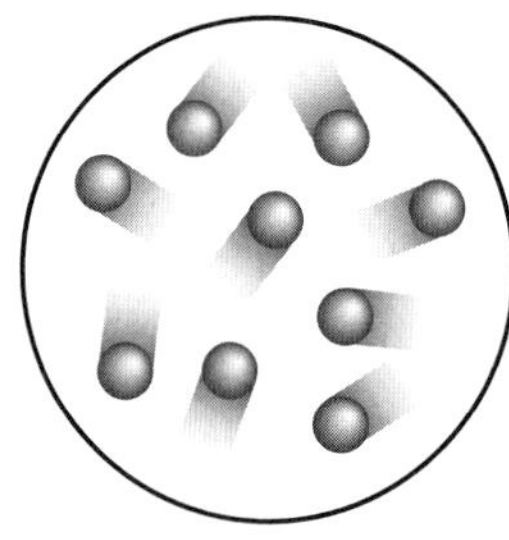

2 Which state of matter is represented above?

A amorphous solid

B crystalline solid

C gas

D liquid

3 Which best describes the attractive forces between particles shown in the figure?

A The attractive forces keep the particles vibrating in place.

B The particles hardly are affected by the attractive forces.

C The attractive forces keep the particles close together but still allow movement.

D The particles are locked in their positions because of the attractive forces between them.

4 What happens to matter as its temperature increases?

A The average kinetic energy of its particles decreases.

B The average thermal energy of its particles decreases.

C The particles gain kinetic energy.

D The particles lose potential energy.

Use the figure to answer question 5.

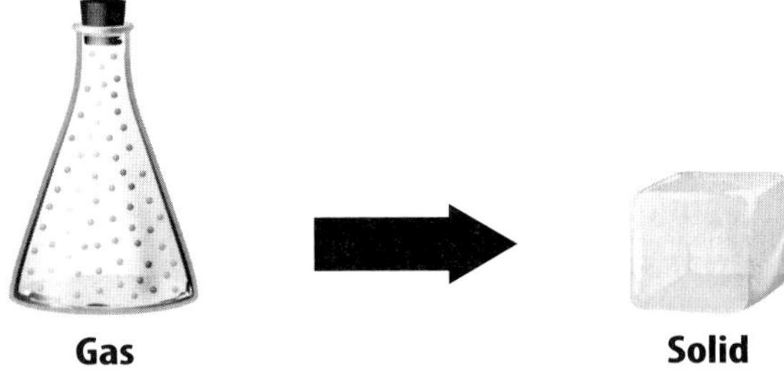

5 Which process is represented in the figure?

A deposition

B freezing

C sublimation

D vaporization

6 Which is a fundamental assumption of the kinetic molecular theory?

A All atoms are composed of subatomic particles.

B The particles of matter move in predictable paths.

C No energy is lost when particles collide with one another.

D Particles of matter never come into contact with one another.

7 Which is true of the thermal energy of particles?

A Thermal energy includes the potential and the kinetic energy of the particles.

B Thermal energy is the same as the average kinetic energy of the particles.

C Thermal energy is the same as the potential energy of particles.

D Thermal energy is the same as the temperature of the particles.

Use the graph below to answer question 8.

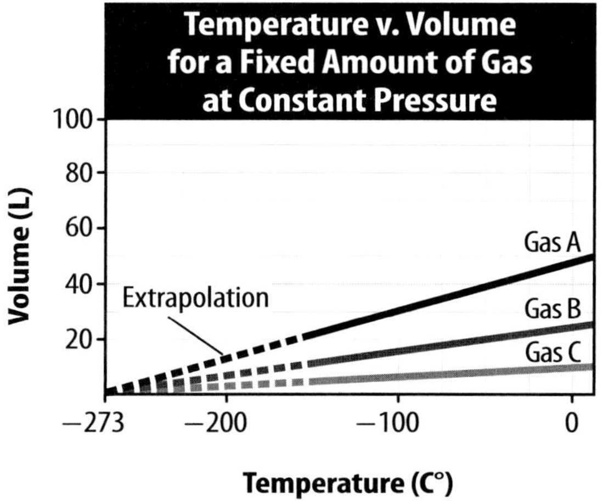

8 Which relationship is shown in the graph?

A Boyle's law

B Charles's law

C kinetic molecular theory

D definition of thermal energy

Constructed Response

9 Some people say that something that does not move very quickly is "as slow as molasses in winter." What property of molasses is described by the saying? Based on the saying, how do you think this property changes with temperature?

Use the graph to answer questions 10 and 11.

A scientist measured the temperature of a sample of frozen mercury as thermal energy is added to the sample. The graph below shows the results.

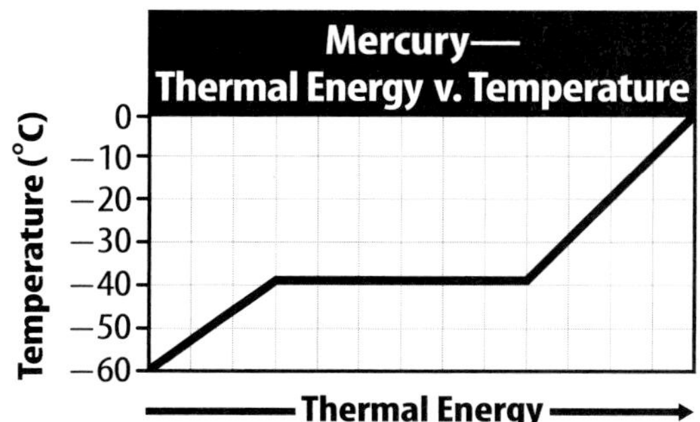

10 At what temperature does mercury melt? How do you know?

11 Describe the motion and arrangement of mercury atoms while the temperature is constant.

12 Atmospheric pressure is greater at the base of a mountain than at its peak. A hiker drinks from a water bottle at the top of a mountain. The bottle is capped tightly. At the base of the mountain, the water bottle has collapsed slightly. What happened to the gas inside the bottle? Assume constant temperature. Explain.

NEED EXTRA HELP?												
If You Missed Question...	1	2	3	4	5	6	7	8	9	10	11	12
Go to Lesson...	1	1	1	2	2	3	2	3	1	1	2	3

Unit 3
Properties of Matter

Sent to her room, Molly Cool dreams of escaping.
If only she could change state and become a liquid, she could flow under the bedroom door and down the stairs...
...then flow to the fireplace where the heat would turn her into vapor and she could escape up the chimney.
I'm free!
Hello birds!

1000 B.C. — 1700 — 1800

350 B.C.
Greek philosopher Aristotle defines an element as "one of those bodies into which other bodies can decompose, and that itself is not capable of being divided into another."

1704
Isaac Newton proposes that atoms attach to each other by some type of force.

1869
Dmitri Mendeleev publishes the first version of the periodic table.

1874
G. Johnstone Stoney proposes the existence of the electron, a subatomic particle that carries a negative electric charge, after experiments in electrochemistry.

1897
J.J. Thompson demonstrates the existence of the electron, proving Stoney's claim.

1900

1950

1907
Physicists Hans Geiger and Ernest Marsden, under the direction of Ernest Rutherford, conduct the famous gold foil experiment. Rutherford concludes that the atom is mostly empty space and that most of the mass is concentrated in the atomic nucleus.

1918
Ernest Rutherford reports that the hydrogen nucleus has a positive charge, and he names it the proton.

1932
James Chadwick discovers the neutron, a subatomic particle with no electric charge and a mass slightly larger than a proton.

? Inquiry

Visit ConnectED for this unit's STEM activity.

Patterns

It's a bird! It's a plane! No, it's Venus! Besides the Sun, Venus is brighter than any other star or planet in the sky. It is often seen from Earth without the aid of a telescope, as shown in **Figure 1.** At certain times of the year, Venus can be seen in the early evening. At other times of the year, Venus is best be seen in the morning or even during daylight hours.

▲ **Figure 1** Venus is often so bright in the morning sky that it has been nicknamed the morning star.

Astronomers study the patterns of each planet's orbit and rotation. A **pattern** is a consistent plan or model used as a guide for understanding and predicting things. Studying the orbital patterns of planets allows scientists to predict the future position of each planet. By studying the pattern of Venus's orbit, astronomers can predict when Venus will be most visible from Earth. Astronomers also can predict when Venus will travel between Earth and the Sun, and be visible from Earth, as shown in **Figure 2.** This event is so rare that it has only occurred seven or eight times since the mid-1600s. Using patterns, scientists are able to predict the date when you will be able to see this event in the future.

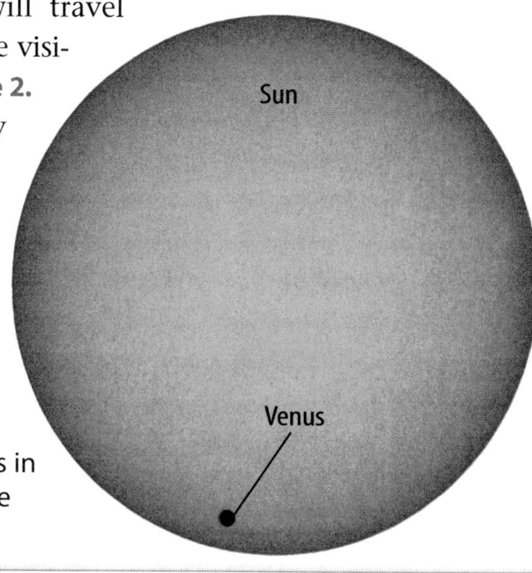

Figure 2 On June 8, 2004, observers around the world watched Venus pass in front of the Sun. This was the first time this event took place since 1882. ▶

Types of Patterns

Physical Patterns

A pattern that you can see and touch is a physical pattern. The crystalline structures of minerals are examples of physical patterns. When atoms form crystals, they produce structural, or physical, patterns. The crystal structure of the Star of India sapphire creates a pattern that reflects light in a stunning star shape.

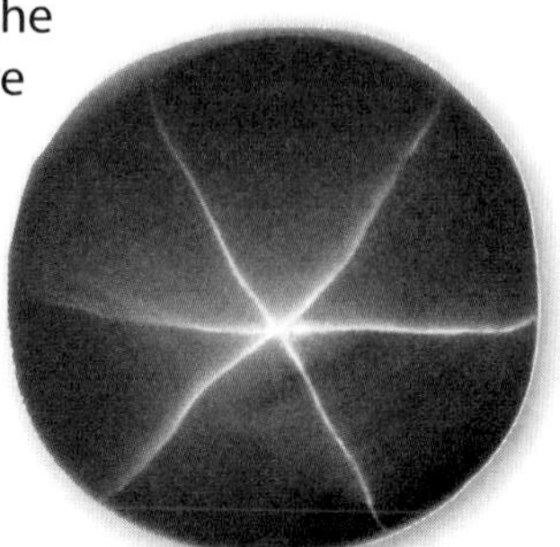

Cyclic Patterns

An event that repeats many times again in a predictable order has a cyclic pattern. Since Earth's axis is tilted, the angle of the Sun's rays on your location on Earth changes as Earth orbits the Sun. This causes the seasons—winter, spring, summer, and fall— to occur in the same pattern every year.

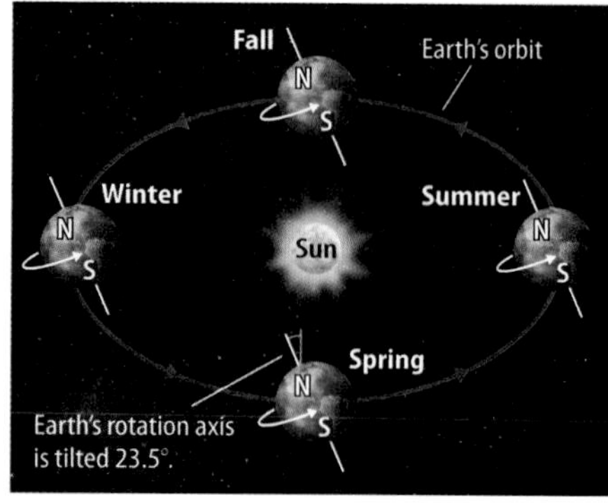

Patterns in Engineering

Engineers study patterns for many reasons, including to understand the physical properties of materials or to optimize the performance of their designs. Have you ever seen bricks with a pattern of holes through them? Clay bricks used in construction are fired, or baked, to make them stronger. Ceramic engineers understand that a regular pattern of holes in a brick assures that the brick is evenly fired and will not easily break.

Maybe you have seen a bridge constructed with a repeating pattern of large, steel triangles. Civil engineers, who design roads and bridges, know that the triangle is one of the strongest shapes in geometry. Engineers often use patterns of triangles in the structure of bridges to make them withstand heavy traffic and high winds.

Patterns in Physical Science

Scientists use patterns to explain past events or predict future events. At one time, only a few chemical elements were known. Chemists arranged the information they knew about these elements in a pattern according to the elements' properties. Scientists predicted the atomic numbers and the properties of elements that had yet to be discovered. These predictions made the discovery of new elements easier because scientists knew what properties to look for.

Look around. There are patterns everywhere—in art and nature, in the motion of the universe, in vehicles traveling on the roads, and in the processes of plant and animal growth. Analyzing patterns helps to understand the universe.

Patterns in Graphs

Scientists often graph their data to help identify patterns. For example, scientists might plot data from experiments on parachute nylon in graphs, such as the one below. Analyzing patterns on graphs then gives engineers information about how to design the strongest parachutes.

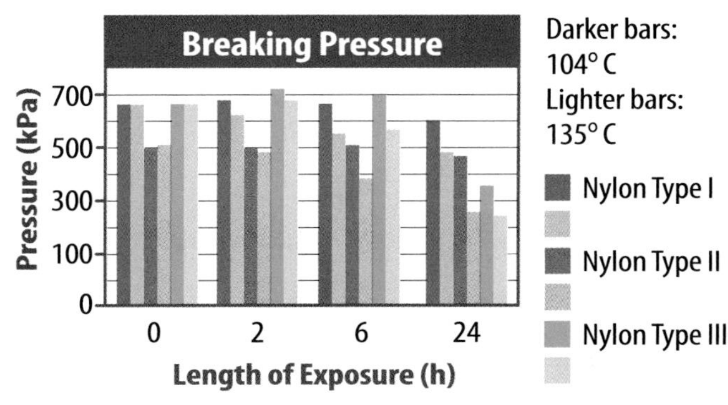

How strong is your parachute?

Suppose you need to design a parachute. The graph to the left shows data for three types of parachute nylon. Each was tested to see how it weakens when exposed to different temperatures for different lengths of time. How would you use the patterns in the graph to design your parachute?

1. Write down the different experiments performed and how the variables changed in your Science Journal.
2. Write down all the patterns that you notice in the graph.

Analyze and Conclude

1. **Compare** Which nylon is weakest? What pattern helps you make this comparison?
2. **Identify** Which nylon is most affected by length of exposure to heat? What is its pattern on the graph?
3. **Select** Which nylon would you choose for your parachute? What pattern helped you make your decision?

Chapter 9

Understanding the Atom

THE BIG IDEA **What are atoms, and what are they made of?**

Inquiry All This to Study Tiny Particles?

This huge machine is called the Large Hadron Collider (LHC). It's like a circular racetrack for particles and is about 27 km long. The LHC accelerates particle beams to high speeds and then smashes them into each other. The longer the tunnel, the faster the beams move and the harder they smash together. Scientists study the tiny particles produced in the crash.

- How might scientists have studied matter before colliders were invented?
- What do you think are the smallest parts of matter?
- What are atoms, and what are they made of?

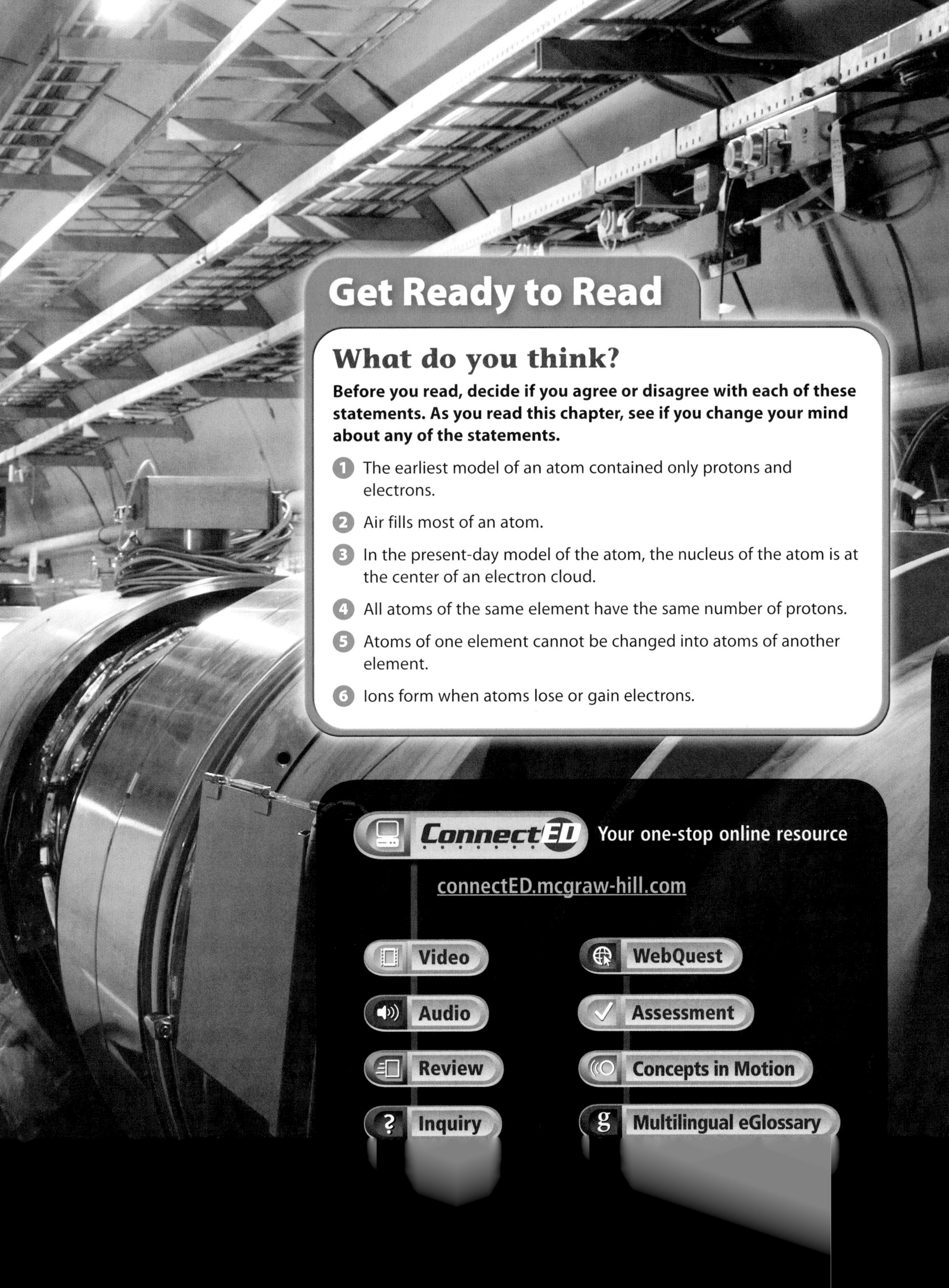

Get Ready to Read

What do you think?

Before you read, decide if you agree or disagree with each of these statements. As you read this chapter, see if you change your mind about any of the statements.

1. The earliest model of an atom contained only protons and electrons.
2. Air fills most of an atom.
3. In the present-day model of the atom, the nucleus of the atom is at the center of an electron cloud.
4. All atoms of the same element have the same number of protons.
5. Atoms of one element cannot be changed into atoms of another element.
6. Ions form when atoms lose or gain electrons.

ConnectED Your one-stop online resource

connectED.mcgraw-hill.com

- Video
- Audio
- Review
- Inquiry
- WebQuest
- Assessment
- Concepts in Motion
- Multilingual eGlossary

Lesson 1

Reading Guide

Key Concepts
ESSENTIAL QUESTIONS

- What is an atom?
- How would you describe the size of an atom?
- How has the atomic model changed over time?

Vocabulary

atom p. 315
electron p. 317
nucleus p. 320
proton p. 320
neutron p. 321
electron cloud p. 322

Multilingual eGlossary

Video BrainPOP®

Discovering Parts of an Atom

Inquiry A Microscopic Mountain Range?

This photo shows a glimpse of the tiny particles that make up matter. A special microscope, invented in 1981, made this image. However, scientists knew these tiny particles existed long before they were able to see them. What are these tiny particles? How small do you think they are? How might scientists have learned so much about them before being able to see them?

10 minutes

What's in there?

When you look at a sandy beach from far away, it looks like a solid surface. You can't see the individual grains of sand. What would you see if you zoomed in on one grain of sand?

1. Read and complete a lab safety form.
2. Have your partner hold a **test tube** of **a substance,** filled to a height of 2–3 cm.
3. Observe the test tube from a distance of at least 2 m. Write a description of what you see in your Science Journal.
4. Pour about 1 cm of the substance onto a piece of **waxed paper.** Record your observations.
5. Use a **toothpick** to separate out one particle of the substance. Suppose you could zoom in. What do you think you would see? Record your ideas in your Science Journal.

Think About This

1. Do you think one particle of the substance is made of smaller particles? Why or why not?
2. **Key Concept** Do you think you could use a microscope to see what the particles are made of? Why or why not?

Early Ideas About Matter

Look at your hands. What are they made of? You might answer that your hands are made of things such as skin, bone, muscle, and blood. You might recall that each of these is made of even smaller structures called cells. Are cells made of even smaller parts? Imagine dividing something into smaller and smaller parts. What would you end up with?

Greek philosophers discussed and debated questions such as these more than 2,000 years ago. At the time, many thought that all matter is made of only four elements—fire, water, air, and earth, as shown in **Figure 1.** However, they weren't able to test their ideas because scientific tools and methods, such as experimentation, did not exist yet. The ideas proposed by the most influential philosophers usually were accepted over the ideas of less influential philosophers. One philosopher, Democritus (460–370 B.C.), challenged the popular idea of matter.

Figure 1 Most Greek philosophers believed that all matter is made of only four elements—fire, water, air, and earth.

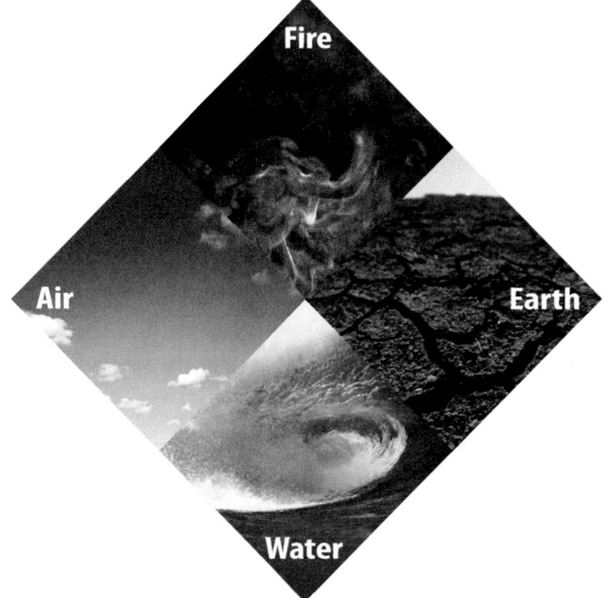

Democritus

Democritus believed that matter is made of small, solid objects that cannot be divided, created, or destroyed. He called these objects *atomos,* from which the English word *atom* is derived. Democritus proposed that different types of matter are made from different types of atoms. For example, he said that smooth matter is made of smooth atoms. He also proposed that nothing is between these atoms except empty space. **Table 1** summarizes Democritus's ideas.

Although Democritus had no way to test his ideas, many of his ideas are similar to the way scientists describe the atom today. Because Democritus's ideas did not conform to the popular opinion and because they could not be tested scientifically, they were open for debate. One philosopher who challenged Democritus's ideas was Aristotle.

Reading Check According to Democritus, what might atoms of gold look like?

Aristotle

Aristotle (384–322 B.C.) did not believe that empty space exists. Instead, he favored the more popular idea—that all matter is made of fire, water, air, and earth. Because Aristotle was so influential, his ideas were accepted. Democritus's ideas about atoms were not studied again for more than 2,000 years.

Dalton's Atomic Model

In the late 1700s, English schoolteacher and scientist John Dalton (1766–1844) revisited the idea of atoms. Since Democritus's time, advancements had been made in technology and scientific methods. Dalton made careful observations and measurements of chemical reactions. He combined data from his own scientific research with data from the research of other scientists to propose the atomic theory. **Table 1** lists ways that Dalton's atomic theory supported some of the ideas of Democritus.

Table 1 Similarities Between Democritus's and Dalton's Ideas

Democritus

1. Atoms are small solid objects that cannot be divided, created, or destroyed.
2. Atoms are constantly moving in empty space.
3. Different types of matter are made of different types of atoms.
4. The properties of the atoms determine the properties of matter.

John Dalton

1. All matter is made of atoms that cannot be divided, created, or destroyed.
2. During a chemical reaction, atoms of one element cannot be converted into atoms of another element.
3. Atoms of one element are identical to each other but different from atoms of another element.
4. Atoms combine in specific ratios.

◀ **Figure 2** If you could keep dividing a piece of aluminum, you eventually would have the smallest possible piece of aluminum—an aluminum atom.

The Atom

Today, scientists agree that matter is made of atoms with empty space between and within them. What is an atom? Imagine dividing the piece of aluminum shown in **Figure 2** into smaller and smaller pieces. At first you would be able to cut the pieces with scissors. But eventually you would have a piece that is too small to see—much smaller than the smallest piece you could cut with scissors. This small piece is an aluminum atom. An aluminum atom cannot be divided into smaller aluminum pieces. *An* **atom** *is the smallest piece of an element that still represents that element.*

 **Key Concept Check** What is a copper atom?

The Size of Atoms

Just how small is an atom? Atoms of different elements are different sizes, but all are very, very small. You cannot see atoms with just your eyes or even with most microscopes. Atoms are so small that about 7.5 trillion carbon atoms could fit into the period at the end of this sentence.

 **Key Concept Check** How would you describe the size of an atom?

Seeing Atoms

Scientific experiments verified that matter is made of atoms long before scientists were able to see atoms. However, the 1981 invention of a high-powered microscope, called a scanning tunneling microscope (STM), enabled scientists to see individual atoms for the first time. **Figure 3** shows an STM image. An STM uses a tiny, metal tip to trace the surface of a piece of matter. The result is an image of atoms on the surface.

Even today, scientists still cannot see inside an atom. However, scientists have learned that atoms are not the smallest particles of matter. In fact, atoms are made of much smaller particles. What are these particles, and how did scientists discover them if they could not see them?

Figure 3 A scanning tunneling microscope created this image. The yellow sphere is a manganese atom on the surface of gallium arsenide. ▼

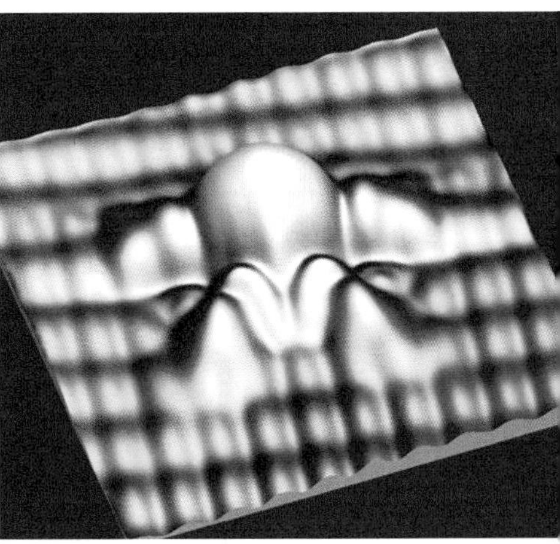

Thomson—Discovering Electrons

Not long after Dalton's findings, another English scientist, named J.J. Thomson (1856–1940), made some important discoveries. Thomson and other scientists of that time worked with cathode ray tubes. If you ever have seen a neon sign, an older computer monitor, or the color display on an ATM screen, you have seen a cathode ray tube. Thomson's cathode ray tube, shown in **Figure 4,** was a glass tube with pieces of metal, called electrodes, attached inside the tube. The electrodes were connected to wires, and the wires were connected to a battery. Thomson discovered that if most of the air was removed from the tube and electricity was passed through the wires, greenish-colored rays traveled from one electrode to the other end of the tube. What were these rays made of?

Negative Particles

Scientists called these rays cathode rays. Thomson wanted to know if these rays had an electric charge. To find out, he placed two plates on opposite sides of the tube. One plate was positively charged, and the other plate was negatively charged, as shown in **Figure 4.** Thomson discovered that these rays bent toward the positively charged plate and away from the negatively charged plate. Recall that opposite charges attract each other, and like charges repel each other. Thomson concluded that cathode rays are negatively charged.

Reading Check If the rays were positively charged, what would Thomson have observed as they passed between the plates?

Figure 4 As the cathode rays passed between the plates, they were bent toward the positive plate. Because opposite charges attract, the rays must be negatively charged.

Thomson's Cathode Ray Tube Experiment

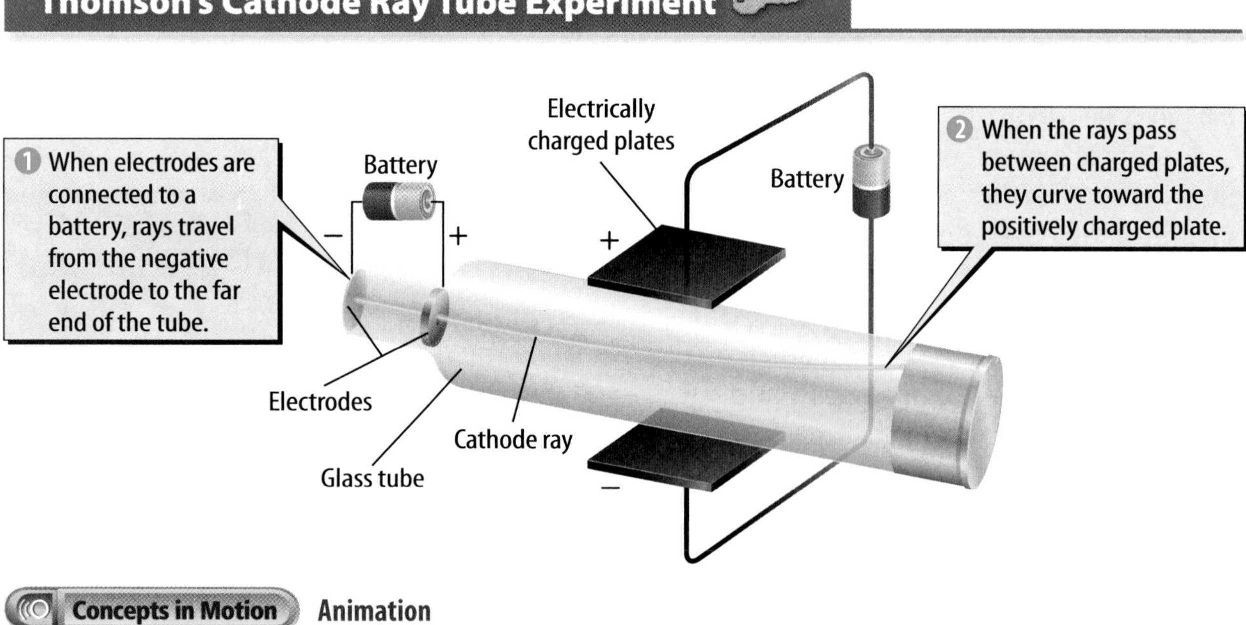
1 When electrodes are connected to a battery, rays travel from the negative electrode to the far end of the tube.
Battery
−
+
Electrically charged plates
+
Battery
2 When the rays pass between charged plates, they curve toward the positively charged plate.
Electrodes
Cathode ray
Glass tube
−

Concepts in Motion Animation

Parts of Atoms

Through more experiments, Thomson learned that these rays were made of particles that had mass. The mass of one of these particles was much smaller than the mass of the smallest atoms. This was surprising information to Thomson. Until then, scientists understood that the smallest particle of matter is an atom. But these rays were made of particles that were even smaller than atoms.

Where did these small, negatively charged particles come from? Thomson proposed that these particles came from the metal atoms in the electrode. Thomson discovered that identical rays were produced regardless of the kind of metal used to make the electrode. Putting these clues together, Thomson concluded that cathode rays were made of small, negatively charged particles. He called these particles electrons. *An* **electron** *is a particle with one negative charge (1−).* Because atoms are neutral, or not electrically charged, Thomson proposed that atoms also must contain a positive charge that balances the negatively charged electrons.

FOLDABLES®

Use two sheets of paper to make a layered book. Label it as shown. Use it to organize your notes and diagrams on the parts of an atom.

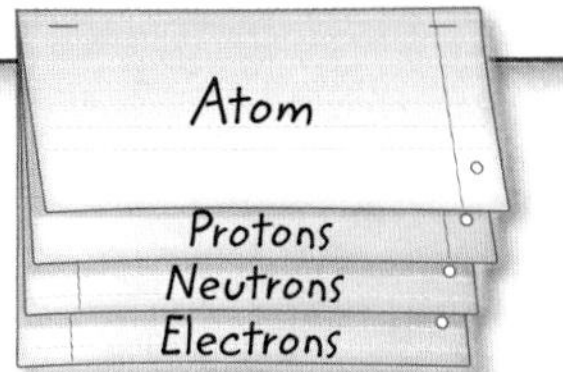

WORD ORIGIN

electron
from Greek *electron,* means "amber," the physical force so called because it first was generated by rubbing amber. Amber is a fossilized substance produced by trees.

Thomson's Atomic Model

Thomson used this information to propose a new model of the atom. Instead of a solid, neutral sphere that was the same throughout, Thomson's model of the atom contained both positive and negative charges. He proposed that an atom was a sphere with a positive charge evenly spread throughout. Negatively charged electrons were mixed through the positive charge, similar to the way chocolate chips are mixed in cookie dough. **Figure 5** shows this model.

Reading Check How did Thomson's atomic model differ from Dalton's atomic model?

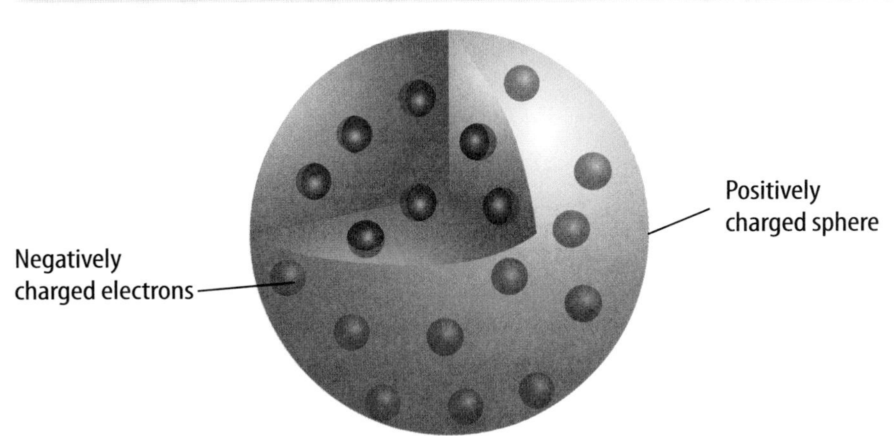

Figure 5 Thomson's model of the atom contained a positively charged sphere with negatively charged electrons within it.

Rutherford—Discovering the Nucleus

The discovery of electrons stunned scientists. Ernest Rutherford (1871–1937) was a student of Thomson's who eventually had students of his own. Rutherford's students set up experiments to test Thomson's atomic model and to learn more about what atoms contain. They discovered another surprise.

Rutherford's Predicted Result

Imagine throwing a baseball into a pile of table tennis balls. The baseball likely would knock the table tennis balls out of the way and continue moving in a relatively straight line. This is similar to what Rutherford's students expected to see when they shot alpha particles into atoms. Alpha particles are dense and positively charged. Because they are so dense, only another dense particle could deflect the path of an alpha particle. According to Thomson's model, the positive charge of the atom was too spread out and not dense enough to change the path of an alpha particle. Electrons wouldn't affect the path of an alpha particle because electrons didn't have enough mass. The result that Rutherford's students expected is shown in **Figure 6.**

 **Reading Check** Explain why Rutherford's students did not think an atom could change the path of an alpha particle.

Figure 6 The Thomson model of the atom did not contain a charge that was dense enough to change the path of an alpha particle. Rutherford expected the positive alpha particles to travel straight through the foil without changing direction.

Rutherford's Predicted Result

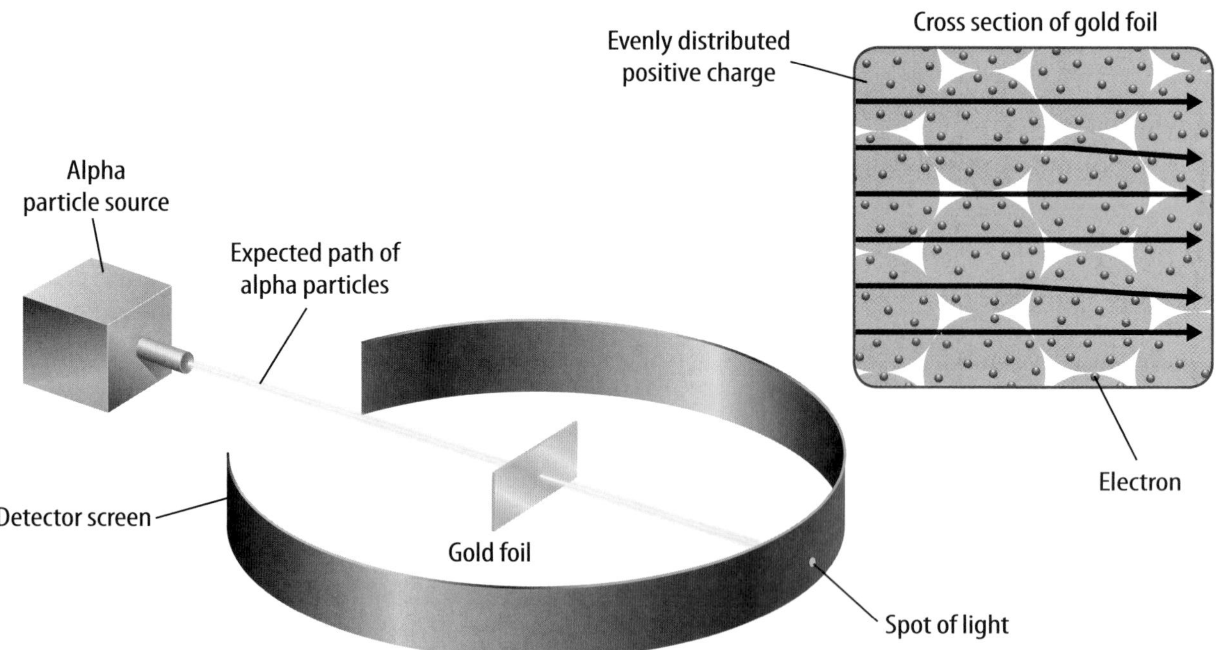

The Gold Foil Experiment

Rutherford's students went to work. They placed a source of alpha particles near a very thin piece of gold foil. Recall that all matter is made of atoms. Therefore, the gold foil was made of gold atoms. A screen surrounded the gold foil. When an alpha particle struck the screen, it created a spot of light. Rutherford's students could determine the path of the alpha particles by observing the spots of light on the screen.

The Surprising Result

Figure 7 shows what the students observed. Most of the particles did indeed travel through the foil in a straight path. However, a few particles struck the foil and bounced off to the side. And one particle in 10,000 bounced straight back! Rutherford later described this surprising result, saying it was almost as incredible as if you had fired a 38-cm shell at a piece of tissue paper and it came back and hit you. The alpha particles must have struck something dense and positively charged inside the nucleus. Thomson's model had to be refined.

Key Concept Check Given the results of the gold foil experiment, how do you think an actual atom differs from Thomson's model?

Figure 7 Some alpha particles traveled in a straight path, as expected. But some changed direction, and some bounced straight back.

Visual Check What do the dots on the screen indicate?

The Surprising Result

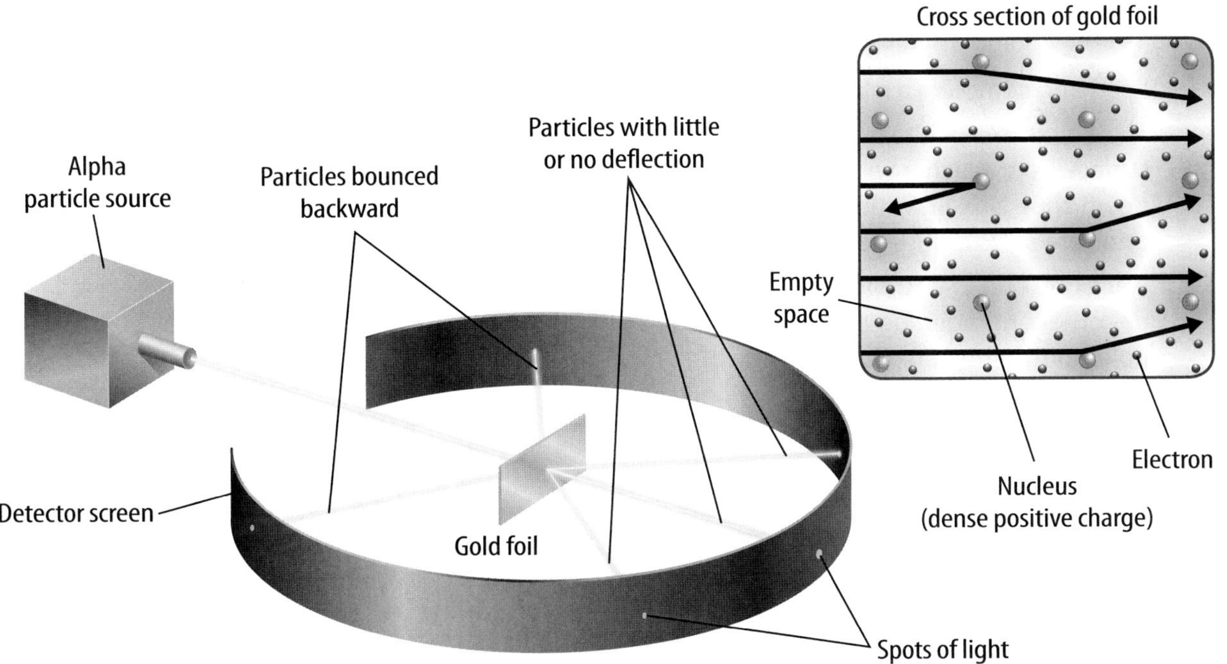

Concepts in Motion Animation

Rutherford's Atomic Model

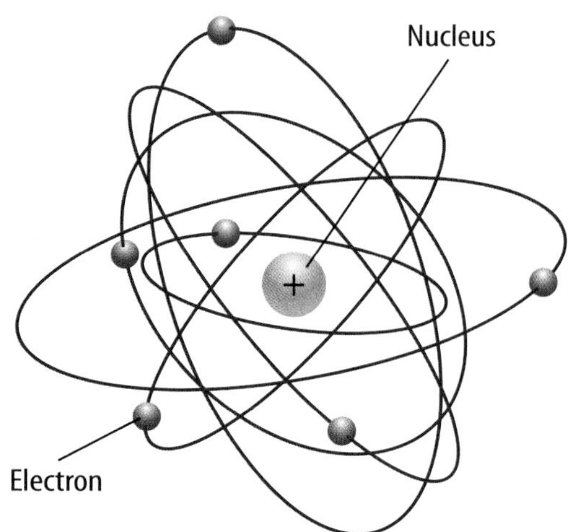

Figure 8 Rutherford's model contains a small, dense, positive nucleus. Tiny, negatively charged electrons travel in empty space around the nucleus.

Rutherford's Atomic Model

Because most alpha particles traveled through the foil in a straight path, Rutherford concluded that atoms are made mostly of empty space. The alpha particles that bounced backward must have hit a dense, positive mass. Rutherford concluded that *most of an atom's mass and positive charge is concentrated in a small area in the center of the atom called the* **nucleus.** Figure 8 shows Rutherford's atomic model. Additional research showed that the positive charge in the nucleus was made of positively charged particles called protons. *A* **proton** *is an atomic particle that has one positive charge (1+).* Negatively charged electrons move in the empty space surrounding the nucleus.

Reading Check How did Rutherford explain the observation that some of the alpha particles bounced directly backward?

Inquiry MiniLab

20–30 minutes

How can you gather information about what you can't see?

Rutherford did his gold foil experiment to learn more about the structure of the atom. What can you learn by doing a similar investigation?

1. Read and complete a lab safety form.
2. Place a piece of white **newsprint** on a flat surface. Your teacher will place an upside-down **shoe box lid** with holes cut on opposite sides on the newsprint.
3. Place one end of a **ruler** on a **book,** with the other end pointing toward one of the holes in the shoe box lid. Roll a **marble** down the ruler and into one of the holes.
4. Team members should use **markers** to draw the path of the marble on the newsprint as it enters and leaves the lid. Predict the path of the marble under the lid. Draw it on the lid. Number the path *1*.
5. Take turns repeating steps 3 and 4 eight to ten times, numbering each path *2, 3, 4*, etc. Move the ruler and aim it in a slightly different direction each time.

Analyze and Conclude

1. **Recognize Cause and Effect** What caused the marble to change its path during some rolls and not during others?
2. **Draw Conclusions** How many objects are under the lid? Where are they located? Draw your answer.
3. **Key Concept** If the shoe box lid were an accurate model of the atom, what hypothesis would you make about the atom's structure?

Discovering Neutrons

The modern model of the atom was beginning to take shape. Rutherford's colleague, James Chadwick (1891–1974), also researched atoms and discovered that, in addition to protons, the nucleus also contained neutrons. *A* **neutron** *is a neutral particle that exists in the nucleus of an atom.*

Bohr's Atomic Model

Rutherford's model explained much of his students' experimental evidence. However, there were several observations that the model could not explain. For example, scientists noticed that if certain elements were heated in a flame, they gave off specific colors of light. Each color of light had a specific amount of energy. Where did this light come from? Niels Bohr (1885–1962), another student of Rutherford's, proposed an answer. Bohr studied hydrogen atoms because they contain only one electron. He experimented with adding electric energy to hydrogen and studying the energy that was released. His experiments led to a revised atomic model.

Electrons in the Bohr Model

Bohr's model is shown in **Figure 9.** Bohr proposed that electrons move in circular orbits, or energy levels, around the nucleus. Electrons in an energy level have a specific amount of energy. Electrons closer to the nucleus have less energy than electrons farther away from the nucleus. When energy is added to an atom, electrons gain energy and move from a lower energy level to a higher energy level. When the electrons return to the lower energy level, they release a specific amount of energy as light. This is the light that is seen when elements are heated.

Limitations of the Bohr Model

Bohr reasoned that if his model were accurate for atoms with one electron, it would be accurate for atoms with more than one electron. However, this was not the case. More research showed that, although electrons have specific amounts of energy, energy levels are not arranged in circular orbits. How do electrons move in an atom?

Key Concept Check How did Bohr's atomic model differ from Rutherford's?

Figure 9 In Bohr's atomic model, electrons move in circular orbits around the atom. When an electron moves from a higher energy level to a lower energy level, energy is released—sometimes as light. Further research showed that electrons are not arranged in orbits.

Concepts in Motion **Animation**

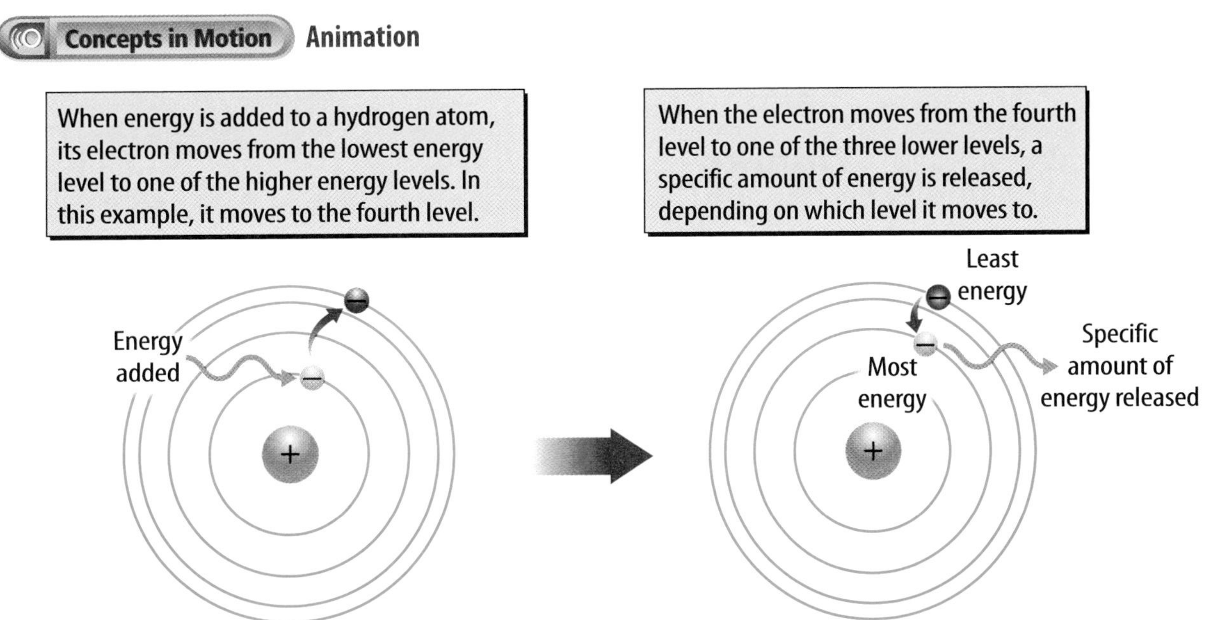

When energy is added to a hydrogen atom, its electron moves from the lowest energy level to one of the higher energy levels. In this example, it moves to the fourth level.
When the electron moves from the fourth level to one of the three lower levels, a specific amount of energy is released, depending on which level it moves to.
Energy added
Least energy
Most energy
Specific amount of energy released

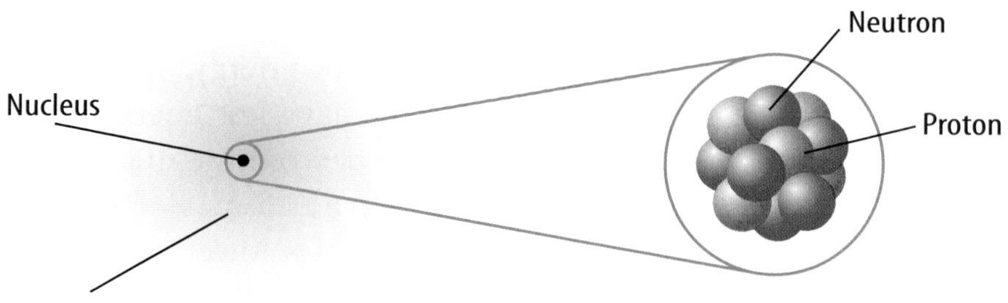

Figure 10 In this atom, electrons are more likely to be found closer to the nucleus than farther away.

Visual Check Why do you think this model of the atom doesn't show the electrons?

The Modern Atomic Model

In the modern atomic model, electrons form an electron cloud. *An* **electron cloud** *is an area around an atomic nucleus where an electron is most likely to be located.* Imagine taking a time-lapse photograph of bees around a hive. You might see a blurry cloud. The cloud might be denser near the hive than farther away because the bees spend more time near the hive.

In a similar way, electrons constantly move around the nucleus. It is impossible to know both the speed and exact location of an electron at a given moment in time. Instead, scientists only can predict the likelihood that an electron is in a particular location. The electron cloud shown in **Figure 10** is mostly empty space but represents the likelihood of finding an electron in a given area. The darker areas represent areas where electrons are more likely to be.

Key Concept Check How has the model of the atom changed over time?

Quarks

You have read that atoms are made of smaller parts—protons, neutrons, and electrons. Are these particles made of even smaller parts? Scientists have discovered that electrons are not made of smaller parts. However, research has shown that protons and neutrons are made of smaller particles called quarks. Scientists theorize that there are six types of quarks. They have named these quarks up, down, charm, strange, top, and bottom. Protons are made of two up quarks and one down quark. Neutrons are made of two down quarks and one up quark. Just as the model of the atom has changed over time, the current model might also change with the invention of new technology that aids the discovery of new information.

Lesson 1 Review

Assessment Online Quiz

Visual Summary

If you were to divide an element into smaller and smaller pieces, the smallest piece would be an atom.

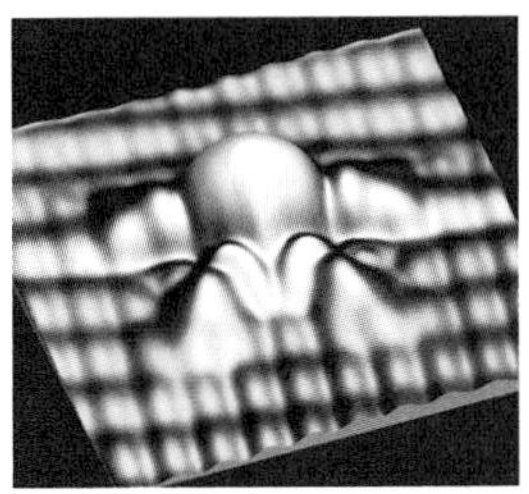

Atoms are so small that they can be seen only by using very powerful microscopes.

Scientists now know that atoms contain a dense, positive nucleus surrounded by an electron cloud.

FOLDABLES®

Use your lesson Foldable to review the lesson. Save your Foldable for the project at the end of the chapter.

What do you think NOW?

You first read the statements below at the beginning of the chapter.

1. The earliest model of an atom contained only protons and electrons.
2. Air fills most of an atom.
3. In the present-day model of the atom, the nucleus of the atom is at the center of an electron cloud.

Did you change your mind about whether you agree or disagree with the statements? Rewrite any false statements to make them true.

Use Vocabulary

1. The smallest piece of the element gold is a gold ________.
2. **Write** a sentence that describes the nucleus of an atom.
3. **Define** *electron cloud* in your own words.

Understand Key Concepts 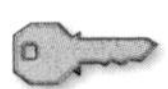

4. What is an atom mostly made of?
   - **A.** air
   - **B.** empty space
   - **C.** neutrons
   - **D.** protons
5. Why have scientists only recently been able to see atoms?
   - **A.** Atoms are too small to see with ordinary microscopes.
   - **B.** Early experiments disproved the idea of atoms.
   - **C.** Scientists didn't know atoms existed.
   - **D.** Scientists were not looking for atoms.
6. **Draw** Thomson's model of the atom, and label the parts of the drawing.
7. **Explain** how Rutherford's students knew that Thomson's model of the atom needed to change.

Interpret Graphics

8. **Contrast** Copy the graphic organizer below and use it to contrast the locations of electrons in Thomson's, Rutherford's, Bohr's, and the modern models of the atom.

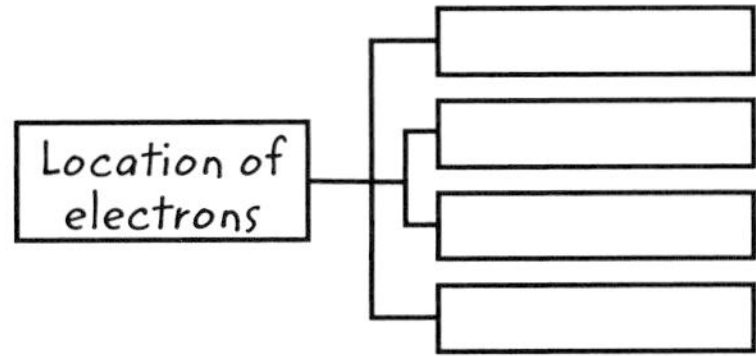

Critical Thinking

9. **Explain** what might have happened in Rutherford's experiment if he had used a thin sheet of copper instead of a thin sheet of gold.

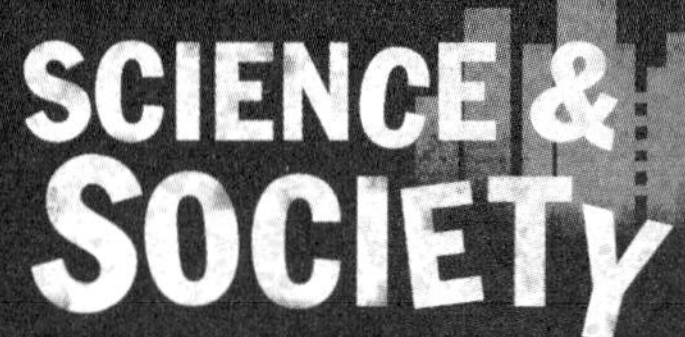

Subatomic Particles

Welcome To The Particle Zoo

QUARKS

BOSONS

LEPTONS

Much has changed since Democritus and Aristotle studied atoms.

When Democritus and Aristotle developed ideas about matter, they probably never imagined the kinds of research being performed today! From the discovery of electrons, protons, and neutrons to the exploration of quarks and other particles, the atomic model continues to change.

You've learned about quarks, which make up protons and neutrons. But quarks are not the only kind of particles! In fact, some scientists call the collection of particles that have been discovered the particle zoo, because different types of particles have unique characteristics, just like the different kinds of animals in a zoo.

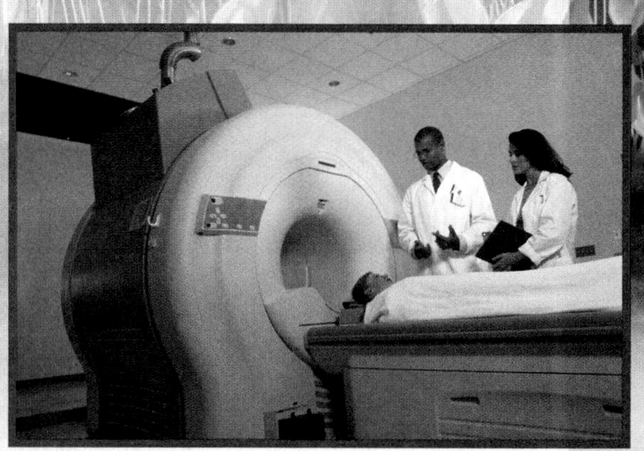

▲ MRIs are just one way in which particle physics technology is applied.

In addition to quarks, scientists have discovered a group of particles called leptons, which includes the electron. Gluons and photons are examples of bosons—particles that carry forces. Some particles, such as the Higgs Boson, have been predicted to exist but have yet to be observed in experiments.

Identifying and understanding the particles that make up matter is important work. However, it might be difficult to understand why time and money are spent to learn more about tiny subatomic particles. How can this research possibly affect everyday life? Research on subatomic particles has changed society in many ways. For example, magnetic resonance imaging (MRI), a tool used to diagnose medical problems, uses technology that was developed to study subatomic particles. Cancer treatments using protons, neutrons, and X-rays are all based on particle physics technology. And, in the 1990s, the need for particle physicists to share information with one another led to the development of the World Wide Web!

It's Your Turn

RESEARCH AND REPORT Learn more about research on subatomic particles. Find out about one recent discovery. Make a poster to share what you learn with your classmates.

Lesson 2

Protons, Neutrons, and Electrons—How Atoms Differ

Reading Guide

Key Concepts

ESSENTIAL QUESTIONS

- What happens during nuclear decay?
- How does a neutral atom change when its number of protons, electrons, or neutrons changes?

Vocabulary

atomic number p. 327

isotope p. 328

mass number p. 328

average atomic mass p. 329

radioactive p. 330

nuclear decay p. 331

ion p. 332

Video BrainPOP®

Inquiry Is this glass glowing?

Under natural light, this glass vase is yellow. But when exposed to ultraviolet light, it glows green! That's because it is made of uranium glass, which contains small amounts of uranium, a radioactive element. Under ultraviolet light, the glass emits radiation.

Launch Lab

15 minutes

How many different things can you make?

Many buildings are made of just a few basic building materials, such as wood, nails, and glass. You can combine those materials in many different ways to make buildings of various shapes and sizes. How many things can you make from three materials?

1. Read and complete a lab safety form.
2. Use **colored building blocks** to make as many different objects as you can with the following properties:
   - Each object must have a different number of red blocks.
   - Each object must have an equal number of red and blue blocks.
   - Each object must have at least as many yellow blocks as red blocks but can have no more than two extra yellow blocks.
3. As you complete each object, record in your Science Journal the number of each color of block used to make it. For example, R = 1; B = 1; Y = 2.
4. When time is called, compare your objects with others in the class.

Think About This

1. How many different objects did you make? How many different objects did the class make?
2. How many objects do you think you could make out of the three types of blocks?
3. **Key Concept** In what ways does changing the number of building blocks change the properties of the objects?

The Parts of the Atom

If you could see inside any atom, you probably would see the same thing—empty space surrounding a very tiny nucleus. A look inside the nucleus would reveal positively charged protons and neutral neutrons. Negatively charged electrons would be whizzing by in the empty space around the nucleus.

Table 2 compares the properties of protons, neutrons, and electrons. Protons and neutrons have about the same mass. The mass of electrons is much smaller than the mass of protons or neutrons. That means most of the mass of an atom is found in the nucleus. In this lesson, you will learn that, while all atoms contain protons, neutrons, and electrons, the numbers of these particles are different for different types of atoms.

Table 2 Properties of Protons, Neutrons, and Electrons

	Electron	Proton	Neutron
Symbol	e–	p	n
Charge	1–	1+	0
Location	electron cloud around the nucleus	nucleus	nucleus
Relative mass	1/1,840	1	1

Different Elements—Different Numbers of Protons

Look at the periodic table on the inside back cover of this book. Notice that there are more than 115 different elements. Recall that an element is a substance made from atoms that all have the same number of protons. For example, the element carbon is made from atoms that all have six protons. Likewise, all atoms that have six protons are carbon atoms. *The number of protons in an atom of an element is the element's* **atomic number.** The atomic number is the whole number listed with each element on the periodic table.

What makes an atom of one element different from an atom of another element? Atoms of different elements contain different numbers of protons. For example, oxygen atoms contain eight protons; nitrogen atoms contain seven protons. Different elements have different atomic numbers. **Figure 11** shows some common elements and their atomic numbers.

Neutral atoms of different elements also have different numbers of electrons. In a neutral atom, the number of electrons equals the number of protons. Therefore, the number of positive charges equals the number of negative charges.

Reading Check What two numbers can be used to identify an element?

FOLDABLES®

Create a three-tab book and label it as shown. Use it to organize the three ways that atoms can differ.

Different Numbers of:
Protons | Neutrons | Electrons

Figure 11 Atoms of different elements contain different numbers of protons.

Visual Check Explain the difference between an oxygen atom and a carbon atom.

Different Elements Review Personal Tutor

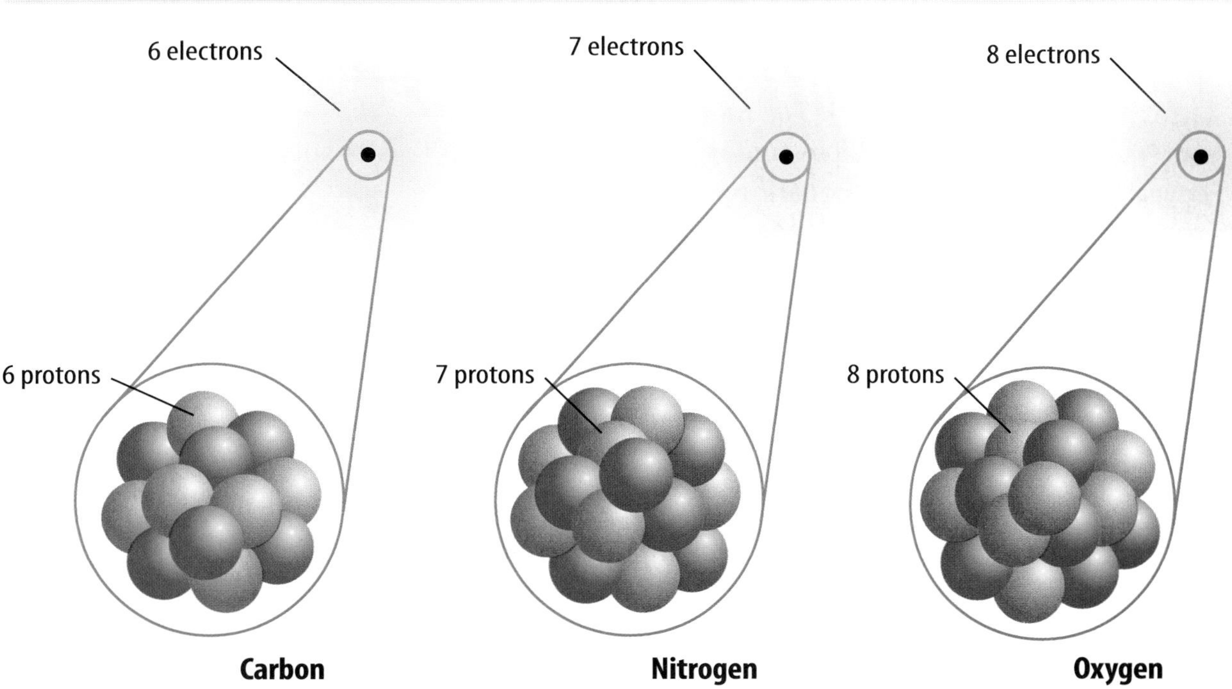

Math Skills 

Use Percentages

You can calculate the average atomic mass of an element if you know the percentage of each isotope in the element. Lithium (Li) contains 7.5% Li-6 and 92.5% Li-7. What is the average atomic mass of Li?

1. Divide each percentage by 100 to change to decimal form.
   $\frac{7.5\%}{100} = 0.075$
   $\frac{92.5\%}{100} = 0.925$
2. Multiply the mass of each isotope by its decimal percentage.
   $6 \times 0.075 = 0.45$
   $7 \times 0.925 = 6.475$
3. Add the values together to get the average atomic mass.
   $0.45 + 6.475 = 6.93$

Practice

Nitrogen (N) contains 99.63% N-14 and 0.37% N-15. What is the average atomic mass of nitrogen?

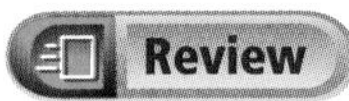

- **Math Practice**
- **Personal Tutor**

Word Origin

isotope
from Greek *isos*, means "equal"; and *topos*, means "place"

Table 3 Naturally Occurring Isotopes of Carbon

Isotope	Carbon-12 Nucleus	Carbon-13 Nucleus	Carbon-14 Nucleus
Abundance	98.89%	<1.11%	<0.01%
Protons	6	6	6
Neutrons	+6	+7	+8
Mass Number	12	13	14

Neutrons and Isotopes

You have read that atoms of the same element have the same numbers of protons. However, atoms of the same element can have different numbers of neutrons. For example, carbon atoms all have six protons, but some carbon atoms have six neutrons, some have seven neutrons, and some have eight neutrons. These three different types of carbon atoms, shown in **Table 3**, are called isotopes. **Isotopes** *are atoms of the same element that have different numbers of neutrons.* Most elements have several isotopes.

Protons, Neutrons, and Mass Number

The **mass number** *of an atom is the sum of the number of protons and neutrons in an atom.* This is shown in the following equation.

$$\text{Mass number} = \text{number of protons} + \text{number of neutrons}$$

Any one of these three quantities can be determined if you know the value of the other two quantities. For example, to determine the mass number of an atom, you must know the number of neutrons and the number of protons in the atom.

The mass numbers of the isotopes of carbon are shown in **Table 3.** An isotope often is written with the element name followed by the mass number. Using this method, the isotopes of carbon are written carbon-12, carbon-13, and carbon-14.

 **Reading Check** How do two different isotopes of the same element differ?

Average Atomic Mass

You might have noticed that the periodic table does not list mass numbers or the numbers of neutrons. This is because a given element can have several isotopes. However, you might notice that there is a decimal number listed with most elements, as shown in **Figure 12.** This decimal number is the average atomic mass of the element. The **average atomic mass** *of an element is the average mass of the element's isotopes, weighted according to the abundance of each isotope.*

Table 3 shows the three isotopes of carbon. The average atomic mass of carbon is 12.01. Why isn't the average atomic mass 13? After all, the average of the mass numbers 12, 13, and 14 is 13. The average atomic mass is weighted based on each isotope's abundance—how much of each isotope is present on Earth. Almost 99 percent of Earth's carbon is carbon-12. That is why the average atomic mass is close to 12.

 **Reading Check** What does the term *weighted average* mean?

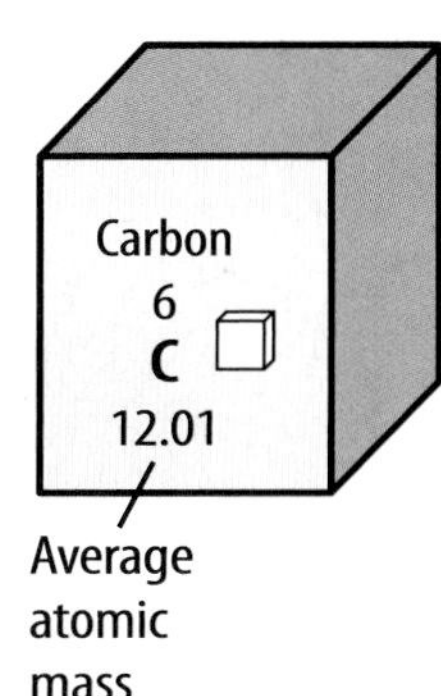

Figure 12 The element carbon has several isotopes. The decimal number 12.01 is the average atomic mass of these isotopes.

Inquiry MiniLab

20 minutes

How many penny isotopes do you have?

All pennies look similar, and all have a value of one cent. But do they have the same mass? Let's find out.

1. Read and complete a lab safety form.
2. Copy the data table into your Science Journal.
3. Use a **balance** to find the mass of **10 pennies minted before 1982.** Record the mass in the data table.
4. Divide the mass by 10 to find the average mass of one penny. Record the answer.
5. Repeat steps 3 and 4 with **10 pennies minted after 1982.**
6. Have a team member combine pre- and post-1982 pennies for a total of 10 pennies. Find the mass of the ten pennies and the average mass of one penny. Record your observations.

Penny Sample	Mass of 10 pennies (g)	Average mass of 1 penny (g)
Pre-1982		
Post-1982		
Unknown mix		

Analyze and Conclude

1. **Compare and Contrast** How did the average mass of pre- and post-1982 pennies compare?
2. **Draw Conclusions** How many pennies of each type were in the 10 pennies assembled by your partner? How do you know?
3. **Key Concept** How does this activity relate to the way in which scientists calculate the average atomic mass of an element?

Radioactivity

More than 1,000 years ago, people tried to change lead into gold by performing chemical reactions. However, none of their reactions were successful. Why not? Today, scientists know that a chemical reaction does not change the number of protons in an atom's nucleus. If the number of protons does not change, the element does not change. But in the late 1800s, scientists discovered that some elements change into other elements spontaneously. How does this happen?

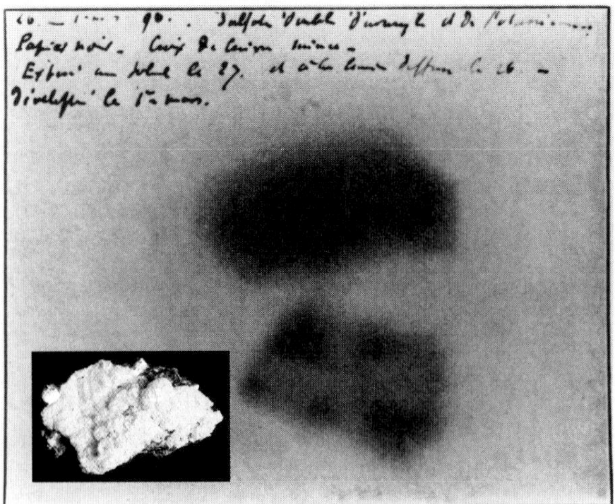

▲ **Figure 13** The black and white photo shows Henri Becquerel's photographic plate. The dark area on the plate was exposed to radiation given off by uranium in the mineral even though the mineral was not exposed to sunlight.

ACADEMIC VOCABULARY

spontaneous
(adjective) occurring without external force or cause

An Accidental Discovery

In 1896, a scientist named Henri Becquerel (1852–1908) studied minerals containing the element uranium. When these minerals were exposed to sunlight, they gave off a type of energy that could pass through paper. If Becquerel covered a photographic plate with black paper, this energy would pass through the paper and expose the film. One day, Becquerel left the mineral next to a wrapped, unexposed plate in a drawer. Later, he opened the drawer, unwrapped the plate, and saw that the plate contained an image of the mineral, as shown in **Figure 13.** The mineral spontaneously emitted energy, even in the dark! Sunlight wasn't required. What was this energy?

Figure 14 Marie Curie studied radioactivity and discovered two new radioactive elements—polonium and radium. ▼

Radioactivity

Becquerel shared his discovery with fellow scientists Pierre and Marie Curie. Marie Curie (1867–1934), shown in **Figure 14,** called *elements that spontaneously emit radiation* **radioactive.** Becquerel and the Curies discovered that the radiation released by uranium was made of energy and particles. This radiation came from the nuclei of the uranium atoms. When this happens, the number of protons in one atom of uranium changes. When uranium releases radiation, it changes to a different element!

Types of Decay

Radioactive elements contain unstable nuclei. **Nuclear decay** *is a process that occurs when an unstable atomic nucleus changes into another more stable nucleus by emitting radiation.* Nuclear decay can produce three different types of radiation—alpha particles, beta particles, and gamma rays. **Figure 15** compares the three types of nuclear decay.

Alpha Decay An alpha particle is made of two protons and two neutrons. When an atom releases an alpha particle, its atomic number decreases by two. Uranium-238 decays to thorium-234 through the process of alpha decay.

Beta Decay When beta decay occurs, a neutron in an atom changes into a proton and a high-energy electron called a beta particle. The new proton becomes part of the nucleus, and the beta particle is released. In beta decay, the atomic number of an atom increases by one because it has gained a proton.

Gamma Decay Gamma rays do not contain particles, but they do contain a lot of energy. In fact, gamma rays can pass through thin sheets of lead! Because gamma rays do not contain particles, the release of gamma rays does not change one element into another element.

Key Concept Check What happens during radioactive decay?

Uses of Radioactive Isotopes

The energy released by radioactive decay can be both harmful and beneficial to humans. Too much radiation can damage or destroy living cells, making them unable to function properly. Some organisms contain cells, such as cancer cells, that are harmful to the organism. Radiation therapy can be beneficial to humans by destroying these harmful cells.

Figure 15 Alpha and beta decay change one element into another element.

Visual Check Explain the change in atomic number for each type of decay.

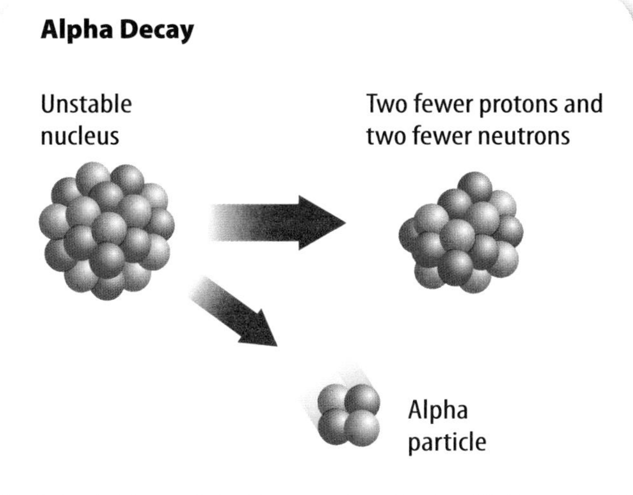

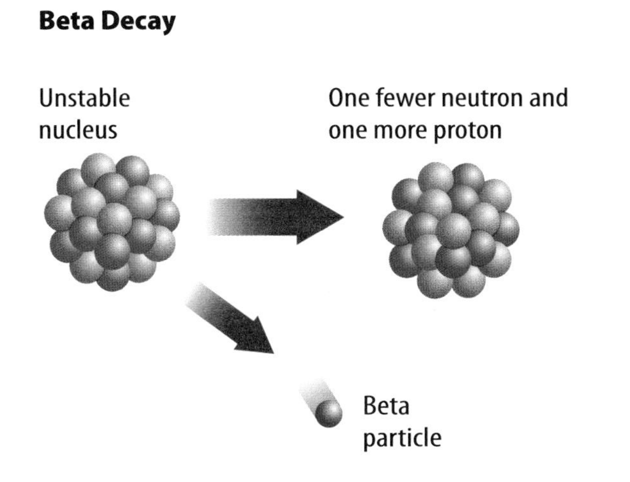

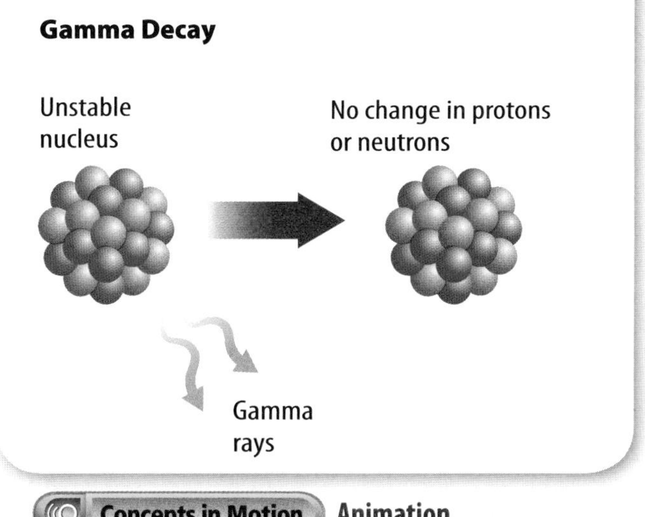

Concepts in Motion Animation

Ions—Gaining or Losing Electrons

What happens to a neutral atom if it gains or loses electrons? Recall that a neutral atom has no overall charge. This is because it contains equal numbers of positively charged protons and negatively charged electrons. When electrons are added to or removed from an atom, that atom becomes an ion. *An* **ion** *is an atom that is no longer neutral because it has gained or lost electrons.* An ion can be positively or negatively charged depending on whether it has lost or gained electrons.

Positive Ions

When a neutral atom loses one or more electrons, it has more protons than electrons. As a result, it has a positive charge. An atom with a positive charge is called a positive ion. A positive ion is represented by the element's symbol followed by a superscript plus sign ($^+$). For example, **Figure 16** shows how sodium (Na) becomes a positive sodium ion (Na^+).

Negative Ions

When a neutral atom gains one or more electrons, it now has more electrons than protons. As a result, the atom has a negative charge. An atom with a negative charge is called a negative ion. A negative ion is represented by the element's symbol followed by a superscript negative sign ($^-$). **Figure 16** shows how fluorine (F) becomes a fluoride ion (F^-).

Key Concept Check How does a neutral atom change when its number of protons, electrons, or neutrons changes?

Figure 16 An ion is formed when a neutral atom gains or loses an electron.

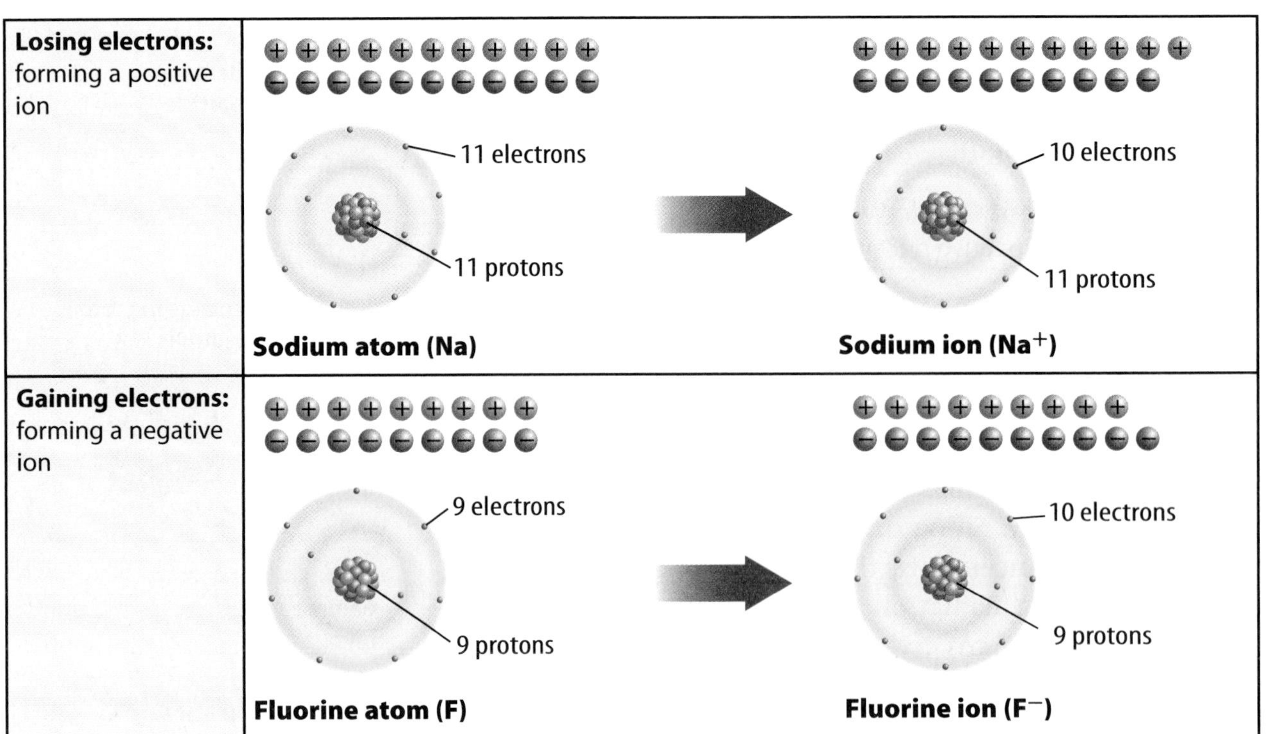

Lesson 2 Review

Assessment Online Quiz
Inquiry Virtual Lab

Visual Summary

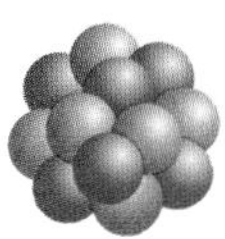
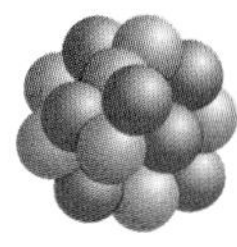

Carbon Nitrogen

Different elements contain different numbers of protons.

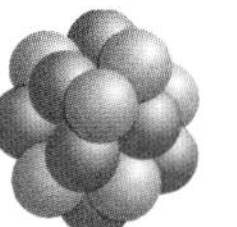

Isotopes

Two isotopes of a given element contain different numbers of neutrons.

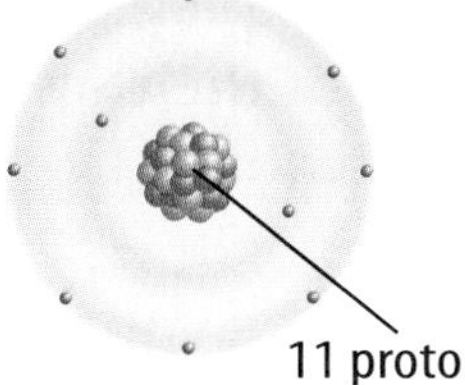

Sodium ion (Na^+)

When a neutral atom gains or loses an electron, it becomes an ion.

FOLDABLES®

Use your lesson Foldable to review the lesson. Save your Foldable for the project at the end of the chapter.

What do you think NOW?

You first read the statements below at the beginning of the chapter.

4. All atoms of the same element have the same number of protons.
5. Atoms of one element cannot be changed into atoms of another element.
6. Ions form when atoms lose or gain electrons.

Did you change your mind about whether you agree or disagree with the statements? Rewrite any false statements to make them true.

Use Vocabulary

1. The number of protons in an atom of an element is its ________.
2. Nuclear decay occurs when an unstable atomic nucleus changes into another nucleus by emitting ________.
3. **Describe** how two isotopes of nitrogen differ from two nitrogen ions.

Understand Key Concepts

4. An element's average atomic mass is calculated using the masses of its
   - **A.** electrons.
   - **B.** isotopes.
   - **C.** neutrons.
   - **D.** protons.
5. **Compare and contrast** oxygen-16 and oxygen-17.
6. **Show** what happens to the electrons of a neutral calcium atom (Ca) when it is changed into a calcium ion (Ca^{2+}).

Interpret Graphics

7. **Contrast** Copy and fill in this graphic organizer to contrast how different elements, isotopes, and ions are produced.

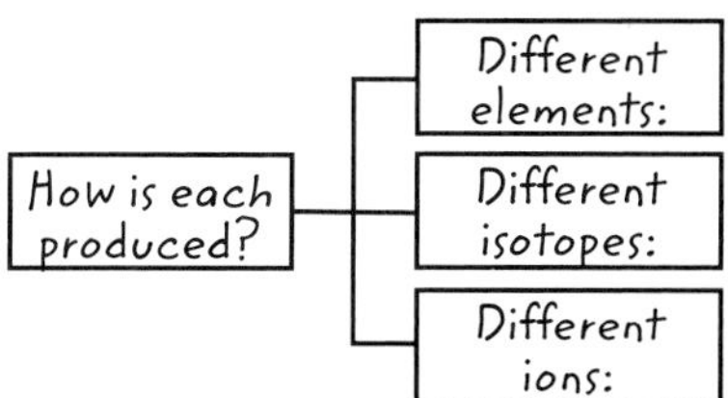

Critical Thinking

8. **Consider** Find two neighboring elements on the periodic table whose positions would be reversed if they were arranged by atomic mass instead of atomic number.
9. **Infer** Can an isotope also be an ion?

Math Skills

Review
Math Practice

10. A sample of copper (Cu) contains 69.17% Cu-63. The remaining copper atoms are Cu-65. What is the average atomic mass of copper?

Communicate Your Knowledge About the Atom

Materials

computer

creative building materials

drawing and modeling materials

office supplies

Also needed: recording devices, software, or other equipment for multimedia presentations

In this chapter, you have learned many things about atoms. Suppose that you are asked to take part in an atom fair. Each exhibit in the fair will help visitors understand something new about atoms in an exciting and interesting way. What will your exhibit be like? Will you hold a mock interview with Democritus or Rutherford? Will you model a famous experiment and explain its conclusion? Can visitors assemble or make models of their own atoms? Is there a multimedia presentation? Will your exhibit be aimed at children or adults? The choice is yours!

Question

Which concepts about the atom did you find most interesting? How can you present the information in exciting, creative, and perhaps unexpected ways? Think about whether you will present the information yourself or have visitors interact with the exhibit.

Procedure

1. In your Science Journal, write your ideas about the following questions:
   - What specific concepts about the atom do you want your exhibit to teach?
   - How will you present the information to your visitors?
   - How will you make the information exciting and interesting to keep your visitors' attention?
2. Outline the steps in preparing for your exhibit.
   - What materials and equipment will you need?
   - How much time will it take to prepare each part of your exhibit?
   - Will you involve anyone else? For example, if you are going to interview a scientist about an early model of the atom, who will play the scientist? What questions will you ask?

3 Have your teacher approve your plan.
4 Follow the steps you outlined, and prepare your exhibit.
5 Ask family members and/or several friends to view your exhibit. Invite them to tell you what they've learned from your exhibit. Compare this with what you had expected to teach in your exhibit.
6 Ask your friends for feedback about what could be more effective in teaching the concepts you intend to teach.
7 Modify your exhibit to make it more effective.
8 If you can, present your exhibit to visitors of various ages, including teachers and students from other classes. Have visitors fill out a comment form.

Analyze and Conclude

9 **Infer** What did visitors to your exhibit find the most and least interesting? How do you know?
10 **Predict** What would you do differently if you had a chance to plan your exhibit again? Why?
11 THE BIG IDEA **The Big Idea** In what ways did your exhibit help visitors understand the current model of the atom?

Communicate Your Results

After the fair is over, discuss the visitors' comments and how you might improve organization or individual experiences if you were to do another fair.

Design and describe an interactive game or activity that would teach the same concept you used for your exhibit. Invite other students to comment on your design.

Lab Tips

- ☑ For very young visitors, you might draw a picture book about the three parts of the atom and make them into cartoon characters.
- ☑ Plan your exhibit so visitors spend about 5 minutes there. Don't try to present too much or too little information.
- ☑ Remember that a picture is worth a thousand words!

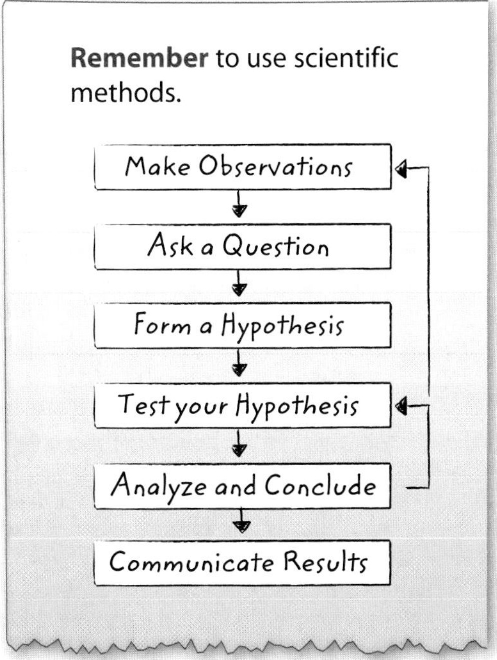

Chapter 9 Study Guide

An atom is the smallest unit of an element and is made mostly of empty space. It contains a tiny nucleus surrounded by an electron cloud.

Key Concepts Summary

Lesson 1: Discovering Parts of the Atom

- If you were to divide an element into smaller and smaller pieces, the smallest piece would be an **atom.**
- Atoms are so small that they can be seen only by powerful scanning microscopes.
- The first model of the atom was a solid sphere. Now, scientists know that an atom contains a dense positive **nucleus** surrounded by an **electron cloud.**

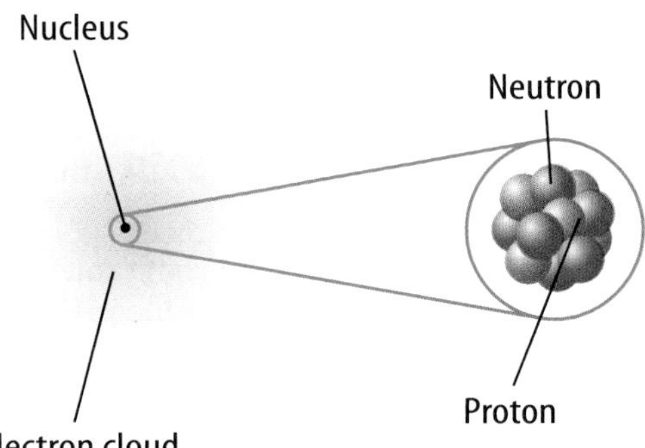

Vocabulary

atom p. 315
electron p. 317
nucleus p. 320
proton p. 320
neutron p. 321
electron cloud p. 322

Lesson 2: Protons, Neutrons, and Electrons—How Atoms Differ

- **Nuclear decay** occurs when an unstable atomic nucleus changes into another more stable nucleus by emitting radiation.
- Different elements contain different numbers of protons. Two **isotopes** of the same element contain different numbers of neutrons. When a neutral atom gains or loses an electron, it becomes an **ion.**

Nuclear Decay

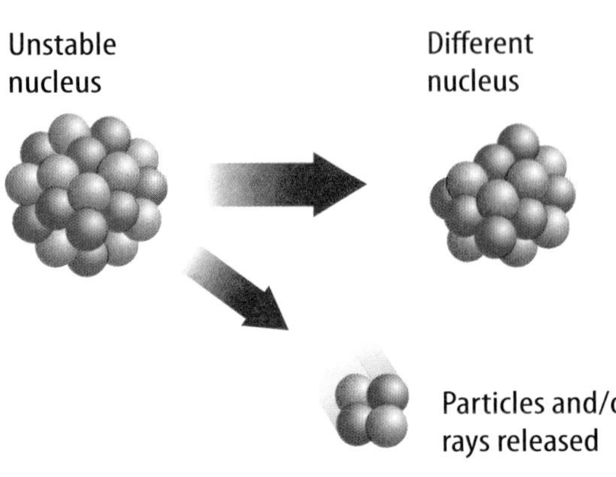

Vocabulary

atomic number p. 327
isotope p. 328
mass number p. 328
average atomic mass p. 329
radioactive p. 330
nuclear decay p. 331
ion p. 332

- Personal Tutor
- Vocabulary eGames
- Vocabulary eFlashcards

FOLDABLES® Chapter Project

Assemble your lesson Foldables as shown to make a Chapter Project. Use the project to review what you have learned in this chapter.

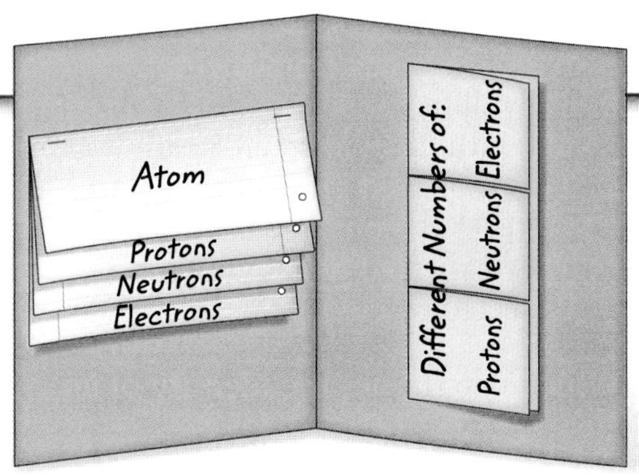

Use Vocabulary

1. A(n) ________ is a very small particle that is the basic unit of matter.
2. Electrons in an atom move throughout the ________ surrounding the nucleus.
3. ________ is the weighted average mass of all of an element's isotopes.
4. All atoms of a given element have the same number of ________.
5. When ________ occurs, one element is changed into another element.
6. Isotopes have the same ________, but different mass numbers.

Interactive Concept Map

Link Vocabulary and Key Concepts

Copy this concept map, and then use vocabulary terms from the previous page to complete the concept map.

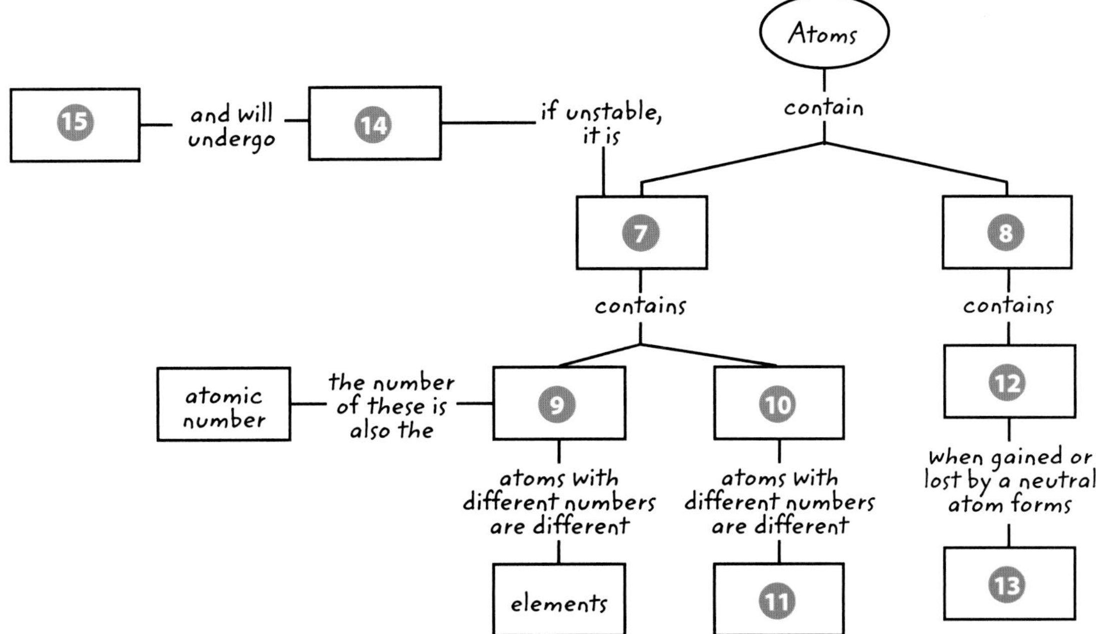

Chapter 9 Review

Understand Key Concepts

1 Which part of an atom makes up most of its volume?

A. its electron cloud
B. its neutrons
C. its nucleus
D. its protons

2 What did Democritus believe an atom was?

A. a solid, indivisible object
B. a tiny particle with a nucleus
C. a nucleus surrounded by an electron cloud
D. a tiny nucleus with electrons surrounding it

3 If an ion contains 10 electrons, 12 protons, and 13 neutrons, what is the ion's charge?

A. 2−
B. 1−
C. 2+
D. 3+

4 J.J. Thomson's experimental setup is shown below.

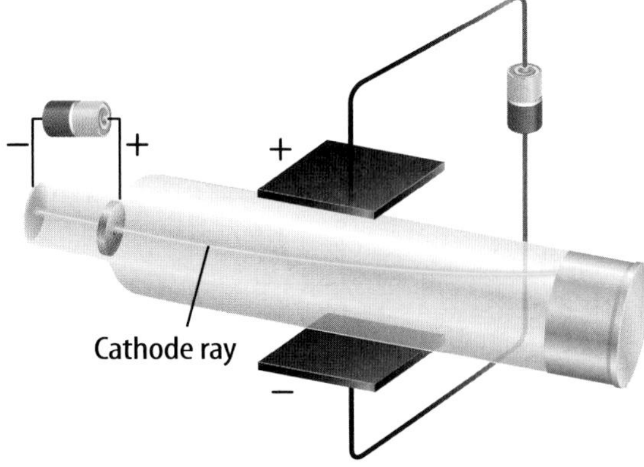

What is happening to the cathode rays?

A. They are attracted to the negative plate.
B. They are attracted to the positive plate.
C. They are stopped by the plates.
D. They are unaffected by either plate.

5 How many neutrons does iron-59 have?

A. 30
B. 33
C. 56
D. 59

6 Why were Rutherford's students surprised by the results of the gold foil experiment?

A. They didn't expect the alpha particles to bounce back from the foil.
B. They didn't expect the alpha particles to continue in a straight path.
C. They expected only a few alpha particles to bounce back from the foil.
D. They expected the alpha particles to be deflected by electrons.

7 Which determines the identity of an element?

A. its mass number
B. the charge of the atom
C. the number of its neutrons
D. the number of its protons

8 The figure below shows which of the following?

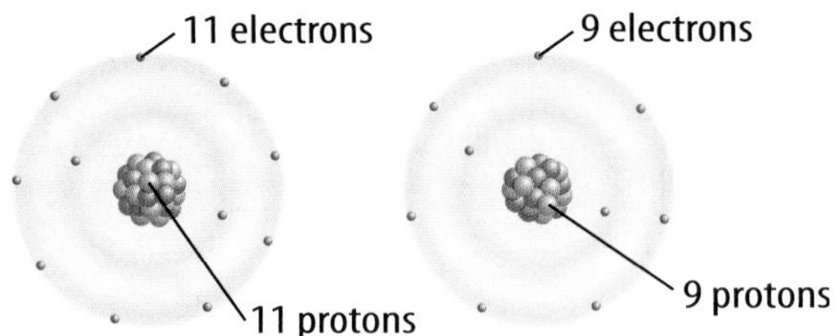

A. two different elements
B. two different ions
C. two different isotopes
D. two different protons

9 How is Bohr's atomic model different from Rutherford's model?

A. Bohr's model has a nucleus.
B. Bohr's model has electrons.
C. Electrons in Bohr's model are located farther from the nucleus.
D. Electrons in Bohr's model are located in circular energy levels.

Assessment
Online Test Practice

Critical Thinking

10 **Consider** what would have happened in the gold foil experiment if Dalton's theory had been correct.

11 **Contrast** How does Bohr's model of the atom differ from the present-day atomic model?

12 **Describe** the electron cloud using your own analogy.

13 **Summarize** how radioactive decay can produce new elements.

14 **Hypothesize** What might happen if a negatively charged ion comes into contact with a positively charged ion?

15 **Infer** Why isn't mass number listed with each element on the periodic table?

16 **Explain** How is the average atomic mass calculated?

17 **Infer** Oxygen has three stable isotopes.

Isotope	Average Atomic Mass
Oxygen-16	0.99757
Oxygen-17	0.00038
Oxygen-18	0.00205

What can you determine about the average atomic mass of oxygen without calculating it?

Writing in Science

18 **Write** a newspaper article that describes how the changes in the atomic model provide an example of the scientific process in action.

REVIEW THE BIG IDEA

19 **Describe** the current model of the atom. Explain the size of atoms. Also explain the charge, the location, and the size of protons, neutrons, and electrons.

20 **Summarize** The Large Hadron Collider, shown below, is continuing the study of matter and energy. Use a set of four drawings to summarize how the model of the atom changed from Thomson, to Rutherford, to Bohr, to the modern model.

Math Skills

Math Practice

Use Percentages

Use the information in the table to answer questions 21 and 22.

Magnesium (Mg) Isotope	Percent Found in Nature
Mg-24	78.9%
Mg-25	10.0%
Mg-26	

21 What is the percentage of Mg-26 found in nature?

22 What is the average atomic mass of magnesium?

Standardized Test Practice

Record your answers on the answer sheet provided by your teacher or on a sheet of paper.

Multiple Choice

1 Which best describes an atom?

A a particle with a single negative charge

B a particle with a single positive charge

C the smallest particle that still represents a compound

D the smallest particle that still represents an element

Use the figure below to answer questions 2 and 3.

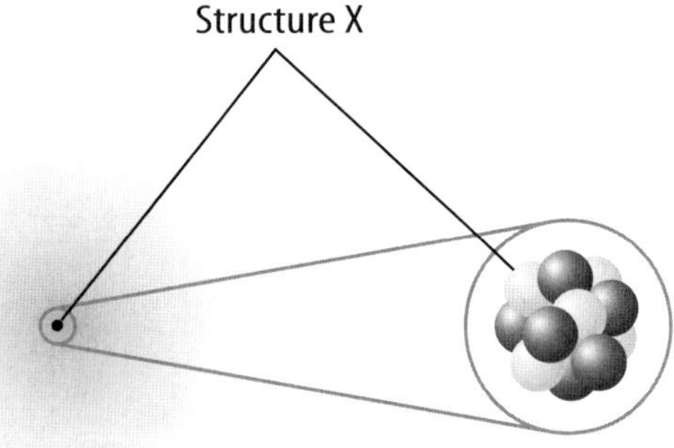

2 What is Structure X?

A an electron

B a neutron

C a nucleus

D a proton

3 Which best describes Structure X?

A most of the atom's mass, neutral charge

B most of the atom's mass, positive charge

C very small part of the atom's mass, negative charge

D very small part of the atom's mass, positive charge

4 Which is true about the size of an atom?

A It can only be seen using a scanning tunneling microscope.

B It is about the size of the period at the end of this sentence.

C It is large enough to be seen using a magnifying lens.

D It is too small to see with any type of microscope.

Use the figure below to answer question 5.

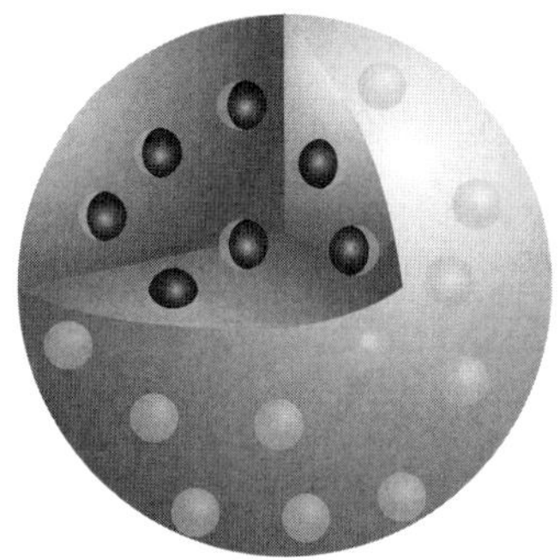

5 Whose model for the atom is shown?

A Bohr's

B Dalton's

C Rutherford's

D Thomson's

6 What structure did Rutherford discover?

A the atom

B the electron

C the neutron

D the nucleus

Assessment
Online Standardized Test Practice

Use the table below to answer questions 7–9.

Particle	Number of Protons	Number of Neutrons	Number of Electrons
1	4	5	2
2	5	5	5
3	5	6	5
4	6	6	6

7 What is atomic number of particle 3?

A 3

B 5

C 6

D 11

8 Which particles are isotopes of the same element?

A 1 and 2

B 2 and 3

C 2 and 4

D 3 and 4

9 Which particle is an ion?

A 1

B 2

C 3

D 4

10 Which reaction starts with a neutron and results in the formation of a proton and a high-energy electron?

A alpha decay

B beta decay

C the formation of positive ion

D the formation of negative ion

Constructed Response

Use the figure below to answer questions 11 and 12.

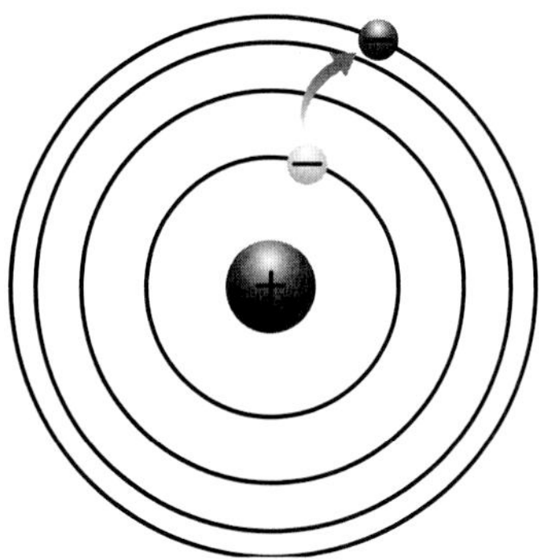

11 Identify the atomic model shown in the figure, and describe its characteristics.

12 How does this atomic model differ from the modern atomic model?

13 Compare two different neutral isotopes of the same element. Then compare two different ions of the same element. What do all of these particles have in common?

14 How does nuclear decay differ from the formation of ions? What parts of the atom are affected in each type of change?

NEED EXTRA HELP?														
If You Missed Question...	1	2	3	4	5	6	7	8	9	10	11	12	13	14
Go to Lesson...	1	1	1	1	1	1	2	2	2	2	1	1	2	2

Chapter 10

The Periodic Table

How is the periodic table used to classify and provide information about all known elements?

Inquiry What makes this balloon so special?

Things are made out of specific materials for a reason. A weather balloon can rise high in the atmosphere and gather weather information. The plastic that forms this weather balloon and the helium gas that fills it were chosen after scientists researched and studied the properties of these materials.

- What property of helium makes the balloon rise through the air?
- How is the periodic table a useful tool when determining properties of different materials?

Get Ready to Read

What do you think?

Before you read, decide if you agree or disagree with each of these statements. As you read this chapter, see if you change your mind about any of the statements.

1. The elements on the periodic table are arranged in rows in the order they were discovered.
2. The properties of an element are related to the element's location on the periodic table.
3. Fewer than half of the elements are metals.
4. Metals are usually good conductors of electricity.
5. Most of the elements in living things are nonmetals.
6. Even though they look very different, oxygen and sulfur share some similar properties.

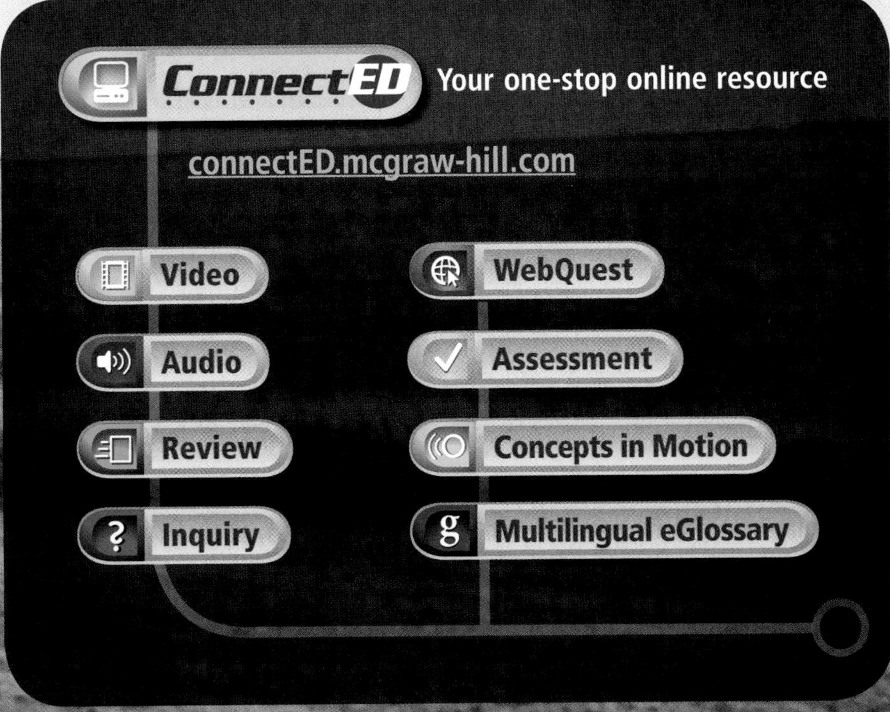

Lesson 1

Using the Periodic Table

Reading Guide

Key Concepts

ESSENTIAL QUESTIONS

- How are elements arranged on the periodic table?
- What can you learn about elements from the periodic table?

Vocabulary

periodic table p. 345

group p. 350

period p. 350

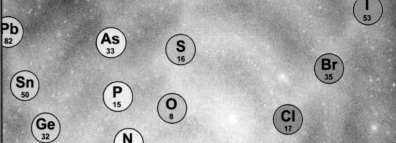

Multilingual eGlossary

Video BrainPOP®

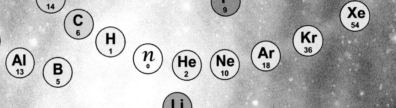

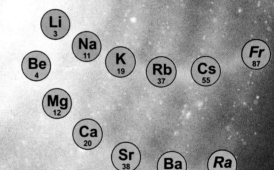

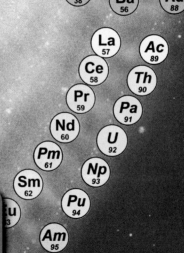

Inquiry Same Information?

You probably have seen a copy of a table that is used to organize the elements. Does it look like this chart? There is no specific shape that a chart of elements must have. However, the relationships among the elements in the chart are important.

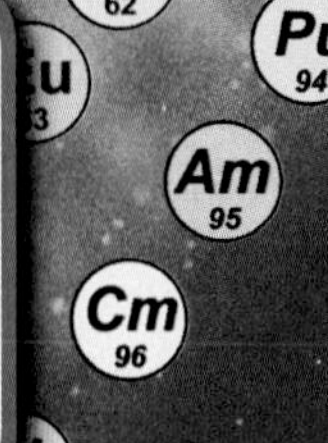

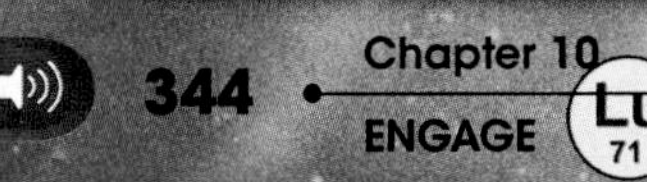

Launch Lab

15 minutes

How can objects be organized?

What would it be like to shop at a grocery store where all the products are mixed up on the shelves? Maybe cereal is next to the dish soap and bread is next to the canned tomatoes. It would take a long time to find the groceries that you needed. How does organizing objects help you to find and use what you need?

1. Read and complete a lab safety form.
2. Empty the **interlocking plastic bricks** from the **plastic bag** onto your desk and observe their properties. Think about ways you might group and sequence the bricks so that they are organized.
3. Organize the bricks according to your plan.
4. Compare your pattern of organization with those used by several other students.

Think About This

1. Describe in your Science Journal the way you grouped your bricks. Why did you choose that way of grouping?
2. Describe how you sequenced the bricks.
3. **Key Concept** How does organizing things help you to use them more easily?

What is the periodic table?

The "junk drawer" in **Figure 1** is full of pens, notepads, rubber bands, and other supplies. It would be difficult to find a particular item in this messy drawer. How might you organize it? First, you might dump the contents onto the counter. Then you could sort everything into piles. Pens and pencils might go into one pile. Notepads and paper go into another. Organizing the contents of the drawer makes it easier to find the things you need, also shown in **Figure 1.**

Just as sorting helped to organize the objects in the junk drawer, sorting can help scientists organize information about the elements. Recall that there are more than 100 elements, each with a unique set of physical and chemical properties.

Scientists use a table called the periodic (pihr ee AH dihk) table to organize elements. *The* **periodic table** *is a chart of the elements arranged into rows and columns according to their physical and chemical properties.* It can be used to determine the relationships among the elements.

In this chapter, you will read about how the periodic table was developed. You will also read about how you can use the periodic table to learn about the elements.

Figure 1 Sorting objects by their similarities makes it easier to find what you need.

Developing a Periodic Table

In 1869 a Russian chemist and teacher named Dimitri Mendeleev (duh MEE tree • men duh LAY uf) was working on a way to classify elements. At that time, more than 60 elements had been discovered. He studied the physical properties such as density, color, melting point, and atomic mass of each element. Mendeleev also noted chemical properties such as how each element reacted with other elements. Mendeleev arranged the elements in a list using their atomic masses. He noticed that the properties of the elements seemed to repeat in a pattern.

When Mendeleev placed his list of elements into a table, he arranged them in rows of increasing atomic mass. Elements with similar properties were grouped the same column. The columns in his table are like the piles of sorted objects in your junk drawer. Both contain groups of things with similar properties.

Reading Check What physical property did Mendeleev use to place the elements in rows on the periodic table?

Patterns in Properties

The term *periodic* means "repeating pattern." For example, seasons and months are periodic because they follow a repeating pattern every year. The days of the week are periodic since they repeat every seven days.

What were some of the repeating patterns Mendeleev noticed in his table? Melting point is one property that shows a repeating pattern. Recall that melting point is the temperature at which a solid changes to a liquid. The blue line in **Figure 2** represents the melting points of the elements in row 2 of the periodic table. Notice that the melting point of carbon is higher than the melting point of lithium. However, the melting point of fluorine, at the far right of the row, is lower than that of carbon. How do these melting points show a pattern? Look at the red line in **Figure 2.** This line represents the melting points of the elements in row 3 of the periodic table. The melting points follow the same increasing and then decreasing pattern as the blue line, or row 2. Boiling point and reactivity also follow a periodic pattern.

A Periodic Property

Figure 2 Melting points increase, then decrease, across a period on the periodic table.

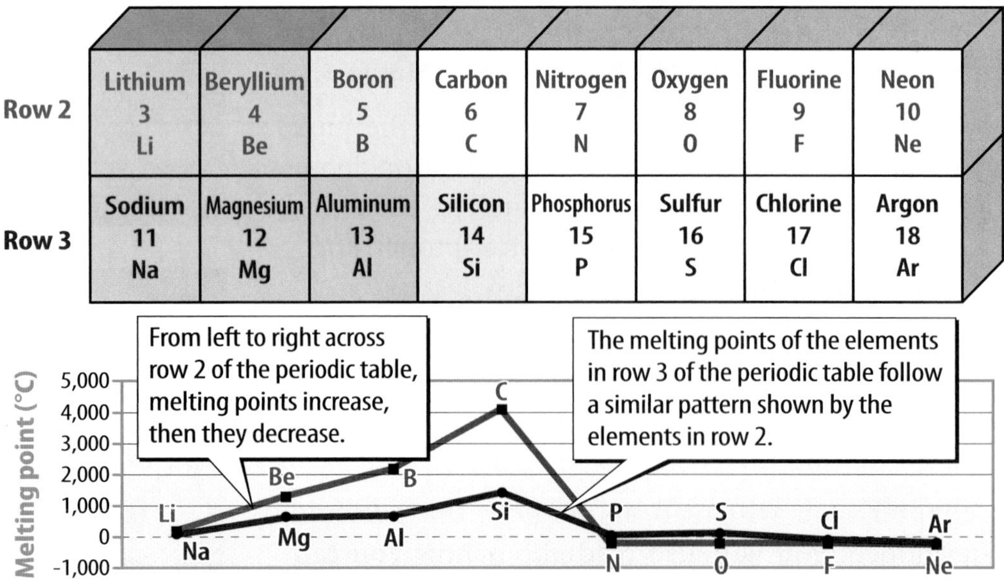

Predicting Properties of Undiscovered Elements

When Mendeleev arranged all known elements by increasing atomic mass, there were large gaps between some elements. He predicted that scientists would discover elements that would fit into these spaces. Mendeleev also predicted that the properties of these elements would be similar to the known elements in the same columns. Both of his predictions turned out to be true.

Changes to Mendeleev's Table

Mendeleev's periodic table enabled scientists to relate the properties of the known elements to their position on the table. However, the table had a problem—some elements seemed out of place. Mendeleev believed that the atomic masses of certain elements must be invalid because the elements appeared in the wrong place on the periodic table. For example, Mendeleev placed tellurium before iodine despite the fact that tellurium has a greater atomic mass than iodine. He did so because iodine's properties more closely resemble those of fluorine and chlorine, just as copper's properties are closer to those of silver and gold, as shown in **Figure 3.**

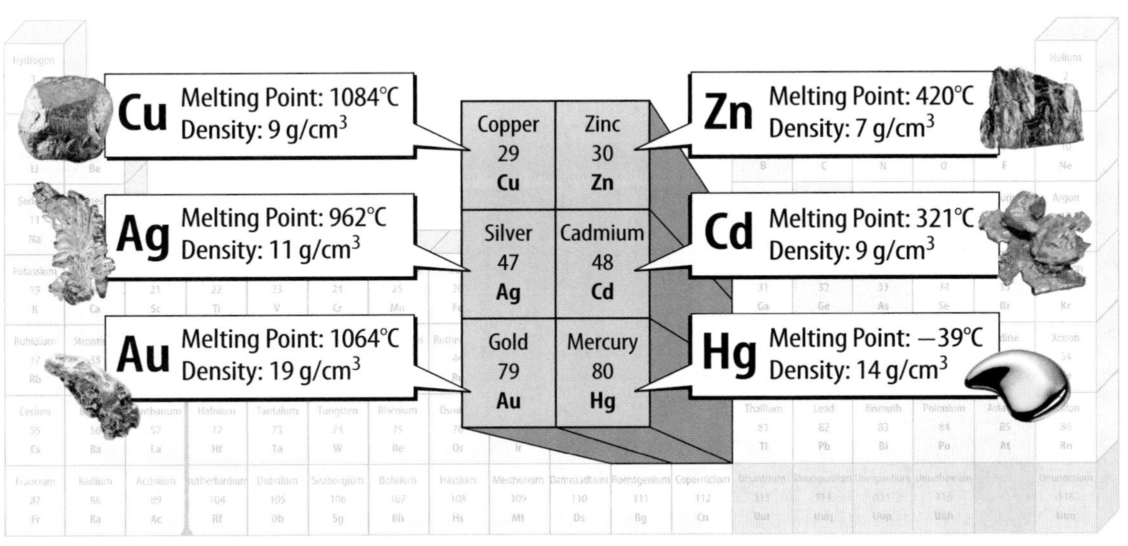

Figure 3 On today's periodic table, copper is in the same column as silver and gold. Zinc is in the same column as cadmium and mercury.

The Importance of Atomic Number

In the early 1900s, the scientist Henry Moseley solved the problem with Mendeleev's table. Moseley found that if elements were listed according to increasing atomic number instead of listing atomic mass, columns would contain elements with similar properties. Recall that the atomic number of an element is the number of protons in the nucleus of each of that element's atoms.

Key Concept Check What determines where an element is located on the periodic table you use today?

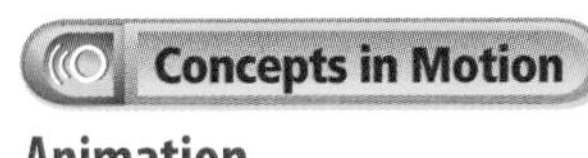

Animation

SCIENCE USE V. COMMON USE

period

Science Use the completion of a cycle; a row on the periodic table

Common Use a point used to mark the end of a sentence; a time frame

Today's Periodic Table

You can identify many of the properties of an element from its placement on the **periodic** table. The table, as shown in **Figure 4,** is organized into columns, rows, and blocks, which are based on certain patterns of properties. In the next two lessons, you will learn how an element's position on the periodic table can help you interpret the element's physical and chemical properties.

Figure 4 The periodic table is used to organize elements according to increasing atomic number and properties.

PERIODIC TABLE OF THE ELEMENTS

Element — Hydrogen; Atomic number — 1; Symbol — H; Atomic mass — 1.01; State of matter

Gas · Liquid · Solid · Synthetic

A column in the periodic table is called a **group.**

A row in the periodic table is called a **period.**

Period	1	2	3	4	5	6	7	8	9
1	Hydrogen 1 H 1.01								
2	Lithium 3 Li 6.94	Beryllium 4 Be 9.01							
3	Sodium 11 Na 22.99	Magnesium 12 Mg 24.31							
4	Potassium 19 K 39.10	Calcium 20 Ca 40.08	Scandium 21 Sc 44.96	Titanium 22 Ti 47.87	Vanadium 23 V 50.94	Chromium 24 Cr 52.00	Manganese 25 Mn 54.94	Iron 26 Fe 55.85	Cobalt 27 Co 58.93
5	Rubidium 37 Rb 85.47	Strontium 38 Sr 87.62	Yttrium 39 Y 88.91	Zirconium 40 Zr 91.22	Niobium 41 Nb 92.91	Molybdenum 42 Mo 95.96	Technetium 43 Tc (98)	Ruthenium 44 Ru 101.07	Rhodium 45 Rh 102.91
6	Cesium 55 Cs 132.91	Barium 56 Ba 137.33	Lanthanum 57 La 138.91	Hafnium 72 Hf 178.49	Tantalum 73 Ta 180.95	Tungsten 74 W 183.84	Rhenium 75 Re 186.21	Osmium 76 Os 190.23	Iridium 77 Ir 192.22
7	Francium 87 Fr (223)	Radium 88 Ra (226)	Actinium 89 Ac (227)	Rutherfordium 104 Rf (267)	Dubnium 105 Db (268)	Seaborgium 106 Sg (271)	Bohrium 107 Bh (272)	Hassium 108 Hs (270)	Meitnerium 109 Mt (276)

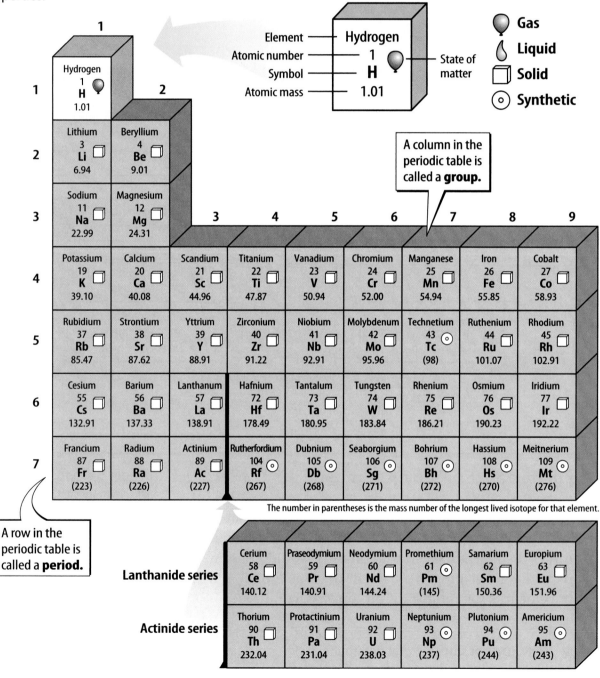

The number in parentheses is the mass number of the longest lived isotope for that element.

Lanthanide series	Cerium 58 Ce 140.12	Praseodymium 59 Pr 140.91	Neodymium 60 Nd 144.24	Promethium 61 Pm (145)	Samarium 62 Sm 150.36	Europium 63 Eu 151.96
Actinide series	Thorium 90 Th 232.04	Protactinium 91 Pa 231.04	Uranium 92 U 238.03	Neptunium 93 Np (237)	Plutonium 94 Pu (244)	Americium 95 Am (243)

What is on an element key?

The element key shows an element's chemical symbol, atomic number, and atomic mass. The key also contains a symbol that shows the state of matter at room temperature. Look at the element key for helium in **Figure 5.** Helium is a gas at room temperature. Some versions of the periodic table give additional information, such as density, conductivity, or melting point.

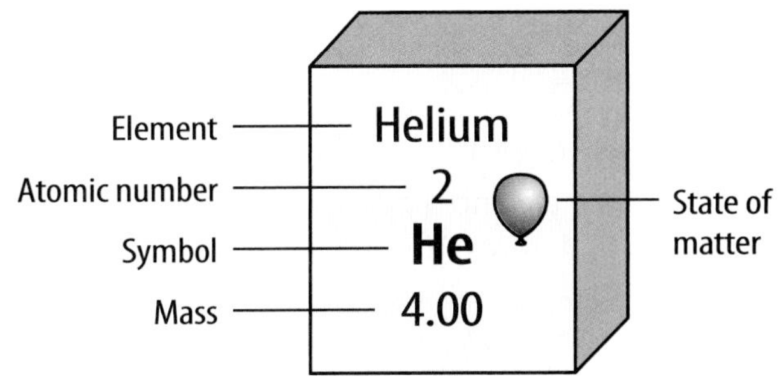

Figure 5 An element key shows important information about each element.

Visual Check What does this key tell you about helium?

Metal
Metalloid
Nonmetal
Recently discovered

10	11	12	13	14	15	16	17	18
								Helium 2 He 4.00
			Boron 5 B 10.81	Carbon 6 C 12.01	Nitrogen 7 N 14.01	Oxygen 8 O 16.00	Fluorine 9 F 19.00	Neon 10 Ne 20.18
			Aluminum 13 Al 26.98	Silicon 14 Si 28.09	Phosphorus 15 P 30.97	Sulfur 16 S 32.07	Chlorine 17 Cl 35.45	Argon 18 Ar 39.95
Nickel 28 Ni 58.69	Copper 29 Cu 63.55	Zinc 30 Zn 65.38	Gallium 31 Ga 69.72	Germanium 32 Ge 72.64	Arsenic 33 As 74.92	Selenium 34 Se 78.96	Bromine 35 Br 79.90	Krypton 36 Kr 83.80
Palladium 46 Pd 106.42	Silver 47 Ag 107.87	Cadmium 48 Cd 112.41	Indium 49 In 114.82	Tin 50 Sn 118.71	Antimony 51 Sb 121.76	Tellurium 52 Te 127.60	Iodine 53 I 126.90	Xenon 54 Xe 131.29
Platinum 78 Pt 195.08	Gold 79 Au 196.97	Mercury 80 Hg 200.59	Thallium 81 Tl 204.38	Lead 82 Pb 207.20	Bismuth 83 Bi 208.98	Polonium 84 Po (209)	Astatine 85 At (210)	Radon 86 Rn (222)
Darmstadtium 110 Ds (281)	Roentgenium 111 Rg (280)	Copernicium 112 Cn (285)	Ununtrium * 113 Uut (284)	Ununquadium * 114 Uuq (289)	Ununpentium * 115 Uup (288)	Ununhexium * 116 Uuh (293)		Ununoctium * 118 Uuo (294)

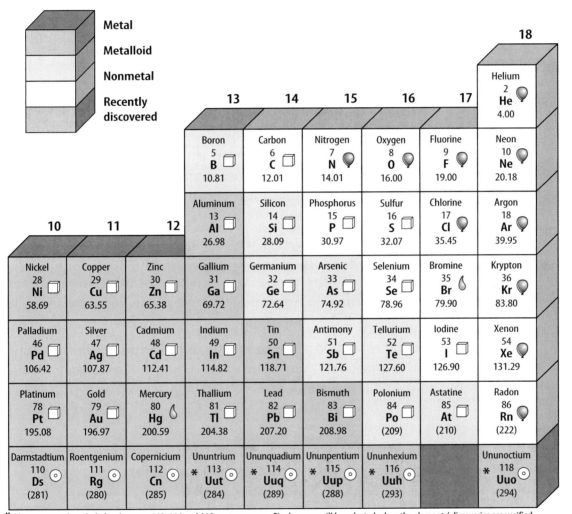

* The names and symbols for elements 113-116 and 118 are temporary. Final names will be selected when the elements' discoveries are verified.

Gadolinium 64 Gd 157.25	Terbium 65 Tb 158.93	Dysprosium 66 Dy 162.50	Holmium 67 Ho 164.93	Erbium 68 Er 167.26	Thulium 69 Tm 168.93	Ytterbium 70 Yb 173.05	Lutetium 71 Lu 174.97
Curium 96 Cm (247)	Berkelium 97 Bk (247)	Californium 98 Cf (251)	Einsteinium 99 Es (252)	Fermium 100 Fm (257)	Mendelevium 101 Md (258)	Nobelium 102 No (259)	Lawrencium 103 Lr (262)

Use Geometry

The distance around a circle is the circumference (C). The distance across the circle, through its center, is the diameter (d). The radius (r) is half of the diameter. The circumference divided by the diameter for any circle is equal to π (pi), or 3.14. The formula for determining the circumference is:

$C = \pi d$ or $C = 2\pi r$

For example, an iron (Fe) atom has a radius of 126 pm (picometers; 1 picometer = one-trillionth of a meter) The circumference of an iron atom is:

$C = 2 \times 3.14 \times 126$ pm

$C =$ **791 pm**

Practice

The radius of a uranium (U) atom is 156 pm. What is its circumference?

- **Math Practice**
- **Personal Tutor**

Groups

A **group** *is a column on the periodic table.* Elements in the same group have similar chemical properties and react with other elements in similar ways. There are patterns in the physical properties of a group such as density, melting point, and boiling point. The groups are numbered 1–18, as shown in **Figure 4.**

Key Concept Check What can you infer about the properties of two elements in the same group?

Periods

The rows on the periodic table are called **periods.** The atomic number of each element increases by one as you read from left to right across each period. The physical and chemical properties of the elements also change as you move left to right across a period.

Metals, Nonmetals, and Metalloids

Almost three-fourths of the elements on the periodic table are metals. Metals are on the left side and in the middle of the table. Individual metals have some properties that differ, but all metals are shiny and conduct thermal energy and electricity.

With the exception of hydrogen, nonmetals are located on the right side of the periodic table. The properties of nonmetals differ from the properties of metals. Many nonmetals are gases, and they do not conduct thermal energy or electricity.

Between the metals and the nonmetals on the periodic table are the metalloids. Metalloids have properties of both metals and nonmetals. **Figure 6** shows an example of a metal, a metalloid, and a nonmetal.

Figure 6 In period 3, magnesium is a metal, silicon is a metalloid, and sulfur is a nonmetal.

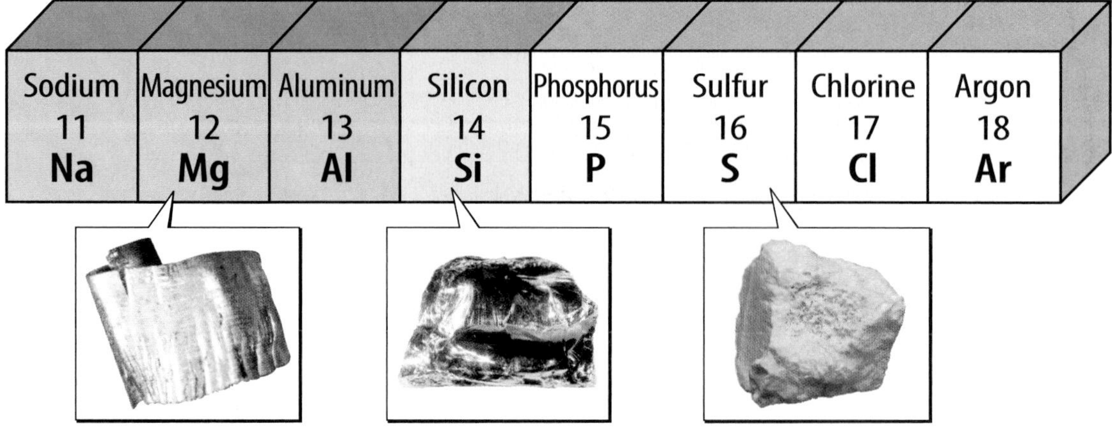

Seaborgium	Bohrium	Hassium	Meitnerium
106	107	108	109
Sg	Bh	Hs	Mt

Figure 7 Three of these synthetic elements are named to honor important scientists.

How Scientists Use the Periodic Table

Even today, new elements are created in laboratories, named, and added to the present-day periodic table. Four of these elements are shown in **Figure 7.** These elements are all synthetic, or made by people, and do not occur naturally on Earth. Sometimes scientists can create only a few atoms of a new element. Yet scientists can use the periodic table to predict the properties of new elements they create. Look back at the periodic table in **Figure 4.** What group would you predict to contain element 117? You would probably expect element 117 to be in group 17 and to have similar properties to other elements in the group. Scientists hope to one day synthesize element 117.

The periodic table contains more than 100 elements. Each element has unique properties that differ from the properties of other elements. But each element also shares similar properties with nearby elements. The periodic table shows how elements relate to each other and fit together into one organized chart. Scientists use the periodic table to understand and predict elements' properties. You can, too.

Reading Check How is the periodic table used to predict the properties of an element?

Inquiry MiniLab — 20 minutes

How does atom size change across a period?

One pattern seen on the periodic table is in the radius of different atoms. The figure below shows how atomic radius is measured.

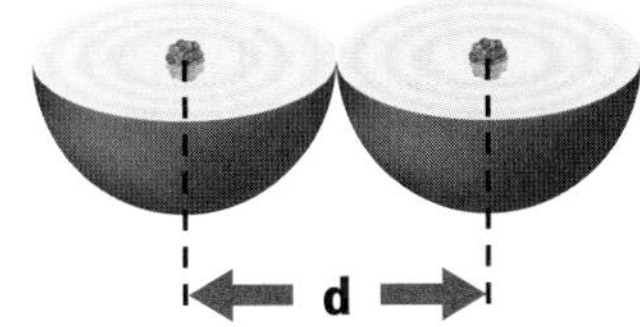

1. Read and complete a lab safety form.
2. Using **scissors** and **card stock paper,** cut seven 2-cm × 4-cm rectangles. Using a **marker,** label each rectangle with the atomic symbol of each of the first seven elements in period 2. Obtain the radius for each atom from your teacher.
3. Using a **ruler,** cut **plastic straws** to the same number of millimeters as each atomic radius given in picometers. For example, if the atomic radius is 145 pm, cut a straw 145 mm long.
4. **Tape** each of the labeled rectangles to the top of its appropriate straw.
5. Insert the straws into **modeling clay** according to increasing atomic number.

Analyze and Conclude

1. **Describe** the pattern you see in your model.
2. **Key Concept** Predict the pattern of atomic radii of the elements in period 4.

Lesson 1 Review

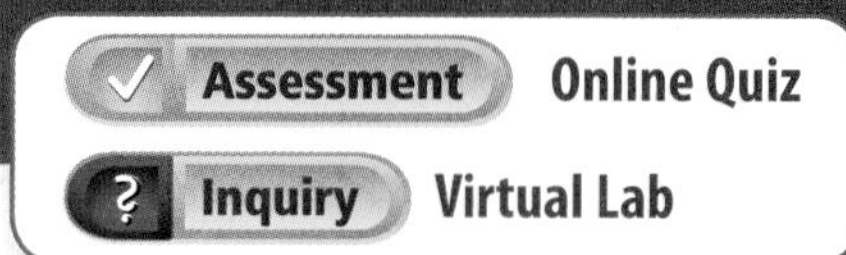

Visual Summary

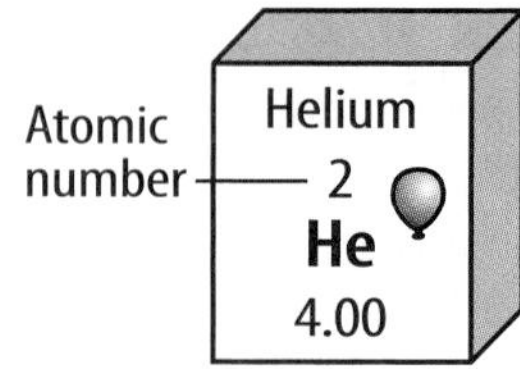

On the periodic table, elements are arranged according to increasing atomic number and similar properties.

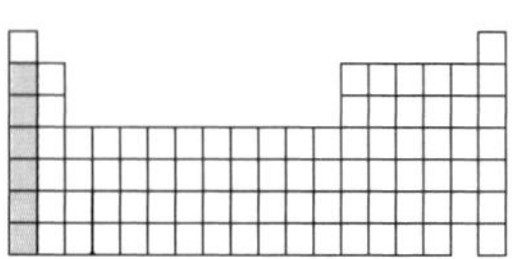

A column of the periodic table is called a group. Elements in the same group have similar properties.

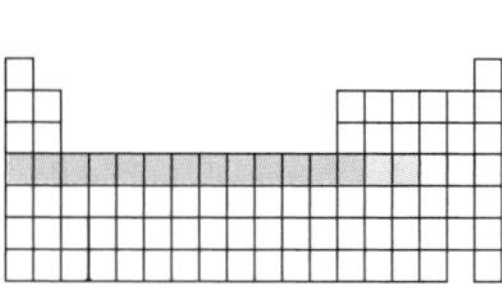

A row of the periodic table is called a period. Properties of elements repeat in the same pattern from left to right across each period.

FOLDABLES®

Use your lesson Foldable to review the lesson. Save your Foldable for the project at the end of the chapter.

What do you think 

You first read the statements below at the beginning of the chapter.

1. The elements on the periodic table are arranged in rows in the order they were discovered.
2. The properties of an element are related to the element's location on the periodic table.

Did you change your mind about whether you agree or disagree with the statements? Rewrite any false statements to make them true.

Use Vocabulary

1. **Identify** the scientific term used for rows on the periodic table.
2. **Name** the scientific term used for columns on the periodic table.

Understand Key Concepts 

3. The ________ increases by one for each element as you move left to right across a period.
4. What does the decimal number in an element key represent?
   A. atomic mass
   B. atomic number
   C. chemical symbol
   D. state of matter

Interpret Graphics

5. **Classify** each marked element, 1 and 2, as a metal, a nonmetal, or a metalloid.

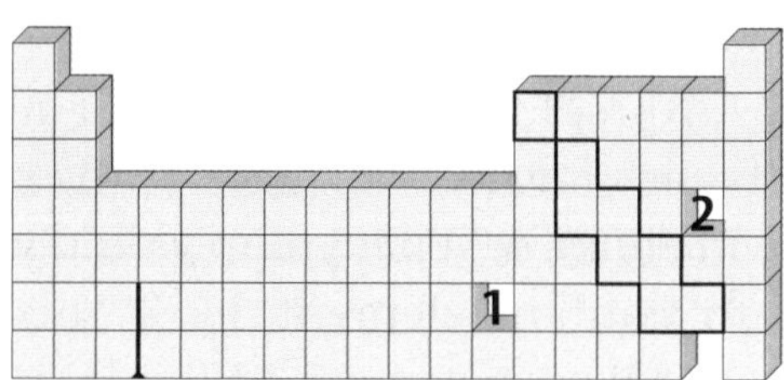

6. **Identify** Copy and fill in the graphic organizer below to identify the color-coded regions of the periodic table.

Critical Thinking

7. **Predict** Look at the perioidic table and predict three elements that have lower melting points than calcium (Ca).

Math Skills

Math Practice

8. Carbon (C) and silicon (Si) are in group 4 of the periodic table. The atomic radius of carbon is 77 pm and sulfur is 117 pm. What is the circumference of each atom?

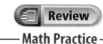

Inquiry **Skill Practice** **Identify Patterns** 25 minutes

How is the periodic table arranged?

Materials

20 cards

What would happen if schools did not assign students to grades or classes? How would you know where to go on the first day of school? What if your home did not have an address? How could you tell someone where you live? Life becomes easier with organization. The following activity will help you discover how elements are organized on the periodic table.

Learn It

Patterns help you make sense of the world around you. The days of the week follow a pattern, as do the months of the year. **Identifying a pattern** involves organizing things into similar groups and then sequencing the things in the same way in each group.

Try It

1. Obtain cards from your teacher. Turn the cards over so the sides with numbers are facing up.
2. Separate the cards into three or more piles. All of the cards in a pile should have a characteristic in common.
3. Organize each pile into a pattern. Use all of the cards.
4. Lay out the cards into rows and columns based on their characteristics and patterns.

Apply It

5. Describe in your Science Journal the patterns you used to organize your cards. Do other patterns exist in your arrangement?
6. Are there gaps in your arrangement? Can you describe what a card in one of those gaps would look like?
7. **Key Concept** What characteristics of elements might you use to organize them in a similar pattern?

Lesson 2

Metals

Reading Guide

Key Concepts
ESSENTIAL QUESTIONS

- What elements are metals?
- What are the properties of metals?

Vocabulary

metal p. 355
luster p. 355
ductility p. 356
malleability p. 356
alkali metal p. 357
alkaline earth metal p. 357
transition element p. 358

g Multilingual eGlossary

Inquiry Where does it strike?

Lightning strikes the top of the Empire State Building approximately 100 times a year. Why does lightning hit the top of this building instead of the city streets or buildings below? Metal lightning rods allow electricity to flow through them more easily than other materials do. Lightning moves through these materials and the building is not harmed.

Launch Lab

20 minutes

What properties make metals useful?

The properties of metals determine their uses. Copper conducts thermal energy, which makes it useful for cookware. Aluminum has low density, so it is used in aircraft bodies. What other properties make metals useful?

1. Read and complete a lab safety form.
2. With your group, observe the **metal objects** in your **container.** For each object, discuss what properties allow the metal to be used in that way.
3. Observe the **photographs of gold and silver jewelry.** What properties make these two metals useful in jewelry?
4. Examine **other objects around the room** that you think are made of metal. Do they share the same properties as the objects in your container? Do they have other properties that make them useful?

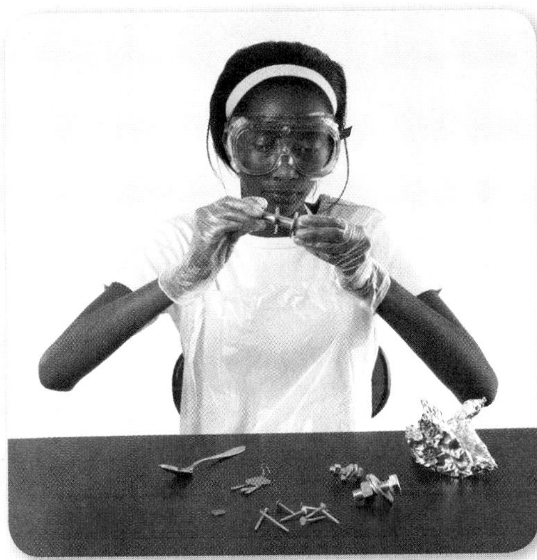

Think About This

1. What properties do all the metals share? What properties are different?
2. **Key Concept** In your Science Journal, list at least four properties of metals that determine their uses.

What is a metal?

What do stainless steel knives and forks, copper wire, aluminum foil, and gold jewelry have in common? They are all made from metals.

As you read in Lesson 1, most of the elements on the periodic table are metals. In fact, of all the known elements, more than three-quarters are metals. With the exception of hydrogen, all of the elements in groups 1–12 on the periodic table are metals. In addition, some of the elements in groups 13–15 are metals. To be a metal, an element must have certain properties.

Key Concept Check How does the position of an element on the periodic table allow you to determine if the element is a metal?

Physical Properties of Metals

Recall that physical properties are characteristics used to describe or identify something without changing its makeup. All metals share certain physical properties.

A **metal** *is an element that is generally shiny. It is easily pulled into wires or hammered into thin sheets. A metal is a good conductor of electricity and thermal energy.* Gold exhibits the common properties of metals.

Luster and Conductivity People use gold for jewelry because of its beautiful color and metallic luster. **Luster** *describes the ability of a metal to reflect light.* Gold is also a good conductor of thermal energy and electricity. However, gold is too expensive to use in normal electrical wires or metal cookware. Copper is often used instead.

Properties of Metals

Figure 8 Gold has many uses based on its properties.

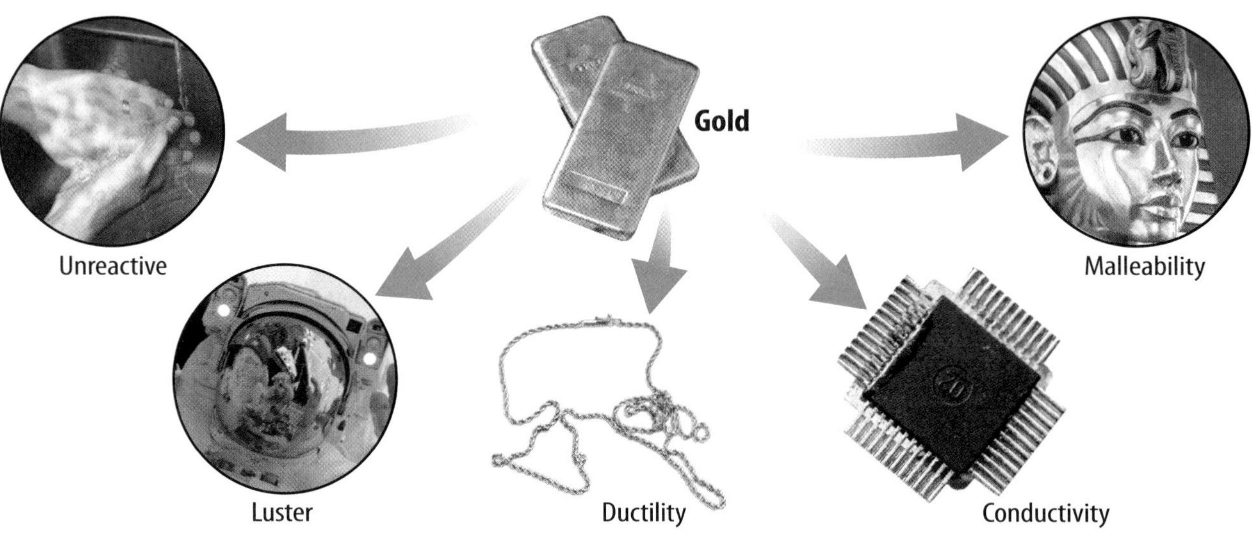

Visual Check Analyze why the properties shown in each photo are an advantage to using gold.

WORD ORIGIN

ductility
from Latin *ductilis*, means "may be led or drawn"

Ductility and Malleability Gold is the most ductile metal. **Ductility** (duk TIH luh tee) *is the ability to be pulled into thin wires.* A piece of gold with the mass of a paper clip can be pulled into a wire that is more than 3 km long.

Malleability (ma lee uh BIH luh tee) *is the ability of a substance to be hammered or rolled into sheets.* Gold is so malleable that it can be hammered into thin sheets. A pile of a million thin sheets would be only as high as a coffee mug.

REVIEW VOCABULARY

density
the mass per unit volume of a substance

Other Physical Properties of Metals In general the **density**, strength, boiling point, and melting point of a metal are greater than those of other elements. Except for mercury, all metals are solid at room temperature. Many uses of a metal are determined by the metal's physical properties, as shown in **Figure 8.**

Key Concept Check What are some physical properties of metals?

FOLDABLES®

Make a two-tab book. Label it as shown. Use it to record information about the properties of metals.

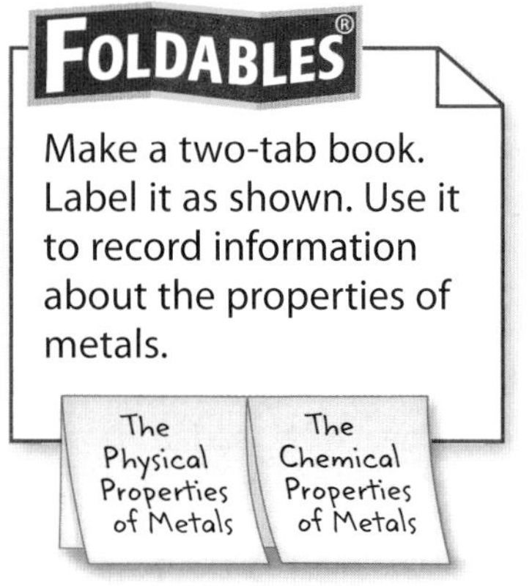

Chemical Properties of Metals

Recall that a chemical property is the ability or inability of a substance to change into one or more new substances. The chemical properties of metals can differ greatly. However, metals in the same group usually have similar chemical properties. For example, gold and other elements in group 11 do not easily react with other substances.

Group 1: Alkali Metals

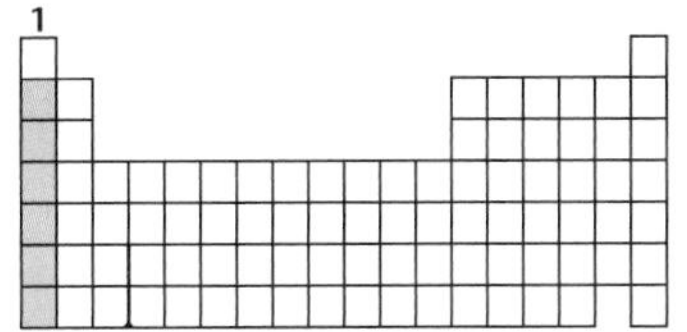

The elements in group 1 are called **alkali** (AL kuh li) **metals.** The alkali metals include lithium, sodium, potassium, rubidium, cesium, and francium.

Because they are in the same group, alkali metals have similar chemical properties. Alkali metals react quickly with other elements, such as oxygen. Therefore, in nature, they occur only in compounds. Pure alkali metals must be stored so that they do not come in contact with oxygen and water vapor in the air. **Figure 9** shows potassium and sodium reacting with water.

Alkali metals also have similar physical properties. Pure alkali metals have a silvery appearance. As shown in **Figure 9,** they are soft enough to cut with a knife. The alkali metals also have the lowest densities of all metals. A block of pure sodium metal could float on water because of its very low density.

Figure 9 Alkali metals react violently with water. They are also soft enough to be cut with a knife.

Concepts in Motion

Animation

Potassium

Sodium

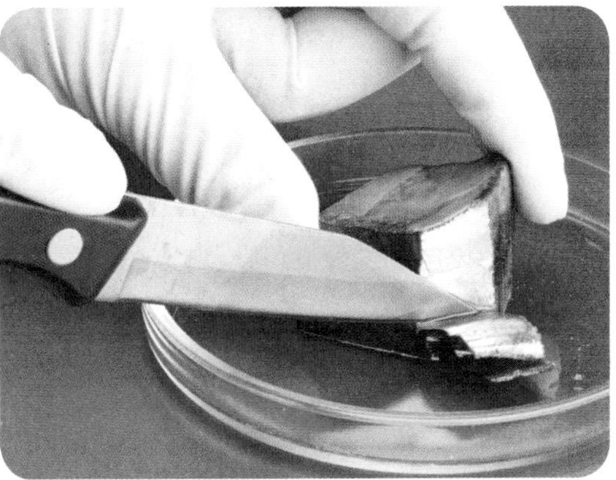

Lithium

Group 2: Alkaline Earth Metals

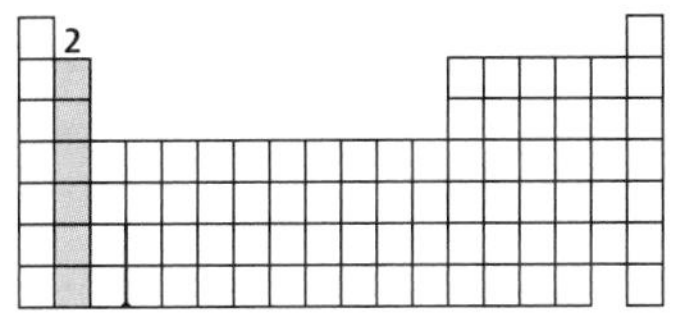

The elements in group 2 on the periodic table are called **alkaline** (AL kuh lun) **earth metals.** These metals are beryllium, magnesium, calcium, strontium, barium, and radium.

Alkaline earth metals also react quickly with other elements. However, they do not react as quickly as the alkali metals do. Like the alkali metals, pure alkaline earth metals do not occur naturally. Instead, they combine with other elements and form compounds. The physical properties of the alkaline earth metals are also similar to those of the alkali metals. Alkaline earth metals are soft and silvery. They also have low density, but they have greater density than alkali metals.

Reading Check Which element reacts faster with oxygen—barium or potassium?

Transition Elements

Concepts in Motion Animation

Figure 10 Transition elements are in blocks at the center of the periodic table. Many colorful materials contain small amounts of transition elements.

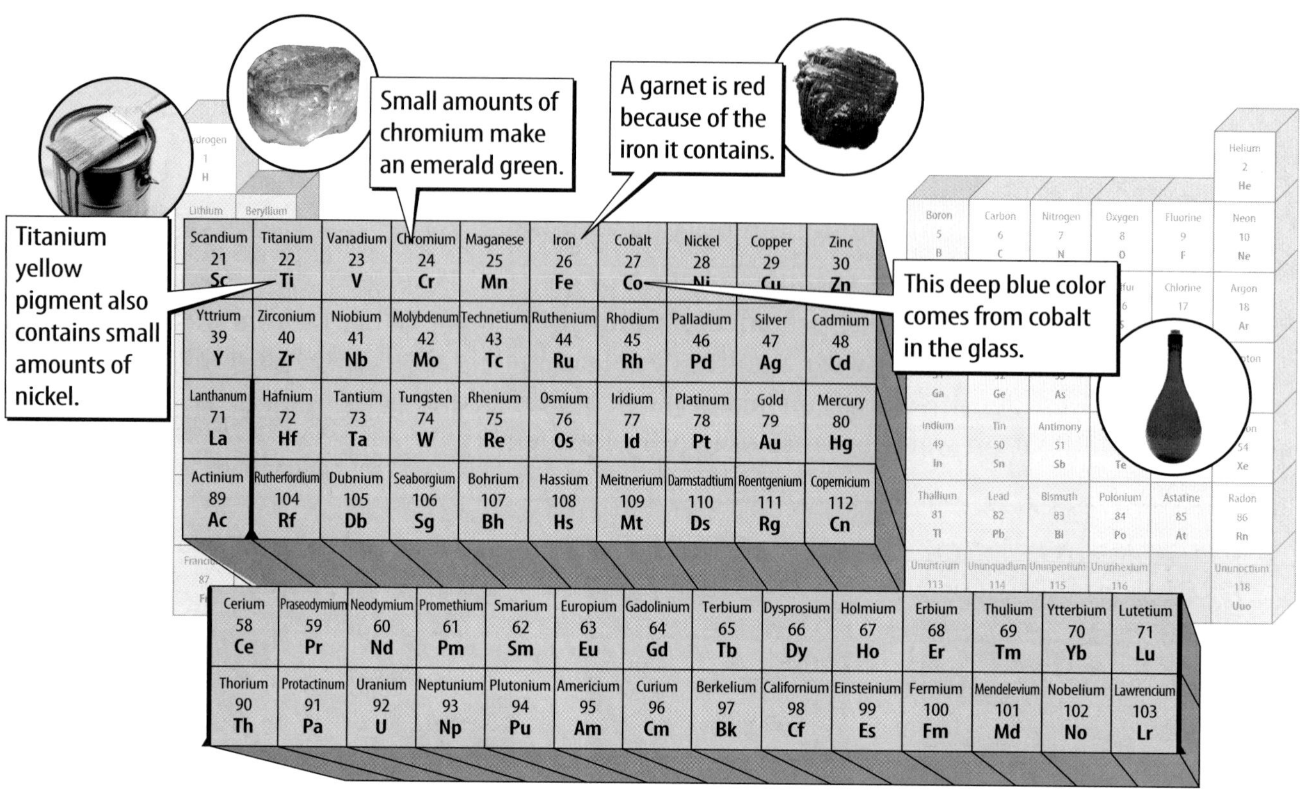

Groups 3–12: Transition Elements

The elements in groups 3–12 are called **transition elements.** The transition elements are in two blocks on the periodic table. The main block is in the center of the periodic table. The other block includes the two rows at the bottom of the periodic table, as shown in **Figure 10.**

Properties of Transition Elements

All transition elements are metals. They have higher melting points, greater strength, and higher densities than the alkali metals and the alkaline earth metals. Transition elements also react less quickly with oxygen. Some transition elements can exist in nature as free elements. An element is a free element when it occurs in pure form, not in a compound.

Uses of Transition Elements

Transition elements in the main block of the periodic table have many important uses. Because of their high densities, strength, and resistance to corrosion, transition elements such as iron make good building materials. Copper, silver, nickel, and gold are used to make coins. These metals are also used for jewelry, electrical wires, and many industrial applications.

Main-block transition elements can react with other elements and form many compounds. Many of these compounds are colorful. Artists use transition-element compounds in paints and pigments. The color of many gems, such as garnets and emeralds, comes from the presence of small amounts of transition elements, as illustrated in **Figure 10.**

Lanthanide and Actinide Series

Two rows of transition elements are at the bottom of the periodic table, as shown in **Figure 10.** These elements were removed from the main part of the table so that periods 6 and 7 were not longer than the other periods. If these elements were included in the main part of the table, the first row, called the lanthanide series, would stretch between lanthanum and halfnium. The second row, called the actinide series, would stretch between actinium and rutherfordium.

Some lanthanide and actinide series elements have valuable properties. For example, lanthanide series elements are used to make strong magnets. Plutonium, one of the actinide series elements, is used as a fuel in some nuclear reactors.

Patterns in Properties of Metals

Recall that the properties of elements follow repeating patterns across the periods of the periodic table. In general, elements increase in metallic properties such as luster, malleability, and electrical conductivity from right to left across a period, as shown in **Figure 11.** The elements on the far right of a period have no metallic properties at all. Potassium (K), the element on the far left in period 4, has the highest luster, is the most malleable, and conducts electricity better than all the elements in this period.

There are also patterns within groups. Metallic properties tend to increase as you move down a group, also shown in **Figure 11.** You could predict that the malleability of gold is greater than the malleability of either silver or copper because it is below these two elements in group 11.

Reading Check Where would you expect to find elements on the periodic table with few or no metallic properties?

MiniLab **20 minutes**

How well do materials conduct thermal energy?

How well a material conducts thermal energy can often determine its use.

1. Read and complete a lab safety form.
2. Have your teacher add about 200 mL of very **hot water** to a **250-mL beaker.**
3. Place **rods of metal, plastic, glass, and wood** in the water for 30 seconds.
4. Set four large **ice cubes** on a sheet of **paper towel.** Use **tongs** to quickly remove each rod from the hot water. Place the heated end of the rod on an ice cube.
5. After 30 seconds, remove the rods and examine the ice cubes.

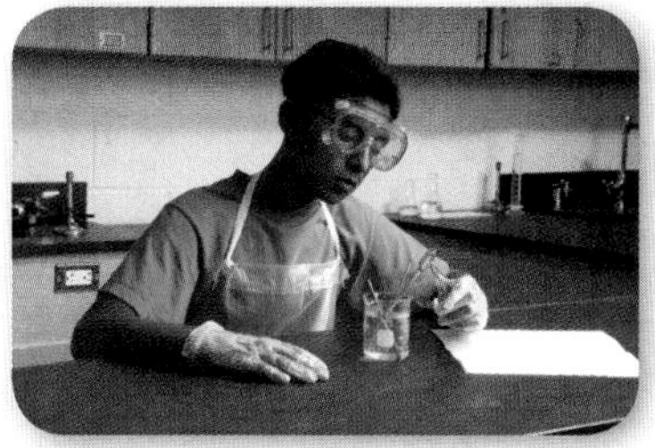

Analyze and Conclude

1. **Conclude** What can you conclude about how well metals conduct thermal energy?
2. **Key Concept** Cookware is often made of metal. What property of metals makes them useful for this purpose?

Figure 11 Metallic properties of elements increase as you move to the left and down on the periodic table.

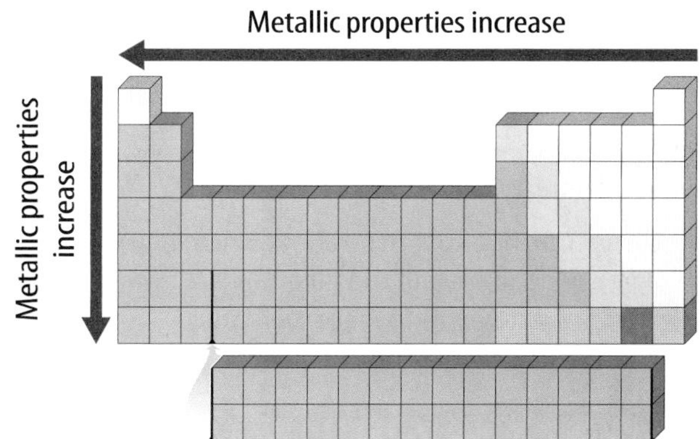

Lesson 2 Review

Assessment Online Quiz

Visual Summary

Properties of metals include conductivity, luster, malleability, and ductility.

Alkali metals and alkaline earth metals react easily with other elements. These metals make up groups 1 and 2 on the periodic table.

Transition elements make up groups 3–12 and the lanthanide and actinide series on the periodic table.

Use your lesson Foldable to review the lesson. Save your Foldable for the project at the end of the chapter.

What do you think NOW?

You first read the statements below at the beginning of the chapter.

3. Fewer than half of the elements are metals.

4. Metals are usually good conductors of electricity.

Did you change your mind about whether you agree or disagree with the statements? Rewrite any false statements to make them true.

Use Vocabulary

1. **Use the term** *luster* in a sentence.
2. **Identify** the property that makes copper metal ideal for wiring.
3. Elements that have the lowest densities of all the metals are called ________.

Understand Key Concepts 

4. **List** the physical properties that most metals have in common.
5. Which is a chemical property of transition elements?
   - **A.** brightly colored
   - **B.** great ductility
   - **C.** denser than alkali metals
   - **D.** reacts little with oxygen
6. **Organize** the following metals from least metallic to most metallic: barium, zinc, iron, and strontium.

Interpret Graphics

7. **Examine** this section of the periodic table. What metal will have properties most similar to those of chromium (Cr)? Why?

Vanadium 23 **V**	Chromium 24 **Cr**	Maganese 25 **Mn**
Niobium 41 **Nb**	Molybdenum 42 **Mo**	Technetium 43 **Tc**

Critical Thinking

8. **Investigate** your classroom and locate five examples of materials made from metal.
9. **Evaluate** the physical properties of potassium, magnesium, and copper. Select the best choice to use for a building project. Explain why this metal is the best building material to use.

Fireworks

SCIENCE & SOCIETY

Metals add variety to color.

About 1,000 years ago, the Chinese discovered the chemical formula for gunpowder. Using this formula, they invented the first fireworks. One of the primary ingredients in gunpowder is saltpeter, or potassium nitrate. Find potassium on the periodic table. Notice that potassium is a metal. How does the chemical behavior of a metal contribute to a colorful fireworks show?

Purple: mix of strontium and copper compounds

Blue: copper compounds

Yellow: sodium compounds

Gold: iron burned with carbon

White-hot: barium-oxygen compounds or aluminum or magnesium burn

Orange: calcium compounds

Green: barium compounds

Red: strontium and lithium compounds

Metal compounds contribute to the variety of colors you see at a fireworks show. Recall that metals have special chemical and physical properties. Compounds that contain metals also have special properties. For example, each metal turns a characteristic color when burned. Lithium, an alkali metal, forms compounds that burn red. Copper compounds burn blue. Aluminum and magnesium burn white.

It's Your Turn

FORM AN OPINION Fireworks contain metal compounds. Are they bad for the environment or for your health? Research the effects of metals on human health and on the environment. Decide if fireworks are safe to use a holiday celebrations.

Lesson 3

Reading Guide

Key Concepts

ESSENTIAL QUESTIONS

- Where are nonmetals and metalloids on the periodic table?
- What are the properties of nonmetals and metalloids?

Vocabulary

nonmetal p. 363
halogen p. 365
noble gas p. 366
metalloid p. 367
semiconductor p. 367

g Multilingual eGlossary

Nonmetals and Metalloids

Inquiry Why don't they melt?

What do you expect to happen to something when a flame is placed against it? As you can see, the nonmetal material this flower sits on protects the flower from the flame. Some materials conduct thermal energy. Other materials, such as this one, do not.

20 minutes

What are some properties of nonmetals?

You now know what the properties of metals are. What properties do nonmetals have?

1. Read and complete a lab safety form.
2. Examine pieces of **copper, carbon, aluminum,** and **sulfur.** Describe the appearance of these elements in your Science Journal.
3. Use a **conductivity tester** to check how well these elements conduct electricity. Record your observations.
4. Wrap each element sample in a **paper towel.** Carefully hit the sample with a **hammer.** Unwrap the towel and observe the sample. Record your observations.

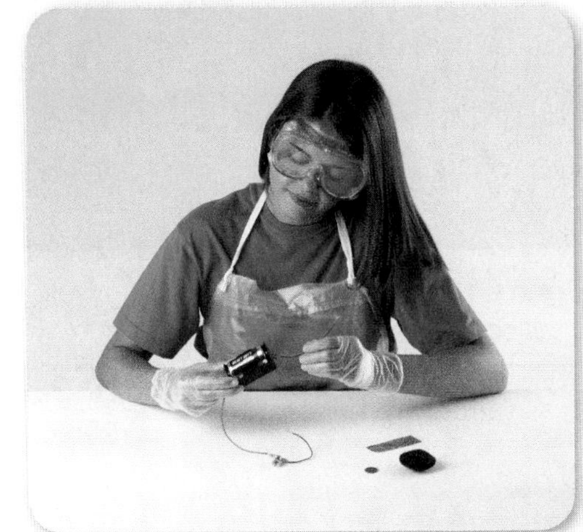

Think About This

1. Locate these elements on the periodic table. From their locations, which elements are metals? Which elements are nonmetals?
2. **Key Concept** Using your results, compare the properties of metals and nonmetals.
3. **Key Concept** What property of a nonmetal makes it useful to insulate electrical wires?

The Elements of Life

Would it surprise you to learn that more than 96 percent of the mass of your body comes from just four elements? As shown in **Figure 12,** all four of these elements—oxygen, carbon, hydrogen, and nitrogen—are nonmetals. **Nonmetals** *are elements that have no metallic properties.*

Of the remaining elements in your body, the two most common elements also are nonmetals—phosphorus and sulfur. These six elements form the compounds in proteins, fats, nucleic acids, and other large molecules in your body and in all other living things.

Reading Check What are the six most common elements in the human body?

Figure 12 Like other living things, this woman's mass comes mostly from nonmetals.

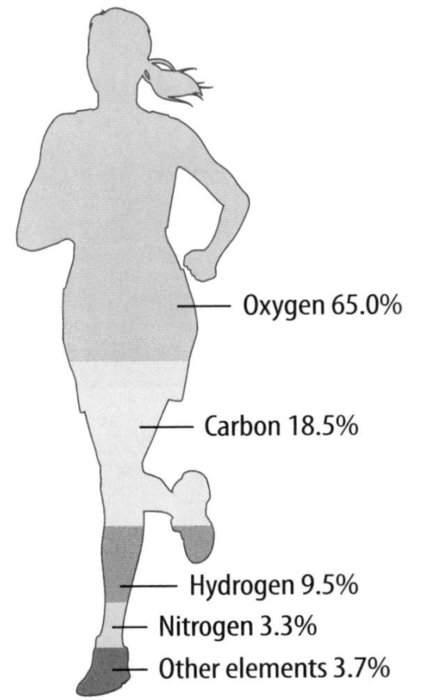

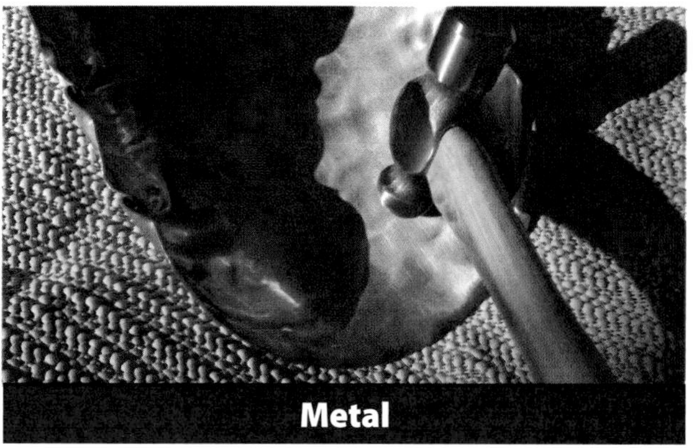

▲ **Figure 13** Solid metals, such as copper, are malleable. Solid nonmetals, such as sulfur, are brittle.

How are nonmetals different from metals?

Recall that metals have luster. They are ductile, malleable, and good conductors of electricity and thermal energy. All metals except mercury are solids at room temperature.

The properties of nonmetals are different from those of metals. Many nonmetals are gases at room temperature. Those that are solid at room temperature have a dull surface, which means they have no luster. Because nonmetals are poor conductors of electricity and thermal energy, they are good insulators. For example, nose cones on space shuttles are insulated from the intense thermal energy of reentry by a material made from carbon, a nonmetal. **Figure 13** and **Figure 14** show several properties of nonmetals.

 **Key Concept Check** What properties do nonmetals have?

Figure 14 Nonmetals have properties that are different from those of metals. Phosphorus and carbon are dull, brittle solids that do not conduct thermal energy or electricity. ▼

Properties of Nonmetals

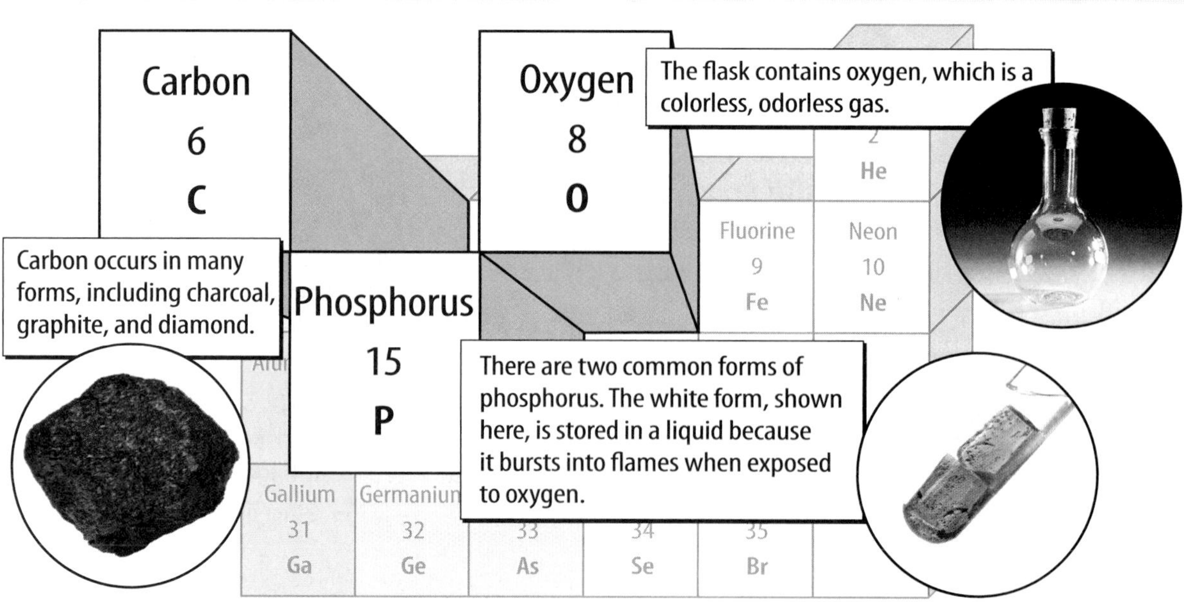

Carbon
6
C
Carbon occurs in many forms, including charcoal, graphite, and diamond.
Oxygen
8
O
The flask contains oxygen, which is a colorless, odorless gas.
He
Fluorine
9
Fe
Neon
10
Ne
Phosphorus
15
P
There are two common forms of phosphorus. The white form, shown here, is stored in a liquid because it bursts into flames when exposed to oxygen.
Gallium
31
Ga
32
Ge
33
As
34
Se
35
Br

Visual Check Compare the properties of oxygen to those of carbon and phosphorus.

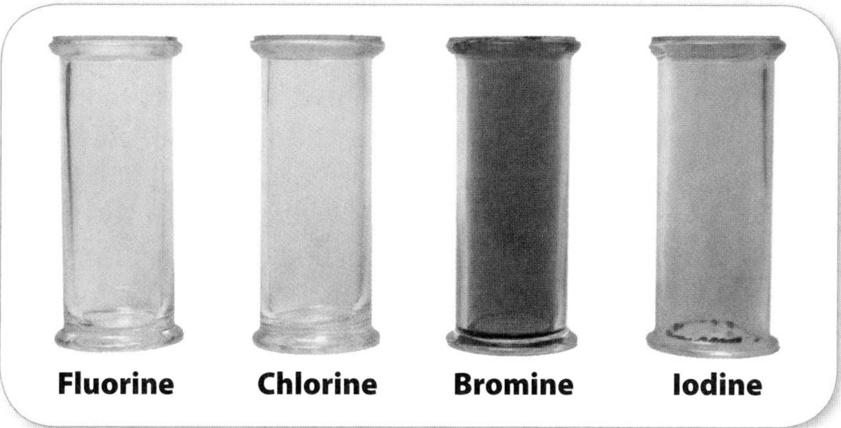

Figure 15 These glass containers each hold a halogen gas. Although they are different colors in their gaseous state, they react similarly with other elements.

Visual Check Compare the colors of these halogens.

FOLDABLES®

Fold a sheet of paper to make a table with three columns and three rows. Label it as shown. Use it to organize information about nonmetals and metalloids.

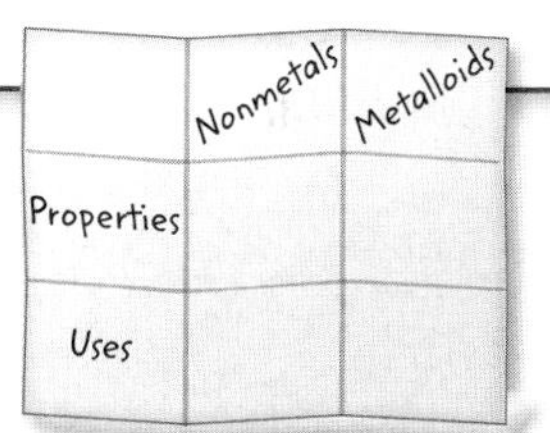

Nonmetals in Groups 14–16

Look back at the periodic table in **Figure 4.** Notice that groups 14–16 contain metals, nonmetals, and metalloids. The chemical properties of the elements in each group are similar. However, the physical properties of the elements can be quite different.

Carbon is the only nonmetal in group 14. It is a solid that has different forms. Carbon is in most of the compounds that make up living things. Nitrogen, a gas, and phosphorus, a solid, are the only nonmetals in group 15. These two elements form many different compounds with other elements, such as oxygen. Group 16 contains three nonmetals. Oxygen is a gas that is essential for many organisms. Sulfur and selenium are solids that have the physical properties of other solid nonmetals.

Group 17: The Halogens

An element in group 17 of the periodic table is called a **halogen** (HA luh jun). **Figure 15** shows the halogens fluorine, chlorine, bromine, and iodine. The term *halogen* refers to an element that can react with a metal and form a salt. For example, chlorine gas reacts with solid sodium and forms sodium chloride, or table salt. Calcium chloride is another salt often used on icy roads.

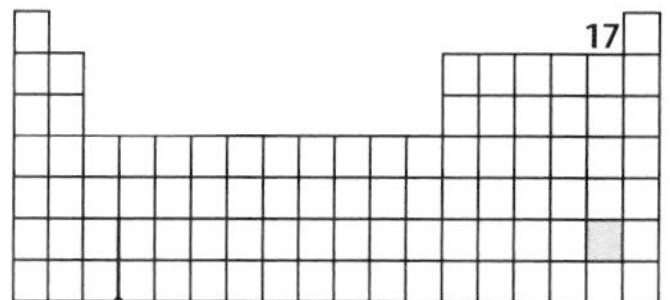

Halogens react readily with other elements and form compounds. They react so readily that halogens only can occur naturally in compounds. They do not exist as free elements. They even form compounds with other nonmetals, such as carbon. In general, the halogens are less reactive as you move down the group.

WORD ORIGIN

halogen
from Greek *hals,* means "salt"; and *-gen,* means "to produce"

Reading Check Will bromine react with sodium? Explain your answer.

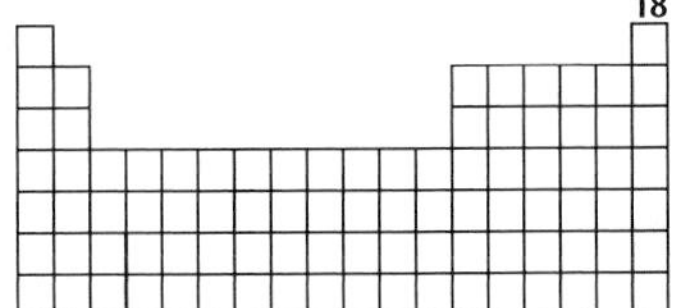

Group 18: The Noble Gases

The elements in group 18 are known as the **noble gases.** The elements helium, neon, argon, krypton, xenon, and radon are the noble gases. Unlike the halogens, the only way elements in this group react with other elements is under special conditions in a laboratory. These elements were not yet discovered when Mendeleev constructed his periodic table because they do not form compounds naturally. Once they were discovered, they fit into a group at the far right side of the table.

ACADEMIC VOCABULARY

construct
(verb) to make by combining and arranging parts

Hydrogen

Figure 16 shows the element key for hydrogen. Of all the elements, hydrogen has the smallest atomic mass. It is also the most common element in the universe.

Is hydrogen a metal or a nonmetal? Hydrogen is most often classified as a nonmetal because it has many properties like those of nonmetals. For example, like some nonmetals, hydrogen is a gas at room temperature. However, hydrogen also has some properties similar to those of the group 1 alkali metals. In its liquid form, hydrogen conducts electricity just like a metal does. In some chemical reactions, hydrogen reacts as if it were an alkali metal. However, under conditions on Earth, hydrogen usually behaves like a nonmetal.

Reading Check Why is hydrogen usually classified as a nonmetal?

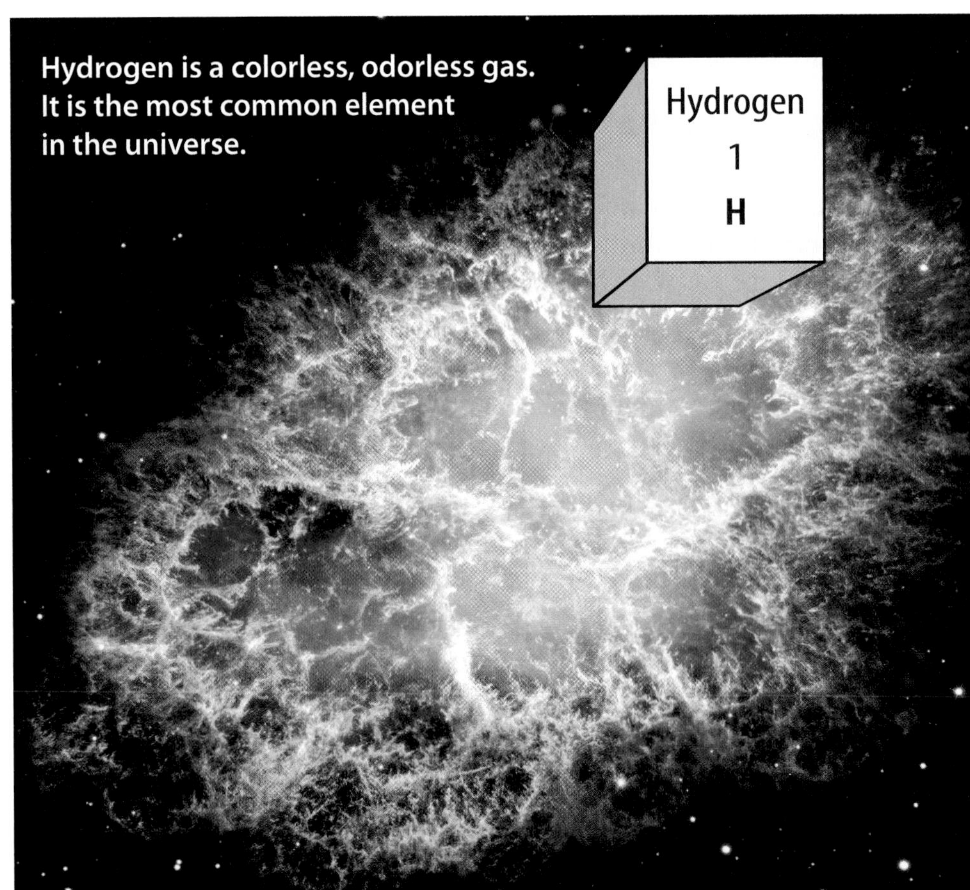

Hydrogen is a colorless, odorless gas. It is the most common element in the universe.
Hydrogen
1
H

Figure 16 More than 90 percent of all the atoms in the universe are hydrogen atoms. Hydrogen is the main fuel for the nuclear reactions that occur in stars.

Metalloids

Between the metals and the nonmetals on the periodic table are elements known as metalloids. *A* **metalloid** (MEH tul oyd) *is an element that has physical and chemical properties of both metals and nonmetals.* The elements boron, silicon, germanium, arsenic, antimony, tellurium, polonium, and astatine are metalloids. Silicon is the most abundant metalloid in the universe. Most sand is made of a compound containing silicon. Silicon is also used in many different products, some of which are shown in **Figure 17.**

Key Concept Check Where are metalloids on the periodic table?

Semiconductors

Recall that metals are good conductors of thermal energy and electricity. Nonmetals are poor conductors of thermal energy and electricity but are good insulators. A property of metalloids is the ability to act as a semiconductor. *A* **semiconductor** *conducts electricity at high temperatures, but not at low temperatures.* At high temperatures, metalloids act like metals and conduct electricity. But at lower temperatures, metalloids act like nonmetals and stop electricity from flowing. This property is useful in electronic devices such as computers, televisions, and solar cells.

WORD ORIGIN

semiconductor
from Latin *semi-*, means "half"; and *conducere*, means "to bring together"

Uses of Silicon

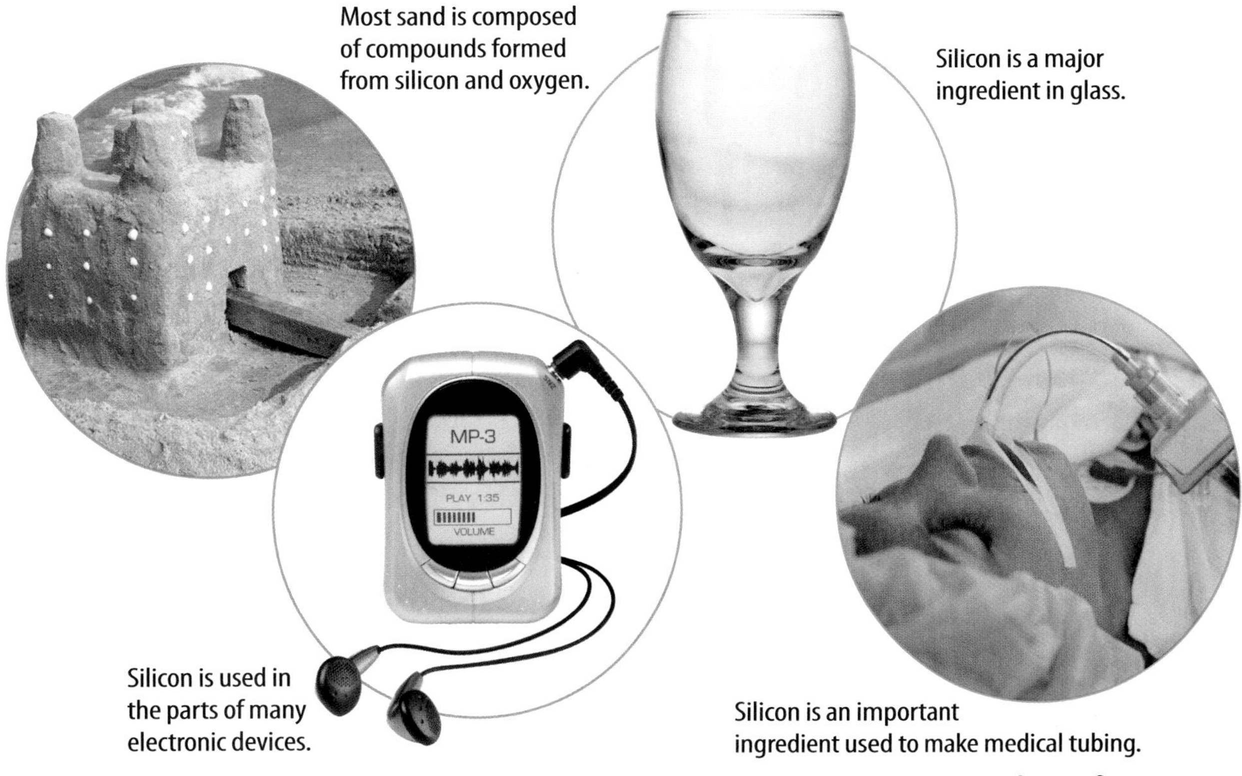

Most sand is composed of compounds formed from silicon and oxygen.
Silicon is a major ingredient in glass.
MP-3
PLAY 1:35
VOLUME
Silicon is used in the parts of many electronic devices.
Silicon is an important ingredient used to make medical tubing.

Figure 17 The properties of silicon make it useful for many different products.

Figure 18 This microchip conducts electricity at high temperatures using a semiconductor.

 Personal Tutor

Inquiry MiniLab 15 minutes

Which insulates better?

In this lab, you will compare how well a metal bowl and a nonmetal ball containing a mixture of nonmetals conduct thermal energy.

1. Read and complete a lab safety form.
2. Pour very **warm water** into a **pitcher.**
3. Pour half of the warm water into a **metal bowl.** In your Science Journal, describe how the outside of the bowl feels.
4. Inflate a **beach ball** until it is one-third full. Mold the partially filled beach ball into the shape of a bowl. Pour the remaining warm water into your beach ball bowl. Feel the outside of the bowl. Describe how it feels.

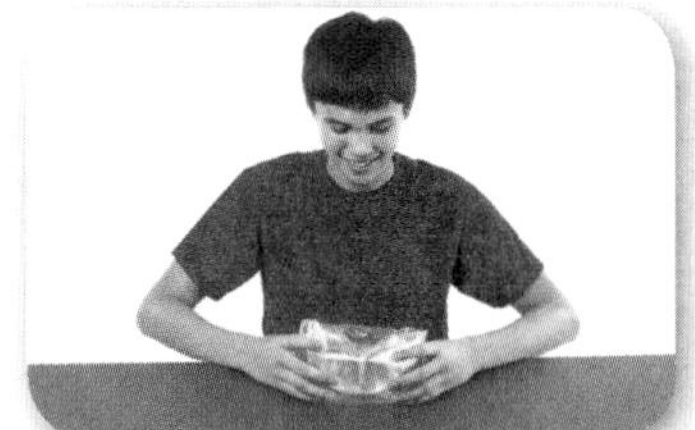

Analyze and Conclude

1. **Explain** the difference in the outside temperatures of the two bowls.
2. **Predict** the results of putting ice in each of the bowls.
3. **Key Concept** Make a statement about how well a nonmetal conducts thermal energy.

Properties and Uses of Metalloids

Pure silicon is used in making semiconductor devices for computers and other electronic products. Germanium is also used as a semiconductor. However, metalloids have other uses, as shown in **Figure 18.** Pure silicon and Germanium are used in semiconductors. Boron is used in water softeners and laundry products. Boron also glows bright green in fireworks. Silicon is one of the most abundant elements on Earth. Sand, clay, and many rocks and minerals are made of silicon compounds.

Metals, Nonmetals, and Metalloids

You have read that all metallic elements have common characteristics, such as malleability, conductivity, and ductility. However, each metal has unique properties that make it different from other metals. The same is true for nonmetals and metalloids. How can knowing the properties of an element help you evaluate its uses?

Look again at the periodic table. An element's position on the periodic table tells you a lot about the element. By knowing that sulfur is a nonmetal, for example, you know that it breaks easily and does not conduct electricity. You would not choose sulfur to make a wire. You would not try to use oxygen as a semiconductor or sodium as a building material. You know that transition elements are strong, malleable, and do not react easily with oxygen or water. These metals make good building materials because they are strong, malleable, and less reactive than other elements. Understanding the properties of elements can help you decide which element to use in a given situation.

 **Reading Check** Why would you not use an element on the right side of the periodic table as a building material?

Lesson 3 Review

Assessment Online Quiz

Visual Summary

A nonmetal is an element that has no metallic properties. Solid nonmetals are dull, brittle, and do not conduct thermal energy or electricity.

Halogens and noble gases are nonmetals. These elements are found in group 17 and group 18 of the periodic table.

Metalloids have some metallic properties and some nonmetallic properties. The most important use of metalloids is as semiconductors.

FOLDABLES®

Use your lesson Foldable to review the lesson. Save your Foldable for the project at the end of the chapter.

What do you think NOW?

You first read the statements below at the beginning of the chapter.

5. Most of the elements in living things are nonmetals.

6. Even though they look very different, oxygen and sulfur share some similar properties.

Did you change your mind about whether you agree or disagree with the statements? Rewrite any false statements to make them true.

Use Vocabulary

1. **Distinguish** between a nonmetal and a metalloid.
2. An element in group 17 of the periodic table is called a(n) ________.
3. An element in group 18 of the periodic table is called a(n) ________.

Understand Key Concepts 

4. The ability of a halogen to react with a metal to form a salt is an example of a ________ property.

   A. chemical **C.** periodic
   B. noble gas **D.** physical

5. **Classify** each of the following elements as a metal, a nonmetal, or a metalloid: boron, carbon, aluminum, and silicon.
6. **Infer** which group you would expect to contain element 117. Use the periodic table to help you answer this question.

Interpret Graphics

7. **Sequence** nonmetals, metals, and metalloids in order from left to right across the periodic table by copying and completing the graphic organizer below.

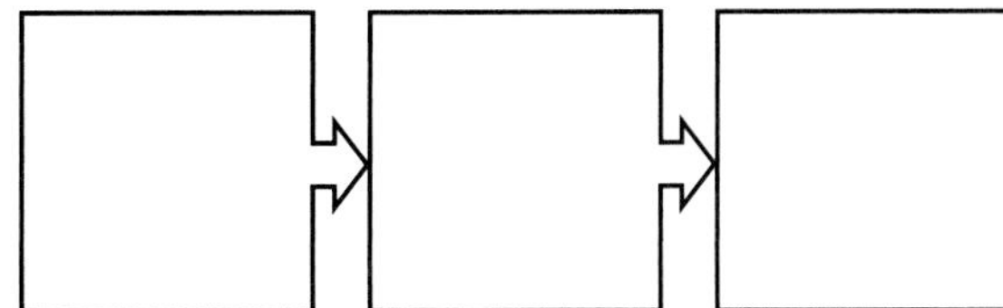

Critical Thinking

8. **Hypothesize** how your classroom would be different if there were no metalloids.
9. **Analyze** why hydrogen is sometimes classified as a metal.
10. **Determine** whether there would be more nonmetals in group 14 or in group 16. Explain your answer.

40 minutes

Alien Insect Periodic Table

Materials

cards

The periodic table classifies elements according to their properties. In this lab, you will model the procedure used to develop the periodic table. Your model will include developing patterns using pictures of alien insects. You will then use your patterns to predict what missing alien insects look like.

Question

How can I arrange objects into patterns by using their properties?

Procedure

1. Obtain a set of alien insect pictures. Spread them out so you can see all of them. Observe the pictures with a partner. Look for properties that you might use to organize the pictures.
2. Make a list of properties you might use to group the alien insects. These properties are those that a number of insects have in common.
3. Make a list of properties you might use to sequence the insects. These properties change from one insect to the next in some pattern.
4. With your partner, decide what pattern you will use to arrange the alien insects in an organized rectangular block. All the insects in a vertical column, or group, must be the same in some way. They must also share some feature that changes regularly as you move down the group. All the aliens in a horizontal row, or period, must be the same in some way and must also share some feature that changes regularly as you move across the period.

4

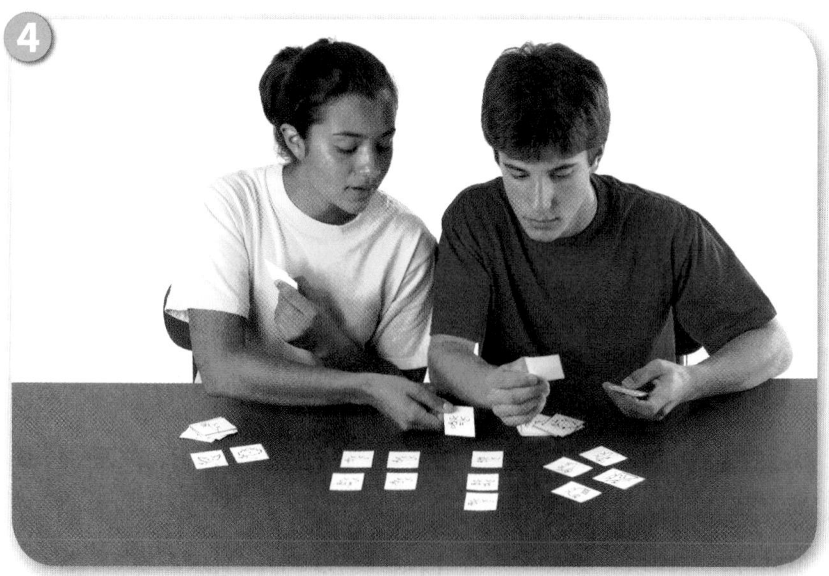

5. Arrange your insects as you planned. Two insects are missing from your set. Leave empty spaces in your rectangular block for these pictures. When you have finished arranging your insects, have the teacher check your pattern.
6. Write a description of the properties you predict each missing alien insect will have. Then draw a picture of each missing insect.

Analyze and Conclude

7. **Explain** Could you have predicted the properties of the missing insects without placing the others in a pattern? Why or why not?
8. **The Big Idea** How is your arrangement similar to the one developed by Mendeleev for elements? How is it different?
9. **Infer** What properties can you use to predict the identity of one missing insect? What do you not know about that insect?

Communicate Your Results

Create a slide show presentation that demonstrates, step by step, how you grouped and sequenced your insects and predicted the properties of the missing insects. Show your presentation to students in another class.

Inquiry Extension

How could you change the insects so that they better represent the properties of elements, such as atomic mass?

Lab Tips

- ☑ A property is any observable characteristic that you can use to distinguish between objects.
- ☑ A pattern is a consistent plan or model used as a guide for understanding or predicting something.

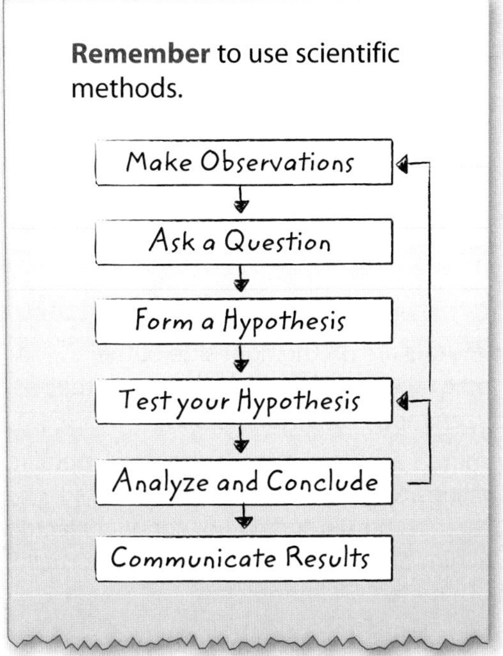

Chapter 10 Study Guide

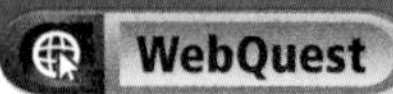

Elements are organized on the periodic table according to increasing atomic number and similar properties.

Key Concepts Summary

Lesson 1: Using the Periodic Table

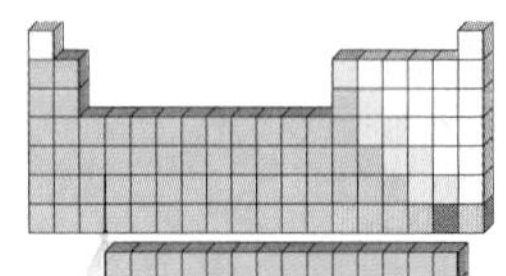

- Elements are organized on the **periodic table** by increasing atomic number and similar properties.
- Elements in the same **group,** or column, of the periodic table have similar properties.
- Elements' properties change across a **period,** which is a row of the periodic table.
- Each element key on the periodic table provides the name, symbol, atomic number, and atomic mass for an element.

Vocabulary

periodic table p. 345
group p. 350
period p. 350

Lesson 2: Metals

- **Metals** are located on the left and middle side of the periodic table.
- Metals are elements that have **ductility, malleability, luster,** and conductivity.
- The **alkali metals** are in group 1 of the periodic table, and the **alkaline earth metals** are in group 2.
- **Transition elements** are metals in groups 3–12 of the periodic table, as well as the lanthanide and actinide series.

Vocabulary

metal p. 355
luster p. 355
ductility p. 356
malleability p. 356
alkali metal p. 357
alkaline earth metal p. 357
transition element p. 358

Lesson 3: Nonmetals and Metalloids

- **Nonmetals** are on the right side of the periodic table, and **metalloids** are located between metals and nonmetals.
- Nonmetals are elements that have no metallic properties. Solid nonmetals are dull in appearance, brittle, and do not conduct electricity. Metalloids are elements that have properties of both metals and nonmetals.
- Some metalloids are **semiconductors.**
- Elements in group 17 are called **halogens,** and elements in group 18 are **noble gases.**

Vocabulary

nonmetal p. 363
halogen p. 365
noble gas p. 366
metalloid p. 367
semiconductor p. 367

Study Guide

- Personal Tutor
- Vocabulary eGames
- Vocabulary eFlashcards

FOLDABLES® Chapter Project

Assemble your lesson Foldables as shown to make a Chapter Project. Use the project to review what you have learned in this chapter.

History | Why It Changed | Today's Table
The Periodic Table
The Physical Properties of Metals
The Chemical Properties of Metals
Uses | Properties | Nonmetals | Metalloids

Use Vocabulary

1. The element magnesium (Mg) is in ________ 3 of the periodic table.
2. An element that is shiny, is easily pulled into wires or hammered into thin sheets, and is a good conductor of electricity and heat is a(n) ________.
3. Copper is used to make wire because it has the property of ________.
4. An element that is sometimes a good conductor of electricity and sometimes a good insulator is a(n) ________.
5. An element that is a poor conductor of heat and electricity but is a good insulator is a(n) ________.

Interactive Concept Map

Link Vocabulary and Key Concepts

Copy this concept map, and then use vocabulary terms from the previous page to complete the concept map.

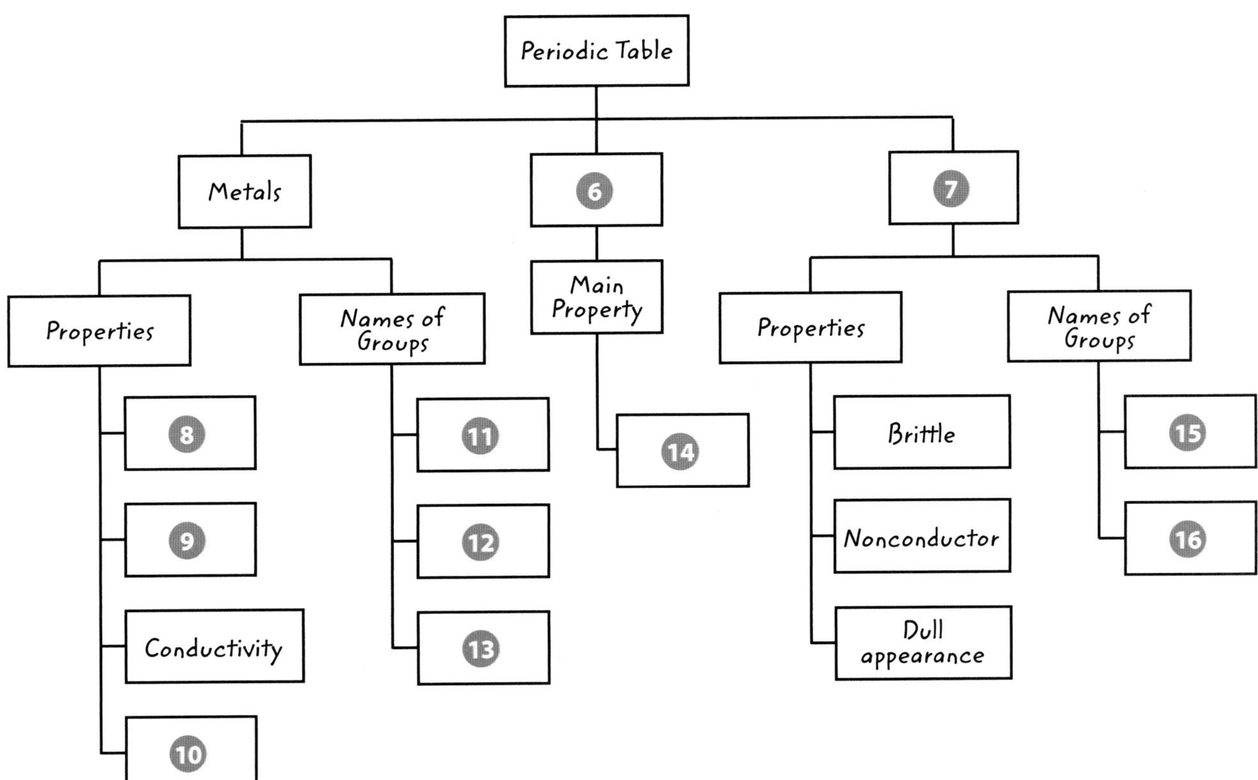

Chapter 10 Review

Understand Key Concepts

1. What determines the order of elements on today's periodic table?
   A. increasing atomic mass
   B. decreasing atomic mass
   C. increasing atomic number
   D. decreasing atomic number

2. The element key for nitrogen is shown below.

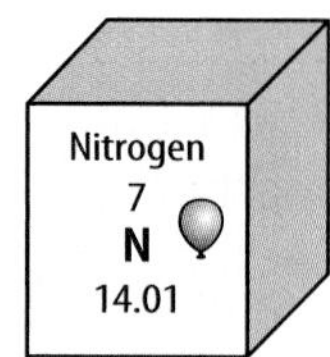

   From this key, determine the atomic mass of nitrogen.
   A. 7
   B. 7.01
   C. 14.01
   D. 21.01

3. Look at the periodic table in Lesson 1. Which of the following lists of elements forms a group on the periodic table?
   A. Li, Be, B, C, N, O, F, and Ne
   B. He, Ne, Ar, Kr, Xe, and Rn
   C. B, Si, As, Te, and At
   D. Sc, Ti, V, Cr, Mn, Fe, Co, Cu, Ni, and Zn

4. Which is NOT a property of metals?
   A. brittleness
   B. conductivity
   C. ductility
   D. luster

5. What are two properties that make a metal a good choice to use as wire in electronics?
   A. conductivity, malleability
   B. ductility, conductivity
   C. luster, malleability
   D. malleability, high density

6. Where are most metals on the periodic table?
   A. on the left side only
   B. on the right side only
   C. in the middle only
   D. on the left side and in the middle

7. Look at the periodic table in Lesson 1 and determine which element is a metalloid.
   A. carbon
   B. silicon
   C. oxygen
   D. aluminum

8. Iodine is a solid nonmetal. What is one property of iodine?
   A. conductivity
   B. dull appearance
   C. malleability
   D. ductility

9. The following table lists some information about certain elements in group 17.

Element Symbol	Atomic Number	Melting Point (°C)	Boiling Point (°C)
F	9	−233	−187
Cl	17	−102	−35
Br	35	−7.3	59
I	53	114	183

   Which statement describes what happens to these elements as atomic number increases?
   A. Both melting point and boiling point decrease.
   B. Melting point increases and boiling point decreases.
   C. Melting point decreases and boiling point increases.
   D. Both melting point and boiling point increase.

Assessment
Online Test Practice

Critical Thinking

10 **Recommend** an element to use to fill bottles that contain ancient paper. The element should be a gas at room temperature, should be denser than helium, and should not easily react with other elements.

11 **Apply** Why is mercury the only metal to have been used in thermometers?

12 **Evaluate** the following types of metals as a choice to make a Sun reflector: alkali metals, alkaline earth metals, or transition metals. The metal cannot react with water or oxygen and must be shiny and strong.

13 The figure below shows a pattern of densities.

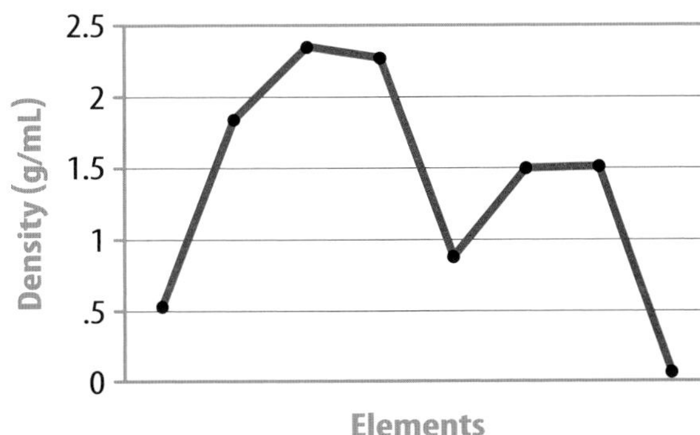

Infer whether you are looking at a graph of elements within a group or across a period. Explain your answer.

14 **Contrast** aluminum and nitrogen. Show why aluminum is a metal and nitrogen is not.

15 **Classify** A student sorted six elements. He placed iron, silver, and sodium in group A. He placed neon, oxygen, and nitrogen in group B. Name one other element that fits in group A and another element that belongs in group B. Explain your answer.

Writing in Science

16 **Write** a plan that shows how a metal, a nonmetal, and a metalloid could be used when constructing a building.

REVIEW THE BIG IDEA

17 Explain how atomic number and properties are used to determine where element 115 is placed on the periodic table.

18 The photo below shows how the properties of materials determine their uses. How can the periodic table be used to help you find elements with properties similar to that of helium?

Math Skills

Math Practice

Use Geometry

19 The table below shows the atomic radii of three elements in group 1 on the periodic table.

Element	Atomic radius
Li	152 pm
Na	186 pm
K	227 pm

a. What is the circumference of each atom?

b. Rubidium (Rb) is the next element in Group 1. What would you predict about the radius and circumference of a rubidium atom?

Standardized Test Practice

Record your answers on the answer sheet provided by your teacher or on a sheet of paper.

Multiple Choice

1 Where are most nonmetals located on the periodic table?

A in the bottom row

B on the left side and in the middle

C on the right side

D in the top row

Use the figure below to answer question 2.

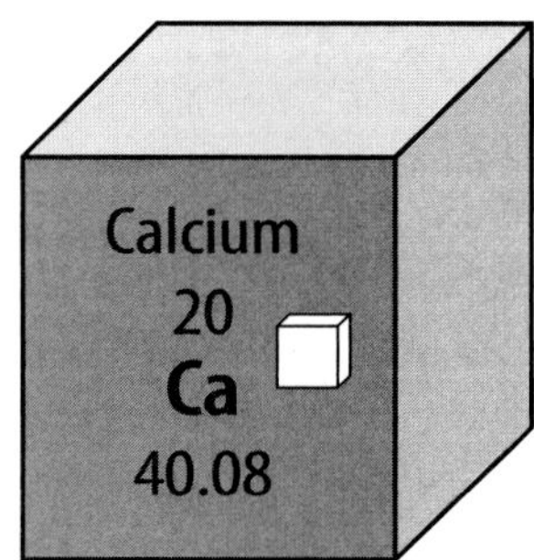

2 What is the atomic mass of calcium?

A 20

B 40.08

C 40.08 ÷ 20

D 40.08 + 20

3 Which element is most likely to react with potassium?

A bromine

B calcium

C nickel

D sodium

4 Which group of elements can act as semiconductors?

A halogens

B metalloids

C metals

D noble gases

Use the table below about group 13 elements to answer question 5.

Element Symbol	Atomic Number	Density (g/cm^3)	Atomic Mass
B	5	2.34	10.81
Al	13	2.70	26.98
Ga	31	5.90	69.72
In	49	7.30	114.82

5 How do density and atomic mass change as atomic number increases?

A Density and atomic mass decrease.

B Density and atomic mass increase.

C Density decreases and atomic mass increases.

D Density increases and atomic mass decreases.

6 Which elements have high densities, strength, and resistance to corrosion?

A alkali metals

B alkaline earth metals

C metalloids

D transition elements

7 Which is a property of a metal?

A It is brittle.

B It is a good insulator.

C It has a dull appearance.

D It is malleable.

Use the figure below to answer questions 8 and 9.

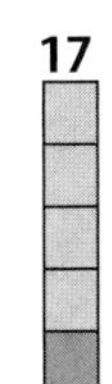

8 The figure shows a group in the periodic table. What is the name of this group of elements?

A halogens

B metalloids

C metals

D noble gases

9 Which is a property of these elements?

A They are conductors.

B They are semiconductors.

C They are nonreactive with other elements.

D They react easily with other elements.

10 What is one similarity among elements in a group?

A atomic mass

B atomic weight

C chemical properties

D practical uses

Constructed Response

Use the figure below to answer questions 11 and 12.

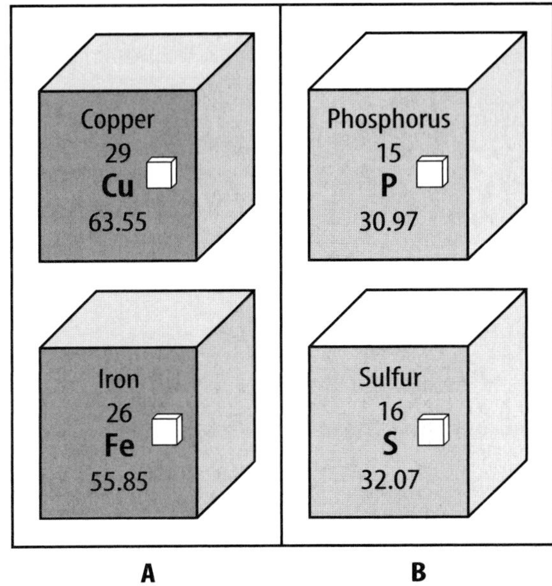

11 Groups A and B each contain two elements. Identify each group as metals, nonmetals, or metalloids. Would silicon belong to one of these groups? Why or why not?

12 Which group in the figure above yields the strongest building elements? Why?

13 How does the periodic table of elements help scientists today?

14 What connection does the human body have with the elements on the periodic table?

NEED EXTRA HELP?														
If You Missed Question...	1	2	3	4	5	6	7	8	9	10	11	12	13	14
Go to Lesson...	1	1	3	3	1	2	2	3	3	1	2, 3	2	1	3

Chapter 11

Elements and Chemical Bonds

THE BIG IDEA How do elements join together to form chemical compounds?

Inquiry How do they combine?

How many different words could you type using just the letters on a keyboard? The English alphabet has only 26 letters, but a dictionary lists hundreds of thousands of words using these letters! Similarly only about 115 different elements make all kinds of matter.

- How do so few elements form so many different kinds of matter?
- Why do you think different types of matter have different properties?
- How are atoms held together to produce different types of matter?

Get Ready to Read

What do you think?

Before you read, decide if you agree or disagree with each of these statements. As you read this chapter, see if you change your mind about any of the statements.

1. Elements rarely exist in pure form. Instead, combinations of elements make up most of the matter around you.
2. Chemical bonds that form between atoms involve electrons.
3. The atoms in a water molecule are more chemically stable than they would be as individual atoms.
4. Many substances dissolve easily in water because opposite ends of a water molecule have opposite charges.
5. Losing electrons can make some atoms more chemically stable.
6. Metals are good electrical conductors because they tend to hold onto their valence electrons very tightly.

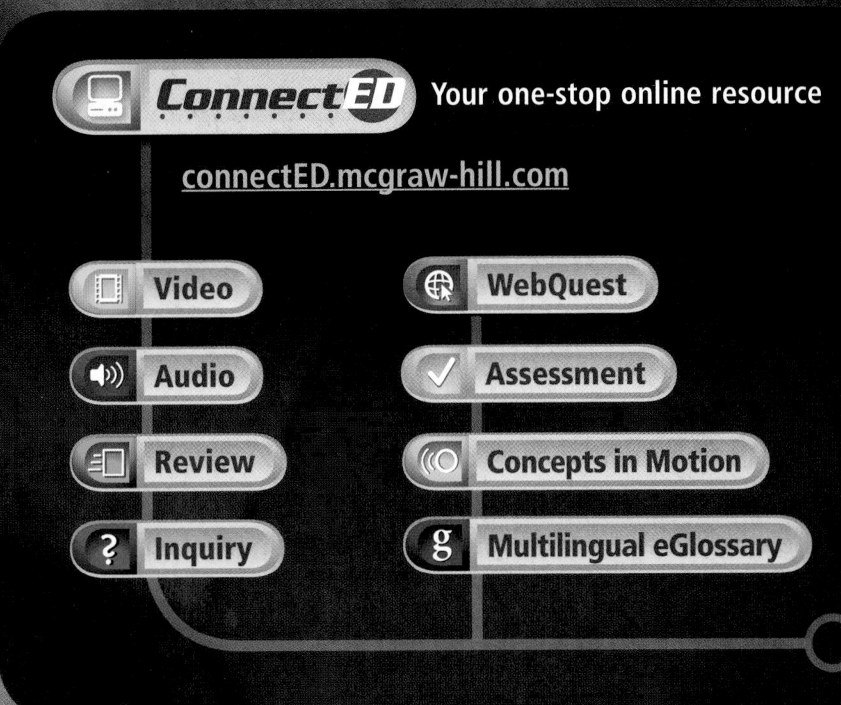

Lesson 1

Electrons and Energy Levels

Reading Guide

Key Concepts
ESSENTIAL QUESTIONS

- How is an electron's energy related to its distance from the nucleus?
- Why do atoms gain, lose, or share electrons?

Vocabulary

chemical bond p. 382
valence electron p. 384
electron dot diagram p. 385

BrainPOP®

Inquiry Are pairs more stable?

Rowing can be hard work, especially if you are part of a racing team. The job is made easier because the rowers each pull on the water with a pair of oars. How do pairs make the boat more stable?

Launch Lab

20 minutes

How is the periodic table organized?

How do you begin to put together a puzzle of a thousand pieces? You first sort similar pieces into groups. All edge pieces might go into one pile. All blue pieces might go into another pile. Similarly, scientists placed the elements into groups based on their properties. They created the periodic table, which organizes information about all the elements.

1. Obtain six **index cards** from your teacher. Using one card for each element name, write the names *beryllium, sodium, iron, zinc, aluminum,* and *oxygen* at the top of a card.
2. Open your textbook to the periodic table printed on the inside back cover. Locate the element key for each element written on your cards.
3. For each element, find the following information and write it on the index card: symbol, atomic number, atomic mass, state of matter, and element type.

Think About This

1. What do the elements in the blue blocks have in common? In the green blocks? In the yellow blocks?
2. **Key Concept** Each element in a column on the periodic table has similar chemical properties and forms bonds in similar ways. Based on this, for each element you listed on a card, name another element on the periodic table that has similar chemical properties.

The Periodic Table

Imagine trying to find a book in a library if all the books were unorganized. Books are organized in a library to help you easily find the information you need. The periodic table is like a library of information about all chemical elements.

A copy of the periodic table is on the inside back cover of this book. The table has more than 100 blocks—one for each known element. Each block on the periodic table includes basic properties of each element such as the element's state of matter at room temperature and its atomic number. The atomic number is the number of protons in each atom of the element. Each block also lists an element's atomic mass, or the average mass of all the different isotopes of that element.

Periods and Groups

You can learn about some properties of an element from its position on the periodic table. Elements are organized in periods (rows) and groups (columns). The periodic table lists elements in order of atomic number. The atomic number increases from left to right as you move across a period. Elements in each group have similar chemical properties and react with other elements in similar ways. In this lesson, you will read more about how an element's position on the periodic table can be used to predict its properties.

Reading Check How is the periodic table organized?

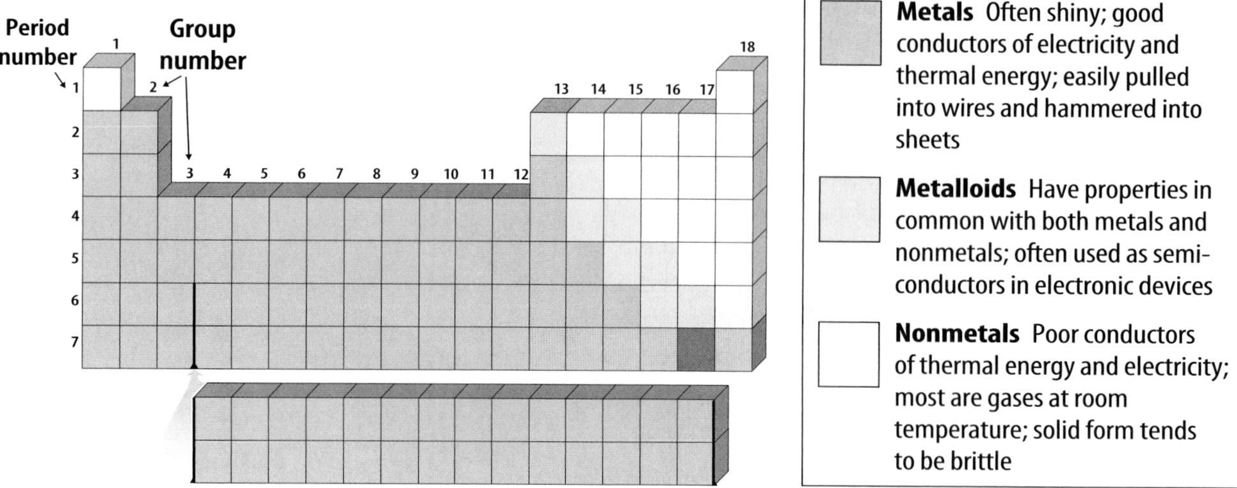

▲ **Figure 1** Elements on the periodic table are classified as metals, nonmetals, or metalloids.

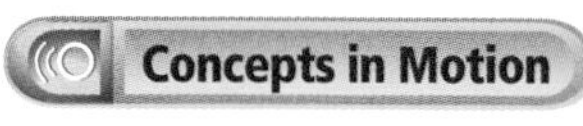

Animation

Metals, Nonmetals, and Metalloids

The three main regions of elements on the periodic table are shown in **Figure 1.** Except for hydrogen, elements on the left side of the table are metals. Nonmetals are on the right side of the table. Metalloids form the narrow stair-step region between metals and nonmetals.

Reading Check Where are metals, nonmetals, and metalloids on the periodic table?

REVIEW VOCABULARY

compound
matter that is made up of two or more different kinds of atoms joined together by chemical bonds

Atoms Bond

In nature, pure elements are rare. Instead, atoms of different elements chemically combine and form **compounds.** Compounds make up most of the matter around you, including living and nonliving things. There are only about 115 elements, but these elements combine and form millions of compounds. Chemical bonds hold them together. *A* **chemical bond** *is a force that holds two or more atoms together.*

Electron Number and Arrangement

Recall that atoms contain protons, neutrons, and electrons, as shown in **Figure 2.** Each proton has a positive charge; each neutron has no charge; and each electron has a negative charge. The atomic number of an element is the number of protons in each atom of that element. In a neutral (uncharged) atom, the number of protons equals the number of electrons.

The exact position of electrons in an atom cannot be determined. This is because electrons are in constant motion around the nucleus. However, each electron is usually in a certain area of space around the nucleus. Some are in areas close to the the nucleus, and some are in areas farther away.

Figure 2 Protons and neutrons are in an atom's nucleus. Electrons move around the nucleus. ▼

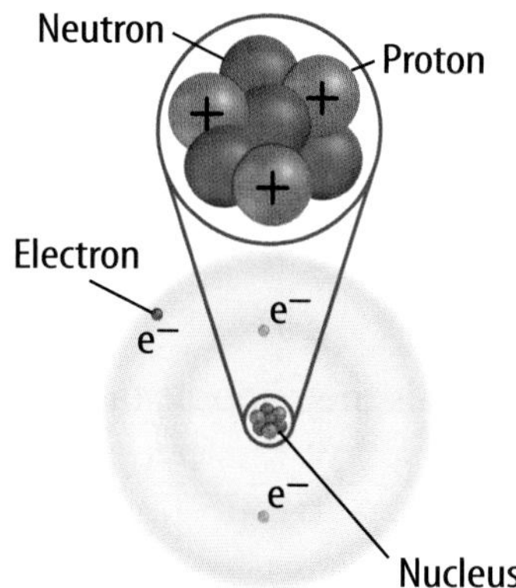

Electrons and Energy Different electrons in an atom have different amounts of energy. An electron moves around the nucleus at a distance that corresponds to its amount of energy. Areas of space in which electrons move around the nucleus are called energy levels. Electrons closest to the nucleus have the least amount of energy. They are in the lowest energy level. Electrons farthest from the nucleus have the greatest amount of energy. They are in the highest energy level. The energy levels of an atom are shown in **Figure 3.** Notice that only two electrons can be in the lowest energy level. The second energy level can hold up to eight.

Key Concept Check How is an electron's energy related to its position in an atom?

Electrons and Bonding Imagine two magnets. The closer they are to each other, the stronger the attraction of their opposite ends. Negatively charged electrons have a similar attraction to the positively charged nucleus of an atom. The electrons in energy levels closest to the nucleus of the same atom have a strong attraction to that nucleus. However, electrons farther from that nucleus are weakly attracted to it. These outermost electrons can easily be attracted to the nucleus of other atoms. This attraction between the positive nucleus of one atom and the negative electrons of another is what causes a chemical bond.

FOLDABLES®

Make two quarter-sheet note cards from a sheet of paper. Use them to organize your notes on valence electrons and electron dot diagrams.

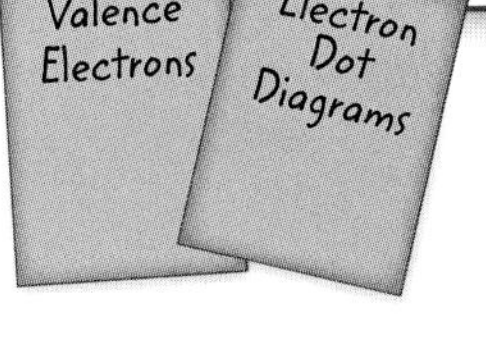

Figure 3 Electrons are in certain energy levels within an atom.

Electron Energy Levels

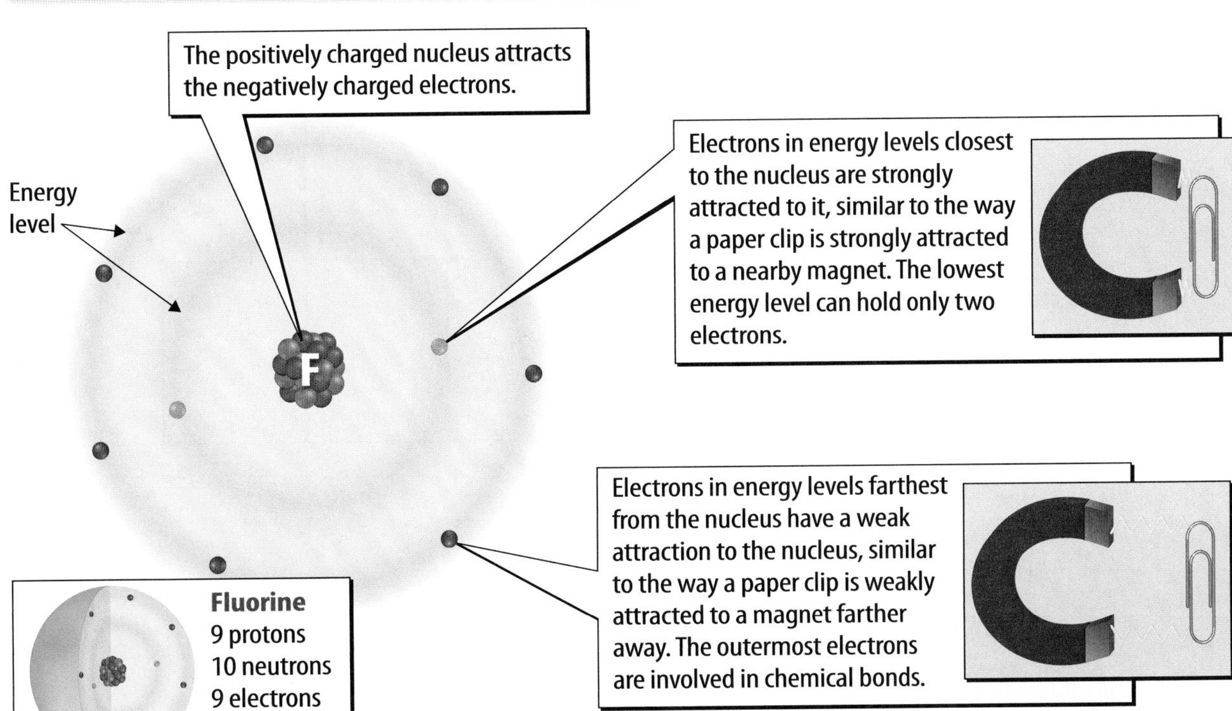

The positively charged nucleus attracts the negatively charged electrons.
Energy level
F
Electrons in energy levels closest to the nucleus are strongly attracted to it, similar to the way a paper clip is strongly attracted to a nearby magnet. The lowest energy level can hold only two electrons.
Electrons in energy levels farthest from the nucleus have a weak attraction to the nucleus, similar to the way a paper clip is weakly attracted to a magnet farther away. The outermost electrons are involved in chemical bonds.
Fluorine
9 protons
10 neutrons
9 electrons

Valence Electrons

You have read that electrons farthest from their nucleus are easily attracted to the nuclei of nearby atoms. These outermost electrons are the only electrons involved in chemical bonding. Even atoms that have only a few electrons, such as hydrogen or lithium, can form chemical bonds. This is because these electrons are still the outermost electrons and are exposed to the nuclei of other atoms. A **valence electron** *is an outermost electron of an atom that participates in chemical bonding.* Valence electrons have the most energy of all electrons in an atom.

Word Origin

valence
from Latin *valentia,* means "strength, capacity"

The number of valence electrons in each atom of an element can help determine the type and the number of bonds it can form. How do you know how many valence electrons an atom has? The periodic table can tell you. Except for helium, elements in certain groups have the same number of valence electrons. **Figure 4** illustrates how to use the periodic table to determine the number of valence electrons in the atoms of groups 1, 2, and 13–18. Determining the number of valence electrons for elements in groups 3–12 is more complicated. You will learn about these groups in later chemistry courses.

Figure 4 You can use the group numbers at the top of the columns to determine the number of valence electrons in atoms of groups 1, 2, and 13–18.

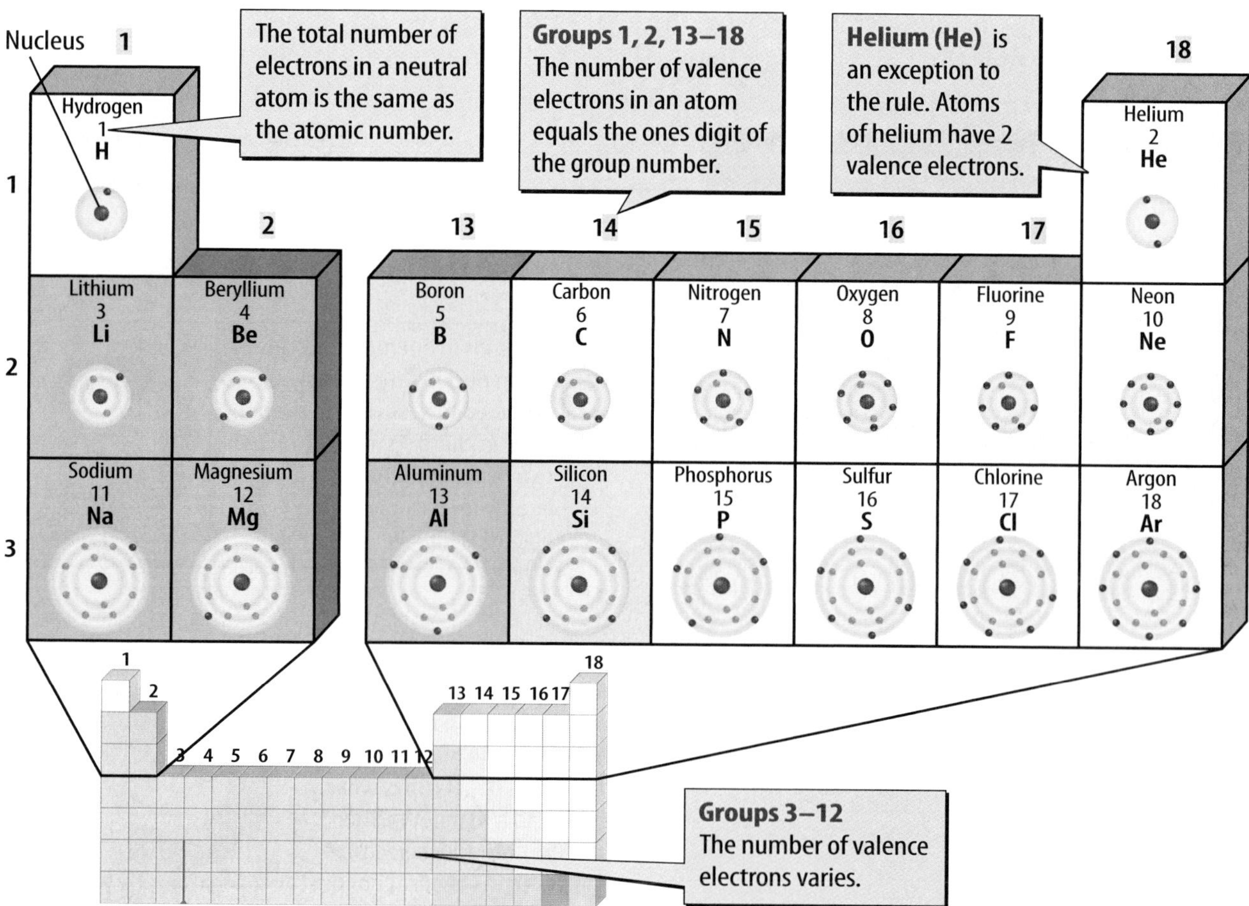

Visual Check How many valence electrons does an atom of phosphorous (P) have?

Writing and Using Electron Dot Diagrams

Figure 5 Electron dot diagrams show the number of valence electrons in an atom.

Steps for writing a dot diagram	Beryllium	Carbon	Nitrogen	Argon
1 Identify the element's group number on the periodic table.	2	14	15	18
2 Identify the number of valence electrons. • This equals the ones digit of the group number.	2	4	5	8
3 Draw the electron dot diagram. • Place one dot at a time on each side of the symbol (top, right, bottom, left). Repeat until all dots are used.	Be·	·C·	·N·	:Ar:
4 Determine if the atom is chemically stable. • An atom is chemically stable if all dots on the electron dot diagram are paired.	Chemically Unstable	Chemically Unstable	Chemically Unstable	Chemically Stable
5 Determine how many bonds this atom can form. • Count the dots that are unpaired.	2	4	3	0

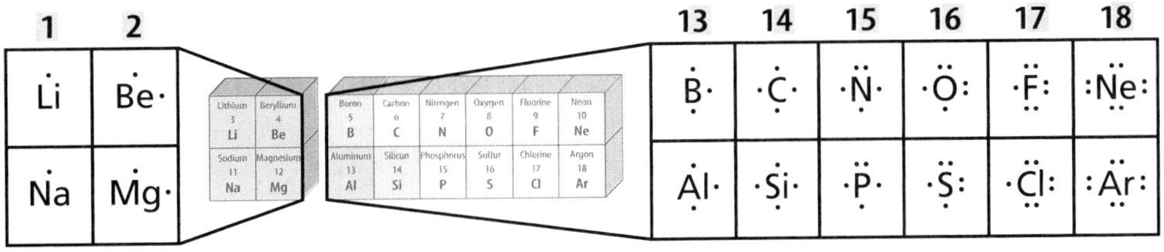

1	2	13	14	15	16	17	18
Li	Be·	B·	·C·	·N·	·O:	·F:	:Ne:
Na	Mg·	Al·	·Si·	·P·	·S:	·Cl:	:Ar:

Electron Dot Diagrams

In 1916 an American Chemist named Gilbert Lewis developed a method to show an element's valence electrons. He developed the **electron dot diagram,** *a model that represents valence electrons in an atom as dots around the element's chemical symbol.*

Electron dot diagrams can help you predict how an atom will bond with other atoms. Dots, representing valence electrons, are placed one-by-one on each side of an element's chemical symbol until all the dots are used. Some dots will be paired up, others will not. The number of unpaired dots is often the number of bonds an atom can form. The steps for writing dot diagrams are shown in **Figure 5.**

Reading Check Why are electron dot diagrams useful?

Recall that each element in a group has the same number of valence electrons. As a result, every element in a group has the same number of dots in its electron dot diagram.

Notice in **Figure 5** that an argon atom, Ar, has eight valence electrons, or four pairs of dots, in the diagram. There are no unpaired dots. Atoms with eight valence electrons do not easily react with other atoms. They are chemically stable. Atoms that have between one and seven valence electrons are reactive, or chemically unstable. These atoms easily bond with other atoms and form chemically stable compounds.

Atoms of hydrogen and helium have only one energy level. These atoms are chemically stable with two valence electrons.

Inquiry MiniLab 20 minutes

How does an electron's energy relate to its position in an atom?

Electrons in energy levels closest to the nucleus are strongly attracted to it. You can use paper clips and a magnet to model a similar attraction.

1. Read and complete a lab safety form.
2. Pick up a **paper clip** with a **magnet.** Use the first paper clip to pick up another one.
3. Continue picking up paper clips in this way until you have a chain of paper clips and no more will attach.
4. Gently pull off the paper clips one by one.

Analyze and Conclude

1. **Observe** Which paper clip was the easiest to remove? Which was the most difficult?
2. **Use Models** In what way do the magnet and the paper clips act as a model for an atom?
3. **Key Concept** How does an electron's position in an atom affect its ability to take part in chemical bonding?

Noble Gases

The elements in Group 18 are called noble gases. With the exception of helium, noble gases have eight valence electrons and are chemically stable. Chemically stable atoms do not easily react, or form bonds, with other atoms. The electron structures of two noble gases—neon and helium—are shown in **Figure 6.** Notice that all dots are paired in the dot diagrams of these atoms.

Stable and Unstable Atoms

Atoms with unpaired dots in their electron dot diagrams are reactive, or chemically unstable. For example, nitrogen, shown in **Figure 6,** has three unpaired dots in its electron dot diagram, and it is reactive. Nitrogen, like many other atoms, becomes more stable by forming chemical bonds with other atoms.

When an atom forms a bond, it gains, loses, or shares valence electrons with other atoms. By forming bonds, atoms become more chemically stable. Recall that atoms are most stable with eight valence electrons. Therefore, atoms with less than eight valence electrons form chemical bonds and become stable. In Lessons 2 and 3, you will read which atoms gain, lose, or share electrons when forming stable compounds.

Key Concept Check Why do atoms gain, lose, or share electrons?

Figure 6 Atoms gain, lose, or share valence electrons and become chemically stable.

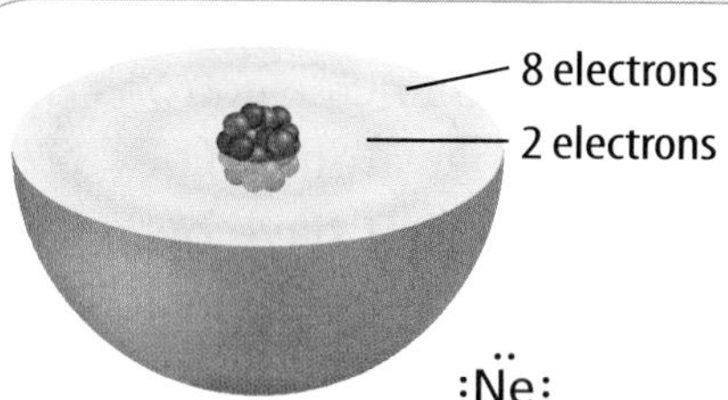

:Ne:

Neon has 10 electrons: 2 inner electrons and 8 valence electrons. A neon atom is chemically stable because it has 8 valence electrons. All dots in the dot diagram are paired.

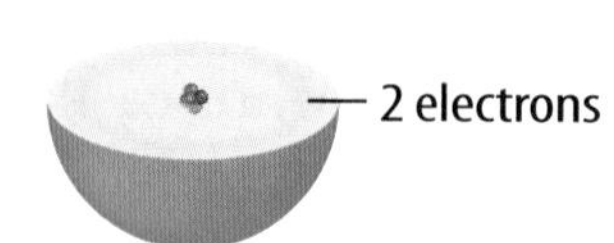

He

Helium has 2 electrons. Because an atom's lowest energy level can hold only 2 electrons, the 2 dots in the dot diagram are paired. Helium is chemically stable.

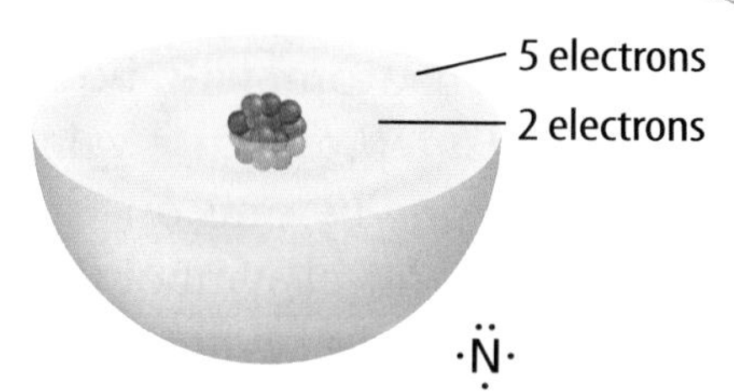

·N·

Nitrogen has 7 electrons: 2 inner electrons and 5 valence electrons. Its dot diagram has 1 pair of dots and 3 unpaired dots. Nitrogen atoms become more stable by forming chemical bonds.

Lesson 1 Review

Assessment Online Quiz
Inquiry Virtual Lab

Visual Summary

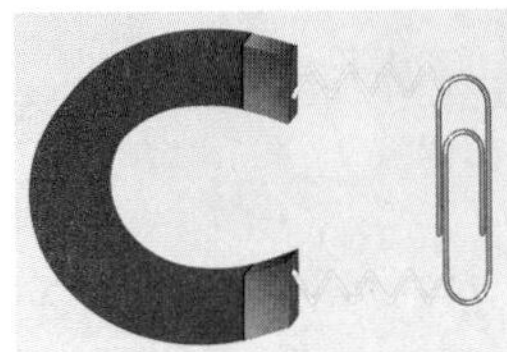

Electrons are less strongly attracted to a nucleus the farther they are from it, similar to the way a magnet attracts a paper clip.

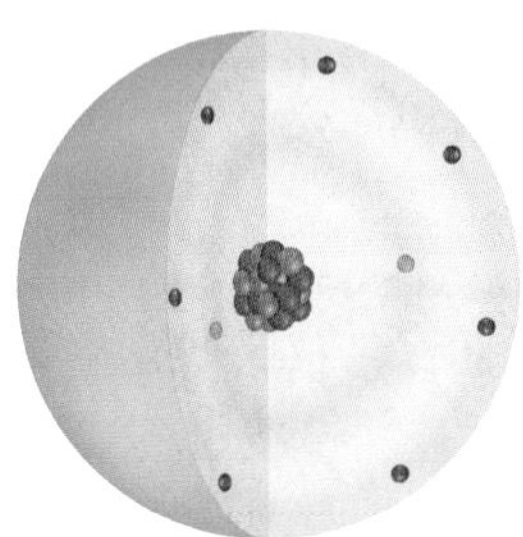

Electrons in atoms are in energy levels around the nucleus. Valence electrons are involved in chemical bonding.

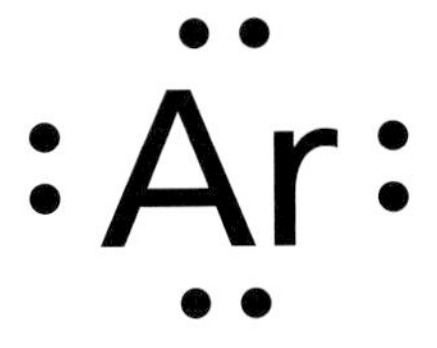

All noble gases, except He, have four pairs of dots in their electron dot diagrams. Noble gases are chemically stable.

FOLDABLES®

Use your lesson Foldable to review the lesson. Save your Foldable for the project at the end of the chapter.

What do you think NOW?

You first read the statements below at the beginning of the chapter.

1. Elements rarely exist in pure form. Instead, combinations of elements make up most of the matter around you.

2. Chemical bonds that form between atoms involve electrons.

Did you change your mind about whether you agree or disagree with the statements? Rewrite any false statements to make them true.

Use Vocabulary

1. **Use the term** *chemical bond* in a complete sentence.
2. **Define** *electron dot diagram* in your own words.
3. The electrons of an atom that participate in chemical bonding are called ________.

Understand Key Concepts 

4. **Identify** the number of valence electrons in each atom: calcium, carbon, and sulfur.
5. Which part of the atom is shared, gained, or lost when forming a chemical bond?
   A. electron **C.** nucleus
   B. neutron **D.** proton
6. **Draw** electron dot diagrams for oxygen, potassium, iodine, nitrogen, and beryllium.

Interpret Graphics

7. **Determine** the number of valence electrons in each diagram shown below.

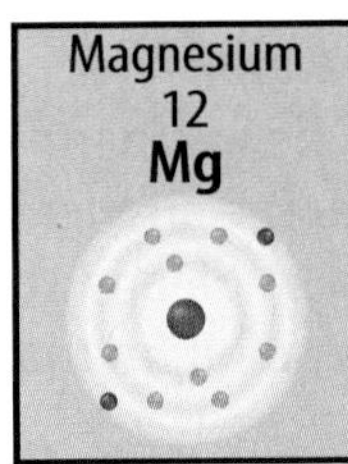

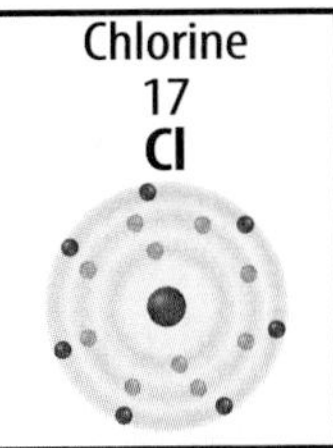

8. **Organize Information** Copy and fill in the graphic organizer below to describe one or more details for each concept: electron energy, valence electrons, stable atoms.

Concept	Description

Critical Thinking

9. **Compare** krypton and bromine in terms of chemical stability.
10. **Decide** An atom of nitrogen has five valence electrons. How could a nitrogen atom become more chemically stable?

GREEN SCIENCE

New Green Airships

The Difference of One Valence Electron

Faster than ocean liners and safer than airplanes, airships used to be the best way to travel. The largest, the *Hindenburg*, was nearly the size of the *Titanic*. To this day, no larger aircraft has ever flown. So, what happened to the giant airship? The answer lies in a valence electron.

The builders of the *Hindenburg* filled it with a lighter-than-air gas, hydrogen, so that it would float. Their plan was to use helium, a noble gas. However, helium was scarce. They knew hydrogen was explosive, but it was easier to get. For nine years, hydrogen airships floated safely back and forth across the Atlantic. But in 1937, disaster struck. Just before it landed, the *Hindenburg* exploded in flames. The age of the airship was over.

Since the *Hindenburg,* airplanes have become the main type of air transportation. A big airplane uses hundreds of gallons of fuel to take off and fly. As a result, it releases large amounts of pollutants into the atmosphere. Some people are looking for other types of air transportation that will be less harmful to the environment. Airships may be the answer. An airship floats and needs very little fuel to take off and stay airborne. Airships also produce far less pollution than other aircraft.

H

Today, however, airships use helium not hydrogen. With two valence electrons instead of one, as hydrogen has, helium is unreactive. Thanks to helium's chemical stability, someday you might be a passenger on a new, luxurious, but not explosive, version of the *Hindenburg*.

▲ A new generation of big airships might soon be hauling freight and carrying passengers.

It's Your Turn

RESEARCH Precious documents deteriorate with age as their surfaces react with air. Parchment turns brown and crumbles. Find out how our founding documents have been saved from this fate by noble gases.

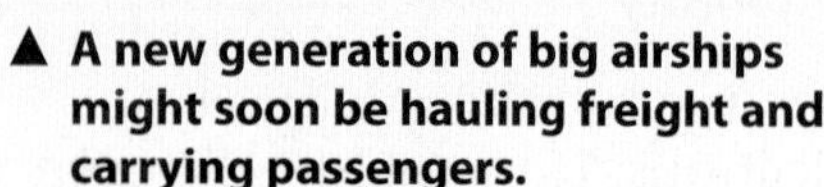

Lesson 2

Reading Guide

Key Concepts
ESSENTIAL QUESTIONS

- How do elements differ from the compounds they form?
- What are some common properties of a covalent compound?
- Why is water a polar compound?

Vocabulary

covalent bond p. 391
molecule p. 392
polar molecule p. 393
chemical formula p. 394

g Multilingual eGlossary

Compounds, Chemical Formulas, and Covalent Bonds

Inquiry How do they combine?

A jigsaw puzzle has pieces that connect in a certain way. The pieces fit together by sharing tabs with other pieces. All of the pieces combine and form a complete puzzle. Like pieces of a puzzle, atoms can join together and form a compound by sharing electrons.

Inquiry Launch Lab

20 minutes

How is a compound different from its elements?

The sugar you use to sweeten foods at home is probably sucrose. Sucrose contains the elements carbon, hydrogen, and oxygen. How does table sugar differ from the elements that it contains?

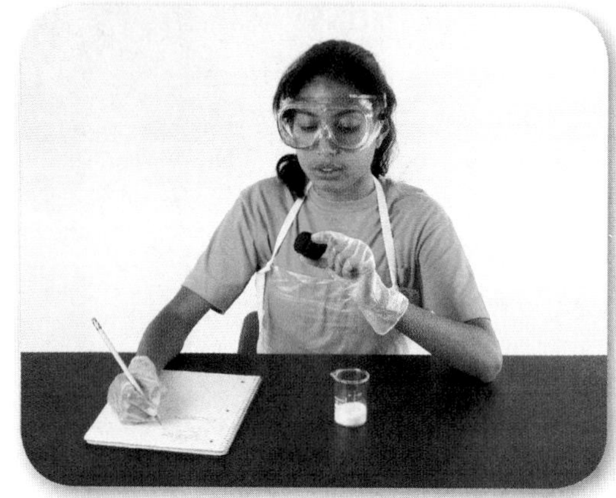

1. Read and complete a lab safety form.
2. Air is a mixture of several gases, including oxygen and hydrogen. Charcoal is a form of carbon. Write some properties of oxygen, hydrogen, and carbon in your Science Journal.
3. Obtain from your teacher a piece of **charcoal** and a **beaker** with **table sugar** in it.
4. Observe the charcoal. In your Science Journal, describe the way it looks and feels.
5. Observe the table sugar in the beaker. What does it look and feel like? Record your observations.

Think About This

1. Compare and contrast the properties of charcoal, hydrogen, and oxygen.
2. **Key Concept** How do you think the physical properties of carbon, hydrogen, and oxygen change when they combined to form sugar?

From Elements to Compounds

Have you ever baked cupcakes? First, combine flour, baking soda, and a pinch of salt. Then, add sugar, eggs, vanilla, milk, and butter. Each ingredient has unique physical and chemical properties. When you mix the ingredients together and bake them, a new product results—cupcakes. The cupcakes have properties that are different from the ingredients.

In some ways, compounds are like cupcakes. Recall that a compound is a substance made up of two or more different elements. Just as cupcakes are different from their ingredients, compounds are different from their elements. An element is made of one type of atom, but compounds are chemical combinations of different types of atoms. Compounds and the elements that make them up often have different properties.

Chemical **bonds** join atoms together. Recall that a chemical bond is a force that holds atoms together in a compound. In this lesson, you will learn that one way that atoms can form bonds is by sharing valence electrons. You will also learn how to write and read a chemical formula.

Key Concept Check How is a compound different from the elements that compose it?

SCIENCE USE V. COMMON USE

bond

Science Use a force that holds atoms together in a compound

Common Use a close personal relationship between two people

Covalent Bonds

Review Personal Tutor

Figure 7 A covalent bond forms when two nonmetal atoms share electrons.

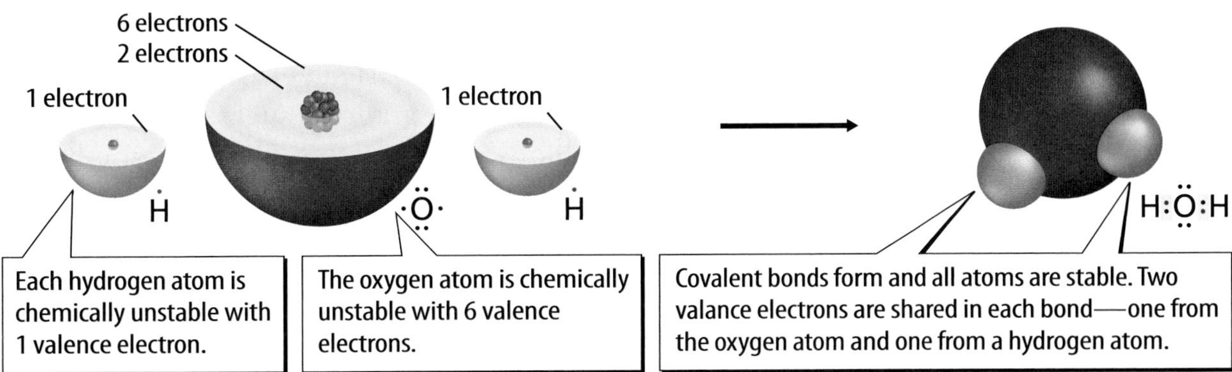

6 electrons
2 electrons
1 electron
1 electron
H
O
H
H:O:H
Each hydrogen atom is chemically unstable with 1 valence electron.
The oxygen atom is chemically unstable with 6 valence electrons.
Covalent bonds form and all atoms are stable. Two valance electrons are shared in each bond—one from the oxygen atom and one from a hydrogen atom.

Covalent Bonds—Electron Sharing

As you read in Lesson 1, one way that atoms can become more chemically stable is by sharing valence electrons. When unstable, nonmetal atoms bond together, they bond by sharing valence electrons. *A* **covalent bond** *is a chemical bond formed when two atoms share one or more pairs of valence electrons.* The atoms then form a stable covalent compound.

A Noble Gas Electron Arrangement

Look at the reaction between hydrogen and oxygen in **Figure 7.** Before the reaction, each hydrogen atom has one valence electron. The oxygen atom has six valence electrons. Recall that most atoms are chemically stable with eight valence electrons—the same electron arrangement as a noble gas. An atom with less than eight valence electrons becomes stable by forming chemical bonds until it has eight valence electrons. Therefore, an oxygen atom forms two bonds to become stable. A hydrogen atom is stable with two valence electrons. It forms one bond to become stable.

Shared Electrons

If the oxygen atom and each hydrogen atom share their unpaired valence electrons, they can form two covalent bonds and become a stable covalent compound. Each covalent bond contains two valence electrons—one from the hydrogen atom and one from the oxygen atom. Since these electrons are shared, they count as valence electrons for both atoms in the bond. Each hydrogen atom now has two valence electrons. The oxygen atom now has eight valence electrons, since it bonds to two hydrogen atoms. All three atoms have the electron arrangement of a noble gas and the compound is stable.

FOLDABLES®

Make three quarter-sheet note cards from a sheet of paper to organize information about single, double, and triple covalent bonds.

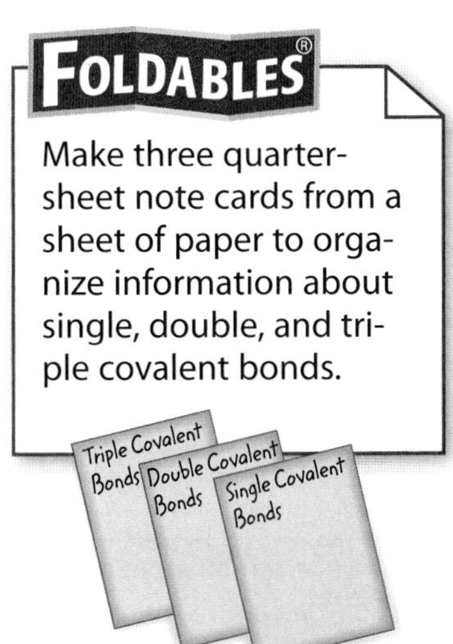

Triple Covalent Bonds
Double Covalent Bonds
Single Covalent Bonds

Double and Triple Covalent Bonds

As shown in **Figure 8**, a single covalent bond exists when two atoms share one pair of valence electrons. A double covalent bond exists when two atoms share two pairs of valence electrons. Double bonds are stronger than single bonds. A triple covalent bond exists when two atoms share three pairs of valence electrons. Triple bonds are stronger than double bonds. Multiple bonds are explained in **Figure 8.**

Covalent Compounds

When two or more atoms share valence electrons, they form a stable covalent compound. The covalent compounds carbon dioxide, water, and sugar are very different, but they also share similar properties. Covalent compounds usually have low melting points and low boiling points. They are usually gases or liquids at room temperature, but they can also be solids. Covalent compounds are poor conductors of thermal energy and electricity.

Molecules

The chemically stable unit of a covalent compound is a molecule. *A* **molecule** *is a group of atoms held together by covalent bonding that acts as an independent unit.* Table sugar ($C_{12}H_{22}O_{11}$) is a covalent compound. One grain of sugar is made up of trillions of sugar molecules. Imagine breaking a grain of sugar into the tiniest microscopic particle possible. You would have a molecule of sugar. One sugar molecule contains 12 carbon atoms, 22 hydrogen atoms, and 11 oxygen atoms all covalently bonded together. The only way to further break down the molecule would be to chemically separate the carbon, hydrogen, and oxygen atoms. These atoms alone have very different properties from the compound sugar.

Key Concept Check What are some common properties of covalent compounds?

Multiple Bonds

Figure 8 The more valence electrons that two atoms share, the stronger the covalent bond is between the atoms.

When two hydrogen atoms bond, they form a single covalent bond.	**One Single Covalent Bond** H· + ·H → H:H	In a single covalent bond, 1 pair of electrons is shared between two atoms. Each H atom shares 1 valence electron with the other.
When one carbon atom bonds with two oxygen atoms, two double covalent bonds form.	**Two Double Covalent Bonds** ·Ö: + ·Ċ· + ·Ö: → :O::C::O:	In a double covalent bond, 2 pairs of electrons are shared between two atoms. One O atom and the C atom each share 2 valence electrons with the other.
When two nitrogen atoms bond, they form a triple covalent bond.	**One Triple Covalent Bond** ·N· + ·N· → :N⋮⋮N:	In a triple covalent bond, 3 pairs of electrons are shared between two atoms. Each N atom shares 3 valence electrons with the other.

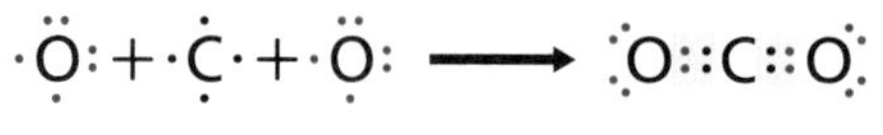

Visual Check Is the bond stronger between atoms in hydrogen gas (H_2) or nitrogen gas (N_2)? Why?

Water and Other Polar Molecules

In a covalent bond, one atom can attract the shared electrons more strongly than the other atom can. Think about the valence electrons shared between oxygen and hydrogen atoms in a water molecule. The oxygen atom attracts the shared electrons more strongly than each hydrogen atom does. As a result, the shared electrons are pulled closer to the oxygen atom, as shown in **Figure 9.** Since electrons have a negative charge, the oxygen atom has a partial negative charge. The hydrogen atoms have a partial positive charge. *A molecule that has a partial positive end and a partial negative end because of unequal sharing of electrons is a* **polar molecule.**

WORD ORIGIN
polar
from Latin *polus,* means "pole"

The charges on a polar molecule affect its properties. Sugar, for example, dissolves easily in water because both sugar and water are polar. The negative end of a water molecule pulls on the positive end of a sugar molecule. Also, the positive end of a water molecule pulls on the negative end of a sugar molecule. This causes the sugar molecules to separate from one another and mix with the water molecules.

 **Key Concept Check** Why is water a polar compound?

Nonpolar Molecules

A hydrogen molecule, H_2, is a nonpolar molecule. Because the two hydrogen atoms are identical, their attraction for the shared electrons is equal. The carbon dioxide molecule, CO_2, in **Figure 9** is also nonpolar. A nonpolar compound will not easily dissolve in a polar compound, but it will dissolve in other nonpolar compounds. Oil is an example of a nonpolar compound. It will not dissolve in water. Have you ever heard someone say "like dissolves like"? This means that polar compounds can dissolve in other polar compounds. Similarly, nonpolar compounds can dissolve in other nonpolar compounds.

Figure 9 Atoms of a polar molecule share their valence electrons unequally. Atoms of a nonpolar molecule share their valence electrons equally.

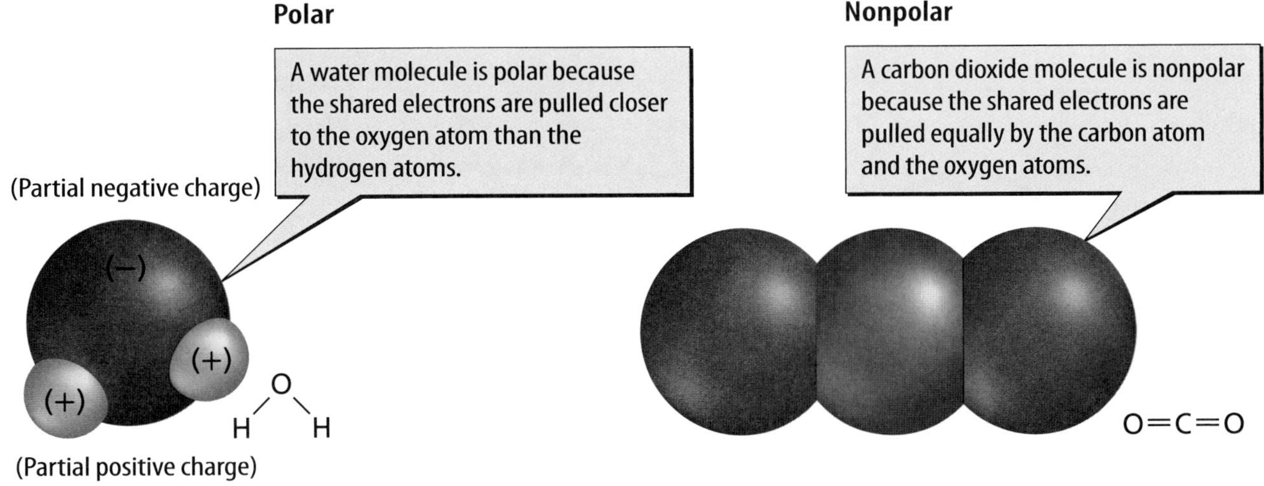
Polar
A water molecule is polar because the shared electrons are pulled closer to the oxygen atom than the hydrogen atoms.
(Partial negative charge)
(−)
(+)
(+)
O
H H
(Partial positive charge)
Nonpolar
A carbon dioxide molecule is nonpolar because the shared electrons are pulled equally by the carbon atom and the oxygen atoms.
O=C=O

Inquiry MiniLab — 20 minutes

How do compounds form?

Use building blocks to model ways in which elements combine to form compounds.

1. Examine various types of **interlocking plastic blocks.** Notice that the blocks have different numbers of holes and pegs. Attaching one peg to one hole represents a shared pair of electrons.

2. Draw the electron dot diagrams for carbon, nitrogen, oxygen, and hydrogen in your Science Journal. Based on the diagrams, decide which block should represent an atom of each element.
3. Use the blocks to make models of H_2, CO_2, NH_3, H_2O, and CH_4. All pegs on the largest block must fit into a hole, and no blocks can stick out over the edge of a block, either above or below it.

Analyze and Conclude

1. **Explain** how you decided which type of block should be assigned to each type of atom.
2. **Key Concept** Name at least one way that your models show the difference between a compound and the elements that combine and form the compound.

Chemical Formulas and Molecular Models

How do you know which elements make up a compound? *A* **chemical formula** *is a group of chemical symbols and numbers that represent the elements and the number of atoms of each element that make up a compound.* Just as a recipe lists ingredients, a chemical formula lists the elements in a compound. For example, the chemical formula for carbon dioxide shown in **Figure 10** is CO_2. The formula uses chemical symbols that show which elements are in the compound. Notice that CO_2 is made up of carbon (C) and oxygen (O). A subscript, or small number after a chemical symbol, shows the number of atoms of each element in the compound. Carbon dioxide (CO_2) contains two atoms of oxygen bonded to one atom of carbon.

A chemical formula describes the types of atoms in a compound or a molecule, but it does not explain the shape or appearance of the molecule. There are many ways to model a molecule. Each one can show the molecule in a different way. Common types of models for CO_2 are shown in **Figure 10.**

Reading Check What information is given in a chemical formula?

Figure 10 Chemical formulas and molecular models provide information about molecules.

Chemical Formula

A carbon dioxide molecule is made up of carbon (C) and oxygen (O) atoms.

$$CO_2$$

A symbol without a subscript indicates one atom. Each molecule of carbon dioxide has one carbon atom.

The subscript 2 indicates two atoms of oxygen. Each molecule of carbon dioxide has two oxygen atoms.

Dot Diagram
- Shows atoms and valence electrons

:Ö::C::Ö:

Structural Formula
- Shows atoms and lines; each line represents one shared pair of electrons

O=C=O

Ball-and-Stick Model
- Balls represent atoms and sticks represent bonds; used to show bond angles

Space-Filling Model
- Spheres represent atoms; used to show three-dimensional arrangement of atoms

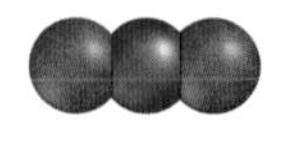

Lesson 2 Review

Assessment Online Quiz

Visual Summary

CO_2

A chemical formula is one way to show the elements that make up a compound.

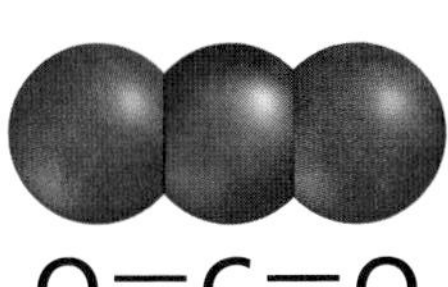

A covalent bond forms when atoms share valence electrons. The smallest particle of a covalent compound is a molecule.

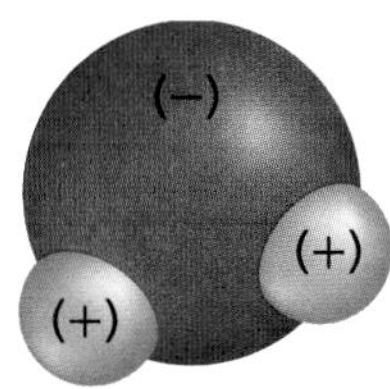

Water is a polar molecule because the oxygen and hydrogen atoms unequally share electrons.

FOLDABLES®

Use your lesson Foldable to review the lesson. Save your Foldable for the project at the end of the chapter.

What do you think NOW?

You first read the statements below at the beginning of the chapter.

3. The atoms in a water molecule are more chemically stable than they would be as individual atoms.

4. Many substances dissolve easily in water because opposite ends of a water molecule have opposite charges.

Did you change your mind about whether you agree or disagree with the statements? Rewrite any false statements to make them true.

Use Vocabulary

1. **Define** *covalent bond* in your own words.
2. The group of symbols and numbers that shows the types and numbers of atoms that make up a compound is a ________.
3. **Use the term** *molecule* in a complete sentence.

Understand Key Concepts 

4. **Contrast** Name at least one way water (H_2O) is different from the elements that make up water.
5. **Explain** why water is a polar molecule.
6. A sulfur dioxide molecule has one sulfur atom and two oxygen atoms. Which is its correct chemical formula?
   A. SO_2
   B. $(SO)_2$
   C. S_2O_2
   D. S_2O

Interpret Graphics

7. **Examine** the electron dot diagram for chlorine below.

In chlorine gas, two chlorine atoms join to form a Cl_2 molecule. How many pairs of valence electrons do the atoms share?

8. **Compare and Contrast** Copy and fill in the graphic organizer below to identify at least one way polar and nonpolar molecules are similar and one way they are different.

Polar and Nonpolar Molecules	
Similarities	
Differences	

Critical Thinking

9. **Develop** an analogy to explain the unequal sharing of valence electrons in a water molecule.

Skill Practice Model

25 minutes

How can you model compounds?

Materials

colored pencils

Chemists use models to explain how electrons are arranged in an atom. Electron dot diagrams are models used to show how many valence electrons an atom has. Electron dot diagrams are useful because they can help predict the number and type of bond an atom will form.

Learn It

In science, **models** are used to help you visualize objects that are too small, too large, or too complex to understand. A model is a representation of an object, idea, or event.

Try It

1. Use the periodic table to write the electron dot diagrams for hydrogen, oxygen, carbon, and silicon.
2. Using your electron dot diagrams from step 1, write electron dot diagrams for the following compounds: H_2O, CO, CO_2, SiO_2, C_2H_2, and CH_4. Use colored pencils to differentiate the electrons for each atom. Remember that all the above atoms, except hydrogen and helium, are chemically stable when they have eight valence electrons. Hydrogen and helium are chemically stable with two valence electrons.

Apply It

3. Based on your model, describe silicon's electron dot diagram and arrangement of valence electrons before and after it forms the compound SiO_2.
4. **Key Concept** Which of the covalent compounds you modeled contain double bonds? Which contain triple bonds?

1	2
Hydrogen 1 H	
Lithium 3 Li	Beryllium 4 Be
Sodium 11 Na	Magnesium 12 Mg
Potassium 19 K	Calcium 20 Ca
Rubidium 37 Rb	Strontium 38 Sr
Cesium 55 Cs	Barium 56 Ba
Francium 87 Fr	Radium 88 Ra

13	14	15	16	17	18
					Helium 2 He
Boron 5 B	Carbon 6 C	Nitrogen 7 N	Oxygen 8 O	Fluorine 9 F	Neon 10 Ne
Aluminum 13 Al	Silicon 14 Si	Phosphorus 15 P	Sulfur 16 S	Chlorine 17 Cl	Argon 18 Ar
Gallium 31 Ga	Germanium 32 Ge	Arsenic 33 As	Selenium 34 Se	Bromine 35 Br	Krypton 36 Kr
Indium 49 In	Tin 50 Sn	Antimony 51 Sb	Tellurium 52 Te	Iodine 53 I	Xenon 54 Xe
Thallium 81 Tl	Lead 82 Pb	Bismuth 83 Bi	Polonium 84 Po	Astatine 85 At	Radon 86 Rn

Lesson 3

Reading Guide

Key Concepts
ESSENTIAL QUESTIONS

- What is an ionic compound?
- How do metallic bonds differ from covalent and ionic bonds?

Vocabulary

ion p. 398

ionic bond p. 400

metallic bond p. 401

g Multilingual eGlossary

Ionic and Metallic Bonds

Inquiry What is this?

This scene might look like snow along a shoreline, but it is actually thick deposits of salt on a lake. Over time, tiny amounts of salt dissolved in river water that flowed into this lake and built up as water evaporated. Salt is a compound that forms when elements form bonds by gaining or losing valence electrons, not sharing them.

Launch Lab

15 minutes

How can atoms form compounds by gaining and losing electrons?

Metals on the periodic table often lose electrons when forming stable compounds. Nonmetals often gain electrons.

1. Read and complete a lab safety form.
2. Make two model atoms of sodium, and one model atom each of calcium, chlorine, and sulfur. To do this, write each element's chemical symbol with a **marker** on a **paper plate.** Surround the symbol with small balls of **clay** to represent valence electrons. Use one color of clay for the metals (groups 1 and 2 elements) and another color of clay for nonmetals (groups 16 and 17 elements).

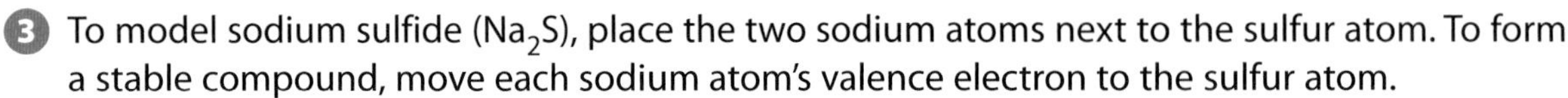

3. To model sodium sulfide (Na_2S), place the two sodium atoms next to the sulfur atom. To form a stable compound, move each sodium atom's valence electron to the sulfur atom.
4. Form as many other compound models as you can by removing valence electrons from the groups 1 and 2 plates and placing them on the groups 16 and 17 plates.

Think About This

1. What other compounds were you able to form?
2. **Key Concept** How do you think your models are different from covalent compounds?

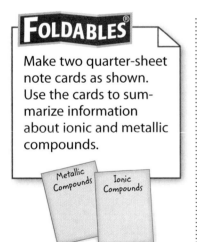

FOLDABLES®

Make two quarter-sheet note cards as shown. Use the cards to summarize information about ionic and metallic compounds.

Metallic Compounds

Ionic Compounds

WORD ORIGIN

ion
from Greek *ienai,* means "to go"

Understanding Ions

As you read in Lesson 2, the atoms of two or more nonmetals form compounds by sharing valence electrons. However, when a metal and a nonmetal bond, they do not share electrons. Instead, one or more valence electrons transfers from the metal atom to the nonmetal atom. After electrons transfer, the atoms bond and form a chemically stable compound. Transferring valence electrons results in atoms with the same number of valence electrons as a noble gas.

When an atom loses or gains a valence electron, it becomes an ion. *An* **ion** *is an atom that is no longer electrically neutral because it has lost or gained valence electrons.* Because electrons have a negative charge, losing or gaining an electron changes the overall charge of an atom. An atom that loses valence electrons becomes an ion with a positive charge. This is because the number of electrons is now less than the number of protons in the atom. An atom that gains valence electrons becomes an ion with a negative charge. This is because the number of protons is now less than the number of electrons.

Reading Check Why do atoms that a gain electrons become an ion with a negative charge?

Losing Valence Electrons

Look at the periodic table on the inside back cover of this book. What information about sodium (Na) can you infer from the periodic table? Sodium is a metal. Its atomic number is 11. This means each sodium atom has 11 protons and 11 electrons. Sodium is in group 1 on the periodic table. Therefore, sodium atoms have one valence electron, and they are chemically unstable.

Metal atoms, such as sodium, become more stable when they lose valence electrons and form a chemical bond with a nonmetal. If a sodium atom loses its one valence electron, it would have a total of ten electrons. Which element on the periodic table has atoms with ten electrons? Neon (Ne) atoms have a total of ten electrons. Eight of these are valence electrons. When a sodium atom loses one valence electron, the electrons in the next lower energy level are now the new valence electrons. The sodium atom then has eight valence electrons, the same as the noble gas neon and is chemically stable.

Gaining Valence Electrons

In Lesson 2, you read that nonmetal atoms can share valence electrons with other nonmetal atoms. Nonmetal atoms can also gain valence electrons from metal atoms. Either way, they achieve the electron arrangement of a noble gas. Find the nonmetal chlorine (Cl) on the periodic table. Its atomic number is 17. Atoms of chlorine have seven valence electrons. If a chlorine atom gains one valence electron, it will have eight valence electrons. It will also have the same electron arrangement as the noble gas argon (Ar).

When a sodium atom loses a valence electron, it becomes a positively charged ion. This is shown by a plus (+) sign. When a chlorine atom gains a valence electron, it becomes a negatively charged ion. This is shown by a negative (−) sign. **Figure 11** illustrates the process of a sodium atom losing an electron and a chlorine atom gaining an electron.

Reading Check Are atoms of a group 16 element more likely to gain or lose valence electrons?

Losing and Gaining Electrons

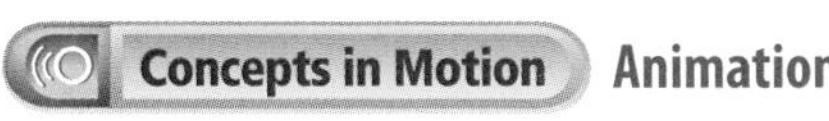

Figure 11 Sodium atoms have a tendency to lose a valence electron. Chlorine atoms have a tendency to gain a valence electron.

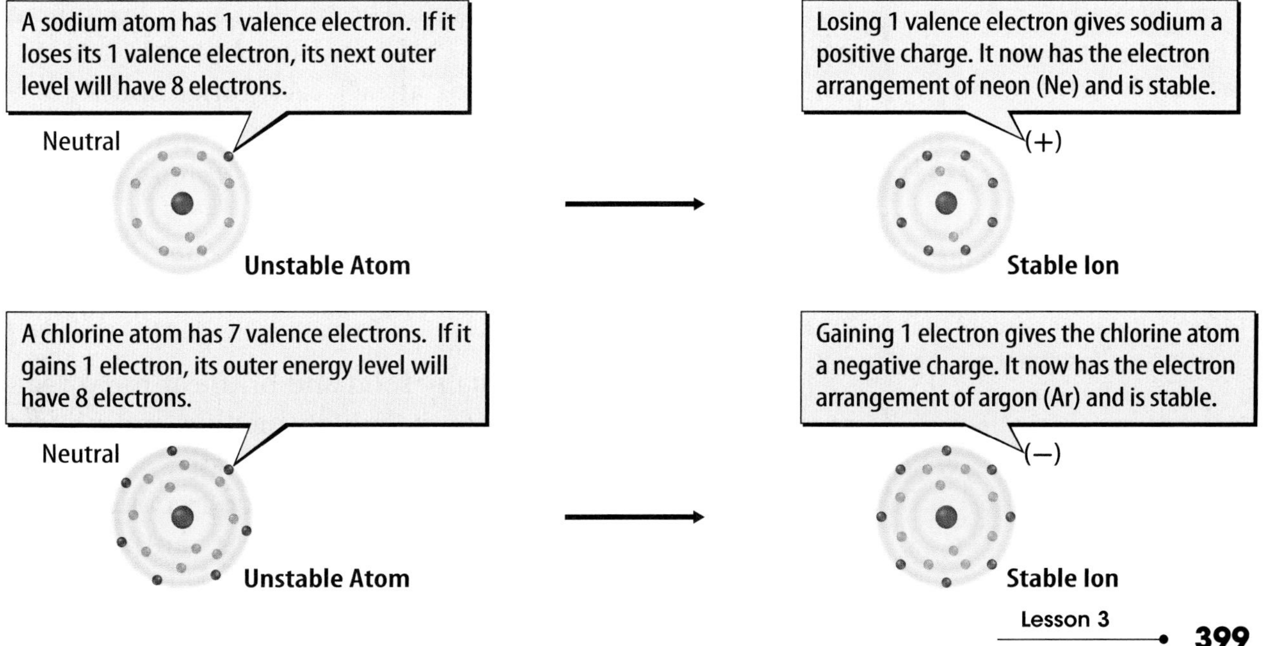

A sodium atom has 1 valence electron. If it loses its 1 valence electron, its next outer level will have 8 electrons.
Neutral
Unstable Atom
Losing 1 valence electron gives sodium a positive charge. It now has the electron arrangement of neon (Ne) and is stable.
(+)
Stable Ion
A chlorine atom has 7 valence electrons. If it gains 1 electron, its outer energy level will have 8 electrons.
Neutral
Unstable Atom
Gaining 1 electron gives the chlorine atom a negative charge. It now has the electron arrangement of argon (Ar) and is stable.
(−)
Stable Ion

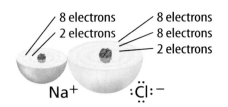

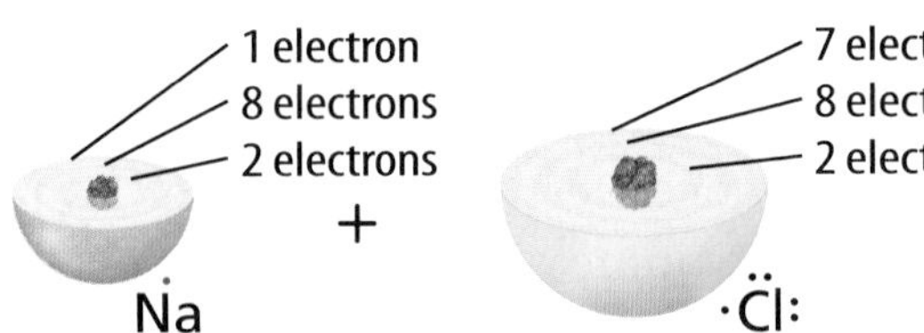

8 electrons
2 electrons
8 electrons
8 electrons
2 electrons
Na^+
Cl^-

Sodium and chlorine atoms are stable when they have eight valence electrons. A sodium atoms loses one valence electron and becomes stable. A chlorine atom gains one valence electron and becomes stable.

The positively charged sodium ion and the negatively charged chlorine ion attract each other. Together they form a strong ionic bond.

Figure 12 An ionic bond forms between Na and Cl when an electron transfers from Na to Cl.

Concepts in Motion Animation

Math Skills

Use Percentage

An atom's radius is measured in picometers (pm), 1 trillion times smaller than a meter. When an atom becomes an ion, its radius increases or decreases. For example, a Na atom has a radius of 186 pm. A Na^+ ion has a radius of 102 pm. By what percentage does the radius change?

Subtract the atom's radius from the ion's radius.

$$102 \text{ pm} - 186 \text{ pm} = -84 \text{ pm}$$

Divide the difference by the atom's radius.

$$-84 \text{ pm} \div 186 \text{ pm} = -0.45$$

Multiply the answer by 100 and add a % sign.

$$-0.45 \times 100 = -45\%$$

A negative value is a decrease in size. A positive value is an increase.

Practice

The radius of an oxygen (O) atom is 73 pm. The radius of an oxygen ion (O^{2-}) is 140 pm. By what percentage does the radius change?

- Math Practice
- Personal Tutor

Determining an Ion's Charge

Atoms are electrically neutral because they have the same number of protons and electrons. Once an atom gains or loses electrons, it becomes a charged ion. For example, the atomic number for nitrogen (N) is 7. Each N atom has 7 protons and 7 electrons and is electrically neutral. However, an N atom often gains 3 electrons when forming an ion. The N ion then has 10 electrons. To determine the charge, subtract the number of electrons in the ion from the number of protons.

$$7 \text{ protons} - 10 \text{ electrons} = -3 \text{ charge}$$

A nitrogen ion has a −3 charge. This is written as N^{3-}.

Ionic Bonds—Electron Transferring

Recall that metal atoms typically lose valence electrons and nonmetal atoms typically gain valence electrons. When forming a chemical bond, the nonmetal atoms gain the electrons lost by the metal atoms. Take a look at **Figure 12.** In NaCl, or table salt, a sodium atom loses a valence electron. The electron is transferred to a chlorine atom. The sodium atom becomes a positively charged ion. The chlorine atom becomes a negatively charged ion. These ions attract each other and form a stable ionic compound. *The attraction between positively and negatively charged ions in an ionic compound is an* **ionic bond.**

 **Key Concept Check** What holds ionic compounds together?

Ionic Compounds

Ionic compounds are usually solid and brittle at room temperature. They also have relatively high melting and boiling points. Many ionic compounds dissolve in water. Water that contains dissolved ionic compounds is a good conductor of electricity. This is because an electrical charge can pass from ion to ion in the solution.

Comparing Ionic and Covalent Compounds

Recall that in a covalent bond, two or more nonmetal atoms share electrons and form a unit, or molecule. Covalent compounds, such as water, are made up of many molecules. However, when nonmetal ions bond to metal ions in an ionic compound, there are no molecules. Instead, there is a large collection of oppositely charged ions. All of the ions attract each other and are held together by ionic bonds.

Metallic Bonds—Electron Pooling

Recall that metal atoms typically lose valence electrons when forming compounds. What happens when metal atoms bond to other metal atoms? Metal atoms form compounds with one another by combining, or pooling, their valence electrons. A **metallic bond** *is a bond formed when many metal atoms share their pooled valence electrons.*

The pooling of valence electrons in aluminum is shown in **Figure 13.** The aluminum atoms lose their valence electrons and become positive ions, indicated by the plus (+) signs. The negative (–) signs indicate the valence electrons, which move from ion to ion. Valence electrons in metals are not bonded to one atom. Instead, a "sea of electrons" surrounds the positive ions.

Key Concept Check How do metal atoms bond with one another?

Figure 13 Valence electrons move among all the aluminum (Al) ions.

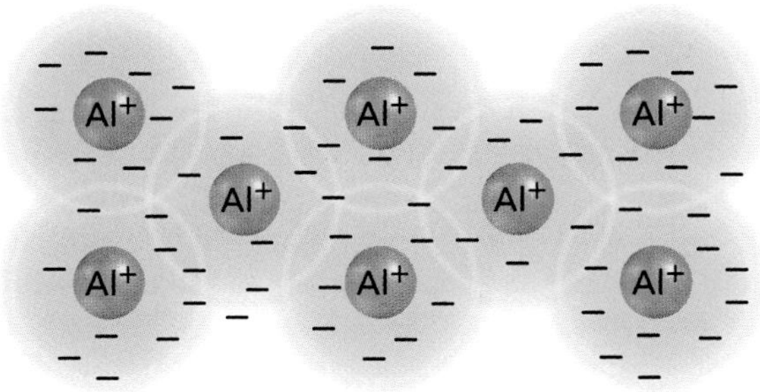

Inquiry MiniLab — 20 minutes

How many ionic compounds can you make?

You have read that in ionic bonding, metal atoms transfer electrons to nonmetal atoms.

1. Copy the table below into your Science Journal.

Group	Elements	Type	Dot Diagram
1	Li, Na, K	Metal	$\dot{X}$
2	Be, Mg, Ca	Metal	
14	C	Nonmetal	
15	N, P	Nonmetal	
16	O, S	Nonmetal	
17	F, Cl	Nonmetal	

2. Fill in the last column with the correct dot diagram for each group. Color the dots of the metal atoms with a **red marker** and the dots of the nonmetal atoms with a **blue marker.**
3. Using the information in your table, create five different ionic bonds. Write (a) the equation for the electron transfer and (b) the formula for each compound. For example:

a. $\dot{Na} + \dot{Na} + \cdot\ddot{\underset{\cdot\cdot}{O}}: \longrightarrow Na^+ + Na^+ + :\ddot{\underset{\cdot\cdot}{O}}:^{2-}$

b. Na_2O

Analyze and Conclude

1. **Explain** What happens to the metal and nonmetal ions after the electrons have been transferred?
2. **Key Concept** Describe the ionic bonds that hold the ions together in your compounds.

ACADEMIC VOCABULARY

conduct
(verb) to serve as a medium through which something can flow

Table 1 Bonds can form when atoms share valence electrons, transfer valence electrons, or pool valence electrons.

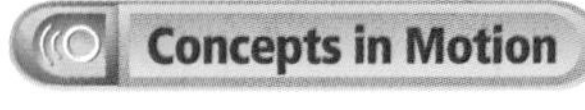

Interactive Table

Properties of Metallic Compounds

Metals are good conductors of thermal energy and electricity. Because the valence electrons can move from ion to ion, they can easily **conduct** an electric charge. When a metal is hammered into a sheet or drawn into a wire, it does not break. The metal ions can slide past one another in the electron sea and move to new positions. Metals are shiny because the valence electrons at the surface of a metal interact with light. **Table 1** compares the covalent, ionic, and metallic bonds that you studied in this chapter.

Reading Check How does valence electron pooling explain why metals can be hammered into a sheet?

Table 1 Covalent, Ionic, and Metallic Bonds

Type of Bond	What is bonding?	Properties of Compounds
Covalent (Water; (−), (+), (+))	nonmetal atoms; nonmetal atoms	• gas, liquid, or solid • low melting and boiling points • often not able to dissolve in water • poor conductors of thermal energy and electricity • dull appearance
Ionic (Salt; Na^+, Cl^-)	nonmetal ions; metal ions	• solid crystals • high melting and boiling points • dissolves in water • solids are poor conductors of thermal energy and electricity • ionic compounds in water solutions conduct electricity
Metallic (Aluminum; Al^+)	metal ions; metal ions	• usually solid at room temperature • high melting and boiling points • do not dissolve in water • good conductors of thermal energy and electricity • shiny surface • can be hammered into sheets and pulled into wires

Lesson 3 Review

Assessment Online Quiz

Visual Summary

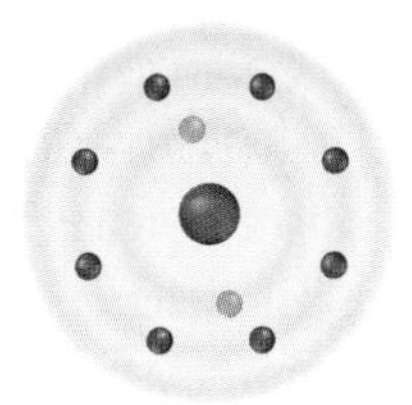

Metal atoms lose electrons and non-metal atoms gain electrons and form stable compounds. An atom that has gained or lost an electron is an ion.

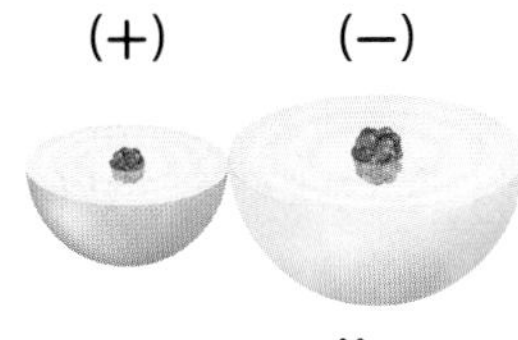

An ionic bond forms between positively and negatively charged ions.

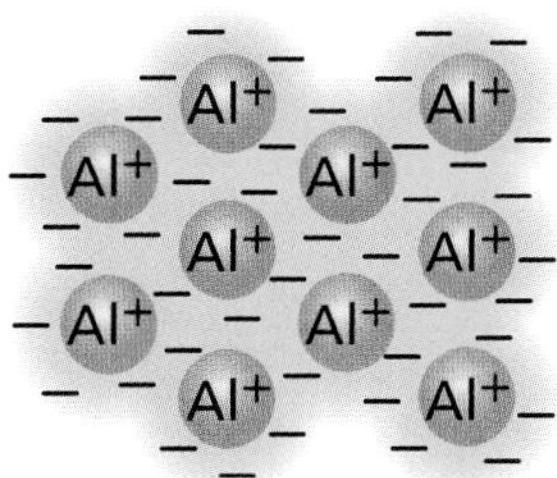

A metallic bond forms when many metal atoms share their pooled valence electrons.

FOLDABLES®

Use your lesson Foldable to review the lesson. Save your Foldable for the project at the end of the chapter.

What do you think NOW?

You first read the statements below at the beginning of the chapter.

5. Losing electrons can make some atoms more chemically stable.

6. Metals are good electrical conductors because they tend to hold onto their valence electrons very tightly.

Did you change your mind about whether you agree or disagree with the statements? Rewrite any false statements to make them true.

Use Vocabulary

1. **Define** *ionic bond* in your own words.
2. An atom that changes so that it has an electrical charge is a(n) ________.
3. **Use the term** *metallic bond* in a sentence.

Understand Key Concepts 

4. **Recall** What holds ionic compounds together?
5. Which element would most likely bond with lithium and form an ionic compound?

   A. beryllium **C.** fluorine
   B. calcium **D.** sodium
6. **Contrast** Why are metals good conductors of electricity while covalent compounds are poor conductors?

Interpret Graphics

7. **Organize** Copy and fill in the graphic organizer below. In each oval, list a common property of an ionic compound.

Critical Thinking

8. **Design** a poster to illustrate how ionic compounds form.
9. **Evaluate** What type of bonding does a material most likely have if it has a high melting point, is solid at room temperature, and easily dissolves in water?

Math Skills

Math Practice

10. The radius of the aluminum (Al) atom is 143 pm. The radius of the aluminum ion (Al^{3+}) is 54 pm. By what percentage did the radius change as the ion formed?

Ions in Solution

Materials

250-mL beaker

plastic spoon

dull pennies (20)

stopwatch

white vinegar

table salt

iron nails (2)

sandpaper

Safety

You know that ions can combine and form stable ionic compounds. Ions can also separate in a compound and dissolve in solution. For example, pennies become dull over time because the copper ions on the surface of the pennies react with oxygen in the air and form copper(II) oxide. When you place dull pennies in a vinegar-salt solution, the copper ions separate from the oxygen ions. These ions dissolve in the solution.

Question

How do elements join together to make chemical compounds?

Procedure

1. Read and complete a lab safety form.
2. Pour 50 mL of white vinegar into a 250-mL beaker. Using a plastic spoon, add a spoonful of table salt to the vinegar. Stir the mixture with the spoon until the salt dissolves.
3. Add 20 dull pennies to the vinegar-salt solution. Leave the pennies in the solution for 10 minutes. Use a stopwatch or a clock with a second hand to measure the time.
4. After 10 minutes, use the plastic spoon to remove the pennies from the solution. Rinse the pennies in tap water. Place them on paper towels to dry. Record the change to the pennies in your Science Journal.

Predict

5. If you place an iron nail in the vinegar-salt solution, predict what changes will occur to the nail.

Test Your Prediction

6. Use sandpaper to clean two nails. Place one nail in the vinegar-salt solution, and place the other nail on a clean paper towel. You will compare the dry nail to the one in the solution and observe changes as they occur.
7. Every 5 minutes observe the nail in the solution and record your observations in your Science Journal. Remember to use the dry nail to help detect changes in the wet nail. Use a stopwatch or a clock with a second hand to measure the time. Keep the nail in the solution for 25 minutes
8. After 25 minutes, use a plastic spoon to remove the nail from the solution. Dispose of all materials as directed by your teacher.

Analyze and Conclude

9. **Compare and Contrast** What changes occurred when you placed the dull pennies in the vinegar-salt solution?
10. **Recognize Cause and Effect** What changes occurred to the nail in the leftover solution? Infer why these changes occurred.
11. **The Big Idea** Give two examples of how elements chemically combine and form compounds in this lab.

Communicate Your Results

Create a chart suitable for display summarizing this lab and your results.

Extension

The Statue of Liberty is made of copper. Research why the statue is green.

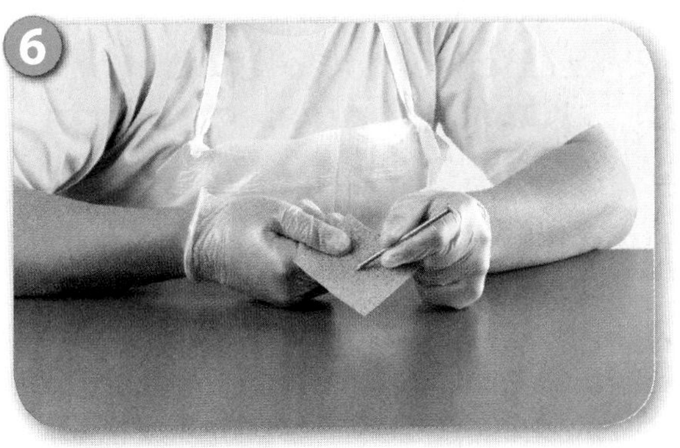

Lab Tips

- ☑ Be sure the pennies are separated when they are in the vinegar-salt solution. You may need to stir them with the plastic spoon.
- ☑ Use the plastic spoon to bring the nail out of the solution when checking for changes.

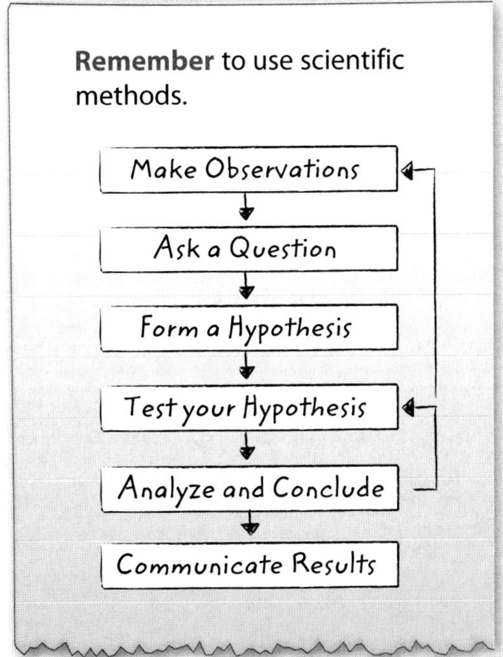

Chapter 11 Study Guide

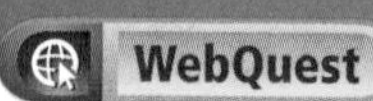

Elements can join together by sharing, transferring, or pooling electrons to make chemical compounds.

Key Concepts Summary

Lesson 1: Electrons and Energy Levels

- Electrons with more energy are farther from the atom's nucleus and are in a higher energy level.
- Atoms with fewer than eight **valence electrons** gain, lose, or share valence electrons and form stable compounds. Atoms in stable compounds have the same electron arrangement as a noble gas.

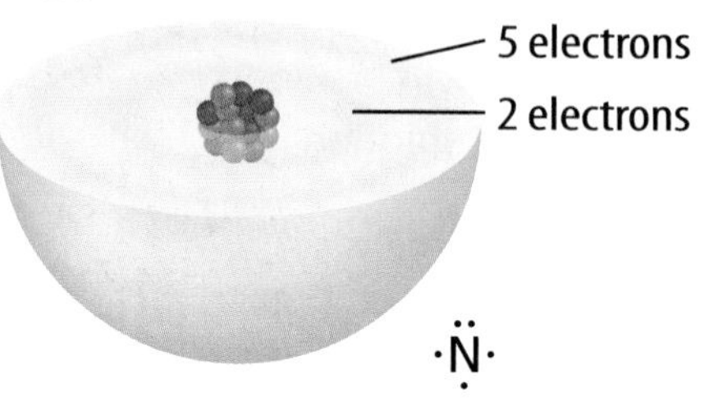

Vocabulary

chemical bond p. 382
valence electron p. 384
electron dot diagram p. 385

Lesson 2: Compounds, Chemical Formulas, and Covalent Bonds

- A compound and the elements it is made from have different chemical and physical properties.
- A **covalent bond** forms when two nonmetal atoms share valence electrons. Common properties of covalent compounds include low melting points and low boiling points. They are usually gas or liquid at room temperature and poor conductors of electricity.
- Water is a polar compound because the oxygen atom pulls more strongly on the shared valence electrons than the hydrogen atoms do.

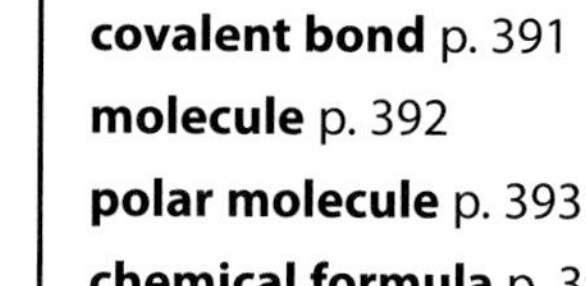

Vocabulary

covalent bond p. 391
molecule p. 392
polar molecule p. 393
chemical formula p. 394

Lesson 3: Ionic and Metallic Bonds

- **Ionic bonds** form when valence electrons move from a metal atom to a nonmetal atom.
- An ionic compound is held together by ionic bonds, which are attractions between positively and negatively charged **ions.**
- A **metallic bond** forms when valence electrons are pooled among many metal atoms.

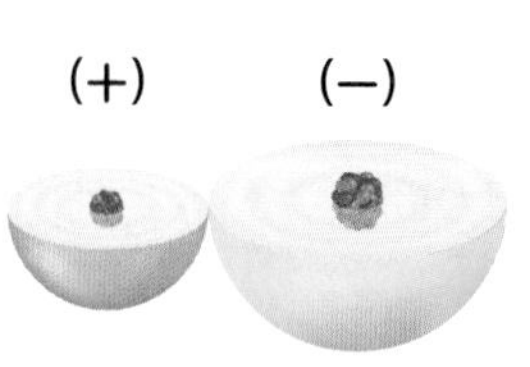

Vocabulary

ion p. 398
ionic bond p. 400
metallic bond p. 401

- Personal Tutor
- Vocabulary eGames
- Vocabulary eFlashcards

FOLDABLES Chapter Project

Assemble your lesson Foldables as shown to make a Chapter Project. Use the project to review what you have learned in this chapter.

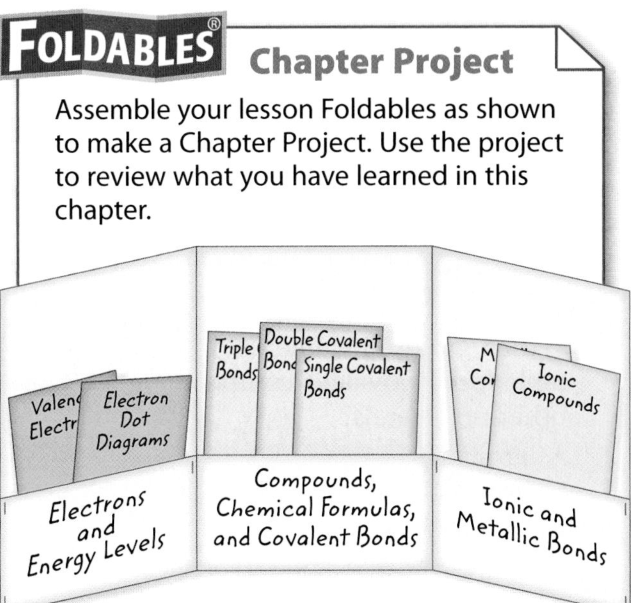

Use Vocabulary

1. The force that holds atoms together is called a(n) ________.
2. You can predict the number of bonds an atom can form by drawing its ________.
3. The nitrogen and hydrogen atoms that make up ammonia (NH_3) are held together by a(n) ________ because the atoms share valence electrons unequally.
4. Two hydrogen atoms and one oxygen atom together are a ________ of water.
5. A positively charged sodium ion and a negatively charged chlorine ion are joined by a(n) ________ to form the compound sodium chloride.

Link Vocabulary and Key Concepts

Copy this concept map, and then use vocabulary terms from the previous page and other terms from the chapter to complete the concept map.

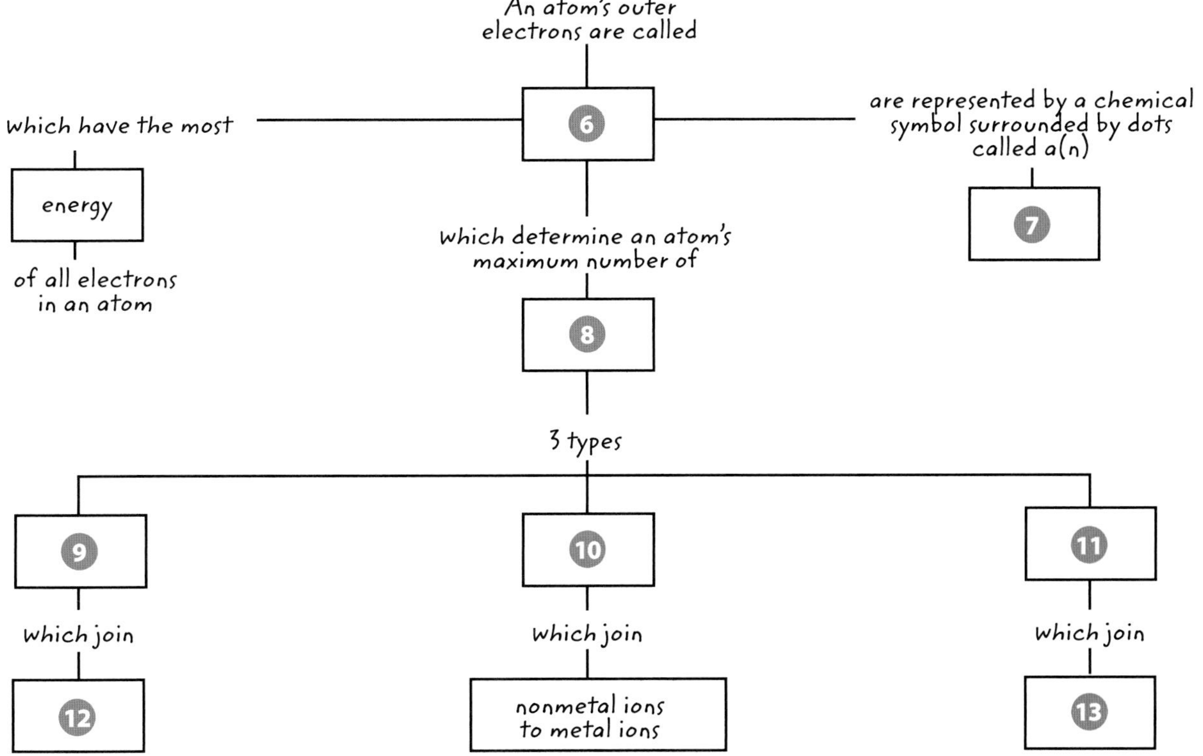

Chapter 11 Review

Understand Key Concepts

1. Atoms lose, gain, or share electrons and become as chemically stable as
   A. an electron.
   B. an ion.
   C. a metal.
   D. a noble gas.

2. Which is the correct electron dot diagram for boron, one of the group 13 elements?
   A. $\overset{\bullet}{\underset{\bullet}{\mathrm{B}}}\bullet$
   B. $\bullet\overset{\bullet\bullet}{\underset{\bullet\bullet}{\mathrm{B}}}:$
   C. $:\overset{\bullet\bullet}{\underset{\bullet\bullet}{\mathrm{B}}}:$
   D. $\bullet\overset{\bullet\bullet}{\underset{\bullet}{\mathrm{B}}}\bullet$

3. If an electron transfers from one atom to another atom, what type of bond will most likely form?
   A. covalent
   B. ionic
   C. metallic
   D. polar

4. What change would make an atom represented by this diagram have the same electron arrangement as a noble gas?

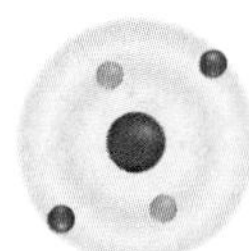

   A. gaining two electrons
   B. gaining four electrons
   C. losing two electrons
   D. losing four electrons

5. What would make bromine, a group 17 element, more similar to a noble gas?
   A. gaining one electron
   B. gaining two electrons
   C. losing one electron
   D. losing two electrons

6. Which would most likely be joined by an ionic bond?
   A. a positive metal ion and a positive nonmetal ion
   B. a positive metal ion and a negative nonmetal ion
   C. a negative metal ion and a positive nonmetal ion
   D. a negative metal ion and a negative nonmetal ion

7. Which group of elements on the periodic table forms covalent compounds with other nonmetals?
   A. group 1
   B. group 2
   C. group 17
   D. group 18

8. Which best describes an atom represented by this diagram?

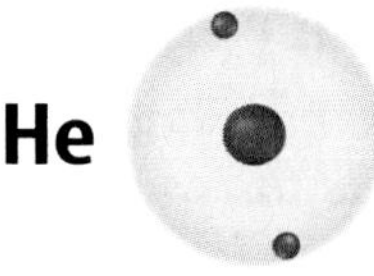

   A. It is likely to bond by gaining six electrons.
   B. It is likely to bond by losing two electrons.
   C. It is not likely to bond because it is already stable.
   D. It is not likely to bond because it has too few electrons.

9. How many dots would a dot diagram for selenium, one of the group 16 elements, have?
   A. 6
   B. 8
   C. 10
   D. 16

Assessment

Online Test Practice

Critical Thinking

10 **Classify** Use the periodic table to classify the elements potassium (K), bromine (Br), and argon (Ar) according to how likely their atoms are to do the following.

a. lose electrons to form positive ions
b. gain electrons to form negative ions
c. neither gain nor lose electrons

11 **Describe** the change that is shown in this illustration. How does this change affect the stability of the atom?

$$\cdot\ddot{\text{N}}\cdot \longrightarrow :\ddot{\text{N}}:^{3-}$$

12 **Analyze** One of your classmates draws an electron dot diagram for a helium atom with two dots. He tells you that these dots mean each helium atom has two unpaired electrons and can gain, lose, or share electrons to have four pairs of valence electrons and become stable. What is wrong with your classmate's argument?

13 **Explain** why the hydrogen atoms in a hydrogen gas molecule (H_2) form nonpolar covalent bonds but the oxygen and hydrogen atoms in water molecules (H_2O) form polar covalent bonds.

14 **Contrast** Why is it possible for an oxygen atom to form a double covalent bond, but it is not possible for a chlorine atom to form a double covalent bond?

Writing in Science

15 **Compose** a poem at least ten lines long that explains ionic bonding, covalent bonding, and metallic bonding.

REVIEW THE BIG IDEA

16 Which types of atoms pool their valence electrons to form a "sea of electrons"?

17 Describe a way in which elements joining together to form chemical compounds is similar to the way the letters on a computer keyboard join together to form words.

Math Skills

Review

Math Practice

Element	Atomic Radius	Ionic Radius
Potassium (K)	227 pm	133 pm
Iodine (I)	133 pm	216 pm

18 What is the percent change when an iodine atom (I) becomes an ion (I^-)?

19 What is the percent change when a potassium atom (K) becomes an ion (K^+)?

Standardized Test Practice

Record your answers on the answer sheet provided by your teacher or on a sheet of paper.

Multiple Choice

1 Which information does the chemical formula CO_2 NOT give you?

A number of valence electrons in each atom

B ratio of atoms in the compound

C total number of atoms in one molecule of the compound

D type of elements in the compound

Use the diagram below to answer question 2.

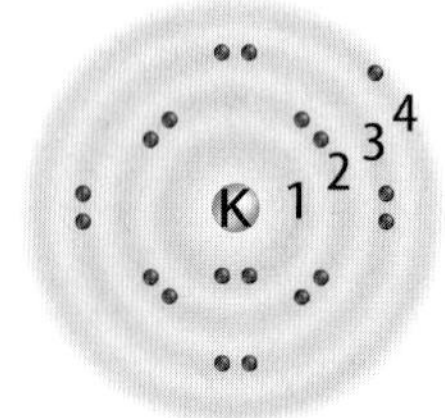

2 The diagram above shows a potassium atom. Which is the second-highest energy level?

A 1

B 2

C 3

D 4

3 What is shared in a metallic bond?

A negatively charged ions

B neutrons

C pooled valence electrons

D protons

4 Which is a characteristic of most nonpolar compounds?

A conduct electricity poorly

B dissolve easily in water

C solid crystals

D shiny surfaces

Use the diagram below to answer question 5.

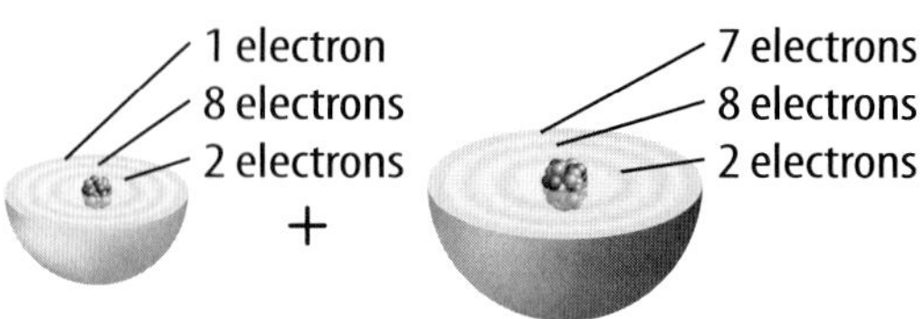

5 The atoms in the diagram above are forming a bond. Which represents that bond?

A

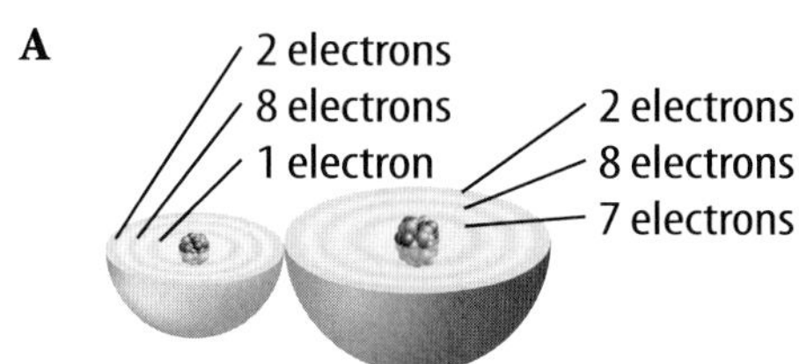

B

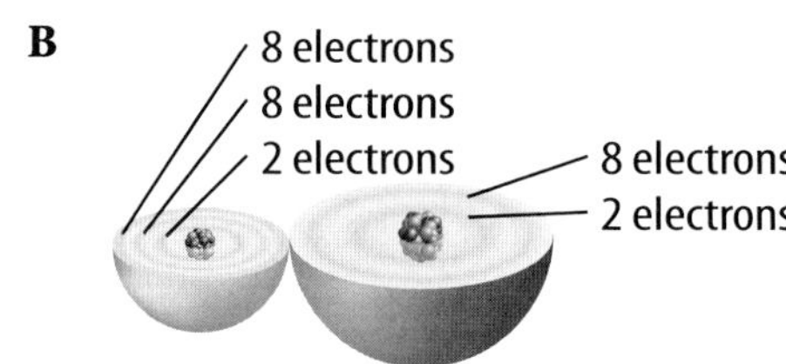

C

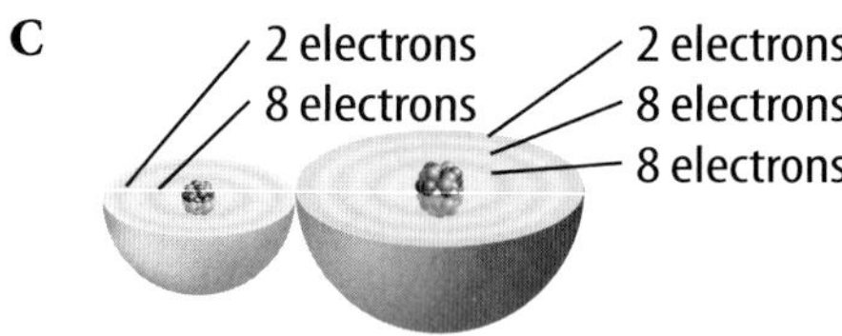

D

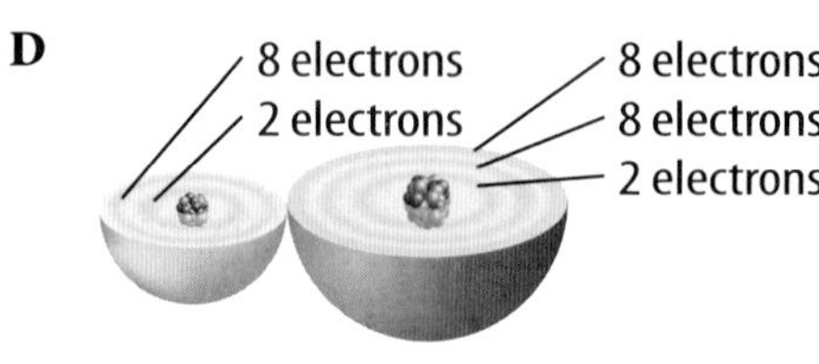

6 Covalent bonds typically form between the atoms of elements that share

A nuclei.

B oppositely charged ions.

C protons.

D valence electrons.

Standardized Test Practice

Assessment

Online Standardized Test Practice

Use the diagram below to answer question 7.

Water Molecule

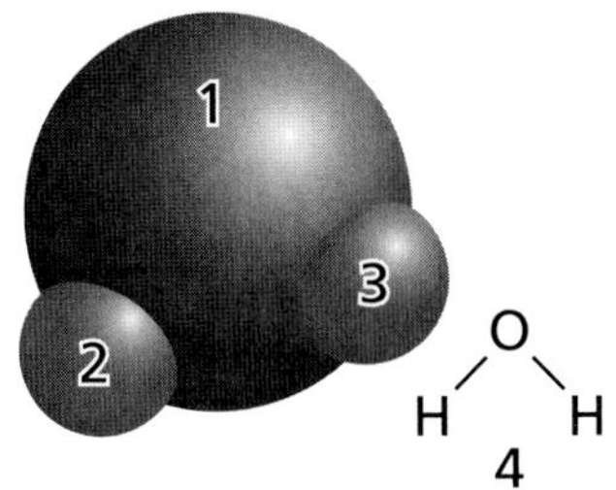

7 In the diagram above, which shows an atom with a partial negative charge?

A 1

B 2

C 3

D 4

8 Which compound is formed by the attraction between negatively and positively charged ions?

A bipolar

B covalent

C ionic

D nonpolar

9 The atoms of noble gases do NOT bond easily with other atoms because their valence electrons are

A absent.

B moving.

C neutral.

D stable.

Constructed Response

Use the table below to answer question 10.

Property	Rust	Iron	Oxygen
Color			Clear
Solid, liquid, or gas			
Strength		Strong	Does NOT apply
Usefulness			

10 Rust is a compound of iron and oxygen. Compare the properties of rust, iron, and oxygen by filling in the missing cells in the table above. What can you conclude about the properties of compounds and their elements?

Use the diagram below to answer questions 11 and 12.

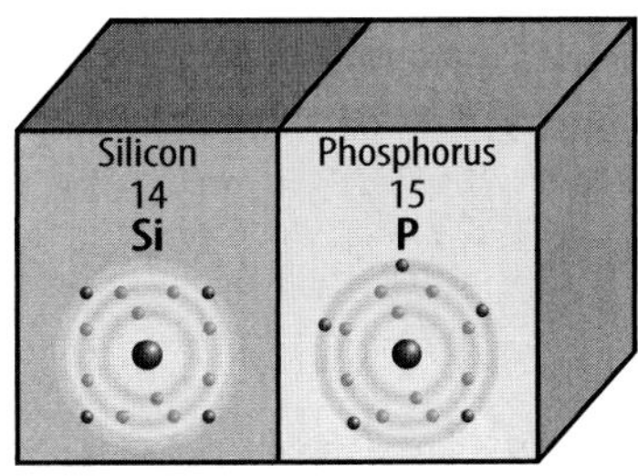

11 In the diagram, how are valence electrons illustrated? How many valence electrons does each element have?

12 Describe a stable electron configuration. For each element above, how many electrons are needed to make a stable electron configuration?

NEED EXTRA HELP?												
If You Missed Question...	1	2	3	4	5	6	7	8	9	10	11	12
Go to Lesson...	2	1	3	3	3	2	2	3	1	2	1	1

Unit 4

INTERACTIONS OF MATTER

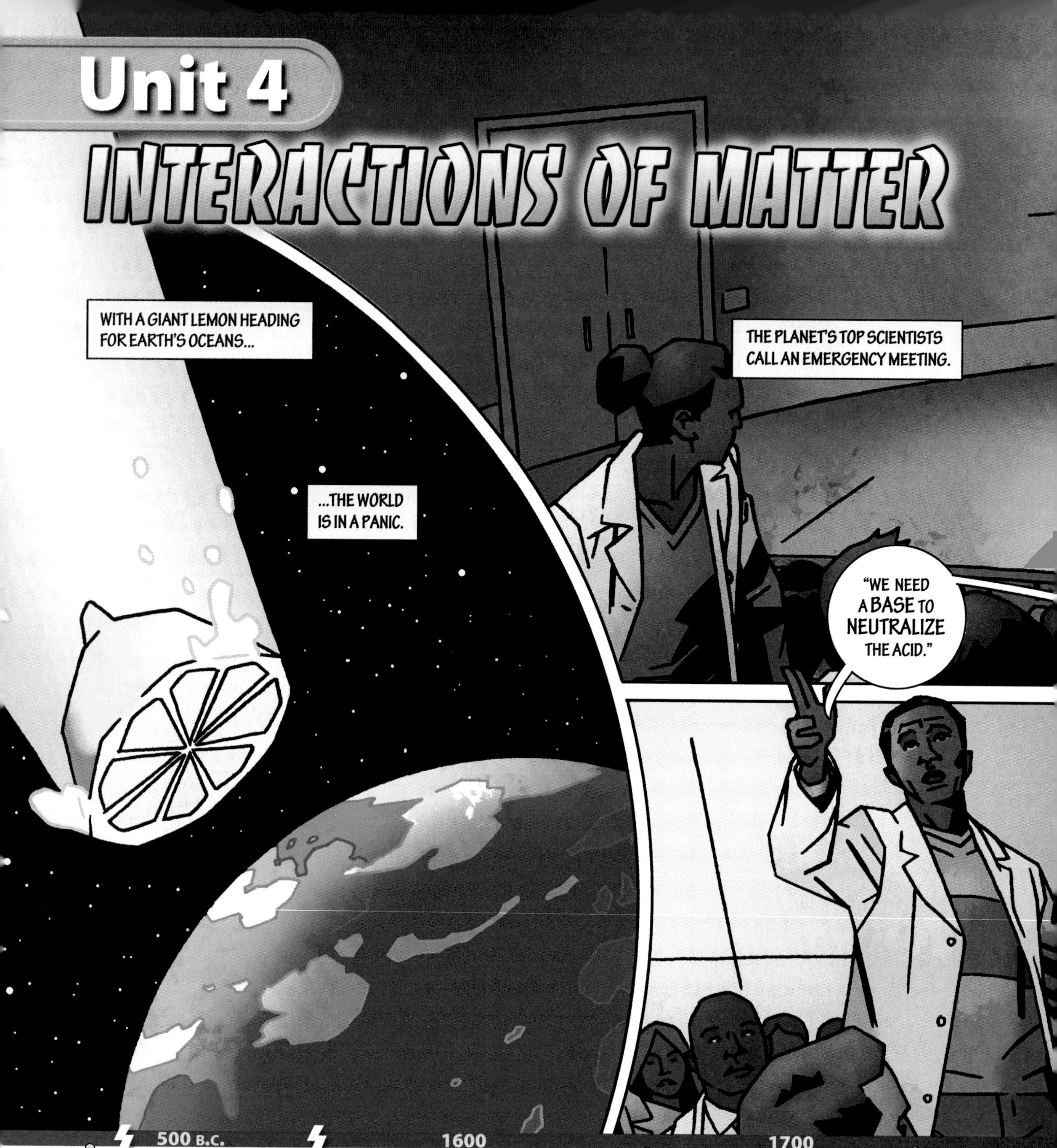

WITH A GIANT LEMON HEADING FOR EARTH'S OCEANS...

...THE WORLD IS IN A PANIC.

THE PLANET'S TOP SCIENTISTS CALL AN EMERGENCY MEETING.

"WE NEED A BASE TO NEUTRALIZE THE ACID."

500 B.C. 1600 1700

1000 B.C.
Chemistry is considered more of an art than a science. Chemical arts include the smelting of metals and the making of drugs, dyes, iron, and bronze.

1661
A clear distinction is made between chemistry and alchemy when *The Sceptical Chymist* is published by Robert Boyle. Modern chemistry begins to emerge.

1789
Antoine Lavoisier, the "father of modern chemistry," clearly outlines the law of conservation of mass.

1803
John Dalton publishes his atomic theory, which states that all matter is composed of atoms, which are small and indivisible and can join together to form chemical compounds. Dalton is considered the originator of modern atomic theory.

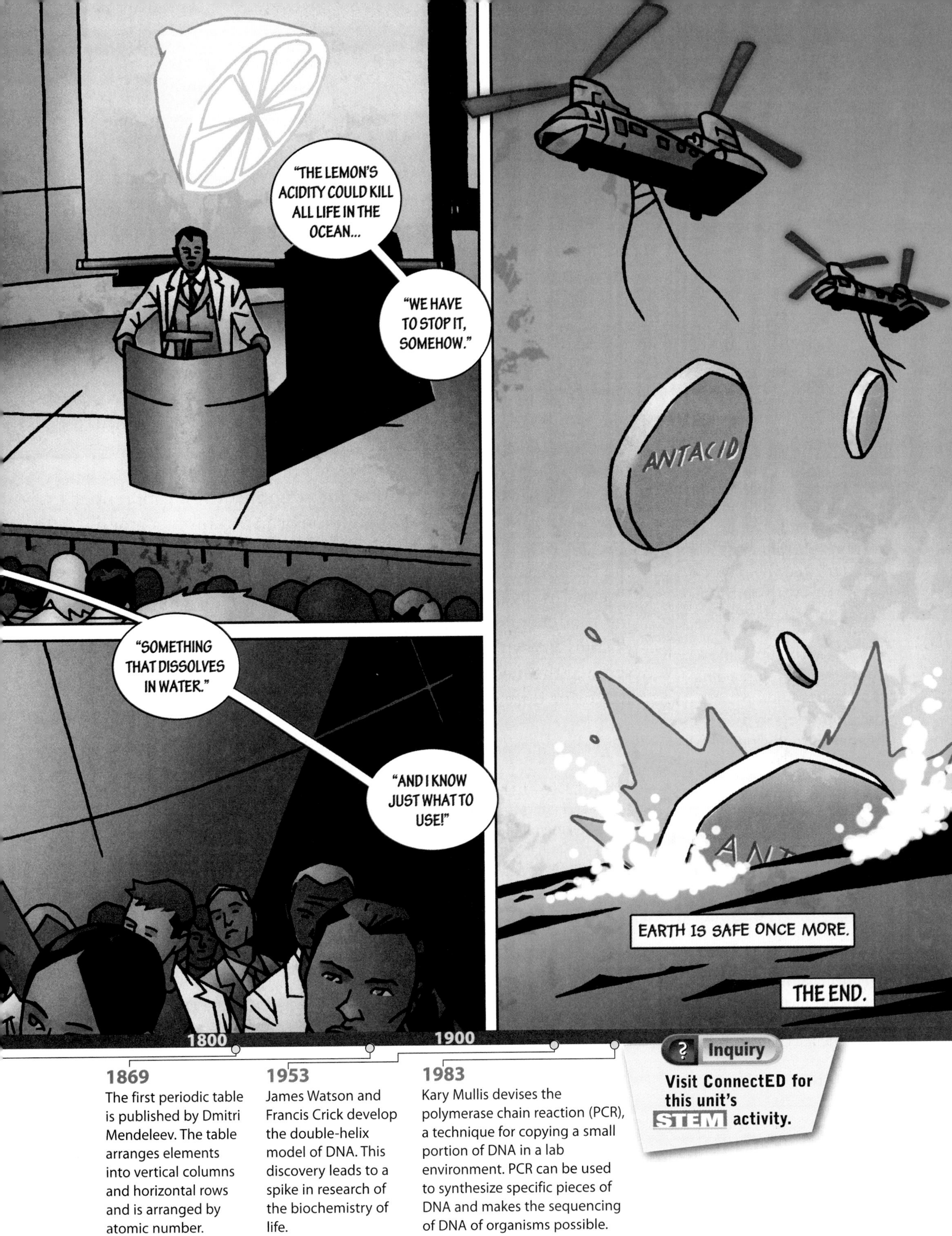

1869
The first periodic table is published by Dmitri Mendeleev. The table arranges elements into vertical columns and horizontal rows and is arranged by atomic number.

1953
James Watson and Francis Crick develop the double-helix model of DNA. This discovery leads to a spike in research of the biochemistry of life.

1983
Kary Mullis devises the polymerase chain reaction (PCR), a technique for copying a small portion of DNA in a lab environment. PCR can be used to synthesize specific pieces of DNA and makes the sequencing of DNA of organisms possible.

Inquiry
Visit ConnectED for this unit's STEM activity.

Health and Science

Have an upset stomach? Chew on some charcoal. Have a headache? Rub a little peppermint oil on your temples. As shown in **Figure 1**, people have used chemicals to fix physical ailments for thousands of years, long before the development of the first medicines. Many cures were discovered by accident. People did not understand why the cures worked, only that they did work.

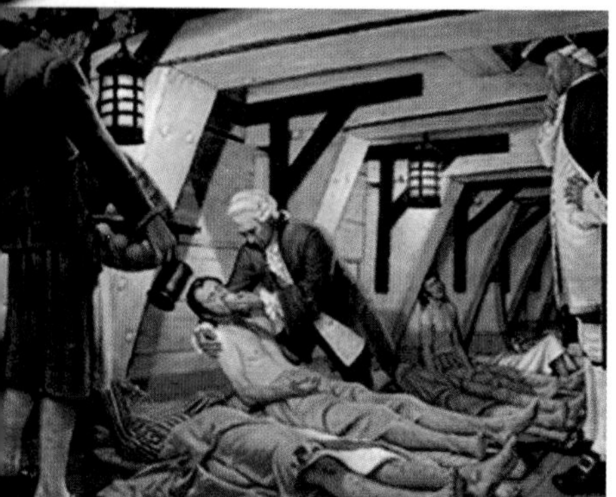

▲ **Figure 1** Thousands of years ago, people believed that evil spirits were responsible for illness. They often treated the physical symptoms with herbs or other natural materials.

Asking Questions About Health

Over time, people asked questions about which cures worked and which cures did not work. They made observations, recorded their findings, and had discussions about cures with other people. This process was the start of the scientific investigation of health. **Health** is the overall condition of an organism or group of organisms at any given time. Early studies focused on treating the physical parts of the body. The study of how chemicals interact in organisms did not come until much later. Recognizing that chemicals can affect health opened a whole new field of study known as biochemistry. The time line in **Figure 2** shows some of the medical and chemical discoveries people made that led to the development of medicines that save lives.

Figure 2 The time line shows several significant discoveries and developments in the history of medicine. ▼

4,200 years ago Clay tablets describe using sesame oil on wounds to treat infection.

3,500 years ago An ancient papyrus described how Egyptians applied moldy bread to wounds to prevent infection.

More than 3,300 years later, scientists found that a chemical in mold broke down the cell membranes of bacteria, killing them. Similar discoveries led to the development of antibiotics.

2,500 years ago Hippocrates, known as the "Father of Medicine," is the first physician known to separate medical knowledge from myth and superstition.

Year 900 The first pharmacy opened in Persia, which is now Iraq.

1740s A doctor found that the disease called scurvy was caused by a lack of Vitamin C.

Early explorers on long sea voyages often lost their teeth or developed deadly sores. Ships could not carry many fruits and vegetables, which contain Vitamin C, because they spoil quickly. Scientists suspect that many early explorers might have died because their diets did not include the proper vitamins.

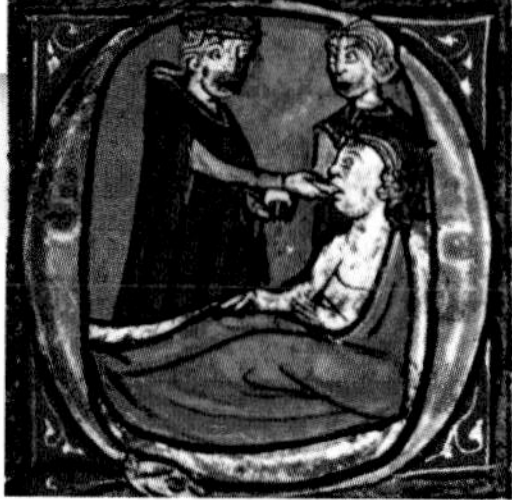

Benefits and Risks of Medicines

Scientists might recognize that a person's body is missing a necessary chemical, but that does not mean they can always fix the problem. For example, people used to get necessary vitamins and minerals by eating natural, whole foods. Today, food processing destroys many nutrients. Foods last longer, but they do not provide all the nutrients the body needs.

Researchers still do not understand the role of many chemicals in the body. Taking a medicine to fix one problem sometimes causes others, called side effects. Some side effects can be worse than the original problem. For example, antibiotics kill some disease-causing bacteria. However, widespread use of antibiotics has resulted in "super bugs"—bacteria that are resistant to treatment.

Histamines are chemicals that have many functions in the body, including regulating sleep and decreasing sensitivity to allergens. However, low levels of histamines have been linked to some serious illnesses. Many medicines have long-term effects on health. Before you take a medicine, you should recognize that you are adding a chemical to your body. You should be as informed as possible about any possible side effects.

Inquiry MiniLab

15 minutes

Is everyone's chemistry the same?

Each person's body is a unique "chemical factory." Why might using the same medicine to treat illness not work exactly the same way in everyone?

1. Read and complete a lab safety form.
2. Place a strip of **pH paper** on your tongue. Immediately place the paper in a **self-sealing plastic bag.**
3. Compare the color of your paper to the **color guide.** Record the pH in your Science Journal.
4. Record your pH on a class chart for comparison.

Analyze and Conclude

1. **Organize Data** What was the range of pH values among your classmates?
2. **Predict** How might differences in pH affect how well a medicine works in different people?

1770s The first vaccination is developed and administered.

1800s Nitrous oxide is first used as an anesthetic by dentists.

1920s Insulin is identified as the missing hormone in people with diabetes.

Scientists studying digestion in dogs noticed that ants were attracted to the urine of a dog whose pancreas had been removed. They determined the dog's urine contained sugar, which attracted ants. Eventually, scientists discovered that diabetes resulted from a lack of insulin, a chemical produced in the pancreas that regulates blood sugar. Today, some people with diabetes wear an insulin pump that monitors their blood sugar and delivers insulin to their bodies.

1920s Penicillin is discovered, but not developed for treatment of disease until the mid-1940s.

2000s First vaccine to target a cause of cancer

Chapter 12

Chemical Reactions and Equations

THE BIG IDEA

What happens to atoms and energy during a chemical reaction?

Inquiry How does it work?

An air bag deploys in less than the blink of an eye. How does the bag open so fast? At the moment of impact, a sensor triggers a chemical reaction between two chemicals. This reaction quickly produces a large amount of nitrogen gas. This gas inflates the bag with a pop.

- A chemical reaction can produce a gas. How is this different from a gas produced when a liquid boils?
- Where do you think the nitrogen gas that is in an air bag comes from? Do you think any of the chemicals in the air bag contain the element nitrogen?
- What do you think happens to atoms and energy during a chemical reaction?

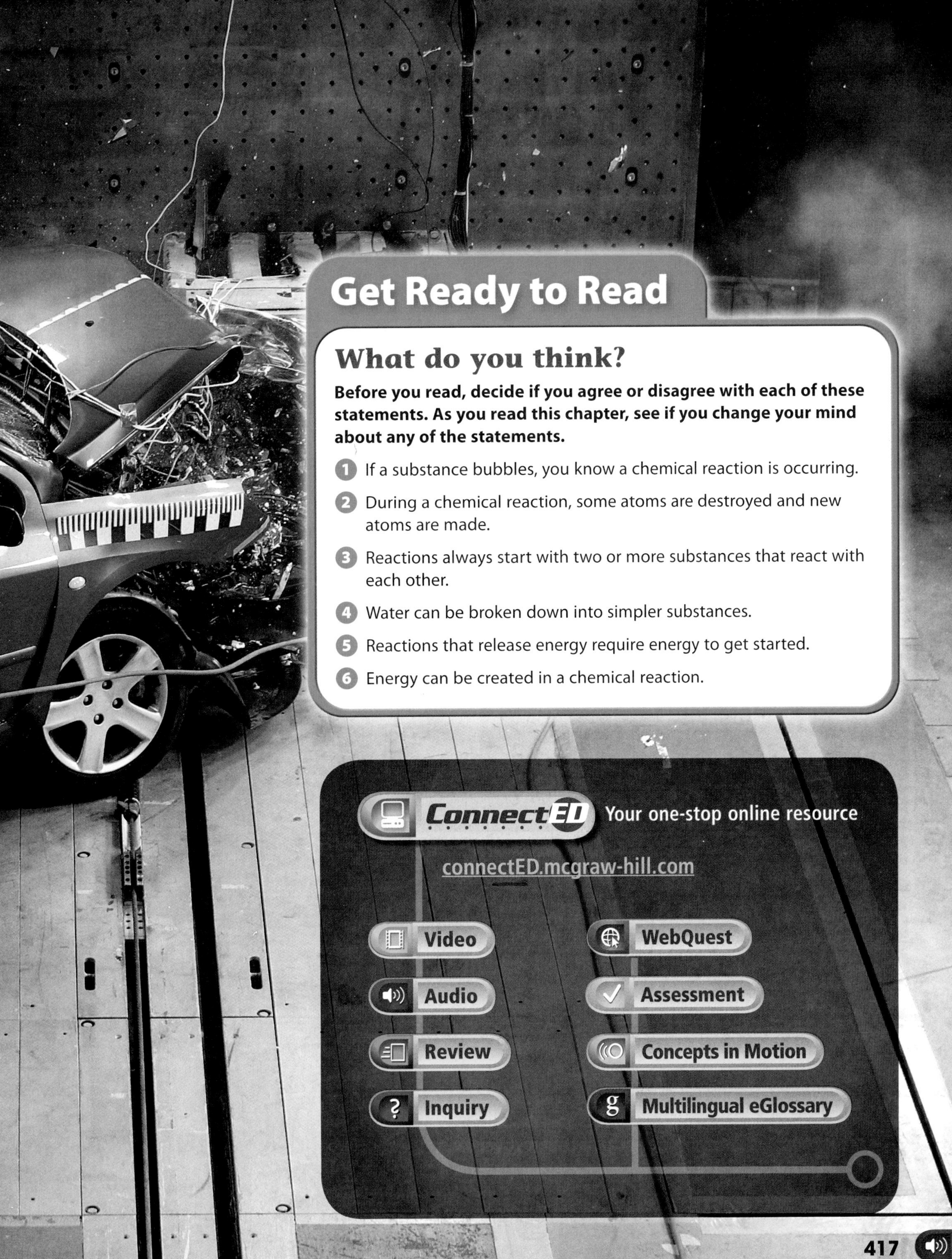

Get Ready to Read

What do you think?

Before you read, decide if you agree or disagree with each of these statements. As you read this chapter, see if you change your mind about any of the statements.

1. If a substance bubbles, you know a chemical reaction is occurring.
2. During a chemical reaction, some atoms are destroyed and new atoms are made.
3. Reactions always start with two or more substances that react with each other.
4. Water can be broken down into simpler substances.
5. Reactions that release energy require energy to get started.
6. Energy can be created in a chemical reaction.

ConnectED Your one-stop online resource

connectED.mcgraw-hill.com

- Video
- Audio
- Review
- Inquiry
- WebQuest
- Assessment
- Concepts in Motion
- Multilingual eGlossary

Lesson 1

Reading Guide

Key Concepts
ESSENTIAL QUESTIONS

- What are some signs that a chemical reaction might have occurred?
- What happens to atoms during a chemical reaction?
- What happens to the total mass in a chemical reaction?

Vocabulary

chemical reaction p. 419
chemical equation p. 422
reactant p. 423
product p. 423
law of conservation of mass p. 424
coefficient p. 426

g Multilingual eGlossary

Understanding Chemical Reactions

Inquiry Does it run on batteries?

Flashes of light from fireflies dot summer evening skies in many parts of the United States. But, firefly light doesn't come from batteries. Fireflies make light using a process called bioluminescence (bi oh lew muh NE cents). In this process, chemicals in the firefly's body combine in a two-step process and make new chemicals and light.

Launch Lab

15–20 minutes

Where did it come from?

Does a boiled egg have more mass than a raw egg? What happens when liquids change to a solid?

1. Read and complete a lab safety form.
2. Use a **graduated cylinder** to add 25 mL of **solution A** to a **self-sealing plastic bag.** Place a **stoppered test tube** containing **solution B** into the bag. Be careful not to dislodge the stopper.
3. Seal the bag completely, and wipe off any moisture on the outside with a **paper towel.** Place the bag on the **balance.** Record the total mass in your Science Journal.
4. Without opening the bag, remove the stopper from the test tube and allow the liquids to mix. Observe and record what happens.
5. Place the sealed bag and its contents back on the balance. Read and record the mass.

Think About This

1. What did you observe when the liquids mixed? How would you account for this observation?
2. Did the mass of the bag's contents change? If so, could the change have been due to the precision of the balance, or did the matter in the bag change its mass? Explain.
3. **Key Concept** Do you think matter was gained or lost in the bag? How can you tell?

Changes in Matter

When you put liquid water in a freezer, it changes to solid water, or ice. When you pour brownie batter into a pan and bake it, the liquid batter changes to a solid, too. In both cases, a liquid changes to a solid. Are these changes the same?

Physical Changes

Recall that matter can undergo two types of changes—chemical or physical. A physical change does not produce new substances. The substances that exist before and after the change are the same, although they might have different physical properties. This is what happens when liquid water freezes. Its physical properties change from a liquid to a solid, but the water, H_2O, does not change into a different substance. Water molecules are always made up of two hydrogen atoms bonded to one oxygen atom regardless of whether they are solid, liquid, or gas.

Chemical Changes

Recall that during a chemical change, one or more substances change into new substances. The starting substances and the substances produced have different physical and chemical properties. For example, when brownie batter bakes, a chemical change occurs. Many of the substances in the baked brownies are different from the substances in the batter. As a result, baked brownies have physical and chemical properties that are different from those of brownie batter.

A chemical change also is called a chemical reaction. These terms mean the same thing. *A* **chemical reaction** *is a process in which atoms of one or more substances rearrange to form one or more new substances.* In this lesson, you will read what happens to atoms during a reaction and how these changes can be described using equations.

Reading Check What types of properties change during a chemical reaction?

Signs of a Chemical Reaction

How can you tell if a chemical reaction has taken place? You have read that the substances before and after a reaction have different properties. You might think that you could look for changes in properties as a sign that a reaction occurred. In fact, changes in the physical properties of color, state of matter, and odor are all signs that a chemical reaction might have occurred. Another sign of a chemical reaction is a change in energy. If substances get warmer or cooler or if they give off light or sound, it is likely that a reaction has occurred. Some signs that a chemical reaction might have occurred are shown in **Figure 1.**

However, these signs are not proof of a chemical change. For example, bubbles appear when water boils. But, bubbles also appear when baking soda and vinegar react and form carbon dioxide gas. How can you be sure that a chemical reaction has taken place? The only way to know is to study the chemical properties of the substances before and after the change. If they have different chemical properties, then the substances have undergone a chemical reaction.

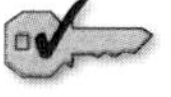

Key Concept Check What are some signs that a chemical reaction might have occurred?

Figure 1 You can detect a chemical reaction by looking for changes in properties and changes in energy of the substances that reacted.

Change in Properties	
Change in color Bright copper changes to green when the copper reacts with certain gases in the air. 	**Formation of bubbles** Bubbles of carbon dioxide form when baking soda is added to vinegar. 
Change in odor When food burns or rots, a change in odor is a sign of chemical change. 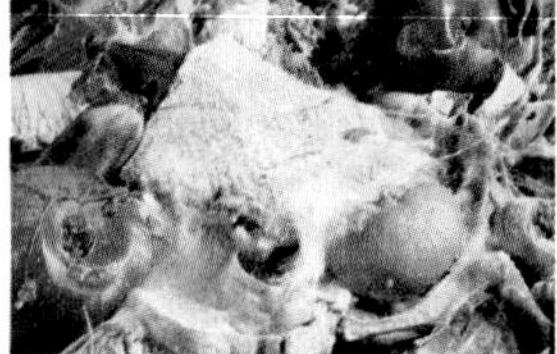	**Formation of a precipitate** A precipitate is a solid formed when two liquids react. 
Change in Energy	
Warming or cooling Thermal energy is either given off or absorbed during a chemical change. 	**Release of light** A firefly gives off light as the result of a chemical change. 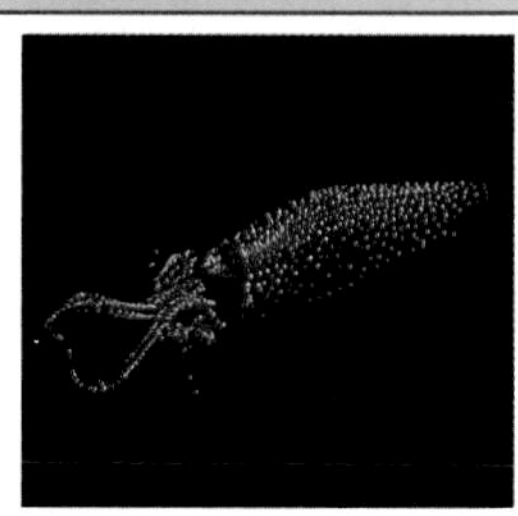

What happens in a chemical reaction?

During a chemical reaction, one or more substances react and form one or more new substances. How are these new substances formed?

Atoms Rearrange and Form New Substances

To understand what happens in a reaction, first review substances. Recall that there are two types of substances—elements and compounds. Substances have a fixed arrangement of atoms. For example, in a single drop of water, there are trillions of oxygen and hydrogen atoms. However, all of these atoms are arranged in the same way—two atoms of hydrogen are bonded to one atom of oxygen. If this arrangement changes, the substance is no longer water. Instead, a different substance forms with different physical and chemical properties. This is what happens during a chemical reaction. Atoms of elements or compounds rearrange and form different elements or compounds.

Bonds Break and Bonds Form

How does the rearrangement of atoms happen? Atoms rearrange when **chemical bonds** between atoms break. Recall that constantly moving particles make up all substances, including solids. As particles move, they collide with one another. If the particles collide with enough energy, the bonds between atoms can break. The atoms separate, rearrange, and new bonds can form. The reaction that forms hydrogen and oxygen from water is shown in **Figure 2.** Adding electric energy to water molecules can cause this reaction. The added energy causes bonds between the hydrogen atoms and the oxygen atoms to break. After the bonds between the atoms in water molecules break, new bonds can form between pairs of hydrogen atoms and between pairs of oxygen atoms.

REVIEW VOCABULARY

chemical bond
an attraction between atoms when electrons are shared, transferred, or pooled

 **Key Concept Check** What happens to atoms during a chemical reaction?

Figure 2 Notice that no new atoms are created in a chemical reaction. The existing atoms rearrange and form new substances.

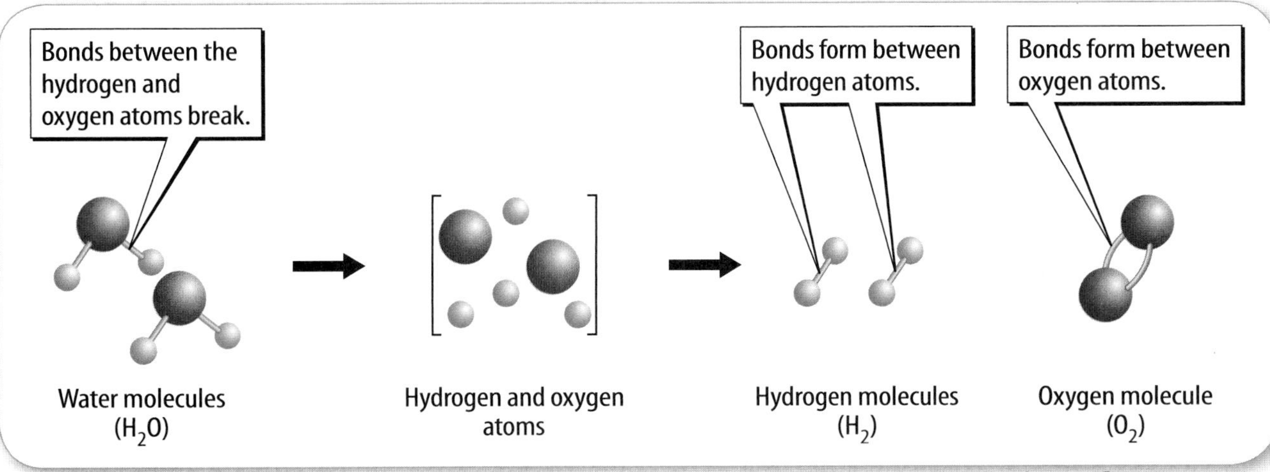
Bonds between the hydrogen and oxygen atoms break.
Bonds form between hydrogen atoms.
Bonds form between oxygen atoms.
Water molecules (H_2O)
Hydrogen and oxygen atoms
Hydrogen molecules (H_2)
Oxygen molecule (O_2)

Table 1 Symbols and Formulas of Some Elements and Compounds

Substance	Formula	# of atoms
Carbon	C	C: 1
Copper	Cu	Cu: 1
Cobalt	Co	Co: 1
Oxygen	O_2	O: 2
Hydrogen	H_2	H: 2
Chlorine	Cl_2	Cl: 2
Carbon dioxide	CO_2	C: 1 O: 2
Carbon monoxide	CO	C: 1 O: 1
Water	H_2O	H: 2 O: 1
Hydrogen peroxide	H_2O_2	H: 2 O: 2
Glucose	$C_6H_{12}O_6$	C: 6 H: 12 O: 6
Sodium chloride	NaCl	Na: 1 Cl: 1
Magnesium hydroxide	$Mg(OH)_2$	Mg: 1 O: 2 H: 2

Table 1 Symbols and subscripts describe the type and number of atoms in an element or a compound.

Visual Check Describe the number of atoms in each element in the following: C, Co, CO, and CO_2.

Interactive Table

Chemical Equations

Suppose your teacher asks you to produce a specific reaction in your science laboratory. How might your teacher describe the reaction to you? He or she might say something such as "react baking soda and vinegar to form sodium acetate, water, and carbon dioxide." It is more likely that your teacher will describe the reaction in the form of a chemical equation. *A* **chemical equation** *is a description of a reaction using element symbols and chemical formulas.* Element symbols represent elements. Chemical formulas represent compounds.

Element Symbols

Recall that symbols of elements are shown in the periodic table. For example, the symbol for carbon is C. The symbol for copper is Cu. Each element can exist as just one atom. However, some elements exist in nature as diatomic molecules—two atoms of the same element bonded together. A formula for one of these diatomic elements includes the element's symbol and the subscript *2*. A subscript describes the number of atoms of an element in a compound. Oxygen (O_2) and hydrogen (H_2) are examples of diatomic molecules. Some element symbols are shown above the blue line in **Table 1.**

Chemical Formulas

When atoms of two or more different elements bond, they form a compound. Recall that a chemical formula uses elements' symbols and subscripts to describe the number of atoms in a compound. If an element's symbol does not have a subscript, the compound contains only one atom of that element. For example, carbon dioxide (CO_2) is made up of one carbon atom and two oxygen atoms. Remember that two different formulas, no matter how similar, represent different substances. Some chemical formulas are shown below the blue line in **Table 1.**

Writing Chemical Equations

A chemical equation includes both the substances that react and the substances that are formed in a chemical reaction. *The starting substances in a chemical reaction are* **reactants.** *The substances produced by the chemical reaction are* **products.** Figure 3 shows how a chemical equation is written. Chemical formulas are used to describe the reactants and the products. The reactants are written to the left of an arrow, and the products are written to the right of the arrow. Two or more reactants or products are separated by a plus sign. The general structure for an equation is:

reactant + reactant → product + product

When writing chemical equations, it is important to use correct chemical formulas for the reactants and the products. For example, suppose a certain chemical reaction produces carbon dioxide and water. The product carbon dioxide would be written as CO_2 and not as CO. CO is the formula for carbon monoxide, which is not the same compound as CO_2. Water would be written as H_2O and not as H_2O_2, the formula for hydrogen peroxide.

Inquiry MiniLab — 10 minutes

How does an equation represent a reaction?

Sulfur dioxide (SO_2) and oxygen (O_2) react and form sulfur trioxide (SO_3). How does an equation represent the reaction?

1. Read and complete a lab safety form.
2. Use **yellow modeling clay** to model two atoms of sulfur. Use **red modeling clay** to model six atoms of oxygen.
3. Make two molecules of SO_2 with a sulfur atom in the middle of each molecule. Make one molecule of O_2. Sketch the models in your Science Journal.
4. Rearrange atoms to form two molecules of SO_3. Place a sulfur atom in the middle of each molecule. Sketch the models in your Science Journal.

Analyze and Conclude

1. **Identify** the reactants and the products in this chemical reaction.
2. **Write** a chemical equation for this reaction.
3. **Explain** What do the letters represent in the equation? The numbers?
4. **Key Concept** In terms of chemical bonds, what did you model by pulling molecules apart and building new ones?

Figure 3 An equation is read much like a sentence. This equation is read as "carbon plus oxygen produces carbon dioxide."

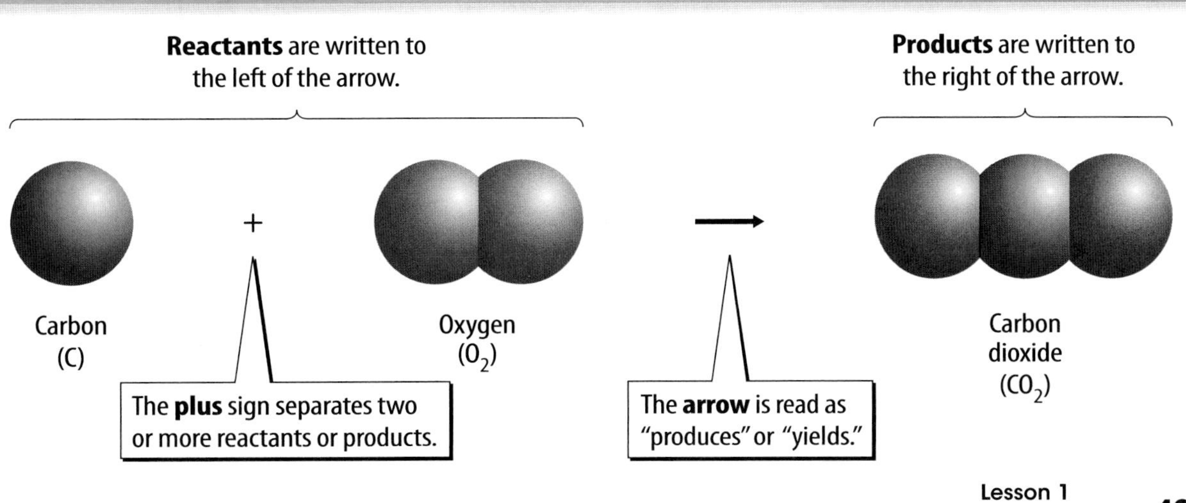
Parts of an Equation
Reactants are written to the left of the arrow.
Products are written to the right of the arrow.
+
Carbon (C)
Oxygen (O2)
Carbon dioxide (CO2)
The plus sign separates two or more reactants or products.
The arrow is read as "produces" or "yields."

WORD ORIGIN

product
from Latin *producere,* means "bring forth"

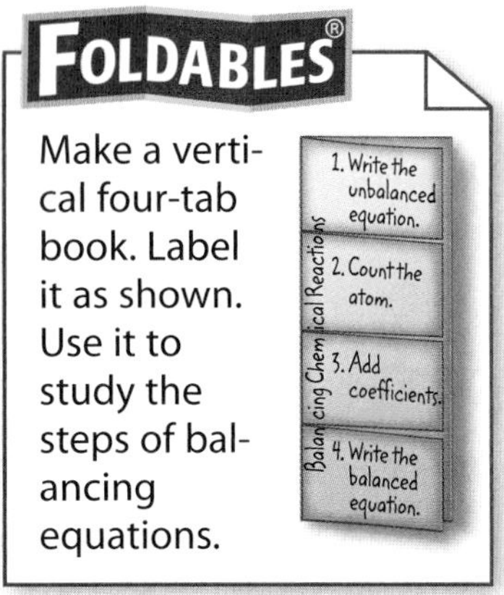
FOLDABLES
1. Write the unbalanced equation.
2. Count the atom.
3. Add coefficients.
4. Write the balanced equation.
Balancing Chemical Reactions

Make a vertical four-tab book. Label it as shown. Use it to study the steps of balancing equations.

Conservation of Mass

A French chemist named Antoine Lavoisier (AN twan • luh VWAH see ay) (1743–1794) discovered something interesting about chemical reactions. In a series of experiments, Lavoisier measured the masses of substances before and after a chemical reaction inside a closed container. He found that the total mass of the reactants always equaled the total mass of the **products.** Lavoisier's results led to the law of conservation of mass. *The* **law of conservation of mass** *states that the total mass of the reactants before a chemical reaction is the same as the total mass of the products after the chemical reaction.*

Atoms are conserved.

The discovery of atoms provided an explanation for Lavoisier's observations. Mass is conserved in a reaction because atoms are conserved. Recall that during a chemical reaction, bonds break and new bonds form. However, atoms are not destroyed, and no new atoms form. All atoms at the start of a chemical reaction are present at the end of the reaction. **Figure 4** shows that mass is conserved in the reaction between baking soda and vinegar.

Key Concept Check What happens to the total mass of the reactants in a chemical reaction?

Figure 4 As this reaction takes place, the mass on the balance remains the same, showing that mass is conserved.

Conservation of Mass

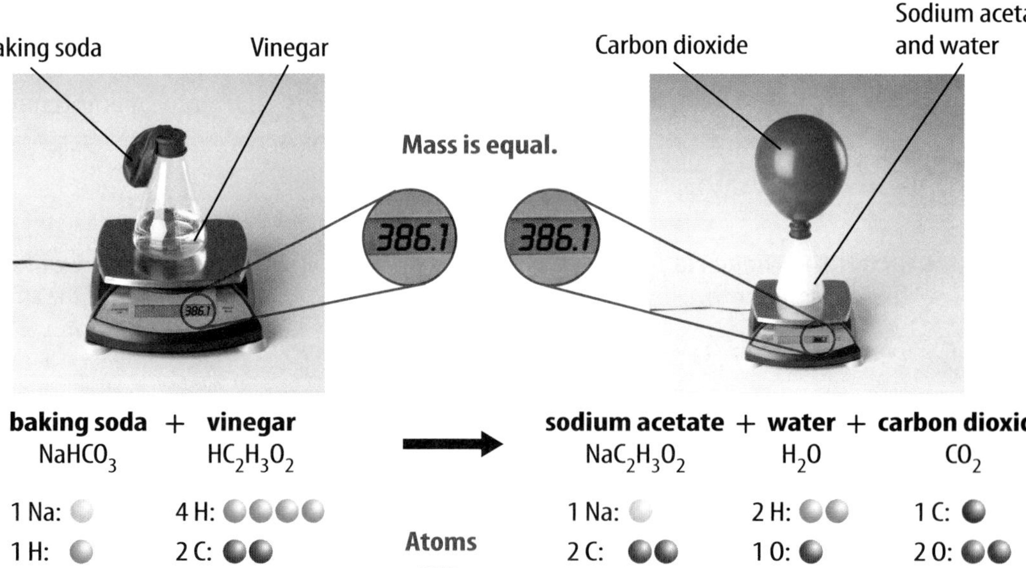
The baking soda is contained in a balloon. The balloon is attached to a flask that contains vinegar.
When the balloon is tipped up, the baking soda pours into the vinegar. The reaction forms a gas that is collected in the balloon.
Baking soda
Vinegar
Carbon dioxide
Sodium acetate and water
Mass is equal.
386.1
386.1
baking soda + vinegar → sodium acetate + water + carbon dioxide
$NaHCO_3$
$HC_2H_3O_2$
$NaC_2H_3O_2$
H_2O
CO_2
1 Na:
1 H:
1 C:
3 O:
4 H:
2 C:
2 O:
Atoms are equal.
1 Na:
2 C:
3 H:
2 O:
2 H:
1 O:
1 C:
2 O:

Is an equation balanced?

How does a chemical equation show that atoms are conserved? An equation is written so that the number of atoms of each element is the same, or balanced, on each side of the arrow. The equation showing the reaction between carbon and oxygen that produces carbon dioxide is shown below. Remember that oxygen is written as O_2 because it is a diatomic molecule. The formula for carbon dioxide is CO_2.

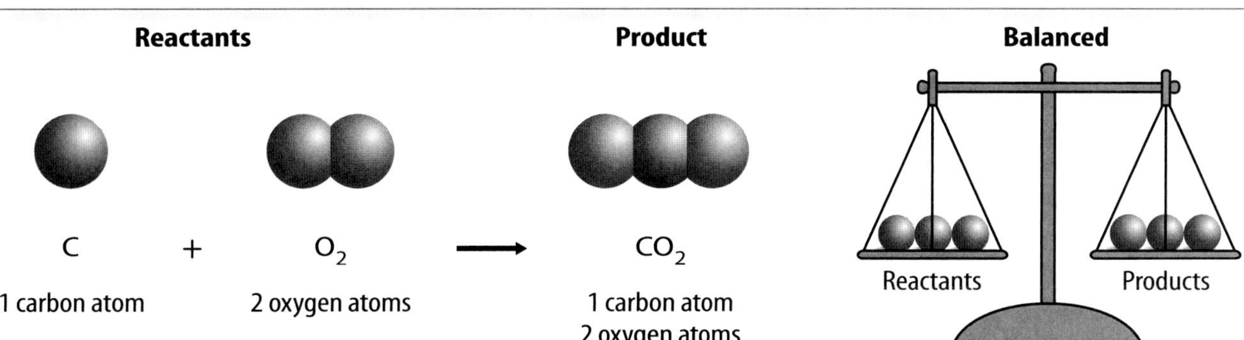

Is there the same number of carbon atoms on each side of the arrow? Yes, there is one carbon atom on the left and one on the right. Carbon is balanced. Is oxygen balanced? There are two oxygen atoms on each side of the arrow. Oxygen also is balanced. The atoms of all elements are balanced. Therefore, the equation is balanced.

You might think a balanced equation happens automatically when you write the symbols and formulas for reactants and products. However, this usually is not the case. For example, the reaction between hydrogen (H_2) and oxygen (O_2) that forms water (H_2O) is shown below.

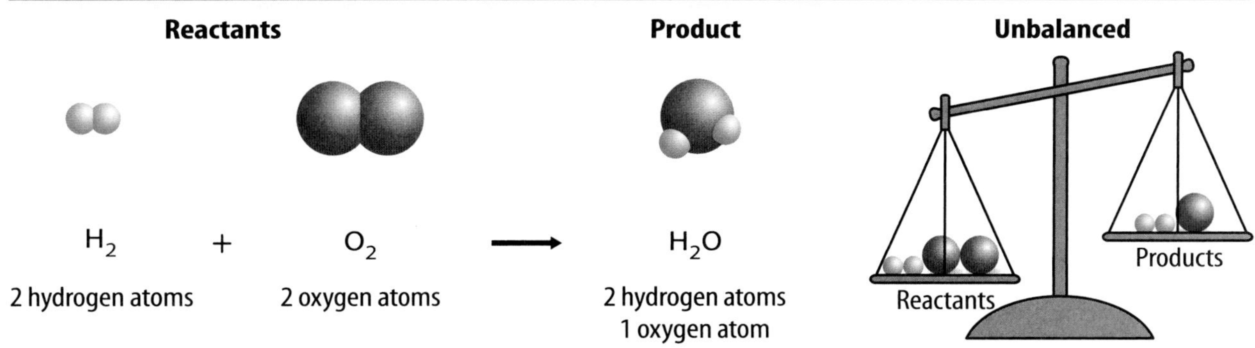

Count the number of hydrogen atoms on each side of the arrow. There are two hydrogen atoms in the product and two in the reactants. They are balanced. Now count the number of oxygen atoms on each side of the arrow. Did you notice that there are two oxygen atoms in the reactants and only one in the product? Because they are not equal, this equation is not balanced. To accurately represent this reaction, the equation needs to be balanced.

Balancing Chemical Equations

When you balance a chemical equation, you count the atoms in the reactants and the products and then add coefficients to balance the number of atoms. A **coefficient** *is a number placed in front of an element symbol or chemical formula in an equation.* It is the number of units of that substance in the reaction. For example, in the formula $2H_2O$, the *2* in front of H_2O is a coefficient, This means that there are two molecules of water in the reaction. Only coefficients can be changed when balancing an equation. Changing subscripts changes the identities of the substances that are in the reaction.

If one molecule of water contains two hydrogen atoms and one oxygen atom, how many H and O atoms are in two molecules of water ($2H_2O$)? Multiply each by 2.

2 × 2 H atoms = 4 H atoms
2 × 1 O atom = 2 O atoms

When no coefficient is present, only one unit of that substance takes part in the reaction. **Table 2** shows the steps of balancing a chemical equation.

Table 2 Balancing a Chemical Equation

Step	Example
1 Write the unbalanced equation. Make sure that all chemical formulas are correct.	$H_2 + O_2 \rightarrow H_2O$ **reactants** / **products**
2 Count atoms of each element in the reactants and in the products. **a.** Note which, if any, elements have a balanced number of atoms on each side of the equation. Which atoms are not balanced? **b.** If all of the atoms are balanced, the equation is balanced.	$H_2 + O_2 \rightarrow H_2O$ **reactants**: H = 2, O = 2 **products**: H = 2, O = 1
3 Add coefficients to balance the atoms. **a.** Pick an element in the equation that is not balanced, such as oxygen. Write a coefficient in front of a reactant or a product that will balance the atoms of that element. **b.** Recount the atoms of each element in the reactants and the products. Note which atoms are not balanced. Some atoms that were balanced before might no longer be balanced. **c.** Repeat step 3 until the atoms of each element are balanced.	$H_2 + O_2 \rightarrow 2H_2O$ **reactants**: H = 2, O = 2 **products**: H = 4, O = 2 $2H_2 + O_2 \rightarrow 2H_2O$ **reactants**: H = 4, O = 2 **products**: H = 4, O = 2
4 Write the balanced chemical equation including the coefficients.	$2H_2 + O_2 = 2H_2O$

Visual Check In row 2 above, which element is not balanced? In the top of row 3, which element is not balanced?

Lesson 1 Review

Assessment Online Quiz

Visual Summary

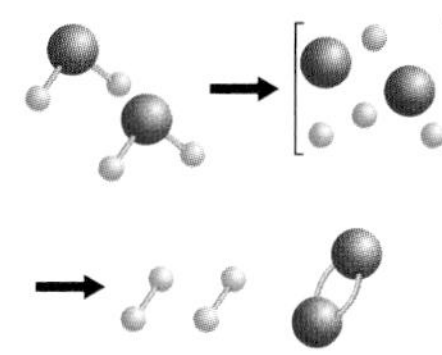

A chemical reaction is a process in which bonds break and atoms rearrange, forming new bonds.

$2H_2 + O_2 \rightarrow 2H_2O$

A chemical equation uses symbols to show reactants and products of a chemical reaction.

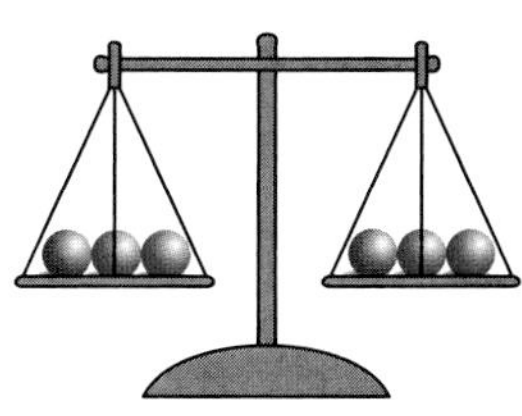

The mass and the number of each type of atom do not change during a chemical reaction. This is the law of conservation of mass.

FOLDABLES®

Use your lesson Foldable to review the lesson. Save your Foldable for the project at the end of the chapter.

What do you think NOW?

You first read the statements below at the beginning of the chapter.

1. If a substance bubbles, you know a chemical reaction is occurring.

2. During a chemical reaction, some atoms are destroyed and new atoms are made.

Did you change your mind about whether you agree or disagree with the statements? Rewrite any false statements to make them true.

Use Vocabulary

1 **Define** *reactants* and *products*.

Understand Key Concepts

2 Which is a sign of a chemical reaction?

- **A.** chemical properties change
- **B.** physical properties change
- **C.** a gas forms
- **D.** a solid forms

3 **Explain** why subscripts cannot change when balancing a chemical equation.

4 **Infer** Is the reaction below possible? Explain why or why not.

$$H_2O + NaOH \rightarrow NaCl + H_2$$

Interpret Graphics

5 **Describe** the reaction below by listing the bonds that break and the bonds that form.

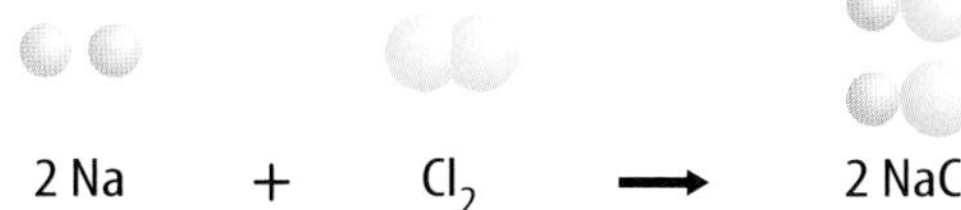

$2\ Na + Cl_2 \rightarrow 2\ NaCl$

6 **Interpret** Copy and complete the table to determine if this equation is balanced:

$$CH_4 + 2O_2 \rightarrow CO_2 + 2H_2O$$

Is this reaction balanced? Explain.

Type of Atom	Number of Atoms in the Balanced Chemical Equation	
	Reactants	Products

Critical Thinking

7 Balance this chemical equation. Hint: Balance Al last and then use a multiple of 2 and 3.

$$Al + HCl \rightarrow AlCl_3 + H_2$$

Inquiry **Skill Practice** **Follow a Procedure** 40 minutes

What can you learn from an experiment?

Observing reactions allows you to compare different types of changes that can occur. You can then design new experiments to learn more about reactions.

Materials

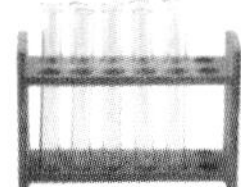

test tubes and rack

ammonium hydroxide (NH_4OH)

aluminum foil

sodium bicarbonate ($NaHCO_3$)

Also needed: copper foil, tongs, salt water, copper sulfate solution ($CuSO_4$), 25-mL graduated cylinder, Bunsen burner, plastic spoon, toothpick, ring stand and clamp, splints, matches, paper towel

Safety

Learn It

If you have never tested for a chemical reaction before, it is helpful to **follow a procedure.** A procedure tells you which materials to use and what steps to take.

Try It

1. Read and complete a safety form.
2. Copy the table into your Science Journal. During each procedure, record observations in the table.

3a. Dip a strip of aluminum foil into salt water in a test tube for about 1 min to remove the coating.

3b. Place 5 mL of copper sulfate solution in a test tube. Lift the aluminum foil from the salt water. Drop it into the test tube of copper sulfate so that the bottom part is in the liquid. Look for evidence of a chemical change. Set the test tube in a rack, and do the other procedures.

4. Use tongs to hold a small piece of copper foil in a flame for 3 min. Set the foil on a heat-proof surface, and allow it to cool. Use a toothpick to examine the product.
5. Place a spoonful of sodium bicarbonate in a dry test tube. Clamp the tube to a ring stand at a 45° angle. Point the mouth of the tube away from people. Move a burner flame back and forth under the tube. Observe the reaction. Test for carbon dioxide with a lighted wood splint.
6. Add 1 drop of ammonium hydroxide to a test tube containing 5 mL of copper sulfate solution.
7. Pour the liquid from the test tube in step 3b into a clean test tube. Dump the aluminum onto a paper towel. Record your observations of both the liquid and the solid.

Apply It

8. Using the table, write a balanced equation for each reaction.
9. Why did the color of the copper sulfate disappear in step 3b?
10. **Key Concept** How can you tell the difference between types of reactions by the number and type of reactants and products?

Step	Reactants	Products	Observations and Evidence of Chemical Reaction
3 + 7	$Al + CuSO_4$	$Cu + Al_2(SO_4)_3$	
4	$Cu + O_2$	CuO	
5	$NaHCO_3$	$CO_2 + Na_2CO_3 + H_2O$	
6	$NH_4OH + CuSO_4$	$(NH_4)_2SO_4 + CU(OH)_2$	

Lesson 2

Types of Chemical Reactions

Reading Guide

Key Concepts
ESSENTIAL QUESTIONS

- How can you recognize the type of chemical reaction by the number or type of reactants and products?
- What are the different types of chemical reactions?

Vocabulary

synthesis p. 431
decomposition p. 431
single replacement p. 432
double replacement p. 432
combustion p. 432

Multilingual eGlossary

Video

What's Science Got to do With It?

Inquiry Where did it come from?

When lead nitrate, a clear liquid, combines with potassium iodide, another clear liquid, a yellow solid appears instantly. Where did it come from? Here's a hint—the name of the solid is lead iodide. Did you guess that parts of each reactant combined and formed it? You'll learn about this and other types of reactions in this lesson.

15 minutes

What combines with what?

The reactants and the products in a chemical reaction can be elements, compounds, or both. In how many ways can these substances combine?

1. Read and complete a lab safety form.
2. Divide a **sheet of paper** into four equal sections labeled *A, B, Y,* and *Z*. Place **red paper clips** in section A, **yellow clips** in section B, **blue clips** in section Y, and **green clips** in section Z.
3. Use another sheet of paper to copy the table shown to the right. Turn the paper so that a long edge is at the top. Print *REACTANTS → PRODUCTS* across the top then complete the table.
4. Using the paper clips, model the equations listed in the table. Hook the clips together to make diatomic elements or compounds. Place each clip model onto your paper over the matching written equation.
5. As you read this lesson, match the types of equations to your paper clip equations.

	REACTANTS	→	PRODUCTS
1	AY	→	A + Y
2	B + Z	→	BZ
3	$2A_2 + Y_2$	→	$2A_2Y$
4	A + BY	→	B + AY
5	Z + BY	→	Y + BZ
6	AY + BZ	→	AZ + BY

Think About This

1. Which equation represents hydrogen combining with oxygen and forming water? How do you know?
2. **Key Concept** How could you use the number and type of reactants to identify a type of chemical reaction?

Patterns in Reactions

If you have ever used hydrogen peroxide, you might have noticed that it is stored in a dark bottle. This is because light causes hydrogen peroxide to change into other substances. Maybe you have seen a video of an explosion demolishing an old building, like in **Figure 5.** How is the reaction with hydrogen peroxide and light similar to a building demolition? In both, one reactant breaks down into two or more products.

The breakdown of one reactant into two or more products is one of four major types of chemical reactions. Each type of chemical reaction follows a unique pattern in the way atoms in reactants rearrange to form products. In this lesson, you will read how chemical reactions are classified by recognizing patterns in the way the atoms recombine.

Figure 5 When dynamite explodes, it chemically changes into several products and releases energy.

Types of Chemical Reactions

There are many different types of reactions. It would be impossible to memorize them all. However, most chemical reactions fit into four major categories. Understanding these categories of reactions can help you predict how compounds will react and what products will form.

Synthesis

A **synthesis** (SIHN thuh sus) *is a type of chemical reaction in which two or more substances combine and form one compound.* In the synthesis reaction shown in **Figure 6,** magnesium (Mg) reacts with oxygen (O_2) in the air and forms magnesium oxide (MgO). You can recognize a synthesis reaction because two or more reactants form only one product.

Decomposition

In a **decomposition** *reaction, one compound breaks down and forms two or more substances.* You can recognize a decomposition reaction because one reactant forms two or more products. For example, hydrogen peroxide (H_2O_2), shown in **Figure 6,** decomposes and forms water (H_2O) and oxygen gas (O_2). Notice that decomposition is the reverse of synthesis.

Key Concept Check How can you tell the difference between synthesis and decomposition reactions?

FOLDABLES®

Make a horizontal four-door book. Label it as shown. Use it to organize your notes about the different types of chemical reactions.

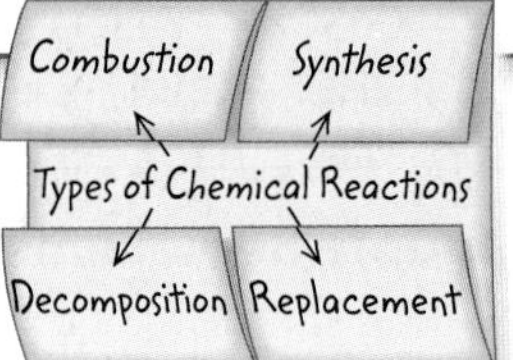

WORD ORIGIN

synthesis
from Greek *syn-*, means "together"; and *tithenai,* means "put"

Figure 6 Synthesis and decomposition reactions are opposites of each other.

Concepts in Motion

Animation

Synthesis and Decomposition Reactions

Synthesis Reactions

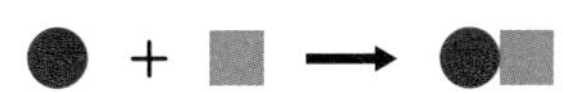

Examples:

$2Na + Cl_2 \rightarrow 2NaCl$

$2H_2 + O_2 \rightarrow 2H_2O$

$H_2O + SO_3 \rightarrow H_2SO_4$

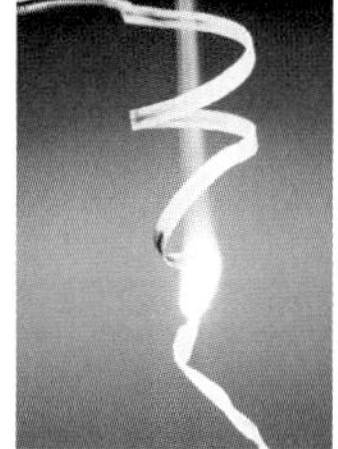

$2Mg$ (magnesium) $+$ O_2 (oxygen) $\rightarrow$ $2MgO$ (magnesium oxide)

Decomposition Reactions

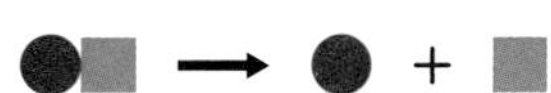

Examples:

$CaCO_3 \rightarrow CaO + CO_2$

$2H_2O \rightarrow 2H_2 + O_2$

$2KClO_3 \rightarrow 2KCl + 3O_2$

$2H_2O_2$ (hydrogen peroxide) $\rightarrow$ $2H_2O$ (water) $+$ O_2 (oxygen)

Replacement Reactions

Single Replacement

Examples:
$Fe + CuSO_4 \rightarrow FeSO_4 + Cu$
$Zn + 2HCl + ZnCl_2 + H_2$

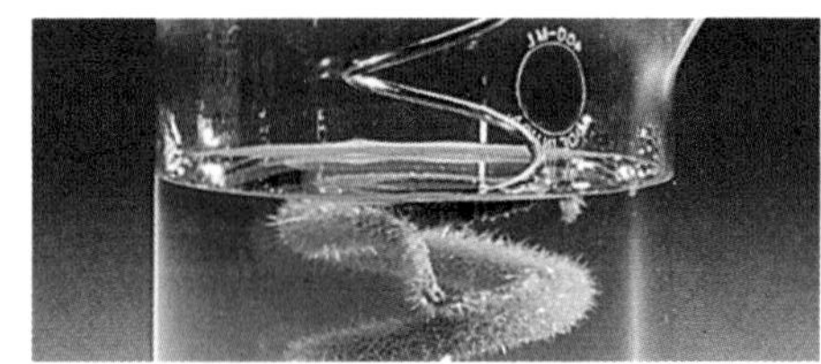

$2AgNO_3$	+	Cu	$\rightarrow$	$Cu(NO_3)_2$	+	$2Ag$
silver nitrate		copper		copper nitrate		silver

Double Replacement

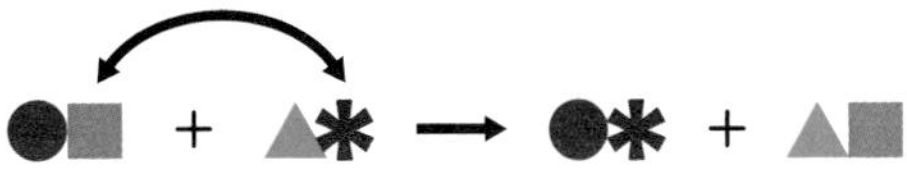

Examples:
$NaCl + AgNO_3 \rightarrow NaNO_3 + AgCl$
$HCl + FeS \rightarrow FeCl_2 + H_2S$

$Pb(NO_3)_2$	+	$2KI$	$\rightarrow$	$2KNO_3$	+	PbI_2
lead nitrate		potassium iodide		potassium nitrate		lead iodide

▲ **Figure 7** In each of these reactions, an atom or group of atoms replaces another atom or group of atoms.

Replacement

In a replacement reaction, an atom or group of atoms replaces part of a compound. There are two types of replacement reactions. *In a* **single-replacement** *reaction, one element replaces another element in a compound.* In this type of reaction, an element and a compound react and form a different element and a different compound. *In a* **double-replacement** *reaction, the negative ions in two compounds switch places, forming two new compounds.* In this type of reaction, two compounds react and form two new compounds. **Figure 7** describes these replacement reactions.

Combustion Reactions

substance + O_2 $\rightarrow$ substance(s)

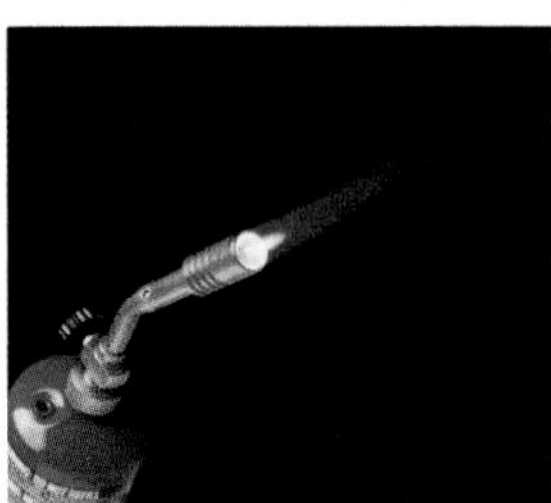

C_3H_8	+	$5O_2$	$\rightarrow$	$3CO_2$	+	$4H_2O$
propane		oxygen		carbon dioxide		water

Example:
$2C_4H_{10} + 13O_2 \rightarrow 8CO_2 + 10H_2O$

▲ **Figure 8** Combustion reactions always contain oxygen (O_2) as a reactant and often produce carbon dioxide (CO_2) and water (H_2O).

Combustion

Combustion *is a chemical reaction in which a substance combines with oxygen and releases energy.* This energy usually is released as thermal energy and light energy. For example, burning is a common combustion reaction. The burning of fossil fuels, such as the propane (C_3H_8) shown in **Figure 8,** produces the energy we use to cook food, power vehicles, and light cities.

Key Concept Check What are the different types of chemical reactions?

Lesson 2 Review

Assessment Online Quiz

? Inquiry Virtual Lab

Visual Summary

Chemical reactions are classified according to patterns seen in their reactants and products.

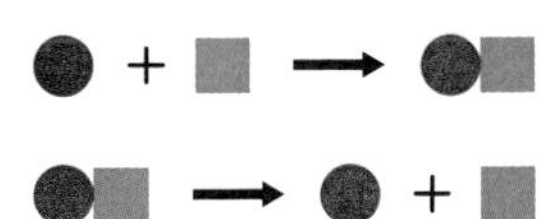

In a synthesis reaction, there are two or more reactants and one product. A decomposition reaction is the opposite of a synthesis reaction.

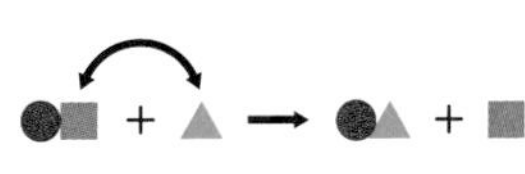

In replacement reactions, an element, or elements, in a compound is replaced with another element or elements.

FOLDABLES®

Use your lesson Foldable to review the lesson. Save your Foldable for the project at the end of the chapter.

What do you think NOW?

You first read the statements below at the beginning of the chapter.

3. Reactions always start with two or more substances that react with each other.

4. Water can be broken down into simpler substances.

Did you change your mind about whether you agree or disagree with the statements? Rewrite any false statements to make them true.

Use Vocabulary

1. **Contrast** synthesis and decomposition reactions using a diagram.
2. A reaction in which parts of two substances switch places and make two new substances is a(n) ________.

Understand Key Concepts 

3. **Classify** the reaction shown below.

$$2Na + Cl_2 \rightarrow 2NaCl$$

A. combustion **C.** single replacement

B. decomposition **D.** synthesis

4. Write a balanced equation that produces Na and Cl_2 from NaCl. Classify this reaction.
5. **Classify** In which two groups of reactions can this reaction be classified?

$$2SO_2 + O_2 \rightarrow 2SO_3$$

Interpret Graphics

6. **Complete** this table to identify four types of chemical reactions and the patterns shown by the reactants and the products.

Type of Reaction	Pattern of Reactants and Products
Synthesis	at least two reactants; one product

Critical Thinking

7. **Design** a poster to illustrate single- and double-replacement reactions.
8. **Infer** The combustion of methane (CH_4) produces energy. Where do you think this energy comes from?

HOW IT WORKS

How does a light stick work?

What makes it glow?

Glowing neon necklaces, bracelets, or sticks—chances are you've worn or used them. Light sticks—also known as glow sticks—come in brilliant colors and provide light without electricity or batteries. Because they are lightweight, portable, and waterproof, they provide an ideal light source for campers, scuba divers, and other activities in which electricity is not readily available. Light sticks also are useful in emergency situations in which an electric current from battery-powered lights could ignite a fire.

Light sticks give off light because of a chemical reaction that happens inside the tube. During the reaction, energy is released as light. This is known as chemiluminescence (ke mee lew muh NE sunts).

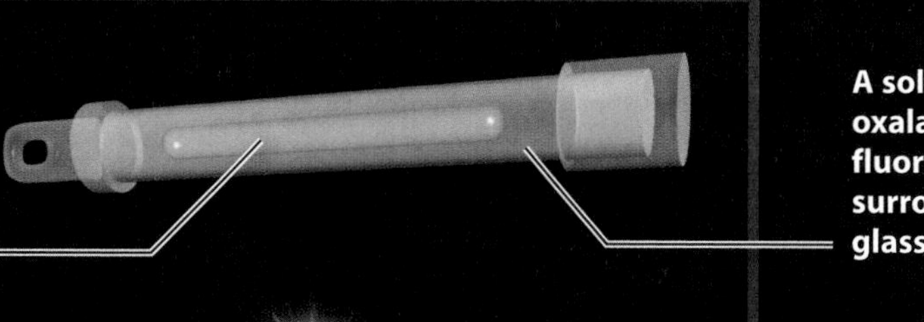

A light stick consists of a plastic tube with a glass tube inside it. Hydrogen peroxide fills the glass tube..

A solution of phenyl oxalate ester and fluorescent dye surround the glass tube.

When you bend the outer plastic tube, the inner glass tube breaks, causing the hydrogen peroxide, ester, and dye to mix together.

When the solutions mix together, they react. Energy produced by the reaction causes the electrons in the dye to produce light.

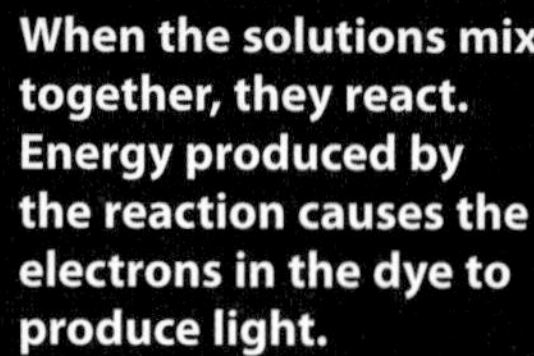

It's Your Turn

RESEARCH AND REPORT Research bioluminescent organisms, such as fireflies and sea animals. How is the reaction that occurs in these organisms similar to or different from that in a glow stick? Work in small groups, and present your findings to the class.

Lesson 3

Reading Guide

Key Concepts
ESSENTIAL QUESTIONS

- Why do chemical reactions always involve a change in energy?
- What is the difference between an endothermic reaction and an exothermic reaction?
- What factors can affect the rate of a chemical reaction?

Vocabulary

endothermic p. 437
exothermic p. 437
activation energy p. 438
catalyst p. 440
enzyme p. 440
inhibitor p. 440

g Multilingual eGlossary

Energy Changes and Chemical Reactions

Inquiry Energy from Bonds?

A deafening roar, a blinding light, and the power to lift 2 million kg—what is the source of all this energy? Chemical bonds in the fuel store all the energy needed to launch a space shuttle. Chemical reactions release the energy in these bonds.

Launch Lab

20 minutes

Where's the heat?

Does a chemical change always produce a temperature increase?

1. Read and complete a lab safety form.
2. Copy the table into your Science Journal.
3. Use a **graduated cylinder** to measure 25 mL of **citric acid solution** into a **foam cup.** Record the temperature with a **thermometer**.
4. Use a **plastic spoon** to add a rounded spoonful of **solid sodium bicarbonate** to the cup. Stir.
5. Use a **clock** or **stopwatch** to record the temperature every 15 s until it stops changing. Record your observations during the reaction.
6. Add 25 mL of **sodium bicarbonate solution** to a **second foam cup.** Record the temperature. Add a spoonful of **calcium chloride**. Repeat step 5.

Time	Temperature (°C)	
	Citric Acid Solution	Sodium Bicarbonate Solution
Starting temp.		
15 s		
30 s		
45 s		
1 min		
1 min, 15 s		
1 min, 30 s		
1 min, 45 s		
2 min		
2 min, 15 sec		

Think About This

1. What evidence do you have that the changes in the two cups were chemical reactions?
2. What happened to the temperature in the two cups? How would you explain the changes?
3. **Key Concept** Based on your observations and past experience, would a change in temperature be enough to convince you that a chemical change had taken place? Why or why not? What else could cause a temperature change?

Energy Changes

What is about 1,500 times heavier than a typical car and 300 times faster than a roller coaster? Do you need a hint? The energy it needs to move this fast comes from a chemical reaction that produces water. If you guessed a space shuttle, you are right!

It takes a large amount of energy to launch a space shuttle. The shuttle's main engines burn almost 2 million L of liquid hydrogen and liquid oxygen. This chemical reaction produces water vapor and a large amount of energy. The energy produced heats the water vapor to high temperatures, causing it to expand rapidly. When the water expands, it pushes the shuttle into orbit. Where does all this energy come from?

Chemical Energy in Bonds

Recall that when a chemical reaction occurs, chemical bonds in the reactants break and new chemical bonds form. Chemical bonds contain a form of energy called chemical energy. Breaking a bond absorbs energy from the surroundings. The formation of a chemical bond releases energy to the surroundings. Some chemical reactions release more energy than they absorb. Some chemical reactions absorb more energy than they release. You can feel this energy change as a change in the temperature of the surroundings. Keep in mind that in all chemical reactions, energy is conserved.

Key Concept Check Why do chemical reactions involve a change in energy?

Endothermic Reactions—Energy Absorbed

Have you ever heard someone say that the sidewalk was hot enough to fry an egg? To fry, the egg must absorb energy. *Chemical reactions that absorb thermal energy are* **endothermic** *reactions.* For an endothermic reaction to continue, energy must be constantly added.

reactants + thermal energy → products

In an endothermic reaction, more energy is required to break the bonds of the reactants than is released when the products form. Therefore, the overall reaction absorbs energy. The reaction on the left in **Figure 9** is an endothermic reaction.

Exothermic Reactions—Energy Released

Most chemical reactions release energy as opposed to absorbing it. *An* **exothermic** *reaction is a chemical reaction that releases thermal energy.*

reactants → products + thermal energy

In an exothermic reaction, more energy is released when the products form than is required to break the bonds in the reactants. Therefore, the overall reaction releases energy. The reaction shown on the right in **Figure 9** is exothermic.

Key Concept Check What is the difference between an endothermic reaction and an exothermic reaction?

FOLDABLES®

Make a vertical three-tab Venn book. Label it as shown. Use it to compare and contrast energy in chemical reactions.

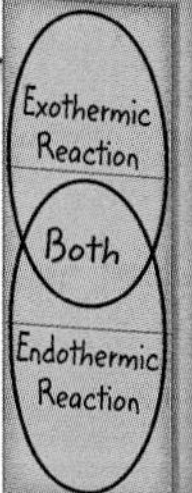

WORD ORIGIN

exothermic
from Greek *exo-*, means "outside"; and *therm,* means "heat"

Figure 9 Whether a reaction is endothermic or exothermic depends on the amount of energy contained in the bonds of the reactants and the products.

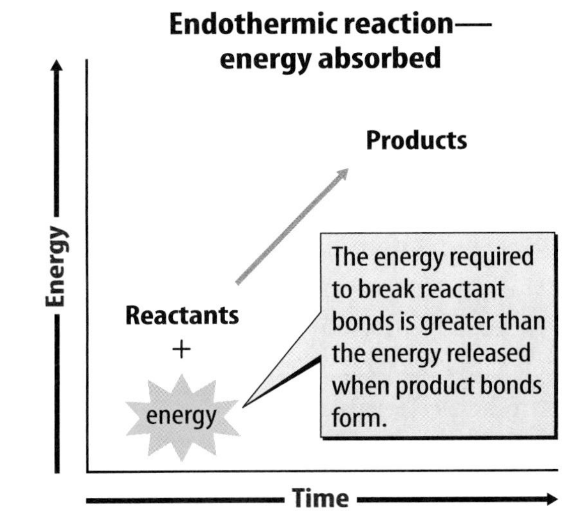

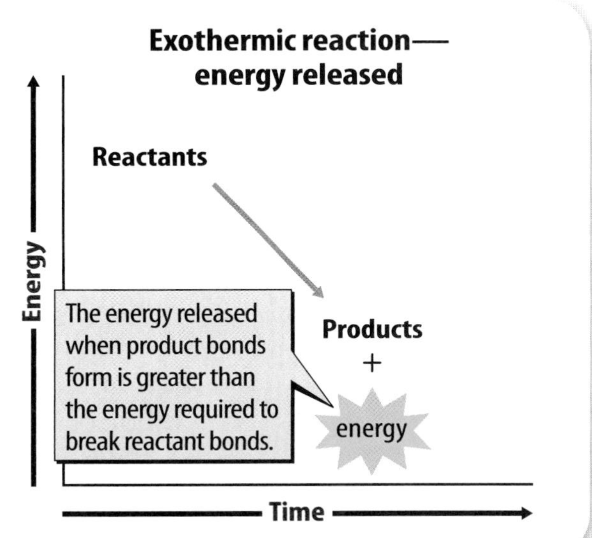

Visual Check Why does one arrow point upward and the other arrow point downward in these diagrams?

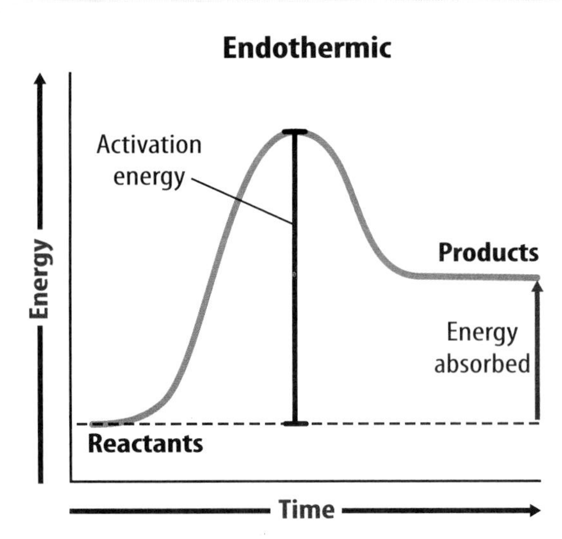

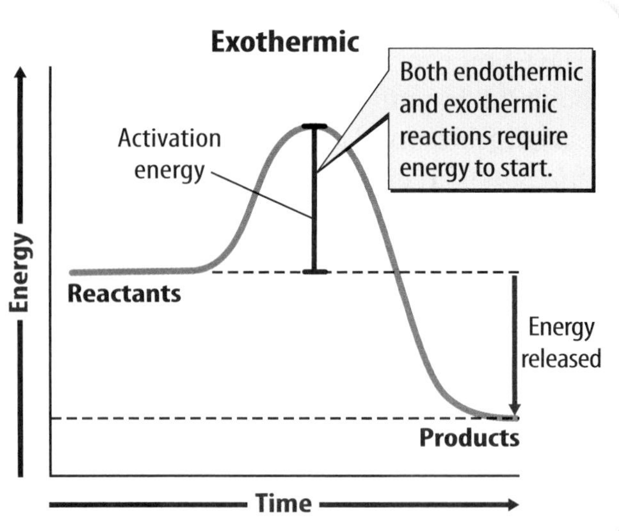

Figure 10  Both endothermic and exothermic reactions require activation energy to start the reaction.

Visual Check How can a reaction absorb energy to start but still be exothermic?

Activation Energy

You might have noticed that some chemical reactions do not start by themselves. For example, a newspaper does not burn when it comes into contact with oxygen in air. However, if a flame touches the paper, it starts to burn.

All reactions require energy to start the breaking of bonds. This energy is called activation energy. **Activation energy** *is the minimum amount of energy needed to start a chemical reaction.* Different reactions have different activation energies. Some reactions, such as the rusting of iron, have low activation energy. The energy in the surroundings is enough to start these reactions. If a reaction has high activation energy, more energy is needed to start the reaction. For example, wood requires the thermal energy of a flame to start burning. Once the reaction starts, it releases enough energy to keep the reaction going. **Figure 10** shows the role activation energy plays in endothermic and exothermic reactions.

Reaction Rates

Some chemical reactions, such as the rusting of a bicycle wheel, happen slowly. Other chemical reactions, such as the explosion of fireworks, happen in less than a second. The rate of a reaction is the speed at which it occurs. What controls how fast a chemical reaction occurs? Recall that particles must collide before they can react. Chemical reactions occur faster if particles collide more often or move faster when they collide. There are several factors that affect how often particles collide and how fast particles move.

Reading Check How do particle collisions relate to reaction rate?

Surface Area

Surface area is the amount of exposed, outer area of a solid. Increased surface area increases reaction rate because more particles on the surface of a solid come into contact with the particles of another substance. For example, if you place a piece of chalk in vinegar, the chalk reacts slowly with the acid. This is because the acid contacts only the particles on the surface of the chalk. But, if you grind the chalk into powder, more chalk particles contact the acid, and the reaction occurs faster.

Temperature

Imagine a crowded hallway. If everyone in the hallway were running, they would probably collide with each other more often and with more energy than if everyone were walking. This is also true when particles move faster. At higher temperatures, the average speed of particles is greater. This speeds reactions in two ways. First, particles collide more often. Second, collisions with more energy are more likely to break chemical bonds.

Concentration and Pressure

Think of a crowded hallway again. Because the concentration of people is higher in the crowded hallway than in an empty hallway, people probably collide more often. Similarly, increasing the concentration of one or more reactants increases collisions between particles. More collisions result in a faster reaction rate. In gases, an increase in pressure pushes gas particles closer together. When particles are closer together, more collisions occur. Factors that affect reaction rate are shown in **Figure 11.**

Math Skills

Use Geometry

The surface area (SA) of one side of a 1-cm cube is 1 cm × 1 cm, or 1 cm^2. The cube has 6 equal sides. Its total SA is 6 × 1 cm^2, or 6 cm^2. What is the total SA of the two solids made when the cube is cut in half?

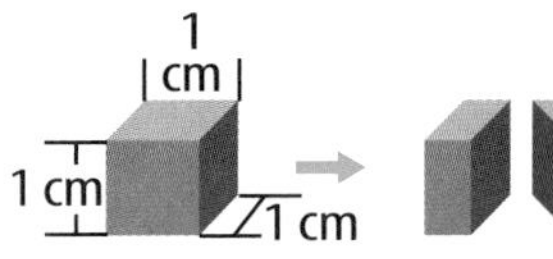

1. The new surfaces made each have an area of 1 cm × 1 cm = 1 cm^2.
2. Multiply the area by the number of new surfaces. 2 × 1 cm^2 = 2 cm^2 The total SA is 8 cm^2.

Practice

Calculate the amount of SA gained when a 2-cm cube is cut in half.

- **Math Practice**
- **Personal Tutor**

Figure 11 Several factors can affect reaction rate.

Slower Reaction Rate

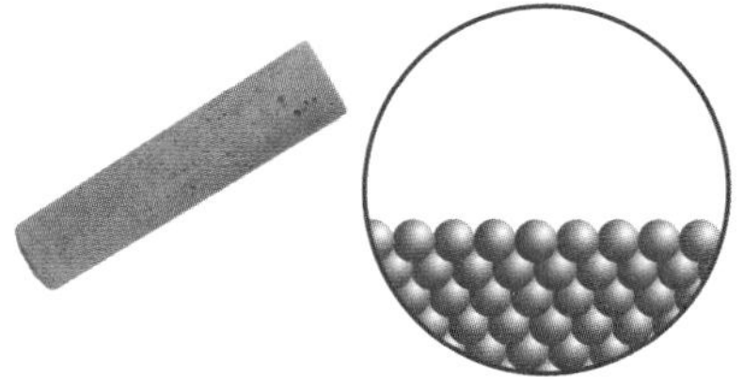

Less surface area

Lower temperature

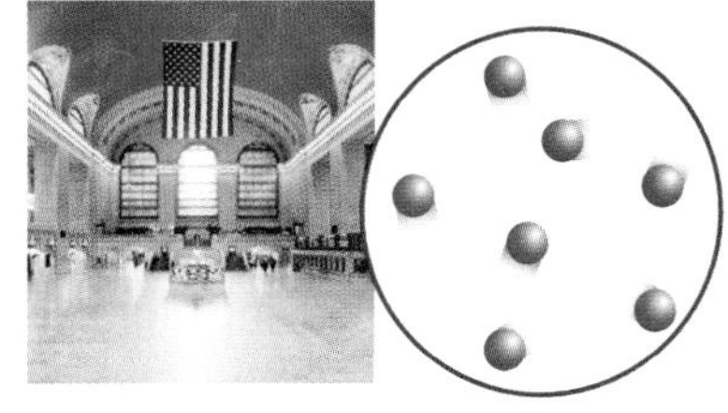

Lower concentration

Faster Reaction Rate

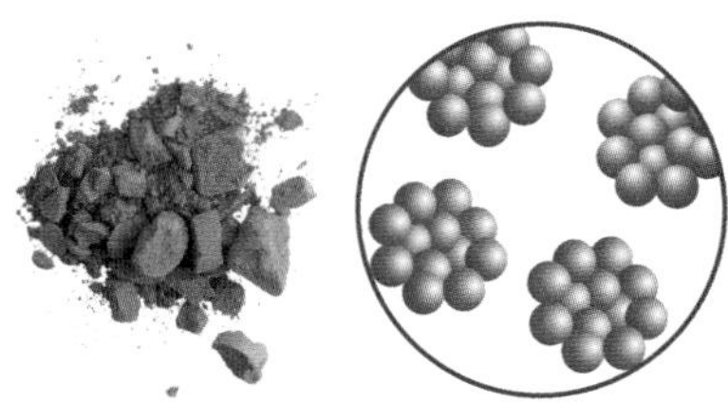

More surface area

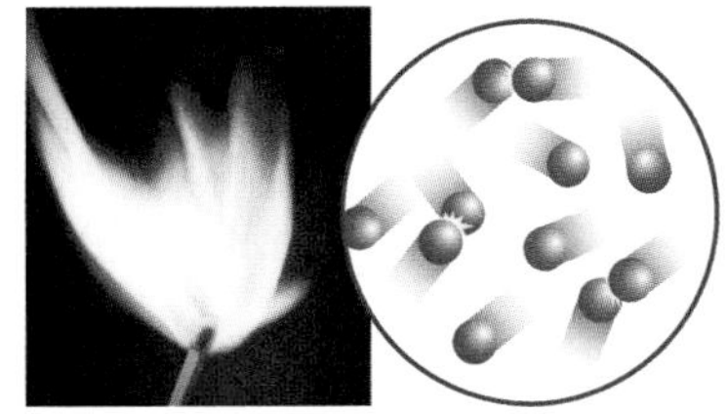

Higher temperature

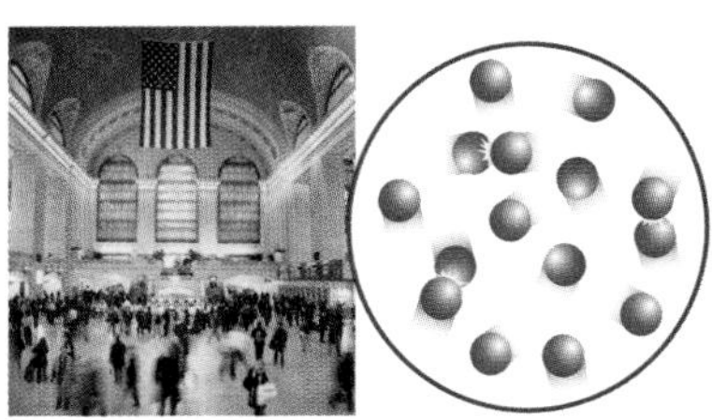

Higher concentration

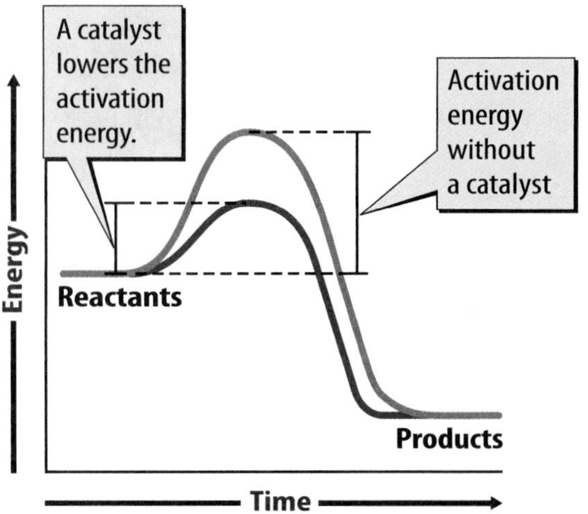

Figure 12 The blue line shows how a catalyst can increase the reaction rate.

Catalysts

A **catalyst** *is a substance that increases reaction rate by lowering the activation energy of a reaction.* One way catalysts speed reactions is by helping reactant particles contact each other more often. Look at **Figure 12.** Notice that the activation energy of the reaction is lower with a catalyst than it is without a catalyst. A catalyst isn't changed in a reaction, and it doesn't change the reactants or products. Also, a catalyst doesn't increase the amount of reactant used or the amount of product that is made. It only makes a given reaction happen faster. Therefore, catalysts are not considered reactants in a reaction.

You might be surprised to know that your body is filled with catalysts called enzymes. *An* **enzyme** *is a catalyst that speeds up chemical reactions in living cells.* For example, the enzyme protease (PROH tee ays) breaks the protein molecules in the food you eat into smaller molecules that can be absorbed by your intestine. Without enzymes, these reactions would occur too slowly for life to exist.

Inhibitors

Recall than an enzyme is a molecule that speeds reactions in organisms. However, some organisms, such as bacteria, are harmful to humans. Some medicines contain molecules that attach to enzymes in bacteria. This keeps the enzymes from working properly. If the enzymes in bacteria can't work, the bacteria die and can no longer infect a human. The active ingredients in these medicines are called inhibitors. *An* **inhibitor** *is a substance that slows, or even stops, a chemical reaction.* Inhibitors can slow or stop the reactions caused by enzymes.

Inhibitors are also important in the food industry. Preservatives in food are substances that inhibit, or slow down, food spoilage.

 **Key Concept Check** What factors can affect the rate of a chemical reaction?

Inquiry MiniLab — 20 minutes

Can you speed up a reaction?

Can you speed up the decomposition of hydrogen peroxide (H_2O_2)? The reaction is $H_2O_2 \rightarrow H_2O$ and O_2.

1. Read and complete a lab safety form.
2. Use **tape** to label three **test tubes** *1, 2,* and *3.* Place the tubes in a **test-tube rack.**
3. Add 10 mL of **hydrogen peroxide** to each test tube.
4. Observe tube 1 for changes. Add a small piece of **raw potato** to tube 2. Record observations in your Science Journal.
5. Add a pinch of **dry yeast** to tube 3. Shake the tube gently. Record observations.
6. Use **matches** to light a **wood splint,** then blow it out, leaving a glowing tip. One at a time, hold each test tube at a 45° angle and insert the glowing splint into the tube just above the liquid. Record your observations.

Analyze and Conclude

1. **Draw Conclusions** What was the chemical reaction when the potato and yeast were added?
2. **Key Concept** Why is the reaction in tube 3 faster than in the other two tubes?

Lesson 3 Review

Assessment Online Quiz

Visual Summary

Endothermic

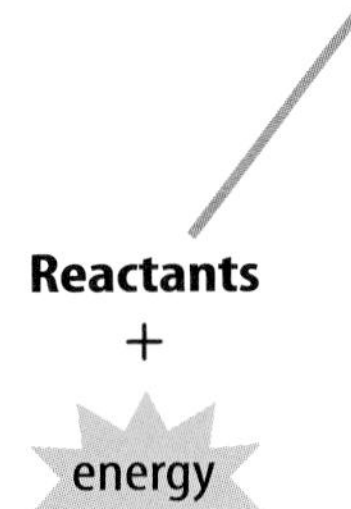

Chemical reactions that release energy are exothermic, and those that absorb energy are endothermic.

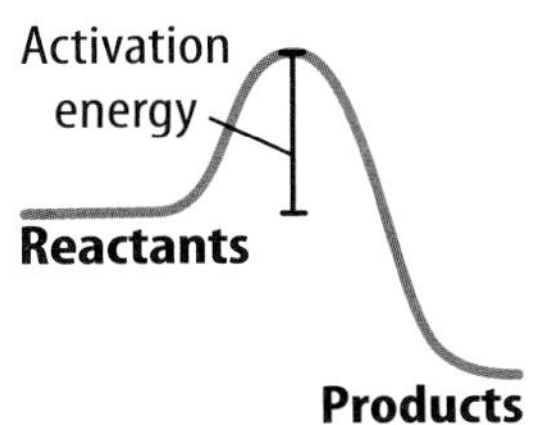

Activation energy must be added to a chemical reaction for it to proceed.

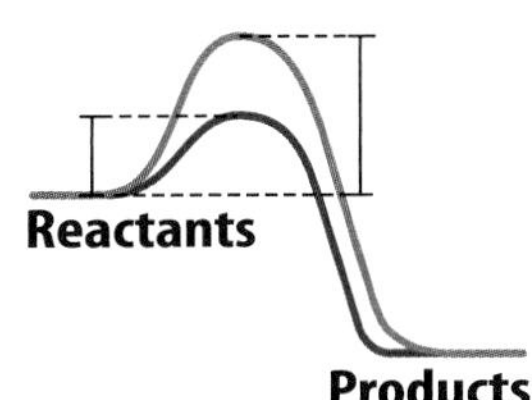

Catalysts, including enzymes, speed up chemical reactions. Inhibitors slow them down.

FOLDABLES®

Use your lesson Foldable to review the lesson. Save your Foldable for the project at the end of the chapter.

What do you think NOW?

You first read the statements below at the beginning of the chapter.

5. Reactions that release energy require energy to get started.

6. Energy can be created in a chemical reaction.

Did you change your mind about whether you agree or disagree with the statements? Rewrite any false statements to make them true.

Use Vocabulary

1 The smallest amount of energy required by reacting particles for a chemical reaction to begin is the ________.

Understand Key Concepts 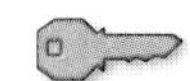

2 How does a catalyst increase reaction rate?

A. by increasing the activation energy
B. by increasing the amount of reactant
C. by increasing the contact between particles
D. by increasing the space between particles

3 **Contrast** endothermic and exothermic reactions in terms of energy.

4 **Explain** When propane burns, heat and light are produced. Where does this energy come from?

Interpret Graphics

5 **List** Copy and complete the graphic organizer to describe four ways to increase the rate of a reaction.

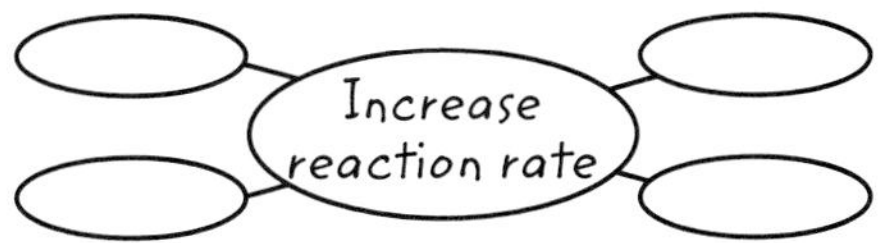

Critical Thinking

6 **Infer** Explain why keeping a battery in a refrigerator can extend its life.

7 **Infer** Explain why a catalyst does not increase the amount of product that can form.

Math Skills

Math Practice

8 An object measures 1 cm × 1 cm × 3 cm.

a. What is the surface area of the object?
b. What is the total surface area if you cut the object into three equal pieces?

2 class periods

Design an Experiment to Test Advertising Claims

Materials

graduated cylinder

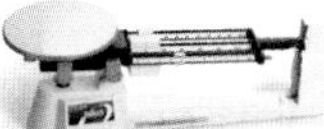

balance

droppers

baking soda

plastic spoon

Also needed: various brands of liquid and solid antacids (both regular and maximum strength), beakers, universal indicator in dropper bottle, 0.1M HCl solution, stirrers

Safety

Antacids contain compounds that react with excess acid in your stomach and prevent a condition called heartburn. Suppose you work for a laboratory that tests advertising claims about antacids. What kinds of procedures would you follow? How would you decide which antacid is the most effective?

Ask a Question

Ask a question about the claims that you would like to investigate. For example: what does *most effective* mean? What would make an antacid the strongest?

Make Observations

1. Read and complete a lab safety form.
2. Study the selection of antacids available for testing. You will use a 0.1M HCl solution to simulate stomach acid. Use the questions below to discuss with your lab partners which advertising claim you might test and how you might test it.
3. In your Science Journal, write a procedure for each variable that you will test to answer your question. Include the materials and steps you will use to test each variable. Place the steps of each procedure in order. Have your teacher approve your procedures.
4. Make a chart or table to record observations during your experiments.

Questions
Which advertising claim will I test? What question am I trying to answer?
What will be the independent and the dependent variables for each test? Recall that the independent variable is the variable that is changed. A dependent variable changes when you change the independent variable.
What variables will be held constant in each test?
How many different procedures will I use, and what equipment will I need?
How much of each antacid will I use? How many antacids will I test?
How will I use the indicator?
How many times will I do each test?
How will I record the data and observations?
What will I analyze to form a conclusion?

Form a Hypothesis

5 Write a hypothesis for each variable. Your hypothesis should identify the independent variable and state why you think changing the variable will alter the effectiveness of an antacid tablet.

Test Your Hypothesis

6 On day 2, use the available materials to perform your experiments. Accurately record all observations and data for each test.

7 Add any additional tests you think you need to answer your questions.

8 Examine the data you have collected. If the data are not conclusive, what other tests can you do to provide more information?

9 Write all your observations and measurements in your Science Journal. Use tables to record any quantitative data.

Analyze and Conclude

10 **Infer** What do you think advertisers mean when they say their product is most effective?

11 **Draw Conclusions** If you needed an antacid, which one would you use, based on the limited information provided from your experiments? Explain your reasoning.

12 **Analyze** Would breaking an antacid tablet into small pieces before using it make it more effective? Why or why not?

13 **The Big Idea** How does understanding chemical reactions enable you to analyze products and their claims?

Communicate Your Results

Combine your data with other teams. Compare the results and conclusions. Discuss the validity of advertising claims for each brand of antacid.

Research over-the-counter antacids that were once available by prescription only. Do they work in the same way as the antacids you tested? Explain.

Lab Tips

- ☑ Think about how you might measure the amount of acid the tablet neutralizes. Would you add the tablet to the acid or the acid to the tablet? What does the indicator show you?
- ☑ Try your tests on a small scale before using the full amounts to see how much acid you might need.
- ☑ Always get your teacher's approval before trying any new test.

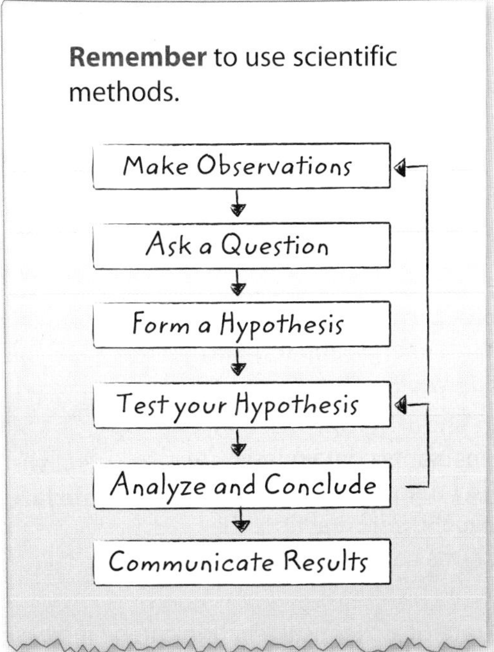

Chapter 12 Study Guide

Atoms are neither created nor destroyed in chemical reactions. Energy can be released when chemical bonds form or absorbed when chemical bonds are broken.

Key Concepts Summary

Lesson 1: Understanding Chemical Reactions

- There are several signs that a **chemical reaction** might have occurred, including a change in temperature, a release of light, a release of gas, a change in color or odor, and the formation of a solid from two liquids.
- In a chemical reaction, atoms of **reactants** rearrange and form **products.**
- The total mass of all the reactants is equal to the total mass of all the products in a reaction.

Reactants			Products		
1 Na: ●	4 H: ●●●●	**Atoms are equal.**	1 Na: ●	2 H: ●●	1 C: ●
1 H: ●	2 C: ●●		2 C: ●●	1 O: ●	2 O: ●●
1 C: ●	2 O: ●●		3 H: ●●●		
3 O: ●●●			2 O: ●●		

Lesson 2: Types of Chemical Reactions

- Most chemical reactions fit into one of a few main categories—synthesis, decomposition, combustion, and single- or double-replacement.
- **Synthesis** reactions create one product. **Decomposition** reactions start with one reactant. **Single-** and **double-replacement** reactions involve replacing one element or group of atoms with another element or group of atoms. **Combustion** reactions involve a reaction between one reactant and oxygen, and they release thermal energy.

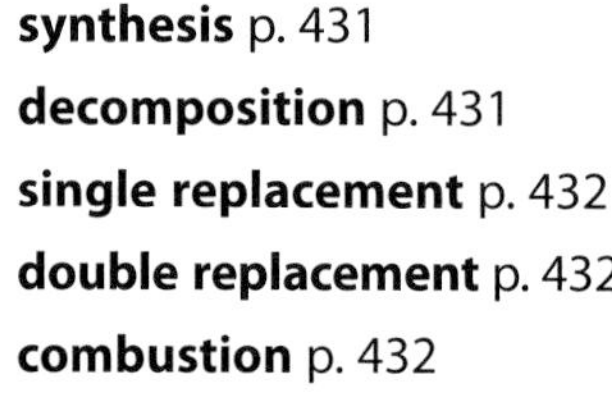

Lesson 3: Energy Changes and Chemical Reactions

- Chemical reactions always involve breaking bonds, which requires energy, and forming bonds, which releases energy.
- In an **endothermic** reaction, the reactants contain less energy than the products. In an **exothermic** reaction, the reactants contain more energy than the products.

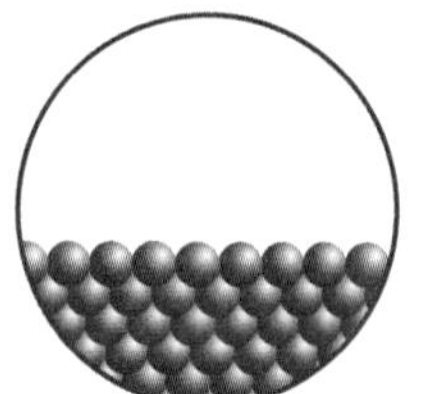

Less surface area

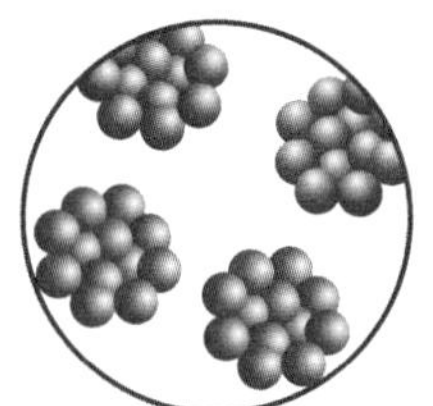

More surface area

- The rate of a chemical reaction can be increased by increasing the surface area, the temperature, or the concentration of the reactants or by adding a **catalyst.**

Vocabulary

chemical reaction p. 419
chemical equation p. 422
reactant p. 423
product p. 423
law of conservation of mass p. 424
coefficient p. 426

synthesis p. 431
decomposition p. 431
single replacement p. 432
double replacement p. 432
combustion p. 432

endothermic p. 437
exothermic p. 437
activation energy p. 438
catalyst p. 440
enzyme p. 440
inhibitor p. 440

Review
- Personal Tutor
- Vocabulary eGames
- Vocabulary eFlashcards

FOLDABLES® Chapter Project

Assemble your lesson Foldables as shown to make a Chapter Project. Use the project to review what you have learned in this chapter.

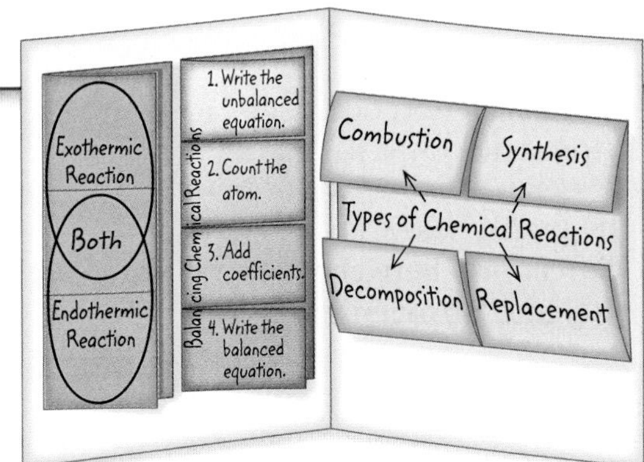

Use Vocabulary

1. When water forms from hydrogen and oxygen, water is the ________.
2. A(n) ________ uses symbols instead of words to describe a chemical reaction.
3. In a(n) ________ reaction, one element replaces another element in a compound.
4. When Na_2CO_3 is heated, it breaks down into CO_2 and Na_2O in a(n) ________ reaction.
5. The chemical reactions that keep your body warm are ________ reactions.
6. Even exothermic reactions require ________ to start.

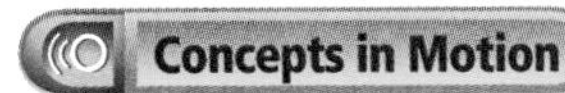

Interactive Concept Map

Link Vocabulary and Key Concepts

Copy this concept map, and then use vocabulary terms from the previous page and other terms from the chapter to complete the concept map.

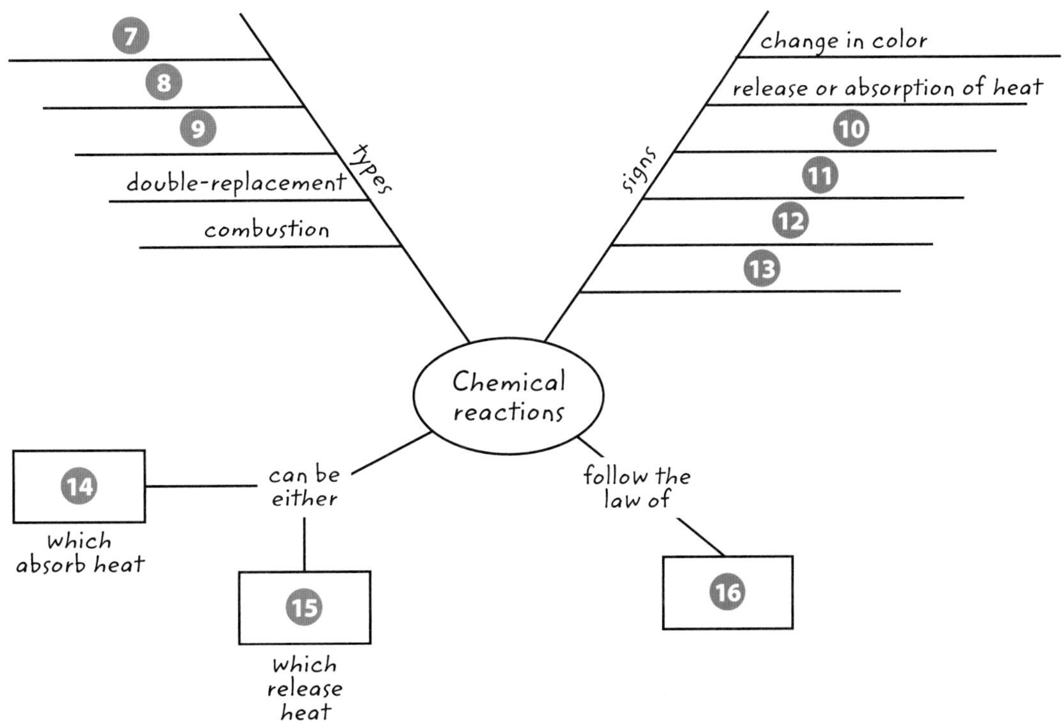

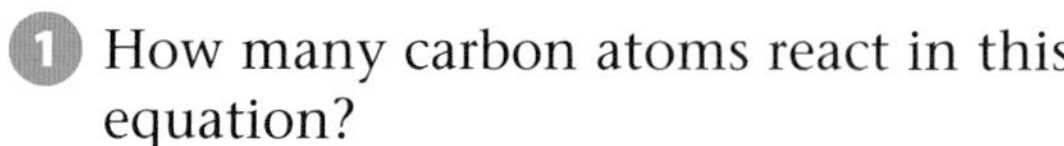

Chapter 12 Review

Understand Key Concepts

1. How many carbon atoms react in this equation?

$$2C_4H_{10} + 13O_2 \rightarrow 8CO_2 + 10H_2O$$

A. 2
B. 4
C. 6
D. 8

2. The chemical equation below is unbalanced.

$$Zn + HCl \rightarrow ZnCl_2 + H_2$$

Which is the correct balanced chemical equation?

A. $Zn + H_2Cl_2 \rightarrow ZnCl_2 + H_2$
B. $Zn + HCl \rightarrow ZnCl + H$
C. $2Zn + 2HCl \rightarrow ZnCl_2 + H_2$
D. $Zn + 2HCl \rightarrow ZnCl_2 + H_2$

3. When iron combines with oxygen gas and forms rust, the total mass of the products

A. depends on the reaction conditions.
B. is less than the mass of the reactants.
C. is the same as the mass of the reactants.
D. is greater than the mass of the reactants.

4. Potassium nitrate forms potassium oxide, nitrogen, and oxygen in certain fireworks.

$$4KNO_3 \rightarrow 2K_2O + 2N_2 + 5O_2$$

This reaction is classified as a

A. combustion reaction.
B. decomposition reaction.
C. single-replacement reaction.
D. synthesis reaction.

5. Which type of reaction is the reverse of a decomposition reaction?

A. combustion
B. synthesis
C. double-replacement
D. single-replacement

6. The compound NO_2 can act as a catalyst in the reaction that converts ozone (O_3) to oxygen (O_2) in the upper atmosphere. Which statement is true?

A. More oxygen is created when NO_2 is present.
B. NO_2 is a reactant in the chemical reaction that converts O_3 to O_2.
C. This reaction is more exothermic in the presence of NO_2 than in its absence.
D. This reaction occurs faster in the presence of NO_2 than in its absence.

7. The graph below is an energy diagram for the reaction between carbon monoxide (CO) and nitrogen dioxide (NO_2).

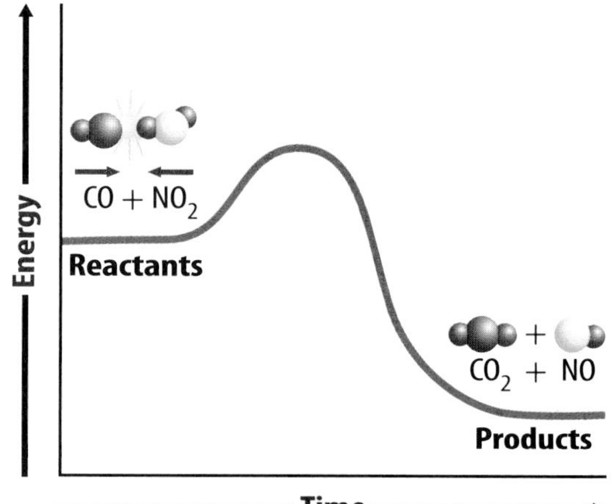

Which is true about this reaction?

A. More energy is required to break reactant bonds than is released when product bonds form.
B. Less energy is required to break reactant bonds than is released when product bonds form.
C. The bonds of the reactants do not require energy to break because the reaction releases energy.
D. The bonds of the reactants require energy to break, and therefore the reaction absorbs energy.

Assessment
Online Test Practice

Critical Thinking

8 **Predict** The diagram below shows two reactions—one with a catalyst (blue) and one without a catalyst (orange).

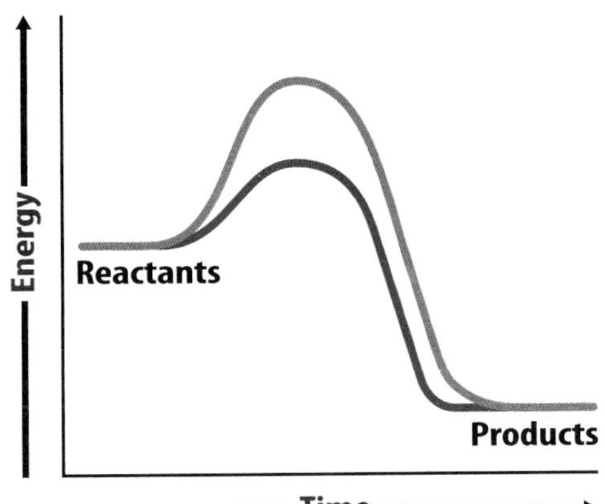

How would the blue line change if an inhibitor were used instead of a catalyst?

9 **Analyze** A student observed a chemical reaction and collected the following data:

Observations before the reaction	A white powder was added to a clear liquid.
Observations during the reaction	The reactants bubbled rapidly in the open beaker.
Mass of reactants	4.2 g
Mass of products	4.0 g

The student concludes that mass was not conserved in the reaction. Explain why this is not a valid conclusion. What might explain the difference in mass?

10 **Explain Observations** How did the discovery of atoms explain the observation that the mass of the products always equals the mass of the reactants in a reaction?

Writing in Science

11 **Write instructions** that explain the steps in balancing a chemical equation. Use the following equation as an example.

$$MnO_2 + HCl \rightarrow MnCl_2 + H_2O + Cl_2$$

REVIEW THE BIG IDEA

12 Explain how atoms and energy are conserved in a chemical reaction.

13 When a car air bag inflates, sodium azide (NaN_3) decomposes and produces nitrogen gas (N_2) and another product. What element does the other product contain? How do you know?

Math Skills

Review
Math Practice

Use Geometry

14 What is the surface area of the cube shown below? What would the total surface area be if you cut the cube into 27 equal cubes?

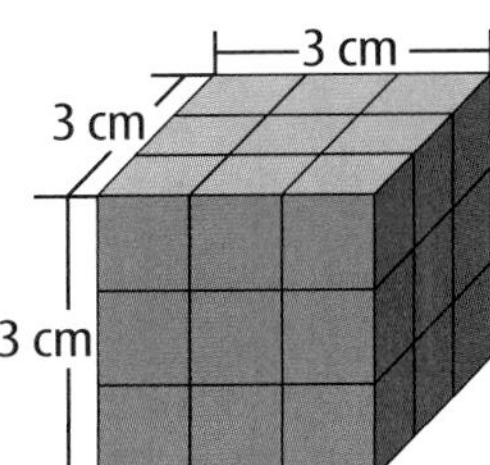

15 Suppose you have ten cubes that measure 2 cm on each side.

a. What is the total surface area of the cubes?
b. What would the surface area be if you glued the cubes together to make one object that is two cubes wide, one cube high, and five cubes long? Hint: draw a picture of the final cube and label the length of each side.

Standardized Test Practice

Record your answers on the answer sheet provided by your teacher or on a sheet of paper.

Multiple Choice

1 How can you verify that a chemical reaction has occurred?

A Check the temperature of the starting and ending substances.

B Compare the chemical properties of the starting substances and ending substances.

C Look for a change in state.

D Look for bubbling of the starting substances.

Use the figure below to answer questions 2 and 3.

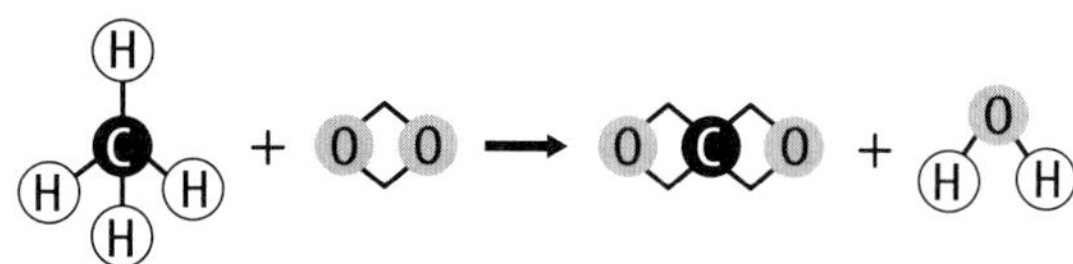

2 The figure above shows models of molecules in a chemical reactions. Which substances are reactants in this reaction?

A CH_4 and CO_2

B CH_4 and O_2

C CO_2 and H_2O

D O_2 and H_2O

3 Which equation shows that atoms are conserved in the reaction?

A $CH_4 + O_2 \longrightarrow CO_2 + H_2O$

B $CH_4 + O_2 \longrightarrow CO_2 + 2H_2O$

C $CH_4 + 2O_2 \longrightarrow CO_2 + 2H_2O$

D $2CH_4 + O_2 \longrightarrow 2CO_2 + H_2O$

4 Which occurs before new bonds can form during a chemical reaction?

A The atoms in the original substances are destroyed.

B The bonds between atoms in the original substances are broken.

C The atoms in the original substances are no longer moving.

D The bonds between atoms in the original substances get stronger.

Use the figure below to answer question 5.

5 The figure above uses shapes to represent a chemical reaction. What kind of chemical reaction does the figure represent?

A decomposition

B double-replacement

C single-replacement

D synthesis

6 Which type of chemical reaction has only one reactant?

A decomposition

B double-replacement

C single-replacement

D synthesis

7 Which element is always a reactant in a combustion reaction?

A carbon

B hydrogen

C nitrogen

D oxygen

Use the figure below to answer question 8.

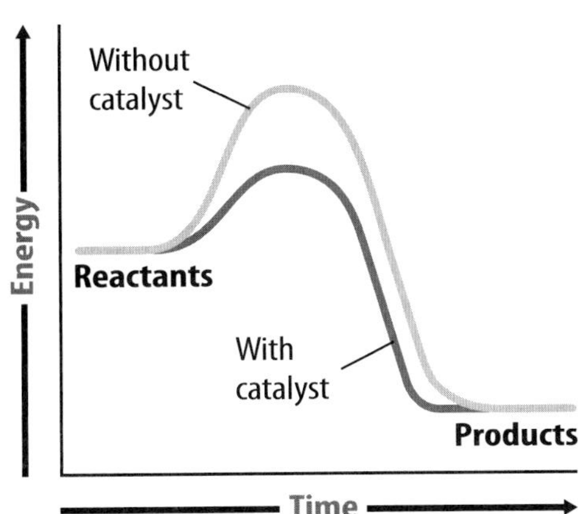

8 The figure above shows changes in energy during a reaction. The lighter line shows the reaction without a catalyst. The darker line shows the reaction with a catalyst. Which is true about these two reactions?

A The reaction with the catalyst is more exothermic than the reaction without the catalyst.

B The reaction with the catalyst requires less activation energy than the reaction without the catalyst.

C The reaction with the catalyst requires more reactants than the reaction without the catalyst.

D The reaction with the catalyst takes more time than the reaction without the catalyst.

Constructed Response

9 Explain the role of energy in chemical reactions.

10 How does a balanced chemical equation illustrate the law of conservation of mass?

11 Many of the reactions that occur when something decays are decomposition reactions. What clues show that this type of reaction is taking place? What happens during a decomposition reaction?

Use the figure below to answer questions 12 and 13.

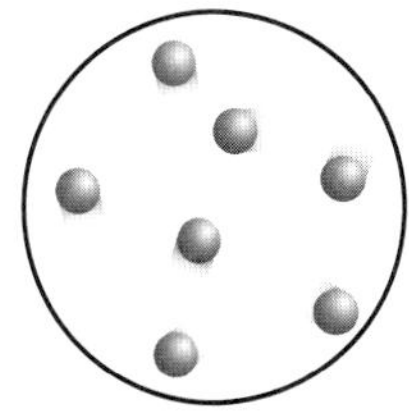

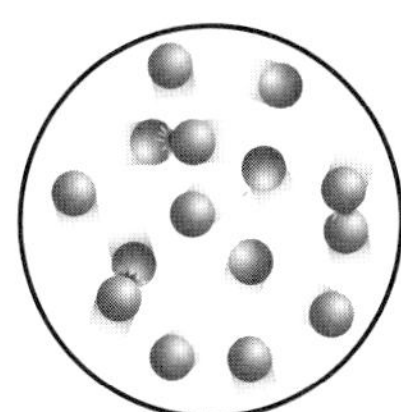

12 Compare the two gas samples represented in the figure in terms of pressure and concentration.

13 Describe the conditions that would increase the rate of a reaction.

NEED EXTRA HELP?													
If You Missed Question...	1	2	3	4	5	6	7	8	9	10	11	12	13
Go to Lesson...	1	1	1	1	2	2	2	3	3	1	2	3	3

Chapter 13

Mixtures, Solubility, and Acid/Base Solutions

THE BIG IDEA What are solutions, and how are they described?

Inquiry Why So Green?

Havasu Falls is located in northern Arizona near the Grand Canyon. The creek that feeds these falls runs through a type of limestone called travertine. Small amounts of travertine are mixed evenly in the water, giving the water its unique blue-green color.

- Can you think of other examples of something mixed evenly in water?
- What do you think of when you hear the word *solution?*
- How would you describe a solution?

Get Ready to Read

What do you think?

Before you read, decide if you agree or disagree with each of these statements. As you read this chapter, see if you change your mind about any of the statements.

1. You can identify a mixture by looking at it without magnification.
2. A solution is another name for a homogeneous mixture.
3. Solutions can be solids, liquids, or gases.
4. A teaspoon of soup is less concentrated than a cup of the same soup.
5. Acids are found in many foods.
6. You can determine the exact pH of a solution by using pH paper.

Lesson 1

Reading Guide

Key Concepts
ESSENTIAL QUESTIONS

- How do substances and mixtures differ?
- How do solutions compare and contrast with heterogeneous mixtures?
- In what three ways do compounds differ from mixtures?

Vocabulary

substance p. 454
mixture p. 454
heterogeneous mixture p. 455
homogeneous mixture p. 455
solution p. 455

g Multilingual eGlossary

Substances and Mixtures

Inquiry What's in the water?

The water in which these fish live is so clear that you might think it is pure water. But if it were pure water (H_2O), the fish could not survive. This is because fish need oxygen. What looks like pure water is actually a mixture of water and other substances, including oxygen. Fish obtain the oxygen from the water.

15 minutes

What makes black ink black?

Many of the products we use every day are mixtures. How can you tell if something is a mixture?

1. Read and complete a lab safety form.
2. Lay a **coffee filter** on your table.
3. Find the center of the coffee filter, and mark it lightly with a **pencil.**
4. Use a **permanent marker** to draw a circle with a diameter of 5 cm around the center of the coffee filter. Do not fill this circle in.
5. Pour **rubbing alcohol** to a depth of 1 cm into a **beaker.**
6. Using the eraser end of your pencil, push the center of the coffee filter down into the beaker until the center, but not the ink, touches the liquid. Keep the ink above the surface of the liquid.
7. Observe the liquid in the bottom of the beaker and the circle on the coffee filter. Record your observations in your Science Journal.
8. Dispose of rubbing alcohol and used coffee filters as instructed by your teacher.

Think About This

1. What happened to the black ink circle on the coffee filter?
2. What was the purpose of the rubbing alcohol?
3. What do you think you would see if you used green ink instead?
4. 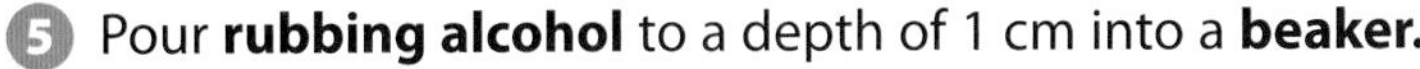 **Key Concept** How do you think this shows that black ink is a mixture?

Matter: Substances and Mixtures

Think about the journey you take to get to school. How many different types of matter do you see? You might see metal, plastic, rocks, concrete, bricks, plants, fabric, water, skin, and hair. So many different types of matter exist around you that it's hard to imagine them all. You might notice that you can group types of matter into categories. For example, keys, coins, and paper clips are all made of metal. Grouping matter into categories helps you understand how some things are similar to each other, but different from other things.

You might be surprised to know that nearly all types of matter can be sorted into just two major categories—substances and mixtures. What are substances and mixtures, and how are they different from each other?

FOLDABLES®

Make horizontal two- and four-tab books. Assemble, staple, and label as shown. Use it to organize your notes on matter.

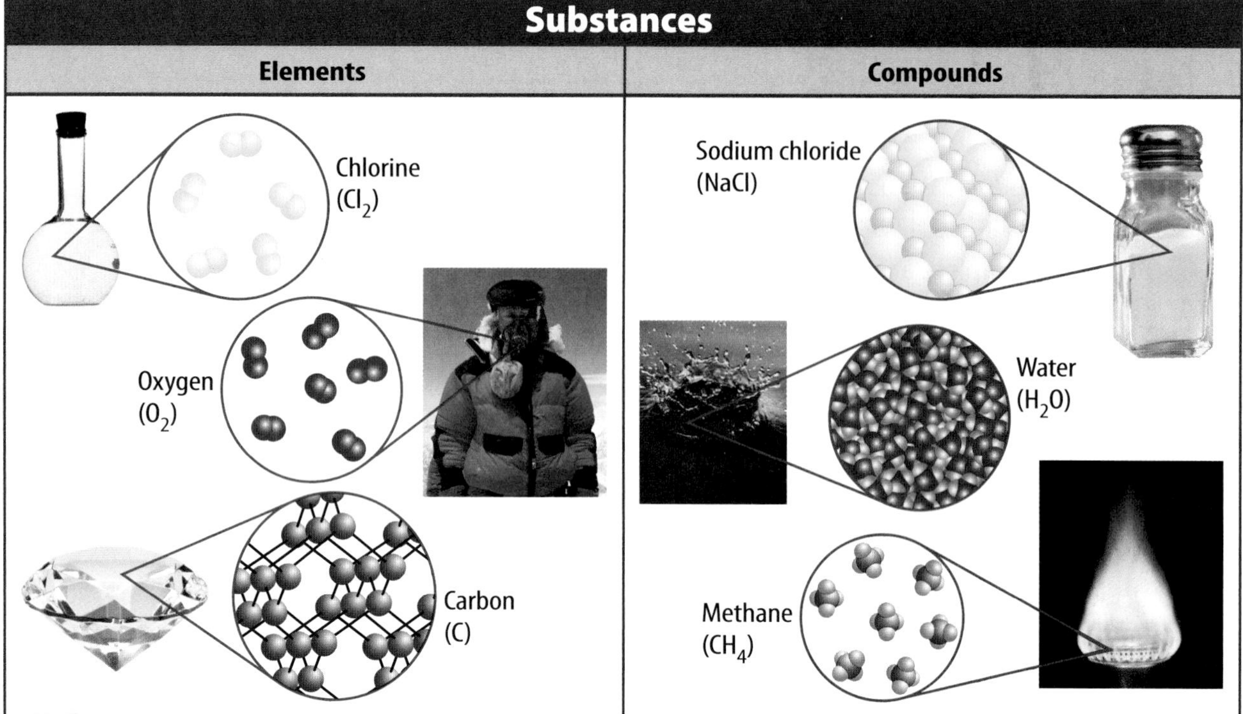

Figure 1 Elements are substances made of only one type of atom. Compounds are made of atoms of two or more elements bonded together.

What is a substance?

A **substance** *is matter that is always made up of the same combination of atoms.* Some substances are shown in **Figure 1.** There are two types of substances—elements and compounds. Recall that an element is matter made of only one type of atom, such as oxygen. A **compound** is matter made of atoms of two or more elements chemically bonded together, such as water (H_2O). Because the compositions of elements and compounds do not change, all elements and compounds are substances.

SCIENCE USE V. COMMON USE

substance

Science Use matter that is always made of the same combination of atoms

Common Use any physical material from which something is made

What is a mixture?

A **mixture** *is two or more substances that are physically blended but are not chemically bonded together.* The amounts of each substance in a mixture can vary. Granite, a type of rock, is a mixture. If you look at a piece of granite, you can see bits of white, black, and other colors. Another piece of granite will have different amounts of each color. The composition of rocks varies.

Air is a mixture, too. Air contains about 78 percent nitrogen, 21 percent oxygen, and 1 percent other substances. However, this composition varies. Air in a scuba tank can have more than 21 percent oxygen and less of the other substances.

REVIEW VOCABULARY

compound

matter made of two or more types of atoms bonded together

It's not always easy to identify a mixture. A rock looks like a mixture, but air does not. Rocks and air are examples of the two different types of mixtures—heterogeneous (he tuh roh JEE nee us) and homogeneous (hoh muh JEE nee us).

 **Key Concept Check** How do substances and mixtures differ?

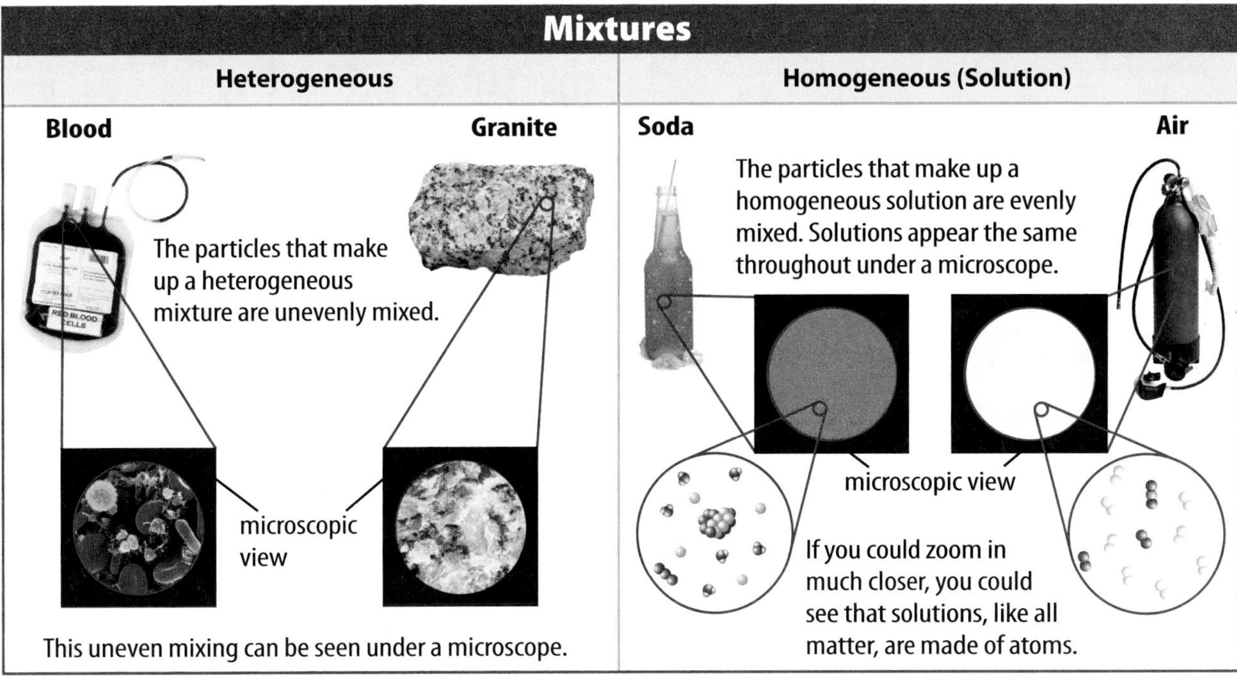

Mixtures
Heterogeneous
Homogeneous (Solution)
Blood
Granite
Soda
Air
The particles that make up a heterogeneous mixture are unevenly mixed.
The particles that make up a homogeneous solution are evenly mixed. Solutions appear the same throughout under a microscope.
microscopic view
microscopic view
If you could zoom in much closer, you could see that solutions, like all matter, are made of atoms.
This uneven mixing can be seen under a microscope.

Figure 2 Solutions look evenly mixed even under a microscope. But if you could zoom in even closer to see individual atoms, you would see that solutions are made of two or more substances mixed evenly but not bonded.

Visual Check Describe what you might see if you were to look at a homogeneous mixture under an ordinary microscope.

Heterogeneous Mixtures *A* **heterogeneous mixture** *is a mixture in which substances are not evenly mixed.* For example, the substances that make up granite, a heterogeneous mixture, are unevenly mixed. When you look at a piece of granite, you can easily see the different parts. Often, you can see the different substances and parts of a heterogeneous mixture with unaided eyes, but sometimes you can only see them with a microscope. For example, blood looks evenly mixed—its color and texture are the same throughout. But, when you view blood with a microscope, as shown in **Figure 2,** you can see areas with more of one component and less of another.

Solutions—Homogeneous Mixtures Many mixtures look evenly mixed even when you view them with a powerful microscope. These mixtures are homogeneous. *A* **homogeneous mixture** *is a mixture in which two or more substances are evenly mixed on the atomic level but not bonded together.* The individual atoms or compounds of each substance are mixed. The mixture looks the same throughout under a microscope because individual atoms and compounds are too small to see.

Air is a homogeneous mixture. If you view air under a microscope, you can't see the individual substances that make it up. This is shown in **Figure 2.** *Another name for a homogeneous mixture is* **solution.** As you read about solutions in this chapter, remember that the term *solution* means "homogeneous mixture."

Word Origin

heterogeneous
from Greek *heteros,* means "different"; and *genos,* means "kind"

homogeneous
from Greek *homos,* means "same"; and *genos,* means "kind"

 **Key Concept Check** How can you determine whether a mixture is homogeneous or heterogeneous?

Figure 3 Some properties of both the sugar and the water are observed in the mixture.

Inquiry MiniLab 20 minutes

Which one is the mixture?

Can you tell the difference between pure water and a mixture of water and salt?

1. Read and complete a lab safety form.
2. Label two **beakers** *A* and *B* as shown.
3. Pour 50 mL of **liquid A** into beaker A.
4. Pour 50 mL of **liquid B** into beaker B.
5. Place both beakers on a **hot plate.**
6. Allow the water to boil until the beakers are nearly dry. Record your observations in your Science Journal.
   ⚠ *Use a hot mitt to handle hot glassware.*

Analyze and Conclude

1. **Compare and Contrast** How did the appearance of beakers A and B change?
2. **Key Concept** Which liquid, A or B, was the mixture? How do you know?
3. **Describe** how the water could be collected and used as a freshwater source.

How do compounds and mixtures differ?

You have read that a compound contains two or more elements chemically bonded together. In contrast, the substances that make up a mixture are not chemically bonded. So, mixing is a physical change. The substances that exist before mixing still exist in the mixture. This leads to two important differences between compounds and mixtures.

Substances keep their properties.

Because substances that make up a mixture are not changed chemically, some of their properties are observed in the mixture. Sugar water, shown in **Figure 3,** is a mixture of two compounds—sugar and water. After the sugar is mixed in, you can't see the sugar in the water, but you can still taste it. Some properties of the water, such as its liquid state, are also observed in the mixture.

In contrast, the properties of a compound can be different from the properties of the elements that make it up. Sodium and chlorine bond to form table salt. Sodium is a soft, opaque, silvery metal. Chlorine is a greenish, poisonous gas. None of these properties are observed in table salt.

Mixtures can be separated.

Because the substances that make up a mixture are not bonded together, they can be separated from each other using physical methods. The physical properties of one substance are different from those of another. These differences can be used to separate the substances. In contrast, the only way compounds can be separated is by a chemical change that breaks the bonds between the elements. **Figure 4** summarizes the characteristics of substances and mixtures.

Key Concept Check In what three ways do compounds differ from mixtures?

Figure 4 This organizational chart shows how different types of matter are classified. All matter can be classified as either a substance or a mixture.

Visual Check Can a mixture be made only of elements? Explain.

Matter

- anything that has mass and takes up space
- Most matter on Earth is made up of atoms.

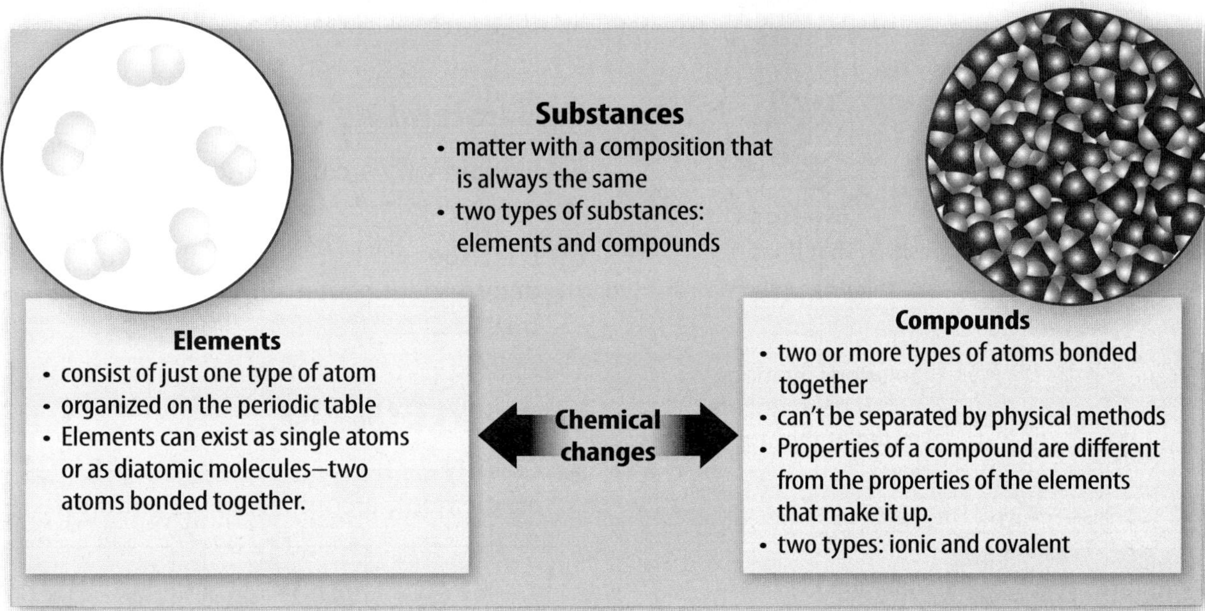

Substances
• matter with a composition that is always the same
• two types of substances: elements and compounds
Elements
• consist of just one type of atom
• organized on the periodic table
• Elements can exist as single atoms or as diatomic molecules—two atoms bonded together.
Chemical changes
Compounds
• two or more types of atoms bonded together
• can't be separated by physical methods
• Properties of a compound are different from the properties of the elements that make it up.
• two types: ionic and covalent

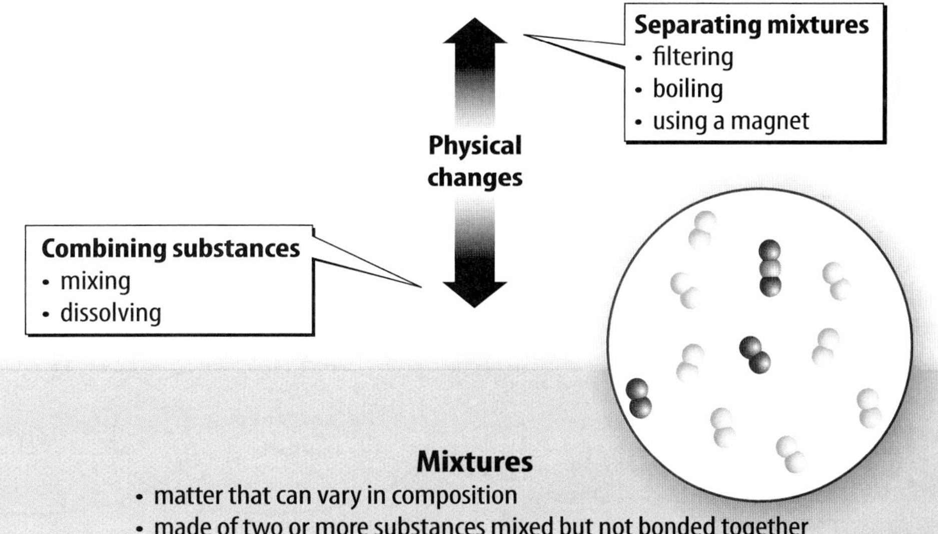

Lesson 1 Review

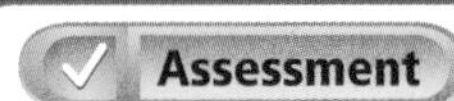

Online Quiz

Visual Summary

Substance **Mixtures**

Substances have a composition that does not change. The composition of mixtures can vary.

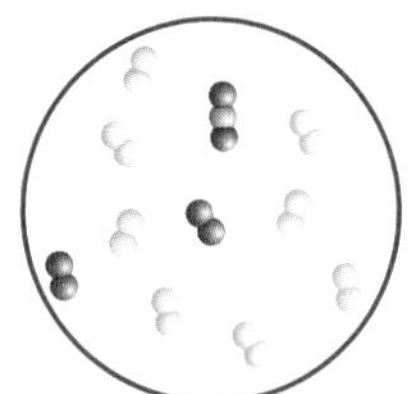

Solutions (homogeneous mixtures) are mixed at the atomic level.

Mixtures contain parts that are not bonded together. These parts can be separated using physical means.

FOLDABLES®

Use your lesson Foldable to review the lesson. Save your Foldable for the project at the end of the chapter.

What do you think NOW?

You first read the statements below at the beginning of the chapter.

1. You can identify a mixture by looking at it without magnification.
2. A solution is another name for a homogeneous mixture.

Did you change your mind about whether you agree or disagree with the statements? Rewrite any false statements to make them true.

Use Vocabulary

1. **Identify** What is another name for a homogeneous mixture?
2. **Contrast** homogeneous and heterogeneous mixtures.

Understand Key Concepts 

3. **Explain** why a compound is classified as a substance.
4. **Describe** two tests that you can run to determine if something is a substance or a mixture.

Interpret Graphics

5. **Classify** each group of particles below as an element, a compound, or a mixture.

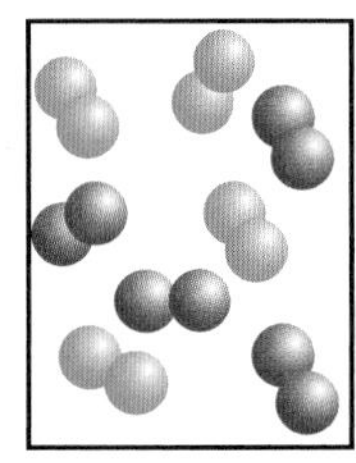

A

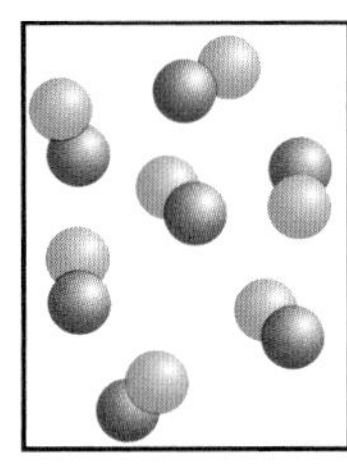

B

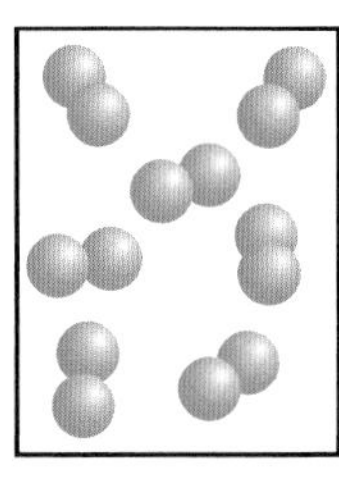

C

6. **Compare and contrast** Copy the graphic organizer below and use it to compare and contrast heterogeneous mixtures and solutions.

Heterogeneous mixtures		Solutions

Critical Thinking

7. **Explain** the following statement: All compounds are substances, but not all substances are compounds.
8. **Suppose** you have found an unknown substance in a laboratory. It has the formula H_2O_2 written on the bottle. Is it water? How do you know?

Sports Drinks and Your Body

HOW NATURE WORKS

Electrolytes and the Conduction of Nerve Impulses

When you exercise, you sweat. When you're thirsty, do you drink water or a sports drink? Chances are you have seen ads that claim sports drinks are better than water because they contain electrolytes. What are electrolytes, and how are they used in your body?

What are electrolytes?

Many elements, including potassium, sodium, chlorine, and calcium form ions. An electrolyte is a charged particle, scientifically known as an ion. Electrolytes can conduct electric charges. Pure water cannot conduct electric charges, but water containing electrolytes can.

Electrolytes in Your Body

Why does your body need electrolytes? Solutions of water and electrolytes surround all of the cells in your body. Electrolytes enable these solutions to carry nerve impulses from one cell to another. Your body's voluntary movements, such as walking, and involuntary movements, such as your heart beating, are caused by nerve impulses. Without electrolytes, these nerve impulses cannot move normally.

Nutrient	Function	Sex/Age	Adequate Intake	Food Sources
Sodium	maintains fluid volume outside of cells and aids in normal cell function	males/females 14–18	1.5 g/day	Sodium is added to many processed foods.
Potassium	maintains fluid volume inside/outside of cells and aids in normal cell function; supports blood and kidney health	males/females 14–18	4.7 g/day	fresh fruits and vegetables, dried peas, dairy products, meats, and nuts
Water	allows transport of nutrients to cells and removal of waste products	males females	3.3 L/day 2.3 L/day	all beverages, including water, as well as moisture in foods

Replenishing Fluids

Have you ever noticed that sweat is salty? Sweat is a solution of water and electrolytes, including sodium. Sports drinks can replace water and electrolytes. Some foods, such as bananas, oranges, and lima beans, also contain electrolytes. However, many sports drinks contain ingredients your body doesn't need, such as caffeine, sugar, and artificial colors. Unless you are sweating for an extended period of time, you don't need to replace electrolytes, but replacing water is always essential.

It's Your Turn

RESEARCH Study three different sports drinks. Create a bar graph that compares ingredients such as water, sugar, electrolytes, and caffeine in each brand. Draw a conclusion about which type of drink is best for your body.

Lesson 2

Reading Guide

Key Concepts
ESSENTIAL QUESTIONS

- Why do some substances dissolve in water and others do not?
- How do concentration and solubility differ?
- How can the solubility of a solute be changed?

Vocabulary

solvent p. 461
solute p. 461
polar molecule p. 462
concentration p. 464
solubility p. 466
saturated solution p. 466
unsaturated solution p. 466

Multilingual eGlossary

Video

What's Science Got to do With It?

Properties of Solutions

Inquiry Stairs?

These stair-like formations, called terraces, are located in Mammoth Hot Springs, a part of Yellowstone National Park in Wyoming. Hot spring water is a mixture of water and dissolved carbon dioxide and limestone. When this mixture reaches Earth's surface, the pressure on the mixture is reduced. This causes the limestone to leave the mixture, forming the terraces.

Launch Lab

15 minutes

How are they different?

If you have ever looked at a bottle of Italian salad dressing, you know that some substances do not easily form solutions. The oil and vinegar do not mix, and the spices sink to the bottom. However, the salt in salad dressing does mix evenly with the other substances and forms a solution. How can we describe the difference quantitatively?

1. Read and complete a lab safety form.
2. Label one **beaker** *A* and **another beaker** *B*.
3. Measure 100 mL of water and pour it into beaker A.
4. Measure 100 mL of water and pour it into beaker B.
5. Add 10 g of **baking soda** to beaker A, and stir with a **plastic spoon** for 2 min or until all the baking soda dissolves, whichever happens first.
6. Add 25 g of **sugar** to beaker B and stir with a plastic spoon for 2 min or until all of the dissolves, whichever happens first.
7. Observe the mixtures in each beaker. Record your observations in your Science Journal.

Think About This

1. What substance dissolved better in water? How do you know?
2. Predict what would happen if you were to use 200 mL of water instead of 100 mL.
3. Do you think more baking soda might dissolve if you stirred the solution longer?
4. **Key Concept** Why do you think one substance dissolved more easily in water than the other substance? What factors do you think contribute to this difference?

Parts of Solutions

You've read that a solution is a homogeneous mixture. Recall that in a solution, substances are evenly mixed on the atomic level. How does this mixing occur? Dissolving is the process of mixing one substance into another to form a solution. Scientists use two terms to refer to the substances that make up a solution. Generally, the **solvent** *is the substance that exists in the greatest quantity in a solution. All other substances in a solution are* **solutes.** Recall that air is a solution of 78 percent nitrogen, 21 percent oxygen, and 1 percent other substances. Which substance is the solvent? In air, nitrogen exists in the greatest quantity. Therefore, it is the solvent. The oxygen and other substances are solutes. In this lesson, you will read the terms *solute* and *solute* often. Refer back to this page if you forget what these terms mean.

 **Reading Check** How do a solute and a solvent differ?

FOLDABLES®

Make a four-tab shutter-fold. Label it as shown. Collect information about which solvents dissolve which solutes.

Polar solvents dissolve:	Nonpolar solvents dissolve:
Like Dissolves Like	
Polar solvents do not dissolve:	Nonpolar solvents do not dissolve:

Table 1 Types of Solutions

State of Solution	Solvent Is:	Solute Can Be:
Solid	solid	**gas or solid (called alloys)** This saxophone is a solid solution of solid copper and solid zinc.
Liquid	liquid	**solid, liquid, and/or gas** Soda is a liquid solution of liquid water, gaseous carbon dioxide, and solid sugar and other flavorings.
Gas	gas	**gas** This lighted sign contains a gaseous mixture of gaseous argon and gaseous mercury.

Interactive Table

The electrons spend more time near the oxygen atom. This makes the end with the oxygen atom slightly negative (–).

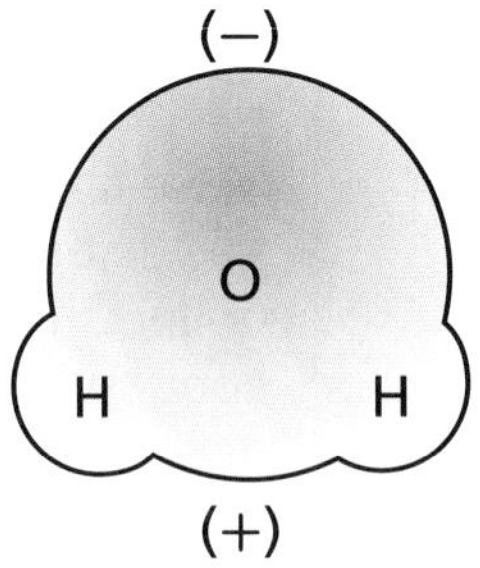

The end with the hydrogen atoms is slightly positive (+).

▲ **Figure 5** Water is a polar molecule. The end with the oxygen atom is slightly negative, and the end with the hydrogen atoms is slightly positive.

Personal Tutor

Types of Solutions

When you think of a solution, you might think of a liquid. However, solutions can exist in all three states of matter—solid, liquid, or gas. The state of the solvent, because it exists in the greatest quantity, determines the state of the solution. **Table 1** contrasts solid, liquid, and gaseous solutions.

Water as a Solvent

Did you know that over 75 percent of your brain and almost 90 percent of your lungs are made of water? Water is one of the few substances on Earth that exists naturally in all three states—solid, liquid, and gas. However, much of this water is not pure water. In nature, water almost always exists as a solution; it contains dissolved solutes. Why does nearly all water on Earth contain dissolved solutes? The answer has to do with the structure of the water molecule.

The Polarity of Water

A water molecule, such as the one shown in **Figure 5,** is a covalent compound. Recall that atoms are held together with covalent bonds when sharing electrons. In a water molecule, one oxygen atom shares electrons with two hydrogen atoms. However, these electrons are not shared equally. The electrons in the oxygen-hydrogen bonds more often are closer to the oxygen atom than they are to the hydrogen atoms. This unequal sharing of electrons gives the end with the oxygen atom a slightly negative charge. And, it gives the end with the hydrogen atoms a slightly positive charge. Because of the unequal sharing of electrons, a water molecule is said to be polar. *A* **polar molecule** *is a molecule with a slightly negative end and a slightly positive end.* Nonpolar molecules have an even distribution of charge. Solutes and solvents can be polar or nonpolar.

Like Dissolves Like

Water is often called the universal solvent because it dissolves many different substances. But water can't dissolve everything. Why does water dissolve some substances but not others? Water is a polar solvent. Polar solvents dissolve polar solutes easily. Nonpolar solvents dissolve nonpolar solutes easily. This is summarized by the phrase "like dissolves like." Because water is a polar solvent, it dissolves most polar and ionic solutes.

Key Concept Check Why do some substances dissolve in water and others do not?

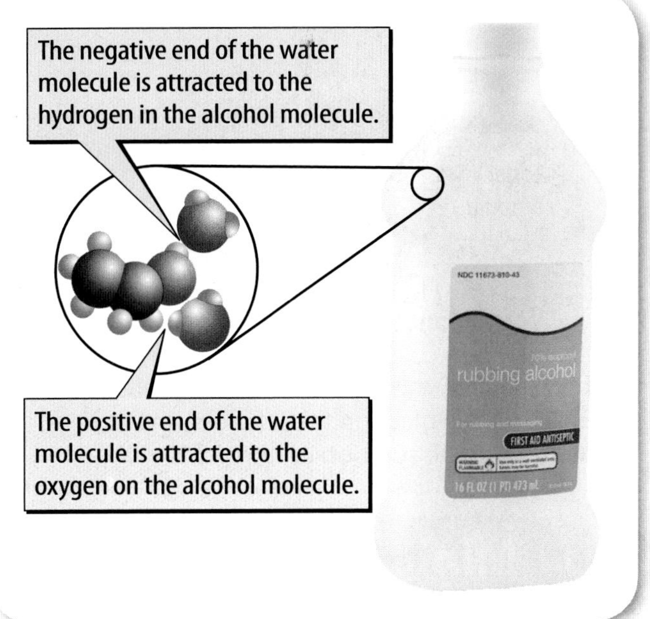

The negative end of the water molecule is attracted to the hydrogen in the alcohol molecule.
The positive end of the water molecule is attracted to the oxygen on the alcohol molecule.
rubbing alcohol
FIRST AID ANTISEPTIC

▲ **Figure 6** When a polar solute, such as rubbing alcohol, dissolves in a polar solvent, such as water, the poles of the solvent are attracted to the oppositely charged poles of the solute.

Polar Solvents and Polar Molecules

Because water molecules are polar, water dissolves groups of other polar molecules. **Figure 6** shows what rubbing alcohol, a substance used as a disinfectant, looks like when it is in solution with water. Molecules of rubbing alcohol also are polar. Therefore, when rubbing alcohol and water mix, the positive ends of the water molecules are attracted to the negative ends of the alcohol molecules. Similarly, the negative ends of the water molecules are attracted to the positive ends of the alcohol. In this way, alcohol molecules dissolve in the solvent.

Polar Solvents and Ionic Compounds

Many ionic compounds are also soluble in water. Recall that ionic compounds are composed of alternating positive and negative ions. Sodium chloride (NaCl) is an ionic compound composed of sodium ions (Na^+) and chloride ions (Cl^-). When sodium chloride dissolves, these ions are pulled apart by the water molecules. This is shown in **Figure 7.** The negative ends of the water molecules attract the positive sodium ions. The positive ends of the water molecules attract the negative chloride ions.

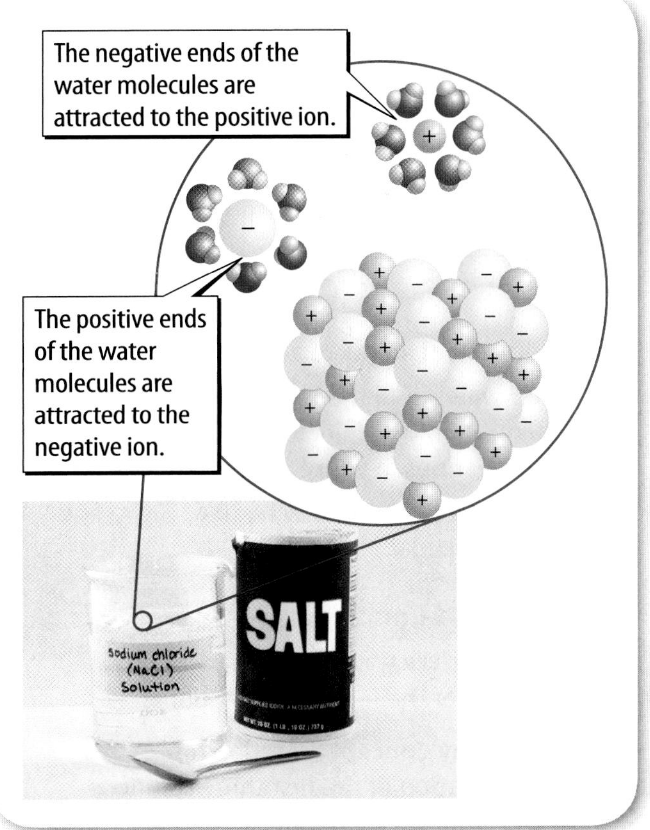

▲ **Figure 7** When ionic solutes dissolve, the positive poles of the solvent are attracted to the negative ions. The negative poles of the solvent are attracted to the positive ions.

Figure 8 The volumes of both drinks are the same, but the glass on the left contains more solute than the solution on the right.

Inquiry MiniLab — 15 minutes

How much is dissolved?

How can you make a glass and a pitcher of lemonade with the same sweetness?

1. Read and complete a lab safety form.
2. Make a table similar to the one below.

Mass of sugar (g)	Volume of solution (mL)	Observation
25 g	50 mL	
50 g	50 mL	
100 g	50 mL	
125 g	50 mL	

3. Add 25 g of **sugar** to a **beaker.**
4. Add water until the volume of solution is 50 mL. Stir for 2 min. Record your observations in your Science Journal.
5. Repeat steps 3 and 4 with 50 g, 100 g, and 125 g of sugar.

Analyze and Conclude

1. **Calculate** What mass of sugar is contained in 25 mL of the first three solutions?
2.  **Key Concept** Describe the concentration of the first three solutions using words and quantities.
3. **Infer** How might you make 100 mL of the first solution so that it has the same concentration as 50 mL of this solution?

Concentration—How much is dissolved?

Have you ever tasted a spoonful of soup and wished it had more salt in it? In a way, your taste buds were measuring the amount, or concentration, of salt in the soup. **Concentration** *is the amount of a particular solute in a given amount of solution.* In the soup, salt is a solute. Saltier soup has a higher concentration of salt. Soup with less salt has a lower concentration of salt. Look at the glasses of fruit drink in **Figure 8.** Which drink has a higher concentration of solute? The darker blue drink has a higher concentration of solute.

Concentrated and Dilute Solutions

One way to describe the saltier soup is to say that it is more concentrated. The less salty soup is more dilute. The terms *concentrated* and *dilute* are one way to describe how much solute is dissolved in a solution. However, these terms don't state the exact amount of solute dissolved. What one person thinks is concentrated might be what another person thinks is dilute. Soup that tastes too salty to you might be perfect for someone else. How can concentration be described more precisely?

Reading Check Why is the term *dilute* not a precise way to describe concentration?

Describing Concentration Using Quantity

A more precise way to describe concentration is to state the quantity of solute in a given quantity of solution. When a solution is made of a solid dissolved in a liquid, such as salt in water, concentration is the mass of solute in a given volume of solution. Mass usually is stated in grams, and volume usually is stated in liters. For example, concentration can be stated as grams of solute per 1 L of solution. However, concentration can be stated using any units of mass or volume.

Calculating Concentration—Mass per Volume

One way that concentration can be calculated is by the following equation:

$$\text{Concentration } (C) = \frac{\text{mass of solute } (m)}{\text{volume of solution } (V)}$$

To calculate concentration, you must know both the mass of solute and the volume of solution that contains this mass. Then divide the mass of solute by the volume of solution.

Reading Check If more solvent is added to a solution, what happens to the concentration of the solution?

Concentration—Percent by Volume

Not all solutions are made of a solid dissolved in a liquid. If a solution contains only liquids or gases, its concentration is stated as the volume of solute in a given volume of solution. In this case, the units of volume must be the same—usually mL or L. Because the units match, the concentration can be stated as a percentage. Percent by volume is calculated by dividing the volume of solute by the total volume of solution and then multiplying the quotient by 100. For example, if a container of orange drink contains 3 mL of acetic acid in a 1,000-mL container, the concentration is 0.3 percent.

Math Skills Calculate Concentration

Solve for Concentration

Suppose you want to calculate the concentration of salt in a **0.4 L** can of soup. The back of the can says it contains **1.6 g** of salt. What is its concentration in g/L? In other words, how much salt would be contained in 1 L of soup?

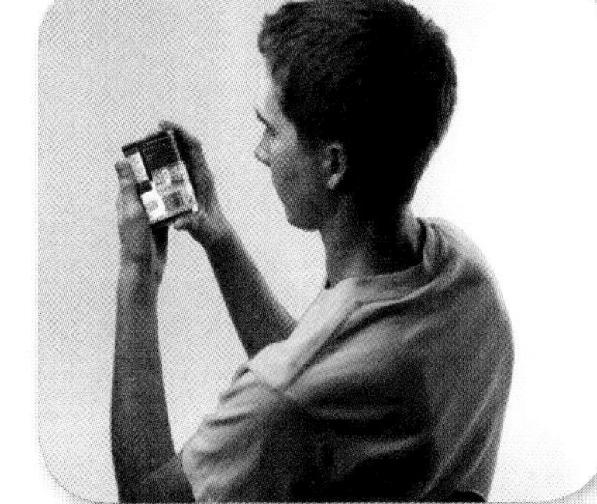

1. **This is what you know:** mass: **1.6 g**; volume: **0.4 L**
2. **This is what you need to find:** concentration: C
3. **Use this formula:** $C = \frac{m}{V}$
4. **Substitute:** the values for m and V into the formula and divide. $C = \frac{1.6 \text{ g}}{0.4 \text{ L}} = 4 \text{ g/L}$

Answer: The concentration is 4 g/L. As you might expect, 0.4 L of soup contains less salt (1.6 g) than 1 L of soup (4 g). However, the concentration of both amounts of soup is the same—4 g/L.

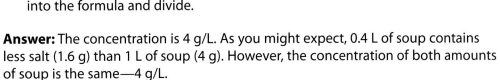

Practice

1. What is the concentration of 5 g of sugar in 0.2 L of solution?
2. How many grams of salt are in 5 L of a solution with a concentration of 3 g/L?
3. Suppose you add water to 6 g of sugar to make a solution with a concentration of 3 g/L. What is the total volume of the solution?

- Math Practice
- Personal Tutor

Solubility—How much can dissolve?

WORD ORIGIN

solubility
from Latin solvere, means "to loosen"

Have you ever put too much sugar into a glass of iced tea? What happens? Not all of the sugar dissolves. You stir and stir, but there is still sugar at the bottom of the glass. That is because there is a limit to how much solute (sugar) can be dissolved in a solvent (water). **Solubility** (sahl yuh BIH luh tee) *is the maximum amount of solute that can dissolve in a given amount of solvent at a given temperature and pressure.* If a substance has a high solubility, more of it can dissolve in a given solvent.

Key Concept Check How do concentration and solubility differ?

Saturated and Unsaturated Solutions

ACADEMIC VOCABULARY

analogous
(adjective) showing a likeness in some ways between things that are otherwise different

If you add water to a dry sponge, the sponge absorbs the water. If you keep adding water, the sponge becomes saturated. It can't hold any more water. This is **analogous** (uh NA luh gus), or similar, to what happens when you stir too much sugar into iced tea. Some sugar dissolves, but the excess sugar does not dissolve. The solution is saturated. *A* **saturated solution** *is a solution that contains the maximum amount of solute the solution can hold at a given temperature and pressure. An* **unsaturated solution** *is a solution that can still dissolve more solute at a given temperature and pressure.*

Factors that Affect How Much Can Dissolve

Can you change the amount of a particular solute that can dissolve in a solvent? Yes. Recall the definition of solubility—the maximum amount of solute that can dissolve in a given amount of solvent at a given temperature and pressure. Changing either temperature or pressure changes how much solute can dissolve in a solvent.

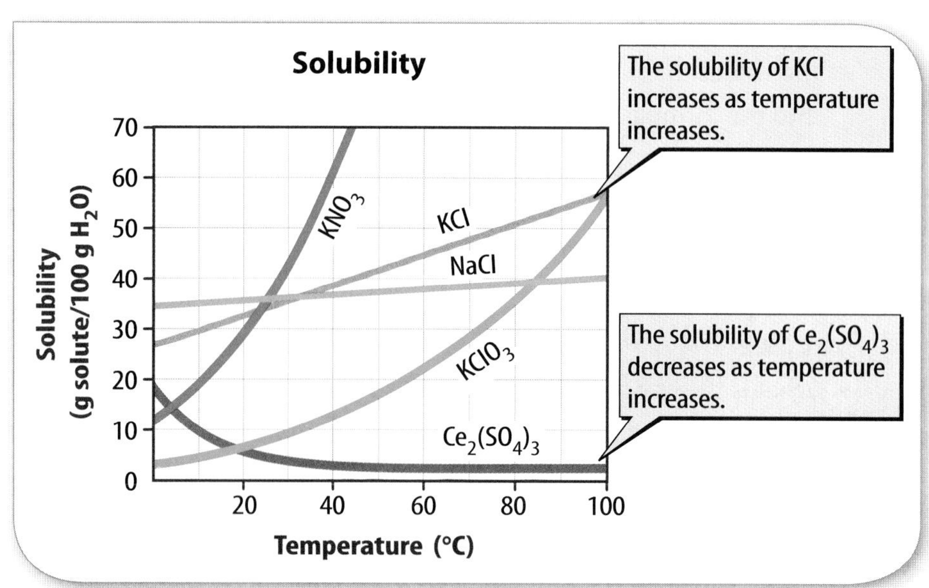

Figure 9 Some solids are more soluble in warmer liquids than cooler ones. Other solids are less soluble in warmer liquids than cooler ones. This difference depends on the chemical structure of the solid.

Visual Check How many grams of KNO_3 will dissolve in 100 g of water at 10°C?

Effect of Temperature Have you noticed that more sugar dissolves in hot tea than in iced tea? The solubility of sugar in water increases with temperature. This is true for many solid solutes, as shown in **Figure 9.** Notice that some solutes become less soluble when temperature is increased.

How does temperature affect the solubility of a gas in a liquid? Recall that soda, or soft drinks, contains carbon dioxide, a gaseous solute, dissolved in liquid water. The bubbles you see in soda are made of undissolved carbon dioxide. Have you ever noticed that more carbon dioxide bubbles out when you open a warm can of soda than when you open a cold can? This is because the solubility of a gas in a liquid decreases when the temperature of the solution increases.

▲ **Figure 10** When the pressure of a gas is increased, it becomes more soluble in a liquid. When the can is opened, this pressure is lowered and the gas leaves the solution.

Effect of Pressure What keeps carbon dioxide dissolved in an unopened can of soda? In a can, the carbon dioxide in the space above the liquid soda is under pressure. This causes the gas to move to an area of lower pressure—the solvent. The gas moves into the solvent, and a solution is formed. When the can is opened, as shown in **Figure 10,** this pressure is released, and the carbon dioxide gas leaves the solution. Pressure does not affect the solubility of a solid solute in a liquid.

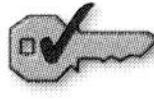

Key Concept Check How can the solubility of a solute be changed?

How Fast a Solute Dissolves

Temperature and pressure can affect how much solute dissolves. If solute and solvent particles come into contact more often, the solute dissolves faster. **Figure 11** shows three ways to increase how often solute particles contact solvent particles. Each of these methods will make a solute dissolve faster. However, it is important to note that stirring the solution or crushing the solute will not make more solute dissolve.

Stirring the solution

Crushing the solute

Increasing the temperature

Figure 11 Several factors can affect how quickly a solute will dissolve in a solution. However, dissolving more quickly won't necessarily make more solute dissolve.

Lesson 2 Review

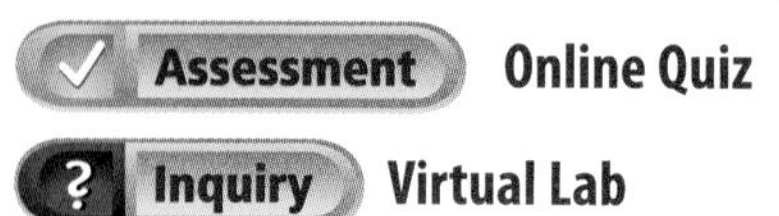

Visual Summary

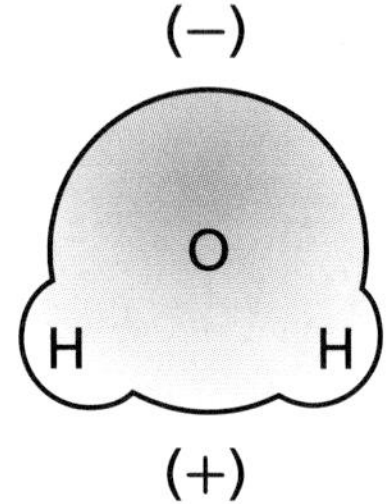

Polar molecule

Substances dissolve in other substances that have similar polarity. In other words, like dissolves like.

Concentration is the amount of substance that is dissolved. Solubility is the maximum amount that can dissolve.

Both temperature and pressure affect the solubility of solutes in solutions.

FOLDABLES®

Use your lesson Foldable to review the lesson. Save your Foldable for the project at the end of the chapter.

What do you think NOW?

You first read the statements below at the beginning of the chapter.

3. Solutions can be solids, liquids, or gases.

4. A teaspoon of soup is less concentrated than a cup of the same soup.

Did you change your mind about whether you agree or disagree with the statements? Rewrite any false statements to make them true.

Use Vocabulary

1. **Define** *polar molecule* in your own words.

Understand Key Concepts 

2. **Explain** how you could use the solubility of a substance to make a saturated solution.

3. **Predict** whether an ionic compound will dissolve in a nonpolar solvent.

Interpret Graphics

4. **Read a Graph** Use the graph to determine what you would observe in a solution of 30 g of $KClO_3$ in 100 g of water at 10°C.

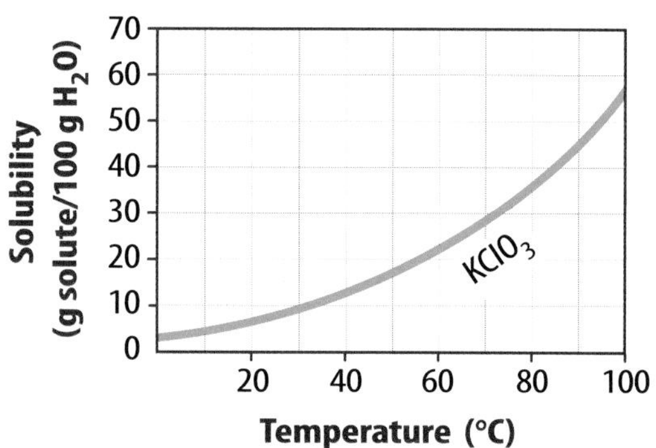

5. **Organize** Copy the graphic organizer, and use it to organize three factors that increase the speed a solute dissolves in a liquid.

Critical Thinking

6. **Explain** A student wants to increase the maximum amount of sugar that can dissolve in water. She crushes the sugar and then stirs it into the water. Does this work?

Math Skills

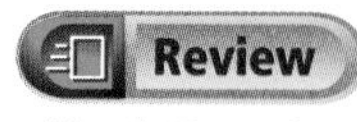

Math Practice

7. Use ratios to explain how a tablespoon of soup and a cup of the same soup have the same concentration.

Manipulate Variables

45 minutes

How does a solute affect the conductivity of a solution?

Materials

triple-beam balance

250-mL beaker

stirring rod

salt

Also needed: 6-V battery, wires, sugar, miniature lightbulb with base

Safety

When some substances dissolve in water, they form ions, or charged particles. Other substances do not. Solutions that contain ions conduct electricity. In this lab, you will determine how one variable affects another.

Learn It

In an experiment, you can manipulate variables or factors. The factor you change is called the independent variable. The factor or factors that change as a result of a change to the independent variable are called the dependent variables. All other factors should be kept constant.

Try It

1. Read and complete a lab safety form.
2. Dissolve 20 g of salt in 100 mL of water to make a saltwater solution with a concentration of 200 g/L. Label the beaker *salt water, 200 g/L.*
3. Prepare a sugar solution with a concentration of 200 g/L. Label the beaker *sugar water, 200 g/L.*
4. Create a circuit as shown in the photograph below using the salt-water solution, wires, a 6-V battery, and a lightbulb with base.

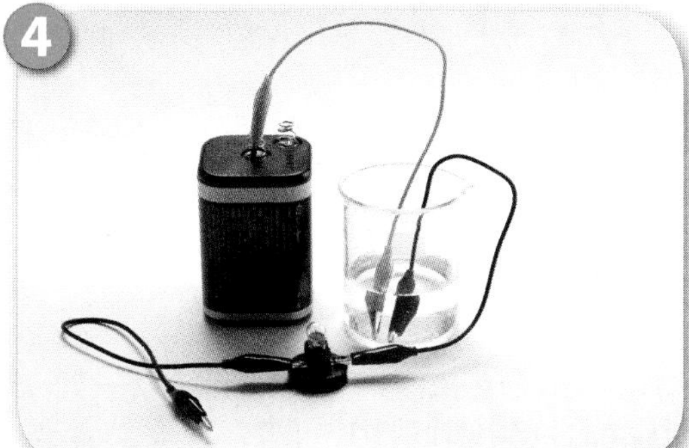

5. Once the circuit is complete, record your observations in your Science Journal.
6. Remove the wires from the solution, and rinse them with plain water.
7. Repeat steps 4–6 using the sugar-water solution.
8. Repeat steps 4–6 using just water.

Apply It

9. Based on your observations, what happened when the circuit was made using each of the solutions?
10. What would happen if the concentration of the solution that conducted the electricity were changed to a more dilute solution? What if it was changed to a more concentrated solution?
11. **Key Concept** Test your hypothesis by creating two more saltwater solutions, one with a concentration of 100 g/L and one with a concentration of 300 g/L. Use them, one at a time, to complete the circuit. Record your observations in your Science Journal.

Lesson 3

Reading Guide

Key Concepts
ESSENTIAL QUESTIONS

- What happens when acids and bases dissolve in water?
- How does the concentration of hydronium ions affect pH?
- What methods can be used to measure pH?

Vocabulary

acid p. 472
hydronium ion p. 472
base p. 472
pH p. 474
indicator p. 476

Multilingual eGlossary

Video BrainPOP®

Acid and Base Solutions

Inquiry What's eating her?

When this statue was first carved, it didn't have any of these odd-shaped marks on its surface. This damage was caused by acid rain—precipitation that contains water and dissolved substances called acids. When acid rain falls on this statue, the acid reacts with the stone, dissolving it and then carrying it away.

15 minutes

What color is it?

Did you know that all rain is naturally acidic? As raindrops fall through the air, they pick up molecules of carbon dioxide. An acid called carbonic acid is formed when the water molecules react with the carbon dioxide molecules. An indicator is a substance that can be used to tell if a solution is acidic, basic, or neutral.

1. Read and complete a lab safety form.
2. Half fill a **beaker** with the **colored solution.**
3. Place one end of a **straw** into the solution.
   ⚠ **Caution:** *Do not suck liquid through the straw.*
4. Blow through the straw, making bubbles in the solution. Continue blowing, and count how many times you have to blow bubbles until you observe a change.
5. Record your observations in your Science Journal.

Think About This

1. Describe what change you saw take place.
2. What do you think made this change occur?
3. How do you think the results would have been different if you had held your breath for several seconds before blowing through the straw?
4. **Key Concept** Using the terms *acidic* and *basic,* explain what you have learned about the colored solution being used.

What are acids and bases?

Would someone ever drink an acid? At first thought, you might answer no. After all, when people think of acids, they often think of acids such as those found in batteries or in acid rain. However, acids are found in other items, including milk, vinegar, fruits, and green leafy vegetables. Some examples of acids that you might eat are shown in **Figure 12.** Along with the word *acid,* you might have heard the word *base.* Like acids, you can also find bases in your home. Detergent, antacids, and baking soda are examples of items that contain bases. But acids and bases are found in more than just household goods. As you will learn in this lesson, they are necessities for our daily life.

Figure 12 You might be surprised to learn that acids are common in the foods you eat. All of the foods pictured here contain acids.

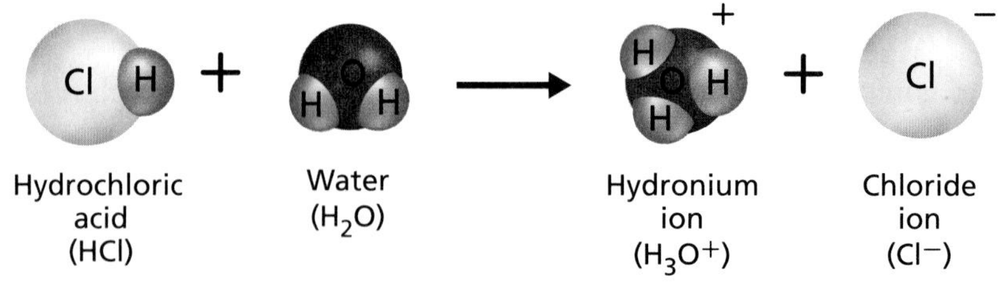

Figure 13 Acids, such as hydrochloric acid, produce hydronium ions when they dissolve in water. ▶

Acids

Have you ever tasted the sourness of a lemon or a grapefruit? This sour taste is due to the acid in the fruit. *An* **acid** *is a substance that produces a hydronium ion (H_3O^+) when dissolved in water.* Nearly all acid molecules contain one or more hydrogen atoms (H). When an acid mixes with water, this hydrogen atom separates from the acid. It quickly combines with a water molecule, resulting in a hydronium ion. This process is shown in **Figure 13.** A **hydronium ion**, H_3O^+, *is a positively charged ion formed when an acid dissolves in water.*

WORD ORIGIN

acid
from Latin *acidus,* means "sour"

Bases

A **base** *is a substance that produces hydroxide ions (OH^-) when dissolved in water.* When a hydroxide compound such as sodium hydroxide (NaOH) mixes with water, hydroxide ions separate from the base and form hydroxide ions (OH^-) in water. Some bases, such as ammonia (NH_3), do not contain hydroxide ions. These bases produce hydroxide ions by taking hydrogen atoms away from water, leaving hydroxide ions (OH^-). This process is shown in **Figure 14.** Some properties and uses of acids and bases are shown in **Table 2.**

Figure 14 Bases, such as sodium hydroxide and ammonia, produce hydroxide ions when they dissolve in water. ▼

Visual Check How is dissolving an acid, shown above, similar to dissolving ammonia, shown below?

Key Concept Check What happens when acids and bases dissolve in water?

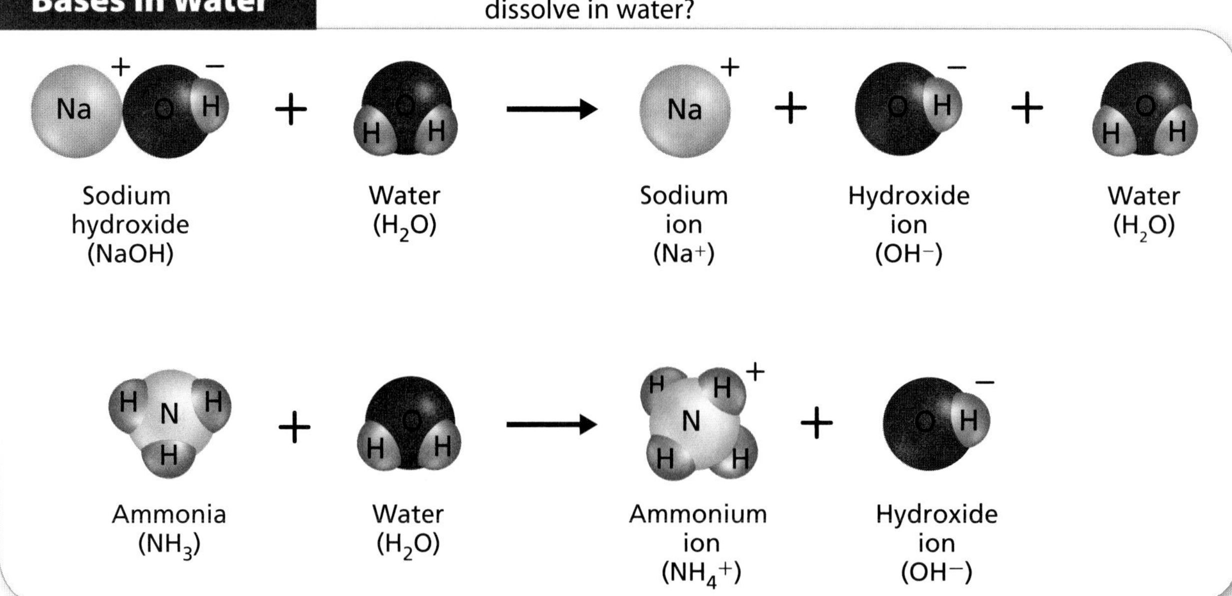

Table 2 Properties and Uses of Acids and Bases

	Acids	Bases
Ions produced	Acids produce H_3O^+ in water.	Bases produce OH^- ions in water.
Examples	• hydrochloric acid, HCl • acetic acid, CH_3COOH • citric acid, $H_3C_6H_5O_7$ • lactic acid, $C_3H_6O_3$	• sodium hydroxide, NaOH • ammonia, NH_3 • sodium carbonate, Na_2CO_3 • calcium hydroxide, $Ca(OH)_2$
Some properties	• Acids provide the sour taste in food (never taste acids in the laboratory). • Most can damage skin and eyes. • Acids react with some metals to produce hydrogen gas. • H_3O^+ ions can conduct electricity in water. • Acids react with bases to form neutral solutions.	• Bases provide the bitter taste in food (never taste bases in the laboratory). • Most can damage skin and eyes. • Bases are slippery when mixed with water. • OH^- ions can conduct electricity in water. • Bases react with acids to form neutral solutions.
Some uses	• Acids are responsible for for natural and artificial flavoring in foods, such as fruits. • Lactic acid is found in milk. • Acid in your stomach breaks down food. • Blueberries, strawberries, and many vegetable crops grow better in acidic soil. • Acids are used to make products such as fertilizers, detergents, and plastics.	• Bases are found in natural and artificial flavorings in food, such as cocoa beans. • Antacids neutralize stomach acid, alleviating heartburn. • Bases are found in cleaners such as shampoo, dish detergent, and window cleaner. • Many flowers grow better in basic soil. • Bases are used to make products such as rayon and paper.

What is pH?

Have you ever seen someone test the water in a swimming pool? It is likely that the person was testing the pH of the water. Swimming pool water should have a pH around 7.4. If the pH of the water is higher or lower than 7.4, the water might become cloudy, burn swimmers' eyes, or contain too many bacteria. What does a pH of 7.4 mean?

Hydronium Ions

The **pH** *is an inverse measure of the concentration of hydronium ions (H_3O^+) in a solution.* What does *inverse* mean? It means that as one thing increases, another thing decreases. In this case, as the concentration of hydronium ions increases, pH decreases. A solution with a lower pH is more acidic. As the concentration of hydronium ions decreases, the pH increases. A solution with a higher pH is more basic. This relationship is shown in **Figure 15.**

Balance of Hydronium and Hydroxide Ions

All acid and base solutions contain both hydronium and hydroxide ions. In a neutral solution, such as water, the concentrations of hydronium and hydroxide ions are equal. What distinguishes an acid from a base is which of the two ions is present in the greater concentration. Acids have a greater concentration of hydronium ions (H_3O^+) than hydroxide ions (OH^-). Bases have a greater concentration of hydroxide ions than hydronium ions. Brackets around a chemical formula mean *concentration.*

Acids	$[H_3O^+] > [OH^-]$
Neutral	$[H_3O^+] = [OH^-]$
Bases	$[H_3O^+] < [OH^-]$

Key Concept Check How does the concentration of hydronium ions affect pH?

pH Scale

Figure 15 Notice that as hydronium concentration increases, the pH decreases.

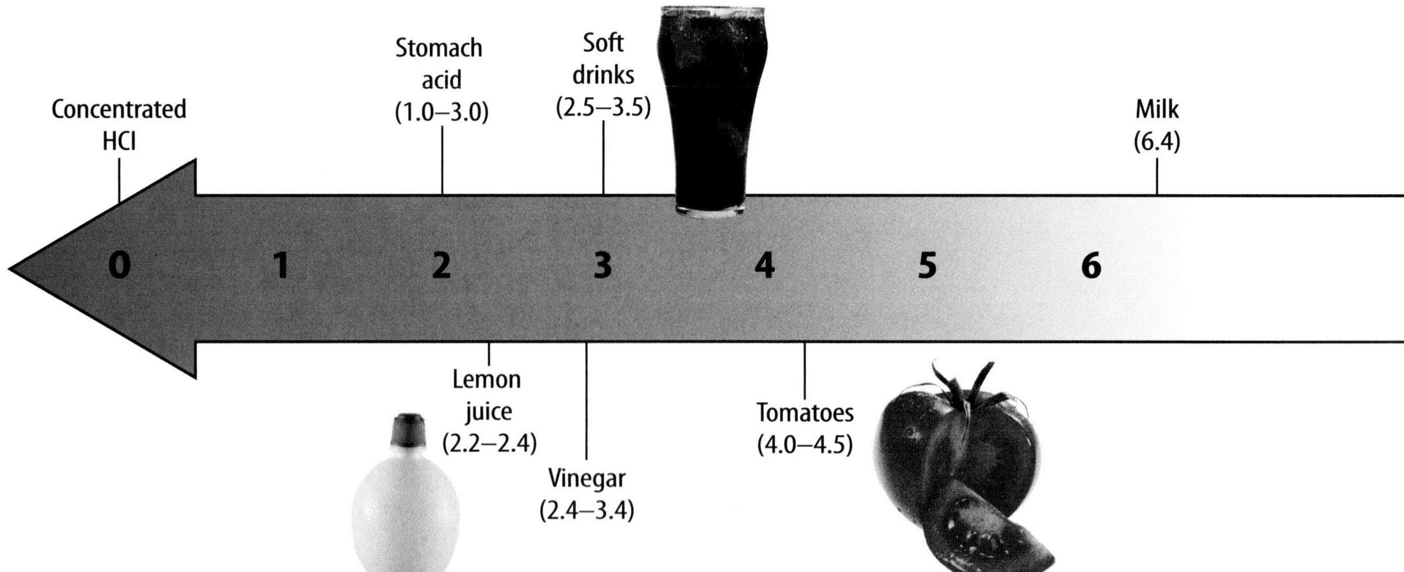

The pH Scale

The pH scale is used to indicate how acidic or basic a solution is. Notice in **Figure 15** that the pH scale contains values that range from below 0 to above 14. Acids have a pH below 7. Bases have a pH above 7. Solutions that are neutral have a pH of 7—they are neither acidic nor basic.

You might be wondering what the numbers on the pH scale mean. How is the concentration of hydronium ions different in a solution with a pH of 1 from the concentration in a solution with a pH of 2? A change in one pH unit represents a tenfold change in the acidity or basicity of a solution. For example, if one solution has a pH of 1 and a second solution has a pH of 2, then the first solution is not twice as acidic as the second solution; it is ten times more acidic.

The difference in acidity or basicity between two solutions is represented by 10^n, where n is the difference between the two pH values. For example, how much more acidic is a solution with a pH of 1 than a solution with pH of 3? First, calculate the difference, n, between the two pH values: $n = 3 - 1 = 2$. Then use the formula, 10^n, to calculate the difference in acidity: $10^2 = 100$. A solution with a pH of 1 is 100 times more acidic than a solution with a pH of 3.

Reading Check How much more acidic is a solution with a pH of 1 than a solution with a pH of 4?

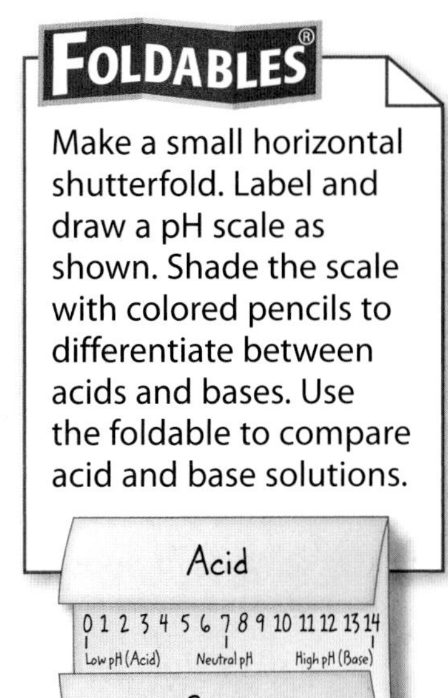

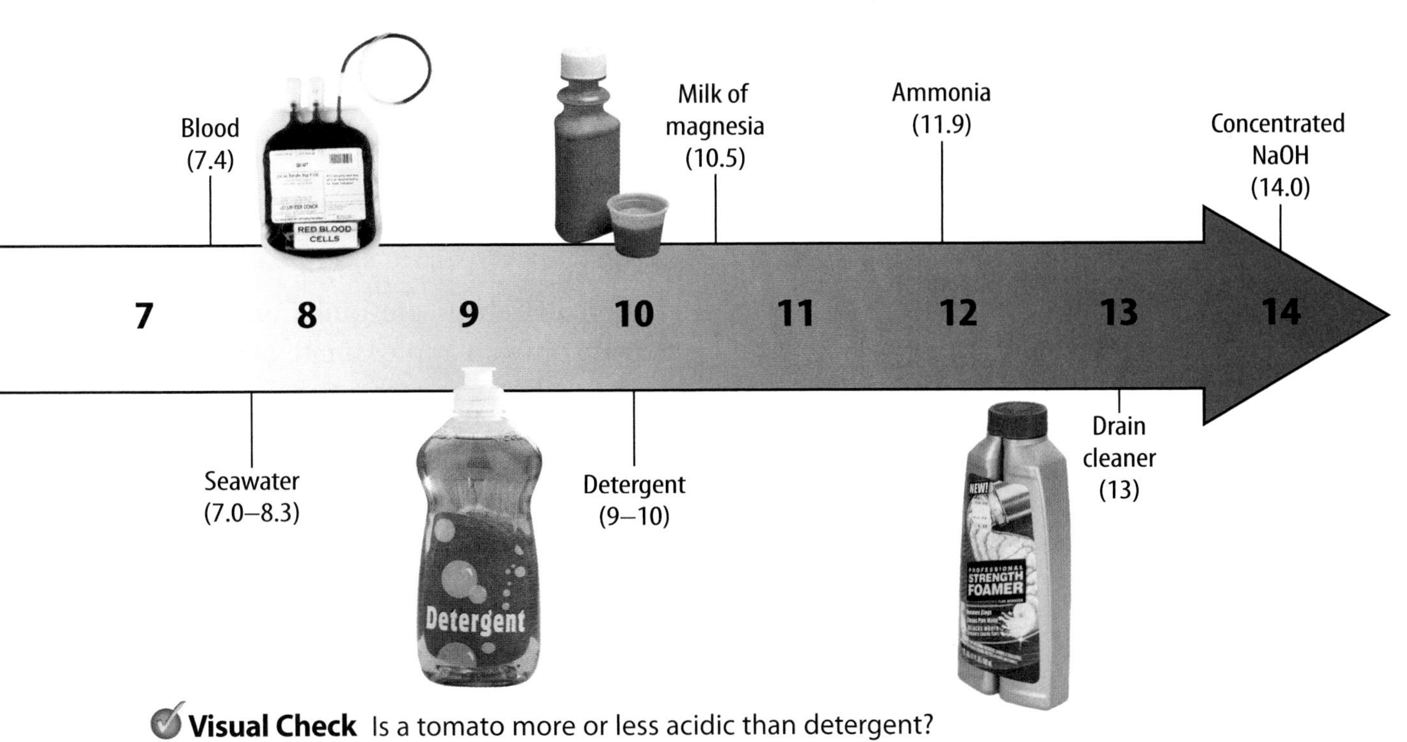

Visual Check Is a tomato more or less acidic than detergent? What is the difference in acidity?

MiniLab

30 minutes

Is it an acid or a base?

Acids and bases are found in many of the products people use in their homes every day. Properties of acids and bases make them desirable for use as cleaners, preservatives, or even flavorings.

1. Read and complete a lab safety form.
2. Get samples of **household products** from your teacher.
3. Dip a **pH strip** into the first product. Blot off excess liquid with a **paper towel.**
4. Compare the color of the pH strip to the **pH strip color chart,** and read the pH value.
5. Make a table in your Science Journal that shows the products and their pH values. Identify each product as acidic, neutral, or basic.

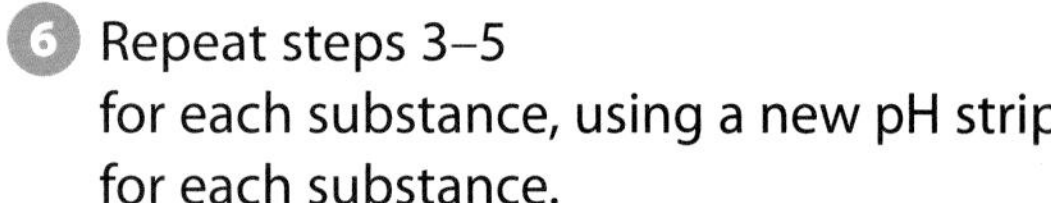

6. Repeat steps 3–5 for each substance, using a new pH strip for each substance.

Analyze and Conclude

1. **Predict** Using your table, think of three other products that could be tested. Predict the pH of each product. Explain your predictions.
2. **Key Concept** Based on your table, what characteristics do some of the acids have in common? What characteristics do some of the bases share?

How is pH measured?

How is the pH of a solution, such as swimming pool water, measured? Water test kits contain chemicals that change color when an acid or a base is added to them. These chemicals are called indicators.

pH Indicators

Indicators can be used to measure the approximate pH of a solution. *An* **indicator** *is a compound that changes color at different pH values when it reacts with acidic or basic solutions.* The pH of a solution is measured by adding a drop or two of the indicator to the solution. When the solution changes color, this color is matched to a set of standard colors that correspond to certain pH values. There are many different indicators—each indicator changes color over a specific range of pH values. For example, bromthymol blue is an indicator that changes from yellow to green to blue between pH 6 and pH 7.6.

pH Testing Strips

pH also can be measured using pH testing strips. The strips contain an indicator that changes to a variety of colors over a range of pH values. To use pH strips, dip the strip into the solution. Then match the resulting color to the list of standard colors that represent specific pH values.

pH Meters

Although pH strips are quick and easy, they provide only an approximate pH value. A more accurate way to measure pH is to use a pH meter. A pH meter is an electronic instrument with an electrode that is sensitive to the hydronium ion concentration in solution.

Key Concept Check What are two methods that can be used to measure the pH of a solution?

Lesson 3 Review

Assessment Online Quiz

Visual Summary

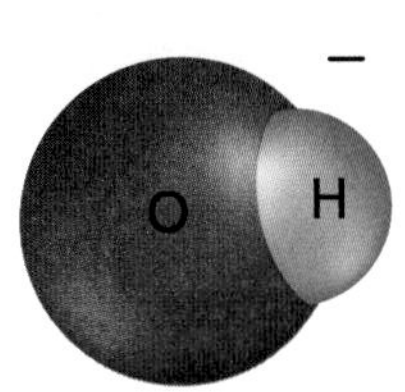

Acids contain hydrogen ions that are released and form hydronium ions in water. Bases are substances that form hydroxide ions when dissolved in water.

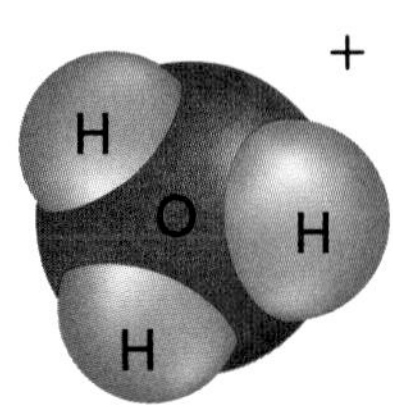

Hydronium ion concentration changes inversely with pH. This means that as hydronium ion concentration increases, the pH decreases.

pH can be measured using indicators or digital pH meters.

FOLDABLES®

Use your lesson Foldable to review the lesson. Save your Foldable for the project at the end of the chapter.

What do you think NOW?

You first read the statements below at the beginning of the chapter.

5. Acids are found in many foods.

6. You can determine the exact pH of a solution using pH paper.

Did you change your mind about whether you agree or disagree with the statements? Rewrite any false statements to make them true.

Use Vocabulary

1 A measure of the concentration of hydronium ions (H_3O^+) in a solution is ____________.

2 A(n) ____________ is used to determine the approximate pH of a solution.

Understand Key Concepts 

3 **Describe** What happens to a hydrogen atom in an acid when the acid is dissolved in water?

4 **Explain** How does pH vary with hydronium ion and hydroxide ion concentrations in water?

5 **Show** Does an acidic solution contain hydroxide ions? Explain your answer with a diagram.

Interpret Graphics

6 **Predict** what is produced when hydrofluoric acid (HF) is dissolved in water in the equation below.

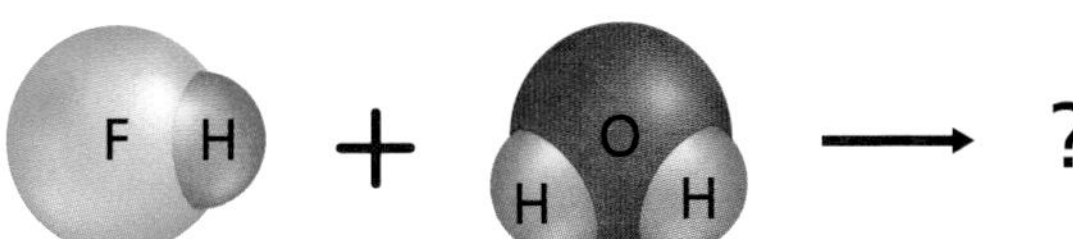

7 **Contrast** Copy the graphic organizer below, and use it to describe and contrast three ways to measure pH. In the organizer, describe which methods are most and least accurate.

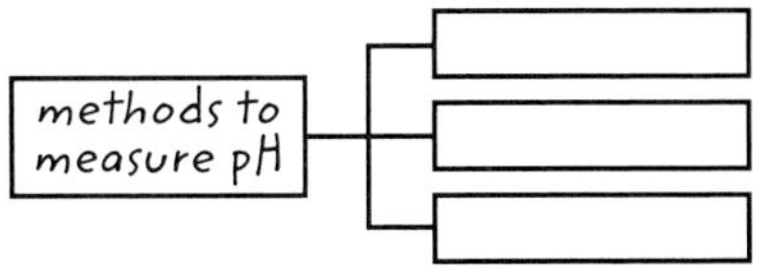

Critical Thinking

8 **Describe** the concentration of hydronium ions and hydroxide ions when a base is added slowly to a white vinegar solution. The pH of white vinegar is 3.1.

Lab 1 class period

Can the pH of a solution be changed?

Materials

100-mL graduated cylinder

triple-beam balance

beakers

dropper

white vinegar

Also needed: baking soda, stirring rod, universal indicator, water

Safety

Many foods that people eat come from plants that grow best in soils that have a pH within a particular range. Potatoes grow best in soils that have a pH between 4.5 and 6.0. Blueberry plants grow best in soils with a pH of 4.0 to 4.5. Strawberry plants grow best in soils with a pH of 5.3 to 6.2. How might someone grow all of these plants in the same garden? The pH of soil often has to be changed to make it suitable to grow these plants. Using the knowledge you have gained from this chapter and the techniques that you have practiced, you will discover whether the pH of a solution can be changed.

Question

Can the pH of a solution be changed? If so, how quickly can it happen, and how does one tell?

Procedure

1. Read and complete a lab safety form.
2. Using the materials provided, make a baking soda solution with a concentration of 5 g/L.
3. Add 15 drops of universal indicator to the solution.
4. In your Science Journal, make a table like the one shown here.

Number of Drops of Vinegar Added	Solution pH According to pH Paper	Color of Solution	Solution pH According to Indicator
0			
10			
20			
30			

5. Determine and record the pH and color of the solution.

6. Add 10 drops of vinegar to the solution and stir.
7. Determine and record the pH and color of the solution.
8. Repeat steps 7 and 8 until you have added 200 drops of vinegar.
9. Dispose of your solutions according to your teacher's instructions.

Analyze and Conclude

10. **Organize Information** Create a line graph that shows the data you have collected. Remember to put your independent variable on the *x*-axis.
11. **Analyze Data** Look at your table and the graph you made and discuss what, if any, relationships you see in your data.
12. **Predict** What do you think would happen if you performed this experiment again using an acid stronger than vinegar?
13. **The Big Idea** How can the pH of a solution or substance be changed?

Communicate Your Results

Compare your results with those of other groups. Look for differences, and discuss possible causes for discrepancies among the data collected.

Inquiry Extension

Perform some additional research in one of the following areas:

- Why is soil pH important?
- Why are acids and bases added to products, such as pH-balanced shampoo, to adjust their pH?
- Why is the pH of blood important to a person's health? How can blood pH become too high or too low? How can it be corrected?

Lab Tips

- ☑ The concentration of the starting solution is important. Zero your balance, and make accurate measurements.
- ☑ Stir the solution thoroughly after adding the vinegar each time.
- ☑ Always get your teacher's approval before trying any new test.

Remember to use scientific methods.

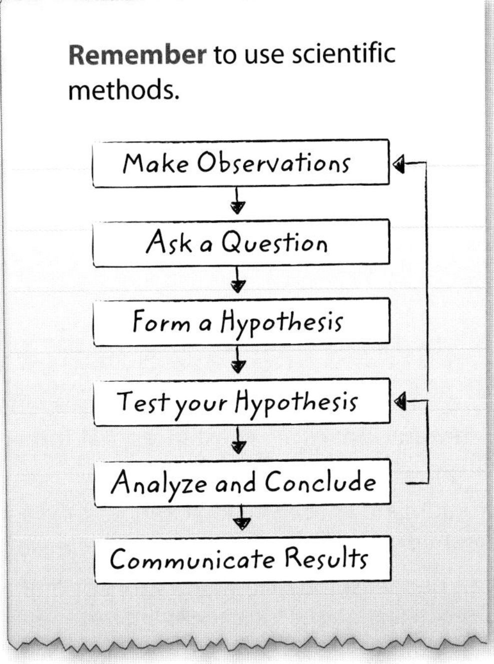

Chapter 13 Study Guide

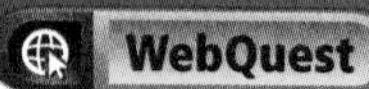

Solutions are homogeneous mixtures. They can be described by the concentration and type of solute they contain.

Key Concepts Summary

Lesson 1: Substances and Mixtures

- **Substances** have a fixed composition. The composition of **mixtures** can vary.
- **Solutions** and **heterogeneous mixtures** are both types of mixtures. Solutions are mixed at the atomic level.
- Mixtures contain parts that are not bonded together. These parts can be separated using physical means, and their properties can be seen in the solution.

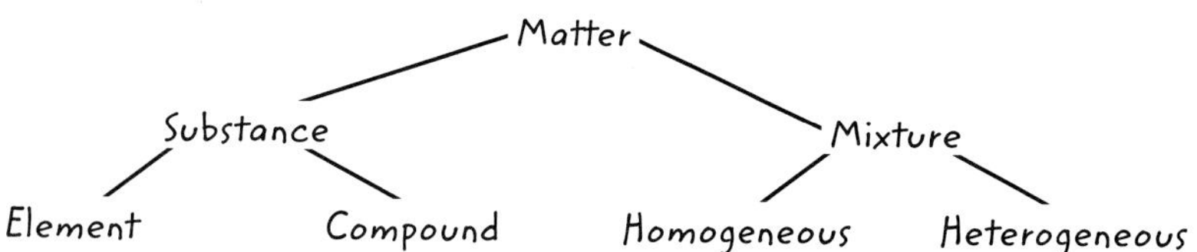

Vocabulary

substance p. 454
mixture p. 454
heterogeneous mixture p. 455
homogeneous mixture p. 455
solution p. 455

Lesson 2: Properties of Solutions

- Substances dissolve other substances that have a similar polarity. In other words, like dissolves like.
- **Concentration** is the amount of a **solute** that is dissolved. **Solubility** is the maximum amount of a solute that can dissolve.
- Both temperature and pressure affect the solubility of solutes in solutions.

Vocabulary

solvent p. 461
solute p. 461
polar molecule p. 462
concentration p. 464
solubility p. 466
saturated solution p. 466
unsaturated solution p. 466

Lesson 3: Acid and Base Solutions

- **Acids** contain hydrogen ions that are released and form **hydronium ions** in water. **Bases** are substances that form hydroxide ions when dissolved in water.
- Hydronium ion concentration changes inversely with **pH.** This means that as hydronium ion concentration increases, the pH decreases.
- pH can be measured using **indicators** or digital pH meters.

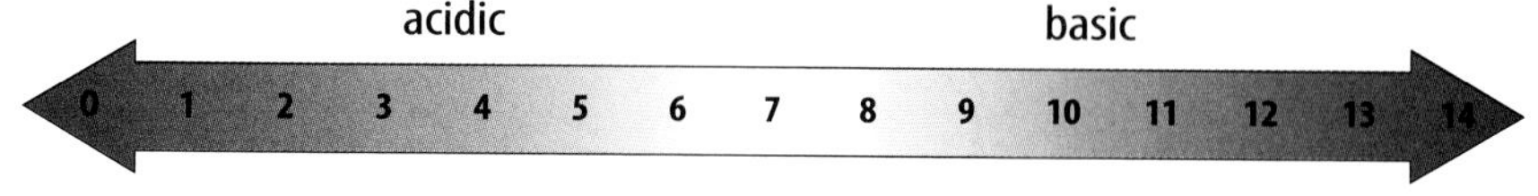

Vocabulary

acid p. 472
hydronium ion p. 472
base p. 472
pH p. 474
indicator p. 476

- Personal Tutor
- Vocabulary eGames
- Vocabulary eFlashcards

FOLDABLES® Chapter Project

Assemble your lesson Foldables as shown to make a Chapter Project. Use the project to review what you have learned in this chapter.

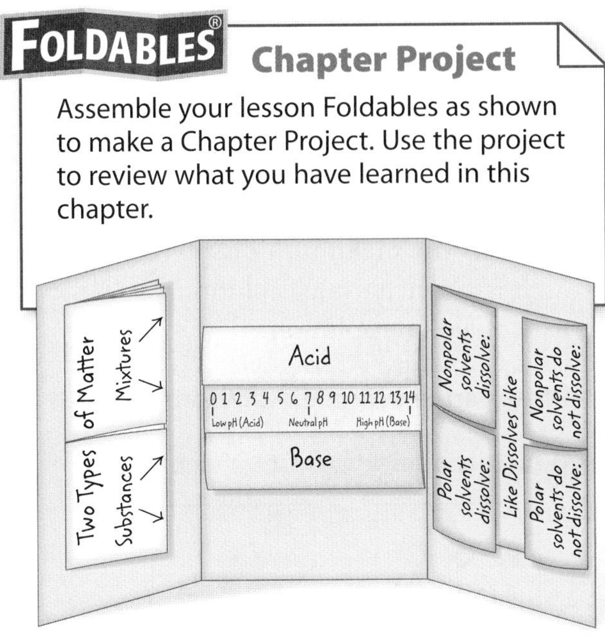

Use Vocabulary

1. The parts of a __________ can be seen with unaided eyes or with a microscope.
2. It is impossible to tell the difference between a solution and a __________ just by looking at them.
3. Water dissolves other __________ easily.
4. Two equal volumes of a solution that contain different amounts of the same solute have a different __________.
5. As __________ concentration decreases, pH increases.
6. A(n) __________ can be added to milk to neutralize it.

Link Vocabulary and Key Concepts

Concepts in Motion Interactive Concept Map

Use vocabulary terms from the previous page to complete the concept map.

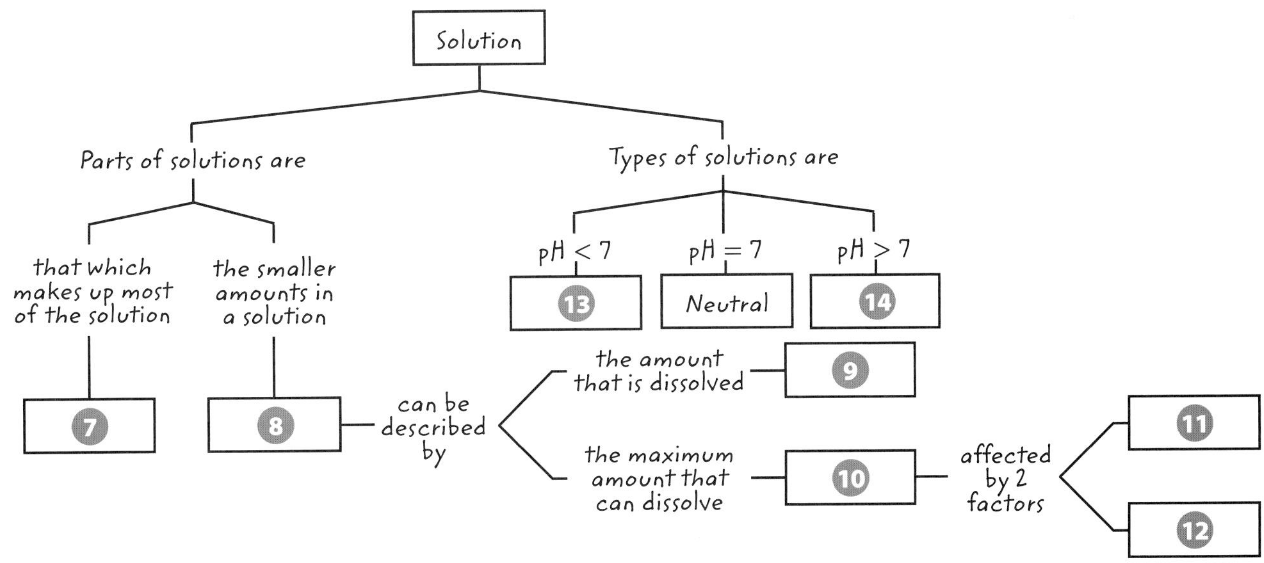

Chapter 13 Review

Understand Key Concepts

1. Which is a solution?
   A. copper
   B. vinegar
   C. pure water
   D. a raisin cookie

2. The graph below shows the solubility of sodium chloride (NaCl) in water.

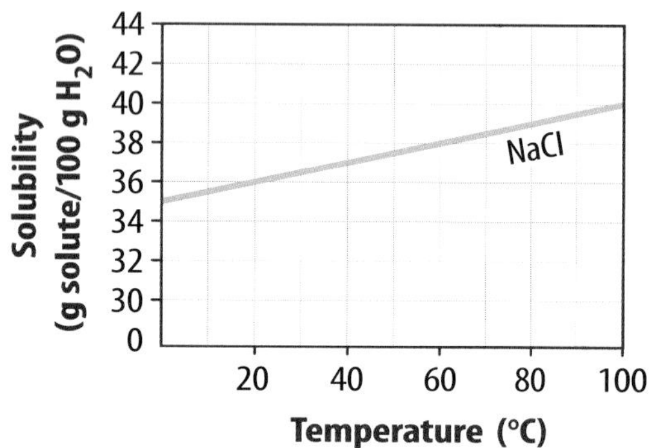

   What mass of sodium chloride is needed to form a saturated solution at 80°C?
   A. 30 g
   B. 40 g
   C. 50 g
   D. 60 g

3. What would you add to a solution with a pH of 1.5 to obtain a solution with a pH of 7?
   A. milk (pH 6.4)
   B. vinegar (pH 3.0)
   C. lye (pH 13.0)
   D. coffee (pH 5.0)

4. Which can change the solubility of a solid in a liquid?
   A. crushing the solute
   B. stirring the solute
   C. increasing the pressure of the solution
   D. increasing the temperature of the solution

5. Which ions are present in the greatest amount in a solution with a pH of 8.5?
   A. hydrogen ions
   B. hydronium ions
   C. hydroxide ions
   D. oxygen ions

6. Which best describes a solution that contains the maximum dissolved solute?
   A. It is a concentrated solution.
   B. It is a dilute solution.
   C. It is a saturated solution.
   D. It is an unsaturated solution.

7. Which is a mixture of two elements?

A.
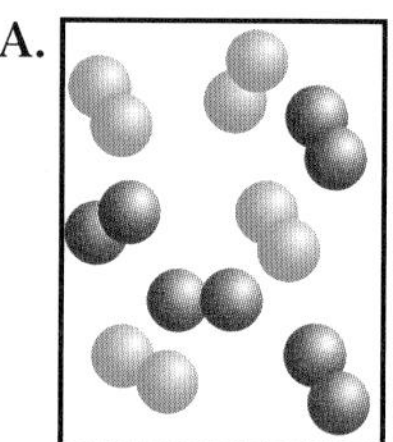

C.
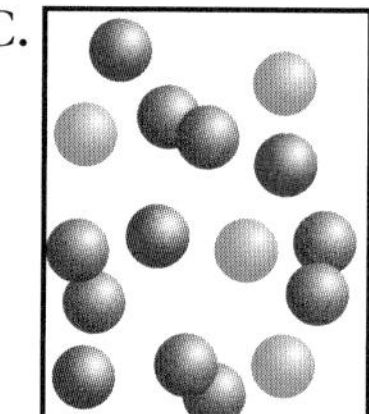

B.
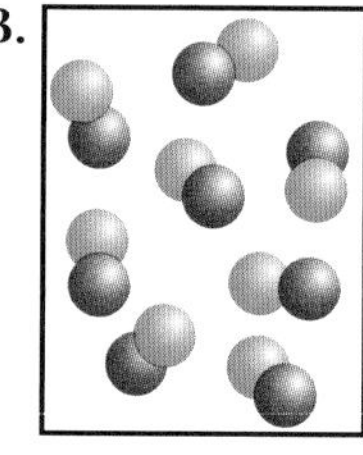

D.
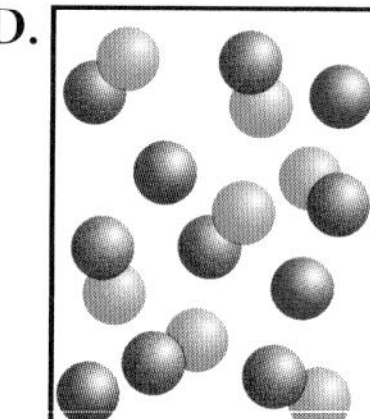

8. Which explains why a soft drink bubbles when the cap is released?
   A. The gas becomes less soluble when temperature decreases.
   B. The gas becomes more soluble when temperature decreases.
   C. The gas becomes less soluble when pressure decreases.
   D. The gas becomes more soluble when pressure decreases.

Assessment

Online Test Practice

Critical Thinking

9 **Infer** How can you tell which component in a solution is the solvent?

10 **Predict** The graph below shows the solubility of potassium chloride (KCl) in water.

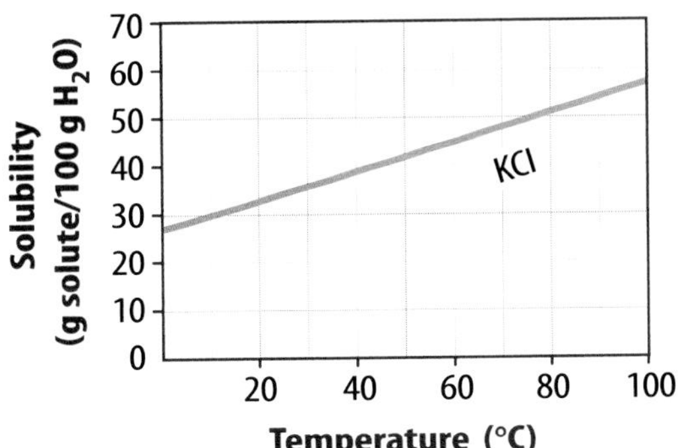

Imagine you have made a solution that contains 50 g of potassium chloride (KCl) in 100 g of solution. Predict what you would observe as you gradually increased the temperature from 0°C to 100°C.

11 **Organize** The pH of three solutions is shown below.

Milk (pH 6.7)
Coffee (pH 5)
Ammonia (pH 11.6)

Place these solutions in order of

a. most acidic to least acidic
b. most basic to least basic
c. highest OH^- concentration to lowest OH^- concentration

12 **Explain** The pH of a solution is inversely related to the concentration of hydronium ions in solution. Explain what this means.

13 **Design** a method to determine the solubility of an unknown substance at 50°C.

Writing in Science

14 **Compose** A haiku is a poem containing three lines of five, seven, and five syllables, respectively. Write a haiku describing what happens when an acid is dissolved in water.

REVIEW THE BIG IDEA

15 What are solutions? List at least three ways a solution can be described.

16 How do solutions differ from other types of matter?

Math Skills

Math Practice

Calculate Concentration

17 Calculate the concentration of sugar in g/L in a solution that contains 40 g of sugar in 100 mL of solution. There are 1,000 mL in 1 L.

18 There are many ways to make a solution of a given concentration. What are two ways you could make a sugar solution with a concentration of 100 g/L?

19 A salt solution has a concentration of 200 g/L. How many grams of salt are contained in 500 mL of this solution? How many grams of salt would be contained in 2 L of this solution?

Standardized Test Practice

Record your answers on the answer sheet provided by your teacher or on a sheet of paper.

Multiple Choice

Use the figures below to answer question 1.

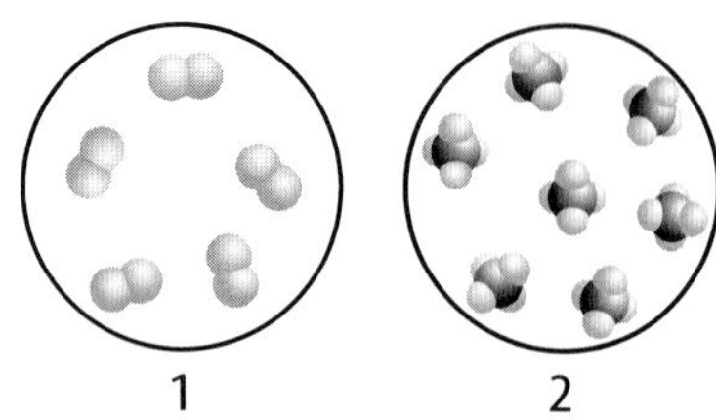

1 Which statement describes the two figures?

A Both 1 and 2 are mixtures.

B Both 1 and 2 are substances.

C 1 is a mixture and 2 is a substance.

D 1 is a substance and 2 is a mixture.

2 Which statement is an accurate comparison of solutions and homogeneous mixtures.

A They are the same.

B They are opposites.

C Solutions are more evenly mixed than homogeneous mixtures.

D Homogeneous mixtures are more evenly mixed than solutions.

3 A worker uses a magnet to remove bits of iron from a powdered sample. Which describes the sample before the worker used the magnet to remove the iron?

A The sample is a compound because the iron was removed using a physical method.

B The sample is a compound because the iron was removed using a chemical change.

C The sample is a mixture because the iron was removed using a chemical change.

D The sample is a mixture because the iron was removed using a physical method.

4 A beaker contains a mixture of sand and small pebbles. What kind of mixture is this?

A compound

B heterogeneous

C homogeneous

D solution

5 Which type of substance would best dissolve in a solvent that was made of nonpolar molecules?

A a water-based solvent

B an ionic compound

C a solute made of polar molecules

D a solute made of nonpolar molecules

Use the figure to answer question 6.

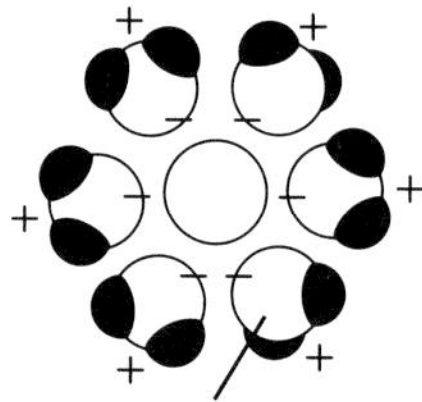

6 The figure shows how water molecules surround an ion in a solution. What can you conclude about the ions?

A It is negative because the negative ends of the water molecule are attracted to it.

B It is negative because the positive ends of the water molecule are attracted to it.

C It is positive because the negative ends of the water molecule are attracted to it.

D It is positive because the positive ends of the water molecule are attracted to it.

7 A girl makes two glasses of lemonade using a powder mix. She pours one cup of water into each glass. She adds one spoonful of powder to the first glass and two spoonfuls of powder to the second glass. How do the solutions in the two glasses compare?

A The first glass has a greater concentration of powder mix.

B The first glass has a greater solubility.

C The second glass has a greater concentration of powder mix.

D The second glass has a greater solubility.

Use the table below to answer question 8.

Sample solution	Change in blue litmus	Change in red litmus
1	turns red	no change
2	no change	turns blue
3	turns red	no change
4	no change	no change

8 A scientist collects the data above using litmus paper. Blue litmus paper is a type of pH indicator that turns red when placed in an acidic solution. Red litmus paper is an indicator that turns blue when placed in an basic solution. Neutral solutions cause no change in either color of litmus paper. Which sample solution must be a base?

A solution 1

B solution 2

C solution 3

D solution 4

Constructed Response

9 Explain how the concentration of hydronium ions and the concentration of hydroxide ions change when a base is dissolved in water.

10 A researcher mixes a solution that is 40 percent helium gas and 60 percent nitrogen gas. Which gas is the solute and which is the solvent? What would the mixture look like through a microscope? What would the mixture look like at the atomic level?

11 A student is dissolving rock salt in water. Describe three ways to increase the rate of dissolving.

Use the figure to answer questions 12 and 13.

12 The figure shows what happens when hydrogen iodide (HI) dissolves in water. Is hydrogen iodide an acid, a base, or a neutral substance? Explain.

13 What can you conclude about the pH of the aqueous solution of hydrogen iodide?

NEED EXTRA HELP?													
If You Missed Question...	1	2	3	4	5	6	7	8	9	10	11	12	13
Go to Lesson...	1	1	1	1	2	2	2	2	3	3	1	2	3

Chapter 14

Carbon Chemistry

THE BIG IDEA

What is carbon's role in the chemistry of living things?

Inquiry Where's the carbon?

When you walk through a forest, you probably do not think about carbon atoms. However, all the living things surrounding you contain carbon atoms.

- Where are these hidden carbon atoms?
- What other elements are in living things?
- What is carbon's role in the chemistry of living things?

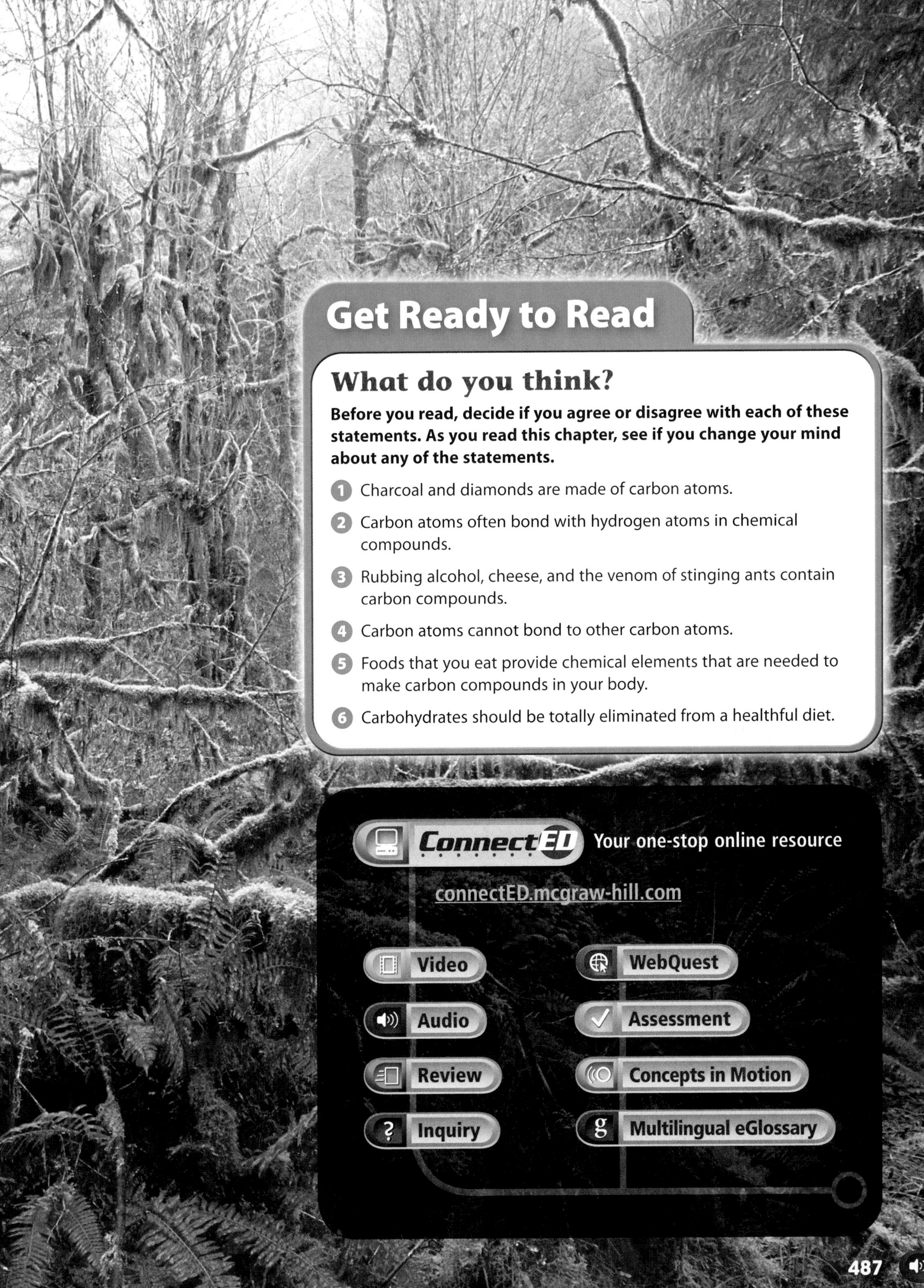

Get Ready to Read

What do you think?

Before you read, decide if you agree or disagree with each of these statements. As you read this chapter, see if you change your mind about any of the statements.

1. Charcoal and diamonds are made of carbon atoms.
2. Carbon atoms often bond with hydrogen atoms in chemical compounds.
3. Rubbing alcohol, cheese, and the venom of stinging ants contain carbon compounds.
4. Carbon atoms cannot bond to other carbon atoms.
5. Foods that you eat provide chemical elements that are needed to make carbon compounds in your body.
6. Carbohydrates should be totally eliminated from a healthful diet.

Lesson 1

Reading Guide

Key Concepts
ESSENTIAL QUESTIONS

- How is carbon unique compared to other elements?
- How does carbon bond with other carbon atoms?

Vocabulary

organic compound p. 490
hydrocarbon p. 492
isomer p. 492
saturated hydrocarbon p. 493
unsaturated hydrocarbon p. 493

Multilingual eGlossary

Elemental Carbon and Simple Organic Compounds

Inquiry A Diamond?

The yellow rock probably does not look like a diamond that you have seen. It lacks sparkle because it is rough and uncut. It formed deep within Earth under intense pressure and heating. Believe it or not, diamonds are pure carbon.

Launch Lab

15 minutes

Why is carbon a unique element? *Do not eat the gumdrops.*

A carbon atom is unique because it can easily form four bonds with other atoms, including other carbon atoms. Because of this property, carbon forms many different compounds.

1. Read and complete a lab safety form.
2. Use **gumdrops** and **toothpicks** to make as many different carbon molecules as you can. Keep these rules in mind.
   - Each gumdrop represents one carbon atom.
   - Each toothpick represents one chemical bond.
   - Each molecule must contain four carbon atoms (gumdrops).
   - Each carbon atom must have four chemical bonds (toothpicks).
   - One carbon atom can share up to three bonds with another carbon atom.
3. Make a sketch of each molecule you make in your Science Journal.

Think About This

1. How many different molecules were you able to build?
2. If you had five gumdrops, would you be able to build more molecules? Explain your answer.
3. **Key Concept** How do you think carbon bonds with other carbon atoms?

Elements in Living Things

What do you have in common with the fish and sea anemones in **Figure 1?** You might be surprised to learn that you, a fish, and a sea anemone have several things in common. Each of you is made of cells that contain carbon, hydrogen, oxygen, nitrogen, and a few other elements. In fact, the mass of all living organisms contain about 18 percent carbon compounds.

Except for water and some salts, most things you put in or on your body—food, clothing, cosmetics, and medicines—consist of compounds that contain carbon. This chapter explores various types of carbon compounds that make up living things.

Reading Check Which four elements are in most living organisms?

Figure 1 All living things are made of similar carbon-containing compounds.

Organic Compounds

Scientists once thought that all carbon compounds came from living or once-living organisms, and they called these compounds organic. Scientists now know that carbon is also in many nonliving things. Today, scientists define an **organic compound** *as a chemical compound that contains carbon atoms usually bonded to at least one hydrogen atom.* Organic compounds can also contain other elements such as oxygen, phosphorus, or sulfur. However, compounds such as carbon dioxide (CO_2) and carbon monoxide (CO) are not organic because they do not have a carbon-hydrogen bond.

SCIENCE USE V. COMMON USE

organic

Science Use a chemical compound that contains carbon and usually contains at least one carbon-hydrogen bond

Common Use pertaining to or grown with natural fertilizers and pesticides; often used to refer to foods, such as fruits, vegetables, and meats

Understanding Carbon

A carbon atom is unique because it can easily combine with other atoms and form millions of compounds. Find carbon on the periodic table on the inside back cover of this book. Carbon has an atomic number of 6. Therefore, a neutral carbon atom has six protons and six electrons. Four of these electrons are valence electrons, or are in the outermost energy level. Recall that many atoms are chemically stable when they have eight valence electrons. Carbon atoms become more chemically stable through **covalent bonding,** as shown in **Figure 2.** In a covalent bond, carbon atoms have eight valence electrons, like a stable, unreactive noble gas.

REVIEW VOCABULARY

covalent bond

a chemical bond formed by sharing one or more pairs of electrons between atoms

The Carbon Group

Look again at the periodic table. Notice that silicon and germanium are in the same group as carbon. They each have four valence electrons. Silicon and germanium atoms also become stable by forming four covalent bonds. However, it takes more energy for them to do this. The more energy it takes, the less likely it is that bonding will occur.

Key Concept Check How is carbon unique compared to other elements?

Figure 2 Carbon often bonds with four hydrogen atoms and forms a stable compound.

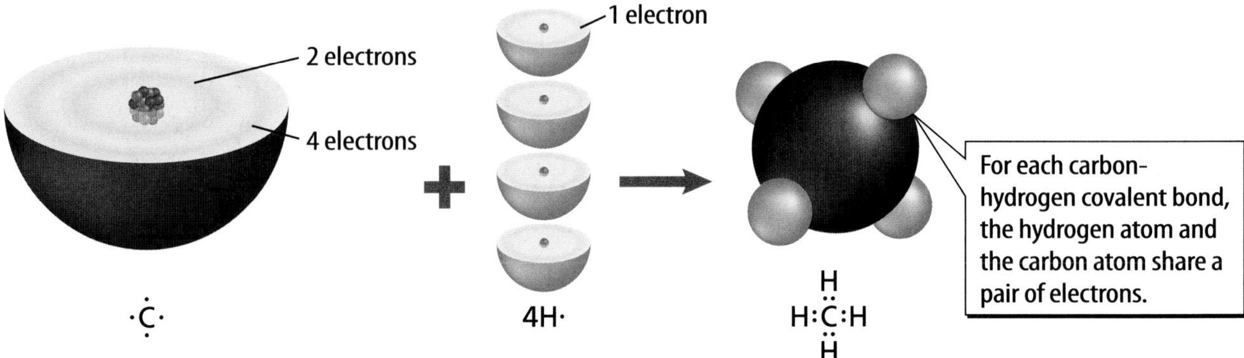
Bonding with Carbon
2 electrons
4 electrons
1 electron
For each carbon-hydrogen covalent bond, the hydrogen atom and the carbon atom share a pair of electrons.
·C·
4H·
H
H:C:H
H

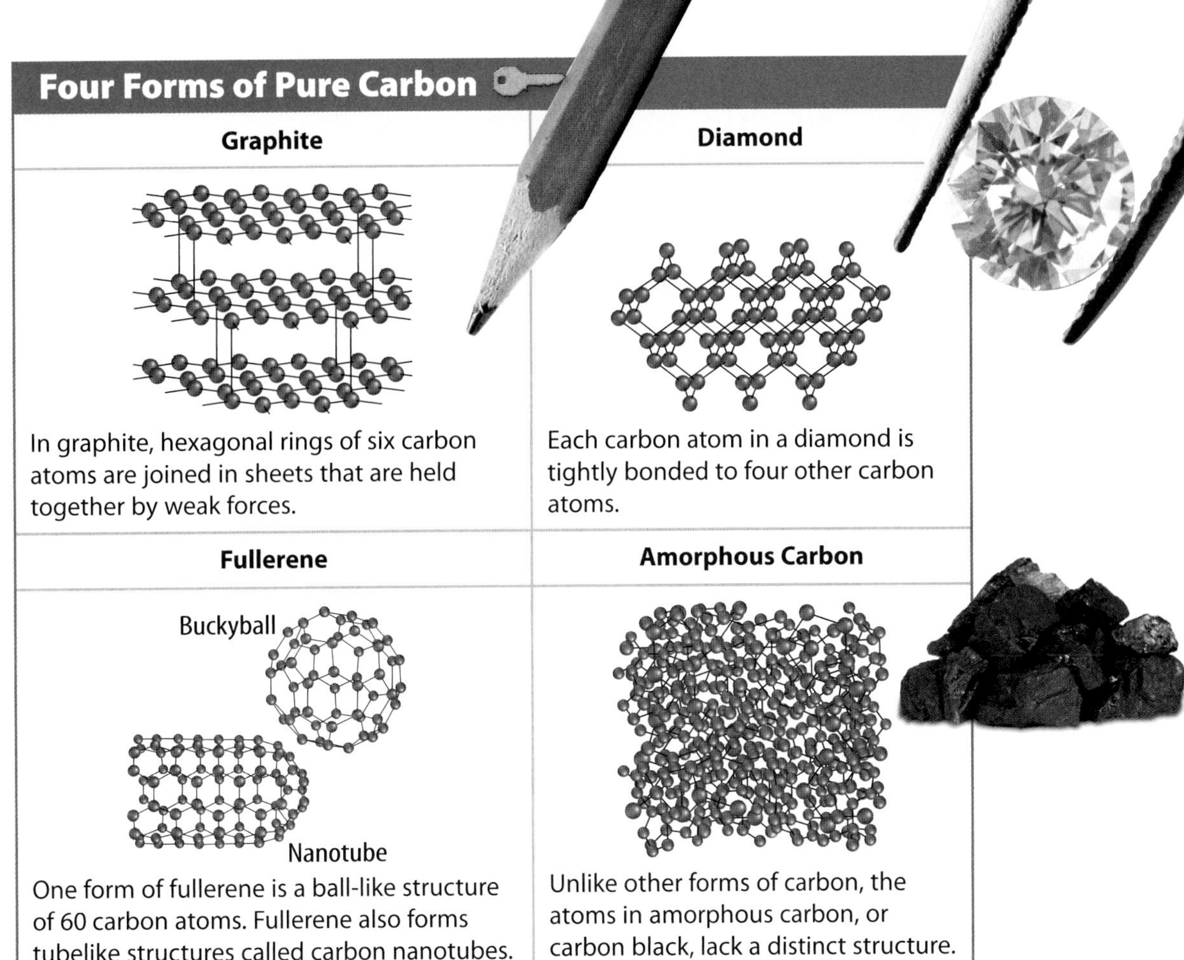

Four Forms of Pure Carbon
Graphite
Diamond
In graphite, hexagonal rings of six carbon atoms are joined in sheets that are held together by weak forces.
Each carbon atom in a diamond is tightly bonded to four other carbon atoms.
Fullerene
Amorphous Carbon
Buckyball
Nanotube
One form of fullerene is a ball-like structure of 60 carbon atoms. Fullerene also forms tubelike structures called carbon nanotubes.
Unlike other forms of carbon, the atoms in amorphous carbon, or carbon black, lack a distinct structure.

Figure 3 Four common forms of pure carbon are graphite, diamond, fullerene, and amorphous.

Visual Check Compare the structures of graphite and diamonds and explain why graphite is used in pencil lead but diamonds are not.

The Forms of Pure Carbon

When carbon atoms bond together, they form one of several different arrangements, such as those shown in **Figure 3.** Forms of carbon are described below.

- One form of carbon is graphite (GRA fite). In graphite, carbon atoms form thin sheets that can slide over one another or bend. Graphite is used as a lubricant and in making golf clubs, tennis rackets, pencil lead, and other items.
- Diamonds, another form of carbon, are used in jewelry, on the tips of drill bits, and on the edges of some saw blades. The carbon atoms bond to one another in a rigid and orderly structure, making diamonds extremely strong. This makes diamonds one of the hardest materials known.
- Carbon atoms in fullerene (FOOL uh reen) form various cagelike structures. Fullerene was discovered late in the twentieth century and uses for fullerene are still being explored. However, future fullerene uses might include the development of faster, smaller electronic components.
- The atoms in **amorphous** (uh MOR fus) carbon lack an orderly arrangement. Amorphous carbon is found in soot, coal, and charcoal.

Word Origin

amorphous
from *a–* and Greek *morphe*, means "without form"

Figure 4 Methane consists of one carbon atom bonded to four hydrogen atoms. ▶

Make a three-quarter book and use it to organize your notes about hydrocarbons and how they bond.

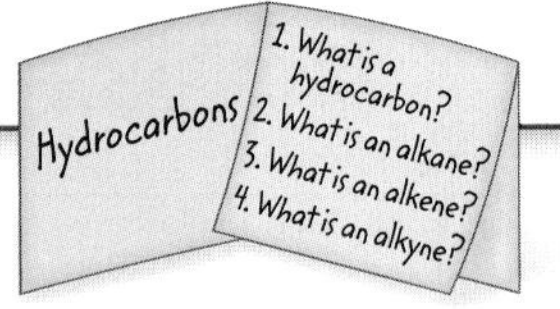

Hydrocarbons

Many organic compounds contain only carbon and hydrogen atoms. *A compound that contains only carbon and hydrogen atoms is called a* **hydrocarbon.** There are many different hydrocarbons. The simplest is methane (CH_4), shown in **Figure 4.**

Hydrocarbon Chains

When carbon atoms form hydrocarbons, the carbon atoms can link together in different ways. They can form straight chains, branched chains, or rings. **Table 1** shows examples of each type of arrangement. Look closely at the molecular formula for each compound. Notice that butane and isobutane have the same molecular formula. They have the same ratio of carbon atoms to hydrogen atoms. *Compounds that have the same molecular formula but different structural arrangements are called* **isomers** (I suh murz). Each isomer is a different molecule with its own unique name and properties.

Table 1 Carbon atoms can form straight chains, branched chains, or rings.

Animation

Table 1 Hydrocarbon Arrangements

Butane	Isobutane	Cyclobutane
Molecular formula: C_4H_{10}	Molecular formula: C_4H_{10}	Molecular formula: C_4H_8
H H H H H—C—C—C—C—H H H H H Structural formula	H H—C—H H H H—C—C—C—H H H H Structural formula	H H H—C—C—H H—C—C—H H H Structural formula
Ball-and-stick model	Ball-and-stick model	Ball-and-stick model

Carbon-to-Carbon Bonding

When a carbon atom bonds to another carbon atom, the two atoms can share two, four, or six electrons, as shown in **Figure 5.** In all cases, the carbon atoms in each molecule have eight valence electrons and are stable. When two carbon atoms share two electrons, it is called a single bond. A hydrocarbon that contains only single bonds is called an alkane. When two carbon atoms share four electrons, it is called a double bond. A hydrocarbon that contains at least one double bond is called an alkene. If two carbon atoms share six electrons, it is called a triple bond. A hydrocarbon that contains at least one triple bond is called an alkyne.

Key Concept Check What kind of bonds do carbon atoms form with other carbon atoms?

Saturated Hydrocarbons

Hydrocarbons are often classified by the type of bonds the carbon atoms share. *A hydrocarbon that contains only single bonds is called a* **saturated hydrocarbon.** It is called saturated because no more hydrogen atoms can be added to the molecule. Look at the top image in **Figure 5.** Notice that three of the valence electrons in each carbon atom bond with hydrogen atoms. The carbon atoms are saturated with hydrogen atoms.

Unsaturated Hydrocarbons

A hydrocarbon that contains one or more double or triple bonds is called an **unsaturated hydrocarbon.** Look at the double bond and triple bond examples in **Figure 5.** If the double and triple bonds are broken, additional hydrogen atoms could bond to the carbon atoms. Therefore, molecules containing double and triple bonds are not saturated with hydrogen atoms.

Reading Check Explain the difference between saturated and unsaturated compounds.

Carbon-to-Carbon Bonding

Alkane

- single bond
- two electrons shared between two carbon atoms

single line indicates a single bond

$CH_3 - CH_3$

H H
H:C:C:H
H H

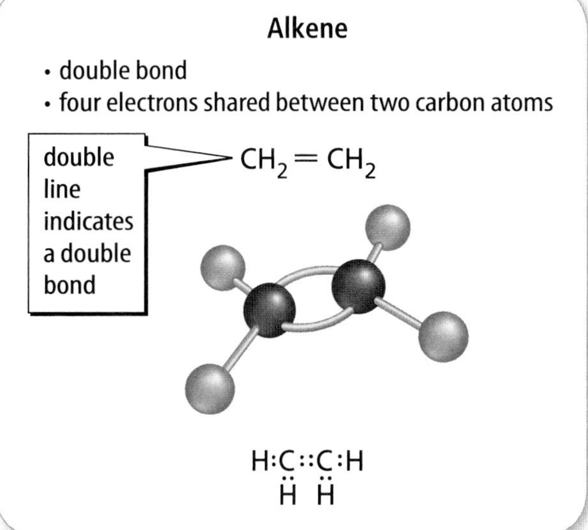

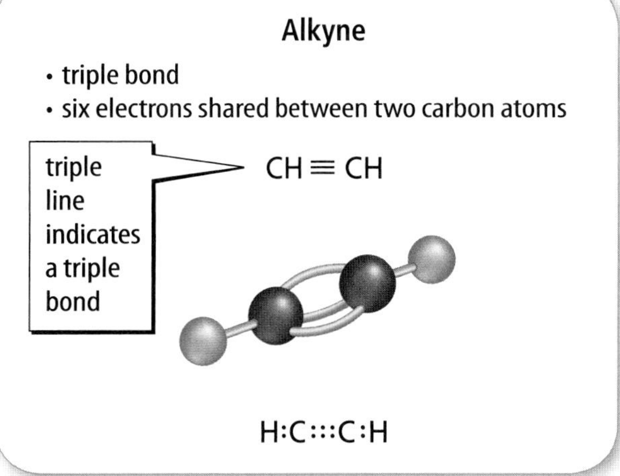

Figure 5 Two carbon atoms can form a single, double, or triple bond.

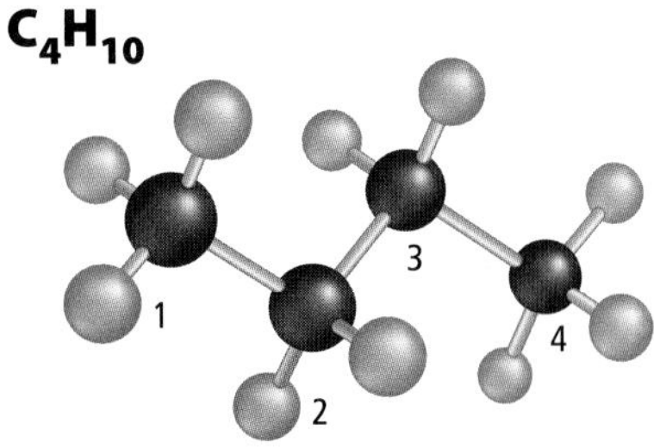

▲ **Figure 6** This hydrocarbon has four carbon atoms in its chain.

Animation

Naming Hydrocarbons

What type of shape is a stop sign? It is an octagon. An octagon is a figure with eight sides. Its name comes from the root *oct–*, which means "eight." Most geometric shapes have names that refer to the number of sides they have, such as a triangle. Similarly, hydrocarbons have names that indicate how many carbon atoms are in each molecule.

Carbon Chains

When naming a hydrocarbon, the first thing you need to do is find the longest carbon chain and count the number of carbon atoms in it. Look at **Figure 6.** Find the carbon chain and count the carbon atoms. In this molecule, there are four carbon atoms. The number of carbon atoms gives you the root word of the name. Now, look at **Table 2.** This table shows the root word for any hydrocarbon that has one through ten carbon atoms. What is the root name for the molecule in **Figure 6?** The root name is *but–* (BYEWT). What would be the root name if the carbon chain had eight carbon atoms like a stop sign? The root name is *oct–*.

Table 2 The number of carbon atoms in the longest continuous chain determines the root word for the name of the hydrocarbon. ▼

Table 2 Root Words

Carbon Atoms	Name	Carbon Atoms	Name
1	meth–	6	hex–
2	eth–	7	hept–
3	prop–	8	oct–
4	but–	9	non–
5	pent–	10	dec–

Inquiry MiniLab

20 minutes

How do carbon atoms bond with carbon and hydrogen atoms?

⚠ *Do not eat the gumdrops or the raisins.*

Models help you see how carbon atoms form so many different compounds.

1. Read and complete a lab safety form.
2. Draw the structural formulas for propane, propene, and propyne in your Science Journal.
3. Use **toothpicks, gumdrops,** and **raisins** to make a ball-and-stick model of each compound. Gumdrops represent carbon atoms. Raisins represent hydrogen atoms. The toothpicks represent bonds between atoms.

Analyze and Conclude

1. **Classify** each model as a saturated or unsaturated hydrocarbon.
2. **Key Concept** Why can carbon atoms form so many different kinds of compounds with other atoms?

Determine the Suffix

Now that you know how to find the root word of a hydrocarbon, you must also find the suffix, or end, of the name. Recall that carbon atoms bond to other carbon atoms by single bonds, double bonds, or triple bonds. **Table 3** shows which suffix to use when the type of bonds in the molecule are determined. Look at **Figure 6** again. The molecule has all single bonds and should have the suffix *–ane*. Put the root and the suffix together and you get *butane*.

Reading Check What is the name of a hydrocarbon with six carbon atoms that contains only single bonds?

Determine the Prefix

Sometimes hydrocarbons have a prefix, and sometimes they do not. Recall that hydrocarbons form chains, branched chains, and rings. If a hydrocarbon contains a ring structure, the prefix *cyclo–* is added before the root name. Hydrocarbons sometimes have other prefixes and numbers added before their name. You might read about this naming system in more advanced chemistry courses. For this lesson, only hydrocarbons in the form of a ring will get a prefix. **Table 4** summarizes the steps used to name a hydrocarbon.

Table 3 The type of bonds in the hydrocarbon chain determines the suffix in the name.

Table 3 Bond Type and Hydrocarbon Suffix

Bond Type	Suffix
All single bonds —C—C—	–ane
At least one double bond —C=C—	–ene
At least one triple bond —C≡C—	–yne

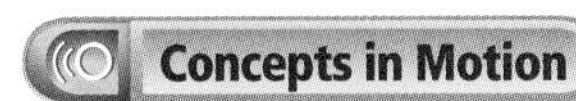

Interactive Table

Table 4 Naming Hydrocarbons

Steps for Naming Hydrocarbons	Example A	Example B
1 Examine the compound.	H—C≡C—C(H_2)—C(H_2)—H (structural formula with H atoms above and below the two single-bonded carbons)	Ring of six CH_2 groups
2 Count the number of carbon atoms in the longest continuous chain.	There are 4 carbon atoms in the longest chain.	There are 6 carbon atoms in the longest chain.
3 Determine the root name of the hydrocarbon using **Table 2.**	The root name is *but–*.	The root name is *hex–*.
4 Determine the type of bonds in the hydrocarbon, then use **Table 3** to find the suffix.	There is a triple bond, so add the suffix *–yne*.	There are only single bonds, to add the suffix *–ane*.
5 Put the root and suffix together to name the hydrocarbon.	Combining the root and suffix gives the name *butyne*.	Combining the root and suffix gives the name *hexane*.
6 If the hydrocarbon is a ring, add *cyclo-* to the beginning of the name.	No prefix is needed because the structure is not a ring. The name of the hydrocarbon is butyne.	The structure is a ring so the prefix *cyclo–* is added to the name. The name of the hydrocarbon is cyclohexane.

Lesson 1 Review

Online Quiz

Visual Summary

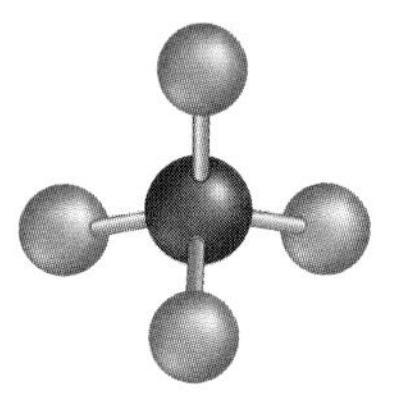

Methane

Chemical compounds that contain carbon and usually at least one hydrogen atom are called organic compounds.

Carbon atoms form several different substances including graphite, diamonds, fullerene, and amorphous carbon.

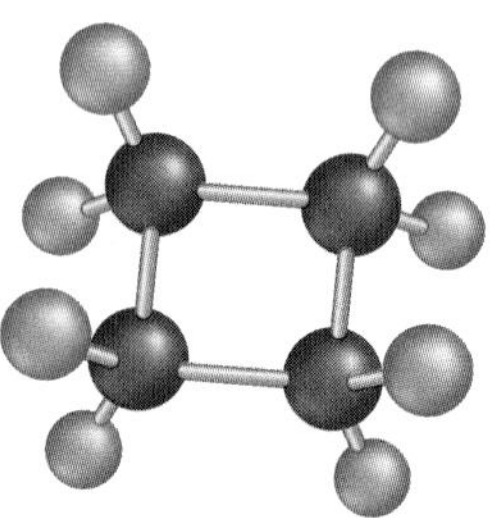

Cyclobutane

A hydrocarbon can be in the form of a straight chain, branched chain, or a ring structure.

FOLDABLES®

Use your lesson Foldable to review the lesson. Save your Foldable for the project at the end of the chapter.

What do you think

You first read the statements below at the beginning of the chapter.

1. Charcoal and diamonds are made of carbon atoms.
2. Carbon atoms often bond with hydrogen atoms in chemical compounds.

Did you change your mind about whether you agree or disagree with the statements? Rewrite any false statements to make them true.

Use Vocabulary

1 **Define** *saturated hydrocarbon* and *unsaturated hydrocarbon.*

2 **Use the term** *isomer* in a complete sentence.

Understand Key Concepts

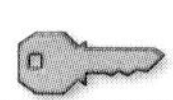

3 **Summarize** how carbon bonds with other carbon atoms.

4 Examine each molecular formula. Which is propane?

A. CH_3 **C.** C_3H_8
B. C_2H_6 **D.** C_4H_{10}

5 **Explain** why carbon is unique compared to other elements.

Interpret Graphics

6 **Identify** Which form of pure carbon is represented in the diagram below?

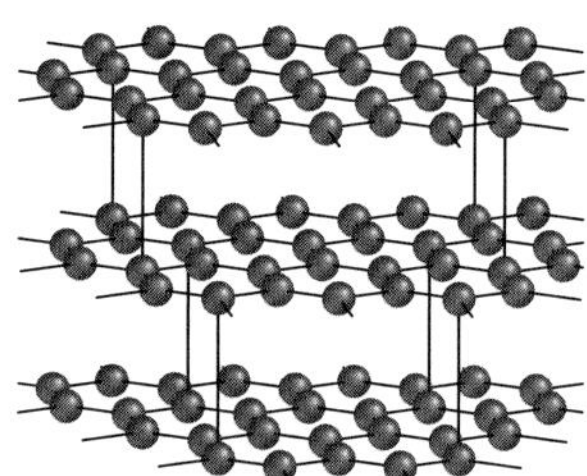

7 **Explain** Copy and fill in the table below to explain how the four forms of carbon differ.

Graphite	
Diamond	
Fullerene	
Amorphous carbon	

Critical Thinking

8 **Explain** why so many different compounds are made from carbon.

9 **Draw** the molecular structure for pentane (C_5H_{12}).

Carbon

SCIENCE & SOCIETY

Will it replace the silicon in your computer?

You have a computer inside your head—your brain. Your brain has 100 billion tiny switches that allow you to process information every day. Those switches are your brain cells. Like your brain, computers have billions of tiny "brain cells" called silicon transistors. These tiny electronic components have been changing society since they were first developed about 50 years ago.

Transistors are devices that can strengthen an electronic signal. Transistors can rapidly turn computer circuits off and on. They are efficient and produce very little heat, which makes them useful in cell phones, radios, and computers. Silicon transistors first were used in computers in 1955. At the time, computers were about the size of three or four adults. Now, of course, computers are much smaller.

Scientists developed the first miniature silicon transistor in 1965. They attached several of the tiny transistors, along with other electronic components, to a piece of plastic. This was the invention of the circuit board. The circuit board allowed designers to fit many more transistors into a computer. This also allowed computers to be made much smaller.

The number of transistors that can be placed on a single circuit board has doubled every two years since 1965. Computers have become smaller, faster, and more powerful. But silicon transistors cannot be made much smaller. How might scientists create smaller transistors? The answer is to use carbon instead of silicon. Carbon nanotubes are cylindrical structures made of pure carbon. They conduct both heat and electricity and are many times faster and more efficient than silicon transistors. And nanotubes are tiny—100,000 of them side by side would be about as thick as a human hair. As if that weren't enough, the tubes are ten times stronger than steel. Once again, society will change as electronics you can't even imagine today become everyday items of the future.

The first silicon transistor, shown here on top of a postage stamp, was small for its time but huge by today's standards.

TRIDAC was the first fully transistorized computer.

By the 1980s, computers had become small enough to fit on a desktop.

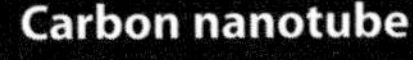
Carbon nanotube

It's Your Turn

RESEARCH AND REPORT What do scientists think is the next step in using carbon nanotube transistors? If nanotube transistors lead to even smaller, faster, and more powerful devices, how might those devices impact your everyday life?

Lesson 2

Other Organic Compounds

Reading Guide

Key Concepts 
ESSENTIAL QUESTIONS

- What are the four common functional groups of organic compounds?
- What are polymers?

Vocabulary

substituted hydrocarbon p. 499
functional group p. 500
hydroxyl group p. 500
halide group p. 501
carboxyl group p. 501
amino group p. 502
polymer p. 503
monomer p. 503
polymerization p. 503

g Multilingual eGlossary

This interesting caterpillar is not feared because of its bite. It shoots an organic compound called formic acid at its enemies. While you might not have caterpillars shooting formic acid at you, you do have other organic compounds all around you.

Launch Lab

20 minutes

How do functional groups affect compounds?

In some hydrocarbons, a hydrogen atom is removed and another atom or group of atoms takes its place. Rubbing alcohol and glycerin are two examples.

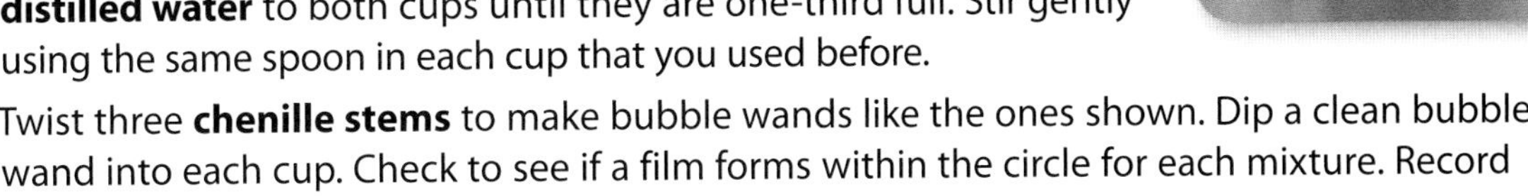

1. Read and complete a lab safety form.
2. Use a **plastic spoon** to measure two spoons of **rubbing alcohol** and pour the liquid into a **clear plastic cup.** Observe the properties of the alcohol. Use the wafting method to check the odor. Record your observations in your Science Journal.
3. Repeat step 2 with **glycerin** using a clean spoon and cup. Add **distilled water** to both cups until they are one-third full. Stir gently using the same spoon in each cup that you used before.
4. Twist three **chenille stems** to make bubble wands like the ones shown. Dip a clean bubble wand into each cup. Check to see if a film forms within the circle for each mixture. Record your observations in your Science Journal.

Think About This

1. Compare and contrast the properties and structural diagrams of rubbing alcohol and glycerin.
2. **Key Concept** Propane is a colorless gas. What changes occur when a hydrogen atom in propane is replaced by an oxygen atom and a hydrogen atom and rubbing alcohol forms?

Substituted Hydrocarbons

Think for a moment about any sports team. Teams often substitute players in and out of a game. In a similar way, other atoms can be substituted for a hydrogen atom in a hydrocarbon. *A* **substituted hydrocarbon** *is an organic compound in which a carbon atom is bonded to an atom, or group of atoms, other than hydrogen.* Just as a team might function differently when new players are substituted into the game, organic compounds function differently when hydrogen atoms are substituted with other atoms.

You might be familiar with the substituted hydrocarbon ethanol. It is often mixed with gasoline and used as a fuel for cars. Ethanol is also in food flavorings, such as vanilla extract. The chemical formula for ethanol is CH_3CH_2OH. In this compound, one hydrogen atom of ethane (CH_3CH_3) has been replaced with –OH. This is just one type of substituted hydrocarbon that you will read about in this lesson.

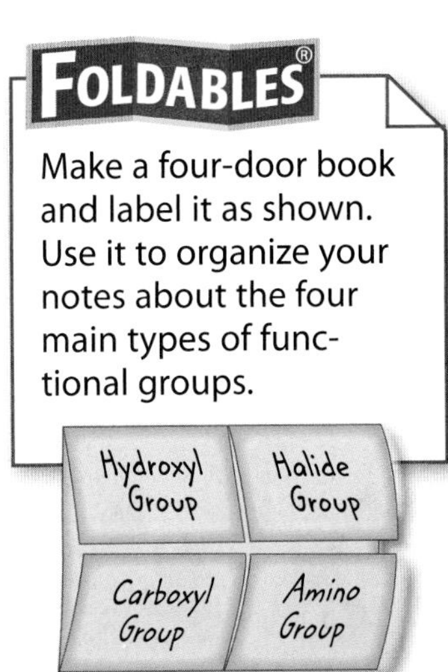

FOLDABLES®
Make a four-door book and label it as shown. Use it to organize your notes about the four main types of functional groups.
Hydroxyl Group
Halide Group
Carboxyl Group
Amino Group

Math Skills

Use Ratios

A ratio expresses the relationship between two or more things. For example, in the formula for methane, CH_4, the ratio of carbon atoms to hydrogen atoms is 1:4, read as "1 to 4."

Ethanol is written CH_3CH_2OH. One molecule contains 2 carbon atoms, **6** hydrogen atoms, and **1** oxygen atom. The ratio is:

C:H:O = 2:**6**:**1**

Practice

What is the ratio of carbon to hydrogen to oxygen atoms in table sugar, $C_{12}H_{22}O_{11}$?

- **Math Practice**
- **Personal Tutor**

Functional Groups

You just read that a hydrogen atom, in an organic compound, can be substituted with other atoms. This causes the substituted hydrocarbon to have new properties. *A* **functional group** *is an atom or group of atoms that determine the function and properties of the compound.* The substituted hydrocarbon is renamed to indicate which functional group has been substituted. There are many functional groups, each with specific characteristics. Four functional groups are discussed in this lesson.

Hydroxyl Group

Have you ever used rubbing alcohol to clean a cut or a scrape? It is often used to soothe and disinfect the skin. Rubbing alcohol is the common name for the compound 2-propanol. 2-Propanol is a substituted hydrocarbon of propane and contains the hydroxyl (hi DRAHK sul) functional group, as shown in **Figure 7.** *The* **hydroxyl group** *contains two atoms—oxygen and hydrogen. Its formula is –OH.* Organic compounds that contain the hydroxyl group are called alcohols. Alcohols are polar compounds and can dissolve in water. They have high melting and boiling points and are commonly used as disinfectants, fuel, and solvents. **Figure 7** also shows how the substituted hydrocarbon ethanol differs from ethane. Larger alcohols form when the hydroxyl group is substituted in larger hydrocarbons.

Figure 7 Substituting a H atom for a functional group in a hydrocarbon changes its properties.

<table>
<tr><th colspan="2">Hydrocarbon</th><th colspan="2">Alcohol</th></tr>
<tr><td>Ethane
H H
| |
H — C — C — H
| |
H H</td><td>• Melting point: −181.7°C
• Boiling point: −88.6°C
• Appearance: colorless, odorless gas
• Uses: automotive fuel; refrigerant in extremely low-temperature systems</td><td>Ethanol
H OH
| |
H — C — C — H
| |
H H</td><td>• Melting point: −117.3°C
• Boiling point: 78.5°C
• Appearance: colorless liquid with mild odor
• Uses: alcohol in alcoholic beverages; solvent in perfumes and paints; automotive fuel; fluid in low-temperature thermometers</td></tr>
<tr><td>Propane
H H H
| | |
H — C — C — C — H
| | |
H H H</td><td>• Melting point: −189.7°C
• Boiling point: −42.1°C
• Appearance: colorless, odorless gas
• Uses: fuel for cooking, hot-air balloons, and some automobiles; raw material for other products</td><td>2-Propanol
H OH H
| | |
H — C — C — C — H
| | |
H H H</td><td>• Melting point: −89.5°C
• Boiling point: 82.4°C
• Appearance: colorless liquid with a strong odor
• Uses: solvent in cleaning fluid; preservative for biological specimens; fuel additive to keep gasoline from freezing</td></tr>
</table>

Halide Group

Group 17 elements—the halogens—can also be substitutions in hydrocarbons. This functional group is called the halide group. *The* **halide group** *contains group 17 halogens—fluorine, chlorine, bromine, and iodine.* Bromomethane (broh moh MEH thayn) is an example of a substance with halide substitution. In bromomethane, a bromine atom replaces one of the hydrogen atoms in methane. This process is similar to the one shown in **Figure 7.** Notice the prefix *bromo–* is before the hydrocarbon name to form the name bromomethane as illustrated in **Figure 8.**

As shown in **Figure 8,** bromomethane is a pesticide. It can be used to kill pests in the soil before strawberries are planted. Its use is strictly regulated because of its environmental hazards. Some countries have banned its use or are phasing out its use.

Carboxyl Group

If you had orange juice for breakfast or a salad for lunch, you ate compounds containing one or more carboxyl (kar BAHK sul) functional groups. *A* **carboxyl group** *consists of a carbon atom with a single bond to a hydroxyl group and a double bond to an oxygen atom. Its formula is -COOH.* When a carboxyl group replaces a hydrogen atom in a hydrocarbon, the result is a carboxylic acid.

Citric acid in citrus fruits, such as oranges, lemons, and limes, is a carboxylic acid. Dairy products such as buttermilk and yogurt also contain a carboxylic acid called lactic acid. Two simple carboxylic acids are methanoic acid and ethanoic acid, as shown in **Figure 9.**

Visual Check How many electrons are shared between each carbon atom and oxygen atom in a carboxyl group?

Figure 8 Bromomethane, sometimes called methyl bromide, contains one carbon atom, three hydrogen atoms, and one bromine atom. It is used for rodent control.

Bromomethane, CH_3Br

WORD ORIGIN

halide
from Greek *hals;* means "salt"

Figure 9 Methanoic and ethanoic acid are simple carboxylic acids.

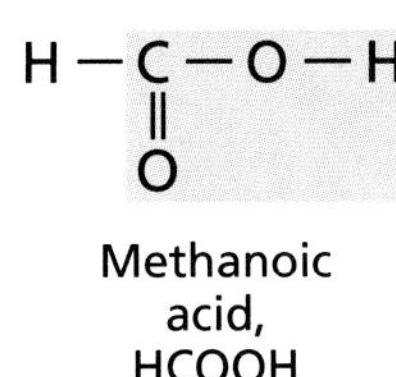

Methanoic acid, HCOOH

Methanoic acid (HCOOH)

Methanoic acid is in the toxin of stinging ants. Methanoic acid is also known as formic acid.

Ethanoic acid, CH_3COOH

Ethanoic acid (CH_3COOH)

Ethanoic acid is in vinegar, which is used in many food items including salad dressings and pickles. Ethanoic acid is also known as acetic acid.

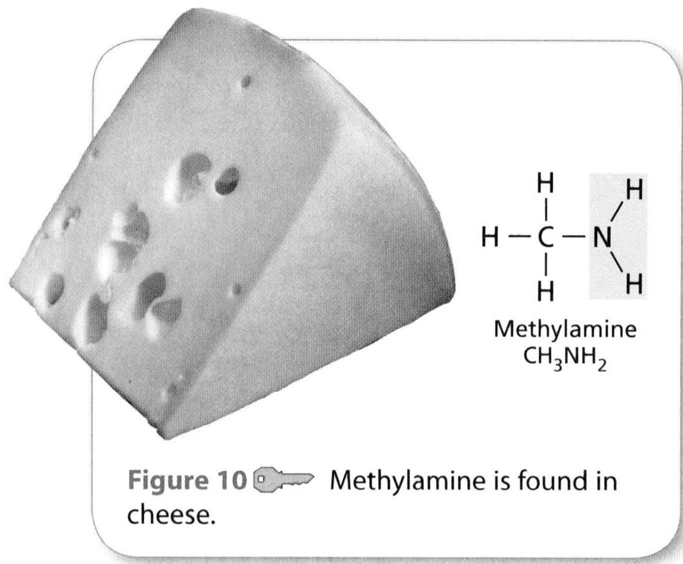

Figure 10  Methylamine is found in cheese.

Amino Group

Have you ever smelled cheese with a strong odor? The strong odor is due to the presence of an amino (uh MEE noh) group. *The* **amino group** *consists of a nitrogen atom covalently bonded to two hydrogen atoms and its formula is* $-NH_2$. The suffix *–amine* is added to the end of each root name to indicate that the amino group is in the compound. The amine, methylamine (meh thuh luh MEEN), forms when an amino group is substituted for a hydrogen in methane. **Figure 10** illustrates the structure and chemical formula of methylamine.

Notice that *–yl* follows the root name *meth*. If a hydrocarbon, such as methane, loses a hydrogen atom, its name changes to methyl. If ethane loses a hydrogen atom, its name becomes ethyl.

Key Concept Check What are four common functional groups of organic compounds?

Table 5 Molecules are not flat. They are three-dimensional. ▼

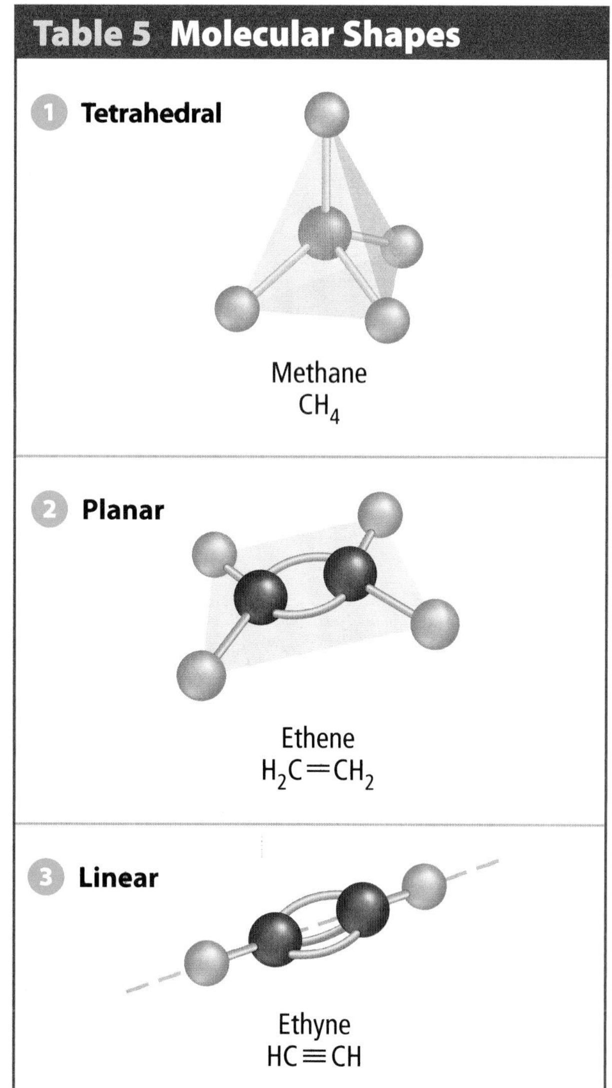

Shapes of Molecules

Molecules come in different shapes and sizes. Scientists often make three-dimensional models of molecules to study their shapes. Knowing a molecule's shape helps scientists understand how it interacts with other molecules, how strong the bonds are between atoms, and what type of bonds are in the molecule. The molecular shapes in **Table 5** show you how some molecules might look in three-dimensions.

1. **Tetrahedral**—Methane is an example of a tetrahedral molecule. The atoms in a tetrahedral molecule form a pyramid.

2. **Planar**—Ethene is an example of a planar molecule. The atoms in a planar molecule are all on the same plane.

3. **Linear**—Ethyne is an example of a linear molecule. The atoms in a linear molecule form a straight line.

Polymers

What do a bottle of water, a toy car, a marker, and a video game have in common? They all contain some amount of plastic. The word *plastic* is a common term that refers to a type of substance called a polymer (PAH luh mur). *A* **polymer** *is a molecule made up of many of the same small organic molecules covalently bonded together, forming a long chain. A* **monomer** *(MAH nuh mur) is one of the small organic molecules that makes up the long chain of a polymer.* Some polymers occur naturally, but many are made in laboratories. Polymers occurring in nature are called natural polymers. Polymers made in laboratories are called synthetic polymers.

Many synthetic polymers are made from simple hydrocarbons by a process called polymerization (pah luh muh ruh ZAY shun). **Polymerization** *is the chemical process in which small organic molecules, or monomers, bond together to form a chain.* Polyethylene is a polymer used to make shampoo bottles, grocery bags, and toys. It is made by the polymerization of ethene—also known as ethylene. As shown in **Figure 11,** first the double bonds are broken in the ethene molecules. Then the carbon atoms bond and form long chains.

 **Key Concept Check** What are polymers?

Inquiry

MiniLab **20 minutes**

How can you make a polymer?

Borax causes organic compounds in white glue to link together. Is this a polymer?

1. Read and complete a lab safety form.
2. Use a **graduated cylinder** to measure 50 mL of **warm water.** Pour the water into a **clear plastic cup.** Stir in **borax** to the water until it no longer dissolves.
3. Measure and pour 5 mL of the water-borax solution into another **clear plastic cup.** Using a **plastic spoon,** add one and one-half spoons of **white glue.** Stir the contents of the cup until completely mixed.
4. Add another 5 mL of the borax mixture to the glue mixture and stir. When the substance is firm, remove it from the cup and knead it until it is like soft clay.

BORAX

Analyze and Conclude

1. **Recognize Cause and Effect** Why did the mixture change as you kneaded it?
2. **Key Concept** Is this mixture a polymer? Why or why not?

Formation of a Polymer

Figure 11 Ethene, also called ethylene, molecules form polyethylene during polymerization.

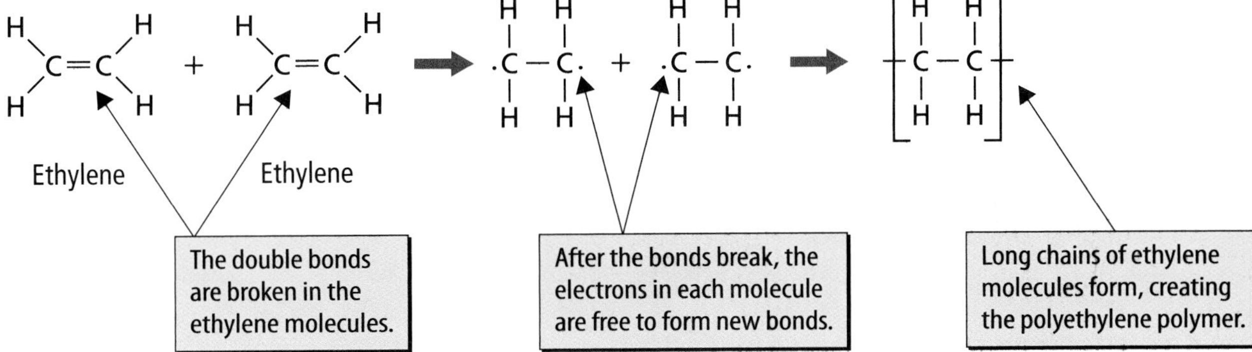
Ethylene
Ethylene
The double bonds are broken in the ethylene molecules.
After the bonds break, the electrons in each molecule are free to form new bonds.
Long chains of ethylene molecules form, creating the polyethylene polymer.

Synthetic Polymers

Polyethylene and many other synthetic polymers are made from petroleum. Petroleum is a thick, oily, flammable mixture of solid, liquid, and gaseous hydrocarbons. Petroleum, an example of a fossil fuel, occurs naturally beneath Earth's surface. It formed from the remains of ancient, microscopic marine organisms.

Examples of polymers and some of their applications are shown in **Table 6.** This is only a small sample of the many polymers and polymer applications that are used today.

Reading Check Why are the polymers in **Table 6** classified as synthetic polymers?

Table 6 Many common objects are made of synthetic polymers.

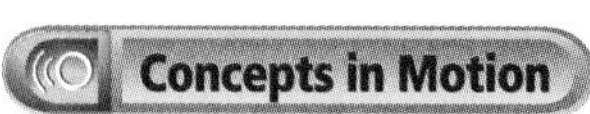

Animation

Table 6 Sample Polymers and Applications

Polymer	Examples
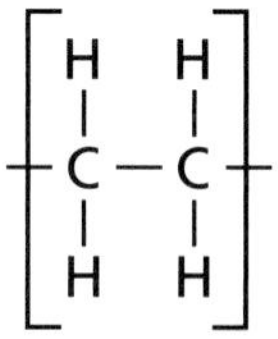 $-[CH_2-CH_2]-$ (H H / C—C / H H) Polyethylene (PE)	 Bales of hay are rolled in polyethylene to protect them from rain. The hay is used to feed farm animals such as cows, horses, and sheep.
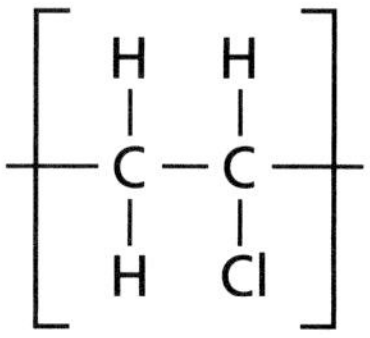 $-[CH_2-CHCl]-$ (H H / C—C / H Cl) Polyvinyl chloride (PVC)	 Pipes made of polyvinyl chloride, or PVC, are used for plumbing. Some rainwear, home siding, and garden hoses are also made of polyvinyl chloride.
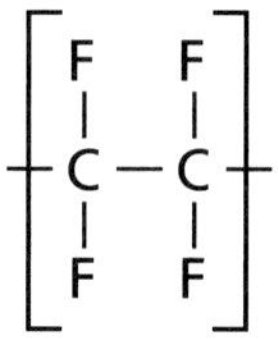 $-[CF_2-CF_2]-$ (F F / C—C / F F) Polytetrafluoroethylene (PTFE)	 Polytetrafluoroethylene (pah lee teh truh flor oh ETH uh leen) is used for nonstick coating on cookware.
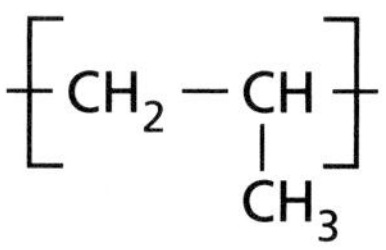 $-[CH_2-CH(CH_3)]-$ Polypropylene (PP)	 Polymer bank notes are made of polypropylene. These bank notes last longer than traditional paper notes. Also, many ropes are made of polypropylene.

Lesson 2 Review

Visual Summary

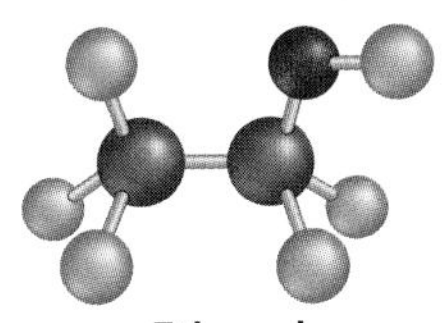

When a functional group replaces a hydrogen atom in a hydrocarbon, it forms a substituted hydrocarbon.

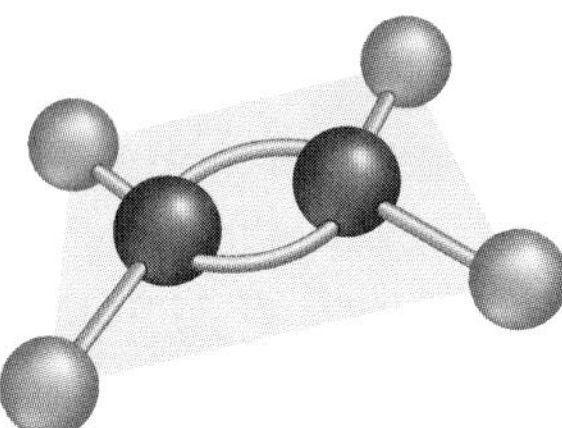

Hydrocarbon molecules form many different three-dimensional shapes such as planar or tetrahedral.

Polymers, made of small, linked, organic molecules, are used to make common items such as money.

FOLDABLES®

Use your lesson Foldable to review the lesson. Save your Foldable for the project at the end of the chapter.

What do you think NOW?

You first read the statements below at the beginning of the chapter.

3. Rubbing alcohol, cheese, and the venom of stinging ants contain carbon compounds.

4. Carbon atoms cannot bond to other carbon atoms.

Did you change your mind about whether you agree or disagree with the statements? Rewrite any false statements to make them true.

Use Vocabulary

1. **Define** *polymerization* in your own words.

Understand Key Concepts

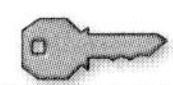

2. **List** the atoms that make up each of the four common functional groups.

3. Select the correct formula for an example of the halide group.

   A. –Cl **C.** –H
   B. –COOH **D.** –OH

4. **Describe** What are polymers?

Interpret Graphics

5. **Determine** the shape of the molecule shown below.

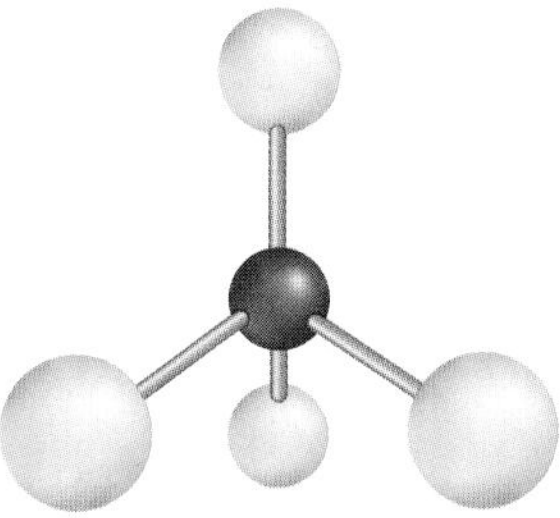

6. **Organize Information** Copy the graphic organizer below. Add additional lines and list the functional groups mentioned in this lesson and their formulas.

Functional Group	Formula

Critical Thinking

7. **Compare and contrast** the structures of butane and butanol.

Math Skills

Math Practice

8. What is the ratio of carbon to hydrogen to oxygen atoms in ethanoic acid, CH_3COOH?

Use Lab Techniques

45 minutes

How do you test for vitamin C?

Materials

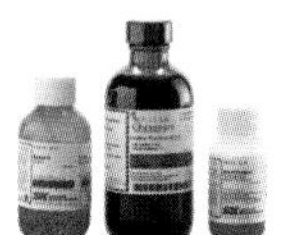

indicator solution

vitamin C solution

distilled water

droppers (10)

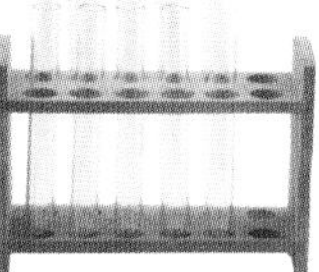

test tubes (10), test tube rack

liquid foods, solid foods

Safety

Vitamin C is a substituted hydrocarbon that is found in many fruits and vegetables.

Learn It

Using an indicator to test for a specific compound is a **lab technique** often used by scientists. When an indicator is added to a mixture, the indicator chemically reacts and changes color if the specific compound is present. In this lab, you will use an indicator to test for the presence of vitamin C.

Try It

1. Complete a lab safety form.

⚠ *Wear protective gloves and do not eat the lab materials.*

2. Copy the data table to the right in your Science Journal. Add additional lines as needed.
3. Obtain 5 mL of the indicator solution in a test tube from your teacher
4. Using a clean dropper, add one drop at a time of the vitamin C solution, obtained from your teacher, to one test tube with the indicator solution. Record the number of drops required for a color change to occur. This sample indicates a strong positive vitamin C reading.
5. Repeat step 4 with distilled water. Notice that the water dilutes the indicator solution, but no distinct color change occurs. This sample demonstrates a negative vitamin C reading.
6. Using a clean test tube and dropper, repeat step 4 for the other samples supplied by your teacher.

Data Table

Food/Beverage Sample	Number of Drops	Results

Apply It

7. Which foods should you include in your diet to ensure that you get enough vitamin C every day?
8. **Key Concept** The structural diagram of vitamin C is shown below. Identify the functional groups discussed in this lesson that are present in a vitamin C molecule.

Vitamin C molecule

Lesson 3

Reading Guide

Key Concepts 
ESSENTIAL QUESTIONS

- What are biological molecules?
- What are some groups of carbon compounds found in living organisms?

Vocabulary

biological molecule p. 508
protein p. 509
amino acid p. 509
carbohydrate p. 510
nucleic acid p. 511
lipid p. 512

 Multilingual eGlossary

Video
What's Science Got to do With It?

Compounds of Life

Inquiry Healthful Diet?

The man and the dog get their energy from the foods they eat. Foods contain carbon compounds that the body uses to function. What carbon compounds are known as the compounds of life?

Inquiry Launch Lab

15 minutes

What does a carbon compound look like when its structure changes?

Proteins are carbon compounds that are in many living things. They are also in products made by living things, such as milk. Like all proteins, the proteins in milk have a three-dimensional structure. You can observe the result of a structural change of one protein when acid is added to milk.

1. Read and complete a lab safety form.
2. Add **skim milk** to a **clear plastic cup** until it is one-third filled. Observe the milk with a **magnifying lens.** Record your observations in your Science Journal.
3. Use a **plastic spoon** to add two spoonfuls of **white vinegar (acetic acid)** to the skim milk. After 1 minute, observe the mixture with the magnifying lens. Record your observations.
4. Place a **coffee filter** in a **funnel.** Hold the funnel over another **clear plastic cup** and pour the milk-vinegar mixture into the filter-lined funnel. Observe the contents of the cup and the funnel. Record your observations.

Think About This

1. Compare and contrast the milk before and after the vinegar was added.
2. What do you think caused the protein to clump when the vinegar was added?
3. **Key Concept** What is one group of carbon compounds in living organisms?

REVIEW VOCABULARY

genetic information
a set of instructions—passed from one generation to the next by genes—that defines how biological processes will occur and determines physical characteristics

FOLDABLES®

Fold a sheet of paper into a four column chart. Label it as shown. Use it to record information about the four biological molecules and their functions.

Biological Molecules: Proteins | Carbohydrates | Nucleic Acids | Lipids

Biological Molecules

Many of the objects you use, such as CDs, DVDs, sandwich bags, plastic bowls, combs, and hairbrushes, are made of synthetic polymers. The bodies of the man and the dog on the previous page also contain polymers. Recall that polymers in nature are called natural polymers. Cells, tissues, and organs in your body and in the bodies of other living things contain natural polymers.

Individual cells of a living thing contain polymers that carry genetic information and pass this information to new cells. The chemical energy stored in your muscles is a polymer too. All of these natural polymers are called biological molecules. *A* **biological molecule** *is a large organic molecule in any living organism.* Biological molecules help determine the structure and function of many different body parts. They also provide the energy needed to run, pedal a bicycle, and the many other activities that you do. The chemical elements that make up these molecules come from the variety of foods you eat and the air you breathe.

Key Concept Check What are biological molecules?

Figure 12 Amino acids are the monomers that form proteins. ▶

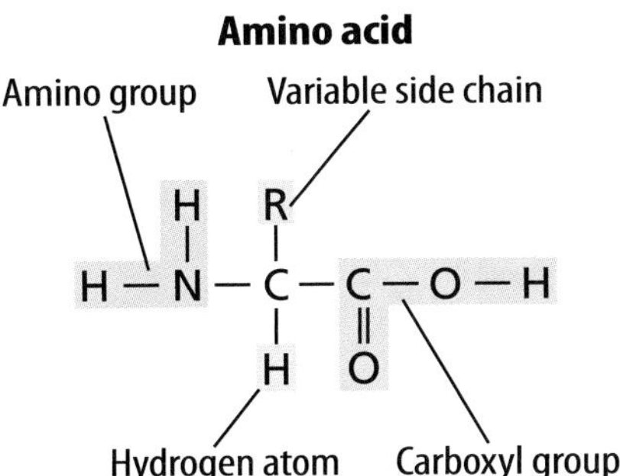
Amino acid
Amino group
Variable side chain
Hydrogen atom
Carboxyl group

Proteins

What do spider webs, plant leaves and roots, and the feathers of a peacock have in common? They contain natural polymers called proteins. Much of your body is made of proteins too, including your hair, muscles, blood, organs, immune system, and fingernails. *A* **protein** (PROH teen) *is a biological polymer made of amino acid monomers. An* **amino acid** (uh MEE noh • A sud) *is a carbon compound that contains the two functional groups—amino and carboxyl.*

Amino Acid Chains

Amino acids link together and form long chains. **Figure 12** shows the basic chemical structure of an amino acid. The *R* represents a side chain of molecules that can differ. There are 20 different side chains. Therefore, 20 different amino acids can link and form proteins, as shown in **Figure 13.**

Proteins and the Human Body

Proteins are important to the body. Of the 20 different amino acids, the human body can make 11 of them. The other nine must be included in the foods that you eat. These nine amino acids are often referred to as essential amino acids. They are in a variety of foods including fish, dairy products, beans, and meat.

 **Reading Check** What are essential amino acids?

Figure 13 Proteins are polymers that contain hundreds of amino acids linked together in a chain. ▼

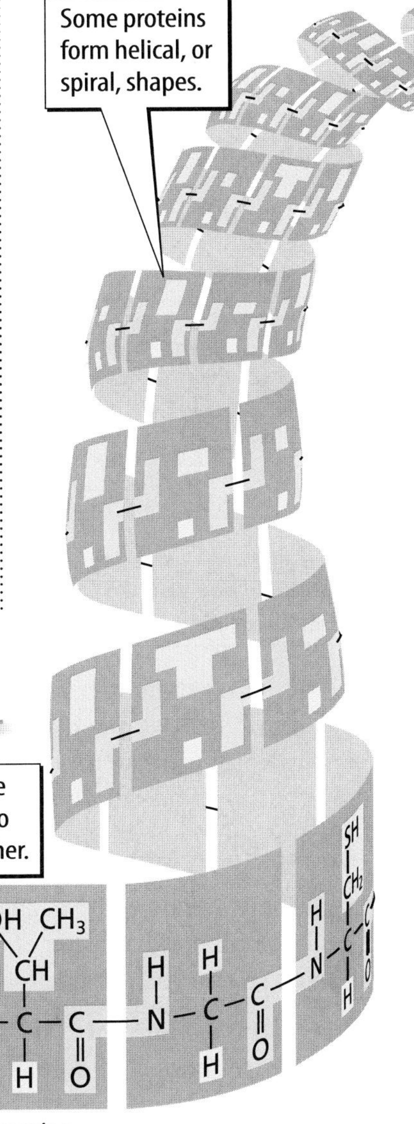
Some proteins form helical, or spiral, shapes.

Protein Formation

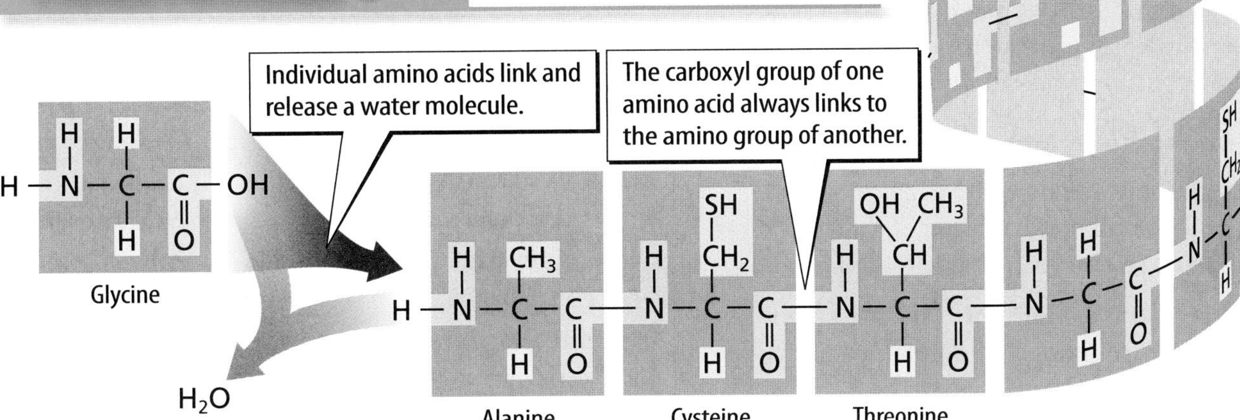
Individual amino acids link and release a water molecule.
The carboxyl group of one amino acid always links to the amino group of another.
Glycine
H_2O
Alanine
Cysteine
Threonine

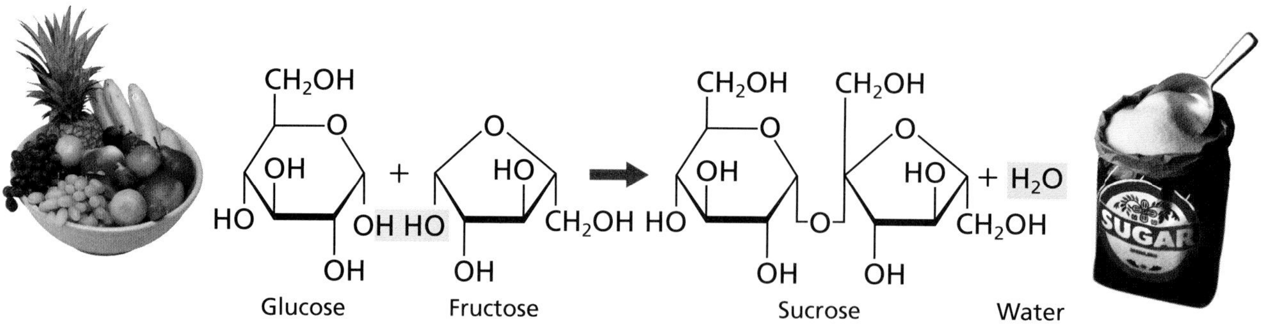

Figure 14 Glucose and fructose are simple sugars in fruits. Sucrose, also known as table sugar, forms from the chemical reaction between glucose and fructose.

Inquiry MiniLab — 20 minutes

How many carbohydrates do you consume?

Nutritionists recommend that the average person get 60 percent of daily calories from carbohydrates. For a 2,000-calorie diet, this is about 300 grams of carbohydrates per day.

1. Copy the table below in your Science Journal. Add lines as needed. List the foods and beverages you consumed yesterday and the quantity of each. Using the **carbohydrate chart** obtained from your teacher, determine the amount of carbohydrates in each type of food.

Food Type	Serving Size	Grams of Carbohydrate per Serving	Total Grams of Carbohydrate
Total carbohydrate consumption			

2. Calculate the total carbohydrates for each item. Then add the grams of carbohydrates for your total carbohydrate consumption that day.

Analyze and Conclude

1. **Infer** how a person can eat the recommended amount of carbohydrates and still not have a healthful diet.
2. **Key Concept** How are carbohydrates important to the overall health of your body?

Carbohydrates

When you eat pasta or sugary snacks, you probably do not think about the carbohydrates in your food. *A* **carbohydrate** (kar boh HI drayt) *is a group of organic molecules that includes sugars, starches, and cellulose.* They are natural polymers that contain carbon, hydrogen, and oxygen atoms. Carbohydrates are a source of energy in cells.

Sugars

Simple sugars, such as glucose and fructose, usually contain five or six carbon atoms. The carbon atoms can be arranged in a ring, as shown in **Figure 14,** or in a straight chain. Your cells can easily break apart simple sugars, which provide quick energy. Glucose is in foods such as fruits and honey. It is also in your blood.

Glucose and fructose combine and form a sugar called sucrose, also shown in **Figure 14.** Sucrose is used to sweeten many foods.

Starch and Cellulose

When simple sugar molecules form chains, they form polymers called complex carbohydrates. Starch and cellulose are complex carbohydrates made of glucose monomers. The chemical bonds in starches take longer to break apart than simple sugars. They provide energy over a longer period. Human digestive systems cannot break the bonds in cellulose, but the digestive systems of animals, such as cows and horses, can.

Nucleotide

Review Personal Tutor

Phosphate group

Sugar

Nitrogen-containing base

Figure 15 Nucleotide monomers make up nucleic acid polymers.

Visual Check In the nitrogen-containing base, how many covalent bonds does each nitrogen atom form?

Nucleic Acids

A biological polymer that stores and **transmits** *genetic information is a* **nucleic acid** (new KLEE ihk • A sud). Genetic information includes instructions for cells on how to make proteins, produce new cells, and transfer genetic information. It is genetic information that determines how you look and how your body functions.

ACADEMIC VOCABULARY

transmit

(verb) to pass, to send, to forward, or to convey

The monomer in a nucleic acid polymer is called a nucleotide, as shown in **Figure 15.** Each nucleotide monomer contains a phosphate group, a sugar, and a nitrogen-containing base. All nucleotides contain the same phosphate group. However, the sugar and nitrogen base can vary in nucleic acids. The elements needed by your body to make nucleic acids come from the foods that you eat.

DNA

Two common nucleic acids, DNA and RNA, control cellular function and heredity. DNA is deoxyribonucleic acid (dee AHK sih rib oh noo klay ihk • A sud). DNA is a spiral-shaped molecule that resembles a twisted zipper, as shown in **Figure 16.** Each DNA monomer contains the five-carbon sugar deoxyribose. Deoxyribose and phosphate groups form the outside of the zipper. Pairs of nitrogen-containing bases—adenine (A) and thymine (T) or cytosine (C) and guanine (G)—form the teeth of the zipper.

RNA

RNA is ribonucleic acid. It contains the five-carbon sugar ribose. RNA is usually single stranded, not double stranded like DNA. It contains the nitrogen bases adenine, cytosine, guanine, and uracil. DNA provides information to make RNA, and then RNA makes the proteins that a cell needs to function.

Key Concept Check What are two groups of carbon compounds found in living organisms?

Figure 16 DNA is a spiral-shaped molecule which is often called a double helix.

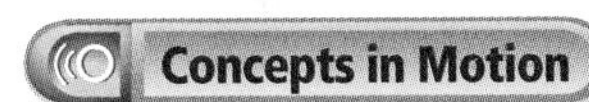

Concepts in Motion

Animation

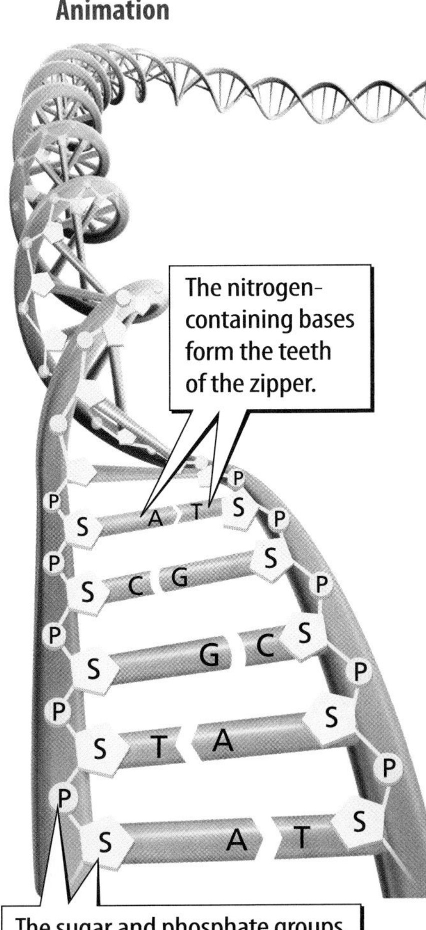

The nitrogen-containing bases form the teeth of the zipper.
The sugar and phosphate groups form the backbone of the zipper.

Oleic acid

$$\mathrm{HO{-}C({=}O)CH_2CH_2CH_2CH_2CH_2CH_2CH_2CH{=}CHCH_2CH_2CH_2CH_2CH_2CH_2CH_2CH_3}$$

Stearic acid

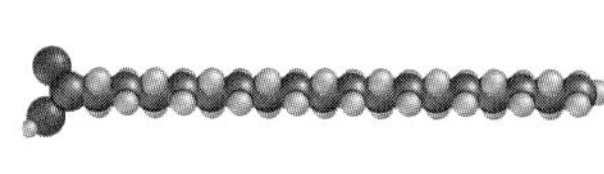

$$\mathrm{HO{-}C({=}O)CH_2CH_2CH_2CH_2CH_2CH_2CH_2CH_2CH_2CH_2CH_2CH_2CH_2CH_2CH_2CH_2CH_3}$$

▲ **Figure 17** Oleic acid is an unsaturated lipid found in olive oil. Stearic acid is a saturated lipid found in bacon.

Lipids

Examples of lipids are shown in **Figure 17.** Lipids are biological molecules, but they are not polymers. *A* **lipid** *is a type of biological molecule that includes fats, oils, hormones, waxes, and components of cellular membranes.* Lipids have two major functions in living organisms. They store energy and make up cellular membranes.

WORD ORIGIN

lipid
from Greek *lipos,* means "fat, grease"

Saturated and Unsaturated Lipids

There are two main groups of lipids—saturated and unsaturated. Saturated lipids contain only single bonds. Just like saturated hydrocarbons, they contain carbon bonded to the maximum number of hydrogen atoms possible. Unsaturated lipids contain at least one double bond. If an unsaturated lipid has one double bond, it is called monounsaturated. If it has more than one double bond, it is called polyunsaturated.

Lipids in Organisms

Lipids, such as fats and oils, store energy for organisms. Another function of lipids is to control what enters and leaves individual cells. These lipids are called phospholipids because they contain a phosphate functional group in their structure. Phospholipids form the cell membrane around individual cells, as shown in **Figure 18.**

Figure 18 Phospholipids form a two-layer cell membrane that controls what enters and leaves cells, such as nutrients, waste, and water. ▼

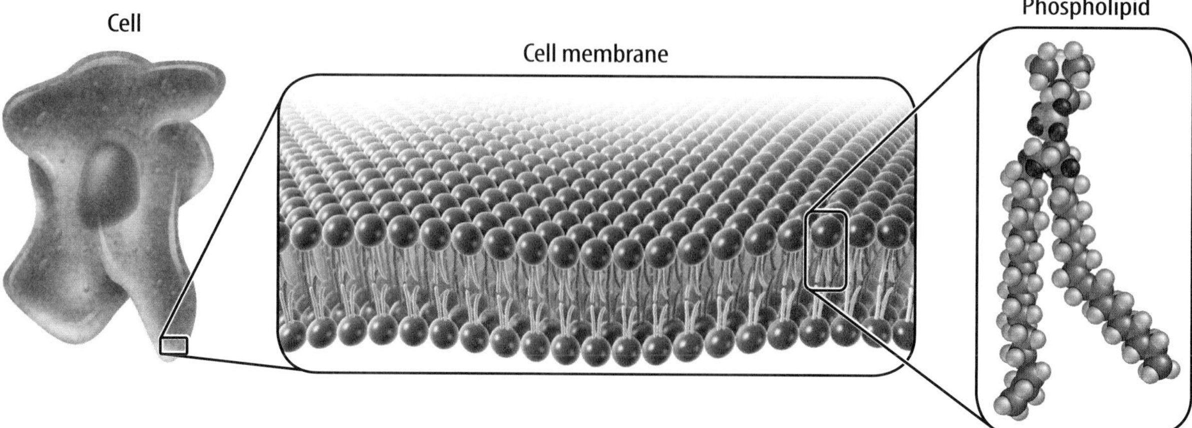

Lesson 3 Review

Assessment — Online Quiz

Inquiry — Virtual Lab

Visual Summary

```
      H   R
      |   |
H — N — C — C — O — H
          |   ||
          H   O
```

Amino acid

A protein is a long chain of amino acids.

The carbohydrate group contains sugars, such as glucose, starches, and cellulose.

Oleic acid

Lipids can be saturated and unsaturated.

FOLDABLES®

Use your lesson Foldable to review the lesson. Save your Foldable for the project at the end of the chapter.

What do you think NOW?

You first read the statements below at the beginning of the chapter.

5. Foods that you eat provide chemical elements that are needed to make carbon compounds in your body.

6. Carbohydrates should be totally eliminated from a healthful diet.

Did you change your mind about whether you agree or disagree with the statements? Rewrite any false statements to make them true.

Use Vocabulary

1. **Write** a complete sentence correctly using the term *biological molecule.*
2. **Define** *lipid* in your own words.

Understand Key Concepts 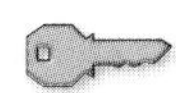

3. **Explain** What are the four types of biological molecules?
4. **Describe** the function of two carbon compounds in living organisms.
5. Which monomer makes up nucleic acids?
   - **A.** amino acids
   - **B.** carbohydrates
   - **C.** glucose
   - **D.** nucleotide

Interpret Graphics

6. **Determine** The molecule below belongs to which group of biological molecules?

```
          H
          |
$H_2N$ — C — C — OH
          |   ||
          H   O
```

7. **Organize Information** Copy and fill in the table below with details about each biological molecule from this lesson.

Molecule	Details

Critical Thinking

8. **Explain** why a meal containing boiled chicken is more healthful than a meal containing fried chicken.
9. **Evaluate** Many weight loss diets stress eliminating carbohydrates and fats from the diet. Explain why eliminating these foods completely from the diet over long periods of time might not be a good idea.

45 minutes

Testing for Carbon Compounds

Many foods come from living things and contain biological molecules. Just as you used an indicator to test for vitamin C in foods, you also can use an indicator to test for other biological molecules.

Ask a Question

How can whole milk be tested to determine if it contains protein, carbohydrates, or lipids?

Make Observations

1. Read and complete a lab safety form. Copy the data table below in your Science Journal to record your results.
2. Place nine clean test tubes in a test-tube rack.
3. To test for starches and complex sugars, pour 2 mL of distilled water in a clean test tube and 2 mL of the starch solution in another clean test tube. Add five drops of Lugol's solution to each test tube. Lugol's solution changes from amber to dark blue in the presence of starch and complex sugars. Record your observations. Remove the used test tubes from the test-tube rack. Place them in a large beaker and set aside.
4. To test for protein, place 2 mL of distilled water in a clean test tube and 2 mL of the protein solution in another clean test tube. Add five drops of Biuret solution to each test tube. Biuret solution changes to purple in the presence of protein. Record your results. Place your used test tubes in the large beaker.
5. To test for lipids, place 2 mL of distilled water in a clean test tube and 2 mL of vegetable oil in another clean test tube. Add five drops of Sudan III to each test tube. Sudan III changes to an orange-pink color if lipids are present. Record your observations and place your used test tubes in the large beaker.

Data Table	
Sample	Results
Starch solution	
Distilled water	
Protein solution	
Distilled water	
Vegetable oil	
Distilled water	
Whole milk	

Materials

indicator solutions

test solutions

whole milk, distilled water

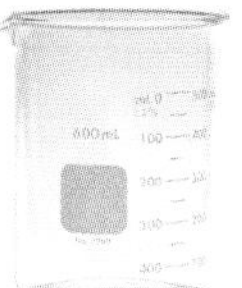

500-mL beakers (2)

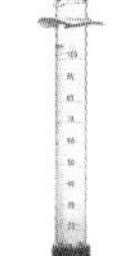

100-mL graduated cylinder

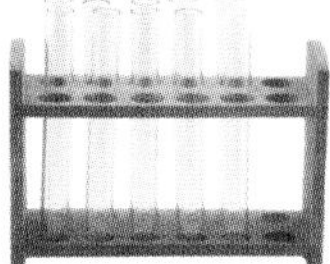

test tubes (9), test-tube rack

Safety

Form a Hypothesis

6. Use your data to form a hypothesis about whether whole milk contains proteins, simple sugars, starches, and lipids.

Test Your Hypothesis

7. Place 2 mL of whole milk in each of the three remaining test tubes.
8. Test the whole milk for starches and complex sugars, proteins, and lipids using the same procedures that you used before. Record your observations. Clean all of the glassware and your lab area. Put the materials where your teacher directs.

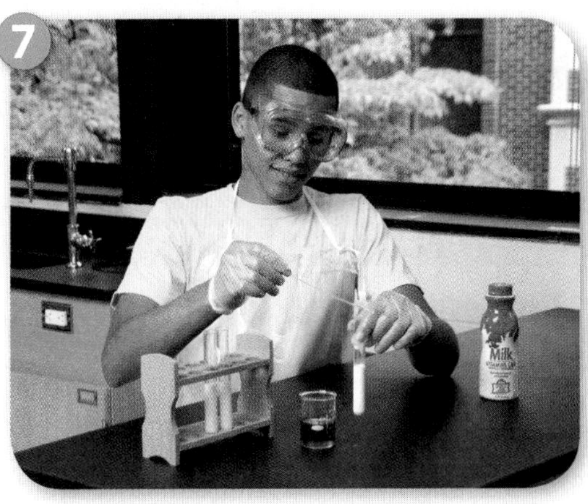

Analyze and Conclude

9. **Interpret Data** Which types of biological molecules did the milk contain?
10. THE BIG IDEA **The Big Idea** What roles do the carbon compounds in whole milk have in human growth and development?

Communicate Your Results

Create a poster that shows the biological molecules found in whole milk and the importance of these molecules for human growth and development.

Indicators provide a quick and easy way to detect the presence of certain molecules. Describe two real-life situations in which an indicator might be used to identify nutrients in different foods.

Lab TIPS

☑ Place a sheet of white paper under the test-tube rack. Use the paper to write down what is in each test tube. The sheet of paper can be used for each trial if there are no spills on it.

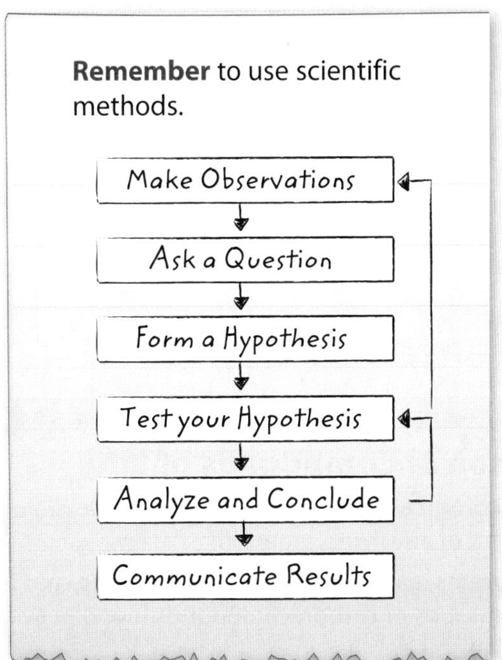

Chapter 14 Study Guide

Carbon is the foundation of life because of its ability to bond easily to other atoms and to other carbon atoms.

Key Concepts Summary

Lesson 1: Elemental Carbon and Simple Organic Compounds

- Carbon atoms can form four covalent bonds using less energy than any other element in group 14. Carbon atoms can also form four different arrangements when bonded to other carbon atoms.
- When carbon bonds with only hydrogen atoms, it forms a **hydrocarbon.**
- A hydrocarbon that contains only single bonds is a **saturated hydrocarbon.** If a hydrocarbon contains at least one double or triple bond, it is an **unsaturated hydrocarbon.**

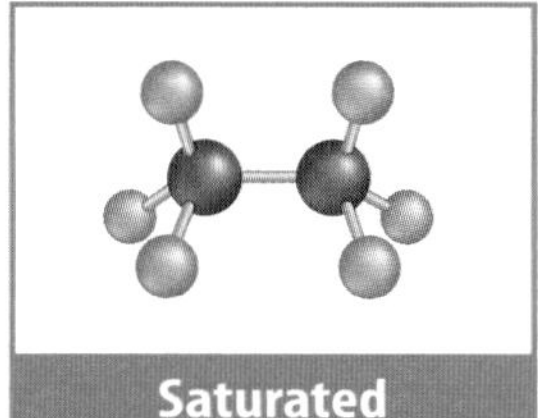
Saturated

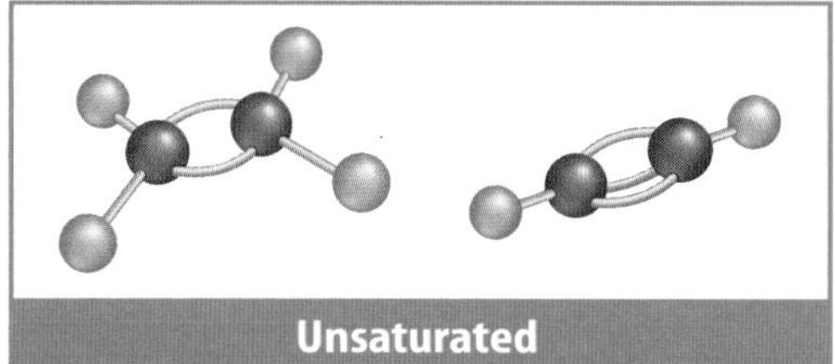
Unsaturated

Vocabulary

organic compound p. 490
hydrocarbon p. 492
isomer p. 492
saturated hydrocarbon p. 493
unsaturated hydrocarbon p. 493

Lesson 2: Other Organic Compounds

- When a hydrogen atom is replaced by a **functional group,** it forms a **substituted hydrocarbon.**
- Alcohols, halides, carboxylic acids, and amines form when specific functional groups are added to a hydrocarbon.
- **Monomers** link to form long chains called **polymers.**

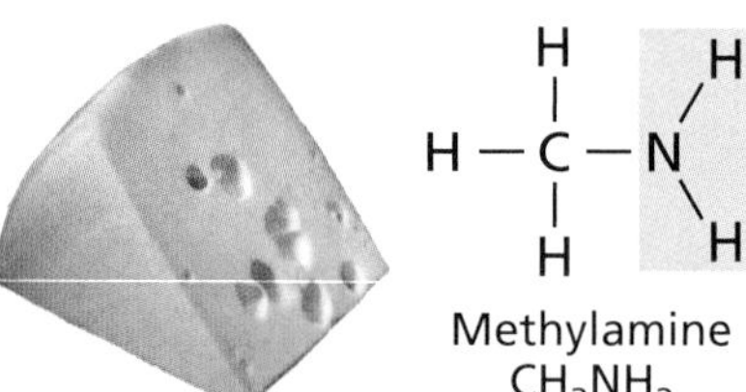

Methylamine
CH_3NH_2

Vocabulary

substituted hydrocarbon p. 499
functional group p. 500
hydroxyl group p. 500
halide group p. 501
carboxyl group p. 501
amino group p. 502
polymer p. 503
monomer p. 503
polymerization p. 503

Lesson 3: Compounds of Life

- A **biological molecule** is a large organic molecule found in any living organism.
- **Proteins, carbohydrates, nucleic acids,** and **lipids** are groups of carbon compounds found in living organisms.

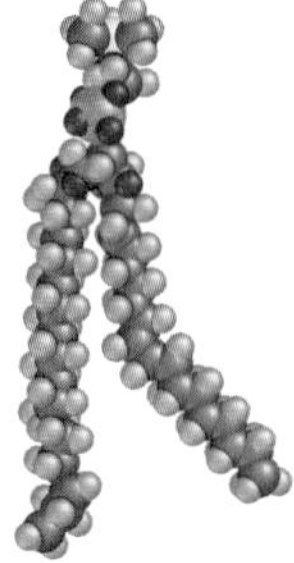
Phospholipid

Vocabulary

biological molecule p. 508
protein p. 509
amino acid p. 509
carbohydrate p. 510
nucleic acid p. 511
lipid p. 512

- **Personal Tutor**
- **Vocabulary eGames**
- **Vocabulary eFlashcards**

FOLDABLES® Chapter Project

Assemble your lesson Foldables as shown to make a Chapter Project. Use the project to review what you have learned in this chapter.

Carbon Chemistry

Use Vocabulary

1. A chemical compound that contains carbon and at least one carbon-hydrogen bond is a(n) ________.
2. A(n) ________ contains the maximum number of hydrogen atoms possible.
3. Compounds that have the same chemical formula but different structural arrangements are called ________.
4. A molecule made up of many small organic molecules covalently bonded together forming a long chain is called a(n) ________.
5. A biological molecule made of amino acids is called a(n) ________.
6. RNA and DNA are two different types of ________.
7. A group of polymers that includes sugars, starches, and cellulose is called a(n) ________.

Link Vocabulary and Key Concepts

Concepts in Motion **Interactive Concept Map**

Copy this concept map, and then use vocabulary terms from the previous page to complete the concept map.

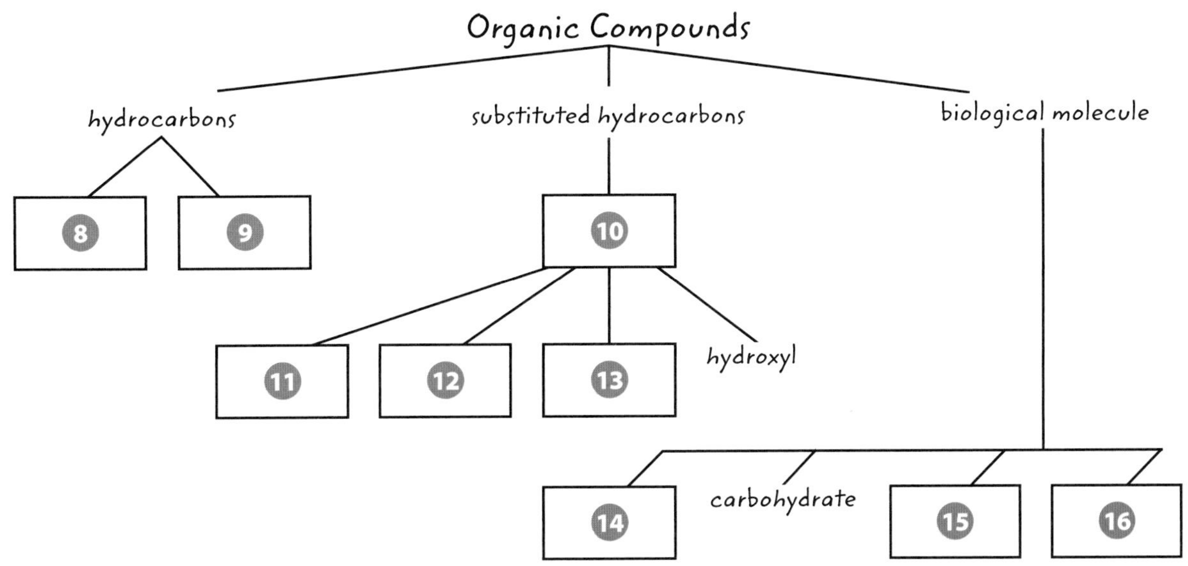

Chapter 14 Review

Understand Key Concepts

1. What is the maximum number of covalent bonds that carbon can form?
   A. 1
   B. 2
   C. 3
   D. 4

2. Which is the correct name for the following carbon compound pictured below?

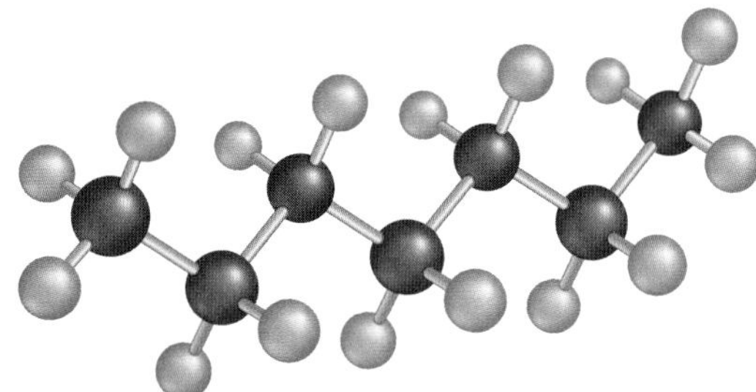

   A. heptane
   B. hexane
   C. pentane
   D. pentene

3. Which is NOT a functional group?
   A. carboxyl
   B. halide
   C. hydroxyl
   D. protein

4. Which functional group contains nitrogen in its structural formula?
   A. amino
   B. carboxyl
   C. halide
   D. hydroxyl

5. For which purpose does the body primarily use carbohydrates?
   A. muscle formation
   B. transmit genetic information
   C. to store energy
   D. cell membrane formation

6. Which item below does not contain lipids?
   A. beeswax
   B. cellular membranes
   C. cooking oil
   D. table sugar

7. Which functional group does this molecule contain?

```
    H
    |
H — C — C — O — H
    |   ‖
    H   O
```

   A. amino
   B. carboxyl
   C. halide
   D. hydroxyl

8. Which biological molecule is single-stranded and contains the instructions for making the proteins that cells need to function?
   A. DNA
   B. RNA
   C. lipid
   D. carbohydrate

9. Which is true about the carbon compound cyclooctane?
   A. It is a straight-chain hydrocarbon with eight carbon atoms and only single bonds.
   B. It is a branched-chain hydrocarbon with eight carbon atoms and at least one double bond.
   C. It is a ringed hydrocarbon with eight carbon atoms and at least one double bond.
   D. It is a ringed hydrocarbon with eight carbon atoms and only single bonds.

10. Which sugar is in DNA?
   A. deoxyribose
   B. fructose
   C. glucose
   D. ribose

11. Which is NOT a form of pure carbon?
   A. diamond
   B. fullerene
   C. graphite
   D. halide

Assessment

Online Test Practice

Critical Thinking

12 **Draw** three isomers for pentane.

13 **Differentiate** between saturated and unsaturated hydrocarbons.

14 **Compare and contrast** the properties of diamond and graphite.

15 **Create** a poster that illustrates why hydrocarbons are present in so many different compounds.

16 **Evaluate** which of these formulas contains at least one double bond: C_7H_{16}, C_4H_8, C_3H_8. Explain your reasoning.

17 **Compare** substituted hydrocarbons with regular hydrocarbons.

18 **Describe** the process used to make most synthetic polymers.

19 **Describe** similarities and differences between proteins and carbohydrates.

20 **Examine** the monomer below. Which biological molecules does it form and what is the function of the biological molecule?

21 **Evaluate** Some diet plans suggest eating only one type of food over a long period of time. Is this a wise practice for maintaining a healthy body? Explain why or why not.

22 **Explain** the function of phospholipids in cells.

Writing in Science

23 **Design** a brochure that explains information about each biological molecule discussed in this chapter.

REVIEW THE BIG IDEA

24 What role does carbon have in the chemistry of living things? Explain a carbon atom's electron structure, how it bonds with other atoms, and how it forms substituted hydrocarbons.

25 The photo below shows a collection of many living things. Describe as many carbon compounds as possible in the living things in the photo.

Math Skills

Math Practice

Use Ratios

26 Many plastics, known as PVCs, are polymers of vinyl chloride, CH_2CHCl. What is the ratio of carbon to hydrogen to chlorine atoms in vinyl chloride?

27 Isopropy alcohol, or rubbing alcohol, has the formula $CH_3CH_2CH_2OH$. What is the ratio of carbon to hydrogen to oxygen atoms in this compound?

Standardized Test Practice

Record your answers on the answer sheet provided by your teacher or on a sheet of paper.

Multiple Choice

Use the diagram below to answer question 1.

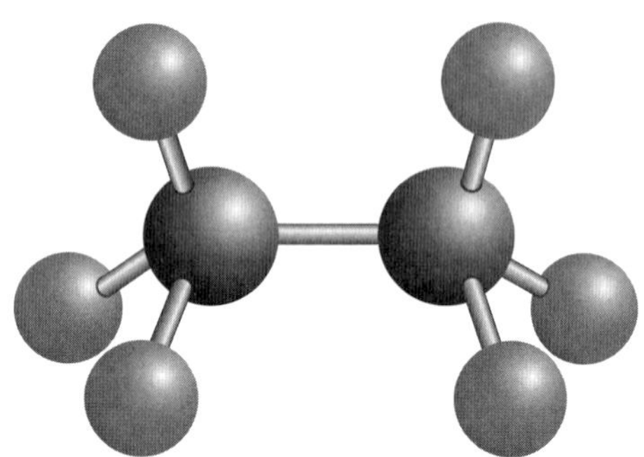

1 Which does the above diagram illustrate?

A an alkane

B an alkyne

C a polymer

D an unsaturated hydrocarbon

2 Which describes the structure of amorphous carbon?

A cage-like

B cross-bonded

C formless

D pyramidal

3 Which group includes a sugar, a phosphate group, and a nitrogen-containing base?

A hydrocarbons

B monomers

C nucleotides

D polymers

4 Which is NOT true about essential amino acids?

A Foods supply them to the body.

B Eleven are known to exist.

C They can link to form proteins.

D They form long chains.

Use the charts below to answer question 5.

Carbon Atoms	Name	Carbon Atoms	Name
1	meth–	6	hex–
2	eth–	7	hept–
3	prop–	8	oct–
4	but–	9	non–
5	pent–	10	dec–

Bond Type	Suffix
All single bonds –C–C–	-ane
At least one double bond –C=C–	-ene
At least one triple bond –C≡C–	-yne

5 Based on the charts, which hydrocarbon has four carbon atoms and only single bonds?

A butane

B butene

C methane

D methyne

6 Which property makes a carbon atom unique?

A It has protons, neutrons, and electrons.

B It has the ability to bond easily.

C It is a nonmetal.

D It is solid at room temperature.

7 Which is characteristic of biological molecules?

A They are very small molecules.

B They are found in living things.

C They are classified as inorganic.

D They form synthetic polymers.

Assessment
Online Standardized Test Practice

Use the diagram below to answer question 8.

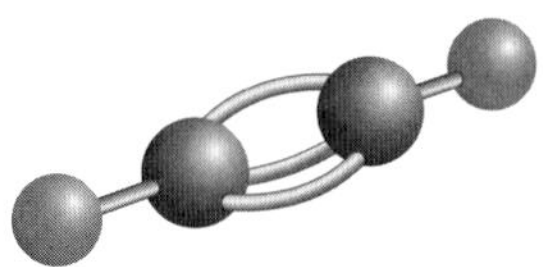

8 How many bonds link the two carbon atoms in the diagram?

A one
B two
C three
D four

9 Which is true about the formation of polymers?

A Laboratories produce most natural polymers.
B Most polymers result from splitting large molecules.
C Many synthetic polymers are products of petroleum.
D Typical polymers result from physical changes in compounds.

10 What is true about an unsaturated hydrocarbon?

A It contains only single bonds.
B It gives off molecules of water.
C It has double or triple bonds.
D Its name ends with the suffix *–ane*.

Constructed Response

Use the table below to answer questions 11 and 12.

Type of Molecule	Example

11 In the table, list the four types of biological molecules and give an example of each.

12 Describe at least one function of each type of biological molecule.

Use the table below to answer question 13.

Functional Group	Example

13 Complete the table above by listing the four common functional groups of organic compounds. Then, provide an example of an organic compound in which each group can be found. In a paragraph, describe one practical use of each group of organic compounds.

NEED EXTRA HELP?													
If You Missed Question...	1	2	3	4	5	6	7	8	9	10	11	12	13
Go to Lesson...	1	1	3	3	1	1	3	1	2	1	3	3	2

Unit 5

WAVES, ELECTRICITY, & MAGNETISM
IT'S 9:30, DID YOU CALL YOUR FATHER?!
WATER?! Nah, I don't need any water.
1600
1700
1800

1660
Robert Hooke publishes the wave theory of light, comparing light's movement to that of waves in water.

1705
Francis Hauksbee experiments with a clock in a vacuum and proves that sound cannot travel without air.

1820
Danish physicist Hans Christian Ørsted publishes his discovery that an electric current passing through a wire produces a magnetic field.

1878
Thomas Edison develops a system to provide electricity to homes and businesses using locally generated and distributed direct current (DC) electricity.

1882
Thomas Edison develops and builds the first electricity-generating plant in New York City, which provides 110 V of direct current to 59 customers in lower Manhattan.

I don't know why I wear these shades, I NEVER get the spotlight!

1900

2000

1883
The first standardized incandescent electric lighting system using overhead wires begins service in Roselle, New Jersey.

1890s
Physicist Nikola Tesla introduces alternating current (AC) by inventing the alternating current generator, allowing electricity to be transmitted at higher voltages over longer distances.

1947
Chuck Yeager becomes the first pilot to travel faster than the speed of sound.

? Inquiry
Visit ConnectED for this unit's STEM activity.

Unit 5 Nature of SCIENCE

Graphs

Have you ever felt a shock from static electricity? The electric energy that you feel is similar to the electric energy you see as a flash of lightning, such as in **Figure 1**, only millions of times smaller. Scientists are still investigating what causes lightning and where it will occur. They use graphs to learn about the risk of lightning in different places and at different times. A **graph** is a type of chart that shows relationships between variables. Graphs organize and summarize data in a visual way. Three of the most common graphs are circle graphs, bar graphs, and line graphs.

Types of Graphs

Line Graphs

A line graph is used when you want to analyze how a change in one variable affects another variable. This line graph shows how the average number of lightning flashes changes over time in Illinois. Time is plotted on the *x*-axis. The average numbers of lightning flashes are plotted the *y*-axis. Each dot, called a data point, indicates the average number of flashes recorded during that hour. A line connects the data points so a trend can be analyzed.

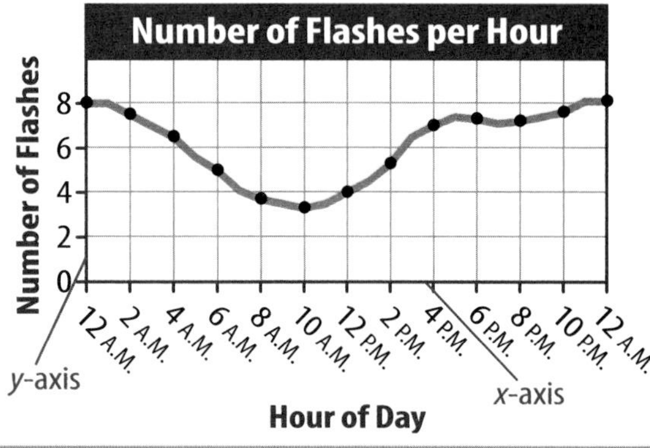

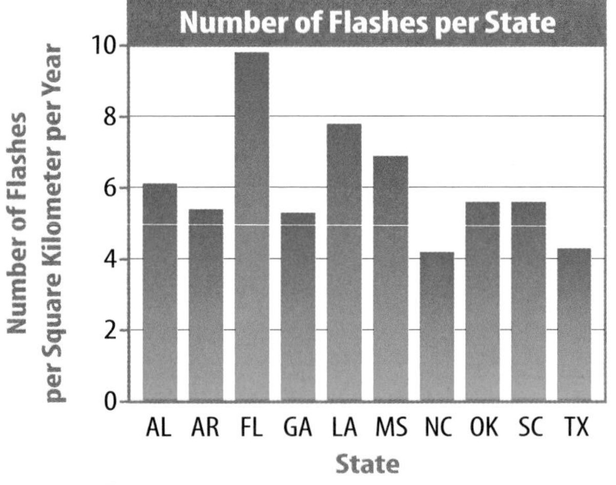

Bar Graphs

When you want to compare amounts in different categories, you use a bar graph. The horizontal axis often contains categories instead of numbers. This bar graph shows the average number of lightning flashes that occur in different states. On average, about 9.8 lightning flashes strike each square kilometer of land in Florida every year. Florida has the more lightning flashes per square kilometer than all other states shown on the graph.

Circle Graphs

If you want to show how the parts of something relate to the whole, use a circle graph. This circle graph shows the average percentage of lightning flashes each U.S. region receives in a year. The graph shows that the northeastern region of the United States receives about 14 percent of all lightning flashes that strike the country each year. From the graph, you can also determine that the southeast receives the most lightning in a given year.

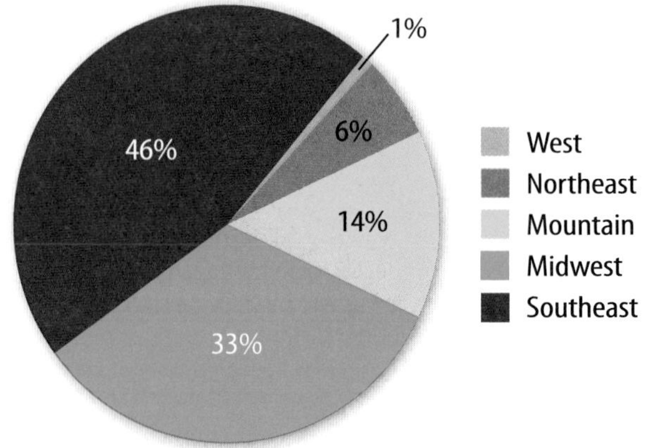

Line Graphs and Trends

Suppose you are planning a picnic in an area that experiences quite a bit of lightning. When would be the safest time to go? First, you gather data about the average number of lightning flashes per hour. Next, you plot the data on a line graph and analyze trends. Trends are patterns in data that help you find relationships among the data and make predictions.

Follow the orange line on the line graph from 12 A.M. to 10 A.M. in **Figure 2.** Notice that the line slopes downward, indicated by the green arrow. A downward slope means that as measurements on the *x*-axis increase, measurements on the *y*-axis decrease. So, as time passes from 12 A.M. to 10 A.M., the number of lightning flashes decreases.

If you follow the orange line on the line graph from 12 P.M. to 5 P.M., you will notice that the line slopes upward. This is indicated by the blue arrow. An upward slope means that as the measurements on the *x*-axis increase, the measurements on the *y*-axis also increase. So, as time passes from 12 P.M. to 5 P.M., the number of lightning flashes increases.

The line graph shows you that between about 8 A.M. and 12 P.M., you would have the least risk of lightning during your picnic.

Figure 2 The slope of a line in a line graph shows the relationship between the variables on the *x*-axis and the variables on the *y*-axis. ▼

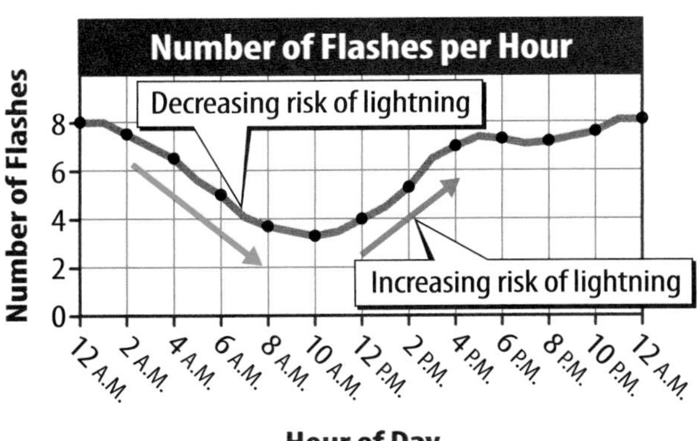

▲ **Figure 1** Scientists study lightning to get a better understanding of what causes it and to predict when it will occur.

Inquiry **MiniLab** 25 minutes

When does lightning strike?

Meteorologists in New Mexico collected data on the number of lightning flashes throughout the day. How can you use a line graph to plan the safest day trip?

1. Make a line graph of the data in the table.
2. Find the trends that show when the risk for lightning is increasing and when it is decreasing.

Hour	# of Flashes
12:00 A.M.	2
3:00 A.M.	2
6:00 A.M.	2
9:00 A.M.	2
12:00 P.M.	5
3:00 P.M.	21
6:00 P.M.	36
9:00 P.M.	22
12:00 A.M.	2

Analyze and Conclude

Decide How could you use your graph to plan a day trip in New Mexico with the least risk of lightning?

Chapter 15

Waves

How do waves travel through matter?

Inquiry What causes waves?

Waves are actually energy moving through matter. Think about the amount of energy this wave must be carrying.

- What do you think caused this giant wave?
- Do you think this is the only large wave in the area?
- How do you think this wave moves through water?

Get Ready to Read

What do you think?

Before you read, decide if you agree or disagree with each of these statements. As you read this chapter, see if you change your mind about any of the statements.

1. Waves carry matter as they travel from one place to another.
2. Sound waves can travel where there is no matter.
3. Waves that carry more energy cause particles in a material to move a greater distance.
4. Sound waves travel fastest in gases, such as those in the air.
5. When light waves strike a mirror, they change direction.
6. Light waves travel at the same speed in all materials.

Your one-stop online resource

connectED.mcgraw-hill.com

WebQuest

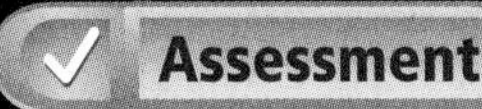

Concepts in Motion

Multilingual eGlossary

Lesson 1

Reading Guide

Key Concepts
ESSENTIAL QUESTIONS

- What is a wave?
- How do different types of waves make particles of matter move?
- Can waves travel through empty space?

Vocabulary

wave p. 529
mechanical wave p. 531
medium p. 531
transverse wave p. 531
crest p. 531
trough p. 531
longitudinal wave p. 532
compression p. 532
rarefaction p. 532
electromagnetic wave p. 535

g Multilingual eGlossary

What are waves?

Inquiry Why Circles?

Have you ever seen raindrops falling on a smooth pool of water? If so, you probably saw a pattern of circles forming. What are these circles? The circles are small waves that spread out from where the raindrops hit the water.

Launch Lab

20 minutes

How can you make waves?

Oceans, lakes, and ponds aren't the only places you can find waves. Can you create waves in a cup of water?

1. Read and complete a lab safety form.
2. Add **water** to a **clear plastic cup** until it is about two-thirds full. Place the cup on a **paper towel.**
3. Explore ways of producing water waves by touching the cup. Do not move the cup.
4. Explore ways of producing water waves without touching the cup. Do not move the cup.

Think About This

1. How did the water's surface change when you produced water waves in the cup?
2. **Key Concept** What did the different ways of producing water waves have in common?

What are waves?

Imagine a warm summer day. You are floating on a raft in the middle of a calm pool. Suddenly, a friend does a cannonball dive into the pool. You probably know what happens next—you are no longer resting peacefully on your raft. Your friend's dive causes you to start bobbing up and down on the water. You might notice that after you stop moving up and down, you haven't moved forward or backward in the pool.

Why did your friend's dive make you move up and down? Your friend created waves by jumping into the pool. *A* **wave** *is a disturbance that transfers energy from one place to another without transferring matter.* You moved up and down because these waves transferred energy.

REVIEW VOCABULARY
energy
the ability to cause change

 **Key Concept Check** What is a wave?

A Source of Energy

The photo on the previous page shows the waves produced when raindrops fall into a pond. The impact of the raindrops on the water is the source of energy for these water waves. Waves transfer energy away from the source of the energy. **Figure 1** shows how light waves spread out in all directions away from a flame. The burning wick is the energy source for these light waves.

Figure 1 All waves, such as light waves, spread out from the energy source that produces the waves.

Figure 2 Water waves transfer energy across the pool, but not matter. As a result, the raft does not move along with the waves. Instead, the raft returns to its initial position.

The raft is at rest in its initial position.

A wave begins to lift the raft upward when it reaches the raft.

The wave transfers energy to the raft as it lifts it upward.

The wave passes the raft and continues to move across the pool.

The raft returns to its initial position after the wave passes.

Visual Check Describe what happens to the raft when the waves transfer energy to it.

Energy Transfer

Think about the waves created by a cannonball dive. When the diver hits the water, the diver's energy transfers to the water. Recall that energy is the ability to cause change. The energy transferred to the water produces waves. The waves transfer energy from the place where the diver hits the water to the place where your raft is floating. **Figure 2** shows that the energy transferred by a wave lifts your raft when the wave reaches it.

Reading Check What do waves transfer from place to place?

The waves created in the pool caused you to move up and then down on your raft. As **Figure 2** shows, however, after the waves passed, you were in the same place in the pool. The waves didn't carry you along after they reached you. Waves transfer energy, but they leave matter in the same place after they pass. Because the water under your raft was not carried along with the waves, you remained in the same place in the pool.

How Waves Transfer Energy

Why wouldn't a water wave carry a raft along with it? How do waves transfer energy without transferring matter? Think about the diver hitting the water. Like all materials, water is made of tiny particles. When the diver hit the water, the impact of the diver exerted a force on water particles. The force of the impact transferred energy to the water by pushing and pulling on water particles. These particles then pushed and pulled on neighboring water particles, transferring energy outward from the point of impact. In this way, the energy of the falling diver is transferred to the water. This energy then travels through the water, from particle to particle, as a wave.

Mechanical Waves

A water wave is an example of a mechanical wave. *A wave that can travel only through matter is a* **mechanical wave.** Mechanical waves can travel through solids, liquids, and gases. They cannot travel through a vacuum. *A material in which a wave travels is called a* **medium.** Two types of mechanical waves are transverse waves and longitudinal waves.

Transverse Waves

You can make a wave on a rope by shaking the end of the rope up and down, as shown in **Figure 3.** A wave on a rope is a transverse wave. *A* **transverse wave** *is a wave in which the disturbance is perpendicular to the direction the wave travels.* **Figure 3** shows that the particles in the rope move up and down while the wave travels horizontally. The up-and-down movement of the rope particles is at right angles to the direction of wave movement.

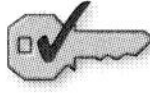
Key Concept Check How do particles move in a transverse wave?

The dotted line in **Figure 3** shows the position of the rope before you start shaking it. This position is called the rest position. The transverse wave on the rope has high points and low points. *The highest points on a transverse wave are* **crests.** *The lowest points on a transverse wave are* **troughs.**

Inquiry MiniLab — 15 minutes

How do waves travel through matter?

When a wave travels through matter, energy transfers from particle to particle. How do particles move in different types of waves?

1. Read and complete a lab safety form.
2. Tie a piece of **yarn** around a **rope.** Stretch the rope on the floor between you and a partner. Make a transverse wave by moving the rope side to side on the floor. Observe how the yarn moves.
3. Tie a piece of yarn around one coil near the middle of a **metal spring toy.** Stretch the toy between you and a partner on the floor. Sharply push one end of the spring toy forward. Observe how the yarn moves.

Analyze and Conclude

1. **Compare** the motion of the yarn on the rope and on the spring in your Science Journal.
2. **Key Concept** How did the particles in each medium move as the wave passed?

Concepts in Motion Animation

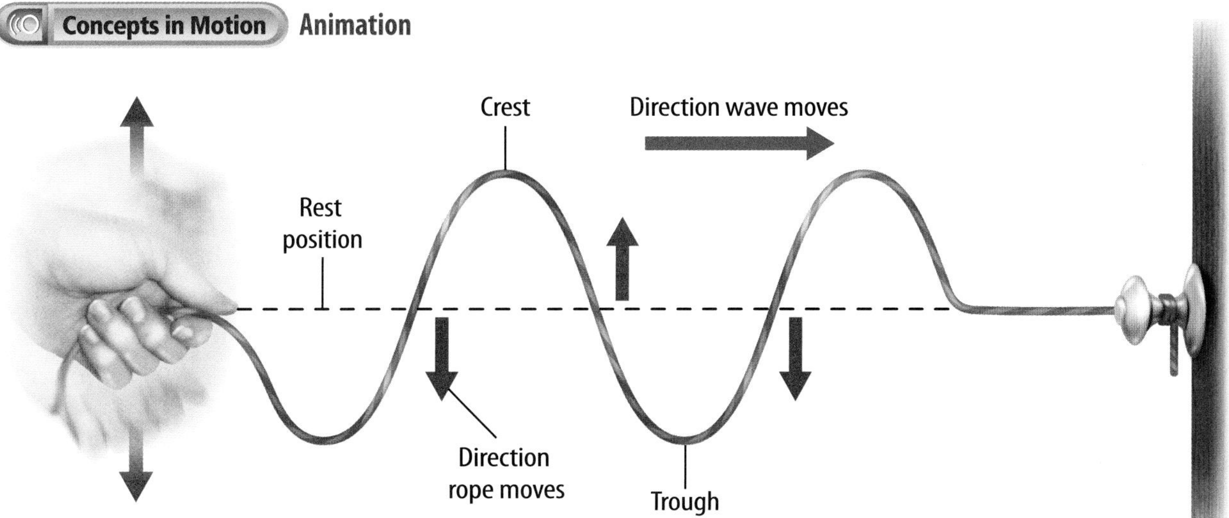

Figure 3 In a transverse wave, particles move at right angles to the direction the wave travels.

Longitudinal Waves

Concepts in Motion Animation

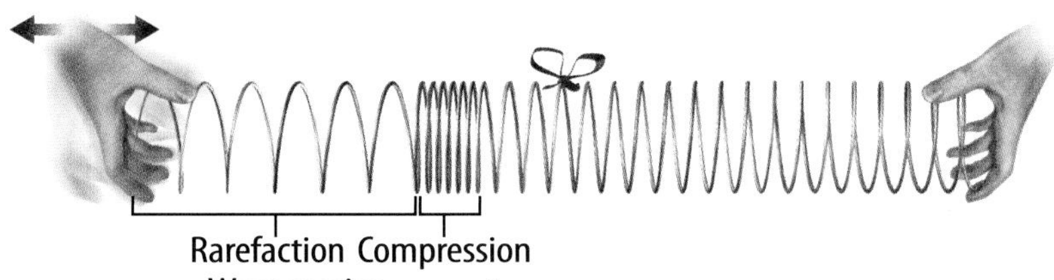

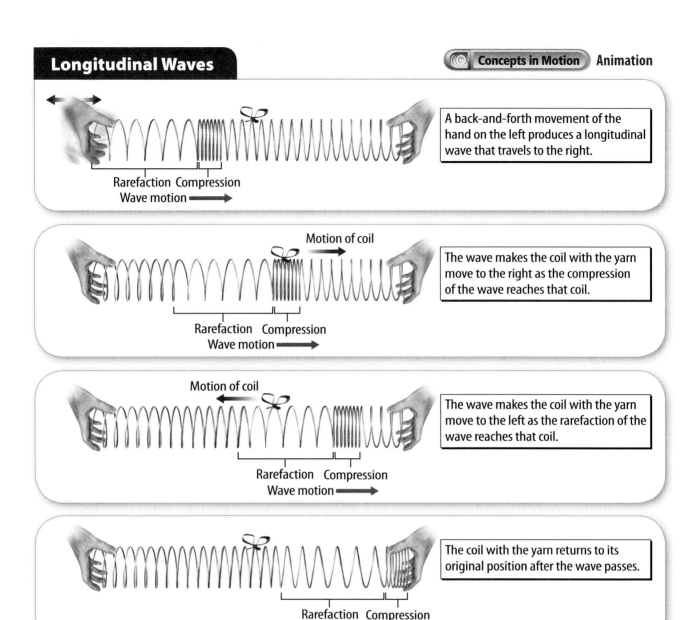

A back-and-forth movement of the hand on the left produces a longitudinal wave that travels to the right.

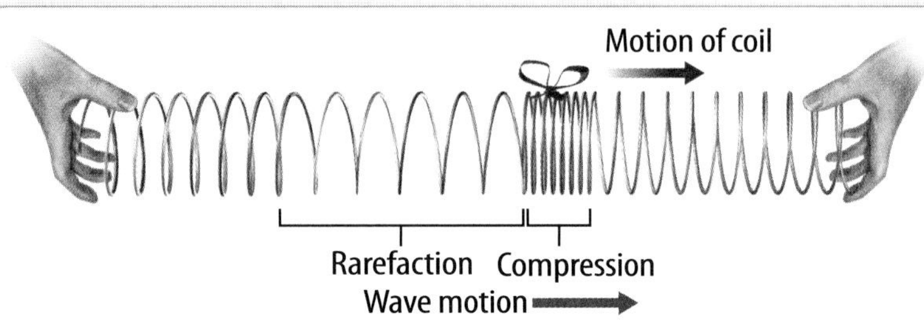

The wave makes the coil with the yarn move to the right as the compression of the wave reaches that coil.

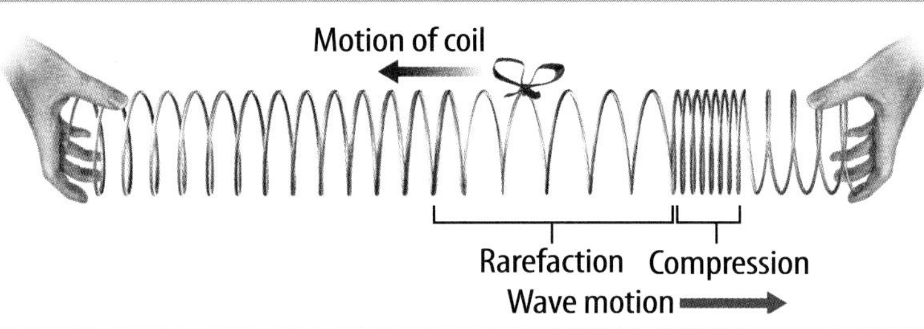

The wave makes the coil with the yarn move to the left as the rarefaction of the wave reaches that coil.

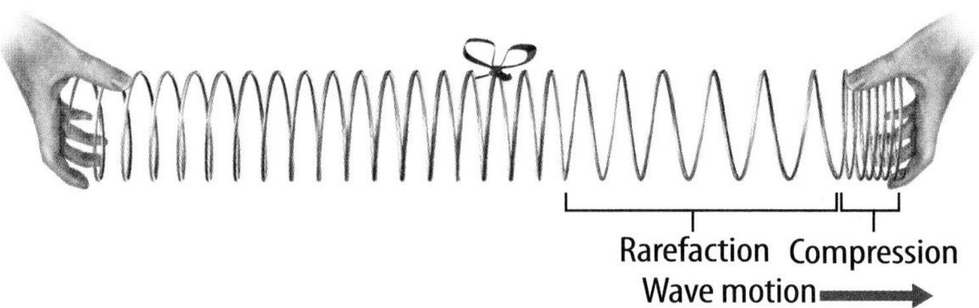

The coil with the yarn returns to its original position after the wave passes.

Figure 4 A longitudinal wave travels along the spring when the hand moves back and forth. The coils of the spring move back and forth along the same direction that the wave travels.

Longitudinal Waves

Another type of mechanical wave is a longitudinal (lahn juh TEWD nul) wave. *A* **longitudinal wave** *makes the particles in a medium move parallel to the direction that the wave travels.* A longitudinal wave traveling along a spring is shown in **Figure 4.** As the wave passes, the coils of the spring move closer together, then farther apart, and then return to their original positions. The coils move back and forth parallel to the direction that the wave moves.

Before a wave is produced in the spring, the coils are the same distance apart. This is the rest position of the spring. **Figure 4** shows that the wave produces regions in the spring where the coils are closer together than they are in the rest position and regions where they are farther apart. *The regions of a longitudinal wave where the particles in the medium are closest together are* **compressions.** *The regions of a longitudinal wave where the particles are farthest apart are* **rarefactions.**

Key Concept Check How do particles move in a longitudinal wave?

Vibrations and Mechanical Waves

If you hit a large drum with a drumstick, the surface of the drum vibrates, or moves up and down. A vibration is a back-and-forth or an up-and-down movement of an object. Vibrating objects, such as a drum or a guitar string, are the sources of energy that produce mechanical waves.

Reading Check What produces mechanical waves?

One Wave per Vibration Suppose you move the end of a rope down, up, and back down to its original position. Then you make a transverse wave with a crest and a trough, as shown in the top of **Figure 5.** The up-and-down movement of your hand is one vibration. One vibration of your hand produces a transverse wave with one crest and one trough.

Similarly, a single back-and-forth movement of the end of a spring produces a longitudinal wave with one compression and one rarefaction. In both cases, the vibration of your hand is the source of energy that produces the transverse or longitudinal wave.

Vibrations Stop—Waves Go Now imagine that you move the end of the rope up and down several times. The motion of your hand transfers energy to the rope and produces several crests and troughs, as the bottom of **Figure 5** shows. As long as your hand keeps moving up and down, energy transfers to the rope and produces waves.

When your hand stops moving, waves no longer are produced. However, as shown in both parts of **Figure 5,** waves produced by the earlier movements of your hand continue to travel along the rope. This is true for any vibrating object. Waves can keep moving even after the object stops vibrating.

Reading Check If your hand makes four vibrations, how many waves are created?

Figure 5 Vibrations produce waves that keep traveling even when the vibrations stop.

Vibrations and Waves

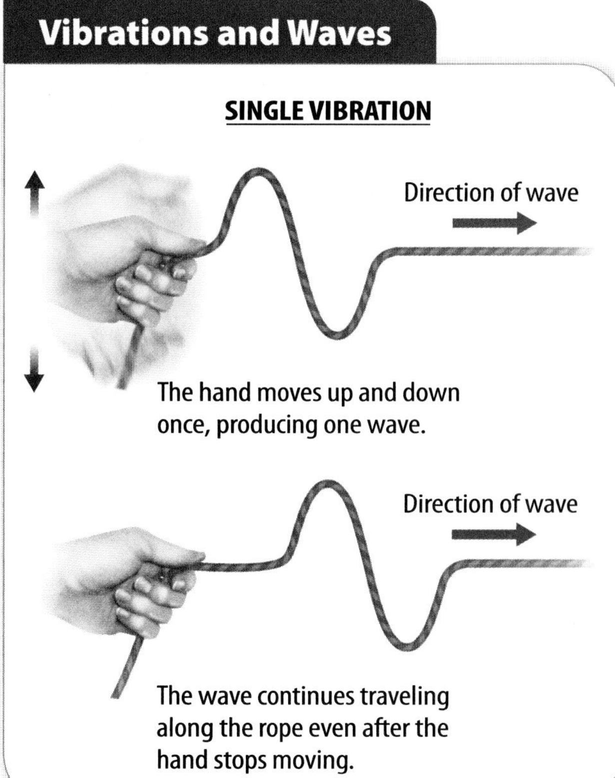

SINGLE VIBRATION
Direction of wave
The hand moves up and down once, producing one wave.
Direction of wave
The wave continues traveling along the rope even after the hand stops moving.

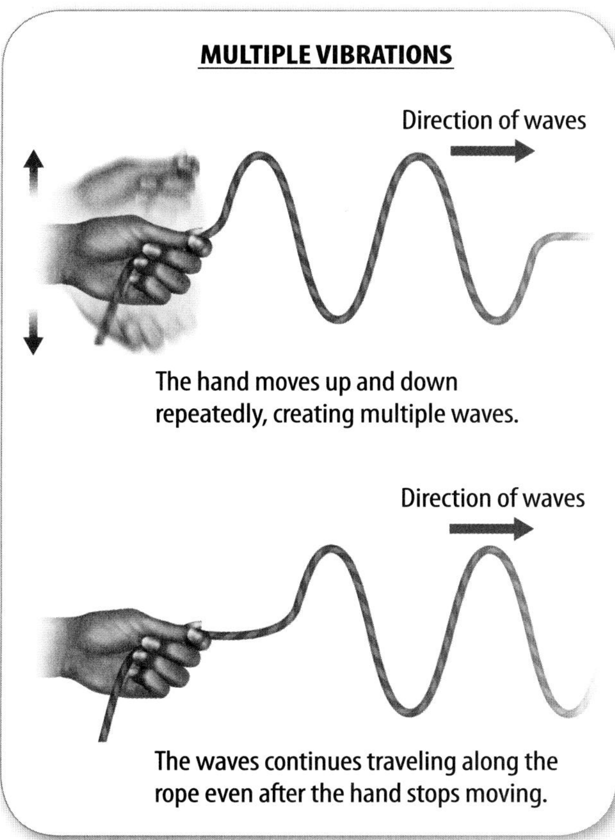

MULTIPLE VIBRATIONS
Direction of waves
The hand moves up and down repeatedly, creating multiple waves.
Direction of waves
The waves continues traveling along the rope even after the hand stops moving.

Types of Mechanical Waves

All mechanical waves travel only in matter. Sound waves, water waves, and waves produced by earthquakes are mechanical waves that travel in different mediums. **Table 1** shows examples of these mechanical waves.

Visual Check What are the mediums for each of the mechanical waves in **Table 1**?

Table 1 Types of Mechanical Waves

Rarefaction Compression Molecules in air The movement of the speaker cone produces compressions and rarefactions.	**Sound Waves** • Sound waves are longitudinal waves that travel in solids, liquids, and gases. • A sound wave is made of a series of compressions and rarefactions. • A paper cone inside the speaker vibrates in and out, producing sound waves. • The speaker makes a compression when the speaker cone pushes air molecules together as it moves outward. • The speaker makes a rarefaction when the speaker cone moves inward and the air molecules spread out.
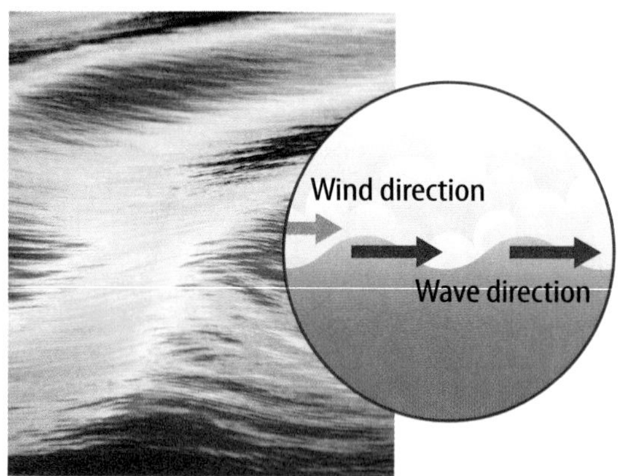	**Water Waves** • Water waves are a combination of transverse waves and longitudinal waves. • Wind produces most waves in oceans and lakes by pushing on the surface of the water.
 Seismic waves travel outward in all directions from their source.	**Seismic Waves** • Waves in Earth's crust, called seismic (SIZE mihk) waves, cause earthquakes. • Seismic waves are mechanical waves that travel within Earth and on Earth's surface. • There are both longitudinal and transverse seismic waves. • In some places, parts of Earth's upper layers can move along a crack called a fault. • The movement of Earth's upper layers along a fault produces seismic waves.

Electromagnetic Waves

The Sun gives off light that travels through space to Earth. Like sound waves and water waves, light is also a type of wave. However, light is not a mechanical wave. A mechanical wave cannot travel through the space between the Sun and Earth. Light is an electromagnetic wave. *An* **electromagnetic wave** *is a wave that can travel through empty space and through matter.*

Key Concept Check Identify a type of wave that can travel through a vacuum.

WORD ORIGIN

electromagnetic
from Greek *elektron,* means "amber" and *magnes,* means "lodestone"

Types of Electromagnetic Waves

In addition to light waves, other types of electromagnetic waves include radio waves, microwaves, infrared waves, and ultraviolet waves. Cell phones use microwaves that carry sounds from one phone to another. When you stand by a fire, infrared waves striking your skin cause the warmth you feel. Ultraviolet waves from the Sun cause sunburns.

FOLDABLES®
Make a two-tab book and label it as shown. Use your book to organize information about mechanical and electromagnetic waves.
Mechanical Waves
Electromagnetic Waves

Electromagnetic Waves and Objects

Every object, including you, gives off electromagnetic waves. The type of electromagnetic waves that an object gives off depends mainly on the temperature of the object. For example, you give off mostly infrared waves. Other objects near human body temperature also give off mostly infrared waves. Hotter objects, such as a piece of glowing metal, give off visible light waves as well as infrared waves. Some animals, such as the copperhead in **Figure 6,** have specialized detectors for perceiving the infrared waves given off by their prey.

Electromagnetic Waves from the Sun

Like all waves, electromagnetic waves carry energy. Scientists often call this radiant energy. Infrared and visible light waves carry about 92 percent of the radiant energy that reaches Earth from the Sun. Ultraviolet waves carry about 7 percent of the Sun's energy.

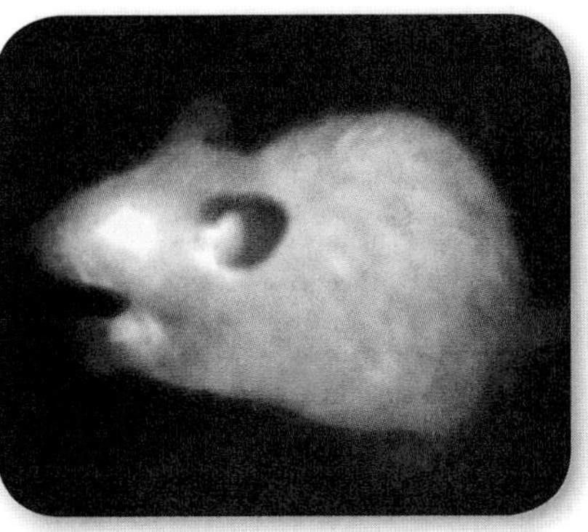

Figure 6 Some snakes have special organs on their heads that allow them to detect infrared waves from prey, such as mice. The right photo shows the infrared waves that a mouse gives off.

Lesson 1 Review

Visual Summary

Waves, such as those from a burning candle, the Sun, or a loudspeaker, transfer energy away from the source of the wave.

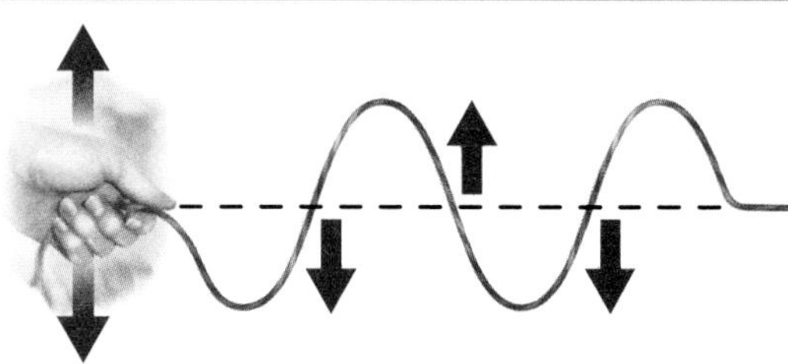

A transverse wave makes particles in a medium move perpendicular to the direction of the wave.

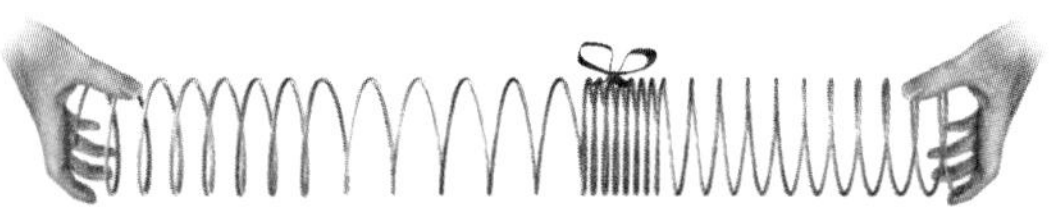

A longitudinal wave makes the particles in a medium move in a direction that is parallel to the direction the wave travels.

FOLDABLES®

Use your lesson Foldable to review the lesson. Save your Foldable for the project at the end of the chapter.

What do you think NOW?

You first read the statements below at the beginning of the chapter.

1. Waves carry matter as they travel from one place to another.
2. Sound waves can travel where there is no matter.

Did you change your mind about whether you agree or disagree with the statements? Rewrite any false statements to make them true.

Use Vocabulary

1. **Define** *wave* in your own words.
2. **Distinguish** between a transverse wave and a longitudinal wave.

Understand Key Concepts 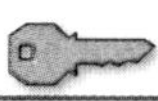

3. What causes a wave?
   A. a crest
   B. a rarefaction
   C. a rope
   D. a vibration
4. **Differentiate** How are sound waves and water waves similar? How are they different?

Interpret Graphics

5. **Describe** what happens to the wave shown below when the hand stops vibrating.

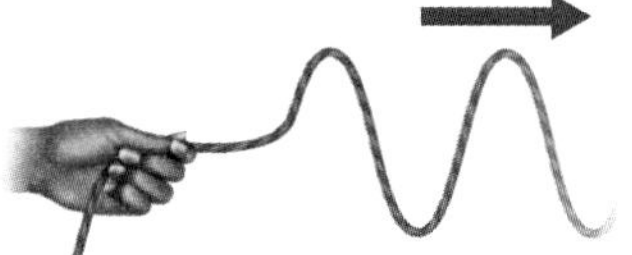

6. **Compare and Contrast** Copy and fill in the graphic organizer below to compare and contrast mechanical waves and electromagnetic waves.

Similarities	Differences

Critical Thinking

7. **Analyze** A scientist wants to analyze signals from outer space that tell her about the age of the universe. Her lab is equipped with advanced sensors that detect sound waves, radio waves, and seismic waves. Which of these sensors will provide the best information? Explain your reasoning.
8. **Predict** You are floating motionless on a rubber raft in the middle of a pool. A friend forms a wave by slapping the water every second. Will the wave carry you to the edge of the pool? Explain your answer.

Making a Computer Tsunami

CAREERS in SCIENCE

Meet Vasily Titov, a scientist working to predict the next big wave.

On the morning of December 26, 2004, a magnitude 9.3 earthquake in the Indian Ocean caused an enormous tsunami. On the other side of the world, a scientist in Seattle, Washington, sprang into action. Vasily Titov worked through the night to develop the first-ever computer model of the tsunami.

Titov is a mathematician for the National Oceanic and Atmospheric Administration (NOAA). His computer model solves equations that describe how tsunamis are produced and how they move. The model includes data about an earthquake's energy, the shape of the ocean floor, and the ocean's depth. When there is an earthquake on the ocean floor, Titov's model predicts the properties of the resulting tsunami.

A tsunami moves about as fast as a jet airplane. By the time Titov's computer model was ready, the 2004 Indian Ocean tsunami had already hit the coast and killed more than 280,000 people. But Titov hopes this new tool will help communities prepare for future tsunamis.

Now Titov is applying his computer model to the Pacific Northwest. His model shows how tsunamis could affect the coastlines of Washington, Oregon, and northern California. Emergency managers can use these results to predict when tsunami waves will reach different parts of the coast. Warnings could be issued to communities just minutes after an earthquake triggers a tsunami.

What's a tsunami?

Wind causes most ordinary water waves, but tsunamis are not ordinary waves. Underwater earthquakes usually cause these giant, destructive waves. As a tectonic plate moves upward, it raises the water above it. As this water falls back down, it forms a massive wave that spreads in all directions.

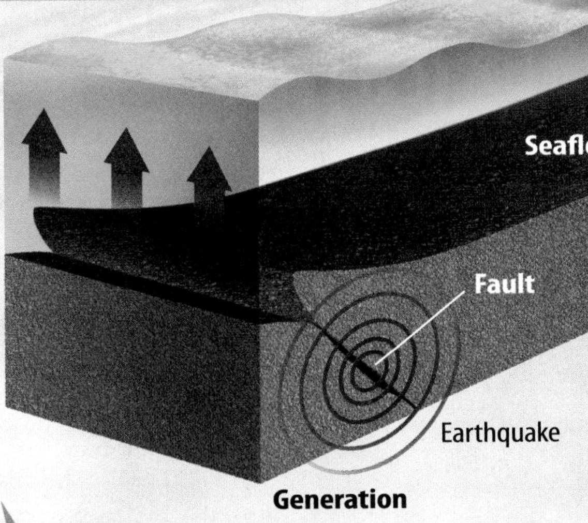

❶ **In deep water, tsunamis usually travel faster than 250 m/s, but they are only a few centimeters high. Most ships in deep water cannot even detect a passing tsunami.**

❷ **As the wave approaches shore, much of its kinetic energy is transformed into gravitational potential energy. The water piles up, forming enormous walls of water that flood coastal areas.**

It's Your Turn

WRITE Suppose you were a witness to the 2004 tsunami. Write a diary entry recording your experience. What was your first hint that something was wrong? What was happening around you? What did you see and hear?

Lesson 2

Reading Guide

Key Concepts 
ESSENTIAL QUESTIONS

- What are properties of waves?
- How are the frequency and the wavelength of a wave related?
- What affects wave speed?

Vocabulary

amplitude p. 539
wavelength p. 541
frequency p. 542

Video

- BrainPOP®
- Science Video

Wave Properties

Inquiry Why So Big?

Have you ever watched a surfer on a huge wave? If so, you might have wondered why some waves are huge and why others are small. The size of a water wave depends on the energy it carries.

Launch Lab

20 minutes

Which sounds have more energy?

Some sounds are loud and others are soft. What is the difference between loud and soft sounds?

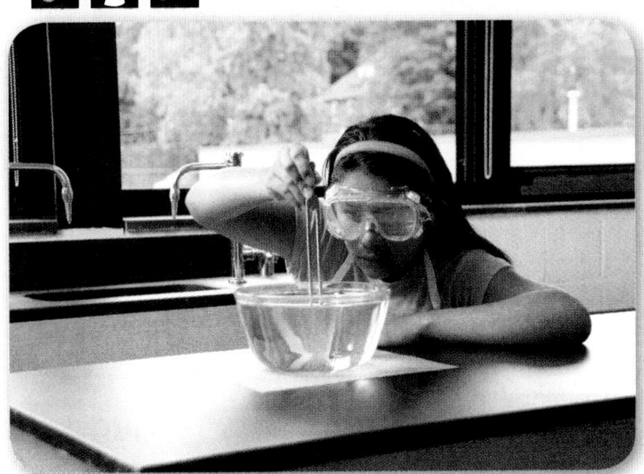

1. Read and complete a lab safety form.
2. Place a **bowl of water** over a sheet of **white paper.**
3. Strike a **tuning fork** gently on your hand so it makes a soft sound and then quickly place its prongs in the bowl of water. Remove the tuning fork.
4. Strike the tuning fork sharply on your hand so it makes a loud sound and then quickly place its prongs in the bowl of water.

Think About This

1. **Contrast** the waves made by the tuning fork in steps 3 and 4 in your Science Journal.
2. In which step did the tuning fork transfer more energy to the water? Explain your answer.
3. **Key Concept** How are the loudness of the sounds and the vibrations of the tuning fork related?

Amplitude and Energy

Imagine you are floating on a raft in a pool and someone gently splashes the water near you, creating waves. You might barely feel these waves lift you up and down as they pass. If someone dives off the diving board, this makes waves that bounce you up and down. What's the difference between these waves?

Initially, the surface of the water you are floating on is nearly flat. This is the rest position for the water. The dive produced waves with higher crests and deeper troughs than those of the waves produced by gently splashing the water. This means that the dive caused water to move a greater distance from its rest position, producing a wave with a greater amplitude. *The* **amplitude** *of a wave is the maximum distance that the wave moves from its rest position.*

For any wave, the larger the amplitude, the more energy the wave carries. The wave produced by the diver hitting the water caused a greater change than the wave produced by the gentle splash. The first wave had more energy.

Reading Check What is the amplitude of a wave? How are the amplitude of a wave and energy related?

WORD ORIGIN

amplitude
from Latin *amplitudinem*, means "width"

FOLDABLES®

Make a layered book from two half sheets of paper. Label it as shown. Use your book to organize your notes about the properties of waves.

Properties of Waves
Amplitude
Wavelength
Frequency

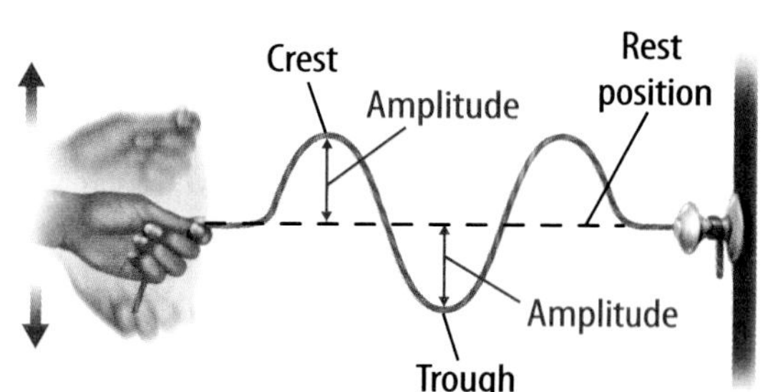

This wave has a smaller amplitude and carries less energy.

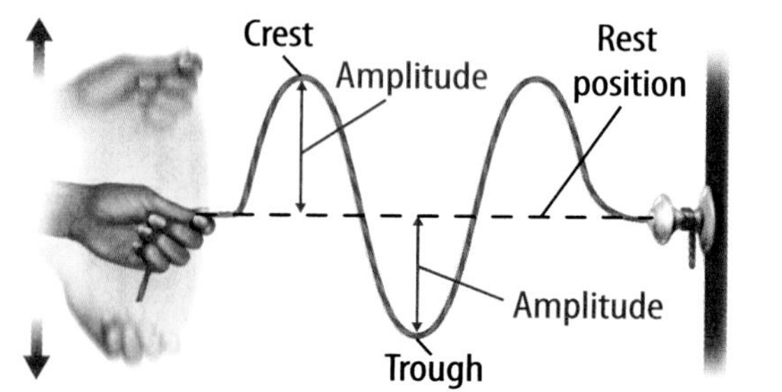

This wave has a greater amplitude and carries more energy.

▲ **Figure 7** The amplitude of a transverse wave is the distance from the resting position to a crest or a trough.

Amplitude and Energy of Transverse Waves

You produce a transverse wave on a rope when you move the rope up and down. For a transverse wave, the greatest distance a particle moves from the rest position is to the top of a crest or to the bottom of a trough. This distance is the amplitude of a transverse wave, as shown in **Figure 7.** The energy carried by a transverse wave increases as the amplitude of the wave increases as shown in **Figure 8.**

Amplitude and Energy of Longitudinal Waves

The amplitude of a longitudinal wave depends on the distance between particles in the compressions and rarefactions. When the amplitude of a longitudinal wave increases, the particles in the medium get closer together in the compressions and farther apart in the rarefactions, as shown in **Figure 9** on the next page. In a longitudinal wave, you transfer more energy when you push and pull the end of the spring a greater distance. Just as for transverse waves, the energy carried by a longitudinal wave increases as its amplitude increases.

Reading Check When comparing two longitudinal waves that are traveling through the same medium, how can you tell which has the greater amount of energy?

Figure 8 The wave with the larger amplitude carries more energy and makes the ball bounce higher. ▼

Visual Check What is the source of energy for these waves?

Lower Amplitude Wave
The parachute transfers less energy to the ball.

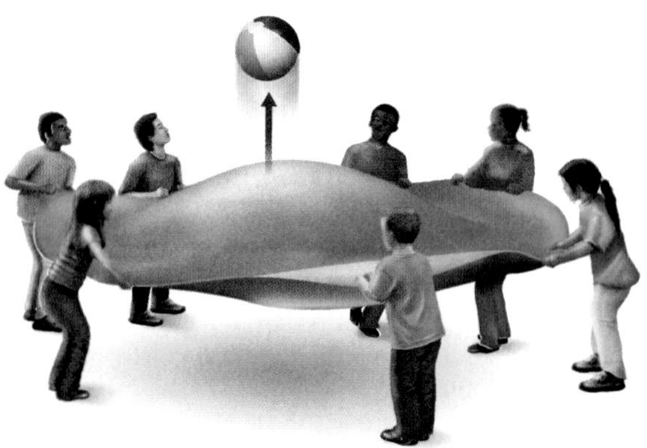

Higher Amplitude Wave
The parachute transfers more energy to the ball.

Lower-Amplitude Wave

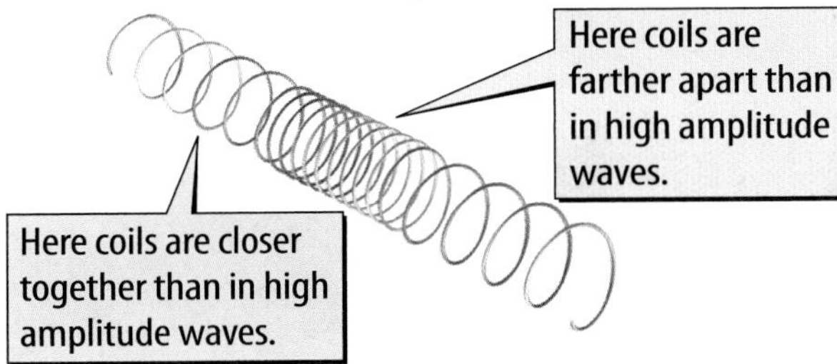

Here coils are farther apart than in high amplitude waves.
Here coils are closer together than in high amplitude waves.

Higher-Amplitude Wave

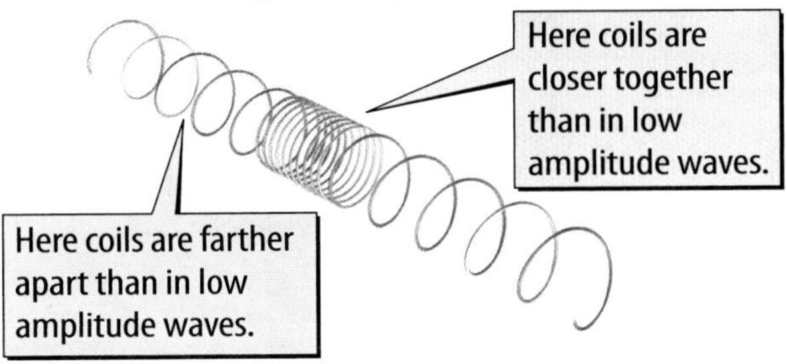

Here coils are closer together than in low amplitude waves.
Here coils are farther apart than in low amplitude waves.

▲ **Figure 9** Amplitude depends on the spacings in the compressions and rarefactions.

Animation

Wavelength

The **wavelength** *of a wave is the distance from one point on a wave to the same point on the next wave.* The wavelengths of a transverse wave and a longitudinal wave are shown in **Figure 10.** To measure the wavelength of a transverse wave, you can measure the distance from one crest to the next crest or from one trough to the next trough. To measure the wavelength of a longitudinal wave, you can measure the distance from one compression to the next compression or from one rarefaction to the next rarefaction. Wavelength is measured in units of distance, such as meters.

Inquiry MiniLab **20 minutes**

How are wavelength and frequency related?

Waves traveling in a material can have different frequencies and wavelengths. Is there a relationship between the wavelength and frequency of a wave?

1. Read and complete a lab safety form.
2. With a partner, stretch a **piece of rope,** approximately 2–3 m long, across a lab table or the floor. Move your hand side to side while your partner holds the other end of the rope in place. Observe the wavelength.
3. Move your hand side to side faster. Observe the wavelength.

Analyze and Conclude

1. **Explain** When was the frequency of the wave higher? Lower?
2. **Key Concept** How are wavelength and frequency related?

Figure 10 Wavelength is the distance from one point on a wave to the nearest point just like it. ▼

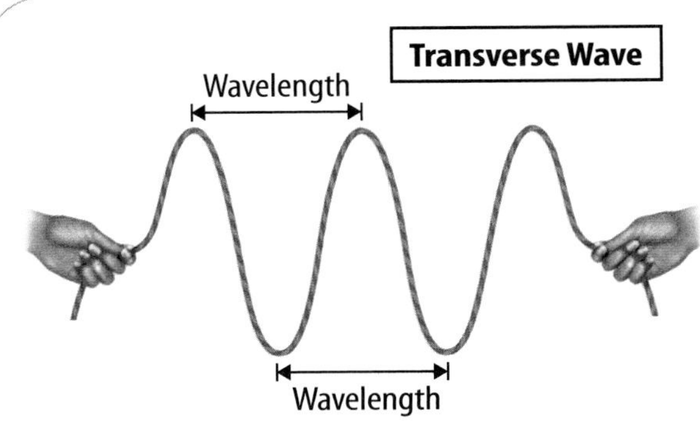

Wavelength is the distance from one crest to the next crest or from one trough to the next trough.

Longitudinal Wave

Wavelength

Wavelength

Wavelength is the distance from one compression to the next compression or from one rarefaction to the next rarefaction.

Frequency and Wavelength

Wave direction →

- Longer wavelength
- Lower frequency
- One complete wave passes in four seconds.

Wave direction →

- Shorter wavelength
- Higher frequency
- Two complete waves pass in four seconds.

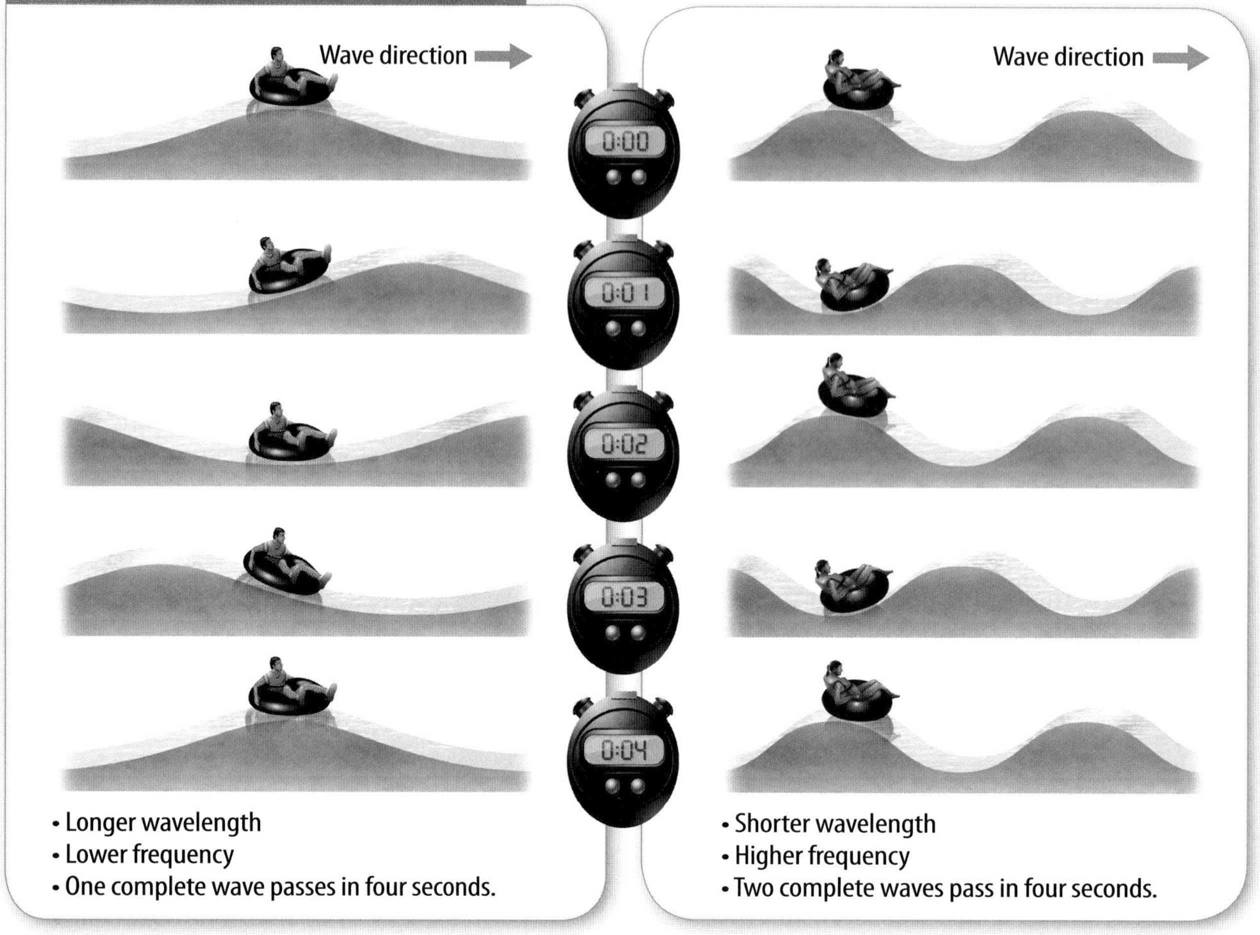

Figure 11 Frequency is the number of wavelengths that pass a point each second. When the frequency increases, the wavelength decreases.

Frequency

Waves have another property called frequency. *The* **frequency** *of a wave is the number of wavelengths that pass by a point each second.* Frequency is related to how rapidly the object or material producing the wave vibrates. Each vibration of the object produces one wavelength. The frequency of a wave is the same as the number of vibrations the vibrating object makes each second.

Key Concept Check What are three properties of waves?

The Unit for Frequency

The SI unit for frequency is hertz (Hz). A wave with a frequency of 2 Hz means that two wavelengths pass a point each second. The unit Hz is the same unit as 1/s.

Wavelength and Frequency

Figure 11 shows how frequency and wavelength are related. The wavelength of the waves in the left column is longer than that of the wave in the right column. To calculate the frequency of waves, divide the number of wavelengths by the time. For the wave on the left, the frequency is 1 wavelength divided by 4 s, which is 0.25 Hz. The wave on the right has a frequency of 2 wavelengths divided by 4 s, which is 0.5 Hz. The wave on the right has a shorter wavelength and a higher frequency. As the frequency of a wave increases, the wavelength decreases.

Key Concept Check How does the wavelength change if the frequency of a wave decreases? What if the frequency increases?

Wave Speed

Different types of waves travel at different speeds. For example, light waves from a lightning flash travel almost 1 million times faster than the sound waves you hear as thunder.

Wave Speed Through Different Materials

The same type of waves travel at different speeds in different materials. Mechanical waves, such as sound waves, usually travel fastest in solids and slowest in gases, as shown in **Table 2.** Mechanical waves also usually travel faster as the temperature of the medium increases. Unlike mechanical waves, electromagnetic waves move fastest in empty space and slowest in solids.

Key Concept Check What does wave speed depend on?

Calculating Wave Speed

You can calculate the speed of a wave by multiplying its wavelength and its frequency together, as shown below. The symbol for wavelength is λ, which is the Greek letter *lambda*.

Wave Speed Equation

wave speed (in m/s) = **frequency** (in Hz) × **wavelength** (in m)

$$s = f\lambda$$

When you multiply wavelength and frequency, the result has units of m × Hz. This equals m/s—the unit for speed.

Table 2 Speed of Sound Waves in Different Materials

Material	Wave Speed (m/s)
Gases (0°C)	
Oxygen	316
Dry air	331
Liquids (25°C)	
Ethanol	1,207
Water	1,500
Solids	
Ice	3,850
Aluminum	6,420

Math Skills Use a Simple Equation

Solve for Wave Speed A mosquito beating its wings produces sound waves with a frequency of 700 Hz and a wavelength of 0.5 m. How fast are the sound waves traveling?

1. **This is what you know:** frequency: $f = 700$ Hz; wavelength: $\lambda = 0.5$ m
2. **This is what you need to find:** wave speed: s
3. **Use this formula:** $s = f\lambda$
4. **Substitute:** the values for f and λ into the formula and multiply: $s = 700 \text{ Hz} \times 0.5 \text{ m} = 350 \text{ Hz} \times \text{m}$
5. **Convert units:** $(\text{Hz}) \times (\text{m}) = (1/\text{s}) \times (\text{m}) = \text{m/s}$

Answer: The wave speed is 350 m/s.

Practice

What is the speed of a wave that has a frequency of 8,500 Hz and a wavelength of 1.5 m?

- Math Practice
- Personal Tutor

Lesson 2 Review

Assessment Online Quiz

Inquiry Virtual Lab

Visual Summary

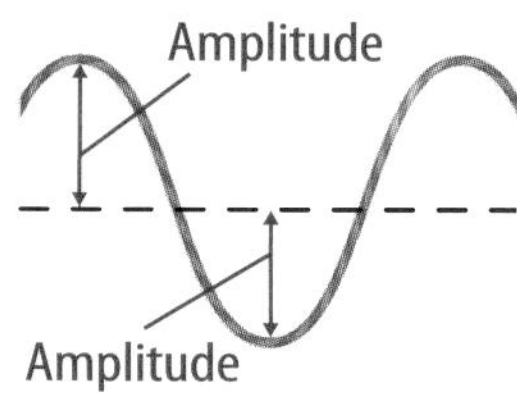

The amplitude of a transverse wave is the maximum distance that the wave moves from its rest position.

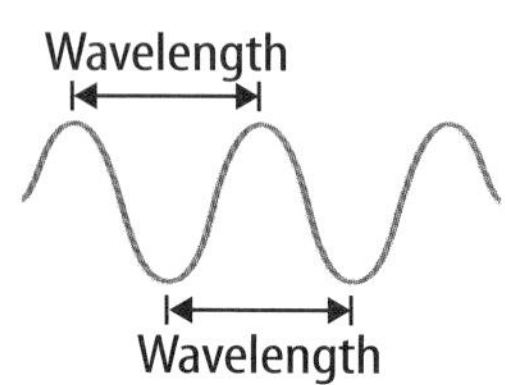

The wavelength of a transverse wave is the distance from one point on a wave to the same point on the next wave, such as from crest to crest or from trough to trough.

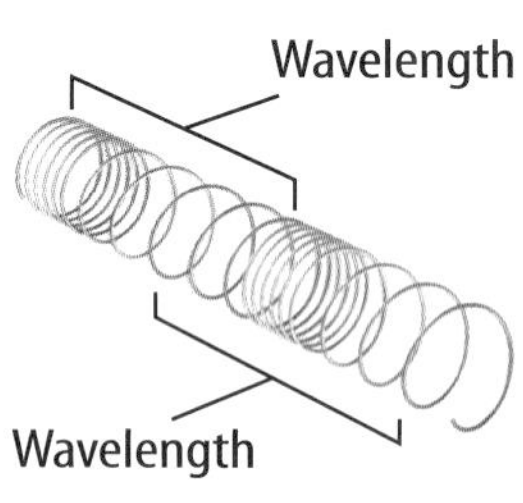

The wavelength of a longitudinal wave is the distance from one point on a wave to the nearest point just like it, such as from compression to compression or from rarefaction to rarefaction.

FOLDABLES®

Use your lesson Foldable to review the lesson. Save your Foldable for the project at the end of the chapter.

What do you think NOW?

You first read the statements below at the beginning of the chapter.

3. Waves that carry more energy cause particles in a material to move a greater distance.

4. Sound waves travel fastest in gases, such as those in the air.

Did you change your mind about whether you agree or disagree with the statements? Rewrite any false statements to make them true.

Use Vocabulary

1. For a transverse wave, the ________ depends on the distance from the rest position to a crest or a trough.

2. The unit for the ________ of a wave is the Hz, which means "per second."

Understand Key Concepts 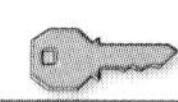

3. **Compare** Which wave would have the greatest wave speed, a wave from a vibrating piano string in an auditorium or a sound wave created by a boat anchor striking an underwater rock? Explain.

4. In which medium would an electromagnetic wave travel the fastest?

   A. air **C.** vacuum
   B. granite **D.** water

Interpret Graphics

5. **Determine** which wave carries the greater amount of energy. Explain.

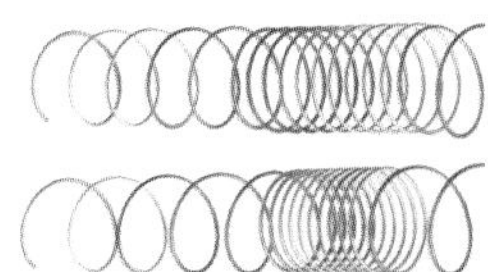

6. **Determine Cause and Effect** Copy and fill in the graphic organizer below.

Critical Thinking

7. **Infer** A loudspeaker produces sound waves that change in wavelength from 1.0 m to 1.5 m. If the wave speed is constant, how did the vibration of the loudspeaker change? Explain.

Math Skills

Math Practice

8. **Use a Simple Equation** A water wave has a frequency of 10 Hz and a wavelength of 150 m. What is the wave speed?

Inquiry **Skill Practice** **Model** 25 minutes

How are the properties of waves related?

Materials

meter tapes (2)

masking tape

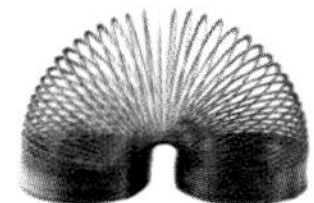

coiled spring toy

stopwatch or clock with second hand

Safety

All waves have amplitude, wavelength, and frequency. The properties of a wave are related and also determine the amount of energy the wave carries.

Learn It

Scientists create **models** to study many objects and concepts that are difficult to observe directly. You used a coiled spring toy in previous labs to model a longitudinal wave. In this lab, you will use the same toy to model transverse waves.

Try It

1. Read and complete a lab safety form.
2. With a partner, use masking tape to secure the meter tapes to the floor, creating *x*- and *y*-axes.
3. With your partner, stretch the spring toy across the tape representing the *x*-axis. Generate a transverse wave by moving the toy back and forth on the floor. Try to be as steady and even as possible to generate consistent waves.
4. Using the tape on the *y*-axis, measure and record the amplitude of the wave in your Science Journal.
5. Using the tape representing the *x*-axis, measure the wavelength of the wave. Record your measurement.
6. Using a stopwatch, count and record the number of crests or troughs that cross the *y*-axis in 10 seconds.
7. Repeat steps 3–6 for waves with different properties by moving the toy faster and slower.

Apply It

8. Calculate the frequency of each wave generated.
9. Which wave transferred the most energy? Explain.
10. **Key Concept** What happened to the frequency and wavelength of the waves when you moved the spring toy faster and slower? How are frequency and wavelength related?

2

Lesson 3

Wave Interactions

Reading Guide

Key Concepts 
ESSENTIAL QUESTIONS

- How do waves interact with matter?
- What are reflection, refraction, and diffraction?
- What is interference?

Vocabulary

absorption p. 548
transmission p. 548
reflection p. 548
law of reflection p. 549
refraction p. 550
diffraction p. 550
interference p. 551

g Multilingual eGlossary

Video Science Video

Inquiry Can waves change?

Have you ever watched two waves bump into each other? If so, you might have noticed that the shapes of the waves changed. How do the shapes of waves change when they interact?

Launch Lab

20 minutes

What happens in wave collisions?

You might have seen ripples on a water surface spreading out from different points. As the water waves reach each other, they collide. Do waves change after they collide?

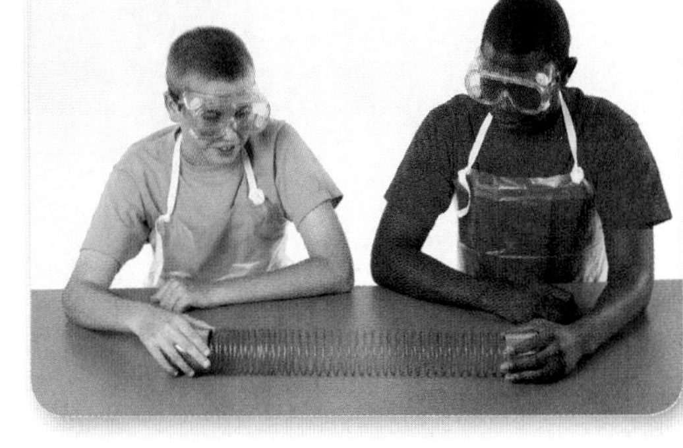

1. Read and complete a lab safety form.
2. Stretch a **metal coiled spring toy** about 30–40 cm between you and a partner.
3. Make a wave by grabbing about five coils at one end and then releasing them. Record your observations in your Science Journal.
4. Make waves at both ends of the spring with your partner. Make waves that appear much different from each other so you can distinguish them easily. Then release them at the same time. Observe and record how each wave moves before, during, and after the collision.

Think About This

1. **Describe** how the two waves moved after the coils were released.

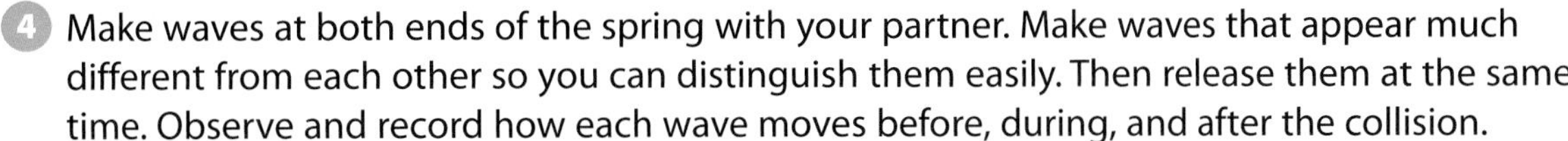

2. **Key Concept** How were the two waves affected by their collision?

Interaction of Waves with Matter

Have you seen photos, like the one shown in **Figure 12,** of objects in space taken with the *Hubble Space Telescope?* The *Hubble* orbits Earth collecting light waves before they enter Earth's atmosphere. Photos taken with the *Hubble* are clearer than photos taken with telescopes on Earth's surface. This is because light waves strike the telescope before they interact with matter in Earth's atmosphere.

Waves interact with matter in several ways. Waves can be reflected by matter or they can change direction when they travel from one material to another. In addition, as waves pass through matter, some of the energy they carry can be transferred to matter. For example, the energy from sound waves can be transferred to soft surfaces, such as the padded walls in movie theaters. Waves also interact with each other. When two different waves overlap, a new wave forms. The new wave has different properties from either original wave.

Figure 12 This *Hubble Space Telescope* photo shows a giant cloud of dust and gas called NGC 3603. Many stars are forming in this cloud.

Waves Interact With Matter

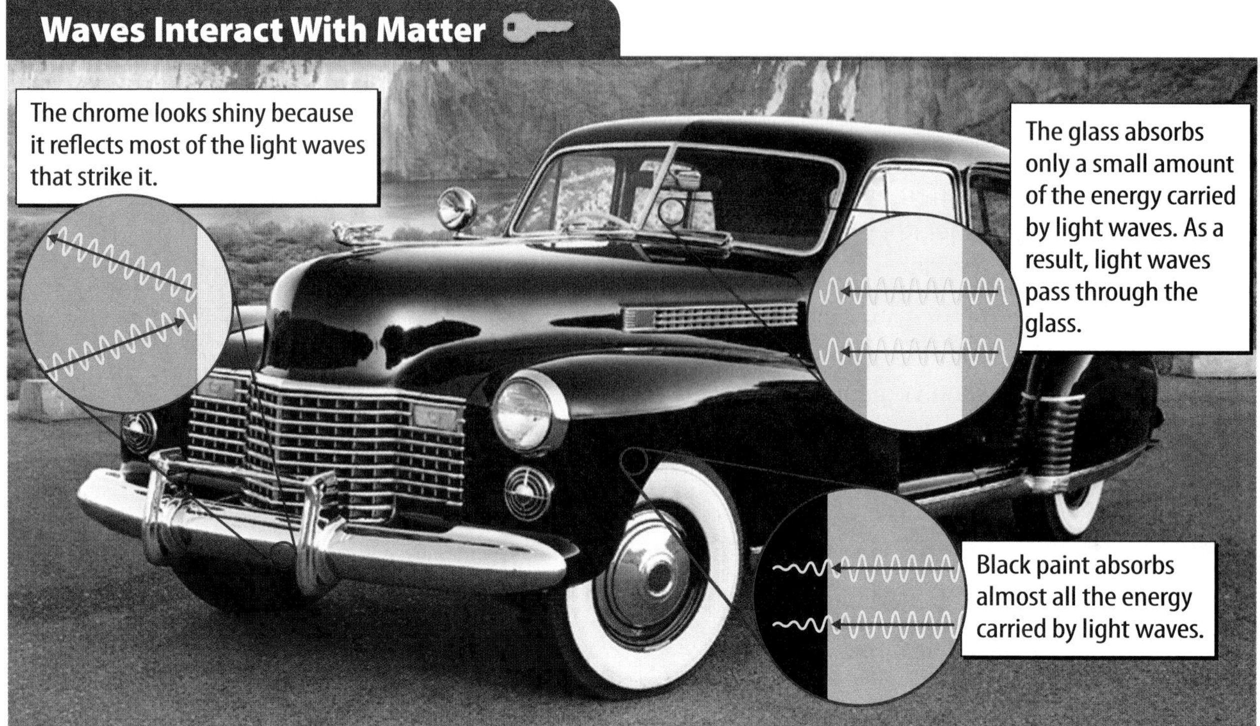

The chrome looks shiny because it reflects most of the light waves that strike it.
The glass absorbs only a small amount of the energy carried by light waves. As a result, light waves pass through the glass.
Black paint absorbs almost all the energy carried by light waves.

Figure 13 Waves can be absorbed, transmitted, or reflected by matter.

Visual Check Do tires usually absorb, transmit, or reflect light waves? Explain your answer.

Absorption

When you shout, you create sound waves. As these waves travel in air, some of their energy transfers to particles in the air. As a result, the energy the waves carry decreases as they travel through matter. **Absorption** *is the transfer of energy by a wave to the medium through which it travels.* The amount of energy absorbed depends on the type of wave and the material in which it moves.

Reading Check Give one reason why the energy carried by sound waves decreases as those sound waves travel through air.

Absorption also occurs for electromagnetic waves. All materials absorb electromagnetic waves, although some materials absorb more electromagnetic waves than others. Darker materials, such as tinted glass, absorb more visible light waves than lighter materials, such as glass that is not tinted.

Absorption occurs at the surface of the car in **Figure 13.** The car's black paint absorbs much of the energy carried by light waves.

Transmission

Why can you see through a sheet of clear plastic wrap but not through a sheet of black construction paper? When visible light waves reach the paper, almost all their energy is absorbed. As a result, no light waves pass through the paper. However, light waves pass through the plastic wrap because the plastic absorbs only a small amount of the wave's energy. **Transmission** *is the passage of light through an object,* such as the windows in **Figure 13.**

Reflection

When waves reach the surface of materials, they can also be reflected. **Reflection** *is the bouncing of a wave off a surface.* Reflection causes the chrome on the car in **Figure 13** to appear grey instead of black. An object that reflects all visible light appears white, while an object that reflects no visible light appears black.

Key Concept Check What are three ways that waves interact with matter?

All types of waves, including sound waves, light waves, and water waves, can reflect when they hit a surface. Light waves reflect when they reach a mirror. Sound waves reflect when they reach a wall. Reflection causes waves to change direction. When you drop a basketball at an angle, it bounces up at the same angle but in the opposite direction. When waves reflect from a surface, they change direction like a basketball bouncing off a surface.

The Law of Reflection

The direction of a wave that hits a surface and the reflected wave are related. As shown in **Figure 14,** an imaginary line, perpendicular to a surface, is called a **normal.** The angle between the direction of the incoming wave and the normal is the angle of incidence. The angle between the direction of the reflected wave and the normal is the angle of reflection. According to the **law of reflection,** *when a wave is reflected from a surface, the angle of reflection is equal to the angle of incidence.*

 **Reading Check** What is the law of reflection?

Reflection of Waves

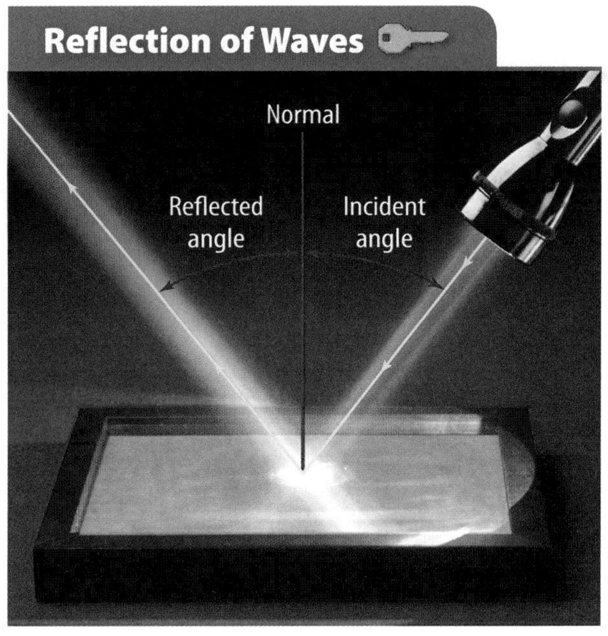

Figure 14 All waves obey the law of reflection. According to the law of reflection, the incident angle equals the reflected angle.

SCIENCE USE V. COMMON USE

normal

Science Use perpendicular to or forming a right angle with a line or plane

Common Use conforming to a standard or common

Inquiry MiniLab

20 minutes

How can reflection be used? 

Light waves, like all waves, obey the law of reflection. By using mirrors, you can see around corners.

1. Read and complete a lab safety form.
2. Place a **small object** on a table and stand a **book** vertically about 30 cm in front of the object.
3. Position a **mirror** vertically so an observer on the opposite side of the book from the object can see the object. Use **modeling clay** to prop up the mirror.
4. Use **string** to represent the path light waves travel from the observer to the mirror and then to the object. Draw the outlined path in your Science Journal.
5. Repeat steps 3–4 with two mirrors.

Analyze and Conclude

 **Key Concept** How could three mirrors be used to see the object behind the book?

Refraction of Waves

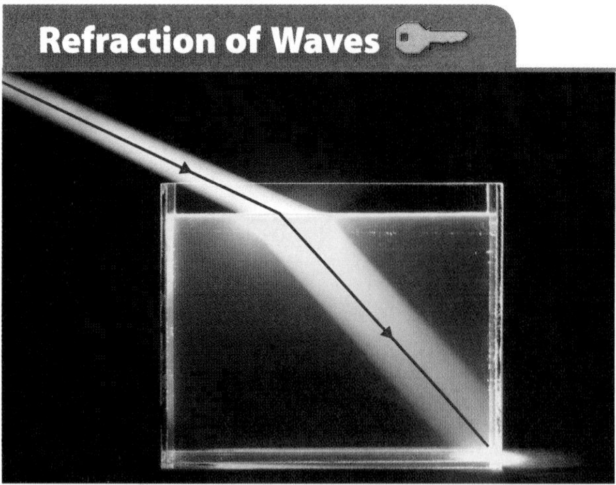

▲ **Figure 15** Refraction occurs when a wave changes speed. The beam of light changes direction because light waves slow down as they move from air into water.

Personal Tutor

WORD ORIGIN
refraction
from Latin *refractus,* means "to break up"

Refraction

Sometimes waves change direction even if they are not reflected from a surface. The light beam in **Figure 15** changes direction as it travels from air into water. The speed of light waves in water is about three-fourths the speed of light waves in air. When light waves slow down, they change direction. **Refraction** *is the change in direction of a wave that occurs as the wave changes speed when moving from one medium to another.* The greater the change in speed, the more the wave changes direction.

Diffraction

Waves can also change direction as they travel by objects. Have you ever been walking down a hallway and heard people talking in a room before you got to the open door of the room? You heard some of the sound waves because they changed direction and spread out as they traveled through the doorway.

What is diffraction?

The change in direction of a wave when it travels by the edge of an object or through an opening is called **diffraction.** Examples of diffraction are shown in **Figure 16.** Diffraction causes the water waves to travel around the edges of the object and to spread out after they travel through the opening. More diffraction occurs as the size of the object or opening becomes similar in size to the wavelength of the wave.

Figure 16 Waves diffract as they pass by an object or pass through an opening. ▼

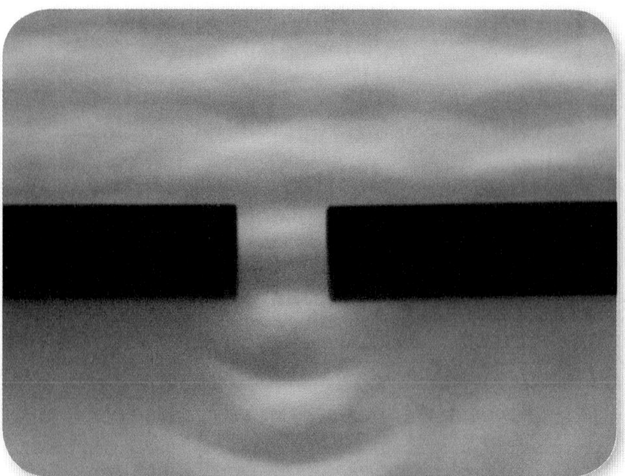

Diffraction of Sound Waves and Light Waves

The wavelengths of sound waves are similar in size to many common objects. Because of this size similarity, you often hear sound from sources that you can't see. For example, the wavelengths of sound waves are roughly the same size as the width of the doorway. Therefore, sound waves spread out as they travel through the doorway. The wavelengths of light waves are more than a million times smaller than the width of a doorway. As a result, light waves do not spread out as they travel through the doorway. Because the wavelengths of light waves are so much smaller than sound waves, you can't see into the room until you reach the doorway. However, you can hear the sounds much sooner.

Key Concept Check Compare and contrast reflection, refraction, and diffraction.

FOLDABLES®

Make a tri-fold book and label the columns as shown. Use your book to record information about how waves interact with matter.

Absorption | Transmission | Reflection

Interference

Waves not only interact with matter. They also interact with each other. Suppose you throw two pebbles into a pond. Waves spread out from the impact of each pebble and move toward each other. When the waves meet, they overlap for a while as they travel through each other. **Interference** *occurs when waves that overlap combine, forming a new wave,* as shown in **Figure 17.** However, after the waves travel through each other, they keep moving without having been changed.

Wave Interference

Two waves approach each other from opposite directions.

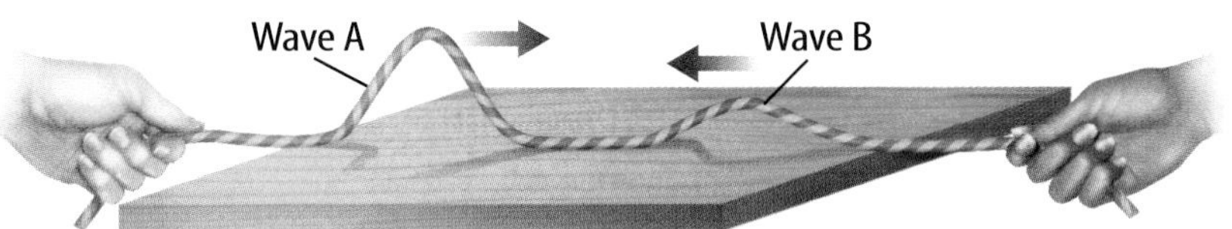

The waves interfere with each other and form a large amplitude wave.

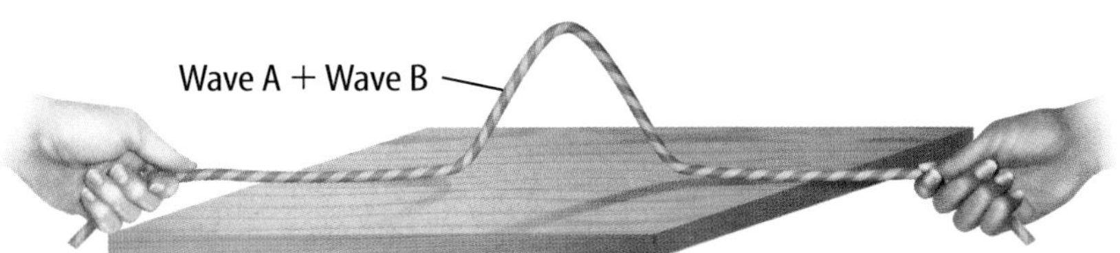

The waves keep traveling in opposite directions after they move through each other.

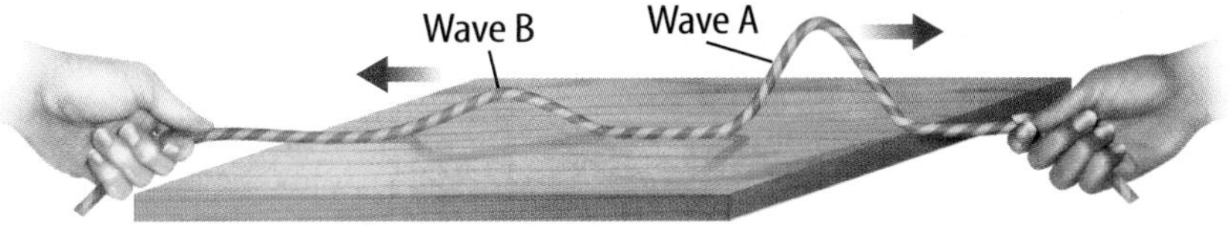

Figure 17 When waves interfere with each other, they create a new wave that has a different amplitude than either original wave.

Visual Check Which wave has the larger amplitude?

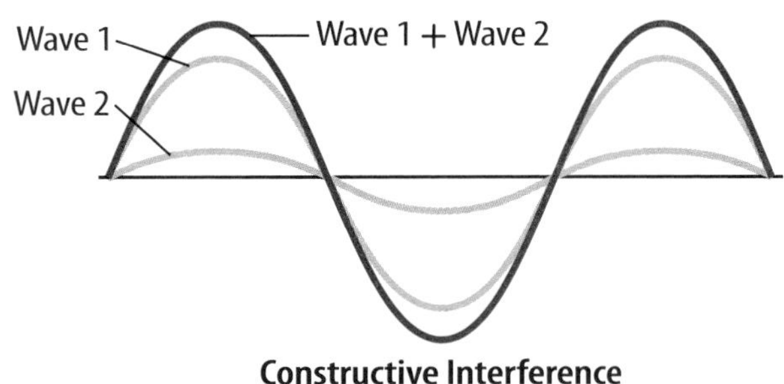

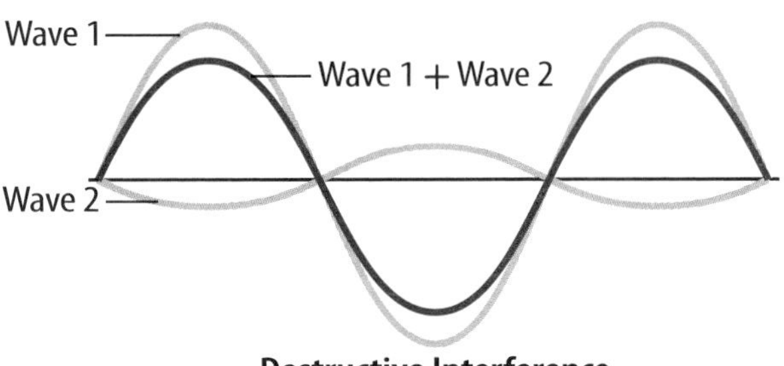

▲ **Figure 18** When constructive interference occurs, the new wave has a greater amplitude than either original wave. When destructive interference occurs, the new wave has a smaller amplitude than the sum of the amplitudes of the original waves.

Animation

ACADEMIC VOCABULARY

constructive
(adjective) pertaining to building or putting parts together to make a whole

Constructive and Destructive Interference

As waves travel through each other, sometimes the crests of both waves overlap, as shown in the top image of **Figure 18.** A new wave forms with greater amplitude than either of the original waves. This type of interference is called **constructive** interference. It occurs when crests overlap with crests and troughs overlap with troughs.

Destructive interference occurs when a crest of one wave overlaps the trough of another wave. The new wave that forms has a smaller amplitude than the sum of the amplitudes of the original waves, as shown in the bottom image of **Figure 18.** If the two waves have the same amplitude, they cancel each other when their crests and troughs overlap.

 **Key Concept Check** Describe two types of wave interference.

Standing Waves

Suppose you shake one end of a rope that has the other end attached to a wall. You create a wave that travels away from you and then reflects off the wall. As the wave you create and the reflected wave interact, interference occurs. For some values of the wavelength, the wave that forms from the combined waves seems to stand still. This wave is called a standing wave. An example is shown in **Figure 19.**

Figure 19 A standing wave can occur when two waves with the same wavelength travel in opposite directions and overlap. The wave that forms seems to be standing still. ▼

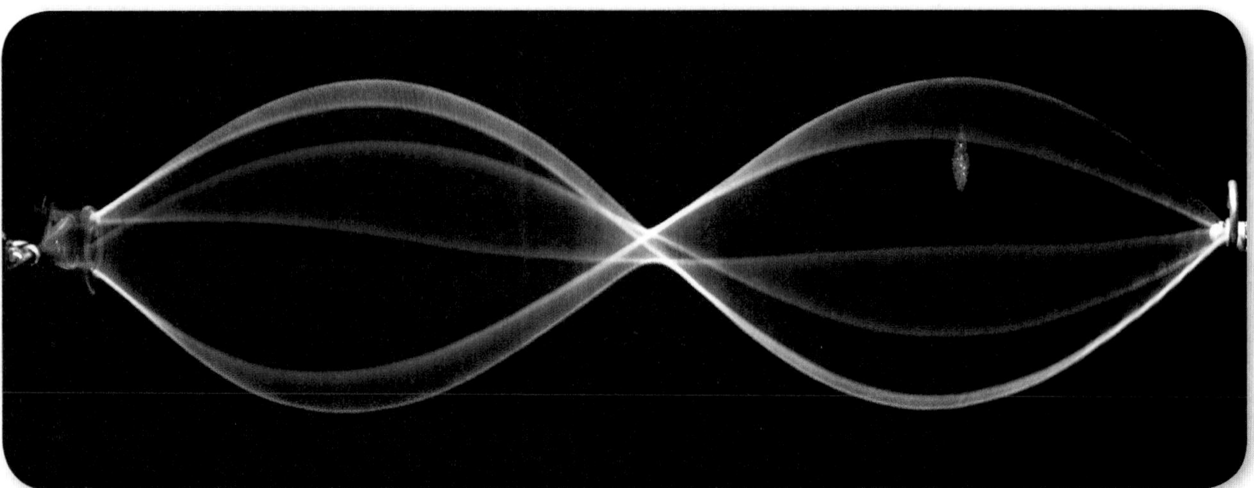

Lesson 3 Review

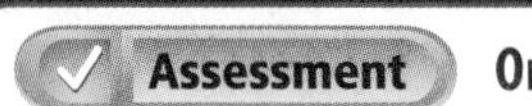

Visual Summary

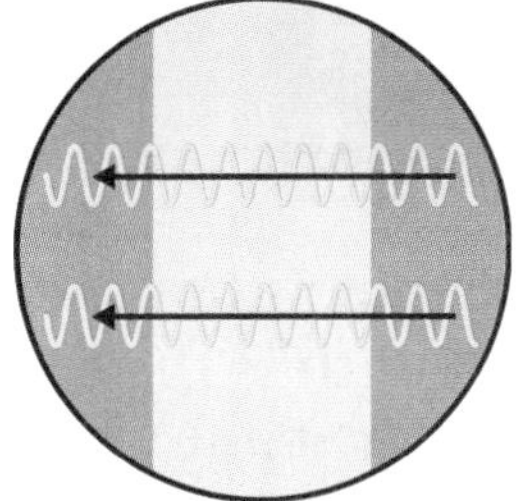

Transmission occurs when waves travel through a material.

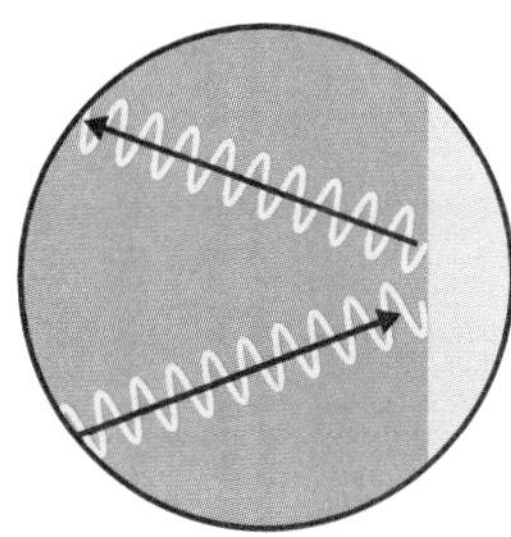

Reflection occurs when waves bounce off the surface of a material.

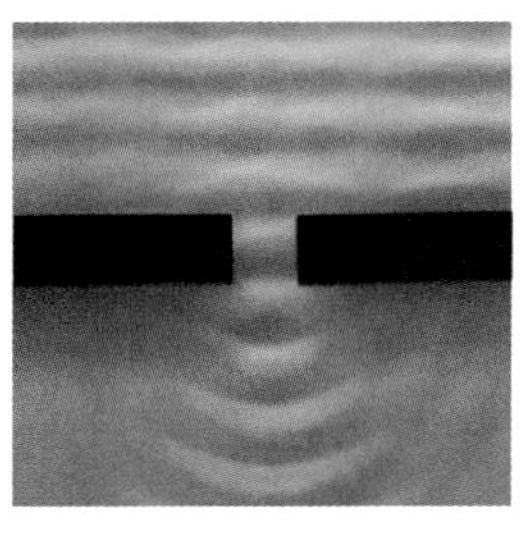

The change in direction of a wave when it travels through an opening is diffraction.

FOLDABLES®

Use your lesson Foldable to review the lesson. Save your Foldable for the project at the end of the chapter.

What do you think NOW?

You first read the statements below at the beginning of the chapter.

5. When light waves strike a mirror, they change direction.

6. Light waves travel at the same speed in all materials.

Did you change your mind about whether you agree or disagree with the statements? Rewrite any false statements to make them true.

Use Vocabulary

1. **Explain** the law of reflection.
2. **Distinguish** between refraction and diffraction.

Understand Key Concepts 

3. **Contrast** the behavior of a water wave that travels by a stone barrier to a sound wave that travels through a door.
4. Which will NOT occur when a light ray interacts with a smooth pane of glass?
   A. absorption
   B. diffraction
   C. reflection
   D. transmission

Interpret Graphics

5. **Describe** what is occurring in the figure below.

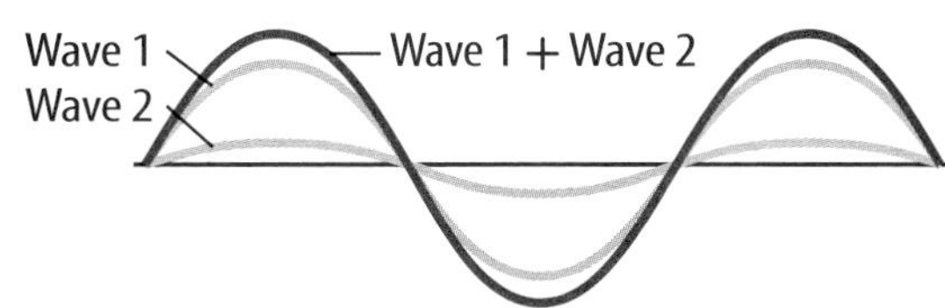

6. **Organize** Copy and fill in the graphic organizer below. In each oval, list something that can happen to a wave when it interacts with matter.

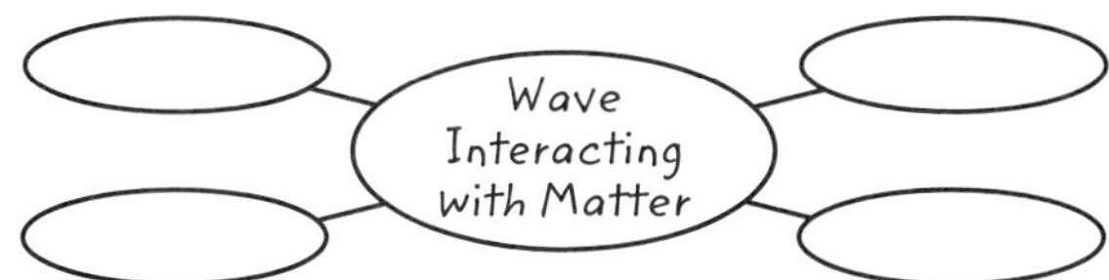

Critical Thinking

7. **Construct** Biologists know that chlorophyll, the pigment responsible for photosynthesis in plants, absorbs red light. Design a machine that you could use to test for the presence of chlorophyll.
8. **Recommend** An architect wants to design a conference room that reduces noise coming from outside the room. Suggest some design features that should be considered in this project.

55 minutes

Measuring Wave Speed

Materials

meter tapes (2)

masking tape

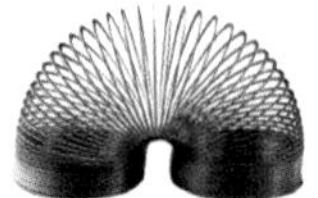

coiled spring toy

twine, 0.25 m

stopwatch or clock with second hand

Safety

When you make a wave on a spring toy, the frequency is how many wavelengths pass a point per second. Wavelength is the distance between one point on the wave and the nearest point just like it. If you can measure the frequency and wavelength of a wave, you can determine the wave speed.

Ask a Question

How can you determine the speed of a wave?

Make Observations

1. Read and complete a lab safety form.
2. Lay the meter tapes on the floor perpendicular to each other to make an *x*- and *y*-axis. Fasten them in place with masking tape.
3. Tie a piece of twine around the last coil of the spring toy.
4. With a partner, stretch the spring toy along the *x*-axis. One person should hold one end at the *y*-axis. The other person should hold the twine at the end of the outstretched spring.
5. One student creates a transverse wave by moving his or her hand up and down along the *y*-axis at a constant rate. When the wave is consistent, another student times a 10-second period while the third person counts the number of vibrations in 10 seconds. Record the number of vibrations in your Science Journal in a data table like the one shown below.
6. As the student continues making the wave, another student should estimate the wavelength along the *x*-axis using the meter tape.
7. Calculate the frequency of the wave. Then calculate the wave speed using the equation, wave speed = frequency × wavelength.
8. Repeat steps 5 through 7 using a different frequency.

Trial	Number of Vibrations in 10 s	Frequency (Hz)	Wavelength (cm)	Wave Speed (cm/s)
1				
2				

Form a Hypothesis

9. Form a hypothesis about the relationship between frequency and wavelength.

Test Your Hypothesis

10. Choose a frequency that you did not use during **Make Observations.** Predict the wavelength for a wave with this frequency.
11. Practice making a wave on the toy spring with your chosen frequency. Repeat steps 4–7 for this wave. Did your prediction of wavelength support your hypothesis? If not, revise your hypothesis and repeat steps 4–7.

Analyze and Conclude

12. **Conclude** How did your prediction of wavelength compare to your measurement?
13. **Think Critically** What measurements were the most difficult to make accurately? Suggest ways to improve on the method.
14. **The Big Idea** How did the wavelength, frequency, and wave speed change for the different waves that you created?

Communicate Your Results

Write a report explaining the steps you took in this lab. Include a table of the measurements you made. Be sure to describe sources of error in your measurements and ways that you might improve the accuracy of your experiment.

Try measuring the wave speed of other waves. Try stretching your spring toy to different lengths or try measuring the wave speed of longitudinal waves. You also might try working with ropes of different thicknesses, different spring toys, or even water in a wave tank.

Lab Tips

- ☑ Keep the amplitude constant by moving the same distance on the *y*-axis in each vibration.
- ☑ Twenty vibrations in 10 s make a wave with a frequency of 2 Hz.

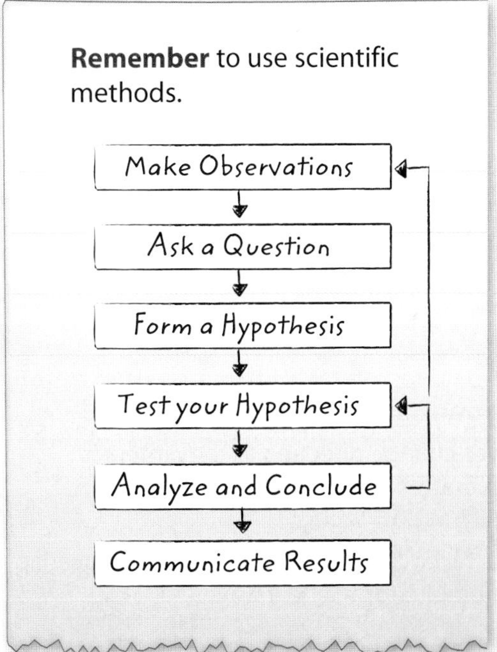

Chapter 15 Study Guide

Waves transfer energy but not matter as they travel.

Key Concepts Summary

Lesson 1: What are waves?

- Vibrations cause **waves.**
- **Transverse waves** make particles in a **medium** move at right angles to the direction that the wave travels. **Longitudinal waves** make particles in a medium move parallel to the direction that the wave travels.
- **Mechanical waves** cannot move through empty space, but **electromagnetic waves** can.

Direction wave moves

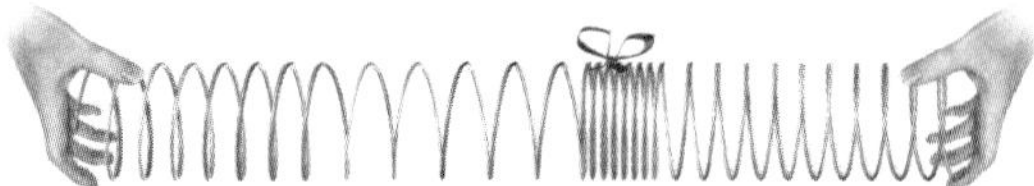

Vocabulary

wave p. 529
mechanical wave p. 531
medium p. 531
transverse wave p. 531
crest p. 531
trough p. 531
longitudinal wave p. 532
compression p. 532
rarefaction p. 532
electromagnetic wave p. 535

Lesson 2: Wave Properties

- All waves have the properties of **amplitude, wavelength,** and **frequency.**
- Increasing the frequency of a wave decreases the wavelength, and decreasing the frequency increases the wavelength.
- The speed of a wave depends on the type of material in which it is moving and the temperature of the material.

Amplitude
Wavelength
Amplitude

Vocabulary

amplitude p. 539
wavelength p. 541
frequency p. 542

Lesson 3: Wave Interactions

- When waves interact with matter, **absorption** and **transmission** can occur.
- Waves change direction as they interact with matter when **reflection, refraction,** or **diffraction** occurs.
- **Interference** occurs when waves that overlap combine to form a new wave.

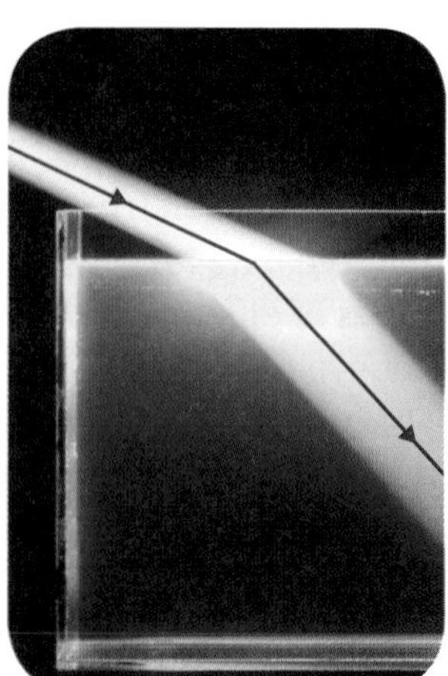

Vocabulary

absorption p. 548
transmission p. 548
reflection p. 548
law of reflection p. 549
refraction p. 550
diffraction p. 550
interference p. 551

- Personal Tutor
- Vocabulary eGames
- Vocabulary eFlashcards

FOLDABLES® Chapter Project

Assemble your lesson Foldables as shown to make a Chapter Project. Use the project to review what you have learned in this chapter.

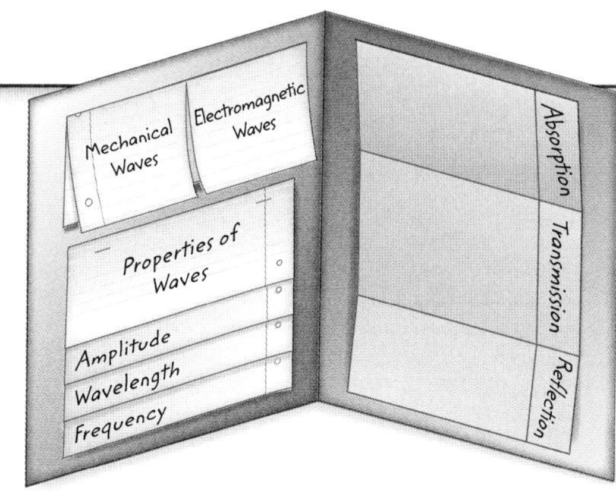

Use Vocabulary

1. A material though which a wave travels is a(n) ________.
2. A(n) ________ is a region where matter is more closely spaced in a longitudinal wave.
3. The Sun gives off energy that travels through space in the form of ________.
4. The product of ________ and wavelength is the speed of the wave.
5. ________ is a property of waves that is measured in hertz.
6. The highest point on a transverse wave is a(n) ________.
7. ________ is when two waves pass through each other and keep going.

Link Vocabulary and Key Concepts

Copy this concept map, and then use vocabulary terms from the previous page to complete the concept map.

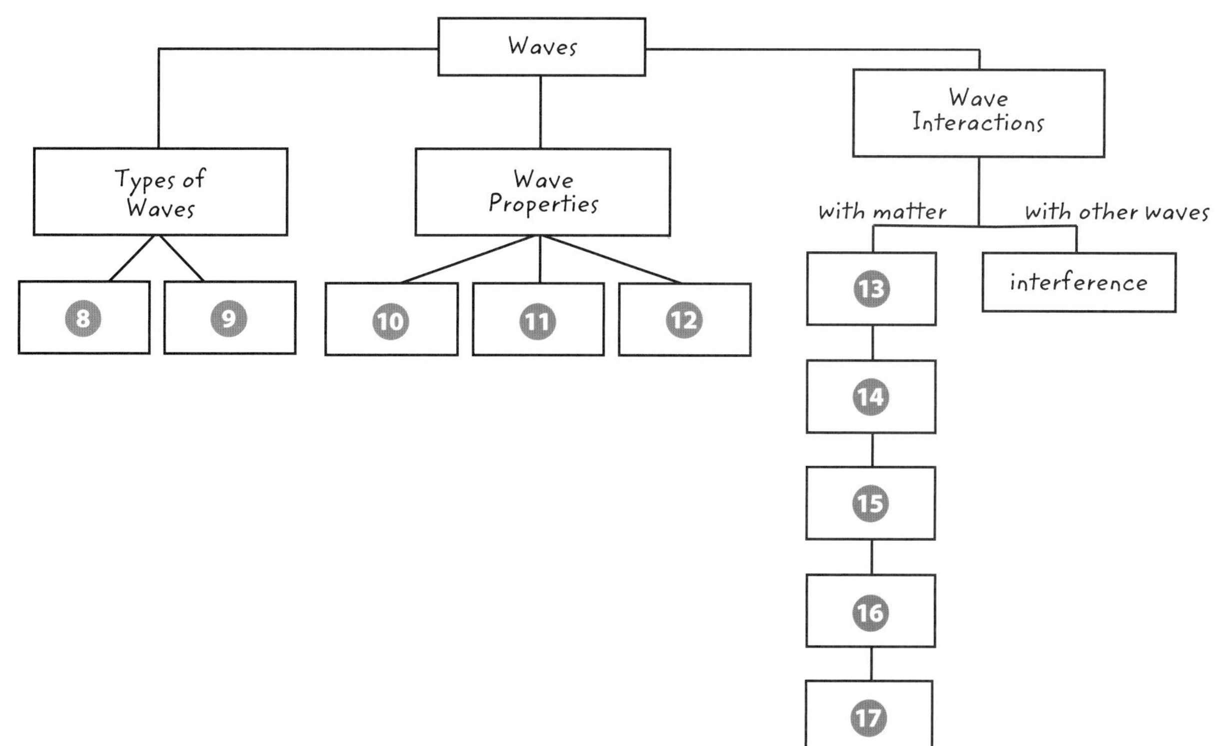

Chapter 15 Review

Understand Key Concepts 

1. What is transferred by a radio wave?
   A. air
   B. energy
   C. matter
   D. space

2. In a longitudinal wave, where are the particles most spread out?
   A. compression
   B. crest
   C. rarefaction
   D. trough

3. Which would produce mechanical waves?
   A. burning a candle
   B. hitting a wall with a hammer
   C. turning on a flashlight
   D. tying a rope to a doorknob

4. Which is an electromagnetic wave?
   A. a flag waving in the wind
   B. a vibrating guitar string
   C. the changes in the air that result from blowing a horn
   D. the waves that heat a cup of water in a microwave oven

5. Identify the crest of the wave in the illustration below.

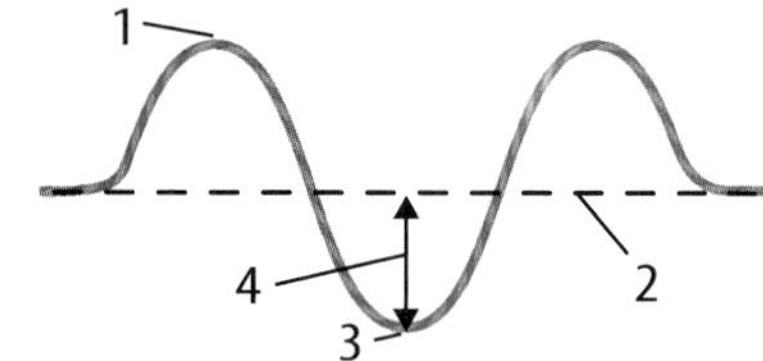

   A. 1
   B. 2
   C. 3
   D. 4

6. If the energy carried by a wave increases, which other wave property also increases?
   A. amplitude
   B. medium
   C. wavelength
   D. wave speed

7. In which medium is the speed of a sound wave the greatest?
   A. air in your classroom
   B. iron railroad track
   C. pool of water
   D. vacuum in space

8. A vibration that produces a wave takes 0.5 seconds to complete. What is the frequency of the wave?
   A. 0.25 Hz
   B. 0.5 Hz
   C. 2 Hz
   D. 4 Hz

9. What does the amount of refraction of a wave depend on?
   A. change in wave speed
   B. location of the normal line
   C. size of the object
   D. size of the opening between objects

10. Two waves travel through each other, and a crest forms with an amplitude smaller than either original wave. What has happened?
    A. constructive interference
    B. destructive interference
    C. reflection
    D. refraction

11. According to the table below, which material is probably a solid?

The Speed of Light in Different Materials

Material	Speed (km/s)
1	300,000
2	298,600
3	225,000
4	125,000

    A. material 1
    B. material 2
    C. material 3
    D. material 4

Assessment

Online Test Practice

Critical Thinking

12 **Assess** A student sets up a line of dominoes so that each is standing vertically next to another. He then pushes the first one and each falls down in succession. How does this demonstration represent a wave? How is it different?

13 **Infer** In the figure below, suppose wave 1 and wave 2 have the same amplitude. Describe the wave that forms when destructive interference occurs.

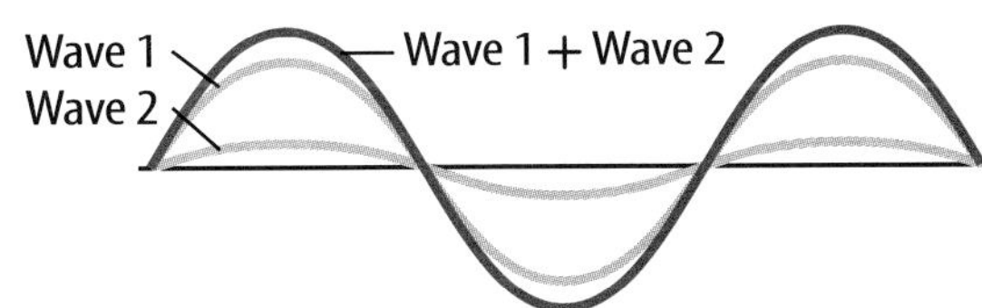

14 **Compare** A category 5 hurricane has more energy than a category 3 hurricane. Which hurricane will create water waves with greater amplitude? Why?

15 **Infer** At a baseball game when you are far from the batter, you might see the batter hit the ball before you hear the sound of the bat hitting the ball. Explain why this happens.

16 **Evaluate** Geologists measure the amplitude of seismic waves using the Richter scale. If an earthquake of 7.3 has a greater amplitude than an earthquake of 4.4, which one carries more energy? Explain your answer.

17 **Recommend** Some medicines lose their potency when exposed to ultraviolet light. Recommend the type of container in which these medicines should be stored.

18 **Explain** why the noise level rises in a room full of many talking people.

Writing in Science

19 **Write** a short essay explaining how an earthquake below the ocean floor can affect the seas near the earthquake area.

REVIEW THE BIG IDEA

20 What are waves and how do they travel? Describe the movement of particles from their resting positions for transverse and longitudinal waves.

21 The photo below shows waves in the ocean. Describe the waves using vocabulary terms from the chapter.

Math Skills

Math Practice

Use Numbers

22 A hummingbird can flap its wings 200 times per second. If the hummingbird produces waves that travel at 340 m/s by flapping its wings, what is the wavelength of these waves?

23 A student did an experiment in which she collected the data shown in the table. What can you conclude about the wave speed and rope diameter in this experiment? What can you conclude about frequency and wavelength?

Wave Speed and Diameter

Trial	Rope Diameter (cm)	Frequency (Hz)	Wavelength (m)
1	2.0	2.0	8.0
2	2.0	8.0	2.0
3	4.0	2.0	10.0
4	4.0	4.0	5.0

Standardized Test Practice

Record your answers on the answer sheet provided by your teacher or on a sheet of paper.

Multiple Choice

1 Through which medium would sound waves move most slowly?

A air

B aluminum

C glass

D water

Use the diagram below to answer question 2.

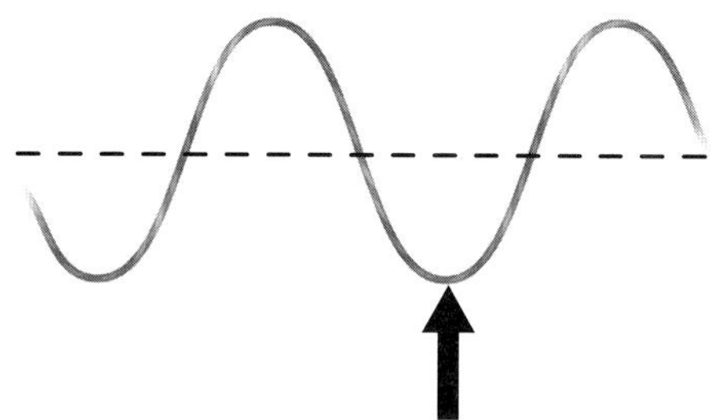

2 The diagram illustrates a mechanical wave. To which does the arrow point?

A compression

B crest

C rarefaction

D trough

3 Which is an electromagnetic wave?

A light

B seismic

C sound

D water

4 Which statement about waves is false?

A Waves transfer matter.

B Waves can change direction.

C Waves can interact with each other.

D Waves can transfer energy to matter.

Use the diagram below to answer question 5.

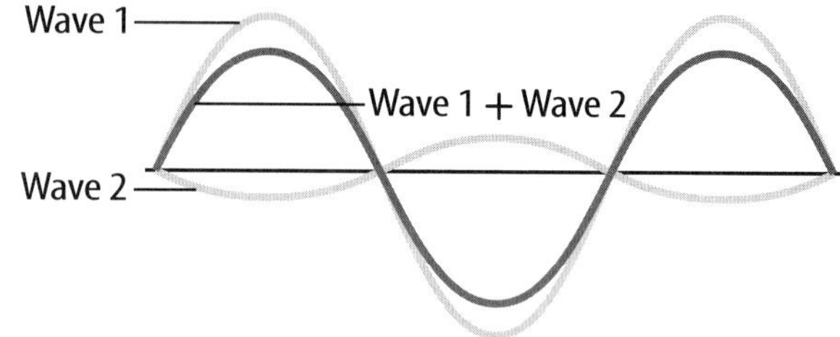

5 Which does the figure illustrate?

A constructive interference

B destructive interference

C diffraction

D reflection

6 In what region of a longitudinal wave are particles closest together?

A compression

B crest

C rarefaction

D trough

7 What happens to most of the light waves that strike a transparent pane of glass?

A absorption

B diffraction

C reflection

D transmission

8 Which wave can travel in both empty space and matter?

A radio

B seismic

C sound

D water

Online Standardized Test Practice

Use the figure below to answer question 9.

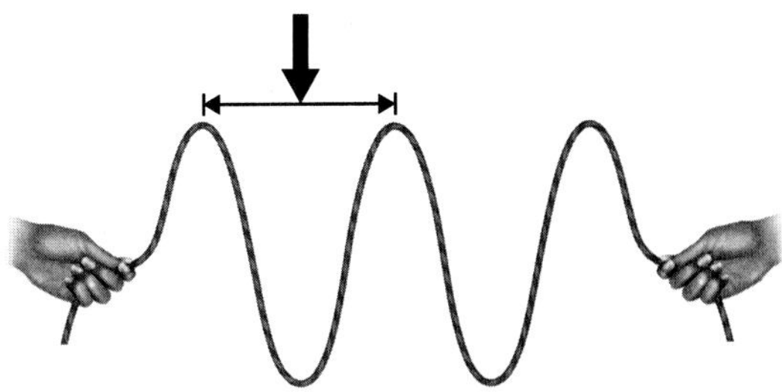

9 To which does the arrow point in the transverse wave diagram?

A amplitude

B crest

C frequency

D wavelength

10 Which is an example of diffraction?

A a flashlight beam hitting a mirror

B a shout crossing a crowded room

C a sunbeam striking a window

D a water wave bending around a rock

11 Which property of waves helps explain why a human shout cannot be heard a mile away?

A absorption

B diffraction

C reflection

D transmission

Constructed Response

Use the table below to answer questions 12 and 13.

	Wave A	Wave B
Number of Wavelengths that pass a point	5	8
Time for wavelengths to pass a point(s)	10	10

12 Which wave has a higher frequency? Why? How does wavelength change as frequency increases?

13 Write and solve an equation to find the speed of wave B if its wavelength is 2 m.

14 Two ocean waves approach a floating beach ball at different times. The second wave has more energy than the first wave. Which wave will have the higher amplitude? Explain your reasoning.

15 Explain why and how the waves from a passing speedboat rock a rowboat. Was the rowboat moved from its original location? Why or why not? Include the definition of a wave in your explanation.

NEED EXTRA HELP?															
If You Missed Question...	1	2	3	4	5	6	7	8	9	10	11	12	13	14	15
Go to Lesson...	2	1	1	1	3	1	3	1	2	3	3	2	2	2	1

Chapter 16

Sound

THE BIG IDEA **How can you produce, describe, and use sound?**

Inquiry How does it sound?

Recording a song involves more than just a band playing music. Often, each instrument records alone. Then, a singer records the vocals. Next, an audio engineer blends together the vocals and the sounds from the instruments. Some sounds may need to be adjusted to sound louder, softer, higher, or lower.

- What other properties of sound do you think the soundboard knobs control?
- Why do you think musicians wear headphones?
- How can you produce, describe, and use sound?

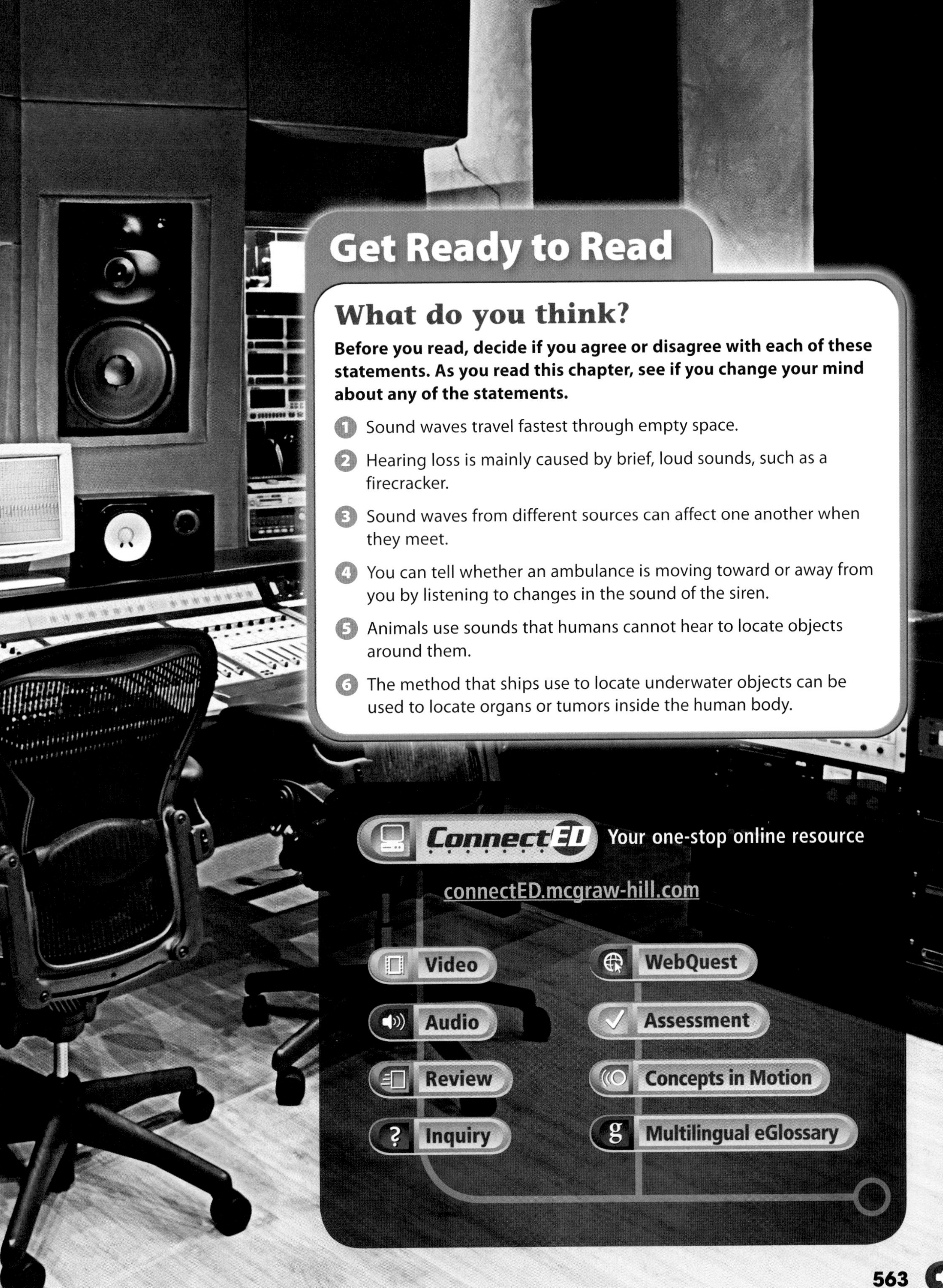

Get Ready to Read

What do you think?

Before you read, decide if you agree or disagree with each of these statements. As you read this chapter, see if you change your mind about any of the statements.

1. Sound waves travel fastest through empty space.
2. Hearing loss is mainly caused by brief, loud sounds, such as a firecracker.
3. Sound waves from different sources can affect one another when they meet.
4. You can tell whether an ambulance is moving toward or away from you by listening to changes in the sound of the siren.
5. Animals use sounds that humans cannot hear to locate objects around them.
6. The method that ships use to locate underwater objects can be used to locate organs or tumors inside the human body.

ConnectED Your one-stop online resource

connectED.mcgraw-hill.com

- Video
- Audio
- Review
- Inquiry
- WebQuest
- Assessment
- Concepts in Motion
- Multilingual eGlossary

Lesson 1

Producing and Detecting Sound

Reading Guide

Key Concepts
ESSENTIAL QUESTIONS

- How is sound produced?
- How does sound move from one place to another?
- Why does sound travel at different speeds through various materials?
- What are the functions of the different parts of the human ear?

Vocabulary

sound wave p. 565
longitudinal wave p. 565
vibration p. 565
medium p. 566
compression p. 566
rarefaction p. 566

Multilingual eGlossary

Video Science Video

Inquiry What is he listening to?

Think about the different ways people use sounds. Like the boy in the picture, many people use headphones to listen to their favorite songs. Others enjoy attending concerts or playing musical instruments. The sound of a car horn or a person shouting can alert people to danger. How are the different types of sounds produced? How are your ears able to detect the sounds?

10 minutes

What causes sound?

Sound travels through air as vibrations. When those vibrations reach the ear, sound is heard.

1. Read and complete a lab safety form.
2. Stretch **waxed paper** over the top of a **beaker.** Wrap a **rubber band** around the top to hold it tight.
3. Strike the center of the waxed paper gently with the eraser end of a **pencil.** Then strike it harder. How did the sound change? Write your observations in your Science Journal.
4. Sprinkle a few grains of **rice** onto the waxed paper. Strike the paper gently and then harder. Observe how the rice moves each time. Record your observations.

Think About This

1. How was the change in the rice's motion related to the change in sound?
2. **Key Concept** Based on your results, what do you think causes sound?

What is sound?

Everywhere you look, it seems that people have something on their ears! Some are talking on cell phones or listening to music. Some, such as people who work around airplanes, wear ear protection to prevent damage to their hearing. All of these devices have something to do with sound. The sounds you hear are produced by **sound waves**—*longitudinal waves that can only travel through matter.* A **longitudinal wave** *is a wave that makes the particles in the material that carries the wave move back and forth along the direction the wave travels.*

Sources of Sound

Every sound, from the buzzing of a bee to a loud siren, is the result of a vibration. *A* **vibration** *is a rapid, back-and-forth motion that can occur in solids, liquids, or gases.* The energy carried by a sound wave is caused by vibration. For example, as you pull on a guitar string, you transfer energy to the string. When you let go, the string snaps back and vibrates. As the string vibrates, it collides with nearby air particles. The string transfers energy to these particles. The air particles collide with other air particles and pass on energy, as shown in **Figure 1.**

As the string vibrates, its back-and-forth motion causes a disturbance in the air that carries energy outward from the source of the sound. The disturbance is a sound wave.

 **Key Concept Check** How is sound produced?

Figure 1 As the guitar string vibrates, it transfers energy to nearby air particles.

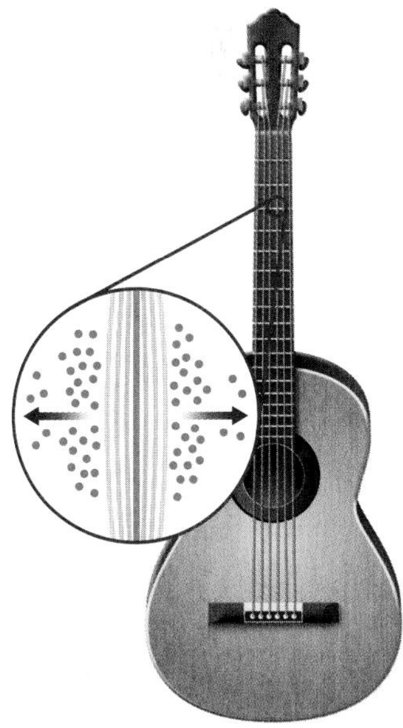

Visual Check How would the picture of the particles be different if the string vibrated faster?

Figure 2 Sound waves move away from a source as compressions and rarefactions.

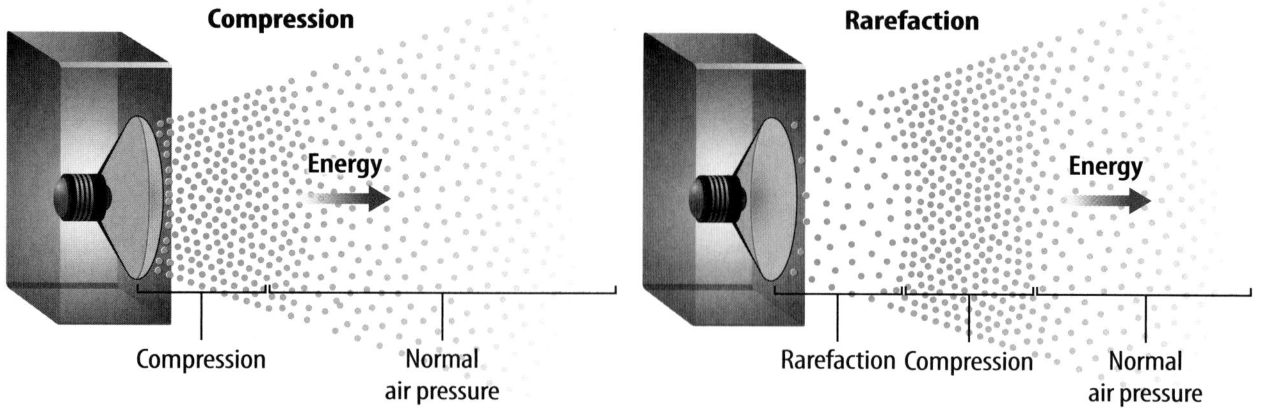

Compression

Energy

Compression

Normal air pressure

When the speaker cone moves out, it forces particles in the air closer together. This produces a high-pressure area, or compression.

Rarefaction

Energy

Rarefaction Compression

Normal air pressure

When the speaker cone moves back, it leaves behind an area with fewer particles. This is a low-pressure area called a rarefaction.

Science Use v. Common Use

media

Science Use plural form of *medium;* forms of matter through which sound travels

Common Use a type of mass communication, such as radio

Word Origin

rarefaction

from Latin *rarefacere,* means "to make rare or less dense"

How Sound Waves Travel

Vibrating objects cause sound waves that occur only in matter. For this reason, sound waves must travel through a solid, a liquid, or a gas. *A material in which a wave travels is a* **medium.** You usually hear sound through the medium of air, but sound waves also can travel through other **media,** such as water, wood, and metal. Sound waves cannot travel through empty space because there is no medium to carry the energy.

From a Sound Source to Your Ear If you touch a speaker, like the one in **Figure 2,** you can feel it vibrate as it produces sound waves. Air particles fill a room. Each time the speaker cone moves forward, it pushes air particles ahead of it in the room. This push forces the particles closer together, increasing air pressure in that area. *A region of a longitudinal wave where the particles in the medium are closest together is a* **compression.** With each vibration, the speaker cone moves forward and then back. This motion leaves behind a low-pressure region with fewer air particles, as shown on the right in **Figure 2.** *A* **rarefaction** (rayr uh FAK shun) *is a region of a longitudinal wave where the particles are farthest apart.*

Figure 3 Sound waves carry energy in the direction of the vibrations. ▼

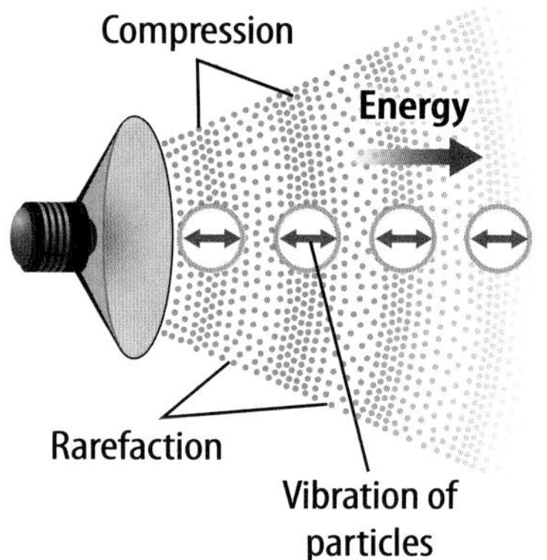

Energy in Sound Waves Suppose you are in the ticket line at the movies. Someone at the back of the line bumps into the next person in line. That person stumbles and bumps the next person before returning to his or her place in line. The energy of the bump continues down the line as each person bumps the next. In the same way, particles of a medium vibrate back and forth as a sound wave carries energy away from a source. This process is shown in **Figure 3.**

 **Key Concept Check** How do sound waves travel?

Speed of Sound

You are swimming in a pool when someone taps on the side nearby. Would you hear the sound if your head was under water? Yes, and you would probably hear the sound better than if your head were above water. Sound waves travel faster in water than in air. **Table 1** compares the speed of sound in different media.

Material Two factors that affect the speed of sound waves are the density and the stiffness of the material, or medium. Density is how closely the particles of a medium are packed. Gas particles are far apart and do not collide as often as do the particles of a liquid or a solid. Therefore, sound energy transfers more slowly in a gas than in a liquid or a solid. Notice in **Table 1** that sound travels fastest in solids. In a stiff or rigid solid where particles are packed very close together, the particles collide and transfer energy very quickly.

Sound waves also travel faster in seawater than in freshwater. Seawater contains dissolved salts and has a higher density than freshwater. The fin whale in **Figure 4** emits sounds heard by other whales hundreds of kilometers away.

Table 1 The Speed of Sound

Material	Speed (m/s)
Air (0°C)	331
Air (20°C)	343
Water (20°C)	1,481
Water (0°C)	1,500
Seawater (25°C)	1,533
Ice (0°C)	3,500
Iron	5,130
Glass	5,640

Table 1 The speed at which sound waves travels through different materials depends on factors such as density and temperature.

◀ **Figure 4** The low-pulse sounds of a fin whale travel more than four times faster through water than they would through air.

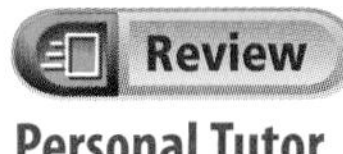

Personal Tutor

Temperature The temperature of a medium also affects the speed of sound. As the temperature of a gas increases, the particles move faster and collide more often. This increase in collisions transfers more energy in less time. Notice that sound waves travel faster in air at 20°C than in air at 0°C.

In liquids and solids, temperature has the opposite effect. Why do sound waves travel faster in water at 0°C than in water at 20°C? As water cools, the molecules move closer together, so they collide more often. Sound waves travel even faster when the water freezes into ice because it is rigid.

Key Concept Check Why does sound travel at different speeds through various materials?

FOLDABLES®

Make a vertical three-tab book and label it as shown. Use it to organize your notes about sound.

▲ **Figure 5** The large ears of the fennec fox can detect prey moving underground.

Detecting Sound

How do you think having large ears helps the fennec fox in **Figure 5**? Sound waves fill the air, and the large outer ear helps funnel sound waves to the inner ear, where sound is detected. Ears also tell you the direction a sound comes from. With its large ears, the fennec fox is better able to hear predators approach from greater distances.

The Human Ear

Have you ever cupped your hand around the back of your ear so you could better hear? Why does that work? The human ear has three main parts. The outer ear collects sound waves. By cupping your hand around your ear, you extend the outer ear and therefore collect more waves. The middle ear amplifies sound. The inner ear sends signals about sound to the brain. As shown in **Figure 6**, each part of the ear has a special shape with different parts that help it perform its function.

Key Concept Check What are the functions of the different parts of the human ear?

Figure 6 Each part of the ear has an important job that helps you hear. ▼

Visual Check How does the eardrum help you hear?

Functions of Human Ear Parts

Concepts in Motion Animation

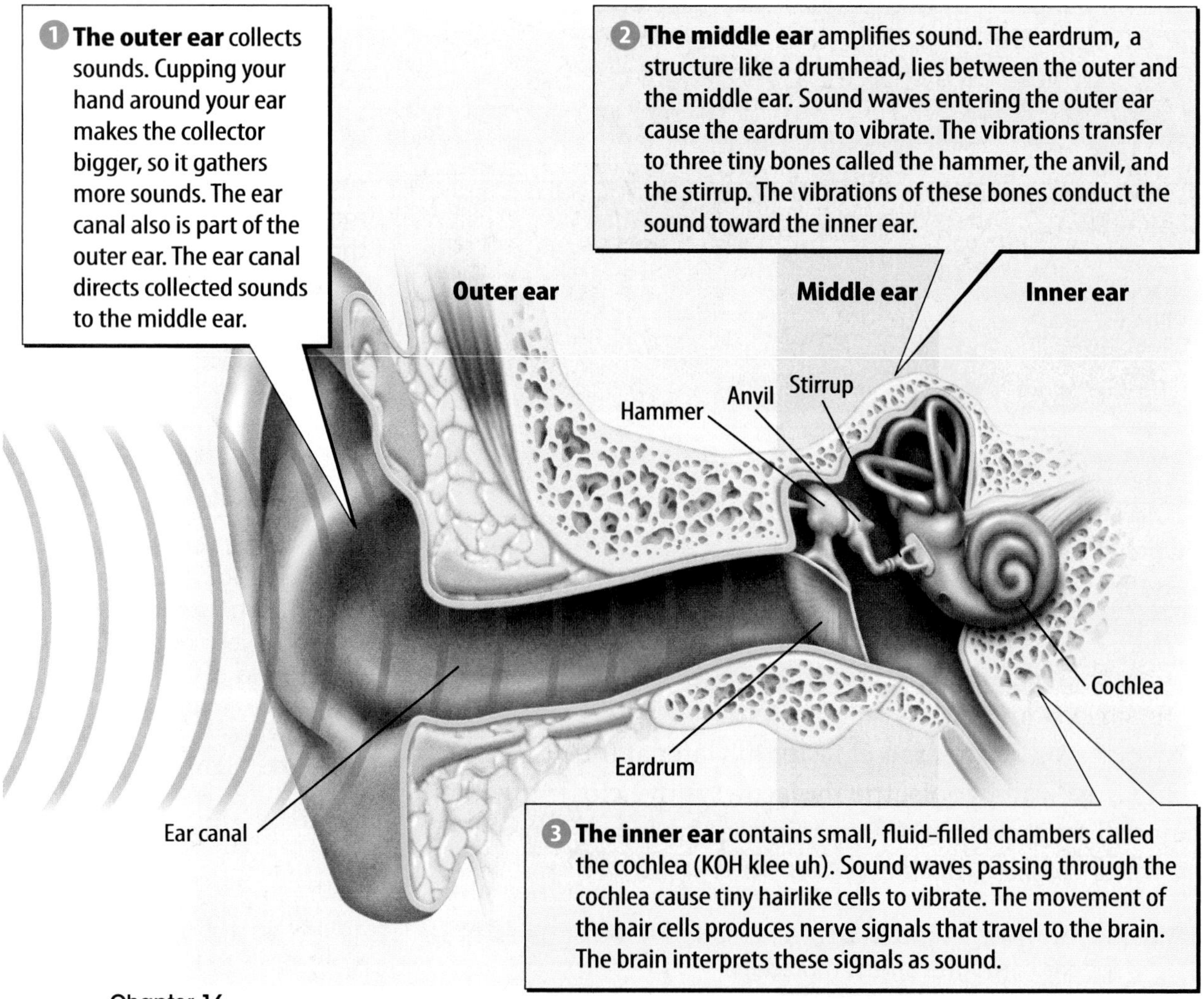

① **The outer ear** collects sounds. Cupping your hand around your ear makes the collector bigger, so it gathers more sounds. The ear canal also is part of the outer ear. The ear canal directs collected sounds to the middle ear.

② **The middle ear** amplifies sound. The eardrum, a structure like a drumhead, lies between the outer and the middle ear. Sound waves entering the outer ear cause the eardrum to vibrate. The vibrations transfer to three tiny bones called the hammer, the anvil, and the stirrup. The vibrations of these bones conduct the sound toward the inner ear.

Outer ear
Middle ear
Inner ear
Hammer
Anvil
Stirrup
Cochlea
Eardrum
Ear canal

③ **The inner ear** contains small, fluid-filled chambers called the cochlea (KOH klee uh). Sound waves passing through the cochlea cause tiny hairlike cells to vibrate. The movement of the hair cells produces nerve signals that travel to the brain. The brain interprets these signals as sound.

MiniLab

10 minutes

How do you know a sound's direction?

One way an animal can determine the location of a predator is by listening for the sounds the predator makes as it moves. How can ears help determine the direction of sound?

1. Read and complete a lab safety form.
2. Hold the ends of a 1-m piece of **flexible tubing** over your ears. Have your partner use a **pencil** to tap the tubing close to one ear and then the other. Next, have your partner tap at the midpoint of the tubing. Did you notice a change in the sound? Record your observations in your Science Journal.
3. Close your eyes. Have your partner tap at random places along the tubing. Try to guess whether each tap was closest to your left ear, your right ear, or in the middle of the tubing.
4. Switch roles, and tap the tubing as your partner holds the ends to his or her ears.

Analyze and Conclude

1. **Compare** the sounds you heard when your partner tapped the tubing at different locations.
2. **Key Concept** Why is it useful to have two ears instead of one?

Hearing Loss

The harder you pound on a drum, the farther the drumhead travels as it vibrates. What would a drum sound like if the drumhead had a big tear in it? Like a drumhead, the eardrum can be damaged. The eardrum vibrates as pressure changes in the ear. The louder the sound, the farther the eardrum moves in and out as it vibrates.

A very loud sound can make the eardrum vibrate so hard that it tears. Damage to the eardrum can cause hearing loss. Also, the tear can allow bacteria into the ear, causing infection. The tear may heal, but thick, uneven scar tissue can make the eardrum less sensitive to sounds.

Listening to loud music over a long period of time also can damage the ears. Look again at **Figure 6** and locate the cochlea. Infection or loud sounds can damage tiny hair cells in the cochlea. Cells that are damaged or die do not grow back, so hearing becomes less sensitive. Many people who work around loud machines, construction, or traffic wear ear protection to prevent damage, as shown in **Figure 7.** Wearing a headset while listening to loud music, however, traps the pressure changes in the ear; this can lead to permanent hearing loss.

Reading Check What are some things that might happen to your ear to cause hearing loss?

Figure 7 This woman is protecting two of her most important senses—sight and hearing.

Lesson 1 Review

Visual Summary

Sound waves carry energy away from a sound source.

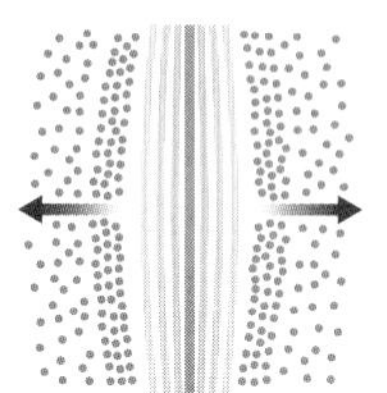

Sound waves move out from a source as a series of compressions and rarefactions.

Ears detect vibrations and interpret them as different sounds.

FOLDABLES®

Use your lesson Foldable to review the lesson. Save your Foldable for the project at the end of the chapter.

What do you think NOW?

You first read the statements below at the beginning of the chapter.

1. Sound waves travel fastest through empty space.
2. Hearing loss is mainly caused by brief, loud sounds, such as a firecracker.

Did you change your mind about whether you agree or disagree with the statements? Rewrite any false statements to make them true.

Use Vocabulary

1. **Define** *longitudinal wave* in your own words.
2. **Identify** the region in a sound wave where the particles are farthest apart.
3. The energy carried by a sound wave comes from a(n) ________ in a medium.

Understand Key Concepts 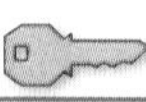

4. Through which medium would sound travel fastest?

   A. air **C.** cold water
   B. iron **D.** warm water

5. **Summarize** how you produce sound when you tap a pencil against your desk.
6. **Compare and contrast** the functions of the three main parts of the human ear.
7. **Describe** the motion of an air particle as a sound wave passes through it.

Interpret Graphics

8. **Identify** the regions labeled *A* and *B* in the image below. What causes these regions?

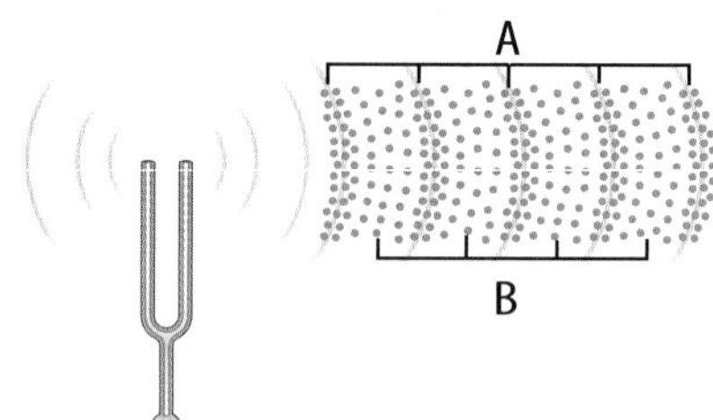

9. **Sequence** Copy and fill in a graphic organizer like the one below that identifies the path of a sound wave as it enters the ear.

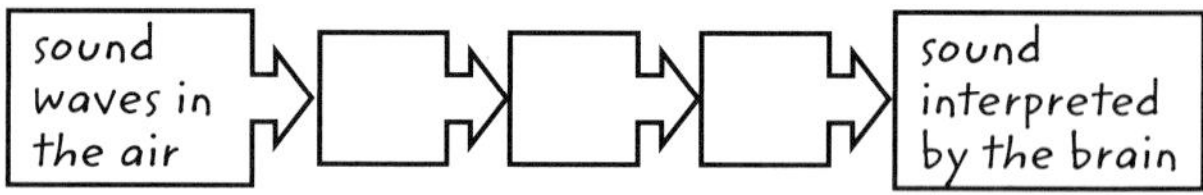

Critical Thinking

10. **Evaluate** A spaceship in a science fiction movie explodes. People in a nearby spaceship hear a loud sound. Is this realistic? Explain.

Cochlear Implants

HOW IT WORKS

Helping Damaged Ears Hear Again

Is there any way to hear again after hair cells in your cochlea are destroyed? Not long ago, the answer was no. But over the last 20 to 30 years, scientists developed a way to bypass the damaged cells. It is called a cochlear implant—a device that uses electrical signals to stimulate the nerves that go from the ear to the brain. How does it work?

First a surgeon implants the interior part of the device under the scalp and into the inner ear. Then the exterior part of the device is put to work. With hearing restored, you are once again connected to the sounds in the world around you!

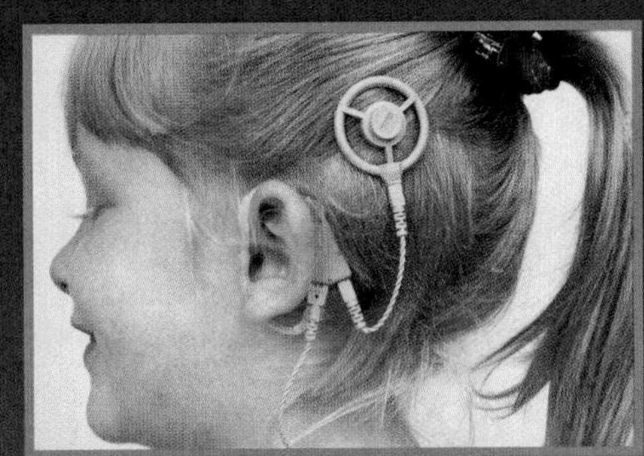

1. **A microphone receives sound waves from the environment. A speech processor changes sounds into electrical signals.**
2. **The electrical signals are then sent across the scalp from the transmitter to a receiver.**
3. **The receiver sends the signals through a wire to electrodes implanted in the cochlea.**
4. **Nerves in the inner ear pick up the signals and send them to the brain. The brain interprets the signals as sound.**

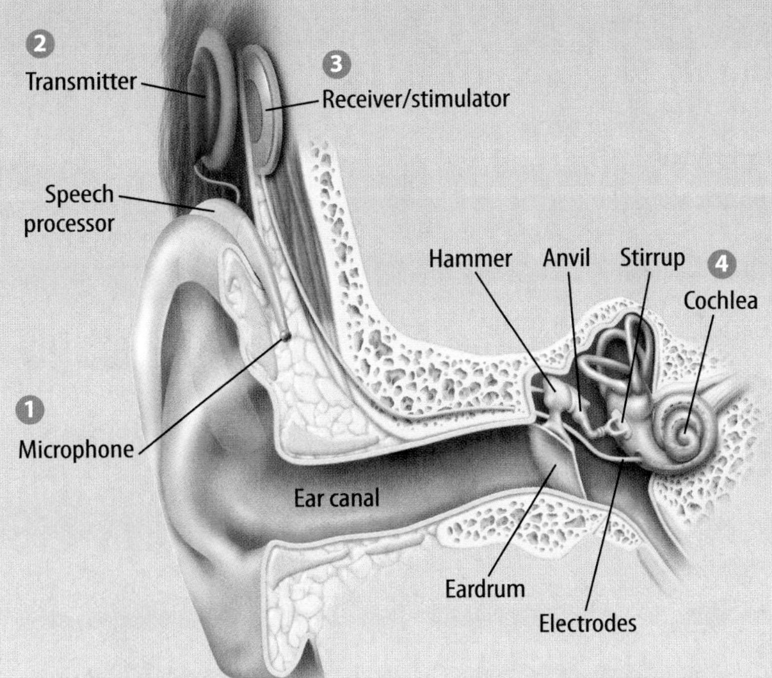

It's Your Turn

MAKE A POSTER How does a hearing aid work? How is it different from a cochlear implant? Research these questions and create a poster from what you discover.

Lesson 2

Properties of Sound Waves

Reading Guide

Key Concepts
ESSENTIAL QUESTIONS

- How are amplitude and intensity related to energy?
- What is the relationship among frequency, pitch, and wavelength?
- How can you recognize sounds from different sources?
- In what ways are musical sounds produced?

Vocabulary

amplitude p. 573
intensity p. 574
wavelength p. 575
frequency p. 575
pitch p. 575
Doppler effect p. 576
interference p. 576
resonance p. 578

Multilingual eGlossary

Video Science Video

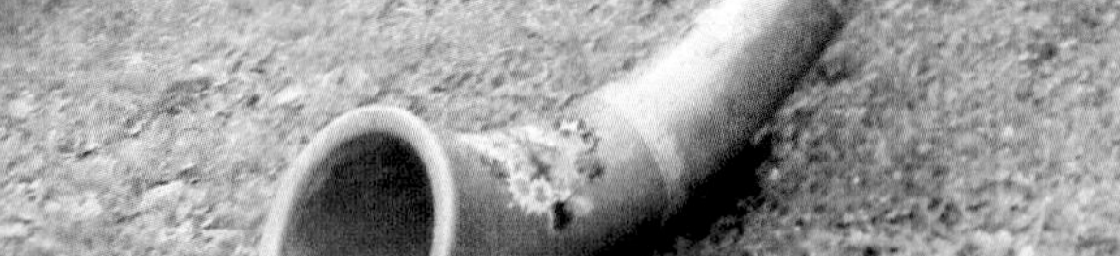

Inquiry What do they sound like?

High in the mountains of Switzerland, you might hear the clear, mellow tones of an alphorn. At one time, herders used the horns to call or soothe cows. How do you think the shape of the horn affects the sound? Why do you think the sound is able to travel so far?

15 minutes

How can sound blow out a candle?

Sound waves carry energy. Can the energy in a sound wave affect a nearby candle flame?

1. Read and complete a lab safety form.
2. Cut off the neck of a **balloon** with **scissors.** Stretch the remaining part of the balloon over the wide end of a **small funnel.**
3. Set a ball of **modeling clay** on a table. Insert a small **candle** in the clay, and use a **safety match** to light it.
4. Hold the funnel with the narrow end pointing toward the lit candle, about 2 cm away.
   ⚠ *Do not let the funnel touch the flame.*
5. Sharply strike the rim of the funnel several times with a **ruler** so that it makes a loud sound. What happens to the candle flame? Record your observations in your Science Journal.
6. Strike the funnel several more times. Each time, vary the amount of energy you use. Record your observations.

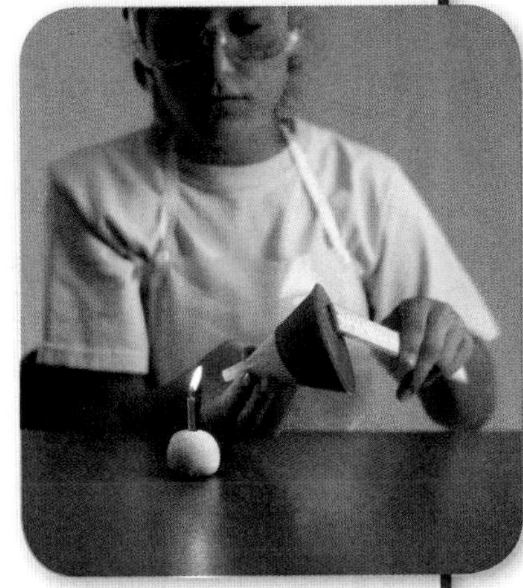

Think About This

1. What was different about the flame when you made a soft sound compared to when you made a loud sound? Why do you think this happened?
2. **Key Concept** How did the amount of energy you used to strike the funnel affect the sound? How did it affect the flame?

Energy of Sound Waves

Shhhhh! How do you change your voice from a yell to a whisper? The energy a sound wave carries depends on the amount of energy that caused the original vibration. To speak softly, just use less energy!

Amplitude

How do the sound waves produced when you yell and when you whisper differ? The more energy you put into your voice, the farther the air particles move as they vibrate back and forth. *For a longitudinal wave,* **amplitude** *is the maximum distance the particles in a medium move from their rest positions as the wave passes through the medium.* As the energy in a sound wave increases, its amplitude increases. Sound waves with small and large amplitudes are shown in **Figure 8.**

Key Concept Check How is the amplitude of sound related to energy?

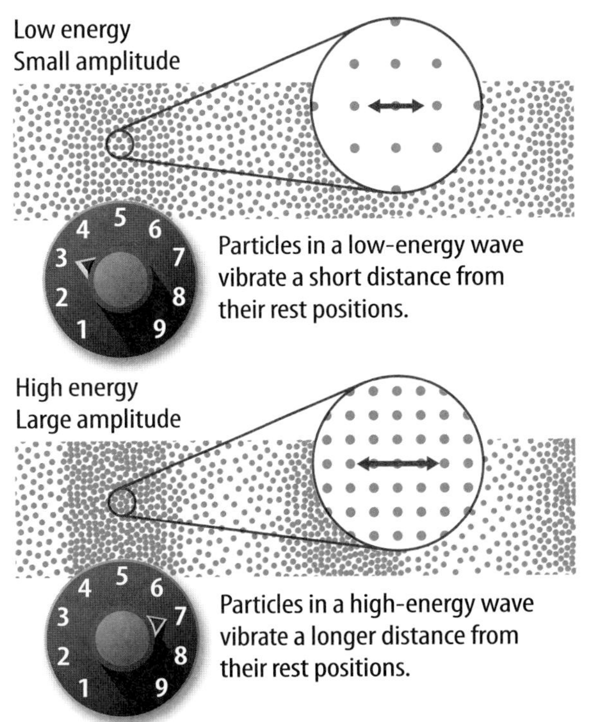

Low energy
Small amplitude
Particles in a low-energy wave vibrate a short distance from their rest positions.
High energy
Large amplitude
Particles in a high-energy wave vibrate a longer distance from their rest positions.
1 2 3 4 5 6 7 8 9

Figure 8 Particle spacing differs in high-energy and low-energy waves.

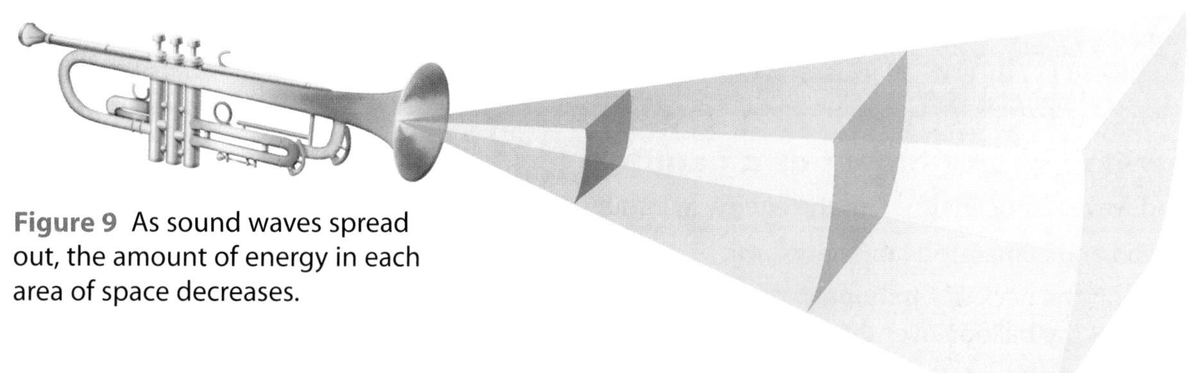
Figure 9 As sound waves spread out, the amount of energy in each area of space decreases.

Amplitude, Intensity and Loudness

Imagine blowing into a trumpet, like the one in **Figure 9.** Sound waves leave the horn with a certain amount of energy and, therefore, a certain amplitude. Recall that amplitude is the distance the particles of air vibrate back and forth. Loudness is how you perceive the energy of a sound wave. So, a wave with a greater amplitude will produce a louder sound.

Why do sounds get quieter as you get farther from the source? Think of what happens as a sound wave travels away from the horn. As the particles of air in front of the horn vibrate back and forth, they collide with, and transfer energy to, surrounding particles of air. As the energy spreads out among more and more particles, the intensity of the wave decreases. **Intensity** *is the amount of sound energy that passes through a square meter of space in one second.*

 **Key Concept Check** How is the intensity of sound related to energy?

As a sound wave travels farther from the horn, there is a larger area of particles sharing the same amount of energy that left the horn. Therefore, the farther you are from the horn, the less energy passing trough one square meter. This results in less intensity of the wave. As intensity decreases, amplitude decreases. Therefore, loudness decreases.

The Decibel Scale

The unit decibel (dB) describes the intensity and, in turn, the loudness of sound. Decibel levels of common sounds are shown in **Figure 10.** Each increase of 10 dB indicates the sound is about twice as loud and has about 10 times the energy. For example, the decibel level of city traffic is about 85 dB, and the level of a rock concert is about 105 dB. This means a concert, which is 20 dB higher, has about 10 × 10, or 100 times, more energy than traffic. Recall that a loud sound can make the eardrum vibrate so hard that it tears. As sounds get louder, the amount of time you can listen without hearing loss gets shorter.

Decibel Scale

Figure 10 The decibel scale rates the loudness of some common sounds.

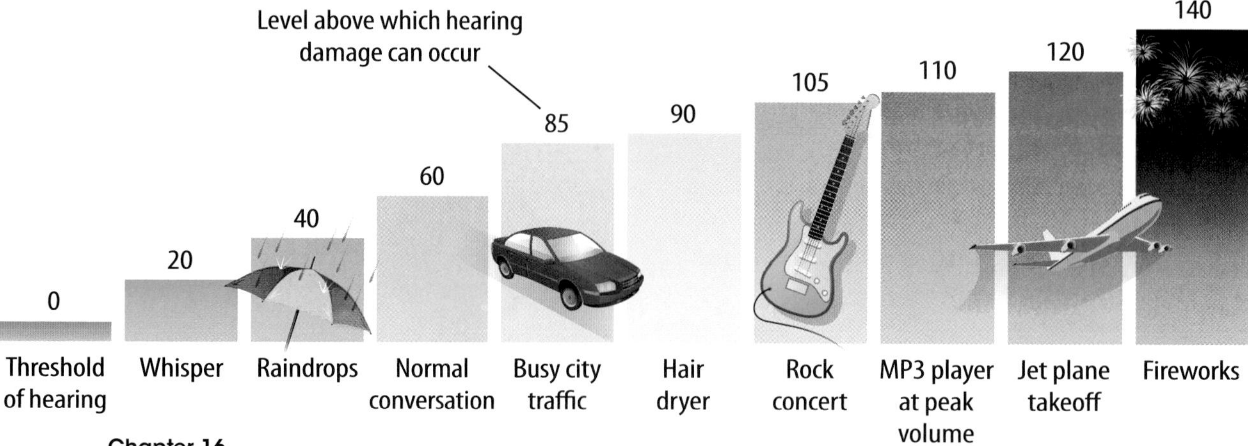

Wavelength and Frequency

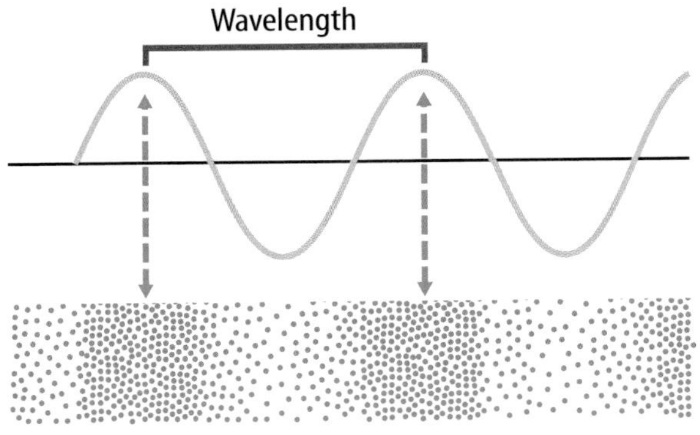

Low frequency; long wavelength

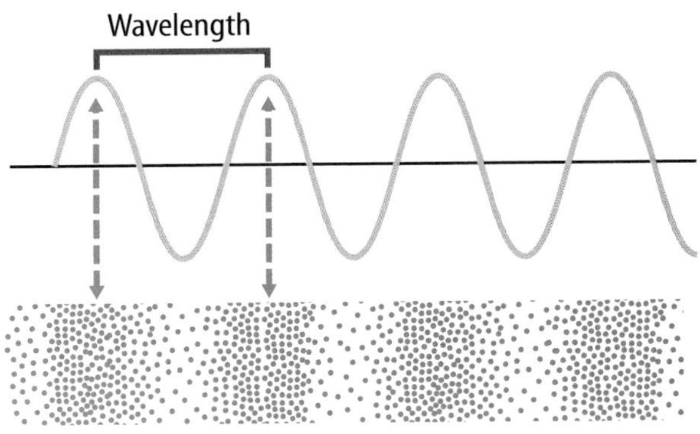

High frequency; short wavelength

Figure 11 If you compare two sound waves, the wave with the longer wavelength has a lower frequency. ▼

Visual Check How can you tell that the sounds have the same intensity?

Describing Sound Waves

Sounds depend on many properties of the sound waves that enter your ear. Loudness or softness depends on the amplitude of the wave. You also might describe sound according to how frequently the waves occur or how long the waves are.

Wavelength

One property of a sound wave is its wavelength. *The distance between a point on one wave and the nearest point just like it is called* **wavelength.** For example, you could measure a wavelength as the distance between the midpoint of one compression, or rarefaction and the midpoint of the next compression, or rarefaction, as shown in **Figure 11.**

Frequency and Pitch

Suppose you could count sound waves produced by playing middle C on a piano. You would find that 262 wavelengths pass you each second. *The* **frequency** *of sound is the number of wavelengths that pass by a point each second.* Notice in **Figure 11,** that as the wavelength of a sound wave decreases, its frequency increases. The frequency of one vibration, or wavelength, per second is called a hertz (Hz). The frequency of middle C on a piano is 262 Hz.

The perception of how high or low a sound seems is **pitch.** A higher frequency produces a higher pitch. For example, an adult male voice might range from 85 Hz to 155 Hz. An adult female voice might range from about 165 Hz to 255 Hz.

The human ear can detect sounds with frequencies between about 20 Hz and 20,000 Hz. Frequencies above this range are called ultrasound. The range of sounds heard by various animals is shown in **Figure 12.**

Key Concept Check What is the relationship among frequency, pitch, and wavelength?

▲ **Figure 12** Many animals can hear sounds outside the range of human hearing.

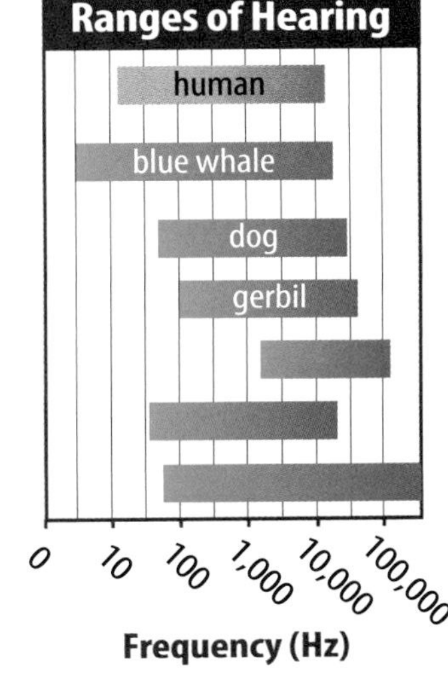

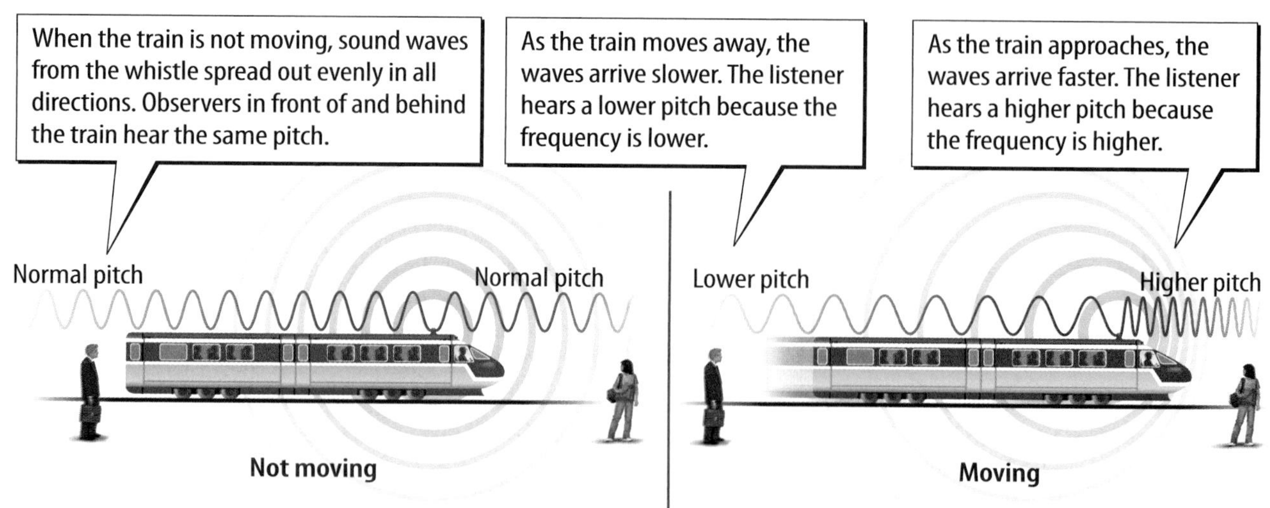

When the train is not moving, sound waves from the whistle spread out evenly in all directions. Observers in front of and behind the train hear the same pitch.
As the train moves away, the waves arrive slower. The listener hears a lower pitch because the frequency is lower.
As the train approaches, the waves arrive faster. The listener hears a higher pitch because the frequency is higher.
Normal pitch
Normal pitch
Lower pitch
Higher pitch
Not moving
Moving

▲ **Figure 13** Sound waves bunch together ahead of a moving source.

The Doppler Effect

You might have heard the high pitch of a train whistle as the train approaches. As the train passes, the pitch drops. Sound frequency depends on the motions of the source and the listener. Compare the wave frequencies in front of and behind the moving train in **Figure 13.** The frequency increases if the distance between the listener and the source is decreasing. The frequency decreases if the distance between the listener and the source increases. *The change of pitch when a sound source is moving in relation to an observer is the* **Doppler effect.**

Sound Interference

If you walk through a room with stereo speakers at each end, the sound might seem louder in some places and softer in others. The waves from each speaker interact with one another. **Interference** *occurs when waves that overlap combine, forming a new wave.*

Figure 14 shows why this happens. When compressions meet, they join to form a wave with higher intensity and greater amplitude. This is called constructive interference. However, when a compression meets a rarefaction, the intensity and amplitude decrease. This is destructive interference.

▼ **Figure 14** Waves' interference can result in an increase or a decrease in amplitude.

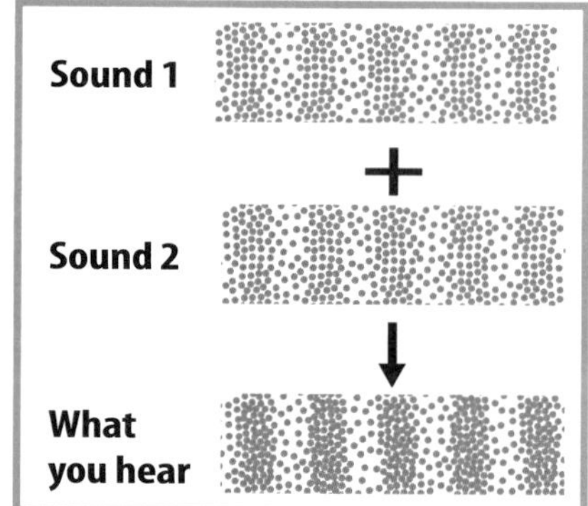

Constructive Interference
When the compressions and rarefactions of waves overlap, the combined compressions have greater intensity.

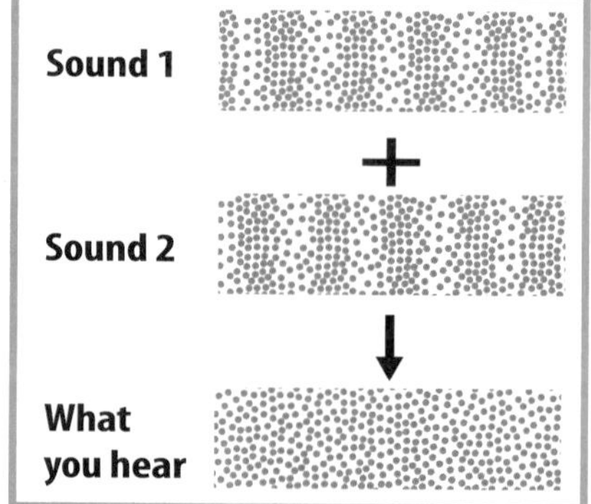

Destructive Interference
When the compressions of one wave overlap the rarefactions of another wave, the waves cancel and the result is no sound.

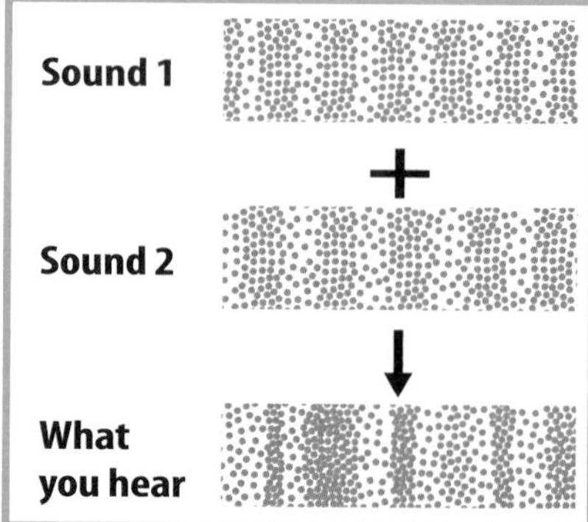

Beats
When the compressions of two waves are slightly offset, a pattern of increasing and decreasing compressions, called beats, occurs.

10 minutes

How can you hear beats?

When musicians play tones with slightly different pitches, the combined sound creates a beat.

1. Read and complete a lab safety form.
2. Wrap two **rubber bands** around a **box** so that they stretch across the open side. Listen as you pluck the rubber bands.
3. Loosen or tighten the rubber bands until their pitches are low and very close. Put your ear near the box, and pluck both rubber bands at the same time. Listen carefully for slight variations in the loudness of the combined sound.
4. Repeat several times. Adjust the rubber bands until you hear beats. Record your observations in your Science Journal.

Analyze and Conclude

1. Describe how you produced the beats in the sound.
2. **Key Concept** How do you think musicians playing together can avoid beats?

Beats

Have you ever been to a concert where the musicians start by all playing the same note? Why do they do that? If the pitches of the notes are slightly different, the sounds will interfere. The audience might hear the notes get louder and softer several times a second. The repeating increases and decreases in amplitude are beats. Look back at **Figure 14.** The difference in frequencies determines how often beats occur. If one musician plays a note with a pitch of 392 Hz and another plays a note with a pitch of 395 Hz, the difference is 3 Hz. Beats will occur 3 times each second. Musicians can avoid beats by playing notes at the correct pitch on their instruments.

Fundamental and Overtones

When a musician plucks a guitar string, the string vibrates with a certain sound. If the musician plucks it again in the exact same way, the sound will be the same. All objects tend to vibrate with a certain frequency that depends on the object's properties.

The lowest frequency at which a material naturally vibrates is called its fundamental. Higher frequencies at which the material vibrates are called overtones. Objects vibrate with both a fundamental and overtones, as shown in **Figure 15.** The interference of these waves produces the sound you hear.

Reading Check What is the difference between a fundamental and overtones?

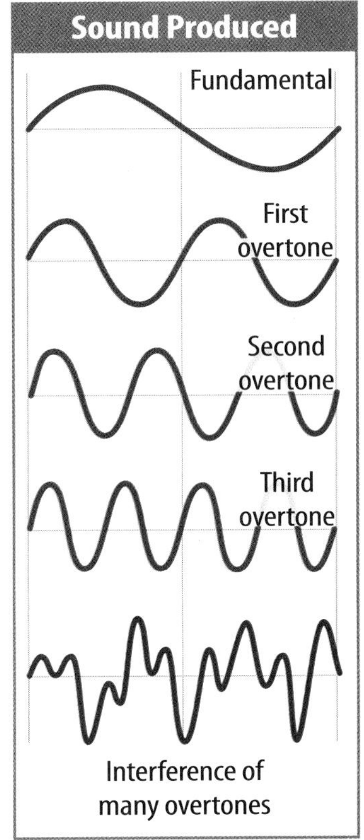

Figure 15 A fundamental and overtones combine to produce an object's sound.

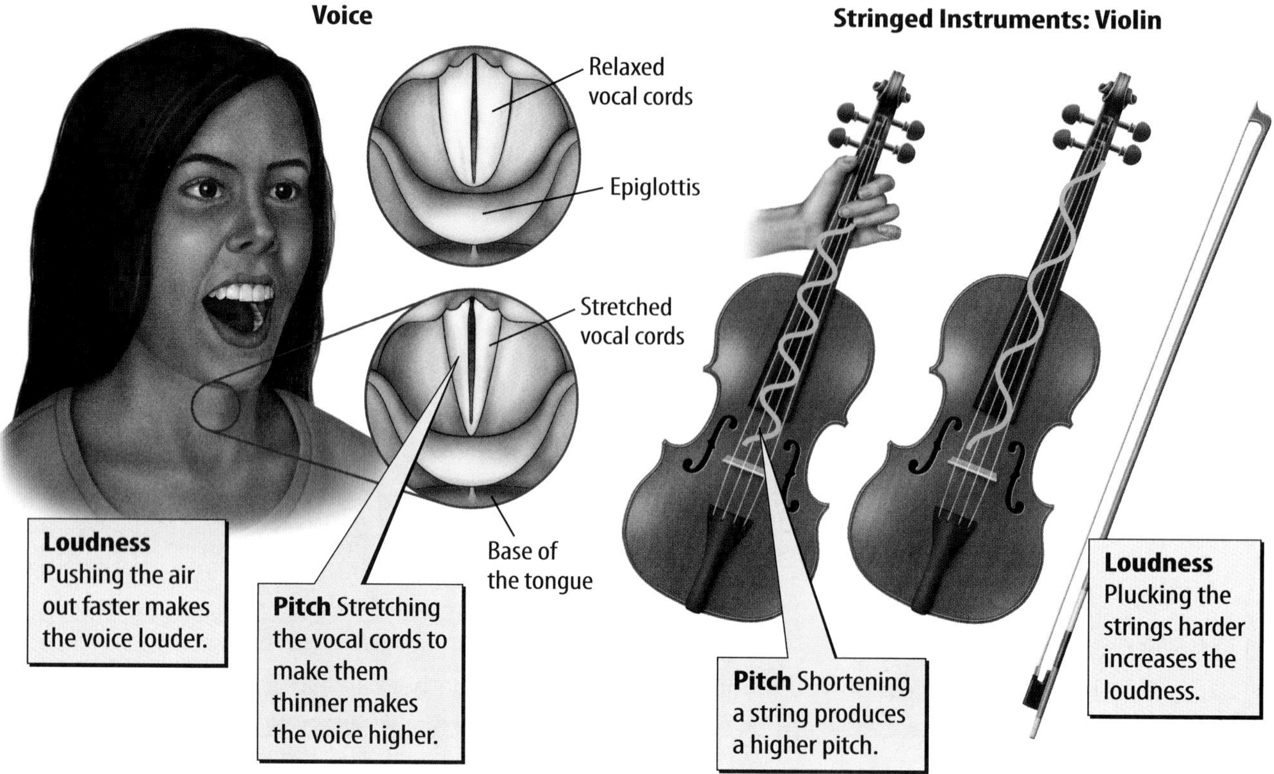

Voice
Relaxed vocal cords
Epiglottis
Stretched vocal cords
Base of the tongue
Loudness Pushing the air out faster makes the voice louder.
Pitch Stretching the vocal cords to make them thinner makes the voice higher.
Stringed Instruments: Violin
Pitch Shortening a string produces a higher pitch.
Loudness Plucking the strings harder increases the loudness.

Figure 16 Musicians change the pitch and loudness of sound in different ways.

FOLDABLES®

Make a vertical two-tab book and label it as shown. Use it to explain what you have learned about the relationships between sound properties and energy.

WORD ORIGIN

resonance
from Latin *resonare*, means "sound again"

Music

Have you ever been listening to your favorite music and someone tells you "Turn off that noise!" How are music and noise different? Unlike noise, music is sound with a pleasing pattern.

Sound Quality

The unique sound of a musical instrument is a mix of its fundamental and overtones. These waves interact to form a distinct sound, or timbre. Suppose, for example, a clarinet player and a piano player both play the same note. The fundamental of both instruments is the same. The number and intensity of the overtones, however, differs. Overtones produce the complex sound waves that let you distinguish the unique sound quality of each instrument.

 **Key Concept Check** How can you recognize sounds from different sources?

Resonance

If you hold a guitar string by its ends and have a friend pluck the string, the sound is almost too low to hear. Instruments use resonance to amplify sound. **Resonance** *is an increase in amplitude that occurs when an object vibrating at its natural frequency absorbs energy from a nearby object vibrating at the same frequency.* The vibrating string causes the back of the guitar and the air inside to vibrate by resonance. The sound is then much louder.

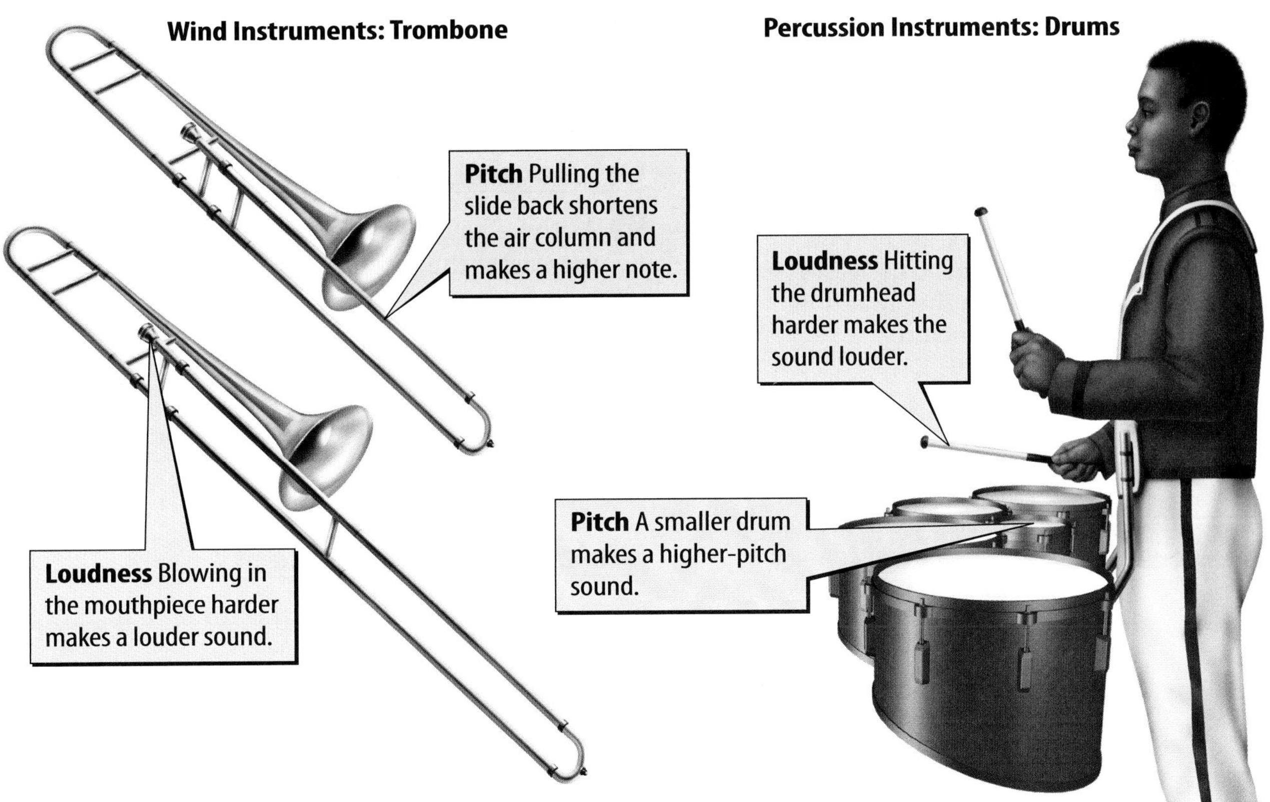

Wind Instruments: Trombone
Pitch Pulling the slide back shortens the air column and makes a higher note.
Loudness Blowing in the mouthpiece harder makes a louder sound.
Percussion Instruments: Drums
Loudness Hitting the drumhead harder makes the sound louder.
Pitch A smaller drum makes a higher-pitch sound.

Types of Musical Instruments

As shown in **Figure 16**, different types of instruments control the pitch and the loudness of sound in different ways.

String Instruments Musical instruments that have strings, such as a guitar, a violin, a harp, and a piano, produce sound when the string vibrates. A player plucks the strings of a guitar or a harp. The motion of a bow vibrates the strings of a violin. Pressing a piano key causes a felt-covered hammer to strike a particular string inside the piano. When the string inside the piano vibrates, you hear a tone. The pitch a string makes depends on its length and its thickness.

Pitch also depends on how tightly the string is stretched and the material it is made from. You can hear sounds a stringed instrument makes because of resonance between the string and the instrument's hollow body. Plucking or pressing the strings harder increases the volume.

Wind Instruments The vibrating medium in wind instruments, such as a saxophone or a trumpet, is air. Either your lips or a thin piece of wood, called a reed, vibrates. An air column then vibrates by resonance. The length of the air column determines pitch. Blowing harder increases the volume.

Percussion Instruments You make sound with a percussion instrument by striking it. Examples include a drum, cymbals, and a bell. The pitch depends on the instrument's size, its thickness, and the material from which it is made. Resonance can make the sound of a percussion instrument louder.

Voice The source of sound in your voice is vocal cords. Muscles in your throat allow you to increase the pitch of your voice by pulling the cords tighter. Other parts of your mouth and throat also affect the sounds your voice makes. You can make your voice louder by pushing out the air with a greater force.

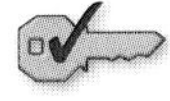
Key Concept Check In what ways are musical sounds produced?

Lesson 2 Review

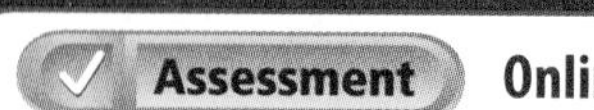

Visual Summary

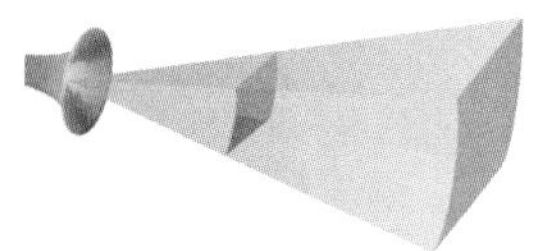

The intensity of sound decreases farther away from a source.

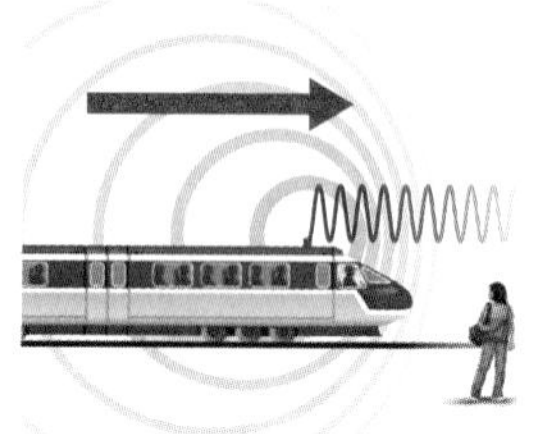

The pitch of sound increases if the distance between a sound source and a listener decreases.

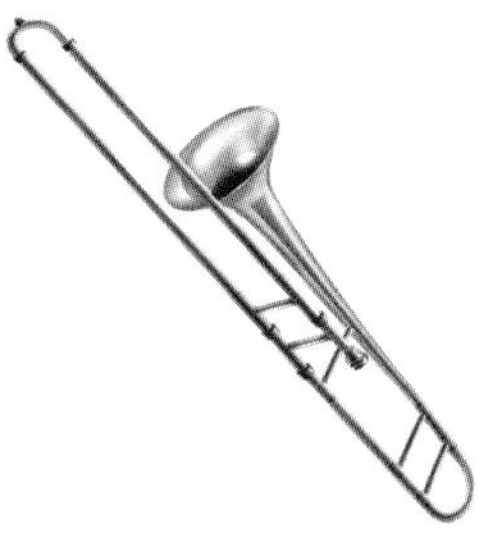

An instrument's sound is a mix of an its fundamental and overtones.

Use your lesson Foldable to review the lesson. Save your Foldable for the project at the end of the chapter.

What do you think NOW?

You first read the statements below at the beginning of the chapter.

3. Sound waves from different sources can affect one another when they meet.

4. You can tell whether an ambulance is moving toward or away from you by listening to changes in the sound of the siren.

Did you change your mind about whether you agree or disagree with the statements? Rewrite any false statements to make them true.

Use Vocabulary

1. As the ________ of a sound wave increases, the pitch of the sound gets lower.
2. **Describe** the Doppler effect in your own words.
3. **Use** the term *resonance* in a complete sentence.

Understand Key Concepts 

4. Which increases if the amplitude of a sound wave increases?
   - **A.** intensity
   - **B.** interference
   - **C.** pitch
   - **D.** sound quality
5. **Identify** the vibrating media in three different types of musical instruments.
6. **Explain** how you can recognize the sound of a flute.
7. **Analyze** If the frequency of a sound wave increases, what happens to the wavelength?

Interpret Graphics

8. **Identify** Copy and fill in a graphic organizer like the one below. Use it to identify and briefly explain four properties that distinguish sounds.

Critical Thinking

9. **Decide** If you double the amplitude of a string vibrating at 10 Hz, will it sound louder? Why or why not?
10. **Construct** a drawing of two waves that represent a violin and a tuba each playing a note with the same frequency and loudness.

Manipulate Variables

40 minutes

How can you use a wind instrument to play music?

Materials

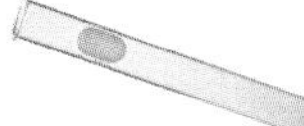

6 test tubes

test-tube rack

beaker

ruler

straw

Safety

Wind instruments use a vibrating column of air to create a sound. By changing the length of the column of air vibrating inside the instrument, you change the frequency of the vibrating air.

Learn It

When you **manipulate variables,** you change only one factor, called the independent variable. A factor that changes as a result of the independent variable is a dependent variable. By changing only one variable, you know what causes any effects you observe.

Try It

1. Read and complete a lab safety form.
2. Place 6 test tubes in a rack. Add just enough water to cover the bottom of the first tube. Then, measure the length of the air column inside the test tube. Record your measurement in your Science Journal.
3. Blow across the top of the tube with a straw. Record your observations about the sound.
4. Add a greater amount of water to another test tube. Observe the difference in the sound as you blow across the tube. Measure the length of the air column, and again record your observations.
5. Think about how the pitch of the sound relates to the length of the air column. Choose a simple song to play by blowing on the test tubes.
6. Add different amounts of water to the test tubes to make different pitches. The pitches should correspond to the notes of your song.
7. Measure and record the air column length for each pitch. Write the pattern of lengths needed for your song.

Apply It

8. Identify the independent and the dependent variables.
9. Describe the relationship between pitch and the length of the air column.
10. **Key Concept** Based on what you have learned from this lab, explain how musicians use wind instruments to play music.

Lesson 3

Reading Guide

Key Concepts
ESSENTIAL QUESTIONS

- In what ways does sound interact with matter?
- How can people control sound?
- What are some ways to use ultrasound?

Vocabulary
absorption p. 584
reflection p. 584
echo p. 584
reverberation p. 585
acoustics p. 585
echolocation p. 587
sonar p. 587

g Multilingual eGlossary

Using Sound Waves

Inquiry Where is it?

These workers are preparing to lower a special device into the water that uses sound to search for underwater objects. Sometimes they locate sunken ships, such as the oil tanker in the large image. What properties of sound waves are useful for finding things under water?

Launch Lab

15 minutes

Why didn't you hear the phone ringing?

A cell phone ringing on a countertop can be heard from far away, but the same phone in your jacket pocket is not so easy to hear. What explains this difference?

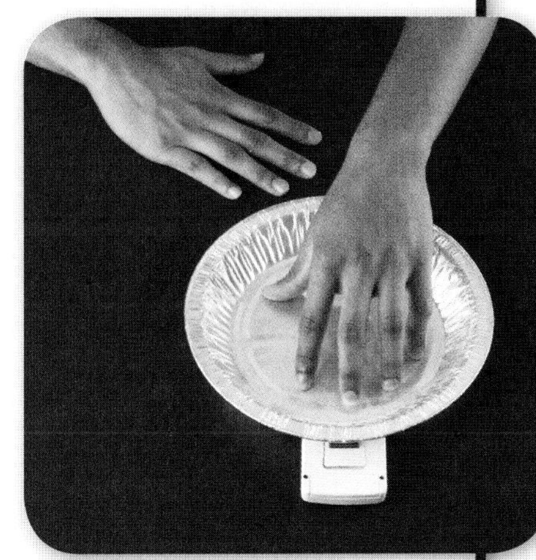

1. Read and complete a lab safety form.
2. Set a **kitchen timer** for 1 s. Hold it about 15 cm from your ear, and listen to the sound.
3. Locate the exit hole for sound waves on the back of the timer. Set the timer to a 5-s delay. Hold a **foil pie pan** flat against the exit hole. Move your ear about 15 cm from the timer and pie pan. Observe the difference in the sound. Record your observations in your Science Journal.
4. Set the timer with a 5-s delay. Place the timer in a **shoe box,** and cover it with several crumpled **paper towels.** Listen with your ear 15 cm from the box. Record your observations.

Think About This

1. Compare the movement of sound through the foil pan and through air.
2. What do you think changed the sound when you covered the timer with paper towels?
3. **Key Concept** Based on your results, why do you think a cell phone is harder to hear in a jacket pocket than on a countertop?

Sound Waves and Matter

Why do the cheers of a crowd in an indoor gymnasium sound so different from a crowd yelling at an outdoor football game? Sound waves at the football game spread out with few barriers. What happens to sound waves when they strike a different medium, such as the walls of a building?

Transmission

Have you ever heard someone talking in the next room? This is possible because of transmission—the movement of sound waves through a medium. When sound waves move from air into a wall, the vibrations of air particles cause particles in the solid wall to vibrate. Even though solids transmit sound waves better than gases, most sound waves do not move easily from gases into a solid. However, loud sounds, which have a lot of energy, will move into and through the solid wall. Waves of quieter sounds, with less energy, may be partially or completely blocked. These waves don't carry enough energy to cause much vibration in the wall. As the vibrations reach the next room, they transfer the remaining energy to the air particles, and you hear the sound. The amplitude of the sound is lower because the wall could not transmit all of the energy.

FOLDABLES®

Make a vertical five-tab book. Label it as shown. Use it to organize your notes about the different ways that sound interacts with matter.

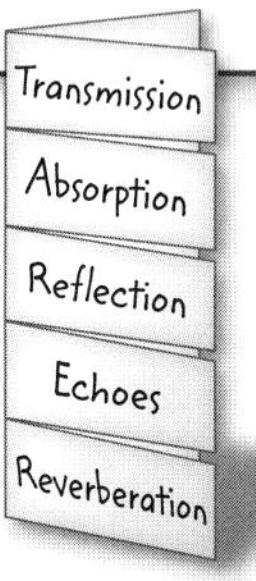

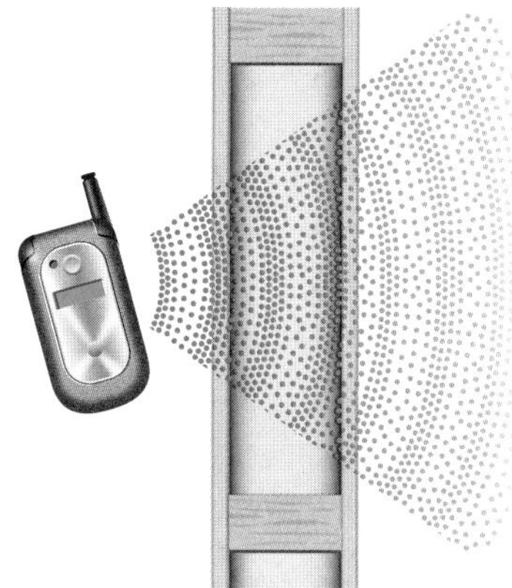

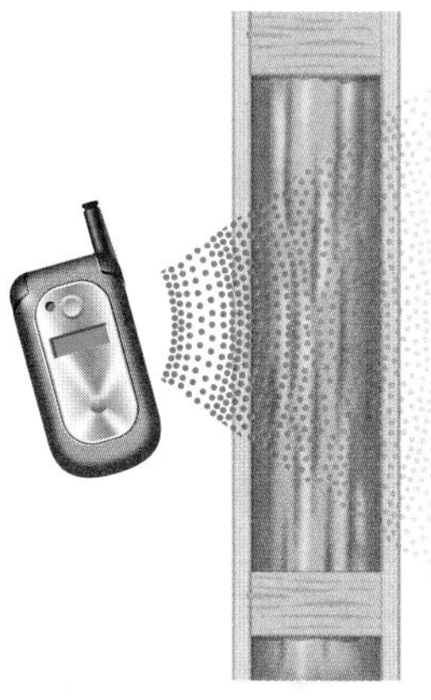

▲ **Figure 17** Insulation absorbs much of the energy in a sound wave and converts it to a small amount of heat.

Absorption

If you throw a tennis ball at a pillow, it will not bounce back. Most of the ball's energy goes into the pillow. Some materials, such as the wall insulation in **Figure 17,** act like the pillow when sound waves strike them. *The transfer of energy by a wave to the medium through which it travels is called* **absorption.** How well a material absorbs the energy of a sound wave depends on various factors, such as its inner structure and the amount of air in it. Rather than passing from one particle to another, some of the sound energy changes to heat due to friction.

Reflection

What happens if you throw a tennis ball at a hard surface? The ball probably bounces back at you. Similarly, a sound wave might bounce back when it strikes a different medium. *The bouncing of a wave off a surface is called* **reflection.** The way in which sound waves reflect is shown in **Figure 18.** The angle at which a sound wave strikes a surface is always equal to the angle at which the sound wave is reflected off of the surface.

Key Concept Check What are some ways in which sound waves interact with matter?

Echoes

Have you ever yelled a name in a gym and heard the same voice yell back? That was you, of course! As you yelled, you heard the original sound of your own voice. Then you heard the sound again after the sound waves reflected off the walls of the gymnasium and traveled back to your ears. *A reflected sound wave is an* **echo.**

WORD ORIGIN

echo
from Greek *ekhe,* means "sound"

Sound waves travel at about 343 m/s in air, or 34.3 m in 0.1 s And, the brain holds onto a sound for about 0.1 s. When the reflecting surface is far enough away that the sound waves take more than 0.1 s to return, the listener hears the original sound followed by the reflected sound. So, if you clap your hands at one end of a long room, you hear an echo only if the sound wave returns more than 0.1 s later.

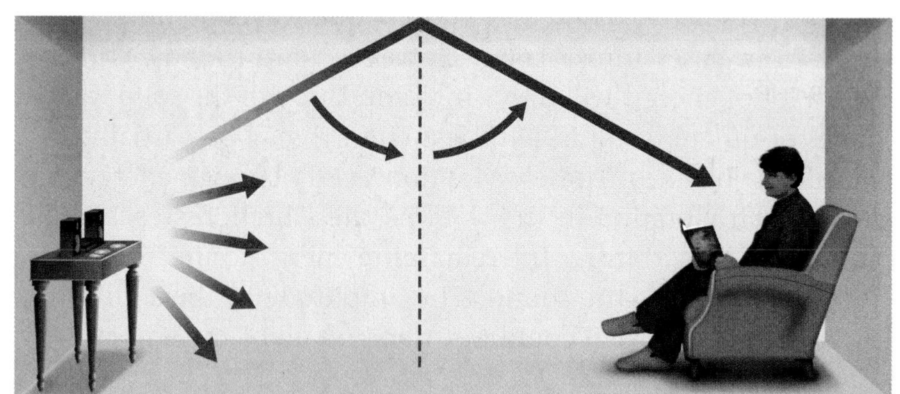

Figure 18 Sound waves reflect from a surface at the same angle at which they strike the surface.

Reverberation

In many closed spaces, sound waves reflect from surfaces that are different distances from the listener. Because some waves travel farther than other waves, reflected waves reach the listener at different times. *The collection of reflected sounds from the surfaces in a closed space is called* **reverberation.**

Sound waves that reach the listener directly are heard sooner than reflected sound waves. If each reflected wave reaches the ear before the previous sound fades, the original sound seems to last longer. However, too much reverberation can make words hard to understand because the echoes of old sounds can interfere with new ones.

Acoustics

In a room with no furniture, rugs, or drapes, the sound waves of footsteps and speech bounce around the room and sound loud. Soft or fuzzy materials, such as the rug, curtains, and padded furniture on the right of **Figure 19,** absorb much of the energy of sound waves. Footsteps are almost silent. Voices are softer. *The study of how sound interacts with structures is called* **acoustics.** Acoustical engineers use their knowledge of sound transmission, absorption, and reflection to control sounds.

MiniLab **25 minutes**

How fast is sound?

The speed of sound in air is about 343 m/s. You can use echoes to measure this speed.

1. Read and complete a lab safety form.
2. Use a **meterstick** to measure a spot 30 m from a wall. Standing at this spot, clap your hands once and listen for the echo.
3. Clap in time with the echo. If you hear the echo after each clap, clap slightly faster.
4. When your clapping matches the echo speed, have your partner use a **stopwatch** to measure the time as you clap 25 times. Record the time in your Science Journal.
5. Repeat steps 2–4 at distances of 40 m and 50 m from the wall.

Analyze and Conclude

1. **Calculate** the speed of sound for each measurement: *speed* = *distance* ÷ *time.*
2. **Key Concept** Describe how echoes helped you measure the speed of sound.

Figure 19 Soft materials reduce the reverberation in a room.

Visual Check Why do you hear reverberation only in a closed space?

▲ **Figure 20** Noise-canceling earphones protect the worker from noises that otherwise would damage his hearing.

Noise Pollution and Control

One way acoustical engineers control sound is by developing methods to protect people from noise pollution. Think about the noises people might hear during a typical day. Trucks rumble by, cars honk their horns, and claps of thunder boom during a storm. At home, you might use noisy appliances such as a hair dryer, a vacuum cleaner, or a dishwasher. Severe noise pollution can result in hearing loss, stress, and other types of health problems. Laws limit the noise that might be produced by machinery or landing aircraft. The government requires ear protection for workers in many jobs.

Noise-canceling earphones, such as those worn by the person in **Figure 20**, work in several ways. They cover the ears and can block incoming sound waves. Other types of earphones reduce the sound by analyzing incoming sound waves and then producing waves that create destructive interference.

Designing Spaces

Acoustical engineers develop ways to control sound in buildings. When you enter a concert hall such as the one in **Figure 21**, you might think that the unusual appearance is just for looks. However, engineers have carefully chosen shapes and materials to control sound waves. The stage has a wooden floor to improve vibrations. The curved panels on the ceiling reflect sound waves in different directions to fill the space. However, the recording studio has foam panels on the walls. These soft materials absorb sound waves to **prevent** reverberation as the musician performs.

ACADEMIC VOCABULARY
prevent
(verb) to keep from happening

 **Key Concept Check** What are some ways in which people can control sound?

Figure 21 Engineers design concert halls and recording studios to control sounds. ▼

Visual Check What methods have designers used to control sound in each of these spaces?

1 A dolphin sends out a series of high-pitched clicks.
2 When a sound wave strikes an object or another dolphin, some of the sound reflects back.

Figure 22 Organs in the dolphin's head send out ultrasonic waves and analyze the echoes.

Ultrasound

You have read that the waves of ultrasound have a higher frequency than humans can hear. Both animals and humans use these high-frequency sound waves.

Echolocation

Recall that many animals, such as bats and whales, can hear sounds above the range of human hearing. Animals use some of these ultrasound frequencies to communicate. They use other frequencies to locate and identify objects. *The process an animal uses to locate an object by means of reflected sounds is* **echolocation** (e koh loh KAY shun). The dolphins in **Figure 22** might use an echo to determine an object's distance, shape, and speed or to find other dolphins.

Sonar

If you go fishing often, you probably wish you could easily discover exactly where the fish are located. A device called a fish-finder does just that by using a technology similar to echolocation. A fish-finder is an example of **sonar**, *a system that uses the reflection of sound waves to find underwater objects.* The word *sonar* is an acronym for Sound Navigation and Ranging. Ships use sonar to send a high frequency sound wave into the water. As the sound wave moves deeper, it spreads out, forming a cone, or beam. When the sound wave strikes something within this beam, it bounces back to the ship. Sonar contrasts signals that strike the ocean floor with signals from other objects. It measures the amount of time between when the sound wave leaves and when it bounces back. The sonar system then calculates the distance and draws an image on a screen. The picture of the sunken oil tanker on the first page of this lesson shows what a sonar image might look like.

 **Reading Check** How can people "see" under water using sonar?

Math Skills

Use a Formula

You can use a formula to calculate sonar distances. For example, sound travels at **1,530 m/s** in seawater. A ship sends out a signal to find the water's depth. The signal returns in **5.0 s**. How deep is the water?

1. Use the speed = distance/time formula to calculate distance.

   distance = *speed* × *time*

2. The signal traveled to the ocean's bottom and back to the ship. Divide by 2 to get the time it took to reach the bottom.

   $5.0\ s/2 = 2.5\ s$

3. Replace the terms with the values and multiply.

   $d = (1{,}530\ m/s)(2.5\ s)$
   $= 3{,}825\ m$

Practice

A sonar signal traveling at 1,490 m/s returns to the ship in 4.0 s. What is the distance to the target?

 Review

- Math Practice
- Personal Tutor

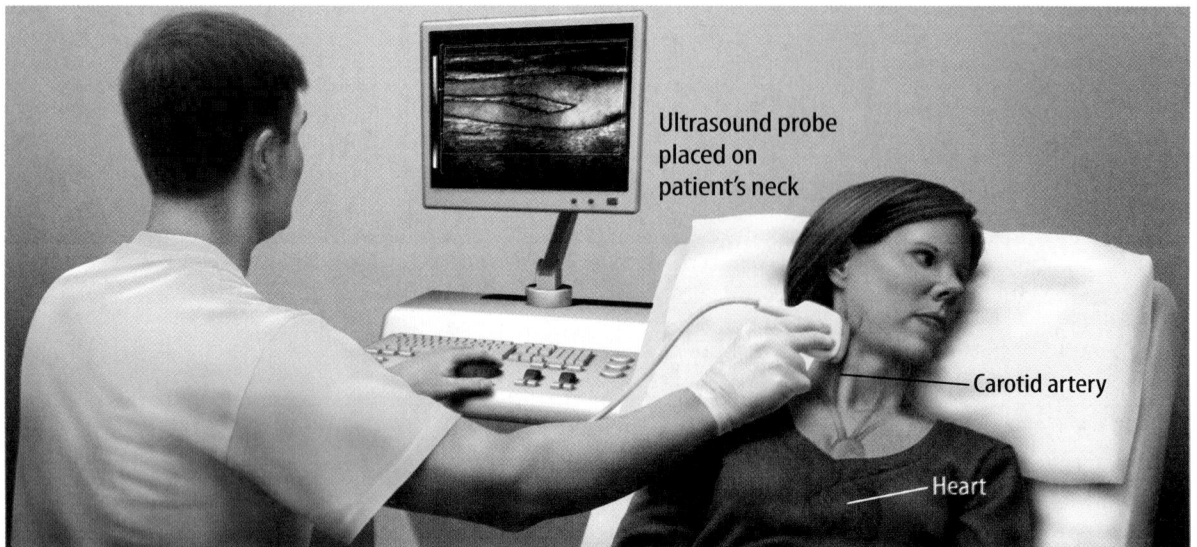

▲ **Figure 23** This ultrasound scanner produces an image of the patient's artery.

Medical Uses of Ultrasound

Suppose a doctor could see your heart beating or watch blood flowing through your arteries. He or she could determine whether something was wrong without performing surgery. Believe it or not, as **Figure 23** shows, these things are possible using ultrasound.

REVIEW VOCABULARY
artery
a vessel that takes blood away from the heart

Ultrasound Imaging Ultrasound scanners work much like sonar. A doctor moves the scanner over different parts of the body. The scanner emits safe, high-frequency sound waves. Body structures such as muscle, fat, blood, and bone reflect sound at different rates. Based on the reflected waves, the scanner produces an image called a sonogram. It is even possible to analyze motion with ultrasound. The scanner in **Figure 23** uses the Doppler effect. By determining how much the frequency of the reflected wave changes, the scanner can determine the blood's speed and direction. Doctors often use sonograms to check the health of unborn babies. Ultrasound is safer than X-rays because it doesn't damage cells.

Figure 24 Ultrasound can break large kidney stones into tiny pieces. ▼

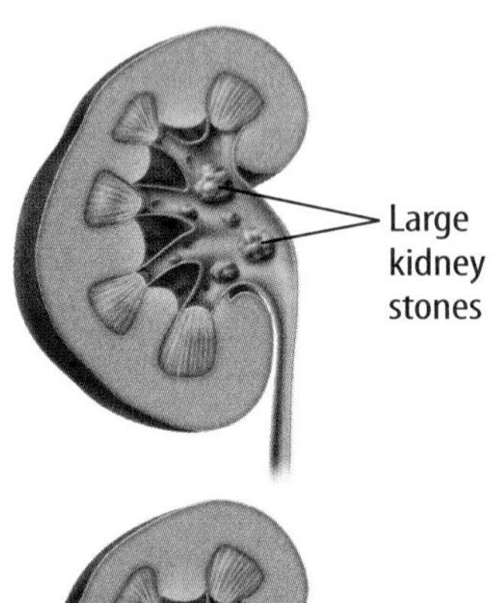

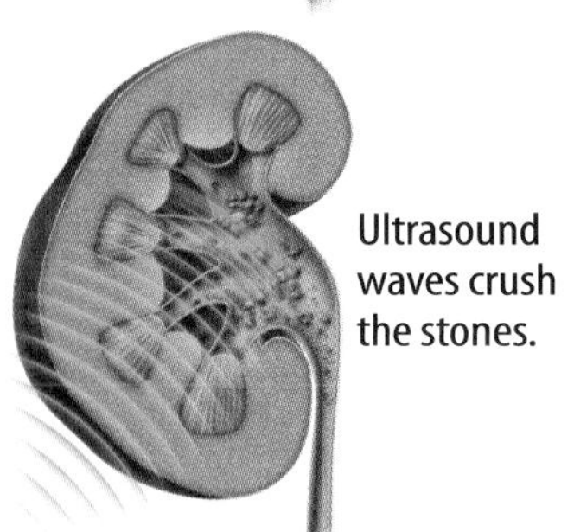

Treating Medical Problems Has anyone ever massaged your neck or back when it was sore? It not only feels good, but it also relaxes stiff muscles. Many physical therapists use ultrasound to treat joint and muscle sprains or to ease muscle spasms. The vibrations travel through the skin and soft tissue and act like hundreds of tiny fingers massaging the area. Short pulses of high-frequency sound waves can even break apart kidney stones, as shown in **Figure 24.**

Key Concept Check What are some ways in which people use ultrasound?

Lesson 3 Review

Visual Summary

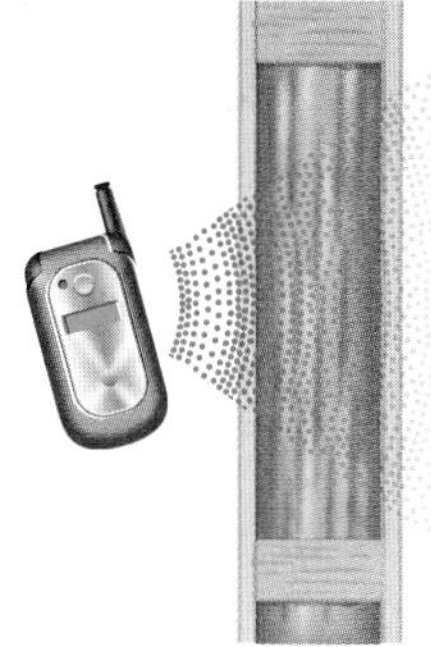

A medium might transmit, absorb, or reflect sound waves that strike it.

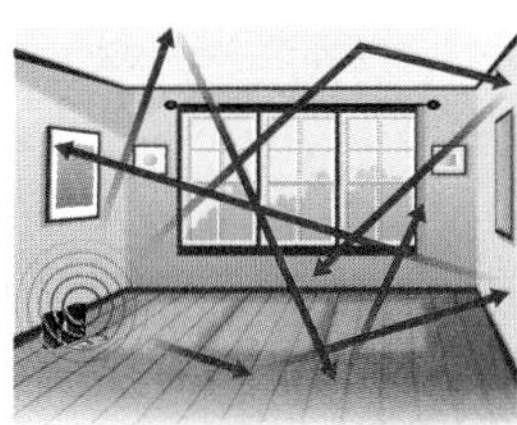

Sound waves reflect off hard surfaces in an enclosed space.

Acoustical engineers develop ways to control sound.

FOLDABLES®

Use your lesson Foldable to review the lesson. Save your Foldable for the project at the end of the chapter.

What do you think NOW?

You first read the statements below at the beginning of the chapter.

5. Animals use sounds that humans cannot hear to locate objects around them.

6. The method that ships use to locate underwater objects can be used to locate organs or tumors inside the human body.

Did you change your mind about whether you agree or disagree with the statements? Rewrite any false statements to make them true.

Use Vocabulary

1 **Describe** reverberation in your own words.

Understand Key Concepts 

2 Which is the use of sound to identify the distances and positions of objects?

A. echolocation
B. reflection
C. reverberation
D. timbre

3 **Suggest** a way to prevent sounds from echoing in a large gymnasium.

4 **Describe** several ways to limit sound transmission through walls.

Interpret Graphics

5 **Identify** conditions that would enable you to hear an echo from the reflected sound waves in the diagram below.

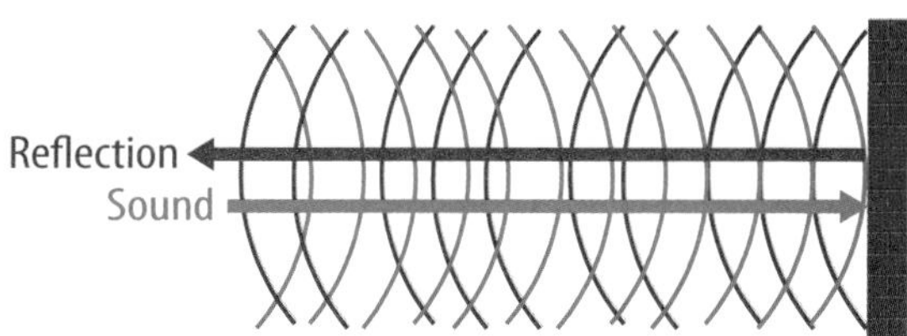

6 **Summarize** Copy and fill in a graphic organizer like the one below. Use it to identify and briefly explain three ways sound can interact with matter.

Critical Thinking

7 **Assess** the value of echolocation to animals such as bats and dolphins.

Math Skills

Math Practice

8 A ship's sonar signal travels at 1,500 m/s and returns 0.40 s after it is sent. How deep is the shipwreck the ship locates?

Inquiry Lab

90 minutes

Make Your Own Musical Instrument

Materials

ruler

rubber bands

scissors

balloons

drinking straws

string

Also needed: tape, pencil, bottles, boxes, plastic wrap, aluminum pans, paper, metal cans, tubing, cardboard tubes

Safety

Suppose you are a professional musician and want to design a new instrument. What type of instrument will you make? Your instrument must be able to produce a range of pitches. It also must be able to play loudly or quietly. The instrument must be durable enough to play over and over. In this lab, you will make a model of your instrument. Then you will modify it, one part at a time, until you are satisfied with the result.

Question

What factors should you consider when making a musical instrument?

Procedure

1. Read and complete a lab safety form.
2. Design a musical instrument made from common materials. Your instrument must be able to play a range of frequencies and a range of amplitudes. Record your design in your Science Journal.
3. Have your teacher approve your design before you build your instrument.
4. Build your instrument, and then play it.
5. Think about how you could improve the design of your instrument. Record your observations and ideas in your Science Journal.

6 Make a modification to your instrument. Remember that in any investigation you should change only one variable at a time—the independent variable. In your Science Journal, record what you changed and why you changed it.

7 Play the instrument. Did your modification produce the change in sound you intended? Why or why not? Record your observations.

8 Continue to improve your instrument as time permits. You may change the same variable or a different one, but be sure to change only one variable at a time. Record your observations and details about all changes.

Analyze and Conclude

9 **Classify** your instrument as a string instrument, a wind instrument, or a percussion instrument. Identify the properties you used to classify the instrument.

10 **Draw** a sketch of your instrument. Label the parts of the instrument that vibrate to produce sounds.

11 **Compare** your instrument to a similar common musical instrument.

12 **Critique** your design, your changes, and the results.

13 **The Big Idea** Describe how your instrument produces sounds.

Communicate Your Results

Present your musical instrument to the class, and explain your design. Demonstrate your instrument by playing a simple song.

Extension

Common musical instruments are both durable and able to be tuned. Explore what it means to tune an instrument. In what ways can you make your instrument playable by a professional musician.

Lab Tips

- ☑ Think about how each type of instrument changes the frequency and amplitude of sound.
- ☑ When you decide to change something about your instrument, consider whether the change might affect frequency, amplitude, and durability.

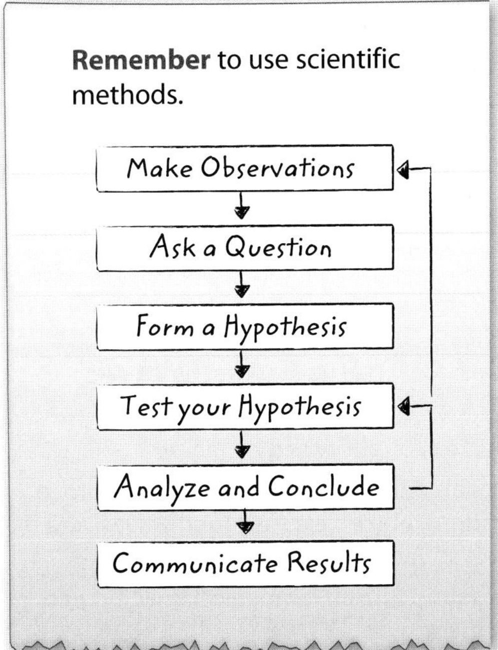

Chapter 16 Study Guide

Vibrations in matter produce sound waves. Each sound source creates a unique wave form that the ear can identify. People use the way sound waves are absorbed, transmitted, and reflected by matter in many ways.

Key Concepts Summary

Lesson 1: Producing and Detecting Sound

- **Vibrations** in a medium produce **sound waves.**
- Sound waves travel as **compressions** and **rarefactions.**
- Sound waves travel faster through materials in which the particles are closer together.
- The outer ear collects sound. The middle ear amplifies sound. The inner ear converts vibrations to nerve signals.

Vocabulary

sound wave p. 565
longitudinal wave p. 565
vibration p. 565
medium p. 566
compression p. 566
rarefaction p. 566

Lesson 2: Properties of Sound Waves

- The greater a sound wave's energy, the larger the **amplitude** and the greater the wave's **intensity.**
- Sound waves with a longer **wavelength** have a lower **frequency** and a lower **pitch.**
- Different frequencies of sound waves combine and form a complex wave the brain recognizes.
- Strings, air columns, or surfaces of instruments vibrate and produce music.

Vocabulary

amplitude p. 573
intensity p. 574
wavelength p. 575
frequency p. 575
pitch p. 575
Doppler effect p. 576
interference p. 576
resonance p. 578

Lesson 3: Using Sound Waves

- Sound waves can be transmitted, **reflected,** or **absorbed** by matter.
- Materials and shapes in a room can improve vibrations, absorb excess sound waves, and reflect sound waves to fill the room.
- Ultrasound is used for medical imaging and treatment.

Vocabulary

absorption p. 584
reflection p. 584
echo p. 584
reverberation p. 585
acoustics p. 585
echolocation p. 587
sonar p. 587

Study Guide

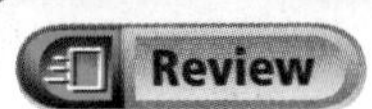

- Personal Tutor
- Vocabulary eGames
- Vocabulary eFlashcards

FOLDABLES® Chapter Project

Assemble your lesson Foldables as shown to make a Chapter Project. Use the project to review what you have learned in this chapter.

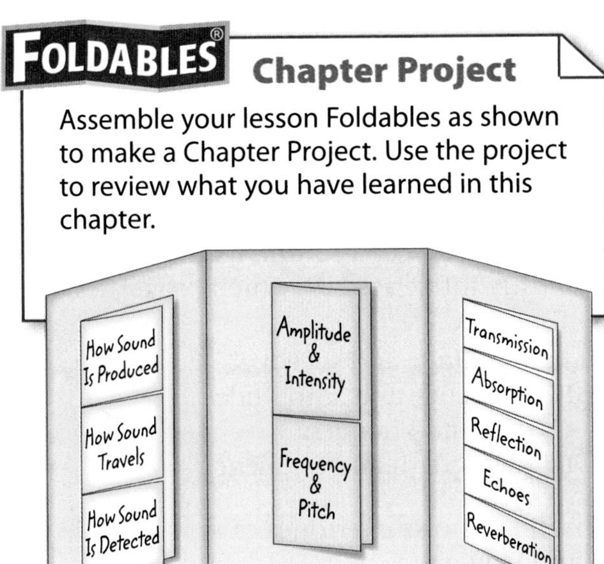

Use Vocabulary

1. Describe a longitudinal wave.
2. The part of a sound wave in which the particles are most spread out is called a(n) ________.
3. Explain the relationship between frequency and pitch.
4. A change in frequency called the ________ depends on the motion of the sound source and the position of the listener.
5. Describe how resonance works.
6. The collection of sound reflections in a closed space is called ________.
7. An example of ________ is when a dolphin emits loud clicks and listens for the echo.

Link Vocabulary and Key Concepts

Interactive Concept Map

Copy this concept map, and then use vocabulary terms from the previous page to complete the concept map.

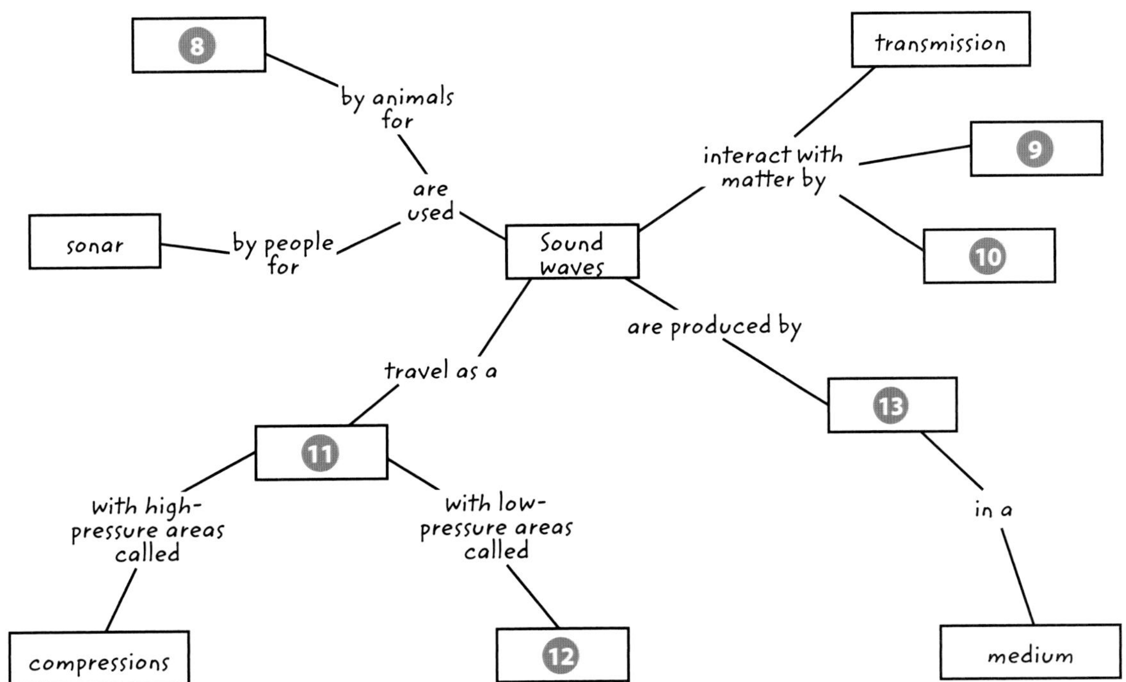

Chapter 16 Review

Understand Key Concepts 

1. Which type of matter would transmit sound waves fastest?
   A. air at 5°C
   B. air at 20°C
   C. ice at 0°C
   D. seawater at 20°C

2. A large windowpane on a storefront vibrates as a large truck drives by. Which type of interaction between sound and matter best explains this vibration?
   A. absorption
   B. resonance
   C. reverberation
   D. transmission

3. As ________ decreases, sound intensity decreases.
   A. amplitude
   B. quality
   C. wave speed
   D. wavelength

4. Wave frequency is measured in which unit?
   A. decibel
   B. hertz
   C. meter
   D. second

5. What causes the sound waves's intensity to decrease in the picture below?

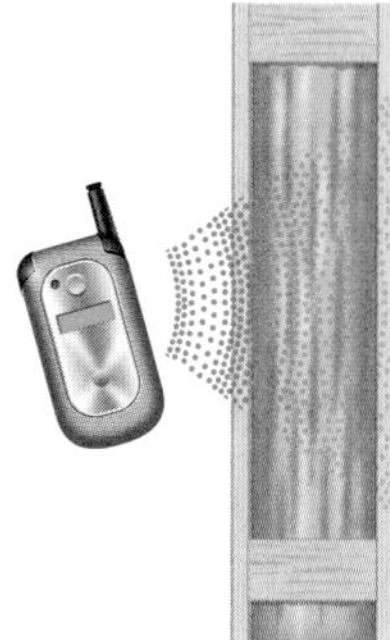

   A. insulation
   B. interference
   C. rarefaction
   D. resonance

6. Which property describes the distance between two identical points on a sound wave?
   A. amplitude
   B. frequency
   C. pitch
   D. wavelength

7. When you hear a sound, its pitch is mostly influenced by which properties of a sound wave?
   A. amplitude and speed
   B. frequency and amplitude
   C. speed and frequency
   D. wavelength and frequency

8. Which shows overtones of a fundamental?

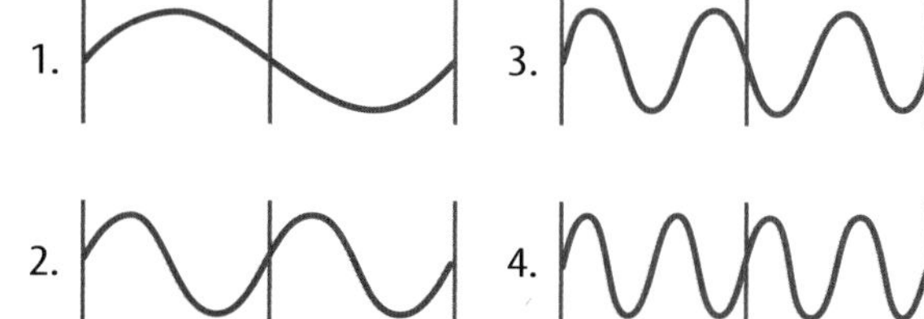

   A. only 1
   B. 1 and 2
   C. 2, 3, and 4
   D. 1, 2, 3, and 4

9. Which property of sound waves changes when you increase the volume on a car radio?
   A. amplitude
   B. frequency
   C. speed
   D. wavelength

10. Which is the region of a sound wave where particles of a medium are most spread out?
    A. compression
    B. rarefaction
    C. reverberation
    D. transmission

Assessment
Online Test Practice

Critical Thinking

11 **Visualize** A bat in a dark cave sends out a high-frequency sound wave. The bat detects an increase in frequency after the sound bounces off its prey. Describe the possible motion of the bat and its prey.

12 **Synthesize** An MP3 player at maximum volume produces sound at 110 dB. The table below shows the recommended time exposure before risk of damage to the ear. How many hours a day could you listen to your MP3 player at full volume without risking hearing loss? Explain.

Recommended Noise Exposure Limits

Sound Level (dB)	Time Permitted (hr)
90	8
95	4
100	2
105	1

13 **Construct** a diagram of two sound waves of equal amplitude and frequency that experience destructive interference.

14 **Create** a diagram that includes drawings of sound waves to show how the ear distinguishes among sounds, such as different voices or musical instruments.

15 **Compare** the effects on the ear of a single very loud sound of 150 dB with prolonged exposure to sounds of 90 dB.

Writing in Science

16 **Write** A friend tells you that his family always complains about the noise when he practices his electric guitar in his room. Write at least four recommendations for how your friend might reduce the sound that escapes from his room.

REVIEW THE BIG IDEA

17 Explain how sound is produced, travels from one place to another, and is detected by the ear.

18 Look at all the equipment in the music control room shown in the photograph. Think about what you have learned about sound in the chapter. How can musicians produce, describe, and use sound?

Math Skills

Review
Math Practice

Use a Formula

19 A ship sends out a sonar signal to locate other ships. Traveling at 1,530 m/s, a signal returns to the ship 3.6 s after it is sent. What is the distance from the other ship?

20 A sonar signal takes 3 s to return from a sunken ship directly below. Find the depth of the sunken ship if the speed of the signal is 1,440 m/s.

21 A ship is sailing in water that is 500 m deep. Sound travels at 1,520 m/s in the body of water. A sonar echo returns to the ship in 0.6 s. Did the signal bounce off the bottom? Explain.

Standardized Test Practice

Record your answers on the answer sheet provided by your teacher or on a sheet of paper.

Multiple Choice

Use the figure below to answer questions 1–3.

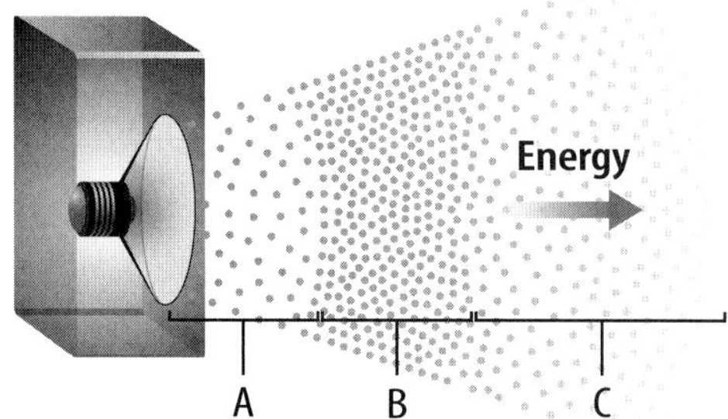

1 How does this speaker cause sound that can be heard?

A It produces echoes.
B It produces vibrations.
C It releases light.
D It releases heat.

2 What type of wave does the speaker produce for sound to be heard?

A electromagnetic
B longitudinal
C surface
D transverse

3 Point A in the figure is in a rarefaction. Point B is in a compression. Point C is a location in the air at normal air pressure before the sound wave has arrived. Which statement is true of the pressure at points A, B, and C?

A The pressure at point A is the greatest.
B The pressure at point B is the greatest.
C The pressure at point C is the greatest.
D The pressure at all three points is equal.

4 Which two factors can affect the speed of sound?

A size and density of the medium
B mass and density of the medium
C temperature and mass of the medium
D temperature and density of the medium

Use the table below to answer question 5.

Speed of Sound in Different Materials	
Material (at 20°C)	**Speed (m/s)**
Air	343
Glass	5,640
Iron	5,130
Water	1,481

5 A sound wave takes about 0.03 s to move through a material that is 10.3 m long. What is the material?

A air
B glass
C iron
D water

6 Which structure in the human ear transfers sound vibrations to the stirrup?

A anvil
B cochlea
C ear canal
D eardrum

7 Which type of sound wave has the greatest energy?

A a wave that has a very high intensity
B a wave that has a very low amplitude
C a wave with particles that do not vibrate from their rest positions
D a wave with particles that vibrate a very short distance from their rest positions

Assessment

Online Standardized Test Practice

Use the figure below to answer questions 8 and 9.

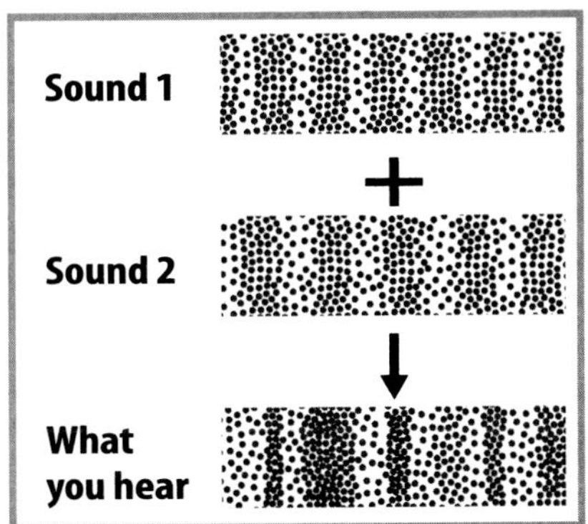

8 What can you tell about sounds 1 and 2 by looking at the figure?

A Sound 1 has a lower pitch.

B Sound 1 has a greater frequency.

C Sound 1 has a smaller amplitude.

D Sound 1 has a greater wavelength.

9 What does the bottom wave in the figure represent?

A silence

B interference

C the Doppler effect

D a fundamental frequency

10 What differs when the same note is played with the same amplitude by two different kinds of instruments?

A energy

B fundamentals

C intensity

D overtones

Constructed Response

Use the figure below to answer questions 11 and 12.

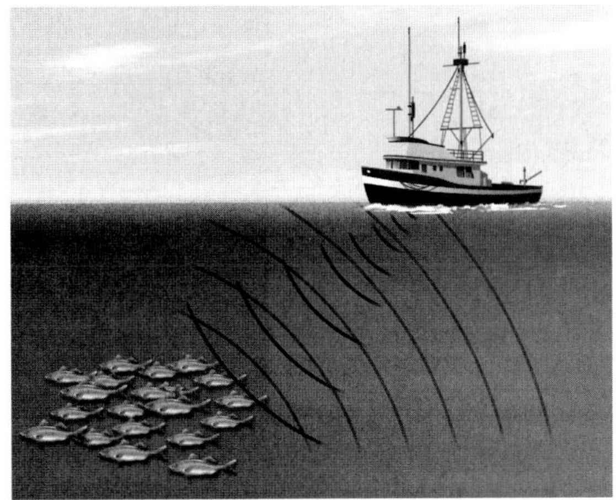

11 Describe the type of sound technology being used in the figure. How does this type of technology differ from echolocation?

12 The ship in this figure sends out a signal that returns to the ship in 4 s. If sound travels through seawater at 1,530 m/s, how far away from the ship are the fish?

13 Why would a sound studio have soft foam squares on the walls?

14 Name two ways that ultrasound is used to detect medical problems. Name two ways that ultrasound is used to treat medical problems.

NEED EXTRA HELP?														
If You Missed Question...	1	2	3	4	5	6	7	8	9	10	11	12	13	14
Go to Lesson...	1	1	1	1	1	1	2	2	2	2	3	3	3	3

Chapter 17

Electromagnetic Waves

How can you describe and use electromagnetic waves?

Invisible Waves?

Each of these photos shows an image of the Sun taken with a camera that is sensitive to different temperatures. Color has been added with the help of a computer because the waves are invisible to the human eye. In this chapter, you will read that all objects emit electromagnetic waves. The wavelengths depend on the temperature of the object.

- Why do you think all objects give off electromagnetic waves?
- How do photographs such as these help scientists study the Sun?
- How can you describe and use electromagnetic waves?

Get Ready to Read

What do you think?

Before you read, decide if you agree or disagree with each of these statements. As you read this chapter, see if you change your mind about any of the statements.

1. Warm objects emit radiation, but cool objects do not.
2. Light always travels at a speed of 300,000 km/s.
3. Red light has the least amount of energy of all colors of light.
4. A television remote control emits radiation.
5. Thermal images show differences in the amount of energy people or objects give off.
6. When you call a friend on a cell phone, a signal travels directly from your phone to your friend's phone.

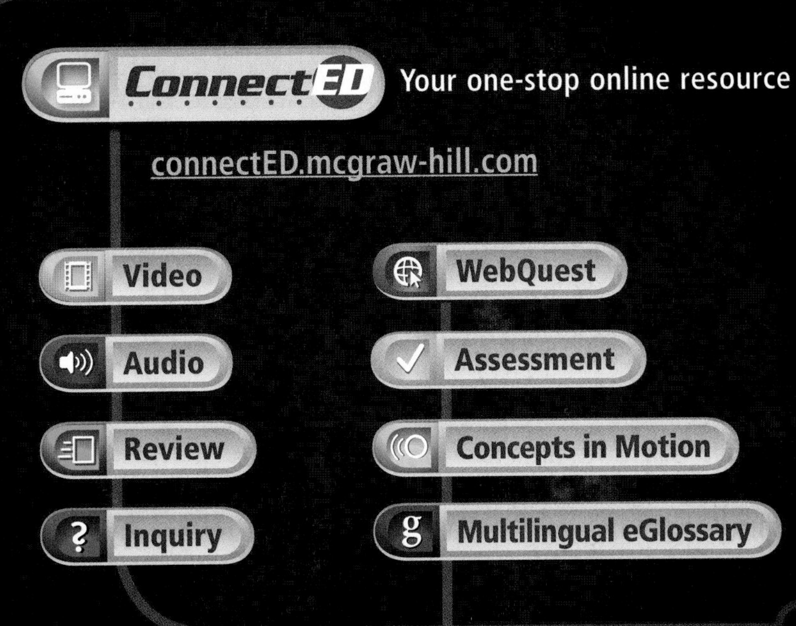

Lesson 1

Reading Guide

Key Concepts
ESSENTIAL QUESTIONS

- How do electromagnetic waves form?
- What are some properties of electromagnetic waves?

Vocabulary

electromagnetic wave p. 601
radiant energy p. 601

Multilingual eGlossary

Video

What's Science Got to do With It?

Electromagnetic Radiation

Inquiry Catching Waves?

Camping in a remote area far from electric lines can be a challenge. Is it possible to get energy for things such as lights and heating? This camper can use these solar panels to capture energy from the Sun. Because electromagnetic waves can travel through space, the Sun is Earth's most important source of energy, both in cities and in remote areas.

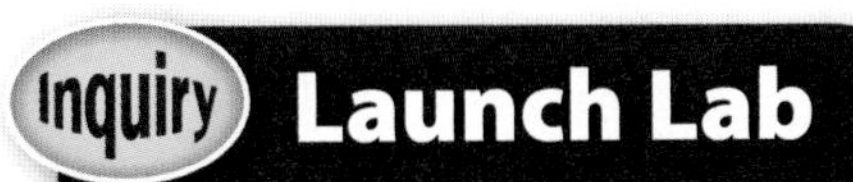

Inquiry Launch Lab

15 minutes

How can you detect invisible waves?

You can see light from the Sun, but other forms of the Sun's energy are invisible. One way to detect invisible waves is to observe their effects on things.

1. Read and complete a lab safety form.
2. With a **marker,** label a **clear plastic cup** *TAP* near the bottom of the cup. Fill it with tap water.
3. Label another cup *TONIC,* and fill it with **tonic water.**
4. Hold the cup with tap water near a **lamp.** Hold a sheet of **black construction paper** behind the cup. Observe the water as you slowly move the cup and paper away from and then closer to the lamp several times. Do you notice any change? Record observations in your Science Journal.
5. Repeat step 4 with the tonic water. Record your observations.
6. Repeat steps 4 and 5, but this time move each cup and paper closer to and then away from bright sunlight instead of a lamp. Record your observations.

Think About This

1. How was the effect of sunlight different from the effect of lamplight on each type of water?
2. **Key Concept** How did your results show that sunlight emits invisible waves?

What are electromagnetic waves?

Suppose you live in a remote location. Would you be able to have electric lights? Could you watch television? Thanks to the Sun, the answer is yes! Like the camper on the previous page, you could use solar panels to capture the Sun's energy and transform it into electricity.

Energy from the Sun reaches Earth by traveling in waves. Many types of waves can travel only through a medium, or matter. Waves in a pond, for example, require water to travel. Waves from the Sun are different because they can travel through empty space. *A wave that can travel through empty space and through matter is called an* **electromagnetic wave.** These waves radiate, or spread out, in all directions from a source. *Energy carried by an electromagnetic wave is called* **radiant energy.** This energy also is known as electromagnetic radiation.

You will read in this lesson that the Sun is not the only source of radiant energy. However, the Sun is the source that provides most of Earth's radiant energy. You also will read about how electromagnetic waves form, and learn about some properties of electromagnetic waves.

Reading Check What is electromagnetic radiation?

Create a vertical three-tab book. Label it as shown. Use it to organize your notes about electromagnetic waves.

What are electromagnetic waves?
How are they produced?
What are some of their properties?

Word Origin

radiant
from Latin *radiantem,* means "shining"

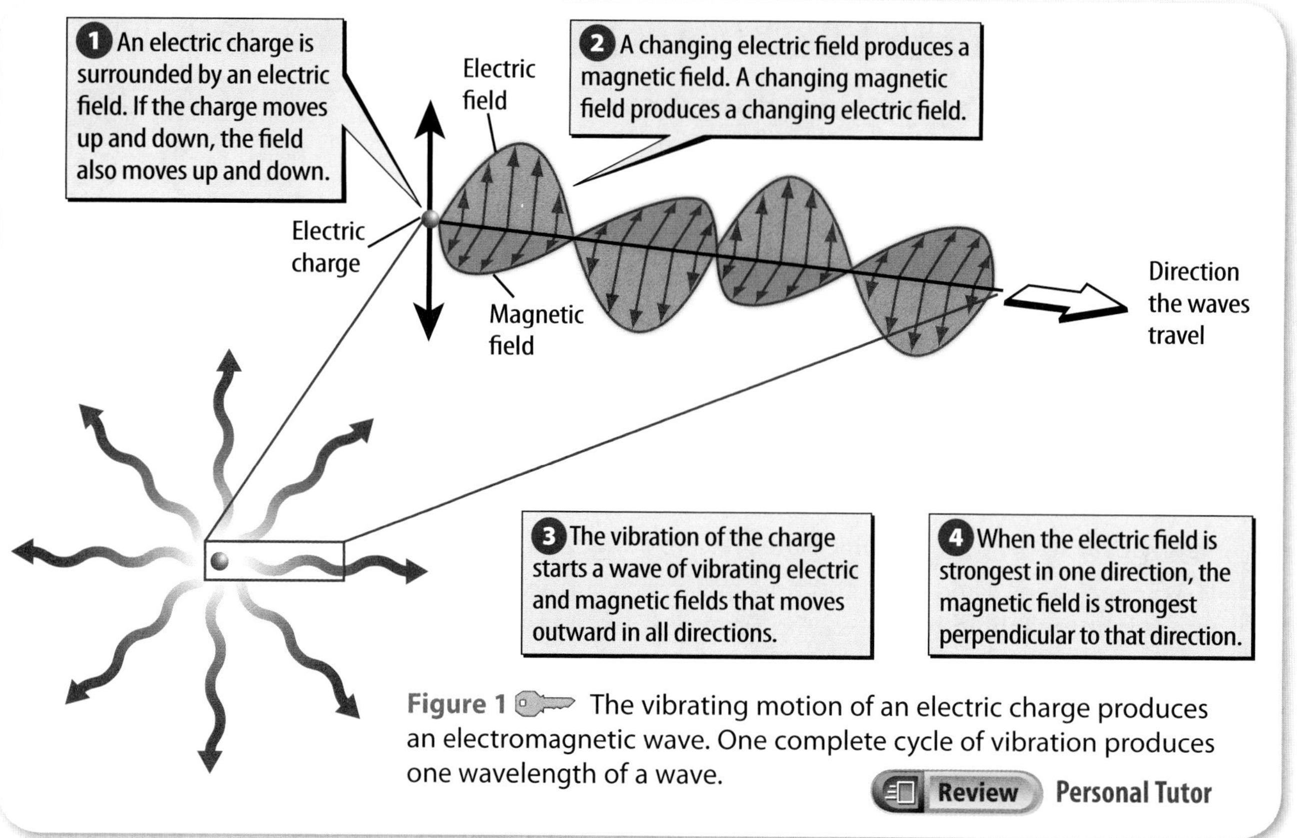

1 An electric charge is surrounded by an electric field. If the charge moves up and down, the field also moves up and down.
Electric field
2 A changing electric field produces a magnetic field. A changing magnetic field produces a changing electric field.
Electric charge
Magnetic field
Direction the waves travel
3 The vibration of the charge starts a wave of vibrating electric and magnetic fields that moves outward in all directions.
4 When the electric field is strongest in one direction, the magnetic field is strongest perpendicular to that direction.

Figure 1 The vibrating motion of an electric charge produces an electromagnetic wave. One complete cycle of vibration produces one wavelength of a wave.

Review Personal Tutor

How Electromagnetic Waves Form

Electromagnetic waves form when an electric charge accelerates by either speeding up, slowing down, or changing direction. This happens when a charged particle vibrates, as shown in **Figure 1.**

Force Fields If you ever have played with a magnet, you know that it is surrounded by a field, or area, where the force of the magnet is present. The same is true for a charged particle, such as an electron. An electric field surrounds the charged particle.

Connected Fields Scientists have found that electric fields and magnetic fields are related.

- A changing electric field produces a magnetic field.
- A changing magnetic field produces a changing electric field.

As a charged particle vibrates, the electric field around it vibrates. This changing electric field produces a magnetic field. As the magnetic field changes, it produces a changing electric field. These connected fields spread out in all directions as electromagnetic waves. Just as shaking the end of the rope in **Figure 2** produces a wave, a vibrating charge produces a wave.

Key Concept Check How do electromagnetic waves form?

Figure 2 Shaking the rope up and down one time produces one wavelength. A charged particle also produces one wavelength when it moves up and down one time. ▼

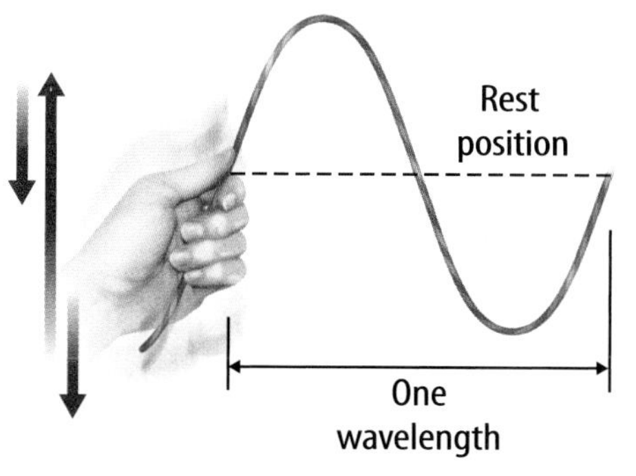

Properties of Electromagnetic Waves

Electromagnetic waves usually are drawn as a single curve, like the rope in **Figure 2.** The waves are called transverse because the disturbance is perpendicular to the direction they travel.

Wavelength and Frequency As with all waves, the wavelength and frequency of electromagnetic waves are related. As shown in **Figure 2,** wavelength is the distance between one point on a wave to the nearest point just like it. Frequency is the number of wavelengths that pass by a point in a certain period of time, such as a second. As frequency decreases, wavelength increases. You will read in Lesson 2 that electromagnetic waves are grouped according to wavelength and frequency.

Wave Speed Electromagnetic waves travel through space at 300,000 km/s, or the speed of light (c). A wave's speed (s) is its frequency (f) multiplied by its wavelength (λ). To determine the wavelength of a wave moving through space, divide the speed of light (c) by the frequency of the wave, as shown in the Math Skills box.

What happens when electromagnetic waves move through matter? Suppose you run across a beach toward a lake or an ocean. You move quickly over the sand, but you slow down in the water. Electromagnetic waves behave similarly. When they encounter matter, electromagnetic waves slow down.

Key Concept Check What are some properties of electromagnetic waves?

Math Skills 

Solve One-Step Equations

Use inverse operations to keep sides of an equation equal. What is the wavelength of an electromagnetic wave in space that has a frequency of 500,000 Hz?

a. Use the wave-speed equation: **wave speed** = **frequency** × **wavelength**:

$$s = f\lambda$$

b. Divide by frequency. Then substitute known values. (Hint: Hz = 1/s)

$$\lambda = \frac{s}{f} = \frac{300{,}000 \text{ km/s}}{500{,}000 \text{ Hz}} = 0.6 \text{ km}$$

Practice

What is the wavelength of an electromagnetic wave with a frequency of 125,000 Hz?

- Math Practice
- Personal Tutor

Inquiry MiniLab

10 minutes

How are electric fields and magnetic fields related?

A compass can show the effect of an electric field on a magnetic field.

1. Read and complete a lab safety form.
2. Place a **battery** on a table, as shown in the photo. Place a **compass** a few centimeters from one end of the battery.
3. Tap the ends of a **20-cm wire** to each end of the battery and quickly remove them. What happens to the compass? Record your observations in your Science Journal.

   ⚠ **Do not leave the wire attached to the battery. It can get very hot.**
4. Repeat step 3 several times. Then hold the wire to the battery for only 2 seconds. Record your observations of the compass.

Analyze and Conclude

1. **Contrast** the effect on the compass when you touched the wire quickly and slowly.
2. **Key Concept** How do your results relate electric fields and magnetic fields?

Figure 3 Most of the Sun's energy is carried by infrared waves. ▶

Visual Check What percentage of the Sun's electromagnetic waves are either light or infrared waves?

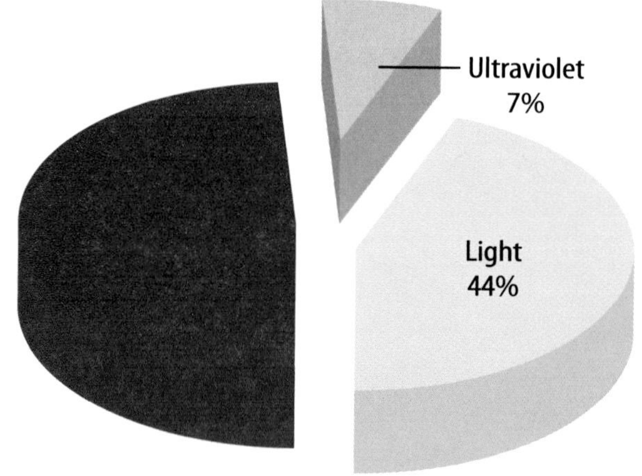

Sources of Electromagnetic Waves

When you hear the term *electromagnetic radiation, or radiant energy,* you might imagine dangerous rays that you should avoid. It might surprise you to learn, however, that you are a source of electromagnetic waves! All matter contains charged particles that constantly vibrate. As a result, all matter—including you—produces electromagnetic waves, and therefore radiant energy. As an object's temperature increases, its particles vibrate faster. Therefore, the object produces electromagnetic waves with a greater frequency, or shorter wavelengths.

REVIEW VOCABULARY
temperature
a measure of the average kinetic energy of the particles that make up an object

The Sun

Earth's most important energy source is the Sun. The Sun emits energy by giving off electromagnetic waves. Only a tiny amount of these waves reach Earth. The Sun has many areas with different temperatures and produces electromagnetic waves with many different wavelengths. As shown in **Figure 3,** almost all of the Sun's energy is carried by three types of waves—ultraviolet waves, light waves, and infrared waves. Light waves are the only type of electromagnetic waves you can see. Infrared waves have wavelengths longer than light. Ultraviolet waves have wavelengths shorter than light. You will learn more about the different types of electromagnetic waves in Lesson 2.

Figure 4 A supernova is the result of an exploding star. Colors were added with a computer to show the different wavelengths of electromagnetic waves emitted by this supernova. ▼

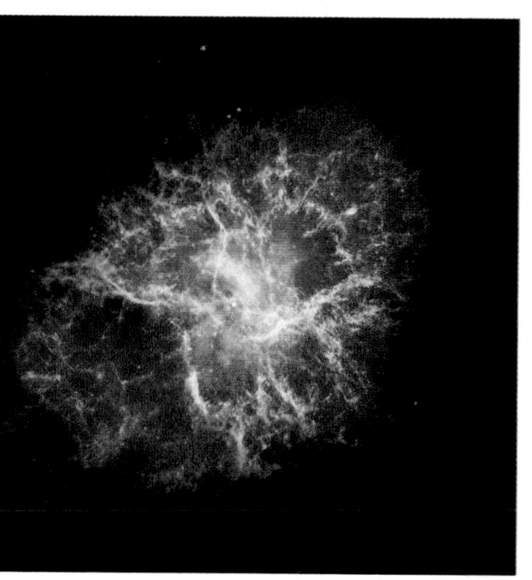

Other Sources of Electromagnetic Waves

Even though the Sun is Earth's most important source of electromagnetic radiation, it is not the only one. All matter, both in space and on Earth, produces electromagnetic waves.

Sources in Space If you look up at the night sky, you see the Moon, stars, and planets. Telescopes on Earth and on satellites above Earth's atmosphere produce images of radiation emitted by these objects. Some of the radiation is visible, but most of it is not. **Figure 4** shows what radiation emitted by an exploding star might look like if your eyes could detect it.

Sources on Earth What do a campfire, a lightbulb, and a burner on an electric stove have in common? They all are hot enough to produce electromagnetic waves that carry energy you can feel. Look around you right now. Your book, the wall, people, and everything else you see produce electromagnetic waves, too. Some waves you detect, but others you do not. Telescopes produce visible images of radiation from space. Also, special cameras produce visible images of invisible waves on Earth. As shown in **Figure 5,** the ultraviolet waves from a flower produce an image very different from the waves you see with your eyes.

▲ **Figure 5** The image on the bottom shows what the dandelions would look like if your eyes could detect the ultraviolet waves the flowers emit.

The Energy of Electromagnetic Waves

Have you ever seen someone with a terrible sunburn, such as the boy in **Figure 6?** Some of the Sun's waves have enough energy to damage your skin. The energy of electromagnetic waves is related to their frequency. Waves that have a higher frequency, such as ultraviolet waves, have higher energy. Waves that have a lower frequency, such as light waves and infrared waves, have lower energy. Light from the girl's flashlight in **Figure 6** can never damage her skin because the light waves do not have enough energy.

The relationship of energy to other wave properties is different for mechanical waves and electromagnetic waves. The energy of a mechanical wave is related to its amplitude. A water wave, for example, with a high amplitude has a lot of energy. The energy of electromagnetic waves is related to their frequency, not amplitude. As the frequency of an electromagnetic wave increases, the energy of the wave increases.

Reading Check Why can the Sun's rays cause a burn, but the light from a flashlight cannot harm your skin?

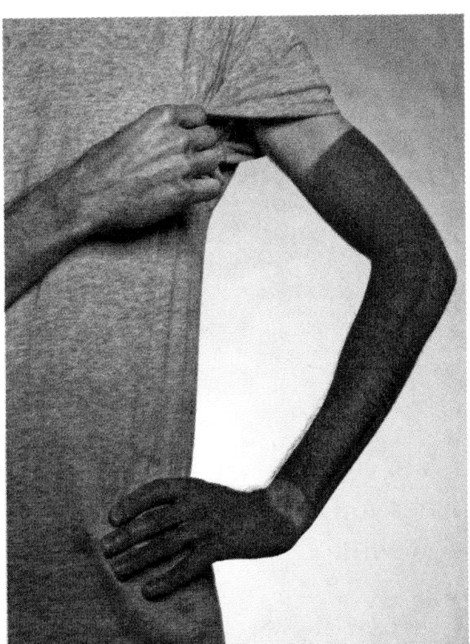

◀ **Figure 6** Ultraviolet waves from the Sun carry enough energy to damage skin cells. Light waves from a flashlight do not carry enough energy to cause damage.

Lesson 1 Review

Online Quiz

Visual Summary

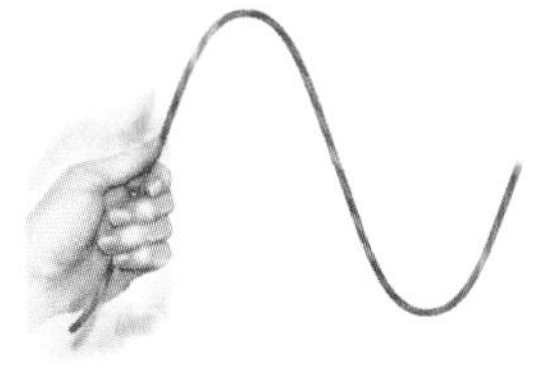

An accelerating charge produces an electromagnetic wave similar to the wave produced by shaking the end of a rope.

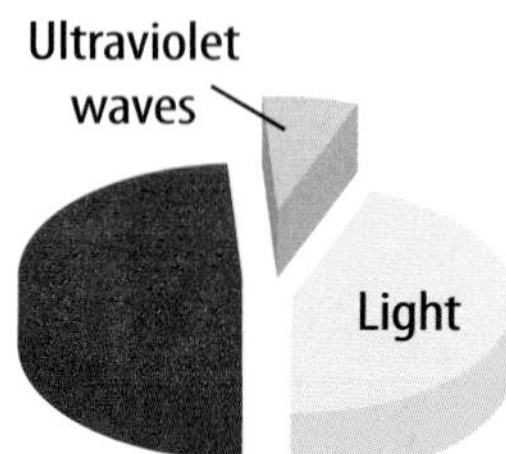

Almost all of the Sun's energy travels by infrared waves, light waves, or ultraviolet waves.

Electromagnetic waves transfer energy from one place to another, even through empty space.

FOLDABLES®

Use your lesson Foldable to review the lesson. Save your Foldable for the project at the end of the chapter.

What do you think NOW?

You first read the statements below at the beginning of the chapter.

1. Warm objects emit radiation, but cool objects do not.
2. Light always travels at a speed of 300,000 km/s.

Did you change your mind about whether you agree or disagree with the statements? Rewrite any false statements to make them true.

Use Vocabulary

1. **Use the term** *radiant energy* in a sentence.
2. A wave that can travel though space and through matter is a(n) ________.

Understand Key Concepts 

3. What is the speed of electromagnetic waves in space?
   - **A.** 30,000 m/s
   - **B.** 30,000 km/s
   - **C.** 300,000 m/s
   - **D.** 300,000 km/s
4. **Identify** What must happen in order for an electromagnetic wave to form?

Interpret Graphics

5. **Examine** the electromagnetic wave below.

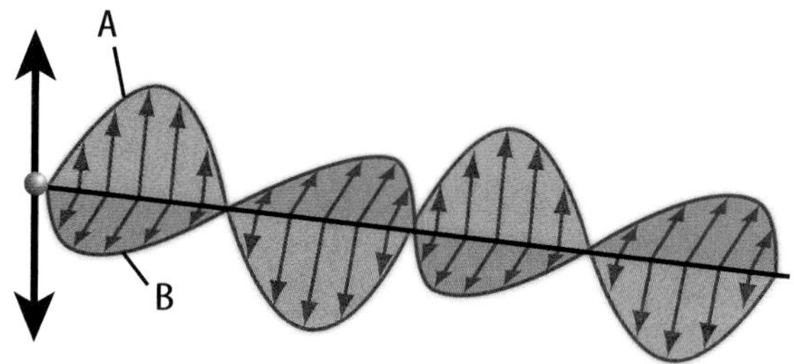

Identify regions A and B in the figure. Explain what causes these regions.

6. **Sequence** Copy and fill in a graphic organizer like the one below to describe how an electromagnetic wave forms and travels away from a source.

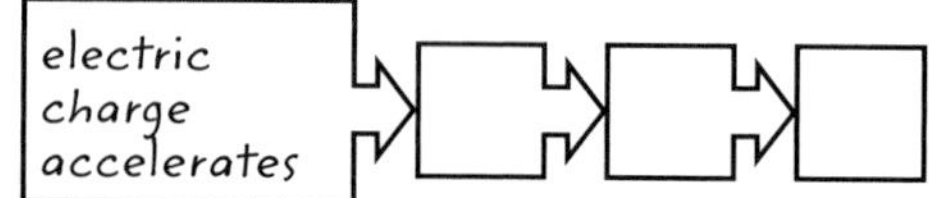

Critical Thinking

7. **Create** a poster that explains how electromagnetic waves sometimes behave like a stream of particles.

Math Skills

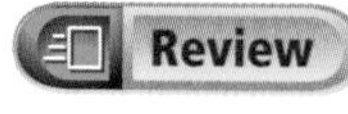

Math Practice

8. What is the wavelength of an electromagnetic wave with a frequency of 135,000 Hz?

HOW IT WORKS

Solar Sails

Using Light to Sail Through Space

How far and how fast can a spacecraft travel? Scientists are studying an exciting way to travel beyond the solar system. A technology called solar sails would allow a spacecraft to travel faster and farther than ever before possible.

Johannes Kepler first proposed solar sails about 400 years ago. He observed comets' tails facing away from the Sun and concluded that sunlight caused pressure—a force that might be harnessed with sails.

The Power of the Sun

Solar sails use light, not matter, to propel spacecraft. When reflected off a solar sail, light transfers its momentum to the surface of the sail, giving it a slight push. The solar sail moves a spacecraft very slowly at first. However, as long as there is a steady stream of light, the sail can eventually accelerate a spacecraft to incredibly high speeds.

The future of solar sail technology is unknown. So far, the use of solar sails has been successful only in a laboratory test chamber. The image below shows how scientists believe a solar sail could work if used in space.

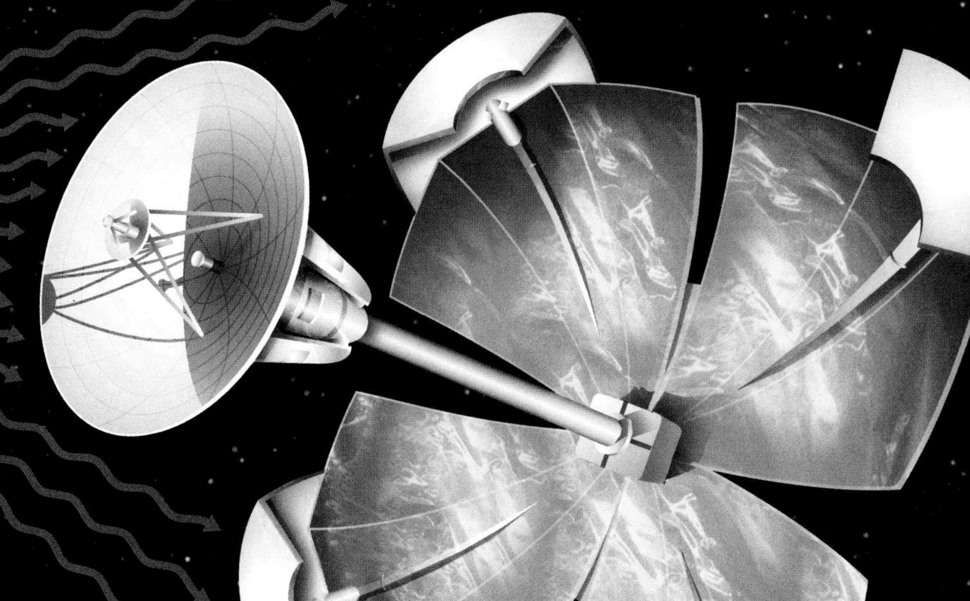

Sun

Earth

Characteristics of a Solar Sail

- Large: at least as big as a football field
- Lightweight: 40–100 times thinner than a sheet of paper
- Very shiny: push provided by reflected light
- Rigid and durable: lasts many years in deep space

It's Your Turn

CALCULATE After traveling continuously for 3 years, a solar sail could reach a speed of 160,000 km/h. The speed of light is about 300,000 km/s. First calculate the kilometers per second a solar sail could reach. Then compare that number to the speed of light. What comparison can you make?

Lesson 2

Reading Guide

Key Concepts 
ESSENTIAL QUESTIONS

- What is the electromagnetic spectrum?
- How do electromagnetic waves differ?

Vocabulary

electromagnetic spectrum p. 609
radio wave p. 610
microwave p. 610
infrared wave p. 610
ultraviolet wave p. 611
X-ray p. 611
gamma ray p. 611

g Multilingual eGlossary

 Video
What's Science Got to do With It?

The Electromagnetic Spectrum

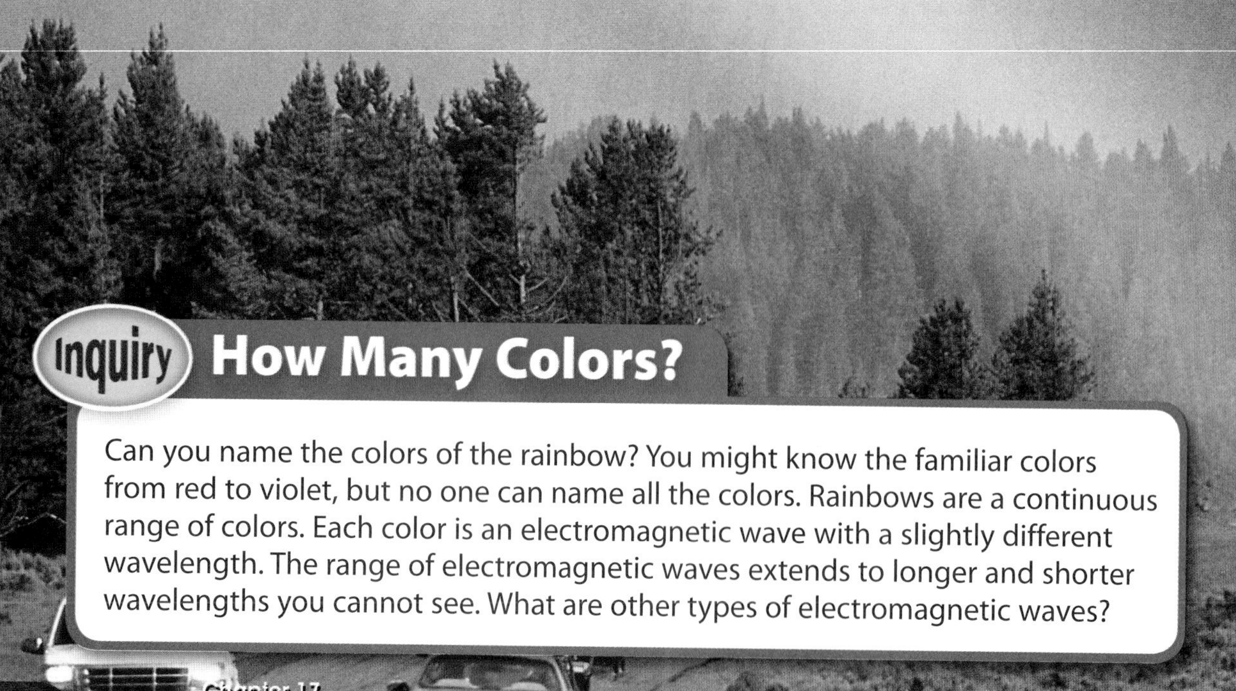

Inquiry How Many Colors?

Can you name the colors of the rainbow? You might know the familiar colors from red to violet, but no one can name all the colors. Rainbows are a continuous range of colors. Each color is an electromagnetic wave with a slightly different wavelength. The range of electromagnetic waves extends to longer and shorter wavelengths you cannot see. What are other types of electromagnetic waves?

Launch Lab

20 minutes

How do electromagnetic waves differ?

Sunlight contains many types of invisible electromagnetic waves. How are these waves different from light you can see?

1. Read and complete a lab safety form.
2. Obtain **four beads.** Place each in a separate **small, self-sealing plastic bag.**
3. Obtain **three types of sunscreen.** Each type should have a different sun protection factor (SPF).
4. Use a **permanent marker** to write the SPF rating on one side of each bag. On the other side, apply a thin layer of sunscreen that has that SPF. Label the fourth bag *No Sunscreen.*
5. Place the bags and beads near a **lamp** with the sunscreen side up. Observe the beads for several minutes. Record your observations in your Science Journal.
6. Place the bags outside in sunlight. Observe the beads again for several minutes, and record your observations.

Think About This

1. Contrast what happened to the beads when you placed them near lamplight and in sunlight.
2. Describe the relationship between the SPF numbers and what happened to the beads.
3. **Key Concept** What do you think could have caused your results?

What is the electromagnetic spectrum?

The waves that carry voices to your cell phone, the waves of energy that toast your bread, and the X-rays that a dentist uses to check the health of your teeth are all electromagnetic waves. The changing motion of an electric charge produces each type of electromagnetic wave. Each type of wave has a different frequency and wavelength, and each carries a different amount of energy. Electromagnetic waves might vibrate from a thousand times a second to trillions of times a second. They might be as large as a house or as small as an atom's nucleus. *The* **electromagnetic spectrum** *is the entire range of electromagnetic waves with different frequencies and wavelengths.*

Key Concept Check What is the electromagnetic spectrum, and how do electromagnetic waves differ?

Classifying Electromagnetic Waves

There are many shades among the familiar colors of a rainbow. Each color gradually becomes another. However, the electromagnetic spectrum is organized into groups based on the wavelengths and frequencies of the waves. Like the colors of a rainbow, each group blends into the next. You will read about how electromagnetic waves are grouped on the next pages.

WORD ORIGIN

spectrum
from Latin *spectrum*, means "appearance"

FOLDABLES®

Use four sheets of paper to make an eight-layer book. Label it as shown. Use it to compare and contrast the different types of electromagnetic waves.

Electromagnetic Waves
Radio Waves
Microwaves
Infrared Waves
Light Waves
Ultraviolet Waves
X-Rays
Gamma Rays

Electromagnetic Spectrum

A **radio wave** *is a low-frequency, low-energy electromagnetic wave that has a wavelength longer than about 30 cm.* Some radio waves have wavelengths as long as a kilometer or more. Radio waves often are used for communication. The wavelengths are long enough to move around many objects, but the energy is low enough that they aren't harmful. On Earth, radio waves usually are produced by an electric charge moving in an antenna, but the Sun and other objects in space also produce radio waves.

1 **Radio waves** 2 **Microwaves** 3 **Infrared waves**

Longer wavelengths, Lower frequencies, Lower energy

A **microwave** *is a low-frequency, low-energy electromagnetic wave that has a wavelength between about 1 mm and 30 cm.* Like radio waves, microwaves are used for communication, such as cell phone signals. With shorter wavelengths than radio waves, microwaves are less often scattered by particles in the air. Microwaves are useful for satellite communications because they can pass through Earth's upper atmosphere. Because of the frequency range of microwaves, food molecules such as water and sugar can absorb their energy. This makes microwaves useful for cooking.

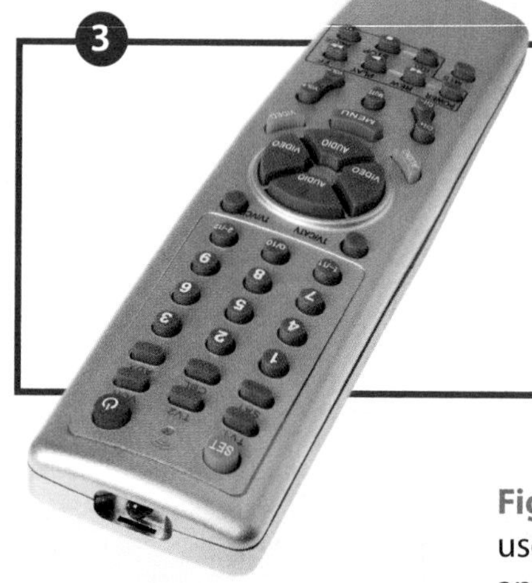

An **infrared wave** *is an electromagnetic wave that has a wavelength shorter than a microwave but longer than light.* Vibrating molecules in any matter emit infrared waves. Even your body emits infrared radiation. You cannot see infrared waves, but if you warm your hands near a campfire, you can feel them. Your skin senses infrared waves with longer wavelengths as warmth. Infrared waves with shorter wavelengths do not feel warm. Your television remote control, for example, sends out these waves.

Figure 7 Electromagnetic waves have a wide range of uses because of their different wavelengths, frequencies, and energy.

Visual Check Which types of electromagnetic waves have wavelengths too long for your eyes to see?

4

Light is electromagnetic waves that your eyes can see. You might describe light as red, orange, yellow, green, blue, indigo, and violet. Red light has the longest wavelength and the lowest frequency. Violet light has the shortest wavelength and the highest frequency. Each name represents a family of colors, each with a range of wavelengths.

4 Light waves | 5 Ultraviolet waves | 6 X-rays | 7 Gamma rays

Shorter wavelengths, Higher frequencies, Higher energy

5

An **ultraviolet wave** *is an electromagnetic wave that has a slightly shorter wavelength and higher frequency than light and carries enough energy to cause chemical reactions.* Earth's atmosphere prevents most of the Sun's ultraviolet rays from reaching Earth. But did you know that you can get a sunburn on a cloudy day? This is because ultraviolet waves carry enough energy to move through clouds and to penetrate the skin. They can damage or kill cells, causing sunburn or even skin cancer.

6

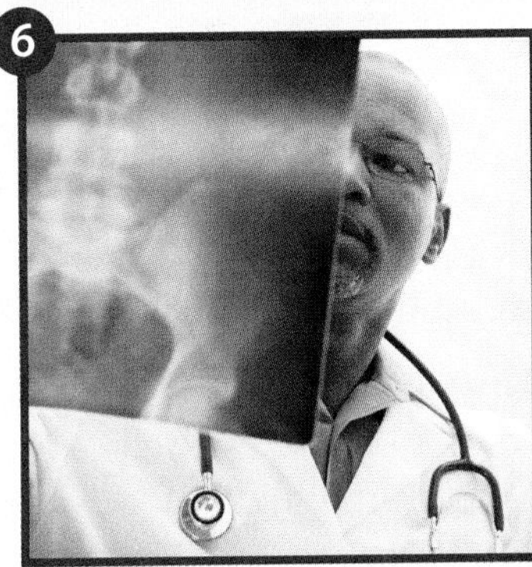

An **X-ray** *is a high-energy electromagnetic wave that has a slightly shorter wavelength and higher frequency than an ultraviolet wave.* Have you ever had an X-ray taken to see if you had a broken bone? X-rays have enough energy to pass through skin and muscle, but the calcium in bone can stop them. Scientists learn about objects and events in space, such as black holes and star explosions, by studying the X-rays they emit.

7

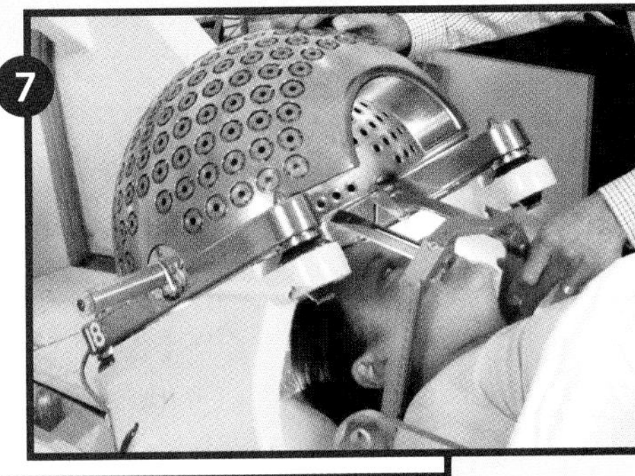

A **gamma ray** *is a high-energy electromagnetic wave with a shorter wavelength and higher frequency than all other types of electromagnetic waves.* Gamma rays are produced when the nucleus of an atom breaks apart or changes. As shown in the photo, gamma rays have enough energy that physicians can use them to destroy cancerous cells. Like X-rays, gamma rays form in space during violent events, such as the explosion of stars.

Lesson 2 Review

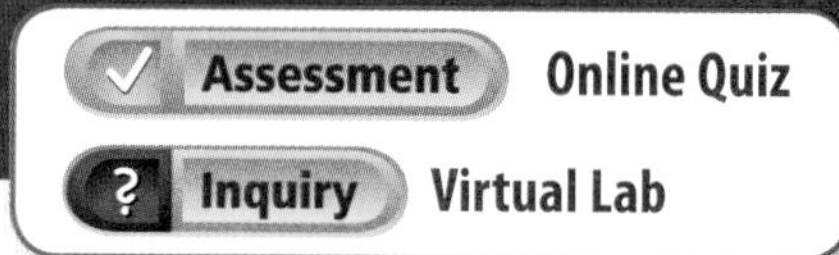

Visual Summary

Light represents only a small portion of the total electromagnetic spectrum.

Because of differences in wavelength, frequency, and energy, electromagnetic waves have many common uses.

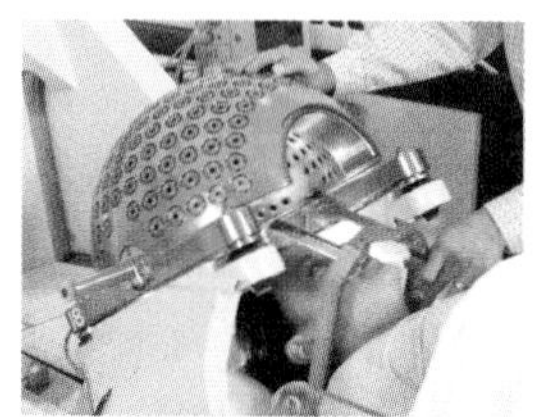

High-frequency electromagnetic waves carry so much energy that they often are used for medical imaging and treatment.

FOLDABLES®

Use your lesson Foldable to review the lesson. Save your Foldable for the project at the end of the chapter.

What do you think NOW?

You first read the statements below at the beginning of the chapter.

3. Red light has the least amount of energy of all colors of light.

4. A television remote control emits radiation.

Did you change your mind about whether you agree or disagree with the statements? Rewrite any false statements to make them true.

Use Vocabulary

1. **Explain** the difference between a microwave and a radio wave.
2. **Explain** the difference between an infrared wave and an ultraviolet wave.

Understand Key Concepts 

3. Which electromagnetic waves have the highest energy?
   A. gamma rays
   B. light waves
   C. radio waves
   D. X-rays
4. **Name** the types of waves that make up the electromagnetic spectrum, from longest wavelength to shortest wavelength.
5. **Compare** the frequency of gamma waves with the frequency of light waves.

Interpret Graphics

6. **Identify** The diagram below shows three types of electromagnetic waves approaching a human body.

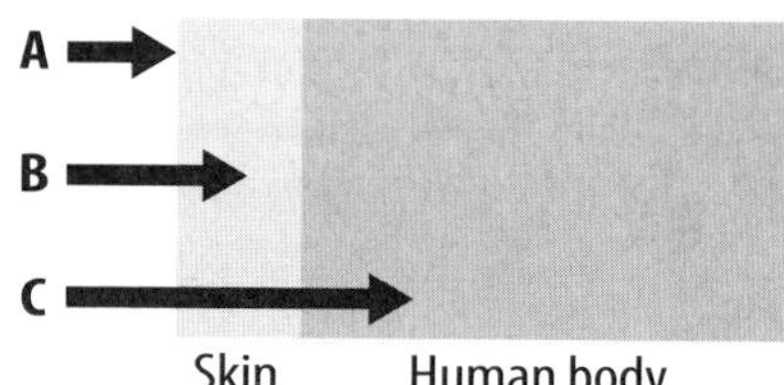

Which represents X-rays? Which represents ultraviolet waves? Explain.

7. **Sequence** Draw a graphic organizer like the one below. In the boxes, write the types of electromagnetic waves from lowest to highest energy. Add boxes as necessary.

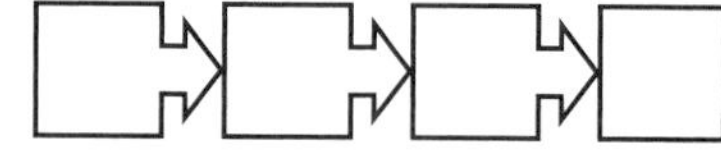

Critical Thinking

8. **Compare and contrast** infrared waves, light, and ultraviolet waves.
9. **Infer** Why are X-rays and gamma rays able to penetrate the body, but radio waves are not?

What's at the edge of a rainbow?

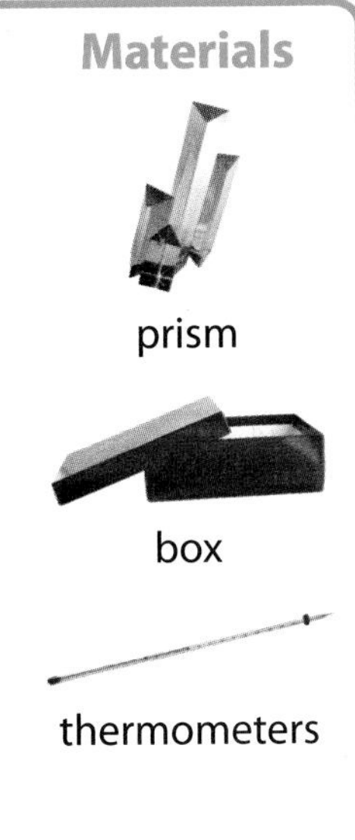

Materials

prism

box

thermometers

Safety

If you have been to a fast food restaurant or cafeteria, you might have noticed special lights hanging over the french fries to keep them warm. Why are the lightbulbs usually red?

Learn It

A hypothesis is a possible explanation for an observation. It might be based on things you have seen, as well as knowledge you already have. You **test a hypothesis** with an experiment. You make more observations and collect data. Your data either support or do not support your hypothesis.

Try It

1. Read and complete a lab safety form.
2. Mount a prism on the end of a large box, as shown in the photo.
3. Adjust the prism and box so that the prism catches the Sun's rays and produces a spectrum on the bottom of the box. You might need to tilt the box.
4. Think about the lights that keep food warm at a restaurant. Based on your observations, form a hypothesis about what part of the electromagnetic spectrum is most useful for keeping food warm and why.

5. Place two thermometers in the box. One should be in the visible part of the spectrum, and the other should be just past the red end of the spectrum.
6. Create a chart like the one below in your Science Journal. Monitor and record the temperatures shown on the thermometers each minute for 5 minutes.

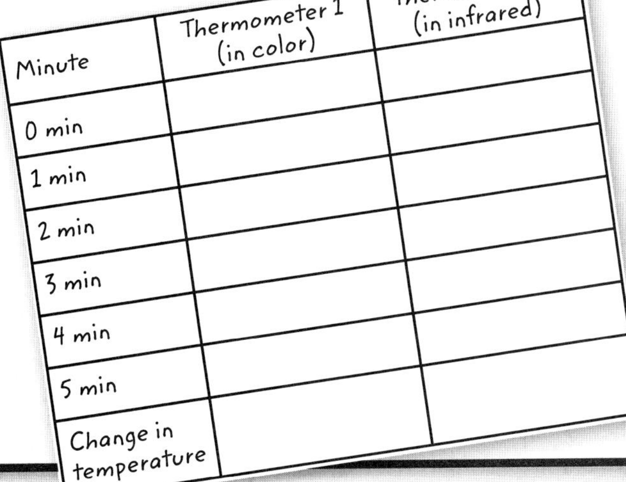

Minute	Thermometer 1 (in color)	Thermometer 2 (in infrared)
0 min		
1 min		
2 min		
3 min		
4 min		
5 min		
Change in temperature		

Apply It

7. **Design** a double-line graph to compare the temperature readings of each thermometer.
8. **Conclude** Did your data support your hypothesis? Why or why not?
9. **Decide** How do you think a thermometer would respond if you placed it just past the violet portion of the spectrum? Explain.
10. **Key Concept** Describe the differences between the types of electromagnetic waves that reach each thermometer.

Lesson 3

Using Electromagnetic Waves

Reading Guide

Key Concepts
ESSENTIAL QUESTIONS

- How are different types of electromagnetic waves used for communication?
- What are some everyday applications of electromagnetic waves?
- What are some medical uses of electromagnetic waves?

Vocabulary

broadcasting p. 615
carrier wave p. 616
amplitude modulation p. 616
frequency modulation p. 616
Global Positioning System p. 618

BrainPOP®

Inquiry Glowing Fingerprint?

A glowing fingerprint might look strange, but to a crime-scene detective, it's a clue. Detectives dust the scene with a powder that glows when ultraviolet waves strike it. If the powder sticks to a handprint, using ultraviolet waves might solve the crime. What are some other uses of electromagnetic waves?

Launch Lab

15 minutes

How do X-rays see inside your teeth?

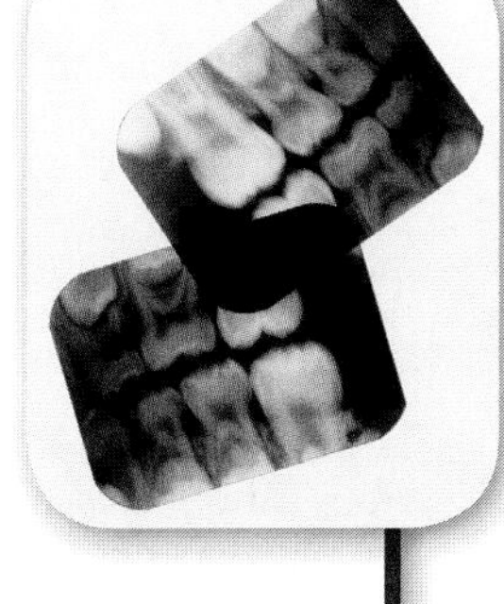

Dentists take X-rays of your mouth to see if you have any problems. How does an X-ray produce images of hard materials such as teeth?

1. Read and complete a lab safety form.
2. Use **scissors** to cut an **index card** into four teeth shapes. Put the paper teeth in different places on a **piece of black paper** on the bottom of a **shoe box.**
3. Cover the shoebox opening with a piece of **screen.** Use **rubber bands** to attach the screen to the top of the shoe box.
4. Sprinkle **flour** evenly across the surface of the screen. Use a **toothbrush** to gently brush the flour on the screen until all of it falls through the screen.
5. Carefully remove the screen and remove the card parts from the box.
6. Observe the paper "X-ray." Record your observations in your Science Journal.

Think About This

1. What did you observe in the bottom of the box?
2. **Key Concept** If the screen represents your skin, muscle, and fatty tissues, what do you think this activity tells you about how X-ray machines form images?

How do you use electromagnetic waves?

There are few things you do that don't involve some type of electromagnetic wave. Think about a typical day. Perhaps a clock radio wakes you in the morning with your favorite music. You are warm under your blanket because infrared radiation from your body changed into thermal energy. When you finally crawl out of bed, you turn on a lamp and use light to see which clothes to wear. These are just a few of the ways you use electromagnetic waves. In this lesson, you will read about common uses for all types of electromagnetic waves.

Radio Waves

It probably is no surprise to you that radio and television stations use radio waves to send out signals. Radio waves can pass through many buildings, yet they don't carry enough energy to harm humans. The wavelengths of radio waves are long enough to go around many obstacles. They travel through air at the same speed as all electromagnetic waves.

In the past, it might have taken minutes, hours, or even days for news to travel from one part of the United States to another. Today, you can watch and listen to events around the world as they happen! **Broadcasting** *is the use of electromagnetic waves to send information in all directions.* Broadcasting became possible when scientists learned to use radio waves.

FOLDABLES®

Make a vertical seven-tab book. Label it as shown. Use it to organize your notes on uses of the different types of electromagnetic waves.

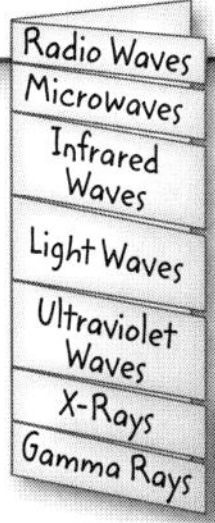

Radio and Television

How do your radio and television receive information to produce sounds or images? A radio or TV station uses radio waves to carry information. Each radio and TV station sends out radio waves at a certain frequency. The station converts sounds or images into an electric signal. Then, the station produces a **carrier wave**—*an electromagnetic wave that a radio or television station uses to carry its sound or image signals.* The station **modulates,** or varies, the carrier wave to match the electric signal. The signal then gets converted back into images or sounds when it reaches your radio or television.

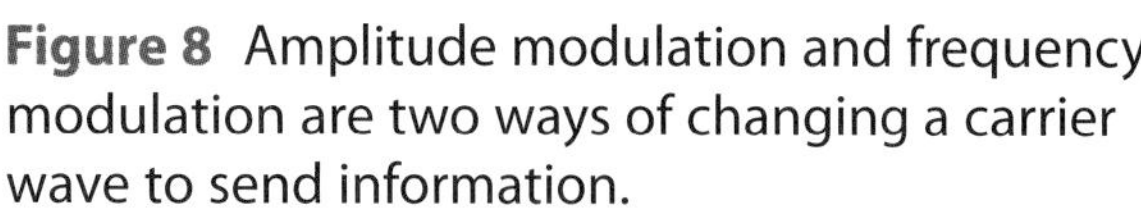

WORD ORIGIN
modulate
from Latin *modulari,* means "to play or sing"

Figure 8 Amplitude modulation and frequency modulation are two ways of changing a carrier wave to send information.

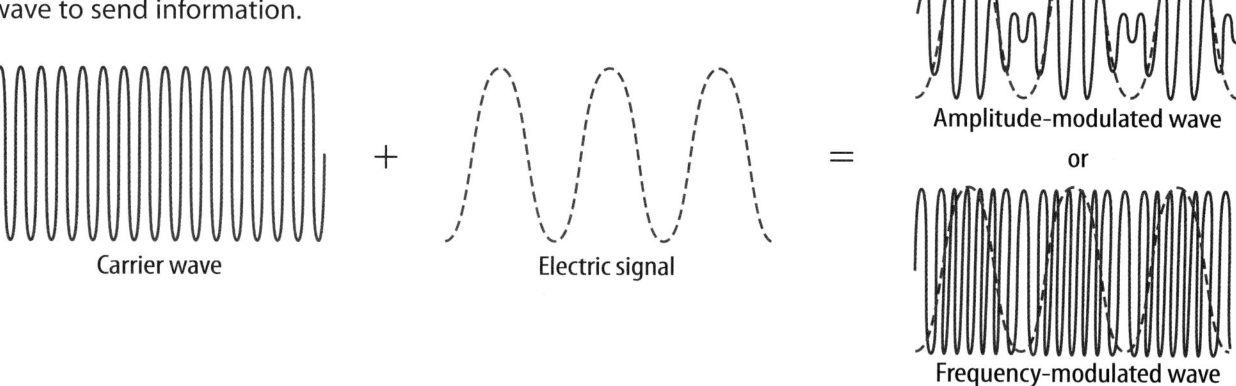

Two ways a station might change its carrier wave are shown in **Figure 8.** **Amplitude modulation** *(AM) is a change in the amplitude of a carrier wave.* Recall that amplitude is the maximum distance a wave varies from its rest position. **Frequency modulation** *(FM) is a change in the frequency of a carrier wave.* Changes in frequency match changes in sounds or images.

Key Concept Check How are radio waves used for communication?

Digital Signals

Television stations and some radio stations broadcast digital signals. To understand types of signals, look at the amplitude-modulated wave in **Figure 8.** Notice how the wave changes smoothly from high to low. This type of change is called analog. A digital signal, however, changes in steps. Stations can produce a digital signal by changing different properties of a carrier wave. For example, the station might send the signal as pulses, or a pattern of starting and stopping. The wave might have a code of high and low amplitudes. Sounds and images sent by digital signals are usually clearer than analog signals.

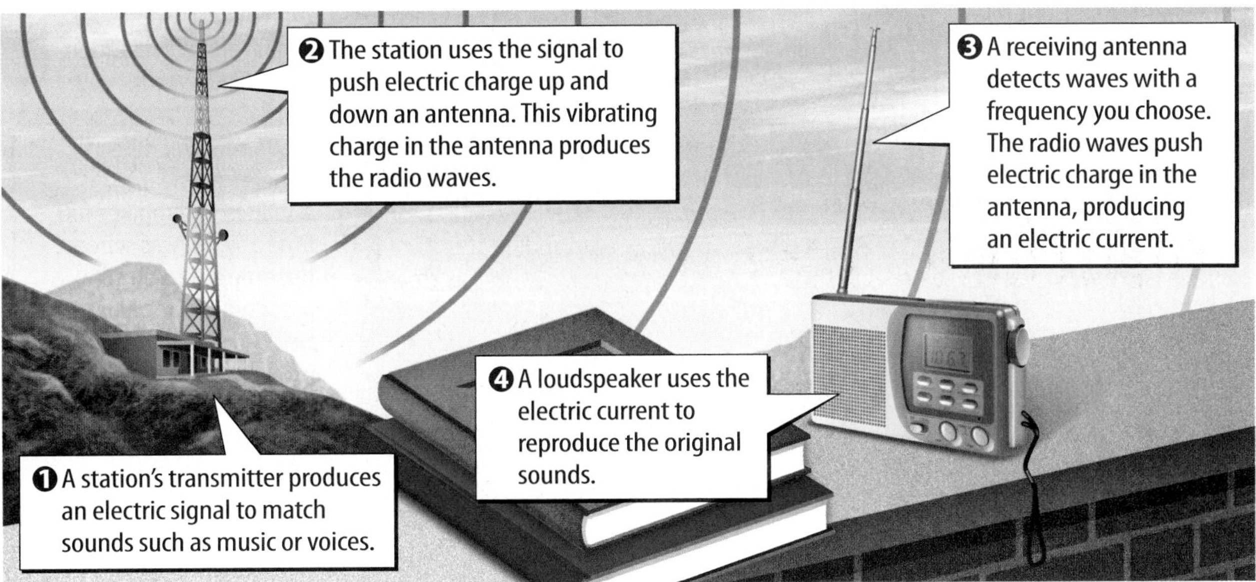

❷ The station uses the signal to push electric charge up and down an antenna. This vibrating charge in the antenna produces the radio waves.
❸ A receiving antenna detects waves with a frequency you choose. The radio waves push electric charge in the antenna, producing an electric current.
❹ A loudspeaker uses the electric current to reproduce the original sounds.
❶ A station's transmitter produces an electric signal to match sounds such as music or voices.

▲ **Figure 9** Radio waves travel from a transmitting station to a radio antenna.

Visual Check What types of energy changes take place during this process?

Transmission and Reception

The way a radio station broadcasts a signal is shown in **Figure 9.** A transmitter produces and sends out radio waves. An antenna receives the waves and changes them back to sound. Television transmission is similar, but for cable television, the waves travel through cables designed to carry radio waves.

The wavelength of an AM carrier wave is typically a lot longer than the wavelength of an FM carrier wave. This means AM signals can pass around obstacles and travel farther than FM waves. AM waves also are long enough to reach Earth's upper atmosphere without scattering. FM waves typically have higher frequencies and more energy than AM waves. Therefore, FM waves usually produce better sound quality than AM waves.

Microwaves

You've probably used a microwave oven to heat food. Did you know you also use microwaves when you watch a television show that uses satellite transmission? Like radio waves, microwaves are useful for sending and receiving signals. However, microwaves can carry more information than radio waves because their wavelength is shorter. Microwaves also easily pass through smoke, light rain, snow, and clouds.

Cell Phones

A cell phone company sets up small regions of service called cells. Each cell has a base station with antennae, as shown in **Figure 10.** When you make a call, your phone sends and receives signals to and from the cell tower. Electric circuits in the base station direct your signal to other towers. A signal passes from cell to cell until it reaches the phone of the person you called.

Figure 10 Cell towers hold antennae from one or more service providers. Each provider positions its antennae toward a different direction to send and receive signals. ▼

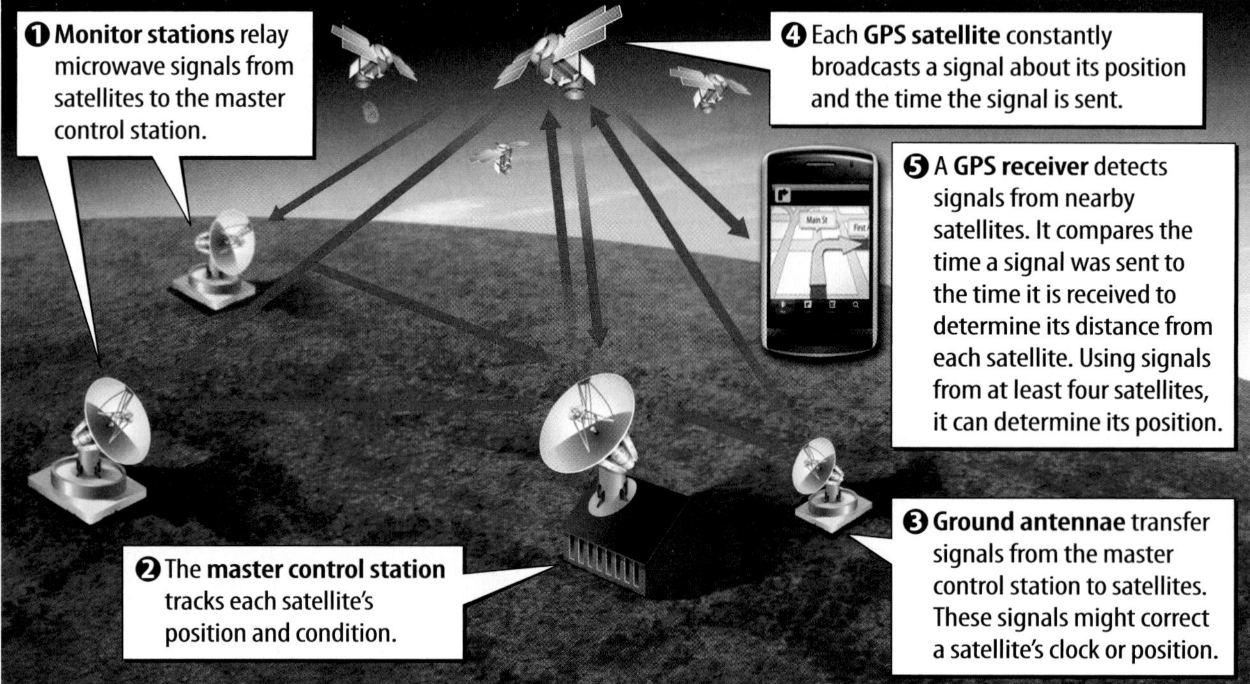

1 Monitor stations relay microwave signals from satellites to the master control station.
4 Each GPS satellite constantly broadcasts a signal about its position and the time the signal is sent.
5 A GPS receiver detects signals from nearby satellites. It compares the time a signal was sent to the time it is received to determine its distance from each satellite. Using signals from at least four satellites, it can determine its position.
3 Ground antennae transfer signals from the master control station to satellites. These signals might correct a satellite's clock or position.
2 The master control station tracks each satellite's position and condition.

Figure 11 The GPS system includes more than 24 satellites evenly spaced in orbit around Earth. To calculate its position, a receiver needs signals from at least four satellites.

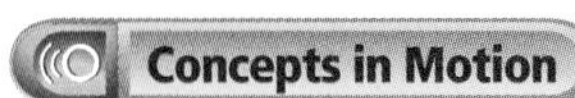

Animation

SCIENCE USE V. COMMON USE

microwave
Science Use a low-frequency, low-energy electromagnetic wave

Common Use appliance used to cook food quickly

ACADEMIC VOCABULARY

analyze
(verb) to study how parts work together

Communication Satellites

Have you seen a sports event or heard a news story from around the world on your TV? Information about the sounds and the images probably traveled by **microwaves** from a satellite to your TV. Some homes have a satellite dish that receives signals directly from satellites. Satellites send and receive signals similar to the way antennae on Earth do. A transmitter sends microwave signals to the satellite. The satellite can pass the signal to other satellites or send it to another place on Earth.

You can use satellite signals to find directions. *The* **Global Positioning System** (GPS) *is a worldwide navigation system that uses satellite signals to determine a receiver's location.* As shown in **Figure 11,** at least 24 GPS satellites continually orbit Earth, sending out signals about their orbits and the time. Receivers **analyze** this information to calculate their location.

Key Concept Check How are microwaves used for communication?

Infrared Waves

You read in Lesson 2 that vibrating molecules in any matter emit infrared waves. The wavelength of the infrared waves depends on the object's temperature. Hotter objects emit infrared waves with a greater frequency and shorter wavelength than do cooler objects. Scientists have developed technology that produces images showing how much thermal energy a person or thing emits.

Infrared Imaging

Thermal cameras take pictures by detecting infrared waves rather than light waves. They convert invisible infrared waves, or different temperatures, to different colors so that your eyes can interpret the information. Medical professionals use thermal imaging to locate areas of poor circulation in the body. Thermal imaging of a building can identify areas where excessive thermal energy loss occurs.

Can you see in the dark? Night-vision goggles, such as those in **Figure 12,** use infrared waves similar to the way a normal camera uses light. Using the small amount of infrared light present, the goggles produce enough light that objects can be seen easily in the dark.

▲ **Figure 12** Infrared waves detected by night-vision goggles make buildings and other objects in a city visible at night.

Imaging Earth

Scientists launch satellites that detect and photograph infrared waves coming from Earth. These infrared images might show variations in vegetation or snowfall. Some images can even show a fire smoldering in a forest before it bursts into flame. The thermal image in **Figure 13** shows the lava flow from a volcano.

▲ **Figure 13** This infrared image of the Colima Volcano in Mexico shows recent lava flow as red. Notice the volcano's two snow-capped peaks.

Reading Check What are some ways in which infrared images help humans?

Light

The Sun provides most of Earth's light, but light emitted on Earth also has important uses. Think about how difficult driving in a city would be without automobile headlights, street lights, and traffic signals, such as the one in **Figure 14.** Without electric lights, homes and businesses would be dark at night. Televisions, computers, and movie theaters also rely on light. Some forms of communication rely on light that travels through optical fibers—thin strands of glass or plastic that transmit a beam of light over long distances.

Figure 14 A traffic signal is an example of how light is useful. ▶

Inquiry MiniLab 10 minutes

How does infrared imaging work?

Like light waves, infrared waves can be either emitted or reflected by an object. Infrared photography can be used to detect waves that reflect off vegetation.

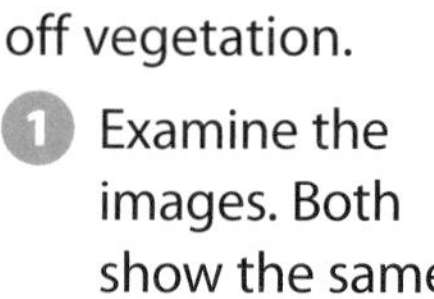

1. Examine the images. Both show the same location. Compare the true-color visual light image at the top with the false-color infrared image. In the false-color image, red indicates the presence of infrared light.
2. Interpret the images to answer the following questions.

Analyze and Conclude

1. **Analyze Data** What parts of the false-color image show the most infrared light?
2. **Observe** What does the false-color image show about the land in the background?
3. **Key Concept** How could infrared photography be used to identify unhealthy patches of trees in a forest?

Figure 15 A dentist can use an ultraviolet wand to quickly harden adhesives.

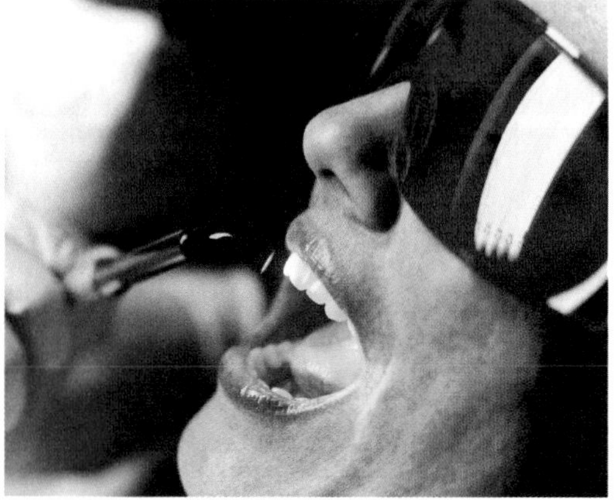

Ultraviolet Waves

Too much exposure to ultraviolet waves can damage your skin, but these same waves also can be useful. Certain materials glow when ultraviolet waves strike them. The materials absorb the energy of the waves and reemit it as light. Credit cards, for example, have invisible symbols stamped on them using this material. If a store clerk holds the card under an ultraviolet lamp, the symbol appears. Scientists also can identify some minerals in rocks by the way they glow under an ultraviolet lamp.

Ultraviolet waves also are useful for killing germs. Just as your skin can be damaged by absorbing the energy of ultraviolet waves, germs also can be damaged. Food manufacturers might use ultraviolet lamps to kill germs in some foods. Campers might use ultraviolet lamps to purify lake water.

Key Concept Check What are some everyday applications of electromagnetic waves?

Medical Uses

In hospitals and clinics, having a germ-free environment might be the difference between life and death. Medical facilities sometimes clean tools and surfaces by bringing them near an ultraviolet lamp. Some air and water purifiers use ultraviolet light to kill germs and reduce the spread of disease.

Ultraviolet light has other medical uses. For example, exposure to ultraviolet waves can help control or cure certain skin problems, such as psoriasis. Several times a week, the patient's skin is exposed to ultraviolet light. The waves carry enough energy to slow the growth of the diseased skin cells.

Another use of ultraviolet waves is shown in **Figure 15.** The dentist is using ultraviolet light to harden an adhesive in just a few seconds. Without ultraviolet light, adhesives might take several minutes to harden.

Fluorescent Lightbulbs

Lightbulbs like the one shown in **Figure 16** use the energy from ultraviolet waves to produce light. Similar to the way a symbol on a credit card glows under ultraviolet light, certain chemicals inside the lightbulb glow when they are exposed to ultraviolet light. When you flip a switch to provide electricity to the bulb, an electric current flows through a gas inside the bulb. The heated gas emits ultraviolet light. This light strikes a chemical that coats the inside of the bulb. The chemical absorbs the energy of the ultraviolet waves and reemits it as light.

Reading Check How does a fluorescent lightbulb use ultraviolet waves?

▲ **Figure 16** A fluorescent lightbulb uses ultraviolet waves to produce light.

X-Rays

Ultraviolet waves can be dangerous because they can pass through the upper layer of your skin. X-rays have even more energy than ultraviolet waves and can pass through skin and muscle. Although this property makes X-rays dangerous, it also makes them useful. Two important uses of X-rays are security and medical imaging.

Detection and Security

Have you ever had your luggage scanned at an airport, as shown in **Figure 17**? X-rays are useful for security scanning because they can pass through many materials, but metal objects block them. Computers can form images of the contents of luggage by measuring how different materials transmit the X-rays.

Figure 17 Security workers use X-ray scanners to form images of the contents of passenger luggage. ▼

Airport Security

Medical Detection

You might have had an X-ray taken by a dentist or a doctor, similar to that shown in **Figure 18.** X-rays pass through soft parts of your body, but bone stops them. When X-rays strikes photographic film, the film turns dark. Light parts of the film show where bone absorbed the X-rays.

A doctor can obtain even more-detailed images using a computed tomography (CT) scanner. The scanner is an X-ray machine that rotates around a patient, producing a three-dimensional view of the body.

 **Reading Check** How are a CT scan and a normal X-ray different?

Gamma Rays

Because of their extremely high energy, gamma rays can be used to destroy diseased tissue in a patient. Gamma rays also can be used to diagnose medical conditions. Recall that gamma rays are produced when an atom's nucleus breaks apart or changes. In a procedure called a positron emission tomography (PET) scan, a detector monitors the breakdown of a chemical injected into a patient. The chemical is chosen because it is attracted to diseased parts of the body. The detector can find the location of the disease by detecting the gamma rays emitted by the chemical.

 **Key Concept Check** What are some medical uses of electromagnetic waves?

Figure 18 Both X-rays and gamma rays are used in medicine to diagnose and treat diseases.

X-Rays and Gamma Rays

Lesson 3 Review

Assessment | Online Quiz

Visual Summary

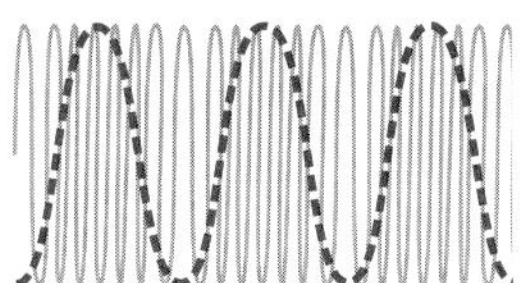

Radio and television stations can broadcast their programming with modulated radio waves.

Satellites send and receive microwave signals to communicate information over long distances.

Because warmer objects emit more infrared waves, scientists can study heat-related changes on Earth's surface, such as weather and volcanic activity.

FOLDABLES®

Use your lesson Foldable to review the lesson. Save your Foldable for the project at the end of the chapter.

What do you think NOW?

You first read the statements below at the beginning of the chapter.

5. Thermal images show differences in the amount of energy people or objects give off.

6. When you call a friend on a cell phone, a signal travels directly from your phone to your friend's phone.

Did you change your mind about whether you agree or disagree with the statements? Rewrite any false statements to make them true.

Use Vocabulary

1. The wave that is changed to carry a radio or television signal is a(n) ________.
2. **Use the term** *broadcasting* in a complete sentence.

Understand Key Concepts 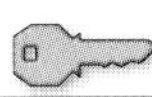

3. Which of the following changes in an AM radio wave?
   A. amplitude
   B. frequency
   C. speed
   D. wavelength
4. **Explain** why X-rays are less harmful than infrared waves in medical imaging.
5. **Describe** how microwaves are useful for communication.
6. **Identify** an everyday application of ultraviolet waves.

Interpret Graphics

7. **Interpret** Which type of modulation does the radio wave below have? Do you think a station could broadcast this signal for hundreds of kilometers? Explain.

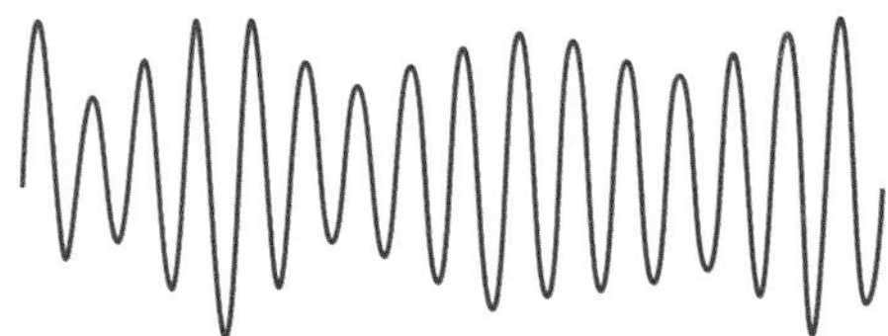

8. **Identify** Copy and fill in a graphic organizer like the one below to identify several uses of electromagnetic waves.

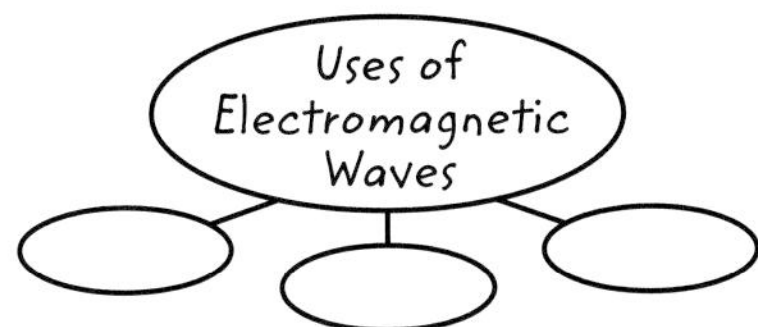

Critical Thinking

9. **Predict** the effect on human life if the ozone layer, which blocks harmful ultraviolet waves, becomes thinner or develops more holes.

Lab

1 class period

Design an Exhibit for a Science Museum

Materials

miscellaneous supplies for making an exhibit—poster board, markers, balsa wood, masking tape, etc.

Safety

Scientists learn about distant regions of the universe by detecting different types of electromagnetic waves. Doctors use electromagnetic waves to find and treat diseases. Anytime you use a cell phone or watch a television program, you use electromagnetic waves for communication. Even though people use electromagnetic waves every day, many people don't know much about the electromagnetic spectrum. Your assignment is to help change that.

Question

What sort of science museum exhibit can you design to educate others about electromagnetic waves?

Procedure

1. Read and complete a lab safety form.
2. Brainstorm with your partner(s) about different types of exhibits you enjoy when visiting museums (sports, science, history, music) or zoos. What makes them fun? Make a list of your ideas titled *Exhibits* in your Science Journal.
3. Discuss with your partner(s) ideas, facts, or concepts you remember from your visits to museums or zoos. What was it about your experience that made this information memorable? Make a list of these qualities titled *Memorable*.
4. Determine with your class which part of the electromagnetic spectrum your exhibit will describe.
5. Research your assigned part of the electromagnetic spectrum using your textbook, reliable sources on the Internet, and other reference materials. You also might interview people. In addition to specifics about your topic, your exhibit should provide a general overview of the electromagnetic spectrum.

6. Plan your exhibit. Where will the information portion be? Is the exhibit interactive? Will people do something when they visit your exhibit? Will any materials be used while people visit? Are your ideas safe?
7. Gather the supplies you need, and then construct your exhibit.
8. Invite your classmates to evaluate your exhibit. Have them write on an index card what they enjoyed and what they remembered after leaving your exhibit. Ask them to write questions or suggestions they might have about your topic or exhibit.

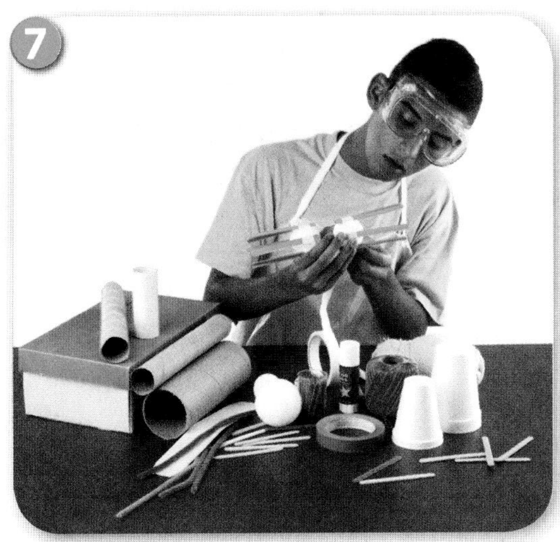

Analyze and Conclude

9. **Analyze Data** Read the comments made on the index cards by your classmates. Organize this information into three lists: things people liked, things people learned, and questions or suggestions people made.
10. **Draw Conclusions** Use the feedback lists to evaluate your display. Were you able to create an informative and fun exhibit? Is there anything you would change or improve based on the feedback?
11. **The Big Idea** How did you describe and use electromagnetic waves in your exhibit?

Communicate Your Results

Work with your classmates to place your exhibits in a public area of the school. Invite other students to visit your class museum.

Inquiry Extension

Research a topic you found interesting when you visited your classmates' exhibits. Design an experiment, write a paper, or make a poster to explore or explain the topic.

Lab Tips

- ☑ Consider your lists of exhibits and memorable qualities carefully.
- ☑ When preparing your exhibit, think about your audience. Make your exhibit understandable and entertaining for someone your age.
- ☑ Quality counts. Take time to make a neat and error-free exhibit.

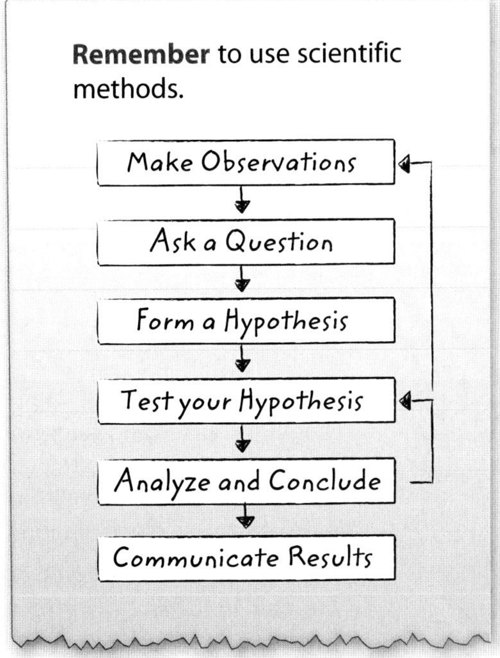

Chapter 17 Study Guide

Electromagnetic waves move through matter and space. They have different frequencies, wavelengths, and energy. Uses include communication and medical imaging.

Key Concepts Summary

Lesson 1: Electromagnetic Radiation

- An accelerating electric charge produces an **electromagnetic wave.**
- Electromagnetic waves travel through space at 300,000 km/s. Properties include their wavelengths, their frequencies, and the amount of **radiant energy** each wave carries.

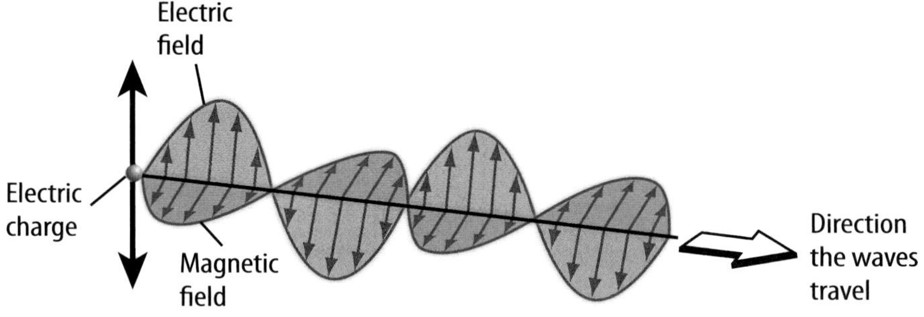

Vocabulary

electromagnetic wave p. 601

radiant energy p. 601

Lesson 2: The Electromagnetic Spectrum

- The **electromagnetic spectrum** is the entire range of wavelengths and frequencies of electromagnetic waves.
- Electromagnetic waves differ in their wavelengths, frequencies, and energy. **Radio waves** have the longest wavelengths, the lowest frequencies, and the lowest energy. **Gamma rays** have the shortest wavelengths, the highest frequencies, and the highest energy.

Vocabulary

electromagnetic spectrum p. 609

radio wave p. 610

microwave p. 610

infrared wave p. 610

ultraviolet wave p. 611

X-ray p. 611

gamma ray p. 611

Lesson 3: Using Electromagnetic Waves

- Electromagnetic waves are used in radio and television **broadcasting,** and in cell phone communication.
- Using light to see and microwaves to cook food are everyday applications of electromagnetic waves. Listening to radio and watching television require devices that use electromagnetic waves.
- Doctors use X-rays to identify broken bones or damaged teeth. Gamma rays can be used to diagnose or treat diseases.

Vocabulary

broadcasting p. 615

carrier wave p. 616

amplitude modulation p. 616

frequency modulation p. 616

Global Positioning System p. 618

Review
- Personal Tutor
- Vocabulary eGames
- Vocabulary eFlashcards

FOLDABLES® Chapter Project

Assemble your lesson Foldables as shown to make a Chapter Project. Use the project to review what you have learned in this chapter.

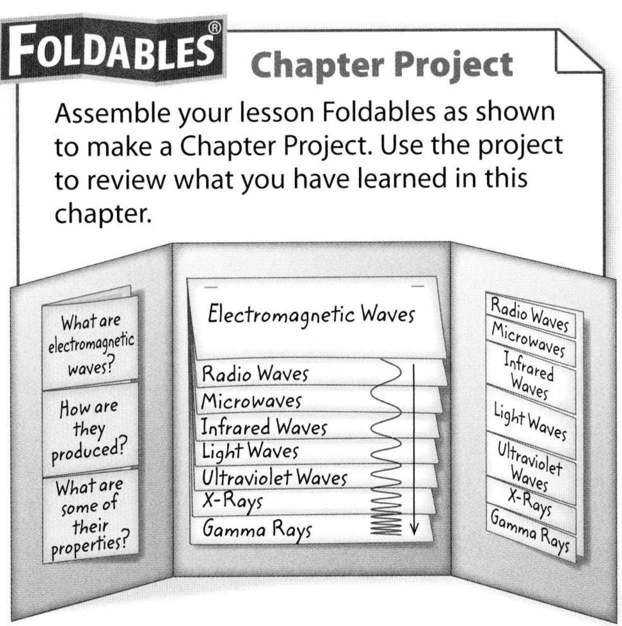

Use Vocabulary

1. The energy carried by electromagnetic waves is ________.
2. A transverse wave that is able to move through space is a(n) ________.
3. The type of electromagnetic wave that causes sunburn is a(n) ________.
4. The electromagnetic wave that has the highest energy and is therefore most dangerous to humans is a(n) ________.
5. A television signal travels from one place to another by a(n) ________ that is modulated.
6. An FM station changes its carrier wave by ________.

Concepts in Motion Interactive Concept Map

Link Vocabulary and Key Concepts

Copy this concept map, and then use vocabulary terms from the previous page to complete the concept map.

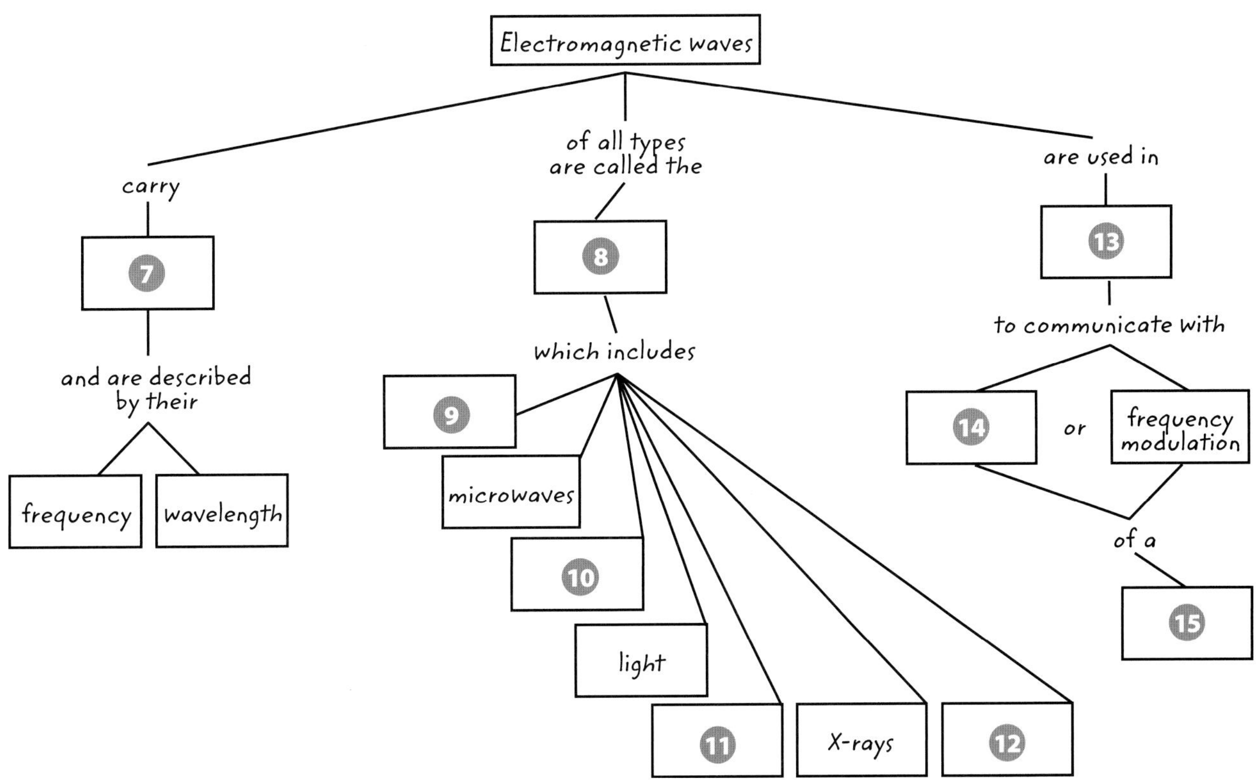

Chapter 17 Review

Understand Key Concepts 

1. What happens to an electromagnetic wave as it passes from space to matter?
   A. It changes frequency.
   B. It changes wavelength.
   C. It slows down.
   D. It speeds up.

2. Which property of the carrier wave shown below has been modulated?

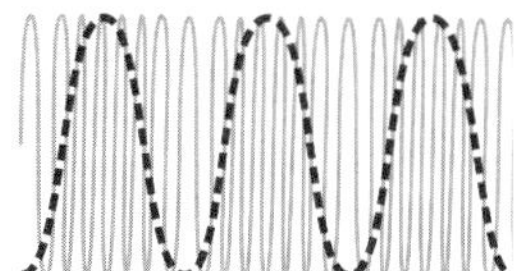

   A. amplitude
   B. energy
   C. frequency
   D. speed

3. Which type of electromagnetic waves has the longest wavelengths?
   A. infrared waves
   B. microwaves
   C. radio waves
   D. ultraviolet waves

4. Which type of electromagnetic waves does your body emit?
   A. infrared waves
   B. microwaves
   C. radio waves
   D. ultraviolet waves

5. Which three types of electromagnetic waves carry most of the Sun's energy that strikes Earth?
   A. infrared waves, light, ultraviolet waves
   B. ultraviolet waves, light, microwaves
   C. light, ultraviolet waves, radio waves
   D. X-rays, infrared waves, gamma waves

6. What happens to a wave if its frequency decreases?
   A. Its amplitude increases.
   B. Its energy decreases.
   C. Its speed increases.
   D. Its wavelength decreases.

7. Which list of electromagnetic waves is in the correct order from highest to lowest energy?
   A. gamma rays, radio waves, infrared waves, microwaves
   B. ultraviolet waves, gamma rays, X-rays, light
   C. light, infrared waves, microwaves, radio waves
   D. X-rays, gamma rays, ultraviolet waves, light

8. The image below most likely was produced using which type of electromagnetic wave?

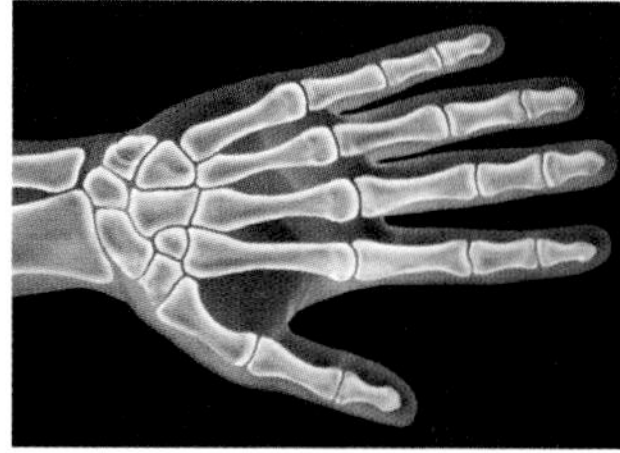

   A. gamma rays
   B. light
   C. microwaves
   D. X-rays

9. Which type of electromagnetic wave causes a chemical to glow in a fluorescent lightbulb?
   A. gamma waves
   B. infrared waves
   C. radio waves
   D. ultraviolet waves

Chapter Review

Assessment

Online Test Practice

Critical Thinking

10 **Classify** Look around your home, school, and community. Make a list of different devices that use electromagnetic waves. Beside each device, write the type of electromagnetic wave the device uses.

11 **Sequence** the types of electromagnetic waves according to their ability to penetrate matter.

12 **Synthesize** The waves in the diagram below represent the colors yellow, blue, and red. Describe which color is represented by waves A, B, and C, and explain how you made your choice.

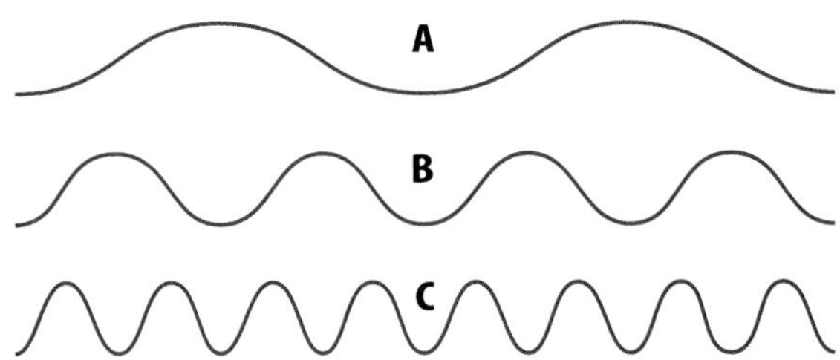

13 **Apply** You are in a dark room. Are there any electromagnetic waves present? Why or why not?

14 **Explain** how you can determine the frequency of an electromagnetic wave moving through space if you know the wavelength.

15 **Assess** New Zealand has one of the highest rates of skin cancer in the world. In terms of electromagnetic waves, what might be causing the disease?

Writing in Science

16 **Write** A classmate can't understand what electromagnetic waves have to do with his or her everyday life. Write a note at least five sentences long to your friend explaining why understanding electromagnetic waves is important and how the information might apply to everyday life.

REVIEW THE BIG IDEA

17 What are electromagnetic waves? How do they travel outward from their source?

18 The photos below show that the Sun emits different types of electromagnetic waves. How can you describe the main types of electromagnetic waves the Sun emits? What are some ways in which people use electromagnetic waves?

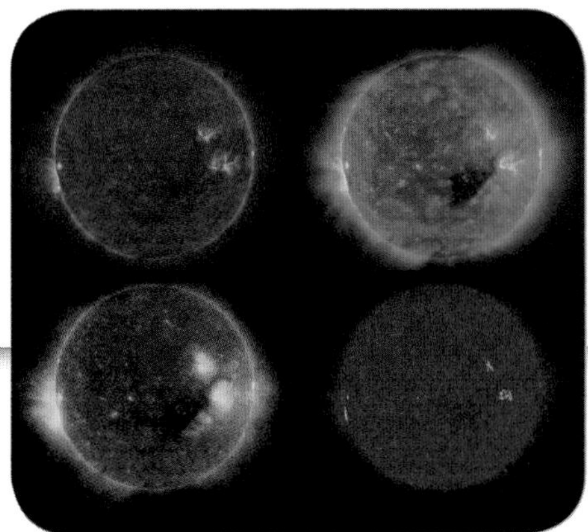

Math Skills

Math Practice

Solve One-Step Equations

19 What is the wavelength of an electromagnetic wave with a frequency of 145,000 Hz?

20 What is the wavelength of an electromagnetic wave with a frequency of 165,000 Hz?

21 An electromagnetic wave has a wavelength of 0.65 km. What is its frequency?

22 An electromagnetic wave has a wavelength of 0.45 km. What is its frequency?

23 What is the wavelength of an electromagnetic wave with a frequency of 225,000 Hz?

Standardized Test Practice

Record your answers on the answer sheet provided by your teacher or on a sheet of paper.

Multiple Choice

Use the figures to answer questions 1 and 2.

Electromagnetic wave

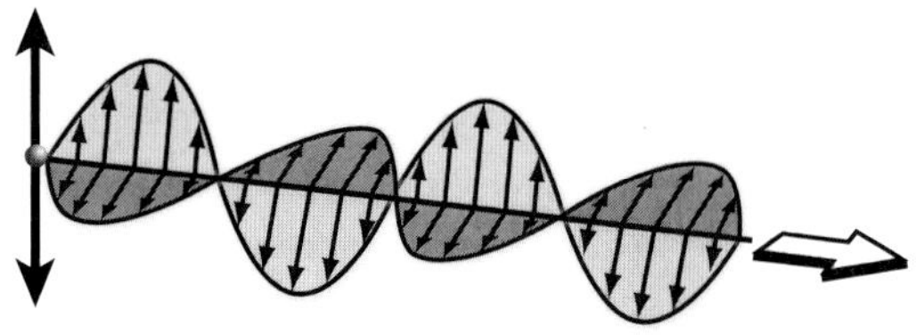

Rope wave

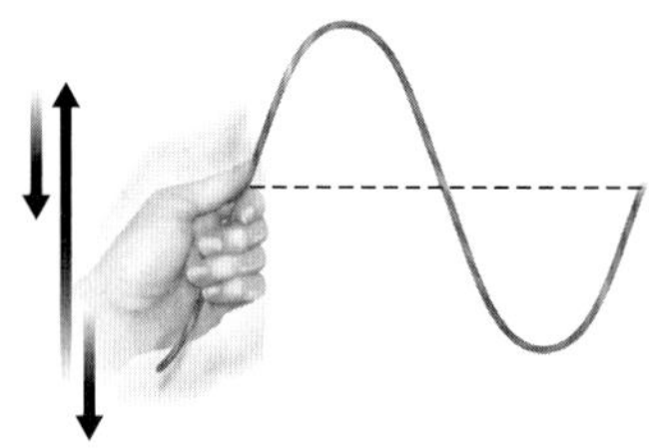

1 The figures show an electromagnetic wave and a wave in a rope. Which characteristic do these two waves have in common?

A They are both transverse waves.

B They are both longitudinal waves.

C They both have electric and magnetic fields.

D They both have compressions and rarefactions.

2 How do these waves differ?

A One has crests, while the other has troughs.

B One can travel through empty space, while the other cannot.

C One transmits energy, while the other transmits matter.

D One has an electric field, while the other has a magnetic field.

3 Which forms an electromagnetic wave when it vibrates?

A an electron

B a carrier wave

C a photon

D an X-ray

4 Which shows the correct order from lowest to highest energy for different electromagnetic waves?

A light → X-ray → microwave

B gamma ray → light → radio wave

C X-ray → microwave → ultraviolet wave

D radio wave → microwave → gamma ray

Use the figure to answer questions 5 and 6.

Violet Blue Green Yellow Orange Red

5 The figure lists different colors of light. Based on the sequence shown, what could the arrow represent?

A increasing energy

B increasing frequency

C increasing speed

D increasing wavelength

6 What do all these colors of light have in common?

A their energy

B their frequency

C their speed

D their wavelength

Assessment
Online Standardized Test Practice

Use the figure to answer questions 7 and 8.

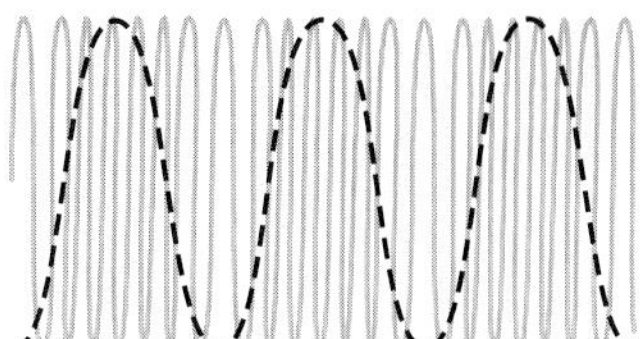

7 The figure shows a carrier wave used to broadcast radio signals. What is modulated in this wave in order to carry information?

A amplitude

B crest height

C frequency

D loudness

8 Which is true of this type of signal wave?

A It can easily pass around obstacles.

B It can be seen with the unaided eye.

C It can travel far by bouncing off particles in the atmosphere.

D It can have wavelengths that are less than a meter long.

9 Which is the number of satellites needed to calculate the position of a Global Positioning System (GPS) receiver on Earth?

A 2

B 4

C 6

D 8

Constructed Response

10 How does the speed of an electromagnetic wave relate to its frequency and its wavelength? If the speed of light is 300,000 km/s through empty space, how would you calculate the wavelength of light that has a frequency of 150,000 Hz?

Use the figure to answer questions 11 and 12.

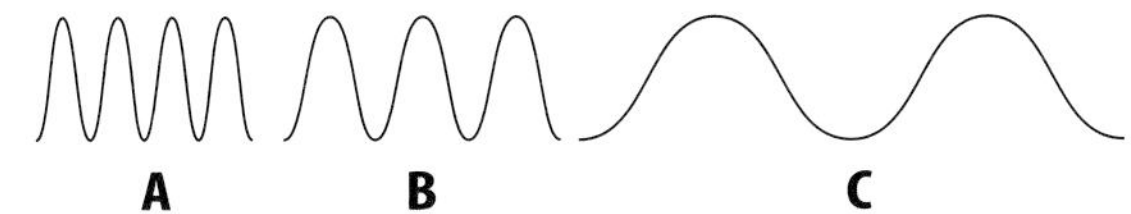

11 Describe how electromagnetic waves A, B, and C differ (if at all) in amplitude, wavelength, frequency, and energy?

12 Choose three types of electromagnetic radiation which would fit the characteristics of the waves shown. Identify which wave could represent each type and why.

13 What kind of information does an infrared image taken from a satellite give about Earth? How might this kind of information be used?

NEED EXTRA HELP?														
If You Missed Question...	1	2	3	4	5	6	7	8	9	10	11	12	13	14
Go to Lesson...	1	1	1	2	2	2	3	3	3	3	1	2	2	3

Chapter 18

Light

THE BIG IDEA **How does matter affect the way you perceive and use light?**

Are they real?

The tiny flowers seem real, but they are not. This is a close-up photo of a chrysanthemum flower covered with water drops. The tiny flower images are reflections in the water drops.

- What causes images of the flower to appear in the water drops?
- Why do reflections appear in the water but not on the flower?
- How does matter affect the way you perceive light?

Get Ready to Read

What do you think?

Before you read, decide if you agree or disagree with each of these statements. As you read this chapter, see if you change your mind about any of the statements.

1. Both the Sun and the Moon produce their own light.
2. The color of an object depends on the light that strikes it.
3. Mirrors are the only surfaces that reflect light.
4. The image in a mirror is always right side up.
5. A prism adds color to light.
6. When light moves from air into water or glass, it travels more slowly.
7. A laser will burn a hole through your skin.
8. Telephone conversations can travel long distances as light waves.

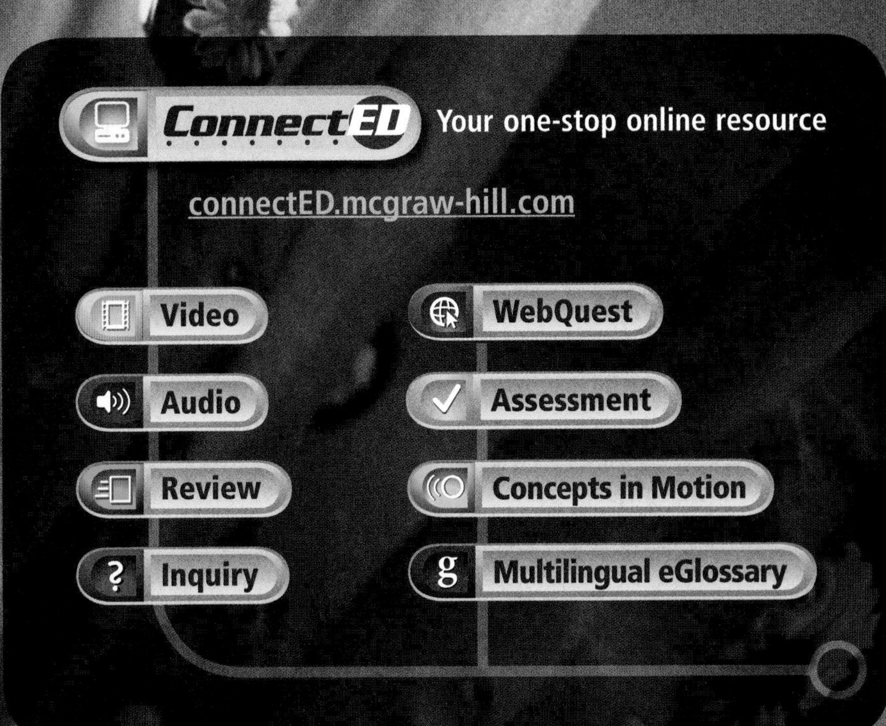

Lesson 1

Light, Matter, and Color

Reading Guide

Key Concepts ESSENTIAL QUESTIONS

- What are some sources of light, and how does light travel?
- What can happen to light that strikes matter?
- Why do objects appear to have different colors?

Vocabulary

light p. 635
reflection p. 635
transparent p. 636
translucent p. 636
opaque p. 636
transmission p. 637
absorption p. 637

Multilingual eGlossary

Video BrainPOP®

Inquiry Why So Colorful?

On a clear day, it's exciting to watch colorful hot-air balloons launch into the blue sky. What makes the balloons so colorful? Did you ever wonder how you are able to see all the different colors?

Launch Lab

10 minutes

How can you make a rainbow?

After it rains, you might see a rainbow stretching across the sky. How can sunlight change into so many different colors? How can you make a rainbow of colors by shining light through a prism?

1. Read and complete a lab safety form.
2. Use **tape** to attach a sheet of **white paper** to a wall.
3. In a darkened room, stand about 1 m from the paper. Hold a **prism** in front of a **flashlight,** as shown in the photo. Turn on the flashlight. Shine the light through the prism onto the paper.
4. Adjust the prism until you see bands of color on the paper. Turn the prism, and have your partner trace the bands of color on the paper with **colored pencils.** Record your observations in your Science Journal. Turn the prism two more times, and trace the colors.
5. Switch places with your partner, and repeat steps 3–4.

Think About This

1. What is the sequence of colors in the bands of light in your drawings? Is it the same in all the drawings?
2.  **Key Concept** How do you think the prism affected the white light from the flashlight?

What is light?

Suppose you are taking a tour of a cave below Earth's surface and the lights go out. Would you see anything? No! You could not see anything because there would be no light for your eyes to perceive. **Light** *is electromagnetic radiation that you can see.* Electromagnetic radiation has wave properties and particle properties. A particle of electromagnetic radiation is called a photon. The frequency of a light wave depends on the amount of energy carried by a photon of light. Light waves can carry this energy through space and some matter.

Review Vocabulary

electromagnetic radiation
energy carried through space or matter by electromagnetic waves

Sources of Light

You can see an object if it is luminous or if it is illuminated. Luminous objects, such as the campfire in **Figure 1,** release, or emit, light. The Sun is luminous because it is made of hot, glowing gases. A traffic light and a firefly also are luminous objects. Luminous objects are sources of light. We also can see illuminated objects, such as the camper and trees in **Figure 1.** However, they do not emit light and are not sources of light. The Moon might appear bright, but it is just a rocky sphere. You see the Moon because it reflects light from the Sun. **Reflection** *is the bouncing of a wave off a surface.* In a dark cave, there is no light to reflect off objects so you cannot see anything.

Key Concept Check What are some sources of light?

Figure 1 You see the campfire because it is luminous. You see other objects in the picture because light from the fire illuminates them.

▲ **Figure 2** Light spreads out as it moves away from a source. An object in the path of light forms shadows where there is less light.

Figure 3 Different amounts of light can pass through the transparent and translucent parts of this window. The opaque walls and table do not allow light to pass through them. ▼

How Light Travels

Light travels as waves moving away from a source. Scientists often describe these waves as countless numbers of light rays spreading out in all directions from a source. You can model light rays by shining a light through a comb, as shown in **Figure 2.**

What are the spaces between the rays in **Figure 2?** They are shadows or places with less light. The shadows show that light normally travels in a straight line. However, objects in its path can cause light to change direction. It also can spread out slightly as it moves through a small opening. Sometimes, the effect is difficult to see because light waves are so small.

 **Key Concept Check** How does light travel?

Light and Matter

What can you see through your classroom window? Would you still be able to see anything if the blinds were closed? How do different types of matter affect light?

You can see objects clearly through air, clean water, plain glass, and some plastics. *A material that allows almost all the light that strikes it to pass through and form a clear image is* **transparent.** The unfrosted parts of the window in **Figure 3** are transparent.

Light also passes through the frosted parts of the window, but clear images do not form. *A material that allows most of the light that strikes it to pass through and form a blurry image is* **translucent.** Plastics with textured surfaces also are examples of translucent materials.

A material through which light does not pass is **opaque.** No light passes through the chairs or the table in **Figure 3.** Wood and metal are examples of opaque materials.

 **Key Concept Check** What can happen to light that strikes matter?

Figure 4 The window transmits, absorbs, and reflects light.

Visual Check Which materials in the photograph are transparent? Which are opaque?

Transmission of Light

You just read that light passes through transparent and translucent objects. *The passage of light through an object is called* **transmission.** The girls and you can see objects through the window in **Figure 4** because the window transmits light. A luminous object or an illuminated object on one side of the window is visible on the other side of the window. Also, the energy carried by the light waves from these objects can pass through the window.

Absorption of Light

Imagine standing near a window on a spring day. The transparent window transmits some sunlight, and it lands on you. If you touch the window, it might feel warm. Some of the energy in sunlight stays inside the window. *The transfer of energy by a wave to the medium through which it travels is called* **absorption.** The energy causes atoms in the material to vibrate faster, increasing the temperature of the material. All materials absorb some of the light that strikes them. The window feels warm because it absorbs some of the sunlight's energy.

Reflection of Light

When you look at a pane of glass, you sometimes can see an image of yourself. Light bounces off you, strikes the glass, and bounces back to your eye. Recall that the bouncing of a wave off a surface is called reflection.

Look again at the window in **Figure 4.** Notice that you can see the transmission and reflection of light. You cannot see it, but some of the light also is absorbed. Most types of matter interact with light in a combination of ways.

WORD ORIGIN

transmission
from Latin *transmissionem,* means "sending over or across"

FOLDABLES®

Make a vertical trifold book and label it as shown. Use it to organize your notes on the sources of light, light and matter, and the relationship between light and color.

Light Sources

Light & Matter

Light & Color

ACADEMIC VOCABULARY

image
(noun) a figure, picture, or representation of an object

Inquiry MiniLab

15 minutes

What color is that?

Have you ever noticed that objects seem to change color when you put on sunglasses? What causes this to happen?

Object	Filter			
	None	Red	Blue	Green

1. Read and complete a lab safety form.
2. Copy the data table into your Science Journal. You can add lines as needed.
3. Write *PHYSICS!* on **white paper** as shown here using **red** and **blue markers.**
4. Choose at least **4 different-colored objects.** Observe the word *PHYSICS!* and the objects with no filter and then through **red, blue,** and **green filters.** Record your observations in your Science Journal.

Analyze and Conclude

1. **Contrast** your observations with and without filters.
2. **Key Concept** Explain why the objects you observed appeared to change color when you used a filter.

Light and Color

Recall that visible light is electromagnetic radiation with wavelengths of all colors of the rainbow. The longest wavelengths of light appear red. Violet has the shortest wavelengths. Other colors of light have different wavelengths. White light is a mixture of all wavelengths of light. How does this account for the colors you might see?

Colors you see result from wavelengths of light that enter your eyes. When you look at a luminous object such as a campfire, you see the colors emitted by the fire and glowing logs. What happens when you look at an illuminated object? That depends on whether the object is opaque, transparent, or translucent.

Opaque Objects

Suppose white light strikes a box of crayons, as shown in **Figure 5.** Each crayon absorbs all wavelengths of light except its color. For example, the green crayon absorbs all colors except green. The green wavelengths reflect back into your eyes, and you see green. The red crayon absorbs all colors except red, and it reflects red. The black crayon absorbs all colors. The color of an opaque object is the color of light it reflects.

What do you think would happen to the colors reflected by the crayons if red light, instead of white light, were shining on them? Would the green crayon still appear green? No. It would absorb the red light, but there would be no green light to reflect. The green crayon would appear black. The blue crayon also would appear black. The red crayon and the white crayon would appear red because they would reflect red light. The color you see always depends on the color of light that the object reflects.

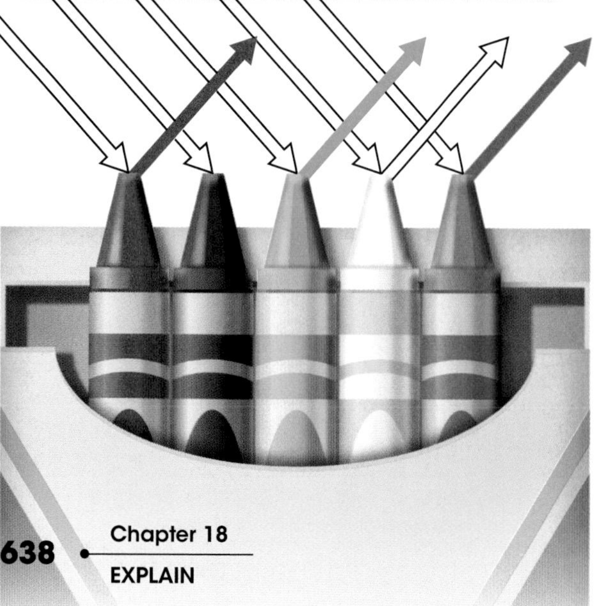

Figure 5 The color of an object is the color of light that reflects off the object.

Transparent and Translucent Objects

Absorption, transmission, and reflection also explain the color of transparent and translucent objects. For example, suppose white light, such as sunlight, shines through a piece of blue glass. The glass absorbs all wavelengths of light except blue. The blue wavelengths pass through the glass to your eyes. If the blue glass is translucent, it still only transmits blue light, but the image is blurry. The color of a transparent or translucent object is the color it transmits.

What color would you see if you held a piece of red cellophane, or a red filter, in front of each crayon in **Figure 5?** Remember that the color you see depends on the colors in the light source, the absorbed colors, and the colors that reach your eyes.

Combining Colors

Have you ever mixed several colors of watercolor paints to get just the shade you want? You know that you can make many different shades from a few basic colors. If you mix too many colors, you get black! Why does that happen?

Combining Pigments Each color of paint in a set of watercolors contains different pigments, or dyes. Each pigment absorbs some colors of light and reflects other colors. Mixing pigments produces many different shades as certain wavelengths are absorbed and fewer colors are reflected to your eyes. As you add each color of pigment, the mixture gets darker and darker because more colors are absorbed. Cyan, magenta, and yellow are the primary pigments. Combining equal amounts of these pigments makes black, as shown in the center of the artist's palette in **Figure 6.**

Combining Colors of Light Red, green, and blue are the primary light colors. If you shine equal amounts of red light, green light, and blue light at a white screen, each color reflects to your eyes. Where two of the colors overlap, both wavelengths reflect to your eyes and you see a third color. Where the three colors overlap, all colors reflect and you see white light, also shown in **Figure 6.**

Key Concept Check Why do objects appear to have different colors?

Figure 6 Colors of pigment combine by subtracting wavelengths. Colors of light combine by adding wavelengths.

Visual Check What is the result of adding any two primary pigments?

Combining Pigments

Each primary pigment subtracts color by absorption. Mixing all three pigments equally produces black.

Combining Light

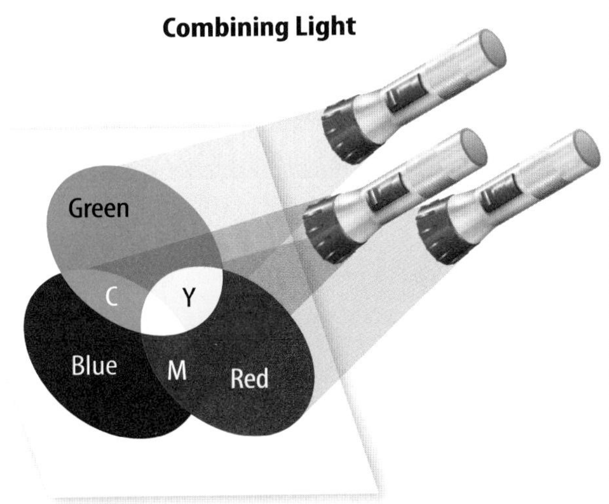

Adding the three primary colors of light produces the colors of the primary pigments and white.

Lesson 1 Review

Assessment Online Quiz

Inquiry Virtual Lab

Visual Summary

The amount of light that can pass through a material determines whether it is transparent, translucent, or opaque.

Reflection is the bouncing of waves off a surface.

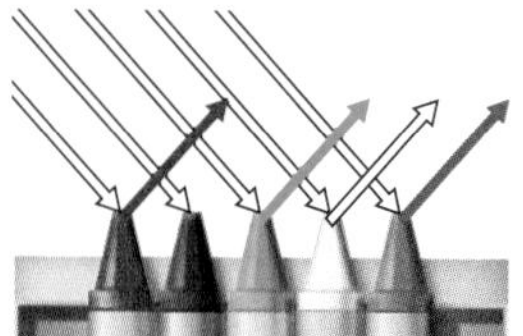

The color of an opaque object is the color of light that reflects off the object.

FOLDABLES®

Use your lesson Foldable to review the lesson. Save your Foldable for the project at the end of the chapter.

What do you think NOW?

You first read the statements below at the beginning of the chapter.

1. Both the Sun and the Moon produce their own light.
2. The color of an object depends on the light that strikes it.

Did you change your mind about whether you agree or disagree with the statements? Rewrite any false statements to make them true.

Use Vocabulary

1. **Compare and contrast** how an opaque object and a transparent object affect light.
2. The passage of light through an object is called ________.
3. An electromagnetic wave that you can see is ________.

Understand Key Concepts 

4. Which indicates that light travels in straight lines?
   - **A.** colors
   - **B.** pigments
   - **C.** shadows
   - **D.** waves
5. **Contrast** how white light interacts with a blue book and a blue stained glass window.
6. **Describe** three things that can happen when light strikes an object.

Interpret Graphics

7. **Explain** why the overlap region of the filters at the right looks black when you look through the filters toward a white light.

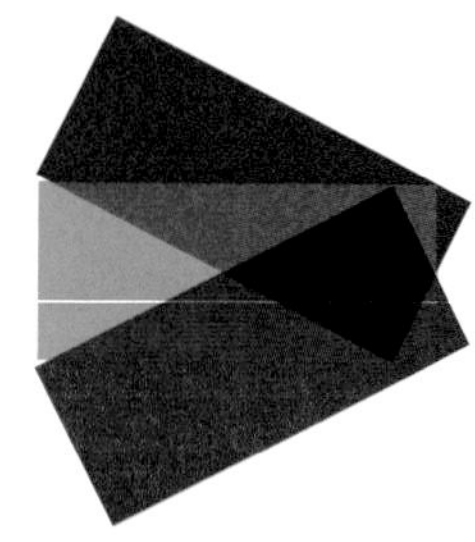

8. **Summarize** Copy the graphic organizer below. Fill in three ways you can describe a material according to whether light can pass through it.

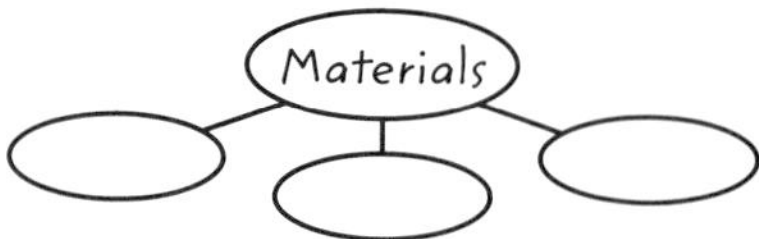

Critical Thinking

9. **Construct** a diagram showing what happens when rays of white light strike a yellow, opaque object.

Color!

There's more to it than meets the eye.

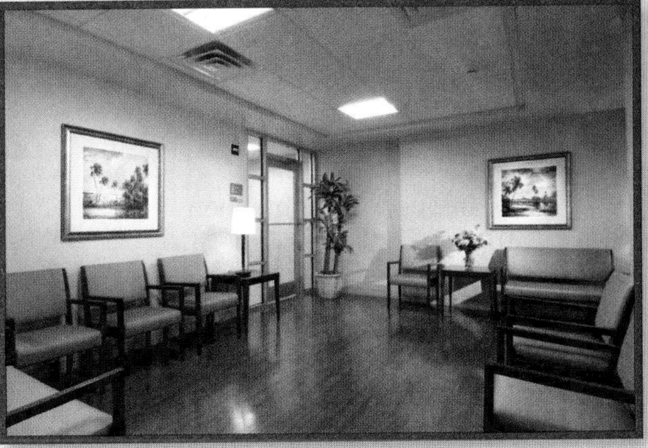

▲ **This waiting room is designed to be soothing.**

Do you have a favorite color? If you had to explain why it's your favorite, what would you say? You know that an object's color is due to the wavelengths of light it reflects, but you might be surprised to learn that color is more than something you see. Color can affect your mood, your emotions, and even your creativity at school or work!

Color psychology is the study of the effect of color on the way people think and behave. For example, the color blue is associated with peacefulness. It tends to make people feel calm. Studies have shown that blue rooms promote creativity.

What words would you use to describe a bright yellow room? Would you say the room is sunny and cheerful? People often connect happiness with yellow and nature with green. Green, like blue, tends to calm people. Hospitals often use green in waiting rooms to reduce people's anxiety. What about orange? Many people connect orange with energy and warmth.

Facts about how color affects moods can be fun to learn. However, for some people, it is their business. Industrial psychologists study ways to maximize workers' productivity and safety. Many industrial psychologists pay close attention to the colors used in different areas of the workplace. Those who work in advertising and merchandising choose specific colors for product packages that connect the product inside with certain ideas or emotions.

So, what does your favorite color mean to you? Perhaps it changes depending on your mood or experiences. Regardless of what your favorite color is, it is interesting to know that there is science behind colors chosen for many places and products used daily.

▲ **Do you think that blue calms people because it makes them think about the ocean or the sky?**

Red makes people think of energy and excitement. That's why it's a popular color for sports cars.

It's Your Turn

RESEARCH AND REPORT Find out more about how a person's culture might influence the way the person perceives colors. Make a poster to share what you learn with your classmates.

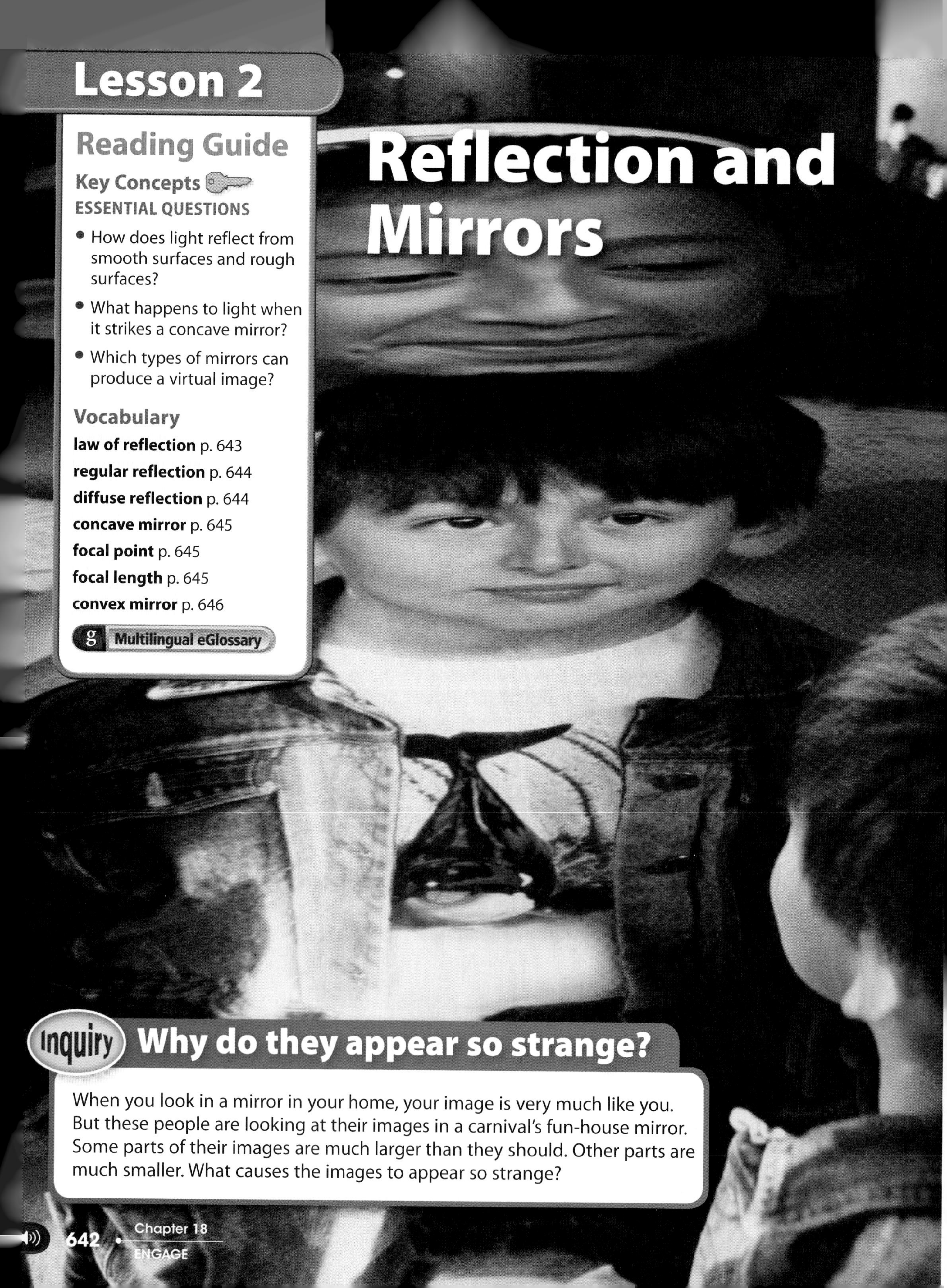

Lesson 2

Reflection and Mirrors

Reading Guide

Key Concepts
ESSENTIAL QUESTIONS

- How does light reflect from smooth surfaces and rough surfaces?
- What happens to light when it strikes a concave mirror?
- Which types of mirrors can produce a virtual image?

Vocabulary

law of reflection p. 643
regular reflection p. 644
diffuse reflection p. 644
concave mirror p. 645
focal point p. 645
focal length p. 645
convex mirror p. 646

g Multilingual eGlossary

Inquiry Why do they appear so strange?

When you look in a mirror in your home, your image is very much like you. But these people are looking at their images in a carnival's fun-house mirror. Some parts of their images are much larger than they should. Other parts are much smaller. What causes the images to appear so strange?

10 minutes

How can you read a sign behind you?

Mirrors aren't just for looking at your reflection. You can use them to see around corners, over and under obstacles, and inside things.

1. Read and complete a lab safety form.
2. Use a **marker** to write *CAT* about 5 cm tall on an **index card.** Use **tape** to attach the card to a wall.
3. Stand with your back toward the wall. Use a **mirror** to look at the letters on the card. Sketch what you see in your Science Journal.
4. Repeat step 3, but this time use **an additional mirror** to look at the letters in the first mirror.

Think About This

1. Draw a simple diagram showing the path of light from the card to your eye for each setup.
2.  **Key Concept** How did one reflection in a mirror affect the letters? How did two reflections affect the letters?

Reflection of Light

In Lesson 1, you read that reflection is the bouncing of a wave off a surface. Suppose you toss a tennis ball against a wall. If you throw the ball straight toward the wall, it will bounce back to you. Where on the wall would you throw the ball so that a friend standing to your left could catch it? You would throw it toward a point on the wall halfway between you and your friend.

Law of Reflection

Like the tennis ball, light behaves in predictable ways when it reflects. Scientists often model the path of light using straight arrows called rays. The rays in **Figure 7** show how light reflects. An imaginary line perpendicular to a reflecting surface is called the normal. The light ray moving toward the surface is the incident ray. The light ray moving away is the reflected ray.

Notice the angle formed where an incident ray meets a normal. This is the angle of incidence. A reflected ray forms an identical angle on the other side of the normal. This angle is the angle of reflection. According to the **law of reflection,** *when a wave is reflected from a surface, the angle of reflection is equal to the angle of incidence.*

Reading Check How do the reflected and incident rays differ?

Figure 7 If you know a light ray's angle of incidence, you can predict its angle of reflection.

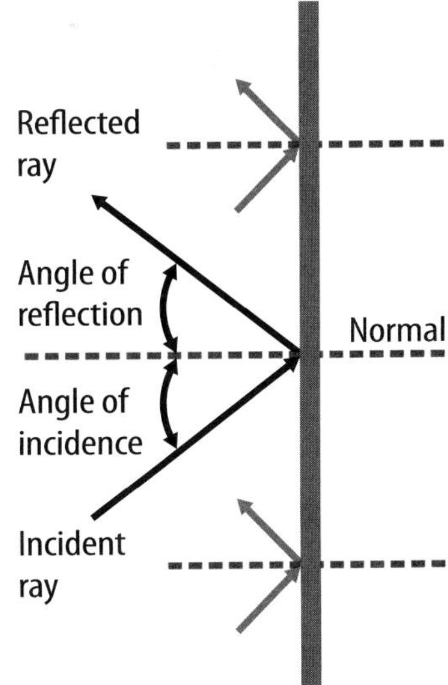

Visual Check If the angle of incidence is 40°, what is the angle of reflection?

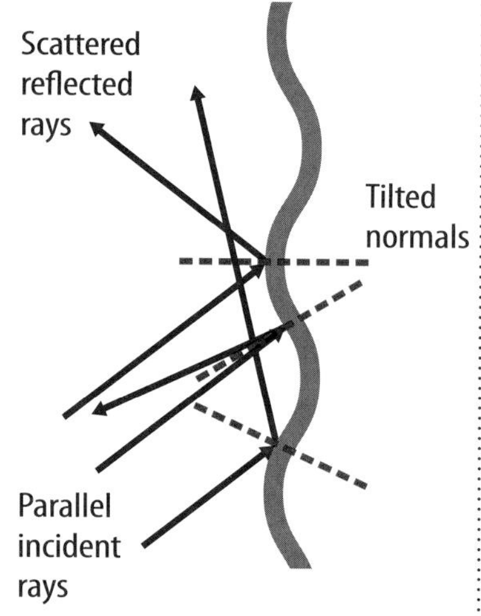

▲ **Figure 8** No clear image forms when light reflects off a rough surface.

Regular and Diffuse Reflection

You see objects when light reflects off them into your eyes. Why can you see your reflection in smooth, shiny surfaces but not in a piece of paper or a painted wall? The law of reflection applies whether the surface is smooth or rough.

Reflection of light from a smooth, shiny surface is called **regular reflection.** Look again at **Figure 7** on the previous page. The three incident rays and the three reflected rays all are parallel. You see a sharp image when parallel rays reflect into your eyes. When light strikes an uneven surface, as in **Figure 8,** the angle of reflection still equals the angle of incidence at each point. However, rays reflect in different directions. *Reflection of light from a rough surface is called* **diffuse reflection.**

Key Concept Check How does light reflect from smooth surfaces and rough surfaces?

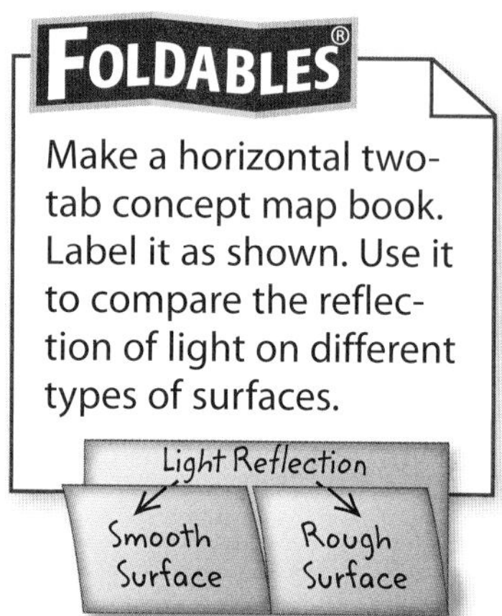

Mirrors

Any surface that reflects light and forms images is a mirror. The type of image depends on whether the reflecting surface is flat or curved.

Plane Mirrors

The word *plane* means "flat," so a plane mirror has a flat reflecting surface. **Figure 9** shows how an image from a plane mirror forms Only a few rays are shown, but the number of rays involved is infinite. Notice that the image formed is the same size as the object. However, it is a virtual image. A virtual image is an image of an object that your brain perceives to be in a place where the object is not.

If you look in a plane mirror and raise your right hand, your image raises its left hand. However, up and down are not reversed.

Figure 9 A plane mirror forms a virtual image.

Your brain knows that light travels in straight lines, so the reflection appears to come from behind the mirror. This is shown by the dashed lines.

Light from a source reflects from each part of the object to the mirror. The mirror reflects the light back to your eye.

Because no object is located at the place where the image appears, the image is called virtual.

Object

Virtual image

15 minutes

Where is the image in a plane mirror?

A plane mirror forms a virtual image. Where is it, compared to the real object?

1. Read and complete a lab safety form.
2. Prop a sheet of **clear acrylic** vertically between **two books.**
3. Place a **candle** 15 cm in front of the acrylic. Use a **safety match** to light the candle in a slightly darkened room.
4. Place an **unlit candle** behind the acrylic. Look through the acrylic. Move the unlit candle to align with the reflection of the lit candle. Observe from several locations to the right and the left of center. Record your observations in your Science Journal.
5. Using a **ruler,** measure from the unlit candle to the acrylic. Record the distance.
6. Repeat steps 4–5 several times with the lit candle different distances from the acrylic.

Analyze and Conclude

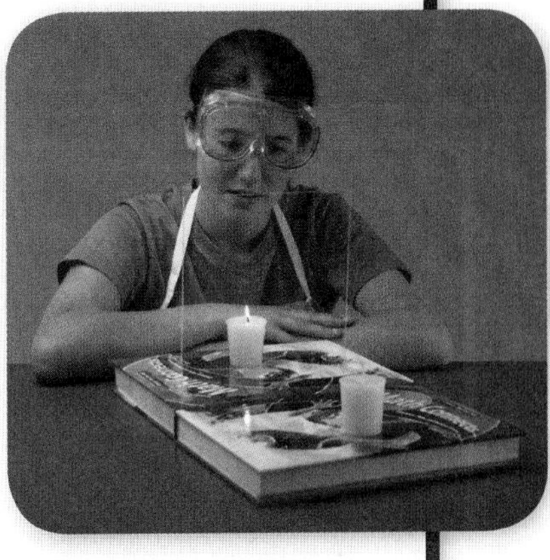

1. **Compare** the distances of the unlit candle and the lit candle from the acrylic.
2. **Draw** a diagram of your setup. Use arrows and labels to show reflection of light and how your brain perceives a virtual image.
3. **Key Concept** Describe and explain the reflection you saw.

Concave Mirrors

Not all mirrors are flat. *A mirror that curves inward is called a* **concave mirror,** like the mirror in **Figure 10.** A line perpendicular to the center of the mirror is the optical axis. The law of reflection determines the direction of reflected rays in **Figure 10.** When rays parallel to the optical axis strike a concave mirror, the reflected rays converge, or come together.

Focal Point Look again at **Figure 10.** Notice the point where the rays converge. *The point where light rays parallel to the optical axis converge after being reflected by a mirror or refracted by a lens is the* **focal point.** In the next lesson, you will read about lenses. Imagine that a concave mirror is part of a hollow sphere. The focal point is halfway between the mirror and the center of the sphere. *The distance along the optical axis from the mirror to the focal point is the* **focal length.** The lesser the curve of a mirror, the longer its focal length. The position of an object compared to the focal point determines the type of image formed by a concave mirror.

If you reverse the direction of the arrows in **Figure 10,** you can see how a flashlight works. The reflector behind the bulb is a concave mirror. The bulb is at the focal point. Light rays from the bulb strike the mirror and reflect as parallel rays.

 **Key Concept Check** How does a concave mirror reflect light?

WORD ORIGIN

concave
from Latin *concavus,* means "hollow"

Figure 10 Light rays that strike a concave mirror converge.

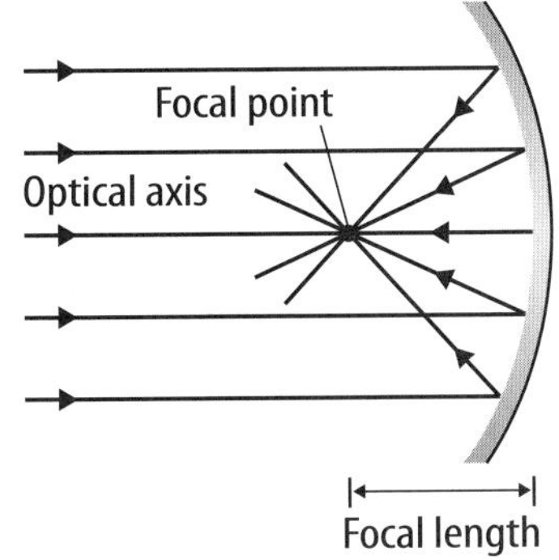

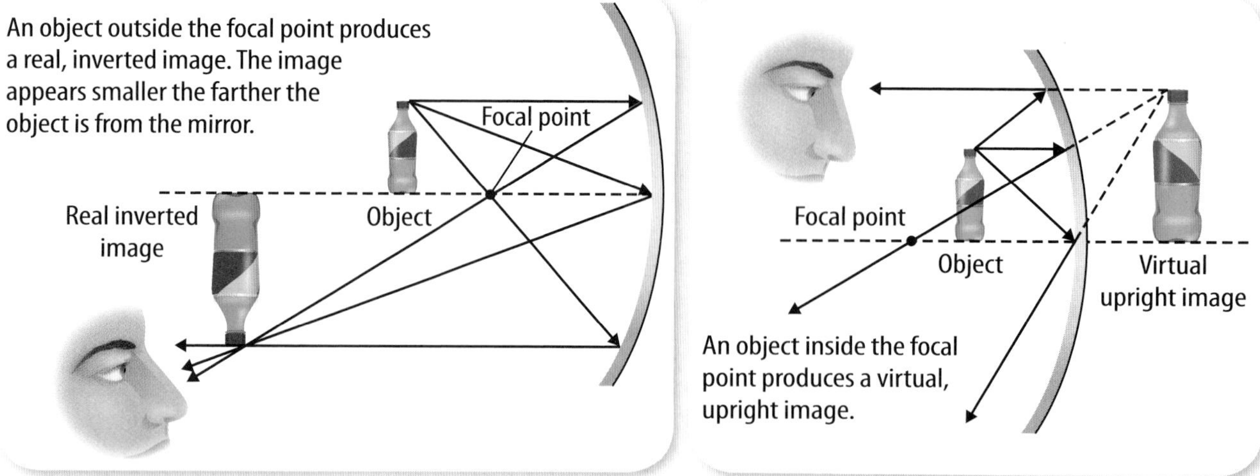
An object outside the focal point produces a real, inverted image. The image appears smaller the farther the object is from the mirror.
Focal point
Real inverted image
Object
Focal point
Object
Virtual upright image
An object inside the focal point produces a virtual, upright image.

▲ **Figure 11** A concave mirror can produce a real or a virtual image.

▲ **Figure 12** The image flips as the spoon moves away from the object.

Figure 13 The image in a convex mirror looks smaller than the object. ▼

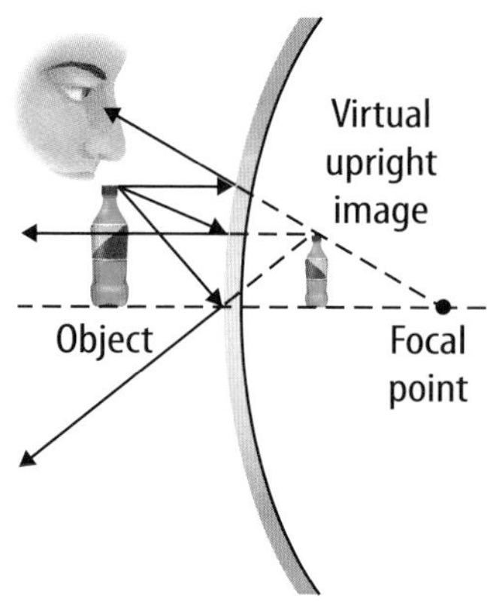
Virtual upright image
Object
Focal point

Types of Images When you look at ray diagrams, think about what you would see if the rays reflected into your eyes. Remember that your brain only perceives where the rays appear to come from, not their actual paths. As shown in **Figure 11**, the image a concave mirror forms depends on the object's location relative to the focal point. The image is virtual if the object is between the focal point and the mirror. The image is real if the object is beyond the focal point. A real image is one that forms where rays converge. No image forms if the object is at the focal point.

Suppose you look at your reflection in the bowl of a shiny spoon. If your face is outside the spoon's focal point, your image is inverted, or upside down, as shown in **Figure 12.** Your image disappears when your face is at the focal point. When your face is inside the focal point, the image is upright.

Convex Mirrors

Have you ever seen a large, round mirror high in the corner of a store? The mirror enables store personnel to see places they cannot see with a plane mirror. *A mirror that curves outward, like the back of a spoon, is called a* **convex mirror.** Light rays diverge, or spread apart, after they strike the surface of a convex mirror. The dashed lines in **Figure 13** model how your brain interprets these rays as coming from a smaller object behind the mirror. Therefore, a convex mirror always produces a virtual image that is upright and smaller than the object.

As you just read, convex mirrors and plane mirrors form only virtual images. However, concave mirrors can form both virtual images and real images.

Key Concept Check Which types of mirrors can form virtual images?

Lesson 2 Review

Assessment Online Quiz

Visual Summary

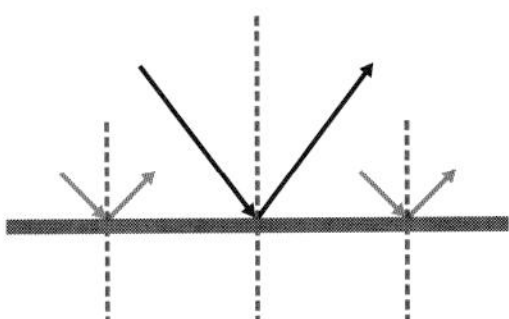

The angle of reflection equals the angle of incidence, even if a surface is rough.

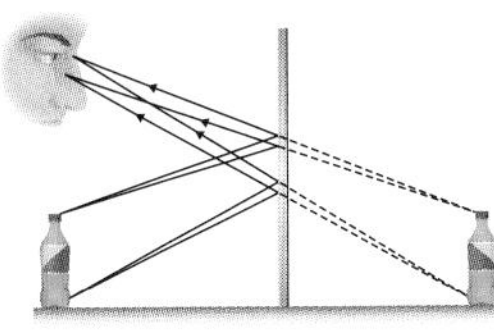

The reflection of an object in a plane mirror is a virtual image the same size as the object.

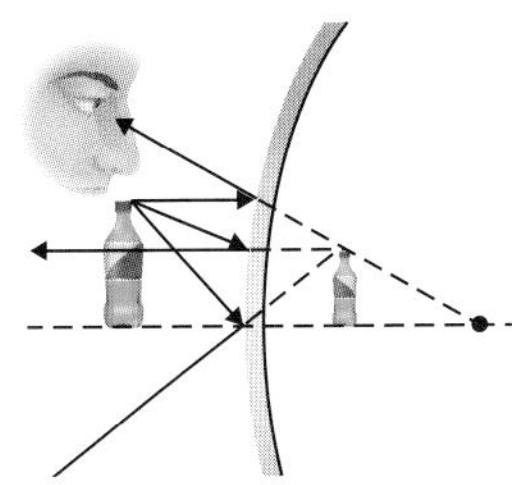

The image formed by a curved mirror might have a different size and might be inverted.

Use your lesson Foldable to review the lesson. Save your Foldable for the project at the end of the chapter.

What do you think NOW?

You first read the statements below at the beginning of the chapter.

3. Mirrors are the only surfaces that reflect light.

4. The image in a mirror is always right side up.

Did you change your mind about whether you agree or disagree with the statements? Rewrite any false statements to make them true.

Use Vocabulary

1. **Distinguish** between the focal point and the focal length of a mirror.
2. The reflection in a clear window of a store is a(n) ________.
3. A mirror that converges light rays is a ________ mirror.

Understand Key Concepts 

4. Which term describes reflection from a rough surface?
   - **A.** diffuse
   - **B.** irregular
   - **C.** regular
   - **D.** translucent
5. Which terms describe the image of an object located between a concave mirror and its focal point?
   - **A.** real, inverted
   - **B.** real, upright
   - **C.** virtual, inverted
   - **D.** virtual, upright
6. **Contrast** your reflection from the back of a shiny spoon and from a plane mirror.
7. **Explain** why the image in a plane mirror is a virtual image.

Interpret Graphics

8. **Organize Information** Copy and fill in the graphic organizer below to describe how real and virtual images are produced by different types of mirrors.

Type of Mirror	How Images Form

Critical Thinking

9. **Evaluate** The rearview mirror on a car is often slightly convex. Explain why this is more practical than a plane mirror.
10. **Construct** a diagram showing the image that forms when an object is inside the focal point of a concave mirror.

Inquiry **Skill Practice** **Measure** 40 minutes

How can you demonstrate the law of reflection?

Materials

modeling clay

plane mirror

ruler

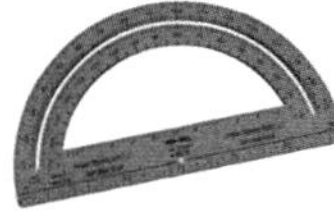

protractor

Also needed: paper, corrugated cardboard, pin

Safety

When light strikes a mirror, the angle of reflection equals the angle of incidence. You can demonstrate the law of reflection by measuring the angles of incident and reflected light rays between an object and a mirror.

Learn It

One method of collecting data is to use a scientific tool and **measure** a value. You can use a protractor to measure the angles at which light strikes and reflects off a surface. You can use a ruler to measure distances of an object and its image from a mirror.

Try It

1. Read and complete a lab safety form.
2. Draw a line across a piece of paper. Label the line *Mirror*. Place the paper on cardboard so it will support pins.
3. Use clay to hold a mirror upright along the line, as shown in the picture.
4. Push a pin into the paper 5 cm in front of the mirror. Label this point on the paper *Object.*
5. With your eyes to the right of the pin, look toward the pin's image in the mirror. Draw a line on the paper from you to the image.

5

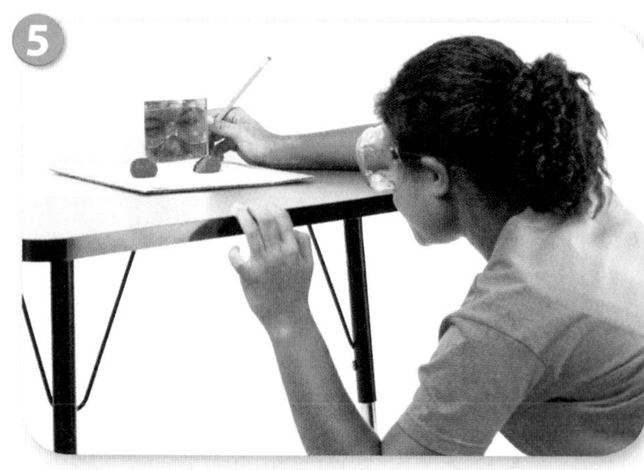

6. Have your partner repeat step 5 by looking at eye-level from the left of the pin.
7. Remove the pin and the mirror.

Apply It

8. First determine the location of the pin's virtual image. Extend the lines that you and your partner drew so that they meet at a point behind the mirror line. Label this point *Virtual image.*
9. Draw and label the incident ray from the object to the point where the line you drew in step 5 crossed the mirror. Draw a normal from the mirror at this point.
10. Light traveled from the object, toward this point, and back to your eye. Label the reflected ray.
11. Have your partner repeat steps 9–10 for the line drawn in step 6.
12. Draw your setup in your Science Journal. Measure and record the distance of the virtual image to the mirror.
13. Use a protractor to measure the angles of incidence and reflection for the light rays you and your partner saw. Record your measurements.
14. **Key Concept** How did your measurements support what you have learned about virtual images formed by a plane mirror? How did they demonstrate the law of reflection?

Lesson 3

Refraction and Lenses

Reading Guide

Key Concepts
ESSENTIAL QUESTIONS

- What happens to light as it moves from one transparent substance to another?
- How do convex lenses and concave lenses affect light?
- How do eyes detect light and color?

Vocabulary

refraction p. 650
lens p. 652
convex lens p. 652
concave lens p. 652
rod p. 657
cone p. 657

 Multilingual eGlossary

 Video BrainPOP®

Inquiry Why is its head so big?

The cat's head might look gigantic compared to its body, but it is just an illusion. Light can produce interesting effects when it moves from one transparent material, such as glass, to another transparent material, such as water. Why do you think this happens?

Inquiry Launch Lab

15 minutes

What happens to light that passes from one transparent substance to another?

Objects look different in water than in air. Light that reflects off objects changes when it moves from one transparent material to another.

1. Read and complete a lab safety form.
2. Pour about 150 mL of water into a **250-mL beaker.** Pour 150 mL of **vegetable oil** into a **second beaker.**
3. Place three **test tubes** in a **test-tube rack.** Leave one test tube empty. Half fill one with water. Half fill another with vegetable oil.
4. One at a time, place each test tube into the water beaker. Look through the side of the beaker at the test tube. Record your observations in your Science Journal.
5. Remove and dry the test tubes with a **paper towel.**
6. Repeat steps 4–5 using the oil beaker. Record your observations.

Think About This

1. Draw a diagram showing the substances light passed through as it moved from the test tube to your eye for each setup. Remember to include the glass of the test tube and the beaker.
2. **Key Concept** Summarize your observations of the test tubes of air, water, and oil.

Refraction of Light

Have you ever tried to pick up something from the bottom of a container of water and the object was deeper than you thought it was? This happens because light waves can change direction. Light always travels through empty space at the same speed—300,000 km/s. Light travels more slowly through a medium (plural, media), such as air, glass, or water. The atoms of the material interact with the light waves and slow them down. Some substances, such as air, only slow light a little. Others, such as glass, slow the light more.

As a light wave moves from one medium into another, its speed changes. If it enters the new medium at an angle, the wave will change direction. *The change in direction of a wave as it changes speed while moving from one medium to another is called* **refraction.**

Each transparent material has a property called the index of refraction, as shown in **Table 1.** A medium that has a high index of refraction is sometimes called slow because light moves more slowly through it. A medium that has a relatively low index of refraction, such as air, is called fast.

Key Concept Check What happens to light as it moves between transparent substances?

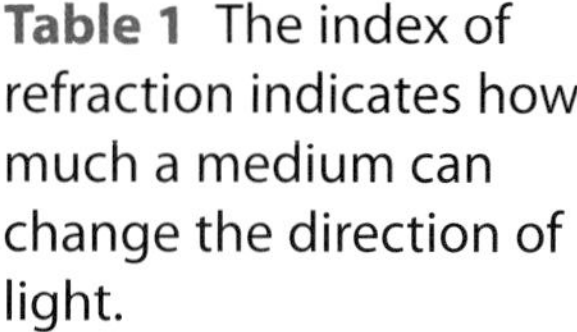

Table 1 The index of refraction indicates how much a medium can change the direction of light.

Table 1

Medium	Index of Refraction
Vacuum	1.0000
Air	1.0003
Ice	1.31
Water	1.333
Oil	1.47
Ovenproof glass	1.47
Diamond	2.417

Moving Into a Slower Medium

Suppose you roll a toy car across a table straight at a piece of fabric. The front tires of the car slow down when they hit the fabric. The car continues to move in a straight line but more slowly. If you roll the car at an angle, one of the front tires will hit the fabric before the other. That side of the car will slow down, but the rest of the car will continue at the same speed until the other tire hits the fabric. This will cause the car to turn and change direction.

As shown in **Figure 14,** a light wave behaves in a similar way when it moves into a slower medium. Recall that a normal is a line perpendicular to a surface. As light moves into a slower medium at an angle, it changes direction toward the normal.

Moving Into a Faster Medium

What happens when light in the figure moves back into the air? Suppose you ride your bike from a sidewalk into a muddy field and then back onto a sidewalk. You use the same energy to pedal the whole time, but you move more slowly in the mud. When you move back onto the sidewalk, you speed up.

Similarly, as light moves into a medium with a lower index of refraction, it speeds up. The wave is still at an angle, so the part that leaves the slower medium first, speeds up sooner. This causes the wave to turn away from the normal, as shown on the right in **Figure 14.**

You see the boundaries between surfaces such as air, glass, and water because of refraction. If transparent substances have the same index of refraction, you do not see the surfaces. The Launch Lab at the start of this lesson shows this effect.

Math Skills

Use Scientific Notation

You can calculate the index of refraction of a material by dividing the speed of light in a vacuum (3.0×10^8 m/s) by the speed of light in that material. When dividing numbers written in scientific notation, divide the coefficients and subtract the exponents. For example, what is the index of refraction of a substance in which light travels at 2.0×10^8 m/s?

a. Set up the problem.

Index of refraction $= \frac{3.0 \times 10^8 \text{ m/s}}{2.0 \times 10^8 \text{ m/s}}$

Note that units cancel.

b. Divide the coefficients.

$3.0 \div 2.0 = 1.5$

c. Subtract the exponents.

$8 - 8 = 0$

The index of refraction is $1.5 \times 10^0 = 1.5 \times 1 = 1.5$.

Practice

The speed of light in a material is 1.56×10^8 m/s. What is its index of refraction?

- **Math Practice**
- **Personal Tutor**

Refraction

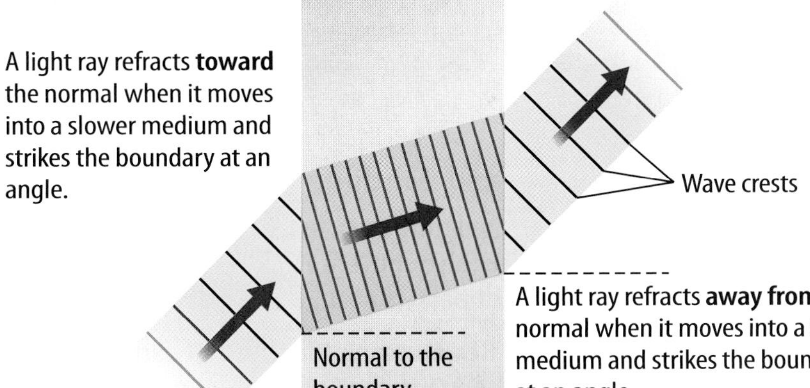
A light ray refracts **toward** the normal when it moves into a slower medium and strikes the boundary at an angle.
Wave crests
A light ray refracts **away from** the normal when it moves into a faster medium and strikes the boundary at an angle.
Normal to the boundary
Air
Glass
Air

Figure 14 The direction of refraction depends each material's index of refraction.

Personal Tutor

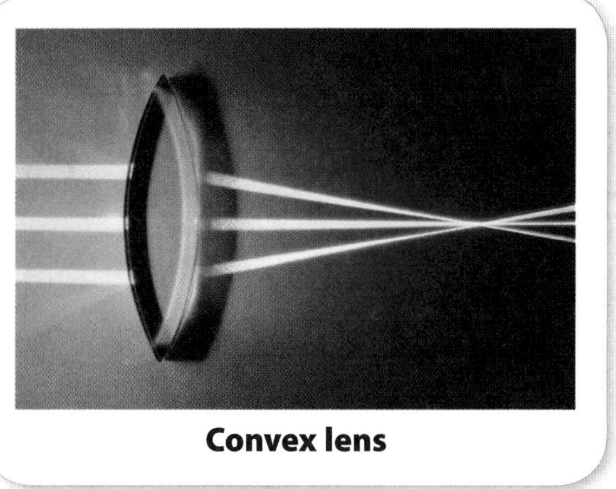
Convex lens

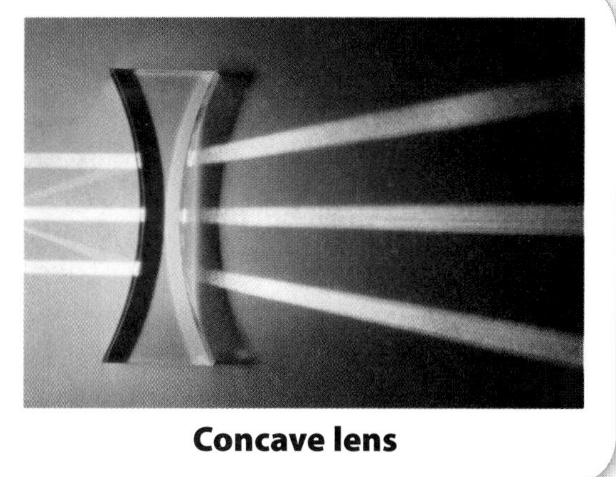
Concave lens

Figure 15 A convex lens often is called a converging lens. A concave lens often is called a diverging lens.

Lenses

What do binoculars, eyeglasses, your eyes, and a camera have in common? Each contains a lens. *A* **lens** *is a transparent object with at least one curved side that causes light to change direction.* Recall that most of the light that strikes a transparent material passes through it. Light refracts as it passes through a lens. The greater the curve of the lens, the more the light refracts. The direction of refraction depends on whether the lens is curved outward or inward.

As shown in **Figure 15,** light passes through two different lenses. *A lens that is thicker in the middle than at the edges is a* **convex lens.** Notice that the light rays that move through a convex lens come together, or converge. *A lens that is thicker at the edges than in the middle is a* **concave lens.** Notice that the light rays spread apart, or diverge, as they move through a concave lens.

Reading Check Describe the shapes of a convex lens and a concave lens.

Inquiry MiniLab

15 minutes

How can water move light? 

You can draw light rays to investigate refraction of light.

1. Read and complete a lab safety form.
2. Place a **sheet of paper** over **corrugated cardboard.** Place a **clear plastic box** on the paper. Use a pencil to trace the long sides of the box. Fill the box with water.
3. Place two **pins** into the paper and cardboard on opposite edges of the box. With one eye, look from one pin to the other, diagonally through the box. Imagine a line through the pins. Draw along this line extending out from each pin and away from the box.
4. Remove the box. Use a **ruler** to draw a line between the pins.

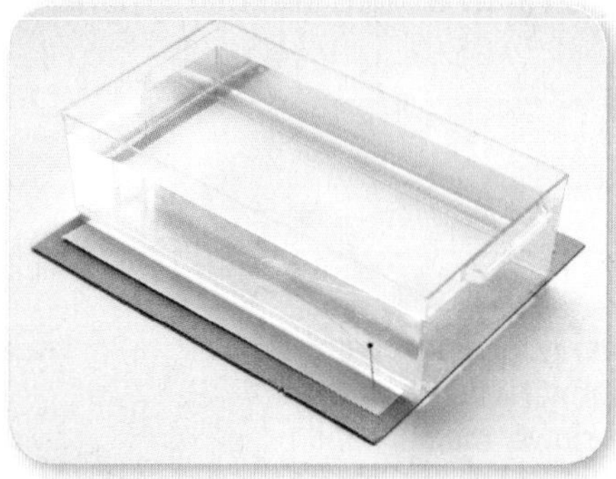

Analyze and Conclude

1. **Draw** the light's path as it moved through the box to your eye. Label each medium the light passed through.
2. **Key Concept** Identify and explain each change in the light's direction as it moved among the different media.

Convex Lenses

Have you ever used a magnifying lens to look at a tiny insect? A magnifying lens makes things appear larger. If you look closely, you can see that a magnifying lens is a convex lens because its center is thicker than its edges.

The refraction of light by a convex lens is shown in **Figure 16.** Notice that a normal to the curved surface slants toward the optical axis. Recall that light moving into a slower **medium** turns toward the normal. As a result, a convex lens refracts light inward and it converges.

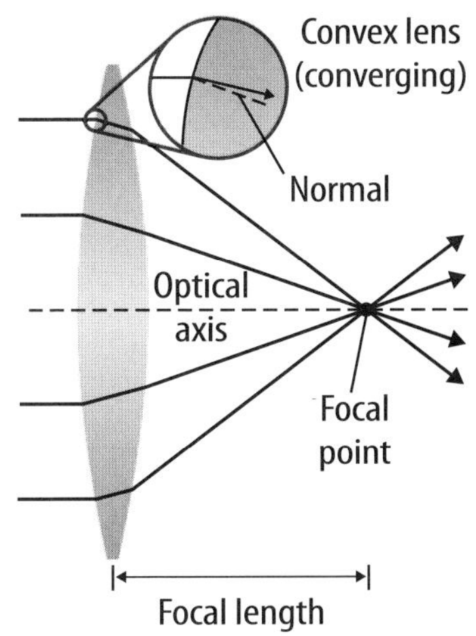

▲ **Figure 16** The outward curve of a convex lens refracts light rays inward.

Focal Point and Focal Length Similar to a mirror, the point where rays parallel to the optical axis converge after passing through a lens is the focal point. The distance along the optical axis between the lens and the focal point is the focal length of the lens. Because you can look through a lens from either side, a focal point is on both sides of the lens. For a lens with the same curve on both sides, the lens's two focal points are the same distance from it.

Key Concept Check What happens to light as it passes through a convex lens?

Science Use v. Common Use

medium

Science Use matter through which a wave travels

Common Use middle or average

Types of Images Like a concave mirror, the type of image a convex lens forms depends on the location of the object. A convex lens can form both real and virtual images, as shown in **Figure 17.** The diagrams show only two rays, but remember that in reality there are an infinite number of rays.

If you look through a magnifying lens at an object more than one focal length from the lens, the image you see is inverted and smaller. If you look at an object less than one focal length from the lens, the image you see is upright and larger. It is virtual because your brain interprets the rays as moving in a straight line, as shown by the dashed lines in **Figure 17.**

Figure 17 You can see either a virtual image or a real image if you look through a convex lens. ▼

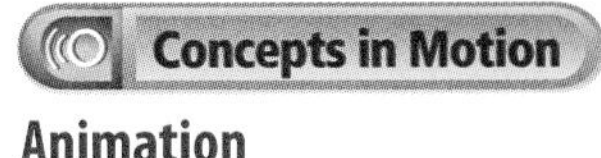

Animation

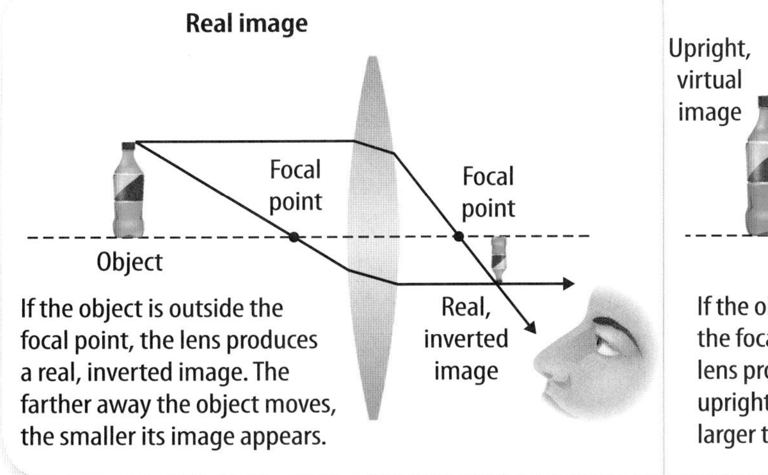

If the object is outside the focal point, the lens produces a real, inverted image. The farther away the object moves, the smaller its image appears.

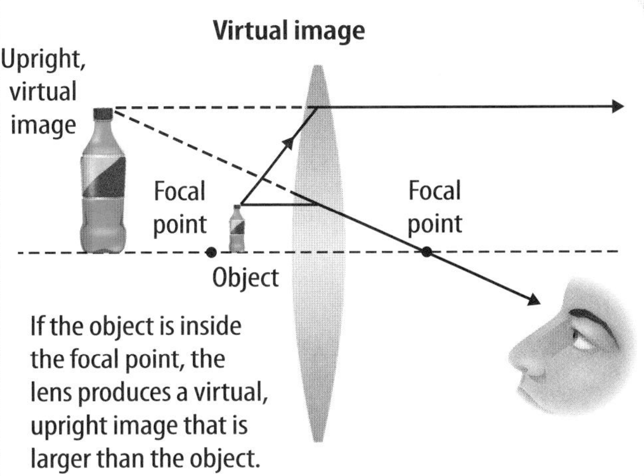

If the object is inside the focal point, the lens produces a virtual, upright image that is larger than the object.

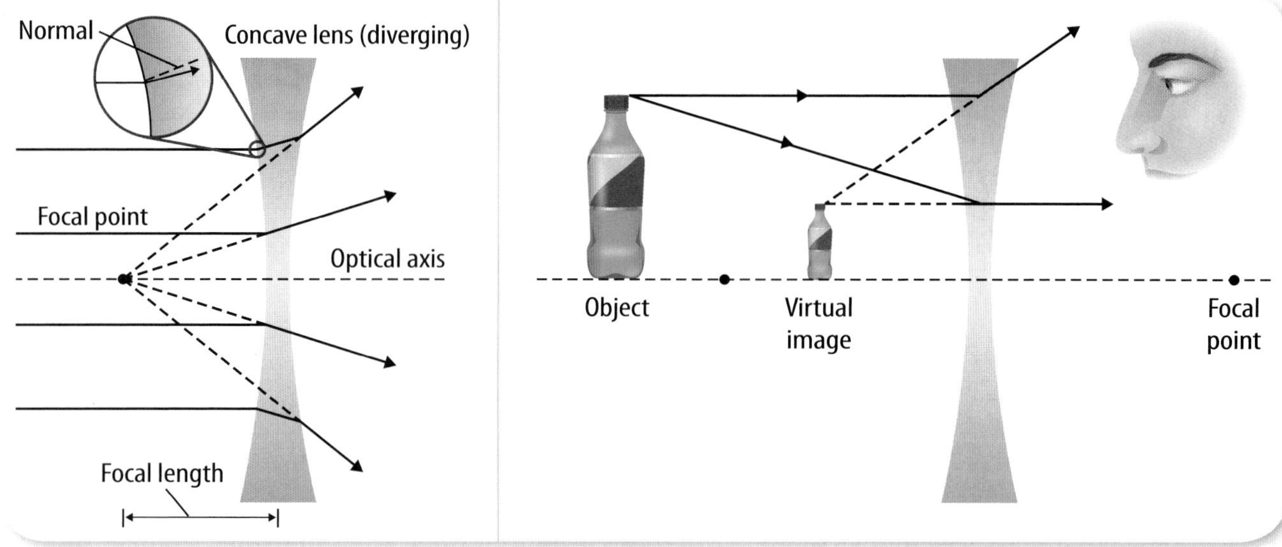

▲ **Figure 18**  The inward curve of a concave lens refracts light rays outward. You see a virtual image if you look through a concave lens.

Concave Lenses

Light rays that pass through a concave lens diverge, or spread apart. The drawing on the left in **Figure 18** shows why. Notice that a normal to the curved surface slants away from the optical axis. Because light entering a slower medium changes direction toward the normal, the lens refracts light outward. The light diverges.

The drawing on the right in **Figure 18** shows the type of image a concave lens forms. Suppose you stand to the right of the lens and look at the object. The refracted rays reach your eyes, but your brain assumes the light traveled the straight dashed lines. You see a virtual image smaller than the object.

Key Concept Check What happens to light as it passes through a concave lens?

Refraction and Wavelength

You have read that the curved surfaces of lenses affect the refraction of light rays. The wavelength of light also affects the amount of refraction. You might have seen sparkling cut-glass ornaments hanging in a window. When sunlight strikes the glass, rainbows appear on the wall. These "sun catchers" work because different wavelengths of light refract different amounts.

Figure 19 Each color of light has a different wavelength and frequency. ▼

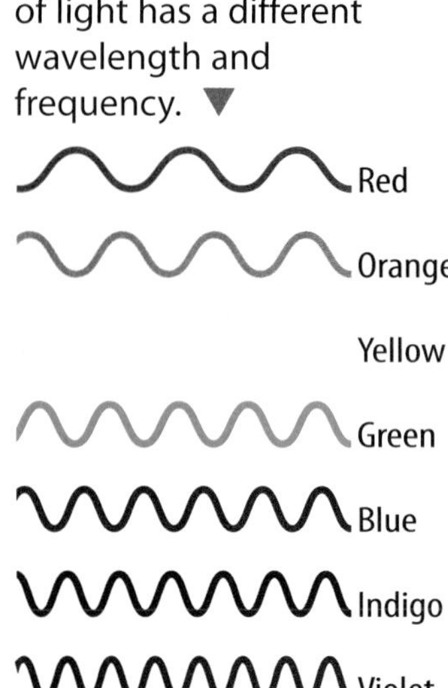

You read in Lesson 1 that white light is made up of different colors. A few colors are shown in **Figure 19.** The speed of a wave in a material is related to its wavelength. Waves with longer wavelengths travel at greater speeds in a material than waves with shorter wavelengths. Therefore, when entering a material, light with longer wavelengths travels faster and refracts less than light with shorter wavelengths. As a result, violet refracts the most because its wavelength is the smallest. Red has the longest wavelength and refracts the least.

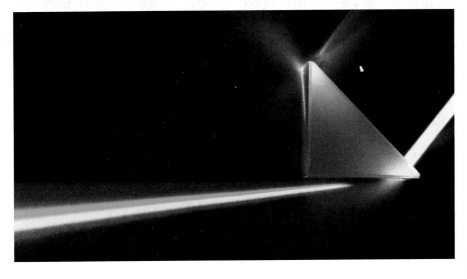

◀ **Figure 20** A prism spreads light into different wavelengths.

Visual Check Compare the refraction of yellow and blue light by a prism.

FOLDABLES®

Make a vertical two-tab book. Label it as shown. Use it to summarize your notes on prisms and rainbows.

Prisms

Rainbows

Prisms

As white light passes through a piece of glass called a prism, each wavelength of light refracts when it enters and again when it leaves. Because each color refracts at a slightly different angle, the colors separate. They spread out into the familiar spectrum, as shown in **Figure 20.**

Rainbows

Why do rainbows appear in the sky only during or after a rain shower? Rainbows form when water droplets in the air refract light like a prism. Each wavelength of light refracts as it enters the droplet, reflects back into the droplet, and refracts again when it leaves the droplet. Notice in **Figure 21** that wavelengths of light near the blue end of the spectrum refract more than wavelengths near the red end of the spectrum. This effect produces the separate colors you see in a rainbow.

Figure 21 Sunlight refracts as it passes into and out of a raindrop. ▼

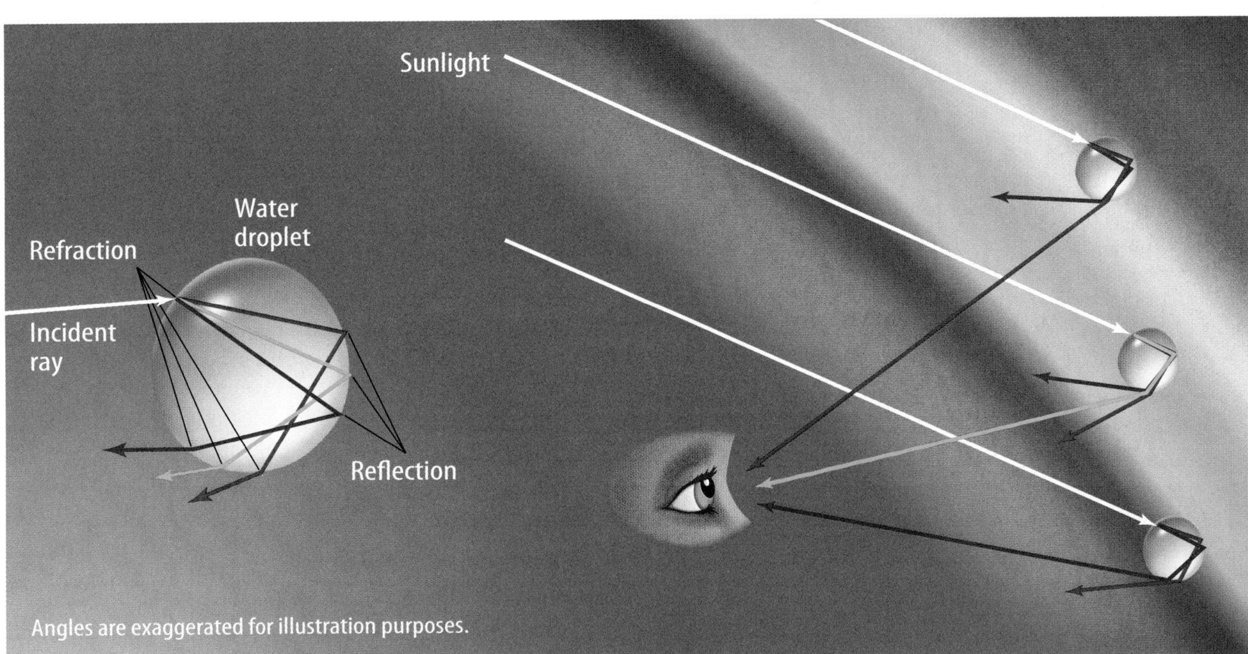

Visual Check Why do you have to stand with the Sun behind you to see a rainbow?

Detecting Light

You receive much information about the world around you when objects either emit light or reflect light into your eyes. You have read that the material that makes up an object reacts to various wavelengths of light in different ways. The material absorbs some wavelengths of light. Other wavelengths of light reflect to your eyes, providing information about shapes and colors. Your brain interprets this light energy so that you can recognize the images you see as people, places, and objects.

The Human Eye

Eyes respond quickly to changing conditions. As shown in **Figure 22,** the iris changes the size of the pupil, which controls the amount of light that enters an eye. The cornea acts as a convex lens and focuses light when it enters the eye. The shape of the cornea cannot change, but the shape of the eye's lens can. Changes to the shape of the lens can alter its focal point. This can enable a person to see objects clearly either far away or near. Follow the steps in the **Figure 22** to learn what each part of the eye does that enables a person to see.

How the Eye Works

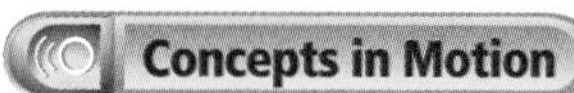

Animation

Figure 22 The parts of an eye work together to focus light and send signals about what you see to the brain.

1. The *cornea* protects the eye. Light refracts as it enters and as it leaves the cornea. Because air and the cornea have very different indexes of refraction, the greatest change in direction of light rays occurs as light moves into the cornea.

2. The *iris* is the colored part of your eye. The *pupil* is an opening in the center of the iris that can change size from 2 mm to 8 mm to let different amounts of light into the eye.

3. The *lens* is convex to help focus incoming light to form an upside-down image on the retina.

4. The *ciliary muscles* change the shape of the lens, making it flatter or rounder so that you can focus clearly on objects at any distance.

5. Cells in the *retina* absorb light and convert information about the amount of light and color into electric signals.

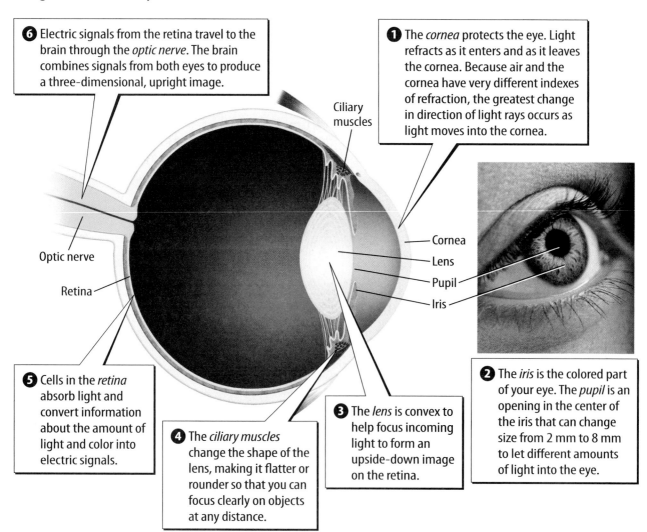

6. Electric signals from the retina travel to the brain through the *optic nerve*. The brain combines signals from both eyes to produce a three-dimensional, upright image.

Ciliary muscles

Cornea

Lens

Pupil

Iris

Optic nerve

Retina

Seeing Color

You might have noticed that it is harder to see colors in dim light. Why does this happen? Different cells on the retina of an eye enable a person to see both colors and shades of gray. *A* **rod** *is one of many cells in the retina of the eye that responds to low light.* There are about 120 million rods in the human retina. Rods enable you to see near the sides of your eyes rather than along the direct line of vision. This type of eyesight is called peripheral [puh RIH fuh rul] vision.

A **cone** *is one of many cells in the retina of the eye that respond to colors.* The human retina contains 6 million to 7 million cones that require more light to produce signals. These cones respond to red, green, and blue wavelengths of light. Recall that these primary colors of light can combine and form all the other colors. When light strikes a cone, it produces an electric signal that depends on the light's intensity and its wavelength. Your brain combines signals from all rods and cones and forms the colors, the shapes, and the brightness of the objects you see, as shown in **Figure 23.**

▲ **Figure 23** The ability to distinguish colors allows you to see more detail in your surroundings.

WORD ORIGIN

cone
from Latin *conus,* means "wedge"

 **Key Concept Check** How do eyes detect light and color?

Correcting Vision

In a person with normal eyesight, the cornea and the lens focus light directly on the retina. This forms a sharp, clear image. Some eyes, however, have an irregular shape. If an eye is too long, light focuses in front of the retina. This person sees near objects clearly, but objects far away are blurred. In other instances, the eye can be too short, and light focuses behind the retina. People with this problem clearly see objects far away, but objects nearby are blurred. **Figure 24** shows how manufactured lenses can correct these problems.

You have read that refraction and lenses enable you to see. Light can refract as it moves from one medium into another. Different colors refract at different angles. A lens causes light to change direction. An example is the focusing of light onto the retina by the cornea and lens of the eye. Eyeglasses or contact lenses may be needed to help focus this light.

Figure 24 Some people have eyes with irregular shapes that require lenses to correct their vision. ▼

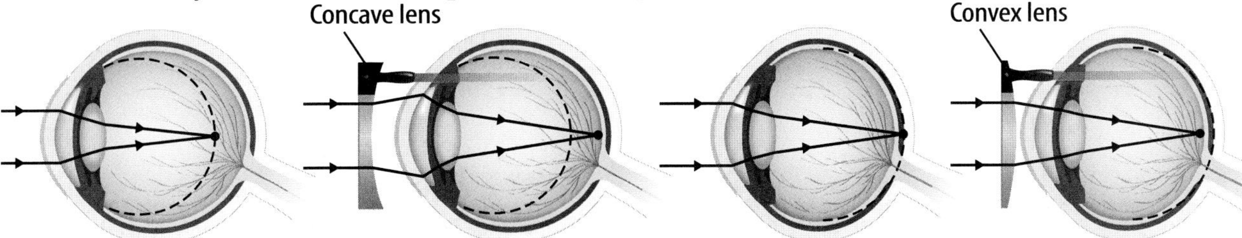

A person is **nearsighted** when light focuses in front of the retina. A **concave** lens corrects the problem.

A person is **farsighted** when light focuses behind the retina. A **convex** lens corrects the problem.

Lesson 3 Review

Visual Summary

Refraction is the change in direction of a wave as it moves from one medium into another at an angle.

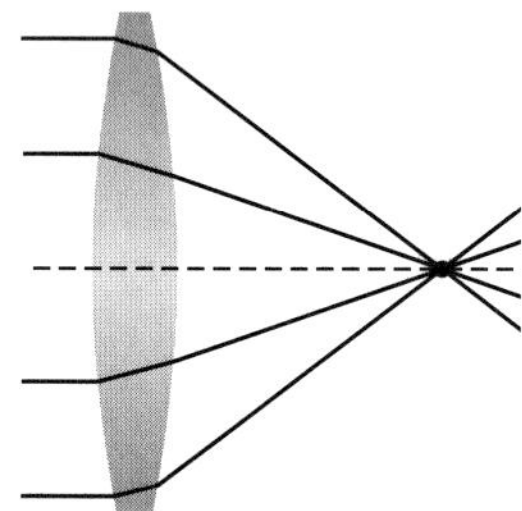

A lens changes the direction of light rays that pass through it.

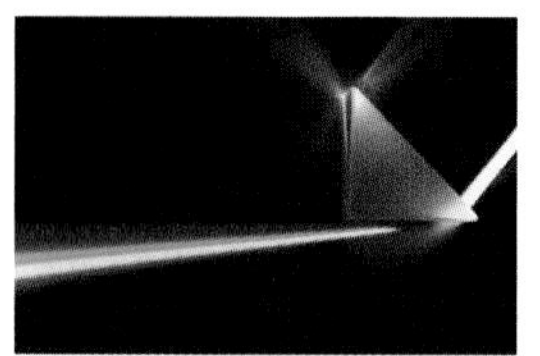

Each color of light refracts differently as it passes through a prism.

FOLDABLES®

Use your lesson Foldable to review the lesson. Save your Foldable for the project at the end of the chapter.

What do you think 

You first read the statements below at the beginning of the chapter.

5. A prism adds color to light.

6. When light moves from air into water or glass, it travels more slowly.

Did you change your mind about whether you agree or disagree with the statements? Rewrite any false statements to make them true.

Use Vocabulary

1. **Identify** two differences in rods and cones.
2. **Describe** a convex lens in your own words.

Understand Key Concepts 

3. Which terms describe the image produced by a concave lens?
   - **A.** real, inverted
   - **B.** real, upright
   - **C.** virtual, inverted
   - **D.** virtual, upright
4. **Contrast** refraction when light moves into an area of higher and lower index of refraction.
5. **Identify** the functions of the pupil, the cornea, the lens, and the retina.

Interpret Graphics

6. **Examine** What does the direction of the light wave tell you about the index of refraction of each of the materials below?

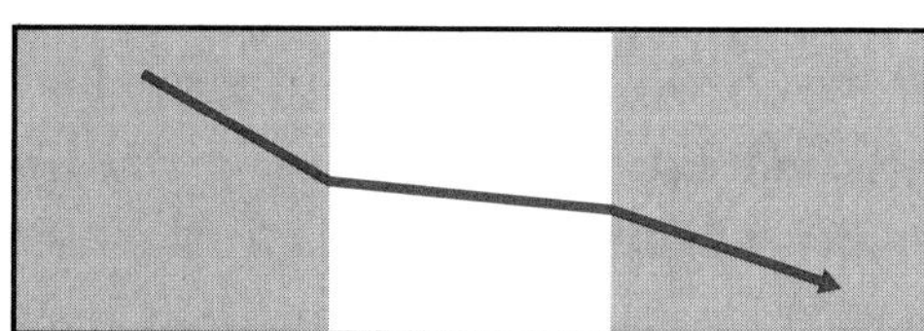

7. **Sequence** Copy and fill in a graphic organizer like the one below that shows what happens as light enters and interacts with parts of the eye and allows you to see an image. Add boxes as necessary.

Critical Thinking

8. **Predict** How would an object appear if you looked at it through a concave lens and a convex lens at the same time? Why?

Math Skills

Math Practice

9. Light travels through a sheet of plastic at 1.8×10^8 m/s. What is the index of refraction of the plastic?

How does a lens affect light?

Materials

index card

convex lens

ruler

meterstick

Also needed: concave lens

Safety

Light changes direction as it passes through a lens. This can make an object appear larger or smaller. It can even make the object appear upside down. The shape of the lens and the distance of the lens from the object affect the image you see.

Learn It

Making observations and measuring are two ways to **gather information** during an investigation. You can **organize information** you gather by recording it in a data table.

Try It

1. Read and complete a lab safety form.
2. Copy the table below into your Science Journal. Add lines as needed.
3. Prop a textbook upright. Hold a convex lens at arm's length and 1.5 m from the book. Look at the book through the lens. Record your observations.
4. Observe the image as you slowly walk toward the book, keeping the lens at arm's length. Have your partner measure the distance from the lens to the book whenever the image changes significantly. Record your data and observations.
5. With the classroom lights dimmed, hold a card in front of a window. Hold the lens between the card and the window, as shown.

6. Try to focus an image of the scene outside onto the card by moving the card closer and farther from the lens. When the image is focused as clearly as possible, have your partner measure the distance between the lens and the card. Record data and observations.
7. Repeat steps 3–6 with a concave lens.

Apply It

8. **Identify** positions of each lens when you saw significant changes in the image. What caused these?
9. **Infer** the focal length of the convex lens.
10. **Key Concept** Summarize your observations of the convex-lens and concave-lens images.

Type of Lens	Distance from the Lens to the Object	Description of the Object
Convex		

Lesson 4

Reading Guide

Key Concepts
ESSENTIAL QUESTIONS

- What do devices such as telescopes, microscopes, and cameras have in common?
- What is laser light, and how is it used?
- How do optical fibers work, and how are they used?

Vocabulary

optical device p. 661

refracting telescope p. 662

reflecting telescope p. 662

microscope p. 662

laser p. 664

hologram p. 665

Multilingual eGlossary

Video

What's Science Got to do With It?

Optical Technology

Inquiry How does the light do that?

A laser-light show is a spectacular site as the beams of light streak through the air. The lights in your home and school fill a room with brightness, but laser lights do not. How do laser lights travel in such a straight line? What makes them different from other lights?

Launch Lab

10 minutes

How do long-distance phone lines work?

Long-distance telephone calls often are sent as light signals along thin strands of glass called optical fibers. Why doesn't light escape out the sides of the fibers?

1. Read and complete a lab safety form.
2. Fill a **clear plastic box** with water.
3. In a darkened room, shine a thin **flashlight** beam through the side of the box into the water. Stir **milk** into the water one drop at a time until you can see the beam in the water.
4. Position the beam of light through the side of the box so that it is almost perpendicular to the underside of the water's surface. Record your observations of the reflected beam and the water in your Science Journal.
5. Hold an **index card** above the box so that any light transmitted through the surface shines on it.
6. Slowly change the angle of the beam until no light is transmitted onto the card. Draw your setup, and record your observations in your Science Journal.

Think About This

1. How did the angle of the incident beam affect the amount of light that appeared on the card?
2. **Key Concept** How do you think your observations show what happens to light in an optical fiber?

Optical Devices That Magnify

Do you wear sunglasses or contact lenses? Have you ever watched an animal through binoculars or looked at a planet through a telescope? If you answered yes to any of these questions, you have used an optical device. *Any instrument or object used to produce or control light is an* **optical device.** Some optical devices use lenses to help you see small or distant objects.

Telescopes

Hundreds of years ago, people could not see details of the Moon, such as those shown in **Figure 25.** However, by using telescopes, people have learned about objects in space. Telescopes are optical devices that produce magnified images of distant objects. A telescope uses a lens or a concave mirror to gather light. You might have used a small telescope to observe the Moon or planets, but some telescopes are much larger. The largest optical telescope in the United States has a mirror that is 10 m across. This enormous size means that the telescope can gather more light from a distant object than your eye can. The *Hubble Space Telescope* orbits about 600 km above Earth. It has a clearer view because it is above Earth's atmosphere. Astronomers use it to see objects billions of light-years away.

Figure 25 A telescope gives a close-up view of the Moon's surface.

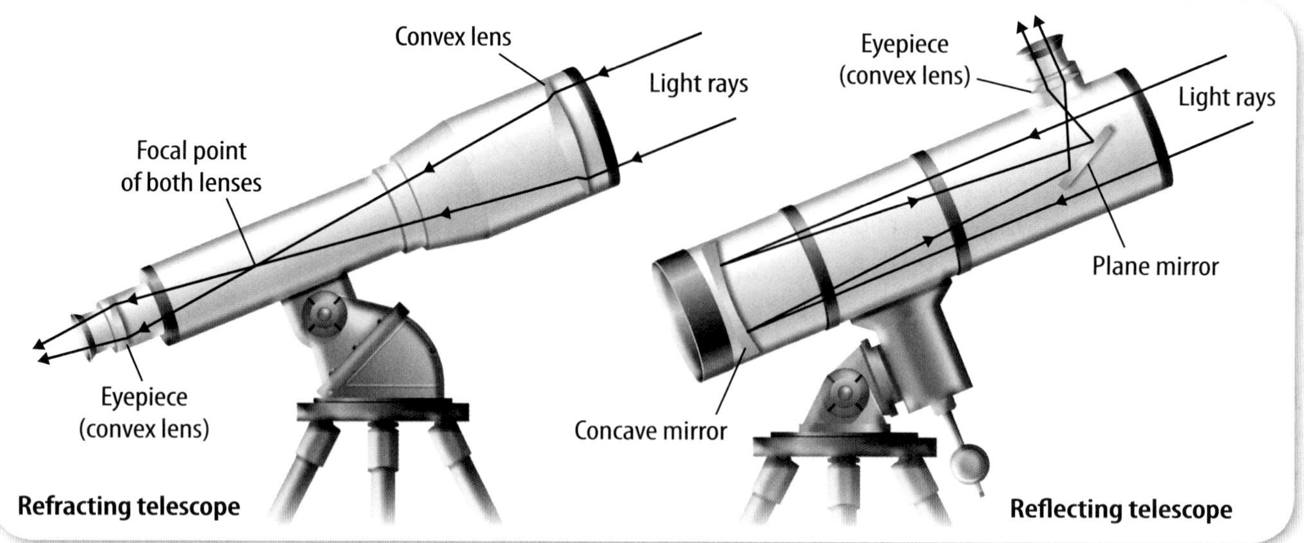

▲ **Figure 26** Telescopes use either a lens or a mirror to gather light from a faraway source.

WORD ORIGIN

microscope
from Latin *microscopium,* means "an instrument used to view small things"

Refracting Telescope *A telescope that uses lenses to gather and focus light from distant objects is a* **refracting telescope.** The convex lens that gathers light is called the objective. As shown on the left in **Figure 26,** light rays that pass through the objective converge and form a real image. The convex lens that a viewer looks through is called the eyepiece. This lens magnifies the image. Recall that refraction differs slightly for different wavelengths of light. Eyepieces often include multiple lenses with different focal lengths so that all wavelengths of light focus clearly.

Reflecting Telescope *A telescope that uses a mirror to gather and focus light from distant objects is a* **reflecting telescope.** A simple reflecting telescope is shown on the right in **Figure 26.** Light enters the tube and reflects off a concave mirror at the far end. The light begins to converge, but before it converges completely, it reflects off a plane mirror toward the eyepiece. A real image forms, and the eyepiece lens magnifies the image.

Figure 27 The main difference in microscopes and refracting telescopes is focal length. ▼

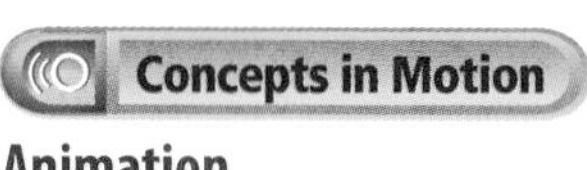

Animation

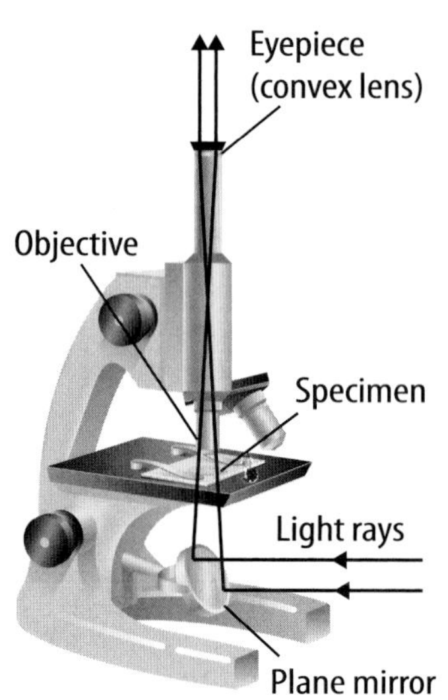

Light Microscopes

Telescopes are useful for observing far away objects, but to observe tiny objects up close, you need a different instrument. *A* **microscope** *is an optical device that forms magnified images of very small, close objects.* A microscope works like a refracting telescope, but its lenses have shorter focal lengths. As shown in **Figure 27,** light from a mirror or a light source near the microscope's base passes through a thin sample of an organism or object. The light then travels through a convex lens, called an objective, that focuses it. The eyepiece magnifies the image. Microscopes often have several objective lenses with different focal lengths for different magnifications.

Key Concept Check How are refracting telescopes and microscopes alike?

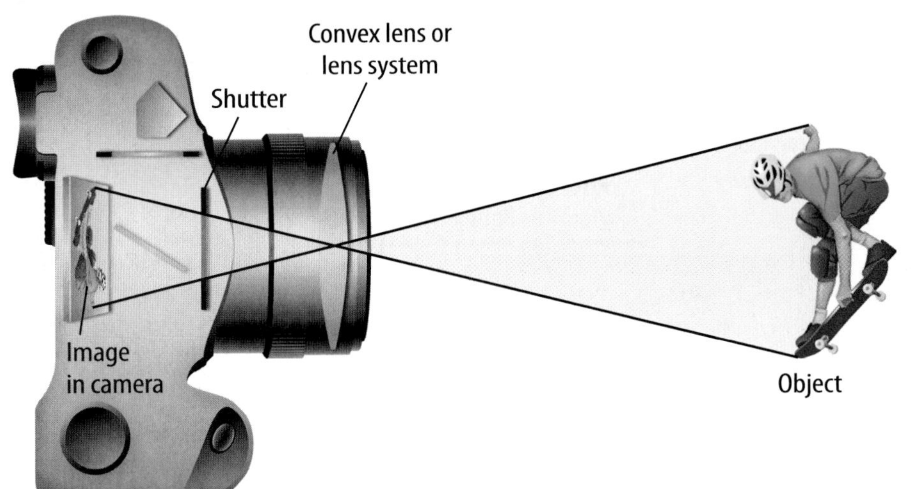

Figure 28 A digital camera uses one or more convex lenses to focus light onto a sensor.

Visual Check Why would the lens system have to move in order to photograph more distant objects?

Cameras

Cameras come in many different designs that use lenses to focus images. The design of a typical digital camera is illustrated in **Figure 28.** Suppose you want to take a photo of your friend on a skateboard. Light reflects off your friend and enters the camera. The light passes through a convex lens or, more often, a lens system. A shutter normally blocks light from entering the camera. However, pressing a button briefly opens the shutter. Light enters and is detected by an electronic sensor in the back of the camera. Recall that a convex lens is a converging lens. Inside the camera, a lens or system of lenses converges the light rays toward the sensor.

The boy you photograph is farther away from the lens than the focal point. For a convex lens, this means the image produced is real, inverted, and smaller than the object. Notice in the figure that this is the type of image that appears on the electronic sensor.

With some cameras, you can zoom, or move the lenses and increase the size of an image. Zooming changes the lenses' focal lengths. This means a camera can take enlarged images of distant or close objects.

Key Concept Check What does a camera have in common with a telescope and a microscope?

Inquiry MiniLab — 15 minutes

How does a zoom lens work?

A zoom lens can focus on distant objects. You can use a convex lens and a concave lens to observe the basic principle behind how zoom lenses work.

1. Read and complete a lab safety form.
2. Hold a **convex lens** at arm's length. Look through the lens at a distant object. Record your detailed observations in your Science Journal.
3. Hold a **concave lens** between you and the convex lens. Look through both lenses at the same object. Record your observations. Note changes in the image.
4. Move the concave lens closer and then farther from the convex lens. Does the image change? Record your observations.

Analyze and Conclude

1. **Contrast** the image with the convex lens with the image using both lenses.
2. **Key Concept** Based on your observations, what do you think is one reason telescopes, microscopes, and cameras often have systems of lenses for focusing rather than just one lens?

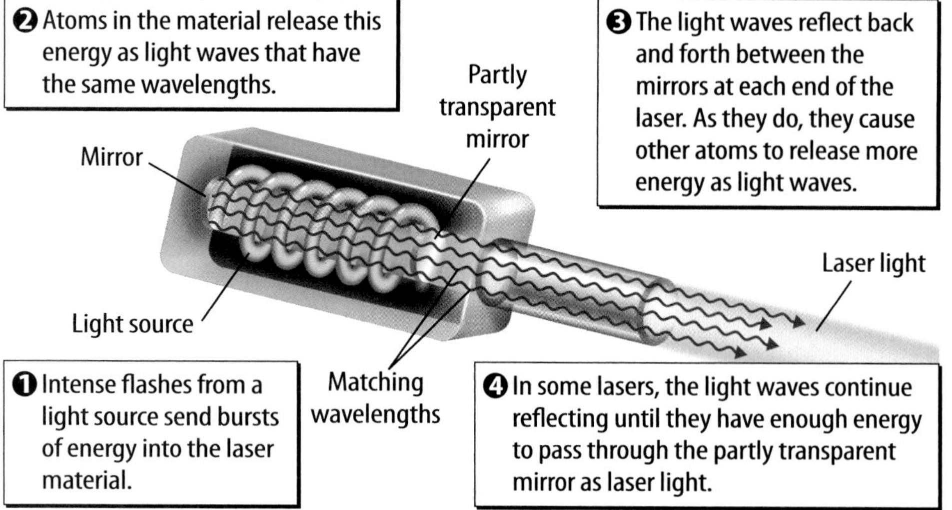

❷ Atoms in the material release this energy as light waves that have the same wavelengths.
Partly transparent mirror
❸ The light waves reflect back and forth between the mirrors at each end of the laser. As they do, they cause other atoms to release more energy as light waves.
Mirror
Laser light
Light source
❶ Intense flashes from a light source send bursts of energy into the laser material.
Matching wavelengths
❹ In some lasers, the light waves continue reflecting until they have enough energy to pass through the partly transparent mirror as laser light.

▲ **Figure 29** Light from the laser spreads much less than light from other sources because the light waves have the same wavelength.

Lasers

Some lasers are gentle enough to operate on the human eye, but some are powerful enough to cut through steel. *A* **laser** *is an optical device that produces a narrow beam of coherent light.* Recall that light from a luminous source, such as a lightbulb, has many wavelengths and spreads out in all directions. Coherent light is different. The light waves of coherent light all have the same wavelength and travel together. The beam is narrow because the waves do not spread apart, unlike light from other sources. Also, the intensity of laser light does not decrease quickly as it moves away from its source.

The word *laser* stands for light amplification through stimulated emission of radiation. One type of laser is shown in **Figure 29.** An energy source causes atoms of a material in the laser to emit light. This is the stimulated emission. This light travels back and forth within a tube, causing other atoms to emit similar waves. In this way, the light intensity increases. This is the light amplification.

 **Key Concept Check** What is laser light?

CDs and DVDs

Whenever you use a CD player, you use a laser. To make a CD, electric signals represent sounds. A laser burns tiny pits in the surface of the CD that correspond to the signals. In your CD player, another laser passes over the pits as the CD spins. Different amounts of light reflect back to a sensor that converts the light back into an electric signal. The signal causes the speakers to produce sound.

DVDs are similar to CDs except the DVD's pits are much smaller and closer. Therefore, more information fits on a DVD than a CD, as shown in **Figure 30.**

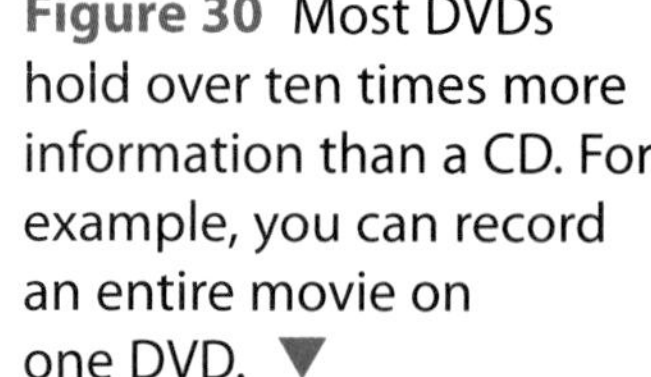

Figure 30 Most DVDs hold over ten times more information than a CD. For example, you can record an entire movie on one DVD. ▼

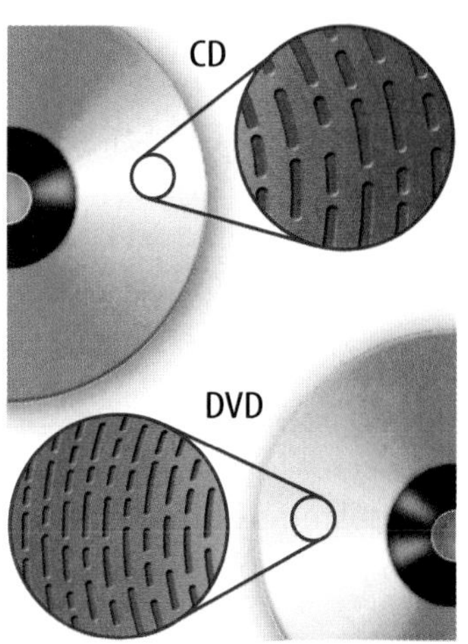

Holograms

Perhaps you've seen images on a book cover or on a credit card that seem to float in space. You can see different sides of the object, just as you would with a real object. *A* **hologram** *is a three-dimensional photograph of an object.*

Figure 31 shows how one type of hologram forms. First, laser light splits into two beams. One beam passes through a convex lens then reflects from the object onto photographic film. The other beam passes through a convex lens and then travels to the film. The combined light from the two beams produces a pattern on the film that shows both the brightness of light and its direction. The pattern appears as swirls, but when a laser shines on the film, a holographic image appears.

Some paper bills and credit cards have a different type of hologram printed on them to prevent counterfeiting. The holograms are made with lasers but can be viewed under regular light. Other uses of lasers are shown in **Figure 32.**

 **Key Concept Check** What are some ways lasers are used?

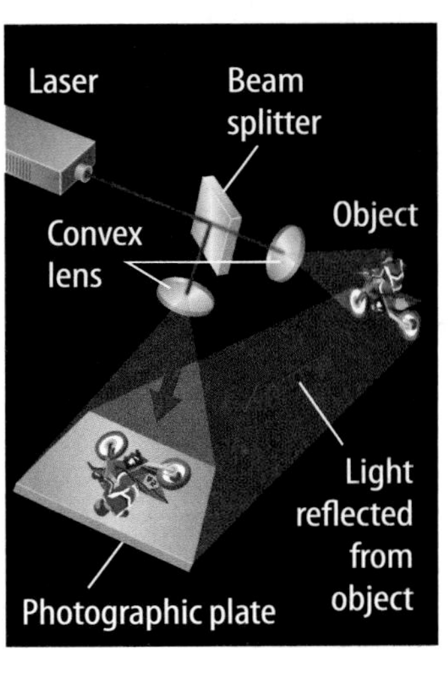

▲ **Figure 31** Interference of two laser-light beams produces a hologram.

Figure 32 Laser light is useful because its intensity does not decrease quickly. ▼

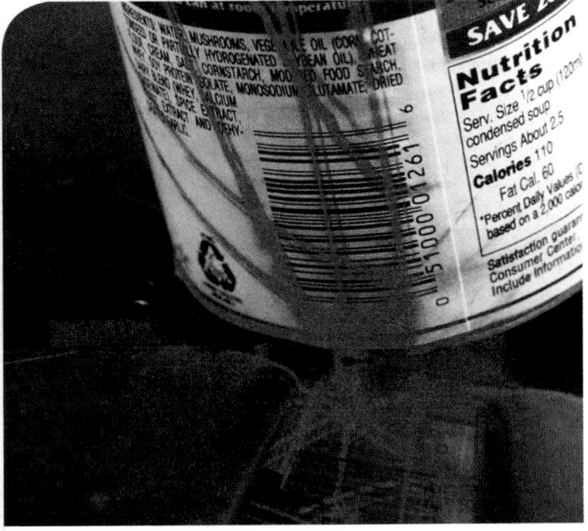

▲ Lasers scan barcodes at stores and libraries.

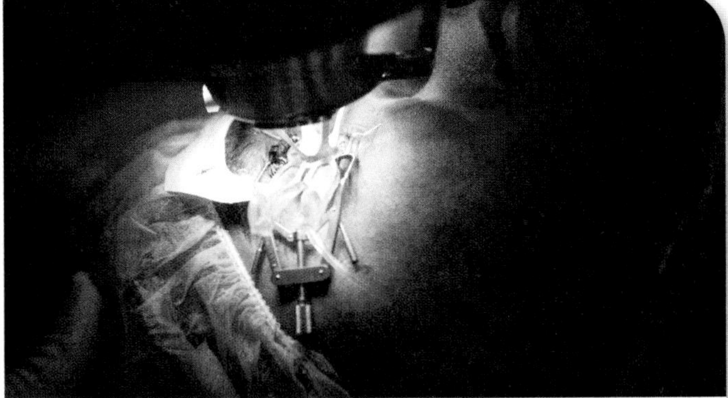

▲ Doctors use low-power lasers to correct problems in the retina.

▲ Lasers guide tunnel-boring machines to keep tunnels straight.

◀ Powerful lasers cut through metal to form machine parts or intricate designs.

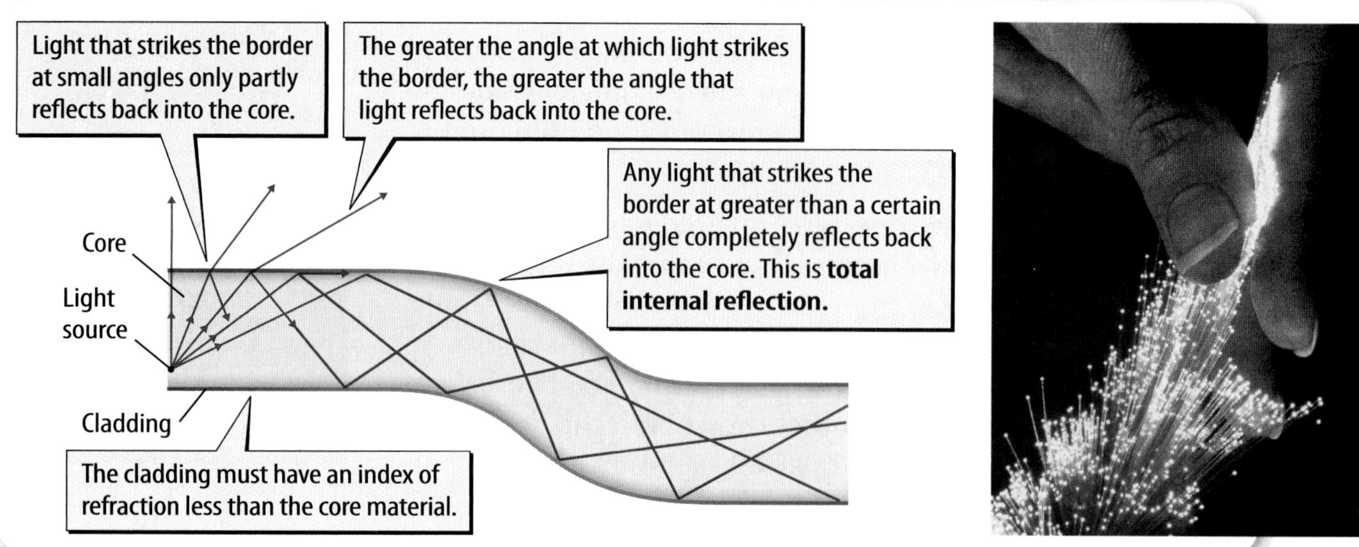

Light that strikes the border at small angles only partly reflects back into the core.
The greater the angle at which light strikes the border, the greater the angle that light reflects back into the core.
Any light that strikes the border at greater than a certain angle completely reflects back into the core. This is total internal reflection.
Core
Light source
Cladding
The cladding must have an index of refraction less than the core material.

Figure 33 Light moves through an optical fiber because of a phenomenon called total internal reflection.

Visual Check Why does light stay inside an optical fiber?

Optical Fibers

Almost every time you make a phone call, use a computer, or watch a television show, you are using light technology. The electronic signals for these devices travel at least part of the way through optical fibers. An optical fiber is a thin strand of glass or plastic that can transmit a beam of light over long distances.

Total Internal Reflection Just as water moves through a pipe, light moves through an optical fiber. **Figure 33** shows how this happens. A core material surrounded by another material called the cladding makes up an optical fiber. As light moves through the core, it strikes the cladding at an angle greater than a certain angle and reflects back into the core. This process is called total internal reflection. Because light stays in the core, a fiber can carry light signals long distances without losing strength.

Uses of Optical Fibers Phone conversations, computer data, and TV signals travel as pulses of light through optical fibers. Using different wavelengths of light, one fiber can carry thousands of phone conversations at a time. Optical fibers also have medical uses, such as providing light for a physician to perform surgery through a small incision in the body.

Key Concept Check How do optical fibers work, and how are they used?

Relating Optical Technology Each of the optical devices that you have read about uses lenses or mirrors to produce or control light. Telescopes and light microscopes produce magnified images. Cameras use lenses to focus light on an image sensor. Lasers produce coherent light. Optical fibers are also a type of optical device because they transmit and reflect light.

FOLDABLES®

Use two sheets of paper to make a bound book and label it as shown. Use it to organize what you have learned about different types of optical devices.

Lesson 4 Review

Assessment Online Quiz

Visual Summary

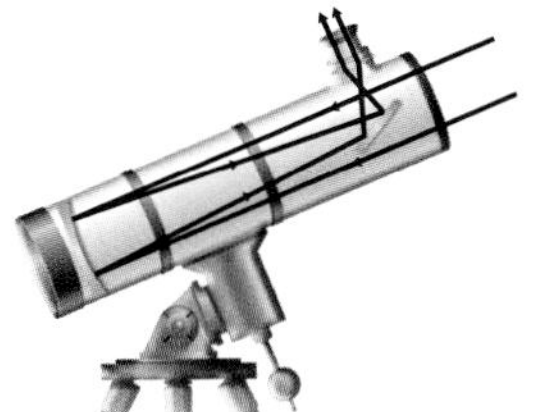

A telescope gathers and focuses light and produce a magnified image of an object far away.

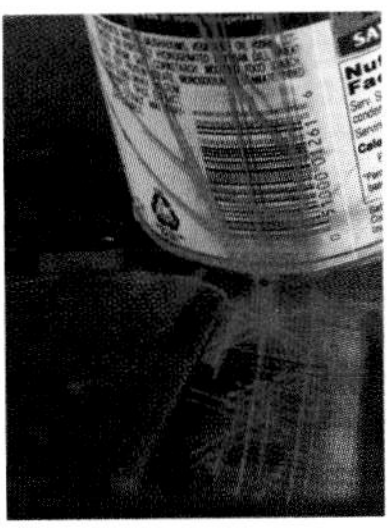

Lasers have many uses because they produce an intense beam of light.

Almost no light is lost as light passes through an optical fiber.

FOLDABLES®

Use your lesson Foldable to review the lesson. Save your Foldable for the project at the end of the chapter.

What do you think NOW?

You first read the statements below at the beginning of the chapter.

7. A laser will burn a hole through your skin.

8. Telephone conversations can travel long distances as light waves.

Did you change your mind about whether you agree or disagree with the statements? Rewrite any false statements to make them true.

Use Vocabulary

1. **Use the term** *microscope* in a sentence.
2. **Define** *hologram* in your own words.
3. An optical device that produces a narrow beam of coherent light is a(n) ________.

Understand Key Concepts 

4. Which collects light in a reflecting telescope?
   A. concave lens
   B. concave mirror
   C. convex lens
   D. convex mirror
5. **Contrast** a microscope and a camera.
6. **Describe** how coherent light is produced in a laser.
7. **Explain** total internal reflection.

Interpret Graphics

8. **Identify** the optical effect that a jeweler achieves when he cuts facets, or faces, on a diamond, as shown here. Why does cutting increase the sparkle?

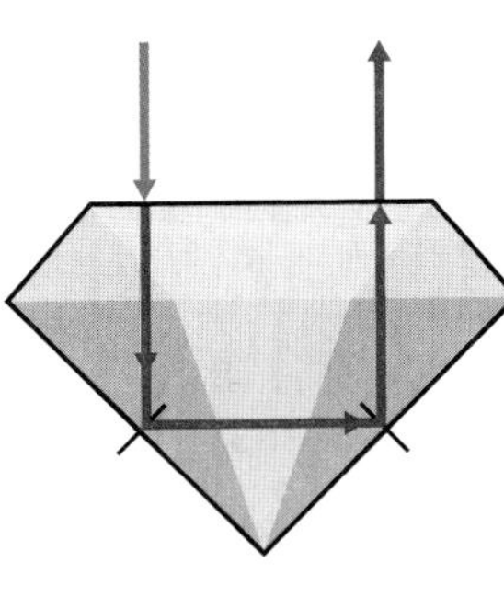

9. **Compare and Contrast** Copy and fill in a graphic organizer like the one below. How are the lenses in a refracting telescope and a microscope alike and different?

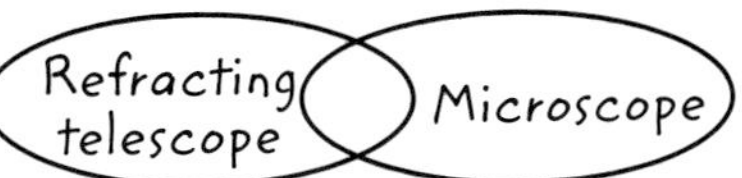

Critical Thinking

10. **Compare and contrast** a hologram and a regular image taken by a camera.
11. **Extend** When you look through binoculars, each eye looks through two convex lenses. Which other optical devices work in this way?

Design an Optical Illusion

Materials

beakers

clay

color filters

flashlights

markers

prisms

Also needed: various lenses, various mirrors, cardboard box, plastic box, scissors, stands, candles, matches, test tubes, water, vegetable oil

Safety

A famous illusionist has hired you to design a new type of optical illusion for a show. How will you do it? To make an optical illusion, you have to use lights to make people see an image that is not really there. Think about how you can use all you have learned about light and optical devices to make an illusion. Here is your challenge:

- You may use only common materials and devices approved by your teacher.
- You must use light in at least two different ways (transmit, reflect, refract, or absorb).
- The image you create must be different from the object(s) in at least two ways (for example, size, orientation, color, or position).

Question

How can you design and build an optical illusion? How can light be used to create an optical illusion? What materials are easily manipulated to construct a simple optical illusion?

Procedure

1. Read and complete a lab safety form.
2. Brainstorm ideas for your illusion with others in your group. Think about light sources and shadows. Can you combine images or show only part of an image? How does each optical device work, and how could you combine them? Record your ideas in your Science Journal.
3. Use your ideas to design an optical illusion. List the materials and steps you will follow to construct it. Include sketches of your setup.
4. Have your teacher approve your design.
5. Collect your materials and build your optical illusion.
6. Test your illusion several times. Record your observations.

5

Analyze and Conclude

7. Evaluate the quality of your illusion. Consider these questions: *Does it meet all the requirements? Is your illusion creative and unique? How easy is it for an audience to see? How different is the image your illusion produces from the object?*
8. Make a modification to your setup. Test and evaluate the effectiveness of your setup in producing a realistic illusion.
9. Record details about your modification. Then record your observations.
10. Continue to improve your illusion as time permits.
11. **Sketch** and label your final setup. Indicate the path of light.
12. **Critique** your design and results. Did your setup work the way you expected? Did it meet all the requirements?
13. **The Big Idea** How did matter and light interact to produce your illusion?

Communicate Your Results

Demonstrate your illusion to the class. Imagine that you are presenting the illusion to the illusionist who hired you. Explain your design. The illusionist needs to understand how all the parts work in order to present the illusion to an audience. Describe how light travels between media. Explain how reflection and refraction affect the path of the light to produce an illusion.

Inquiry Extension

Explain how you could modify your illusion for use on stage. This might include making the setup larger so that an entire theater audience could see the illusion. Also consider whether you would be able to take your setup apart and put it together again as the show travels from one city to another.

8

Lab Tips

- ☑ Think about how you might combine images of objects that are far apart into a single image.
- ☑ Consider how you could use specific colors in your illusion or how you could use filters to block some colors of light.
- ☑ Remember that you can use lenses and mirrors to make images appear in different locations than where the objects are.

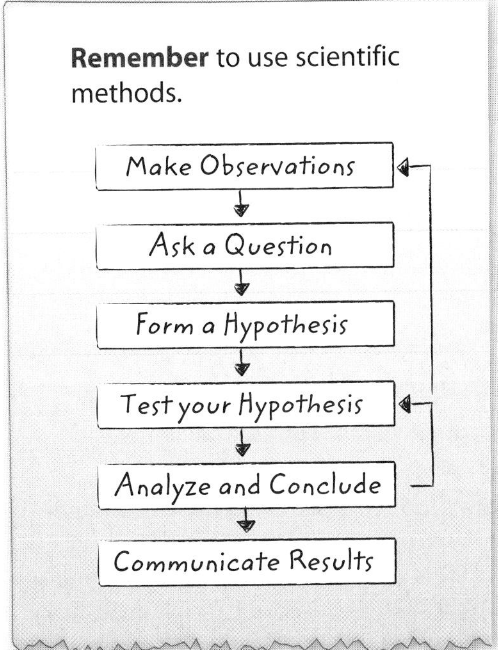

Chapter 18 Study Guide

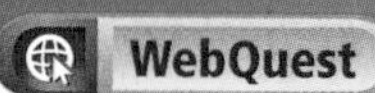

Matter can absorb, reflect, or transmit light. Objects might appear as different colors or sizes because of the way light interacts with matter.

Key Concepts Summary

Lesson 1: Light, Matter, and Color

- Luminous objects, such as a flashlight, are sources that produce **light** that spreads out in straight lines.
- Matter can emit, absorb, reflect, or transmit light.
- An object's color is determined by the wavelengths of light the object reflects or transmits to you.

Vocabulary

light p. 635
reflection p. 635
transparent p. 636
translucent p. 636
opaque p. 636
transmission p. 637
absorption p. 637

Lesson 2: Reflection and Mirrors

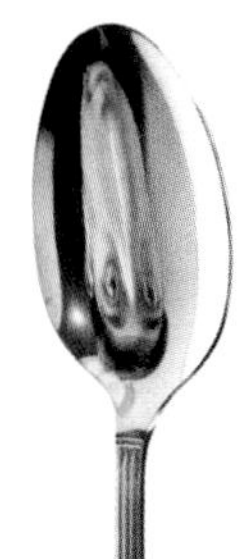

- Light rays reflect parallel from smooth surfaces and in many directions from rough surfaces.
- The shape of a mirror affects the way it reflects light. **Concave mirrors** converge light rays. **Convex mirrors** diverge light rays.
- All mirrors can produce virtual images, but only concave mirrors can produce real images.

Vocabulary

law of reflection p. 643
regular reflection p. 644
diffuse reflection p. 644
concave mirror p. 645
focal point p. 645
focal length p. 645
convex mirror p. 646

Lesson 3: Refraction and Lenses

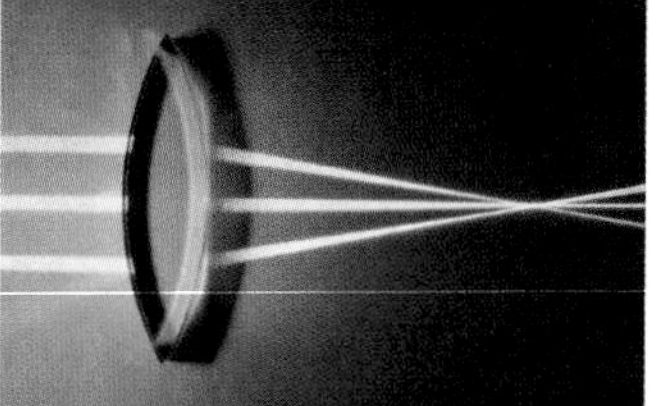

- Light changes direction if it moves into a medium with a different index of **refraction.**
- **Convex lenses** converge light rays. **Concave lenses** diverge light rays.
- **Rods** in the retina of the eye detect the intensity of light. **Cones** detect the colors of objects.

Vocabulary

refraction p. 650
lens p. 652
convex lens p. 652
concave lens p. 652
rod p. 657
cone p. 657

Lesson 4: Optical Technology

- Telescopes, **microscopes,** and cameras use lenses to focus light.
- **Lasers** produce a narrow beam of coherent light. Uses of laser light include detecting information on DVDs, making **holograms,** and cutting metal.
- Light completely reflects back into optical fibers. These fibers can carry light signals over long distances.

Vocabulary

optical device p. 661
refracting telescope p. 662
reflecting telescope p. 662
microscope p. 662
laser p. 664
hologram p. 665

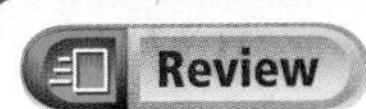
Review

- Personal Tutor
- Vocabulary eGames
- Vocabulary eFlashcards

FOLDABLES® Chapter Project

Assemble your lesson Foldables as shown to make a Chapter Project. Use the project to review what you have learned in this chapter.

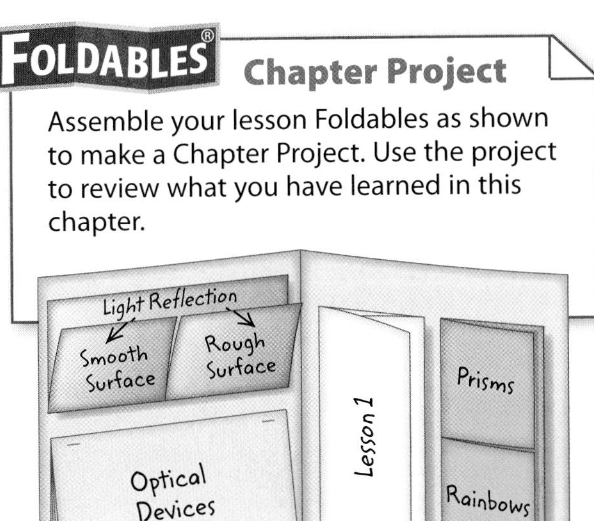
Light Reflection
Smooth Surface
Rough Surface
Optical Devices
Lesson 1
Prisms
Rainbows

Use Vocabulary

1. You can see light but not objects clearly through a(n) ________ object.
2. Contrast opaque and transparent objects.
3. A mirror that causes light waves to converge is a(n) ________.
4. The place where rays parallel to the optical axis cross after reflecting from a mirror is called the ________.
5. Describe refraction in your own words.
6. A three-dimensional image of an object is a(n) ________.
7. Which optical instrument uses at least two convex lenses to magnify small, close objects?

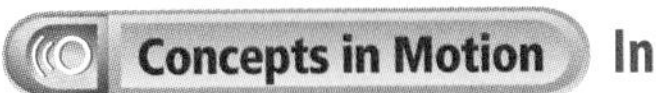
Concepts in Motion

Interactive Concept Map

Link Vocabulary and Key Concepts

Copy this concept map, and then use vocabulary terms from the previous page to complete the concept map.

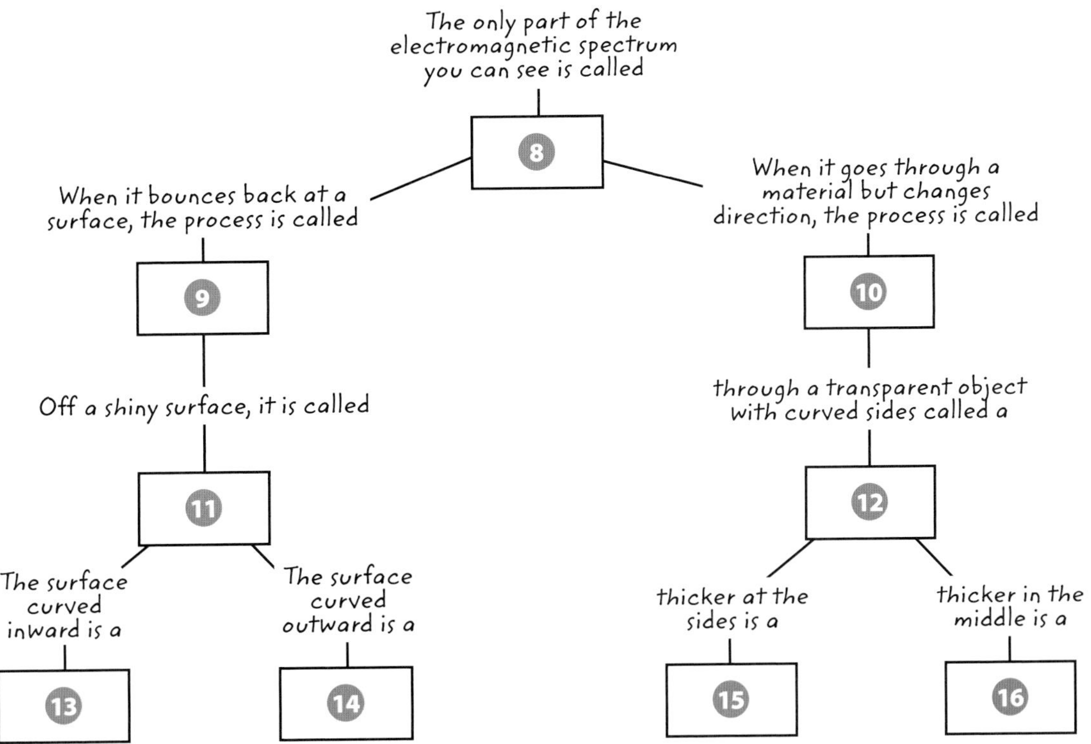
The only part of the electromagnetic spectrum you can see is called
8
When it bounces back at a surface, the process is called
When it goes through a material but changes direction, the process is called
9
10
Off a shiny surface, it is called
through a transparent object with curved sides called a
11
12
The surface curved inward is a
The surface curved outward is a
thicker at the sides is a
thicker in the middle is a
13
14
15
16

Chapter 18 Review

Understand Key Concepts

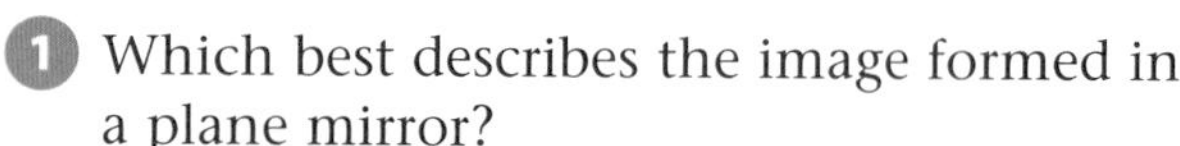

1. Which best describes the image formed in a plane mirror?
   A. a real image behind the mirror
   B. a real image in front of the mirror
   C. a virtual image behind the mirror
   D. a virtual image in front of the mirror

2. Which MOST determines the color of an opaque object?
   A. diffraction
   B. reflection
   C. refraction
   D. transmission

3. Which describes a material that does NOT transmit any light?
   A. opaque
   B. reflective
   C. translucent
   D. transparent

4. If the angle of incidence of a ray striking a plane mirror is 50°, what is the angle of reflection?
   A. 40°
   B. 50°
   C. 100°
   D. 140°

5. Which terms describe the image that forms as light passes through the lens shown below?

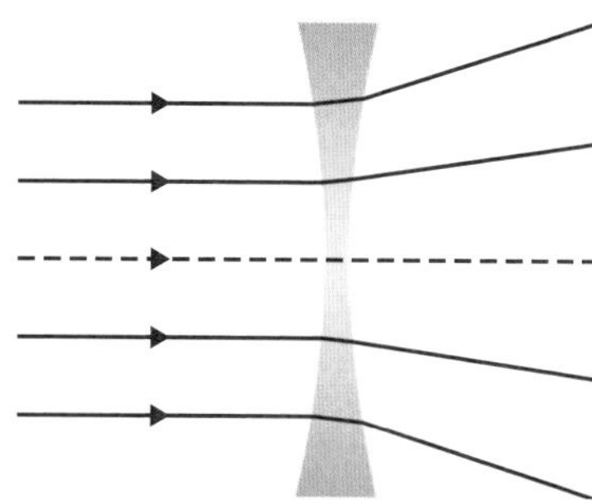

   A. real image; larger than the object
   B. real image; smaller than the object
   C. virtual image; larger than the object
   D. virtual image; smaller than the object

6. What color of light is produced when the three primary colors are combined in equal amounts?
   A. black
   B. blue
   C. red
   D. white

Use the diagram below to answer questions 7–9.

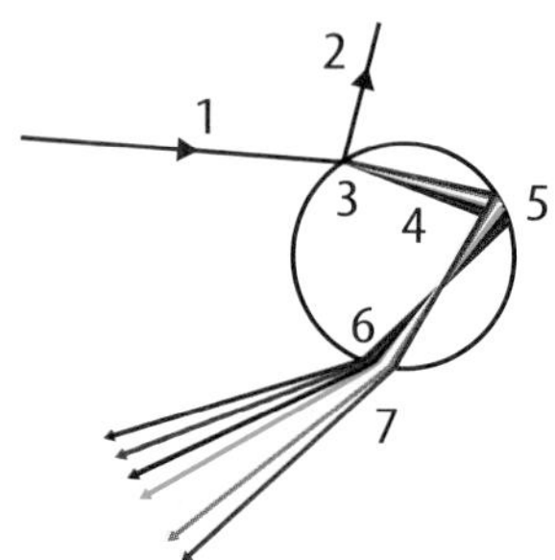

7. The diagram above shows light entering a raindrop. What produced the light represented by ray 2?
   A. It was absorbed by the drop.
   B. It was reflected from the drop.
   C. It was refracted by the drop.
   D. It was transmitted by the drop.

8. What caused the spread of the colors at point 7?
   A. absorption
   B. reflection
   C. refraction
   D. transmission

9. What process took place at position 5?
   A. absorption
   B. holography
   C. coherent light refraction
   D. total internal reflection

10. What term describes the degree to which a material causes light to move more slowly?
   A. focal length
   B. focal point
   C. index of refraction
   D. optical axis

Assessment
Online Test Practice

Critical Thinking

11 **Analyze** A girl wants to take a picture of her reflection in a plane mirror. She stands 1.3 m in front of the mirror. At what distance should she focus the camera to get a clear image of her reflection?

12 **Hypothesize** How might a magician make an object seem to disappear using a concave mirror?

13 **Infer** A solid, transparent object is placed in a container of clear liquid. The object is no longer visible. What could explain this effect?

14 **Deduce** At sunset, the Sun is at position A, but the observer sees the Sun at position B. Why does this happen?

15 **Construct** a diagram showing how light forms an image in a concave mirror when the object is farther from the mirror than the focal point is. Explain the type of image that forms.

16 **Predict** what color a white shirt would appear to be if the light reflected from the shirt passes through a red filter and then a green filter before reaching your eye.

Writing in Science

17 **Write** You are an eye doctor with a patient who has problems seeing things up close. The patient's distance vision is good. Describe the patient's problem in detail. Identify how you would recommend solving the patient's problem.

REVIEW THE BIG IDEA

18 Describe how the interaction of matter and light affects what you see when you look at a window, a lake, and a tree.

19 How does matter affect the way you perceive light in the picture below?

Math Skills

Math Practice

Use Scientific Notation

20 Light travels through a manufactured transparent material at about 1.0×10^8 m/s. What is the index of refraction of the material?

21 The speed of light is 1.83×10^8 m/s through a material. What is the material's index of refraction?

22 Light travels through a material at a speed of 1.42×10^8 m/s. What is the index of refraction of the material?

23 The speed of light in a sodium chloride crystal is 1.95×10^8 m/s. What is the index of refraction of sodium chloride?

24 A diamond's index of refraction is about 2.4. What is the speed of light through a diamond? [*Hint:* The speed of light through a substance equals the speed of light through a vacuum divided by the index of refraction.]

Standardized Test Practice

Record your answers on the answer sheet provided by your teacher or on a sheet of paper.

Multiple Choice

1 Which describes one way concave mirrors and convex mirrors differ?

A Concave mirrors are curved, while convex mirrors are flat.

B Concave mirrors always produce a virtual image, while convex mirrors can produce a real image.

C Convex mirrors are curved, while concave mirrors are flat.

D Convex mirrors always produce a virtual image, while concave mirrors can produce a real image.

2 What happens to most of the light that strikes a flat transparent object?

A It is absorbed by the object.

B It is magnified by the object.

C It is reflected by the object.

D It is transmitted by the object.

Use the figure below to answer question 3.

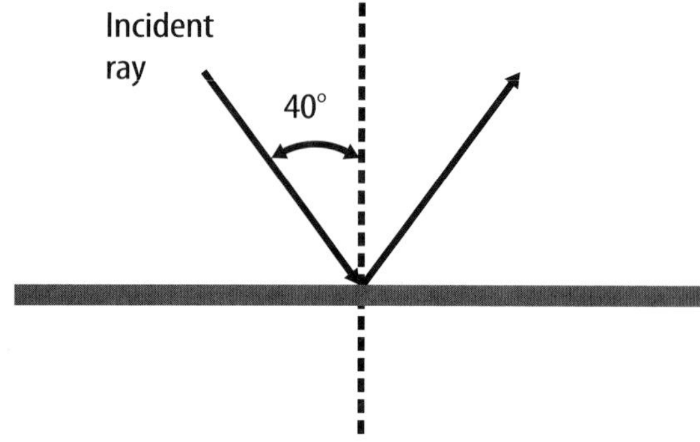

3 The figure shows how light is affected by a plane mirror. What angle does the reflected ray form with the normal?

A 20°

B 40°

C 80°

D 90°

Use the figure below to answer question 4.

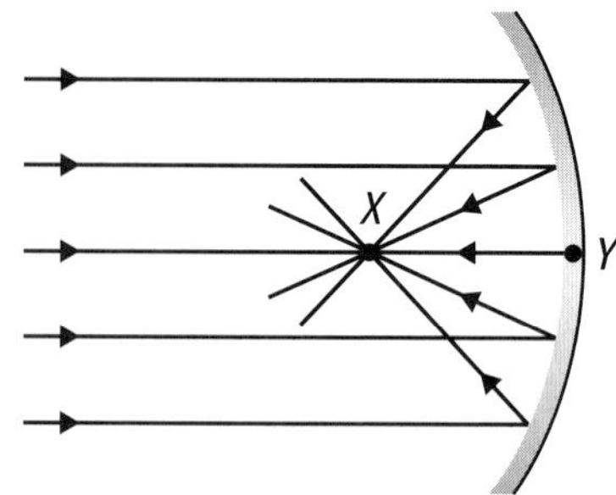

4 The figure shows how light rays are affected by a concave mirror. What is between points X and Y?

A the angle of incidence

B the focal length

C the focal point

D a virtual image

5 Which correctly describes an optical fiber?

A The core is opaque and luminous.

B The core is opaque and illuminates other objects.

C The core is transparent and luminous.

D The core is transparent and transmits light.

6 Which describes how a refracting telescope and a reflecting telescope differ?

A A refracting telescope gathers light using a concave lens, while a reflecting telescope uses a convex lens.

B A refracting telescope gathers light using a concave mirror, while a reflecting telescope uses a convex lens.

C A refracting telescope gathers light using a convex lens, while a reflecting telescope uses a concave mirror.

D A refracting telescope gathers light using a concave mirror, while a reflecting telescope uses a convex mirror.

7 Which describes how concave lenses and convex lenses differ?

A Concave lenses are flat on both sides. Convex lenses are curved on both sides.

B Concave lenses are curved on both sides. Convex lenses are flat on both sides.

C Concave lenses are thicker in the middle. Convex lenses are thinner in the middle.

D Concave lenses are thinner in the middle. Convex lenses are thicker in the middle.

Use the figure below to answer question 8.

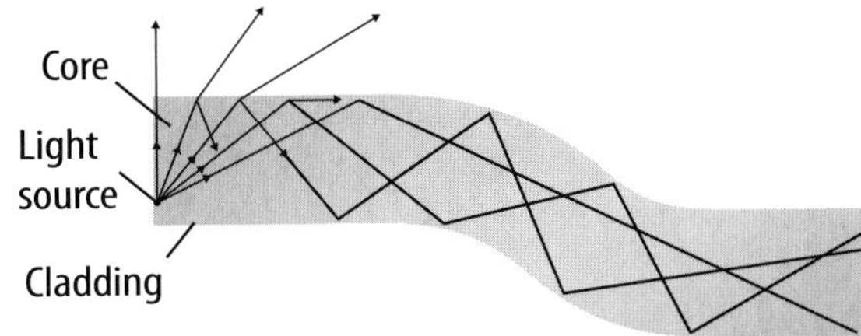

8 What determines whether light that enters the fiber will reflect back into the core?

A the angle at which the light strikes

B the intensity of the light

C the thickness of the fiber

D the wavelength of the light

9 What is unique about light from a laser?

A It is more spread out.

B It is more coherent.

C It has more wavelengths.

D Its intensity decreases faster.

Constructed Response

10 Compare and contrast the interaction of light with the rods and cones in the human eye.

11 An object viewed in white light appears to have blue and white stripes. Describe how those same stripes would appear if red light were shining on the object.

Use the figure below to answer questions 12 and 13.

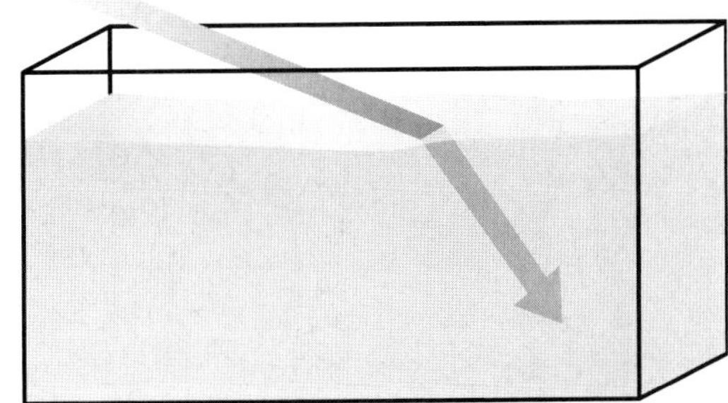

12 The figure shows light shining through air into water. What causes the light to change direction as it enters the water?

13 What would happen to the light if the tank were filled with a liquid that had the same index of refraction as air?

NEED EXTRA HELP?													
If You Missed Question...	1	2	3	4	5	6	7	8	9	10	11	12	13
Go to Lesson...	2	1	3	3	4	3	3	4	4	4	3	3	3

Chapter 19

Electricity

THE BIG IDEA

How do electric circuits and devices transform energy?

Inquiry What did he spark?

You might not know the name Nikola Tesla, despite his great achievements in electrical science. In the past, many people believed electric energy was too dangerous to use in their homes. Tesla made this photograph to demonstrate that his inventions in electricity were safe for everyday use. His work set the stage for the power and lighting systems now used around the world.

- What might happen if any of the electric streaks touch the man?
- What different ways is electricity used in the room?
- How is energy transformed in this scene?

Get Ready to Read

What do you think?

Before you read, decide if you agree or disagree with each of these statements. As you read this chapter, see if you change your mind about any of the statements.

1. Protons and electrons have opposite electric charges.
2. Objects must be touching to exert a force on each other.
3. When electric current flows in a wire, the number of electrons in the wire increases.
4. Electrons flow more easily in metals than in other materials.
5. In any electric circuit, current stops flowing in all parts of the circuit if a connecting wire is removed or cut.
6. The light energy given off by a flashlight comes from the flashlight's batteries.

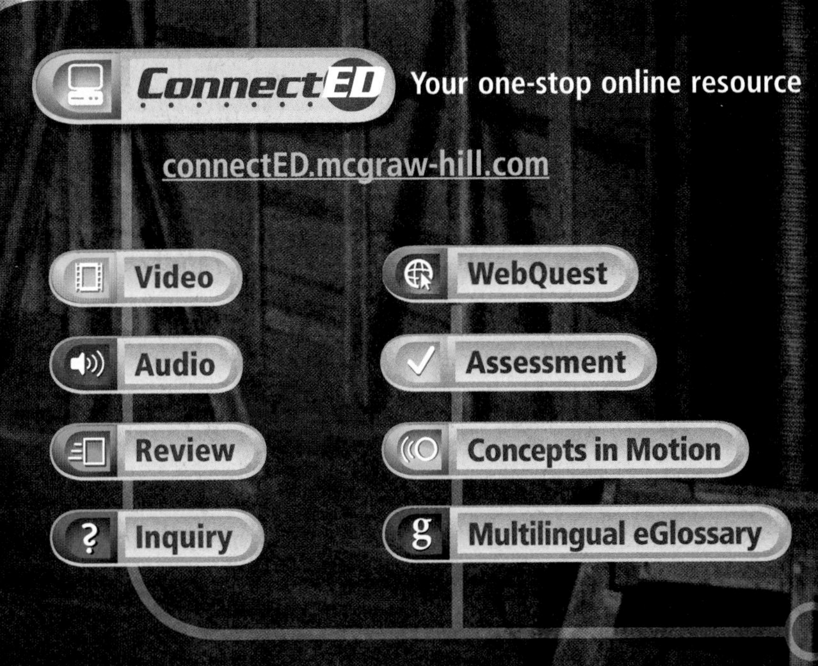

Lesson 1

Electric Charge and Electric Forces

Reading Guide

Key Concepts
ESSENTIAL QUESTIONS

- How do electrically charged objects interact?
- How can objects become electrically charged?
- What is an electric discharge?

Vocabulary

static charge p. 680
electric insulator p. 682
electric conductor p. 682
polarized p. 683
electric discharge p. 685
grounding p. 685

Multilingual eGlossary

Video

- BrainPOP®
- Science Video
- What's Science Got to do With It?

Inquiry Where is the charge?

You are surrounded by electric charges at all times. Without them, you could not exist. However, some electric charges, such as those in lightning, can be dangerous. How can you control electric charges to make your life better?

10 minutes

How can you bend water?

You open the clothes dryer, grab a warm sweater, and pull it on over your head. Then, you look in the mirror and see a sock clinging to your sleeve. A force causes the sock to cling to your sweater. What else could that force do?

1. Read and complete a lab safety form.
2. Inflate a **balloon,** and tie the end. With a **permanent marker,** draw an *X* on one side of the balloon.
3. Your partner holds a **funnel** over a **large bowl** and pours a cup of water through the funnel. As the water gently flows, bring the balloon as close to the stream of water as you can without getting it wet. Record your observations in your Science Journal.
4. Next, rub the X side of the balloon on your **sweater,** and then hold the balloon next to the spot where you rubbed. Observe the interaction between your sweater and the balloon. Record your observations.
5. Rub the X side of the balloon against your sweater again. Immediately repeat step 3.

Think About This

1. **Infer** Why did you rub the balloon on your sweater? Predict what might have happened if you simply touched the balloon to your sweater instead of rubbing it.
2. **Key Concept** Why do you think the balloon interacted the way it did with your sweater and with the stream of water?

Electric Charges

Imagine a hot summer afternoon. Dark clouds fill the sky. Suddenly, a bolt of lightning streaks across the sky. Seconds later, you hear thunder in the distance. The lightning released a tremendous amount of energy. Some of the energy was released as the light that flashed through the sky. Some of the energy was released as the sound you heard as thunder. And, some of the energy was released as thermal energy that heated the air. Where did the energy of the lightning come from? The answer has to do with electric charge.

Charged Particles

Recall that all matter is made of particles called atoms. Also recall that atoms are made of even smaller particles—protons, neutrons, and electrons. Protons and neutrons make up the nucleus of an atom, as shown in **Figure 1.** Electrons move around the nucleus. Protons and electrons have the property of electric charge, neutrons do not. As you read further, you will learn how charged particles interact to affect your everyday life.

Reading Check Which particles found in atoms have the property of electric charge?

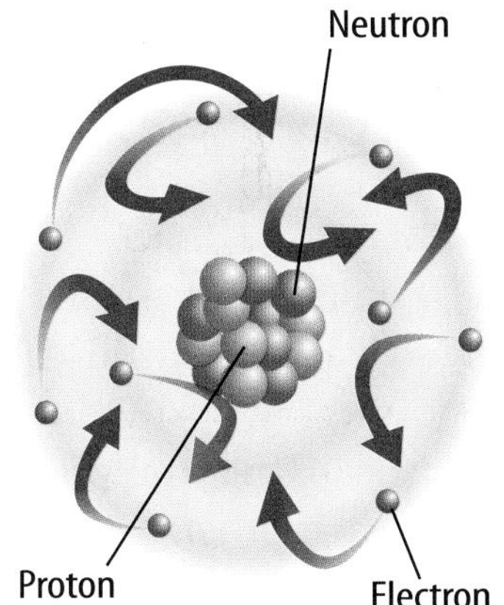

Figure 1 Protons, neutrons, and electrons make up an atom.

Positive Charge and Negative Charge

There are two types of electric charge—positive charge and negative charge. Protons in the nuclei of atoms have positive charge. Electrons moving around a nucleus have negative charge. The amount of positive charge of one proton is equal to the amount of negative charge of one electron.

Recall from previous chapters that oppositely charged particles attract each other. Similarly charged particles repel each other. Therefore, a positively charged proton and a negatively charged electron attract each other. Two protons, or two electrons, repel, or push away from each other.

An atom is electrically neutral—it has equal amounts of positive charge and negative charge. This is because an atom has equal numbers of protons and electrons. Electrons can move from one atom to another. When an atom gains one or more electrons, it becomes negatively charged. If an atom loses one or more electrons, it becomes positively charged. Electrically charged atoms are called ions. It is important to understand that any object can become electrically charged.

Neutral Objects

Similar to electrically neutral atoms, larger electrically neutral objects have equal amounts of positive and negative charge. Electrically neutral objects do not attract or repel each other.

Charged Objects

Just as atoms sometimes gain or lose electrons, larger objects can gain or lose electrons, too. Some materials hold electrons more loosely than other materials. As a result, electrons often move from one object to another. When this happens, the positive charge and negative charge on the objects are unbalanced. *An unbalanced negative or positive electric charge on an object is sometimes referred to as a* **static charge.**

Like an atom, an object that gains electrons has more negative charge than positive charge. It is said to be negatively charged. Likewise, an object that loses electrons has more positive charge than negative charge. It is said to be positively charged.

Electric Forces

The region surrounding a charged object is called an electric field. An electric field applies a force, called an electric force, to other charged objects, even if the objects are not touching.

The electric force applied by an object's electric field will either attract or repel other charged objects. **Figure 2** shows that objects with opposite electric charges attract each other. On the other hand, objects with similar electric charges repel each other.

Key Concept Check How do electrically charged objects interact?

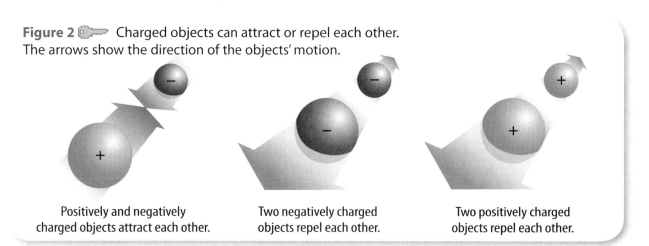

Figure 2 Charged objects can attract or repel each other. The arrows show the direction of the objects' motion.

Positively and negatively charged objects attract each other.

Two negatively charged objects repel each other.

Two positively charged objects repel each other.

Electric Force and Amount of Charge

The strength of the electric force between two charged objects depends on two variables—the total amount of charge on both objects and the distance between the objects. For example, when you brush your hair, electrons move from the brush to your hair. This causes the brush and your hair to each have a static charge. The brush is positively charged. Your hair is negatively charged. Because the brush and your hair have opposite electric charges, they attract each other.

The left portion of **Figure 3** illustrates how the amount of charge affects electric force. If you brush your hair once or twice, some electrons transfer from the brush to your hair. The force of attraction is not very strong. However, if you continue brushing your hair, more electrons transfer from the brush to your hair. This increases the strength of the electric force of attraction between the brush and your hair.

FOLDABLES®

Make a vertical three-tab book, and label it as shown. Use it to organize your notes about electrically charged objects.

Electric Force and Distance

Distance also affects the strength of the electric force between electric charges. As described above, brushing your hair produces an unbalanced positive charge on the brush and an unbalanced negative charge on your hair. Electric fields surround these electric charges. The fields are more intense near the charges. A more intense electric field applies a stronger electric force to other objects. Therefore, when the brush is close to your hair, the force of attraction is very noticeable. The electric force is weaker when the charged objects are far from each other, as shown on the right side of **Figure 3.**

Figure 3 The strength of the electric force between two charged objects depends on the distance between the objects and the total amount of electric charge of the objects.

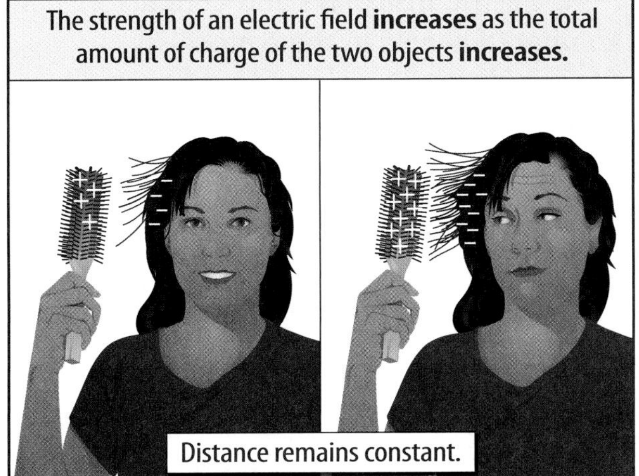

The strength of an electric field **increases** as the total amount of charge of the two objects **increases.**
Distance remains constant.

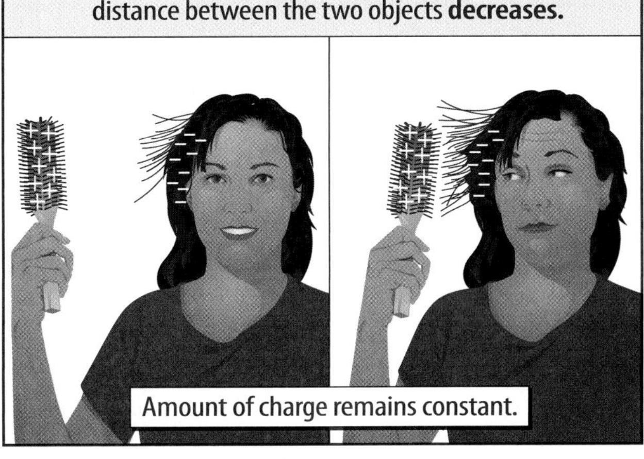

The strength of an electric field **increases** as the distance between the two objects **decreases.**
Amount of charge remains constant.

Inquiry **MiniLab** 10 minutes

How can a balloon push or pull?

When a balloon touches a wool cloth, electrons transfer from the wool onto the balloon.

1. Read and complete a lab safety form.
2. Inflate and tie off two **balloons.** Attach a 10 cm length of **string** to each balloon.
3. Hold the balloons by the strings and slowly bring them together. Record your observations in your Science Journal.
4. Hold one of the balloons 1–3 cm over some **packing peanuts.** Record your observations.
5. Now rub each balloon against a piece of **wool cloth.** Repeat step 3.
6. Rub one of the balloons against the wool again. Repeat step 4.

Analyze and Conclude

1. **Construct** a diagram showing the arrangement of the charges as you are rubbing the balloon against the wool cloth, when the two balloons are interacting, and when the balloon is held near the packing peanuts.
2.  **Key Concept** Summarize how objects with electric charge interact.

Transferring Electrons

You read that a hair brush and strands of hair become electrically charged when electrons transfer from the brush to your hair. How do electrons move from one object to another?

Insulators and Conductors

To understand how electrons move from one object to another, you need to know about two basic types of materials—electric insulators and electric conductors. *A material in which electrons cannot easily move is an* **electric insulator.** Glass, rubber, wood, and even air are good electric insulators. *A material in which electrons can easily move is an* **electric conductor.** Most metals, such as copper and aluminum, are good electric conductors.

Electric insulators and electric conductors are all around you. Electric conductors and electric insulators are used in electrical power cords around your house. In **Figure 4,** copper wire is the conductor of electrons in an extension cord. The copper allows electrons to easily move through the wire. Plastic and rubber are used as protective electric insulators around the metal wire. Electrons cannot easily move in the plastic and rubber.

Electrons transfer between objects by contact, induction, or conduction. As you will read, electric insulators and conductors play an important role in these processes.

Reading Check How do electric insulators and electric conductors differ?

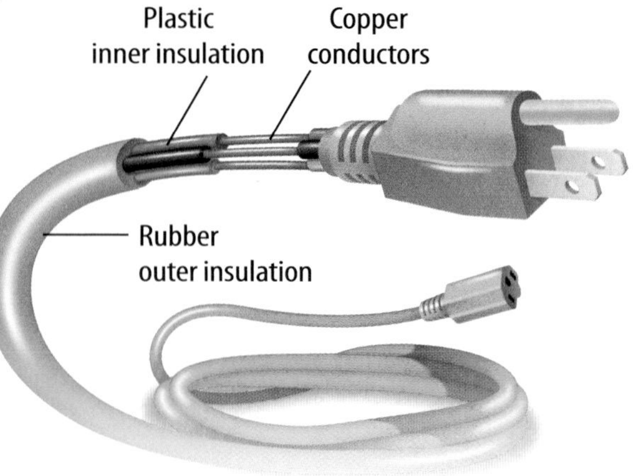

Figure 4 Using conductors and insulators is one way to safely control the flow of electric charges.

 Personal Tutor

Transferring Charge by Contact

Recall that some materials hold their electrons more loosely than other materials. When objects made of different materials touch, electrons tend to collect on the object that holds electrons more tightly. This is called transferring charge by contact.

The wool sweater in **Figure 5** holds electrons more loosely than the rubber balloon. When the balloon comes in contact with the sweater, electrons from the surface of the sweater transfer to the surface of the balloon, creating a static charge on both objects. Because the balloon gained electrons, it has an unbalanced negative charge. Because the sweater lost electrons, it has an unbalanced positive charge. Both insulators and conductors can be charged by contact.

▲ **Figure 5** More loosely held electrons on wool easily transfer onto the surface of a rubber balloon.

Transferring Charge by Induction

Transferring charge by induction is a process by which one object causes two other objects that are conductors to become charged without touching them. **Figure 6** illustrates how this works.

Part 1 of **Figure 6** shows a negatively charged balloon repelling electrons in a metal soda can. Because the can is aluminum and, therefore, a conductor, electrons in the can easily move toward the far end of the can. The can is not charged because it has not gained or lost any electrons. Instead, the can is polarized. *When electrons concentrate at one end of an object, the object is* **polarized.**

In part 2, when a charged balloon is brought near two cans that are touching, the balloon polarizes the cans as if they are one object. Electrons in both cans move toward the far end of the can on the right. Then, the two cans separate. As shown in part 3, the cans that were originally polarized as a group are now individually charged. The can on the right has an unbalanced negative electric charge, and the can on the left has an unbalanced positive electric charge.

Figure 6 Objects that are electric conductors can be charged by induction.▼

1

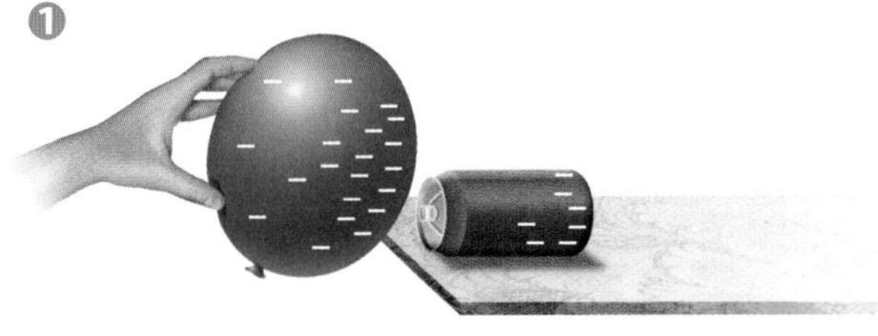

2

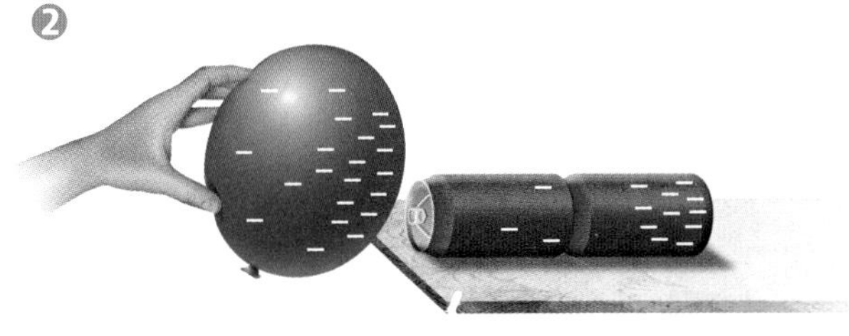

3

Visual Check Why do negative charges in the aluminum cans tend to move away from the balloon?

Transferring Charge by Conduction

Another way that electrons transfer between two conductors is called transferring charge by conduction. As shown in **Figure 7,** when conducting objects with unequal charges touch, electrons flow from the object with a greater concentration of negative charge to the object with a lower concentration of negative charge. This is similar to water flowing from a container with a higher water level to a container with a lower water level. The flow of electrons continues until the concentration on charge of both objects is equal.

Key Concept Check What are three ways by which objects can become electrically charged?

Figure 7 Water flows from the container with more water to the container with less water until the levels are equal. Similarly, negative charges flow between objects until the concentration of charge on both objects is equal.

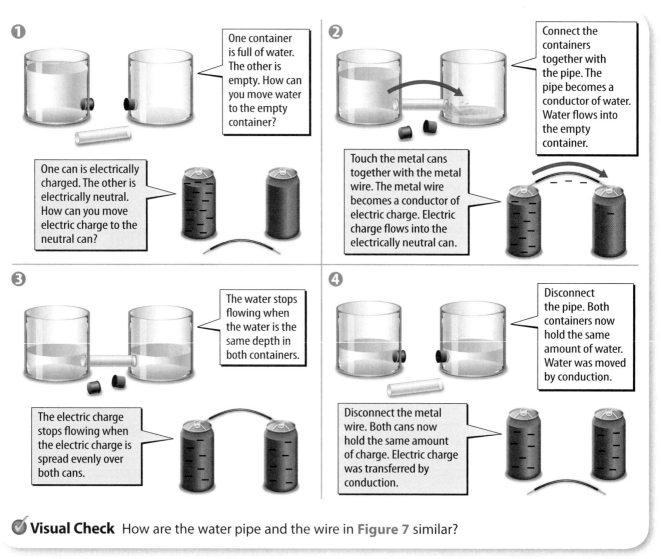

1

One container is full of water. The other is empty. How can you move water to the empty container?

One can is electrically charged. The other is electrically neutral. How can you move electric charge to the neutral can?

2

Connect the containers together with the pipe. The pipe becomes a conductor of water. Water flows into the empty container.

Touch the metal cans together with the metal wire. The metal wire becomes a conductor of electric charge. Electric charge flows into the electrically neutral can.

3

The water stops flowing when the water is the same depth in both containers.

The electric charge stops flowing when the electric charge is spread evenly over both cans.

4

Disconnect the pipe. Both containers now hold the same amount of water. Water was moved by conduction.

Disconnect the metal wire. Both cans now hold the same amount of charge. Electric charge was transferred by conduction.

Visual Check How are the water pipe and the wire in **Figure 7** similar?

Electric Discharge

When you brush your hair, electrons transfer from the brush to your hair. What happens to these excess electrons in your hair? They transfer to objects that come in contact with your hair, such as a hat or your pillow. Electrons even transfer to the air. Gradually, your hair loses its charge. *The process of an unbalanced electric charge becoming balanced is an* **electric discharge.**

Lightning Rods and Grounding

An electric discharge can occur slowly, such as when your hair loses its negative charge and is no longer attracted to a brush. Or, an electric discharge can occur as quickly as a flash. For example, lightning is a powerful electric discharge that occurs in an instant.

A lightning strike can severely damage a building or injure people. Lighting rods help protect people against these dangers. **Figure 8** shows a metal lightning rod attached to the roof of a building. A thick wire connects the lightning rod to the ground. The wire provides a path for the electrons released in a lightning strike to travel into the ground. *Providing a path for electric charges to flow safely into the ground is called* **grounding. Table 1** provides tips on how to protect yourself from a lightning strike.

What causes lightning?

Scientists are not entirely clear on what causes lightning. However, many scientists believe that lightning is related to the electric charges that separate within storm clouds. The large amounts of ice, hail, and partially frozen water droplets that thunderstorms create seem to play a role, too.

Forecasting exactly where and when lightning will strike might never be possible. **Figure 9** on the next page summarizes what is known about how lightning forms.

 **Key Concept Check** What is an electric discharge?

Figure 8 Grounded lightning rods help protect tall buildings from the damaging effects of lightning.

Visual Check What are the two components of a building's lightning protection system?

WORD ORIGIN

electric
from Greek *ēlektron*, means "beaming sun"

Table 1 Lightning Safety Tips

If the time between a lightning flash and thunder is less than 30 seconds, the storm is dangerously close. Protect yourself in the following ways:

- Seek shelter in an enclosed building or a car with a metal top. Never stand under a tree.
- Do not touch metal surfaces.
- Get away from water if swimming, boating, or bathing.
- Wait 30 minutes after the last flash of lightning before leaving the shelter, even if the Sun comes out.

Figure 9 The conditions needed for lightning to form are well known. However, scientists still do not clearly understand how lightning is produced. Lightning is one of the most unpredictable forces of nature.

1

Within a storm cloud, warm air rises past falling cold air. The cold air is filled with hail, ice, and partially frozen water droplets that pick up electrons from the rising air. This causes the bottom of the cloud to become negatively charged.

2

The negatively charged cloud polarizes Earth's surface by repelling negative charges in the ground. Thus, the surface of the ground becomes positively charged.

3

When the bottom of the cloud accumulates enough negative charge, the attraction of the positively charged ground causes electrons in the cloud to begin moving toward the ground.

4

As electrons approach the ground, positive charges quickly flow upward, making an electric connection between the cloud and the ground. You see this electric discharge as lightning.

5

Accumulation of charge does not only occur between clouds and the ground. Charges also separate within or between storm clouds, causing cloud-to-cloud lightning.

Lesson 1 Review

Assessment Online Quiz

Visual Summary

Any object can be positively charged, negatively charged, or neutral.

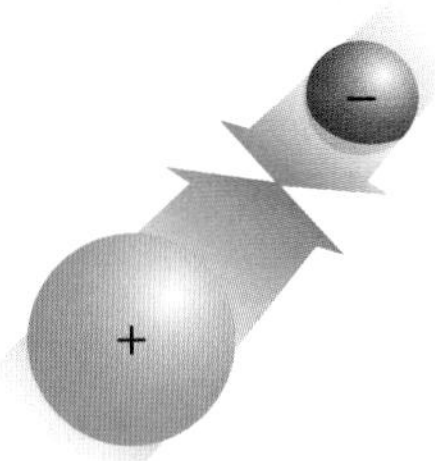

Charged objects exert forces on each other.

Lightning is one form of electric discharge.

FOLDABLES®

Use your lesson Foldable to review the lesson. Save your Foldable for the project at the end of the chapter.

What do you think NOW?

You first read the statements below at the beginning of the chapter.

1. Protons and electrons have opposite electric charges.
2. Objects must be touching to exert a force on each other.

Did you change your mind about whether you agree or disagree with the statements? Rewrite any false statements to make them true.

Use Vocabulary

1. Providing a path for electric charges to flow safely into the ground is called ________.
2. **Distinguish** between an electric conductor and an electric insulator.
3. **Use the term** *static charge* in a sentence.

Understand Key Concepts 

4. Particles that attract each other are
   A. two electrons.
   B. two protons.
   C. one proton and one electron.
   D. one proton and one neutron.
5. **Summarize** How does the electric force between charged objects change as the objects move away from each other?
6. **Describe** Imagine you have to design a safe, useful device that uses an electric discharge to make your life better. How would you describe it to others?

Interpret Graphics

7. **Organize** Copy the graphic organizer below. Fill in the type of charge on protons, neutrons, and electrons.

Particle	Type of Electric Charge
Proton	
Neutron	
Electron	

Critical Thinking

8. **Analyze** How can objects become positively charged?
9. **Hypothesize** A polyester shirt clings to your skin as you put it on. You then learn that skin easily releases electrons. Write a description of what happened between the shirt and your skin.

HOW IT WORKS

Van de Graaff Generator

How can a machine move electrons?

Have you ever seen a Van de Graaff generator? Originally, scientists studied the nuclei of atoms with these machines. Now, they are often seen in science museums, where they are used to demonstrate the effects electric charge.

How does this device generate electric charge? Recall that some materials hold electrons more loosely than other materials. In a Van de Graaff generator, the metal dome holds electrons more loosely than the belt. As the belt travels over the top roller past the top comb, electrons move from the metal dome through the upper comb and onto the belt. This leaves the dome positively charged.

The excess electrons on the belt move from the belt onto the lower comb as the belt moves over the roller at the bottom of the generator. The electrons then travel through a wire out of the machine and into the ground. This process of electrons moving from the dome to the ground continues as long as the generator is running.

Soon, the dome loses so many electrons that it acquires a very large positive charge. The positive charge on the dome becomes so great that electrons create an electric spark as they jump back onto the dome from any object that will release them. That object could be you if you stand close enough.

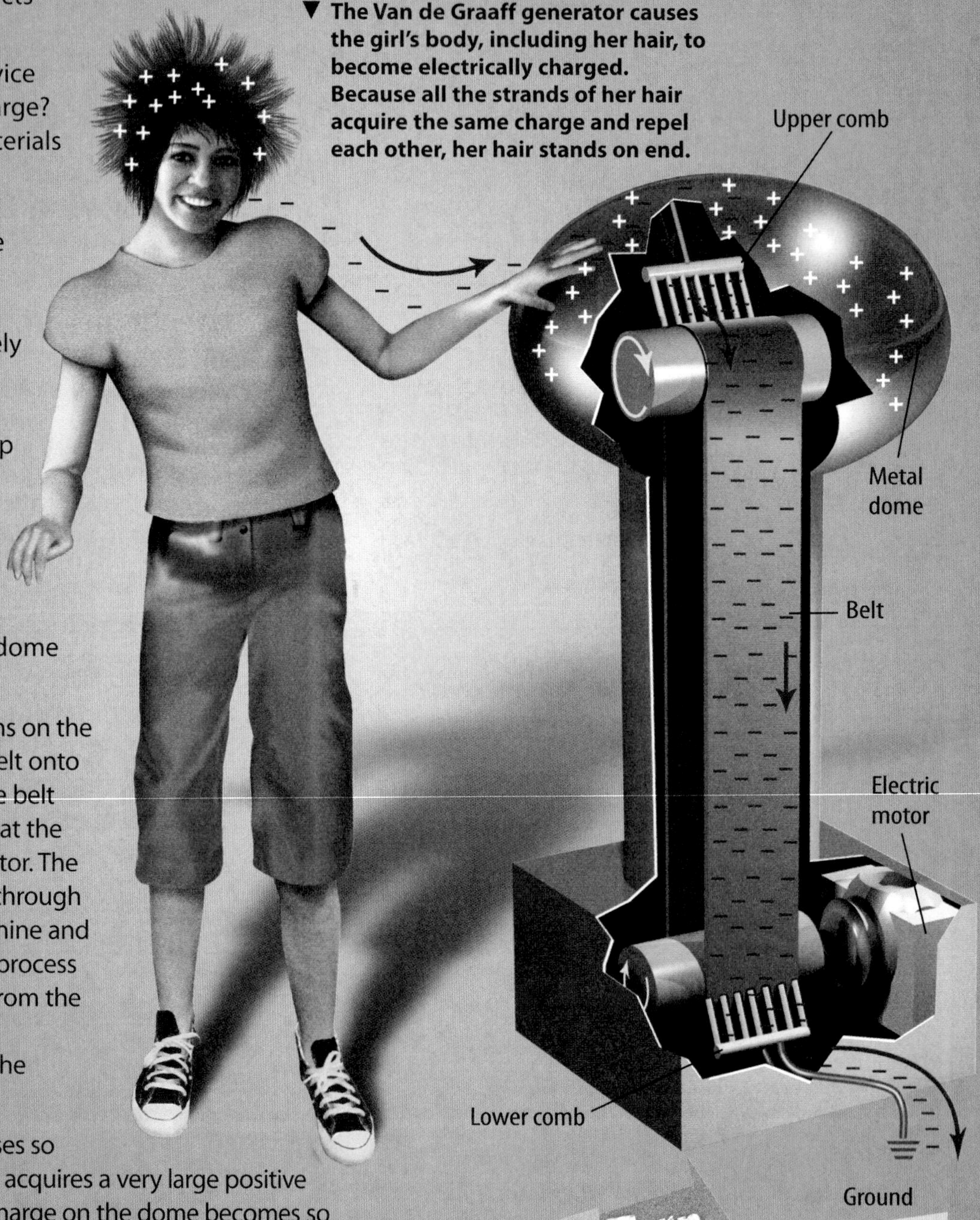

▼ The Van de Graaff generator causes the girl's body, including her hair, to become electrically charged. Because all the strands of her hair acquire the same charge and repel each other, her hair stands on end.

It's Your Turn

REPORT Find out how to build a simple Van de Graaff generator using everyday materials. Share your findings with your classmates.

Lesson 2

Reading Guide

Key Concepts
ESSENTIAL QUESTIONS

- What is the relationship between electric charge and electric current?
- What are voltage, current, and resistance? How do they affect each other?

Vocabulary

electric current p. 690
electric circuit p. 690
electric resistance p. 692
voltage p. 693
Ohm's law p. 694

Multilingual eGlossary

Video
- BrainPOP®
- Science Video

Electric Current and Simple Circuits

Inquiry How far can it go?

Across the country, over 300,000 km of wires carry electric energy from about 500 power companies to homes and businesses. A flashlight operates with only a few centimeters of electric conductors. How are these vastly different systems similar?

Inquiry Launch Lab

20 minutes

What is the path to turn on a light?

How can you make a lightbulb light up?

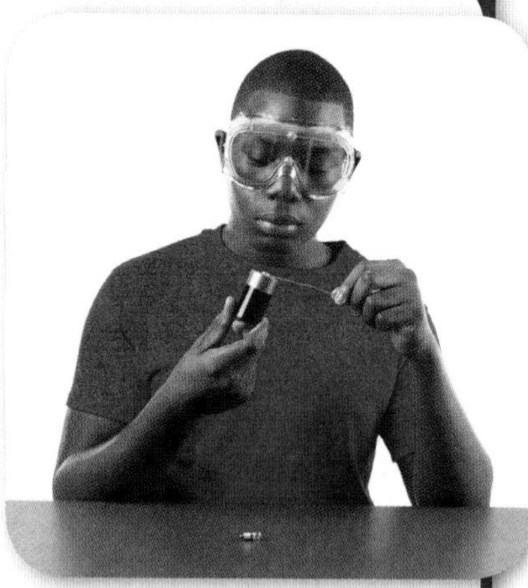

1. Read and complete a lab safety form.
2. Examine a **D-cell battery.** Notice the differences between the two ends. Record your observations in your Science Journal.
3. Using a **paper clip, small lightbulbs,** and the battery, design a way to light the bulb. Draw a diagram of your plan.
   ⚠ *The paperclip can get hot!*
4. Find two other ways to light the bulb using the paper clip and the battery. Draw your plans.
5. Record three configurations that do not light the bulb.

Think About This

1. Why do some arrangements of the materials light the bulb, but other arrangements do not?
2. **Key Concept** How do you think all the pieces interact to light the bulb?

Electric Current and Electric Circuits

In Lesson 1, you read about ways to transfer, or move electric charges. The transferring of electric charges allows you to power the electrical devices you use every day. For example, conduction of electric charges occurs every time you turn on a flashlight or a television. Induction is used for wireless charging of devices, such as electrical toothbrushes. And, as you read in the last lesson, transferring charge by contact occurs every time lightning strikes. How are a flash of lightning and a TV similar? They both transform the energy of moving electrons to light, sound, and thermal energy. *The movement of electrically charged particles is an* **electric current.**

SCIENCE USE V. COMMON USE
current
Science Use the flow of electric charge
Common Use occurring at the present time

A Simple Electric Circuit

The movement of electrons in a lightning strike lasts only a fraction of a second. In a TV, electrons continue moving as long as the TV remains on. An electric current flows in a closed path to and from a source of electric energy. *A closed, or complete, path in which an electric current travels is an* **electric circuit**.

WORD ORIGIN
circuit
from Latin *circuire,* means "to go around"

All electric circuits have one thing in common—they transform electric energy to other forms of energy. For example, electric circuits in a microwave oven transform electric energy to the thermal energy that quickly cooks your food. Circuits in your television transform electric energy to the light and sound that entertains and informs you. How do electric energy transformations improve your life?

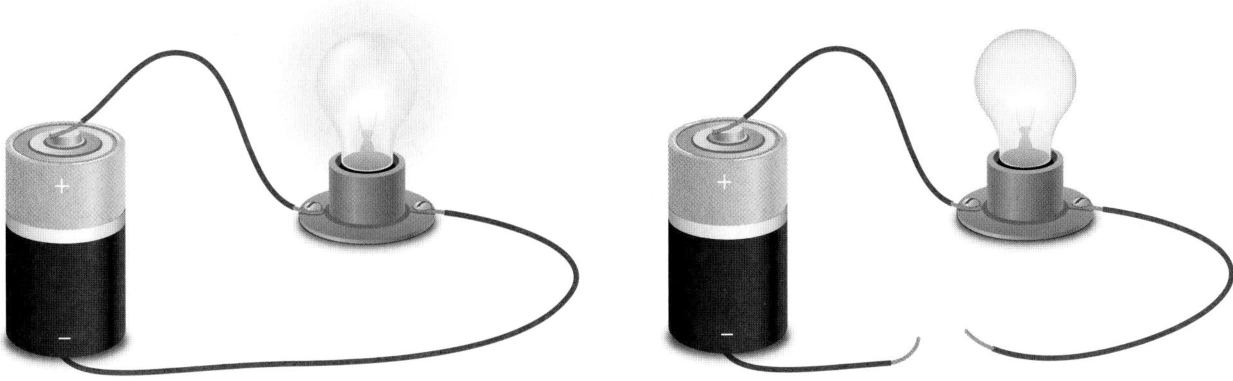

▲ **Figure 10** An electric current flows in a circuit if the circuit is complete, or closed. Current will not flow if the circuit is broken, or open.

How Electric Charges Flow in a Circuit

Figure 10 shows a battery, wires, and a lightbulb connected in a circuit. The lightbulb glows as electrons flow from of one end of the battery, through the wires and lightbulb, and back into the other end of the battery. If the circuit is broken, or open, the electrons stop flowing, and the bulb stops glowing.

A current of electrons in a wire is somewhat like water being pumped through a hose. The amount of water flowing into one end of a hose is the same as the amount of water flowing out the other end. As shown in **Figure 11**, the number of electrons flowing into a wire from a power source equals the number of electrons flowing out of the wire back into the source.

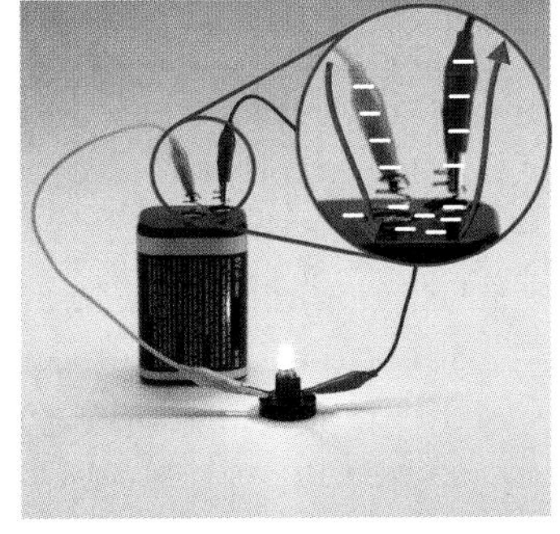

▲ **Figure 11** When current flows in a wire, the number of electrons in the wire does not change.

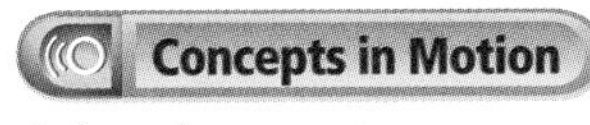

Animation

The Unit For Electric Current

Electric current is approximately measured as the number of electrons that flow past a point every second. However, electrons are so tiny and there are so many in a circuit that you could never count them one at a time. Therefore, just like you count eggs by the dozen, scientists count electrons by a quantity called the coulomb (KEW lahm). A coulomb is a very large number—about 6×10^{18}, or 6,000,000,000,000,000,000! That is 6 quintillion.

The SI unit for electric current is the ampere (AM pihr), commonly called an amp. Its symbol is A. One ampere of current equals about one coulomb of electrons flowing past a point in a circuit every second. The electric current through a 120-W lightbulb is about 1 A. A typical hair dryer uses a current of about 10 A, or 60,000,000,000,000,000,000 electrons per second.

Key Concept Check What is the relationship between electric charge and electric current?

Inquiry MiniLab 20 minutes

When is one more than two?

All parts of a circuit—such as bulbs and wires—add resistance to the circuit.

1. Read and complete a lab safety form.
2. Using **alligator clip wires,** connect a **small bulb in a base** to a **hand-cranked generator.** Draw your circuit in your Science Journal.
3. Crank the generator at different rates. Observe what happens to the brightness of the bulb as the crank turns. Turn the crank at a rate that produces a medium brightness. Record that rate.
4. Using additional wires as needed, add a **second bulb** to your circuit. Reconnect the generator, and turn the crank at the same rate as before. Draw the circuit and record your observations.
5. Repeat the previous step with additional bulbs and wires.

Analyze and Conclude

1. **Describe** how you changed the resistance in the circuit.
2. **Key Concept** Explain how current and resistance are related in this circuit.

Table 2 Electric Resistance of Different Materials

Material (20cm x 1mm)	Resistance (ohms)
Copper	*0.004*
Gold	*0.006*
Iron	*0.025*
Carbon	*8.9*
Rubber	*2,500,000,000,000,000,000*

What is electric resistance?

See **Figure 12** on the next page. Suppose you replaced one of the green wires in the circuit on the left side with a piece of string. The lightbulb would not glow because there would be no current in the circuit. Why do electrons flow easily in metal wire but not in string? The answer is that wire has much less electric resistance than string. **Electric resistance** *is a measure of how difficult it is for an electric current to flow in a material.*

The unit of electric resistance is the ohm (OHM), which is symbolized by the Greek letter Ω (omega). An electric resistance of 20 ohms is written 20 Ω. **Table 2** lists the electric resistances of different materials.

Electric Resistance of Conductors and Insulators

You read that electric conductors are materials, such as copper and aluminum, in which electrons easily move. A good conductor has low electric resistance. Usually electric wires are made of copper because copper is one of the best conductors.

Recall that electrons cannot easily move through insulators such as plastic, wood, or string. A good electric insulator has high electric resistance. Atoms of an insulator hold electrons tightly. This prevents electric charges from easily moving through the material. Therefore, replacing a wire in a circuit with string prevents electrons from flowing in a circuit.

Resistance—Length and Thickness

A material's electric resistance also depends on the material's length and thickness. A thick copper wire has less electric resistance than a thin copper wire of the same length. Because the thick wire has less resistance, it will conduct better. Increasing the length of a conductor also increases its electric resistance.

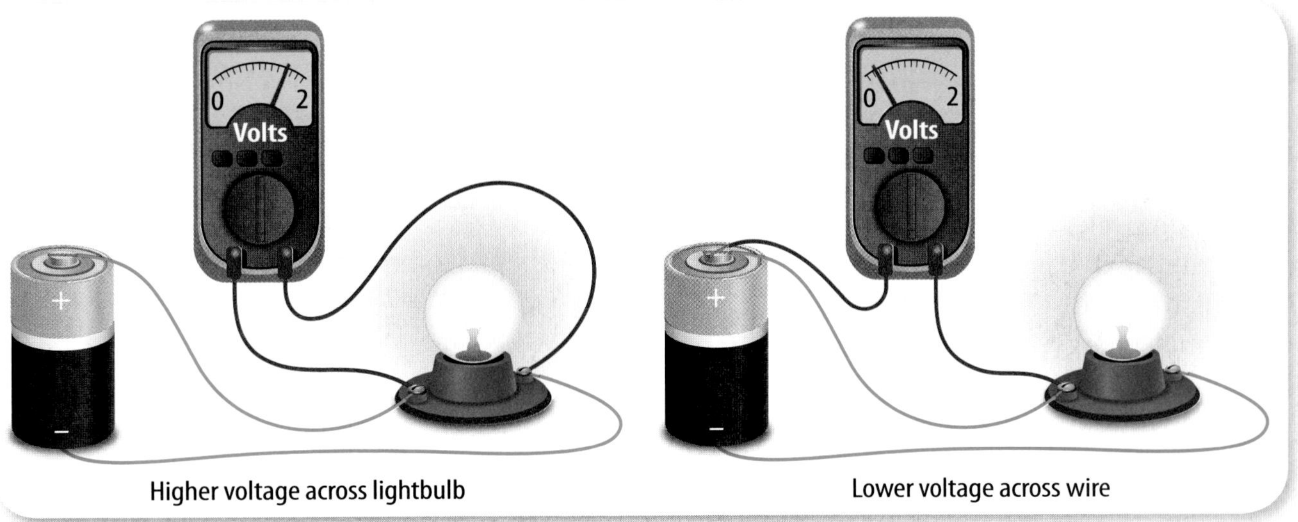

Figure 12 Voltage can be different in different parts of a circuit. The voltmeter shows where most of the battery's energy is used.

Visual Check Why is the voltage reading across the lightbulb higher than across the segment of wire?

What is voltage?

You probably heard the term *volt.* You use 1.5-V batteries in a flashlight. You plug a hair dryer into a 120-V outlet. But, what does this mean?

Battery Voltage

In **Figure 12**, a battery creates an electric current in a closed circuit. Energy stored in the battery moves electrons in the circuit. As the electrons move through the circuit, the amount of energy transformed by the circuit depends on the battery's voltage. **Voltage** *is the amount of energy the source uses to move one coulomb of electrons through the circuit.* A circuit with a high voltage source transforms more electric energy to other energy forms than a circuit with a low voltage source. For example, a lightbulb connected to a 9-V battery produces about six times more light and thermal energy than the same lightbulb connected to a 1.5-V battery.

Voltage in Different Parts of a Circuit

Electric energy transforms to other forms of energy in all parts of a circuit. For example, the lightbulbs in **Figure 12** transform electric energy to light and thermal energy. Even the wires and batteries produce a small amount of thermal energy. In other words, different amounts of energy transform in different parts of a circuit. The voltage measured across a portion of a circuit indicates how much energy transforms in that portion of the circuit. For example, the voltage is greater across the lightbulb in **Figure 12** than across the wire. Therefore, the lightbulb transforms more energy than the wire.

Reading Check What happens to the energy flowing in an electric circuit?

FOLDABLES®

Make a horizontal three-tab book using the labels shown. Use it to organize your notes about the flow of electric charge.

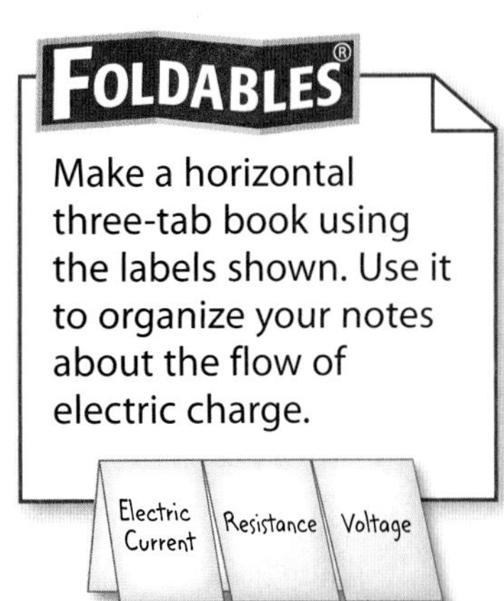

Ohm's Law

When designing electrical devices, engineers choose materials based on their electric resistance. For example, the heating coils in a toaster must be made of a metal with very high electric resistance. As explained on the next page, this allows the coils to transform most of the circuit's energy to thermal energy. But, how much resistance should a conductor have? The answer is found with Ohm's law.

Using Ohm's Law

Named after German physicist Georg Ohm, **Ohm's law** *is a mathematical equation that describes the relationships among voltage, current, and resistance.* The law states that as the voltage of a circuit's electric energy source increases, the current in the circuit increases, too. Also, as the resistance of a circuit increases, the current decreases.

Ohm's Law can be written as the following equation:

Ohm's Law Equation

voltage (V) = **current** (I) × **resistance** (R)

$$V = IR$$

V is the symbol for voltage, measured in volts (V). I is the symbol for electric current, which is measured in amperes (A). And, R is the symbol for electric resistance, measured in ohms (Ω). If you know the value of two of the variables in the equation, you can determine the third, as described in **Figure 13**. You can measure any of these variables with a multimeter, as shown in **Figure 14**.

Key Concept Check How do voltage, current, and resistance affect each other?

Figure 13 Ohm's law can be used to calculate unknown quantities in a circuit.

Calculate the voltage across the lightbulb.

$R = 50\Omega$

To find the voltage, start with Ohm's law:

$$V = IR$$

Substitute the known values into the equation:

$$V = 0.1A \times 50\Omega$$
$$V = 5V$$

Calculate the resistance of the lightbulb.

$R = ?\Omega$

$I = 0.3A$

$V = 6.0V$

To find the resistance, start with this form of Ohm's law:

$$R = \frac{V}{I}$$

Substitute the known values into the equation:

$$R = \frac{6V}{0.3A}$$
$$R = 20\Omega$$

Calculate the current flowing through the lightbulb.

$R = 50\Omega$

To find the current, start with this form of Ohm's law:

$$I = \frac{V}{R}$$

Substitute the known values into the equation:

$$I = \frac{12V}{50\Omega}$$
$$I = 0.2A$$

Math Skills — Use a Simple Equation

Solve for Voltage The current through a lightbulb is **0.5 A**. The resistance of the lightbulb is **220 Ω**. What is the voltage across the lightbulb?

1. **This is what you know:** current: $I = 0.5$ A; resistance: $R = 220\ \Omega$
2. **This is what you need to find:** voltage V
3. **Use this formula:** $V = IR$
4. **Substitute:** the values for I and R into the formula and multiply: $V = (0.5\ \text{A}) \times (220\ \Omega) = 110\ \text{V}$

Answer: The voltage is **110 V**.

Practice

What is the voltage across the ends of a wire coil in a circuit if the current in the wire is 0.1 A and the resistance is 30 Ω?

- Math Practice
- Personal Tutor

Voltage, Resistance, and Energy Transformation

Figure 14 shows two lightbulbs and a battery connected in a circuit. Both bulbs are connected one-after-another in a single loop. The currents through both bulbs are equal. However, the two lightbulbs are not identical. One has greater electric resistance than the other.

You determine which lightbulb has more electric resistance with a voltmeter and an understanding of Ohm's law. Ohm's law states that, with equal current, the voltage is greater across the device with the greater resistance. **Figure 14** also shows that the lightbulb with the greater electric resistance has the greater voltage across it. The higher-resistance lightbulb on the left transforms more electric energy to light.

REVIEW VOCABULARY

energy transformation
the conversion of one form of energy to another

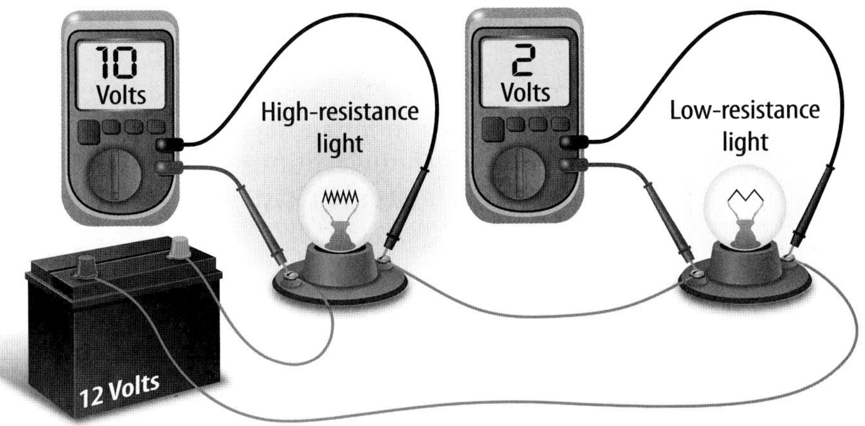

Figure 14 When two devices are in the same circuit, the voltage is greater across the device with higher resistance. The device with higher resistance transforms more electric energy to other forms of energy.

Visual Check Which lightbulb transforms more electric energy?

Lesson 2 Review

Assessment Online Quiz
Inquiry Virtual Lab

Visual Summary

An electric current is the flow of negative electric charges through a conductor.

Electrical devices function only when they are connected in a closed circuit.

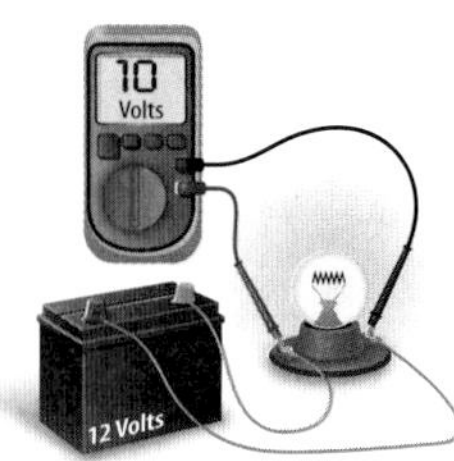

Ohm's law shows the relationship among voltage, current, and resistance.

Use your lesson Foldable to review the lesson. Save your Foldable for the project at the end of the chapter.

What do you think NOW?

At the beginning of this lesson, you read the statements below.

3. When electric current flows in a wire, the number of electrons in the wire increases.
4. Electrons flow more easily in metals than in other materials.

Did you change your mind about whether you agree or disagree with the statements? Rewrite any false statements to make them true.

Use Vocabulary

1. Electrons flow more easily in a material with lower ________.
2. A closed path in which electric charges can flow is a(n) ________.
3. The flow of electric charges is a(n) ______.

Understand Key Concepts 

4. Ohm's law is NOT related to
   A. current. **C.** resistance.
   B. mass. **D.** voltage.
5. **Describe** in your own words the relationship between electric charge and electric current.

Interpret Graphics

6. **Explain** Copy the table below. Fill in the effect of changing voltage and resistance on electric current.

Variable	Effect of Increase in Variable on Current
Voltage	
Resistance	

Critical Thinking

7. **Evaluate** How does the total electric charge of a wire in a circuit change when current stops flowing in the circuit?

Math Skills

Math Practice

8. The current in a hair dryer is 10 A. The hair dryer is plugged into a 110-V outlet. How much electric resistance is there in the hair dryer's circuit?

Inquiry Skill Practice **Identify and Manipulate Variables** **40 minutes**

What effect does voltage have on a circuit?

Materials

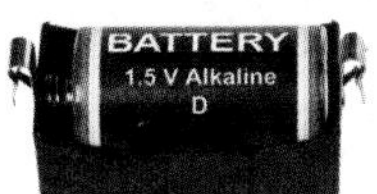

2 batteries in bases

2, 3-V bulbs in bases

alligator clip wires

Safety

In 1827, German scientist Georg Ohm published his work about the relationship among voltage, current, and resistance. At that time many people and other scientists disagreed with him. They did not believe that experiments were necessary to understand the world. We now know that the best way to develop an understanding of nature is through carefully developed and controlled experiments, such as those Ohm performed, in which the researcher **identifies and manipulates the variables.**

Learn It

In an experiment, it is important to keep all factors the same except for the factor you are testing. The factor you manipulate, or change, is called the **independent variable.** The factor that changes as a result is the **dependent variable.**

Try It

1. Read and complete a lab safety form.
2. Notice how the two ends of the battery differ. Record your observations in your Science Journal.
3. Copy the table below in your Science Journal. Record your predictions for each circuit.
4. Construct and observe the circuits described in the table.
5. Compare the predicted brightness of the bulb in each circuit to the observed brightness. Record your findings.

Apply It

6. **Determine** the dependent and the independent variables.
7. **Key Concept** Describe how voltage and current relate in each circuit.

Circuit Description	Sketch of Circuit	Predicted Bulb Brightness	Observed Brightness
One battery and one bulb			
Two batteries and one bulb version #1			
Two batteries and one bulb version #2			

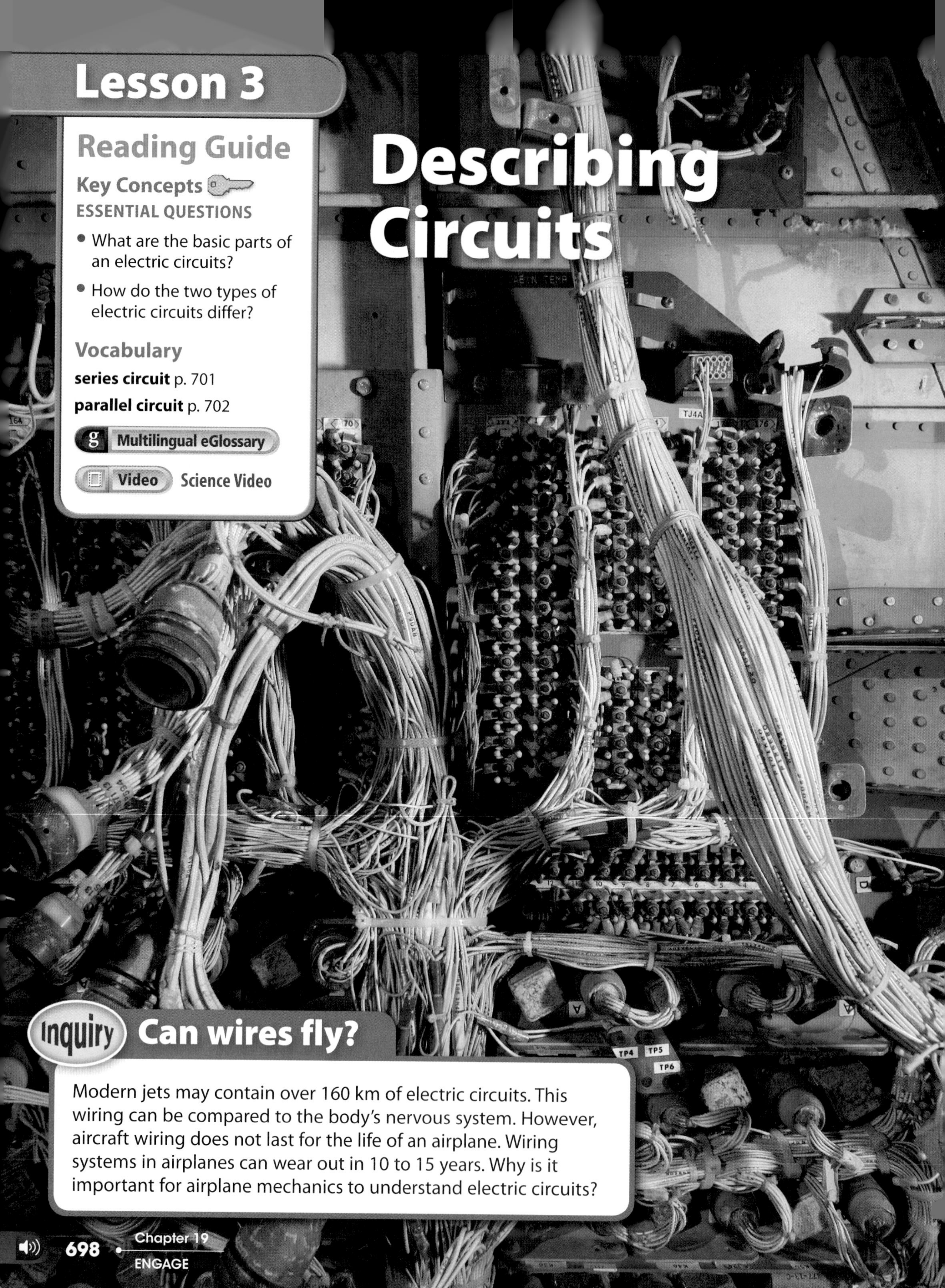

Lesson 3

Reading Guide

Key Concepts
ESSENTIAL QUESTIONS

- What are the basic parts of an electric circuits?
- How do the two types of electric circuits differ?

Vocabulary

series circuit p. 701
parallel circuit p. 702

Multilingual eGlossary

Video Science Video

Describing Circuits

Inquiry Can wires fly?

Modern jets may contain over 160 km of electric circuits. This wiring can be compared to the body's nervous system. However, aircraft wiring does not last for the life of an airplane. Wiring systems in airplanes can wear out in 10 to 15 years. Why is it important for airplane mechanics to understand electric circuits?

Launch Lab

20 minutes

How would you wire a house? 

If you turn off the kitchen light, why does the refrigerator keep running?

1. Read and complete a lab safety form.
2. Using **alligator clip wires,** light a **small bulb in a base** with a **D-cell battery in a base.** Draw the setup in your Science Journal.
3. Disconnect the wires between the battery and the bulb. Add a **second bulb** between the battery and the first bulb. Remove one of the bulbs from its base. Record your observations.
4. Take apart the setup, and reassemble the setup in step 2. Now, add a second bulb by connecting one terminal of the second bulb to one of the terminals of the lit bulb. Connect the other terminal of the second bulb to the remaining terminal of the lit bulb. Both bulbs should be lit. Remove one of the bulbs from its base. Record your observations.

Think About This

1. Which of your assembled circuits would be the best to use when wiring a house?
2. **Key Concept** Which setup is the best way to connect lights in your home? Explain.

Parts of an Electric Circuit

Do you study by an electric lamp? Is there a computer on your desk? These **devices** contain electric circuits.

Three common parts of most electric circuits are a source of electric energy, electrical devices that transform the electric energy, and conductors, such as wires, that connect the other components.

Electric Energy to Kinetic Energy

An energy source, such as the battery in **Figure 15,** produces an electric current in a circuit. Some electrical devices are designed to transform the electric energy of the current to kinetic energy—the energy of motion. For example, the motor in an electric fan transforms electric energy to the kinetic energy of moving air particles that keep you cool.

 **Key Concept Check** What are the three basic parts of an electric circuit?

ACADEMIC VOCABULARY

device
(noun) a piece of equipment

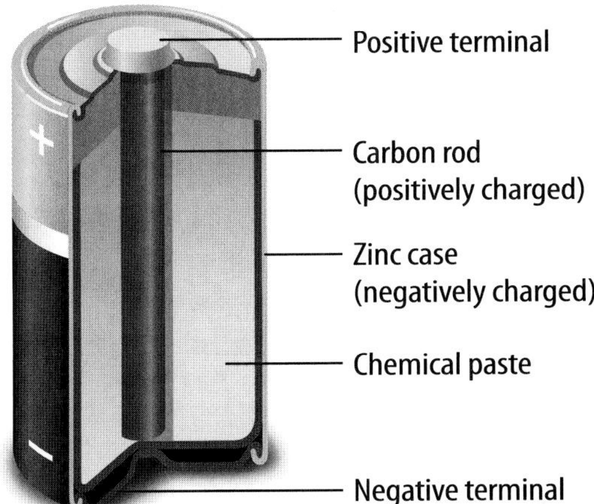

Figure 15 In a battery, chemical reactions in the moist paste cause the carbon rod to become positively charged. The zinc case becomes negatively charged.

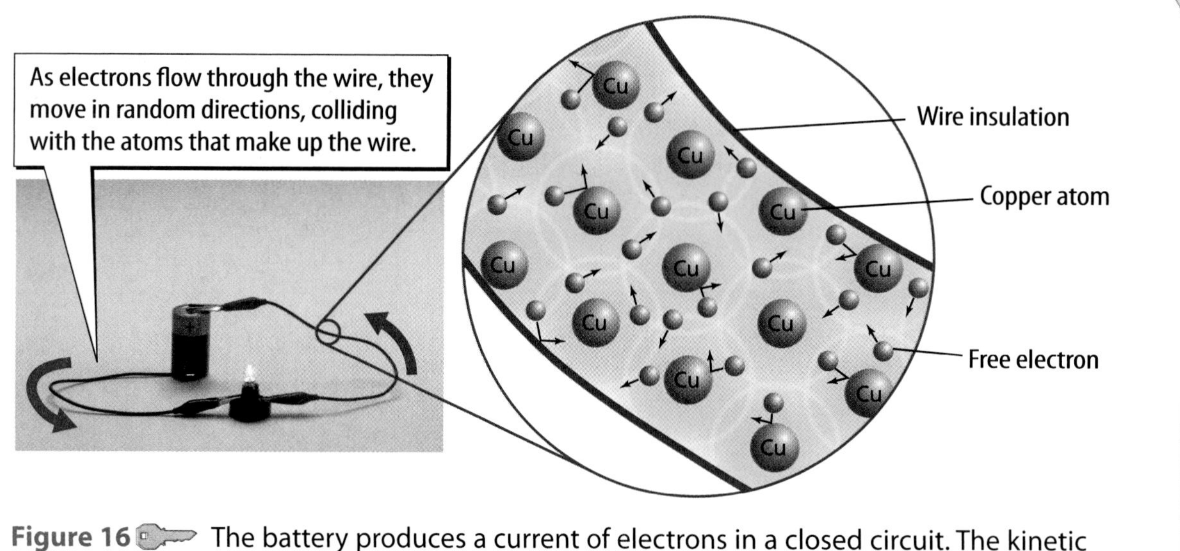
As electrons flow through the wire, they move in random directions, colliding with the atoms that make up the wire.
Wire insulation
Copper atom
Free electron
Cu

Figure 16 The battery produces a current of electrons in a closed circuit. The kinetic energy of the electrons transforms to other forms of energy.

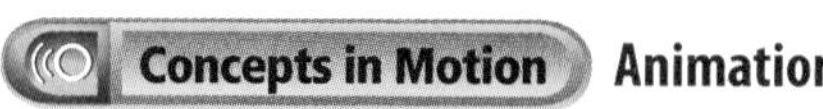
Concepts in Motion
Animation

Batteries supply electric energy.

A battery is often used as a source of electric energy for a circuit. **Figure 15** on the previous page shows a cross section of a common battery. A battery is just a can of chemicals. As the chemicals react, electrons in the battery concentrate at the battery's negative terminal. When a closed circuit connects the terminals, electrons flow in the circuit from the battery's negative terminal to the positive terminal. If the circuit is closed and the chemicals in the battery keep reacting, an electric current continues.

Electric circuits transform energy.

Energy transformations occur in all parts of an electric circuit. For example, a battery transforms stored chemical energy to the electric energy of electrons moving as an electric current. Electrical devices in the circuit transform most of the electric energy to other useful forms of energy. For example, a lightbulb transforms electric energy to light, and stereo speakers transform electric energy to sound. In addition, all parts of a circuit, including the energy source and the connecting wires, transform some of the electric energy to wasted thermal energy.

Figure 16 shows how these energy transformations occur. For example, electrons flowing in the wire filament of a lightbulb collide with the atoms that make up the filament. The collisions transfer the electrons' energy to the atoms. The atoms immediately release the energy in other forms, such as light and thermal energy.

You read that circuits and devices release wasted thermal energy. However, many modern electrical devices, such as compact fluorescent lamps (CFLs), are designed to waste less energy. CFLs conduct an electric current through a gas and do not use wire filaments. More energy transforms to light, and less is wasted as thermal energy.

Wires connect parts of a circuit.

Recall that an electric current flows only in a closed circuit. Therefore, a circuit's energy source and device must be connected with some form of conducting material. Metal wires often are used to complete circuits. Because of their low electric resistance, wires transform only a small amount of electric energy to wasted thermal energy. This leaves more energy available for useful devices in the circuit.

Series and Parallel Circuits

Some strings of holiday lights will not light if any bulbs are missing. Other strings of lights work with or without missing bulbs. These two examples of holiday lights represent the two types of electric circuits—series circuits and **parallel** circuits.

WORD ORIGIN
parallel
from Greek *parallēlos,* means "beside of one another"

Series Circuit—A Single Current Path

A **series circuit** *is an electric circuit with only one path for an electric current to follow.* Many strings of holiday lights are series circuits. In these series circuits all the lightbulbs connect end-to-end in a single conducting loop. As shown in **Figure 17,** breaking the loop at any point stops the current of electrons throughout the entire circuit. If a wire loop in a series circuit is broken, or open, all devices in the circuit will turn off.

Reading Check Why do all devices in a series circuit stop working if the circuit is broken at any point?

The amount of current in a circuit depends on the number of devices in the circuit. **Figure 17** shows that adding devices to a series circuit adds resistance to the total circuit. According to Ohm's law, when resistance increases and voltage remains the same, there is less current in the circuit. Because a string of holiday lights contains so many devices, the current through the circuit is very low. The low current produces very little thermal energy, making these lights safe to use.

Figure 17 A series circuit has only one path for current. The current stops if that path is broken.

Visual Check What happens if one of the wires breaks or is cut?

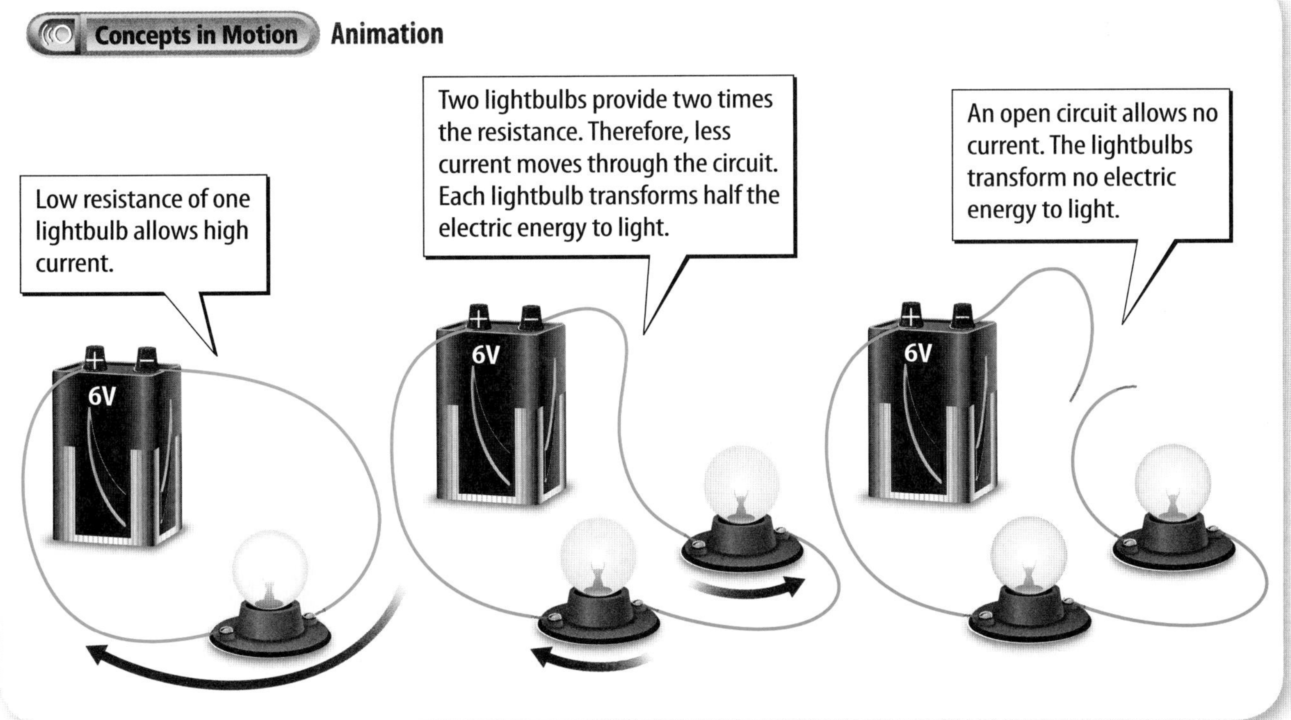

Concepts in Motion
Animation
Low resistance of one lightbulb allows high current.
Two lightbulbs provide two times the resistance. Therefore, less current moves through the circuit. Each lightbulb transforms half the electric energy to light.
An open circuit allows no current. The lightbulbs transform no electric energy to light.
6V
6V
6V

Figure 18 Parallel circuits have more than one path through which electric current can flow.

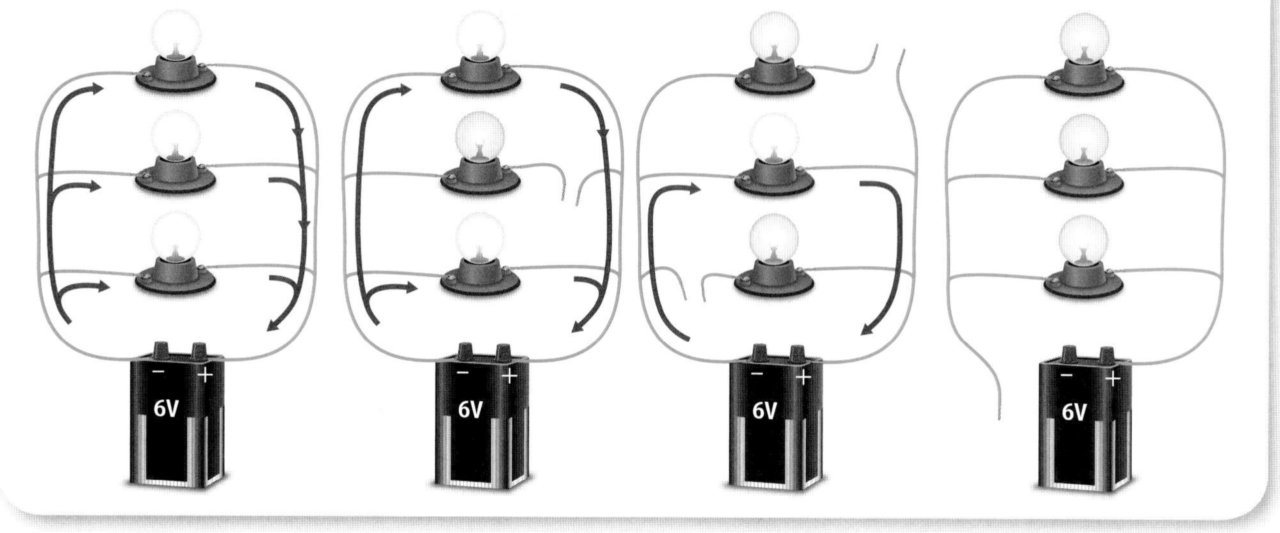

In a parallel circuit, each device has its own path through which current can flow.

If any of the paths of a parallel circuit is broken, current still can flow through the other devices.

However, if the circuit is open at the source of electric energy, no current can flow through any of the devices.

Parallel Circuit—Multiple Current Paths

If the electrical devices in your home were connected as a series circuit, you would need to turn on every electrical device in the house just to watch TV! Luckily, devices in your home are connected to an electric source as a **parallel circuit**—*an electric circuit with more than one path, or branch, for an electric current to follow.*

As shown in **Figure 18,** if you open one branch of a parallel circuit, current continues through the other branches. As a result, you can turn off the TV but keep the kitchen light and the refrigerator on.

You read that a device in a parallel circuit connects to the source with its own branch. The current in one branch has no effect on the current in other branches. However, adding branches to a parallel circuit does increase the total current through the source.

Key Concept Check How do the two types of electric circuits differ?

Electric Circuits in the Home

Most people use electrical devices every day without thinking much about them. However, where does the electric energy that we use come from?

Electric energy used in most homes and businesses is generated at large power plants. These power plants may be many kilometers from your home. A complex system of transmission wires carries the electric energy to all parts of the country.

Figure 19 shows that electric energy enters your home through a main wire. You might see this wire coming from a utility pole outside your home. Sometimes the main wire is underground. Either way, before coming into your house, the main wire travels through an electric meter. The meter measures the energy used in the electric circuits of your home.

From the meter, the wire enters your house and goes to a steel box called the main panel. At the main panel, the main wire divides into the branches of the parallel circuit that carry electric energy to all parts of your house.

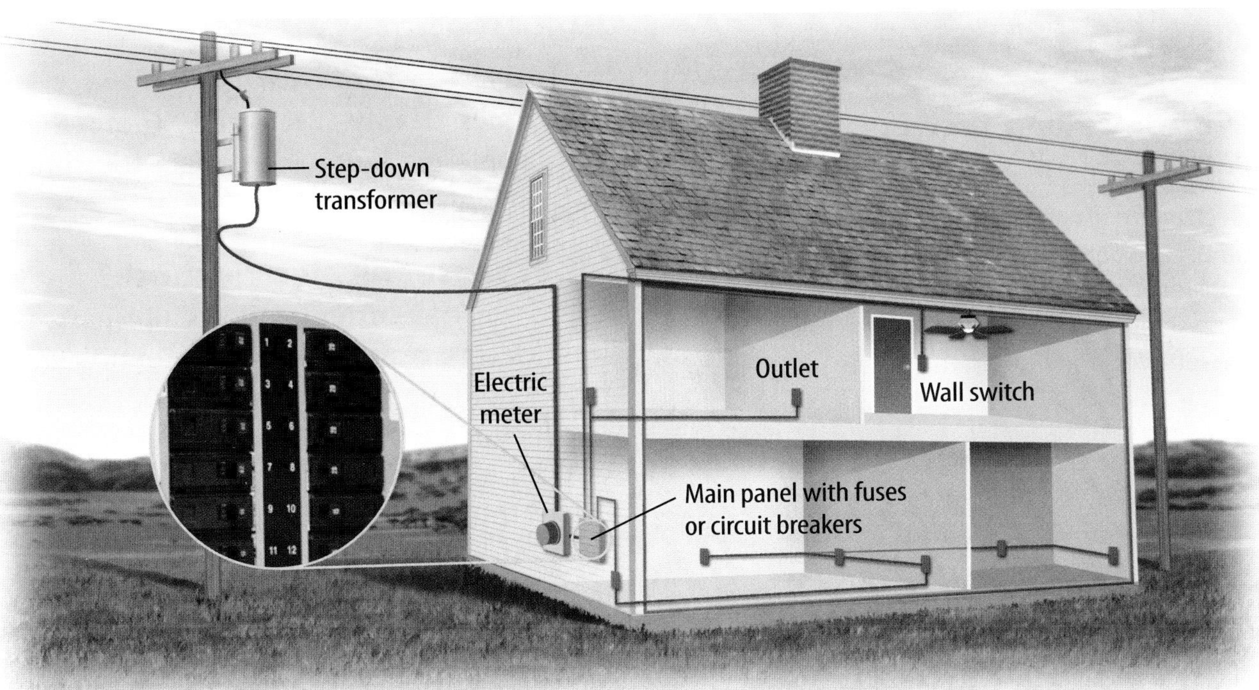

Figure 19 The electric devices in a house are connected in parallel circuit. Each outlet or fixture is connected to a separate branch.

Visual Check Where are circuit breakers located in the home?

Fuses, Circuit Breakers, and GFCI Devices

Recall that adding branches and devices to a parallel circuit increases the current through the source. Too many branches or devices in a circuit can create dangerously high current in the circuit. Excess current can cause the circuit to become hot enough to cause a fire. The branches of your home's parallel circuit connect to the source with a safety mechanism, such as a fuse or a circuit breaker, to help prevent such disasters.

Fuses and circuit breakers automatically open a circuit when the current becomes dangerously high. A fuse is a piece of metal that breaks a circuit by melting from the thermal energy produced by a high current. A circuit breaker is a switch that automatically opens a circuit when the current is too great.

There is another type of automatic safety mechanism found in some circuits. Are there electric outlets in your home with two small buttons labeled *test* and *reset*? Those special outlets are ground-fault circuit interrupters (GFCI). A GFCI is used where an outlet is found near a source of water, such as a sink. A GFCI protects you from a dangerous electric shock.

For example, imagine you are in the bathroom using a hair dryer. Then, accidentally, water is splashed on you and the hair dryer. Since tap water is an electric conductor, some of the electric current from the outlet flows through you and not through the hair dryer. If a current flows through you, it could be fatal. As soon as the GFCI senses that not all of the current flows through the hair dryer, it opens the circuit and stops the current. It is able to react as quickly as 1/30 of a second.

FOLDABLES®

Make a small vertical shutterfold book. Label it as shown. Use it to illustrate and explain the different types of circuits.

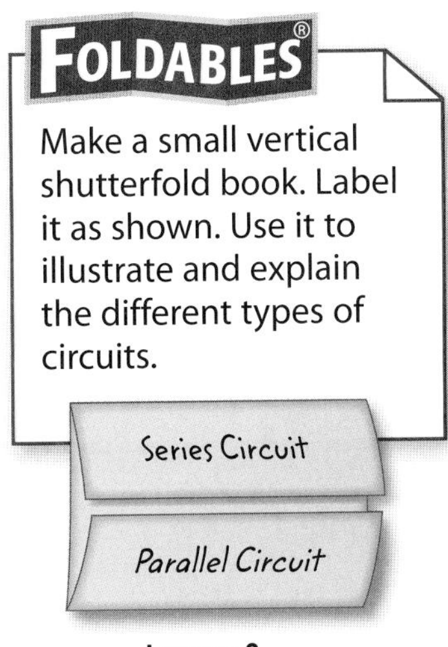

MiniLab

20 minutes

What else can a circuit do?

Electric circuits can do more than light a few bulbs. Look around your kitchen. Circuits transform electric energy to help you do everything from cooking dinner to freezing the leftovers to washing the dishes.

1. Copy the table below into your Science Journal.

Device	What it does	Source of electric energy	Energy transformed into

2. At each lab station, examine the device displayed. Record your observations in your table.

Analyze and Conclude

1. **Select** two devices and compare and contrast their operation.
2. **Key Concept** How do you think the circuits of the devices you examined differ? Explain your reasoning.

Electric Safety

An electric shock can be painful, and sometimes deadly. Each year, more than 500 people die by accidental electric shock in the United States.

What causes an electric shock?

An electric current follows the path of least electric resistance to the ground. That path could be through any good electric conductor, such as metal, wet wood, water, or even you! An electric shock occurs when an electric current passes through the human body. If you touch a bare electric wire or faulty appliance while you are grounded, an electric current could pass through you to the ground, resulting in a dangerous shock.

Current as small as 0.01 A can produce a painful shock. More than 0.1 A of electric current can cause death. The voltage of household electrical devices can cause dangerous amounts of current to pass through the body.

How can you be safe?

Listed below are some ways you can help protect yourself from a deadly electric shock:

- Never use electrical devices with damaged power cords.
- Stay away from water when using electrical devices plugged into an outlet.
- Avoid using extension cords and never plug more than two home appliances into an outlet at once.
- Never allow any object that you are touching to contact electric power lines, such as a kite string or a ladder.
- Do not touch anyone or anything that is touching a downed electric wire.
- Never climb utility poles or play on fences surrounding electricity substations.

Lesson 3 Review

Assessment Online Quiz

Visual Summary

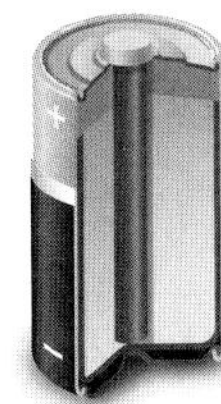

A battery can be the source of electric energy in a circuit.

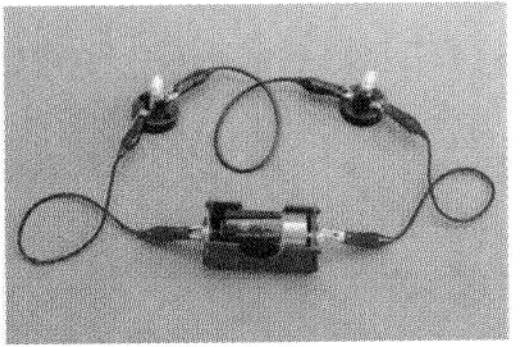

A series circuit has only one path for all devices in the circuit.

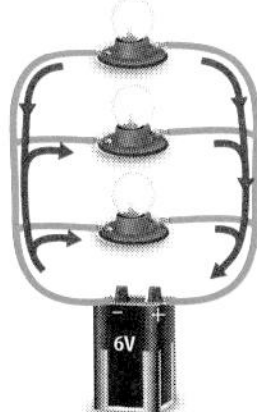

A parallel circuit has a separate path for each device in the circuit.

Use your lesson Foldable to review the lesson. Save your Foldable for the project at the end of the chapter.

Use Vocabulary

1. Electric current decreases as more devices are added to a(n) ________ circuit.
2. Different amounts of current can flow through each device in a(n) ________ circuit.

Understand Key Concepts 

3. One source of electric energy for an electric circuit is a
   A. battery.
   B. lightbulb.
   C. switch.
   D. wire.
4. **List** the basic components of a simple electric circuit. Explain why each of the components is necessary.
5. **Contrast** How does the electric current change when more devices are added to a series circuit? To a parallel circuit?

Interpret Graphics

6. **Organize Information** Copy and fill in the graphic organizer below to show how electric energy is transformed to thermal energy by an electric current. Add additional boxes, if necessary.

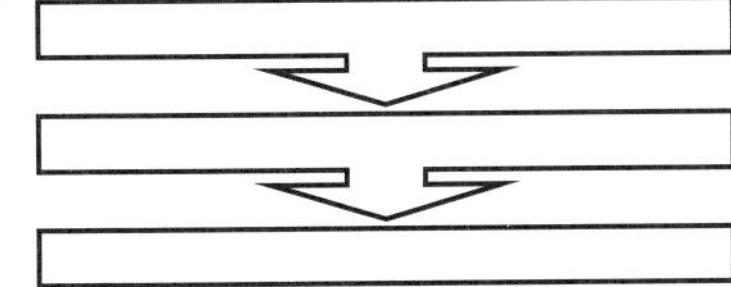

Critical Thinking

7. **Assess** A lightbulb and an electric motor are connected in a series circuit. Describe the amount of current in each device.
8. **Hypothesize** If a lightbulb in a circuit becomes dimmer, how did the energy supplied to the lightbulb change? Explain your answer.
9. **Create** Should a string of holiday lights be connected as a series circuit or as a parallel circuit? Create a graphic organizer to outline the advantages and disadvantages of the two types of circuits.

What do you think NOW?

You first read the statements below at the beginning of the chapter.

5. In any electric circuit, current stops flowing in all parts of the circuit if a connecting wire is removed or cut.

6. The light energy given off by a flashlight comes from the flashlight's batteries.

Did you change your mind about whether you agree or disagree with the statements? Rewrite any false statements to make them true.

2 class periods

Design an Elevator

Materials

thread, 1 m

washers

2 D-cell batteries in plastic bases

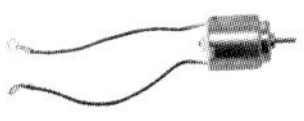

small hobby-type motor

2, 3-V bulbs in bases

alligator clip wires

cork

Also needed: paper clips, masking tape, stopwatch

Safety

There are many things an electric current can do besides making lightbulbs glow. One common electrical device is an electric motor. An electric motor uses an electric current to produce motion. Motors come in all sizes, from 300 times smaller than the diameter of a human hair to as large as a medium-sized house. We use motors for many different jobs—from driving a conveyor belt in a factory to operating a blender in the kitchen. One important use is to lift heavy objects.

Ask a Question

Where could it be important to be able to control the speed of an electric motor?

Make Observations

1. Read and complete a lab safety form.
2. Connect the thread to the cork on the motor's shaft. Attach a paper clip to the end of the thread. Fasten the motor to the edge of the table. Hang several washers on the paper clip over the edge of the table.
3. Draw your setup in your Science Journal.
4. Connect the battery, one lightbulb, and the motor in a series circuit. Observe whether the motor lifts the washers.
5. Adjust the number of washers until the motor lifts the washers as the thread winds up on the cork.
6. Record the time it takes for the motor to lift the washers.
7. Design a new circuit that will lift the washers at a different speed. Your circuit must include at least one battery and one lightbulb with the motor.

6

Form a Hypothesis

8 After observing the behavior of your circuits, formulate a hypothesis about what affects the speed of the motor.

Test Your Hypothesis

9 Design a circuit based on your hypothesis. The circuit should include at least one battery and one lightbulb. The circuit should lift the washers faster than your previous two circuits.

10 Create and test your new circuit. Record your observations.

Analyze and Conclude

11 **Relate** your hypothesis to the time each circuit took to lift the washers.

12 **Compare** your results with those of your classmates. Describe how the fastest time was achieved.

13 **The Big Idea** Explain how energy was transferred and transformed in your circuits.

Communicate Your Results

Create a pamphlet to sell your elevator design to a prospective buyer. Include a diagram and an explanation of how it works.

Design a device that uses an electric motor to provide forward motion instead of vertical motion.

Lab Tips

- Secure the string tightly to the cork on the motor's drive shaft. You want the string to wind around the cork.
- Secure the motor firmly to the table so the weight does not pull it off.
- Think about the voltage used in your motor and the current through flowing in the motor. How can you increase both?

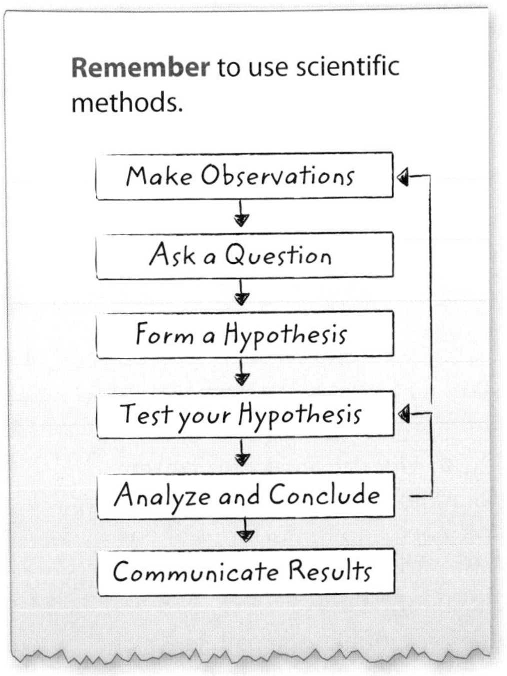

Chapter 19 Study Guide

An energy source, such as a battery, creates an electric current in a circuit. Electrical devices in the circuit transform the electric energy of the current to useful forms of energy.

Key Concepts Summary

Lesson 1: Electric Charge and Electric Forces

- Particles that have the same type of electric charge repel each other. Particles that have different types of electric charge attract each other.
- Objects become negatively charged when they gain electrons, and become positively charged when they lose electrons.
- An **electric discharge** is the loss of **static charge.**

Vocabulary

static charge p. 680
electric insulator p. 682
electric conductor p. 682
polarized p. 683
electric discharge p. 685
grounding p. 685

Lesson 2: Electric Current and Simple Circuits

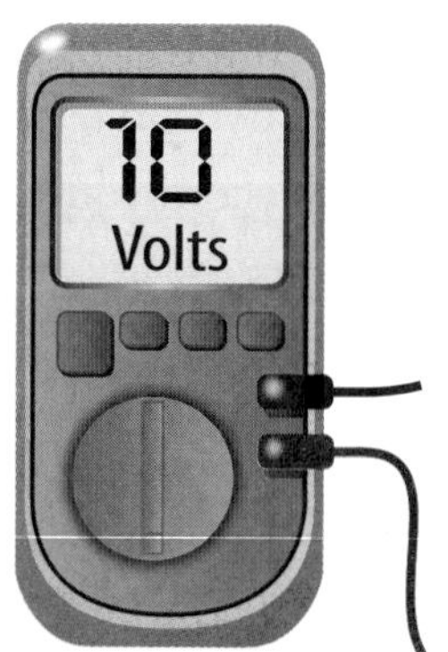

- An **electric current** is the flow of electrically charged particles through a conductor.
- According to Ohm's law, across any portion of an **electric circuit, voltage** (V), current (I), and **electric resistance** (R) are related by the equation $V = IR$.

Vocabulary

electric current p. 690
electric circuit p. 690
electric resistance p. 692
voltage p. 693
Ohm's law p. 694

Lesson 3: Describing Circuits

- Electric circuits have a source of electric energy to produce an electric current, one or more electric devices to transform electric energy to useful forms of energy, and wires to connect the circuit's device(s) to the energy source.
- A **series circuit** has one path in which current flows. A **parallel circuit** has more than one path in which current flows..

Vocabulary

series circuit p. 701
parallel circuit p. 702

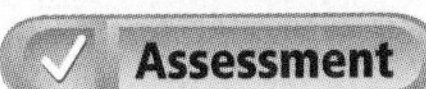

- Personal Tutor
- Vocabulary eGames
- Vocabulary eFlashcards

FOLDABLES® Chapter Project

Assemble your lesson Foldables as shown to make a Chapter Project. Use the project to review what you have learned in this chapter.

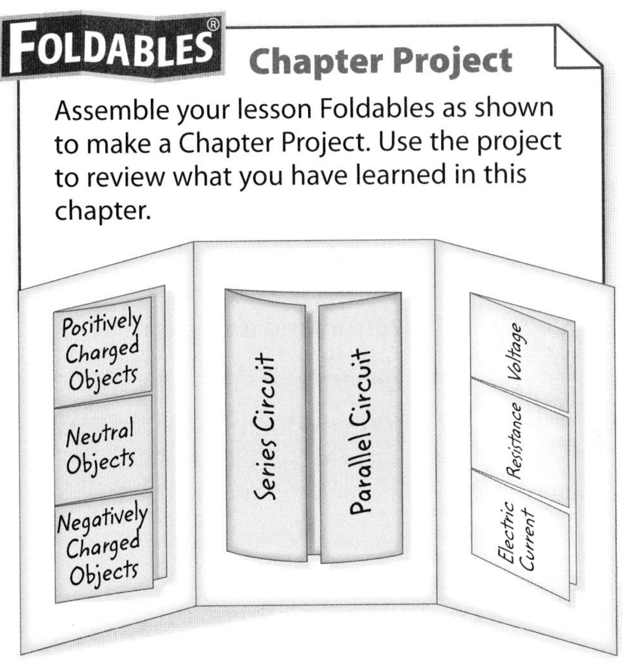

Use Vocabulary

1. An unbalanced electric charge is sometimes called a(n) ________.
2. Ohm's law states that as the voltage of a circuit increases, the current ________.
3. Electric charges flow easily in an electric ________.
4. A measure of the energy transformed in a portion of an electric circuit is ________.
5. Electric current decreases as more devices are added to a(n) ________ circuit.
6. Lightning is a(n) ________ that occurs in a fraction of a second.
7. A closed path in which electric charges flow is an electric ________.
8. Disconnecting one device in a ________ circuit will not cause other devices in the circuit to stop working.

Link Vocabulary and Key Concepts

Concepts in Motion Interactive Concept Map

Copy this concept map, and then use vocabulary terms from the previous page and other terms from the chapter to complete the concept map.

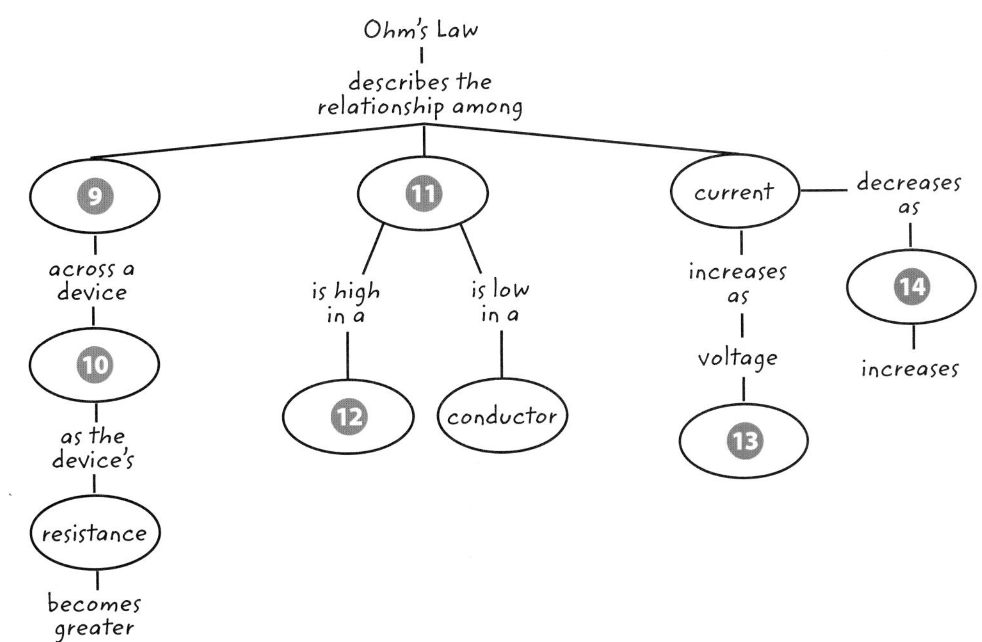

Chapter 19 Review

Understand Key Concepts

1. The measure of the energy transformed between two points in an electric circuit is
   A. resistance.
   B. voltage.
   C. electric current.
   D. electric force.

2. The switch in a circuit breaker opens when which of the following in the circuit becomes too high?
   A. current
   B. resistance
   C. static charge
   D. total charge

3. The electric force between two charges increases when
   A. both charges become negative.
   B. both charges become positive.
   C. the charges get closer together.
   D. the charges get farther apart.

4. Which best describes an electrically charged object?
   A. an electron without a static charge
   B. an object that has gained or lost electrons
   C. an object that has gained or lost neutrons
   D. a proton without a static charge

5. In the figure below, what form of energy is being converted to electric energy?

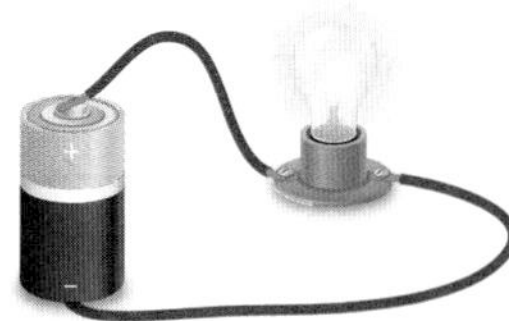

   A. chemical
   B. light
   C. nuclear
   D. thermal

6. What property of a wire changes when the wire is made thicker?
   A. current
   B. resistance
   C. static charge
   D. total electric charge

7. An electric field surrounds
   A. an electron.
   B. a neutron.
   C. both an electron and a neutron.
   D. neither an electron nor a neutron.

8. Which is a good conductor of electricity?
   A. glass
   B. gold
   C. plastic
   D. wood

9. Electric energy transforms into which forms of energy as electrons collide with atoms in a circuit?
   A. gravitational and light
   B. kinetic and nuclear
   C. thermal and light
   D. thermal and nuclear

10. Which lightbulbs in the diagram below will remain lit if the wire is disconnected at point B?

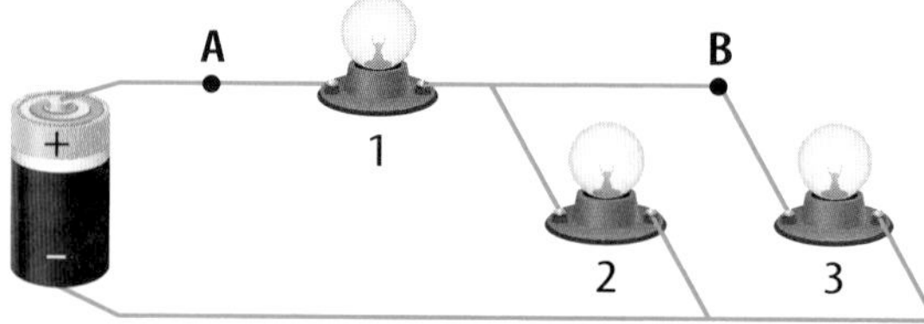

   A. 1 and 2
   B. 1 and 3
   C. 2 and 3
   D. 1, 2, and 3

Assessment
Online Test Practice

Critical Thinking

11 **Suggest** What are two ways the electric force between two charged objects can be increased?

12 **Evaluate** Dry air is a better electric insulator than humid air. Would the electric discharge from a charged balloon happen more slowly in dry air or humid air? Explain your answer.

13 **Recommend** If a metal wire is made thinner, how would you change the length of the wire to keep the electric resistance the same?

14 **Suggest** What are two ways to increase electric current in a simple circuit?

15 **Evaluate** The current in a lightbulb stays the same, but the voltage across the lightbulb decreases. How does the electric energy to the lightbulb change?

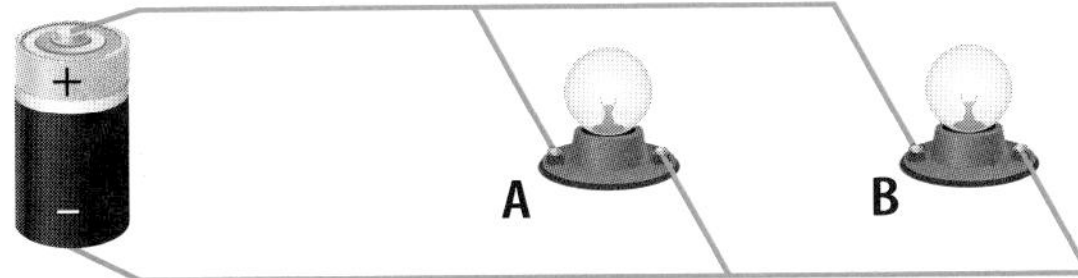

16 **Recommend** Copy the diagram above and show on your drawing where a switch could be placed that would turn only lightbulb B off and on.

17 **Assess** Lightbulb A and lightbulb B are connected in a circuit. The voltage across lightbulb A is higher than the voltage across lightbulb B. Which lightbulb is brighter?

Writing in Science

18 **Write** a short essay describing some of the electrical devices you use every day. Include in your essay the energy transformations that occur in each device.

REVIEW THE BIG IDEA

19 How do electric charges flowing in a circuit regain some of the energy they transfer to atoms in a circuit?

20 The photograph below shows Nikola Tesla sitting and reading in his laboratory. How could he have used this picture to convince people that it was safe to use electricity in their homes?

Math Skills

Math Practice

Use a Simple Equation

21 A current of 20 A flows into a hot water heater. If the electric resistance of the hot water heater is 12 Ω, what is the voltage across the hot water heater?

22 A refrigerator is plugged into an electrical outlet. If the voltage across the refrigerator is 120 V and the current flowing into the refrigerator is 6 A, what is the total electric resistance of the refrigerator?

23 What is the current in a flashlight bulb if the bulb has an electric resistance of 60 Ω and the voltage across the bulb is 6 V?

24 A window fan has a resistance of 80 Ω and the voltage across the fan is 120 V. How much current flows in the fan?

Standardized Test Practice

Record your answers on the answer sheet provided by your teacher or on a sheet of paper.

Multiple Choice

Use the figures to answer questions 1 and 2.

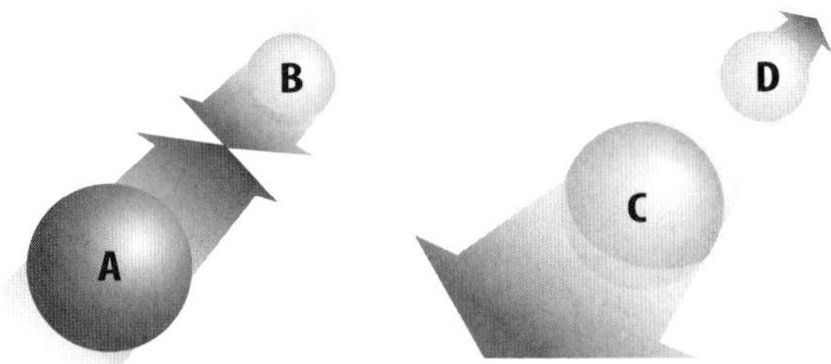

1 What must be true of particles A and B?

A A and B must be like charges.

B A and B must be opposite charges.

C A must be positive, and B must be negative.

D A must be negative, and B must be positive.

2 If the amount of charge on particles A, B C, and D are equal, what must be true of particle pairs AB and CD?

A The attractive force of AB is greater than the repulsive force of CD.

B The attractive force of AB is less than the repulsive force of CD.

C The attractive force of AB equals the repulsive force of CD.

D The repulsive force of AB is less than the attractive force of CD.

3 The movement of negative charges between nonconducting objects that are touching is called transferring charge by

A conduction.

B contact.

C induction.

D polarization

4 Which phenomenon is an example of electric discharge?

A friction

B lightning

C magnetism

D static cling

5 The charged particles that move as an electric current are

A electrons.

B light particles.

C metal atoms.

D neutrons.

Use the table below to answer question 6.

Circuit	Current (A)	Resistance (Ω)
1	0.25	220
2	0.5	220
3	0.5	110

6 Based on the data, use Ohm's law to determine which statement is true.

A Circuits 1 and 2 have the same voltage.

B Circuits 2 and 3 have the same voltage.

C Circuits 1 and 3 have the same voltage.

D Each circuit has a different voltage.

7 When negative charges concentrate at one end of an object that is made of a conducting material, the object is

A inducted.

B insulated.

C polarized.

D undergoing friction.

Standardized Test Practice

Assessment

Online Standardized Test Practice

Use the figure below to answer questions 8-10.

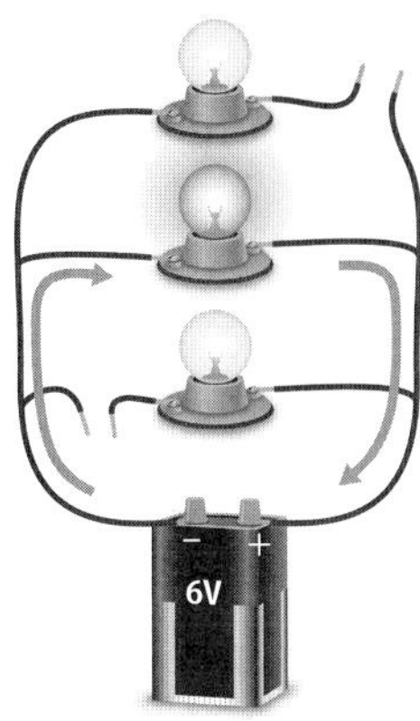

8 When the wires in this circuit are connected and all bulbs are lit, what type of circuit is shown?

A negative

B parallel

C positive

D series

9 If the broken wire of the top branch is connected, what will be true of the lightbulbs?

A All of the bulbs would be lit.

B None of the bulbs would be lit.

C Only the top bulb would be lit.

D Only the top two bulbs would be lit.

10 Connecting then disconnecting the broken ends of the top wire is similar to

A adding then removing a fourth lightbulb.

B adding then removing a second battery.

C turning a switch on and off.

D turning the middle lightbulb on and off.

Constructed Response

Use the figure below to answer questions 11 and 12.

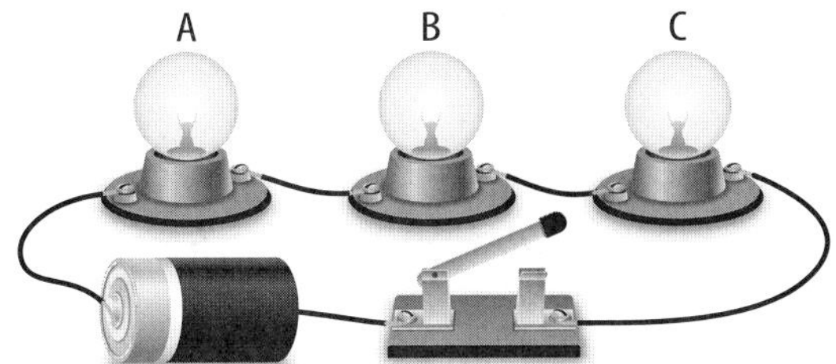

11 Identify the type of circuit shown. Explain what happens if the switch is closed. Explain what happens if the wire connecting lightbulbs A and B is removed.

12 Describe the kind of energy transformation(s) that takes place when the switch is closed.

13 When you turn off a lamp in your home, why do other electrical devices stay on? What does flipping the light switch do? What type of circuit does a house have?

14 What happens to the electric current in a circuit when the circuit's voltage is increased but resistance stays the same? What equation represents this relationship?

15 A string of 120 holiday lights is connected as a series circuit. The voltage across each bulb is 1 V. Another string of lights is connected as a parallel circuit. The voltage across each bulb is 120 V. Explain this voltage difference when both strings are plugged into a 120-V outlet.

NEED EXTRA HELP?															
If You Missed Question...	1	2	3	4	5	6	7	8	9	10	11	12	13	14	15
Go to Lesson...	1	1	1	1	2	2	1	2	2	2	2	3	2	3	3

Chapter 20

Magnetism

THE BIG IDEA

How are electric charges and magnetic fields related?

Inquiry How loud is a magnet?

Since it was first manufactured in 1932, the electric guitar has inspired many new types of music. It is a prominent instrument in rock music and one of the most famous instruments to originate in the United States.

- How does an electric guitar use magnets to produce sound?
- Where are the magnets in the picture?
- How are electric charges and magnetic fields related?

Get Ready to Read

What do you think?

Before you read, decide if you agree or disagree with each of these statements. As you read this chapter, see if you change your mind about any of the statements.

1. All metal objects are attracted to a magnet.
2. Two magnets can attract or repel each other.
3. A magnetic field surrounds a moving electron.
4. Unlike a permanent magnet, an electromagnet has two north magnetic poles or two south magnetic poles.
5. A battery produces an electric current that reverses direction in a regular pattern.
6. An electric generator transforms thermal energy into electric energy.

Lesson 1

Reading Guide

Key Concepts

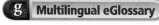

ESSENTIAL QUESTIONS

- What types of forces do magnets apply to other magnets?
- Why are some materials magnetic?
- Why are some magnets temporary while others are permanent?

Vocabulary

magnet p. 717
magnetic pole p. 718
magnetic force p. 718
magnetic material p. 721
ferromagnetic element p. 721
magnetic domain p. 722
temporary magnet p. 723
permanent magnet p. 723

Multilingual eGlossary

Video BrainPOP®

Magnets and Magnetic Fields

Inquiry Is it a magnet?

If you ever played with magnets, you probably noticed that they seemed to have magical powers. Have you ever wondered just what it is that makes magnets so special and necessary for everyone on Earth?

10 minutes

What does *magnetic* mean?

Many years ago, people discovered that some minerals attract each other. The Greeks named these minerals *magnets*, after the ancient city Magnesia. What materials do magnets attract?

	Magnet		Nail		Rubbed Nail		Dropped Nail	
	"N" End	"S" End	Point	Head	Point	Head	Point	Head
Paper clip								
Paper								
Aluminum foil								
Choice #1								
Choice #2								

1. Read and complete a lab safety form.
2. Copy the data table into your Science Journal.
3. Touch each end of a **magnet** to a **paper clip, a piece of paper, a piece of aluminum foil,** and **two objects of your choosing.** Record your observations.
4. Repeat step 3, this time touching the ends of a **nail** to each object.
5. Rub the nail 25 times in the same direction across one end of the magnet. Repeat step 3, using the rubbed nail.
6. Drop the nail several times onto a hard surface. Repeat step 3 using the dropped nail. Record your observations.

Think About This

1. When does the nail behave like the magnet?
2. How do the two ends of the magnet interact with the various materials?
3. **Key Concept** What types of materials does a magnet attract?

Magnets

Did you use a computer or a hair dryer today? Did you listen to a stereo system? It might surprise you to know that all these devices contain magnets. Magnets also are used to produce the electric energy that makes these familiar devices work. *A* **magnet** *is any object that attracts the metal iron.*

As you just read, magnets are used in many ways. Therefore, magnets are manufactured in many shapes and sizes, as shown in **Figure 1.** You might be familiar with common bar magnets and horseshoe magnets. Some of the magnets holding papers on your refrigerator might be disc-shaped or flat and flexible. However, all magnets have certain things in common, regardless of their shape and size.

Figure 1 Magnets come in many sizes and shapes.

Figure 2 When a magnet is broken into pieces, each piece is still a magnet. ▶

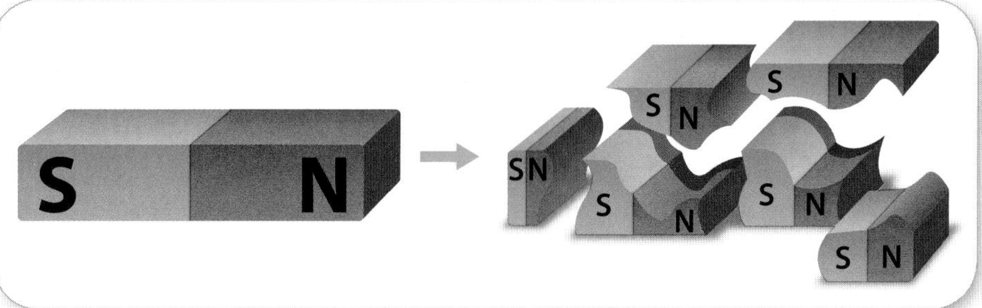

Magnetic Poles

You might have noticed that the ends of some magnets are different colors or labeled *N* and *S*. These colors and labels identify the magnet's magnetic poles. *A* **magnetic pole** *is a place on a magnet where the force it applies is strongest.* There are two magnetic poles on all magnets—a north pole and a south pole. As shown in **Figure 2,** if you break a magnet into pieces, each piece will have a north pole and a south pole.

Word Origin
pole
from Greek *polos,* means "pivot"

The Forces Between Magnetic Poles

A force exists between the poles of any two magnets. If similar poles of two magnets, such as north and north or south and south, are brought near each other, the magnets repel. This means that the magnets will push away from each other. If the north pole of one magnet is brought near the south pole of another magnet, the two magnets attract. This means the magnets will pull together. In other words, as shown in **Figure 3**, similar poles repel, and opposite poles attract.

A force of attraction or repulsion between the poles of two magnets is a **magnetic force.** A magnetic force becomes stronger as magnets move closer together and becomes weaker as the magnets move farther apart.

Foldables®
Make a vertical three-tab book, and label it as shown. Use it to organize your notes about magnetic forces.

Key Concept Check What types of forces do magnets apply to other magnets?

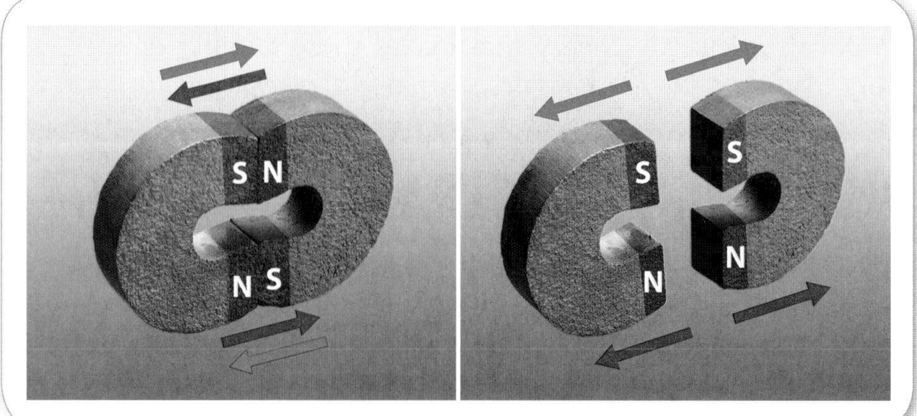

Figure 3 When the opposite poles of two magnets are close to each other, they attract. When similar poles of two magnets are close to each other, they repel. ▶

Concepts in Motion
Animation

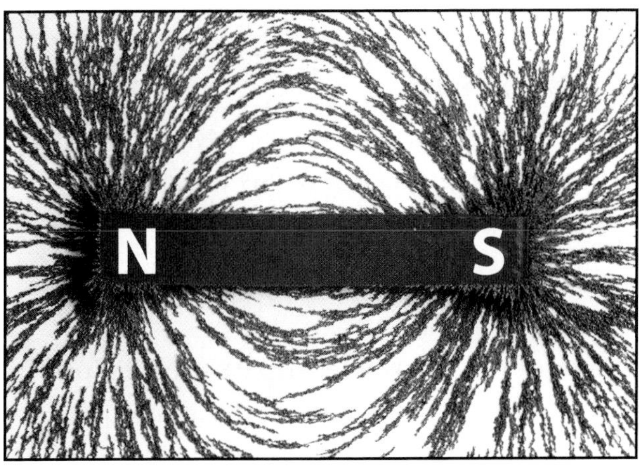

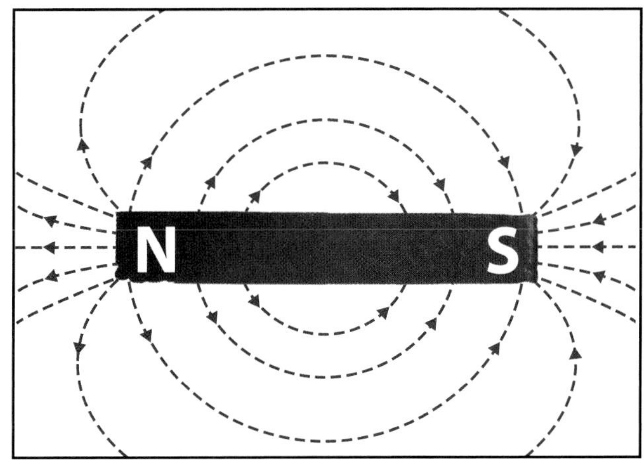

▲ **Figure 4** The magnetic field around a magnet can be shown with iron filings. Drawings of magnetic field lines include arrows to show the north-to-south direction of the magnetic field.

Magnetic Fields

Recall from a previous chapter that charged objects repel or attract each other even when they are not touching. Similarly, magnets can repel or attract each other even when they are not touching. An invisible magnetic **field** surrounds all magnets. It is this magnetic field that applies forces on other magnets.

SCIENCE USE v. COMMON USE

field

Science Use a space in which a given effect (such as magnetism) exists

Common Use an open area free of woods and buildings

Magnetic Field Lines

In **Figure 4,** iron filings have been sprinkled around a bar magnet. The iron filings form a pattern of curved lines that reveal the magnet's magnetic field.

A magnet's magnetic field can be represented by lines, called magnetic field lines. Magnetic field lines have a direction, that is shown on the right in **Figure 4.** Notice that the lines are closest together at the magnet's poles. This is where the magnetic force is strongest. As the field lines become farther apart, the field and the force become weaker.

Combining Magnetic Fields

What happens to the magnetic fields around two bar magnets that are brought together? The two fields combine and form one new magnetic field. The pattern of the new magnetic field lines depends on whether two like poles or two unlike poles are near each other, as shown in **Figure 5.**

Figure 5 When magnetic fields combine, they form field lines with different patterns. The new patterns depend on whether the magnets are attracting or repelling each other. ▼

Attraction

When opposite poles of two magnets are near each other, the resulting magnetic field applies a force of attraction.

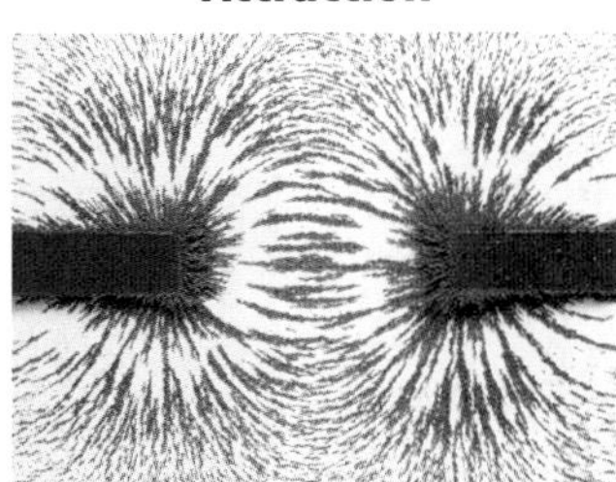

Repulsion

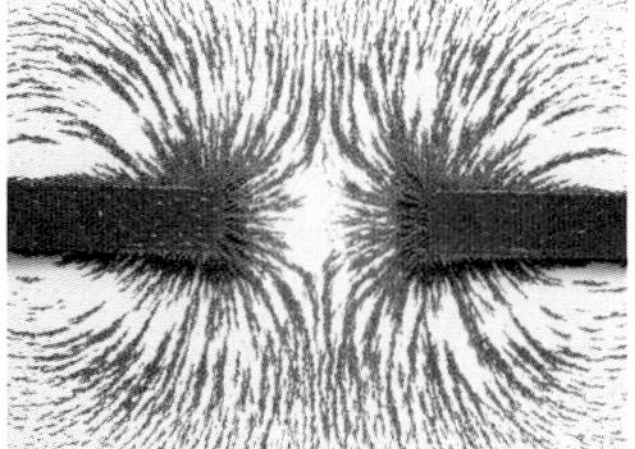

When like poles of two magnets are near each other, the resulting magnetic field applies a force of repulsion.

Earth's Magnetic Field

A magnetic field surrounds Earth similar to the way a magnetic field surrounds a bar magnet. Earth has a magnetic field due to molten iron and nickel in its outer core. Like all magnets, Earth has north and south magnetic poles.

Compasses

Have you ever wondered why a compass needle points toward north? The needle of a compass is a small magnet. Like other magnets, a compass needle has a north pole and a south pole. Earth's magnetic field exerts a force on the needle, causing it to rotate. **Figure 6** shows that if a compass needle is within any magnetic field, including Earth's, it will line up with the magnet's field lines. A compass needle does not point directly toward the poles of a magnet. Instead, the needle aligns with the field lines and points in the direction of the field lines. Earth's magnetic poles and geographic poles are not in the same spot, as shown in **Figure 7.** Therefore, you cannot find your way to the geographic poles with only a compass.

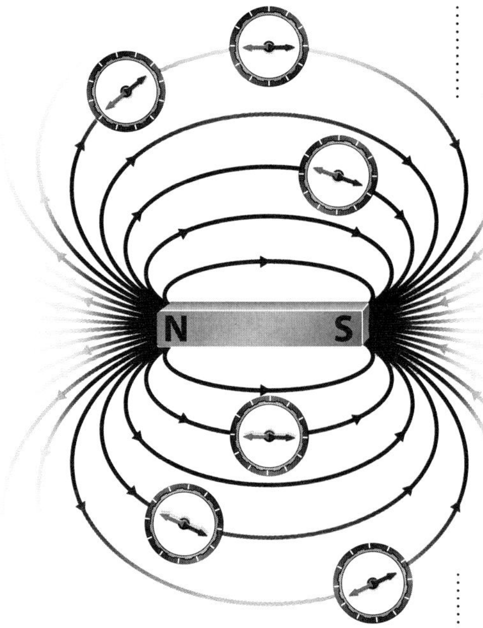

▲ **Figure 6** A compass needle will align with a magnet's magnetic field lines.

Visual Check Why do all the compass needles in **Figure 6** point in different directions?

Auroras

Earth's magnetic field protects Earth from charged particles from by the Sun. These particles can damage living organisms if they reach the surface of Earth. Earth's magnetic field deflects most of these particles. Sometimes, large numbers of particles from the Sun travel along Earth's magnetic field lines and concentrate near the magnetic poles. There, the particles collide with atoms of gases in the atmosphere, causing the atmosphere to glow. The light forms shimmering sheets of color known as auroras. An aurora is shown in **Figure 7.**

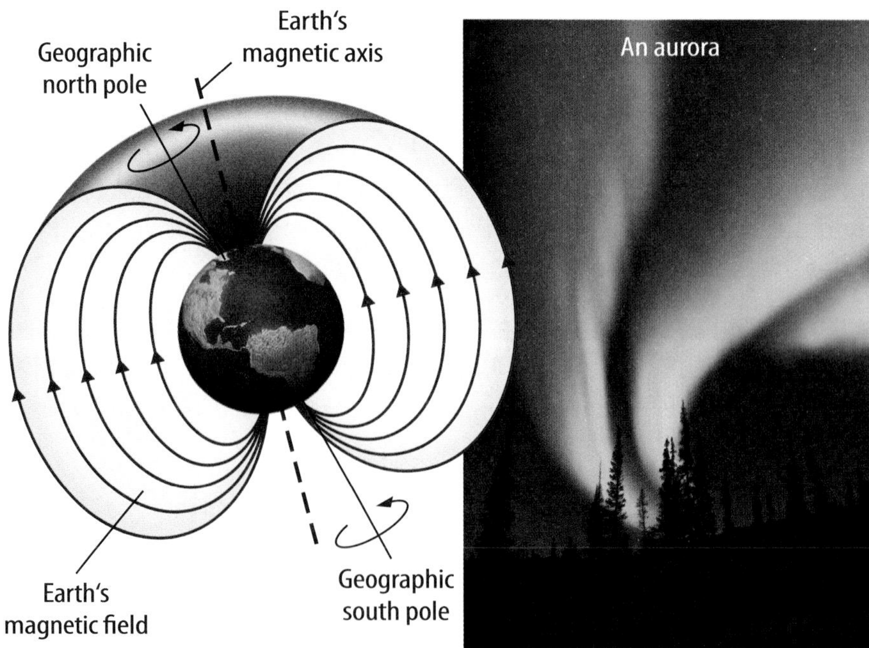

Figure 7 A magnetic field surrounds Earth. This magnetic field can cause auroras to occur near Earth's magnetic north pole and south pole. ▶

Magnetic Materials

You have read that magnets attract other magnets. But, magnets also attract objects, such as nails, which are not magnets. *Any material that is strongly attracted to a magnet is a* **magnetic material**. Magnetic materials often contain ferromagnetic (fer oh mag NEH tik) elements. **Ferromagnetic elements** *are elements, including iron, nickel, and cobalt, that have an especially strong attraction to magnets.*

REVIEW VOCABULARY

element
a substance made up of only one kind of atom

It is important to understand that while not all materials are magnetic, a magnetic field does surround every atom that makes up all materials. The field is created by the atom's constantly moving electrons. The interactions of these individual fields determines whether a material is magnetic. Next, you will read about why magnetic materials make good magnets.

Reading Check Why is nickel a magnetic material?

MiniLab

20 minutes

Where is magnetic north?

Many animals such as birds use Earth's magnetic field for navigation. With the help of a compass, people can find their way, too. What does it mean when a compass points north?

1. Read and complete a lab safety form.
2. On a piece of **paper,** draw a circle with a diameter 2 cm longer than a **bar magnet.** Place the bar magnet in the center of your circle. Label the magnet's north and south poles on the circle.
3. Place a **compass** on the paper just outside the circle. Draw an arrow pointing in the same direction as the north end of the compass.
4. Repeat step 3 in at least eight different places around the circle.
5. In your Science Journal, indicate to which pole of the bar magnet the north end of the compass needle points.
6. Now, step away from your magnet and all metal objects in the room. Hold the compass level in your hand. Face the direction indicated by the compass. Record the direction you are facing in your Science Journal.

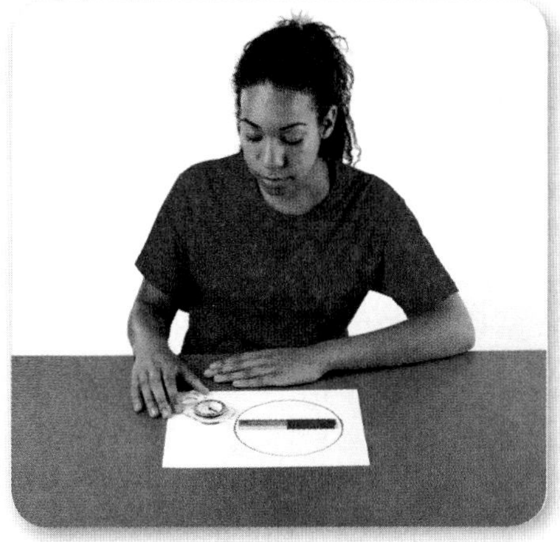

Analyze and Conclude

1. **Describe** the magnetic pole that attracts the north end of a compass needle.
2. **Key Concept** Hypothesize about the interaction of Earth's magnetic field and the compass needle.

Magnetic Domains

If all atoms act like tiny magnets, why do magnets not attract most materials, such as glass and plastic? The answer is that in most materials, the magnetic fields of the atoms point in different directions, as in part a of **Figure 8**. As the fields around the atoms combine, they cancel each other. Thus, most types of matter are nonmagnetic materials and are not attracted to magnets.

In a magnetic material, atoms form groups called magnetic domains. *A* **magnetic domain** *is a region in a magnetic material in which the magnetic fields of the atoms all point in the same direction,* as shown in part b of **Figure 8.** The magnetic fields of the atoms in a domain combine, forming a single field around the domain. Each domain is a tiny magnet with a north pole and a south pole.

Academic Vocabulary
align
(verb) to arrange in a straight line; adjust according to a line

When Domains Don't Line Up

Many magnetic materials are not magnets. This is because the magnetic fields of their domains point in random directions, as shown in part b of **Figure 8.** Similar to the atoms of a nonmagnetic material, the random magnetic fields of the domains cancel each other. Even though the individual domains are magnets, the entire object has no effective magnetic field.

When Domains Line Up

How do a bar magnet and a steel nail that is not a magnet differ? Both are magnetic materials. Both have atoms grouped into magnetic domains. However, for an object to be a magnet, its magnetic domains must **align** as in part c of **Figure 8.** When the domains align, their magnetic fields combine, forming a single magnetic field around the entire material. This causes the object to become a magnet.

Key Concept Check Why are some materials magnetic?

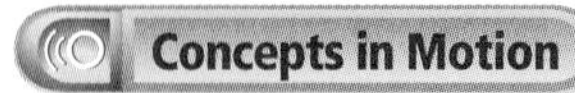

Animation

Figure 8 In many materials, atoms act like tiny magnets. Most materials are nonmagnetic materials (a), but some are magnetic with magnetic domains (b). In a magnet, the poles of the domains line up (c).

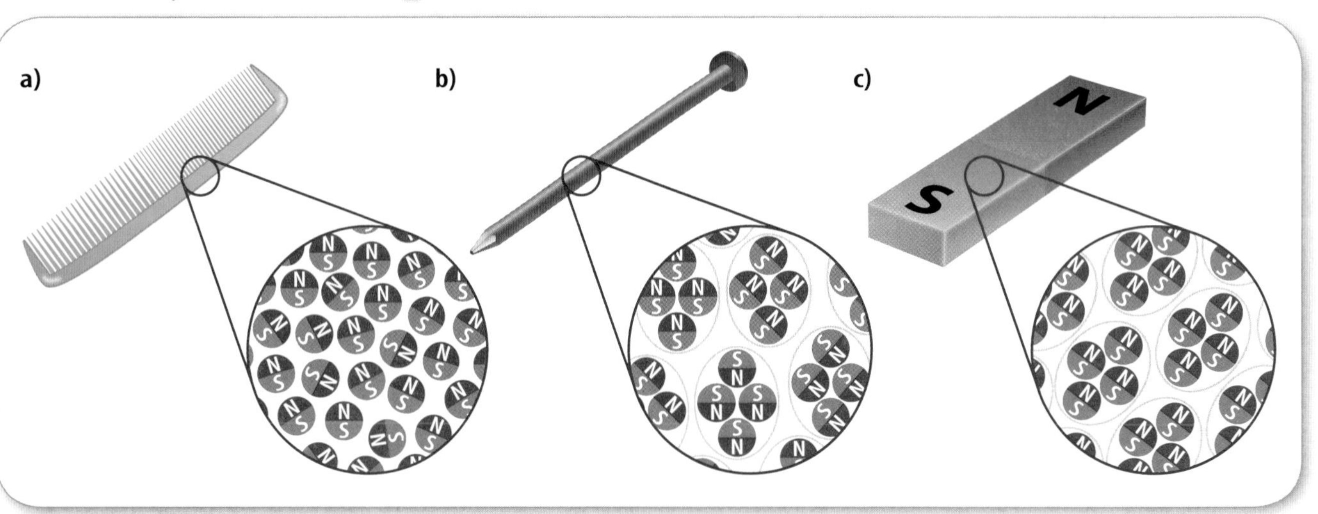

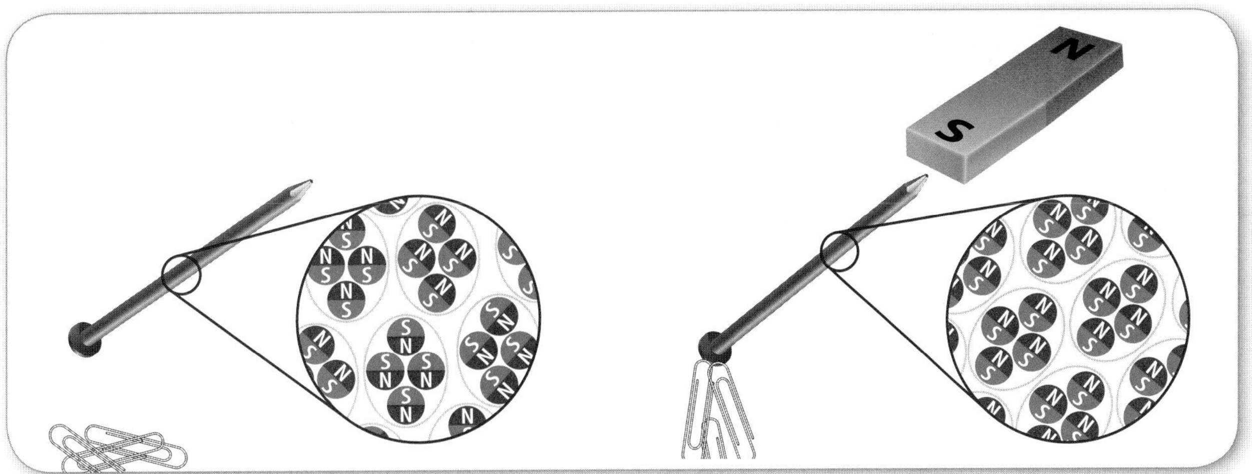

Figure 9 A nail is made of a magnetic material. It becomes a temporary magnet when it is close to another magnet.

Visual Check What happens to the magnetic domains of the nail as it becomes a temporary magnet?

How Magnets Attract Magnetic Materials

Figure 9 shows a nail coming close to one of the poles of a bar magnet. Remember, even though the nail itself is not a magnet, each magnetic domain in the nail is a small magnet. The magnetic field around the bar magnet applies a force to each of the nail's magnetic domains. This force causes the domains in the nail to align along the bar magnet's magnetic field lines. As the poles of the domains of the nail point in the same direction, the nail becomes a magnet. Now the nail can attract other magnetic materials, such as paper clips.

Temporary Magnets

In **Figure 9,** the nail becomes a temporary magnet. *A magnet that quickly loses its magnetic field after being removed from a magnetic field is a* **temporary magnet.** The nail is a magnet only when it is close to the bar magnet. There, the magnetic field of the bar magnet is strong enough to cause the nail's magnetic domains to line up. However, when you move the nail away from the bar magnet, the poles of the domains in the nail return to pointing in different directions. The nail no longer is a magnet and no longer attracts other magnetic materials.

Reading Check Why would the nail in **Figure 9** lose its magnetic field if it is moved away from the bar magnet?

Permanent Magnets

A magnet that remains a magnet after being removed from another magnetic field is a **permanent magnet.** In a permanent magnet, the magnetic domains remain lined up. Some magnetic materials can be made into permanent magnets by placing them in a very strong magnetic field. This causes the magnetic domains to align and stay aligned. The material then remains a magnet after it is removed from the field.

Key Concept Check Why are some magnets temporary while others are permanent?

Lesson 1 Review

Assessment Online Quiz

Visual Summary

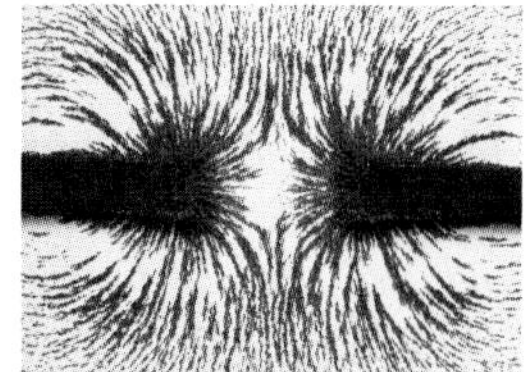

When the magnetic fields of two magnets combine, a magnetic force exists between the magnets.

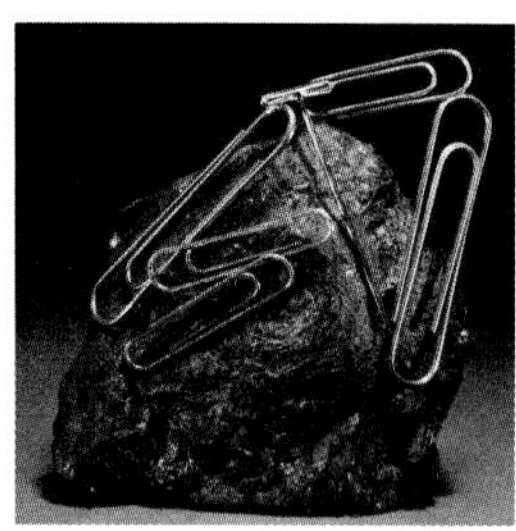

Objects made of magnetic materials are attracted to magnets.

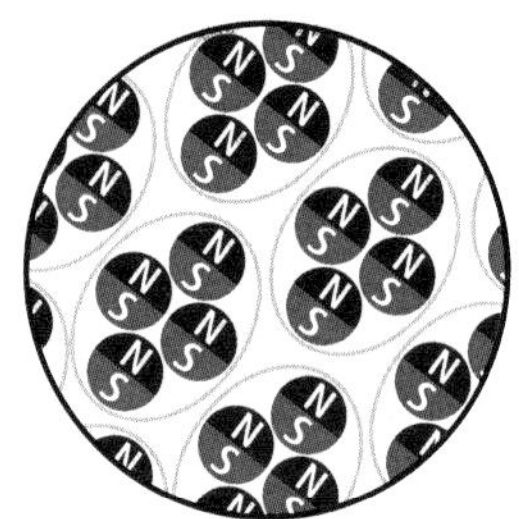

The magnetic domains of a permanent magnet are aligned even when the magnet is not near another magnetic field.

FOLDABLES®

Use your lesson Foldable to review the lesson. Save your Foldable for the project at the end of the chapter.

What do you think NOW?

You first read the statements below at the beginning of the chapter.

1. All metal objects are attracted to a magnet.

2. Two magnets can attract or repel each other.

Did you change your mind about whether you agree or disagree with the statements? Rewrite any false statements to make them true.

Use Vocabulary

1. Elements that are strongly attracted to a magnet are ________.
2. The magnetic poles of the atoms in a(n) ________ point in the same direction.
3. An object that attracts other objects made of magnetic materials is a(n) ________.

Understand Key Concepts 

4. **Identify** Where around a magnet is the magnetic field strongest?
5. Which does NOT contain magnetic domains?
   A. bar magnet
   B. disc magnet
   C. paper airplane
   D. steel nail
6. **Contrast** the magnetic domains in a steel nail that is far from a bar magnet and one that is close to a bar magnet.

Interpret Graphics

7. **Organize** Copy and fill in the table below to describe the type of force between the magnetic poles.

Poles	Type of Force
South pole and south pole	
North pole and north pole	
South pole and north pole	

Critical Thinking

8. **Predict** Where would the north pole of a compass needle point if Earth's magnetic field switched direction?
9. **Evaluate** Two nails have identical sizes and shapes. In one nail, 20 percent of the domains are lined up. In the other nail, 80 percent of the domains are lined up. Which has a stronger magnetic field?

Roller Coasters

SCIENCE & SOCIETY

How Magnets Make a Wild Ride a Safe Ride

Roller coasters certainly have changed throughout their long history! Four hundred years ago, the ancestors of roller coasters appeared in Russia. These rides were long, steep wooden slides covered in ice. Some were over 20 m high. Riders slid down the slope in sleds made of wood or blocks of ice, crashing into a sand pile. Today's roller coasters can reach speeds over 160 km/h! Crashing into a sand pile at this speed is no way to stop a ride safely!

Due to their safety, magnetic brakes are a technology that is gaining popularity with roller-coaster designers. Magnetic brakes rely on the interaction of magnetic fields—these brakes never come in contact with the cars. One design of magnetic brake uses rows of strong magnets built at the side of the track. A metal fin on the car passing between the rows of magnets creates a magnetic field that opposes the fin's motion. Magnetic braking is virtually fail-safe because it relies on the basic properties of magnetism and requires no electricity.

Engineers are designing future roller coasters with magnets to hold the cars to the rails during death-defying drops at extreme speeds never before possible. Computerized sensors and automatic detectors on new generation roller coasters will control the entire roller coaster experience from start to finish. Will science and technology soon make having fun on a roller coaster a process as sophisticated as a rocket launch?

Metal fin: typically copper or a copper/aluminum alloy.

Neodymium magnets: an alloy of neodymium, iron, and boron, currently the strongest type of permanent magnet.

Hydraulic system: High-pressure fluids push against a piston to lower the magnets when braking is not needed.

It's Your Turn

RESEARCH AND REPORT Roller-coaster cars are not the only kind of vehicles that make use of magnets. Research maglev trains to find out more about how these trains use magnets and magnetic fields to provide a safe, smooth ride. Write a short report to share what you learn.

Lesson 2

Reading Guide

Key Concepts
ESSENTIAL QUESTIONS

- Why does a magnet apply a force on an electric current?
- How do electromagnets and permanent magnets differ?
- How do electric motors use magnets?

Vocabulary

electromagnet p. 728

electric motor p. 730

Multilingual eGlossary

Making Magnets with an Electric Current

Inquiry What can magnets find?

What does *metal detector* mean to you? Do you think of searching for buried treasure? Are you reminded of airport security or of the hand-held scanners at a sporting event? Metal-detector technology is a huge part of our lives. How are electricity and magnetism used in these devices?

20 minutes

When is a wire a magnet?

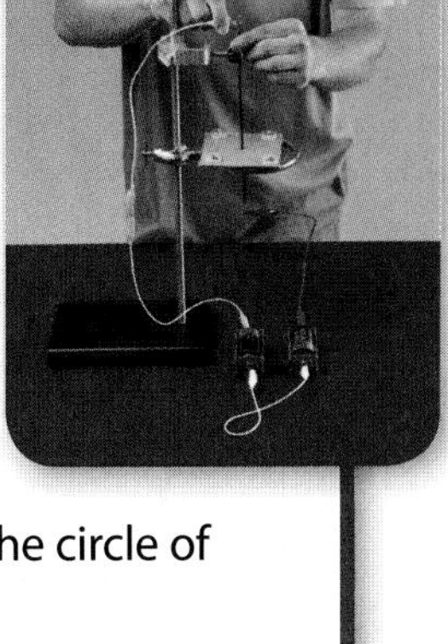

In 1820, Hans Øersted made a remarkable discovery about how magnetism and electricity are parts of the same thing. How do magnetism and electricity relate?

1. Read and complete a lab safety form.
2. With a **clamp attachment,** hang a length of **insulated wire** through the center of an iron ring on a **ring stand.** Make a small hole in the center of a 10-cm square piece of **cardboard.** Slide the wire through the hole. Rest the card on the iron ring. Secure the cardboard to the iron ring with **tape.**
3. Place four **small compasses** on the card in a circle around the wire.
4. In your Science Journal, draw your setup. Include the wire through the card and the circle of compasses in your drawing. Indicate the direction that each compass points.
5. Use **alligator clip wires** to connect two **D-cell batteries in holders** in series with the ends of the wire**.** Draw a second diagram. Show the direction the compasses point with the batteries connected.
   ⚠ *Unhook the wire after a few seconds to prevent it from overheating!*

Think About This

1. What would happen if you reversed the batteries?
2. **Key Concept** Why do you think the compass needles behaved as they did after the batteries were connected?

Moving Charges and Magnetic Fields

Recall that a magnetic field surrounds a magnet. In addition, a magnetic field surrounds an electric current. This is why a compass needle moves when placed near a current-carrying wire. The needle moves because the magnetic field around the wire applies a force to the compass needle.

The Magnetic Field Around a Current

A magnetic field surrounds all moving charged particles. Remember that an electric current is the flow of electric charge. In a current-carrying wire, the magnetic fields of the flowing charges combine to produce a magnetic field around the wire, as shown in Figure 10. The field around the wire becomes stronger as the current in the wire increases, or as more electrons flow in the wire.

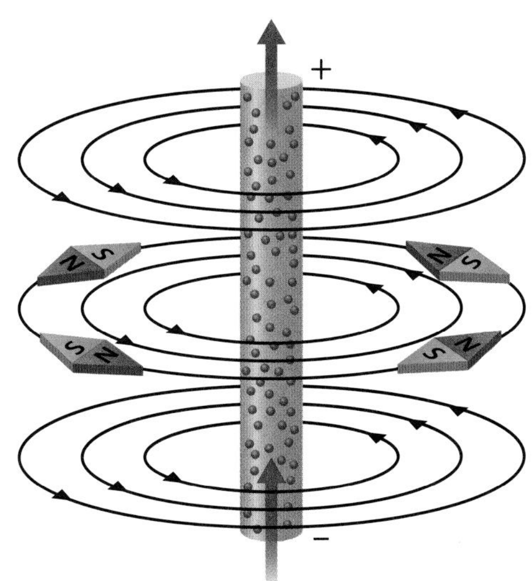

Figure 10 The magnetic field around a current-carrying wire forms closed circles.

Visual Check Why do compass needles around a current-carrying wire point in different directions?

▲ **Figure 11**  A magnet applies a force to a current-carrying wire. The direction of the force is related to the direction of the current.

 **Personal Tutor**

Magnets and Electric Currents

What happens when a magnet comes near a current-carrying wire? Because negatively charged electrons are moving in the wire, the magnet applies a force to the moving charges. This force causes the wire to move, as shown in **Figure 11.** The direction of the force on the wire depends on the direction of the current and the direction of the magnet's field.

 **Key Concept Check** Why does a magnet apply a force to an electric current?

Electromagnets

What do a stereo speaker and a hair dryer have in common? Both use electromagnets. An **electromagnet** *is a magnet created by wrapping a current-carrying wire around a ferromagnetic core.* The core becomes a magnet when an electric current flows through the coil. If a device uses electricity and has moving parts, chances are an electromagnet is inside.

Reading Check What is an electromagnet?

Making an Electromagnet

Recall that an electric current in a wire produces a magnetic field around the wire. The strength of the magnetic field around a current-carrying wire increases when the wire is wound into a coil, as shown in the center of **Figure 12.** The magnetic fields around the individual loops of the coil combine, making the magnetic field around the wire stronger.

An electromagnet is made by placing a ferromagnetic material, such as iron, as a core within the wire coil, as shown on the right in **Figure 12.** The magnetic field of the coil causes the ferromagnetic core to become a magnet. The core greatly increases the strength of the coil's magnetic field. This makes the electromagnet's force stronger. And, as the number of loops in the coil increases, the magnetic field of the electromagnet becomes stronger.

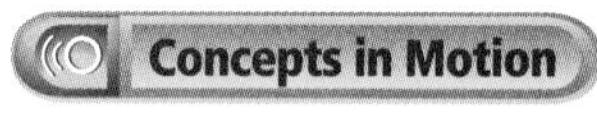 **Animation**

Figure 12 A ferromagnetic core becomes an electromagnet when surrounded by a current-carrying wire. ▼

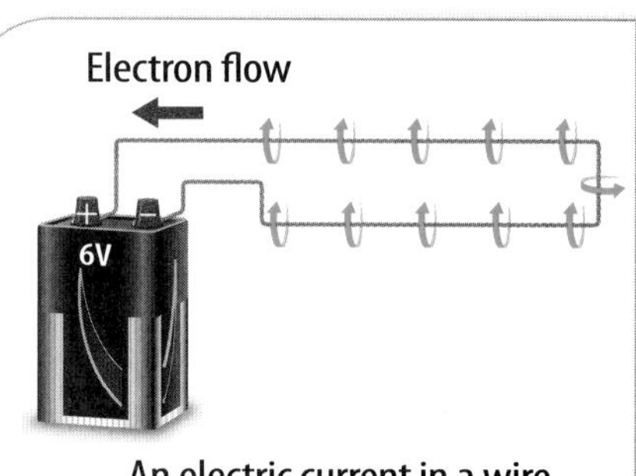

An electric current in a wire produces a magnetic field around the wire.

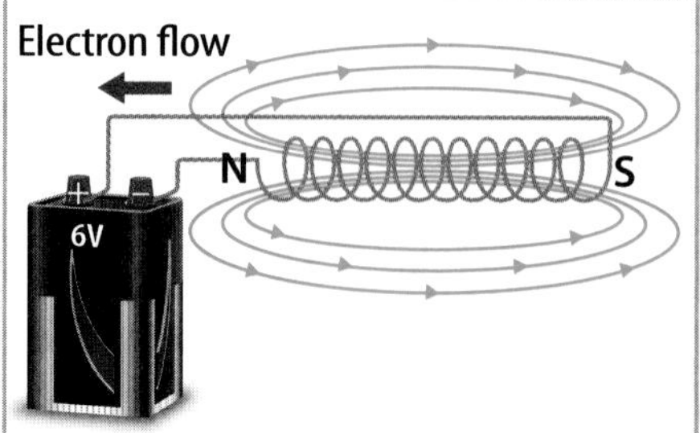

An electric current in a wire coil produces a magnetic field with a north pole and a south pole.

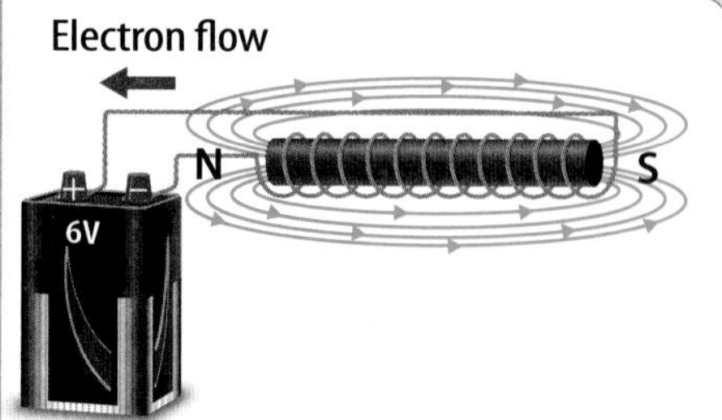

Placing an iron core within the coil greatly intensifies the magnetic field. This device is an electromagnet.

Properties of Electromagnets

An electromagnet is a special type of temporary magnet. An electromagnet differs from other magnets in several ways. First, the magnetic field around an electromagnet can be turned off by turning off the current. Without an electric current in the coil, an electromagnet is just a piece of iron wrapped with wire. But with the current on, it becomes a magnet.

Second, the strength of an electromagnet can be controlled. You read that the strength of the magnetic force of an electromagnet depends on the number of loops in the electromagnet's wire coil. Also, increasing the electric current in the coil strengthens the magnetic force.

Finally, the poles of an electromagnet can be reversed. Changing the direction of the current in the wire coil reverses an electromagnet's north and south magnetic poles. You will read more about this property in the next part of the lesson.

 **Key Concept Check** How do electromagnets and permanent magnets differ?

FOLDABLES®

Make a vertical three-tab Venn book. Use it to compare and contrast the properties of permanent magnets and electromagnets.

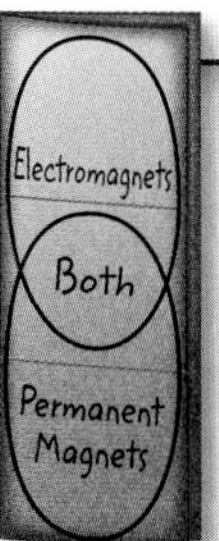

Using Electromagnets

Because an electromagnet has reversible magnetic poles, other magnets can attract and then repel an electromagnet. For example, a loudspeaker produces sound waves when the direction of the electric current in its electromagnetic coil rapidly and repeatedly reverses. **Figure 13** shows the electromagnet in a loudspeaker that connects to a paper or plastic drive cone. The changing direction of the current causes the electromagnet to attract and then repel a permanent magnet. This makes the cone vibrate, producing sound waves.

Figure 13 A rapidly changing electric current in the coil of an electromagnet causes the drive cone of a loudspeaker to produce sound waves.

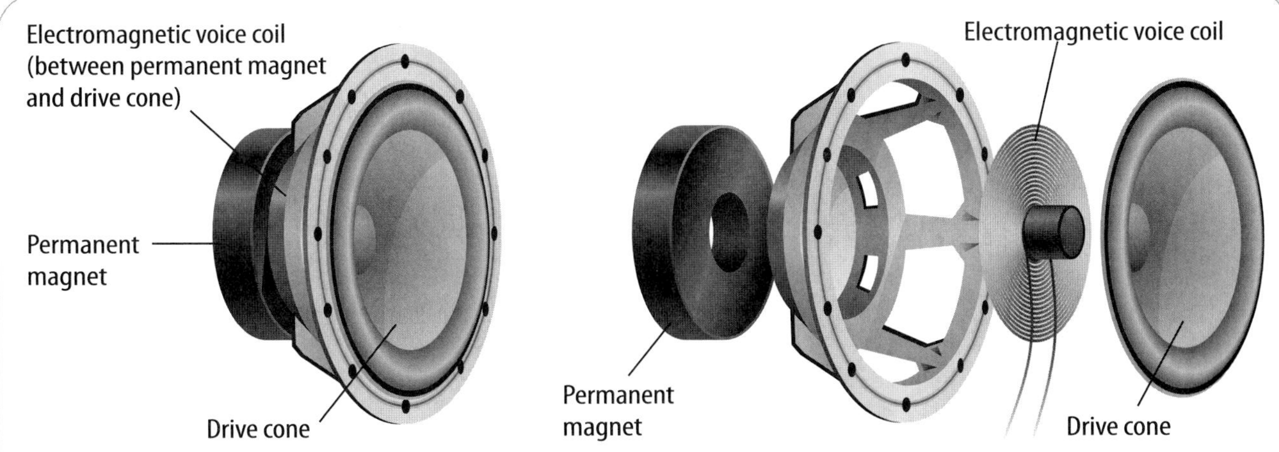

As the electric current in the voice coil changes direction (up to thousands of times per second), the coil's magnetic field changes with the current. The interaction of the coil's changing magnetic field and the permanent magnet causes the drive cone to move with a back-and-forth motion. The vibrating cone produces sound waves in the surrounding air.

20 minutes

What is an electromagnet?

Electromagnets do things that permanent magnets cannot—they can be turned off and their strength altered. As a result, electromagnets are in many modern electrical devices.

1. Read and complete a lab safety form.
2. With **sandpaper,** rub off 2 cm of insulation from both ends of 150 cm of **magnet wire.**
3. Make a coil by wrapping half of the wire around half of a **drinking straw.** Leave a 5-cm tail of wire as you begin wrapping the wire. Approximately 75 cm of wire will remain at the other end of the coil.
4. Count how many **paper clips** the coil picks up and how it interacts with a **permanent magnet.** Record your observations in your Science Journal.
5. Use **alligator clip wires** to connect the tails of the coil to a **D-cell battery in a base.** Repeat step 4. Record your observations. ⚠ *Unhook the wire after a few seconds to avoid overheating!*
6. Slide a **nail** into the center of the straw. Repeat step 4. Record your observations.
7. Disconnect the battery. Remove the nail from the straw. Wrap the rest of the wire around the straw. Repeat steps 4 through 6. Record your observations.

Analyze and Conclude

1. **Explain** your observations of the paper clips lifted by the coil in steps 4, 5, 6, and 7.
2. **Key Concept** Design a graphic organizer that shows possible uses of permanent magnets and electromagnets.

Magnets and Electric Motors

Power tools, electric fans, hair dryers, computers, and even microwave ovens use electric motors. *An* **electric motor** *is a device that uses an electric current to produce motion.* How many devices can you think of that use electric motors?

A Simple Electric Motor

A simple electric motor is shown in Figure 14. The main parts of an electric motor are a coil of wire connected to a rotating shaft, a permanent magnet, and a source of electric energy, such as a battery.

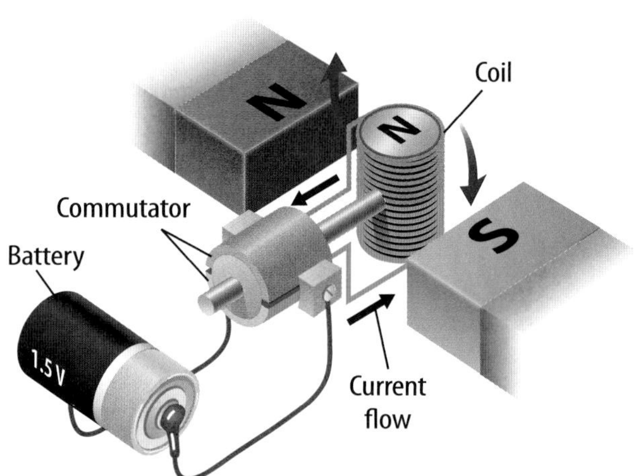

Figure 14 A rotating electromagnet in an electric motor is mounted between the poles of a permanent magnet. Here, a battery supplies a current to the electromagnet.

Animation

Making the Motor Spin

Recall that one useful property of electromagnets is that their magnetic poles easily can be reversed. This property of electromagnets is what makes an electric motor spin.

Unlike Poles Attract When an electric current is supplied to the motor, the unlike poles of the permanent magnet and electromagnet attract each other, causing the motor to begin to turn.

Reversing the Electric Current As the unlike poles of the motor's electromagnet and permanent magnet line up, the attraction between them will stop the motor's spinning. To keep the motor turning, the poles of the electromagnet must reverse.

The poles reverse when the direction of the current in the electromagnet changes, as shown in **Figure 15.** The commutator is the device in the motor that reverses the direction of the current in the electromagnet.

Similar Poles Repel Now, the similar poles of the electromagnet and the permanent magnet are close to each other. Thus, the poles repel each other and the electromagnet keeps spinning.

To keep the electromagnet spinning, the direction of the current in the coiled wire must continue changing direction. The commutator reverses the current when the poles of the electromagnet come near the poles of the permanent magnet. The permanent magnet and the electromagnet continuously attract then repel each other. This makes the electromagnet continue to spin.

Using Electric Motors

You read that an electric motor includes an electromagnet mounted on a shaft. As the electromagnet spins, the shaft spins, too. This spinning motion can be used to create other motions. Can you think of any parts powered with an electric motor in a car?

A system of levers connected to an electric motor produces the back-and-forth motion of the windshield wipers. Electric motors make the power windows go up and down. A compact disk player uses an electric motor to spin a CD and move a small laser across the disk. And now, electric motors are replacing gasoline engines in many cars.

Key Concept Check How are magnets used in electric motors?

Figure 15 The controlled forces of attraction and repulsion between an electromagnet and a permanent magnet make an electric motor spin.

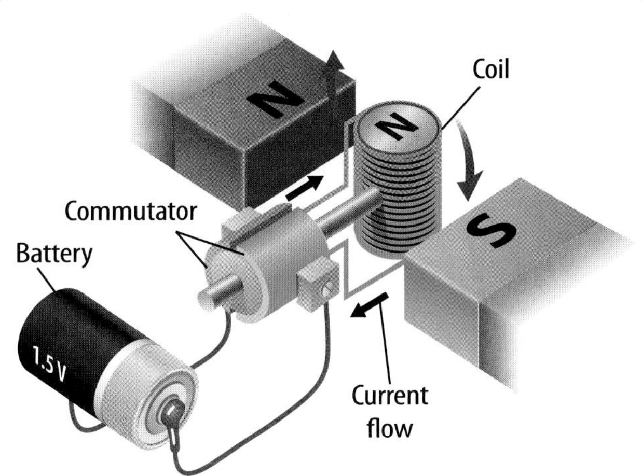

The opposite magnetic poles of the electromagnet and the permanent magnet attract each other and make the electromagnet rotate.

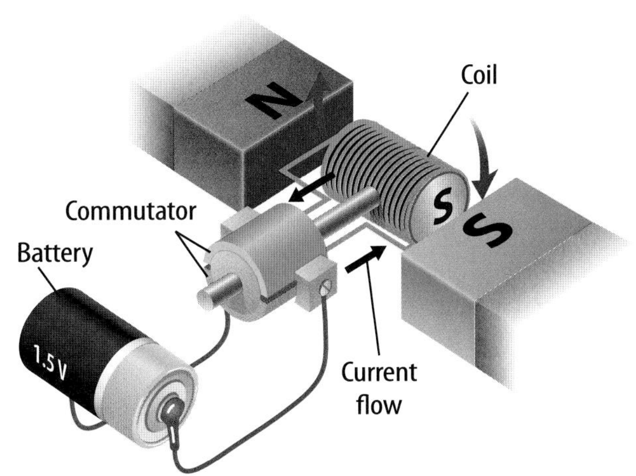

The commutator causes the current in the electromagnet to change direction, which reverses the poles of the electromagnet. The magnets now repel each other and the motor keeps rotating.

Visual Check What happens to the magnetic poles of the permanent magnets in an electric motor as the electromagnet rotates?

Word Origin

motor
from Latin *movēre*, means "to move"

Lesson 2 Review

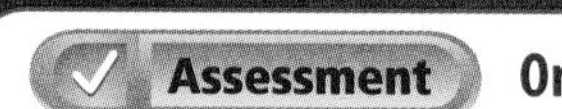

Visual Summary

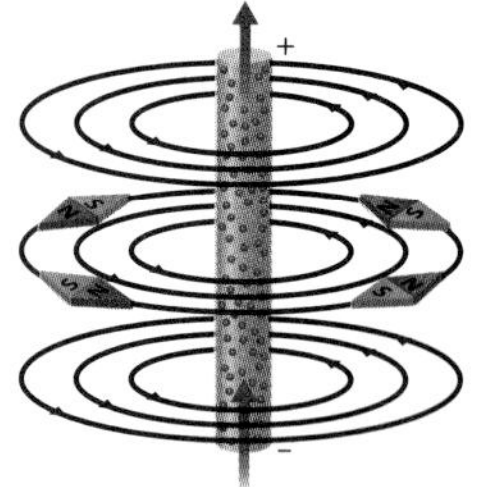

A current-carrying wire is surrounded by a magnetic field.

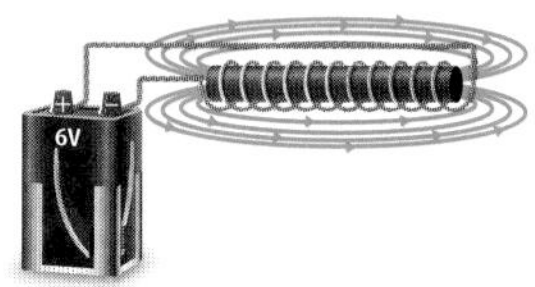

A magnetic core within a curent-carrying wire coil is an electromagnet.

Alternating attraction and repulsion between a permanent magnet and an electromagnet causes an electric motor to rotate.

FOLDABLES®

Use your lesson Foldable to review the lesson. Save your Foldable for the project at the end of the chapter.

What do you think NOW?

You first read the statements below at the beginning of the chapter.

3. A magnetic field surrounds a moving electron.

4. Unlike a permanent magnet, an electromagnet has two north magnetic poles or two south magnetic poles.

Did you change your mind about whether you agree or disagree with the statements? Rewrite any false statements to make them true.

Use Vocabulary

1. A current-carrying wire coil around a ferromagnetic core is a(n) ________.

2. An electric current causes a shaft in a(n) ________ to rotate.

Understand Key Concepts 

3. **Identify** Why does a magnet apply a force to a current-carrying wire?

4. **Summarize** How do electromagnets and permanent magnets differ?

5. In an electric motor, a force is applied to the electromagnet by
   A. a battery.
   B. a commutator.
   C. an electric current.
   D. a permanent magnet.

Interpret Graphics

6. **Compare** How would the magnetic field around the wire at the right change if the electrons flowed in the opposite direction?

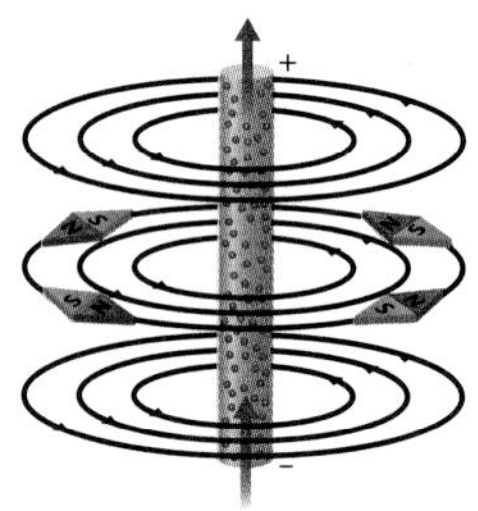

7. **Sequence Events** Copy and fill in the graphic organizer below to show the sequence of events that occur during one revolution of an electric motor.

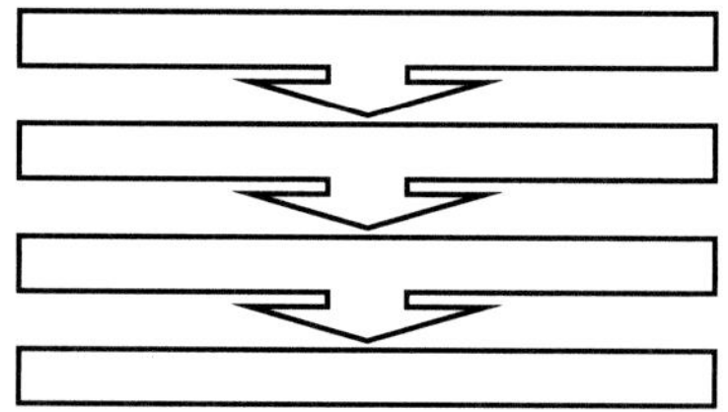

Critical Thinking

8. **Predict** A current-carrying wire made of nonmagnetic copper is attracted to a bar magnet. How will the force between the wire and the magnet change if more current flows in the wire?

Inquiry **Skill Practice** **Measure** **40 minutes**

How can you measure an electric current?

Materials

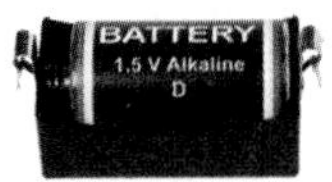

D-cell battery in a plastic holder

3 lightbulbs in screw bases

compass

sandpaper

straws

Also needed: 1.5 m of magnet wire, alligator clip wires

Safety

Moving electric charges create a magnetic field. A compass is a device that shows the direction of a magnetic field. How can you use a magnetic compass to measure an electric current?

Learn It

You **make measurements** every day. Scientists also make measurements. Sometimes the tools they use are familiar, such as rulers or scales. However, researchers must sometimes invent measuring tools based on their specific needs. What could you use to construct your own ammeter—a device that measures electric current?

Try It

1. Read and complete a lab safety form.
2. Copy the data table below into your Science Journal.

Number of Lightbulbs in Circuit	Position of Compass Needle
0	
1	
2	
3	

3. Using sandpaper, strip 1 cm of insulation from both ends of 1.5 m of magnet wire.
4. Wrap the wire around a drinking straw. Do not crush the straw. The coils should be close together and near one end of the straw. Leave 10 cm of wire sticking out from both sides.
5. Tape the compass and the straw on a table so the straw is perpendicular to the compass needle.
6. Using a battery, a lightbulb, and your coil, create a closed-series circuit. Observe the behavior of the compass. Record your observations in your data table. ⚠ *Unhook the wire after a few seconds to avoid overheating!*
7. Add a second lightbulb in series with the first. Record your observations of the compass needle.
8. Add a third lightbulb to your series circuit. Record your observations.

Apply It

9. **Explain** the relationship between the number of lightbulbs and the current.
10. **Key Concept** Explain how the compass measured the amount of an electric current.

Lesson 3

Reading Guide

Key Concepts
ESSENTIAL QUESTIONS

- How can a wire and a magnet produce an electric current?
- How do electric generators create an electric current?
- How are transformers used to bring an electric current into your home?

Vocabulary

electric generator p. 736
direct current p. 737
alternating current p. 737
turbine p. 738
transformer p. 739

g Multilingual eGlossary

Making an Electric Current with Magnets

nquiry How bright is a magnet?

Riding a bicycle safely at night means being visible to other vehicles. A small generator mounted against the bicycle's wheel provides the electric current that powers the lights, which help keep your ride safe. How does this simple device change the motion of your bicycle into a bright light at night?

Launch Lab

20 minutes

Why does the pendulum slow down?

A pendulum swings back and forth. How can magnetism be used to stop a pendulum?

1. Read and complete a lab safety form.
2. Using a **thread,** suspend a **strong magnet** from a **ring stand.** Adjust the height of the ring so that the magnet hangs about 0.5 cm off the table.
3. Pull the pendulum back 5 cm, and gently release it. Count the swings until the pendulum comes to a stop. Record your result in your Science Journal.
4. Fold a **foil square** into a smooth 3-cm-wide strip. With **tape,** secure the foil under the pendulum, parallel to the direction of swing. The magnet should not touch the foil.
5. Repeat step 3. Record your results.

Think About This

1. How might your results differ if you used a marble or a baseball as the pendulum?
2. **Key Concept** How would you describe the interaction of the foil and the magnet?

Magnets and Wire Loops

In the previous lesson, you read that an electric current is surrounded by a magnetic field. You experience a magnetic field surrounding an electric current every time you use a device that has an electric motor. Now, you will read about how magnetic fields can produce electric currents. Electric power plants use powerful magnetic fields to produce the electric energy supplied to homes and businesses.

Generating Electric Current

How can a magnetic field produce an electric current? **Figure 16** shows one way. A magnet is moved through a wire coil that is part of a closed electric circuit. As the magnetic field moves over the wire coil, an electric current is produced in the circuit. When the magnet stops moving, there is no current in the circuit. The direction of the current depends on the direction in which the magnet moves.

Key Concept Check How can a wire and a magnet produce an electric current?

Figure 16 An electric current is produced in a wire coil when a magnet moves within the coil.

Visual Check If the magnet does not touch the wire coil, what causes a current to flow in the wire?

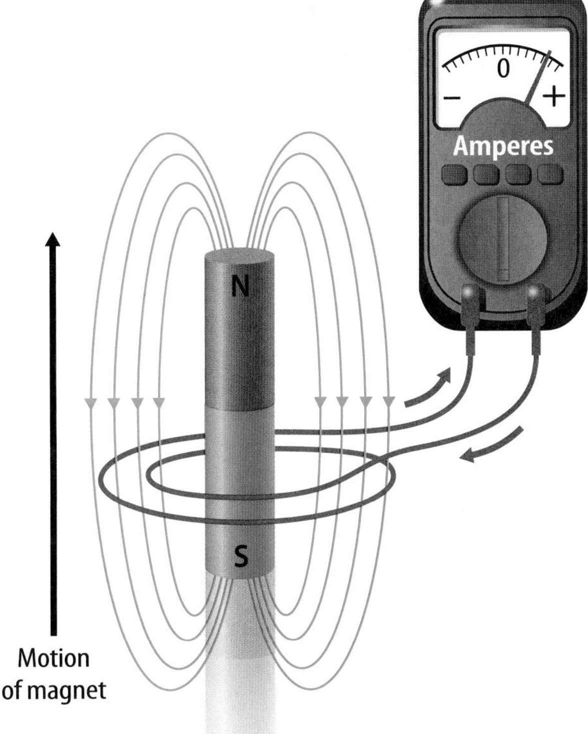

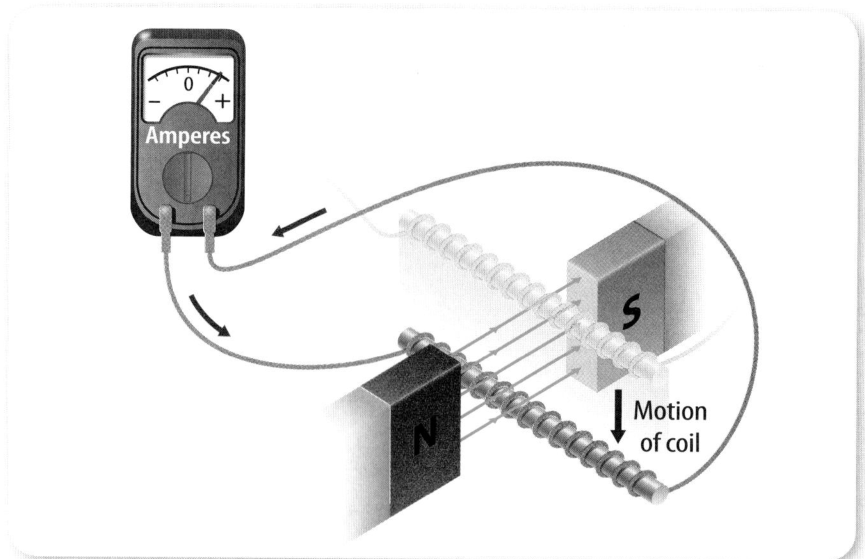

Figure 17 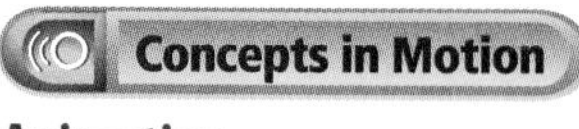 When a wire moves through a magnetic field, an electric current is produced in the circuit. There is no current when the coil is not moving.

Concepts in Motion
Animation

A Moving Wire and an Electric Current.

You just read that a magnetic field passing over a wire coil produces an electric current. **Figure 17** shows moving a wire coil through a magnetic field as another way to generate an electric current. The magnetic force between the magnet and the electrons in the wire causes the electrons in the wire to move as an electric current in the wire.

Either a magnetic field passing over a wire coil or a wire coil passing through a magnetic field produces an electric current. You will now read how this relationship powers the many electric devices you use everyday.

Reading Check How can a magnetic field produce an electric current in a wire coil?

FOLDABLES®

Make a vertical three-tab book. Label it as shown. Use it to organize notes about the electric current in each device.

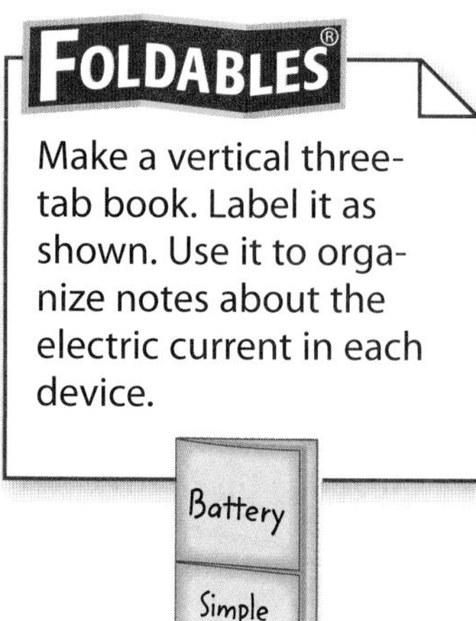

Electric Generators

When you turn on a flashlight, batteries produce an electric current in the lightbulb. However, when you turn on a TV, large electric generators in distant power plants provide the electric current to the TV. *An* **electric generator** *is a device that uses a magnetic field to transform mechanical energy to electric energy.*

A Simple Electric Generator

Figure 18 on the next page shows a hand generator in a circuit. The crank rotates a wire coil through the magnetic field of a small permanent magnet. This produces an electric current in the circuit. The current continues as the crank continues to rotate the coil within the magnetic field.

Unlike the hand-cranked generator that uses the magnetic field of a small permanent magnet, larger generators often use powerful electromagnets to produce a magnetic field.

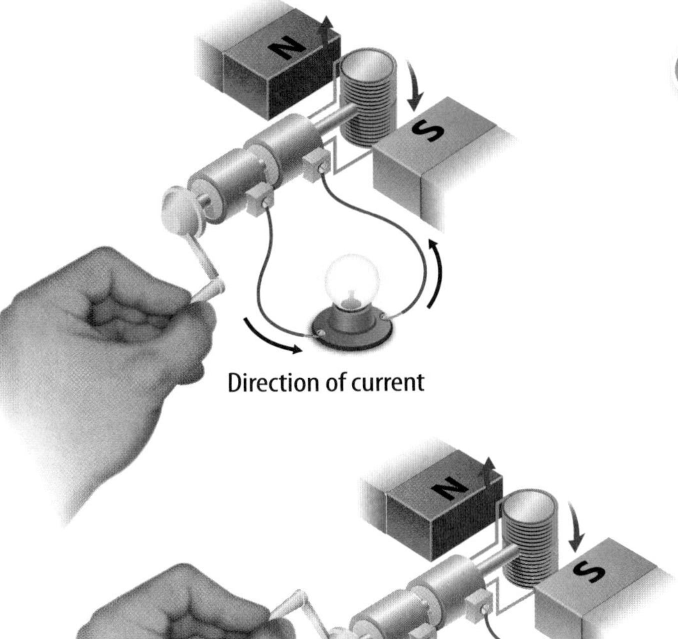

Figure 18 A generator produces an electric current when a wire coil rotates between the poles of two magnets.

Direct Current and Alternating Current

The electric current produced by a battery differs from the current produced by the generator in **Figure 18**. The current produced by a battery flows in a circuit in only one direction. *An electric current that flows in one direction is* **direct current.**

The current produced by a generator changes direction as the poles of the rotating coil line up with the poles of the magnet. *An electric current that changes direction in a regular pattern is* **alternating current.**

Some generators do produce direct current. They use a commutator, such as the one in **Figure 15** in Lesson 2. A commutator is a switch that rotates with the wire coil and prevents the current from changing directions. This results in direct current.

 **Key Concept Check** How do generators produce electric current?

20 minutes

How many paper clips can you lift?

With a hand generator, you provide the mechanical energy that causes a magnet to spin inside a wire coil. How does the speed of the spinning affect the output of a generator?

1. Read and complete a lab safety form.
2. Copy the table into your Science Journal.

Rate of Rotation	Number of Paper Clips	Rate of Rotation (opposite direction)	Number of Paper Clips
Not rotating		Not rotating	
Slow		Slow	
Fast		Fast	

3. Make an electromagnet by wrapping 150 cm of **magnet wire** around a **straw.** Leave 10-cm tails of wire at both ends of the coil.
4. Use **sandpaper** to remove 2 cm of insulation from each end of the wire. Connect the ends to a **hand generator.**
5. Slowly turn the crank of the generator.
6. Have a partner use the electromagnet to pick up **paper clips.** Record how many paper clips the electromagnet picks up.
7. Repeat step 6, turning the generator at a higher speed.
8. Hold a **compass** near one end of the coil. Turn the generator at different speeds. Record your observations.
9. Repeat steps 5–8, turning the generator crank in the opposite direction.

Analyze and Conclude

1. **Explain** how the speed of the generator and the direction it was turned affected your results.
2. **Key Concept** How does the movement of your hand lead to lifting the paper clips?

▲ **Figure 19** Enormous electric generators at Hoover Dam on the Colorado River generate the electric energy used by more than 1 million households in the region.

Generators and Power Plants

Electric power plants use huge generators, such as the ones shown in **Figure 19**. They generate the electric current used in thousands of homes. Instead of one wire coil, these generators have several large coils of wire. Each coil might have thousands of loops. Increasing the number of coils and the number of loops in each coil increases the amount of current a generator produces.

Mechanical Energy to Electric Energy

A useful energy transformation occurs in a generator. A generator produces an electric current as its wire coils rotate through a magnetic field. The rotating coils transform mechanical energy to electric energy. The electric energy is the kinetic energy of the current of electrons in the coil.

Word Origin

turbine
from Latin *turban*, means "confusion"

Supplying Mechanical Energy

As you turn a hand generator, you supply mechanical energy to the generator. The generators in electric power plants also need a source of mechanical energy to keep the coils rotating. Some power plants use the mechanical energy of high-pressure jets of steam from boiling water. Hydroelectric power plants use the mechanical energy of falling water.

In electric power plants, a generator usually is connected to a turbine (TUR bine), as shown in **Figure 20.** *A* **turbine** *is a shaft with a set of blades that spins when a stream of pressurized fluid strikes the blades.* A turbine transfers the mechanical energy of a stream of water, steam, or air to the generator.

Modern turbines that generate electric power are very efficient. They waste little of the mechanical energy used to rotate them.

Figure 20 This turbine will connect to an electric generator. There, it will spin as steam is forced through its blades. ▼

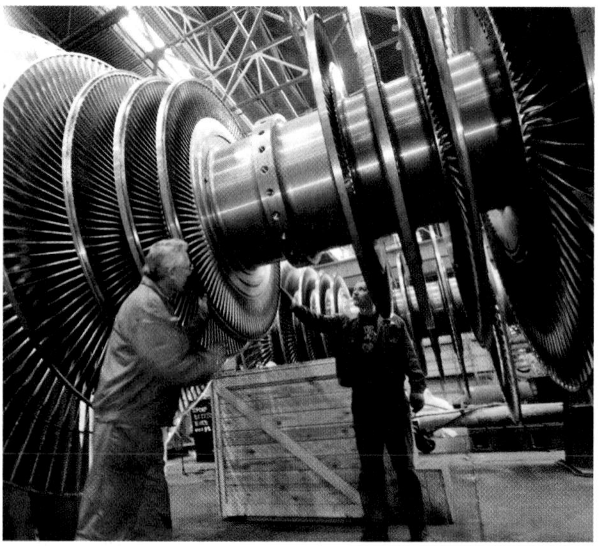

Reading Check How is a turbine used to produce an electric current?

Transformers—Changing Voltage

Recall that voltage is a measure of the energy transformed by a circuit. A high-voltage circuit transforms more energy than a low-voltage circuit with the same electric current.

Household electrical outlets provide an electric current at 120 V—great enough to cause a dangerous electric shock. However, in large transmission wires, voltage can exceed 500,000 V! At electric power plants, transformers raise the voltage for long-distance transmission, Then, other transformers lower the voltage for household use. *A* **transformer** *is a device that changes the voltage of an alternating current.*

How a Transformer Works

A transformer consists of two wire coils wrapped around a single iron core, as shown in **Figure 21.** Alternating current in the primary coil produces a continually reversing magnetic field around the core. This changing magnetic field produces alternating current in the secondary coil. If the input voltage of the primary coil increases, the output voltage of the secondary coil increases. And, if the input voltage of the primary coil decreases, the output voltage of the secondary coil decreases.

Step-Down Transformer The output voltage from a transformer also depends on the number of loops in the transformer's coils. If there are fewer loops in the secondary coil than in the primary coil, the transformer's output voltage is less than the input voltage. This type of transformer is a step-down transformer.

Step-Up Transformer If the number of loops in the secondary coil is greater than the number of loops in the primary coil, the output voltage is greater than the input voltage. This type of transformer is called a step-up transformer.

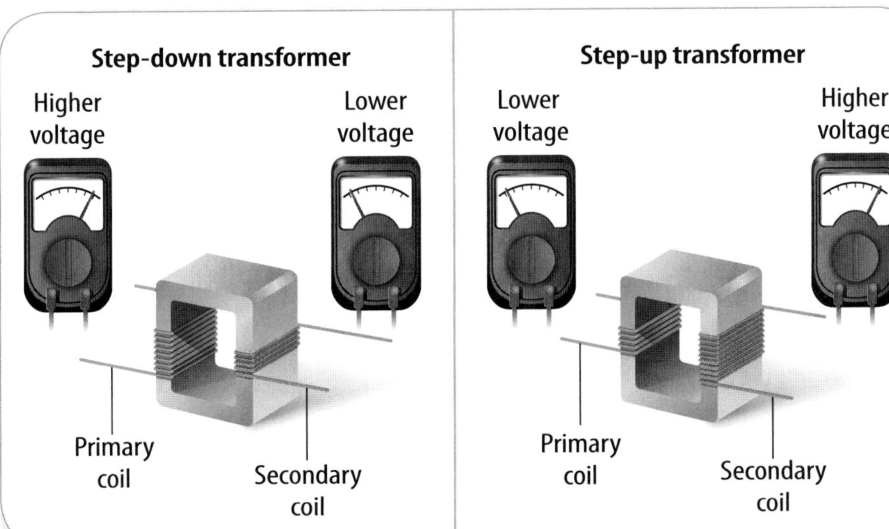

Figure 21 A transformer's output voltage can be more or less than its input voltage.

Math Skills

Solve an Equation

You can use math to find the output voltage in a transformer.

Variable	Notation
Number of Loops on Primary Coil	N_p
Number of Loops on Secondary Coil	N_s
Input Voltage to Primary Coil	V_p
Output Voltage from Secondary Coil	V_s

A transformer's primary coil has **200 loops.** Its secondary coil has **3,000 loops.** If its input voltage is **90 V**, what is its output voltage?

$$V_s = V_p \times (N_s \div N_p)$$

Replace the terms in the equation and solve.

V_s = **90.0 V** × (**3,000 loops /200 loops**) = 1,350 V

Practice

A transformer's primary coil has 50 loops. Its secondary coil has 1,500 loops. If its input voltage is 120 V, what is its output voltage?

Review

- Math Practice
- Personal Tutor

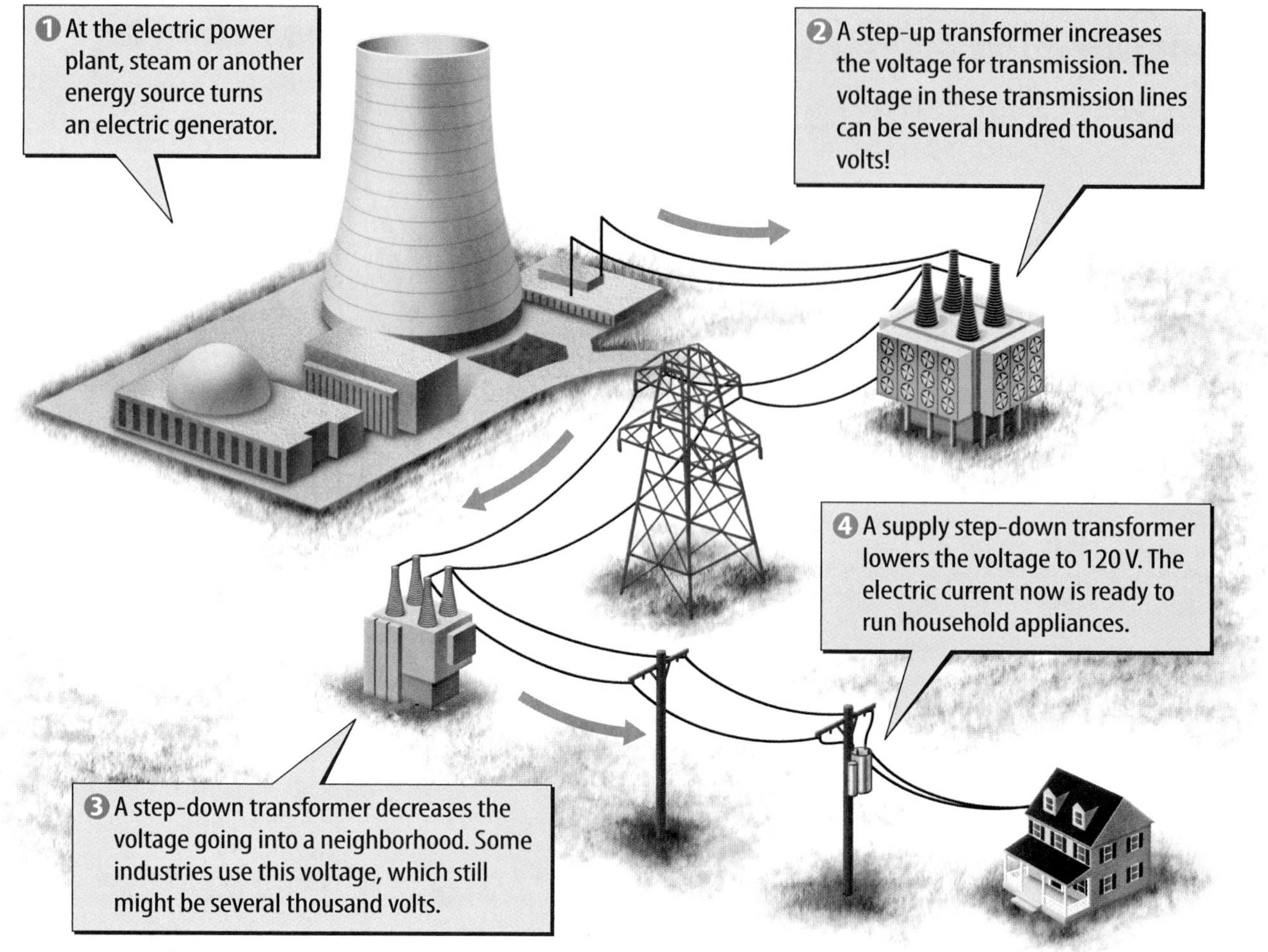

1 At the electric power plant, steam or another energy source turns an electric generator.

2 A step-up transformer increases the voltage for transmission. The voltage in these transmission lines can be several hundred thousand volts!

4 A supply step-down transformer lowers the voltage to 120 V. The electric current now is ready to run household appliances.

3 A step-down transformer decreases the voltage going into a neighborhood. Some industries use this voltage, which still might be several thousand volts.

Figure 22 Power plants move electric energy through long-distance wires. Transformers raise and lower the voltage of the current as needed by homes and businesses.

Visual Check What type of transformer is used to make electric current safe to enter a home?

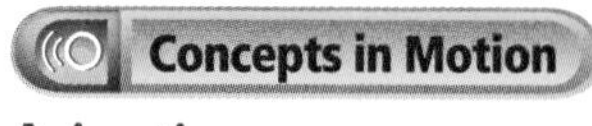

Animation

Electric Energy—From a Power Plant to Your Home

Electric energy is transmitted from an electric power plant to homes, as **Figure 22** shows. Transformers control the voltage of the current as it travels from the power plant to homes and businesses. Step-up transformers increase the voltage of the alternating current produced at electric power plants.

High-voltage current is transmitted through transmission wires, or lines. The electric resistance of the wires causes some electric energy to transform to thermal energy. High voltage current in transmission wires reduces the amount of electric energy released as thermal energy to the environment.

However, the high voltage of the current must be reduced for safe use in homes. At neighborhood substations and on utility poles close to your house, step-down transformers decrease the voltage of the current to 120 V.

Key Concept Check How are transformers used to bring an electric current to your home?

Lesson 3 Review

Assessment Online Quiz
Inquiry Virtual Lab

Visual Summary

An electric current is produced in a wire coil that passes through a magnetic field.

Electric generators transform mechanical energy to electric energy.

A magnetic field can produce an electric current, and an electric current produces a magnetic field.

FOLDABLES®

Use your lesson Foldable to review the lesson. Save your Foldable for the project at the end of the chapter.

What do you think NOW?

You first read the statements below at the beginning of the chapter.

5. A battery produces an electric current that reverses direction in a regular pattern.

6. An electric generator transforms thermal energy into electric energy.

Did you change your mind about whether you agree or disagree with the statements? Rewrite any false statements to make them true.

Use Vocabulary

1. A device that transforms mechanical energy into electric energy is a(n) ________.
2. Electric current that reverses direction in a regular pattern is ________.
3. A device that changes the voltage of an alternating current is a(n) ________.

Understand Key Concepts 

4. **Summarize** how a wire and a magnet can be used to produce an electric current.
5. What do step-up transformers do?
   - **A.** control resistance
   - **B.** decrease voltage
   - **C.** increase current
   - **D.** increase voltage
6. **Explain** how an electric generator produces an electric current.

Interpret Graphics

7. **Sequence Events** Copy the graphic organizer below. Show how transformers are used as electric energy is transmitted from a power plant to a home.

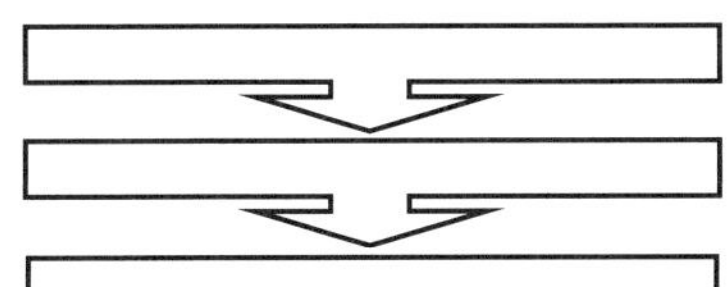

Critical Thinking

8. **Assess** Describe the electric current flowing in the primary coil of a transformer if there is no current flowing in the secondary coil. Explain your answer.

Math Skills

Math Practice

9. A transformer's primary coil has 10,000 loops. The secondary coil has 100 loops. The input voltage is 120 V What is the output voltage?

3 class periods

Design a Wind-Powered Generator

Materials

nails

strong magnet

cardboard

compass

alligator clip wires

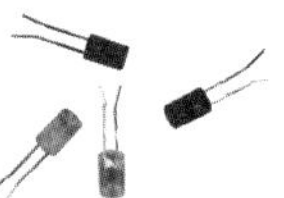
LED lights

Also needed: wood block, 1N34A diode, hammer, metal hangers, enamel-coated wire, drinking straws, small lightbulb (1.5 V), glue or tape, sandpaper

Safety

A generator converts mechanical energy into electric energy. The source of the mechanical energy can be anything from water rushing through a hydroelectric dam to steam produced in a coal-fired or nuclear power plant to wind sweeping across a prairie.

Question

What do you need to design and build a wind-powered generator?

Procedure

1. Read and complete a lab safety form.
2. Build an ammeter: Fold cardboard around a compass to make a base. Wrap 50 loops of wire around the compass. Leave 10-cm tails exposed at each end. Twist the wires to prevent unwinding. Use sandpaper to remove enamel coating from the ends of the wire.
3. Make an electromagnet: Drive a nail 2 cm into a block of wood.
4. Wrap 500 loops of wire around the nail. The coil should be within 1 cm of the head of the nail. Leave 20-cm tails exposed at each end of the coil. Twist the wires to prevent unwinding. Use sandpaper to remove the enamel coating from the ends of the wire.
5. Using alligator clip wires, connect the electromagnet to your ammeter. Position the ammeter at least 1/2 m from the coil. Use additional alligator clip wires if needed.
6. Rotate the ammeter until the needle of the compass is directly under and hidden by the wire wrapping.

7. Glue or tape the magnet to the head of another nail. Position the magnet on the nail so that its north-south axis is perpendicular to the nail.
8. Hold the magnet closely above the electromagnet. Twist the nail with the magnet between your thumb and forefinger so the magnet spins closely over the electromagnet. Do not allow the magnet to touch the nail.
9. Have a lab partner observe the ammeter. Record your team's observations in your Science Journal.
10. Design a structure that captures the wind and will support the nail and the magnet as they spin. Draw your plans in your Science Journal.
11. Have your teacher approve your plans and procedure.
12. Build and test your wind-powered generator.

Analyze and Conclude

13. **Critique** your wind-powered generator. What are the strengths and weaknesses of your generator? How could you revise your design to increase the current produced?
14. **Create** a flow chart showing the transfer of energy in this system.
15. **Explain** how your ammeter works.
16. **The Big Idea** How are moving electric charges and magnetic fields related?

Communicate Your Results

Create a storyboard. Explain the process of making a wind-powered generator and describe how it generates an electric current.

Redesign and improve your wind-powered generator so that it will light a small lightbulb.

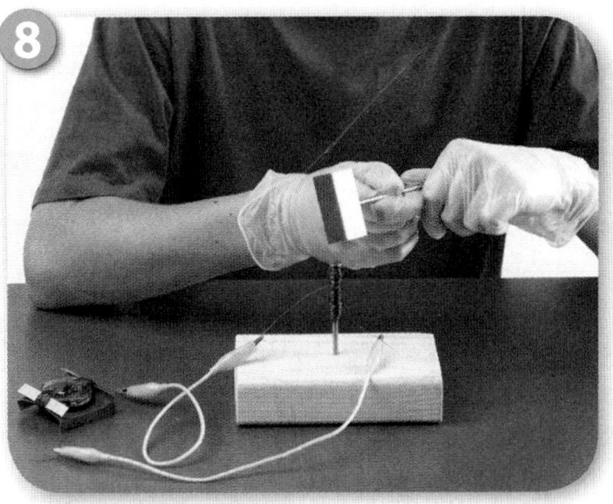

Lab Tips

- The turbine creates AC (alternating current) while the ammeter detects DC (direct current). Watch the ammeter, and carefully observe the needle swinging back and forth.
- The faster the magnet spins, the more current you will generate, but the harder it will be to measure it using your ammeter. Connect a 1N34A diode in series with the ammeter to convert AC from the generator into DC for the ammeter.

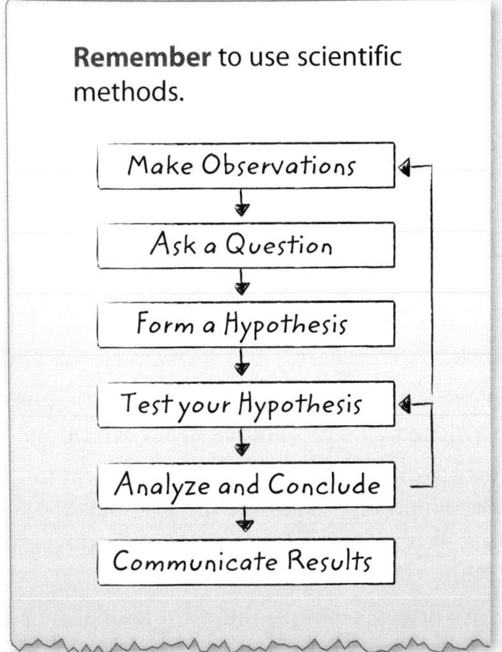

Chapter 20 Study Guide

A moving electric charge is surrounded by a magnetic field, and a magnetic field can make electric charges move.

Key Concepts Summary

Lesson 1: Magnets and Magnetic Fields

- A **magnet** is surrounded by a magnetic field that exerts forces on other magnets.
- **Magnetic materials** have **magnetic domains** that point in the same direction.
- The domains of a **temporary magnet** do not remain aligned when removed from a magnetic field. The domains of a **permanent magnet** remain in alignment, even when they are not near another magnet.

Vocabulary

magnet p. 717
magnetic pole p. 718
magnetic force p. 718
magnetic material p. 721
ferromagnetic element p. 721
magnetic domain p. 722
temporary magnet p. 723
permanent magnet p. 723

Lesson 2: Making Magnets with an Electric Current

- A magnetic field exerts a force on any moving, electrically charged particle, including the charged particles in an electric current.
- The magnetic field around an **electromagnet** can be turned on and off, can reverse direction, and can be made stronger or weaker.
- An **electric motor** contains a permanent magnet and an electromagnet. When an electric current flows in the electromagnet, the forces between the two magnets cause the electromagnet to rotate.

Vocabulary

electromagnet p. 728
electric motor p.730

Lesson 3: Making an Electric Current with Magnets

- An electric current is produced when a magnet and a closed wire loop move past each other.
- An **electric generator** has wire loops that rotate within a magnetic field. The magnetic field causes a current of electric charges to move in the wire.
- A **transformer** changes the voltage of an **alternating current.** Step-up transformers raise voltage for cross-country transmission. Step-down transformers lower voltage for household use.

Vocabulary

electric generator p. 736
direct current p. 737
alternating current p. 737
turbine p. 738
transformer p. 739

- Personal Tutor
- Vocabulary eGames
- Vocabulary eFlashcards

FOLDABLES Chapter Project

Assemble your lesson Foldables as shown to make a Chapter Project. Use the project to review what you have learned in this chapter.

Use Vocabulary

1. A(n) __________ can change the mechanical energy of a turbine into electric energy.
2. Similar poles of the __________ of a magnet point in the same direction.
3. An electric generator operates like a(n) __________ in reverse.
4. The magnetic field around a wire reverses direction in a regular pattern if a(n) __________ flows in the wire.
5. A(n) __________ remains a magnet for a long period of time.
6. A core of a ferromagnetic material placed inside the coil of a wire is a(n) __________.

Link Vocabulary and Key Concepts

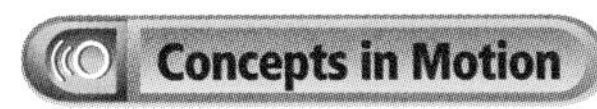

Interactive Concept Map

Copy this concept map, and then use vocabulary terms from the previous page and other terms from the chapter to complete the concept map.

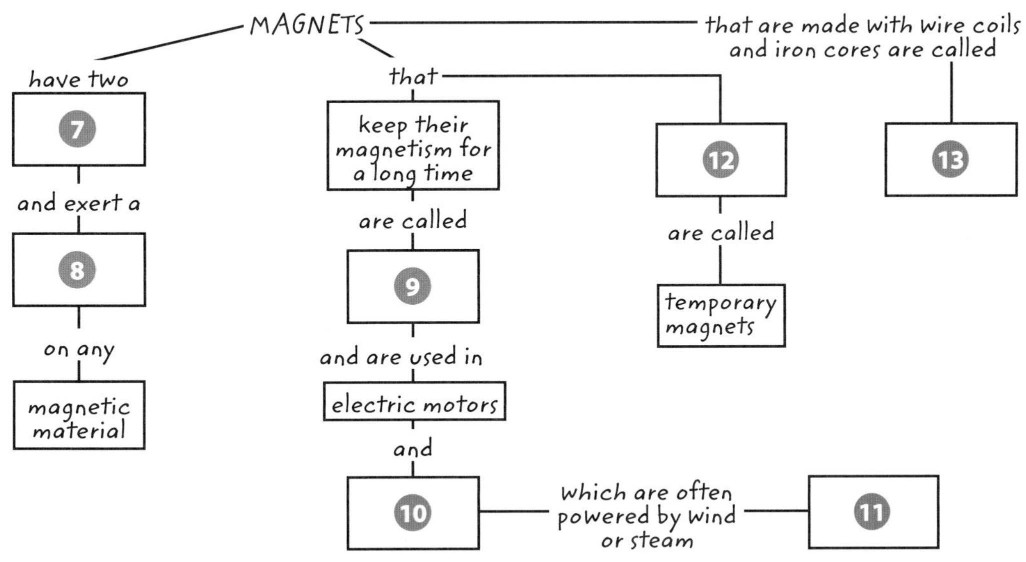

Chapter 20 Review

Understand Key Concepts

1. When the current in an electromagnet increases, what does its magnetic field do?
   A. changes direction
   B. does not change
   C. gets stronger
   D. gets weaker

2. What is the function of the commutator in an electric motor?
   A. It decreases the current in the coil.
   B. It reverses the current in the coil.
   C. It reverses the poles of the permanent magnet.
   D. It weakens the field of the permanent magnet.

3. When a bar magnet moves in a wire loop, what does the wire loop produce?
   A. an electric current
   B. electrical resistance
   C. ferromagnetic elements
   D. magnetic domains

4. Which is NOT a ferromagnetic element?
   A. aluminum
   B. cobalt
   C. iron
   D. nickel

5. Which best describes the force between the two magnets shown below?

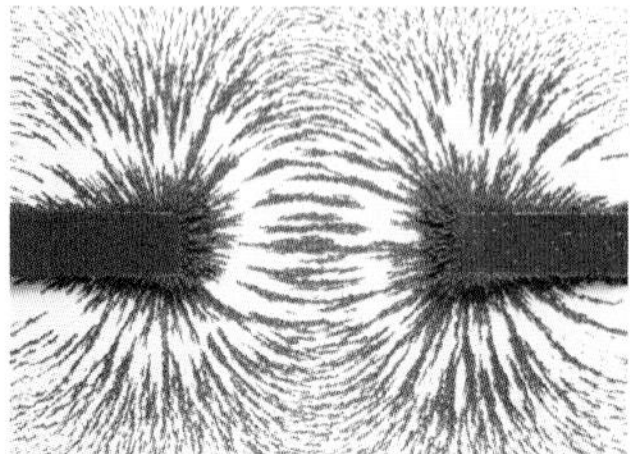

   A. alternating
   B. attractive
   C. direct
   D. repulsive

6. Where on a magnet is the magnetic field the strongest?
   A. both magnetic poles
   B. center of the magnet
   C. magnetic north pole
   D. magnetic south pole

7. In the figure below, how does the magnetic force on the wire change if the current in the wire changes direction?

   A. reverses
   B. strengthens
   C. weakens
   D. remains constant

8. Which energy transformation occurs in a generator?
   A. electric energy to light energy
   B. electric energy to mechanical energy
   C. mechanical energy to electric energy
   D. thermal energy to electric energy

9. Which describes the direction of the magnetic field lines around a magnet?
   A. start at center, end at each pole
   B. start at each pole, end at center
   C. start at north pole, end at south pole
   D. start at south pole, end at north pole

10. Which best describes a nonmagnetic material?
    A. The magnetic poles of its atoms point in different directions.
    B. The magnetic poles of its atoms point in the same direction.
    C. The magnetic poles of its magnetic domains point in different directions.
    D. The magnetic poles of its magnetic domains point in the same direction.

Assessment

Online Test Practice

Critical Thinking

11 **Predict** A step-up transformer is used to connect a lightbulb to a wall outlet. What happens to the brightness of the lightbulb if the transformer is replaced with a step-down transformer?

12 **Defend** One electromagnet has a wood core. Another electromagnet has an iron core. Which has the stronger magnetic field? Explain your answer.

13 **Assess** Compare and contrast the ways magnetic poles interact with the ways electric charges interact.

14 **Suggest** How could you determine a car door is made of a nonmagnetic material, such as plastic?

15 **Suggest** What would you do to a wire to make it attract to a magnet?

16 **Evaluate** The figure below shows the field lines of a magnet. Where is the magnet's north pole? Explain your answer.

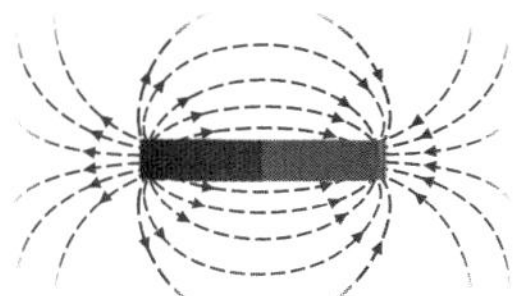

17 **Recommend** How could you increase the number of times per second that an alternating current from a generator changes direction?

18 **Hypothesize** An alternating current flows in an electromagnet. If the electromagnet is placed in a wire coil, will there be an electric current in the wire coil? Explain your answer.

Writing in Science

19 **Write** a short essay describing the different devices you use in a typical day that have electric motors. Include how the electric motors are used to make something move.

REVIEW THE BIG IDEA

20 How can a magnetic field cause a current to flow in a wire coil?

21 How are electric charges and magnetic fields related?

Math Skills

Math Practice

Solve an Equation

22 A transformer has 300 loops on its primary coil and 90,000 loops on its secondary coil. The input voltage is 60 V. What is the output voltage?

23 A transformer has 80 loops on its primary coil and 1,200 loops on its secondary coil. The input voltage is 60 V. What is the output voltage?

24 The primary coil of a transformer has 150 loops and has a 120 V primary source. How many loops should the secondary coil have to produce an output voltage of 600 V?

Standardized Test Practice

Record your answers on the answer sheet provided by your teacher or on a sheet of paper.

Multiple Choice

1 Which statement about magnets is true?

A Two north poles attract each other.

B Two south poles attract each other.

C A north pole and a south pole attract each other.

D A north pole and a south pole repel each other.

Use the images below to answer question 2.

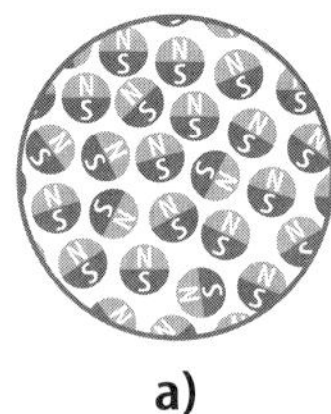

a)

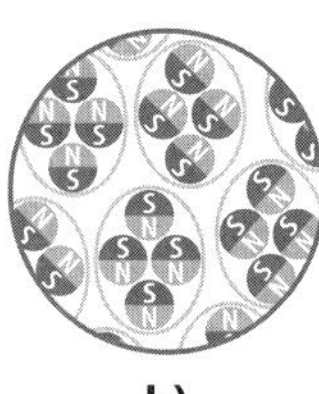

b)

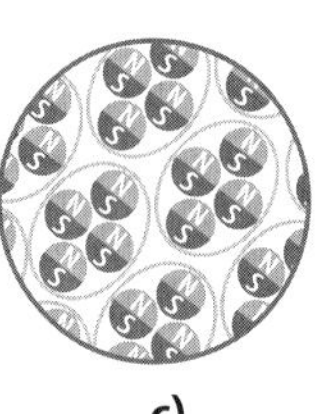

c)

2 Which statement best describes these models?

A Only A is a magnet.

B Only C is a magnet.

C A and B are magnets.

D A and C are magnets.

3 Why does a compass needle move if held near a current-carrying wire?

A A current-carrying wire will cause any object to move.

B A compass needle moves when it is near any piece of metal wire.

C Both the compass needle and the current-carrying wire have magnetic fields.

D The compass needle and the current-carrying wire have the same electric charge.

Use the image below to answer question 4.

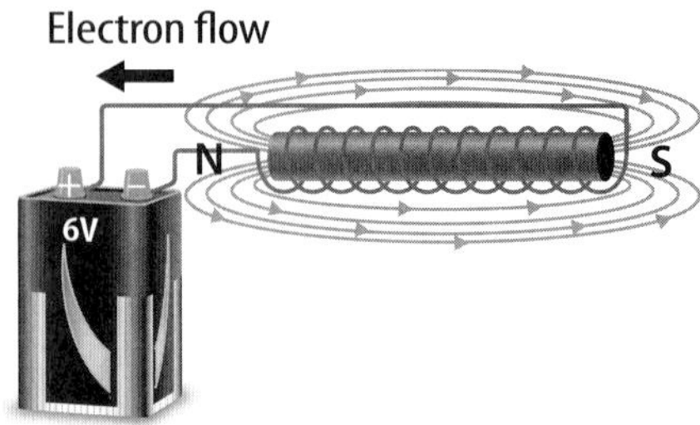

4 Which describes what will happen if the iron core is removed from the electromagnet?

A It is no longer a magnet.

B It is a stronger magnet.

C It is a weaker magnet.

D It loses its north and south poles.

5 What causes the poles of the electromagnet in an electric motor to reverse?

A a change in the direction of the current

B a change in the strength of the magnetic field

C a change in the shape of the current-carrying wire

D a change in the position of the permanent magnet

6 What happens when a magnet passes through a closed loop of wire?

A An electric current is generated in the wire.

B Any nearby magnets gain an electric charge.

C The looped part of the wire no longer is attracted to magnets.

D Any device in the circuit becomes a permanent magnet.

Standardized Test Practice

Assessment

Online Standardized Test Practice

7 How do an electric motor and an electric generator differ?

A One uses a permanent magnet, while the other uses temporary magnets.

B One produces electric energy, while the other produces thermal energy.

C One contains moving parts, while the other has only parts that do not move.

D One uses an electric current to produce motion, while the other uses motion to produce an electric current.

Use the figure below to answer questions 8 and 9.

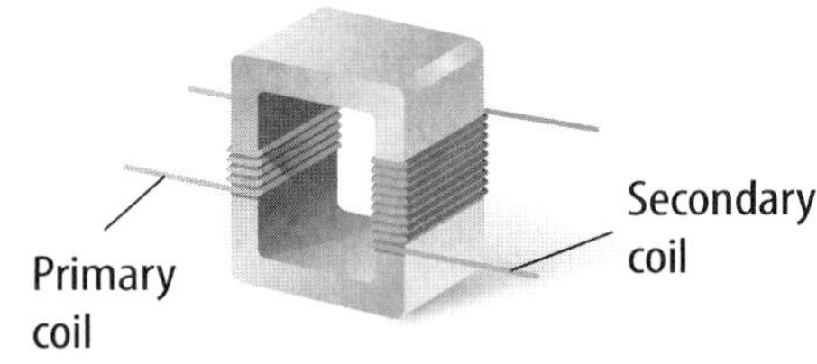

8 What would be the use for this device?

A to generate electric energy

B to increase the voltage

C to make an electric motor

D to reduce the voltage

9 For the device, which statement is true?

A The primary voltage is greater than the secondary voltage.

B The primary voltage is less than the secondary voltage.

C The primary voltage is the same secondary voltage.

D The voltages alternate high to low.

Constructed Response

10 When you touch a magnet to a steel nail, the nail attracts steel paper clips. Explain why this does not happen when you use an aluminum nail.

Use the figure below to answer questions 11 and 12.

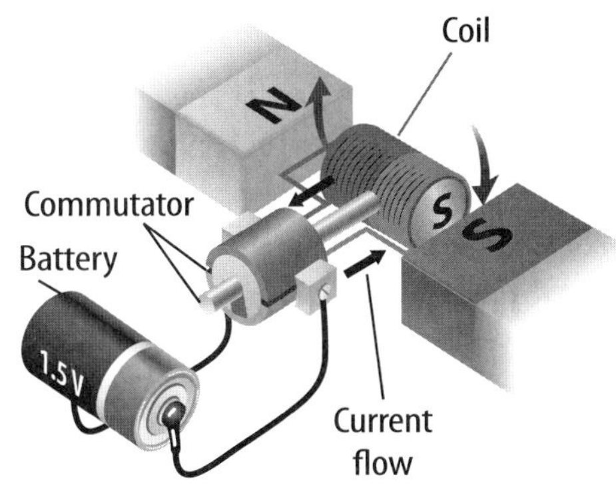

11 Using the terms in the diagram, identify what this device is and describe what it does.

12 What would happen to the device if the wire coil were removed?

13 How are transformers used as an electric current travels between a power plant and a home?

14 Compare and contrast the components and functions of an electric motor and an electric generator.

NEED EXTRA HELP?														
If You Missed Question...	1	2	3	4	5	6	7	8	9	10	11	12	13	14
Go to Lesson...	1	1	2	2	2	3	3	3	3	1	2	2	3	3

Student Resources

For Students and Parents/Guardians

These resources are designed to help you achieve success in science. You will find useful information on laboratory safety, math skills, and science skills. In addition, science reference materials are found in the Reference Handbook. You'll find the information you need to learn and sharpen your skills in these resources.

Table of Contents

Science Skill Handbook **SR-2**

Scientific Methods **SR-2**
- Identify a Question SR-2
- Gather and Organize Information SR-2
- Form a Hypothesis SR-5
- Test the Hypothesis SR-6
- Collect Data SR-6
- Analyze the Data SR-9
- Draw Conclustions SR-10
- Communicate SR-10

Safety Symbols **SR-11**

Safety in the Science Laboratory **SR-12**
- General Safety Rules SR-12
- Prevent Accidents SR-12
- Laboratory Work SR-13
- Emergencies SR-13

Math Skill Handbook **SR-14**

Math Review **SR-14**
- Use Fractions SR-14
- Use Ratios SR-17
- Use Decimals SR-17
- Use Proportions SR-18
- Use Percentages SR-19
- Solve One-Step Equations SR-19
- Use Statistics SR-20
- Use Geometry SR-21

Science Application **SR-24**
- Measure in SI SR-24
- Dimensional Analysis SR-24
- Precision and Significant Digits SR-26
- Scientific Notation SR-26
- Make and Use Graphs SR-27

Foldables Handbook **SR-29**

Reference Handbook **SR-40**
- Periodic Table of the Elements SR-40

Glossary **G-2**

Index **I-2**

Credits **C-2**

Scientific Methods

Scientists use an orderly approach called the scientific method to solve problems. This includes organizing and recording data so others can understand them. Scientists use many variations in this method when they solve problems.

Identify a Question

The first step in a scientific investigation or experiment is to identify a question to be answered or a problem to be solved. For example, you might ask which gasoline is the most efficient.

Gather and Organize Information

After you have identified your question, begin gathering and organizing information. There are many ways to gather information, such as researching in a library, interviewing those knowledgeable about the subject, testing and working in the laboratory and field. Fieldwork is investigations and observations done outside of a laboratory.

Researching Information Before moving in a new direction, it is important to gather the information that already is known about the subject. Start by asking yourself questions to determine exactly what you need to know. Then you will look for the information in various reference sources, like the student is doing in **Figure 1.** Some sources may include textbooks, encyclopedias, government documents, professional journals, science magazines, and the Internet. Always list the sources of your information.

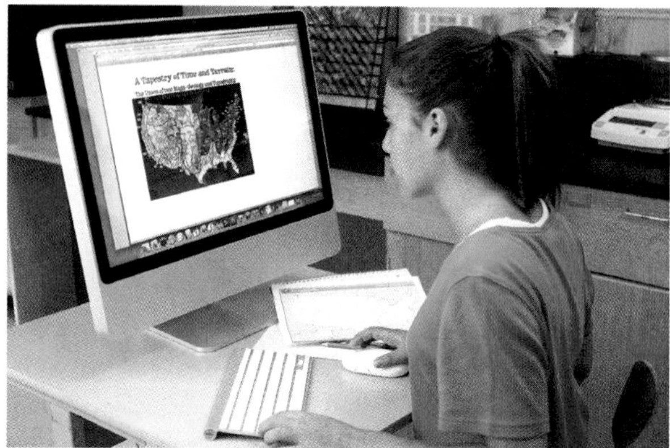

Figure 1 The Internet can be a valuable research tool.

Evaluate Sources of Information Not all sources of information are reliable. You should evaluate all of your sources of information, and use only those you know to be dependable. For example, if you are researching ways to make homes more energy efficient, a site written by the U.S. Department of Energy would be more reliable than a site written by a company that is trying to sell a new type of weatherproofing material. Also, remember that research always is changing. Consult the most current resources available to you. For example, a 1985 resource about saving energy would not reflect the most recent findings.

Sometimes scientists use data that they did not collect themselves, or conclusions drawn by other researchers. This data must be evaluated carefully. Ask questions about how the data were obtained, if the investigation was carried out properly, and if it has been duplicated exactly with the same results. Would you reach the same conclusion from the data? Only when you have confidence in the data can you believe it is true and feel comfortable using it.

Interpret Scientific Illustrations As you research a topic in science, you will see drawings, diagrams, and photographs to help you understand what you read. Some illustrations are included to help you understand an idea that you can't see easily by yourself, like the tiny particles in an atom in **Figure 2.** A drawing helps many people to remember details more easily and provides examples that clarify difficult concepts or give additional information about the topic you are studying. Most illustrations have labels or a caption to identify or to provide more information.

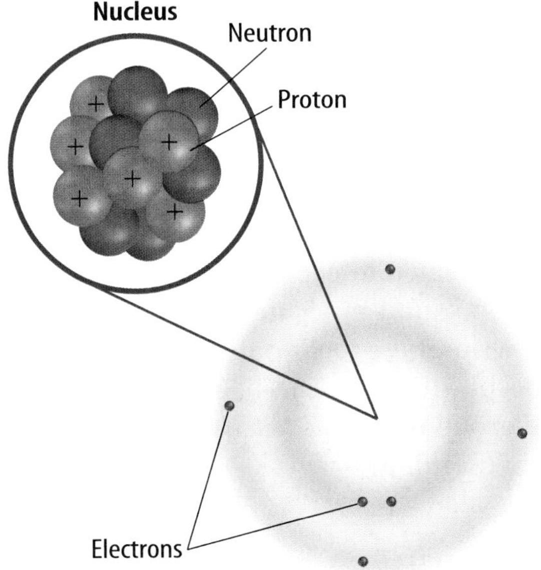

Figure 2 This drawing shows an atom of carbon with its six protons, six neutrons, and six electrons.

Concept Maps One way to organize data is to draw a diagram that shows relationships among ideas (or concepts). A concept map can help make the meanings of ideas and terms more clear, and help you understand and remember what you are studying. Concept maps are useful for breaking large concepts down into smaller parts, making learning easier.

Network Tree A type of concept map that not only shows a relationship, but how the concepts are related is a network tree, shown in **Figure 3.** In a network tree, the words are written in the ovals, while the description of the type of relationship is written across the connecting lines.

When constructing a network tree, write down the topic and all major topics on separate pieces of paper or notecards. Then arrange them in order from general to specific. Branch the related concepts from the major concept and describe the relationship on the connecting line. Continue to more specific concepts until finished.

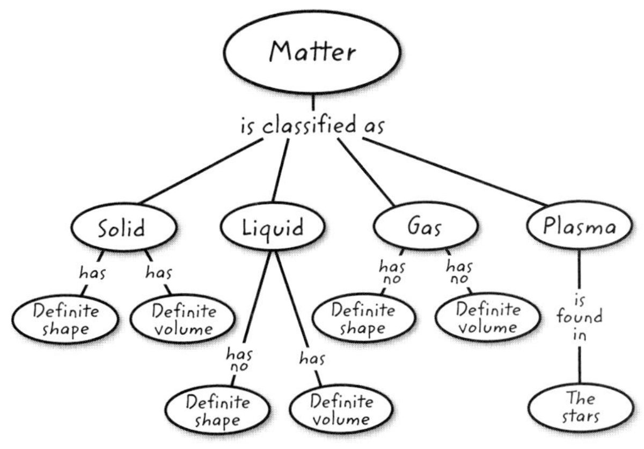

Figure 3 A network tree shows how concepts or objects are related.

Events Chain Another type of concept map is an events chain. Sometimes called a flow chart, it models the order or sequence of items. An events chain can be used to describe a sequence of events, the steps in a procedure, or the stages of a process.

When making an events chain, first find the one event that starts the chain. This event is called the initiating event. Then, find the next event and continue until the outcome is reached, as shown in **Figure 4.**

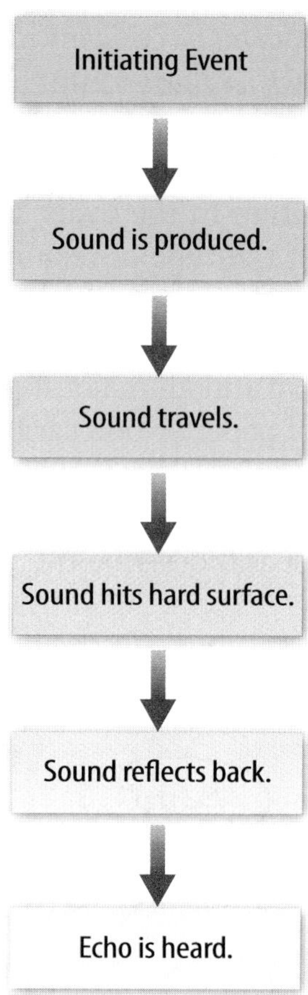

Figure 4 Events-chain concept maps show the order of steps in a process or event. This concept map shows how a sound makes an echo.

Cycle Map A specific type of events chain is a cycle map. It is used when the series of events do not produce a final outcome, but instead relate back to the beginning event, such as in **Figure 5.** Therefore, the cycle repeats itself.

To make a cycle map, first decide what event is the beginning event. This is also called the initiating event. Then list the next events in the order that they occur, with the last event relating back to the initiating event. Words can be written between the events that describe what happens from one event to the next. The number of events in a cycle map can vary, but usually contain three or more events.

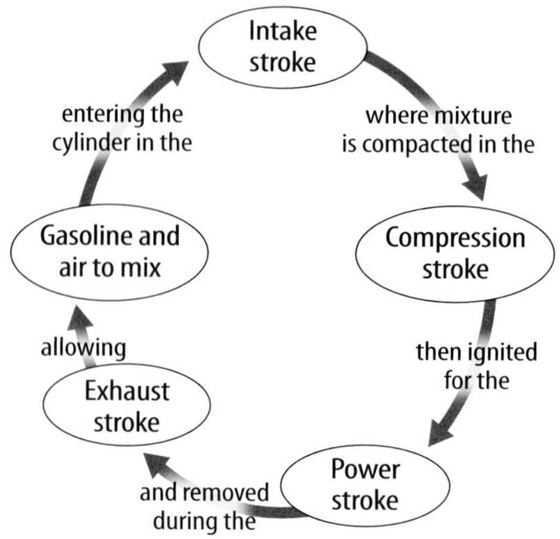

Figure 5 A cycle map shows events that occur in a cycle.

Spider Map A type of concept map that you can use for brainstorming is the spider map. When you have a central idea, you might find that you have a jumble of ideas that relate to it but are not necessarily clearly related to each other. The spider map on sound in **Figure 6** shows that if you write these ideas outside the main concept, then you can begin to separate and group unrelated terms so they become more useful.

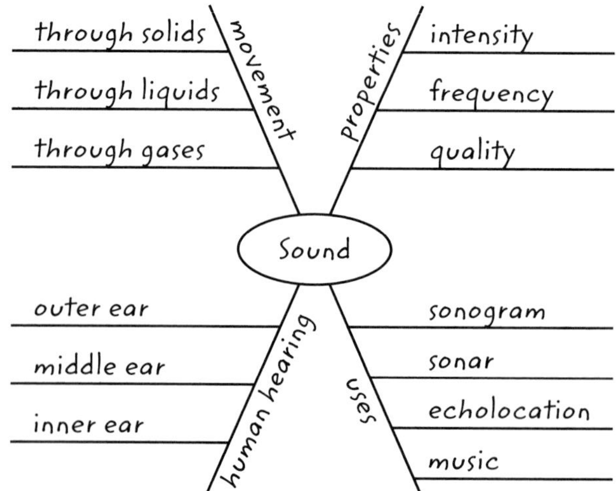

Figure 6 A spider map allows you to list ideas that relate to a central topic but not necessarily to one another.

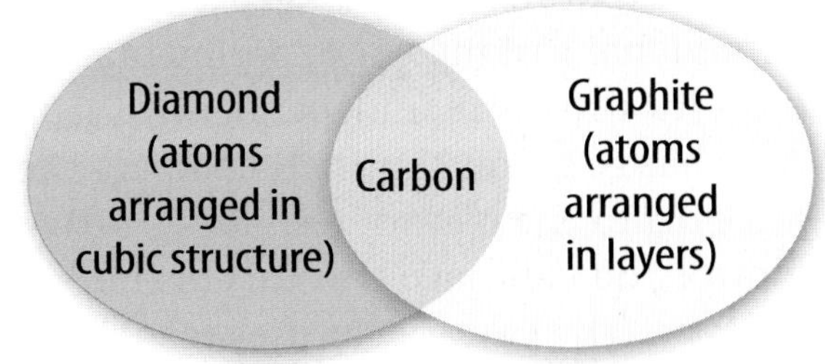

Figure 7 This Venn diagram compares and contrasts two substances made from carbon.

Venn Diagram To illustrate how two subjects compare and contrast you can use a Venn diagram. You can see the characteristics that the subjects have in common and those that they do not, shown in **Figure 7.**

To create a Venn diagram, draw two overlapping ovals that that are big enough to write in. List the characteristics unique to one subject in one oval, and the characteristics of the other subject in the other oval. The characteristics in common are listed in the overlapping section.

Make and Use Tables One way to organize information so it is easier to understand is to use a table. Tables can contain numbers, words, or both.

To make a table, list the items to be compared in the first column and the characteristics to be compared in the first row. The title should clearly indicate the content of the table, and the column or row heads should be clear. Notice that in **Table 1** the units are included.

Table 1 Recyclables Collected During Week

Day of Week	Paper (kg)	Aluminum (kg)	Glass (kg)
Monday	5.0	4.0	12.0
Wednesday	4.0	1.0	10.0
Friday	2.5	2.0	10.0

Make a Model One way to help you better understand the parts of a structure, the way a process works, or to show things too large or small for viewing is to make a model. For example, an atomic model made of a plastic-ball nucleus and pipe-cleaner electron shells can help you visualize how the parts of an atom relate to each other. Other types of models can be devised on a computer or represented by equations.

Form a Hypothesis

A possible explanation based on previous knowledge and observations is called a hypothesis. After researching gasoline types and recalling previous experiences in your family's car you form a hypothesis—our car runs more efficiently because we use premium gasoline. To be valid, a hypothesis has to be something you can test by using an investigation.

Predict When you apply a hypothesis to a specific situation, you predict something about that situation. A prediction makes a statement in advance, based on prior observation, experience, or scientific reasoning. People use predictions to make everyday decisions. Scientists test predictions by performing investigations. Based on previous observations and experiences, you might form a prediction that cars are more efficient with premium gasoline. The prediction can be tested in an investigation.

Design an Experiment A scientist needs to make many decisions before beginning an investigation. Some of these include: how to carry out the investigation, what steps to follow, how to record the data, and how the investigation will answer the question. It also is important to address any safety concerns.

SCIENCE SKILL HANDBOOK · MATH SKILL HANDBOOK · FOLDABLES HANDBOOK · REFERENCE HANDBOOK · GLOSSARY/GLOSARIO · INDEX

Test the Hypothesis

Now that you have formed your hypothesis, you need to test it. Using an investigation, you will make observations and collect data, or information. This data might either support or not support your hypothesis. Scientists collect and organize data as numbers and descriptions.

Follow a Procedure In order to know what materials to use, as well as how and in what order to use them, you must follow a procedure. **Figure 8** shows a procedure you might follow to test your hypothesis.

Procedure

Step 1	Use regular gasoline for two weeks.
Step 2	Record the number of kilometers between fill-ups and the amount of gasoline used.
Step 3	Switch to premium gasoline for two weeks.
Step 4	Record the number of kilometers between fill-ups and the amount of gasoline used.

Figure 8 A procedure tells you what to do step by step.

Identify and Manipulate Variables and Controls In any experiment, it is important to keep everything the same except for the item you are testing. The one factor you change is called the independent variable. The change that results is the dependent variable. Make sure you have only one independent variable, to assure yourself of the cause of the changes you observe in the dependent variable. For example, in your gasoline experiment the type of fuel is the independent variable. The dependent variable is the efficiency.

Many experiments also have a control—an individual instance or experimental subject for which the independent variable is not changed. You can then compare the test results to the control results. To design a control you can have two cars of the same type. The control car uses regular gasoline for four weeks. After you are done with the test, you can compare the experimental results to the control results.

Collect Data

Whether you are carrying out an investigation or a short observational experiment, you will collect data, as shown in **Figure 9.** Scientists collect data as numbers and descriptions and organize it in specific ways.

Observe Scientists observe items and events, then record what they see. When they use only words to describe an observation, it is called qualitative data. Scientists' observations also can describe how much there is of something. These observations use numbers, as well as words, in the description and are called quantitative data. For example, if a sample of the element gold is described as being "shiny and very dense" the data are qualitative. Quantitative data on this sample of gold might include "a mass of 30 g and a density of 19.3 g/cm^3."

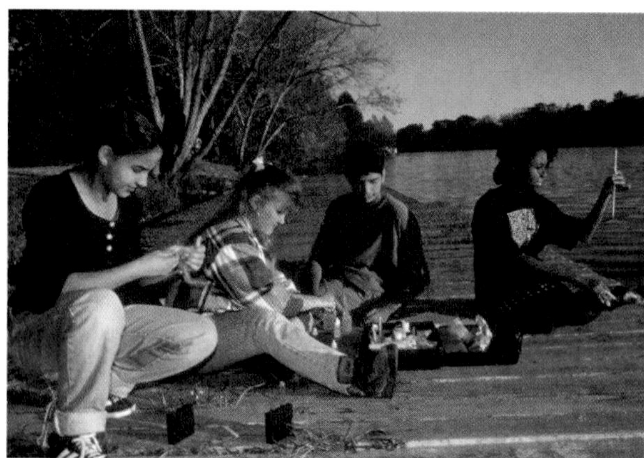

Figure 9 Collecting data is one way to gather information directly.

Figure 10 Record data neatly and clearly so it is easy to understand.

When you make observations you should examine the entire object or situation first, and then look carefully for details. It is important to record observations accurately and completely. Always record your notes immediately as you make them, so you do not miss details or make a mistake when recording results from memory. Never put unidentified observations on scraps of paper. Instead they should be recorded in a notebook, like the one in **Figure 10.** Write your data neatly so you can easily read it later. At each point in the experiment, record your observations and label them. That way, you will not have to determine what the figures mean when you look at your notes later. Set up any tables that you will need to use ahead of time, so you can record any observations right away. Remember to avoid bias when collecting data by not including personal thoughts when you record observations. Record only what you observe.

Estimate Scientific work also involves estimating. To estimate is to make a judgment about the size or the number of something without measuring or counting. This is important when the number or size of an object or population is too large or too difficult to accurately count or measure.

Sample Scientists may use a sample or a portion of the total number as a type of estimation. To sample is to take a small, representative portion of the objects or organisms of a population for research. By making careful observations or manipulating variables within that portion of the group, information is discovered and conclusions are drawn that might apply to the whole population. A poorly chosen sample can be unrepresentative of the whole. If you were trying to determine the rainfall in an area, it would not be best to take a rainfall sample from under a tree.

Measure You use measurements every day. Scientists also take measurements when collecting data. When taking measurements, it is important to know how to use measuring tools properly. Accuracy also is important.

Length To measure length, the distance between two points, scientists use meters. Smaller measurements might be measured in centimeters or millimeters.

Length is measured using a metric ruler or meter stick. When using a metric ruler, line up the 0-cm mark with the end of the object being measured and read the number of the unit where the object ends. Look at the metric ruler shown in **Figure 11.** The centimeter lines are the long, numbered lines, and the shorter lines are millimeter lines. In this instance, the length would be 4.50 cm.

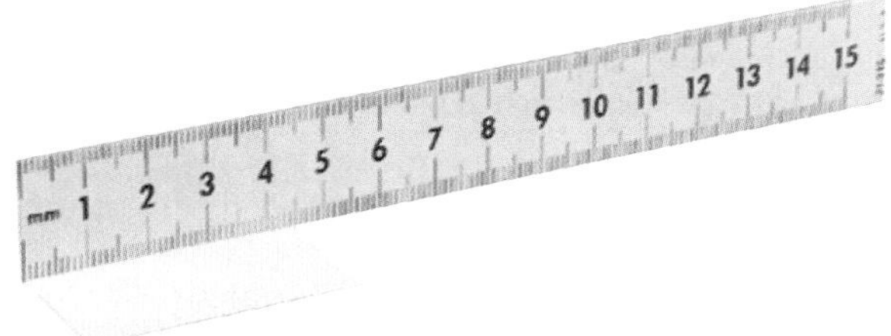

Figure 11 This metric ruler has centimeter and millimeter divisions.

Mass The SI unit for mass is the kilogram (kg). Scientists can measure mass using units formed by adding metric prefixes to the unit gram (g), such as milligram (mg). To measure mass, you might use a triple-beam balance similar to the one shown in **Figure 12.** The balance has a pan on one side and a set of beams on the other side. Each beam has a rider that slides on the beam.

When using a triple-beam balance, place an object on the pan. Slide the largest rider along its beam until the pointer drops below zero. Then move it back one notch. Repeat the process for each rider proceeding from the larger to smaller until the pointer swings an equal distance above and below the zero point. Sum the masses on each beam to find the mass of the object. Move all riders back to zero when finished.

Instead of putting materials directly on the balance, scientists often take a tare of a container. A tare is the mass of a container into which objects or substances are placed for measuring their masses. To mass objects or substances, find the mass of a clean container. Remove the container from the pan, and place the object or substances in the container. Find the mass of the container with the materials in it. Subtract the mass of the empty container from the mass of the filled container to find the mass of the materials you are using.

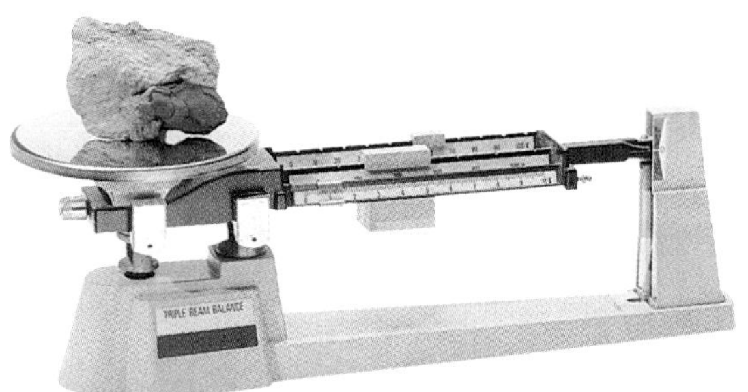

Figure 12 A triple-beam balance is used to determine the mass of an object.

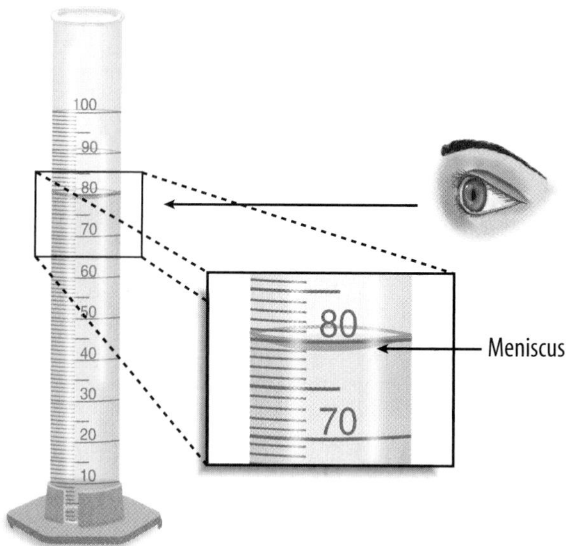

Figure 13 Graduated cylinders measure liquid volume.

Liquid Volume To measure liquids, the unit used is the liter. When a smaller unit is needed, scientists might use a milliliter. Because a milliliter takes up the volume of a cube measuring 1 cm on each side it also can be called a cubic centimeter ($cm^3 = cm \times cm \times cm$).

You can use beakers and graduated cylinders to measure liquid volume. A graduated cylinder, shown in **Figure 13,** is marked from bottom to top in milliliters. In lab, you might use a 10-mL graduated cylinder or a 100-mL graduated cylinder. When measuring liquids, notice that the liquid has a curved surface. Look at the surface at eye level, and measure the bottom of the curve. This is called the meniscus. The graduated cylinder in **Figure 13** contains 79.0 mL, or 79.0 cm^3, of a liquid.

Temperature Scientists often measure temperature using the Celsius scale. Pure water has a freezing point of 0°C and boiling point of 100°C. The unit of measurement is degrees Celsius. Two other scales often used are the Fahrenheit and Kelvin scales.

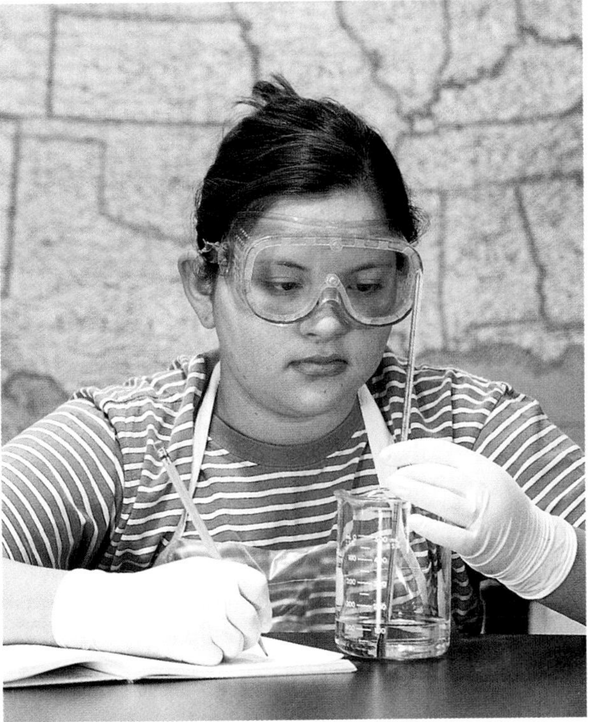

Figure 14 A thermometer measures the temperature of an object.

Scientists use a thermometer to measure temperature. Most thermometers in a laboratory are glass tubes with a bulb at the bottom end containing a liquid such as colored alcohol. The liquid rises or falls with a change in temperature. To read a glass thermometer like the thermometer in **Figure 14,** rotate it slowly until a red line appears. Read the temperature where the red line ends.

Form Operational Definitions An operational definition defines an object by how it functions, works, or behaves. For example, when you are playing hide and seek and a tree is home base, you have created an operational definition for a tree.

Objects can have more than one operational definition. For example, a ruler can be defined as a tool that measures the length of an object (how it is used). It can also be a tool with a series of marks used as a standard when measuring (how it works).

Analyze the Data

To determine the meaning of your observations and investigation results, you will need to look for patterns in the data. Then you must think critically to determine what the data mean. Scientists use several approaches when they analyze the data they have collected and recorded. Each approach is useful for identifying specific patterns.

Interpret Data The word *interpret* means "to explain the meaning of something." When analyzing data from an experiment, try to find out what the data show. Identify the control group and the test group to see whether or not changes in the independent variable have had an effect. Look for differences in the dependent variable between the control and test groups.

Classify Sorting objects or events into groups based on common features is called classifying. When classifying, first observe the objects or events to be classified. Then select one feature that is shared by some members in the group, but not by all. Place those members that share that feature in a subgroup. You can classify members into smaller and smaller subgroups based on characteristics. Remember that when you classify, you are grouping objects or events for a purpose. Keep your purpose in mind as you select the features to form groups and subgroups.

Compare and Contrast Observations can be analyzed by noting the similarities and differences between two or more objects or events that you observe. When you look at objects or events to see how they are similar, you are comparing them. Contrasting is looking for differences in objects or events.

Recognize Cause and Effect A cause is a reason for an action or condition. The effect is that action or condition. When two events happen together, it is not necessarily true that one event caused the other. Scientists must design a controlled investigation to recognize the exact cause and effect.

Draw Conclusions

When scientists have analyzed the data they collected, they proceed to draw conclusions about the data. These conclusions are sometimes stated in words similar to the hypothesis that you formed earlier. They may confirm a hypothesis, or lead you to a new hypothesis.

Infer Scientists often make inferences based on their observations. An inference is an attempt to explain observations or to indicate a cause. An inference is not a fact, but a logical conclusion that needs further investigation. For example, you may infer that a fire has caused smoke. Until you investigate, however, you do not know for sure.

Apply When you draw a conclusion, you must apply those conclusions to determine whether the data supports the hypothesis. If your data do not support your hypothesis, it does not mean that the hypothesis is wrong. It means only that the result of the investigation did not support the hypothesis. Maybe the experiment needs to be redesigned, or some of the initial observations on which the hypothesis was based were incomplete or biased. Perhaps more observation or research is needed to refine your hypothesis. A successful investigation does not always come out the way you originally predicted.

Avoid Bias Sometimes a scientific investigation involves making judgments. When you make a judgment, you form an opinion. It is important to be honest and not to allow any expectations of results to bias your judgments. This is important throughout the entire investigation, from researching to collecting data to drawing conclusions.

Communicate

The communication of ideas is an important part of the work of scientists. A discovery that is not reported will not advance the scientific community's understanding or knowledge. Communication among scientists also is important as a way of improving their investigations.

Scientists communicate in many ways, from writing articles in journals and magazines that explain their investigations and experiments, to announcing important discoveries on television and radio. Scientists also share ideas with colleagues on the Internet or present them as lectures, like the student is doing in **Figure 15.**

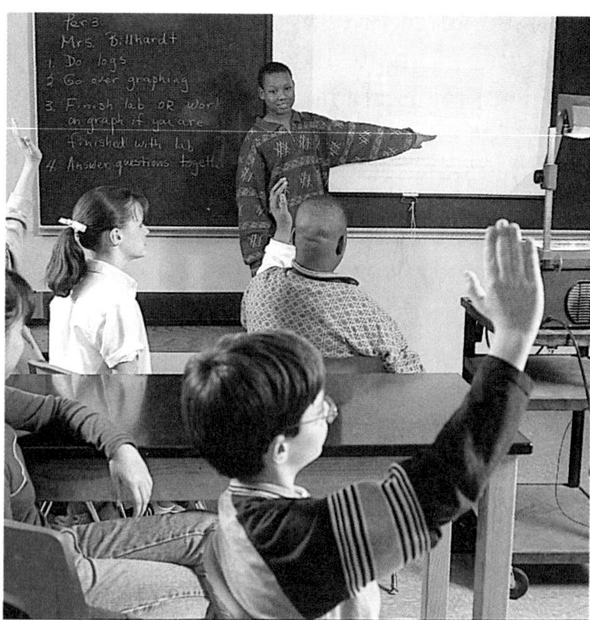

Figure 15 A student communicates to his peers about his investigation.

These safety symbols are used in laboratory and field investigations in this book to indicate possible hazards. Learn the meaning of each symbol and refer to this page often. *Remember to wash your hands thoroughly after completing lab procedures.*

PROTECTIVE EQUIPMENT

Do not begin any lab without the proper protection equipment.

GOGGLES 	Proper eye protection must be worn when performing or observing science activities which involve items or conditions as listed below.	APRON 	Wear an approved apron when using substances that could stain, wet, or destroy cloth.	SOAP 	Wash hands with soap and water before removing goggles and after all lab activities.	GLOVES 	Wear gloves when working with biological materials, chemicals, animals, or materials that can stain or irritate hands.

LABORATORY HAZARDS

Symbols	Potential Hazards	Precaution	Response
DISPOSAL	contamination of classroom or environment due to improper disposal of materials such as chemicals and live specimens	• DO NOT dispose of hazardous materials in the sink or trash can. • Dispose of wastes as directed by your teacher.	• If hazardous materials are disposed of improperly, notify your teacher immediately.
EXTREME TEMPERATURE	skin burns due to extremely hot or cold materials such as hot glass, liquids, or metals; liquid nitrogen; dry ice	• Use proper protective equipment, such as hot mitts and/or tongs, when handling objects with extreme temperatures.	• If injury occurs, notify your teacher immediately.
SHARP OBJECTS	punctures or cuts from sharp objects such as razor blades, pins, scalpels, and broken glass	• Handle glassware carefully to avoid breakage. • Walk with sharp objects pointed downward, away from you and others.	• If broken glass or injury occurs, notify your teacher immediately.
ELECTRICAL	electric shock or skin burn due to improper grounding, short circuits, liquid spills, or exposed wires	• Check condition of wires and apparatus for fraying or uninsulated wires, and broken or cracked equipment. • Use only GFCI-protected outlets	• DO NOT attempt to fix electrical problems. Notify your teacher immediately.
CHEMICAL	skin irritation or burns, breathing difficulty, and/or poisoning due to touching, swallowing, or inhalation of chemicals such as acids, bases, bleach, metal compounds, iodine, poinsettias, pollen, ammonia, acetone, nail polish remover, heated chemicals, mothballs, and any other chemicals labeled or known to be dangerous	• Wear proper protective equipment such as goggles, apron, and gloves when using chemicals. • Ensure proper room ventilation or use a fume hood when using materials that produce fumes. • NEVER smell fumes directly. • NEVER taste or eat any material in the laboratory.	• If contact occurs, immediately flush affected area with water and notify your teacher. • If a spill occurs, leave the area immediately and notify your teacher.
FLAMMABLE	unexpected fire due to liquids or gases that ignite easily such as rubbing alcohol	• Avoid open flames, sparks, or heat when flammable liquids are present.	• If a fire occurs, leave the area immediately and notify your teacher.
OPEN FLAME	burns or fire due to open flame from matches, Bunsen burners, or burning materials	• Tie back loose hair and clothing. • Keep flame away from all materials. • Follow teacher instructions when lighting and extinguishing flames. • Use proper protection, such as hot mitts or tongs, when handling hot objects.	• If a fire occurs, leave the area immediately and notify your teacher.
ANIMAL SAFETY	injury to or from laboratory animals	• Wear proper protective equipment such as gloves, apron, and goggles when working with animals. • Wash hands after handling animals.	• If injury occurs, notify your teacher immediately.
BIOLOGICAL	infection or adverse reaction due to contact with organisms such as bacteria, fungi, and biological materials such as blood, animal or plant materials	• Wear proper protective equipment such as gloves, goggles, and apron when working with biological materials. • Avoid skin contact with an organism or any part of the organism. • Wash hands after handling organisms.	• If contact occurs, wash the affected area and notify your teacher immediately.
FUME	breathing difficulties from inhalation of fumes from substances such as ammonia, acetone, nail polish remover, heated chemicals, and mothballs	• Wear goggles, apron, and gloves. • Ensure proper room ventilation or use a fume hood when using substances that produce fumes. • NEVER smell fumes directly.	• If a spill occurs, leave area and notify your teacher immediately.
IRRITANT	irritation of skin, mucous membranes, or respiratory tract due to materials such as acids, bases, bleach, pollen, mothballs, steel wool, and potassium permanganate	• Wear goggles, apron, and gloves. • Wear a dust mask to protect against fine particles.	• If skin contact occurs, immediately flush the affected area with water and notify your teacher.
RADIOACTIVE	excessive exposure from alpha, beta, and gamma particles	• Remove gloves and wash hands with soap and water before removing remainder of protective equipment.	• If cracks or holes are found in the container, notify your teacher immediately.

Safety in the Science Laboratory

Introduction to Science Safety

The science laboratory is a safe place to work if you follow standard safety procedures. Being responsible for your own safety helps to make the entire laboratory a safer place for everyone. When performing any lab, read and apply the caution statements and safety symbol listed at the beginning of the lab.

General Safety Rules

1. Complete the *Lab Safety Form* or other safety contract BEFORE starting any science lab.
2. Study the procedure. Ask your teacher any questions. Be sure you understand safety symbols shown on the page.
3. Notify your teacher about allergies or other health conditions which can affect your participation in a lab.
4. Learn and follow use and safety procedures for your equipment. If unsure, ask your teacher.

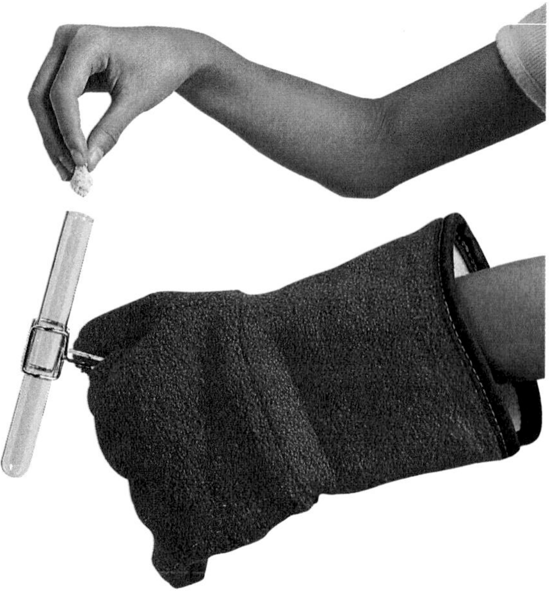

5. Never eat, drink, chew gum, apply cosmetics, or do any personal grooming in the lab. Never use lab glassware as food or drink containers. Keep your hands away from your face and mouth.
6. Know the location and proper use of the safety shower, eye wash, fire blanket, and fire alarm.

Prevent Accidents

1. Use the safety equipment provided to you. Goggles and a safety apron should be worn during investigations.
2. Do NOT use hair spray, mousse, or other flammable hair products. Tie back long hair and tie down loose clothing.
3. Do NOT wear sandals or other open-toed shoes in the lab.
4. Remove jewelry on hands and wrists. Loose jewelry, such as chains and long necklaces, should be removed to prevent them from getting caught in equipment.
5. Do not taste any substances or draw any material into a tube with your mouth.
6. Proper behavior is expected in the lab. Practical jokes and fooling around can lead to accidents and injury.
7. Keep your work area uncluttered.

Laboratory Work

1. Collect and carry all equipment and materials to your work area before beginning a lab.
2. Remain in your own work area unless given permission by your teacher to leave it.

3. Always slant test tubes away from yourself and others when heating them, adding substances to them, or rinsing them.
4. If instructed to smell a substance in a container, hold the container a short distance away and fan vapors towards your nose.
5. Do NOT substitute other chemicals/substances for those in the materials list unless instructed to do so by your teacher.
6. Do NOT take any materials or chemicals outside of the laboratory.
7. Stay out of storage areas unless instructed to be there and supervised by your teacher.

Laboratory Cleanup

1. Turn off all burners, water, and gas, and disconnect all electrical devices.
2. Clean all pieces of equipment and return all materials to their proper places.
3. Dispose of chemicals and other materials as directed by your teacher. Place broken glass and solid substances in the proper containers. Never discard materials in the sink.
4. Clean your work area.
5. Wash your hands with soap and water thoroughly BEFORE removing your goggles.

Emergencies

1. Report any fire, electrical shock, glassware breakage, spill, or injury, no matter how small, to your teacher immediately. Follow his or her instructions.
2. If your clothing should catch fire, STOP, DROP, and ROLL. If possible, smother it with the fire blanket or get under a safety shower. NEVER RUN.
3. If a fire should occur, turn off all gas and leave the room according to established procedures.
4. In most instances, your teacher will clean up spills. Do NOT attempt to clean up spills unless you are given permission and instructions to do so.
5. If chemicals come into contact with your eyes or skin, notify your teacher immediately. Use the eyewash, or flush your skin or eyes with large quantities of water.
6. The fire extinguisher and first-aid kit should only be used by your teacher unless it is an extreme emergency and you have been given permission.
7. If someone is injured or becomes ill, only a professional medical provider or someone certified in first aid should perform first-aid procedures.

Math Review

Use Fractions

A fraction compares a part to a whole. In the fraction $\frac{2}{3}$, the 2 represents the part and is the numerator. The 3 represents the whole and is the denominator.

Reduce Fractions To reduce a fraction, you must find the largest factor that is common to both the numerator and the denominator, the greatest common factor (GCF). Divide both numbers by the GCF. The fraction has then been reduced, or it is in its simplest form.

Example

Twelve of the 20 chemicals in the science lab are in powder form. What fraction of the chemicals used in the lab are in powder form?

Step 1 Write the fraction.

$\frac{\text{part}}{\text{whole}} = \frac{12}{20}$

Step 2 To find the GCF of the numerator and denominator, list all of the factors of each number.

Factors of 12: 1, 2, 3, 4, 6, 12
(the numbers that divide evenly into 12)

Factors of 20: 1, 2, 4, 5, 10, 20
(the numbers that divide evenly into 20)

Step 3 List the common factors.

1, 2, 4

Step 4 Choose the greatest factor in the list. The GCF of 12 and 20 is 4.

Step 5 Divide the numerator and denominator by the GCF.

$\frac{12 \div 4}{20 \div 4} = \frac{3}{5}$

In the lab, $\frac{3}{5}$ of the chemicals are in powder form.

Practice Problem At an amusement park, 66 of 90 rides have a height restriction. What fraction of the rides, in its simplest form, has a height restriction?

Add and Subtract Fractions with Like Denominators To add or subtract fractions with the same denominator, add or subtract the numerators and write the sum or difference over the denominator. After finding the sum or difference, find the simplest form for your fraction.

Example 1

In the forest outside your house, $\frac{1}{8}$ of the animals are rabbits, $\frac{3}{8}$ are squirrels, and the remainder are birds and insects. How many are mammals?

Step 1 Add the numerators.

$\frac{1}{8} + \frac{3}{8} = \frac{(1 + 3)}{8} = \frac{4}{8}$

Step 2 Find the GCF.

$\frac{4}{8}$ (GCF, 4)

Step 3 Divide the numerator and denominator by the GCF.

$\frac{4 \div 4}{8 \div 4} = \frac{1}{2}$

$\frac{1}{2}$ of the animals are mammals.

Example 2

If $\frac{7}{16}$ of the Earth is covered by freshwater, and $\frac{1}{16}$ of that is in glaciers, how much freshwater is not frozen?

Step 1 Subtract the numerators.

$\frac{7}{16} - \frac{1}{16} = \frac{(7 - 1)}{16} = \frac{6}{16}$

Step 2 Find the GCF.

$\frac{6}{16}$ (GCF, 2)

Step 3 Divide the numerator and denominator by the GCF.

$\frac{6 \div 2}{16 \div 2} = \frac{3}{8}$

$\frac{3}{8}$ of the freshwater is not frozen.

Practice Problem A bicycle rider is riding at a rate of 15 km/h for $\frac{4}{9}$ of his ride, 10 km/h for $\frac{2}{9}$ of his ride, and 8 km/h for the remainder of the ride. How much of his ride is he riding at a rate greater than 8 km/h?

Add and Subtract Fractions with Unlike Denominators To add or subtract fractions with unlike denominators, first find the least common denominator (LCD). This is the smallest number that is a common multiple of both denominators. Rename each fraction with the LCD, and then add or subtract. Find the simplest form if necessary.

Example 1

A chemist makes a paste that is $\frac{1}{2}$ table salt (NaCl), $\frac{1}{3}$ sugar ($C_6H_{12}O_6$), and the remainder is water (H_2O). How much of the paste is a solid?

Step 1 Find the LCD of the fractions.

$\frac{1}{2} + \frac{1}{3}$ (LCD, 6)

Step 2 Rename each numerator and each denominator with the LCD.

Step 3 Add the numerators.

$\frac{3}{6} + \frac{2}{6} = \frac{(3 + 2)}{6} = \frac{5}{6}$

$\frac{5}{6}$ of the paste is a solid.

Example 2

The average precipitation in Grand Junction, CO, is $\frac{7}{10}$ inch in November, and $\frac{3}{5}$ inch in December. What is the total average precipitation?

Step 1 Find the LCD of the fractions.

$\frac{7}{10} + \frac{3}{5}$ (LCD, 10)

Step 2 Rename each numerator and each denominator with the LCD.

Step 3 Add the numerators.

$\frac{7}{10} + \frac{6}{10} = \frac{(7 + 6)}{10} = \frac{13}{10}$

$\frac{13}{10}$ inches total precipitation, or $1\frac{3}{10}$ inches.

Practice Problem On an electric bill, about $\frac{1}{8}$ of the energy is from solar energy and about $\frac{1}{10}$ is from wind power. How much of the total bill is from solar energy and wind power combined?

Example 3

In your body, $\frac{7}{10}$ of your muscle contractions are involuntary (cardiac and smooth muscle tissue). Smooth muscle makes $\frac{3}{15}$ of your muscle contractions. How many of your muscle contractions are made by cardiac muscle?

Step 1 Find the LCD of the fractions.

$\frac{7}{10} - \frac{3}{15}$ (LCD, 30)

Step 2 Rename each numerator and each denominator with the LCD.

$\frac{7 \times 3}{10 \times 3} = \frac{21}{30}$

$\frac{3 \times 2}{15 \times 2} = \frac{6}{30}$

Step 3 Subtract the numerators.

$\frac{21}{30} - \frac{6}{30} = \frac{(21 - 6)}{30} = \frac{15}{30}$

Step 4 Find the GCF.

$\frac{15}{30}$ (GCF, 15)

$\frac{1}{2}$

$\frac{1}{2}$ of all muscle contractions are cardiac muscle.

Example 4

Tony wants to make cookies that call for $\frac{3}{4}$ of a cup of flour, but he only has $\frac{1}{3}$ of a cup. How much more flour does he need?

Step 1 Find the LCD of the fractions.

$\frac{3}{4} - \frac{1}{3}$ (LCD, 12)

Step 2 Rename each numerator and each denominator with the LCD.

$\frac{3 \times 3}{4 \times 3} = \frac{9}{12}$

$\frac{1 \times 4}{3 \times 4} = \frac{4}{12}$

Step 3 Subtract the numerators.

$\frac{9}{12} - \frac{4}{12} = \frac{(9 - 4)}{12} = \frac{5}{12}$

$\frac{5}{12}$ of a cup of flour

Practice Problem Using the information provided to you in Example 3 above, determine how many muscle contractions are voluntary (skeletal muscle).

Multiply Fractions To multiply with fractions, multiply the numerators and multiply the denominators. Find the simplest form if necessary.

Example

Multiply $\frac{3}{5}$ by $\frac{1}{3}$.

Step 1 Multiply the numerators and denominators.

$\frac{3}{5} \times \frac{1}{3} = \frac{(3 \times 1)}{(5 \times 3)}\ \frac{3}{15}$

Step 2 Find the GCF.

$\frac{3}{15}$ (GCF, 3)

Step 3 Divide the numerator and denominator by the GCF.

$\frac{3 \div 3}{15 \div 3} = \frac{1}{5}$

$\frac{3}{5}$ multiplied by $\frac{1}{3}$ is $\frac{1}{5}$.

Practice Problem Multiply $\frac{3}{14}$ by $\frac{5}{16}$.

Find a Reciprocal Two numbers whose product is 1 are called multiplicative inverses, or reciprocals.

Example

Find the reciprocal of $\frac{3}{8}$.

Step 1 Inverse the fraction by putting the denominator on top and the numerator on the bottom.

$\frac{8}{3}$

The reciprocal of $\frac{3}{8}$ is $\frac{8}{3}$.

Practice Problem Find the reciprocal of $\frac{4}{9}$.

Divide Fractions To divide one fraction by another fraction, multiply the dividend by the reciprocal of the divisor. Find the simplest form if necessary.

Example 1

Divide $\frac{1}{9}$ by $\frac{1}{3}$.

Step 1 Find the reciprocal of the divisor.

The reciprocal of $\frac{1}{3}$ is $\frac{3}{1}$.

Step 2 Multiply the dividend by the reciprocal of the divisor.

$\frac{\frac{1}{9}}{\frac{1}{3}} = \frac{1}{9} \times \frac{3}{1} = \frac{(1 \times 3)}{(9 \times 1)} = \frac{3}{9}$

Step 3 Find the GCF.

$\frac{3}{9}$ (GCF, 3)

Step 4 Divide the numerator and denominator by the GCF.

$\frac{3 \div 3}{9 \div 3} = \frac{1}{3}$

$\frac{1}{9}$ divided by $\frac{1}{3}$ is $\frac{1}{3}$.

Example 2

Divide $\frac{3}{5}$ by $\frac{1}{4}$.

Step 1 Find the reciprocal of the divisor.

The reciprocal of $\frac{1}{4}$ is $\frac{4}{1}$.

Step 2 Multiply the dividend by the reciprocal of the divisor.

$\frac{\frac{3}{5}}{\frac{1}{4}} = \frac{3}{5} \times \frac{4}{1} = \frac{(3 \times 4)}{(5 \times 1)} = \frac{12}{5}$

$\frac{3}{5}$ divided by $\frac{1}{4}$ is $\frac{12}{5}$ or $2\frac{2}{5}$.

Practice Problem Divide $\frac{3}{11}$ by $\frac{7}{10}$.

Use Ratios

When you compare two numbers by division, you are using a ratio. Ratios can be written 3 to 5, 3:5, or $\frac{3}{5}$. Ratios, like fractions, also can be written in simplest form.

Ratios can represent one type of probability, called odds. This is a ratio that compares the number of ways a certain outcome occurs to the number of possible outcomes. For example, if you flip a coin 100 times, what are the odds that it will come up heads? There are two possible outcomes, heads or tails, so the odds of coming up heads are 50:100. Another way to say this is that 50 out of 100 times the coin will come up heads. In its simplest form, the ratio is 1:2.

Example 1

A chemical solution contains 40 g of salt and 64 g of baking soda. What is the ratio of salt to baking soda as a fraction in simplest form?

Step 1 Write the ratio as a fraction.

$$\frac{\text{salt}}{\text{baking soda}} = \frac{40}{64}$$

Step 2 Express the fraction in simplest form. The GCF of 40 and 64 is 8.

$$\frac{40}{64} = \frac{40 \div 8}{64 \div 8} = \frac{5}{8}$$

The ratio of salt to baking soda in the sample is 5:8.

Example 2

Sean rolls a 6-sided die 6 times. What are the odds that the side with a 3 will show?

Step 1 Write the ratio as a fraction.

$$\frac{\text{number of sides with a 3}}{\text{number of possible sides}} = \frac{1}{6}$$

Step 2 Multiply by the number of attempts.

$\frac{1}{6} \times 6$ attempts $= \frac{6}{6}$ attempts $= 1$ attempt

1 attempt out of 6 will show a 3.

Practice Problem Two metal rods measure 100 cm and 144 cm in length. What is the ratio of their lengths in simplest form?

Use Decimals

A fraction with a denominator that is a power of ten can be written as a decimal. For example, 0.27 means $\frac{27}{100}$. The decimal point separates the ones place from the tenths place.

Any fraction can be written as a decimal using division. For example, the fraction $\frac{5}{8}$ can be written as a decimal by dividing 5 by 8. Written as a decimal, it is 0.625.

Add or Subtract Decimals When adding and subtracting decimals, line up the decimal points before carrying out the operation.

Example 1

Find the sum of 47.68 and 7.80.

Step 1 Line up the decimal places when you write the numbers.

```
  47.68
+  7.80
```

Step 2 Add the decimals.

```
  1 1
  47.68
+  7.80
  55.48
```

The sum of 47.68 and 7.80 is 55.48.

Example 2

Find the difference of 42.17 and 15.85.

Step 1 Line up the decimal places when you write the number.

```
  42.17
 -15.85
```

Step 2 Subtract the decimals.

```
  3 11 1
  42.17
 -15.85
  26.32
```

The difference of 42.17 and 15.85 is 26.32.

Practice Problem Find the sum of 1.245 and 3.842.

Science Skill Handbook
Math Skill Handbook
Foldables Handbook
Reference Handbook
Glossary/Glosario
Index

Multiply Decimals To multiply decimals, multiply the numbers like numbers without decimal points. Count the decimal places in each factor. The product will have the same number of decimal places as the sum of the decimal places in the factors.

Example

Multiply 2.4 by 5.9.

Step 1 Multiply the factors like two whole numbers.

$24 \times 59 = 1416$

Step 2 Find the sum of the number of decimal places in the factors. Each factor has one decimal place, for a sum of two decimal places.

Step 3 The product will have two decimal places.

14.16

The product of 2.4 and 5.9 is 14.16.

Practice Problem Multiply 4.6 by 2.2.

Divide Decimals When dividing decimals, change the divisor to a whole number. To do this, multiply both the divisor and the dividend by the same power of ten. Then place the decimal point in the quotient directly above the decimal point in the dividend. Then divide as you do with whole numbers.

Example

Divide 8.84 by 3.4.

Step 1 Multiply both factors by 10.

$3.4 \times 10 = 34, 8.84 \times 10 = 88.4$

Step 2 Divide 88.4 by 34.

$$\begin{array}{r} 2.6 \\ 34\overline{)88.4} \\ -68 \\ \hline 204 \\ -204 \\ \hline 0 \end{array}$$

8.84 divided by 3.4 is 2.6.

Practice Problem Divide 75.6 by 3.6.

Use Proportions

An equation that shows that two ratios are equivalent is a proportion. The ratios $\frac{2}{4}$ and $\frac{5}{10}$ are equivalent, so they can be written as $\frac{2}{4} = \frac{5}{10}$. This equation is a proportion.

When two ratios form a proportion, the cross products are equal. To find the cross products in the proportion $\frac{2}{4} = \frac{5}{10}$, multiply the 2 and the 10, and the 4 and the 5. Therefore $2 \times 10 = 4 \times 5$, or $20 = 20$.

Because you know that both ratios are equal, you can use cross products to find a missing term in a proportion. This is known as solving the proportion.

Example

The heights of a tree and a pole are proportional to the lengths of their shadows. The tree casts a shadow of 24 m when a 6-m pole casts a shadow of 4 m. What is the height of the tree?

Step 1 Write a proportion.

$$\frac{\text{height of tree}}{\text{height of pole}} = \frac{\text{length of tree's shadow}}{\text{length of pole's shadow}}$$

Step 2 Substitute the known values into the proportion. Let h represent the unknown value, the height of the tree.

$\frac{h}{6} \times \frac{24}{4}$

Step 3 Find the cross products.

$h \times 4 = 6 \times 24$

Step 4 Simplify the equation.

$4h \times 144$

Step 5 Divide each side by 4.

$\frac{4h}{4} \times \frac{144}{4}$

$h = 36$

The height of the tree is 36 m.

Practice Problem The ratios of the weights of two objects on the Moon and on Earth are in proportion. A rock weighing 3 N on the Moon weighs 18 N on Earth. How much would a rock that weighs 5 N on the Moon weigh on Earth?

Use Percentages

The word *percent* means "out of one hundred." It is a ratio that compares a number to 100. Suppose you read that 77 percent of Earth's surface is covered by water. That is the same as reading that the fraction of Earth's surface covered by water is $\frac{77}{100}$. To express a fraction as a percent, first find the equivalent decimal for the fraction. Then, multiply the decimal by 100 and add the percent symbol.

Example 1

Express $\frac{13}{20}$ as a percent.

Step 1 Find the equivalent decimal for the fraction.

$$\begin{array}{r} 0.65 \\ 20\overline{)13.00} \\ \underline{12\ 0} \\ 1\ 00 \\ \underline{1\ 00} \\ 0 \end{array}$$

Step 2 Rewrite the fraction $\frac{13}{20}$ as 0.65.

Step 3 Multiply 0.65 by 100 and add the % symbol.

$0.65 \times 100 = 65 = 65\%$

So, $\frac{13}{20} = 65\%$.

This also can be solved as a proportion.

Example 2

Express $\frac{13}{20}$ as a percent.

Step 1 Write a proportion.

$\frac{13}{20} = \frac{x}{100}$

Step 2 Find the cross products.

$1300 = 20x$

Step 3 Divide each side by 20.

$\frac{1300}{20} = \frac{20x}{20}$

$65\% = x$

Practice Problem In one year, 73 of 365 days were rainy in one city. What percent of the days in that city were rainy?

Solve One-Step Equations

A statement that two expressions are equal is an equation. For example, $A = B$ is an equation that states that A is equal to B.

An equation is solved when a variable is replaced with a value that makes both sides of the equation equal. To make both sides equal the inverse operation is used. Addition and subtraction are inverses, and multiplication and division are inverses.

Example 1

Solve the equation $x - 10 = 35$.

Step 1 Find the solution by adding 10 to each side of the equation.

$$x - 10 = 35$$
$$x - 10 + 10 = 35 - 10$$
$$x = 45$$

Step 2 Check the solution.

$$x - 10 = 35$$
$$45 - 10 = 35$$
$$35 = 35$$

Both sides of the equation are equal, so $x = 45$.

Example 2

In the formula $a = bc$, find the value of c if $a = 20$ and $b = 2$.

Step 1 Rearrange the formula so the unknown value is by itself on one side of the equation by dividing both sides by b.

$$a = bc$$
$$\frac{a}{b} = \frac{bc}{b}$$
$$\frac{a}{b} = c$$

Step 2 Replace the variables a and b with the values that are given.

$$\frac{a}{b} = c$$
$$\frac{20}{2} = c$$
$$10 = c$$

Step 3 Check the solution.

$$a = bc$$
$$20 = 2 \times 10$$
$$20 = 20$$

Both sides of the equation are equal, so $c = 10$ is the solution when $a = 20$ and $b = 2$.

Practice Problem In the formula $h = gd$, find the value of d if $g = 12.3$ and $h = 17.4$.

SCIENCE SKILL HANDBOOK
MATH SKILL HANDBOOK
FOLDABLES HANDBOOK
REFERENCE HANDBOOK
GLOSSARY/GLOSARIO
INDEX

Use Statistics

The branch of mathematics that deals with collecting, analyzing, and presenting data is statistics. In statistics, there are three common ways to summarize data with a single number—the mean, the median, and the mode.

The **mean** of a set of data is the arithmetic average. It is found by adding the numbers in the data set and dividing by the number of items in the set.

The **median** is the middle number in a set of data when the data are arranged in numerical order. If there were an even number of data points, the median would be the mean of the two middle numbers.

The **mode** of a set of data is the number or item that appears most often.

Another number that often is used to describe a set of data is the range. The **range** is the difference between the largest number and the smallest number in a set of data.

Example

The speeds (in m/s) for a race car during five different time trials are 39, 37, 44, 36, and 44.

To find the mean:

Step 1 Find the sum of the numbers.

$39 + 37 + 44 + 36 + 44 = 200$

Step 2 Divide the sum by the number of items, which is 5.

$200 \div 5 = 40$

The mean is 40 m/s.

To find the median:

Step 1 Arrange the measures from least to greatest.

36, 37, 39, 44, 44

Step 2 Determine the middle measure.

36, 37, $\underline{39}$, 44, 44

The median is 39 m/s.

To find the mode:

Step 1 Group the numbers that are the same together.

44, 44, 36, 37, 39

Step 2 Determine the number that occurs most in the set.

$\underline{44, 44}$, 36, 37, 39

The mode is 44 m/s.

To find the range:

Step 1 Arrange the measures from greatest to least.

44, 44, 39, 37, 36

Step 2 Determine the greatest and least measures in the set.

$\underline{44}$, 44, 39, 37, $\underline{36}$

Step 3 Find the difference between the greatest and least measures.

$44 - 36 = 8$

The range is 8 m/s.

Practice Problem Find the mean, median, mode, and range for the data set 8, 4, 12, 8, 11, 14, 16.

A **frequency table** shows how many times each piece of data occurs, usually in a survey. **Table 1** below shows the results of a student survey on favorite color.

Table 1 Student Color Choice

Color	Tally	Frequency
red	IIII	4
blue	~~IIII~~	5
black	II	2
green	III	3
purple	~~IIII~~ II	7
yellow	~~IIII~~ I	6

Based on the frequency table data, which color is the favorite?

Use Geometry

The branch of mathematics that deals with the measurement, properties, and relationships of points, lines, angles, surfaces, and solids is called geometry.

Perimeter The **perimeter** (P) is the distance around a geometric figure. To find the perimeter of a rectangle, add the length and width and multiply that sum by two, or $2(l + w)$. To find perimeters of irregular figures, add the length of the sides.

Example 1

Find the perimeter of a rectangle that is 3 m long and 5 m wide.

Step 1 You know that the perimeter is 2 times the sum of the width and length.

$P = 2(3 \text{ m} + 5 \text{ m})$

Step 2 Find the sum of the width and length.

$P = 2(8 \text{ m})$

Step 3 Multiply by 2.

$P = 16 \text{ m}$

The perimeter is 16 m.

Example 2

Find the perimeter of a shape with sides measuring 2 cm, 5 cm, 6 cm, 3 cm.

Step 1 You know that the perimeter is the sum of all the sides.

$P = 2 + 5 + 6 + 3$

Step 2 Find the sum of the sides.

$P = 2 + 5 + 6 + 3$

$P = 16$

The perimeter is 16 cm.

Practice Problem Find the perimeter of a rectangle with a length of 18 m and a width of 7 m.

Practice Problem Find the perimeter of a triangle measuring 1.6 cm by 2.4 cm by 2.4 cm.

Area of a Rectangle The **area** (A) is the number of square units needed to cover a surface. To find the area of a rectangle, multiply the length times the width, or $l \times w$. When finding area, the units also are multiplied. Area is given in square units.

Example

Find the area of a rectangle with a length of 1 cm and a width of 10 cm.

Step 1 You know that the area is the length multiplied by the width.

$A = (1 \text{ cm} \times 10 \text{ cm})$

Step 2 Multiply the length by the width. Also multiply the units.

$A = 10 \text{ cm}^2$

The area is 10 cm^2.

Practice Problem Find the area of a square whose sides measure 4 m.

Area of a Triangle To find the area of a triangle, use the formula:

$A = \frac{1}{2}(\text{base} \times \text{height})$

The base of a triangle can be any of its sides. The height is the perpendicular distance from a base to the opposite endpoint, or vertex.

Example

Find the area of a triangle with a base of 18 m and a height of 7 m.

Step 1 You know that the area is $\frac{1}{2}$ the base times the height.

$A = \frac{1}{2}(18 \text{ m} \times 7 \text{ m})$

Step 2 Multiply $\frac{1}{2}$ by the product of 18×7. Multiply the units.

$A = \frac{1}{2}(126 \text{ m}^2)$

$A = 63 \text{ m}^2$

The area is 63 m^2.

Practice Problem Find the area of a triangle with a base of 27 cm and a height of 17 cm.

Circumference of a Circle The **diameter** (d) of a circle is the distance across the circle through its center, and the **radius** (r) is the distance from the center to any point on the circle. The radius is half of the diameter. The distance around the circle is called the **circumference** (C). The formula for finding the circumference is:

$C = 2\pi r$ or $C = \pi d$

The circumference divided by the diameter is always equal to 3.1415926... This nonterminating and nonrepeating number is represented by the Greek letter π (pi). An approximation often used for π is 3.14.

Example 1

Find the circumference of a circle with a radius of 3 m.

Step 1 You know the formula for the circumference is 2 times the radius times π.

$C = 2\pi(3)$

Step 2 Multiply 2 times the radius.

$C = 6\pi$

Step 3 Multiply by π.

$C \approx 19$ m

The circumference is about 19 m.

Example 2

Find the circumference of a circle with a diameter of 24.0 cm.

Step 1 You know the formula for the circumference is the diameter times π.

$C = \pi(24.0)$

Step 2 Multiply the diameter by π.

$C \approx 75.4$ cm

The circumference is about 75.4 cm.

Practice Problem Find the circumference of a circle with a radius of 19 cm.

Area of a Circle The formula for the area of a circle is: $A = \pi r^2$

Example 1

Find the area of a circle with a radius of 4.0 cm.

Step 1 $A = \pi(4.0)^2$

Step 2 Find the square of the radius.

$A = 16\pi$

Step 3 Multiply the square of the radius by π.

$A \approx 50\ \text{cm}^2$

The area of the circle is about 50 cm^2.

Example 2

Find the area of a circle with a radius of 225 m.

Step 1 $A = \pi(225)^2$

Step 2 Find the square of the radius.

$A = 50625\pi$

Step 3 Multiply the square of the radius by π.

$A \approx 159043.1$

The area of the circle is about 159043.1 m^2.

Example 3

Find the area of a circle whose diameter is 20.0 mm.

Step 1 Remember that the radius is half of the diameter.

$A = \pi\left(\frac{20.0}{2}\right)^2$

Step 2 Find the radius.

$A = \pi(10.0)^2$

Step 3 Find the square of the radius.

$A = 100\pi$

Step 4 Multiply the square of the radius by π.

$A \approx 314\ \text{mm}^2$

The area of the circle is about 314 mm^2.

Practice Problem Find the area of a circle with a radius of 16 m.

Volume The measure of space occupied by a solid is the **volume** (V). To find the volume of a rectangular solid multiply the length times width times height, or $V = l \times w \times h$. It is measured in cubic units, such as cubic centimeters (cm^3).

Example

Find the volume of a rectangular solid with a length of 2.0 m, a width of 4.0 m, and a height of 3.0 m.

Step 1 You know the formula for volume is the length times the width times the height.

$V = 2.0\text{ m} \times 4.0\text{ m} \times 3.0\text{ m}$

Step 2 Multiply the length times the width times the height.

$V = 24\text{ m}^3$

The volume is 24 m^3.

Practice Problem Find the volume of a rectangular solid that is 8 m long, 4 m wide, and 4 m high.

To find the volume of other solids, multiply the area of the base times the height.

Example 1

Find the volume of a solid that has a triangular base with a length of 8.0 m and a height of 7.0 m. The height of the entire solid is 15.0 m.

Step 1 You know that the base is a triangle, and the area of a triangle is $\frac{1}{2}$ the base times the height, and the volume is the area of the base times the height.

$V = \left[\frac{1}{2}(b \times h)\right] \times 15$

Step 2 Find the area of the base.

$V = \left[\frac{1}{2}(8 \times 7)\right] \times 15$

$V = \left(\frac{1}{2} \times 56\right) \times 15$

Step 3 Multiply the area of the base by the height of the solid.

$V = 28 \times 15$

$V = 420\text{ m}^3$

The volume is 420 m^3.

Example 2

Find the volume of a cylinder that has a base with a radius of 12.0 cm, and a height of 21.0 cm.

Step 1 You know that the base is a circle, and the area of a circle is the square of the radius times π, and the volume is the area of the base times the height.

$V = (\pi r^2) \times 21$

$V = (\pi 12^2) \times 21$

Step 2 Find the area of the base.

$V = 144\pi \times 21$

$V = 452 \times 21$

Step 3 Multiply the area of the base by the height of the solid.

$V \approx 9{,}500\text{ cm}^3$

The volume is about 9,500 cm^3.

Example 3

Find the volume of a cylinder that has a diameter of 15 mm and a height of 4.8 mm.

Step 1 You know that the base is a circle with an area equal to the square of the radius times π. The radius is one-half the diameter. The volume is the area of the base times the height.

$V = (\pi r^2) \times 4.8$

$V = \left[\pi\left(\frac{1}{2} \times 15\right)^2\right] \times 4.8$

$V = (\pi 7.5^2) \times 4.8$

Step 2 Find the area of the base.

$V = 56.25\pi \times 4.8$

$V \approx 176.71 \times 4.8$

Step 3 Multiply the area of the base by the height of the solid.

$V \approx 848.2$

The volume is about 848.2 mm^3.

Practice Problem Find the volume of a cylinder with a diameter of 7 cm in the base and a height of 16 cm.

Science Applications

Measure in SI

The metric system of measurement was developed in 1795. A modern form of the metric system, called the International System (SI), was adopted in 1960 and provides the standard measurements that all scientists around the world can understand.

The SI system is convenient because unit sizes vary by powers of 10. Prefixes are used to name units. Look at **Table 2** for some common SI prefixes and their meanings.

Table 2 Common SI Prefixes

Prefix	Symbol	Meaning	
kilo-	k	1,000	thousandth
hecto-	h	100	hundred
deka-	da	10	ten
deci-	d	0.1	tenth
centi-	c	0.01	hundreth
milli-	m	0.001	thousandth

Example

How many grams equal one kilogram?

Step 1 Find the prefix *kilo-* in **Table 2.**

Step 2 Using **Table 2,** determine the meaning of *kilo-*. According to the table, it means 1,000. When the prefix *kilo-* is added to a unit, it means that there are 1,000 of the units in a "kilounit."

Step 3 Apply the prefix to the units in the question. The units in the question are grams. There are 1,000 grams in a kilogram.

Practice Problem Is a milligram larger or smaller than a gram? How many of the smaller units equal one larger unit? What fraction of the larger unit does one smaller unit represent?

Dimensional Analysis

Convert SI Units In science, quantities such as length, mass, and time sometimes are measured using different units. A process called dimensional analysis can be used to change one unit of measure to another. This process involves multiplying your starting quantity and units by one or more conversion factors. A conversion factor is a ratio equal to one and can be made from any two equal quantities with different units. If 1,000 mL equal 1 L then two ratios can be made.

$$\frac{1{,}000 \text{ mL}}{1 \text{ L}} = \frac{1 \text{ L}}{1{,}000 \text{ mL}} = 1$$

One can convert between units in the SI system by using the equivalents in **Table 2** to make conversion factors.

Example

How many cm are in 4 m?

Step 1 Write conversion factors for the units given. From **Table 2,** you know that 100 cm = 1 m. The conversion factors are

$$\frac{100 \text{ cm}}{1 \text{ m}} \text{ and } \frac{1 \text{ m}}{100 \text{ cm}}$$

Step 2 Decide which conversion factor to use. Select the factor that has the units you are converting from (m) in the denominator and the units you are converting to (cm) in the numerator.

$$\frac{100 \text{ cm}}{1 \text{ m}}$$

Step 3 Multiply the starting quantity and units by the conversion factor. Cancel the starting units with the units in the denominator. There are 400 cm in 4 m.

$$4 \cancel{\text{m}} = \frac{100 \text{ cm}}{1 \cancel{\text{m}}} = 400 \text{ cm}$$

Practice Problem How many milligrams are in one kilogram? (Hint: You will need to use two conversion factors from **Table 2.**)

Table 3 Unit System Equivalents

Type of Measurement	Equivalent
Length	1 in = 2.54 cm 1 yd = 0.91 m 1 mi = 1.61 km
Mass and weight*	1 oz = 28.35 g 1 lb = 0.45 kg 1 ton (short) = 0.91 tonnes (metric tons) 1 lb = 4.45 N
Volume	1 in^3 = 16.39 cm^3 1 qt = 0.95 L 1 gal = 3.78 L
Area	1 in^2 = 6.45 cm^2 1 yd^2 = 0.83 m^2 1 mi^2 = 2.59 km^2 1 acre = 0.40 hectares
Temperature	$°C = \frac{(°F - 32)}{1.8}$ K = °C + 273

*Weight is measured in standard Earth gravity.

Convert Between Unit Systems **Table 3** gives a list of equivalents that can be used to convert between English and SI units.

Example

If a meterstick has a length of 100 cm, how long is the meterstick in inches?

Step 1 Write the conversion factors for the units given. From **Table 3,** 1 in = 2.54 cm.

$$\frac{1 \text{ in}}{2.54 \text{ cm}} \textit{ and } \frac{2.54 \text{ cm}}{1 \text{ in}}$$

Step 2 Determine which conversion factor to use. You are converting from cm to in. Use the conversion factor with cm on the bottom.

$$\frac{1 \text{ in}}{2.54 \text{ cm}}$$

Step 3 Multiply the starting quantity and units by the conversion factor. Cancel the starting units with the units in the denominator. Round your answer to the nearest tenth.

$$100 \cancel{\text{cm}} \times \frac{1 \text{ in}}{2.54 \cancel{\text{cm}}} = 39.37 \text{ in}$$

The meterstick is about 39.4 in long.

Practice Problem 1 A book has a mass of 5 lb. What is the mass of the book in kg?

Practice Problem 2 Use the equivalent for in and cm (1 in = 2.54 cm) to show how 1 $in^3 \approx$ 16.39 cm^3.

Precision and Significant Digits

When you make a measurement, the value you record depends on the precision of the measuring instrument. This precision is represented by the number of significant digits recorded in the measurement. When counting the number of significant digits, all digits are counted except zeros at the end of a number with no decimal point such as 2,050, and zeros at the beginning of a decimal such as 0.03020. When adding or subtracting numbers with different precision, round the answer to the smallest number of decimal places of any number in the sum or difference. When multiplying or dividing, the answer is rounded to the smallest number of significant digits of any number being multiplied or divided.

Example

The lengths 5.28 and 5.2 are measured in meters. Find the sum of these lengths and record your answer using the correct number of significant digits.

Step 1 Find the sum.

5.28 m	2 digits after the decimal
+ 5.2 m	1 digit after the decimal
10.48 m	

Step 2 Round to one digit after the decimal because the least number of digits after the decimal of the numbers being added is 1.

The sum is 10.5 m.

Practice Problem 1 How many significant digits are in the measurement 7,071,301 m? How many significant digits are in the measurement 0.003010 g?

Practice Problem 2 Multiply 5.28 and 5.2 using the rule for multiplying and dividing. Record the answer using the correct number of significant digits.

Scientific Notation

Many times numbers used in science are very small or very large. Because these numbers are difficult to work with scientists use scientific notation. To write numbers in scientific notation, move the decimal point until only one non-zero digit remains on the left. Then count the number of places you moved the decimal point and use that number as a power of ten. For example, the average distance from the Sun to Mars is 227,800,000,000 m. In scientific notation, this distance is 2.278×10^{11} m. Because you moved the decimal point to the left, the number is a positive power of ten.

The mass of an electron is about 0.000 000 000 000 000 000 000 000 000 000 911 kg. Expressed in scientific notation, this mass is 9.11×10^{-31} kg. Because the decimal point was moved to the right, the number is a negative power of ten.

Example

Earth is 149,600,000 km from the Sun. Express this in scientific notation.

Step 1 Move the decimal point until one non-zero digit remains on the left.

1.496 000 00

Step 2 Count the number of decimal places you have moved. In this case, eight.

Step 2 Show that number as a power of ten, 10^8.

Earth is 1.496×10^8 km from the Sun.

Practice Problem 1 How many significant digits are in 149,600,000 km? How many significant digits are in 1.496×10^8 km?

Practice Problem 2 Parts used in a high performance car must be measured to 7×10^{-6} m. Express this number as a decimal.

Practice Problem 3 A CD is spinning at 539 revolutions per minute. Express this number in scientific notation.

Make and Use Graphs

Data in tables can be displayed in a graph—a visual representation of data. Common graph types include line graphs, bar graphs, and circle graphs.

Line Graph A line graph shows a relationship between two variables that change continuously. The independent variable is changed and is plotted on the x-axis. The dependent variable is observed, and is plotted on the *y*-axis.

Example

Draw a line graph of the data below from a cyclist in a long-distance race.

Table 4 Bicycle Race Data

Time (h)	Distance (km)
0	0
1	8
2	16
3	24
4	32
5	40

Step 1 Determine the *x*-axis and *y*-axis variables. Time varies independently of distance and is plotted on the *x*-axis. Distance is dependent on time and is plotted on the *y*-axis.

Step 2 Determine the scale of each axis. The *x*-axis data ranges from 0 to 5. The *y*-axis data ranges from 0 to 50.

Step 3 Using graph paper, draw and label the axes. Include units in the labels.

Step 4 Draw a point at the intersection of the time value on the *x*-axis and corresponding distance value on the *y*-axis. Connect the points and label the graph with a title, as shown in **Figure 8.**

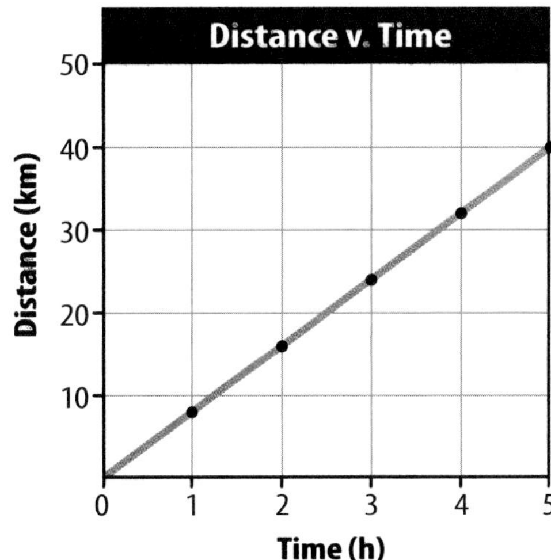

Figure 8 This line graph shows the relationship between distance and time during a bicycle ride.

Practice Problem A puppy's shoulder height is measured during the first year of her life. The following measurements were collected: (3 mo, 52 cm), (6 mo, 72 cm), (9 mo, 83 cm), (12 mo, 86 cm). Graph this data.

Find a Slope The slope of a straight line is the ratio of the vertical change, rise, to the horizontal change, run.

$$\text{Slope} = \frac{\text{vertical change (rise)}}{\text{horizontal change (run)}} = \frac{\text{change in } y}{\text{change in } x}$$

Example

Find the slope of the graph in **Figure 8**.

Step 1 You know that the slope is the change in *y* divided by the change in *x*.

$$\text{Slope} = \frac{\text{change in } y}{\text{change in } x}$$

Step 2 Determine the data points you will be using. For a straight line, choose the two sets of points that are the farthest apart.

$$\text{Slope} = \frac{(40 - 0)\text{ km}}{(5 - 0)\text{ h}}$$

Step 3 Find the change in *y* and *x*.

$$\text{Slope} = \frac{40\text{ km}}{5\text{ h}}$$

Step 4 Divide the change in *y* by the change in *x*.

$$\text{Slope} = \frac{8\text{ km}}{\text{h}}$$

The slope of the graph is 8 km/h.

Bar Graph To compare data that does not change continuously you might choose a bar graph. A bar graph uses bars to show the relationships between variables. The x-axis variable is divided into parts. The parts can be numbers such as years, or a category such as a type of animal. The y-axis is a number and increases continuously along the axis.

Example

A recycling center collects 4.0 kg of aluminum on Monday, 1.0 kg on Wednesday, and 2.0 kg on Friday. Create a bar graph of this data.

Step 1 Select the x-axis and y-axis variables. The measured numbers (the masses of aluminum) should be placed on the y-axis. The variable divided into parts (collection days) is placed on the x-axis.

Step 2 Create a graph grid like you would for a line graph. Include labels and units.

Step 3 For each measured number, draw a vertical bar above the x-axis value up to the y-axis value. For the first data point, draw a vertical bar above Monday up to 4.0 kg.

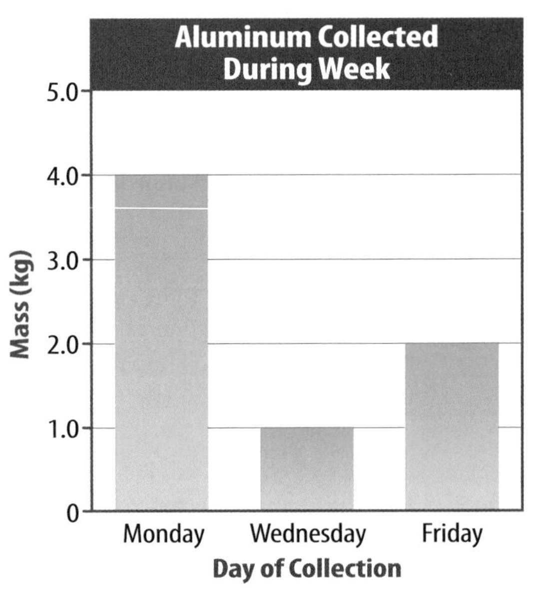

Practice Problem Draw a bar graph of the gases in air: 78% nitrogen, 21% oxygen, 1% other gases.

Circle Graph To display data as parts of a whole, you might use a circle graph. A circle graph is a circle divided into sections that represent the relative size of each piece of data. The entire circle represents 100%, half represents 50%, and so on.

Example

Air is made up of 78% nitrogen, 21% oxygen, and 1% other gases. Display the composition of air in a circle graph.

Step 1 Multiply each percent by 360° and divide by 100 to find the angle of each section in the circle.

$$78\% \times \frac{360°}{100} = 280.8°$$

$$21\% \times \frac{360°}{100} = 75.6°$$

$$1\% \times \frac{360°}{100} = 3.6°$$

Step 2 Use a compass to draw a circle and to mark the center of the circle. Draw a straight line from the center to the edge of the circle.

Step 3 Use a protractor and the angles you calculated to divide the circle into parts. Place the center of the protractor over the center of the circle and line the base of the protractor over the straight line.

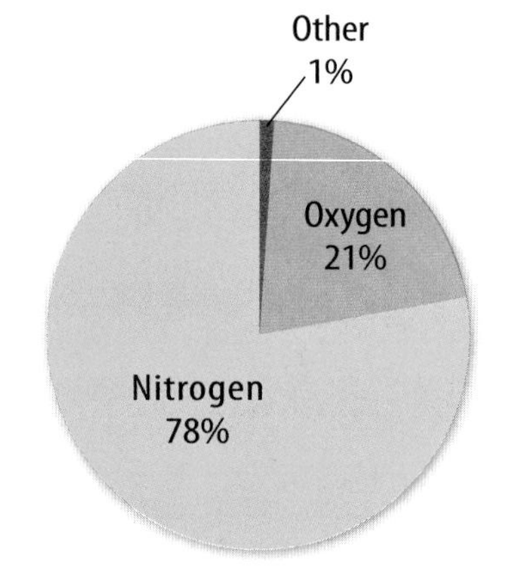

Practice Problem Draw a circle graph to represent the amount of aluminum collected during the week shown in the bar graph to the left.

Student Study Guides & Instructions

By Dinah Zike

1. You will find suggestions for Study Guides, also known as Foldables or books, in each chapter lesson and as a final project. Look at the end of the chapter to determine the project format and glue the Foldables in place as you progress through the chapter lessons.

2. Creating the Foldables or books is simple and easy to do by using copy paper, art paper, and Internet printouts. Photocopies of maps, diagrams, or your own illustrations may also be used for some of the Foldables. Notebook paper is the most common source of material for study guides and 83% of all Foldables are created from it. When folded to make books, notebook paper Foldables easily fit into 11″ × 17″ or 12″ × 18″ chapter projects with space left over. Foldables made using photocopy paper are slightly larger and they fit into Projects, but snugly. Use the least amount of glue, tape, and staples needed to assemble the Foldables.

3. Seven of the Foldables can be made using either small or large paper. When 11″ × 17″ or 12″ × 18″ paper are used, these become projects for housing smaller Foldables. Project format boxes are located within the instructions to remind you of this option.

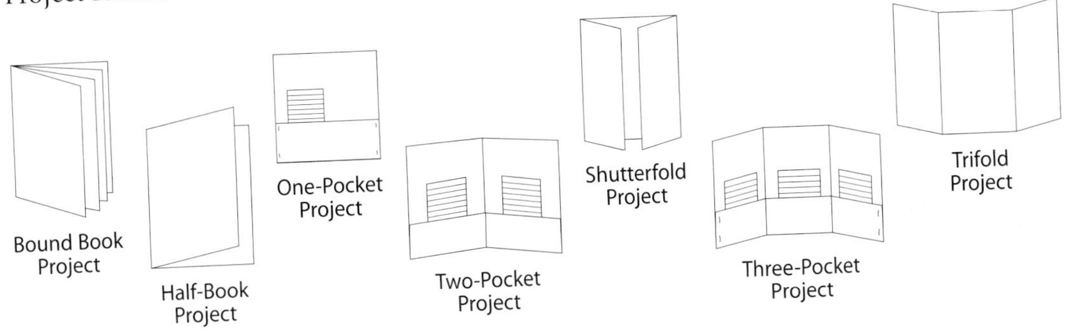

4. Use one-gallon self-locking plastic bags to store your projects. Place strips of two-inch clear tape along the left, long side of the bag and punch holes through the taped edge. Cut the bottom corners off the bag so it will not hold air. Store this Project Portfolio inside a three-hole binder. To store a large collection of project bags, use a giant laundry-soap box. Holes can be punched in some of the Foldable Projects so they can be stored in a three-hole binder without using a plastic bag. Punch holes in the pocket books before gluing or stapling the pocket.

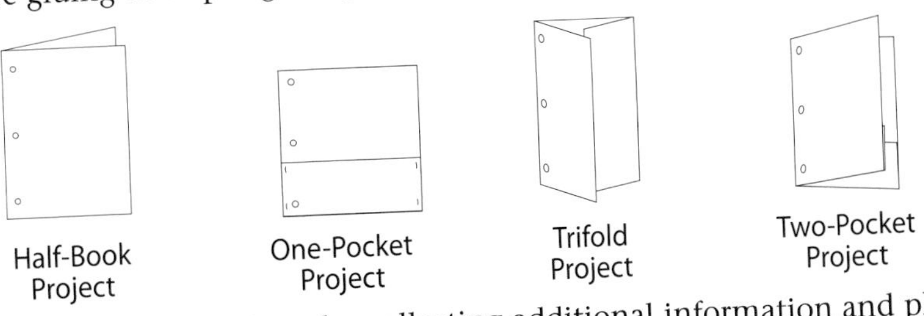

5. Maximize the use of the projects by collecting additional information and placing it on the back of the project and other unused spaces of the large Foldables.

Half-Book Foldable® By Dinah Zike

Step 1 Fold a sheet of notebook or copy paper in half.

Label the exterior tab and use the inside space to write information.

PROJECT FORMAT
Use 11″ × 17″ or 12″ × 18″ paper on the horizontal axis to make a large project book.

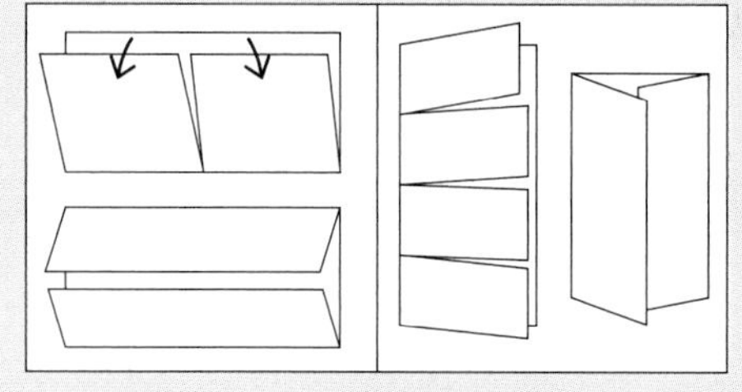

Variations

Paper can be folded vertically, like a *hamburger* or horizontally, like a *hotdog.*

A 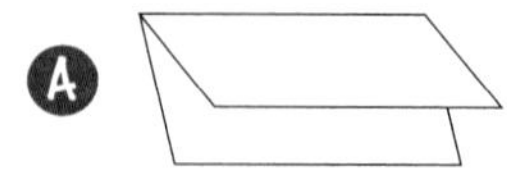

B 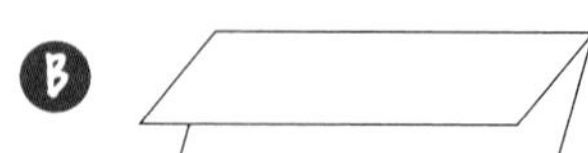

C Half-books can be folded so that one side is ½ inch longer than the other side. A title or question can be written on the extended tab.

Worksheet Foldable or Folded Book® By Dinah Zike

Step 1 Make a half-book (see above) using work sheets, Internet print-outs, diagrams, or maps.

Step 2 Fold it in half again.

Variations

A This folded sheet as a small book with two pages can be used for comparing and contrasting, cause and effect, or other skills.

B When the sheet of paper is open, the four sections can be used separately or used collectively to show sequences or steps.

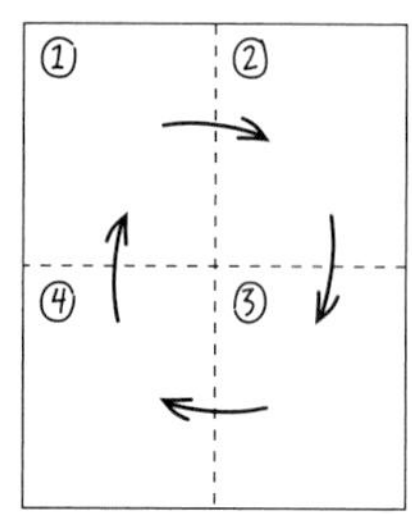

Two-Tab and Concept-Map Foldable® By Dinah Zike

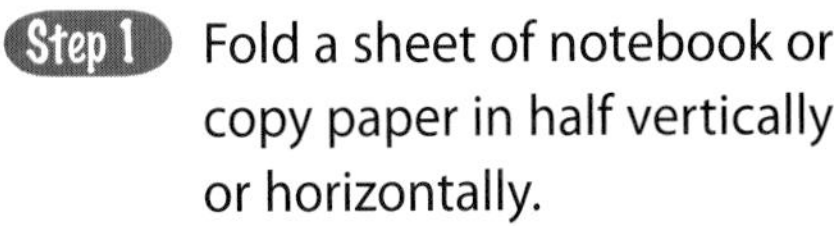

Step 1 Fold a sheet of notebook or copy paper in half vertically or horizontally.

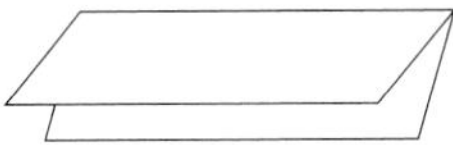

Step 2 Fold it in half again, as shown.

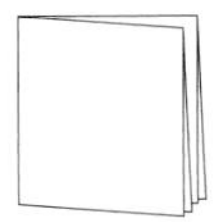

Step 3 Unfold once and cut along the fold line or valley of the top flap to make two flaps.

Variations

A Concept maps can be made by leaving a ½ inch tab at the top when folding the paper in half. Use arrows and labels to relate topics to the primary concept.

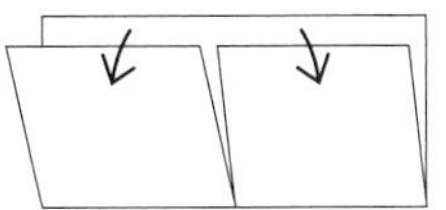

B Use two sheets of paper to make multiple page tab books. Glue or staple books together at the top fold.

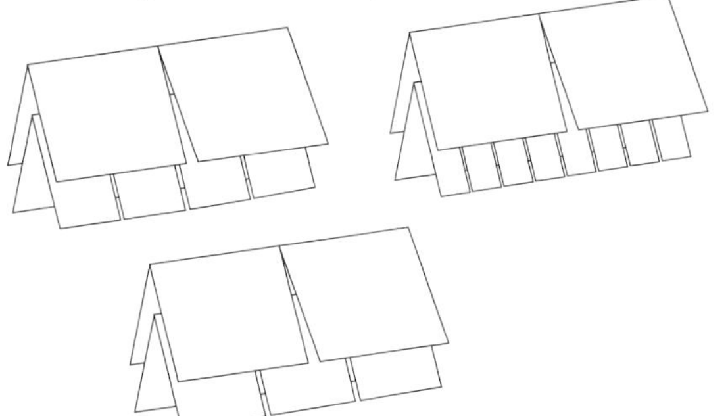

Three-Quarter Foldable® By Dinah Zike

Step 1 Make a two-tab book (see above) and cut the left tab off at the top of the fold line.

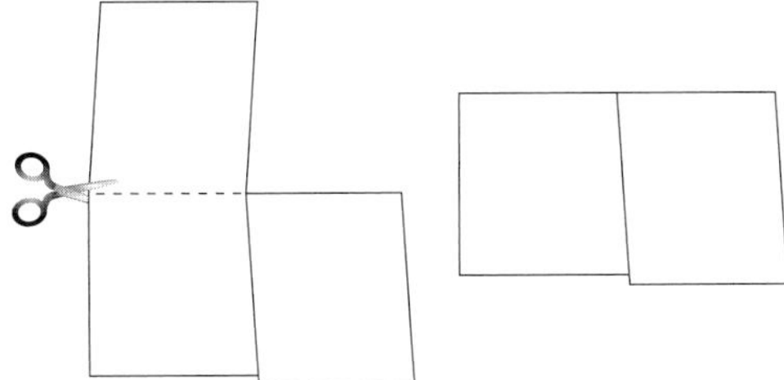

Variations

A Use this book to draw a diagram or a map on the exposed left tab. Write questions about the illustration on the top right tab and provide complete answers on the space under the tab.

B Compose a self-test using multiple choice answers for your questions. Include the correct answer with three wrong responses. The correct answers can be written on the back of the book or upside down on the bottom of the inside page.

Three-Tab Foldable® By Dinah Zike

Step 1 Fold a sheet of paper in half horizontally.

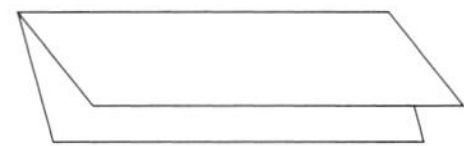

Step 2 Fold into thirds.

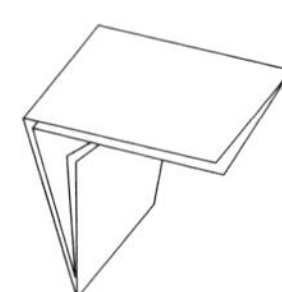

Step 3 Unfold and cut along the folds of the top flap to make three sections.

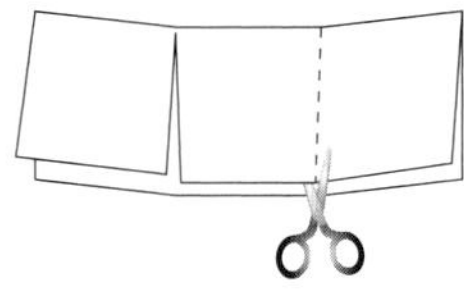

Variations

A Before cutting the three tabs draw a Venn diagram across the front of the book.

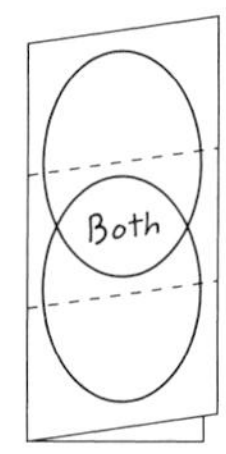

B Make a space to use for titles or concept maps by leaving a ½ inch tab at the top when folding the paper in half.

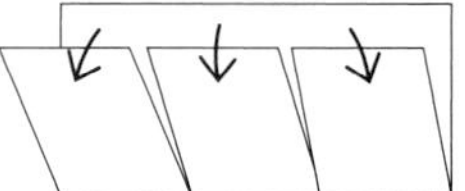

Four-Tab Foldable® By Dinah Zike

Step 1 Fold a sheet of paper in half horizontally.

Step 2 Fold in half and then fold each half as shown below.

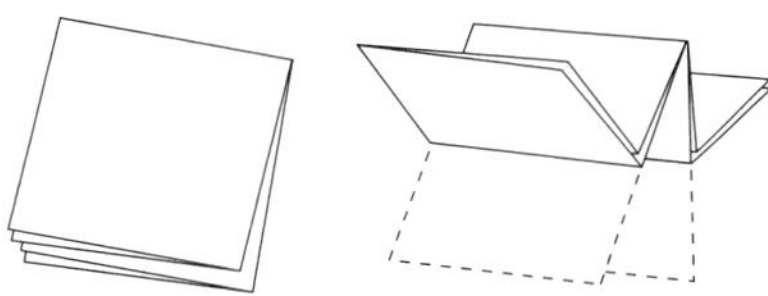

Step 3 Unfold and cut along the fold lines of the top flap to make four tabs.

Variations

A Make a space to use for titles or concept maps by leaving a ½ inch tab at the top when folding the paper in half.

B Use the book on the vertical axis, with or without an extended tab.

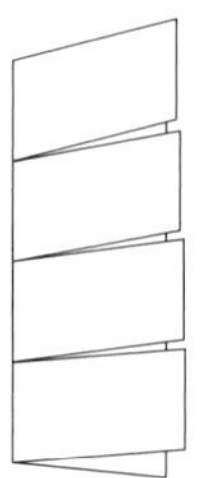

Folding Fifths for a Foldable® By Dinah Zike

Step 1 Fold a sheet of paper in half horizontally.

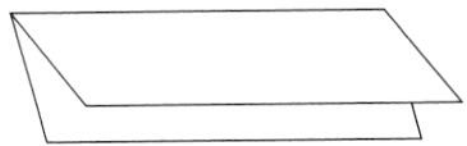

Step 2 Fold again so one-third of the paper is exposed and two-thirds are covered.

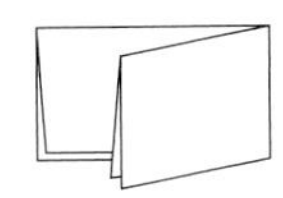

Step 3 Fold the two-thirds section in half.

Step 4 Fold the one-third section, a single thickness, backward to make a fold line.

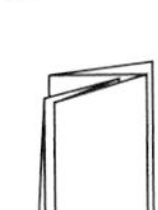

Variations

A Unfold and cut along the fold lines to make five tabs.

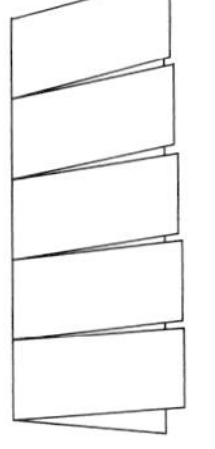

B Make a five-tab book with a ½ inch tab at the top (see two-tab instructions).

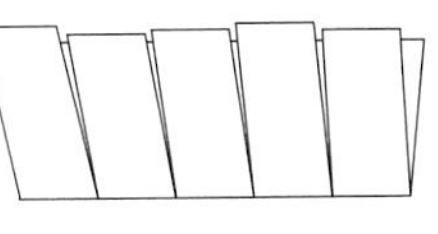

C Use 11″ × 17″ or 12″ × 18″ paper and fold into fifths for a five-column and/or row table or chart.

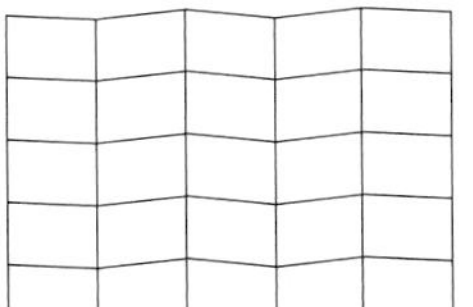

Folded Table or Chart, and Trifold Foldable® By Dinah Zike

Step 1 Fold a sheet of paper in the required number of vertical columns for the table or chart.

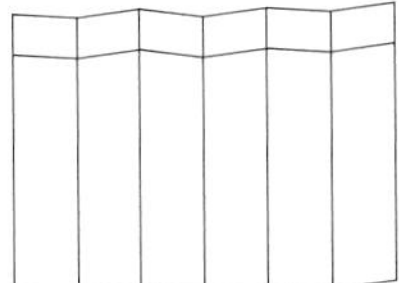

Step 2 Fold the horizontal rows needed for the table or chart.

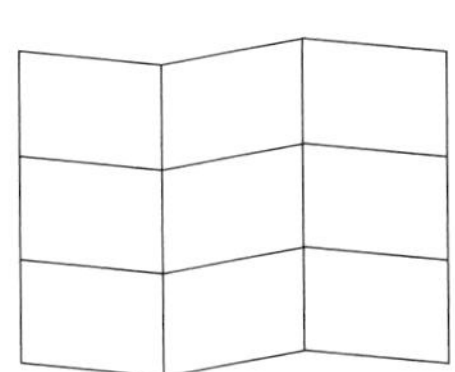

PROJECT FORMAT
Use 11″ × 17″ or 12″ × 18″ paper and fold it to make a large trifold project book or larger tables and charts.

Variations

A Make a trifold by folding the paper into thirds vertically or horizontally.

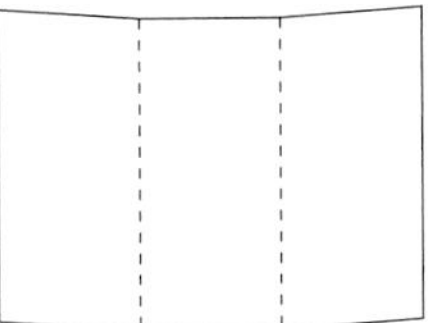

B Make a trifold book. Unfold it and draw a Venn diagram on the inside.

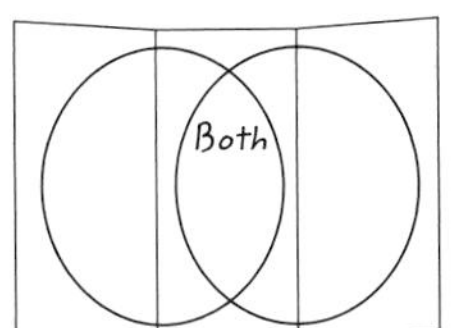

Two or Three-Pockets Foldable® By Dinah Zike

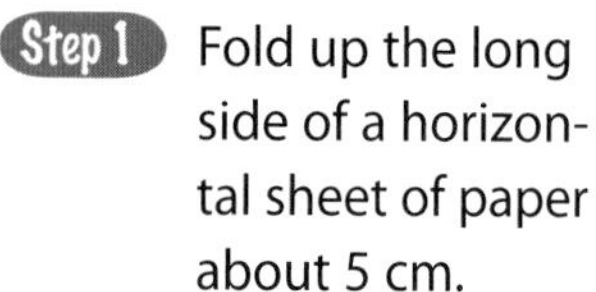

Step 1 Fold up the long side of a horizontal sheet of paper about 5 cm.

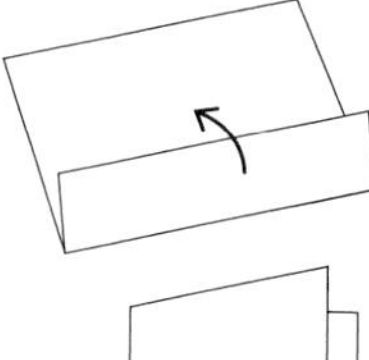

Step 2 Fold the paper in half.

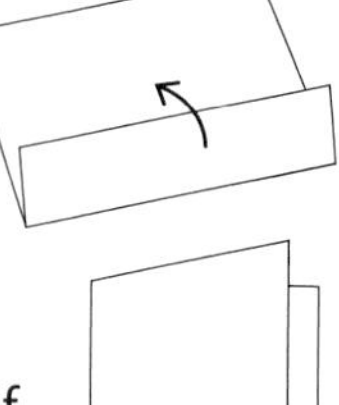

Step 3 Open the paper and glue or staple the outer edges to make two compartments.

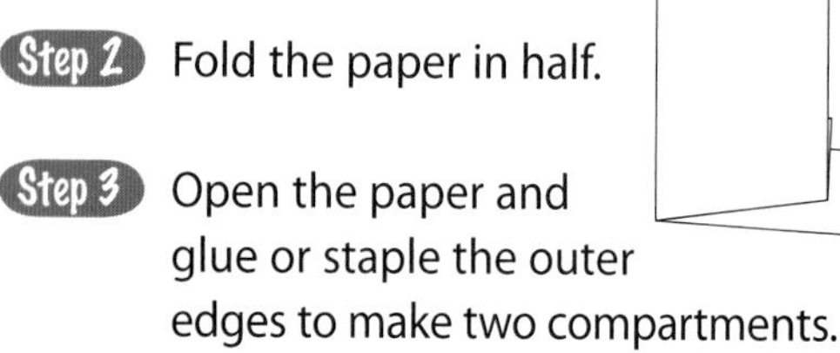

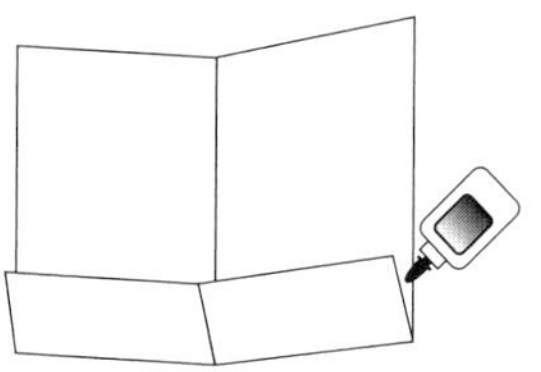

Variations

A Make a multi-page booklet by gluing several pocket books together.

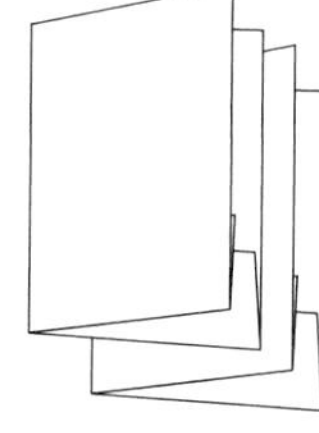

B Make a three-pocket book by using a trifold (see previous instructions).

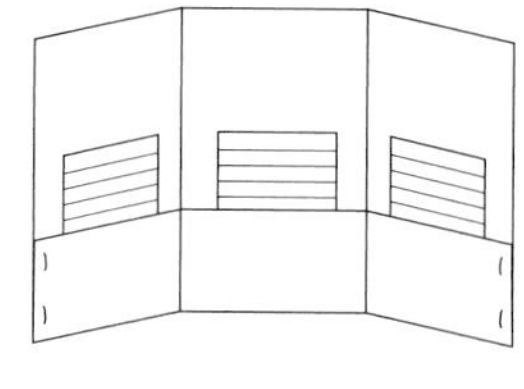

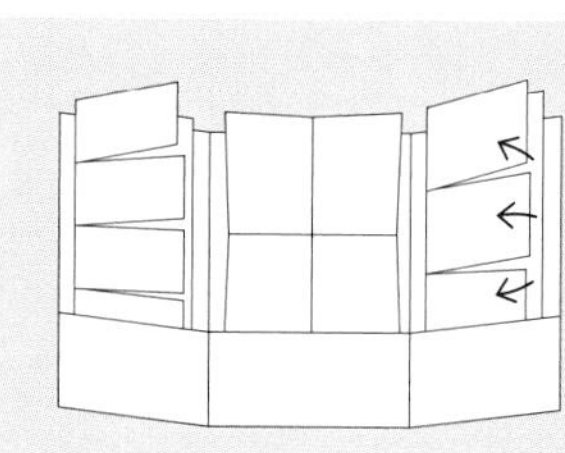

PROJECT FORMAT
Use 11″ × 17″ or 12″ × 18″ paper and fold it horizontally to make a large multi-pocket project.

Matchbook Foldable® By Dinah Zike

Step 1 Fold a sheet of paper almost in half and make the back edge about 1–2 cm longer than the front edge.

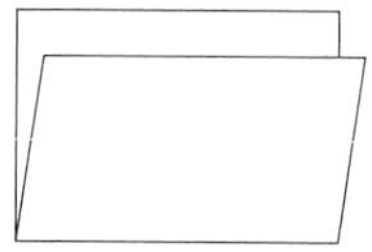

Step 2 Find the midpoint of the shorter flap.

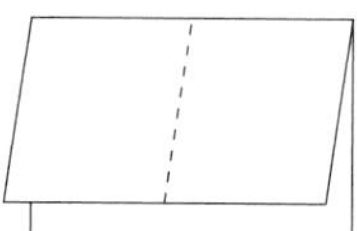

Step 3 Open the paper and cut the short side along the fold lines making two tabs.

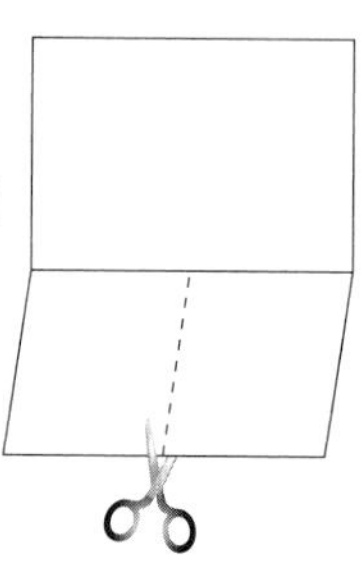

Step 4 Close the book and fold the tab over the short side.

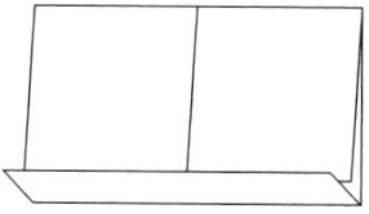

Variations

A Make a single-tab matchbook by skipping Steps 2 and 3.

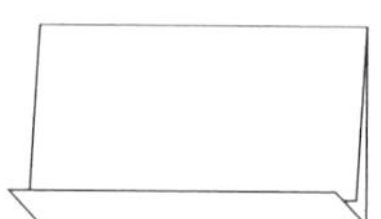

B Make two smaller matchbooks by cutting the single-tab matchbook in half.

Shutterfold Foldable® By Dinah Zike

Step 1 Begin as if you were folding a vertical sheet of paper in half, but instead of creasing the paper, pinch it to show the midpoint.

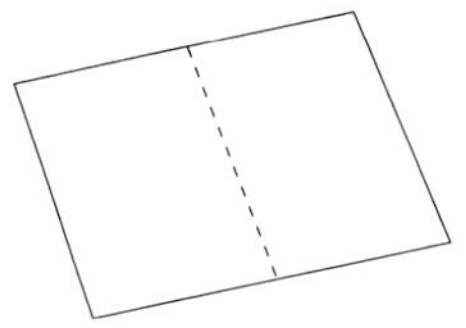

PROJECT FORMAT
Use 11″ × 17″ or 12″ × 18″ paper and fold it to make a large shutterfold project.

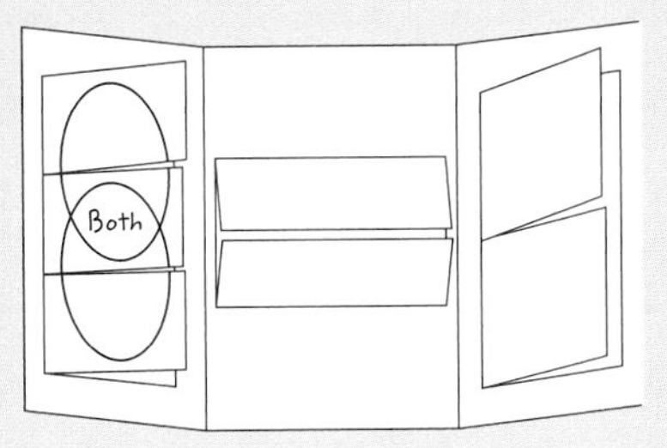

Step 2 Fold the top and bottom to the middle and crease the folds.

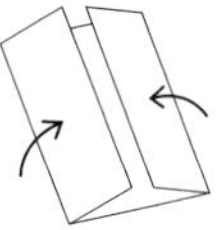

Variations

A Use the shutterfold on the horizontal axis.

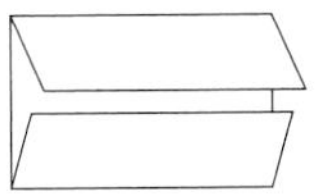

B Create a center tab by leaving .5–2 cm between the flaps in Step 2.

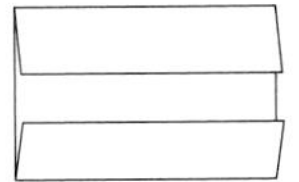

Four-Door Foldable® By Dinah Zike

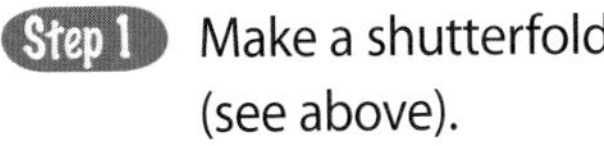

Step 1 Make a shutterfold (see above).

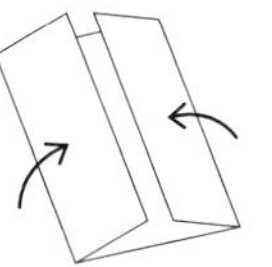

Step 2 Fold the sheet of paper in half.

Step 3 Open the last fold and cut along the inside fold lines to make four tabs.

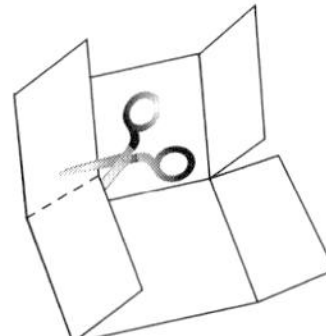

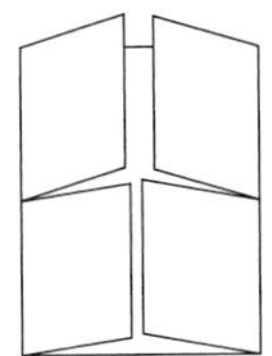

Variations

A Use the four-door book on the opposite axis.

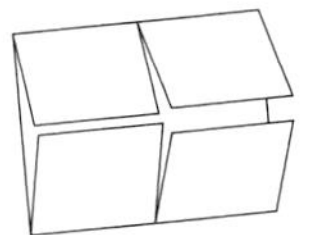

B Create a center tab by leaving .5–2 cm between the flaps in Step 1.

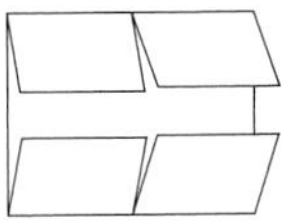

Bound Book Foldable® By Dinah Zike

Step 1 Fold three sheets of paper in half. Place the papers in a stack, leaving about .5 cm between each top fold. Mark all three sheets about 3 cm from the outer edges.

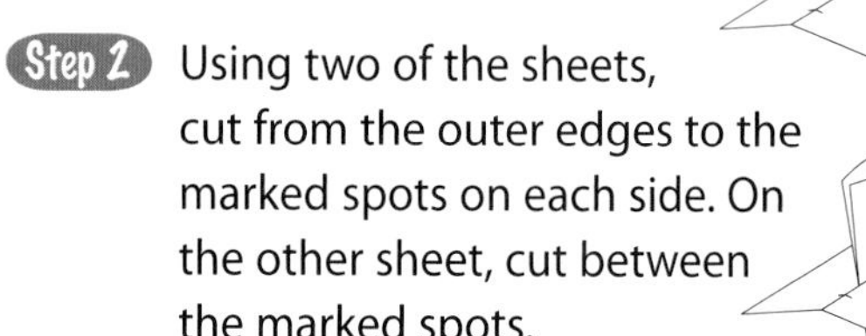

Step 2 Using two of the sheets, cut from the outer edges to the marked spots on each side. On the other sheet, cut between the marked spots.

Step 3 Take the two sheets from Step 1 and slide them through the cut in the third sheet to make a 12-page book.

Step 4 Fold the bound pages in half to form a book.

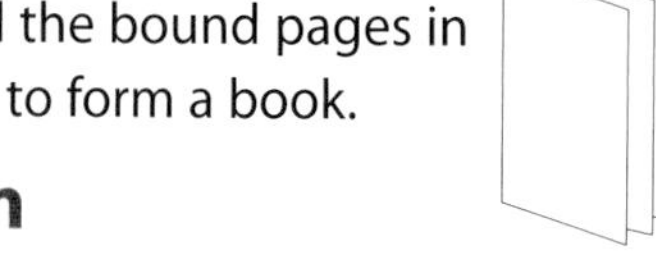

Variation

A Use two sheets of paper to make an eight-page book, or increase the number of pages by using more than three sheets.

PROJECT FORMAT
Use two or more sheets of 11″ × 17″ or 12″ × 18″ paper and fold it to make a large bound book project.

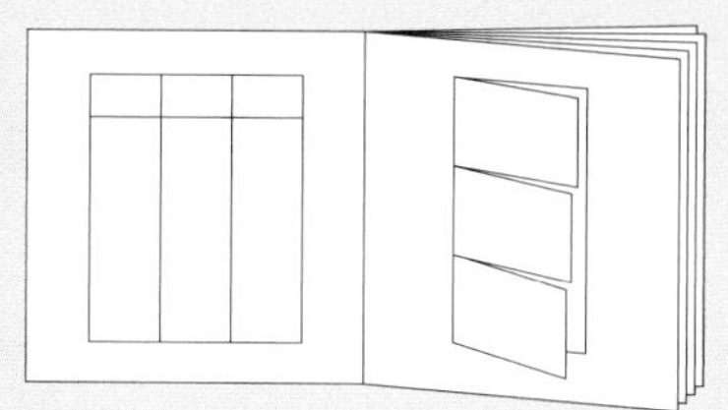

Accordian Foldable® By Dinah Zike

Step 1 Fold the selected paper in half vertically, like a *hamburger*.

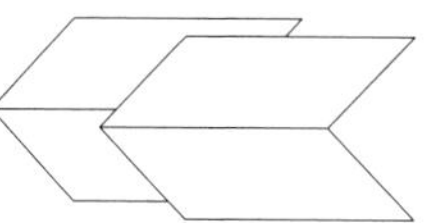

Step 2 Cut each sheet of folded paper in half along the fold lines.

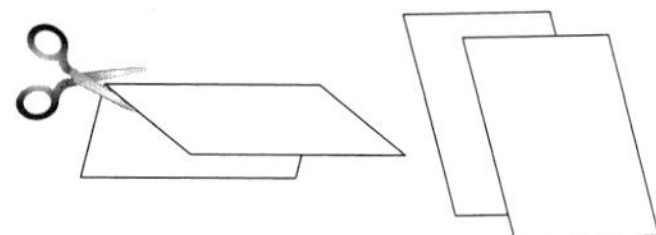

Step 3 Fold each half-sheet almost in half, leaving a 2 cm tab at the top.

Step 4 Fold the top tab over the short side, then fold it in the opposite direction.

Variations

A Glue the straight edge of one paper inside the tab of another sheet. Leave a tab at the end of the book to add more pages.

B Tape the straight edge of one paper to the tab of another sheet, or just tape the straight edges of nonfolded paper end to end to make an accordian.

C Use whole sheets of paper to make a large accordian.

Layered Foldable® By Dinah Zike

Step 1 Stack two sheets of paper about 1–2 cm apart. Keep the right and left edges even.

Step 2 Fold up the bottom edges to to form four tabs. Crease the fold to hold the tabs in place.

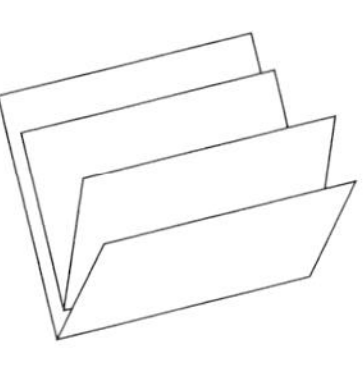

Step 3 Staple along the folded edge, or open and glue the papers together at the fold line.

Variations

A Rotate the book so the fold is at the top or to the side.

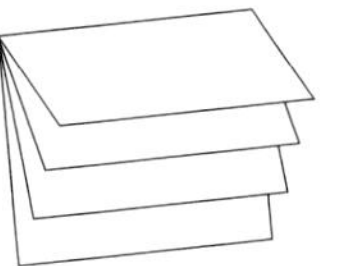

B Extend the book by using more than two sheets of paper.

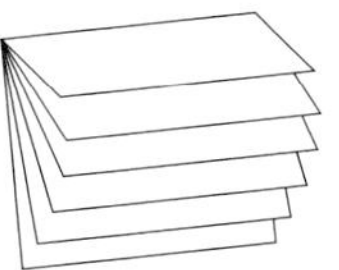

Envelope Foldable® By Dinah Zike

Step 1 Fold a sheet of paper into a *taco*. Cut off the tab at the top.

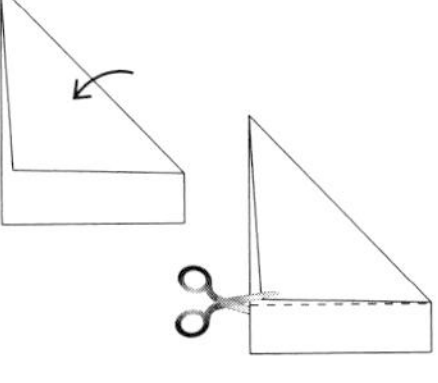

Step 2 Open the *taco* and fold it the opposite way making another *taco* and X-fold pattern on the sheet of paper.

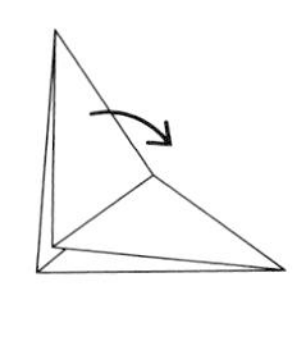

Step 3 Cut a map, illustration or diagram to fit the inside of the envelope.

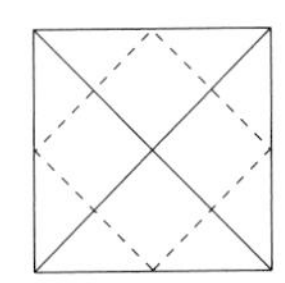

Step 4 Use the outside tabs for labels and inside tabs for writing information.

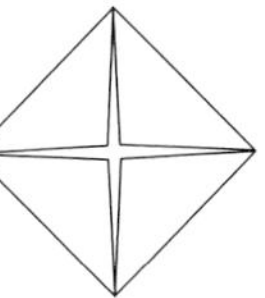

Variations

A Use 11″ × 17″ or 12″ × 18″ paper to make a large envelope.

B Cut off the points of the four tabs to make a window in the middle of the book.

Sentence Strip Foldable® By Dinah Zike

Step 1 Fold two sheets of paper in half vertically, like a *hamburger*.

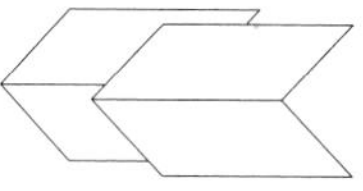

Step 2 Unfold and cut along fold lines making four half sheets.

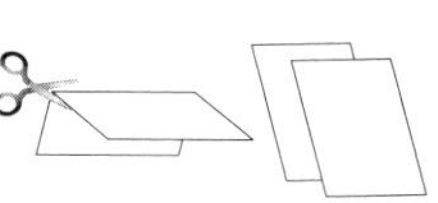

Step 3 Fold each half sheet in half horizontally, like a *hotdog*.

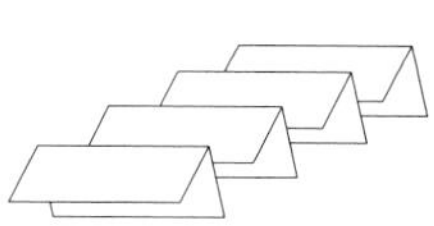

Step 4 Stack folded horizontal sheets evenly and staple together on the left side.

Step 5 Open the top flap of the first sentence strip and make a cut about 2 cm from the stapled edge to the fold line. This forms a flap that can be raisied and lowered. Repeat this step for each sentence strip.

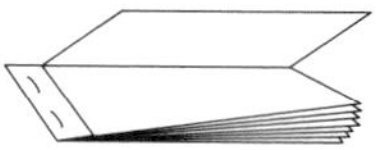

Variations

A Expand this book by using more than two sheets of paper.

B Use whole sheets of paper to make large books.

Pyramid Foldable® By Dinah Zike

Step 1 Fold a sheet of paper into a *taco*. Crease the fold line, but do not cut it off.

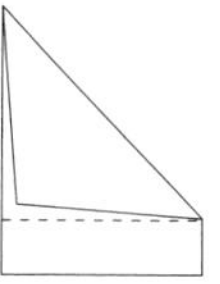

Step 2 Open the folded sheet and refold it like a *taco* in the opposite direction to create an X-fold pattern.

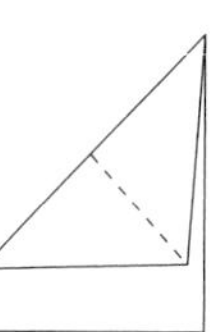

Step 3 Cut one fold line as shown, stopping at the center of the X-fold to make a flap.

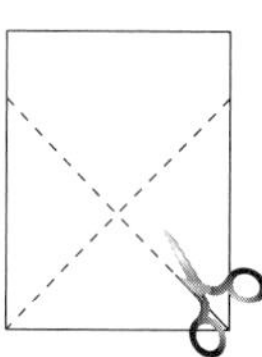

Step 4 Outline the fold lines of the X-fold. Label the three front sections and use the inside spaces for notes. Use the tab for the title.

Step 5 Glue the tab into a project book or notebook. Use the space under the pyramid for other information.

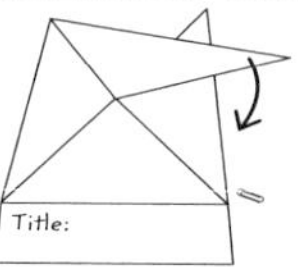

Step 6 To display the pyramid, fold the flap under and secure with a paper clip, if needed.

Single-Pocket or One-Pocket Foldable® By Dinah Zike

Step 1 Using a large piece of paper on a vertical axis, fold the bottom edge of the paper upwards, about 5 cm.

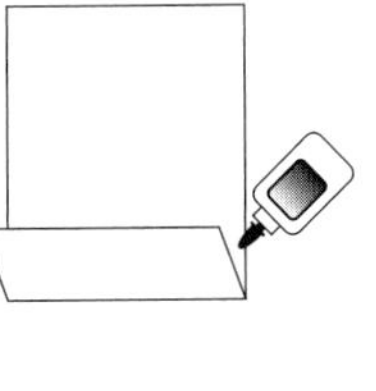

Step 2 Glue or staple the outer edges to make a large pocket.

PROJECT FORMAT
Use 11″ × 17″ or 12″ × 18″ paper and fold it vertically or horizontally to make a large pocket project.

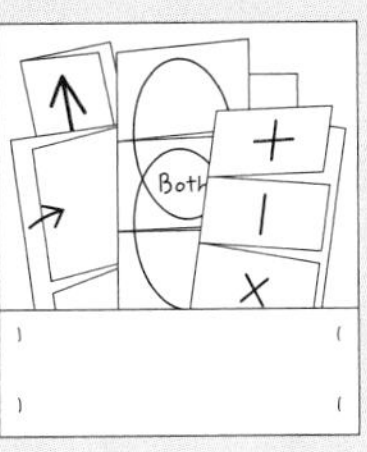

Variations

A Make the one-pocket project using the paper on the horizontal axis.

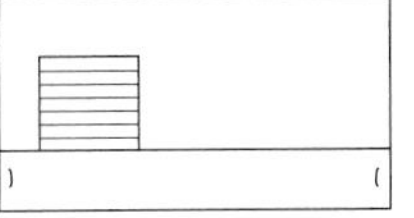

B To store materials securely inside, fold the top of the paper almost to the center, leaving about 2–4 cm between the paper edges. Slip the Foldables through the opening and under the top and bottom pockets.

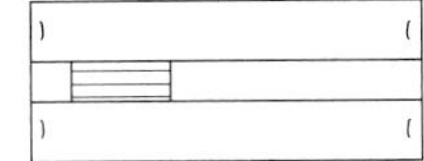

Multi-Tab Foldable® By Dinah Zike

Step 1 Fold a sheet of notebook paper in half like a *hotdog*.

Step 2 Open the paper and on one side cut every third line. This makes ten tabs on wide ruled notebook paper and twelve tabs on college ruled.

Step 3 Label the tabs on the front side and use the inside space for definitions, or other information.

Variation

A Make a tab for a title by folding the paper so the holes remain uncovered. This allows the notebook Foldable to be stored in a three-hole binder.

PERIODIC TABLE OF THE ELEMENTS

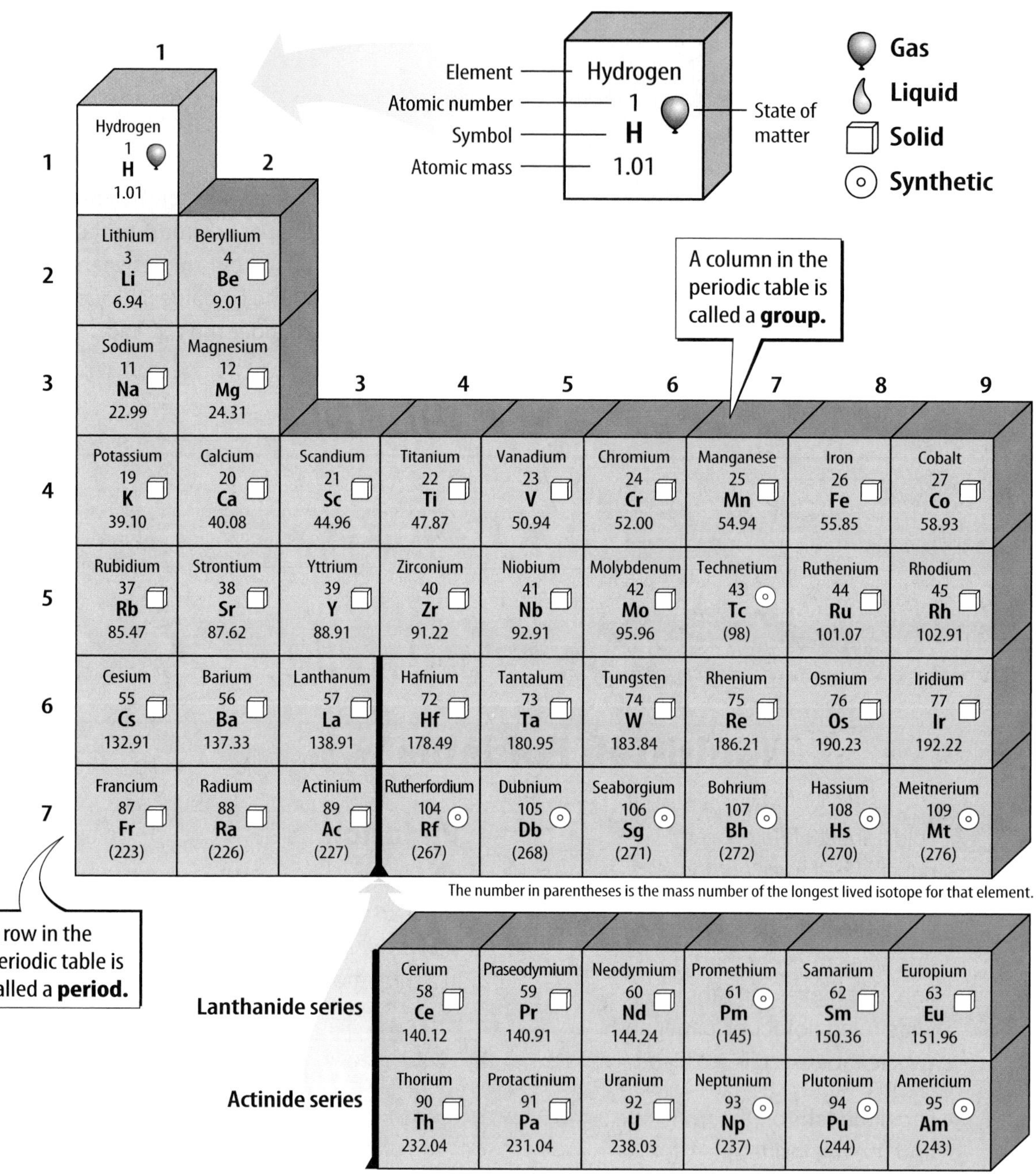

Period	1	2	3	4	5	6	7	8	9
1	Hydrogen 1 H 1.01								
2	Lithium 3 Li 6.94	Beryllium 4 Be 9.01							
3	Sodium 11 Na 22.99	Magnesium 12 Mg 24.31							
4	Potassium 19 K 39.10	Calcium 20 Ca 40.08	Scandium 21 Sc 44.96	Titanium 22 Ti 47.87	Vanadium 23 V 50.94	Chromium 24 Cr 52.00	Manganese 25 Mn 54.94	Iron 26 Fe 55.85	Cobalt 27 Co 58.93
5	Rubidium 37 Rb 85.47	Strontium 38 Sr 87.62	Yttrium 39 Y 88.91	Zirconium 40 Zr 91.22	Niobium 41 Nb 92.91	Molybdenum 42 Mo 95.96	Technetium 43 Tc (98)	Ruthenium 44 Ru 101.07	Rhodium 45 Rh 102.91
6	Cesium 55 Cs 132.91	Barium 56 Ba 137.33	Lanthanum 57 La 138.91	Hafnium 72 Hf 178.49	Tantalum 73 Ta 180.95	Tungsten 74 W 183.84	Rhenium 75 Re 186.21	Osmium 76 Os 190.23	Iridium 77 Ir 192.22
7	Francium 87 Fr (223)	Radium 88 Ra (226)	Actinium 89 Ac (227)	Rutherfordium 104 Rf (267)	Dubnium 105 Db (268)	Seaborgium 106 Sg (271)	Bohrium 107 Bh (272)	Hassium 108 Hs (270)	Meitnerium 109 Mt (276)

The number in parentheses is the mass number of the longest lived isotope for that element.

Lanthanide series	Cerium 58 Ce 140.12	Praseodymium 59 Pr 140.91	Neodymium 60 Nd 144.24	Promethium 61 Pm (145)	Samarium 62 Sm 150.36	Europium 63 Eu 151.96
Actinide series	Thorium 90 Th 232.04	Protactinium 91 Pa 231.04	Uranium 92 U 238.03	Neptunium 93 Np (237)	Plutonium 94 Pu (244)	Americium 95 Am (243)

Metal

Metalloid

Nonmetal

Recently discovered

10	11	12	13	14	15	16	17	18
								Helium 2 **He** 4.00
			Boron 5 **B** 10.81	Carbon 6 **C** 12.01	Nitrogen 7 **N** 14.01	Oxygen 8 **O** 16.00	Fluorine 9 **F** 19.00	Neon 10 **Ne** 20.18
			Aluminum 13 **Al** 26.98	Silicon 14 **Si** 28.09	Phosphorus 15 **P** 30.97	Sulfur 16 **S** 32.07	Chlorine 17 **Cl** 35.45	Argon 18 **Ar** 39.95
Nickel 28 **Ni** 58.69	Copper 29 **Cu** 63.55	Zinc 30 **Zn** 65.38	Gallium 31 **Ga** 69.72	Germanium 32 **Ge** 72.64	Arsenic 33 **As** 74.92	Selenium 34 **Se** 78.96	Bromine 35 **Br** 79.90	Krypton 36 **Kr** 83.80
Palladium 46 **Pd** 106.42	Silver 47 **Ag** 107.87	Cadmium 48 **Cd** 112.41	Indium 49 **In** 114.82	Tin 50 **Sn** 118.71	Antimony 51 **Sb** 121.76	Tellurium 52 **Te** 127.60	Iodine 53 **I** 126.90	Xenon 54 **Xe** 131.29
Platinum 78 **Pt** 195.08	Gold 79 **Au** 196.97	Mercury 80 **Hg** 200.59	Thallium 81 **Tl** 204.38	Lead 82 **Pb** 207.20	Bismuth 83 **Bi** 208.98	Polonium 84 **Po** (209)	Astatine 85 **At** (210)	Radon 86 **Rn** (222)
Darmstadtium 110 **Ds** (281)	Roentgenium 111 **Rg** (280)	Copernicium 112 **Cn** (285)	Ununtrium * 113 **Uut** (284)	Ununquadium * 114 **Uuq** (289)	Ununpentium * 115 **Uup** (288)	Ununhexium * 116 **Uuh** (293)		Ununoctium * 118 **Uuo** (294)

* The names and symbols for elements 113-116 and 118 are temporary. Final names will be selected when the elements' discoveries are verified.

Gadolinium 64 **Gd** 157.25	Terbium 65 **Tb** 158.93	Dysprosium 66 **Dy** 162.50	Holmium 67 **Ho** 164.93	Erbium 68 **Er** 167.26	Thulium 69 **Tm** 168.93	Ytterbium 70 **Yb** 173.05	Lutetium 71 **Lu** 174.97
Curium 96 **Cm** (247)	Berkelium 97 **Bk** (247)	Californium 98 **Cf** (251)	Einsteinium 99 **Es** (252)	Fermium 100 **Fm** (257)	Mendelevium 101 **Md** (258)	Nobelium 102 **No** (259)	Lawrencium 103 **Lr** (262)

Glossary/Glosario

Multilingual eGlossary

A science multilingual glossary is available on the science Web site. The glossary includes the following languages.

Arabic	Hmong	Tagalog
Bengali	Korean	Urdu
Chinese	Portuguese	Vietnamese
English	Russian	
Haitian Creole	Spanish	

Cómo usar el glosario en español:

1. Busca el término en inglés que desees encontrar.
2. El término en español, junto con la definición, se encuentran en la columna de la derecha.

Pronunciation Key

Use the following key to help you sound out words in the glossary.

a	b**a**ck (BAK)	**ew**	f**oo**d (FEWD)
ay	d**ay** (DAY)	**yoo**	p**u**re (PYOOR)
ah	f**ather** (FAH thur)	**yew**	f**ew** (FYEW)
ow	fl**ow**er (FLOW ur)	**uh**	comm**a** (CAH muh)
ar	**car** (CAR)	**u (+ con)**	r**u**b (RUB)
e	**less** (LES)	**sh**	**sh**elf (SHELF)
ee	**leaf** (LEEF)	**ch**	na**t**ure (NAY chur)
ih	**trip** (TRIHP)	**g**	**g**ift (GIHFT)
i (i + com + e)	**i**dea (i DEE uh)	**j**	**g**em (JEM)
oh	**go** (GOH)	**ing**	s**ing** (SING)
aw	**so**ft (SAWFT)	**zh**	vi**si**on (VIH zhun)
or	**orb**it (OR buht)	**k**	ca**k**e (KAYK)
oy	**coin** (COYN)	**s**	**s**eed, **c**ent (SEED, SENT)
oo	**foo**t (FOOT)	**z**	**z**one, rai**s**e (ZOHN, RAYZ)

English — A — Español

absorption: the transfer of energy from a wave to the medium through which it travels. (pp. 548, 584, 637)

acceleration: a measure of the change in velocity during a period of time. (p. 27)

acid: a substance that produces a hydronium ion (H_3O^+) when dissolved in water. (p. 472)

acoustics: the study of how sound interacts with structures. (p. 585)

activation energy: the minimum amount of energy needed to start a chemical reaction. (p. 438)

alkali (AL kuh li) metal: an element in group 1 on the periodic table. (p. 357)

absorción: transferencia de energía desde una onda hacia el medio a través del cual viaja. (pág. 548, 584, 637)

aceleración: medida del cambio de velocidad durante un período de tiempo. (pág. 27)

ácido: sustancia que produce ión hidronio (H_3O^+) cuando se disuelve en agua. (pág. 472)

acústica: estudio de cómo interactúa el sonido con las estructuras. (pág. 585)

energía de activación: cantidad mínima de energía necesaria para iniciar una reacción química. (pág. 438)

metal alcalino: elemento del grupo 1 de la tabla periódica. (pág. 357)

Science Skill Handbook
Math Skill Handbook
Foldables Handbook
Reference Handbook
Glossary/Glosario
Index

alkaline (AL kuh lun) earth metal: an element in group 2 on the periodic table. (p. 357)

metal alcalinotérreo: elemento del grupo 2 de la tabla periódica. (pág. 357)

alternating current: an electric current that changes direction in a regular pattern. (p. 737)

corriente alterna: corriente eléctrica que cambia la dirección en un patrón regular. (pág. 737)

amino acid: a carbon compound that contains the two functional groups amino and carboxyl. (p. 509)

aminoácido: compuesto de carbono que contiene los dos grupos funcionales amino y carboxilo. (pág. 509)

amino group: a group consisting of a nitrogen atom covalently bonded to two hydrogen atoms ($-NH_2$). (p. 502)

grupo amino: grupo que consiste en un átomo de nitrógeno unido de manera covalente a dos átomos de hidrógeno ($-NH_2$). (pág. 502)

amplitude: the maximum distance a wave varies from its rest position. (pp. 539, 573)

amplitud: distancia máxima que varía una onda desde su posición de reposo. (pág. 539, 573)

amplitude modulation (AM): a change in the amplitude of a carrier wave. (p. 616)

amplitud modulada (AM): cambio en la amplitud de una onda portadora. (pág. 616)

Archimedes' (ar kuh MEE deez) principle: principle that states that the buoyant force on an object is equal to the weight of the fluid that the object displaces. (p. 134)

principio de Arquímedes: principio que establece que la fuerza de empuje ejercida sobre un objeto es igual al peso del fluido que el objeto desplaza. (pág. 134)

atmospheric pressure: the ratio of the weight of all the air above you to your surface area. (p. 126)

presión atmosférica: peso del aire sobre una superficie. (pág. 126)

atom: a small particle that is the building block of matter. (p. 299); the smallest piece of an element that still represents that element. (p. 315)

átomo: partícula pequeña que es el componente básico de la materia. (pág. 299); parte más pequeña de un elemento que mantiene la identidad de dicho elemento. (pág. 315)

atomic number: the number of protons in an atom of an element. (p. 327)

número atómico: número de protones en el átomo de un elemento. (pág. 327)

average atomic mass: the average mass of the element's isotopes, weighted according to the abundance of each isotope. (p. 329)

masa atómica promedio: masa atómica promedio de los isótopos de un elemento, ponderado según la abundancia de cada isótopo. (pág. 329)

average speed: the total distance traveled divided by the total time taken to travel that distance. (p. 19)

rapidez promedio: distancia total recorrida dividida por el tiempo usado para recorrerla. (pág. 19)

B

balanced forces: forces acting on an object that combine and form a net force of zero. (p. 56)

fuerzas en equilibrio: fuerzas que actúan sobre un objeto, se combinan y forman una fuerza neta de cero. (pág. 56)

base: a substance that produces hydroxide ions (OH^-) when dissolved in water. (p. 472)

base: sustancia que produce iones hidróxido (OH^-) cuando se disuelve en agua. (pág. 472)

Bernoulli's (ber NEW leez) principle: principle that states that the pressure of a fluid decreases when the speed of that fluid increases. (p. 142)

principio de Bernoulli: principio que establece que la presión de un fluido disminuye cuando la rapidez de dicho fluido aumenta. (pág. 142)

biological molecule: a large organic molecule in any living organism. (p. 508)

Boyle's Law: the law that pressure of a gas increases if the volume decreases and pressure of a gas decreases if the volume increases, when temperature is constant. (p. 294)

broadcasting: the use of electromagnetic waves to send information in all directions. (p. 615)

buoyant (BOY unt) force: an upward force applied by a fluid on an object in the fluid. (p. 132)

molécula biológica: molécula orgánica grande de cualquier organismo vivo. (pág. 508)

Ley de Boyle: ley que afirma que la presión de un gas aumenta si el volumen disminuye y que la presión de un gas disminuye si el volumen aumenta, cuando la temperatura es constante. (pág. 294)

radiodifusión: uso de ondas electromagnéticas para enviar información en todas las direcciones. (pág. 615)

fuerza de empuje: fuerza ascendente que un fluido aplica a un objeto que se encuentra en él. (pág. 132)

C

carbohydrate: a group of organic molecules that includes sugars, starches and cellulose. (p. 510)

carboxyl group: a group consisting of a carbon atom with a single bond to a hydroxyl group and a double bond to an oxygen atom (–COOH). (p. 501)

carrier wave: an electromagnetic wave that a radio or television station uses to carry its sound or image signals. (p. 616)

catalyst: a substance that increases reaction rate by lowering the activation energy of a reaction. (p. 440)

centripetal (sen TRIH puh tuhl) force: in circular motion, a force that acts perpendicular to the direction of motion, toward the center of the curve. (p. 66)

Charles's Law: the law that volume of a gas increases with increasing temperature, if the pressure is constant. (p. 295)

chemical bond: a force that holds two or more atoms together. (p. 382)

chemical change: a change in matter in which the substances that make up the matter change into other substances with different chemical and physical properties. (p. 257)

carbohidrato: grupo de moléculas orgánicas que incluye azúcares, almidones y celulosa. (pág. 510)

grupo carboxilo: grupo que consiste en un átomo de carbono unido a un grupo hidroxilo mediante un enlace sencillo y a un átomo de oxígeno mediante un enlace doble (–COOH). (pág. 501)

onda portadora: onda electromagnética que una estación de radio o televisión usa para transportar señales de sonido o imagen. (pág. 616)

catalizador: sustancia que aumenta la velocidad de reacción al disminuir la energía de activación de una reacción. (pág. 440)

fuerza centrípeta: en movimiento circular, la fuerza que actúa de manera perpendicular a la dirección del movimiento, hacia el centro de la curva. (pág. 66)

Ley de Charles: ley que afirma que el volumen de un gas aumenta cuando la temperatura aumenta, si la presión es constante. (pág. 295)

enlace químico: fuerza que mantiene unidos dos o más átomos. (pág. 382)

cambio químico: cambio de la materia en el cual las sustancias que componen la materia se transforman en otras sustancias con propiedades químicas y físicas diferentes. (pág. 257)

chemical equation: a description of a reaction using element symbols and chemical formulas. (p. 422)

chemical formula: a group of chemical symbols and numbers that represent the elements and the number of atoms of each element that make up a compound. (p. 394)

chemical property: the ability or inability of a substance to combine with or change into one or more new substances. (p. 256)

chemical reaction: a process in which atoms of one or more substances rearrange to form one or more new substances. (p. 419)

circular motion: any motion in which an object is moving along a curved path. (p. 66)

coefficient: a number placed in front of an element symbol or chemical formula in an equation. (p. 426)

combustion: a chemical reaction in which a substance combines with oxygen and releases energy. (p. 432)

compound: a substance containing atoms of two or more different elements chemically bonded together. (p. 234)

compression: region of a longitudinal wave where the particles of the medium are closest together. (pp. 532, 566)

concave lens: a lens that is thicker at the edges than in the middle. (p. 652)

concave mirror: a mirror that curves inward. (p. 645)

concentration: the amount of a particular solute in a given amount of solution. (pp. 260, 464)

condensation: the change of state from a gas to a liquid. (p. 286)

conduction: the transfer of thermal energy due to collisions between particles. (p. 206)

cone: one of many cells in the retina of the eye that respond to colors. (p. 657)

constants: the factors in an experiment that remain the same. (p. NOS 19)

constant speed: the rate of change of position in which the same distance is traveled each second. (p. 18)

ecuación química: descripción de una reacción con símbolos de los elementos y fórmulas químicas. (pág. 422)

fórmula química: grupo de símbolos químicos y números que representan los elementos y el número de átomos de cada elemento que forman un compuesto. (pág. 394)

propiedad química: capacidad o incapacidad de una sustancia para combinarse con o transformarse en una o más sustancias. (pág. 256)

reacción química: proceso en el cual átomos de una o más sustancias se acomodan para formar una o más sustancias nuevas. (pág. 419)

movimiento circular: cualquier movimiento en el cual un objeto se mueve a lo largo de una trayectoria curva. (pág. 66)

coeficiente: número colocado en frente del símbolo de un elemento o de una fórmula química en una ecuación. (pág. 426)

combustión: reacción química en la cual una sustancia se combina con oxígeno y libera energía. (pág. 432)

compuesto: sustancia que contiene átomos de dos o más elementos diferentes unidos químicamente. (pág. 234)

compresión: región de una onda longitudinal donde las partículas del medio están más cerca. (pág. 532, 566)

lente cóncavo: lente que es más grueso en los extremos que en el centro. (pág. 652)

espejo cóncavo: espejo que dobla hacia adentro. (pág. 645)

concentración: cantidad de cierto soluto en una cantidad dada de solución. (pág. 260, 464)

condensación: cambio de estado gaseoso a líquido. (pág. 286)

conducción: transferencia de energía térmica debido a colisiones entre partículas. (pág. 206)

cono: una de muchas células en la retina del ojo que es sensible a los colores. (pág. 657)

constantes: factores que no cambian en un experimento. (pág. NOS 19)

velocidad constante: velocidad a la que se cambia de posición, en la cual se recorre la misma distancia por segundo. (pág. 18)

contact force: a push or a pull on one object by another object that is touching it. (p. 45)

control group: the part of a controlled experiment that contains the same factors as the experimental group, but the independent variable is not changed. (p. NOS 19)

convection: the transfer of thermal energy by the movement of particles from one part of a material to another. (p. 210)

convection current: the movement of fluids in a cycle because of convection. (p. 211)

convex lens: a lens that is thicker in the middle than at the edges. (p. 652)

convex mirror: a mirror that curves outward. (p. 646)

covalent bond: a chemical bond formed when two atoms share one or more pairs of valence electrons. (p. 391)

crest: the highest point on a transverse wave. (p. 531)

critical thinking: comparing what you already know with information you are given in order to decide whether you agree with it. (p. NOS 8)

fuerza de contacto: empuje o arrastre ejercido sobre un objeto por otro que lo está tocando. (pág. 45)

grupo de control: parte de un experimento controlado que contiene los mismos factores que el grupo experimental, pero la variable independiente no se cambia. (pág. NOS 19)

convección: transferencia de energía térmica por el movimiento de partículas de una parte de la materia a otra. (pág. 210)

corriente de convección: movimiento de fluidos en un ciclo debido a la convección. (pág. 211)

lente convexo: lente que es más grueso en el centro que en los extremos. (pág. 652)

espejo convexo: espejo que curva hacia afuera. (pág. 646)

enlace covalente: enlace químico formado cuando dos átomos comparten uno o más pares de electrones de valencia. (pág. 391)

cresta: punto más alto en una onda transversal. (pág. 531)

pensamiento crítico: comparación que se hace cuando se sabe algo acerca de información nueva, y se decide si se está o no de acuerdo con ella. (pág. NOS 8)

D

decomposition: a type of chemical reaction in which one compound breaks down and forms two or more substances. (p. 431)

density: the mass per unit volume of a substance. (p. 249)

dependent variable: the factor a scientist observes or measures during an experiment. (p. NOS 19)

deposition: the process of changing directly from a gas to a solid. (p. 286)

description: a spoken or written summary of an observation. (p. NOS 10)

diffraction: the change in direction of a wave when it travels by the edge of an object or through an opening. (p. 550)

diffuse reflection: reflection of light from a rough surface. (p. 644)

direct current: an electric current that continually flows in one direction. (p. 737)

descomposición: tipo de reacción química en la que un compuesto se descompone y forma dos o más sustancias. (pág. 431)

densidad: cantidad de masa por unidad de volumen de una sustancia. (pág. 249)

variable dependiente: factor que el científico observa o mide durante un experimento. (pág. NOS 19)

deposición: proceso de cambiar directamente de gas a sólido. (pág. 286)

descripción: resumen oral o escrito de una observación. (pág. NOS 10)

difracción: cambio en la dirección de una onda cuando ésta viaja por el borde de un objeto o a través de una abertura. (pág. 550)

reflexión difusa: reflexión de la luz en una superficie rugosa. (pág. 644)

corriente directa: corriente eléctrica que fluye de manera continua en una dirección. (pág. 737)

Science Skill Handbook
Math Skill Handbook
Foldables Handbook
Reference Handbook
Glossary/Glosario
Index

displacement: the difference between the initial, or starting, position and the final position of an object that has moved. (p. 13)
desplazamiento: diferencia entre la posición inicial, o salida, y la final de un objeto que se ha movido. (pág. 13)

dissolve: to form a solution by mixing evenly. (p. 235)
disolver: preparar una solución mezclando de manera homogénea. (pág. 235)

Doppler effect: the change of pitch when a sound source is moving in relation to an observer. (p. 576)
efecto Doppler: cambio de tono cuando una fuente sonora se mueve con relación a un observador. (pág. 576)

double-replacement reaction: a type of chemical reaction in which the negative ions in two compounds switch places, forming two new compounds. (p. 432)
reacción de sustitución doble: tipo de reacción química en la que los iones negativos de dos compuestos intercambian lugares, para formar dos compuestos nuevos. (pág. 432)

drag force: a force that opposes the motion of an object through a fluid. (p. 144)
fuerza de arrastre: fuerza que se opone al movimiento de un objeto a través de un fluido. (pág. 144)

ductility (duk TIH luh tee): the ability to be pulled into thin wires. (p. 356)
ductilidad: capacidad para formar alambres delgados. (pág. 356)

E

echo: a reflected sound wave. (p. 584)
eco: onda sonora reflejada. (pág. 584)

echolocation (e koh loh KAY shun): the process an animal uses to locate an object by means of reflected sounds. (p. 587)
ecocolocación: proceso que un animal hace para ubicar un objeto por medio de sonidos reflejados. (pág. 587)

efficiency: the ratio of output work to input work. (p. 99)
eficiencia: relación entre energía invertida y energía útil. (pág. 99)

electric circuit: a closed, or complete, path in which an electric current flows. (p. 690)
circuito eléctrico: trayectoria cerrada, o completa, por la que fluye corriente eléctrica. (pág. 690)

electric conductor: a material through which electrons easily move. (p. 682)
conductor eléctrico: material a través del cual se mueven los electrones con facilidad. (pág. 682)

electric current: the movement of electrically charged particles. (p. 690)
corriente eléctrica: movimiento de partículas cargadas eléctricamente. (pág. 690)

electric discharge: the process of an unbalanced electric charge becoming balanced. (p. 685)
descarga eléctrica: proceso por el cual una carga eléctrica no balanceada se vuelve balanceada. (pág. 685)

electric energy: energy carried by an electric current. (p. 165)
energía eléctrica: energía transportada por una corriente eléctrica. (pág. 165)

electric generator: a device that uses a magnetic field to transform mechanical energy to electric energy. (p. 736)
generador eléctrico: aparato que usa un campo magnético para transformar energía mecánica en energía eléctrica. (pág. 736)

electric insulator: a material through which electrons cannot easily move. (p. 682)
aislante eléctrico: material por el cual los electrones no pueden fluir con facilidad. (pág. 682)

electric motor: a device that uses an electric current to produce motion. (p. 730)
motor eléctrico: aparato que usa corriente eléctrica para producir movimiento. (pág. 730)

electric resistance: a measure of how difficult it is for an electric current to flow in a material. (p. 692)

electromagnet: a magnet created by wrapping a current-carrying wire around a ferromagnetic core. (p. 728)

electromagnetic spectrum: the entire range of electromagnetic waves with different frequencies and wavelengths. (p. 609)

electromagnetic wave: a transverse wave that can travel through empty space and through matter. (pp. 535, 601)

electron: a negatively charged particle that occupies the space in an atom outside the nucleus. (p. 317)

electron cloud: the region surrounding an atom's nucleus where one or more electrons are most likely to be found. (p. 322)

electron dot diagram: a model that represents valence electrons in an atom as dots around the element's chemical symbol. (p. 385)

element: a substance that consists of only one type of atom. (p. 233)

endothermic reaction: a chemical reaction that absorbs thermal energy. (p. 437)

energy: the ability to cause change. (p. 161)

enzyme: a catalyst that speeds up chemical reactions in living cells. (p. 440)

evaporation: the process of a liquid changing to a gas at the surface of the liquid. (p. 286)

exothermic reaction: a chemical reaction that releases thermal energy. (p. 437)

experimental group: the part of the controlled experiment used to study relationships among variables. (p. NOS 19)

explanation: an interpretation of observations. (p. NOS 10)

resistencia eléctrica: medida de qué tan difícil es para una corriente eléctrica fluir en un material. (pág. 692)

electroimán: imán fabricado al enrollar un alambre que transporta corriente alrededor de un núcleo ferromagnético. (pág. 728)

espectro electromagnético: rango completo de ondas electromagnéticas con frecuencias y longitudes de onda diferentes. (pág. 609)

onda electromagnética: onda transversal que puede viajar a través del espacio vacío y de la materia. (pág. 535, 601)

electrón: partícula cargada negativamente que ocupa el espacio por fuera del núcleo de un átomo. (pág. 317)

nube de electrones: región que rodea el núcleo de un átomo en donde es más probable encontrar uno o más electrones. (pág. 322)

diagrama de puntos de Lewis: modelo que representa electrones de valencia en un átomo a manera de puntos alrededor del símbolo químico del elemento. (pág. 385)

elemento: sustancia que consiste de un sólo tipo de átomo. (pág. 233)

reacción endotérmica: reacción química que absorbe energía térmica. (pág. 437)

energía: capacidad de causar cambio. (pág. 161)

enzima: catalizador que acelera reacciones químicas en las células vivas. (pág. 440)

evaporación: proceso por el cual un líquido cambia a gas en la superficie de dicho líquido. (pág. 286)

reacción exotérmica: reacción química que libera energía térmica. (pág. 437)

grupo experimental: parte del experimento controlado que se usa para estudiar las relaciones entre las variables. (pág. NOS 19)

explicación: interpretación de las observaciones. (pág. NOS 10)

F

ferromagnetic (fer ro mag NEH tik) element: elements, including iron, nickel, and cobalt, that have an especially strong attraction to magnets. (p. 721)

elemento ferromagnético: elementos, incluidos hierro, níquel y cobalto, que tienen una atracción especialmente fuerte hacia los imanes. (pág. 721)

fluid: any substance that can flow and take the shape of the container that holds it. (p. 123)

fluido: cualquier sustancia que puede fluir y toma la forma del recipiente que lo contiene. (pág. 123)

focal length: the distance along the optical axis from the mirror to the focal point. (p. 645)

distancia focal: distancia a lo largo del eje óptico desde el espejo hasta el punto focal. (pág. 645)

focal point: the point where light rays parallel to the optical axis converge after being reflected by a mirror or refracted by a lens. (p. 645)

punto focal: punto donde rayos de luz paralelos al eje óptico convergen después de ser reflejados por un espejo o refractados por un lente. (pág. 645)

force: a push or a pull on an object. (p. 45)

fuerza: empuje o arrastre ejercido sobre un objeto. (pág. 45)

force pair: the forces two objects apply to each other. (p. 71)

par de fuerzas: fuerzas que dos objetos se aplican entre sí. (pág. 71)

fossil fuel: the remains of ancient organisms that can be burned as an energy source. (p. 178)

combustible fósil: restos de organismos antiguos que pueden quemarse como fuente de energía. (pág. 178)

frequency: the number of wavelengths that pass by a point each second. (pp. 542, 575)

frecuencia: número de longitudes de onda que pasan por un punto cada segundo. (pág. 542, 575)

frequency modulation (FM): a change in the frequency of a carrier wave. (p. 616)

frecuencia modulada (FM): cambio en la frecuencia de una onda portadora. (pág. 616)

friction: a contact force that resists the sliding motion of two surfaces that are touching. (pp. 49, 171)

fricción: fuerza de contacto que resiste el movimiento de dos superficies que están en contacto. (pág. 49, 171)

fulcrum: the point about which a lever pivots. (p. 104)

fulcro: punto alrededor del cual gira una palanca. (pág. 104)

functional group: an atom or group of atoms that determines the function or properties of the compound. (p. 500)

grupo funcional: átomo o grupo de átomos que determina la función o las propiedades de un compuesto. (pág. 500)

G

gamma ray: a high-energy electromagnetic wave with a shorter wavelength and higher frequency than all other types of electromagnetic waves. (p. 611)

rayo gamma: onda electromagnética de alta energía con longitud de onda más corta y frecuencia más alta que los demás tipos de ondas electromagnéticas. (pág. 611)

gas: matter that has no definite volume and no definite shape. (p. 278)

gas: materia que no tiene volumen ni forma definidos. (pág. 278)

Global Positioning System (GPS): a worldwide navigation system that uses satellite signals to determine a receiver's location. (p. 618)

sistema de posicionamiento global (SPG): sistema mundial de navegación que usa señales satelitales para determinar la ubicación de un receptor. (pág. 618)

gravity: an attractive force that exists between all objects that have mass. (p. 47)

gravedad: fuerza de atracción que existe entre todos los objetos que tienen masa. (pág. 47)

grounding: providing a path for electric charges to flow safely into the ground. (p. 685)

polo a tierra: suministrar una trayectoria para que las cargas eléctricas fluyan con seguridad hacia el suelo. (pág. 685)

group: a column on the periodic table. (p. 350)

grupo: columna en la tabla periódica. (pág. 350)

H

halide group: a functional group which contains group 17 halogens—fluorine, chlorine, bromine, and iodine. (p. 501)

grupo halide: grupo funcional que contiene 17 halógenos—flúor, cloro, bromo y yodo. (pág. 501)

halogen (HA luh jun): an element in group 17 on the periodic table. (p. 365)

halógeno: elemento del grupo 17 de la tabla periódica. (pág. 365)

heat: the movement of thermal energy from a region of higher temperature to a region of lower temperature. (p. 201)

calor: movimiento de energía térmica de una región de alta temperatura a una región de baja temperatura. (pág. 201)

heat engine: a machine that converts thermal energy into mechanical energy. (p. 218)

motor térmico: máquina que convierte energía térmica en energía mecánica. (pág. 218)

heating appliance: a device that converts electric energy into thermal energy. (p. 215)

calentador: aparato que convierte energía eléctrica en energía térmica. (pág. 215)

heterogeneous mixture: a mixture in which substances are not evenly mixed. (pp. 235, 455)

mezcla heterogénea: mezcla en la cual las sustancias no están mezcladas de manera uniforme. (pág. 235, 455)

hologram: a three-dimensional photograph of an object. (p. 665)

holograma: fotografía tridimensional de un objeto. (pág. 665)

homogeneous mixture: a mixture in which two or more substances are evenly mixed but not bonded together. (pp. 235, 455)

mezcla homogénea: mezcla en la cual dos o más sustancias están mezcladas de manera uniforme, pero no están unidas químicamente. (pág. 235, 455)

hydrocarbon: a compound that contains only carbon and hydrogen atoms. (p. 492)

hidrocarburo: compuesto que contiene solamente átomos de carbono e hidrógeno. (pág. 492)

hydronium ion (H_3O^+): a positively charged ion formed when an acid dissolves in water. (p. 472)

ión hidronio (H_3O^+): ión cargado positivamente que se forma cuando un ácido se disuelve en agua. (pág. 472)

hydroxyl group: the chemical group consisting of one atom of hydrogen and one atom of oxygen (–OH). (p. 500)

grupo hidroxilo: grupo químico que consiste en un átomo de hidrógeno y un átomo de oxígeno (–OH). (pág. 500)

hypothesis: a possible explanation for an observation that can be tested by scientific investigations. (p. NOS 4)

hipótesis: explicación posible para una observación que puede ponerse a prueba en investigaciones científicas. (pág. NOS 4)

I

inclined plane: a simple machine that consists of a ramp, or a flat, sloped surface. (p. 107)

plano inclinado: máquina simple que consiste en una rampa, o superficie plana inclinada. (pág. 107)

independent variable: the factor that is changed by the investigator to observe how it affects a dependent variable. (p. NOS 19)

variable independiente: factor que el investigador cambia para observar cómo afecta la variable dependiente. (pág. NOS 19)

indicator: a compound that changes color at different pH values when it reacts with acidic or basic solutions. (p. 476)

indicador: compuesto que cambia de color a diferentes valores de pH cuando reacciona con soluciones ácidas o básicas. (pág. 476)

inertia (ihn UR shuh): the tendency of an object to resist a change in its motion. (p. 58)

inexhaustible energy resource: an energy resource that cannot be used up. (p. 181)

inference: a logical explanation of an observation that is drawn from prior knowledge or experience. (p. NOS 4)

infrared wave: an electromagnetic wave that has a wavelength shorter than a microwave but longer than visible light. (p. 610)

inhibitor: a substance that slows, or even stops, a chemical reaction. (p. 440)

instantaneous speed: an object's speed at a specific instant in time. (p. 18)

intensity: the amount of energy that passes through a square meter of space in one second. (p. 574)

interference: occurs when waves overlap and combine to form a new wave. (pp. 551, 576)

International System of Units (SI): the internationally accepted system of measurement. (p. NOS 10)

ion (I ahn): an atom that is no longer neutral because it has lost or gained valence electrons. (pp. 332, 398)

ionic bond: the attraction between positively and negatively charged ions in an ionic compound. (p. 400)

isomer: one of two or more compounds that have the same molecular formula but different structural arrangements. (p. 492)

isotopes: atoms of the same element that have different numbers of neutrons. (p. 328)

inercia: tendencia de un objeto a resistirse al cambio en su movimiento. (pág. 58)

recurso energético inagotable: recurso energético que no puede agotarse. (pág. 181)

inferencia: explicación lógica de una observación que se obtiene a partir de conocimiento previo o experiencia. (pág. NOS 4)

onda infrarroja: onda electromagnética que tiene una longitud de onda más corta que la de una microonda, pero más larga que la de la luz visible. (pág. 610)

inhibidor: sustancia que disminuye, o incluso detiene, una reacción química. (pág. 440)

velocidad instantánea: velocidad de un objeto en un instante específico en el tiempo. (pág. 18)

intensidad: cantidad de energía que atraviesa un metro cuadrado de espacio en un segundo. (pág. 574)

interferencia: ocurre cuando las ondas coinciden y combinan para forma una onda nueva. (pág. 551, 576)

Sistema Internacional de Unidades (SI): sistema de medidas aceptado internacionalmente. (pág. NOS 10)

ión: átomo que no es neutro porque ha ganado o perdido electrones de valencia. (pág. 332, 398)

enlace iónico: atracción entre iones cargados positiva y negativamente en un compuesto iónico. (pág. 400)

isómero: uno de dos o más compuestos que tienen la misma fórmula molecular, pero diferentes arreglos estructurales. (pág. 492)

isótopos: átomos del mismo elemento que tienen diferente número de neutrones. (pág. 328)

K

kinetic (kuh NEH tik) energy: energy due to motion. (pp. 162, 282)

kinetic molecular theory: an explanation of how particles in matter behave. (p. 292)

energía cinética: energía debida al movimiento. (pág. 162, 282)

teoría cinética molecular: explicación de cómo se comportan las partículas en la materia. (pág. 292)

L

laser: an optical device that produces a narrow beam of coherent light. (p. 664)

láser: aparato óptico que produce un haz angosto y coherente de luz. (pág. 664)

law of conservation of energy: law that states that energy can be transformed from one form to another, but it cannot be created or destroyed. (p. 170)

ley de la conservación de la energía: ley que plantea que la energía puede transformarse de una forma a otra, pero no puede crearse ni destruirse. (pág. 170)

law of conservation of mass: law that states that the total mass of the reactants before a chemical reaction is the same as the total mass of the products after the chemical reaction. (p. 424)

ley de la conservación de la masa: ley que plantea que la masa total de los reactivos antes de una reacción química es la misma que la masa total de los productos después de la reacción química. (pág. 424)

law of conservation of momentum: a principle stating that the total momentum of a group of objects stays the same unless outside forces act on the objects. (p. 74)

ley de la conservación del momentum: principio que establece que el momentum total de un grupo de objetos permanece constante a menos que fuerzas externas actúen sobre los objetos. (pág. 74)

law of reflection: law that states that when a wave is reflected from a surface, the angle of reflection is equal to the angle of incidence. (pp. 549, 643)

ley de la reflexión: ley que establece que cuando una onda se refleja desde una superficie, el ángulo de reflexión es igual al ángulo de incidencia. (pág. 549, 643)

lever: a simple machine that consists of a bar that pivots, or rotates, around a fixed point. (p. 104)

palanca: máquina simple que consiste en una barra que gira, o rota, alrededor de un punto fijo. (pág. 104)

light: electromagnetic radiation that you can see. (p. 635)

luz: radiación electromagnética que puede verse. (pág. 635)

lipid: a type of biological molecule that includes fats, oils, hormones, waxes, and components of cellular membranes. (p. 512)

lípido: tipo de molécula biológica que incluye grasas, aceites, hormonas, ceras y componentes de las membranas celulares. (pág. 512)

liquid: matter with a definite volume but no definite shape. (p. 276)

líquido: materia con volumen definido y forma indefinida. (pág. 276)

longitudinal (lahn juh TEWD nul) wave: a wave in which the disturbance is parallel to the direction the wave travels. (pp. 532, 565)

onda longitudinal: onda en la que la perturbación es paralela a la dirección en que viaja la onda. (pág. 532, 565)

luster: the way a mineral reflects or absorbs light at its surface. (p. 355)

brillo: forma en que un mineral refleja o absorbe la luz en su superficie. (pág. 355)

M

magnet: an object that attracts iron. (p. 717)

imán: objeto que atrae al hierro. (pág. 717)

magnetic domain: region in a magnetic material in which the magnetic fields of the atoms all point in the same direction. (p. 722)

dominio magnético: región en un material magnético en el que los campos magnéticos de los átomos apuntan en la misma dirección. (pág. 722)

magnetic force: the force that a magnet applies to another magnet. (p. 718)

fuerza magnética: fuerza que un imán aplica a otro imán. (pág. 718)

magnetic material: any material that a magnet attracts. (p. 721)

material magnético: cualquier material que un imán atrae. (pág. 721)

magnetic pole: the place on a magnet where the force it exerts is the strongest. (p. 718)

malleability (ma lee uh BIH luh tee): the ability of a substance to be hammered or rolled into sheets. (p. 356)

mass: the amount of matter in an object. (pp. 47, 242)

mass number: the sum of the number of protons and neutrons in an atom. (p. 328)

matter: anything that has mass and takes up space. (p. 231)

mechanical advantage: the ratio of a machine's output force produced to the input force applied. (p. 98)

mechanical energy: sum of the potential energy and the kinetic energy in a system. (p. 165)

mechanical wave: a wave that can travel only through matter. (p. 531)

medium: a material in which a wave travels. (pp. 531, 566)

metal: an element that is generally shiny, is easily pulled into wires or hammered into thin sheets, and is a good conductor of electricity and thermal energy. (p. 355)

metallic bond: a bond formed when many metal atoms share their pooled valence electrons. (p. 401)

metalloid (MEH tul oyd): an element that has physical and chemical properties of both metals and nonmetals. (p. 367)

microscope: an optical device that forms magnified images of very small, close objects. (p. 662)

microwave: a low-frequency, low-energy electromagnetic wave that has a wavelength between about 1 mm and 30 cm. (p. 610)

mixture: matter that can vary in composition. (pp. 234, 454)

molecule (MAH lih kyewl): two or more atoms that are held together by covalent bonds and act as a unit. (p. 392)

momentum: a measure of how hard it is to stop a moving object. (p. 73)

polo magnético: lugar en un imán donde la fuerza que éste ejerce es la mayor. (pág. 718)

maleabilidad: capacidad de una sustancia de martillarse o laminarse para formar hojas. (pág. 356)

masa: cantidad de materia en un objeto. (pág. 47, 242)

número de masa: suma del número de protones y neutrones de un átomo. (pág. 328)

materia: cualquier cosa que tiene masa y ocupa espacio. (pág. 231)

ventaja mecánica: relación entre la fuerza útil que produce una máquina con la fuerza aplicada. (pág. 98)

energía mecánica: suma de la energía potencial y la energía cinética en un sistema. (pág. 165)

onda mecánica: onda que puede viajar sólo a través de la materia. (pág. 531)

medio: material en el cual viaja una onda. (pág. 531, 566)

metal: elemento que generalmente es brillante, fácilmente puede estirarse para formar alambres o martillarse para formar hojas delgadas y es buen conductor de electricidad y energía térmica. (pág. 355)

enlace metálico: enlace formado cuando muchos átomos metálicos comparten su banco de electrones de valencia. (pág. 401)

metaloide: elemento que tiene las propiedades físicas y químicas de metales y no metales. (pág. 367)

microscopio: aparato óptico que forma imágenes aumentadas de objetos cercanos muy pequeños. (pág. 662)

microonda: onda electromagnética de baja frecuencia y baja energía que tiene una longitud de onda entre 1 mm y 30 cm. (pág. 610)

mezcla: materia cuya composición puede variar. (pág. 234, 454)

molécula: dos o más átomos que están unidos mediante enlaces covalentes y actúan como una unidad. (pág. 392)

momentum: medida de qué tan difícil es detener un objeto en movimiento. (pág. 73)

monomer: one of the small organic molecules that make up the long chain of a polymer. (p. 503)

motion: the process of changing position. (p. 13)

monómero: una de las moléculas orgánicas pequeñas que forman la larga cadena de un polímero. (pág. 503)

movimiento: proceso de cambiar de posición. (pág. 13)

N

net force: the combination of all the forces acting on an object. (p. 55)

neutron: a neutral particle in the nucleus of an atom. (p. 321)

Newton's first law of motion: law that states that if the net force acting on an object is zero, the motion of the object does not change. (p. 57)

Newton's second law of motion: law that states that the acceleration of an object is equal to the net force exerted on the object divided by the object's mass. (p. 65)

Newton's third law of motion: law that states that for every action there is an equal and opposite reaction. (p. 71)

noble gas: an element in group 18 on the periodic table. (p. 366)

noncontact force: a force that one object applies to another object without touching it. (p. 46)

nonmetal: an element that has no metallic properties. (p. 363)

nonrenewable energy resource: an energy resource that is available in limited amounts or that is used faster than it is replaced in nature. (p. 178)

nuclear decay: a process that occurs when an unstable atomic nucleus changes into another more stable nucleus by emitting radiation. (p. 331)

nuclear energy: energy stored in and released from the nucleus of an atom. (p. 165)

nucleic acid (new KLEE ihk • A sud): a biological polymer that stores and transmits genetic information. (p. 511)

nucleus: the region in the center of an atom where most of an atom's mass and positive charge are concentrated. (p. 320)

fuerza neta: combinación de todas las fuerzas que actúan sobre un objeto. (pág. 55)

neutrón: partícula neutra en el núcleo de un átomo. (pág. 321)

primera ley del movimiento de Newton: ley que establece que si la fuerza neta ejercida sobre un objeto es cero, el movimiento de dicho objeto no cambia. (pág. 57)

segunda ley del movimiento de Newton: ley que establece que la aceleración de un objeto es igual a la fuerza neta que actúa sobre él divida por su masa. (pág. 65)

tercera ley del movimiento de Newton: ley que establece que para cada acción hay una reacción igual en dirección opuesta. (pág. 71)

gas noble: elemento del grupo 18 de la tabla periódica. (pág. 366)

fuerza de no contacto: fuerza que un objeto puede aplicar sobre otro sin tocarlo. (pág. 46)

no metal: elemento que tiene propiedades no metálicas. (pág. 363)

recurso energético no renovable: recurso energético disponible en cantidades limitadas o que se usa más rápido de lo que se repone en la naturaleza. (pág. 178)

desintegración nuclear: proceso que ocurre cuando un núcleo atómico inestable cambia a otro núcleo atómico más estable mediante emisión de radiación. (pág. 331)

energía nuclear: energía almacenada en y liberada por el núcleo de un átomo. (pág. 165)

ácido nucleico: polímero biológico que almacena y transmite información genética. (pág. 511)

núcleo: región en el centro de un átomo donde se concentra la mayor cantidad de masa y las cargas positivas. (pág. 320)

O

observation: the act of using one or more of your senses to gather information and take note of what occurs. (p. NOS 4)

Ohm's law: a mathematical equation that describes the relationship between voltage, current, and resistance. (p. 694)

opaque: a material through which light does not pass. (p. 636)

optical device: any instrument or object used to produce or control light. (p. 661)

organic compound: chemical compound that contains carbon atoms usually bonded to at least one hydrogen atom. (p. 490)

observación: acción de usar uno o más sentidos para reunir información y tomar notar de lo que ocurre. (pág. NOS 4)

Ley de Ohm: ecuación matemática que describe la relación entre voltaje, corriente y resistencia. (pág. 694)

opaco: material por el que no pasa la luz. (pág. 636)

aparato óptico: cualquier instrumento usado para producir o controlar luz. (pág. 661)

compuesto orgánico: compuesto químico que contienen átomos de carbono generalmente unidos a, al menos, un átomo de hidrógeno. (pág. 490)

P

parallel circuit: an electric circuit with more than one path, or branch, for an electric current to follow. (p. 702)

Pascal's (pas KALZ) principle: principle that states that when pressure is applied to a fluid in a closed container, the pressure increases by the same amount everywhere in the container. (p. 140)

percent error: the expression of error as a percentage of the accepted value. (p. NOS 13)

period: a row on the periodic table. (p. 350)

periodic table: a chart of the elements arranged into rows and columns according to their physical and chemical properties. (p. 345)

permanent magnet: a magnet that retains its magnetism even after being removed from a magnetic field. (p. 723)

pH: an inverse measure of the concentration of hydronium ions (H_3O^+) in a solution. (p. 474)

photon: a particle of electromagnetic radiation. (p. 605)

physical change: a change in the size, shape, form, or state of matter that does not change the matter's identity. (p. 249)

physical property: a characteristic of matter that you can observe or measure without changing the identity of the matter. (p. 240)

circuito en paralelo: circuito eléctrico con más de una trayectoria, o rama, para que fluya una corriente eléctrica. (pág. 702)

principio de Pascal: principio que establece que cuando se aplica presión a un fluido en un recipiente cerrado, la presión aumenta el mismo valor en todas las partes del recipiente. (pág. 140)

error porcentual: expresión del error como porcentaje del valor aceptado. (pág. NOS 13)

periodo: hilera en la tabla periódica. (pág. 350)

tabla periódica: cuadro en que los elementos están organizados en hileras y columnas según sus propiedades físicas y químicas. (pág. 345)

imán permanente: imán que retiene su magnetismo incluso después de haberlo retirado de un campo magnético. (pág. 723)

pH: medida inversa de la concentración de iones hidronio (H_3O^+) en una solución. (pág. 474)

fotón: partícula de radiación electromagnética. (pág. 605)

cambio físico: cambio en el tamaño, la forma o el estado de la materia en el que no cambia la identidad de la materia. (pág. 249)

propiedad física: característica de la materia que puede observarse o medirse sin cambiar la identidad de la materia. (pág. 240)

pitch: the perception of how high or low a sound is; related to the frequency of a sound wave. (p. 575)

polarized: the condition of having electrons concentrated at one end of an object. (p. 683)

polar molecule: a molecule with a slight negative charge in one area and a slight positive charge in another area. (pp. 393, 462)

polymer: a molecule made up of many of the same small organic molecules covalently bonded together, forming a long chain. (p. 503)

polymerization: the chemical process in which small organic molecules, or monomers, bond together to form a chain. (p. 503)

position: an object's distance and direction from a reference point. (p. 9)

potential (puh TEN chul) energy: stored energy due to the interactions between objects or particles. (p. 162)

power: the rate at which work is done. (p. 91)

prediction: a statement of what will happen next in a sequence of events. (p. NOS 4)

pressure: the amount of force per unit area applied to an object's surface. (pp. 124, 293)

product: a substance produced by a chemical reaction. (p. 423)

protein: a biological polymer made of amino acid monomers. (p. 509)

proton: positively charged particle in the nucleus of an atom. (p. 320)

pulley: a simple machine that consists of a grooved wheel with a rope or cable wrapped around it. (p. 109)

tono: percepción de qué tan alto o bajo es el sonido; relacionado con la frecuencia de la onda sonora. (pág. 575)

polarizado: condición de tener electrones concentrados en un extremo de un objeto. (pág. 683)

molécula polar: molécula con carga ligeramente negativa en una parte y ligeramente positiva en otra. (pág. 393, 462)

polímero: molécula hecha de muchas de las mismas moléculas orgánicas pequeñas unidas por enlaces covalentes, y que forman una cadena larga. (pág. 503)

polimerización: proceso químico en el que moléculas orgánicas pequeñas, o monómeros, se unen para formar una cadena. (pág. 503)

posición: distancia y dirección de un objeto según un punto de referencia. (pág. 9)

energía potencial: energía almacenada debido a las interacciones entre objetos o partículas. (pág. 162)

potencia: velocidad a la que se hace trabajo. (pág. 91)

predicción: afirmación de lo que ocurrirá después en una secuencia de eventos. (pág. NOS 4)

presión: cantidad de fuerza por unidad de área aplicada a la superficie de un objeto. (pág. 124, 293)

producto: sustancia producida por una reacción química. (pág. 423)

proteína: polímero biológico hecho de monómeros de aminoácidos. (pág. 509)

protón: partícula cargada positivamente en el núcleo de un átomo. (pág. 320)

polea: máquina simple que consiste en una rueda acanalada rodeada por una cuerda o cable. (pág. 109)

Q

qualitative data: the use of words to describe what is observed in an experiment. (p. NOS 19)

quantitative data: the use of numbers to describe what is observed in an experiment. (p. NOS 19)

datos cualitativos: uso de palabras para describir lo que se observa en un experimento. (pág. NOS 19)

datos cuantitativos: uso de números para describir lo que se observa en un experimento. (pág. NOS 19)

R

radiant energy: energy carried by an electromagnetic wave. (pp. 165, 601)
radiation: the transfer of thermal energy by electromagnetic waves. (p. 205)
radioactive: any element that spontaneously emits radiation. (p. 330)
radio wave: a low-frequency, low-energy electromagnetic wave that has a wavelength longer than about 30 cm. (p. 610)
rarefaction (rayr uh FAK shun): the region of a longitudinal wave where the particles of the medium are farthest apart. (pp. 532, 566)
reactant: a starting substance in a chemical reaction. (p. 423)
reference point: the starting point you use to describe the motion or the position of an object. (p. 9)
reflecting telescope: a telescope that uses a mirror to gather and focus light from distant objects. (p. 662)
reflection: the bouncing of a wave off a surface. (pp. 548, 584, 635)
refracting telescope: a telescope that uses lenses to gather and focus light from distant objects. (p. 662)
refraction: the change in direction of a wave as it changes speed in moving from one medium to another. (pp. 550, 650)
refrigerator: a device that uses electric energy to pump thermal energy from a cooler location to a warmer location. (p. 216)
regular reflection: reflection of light from a smooth, shiny surface. (p. 644)
renewable energy resource: an energy resource that is replaced as fast as, or faster than, it is used. (p. 180)
resonance: an increase in amplitude that occurs when an object vibrating at its natural frequency absorbs energy from a nearby object vibrating at the same frequency. (p. 578)
reverberation: the collection of reflected sounds from the surfaces in a closed space. (p. 585)
rod: one of many cells in the retina of the eye that responds to low light. (p. 657)

energía radiante: energía que transporta una onda electromagnética. (pág. 165, 601)
radiación: transferencia de energía térmica por ondas electromagnéticas. (pág. 205)
radiactivo: cualquier elemento que emite radiación de manera espontánea. (pág. 330)
onda de radio: onda electromagnética de baja frecuencia y baja energía que tiene una longitud de onda mayor de más o menos 30 cm. (pág. 610)
rarefacción: región de una onda longitudinal donde las partículas del medio están más alejadas. (pág. 532, 566)
reactivo: sustancia inicial en una reacción química. (pág. 423)
punto de referencia: punto que se escoge para describir la ubicación, o posición, de un objeto. (pág. 9)
telescopio de reflexión: telescopio que tiene un espejo para reunir y enfocar luz de objetos lejanos. (pág. 662)
reflexión: rebote de una onda desde una superficie. (pág. 548, 584, 635)
telescopio de refracción: telescopio que tiene lentes para reunir y enfocar luz de objetos lejanos. (pág. 662)
refracción: cambio en la dirección de una onda a medida que cambia de velocidad al moverse de un medio a otro. (pág. 550, 650)
refrigerador: aparato que usa energía eléctrica para bombear energía térmica desde un lugar más frío hacia uno más caliente. (pág. 216)
reflexión especular: reflexión de la luz desde una superficie lisa y brillante. (pág. 644)
recurso energético renovable: recurso energético que se repone tan rápido, o más rápido, de lo que se consume. (pág. 180)
resonancia: aumento en la amplitud que ocurre cuando un objeto que vibra a su frecuencia natural absorbe energía de un objeto cercano y vibran a la misma frecuencia. (pág. 578)
reverberación: colección de sonidos reflejados de superficies en un espacio cerrado. (pág. 585)
bastón: una de muchas células en la retina del ojo sensible a la luminosidad baja. (pág. 657)

S

saturated hydrocarbon: hydrocarbon that contains only single bonds. (p. 493)

saturated solution: a solution that contains the maximum amount of solute the solution can hold at a given temperature and pressure. (p. 466)

science: the investigation and exploration of natural events and of the new information that results from those investigations. (p. NOS 2)

scientific law: a rule that describes a pattern in nature. (p. NOS 7)

scientific literacy: having knowledge of scientific concepts and being able to use that knowledge in your everyday life. (p. NOS 8)

scientific notation: a method of writing or displaying very small or very large numbers. (p. NOS 13)

scientific theory: an explanation of observations or events that is based on knowledge gained from many observations and investigations. (p. NOS 7)

screw: a simple machine that consists of an inclined plane wrapped around a cylinder. (p. 108)

semiconductor: a substance that conducts electricity at high temperatures but not at low temperatures. (p. 367)

series circuit: an electric circuit with only one closed path for an electric current to follow. (p. 701)

simple machine: a machine that does work using one movement. (p. 103)

single-replacement reaction: a type of chemical reaction in which one element replaces another element in a compound. (p. 432)

solid: matter that has a definite shape and a definite volume. (p. 275)

solubility (sahl yuh BIH luh tee): the maximum amount of solute that can dissolve in a given amount of solvent at a given temperature and pressure. (p. 244, 466)

solute: any substance in a solution other than the solvent. (p. 461)

hidrocarburos saturados: hidrocarburo que solamente contiene enlaces sencillos. (pág. 493)

solución saturada: solución que contiene la cantidad máxima de soluto que la solución puede sostener a cierta temperatura y presión. (pág. 466)

ciencia: investigación y exploración de eventos naturales y la información nueva que resulta de dichas investigaciones. (pág. NOS 2)

ley científica: regla que describe un patrón en la naturaleza. (pág. NOS 7)

saber científico: tener conocimiento de conceptos científicos y ser capaz de usarlo en la vida diaria. (pág. NOS 8)

notación científica: método para escribir o expresar números muy pequeños o muy grandes. (pág. NOS 13)

teoría científica: explicación de las observaciones y los eventos basada en conocimiento obtenido en muchas observaciones e investigaciones. (pág. NOS 7)

tornillo: máquina simple que consiste en un plano inclinado incrustado alrededor de un cilindro. (pág. 108)

semiconductor: sustancia que conduce electricidad a altas temperaturas, pero no a bajas temperaturas. (pág. 367)

circuito en serie: circuito eléctrico con sólo una trayectoria cerrada para que fluya una corriente eléctrica. (pág. 701)

máquina simple: máquina que hace trabajo con un movimiento. (pág. 103)

reacción de sustitución sencilla: tipo de reacción química en la que un elemento reemplaza a otro en un compuesto. (pág. 432)

sólido: materia con forma y volumen definidos. (pág. 275)

solubilidad: cantidad máxima de soluto que puede disolverse en una cantidad dada de solvente a temperatura y presión dadas. (pág. 244, 466)

soluto: cualquier sustancia en una solución diferente del solvente. (pág. 461)

solution: another name for a homogeneous mixture. (p. 455)
solución: otro nombre para una mezcla homogénea. (pág. 455)

solvent: the substance that exists in the greatest quantity in a solution. (p. 461)
solvente: sustancia que existe en mayor cantidad en una solución. (pág. 461)

sonar: a system that uses the reflection of sound waves to find underwater objects. (p. 587)
sonar: sistema que usa la reflexión de ondas sonoras para encontrar objetos bajo el agua. (pág. 587)

sound energy: energy carried by sound waves. (p. 165)
energía sonora: energía que transportan las ondas sonoras. (pág. 165)

sound wave: a longitudinal wave that can travel only through matter. (p. 565)
onda sonora: onda longitudinal que sólo viaja a través de la materia. (pág. 565)

specific heat: the amount of thermal energy it takes to increase the temperature of 1 kg of a material by 1°C. (p. 207)
calor específico: cantidad de energía térmica necesaria para aumentar la temperatura de 1 Kg de un material en 1°C. (pág. 207)

speed: the distance an object moves divided by the time it takes to move that distance. (p. 17)
rapidez: distancia que un objeto recorre dividida por el tiempo que éste tarda en recorrer dicha distancia. (pág. 17)

static charge: an unbalanced electric charge on an object. (p. 680)
carga estática: carga eléctrica no balanceada en un objeto. (pág. 680)

sublimation: the process of changing directly from a solid to a gas. (p. 286)
sublimación: proceso de cambiar directamente de sólido a gas. (pág. 286)

substance: matter with a composition that is always the same. (pp. 233, 454)
sustancia: materia cuya composición es siempre la misma. (pág. 233, 454)

substituted hydrocarbon: an organic compound in which a carbon atom is bonded to an atom, or group of atoms, other than hydrogen. (p. 499)
hidrocarburo de sustitución: compuesto orgánico en el cual un átomo de carbono está unido a un átomo, o grupo de átomos, diferente del hidrógeno. (pág. 499)

surface tension: the uneven forces acting on the particles on the surface of a liquid. (p. 277)
tensión superficial: fuerzas desiguales que actúan sobre las partículas en la superficie de un líquido. (pág. 277)

synthesis (SIHN thuh sus): a type of chemical reaction in which two or more substances combine and form one compound. (p. 431)
síntesis: tipo de reacción química en el que dos o más sustancias se combinan y forman un compuesto. (pág. 431)

T

technology: the practical use of scientific knowledge, especially for industrial or commercial use. (p. NOS 6)
tecnología: uso práctico del conocimiento científico, especialmente para empleo industrial o comercial. (pág. NOS 6)

temperature: the measure of the average kinetic energy of the particles in a material. (pp. 199, 282)
temperatura: medida de la energía cinética promedio de las partículas de un material. (pág. 199, 282)

temporary magnet: a magnet that quickly loses its magnetism after being removed from a magnetic field. (p. 723)
imán temporal: imán que rápidamente pierde su magnetismo después de haberlo retirado de un campo magnético. (pág. 723)

thermal conductor: a material through which thermal energy flows quickly. (p. 206)
conductor térmico: material mediante el cual la energía térmica se mueve con rapidez. (pág. 206)

thermal contraction: a decrease in a material's volume when the temperature is decreased. (p. 208)

thermal energy: the sum of the kinetic energy and the potential energy of the particles that make up an object. (pp. 165, 198, 283)

thermal expansion: an increase in a material's volume when the temperature is increased. (p. 208)

thermal insulator: a material through which thermal energy flows slowly. (p. 206)

thermostat: a device that regulates the temperature of a system. (p. 216)

transformer: a device that changes the voltage of an alternating current. (p. 739)

transition element: an element in groups 3–12 on the periodic table. (p. 358)

translucent: a material that allows most of the light that strikes it to pass through, but through which objects appear blurry. (p. 636)

transmission: the passage of light through an object. (pp. 548, 637)

transparent: a material that allows almost all of the light striking it to pass through, and through which objects can be seen clearly. (p. 636)

transverse wave: a wave in which the disturbance is perpendicular to the direction the wave travels. (p. 531)

trough: the lowest point on a transverse wave. (p. 531)

turbine (TUR bine): a shaft with a set of blades that spins when a stream of pressurized fluid strikes the blades. (p. 738)

contracción térmica: disminución del volumen de un material cuando disminuye la temperatura. (pág. 208)

energía térmica: suma de la energía cinética y potencial de las partículas que forman un objeto (pág. 165, 198, 283)

expansión térmica: aumento en el volumen de un material cuando aumenta la temperatura. (pág. 208)

aislante térmico: material en el cual la energía térmica se mueve con lentitud. (pág. 206)

termostato: aparato que regula la temperatura de un sistema. (pág. 216)

transformador: aparato que cambia el voltaje de una corriente alterna. (pág. 739)

elemento de transición: elemento de los grupos 3–12 de la tabla periódica. (pág. 358)

translúcido: material que permite el paso de la mayor cantidad de luz que lo toca, pero a través del cual los objetos se ven borrosos. (pág. 636)

transmisión: paso de la luz a través de un objeto. (pág. 548, 637)

transparente: material que permite el paso de la mayor cantidad de luz que lo toca, y a través del cual los objetos pueden verse con nitidez. (pág. 636)

onda transversal: onda en la que la perturbación es perpendicular a la dirección en que viaja la onda. (pág. 531)

seno: punto más bajo en una onda transversal. (pág. 531)

turbina: eje con una serie de paletas que gira cuando un chorro de fluido a presión golpea las paletas. (pág. 738)

U

ultraviolet wave: an electromagnetic wave that has a slightly shorter wavelength and higher frequency than visible light. (p. 611)

unbalanced forces: forces acting on an object that combine and form a net force that is not zero. (p. 56)

onda ultravioleta: onda electromagnética que tiene una longitud de onda ligeramente menor y mayor frecuencia que la luz visible. (pág. 611)

fuerzas no balanceadas: fuerzas que actúan sobre un objeto, se combinan y forman una fuerza neta diferente de cero. (pág. 56)

unsaturated hydrocarbon: a hydrocarbon that contains one or more double or triple bonds. (p. 493)

unsaturated solution: a solution that can still dissolve more solute at a given temperature and pressure. (p. 466)

hidrocarburos insaturados: hidrocarburo que contiene uno o más enlaces dobles o triples. (pág. 493)

solución insaturada: solución que aún puede disolver más soluto a cierta temperatura y presión. (pág. 466)

valence electron: the outermost electron of an atom that participates in chemical bonding. (p. 384)

vapor: the gas state of a substance that is normally a solid or a liquid at room temperature. (p. 278)

vaporization: the change in state from a liquid to a gas. (p. 285)

variable: any factor that can have more than one value. (p. NOS 19)

velocity: the speed and the direction of a moving object. (p. 23)

vibration: a rapid back-and-forth motion that can occur in solids, liquids, or gases. (p. 565)

viscosity (vihs KAW sih tee): a measurement of a liquid's resistance to flow. (p. 276)

voltage: the amount of energy used to move one coulomb of electrons through an electric circuit. (p. 693)

electrón de valencia: electrón más externo de un átomo que participa en el enlace químico. (pág. 384)

vapor: estado gaseoso de una sustancia que normalmente es sólida o líquida a temperatura ambiente. (pág. 278)

vaporización: cambio de estado líquido a gaseoso. (pág. 285)

variable: cualquier factor que tenga más de un valor. (pág. NOS 19)

velocidad: rapidez y dirección de un objeto en movimiento. (pág. 23)

vibración: movimiento rápido de atrás hacia adelante que puede ocurrir en sólidos, líquidos o gases. (pág. 565)

viscosidad: medida de la resistencia de un líquido a fluir. (pág. 276)

voltaje: cantidad de energía usada para mover un culombio de electrones por un circuito eléctrico. (pág. 693)

wave: a disturbance that transfers energy from one place to another without transferring matter. (p. 529)

wavelength: the distance between one point on a wave and the nearest point just like it. (pp. 541, 575)

wedge: a simple machine that consists of an inclined plane with one or two sloping sides; it is used to split or separate an object. (p. 108)

weight: the gravitational force exerted on an object. (p. 48)

wheel and axle: a simple machine that consists of an axle attached to the center of a larger wheel, so that the shaft and wheel rotate together. (p. 106)

onda: perturbación que transfiere energía de un lugar a otro sin transferir materia. (pág. 529)

longitud de onda: distancia entre un punto de una onda y el punto más cercano similar al primero. (pág. 541, 575)

cuña: máquina simple que consiste en un plano inclinado con uno o dos lados inclinados; se usa para partir o separar un objeto. (pág. 108)

peso: fuerza gravitacional ejercida sobre un objeto. (pág. 48)

rueda y eje: máquina simple que consiste en un eje insertado en el centro de una rueda grande, de manera que el eje y la rueda rotan juntos. (pág. 106)

work: the amount of energy used as a force moves an object over a distance. (pp. 87, 164)

trabajo: cantidad de energía usada como fuerza que mueve un objeto a cierta distancia. (pág. 87, 164)

X-ray: a high-energy electromagnetic wave that has a slightly shorter wavelength and higher frequency than an ultraviolet wave. (p. 611)

rayo X: onda electromagnética de alta energía que tiene una longitud de onda ligeramente más corta y frecuencia más alta que una onda ultravioleta. (pág. 611)

Index

Italic numbers = illustration/photo **Bold numbers = vocabulary term**
lab = indicates entry is used in a lab on this page

A

Absorption, 548
Academic Vocabulary, 11, 48, 91, 366, 402, 511, 552. *See also* **Vocabulary**
Acceleration
- average, 29
- calculation of, 29
- changes in, 28, *28*
- effect of change in mass or force on, 68
- explanation of, **7, 27,** 65
- method to represent, 28, *28*
- Newton's second law of motion and, 65
- speed change and, 30 *lab*
- unbalanced forces and, 64, *64*

Actinide series, 359
Action force, 71, *72*
Adenine, 511
Airships, 388
Alcohols, 500, *500*
Alkali metals, *357,* **357**
Alkaline earth metals, 357
Alkane, 493, *493*
Alkene, 493, *493*
Alkyne, 493, *493*
Amino acids, 509, *509*
Amino group, 502, *502*
Amorphous, 491
Amorphous carbon, 491, *491*
Amplitude
- energy and, 539–540, *540*
- explanation of, **539**
- of longitudinal waves, 540, *540, 541*
- of transverse waves, 540, *540*

Argon, 366, 385
Arms, as levers, 106, *106*
Asteroids, 52
Atomic number, 347
Atoms
- electrons in, *382,* 382–385, *383–385*
- explanation of, 382
- of polar and nonpolar molecules, 393, *393*
- stable and unstable, 386, *386*

Average acceleration, 29
Average acceleration equation, 29
Average speed, 19
Average speed equation, 19
Axle, 106. *See also* **Wheel and axle**

B

Balanced forces
- explanation of, **56**
- illustration of, *63*
- Newton's first law of motion and, 57, 60

Barium, as alkaline earth metal, 357
Beryllium, 357
Big Idea, 36, 78, 114, 188, 372, 406, 516, 556
- Review, 39, 81, 117, 191, 375, 409, 519, 559

Biological molecules, 508
Biomass, 182, *184*
Bohr, Neils, *351*
Boiling point, 346
Bonds, 390. *See also* **Chemical bonds**
Bromine
- explanation of, 501
- properties of, 365, *365*

Bromomethane, 501, *501*
Butane, *492*

C

Cadmium, 347, *347*
Calcium, 357
Carbohydrates
- consumption of, 510 *lab*
- explanation of, **510**
- starch and cellulose as, 510
- sugars as, 510, *510*

Carbon. *See also* **Hydrocarbons; Organic compounds**
- amorphous, 491
- bonding with, 490, *490,* 491, 493, *493,* 494 *lab*
- forms of pure, 491, *491*
- in human body, 363, *363*
- in living things, 489, *489*
- properties of, *364,* 365
- structural change in, 508 *lab*
- testing for, 514–515 *lab*
- uniqueness of, 489 *lab,* 490

Carbon dioxide
- in atmosphere, 167, 179
- chemical formula for, 394, *394*
- explanation of, 490

Carbon monoxide, 490
Carboxylic acid, 501
Careers in Science, 537
Cell phones, 172, *172*
Cellulose, 510
Centripetal force, 66
Cesium, 357
Chapter Review, 38–39, 80–81, 116–117, 190–191, 374–375, 408–409, 518–519, 558–559
Chemical bonds
- covalent, *391,* 391–392, *392*
- explanation of, **382,** 383–386, **390**

Chemical energy, 172, *173*
Chemical formulas, *394,* **394**
Chemical potential energy
- explanation of, 163, *163*
- in fossil fuels, 179

Chlorine
- explanation of, 501
- properties of, 365, *365*

Circular motion, 66, *66*
Coal, 179
Collisions, 74
Common Use. *See* **Science Use v. Common Use**
Compound machines, 110, *110*
Compounds. *See also* **Organic compounds**
- containing metals, 361
- covalent, 392, 393
- elements that make up, 390, 390 *lab*
- explanation of, **382**
- formation of, 394 *lab*
- functional groups and, 499 *lab,* *500,* 500–502, *501, 502*
- halogens in, 365
- ionic, 400, 401
- of life, 508–512, *509–512*
- method to model, 396
- organic, 489–495, *490–495*

Compressions, 532, 540
Conduct, 402
Constant speed, 18, 31
Construct, 366
Constructive, 552
Constructive interference, 552
Contact forces, *45,* **45**
Copper
- on periodic table, 347, *347*
- uses for, 355

Covalent bonds
- double and triple, 392, *392*
- electron sharing and, 391
- explanation of, **391,** *402,* **490**

Covalent compounds
- chemical compounds and, 394
- explanation of, 392
- molecule as unit of, 392
- nonpolar molecules and, 393, *393*
- polar molecules and, 393, *393*

Crests, 531
Critical thinking, 14, 24, 33, 39, 51, 59, 67, 75, 81, 92, 100, 111, 166, 174, 185, 191, 352, 360, 369, 375, 387, 395, 403, 409, 496, 505, 513, 519, 536, 544, 553
Cyclobutane, *492*
Cytosine, 511

D

Dams, 181, *181*
Density, 356

Deoxyribonucleic acid (DNA). *See* **DNA**
Deoxyribose, 511
Destructive interference, 552
Diamonds, 491, *491*
Diffraction, **550,** 551
Dimensions, *12,* **12**
Direction, 28
Displacement, **13**
Distance
displacement and, 13
explanation of, 88
gravitational force and, 47, *47*
Distance-time graphs
changing speed and, 22
comparison of speeds on, 21, *21*
explanation of, 20
used to calculate speed, 21, *21, 22*
DNA, 511, *511*
Ductility, **356**

E

Earth, 47, 52, 66
Earthquakes, 537
Efficiency, **99**
Efficiency equation, 99
Elastic potential energy, 163, *163*
Electrical energy
changed to radiant energy, 169, *169*
explanation of, *165,* **165**
kinetic energy transformed into, 181
nuclear energy transformed to, 180, *180*
sources of, 178, *179,* 183 *lab*
use of, 173
Electric power plants, 178, *178*
Electromechanic waves
absorption in, 548
explanation of, *535,* **535**
from Sun, 535
types of, 535
Electron dot diagrams
explanation of, **385,** *385,* 386
function of, 396
Electron pooling, 401
Electrons
bonding and, 383
energy and, 383, *383*
noble gases and, 386, *386,* 391
number and arrangement of, 382, *383*
shared, 391
valence, 384, *384,* 386, 391, 392, *392,* 401, 490
Elements. *See also specific elements*
in living things, 489
in periodic table, 345–350, *346, 347, 348–350*
synthetic, 351
transition, *358,* 358–359
Energy
amplitude and, 539–540, *540*
changes between forms of, 169–170, 169 *lab,* 177
from electric power plants, 178, *178*
electrons and, 383, *383*
explanation of, **161, 529**
forms of, 164, 165, *165,* 173 *lab*
kinetic, 162
law of conservation of, **170**–171
potential, 162–163
sounds and, 539 *lab*
sources of, 177, *177, 183*
transformations of, 175, *175*
use of, 172–173
waves as source of, *529,* 529–530, *530*
work and, 87, 90, *90,* 164
Energy resources
advantages and disadvantages of, *184*
conservation of, 183
differences between, 177 *lab*
explanation of, **159**
inexhaustible, 181
nonrenewable, 178–180, *179, 180, 183*
renewable, 180–182, *181, 182*
Energy transfer, 530, *530*
Ethanoic acid, 501, *501*
Ethanol, 499
Ethene, 503
Ethyl, 502
Ethylene, 503, *503*

F

Feet, 106, *106*
Fireworks, 361
First-class levers
explanation of, 104, *104*
in human body, 106, *106*
mechanical advantage of, 105, *105*
Fluorine
explanation of, 501
properties of, 365, *365*
Foldables, 10, 22, 29, 37, 46, 58, 65, 72, 79, 88, 98, 107, 115, 162, 172, 178, 189, 347, 356, 365, 373, 383, 391, 398, 407, 492, 499, 508, 517, 535, 539, 551, 557
Force
acceleration and, 68
at angle, 89, *89*
balances, 56, 57, 60, *63*
centripetal, 66
change in size of, 97, *97*
contact, 45, *45*
direction of, 46, *46,* 97, *97*
distance for action of, 97, *97*
explanation of, **45,** 54, 88
friction as, *49,* 49–50, *50*
gravitational, 46–48, *47, 48*
identification of, 54
input to output, 96, *96*
magnetic, 54 *lab*
mass and, 64 *lab*
net, *55,* 55–56, *56*
noncontact, 46, *46*
opposite, 70, *70*
unbalanced, 56, *56,* 62–64, *63, 64*
upward and downward, 90, *90*
work and, 89, *89*
Force pair, 71, *71*
Fossil fuels
advantages and disadvantages of, *184*
CO_2 levels and, 167
explanation of, **178,** 504
formation of, 179, *179*
global warming and, 167, 179
supply of, 183
use of, 179
Francium, 357
Frequency
explanation of, **542**
unit for, 542
wavelength and, 541 *lab,* 542, *542*
of waves, 542, 545 *lab*
Friction
cause of, 50, *50*
explanation of, **49, 159, 171**
fluid, 49
method to decrease, 50, *50*
motion and, 50 *lab,* 58
sliding, 49, *49*
static, 49, *49*
Fructose, 510, *510*
Fulcrum
explanation of, **104,** *104*
location of, 104, 105
Fullerene, 491, *491*
Functional groups
amino, 502, *502*
carboxyl, 501, *501*
effect on compounds, 499 *lab*
explanation of, **500**
halide, 501, *501*
hydroxyl, 500, *500*

G

Gases, noble, 366, 386, *386*
Gears, 110
Genetic information, **508,** 511. *See also* **Heredity**
Geothermal energy, 182, *184*
Germanium, 368, 490
Global Positioning System (GPS), 15
Global warming, 167, 179
Glucose, 510, *510*
Gold
on periodic table, 347, *347*
properties of, 356
uses for, 355, *356*
Graphite, 491, *491*
Gravitational force
distance and, 47, *47*
explanation of, 47
mass and, 47, *47*
weight and, 48, *48*
Gravitational potential energy
explanation of, 163, *163*
transformed into kinetic energy, 175, *175*
Gravity, **47,** 66
Greenhouse gases, 167
Green Science, 167, 388
Groups, in periodic table, **350,** 381, *382*
Guanine, 511
Gunpowder, 361

H

Halide, 501
Halide group, *501*, **501**
Halogens, 365, *365*, 501
Helium
atoms of, 385
electron structure of, 386
on periodic table, 349, *349*
Helium airships, 388
Heredity, 511, *511*. *See also* **Genetic information**
Hertz (Hz), 542
Hindenburg, 388, *388*
Horizontal, 30
Horizontal axis, 30
Horsepower, 93
How It Works, 15
Hubble Space Telescope, 547, *547*
Human body
levers in, 106, *106*
proteins and, 509
Humans, 363, *363*
Hydrocarbons
carbon-to-carbon bonding and, 493, *493*
chains, branched chains, and rings of, 492, *492*, 494, *494*, 495, *495*
explanation of, **492**
method to name, 494, 495, *495*
saturated, 493
substituted, 499, 500, 506
unsaturated, 493
Hydroelectric power plants
advantages and disadvantages of, *184*
explanation of, 181, *181*
Hydrogen
atoms of, 385, 393
in human body, 363, *363*
in hydrocarbons, 492
in methane, *492*
properties of, 366
in universe, *366*
Hydrogen airships, 388, *388*
Hydrogen molecules, 393
Hydroxyl group, *500*, **500**

I

Ideal mechanical advantage
explanation of, 98
of inclined planes, 108
of lever, 105
of pulley, 109
of wheel and axle, 107
Inclined planes
explanation of, *103*, **107,** *107*
mechanical advantage of, 108
Inertia, 58
Inexhaustible energy resources, 181, 182
Infrared waves, 535, *535*
Input force, 96, *96*, 98
Input work, 96, 99
Instantaneous speed, 18
Interference
constructive and destructive, 552
explanation of, *551*, **551**
standing waves and, 552
Interpret Graphics, 14, 24, 33, 51, 59, 67, 75, 92, 100, 111, 166, 174, 185, 352, 360, 369, 387, 395, 403, 496, 505, 513, 536, 544, 553
Iodine
explanation of, 501
properties of, 365, *365*
Ionic bonds, *400*, **400,** *402*
Ionic compounds, 400, 401
Ions
determining charge of, 400
explanation of, **398**
gaining valence electrons and, 399, *399*
losing valence electrons and, 399, *399*
in solution, 404–405 *lab*
Isobutane, *492*
Isomers, 492

J

Joule (J), 88
Jumping, action and reaction forces in, *72*

K

Key Concepts, 8, 16, 26, 44, 53, 61, 69, 86, 94, 102, 160, 168, 176, 344, 354, 362, 380, 389, 397, 488, 498, 507, 528, 538, 546
Check, 10, 12, 13, 17, 21, 23, 28, 32, 46, 47, 50, 57, 58, 65, 66, 71, 74, 88, 90, 91, 97–99, 103, 105, 110, 161–165, 170, 171, 173, 178, 180, 183, 347, 350, 355, 356, 364, 367, 383, 386, 390, 392, 393, 400, 401, 490, 493, 502, 503, 508, 511, 529, 531, 532, 535, 542, 543, 548, 551, 552
Summary, 36, 78, 114, 188, 372, 406, 516, 556
Understand, 14, 24, 33, 38, 51, 59, 67, 75, 80, 92, 100, 111, 116, 166, 174, 190, 352, 360, 369, 387, 395, 403, 408, 496, 505, 513, 518, 536, 544, 553
Kilogram (kg), 47
Kinetic energy
changes between potential energy and, 170, *170*
explanation of, 162, *162*
gravitational potential energy transformed into, 175, *175*
transformed into electrical energy, 181
Krypton, as noble gas, 366

L

Lab, 34–35, 76–77, 112–113, 186–187, 370–371, 404–405, 514–515, 554–555. *See also* **Launch Lab; MiniLab; Skill Practice**
Lanthanide series, 359
Launch Lab, 9, 17, 27, 45, 54, 62, 70, 87, 95, 103, 161, 169, 177, 345, 355, 363, 381, 390, 398, 489, 499, 508, 529, 539, 547
Law of conservation of energy
explanation of, **170,** 177
friction and, 171, *171*
Law of conservation of momentum, 74
Law of reflection, 549
Law of universal gravitation, 47
Lesson Review, 14, 24, 33, 51, 59, 67, 75, 92, 100, 111, 166, 174, 185, 352, 360, 369, 387, 395, 403, 496, 505, 513, 536, 544, 553
Levers
classes of, 104, *104*
explanation of, *103*, **104**
in human body, 106, *106*
mechanical advantages of, 105, *105*
Lewis, Gilbert, 385
Light energy. *See* **Radiant energy**
Light waves. *See also* **Waves**
diffraction of, 551
energy source of, 529, *529*
reflection of, 549
refraction of, 550
transmission of, 548
Linear molecules, 502, *502*
Lipids, *512*, **512**
Lithium, 357, *357*
Longitudinal waves
amplitude and energy of, 540, *540*, *541*
explanation of, 532, *532*
wavelength of, 541, *541*
Lubricants, 50, *50*
Luster, 355

M

Machines
comparison of, 112–113 *lab*
compound, 110, *110*
efficiency of, 99, *99*
explanation of, 95, *95*
function of, 95 *lab*, 96, *96*
mechanical advantage of, 98, 101
methods that ease work by, 96–97, 96 *lab*, *97*
simple, *103*, 103–109, *104–109*
Magnesium, 357
Magnetic forces, 54 *lab*
Malleability, 356, 359
Mass
acceleration and change in, 68
explanation of, **47**
forces and, 64 *lab*
gravitational force and, 47, *47*
kinetic energy and, 162, *162*
relationship between weight and, 48
Math Skills, 33, 39, 65, 67, 73, 75, 81, 88, 91, 92, 98–100, 117, 180, 185, 191, 350, 352, 400, 403, 409, 500, 505, 519, 543, 559

Matter
how waves travel through, 531 *lab*
interaction between waves and, *547*, 547–549, *548*
observations of, 161
Mechanical, 98
Mechanical advantage
effects of, 101
explanation of, **98**
ideal, 105, 107
of inclined planes, 108
of lever, 105, *105*
of pulleys, 109, *109*
of wheel and axle, 107
Mechanical advantage equation, 98
Mechanical energy, *165*, **165**
Mechanical waves
explanation of, **531**
longitudinal waves as, 532, *532*
transverse waves as, 531, *531*
types of, 534
vibrations and, 533, *533*
Medium, 531
Melting point, 346, *346*
Mendeleev, Dimitri, 346, 347, 366
Mercury, *347*
Metallic bonds, 401, *402*
Metalloids
explanation of, **367**
in periodic table, 350, 382
properties and uses of, 368
as semiconductors, 367
Metals
alkali, 357, *357*
alkaline earth, 357
chemical properties of, 356
explanation of, **355**
patterns in properties of, 359, *359*
in periodic table, 350, 355, 382
physical properties of, 355–356, *356*
as transition elements, *358*, 358–359
uses of, 355 *lab*
Meter (m), 88
Methane, 492, *492*, 502
Methanoic acid, 501, *501*
Methyl, 502
Methylamine, 502, *502*
Mettner, Lise, *351*
MiniLab, 11, 20, 30, 50, 57, 64, 74, 89, 96, 109, 164, 173, 183, 351, 359, 368, 386, 394, 401, 494, 503, 510, 531, 541, 549. *See also* **Lab**
Molecular models, 394, *394*
Molecules
biological, 508
explanation of, **392**
nonpolar, 393, *393*
polar, 393, *393*
shapes of, 502, *502*
Momentum
conservation of, 74, 74 *lab*
explanation of, **73**
method to find, 73
Momentum equation, 73
Monomers
amino acid, 509, *509*
explanation of, **503**
nucleic acid polymer, 511
Moon, 47, 52, 66
Moseley, Henry, 347
Motion
along curved path, 62 *lab*
change in, 17 *lab*, 64, *64*
circular, 66, *66*
explanation of, **7,** 13, 32
forces that change, 54, 62–64, *63*, *64*
friction and, 50 *lab*
graphs of, 20 *lab*
Newton's first law of, *57*, 57–58, *58*, 60, 62, 73
Newton's second law of, 65, 72, *73*
Newton's third law of, *71*, 71–72, *72*
unbalanced forces on objects in, 63, *63*, 64, *64*

N

Natural gas, 179
Natural polymers, 503, 508
Neck, 106, *106*
Neon
electron structure of, 386, *386*
as noble gas, 366
Net force
actions by, 63
direction of forces and, *55*, 55–56, *56*
explanation of, **55**
Neutrons, 382, *382*
Newton, Isaac, 47, 57, 65
Newton (N), 46, 88
Newton-meter (N·m), 88
Newton's first law of motion
balanced forces and, 57, *57*
explanation of, **57,** 62, 73, 76 *lab*
inertia and, 58, *58*
modeling of, 60, 76–77 *lab*
momentum and, 73
Newton's second law of motion
explanation of, **65,** 73, 76 *lab*
modeling of, 76–77 *lab*
momentum and, 73
Newton's third law of motion
action and reaction and, 71, *71*
application of, 72, *72*
explanation of, **71,** 76 *lab*
force pairs and, 71, *71*
modeling of, 76–77 *lab*
NGC 3603, *547*
Nitrogen
compounds with, 365
in human body, 363, *363*
Noble gases
electron arrangement in, 391
explanation of, **366,** 386, *386*
Noncontact forces, *46*, **46**
Nonmetals
explanation of, **363**
metals v., *364*, 364–365, *365*
in periodic table, 350, 365, 382
properties of, 363 *lab*, 364, *364*
Nonpolar molecules, 393, *393*
Nonrenewable energy resources
advantages and disadvantages of, *184*
explanation of, **178**
fossil fuels as, 178–179, *179*
nuclear energy as, 180, *180*
percentage of energy from, *183*
Normal, 549
Nuclear energy
advantages and disadvantages of, *184*
explanation of, **165,** *165*, 180
transformed to electrical energy, 180, *180*
Nuclear power plants, 180, *180*
Nuclear waste, 180
Nuclei, 180
Nucleic acids, 511
Nucleotides, 511, *511*
Nucleus, 382, *382*

O

Objects at rest, 30, *30*
Oleic acid, *512*
Opposite forces, 70, *70*, 70 *lab*
Organic, 490
Organic compounds
carbon and, *490*, 490–491, *491*, 508 *lab*
explanation of, **490**
functional groups and, 499 *lab*, *500*, 500–502, *501*, *502*
hydrocarbon, *492*, 492–495, *493–495*
molecule shape and, 502, *502*
polymer, 503, *503*
substituted hydrocarbon, 499
synthetic polymer, 504, *504*
Osprey, 54
Output force, 96, *96*, 98
Output work, 96, 99
Oxygen
atoms of, 391
compounds with, 365
in human body, 363, *363*

P

Periodic, 346
Periodic table. *See also specific elements*
arrangement of, 353
carbon group in, 490
development of, *346*, 346–347, *347*
element key in, 349, *349*
explanation of, **345,** 346
groups in, 350, *350*, 381, *382*
illustration of, *348–349*
modeling procedure used to develop, 370–371 *lab*
organization of, 381 *lab*, 382
periods in, 350, *350*, 381, *382*
use by scientists of, 351
Periods, in periodic table, **350,** 381, *382*
Petroleum, 179, 504
Phospholipids, 512, *512*

Phosphorus
compounds with, 365
in human body, 363
properties of, *364*
Planar molecules, 502, *502*
Plastic, 503. *See also* **Polymers**
Plates, 45
Plutonium, 359
Polar molecules, 393, *393*
Polyethylene, 504, *504*
Polymerization, 503
Polymers
explanation of, **503**
formation of, 503, *503*
making, 503 *lab*
natural, 503
synthetic, 503, 504, *504*, 508
Position
change in, 13, *13*
explanation of, **9**
in two dimensions, 12, *12*
use of reference point to describe, *10*, 10–11, *11*
Potassium
as alkali metal, 357, *357*
properties of, 359
Potassium nitrate, 361
Potential energy
changes between kinetic energy and, 170, *170*, 175, *175*
explanation of, **159, 162**
types of, 163, *163*
work and, *164*
Power, 91
Power equation, 91
Proteins, *509,* **509**
Protons, 382, *382*
Pulleys
explanation of, *103*, **109**
fixed, 109
mechanical advantage of, 109, *109*
movable, 109
systems of, 109

R

Radiant energy
explanation of, *165*, **165, 169,** *169*
use of, 172, *172*
Radium, 357
Radon, 366
Raindrops, 529
Rarefactions, 532, 540
Reaction force, 71, *72*
Reading Check, 9, 10, 12, 18, 20, 22, 27, 28, 31, 47, 48, 50, 55, 57, 63, 64, 70, 72, 89, 90, 95, 97, 104, 108, 163, 170, 173, 179, 180, 346, 351, 357, 359, 363, 365, 366, 368, 381, 382, 385, 394, 398, 399, 402, 489, 493, 495, 504, 509, 530, 533, 539, 540, 548, 549
Reference direction, 11, *11*, **55**
Reference point
change, 10, *10*
to describe position, 10
explanation of, **9**
motion relative to, 13, *13*
use of, 11 *lab*
Reflection
explanation of, **548**
law of, 549, *549*
Refraction, *550*, **550**
Relative, 9
Renewable energy resources
advantages and disadvantages of, *184*
biomass as, 182
explanation of, **180**
geothermal energy as, 182, *182*
hydroelectric power plants and, 181, *181*
solar energy as, 181, *181*
wind energy as, 182, *182*
Reservoir, 181, *181*
Review Vocabulary, 12, 55, 180, 356, 382, 490, 508, 529. *See also* **Vocabulary**
Ribonucleic acid (RNA). *See* **RNA**
RNA, 511
Rockets, *72*
Rubidium, 357

S

Saltpeter, 361
Satellites, 66
Saturated hydrocarbons, 493, *493*
Saturated lipids, 512, *512*
Science Methods, 35, 113, 187, 371, 405, 515, 555
Science & Society, 52, 93, 361
Science Use v. Common Use, 9, 49, 91, 169, 348, 390, 490, 549. *See also* **Vocabulary**
Screws, *103, 108,* **108**
Seaborg, Glen T., *351*
Second-class levers
explanation of, 104, *104*, 105, *105*
in human body, 106, *106*
mechanical advantage of, 105, *105*
Seismic waves, *534*
Selenium, 365
Semiconductors, 367
Significant, 48
Silicon
explanation of, 367, 490
uses of, 367, *367*, 368
Silver, 347, *347*
Simple machines. *See also* **Machines**
comparison of, 112–113 *lab*
explanation of, **103**
Skill Practice, 25, 60, 68, 101, 175, 353, 396, 506, 545. *See also* **Lab**
Sliding friction, 49, *49*
Sodium, 357, *357*
Solar energy
advantages and disadvantages of, *184*
explanation of, 181, *181*
Sound energy, *165*, **165**
Sounds, 539 *lab*
Sound waves. *See also* **Waves**
diffraction of, 551
explanation of, *534*
reflection of, 549
speed of, 543, *543*
Specify, 11
Speed
average, 19
calculation of, 25, 34–35 *lab*
change in, 18, *18*
constant, 18
on distance-time graphs, 21, *21*, 22, *22*
explanation of, **17**
instantaneous, 18
kinetic energy and, 162
units of, 17, *17*, 19
Speed-time graphs
elements of, *30*, 30–32, *31, 32*
explanation of, 30
limitations of, 32
Standardized Test Practice, 40–41, 82–83, 118–119, 192–193, 376–377, 410–411, 520–521, 560–561
Starch, 510
Static, 49
Static friction, *49*, **49**
Steam engine, 93
Stearic acid, *512*
Strontium, 357
Study Guide, 36–37, 78–79, 114–115, 188–189, 372–373, 406–407, 516–517, 556–557
Substituted hydrocarbons
explanation of, 499, 500
Vitamin C as, 506
Sugar, 392, 393
Sugars, *510*, 510
Sulfur
in human body, 363
properties of, 365, 368
Sun
electromagnetic waves from, 535
energy from, 177, *177*
Swimming, *72*
Synthetic polymers
explanation of, 503, 504, *504*
materials made from, 508

T

Tetrahedral molecules, 502, *502*
Thermal energy
chemical potential energy changed to, 179
conduction of, 359 *lab*, 368 *lab*
explanation of, *165*, **165**
friction and, 171, *171*
inside Earth, 182
use of, 172
Third-class levers
explanation of, 104, *104*
in human body, 106, *106*
mechanical advantage of, 105, *105*
Titov, Vasily, 537
Transfer, 91
Transition elements, *358*, **358**
Transmission, 548
Transmit, 511

Transverse waves
- amplitude of, 539, *540*
- energy of, 540
- explanation of, *531*, **531**
- wavelength of, 541, *541*

Troughs, 531
Tsunamis, 537, *537*

Ultraviolet waves, 535
Unbalanced forces
- acceleration and, 64
- change in direction and, 64, *64*
- explanation of, **56,** 60
- objects in motion and, 63, *63*
- objects at rest and, 63, *63*
- velocity and, 57, 62, 64

Unsaturated hydrocarbons, 493, *493*
Unsaturated lipids, 512, *512*
Uracil, 511

Valence electrons
- covalent bond and, 392, *392*
- explanation of, 384, *384*, 386, 401, 490, 493
- metallic bonds and, 401
- shared electrons as, 391

Velocity
- changes in, 23, *23*, 27–28, 27 *lab*, *28*
- explanation of, **7, 23**
- method to represent, 23, *23*
- momentum and, 73
- unbalanced forces and, 57, 62–64

Vertical, 30
Vertical axis, 30
Vibrations, 533, *533*
Visual Check, 10, 12, 18, 20, 21, 23, 28, 32, 46, 48, 55, 56, 58, 63, 66, 71, 72, 89, 90, 105, 110, 170, 178, 181, 349, 356, 364, 365, 384, 392, 491, 501, 511, 530, 534, 540, 548, 551
Vitamin C, testing for, 506
Vocabulary, 8, 16, 26, 43, 44, 53, 61, 69, 78, 85, 86, 94, 102, 114, 159, 160, 168, 176, 343, 344, 354, 362, 372, 379, 380, 389, 397, 406, 487, 488, 498, 507, 516, 527, 528, 538, 546. *See also* **Academic Vocabulary; Review Vocabulary; Science Use v. Common Use; Word Origin**
- Use, 14, 24, 33, 37, 51, 59, 67, 75, 79, 92, 100, 111, 115, 166, 174, 185, 189, 352, 360, 369, 373, 387, 395, 403, 407, 496, 505, 513, 517, 536, 544, 553, 557

Waste energy, 173, *173*
Water molecule, 393, *393*
Water waves. *See also* **Waves**
- energy transfer by, 529, 530, *530*
- explanation of, *534*
- reflection of, 549

Watt, James, 93
Wavelength
- explanation of, *541*, **541,** *542*
- frequency and, 541 *lab*, 542, *542*

Waves
- amplitude and, 539–540, *540*, 545
- collision of, 547 *lab*
- diffraction of, *550*, 550–551
- electromechanic, 535, *535*
- energy transfer by, 530, *530*
- explanation of, **529**
- frequency of, 542, *542*, 545 *lab*
- interaction between matter and, *547*, 547–549, *548*
- interference of, *551*, 551–552, *552*
- longitudinal, 532, *532*
- mechanical, 531–534, *533*, *534*
- method to create, 529 *lab*
- reflection of, 548–549, *549*
- refraction of, 550, *550*
- as source of energy, 529, *529*
- speed of, 543, 554–555 *lab*
- standing, 552, *552*
- transverse, 531, *531*
- tsunamis as, 537

Wave speed equation, 543
Wedges, *103*, **108,** *108*
Weight, *48*, **48**
What do you think?, 7, 14, 24, 33, 43, 51, 59, 67, 75, 85, 92, 100, 111, 166, 174, 185, 343, 352, 360, 369, 379, 387, 395, 403, 487, 496, 505, 513, 527, 536, 544, 553
Wheel and axle
- explanation of, *103*, *106*, **106**
- mechanical advantage of, 107
- use of, 107

Wind energy
- advantages and disadvantages of, *184*
- explanation of, 182, *182*

Wind turbines
- explanation of, 182, *182*
- observation of, 186–187 *lab*

Word Origin, 13, 23, 30, 45, 58, 66, 73, 87, 98, 161, 171, 178, 347, 356, 365, 367, 384, 393, 491, 501, 535, 539. *See also* **Vocabulary**
Work
- energy and, 90, *90*, 164, *164*
- explanation of, **87**
- factors that affect, *89*, 89–90, 89 *lab*, *90*
- input to output, 96, 99
- machines and, 96–97, 96 *lab*, *97*
- method to calculate, 88
- scientific meaning of, 87 *lab*

Work equation, 88
Writing In Science, 39, 81, 117, 409, 497, 519, 559

***x*-axis,** 30
Xenon, 366

***y*-axis,** 30

Z

Zinc, 347, *347*

Credits

Photo Credits

ConnectED (t)Richard Hutchings, (c)Getty Images, (b)Jupiter Images/Thinkstock/Alamy; **ix** (b)Fancy Photography/Veer; **vii** Ransom Studios; **NOS 02–03** NASA—digital version copyright/Science Faction/Corbis; **NOS 04** (inset)SMC Images/Getty Images, (bkgd)Popperfoto/Getty Images; **NOS 05** (t)Maria Stenzel/Getty, (c)Stephen Alvarez/Getty Images, (b)Science Source/Photo Researchers, Inc.; **NOS 06** Martyn Chillmaid/photolibrary.com; **NOS 07** NASA; **NOS 09** (t)Andy Sacks/Getty Images, (c)NASA, H. Ford (JHU), G. Illingworth (UCSC/LO), M.Clampin (STScl), G. Hartig (STScl), the ACS Science Team, and ESA, (b)Michael Rosenfeld/Science Faction/Corbis; **NOS 10** StockShot/Alamy; **NOS 11** Michael Newman/PhotoEdit; **NOS 12** Tim Wright/Corbis; **NOS 13 NOS 14** Hutchings Photography/Digital Light Source; **NOS 16** (tl)Matt Meadows, (tr)ASP/YPP/age fotostock, (bl)Blair Seitz/photolibrary.com, (br)The McGraw-Hill Companies, Inc./Louis Rosenstock, photographer; **NOS 17** (t)The McGraw-Hill Comapanies, (c)photostock1/Alamy, (b)Blend Images/Alamy; **NOS 18** (tl)The McGraw-Hill Companies, (tr)Richard Hutchings (see Digital Light Source), (bl)The McGraw-Hill Companies, (br)The McGraw-Hill Companies; **NOS 19** (t to b)Hutchings Photography, (2)Macmillan/McGraw-Hill, (3)(4)Hutchings Photography/Digital Light Source, (5)Carlos Casariego/Getty Images; **NOS 20** Mandel Ngan/AFP/Getty Images; **NOS 21** Brian Stevenson/Getty Images; **NOS 22** (t)Hisham Ibrahim/Getty Images (bl)Bordner Aerials, AP Photo/NTSB; **NOS 23** AP Photo/The Minnesota Daily, Stacy Bengs; **NOS 24** Larry Downing/Reuters/Landov; **NOS 25** Plus Pix/age fotostock; **NOS 26** Associated Press; **NOS 28** (br) (inset)(t to b)(2)(3)Macmillan/McGraw-Hill, (4)(5)Hutchings Photography/Digital Light Source; **NOS** 29 Hutchings Photography/Digital Light Source; **NOS 31** (l)Andy Sacks/Getty Images, (r)NASA—digital version copyright/Science Faction/Corbis; **4** (t)Wildscape/Alamy, (b)Andy Crawford/BBC Visual Effects—modelmaker/Dorling Kindersley; **5** (t)Miriam Maslo/Science Photo Library/Corbis, (b)The Gravity Group, LLC; **6–7** Andrew Holt/Getty Images; **7** Camille Moirenc/Getty Images; **8** Hutchings Photography/Digital Light Source; **12** Aerial Photos of New Jersey; **14** Camille Moirenc/Getty Images; **15** David J. Green—technology/Alamy; **16** Steve Bloom Images/Alamy; **17** (t to b)Hutchings Photography/Digital Light Source, (2)Department of Defense, (3)Car Culture/Corbis, (4)Andersen Ross/Jupiter Images; **20** Hutchings Photography/Digital Light Source; **22** Michael Dunning/Photo Researchers; **23** 24 Hutchings Photography/Digital Light Source; **25** (r)(bl)(t to b)(2)(3)(4)Richard Hutchings/Digital Light Source, (5)Aaron Haupt; **26** Leo Dennis Productions/Brand X/Corbis; **30** (t)Hutchings Photography/Digital Light Source, (b)Robert Holmes/Corbis; **31** (t)Comstock/PunchStock, (b)Eyecon Images/Alamy Images; **32** (t)NASA, (b)Agence Zoom/Getty Images; **33** Leo Dennis Productions/Brand X/Corbis; **34** Hutchings Photography/Digital Light Source; **36** (t)Camille Moirenc/Getty Images, (c)Steve Bloom Images/Alamy, (b)Agence Zoom/Getty Images; **39** Andrew Holt/Getty Images; **42-43** Jason Horowitz/zefa/Corbis; **44** NASA; **45** (t)Hutchings Photography/Digital Light Source, (b)Terje Rakke/Getty Images; **46** (t)Hutchings Photography/Digital Light Source, (c)Daniel Smith/Getty Images, (b)Ryuhei Shindo/Corbis/Jupiter Images; **48** (l)(r)NASA, (tl)(tr)Horizons Companies; **49** (b)Hutchings Photography/Digital Light Source; **50** Hutchings Photography/Digital Light Source; **51** (t)Terje Rakke/Getty Images, (b)Hutchings Photography/Digital Light Source; **52** (t)NASA/Goddard Space Flight Center Scientific Visualization Studio, (c)VisionsofAmerica/Joe Sohm/Getty Images, (b)Ursula Gahwiler/photolibrary.com, (bkgd)StockTrek/Getty Images; **53** Bryan Mullennix/Getty Images; **54** (t)Hutchings Photography/Digital Light Source, (b)Marvin E. Newman/Getty Images; **55** 56 Hutchings Photography/Digital Light Source; **57** (t)Tim Keatley/Alamy, (b)Hutchings Photography/Digital Light Source; **58** AP Photo/Insurance Institute for Highway Safety; **59** (t)Hutchings Photography/Digital Light Source, (c)Tim Keatley/Alamy, (b)AP Photo/Insurance Institute for Highway Safety; **60** (t)Macmillan/McGraw-Hill, (b)Hutchings Photography/Digital Light Source; **61** Michael Steele/Getty Images; **62** Hutchings Photography/Digital Light Source; **63** (l)Tim Garcha/Corbis, (r)Myrleen Ferguson Cate/PhotoEdit; **64** Hutchings Photography/Digital Light Source; **67** Tim Garcha/Corbis; **68** Hutchings Photography/Digital Light Source; **69** Patrik Giardino/Corbis; **70** Hutchings Photography/Digital Light Source; **71** (t)LIN HUI/Xinhua/Landov, (tl)David Madison/Getty Images, (b)Duncan Soar/Alamy; **72** (r)Design Pics Inc./Alamy, (bl)U.S. Air Force photo by Carleton Bailie; **74** (tl)Richard Megna, Fundamental Photographs, NYC, (r)Hutchings Photography/Digital Light Source, (bl)Richard Megna, Fundamental Photographs, NYC; **75** (t)Hutchings Photography/Digital Light Source, (c)David Madison/Getty Images, (b)Richard Megna, Fundamental Photographs, NYC; **76** (tr)Macmillan/McGraw-Hill, (cr)Hutchings Photography/Digital Light Source, (bl)(br)Macmillan/McGraw-Hill, (inset)C Squared Studios/Getty Images, (t to b)Hutchings Photography/Digital Light Source, (2)Macmillan/McGraw-Hill, (3)Hutchings Photography/Digital Light Source, (4)Macmillan/McGraw-Hill; **77** Hutchings Photography/Digital Light Source; **78** (t to b)Hutchings Photography/Digital Light Source, (2)Marvin E. Newman/Getty Images, (3)Myrleen Ferguson Cate/PhotoEdit, (4)LIN HUI/Xinhua/Landov; **81** Jason Horowitz/zefa/Corbis; **84–85** Neil Duncan/photolibrary.com; **86** Malcolm Fife/Getty Images; **87** (t)Hutchings Photography/Digital Light Source, (bl) (br)Jupiterimages/Brand X/Alamy; **88** 89 (l)Hutchings Photography/Digital Light Source; **90** Hutchings Photography/Digital Light Source; **92** (tl)Jupiterimages/Brand X/Alamy, (cl)Hutchings Photography/Digital Light Source, (r)Jupiterimages/Brand X/Alamy, (bl)Hutchings Photography/Digital Light Source; **93** (t)P.H. Emerson/George Eastman House/Getty Images, (c)Hulton Archive/Getty Images, (b)Bridgeman Art Library/SuperStock, (inset)Eduardo M. Rivero/age fotostock; **94** Philip and Karen Smith/Getty Images; **95** (t)Hutchings Photography/Digital Light Source, (c)Clive Streeter/Getty Images, (b)Corbis; **96** (t)Hutchings Photography/Digital Light Source, (bl)(br)The McGraw-Hill Companies; **97** (t)Steve Gorton/Dorling Kindersley/Getty Images (c)Sean Justice/Getty Images, (b)Dorling Kindersley; **99** Glowimages/Getty Images; **100** (tl)Clive Streeter/Getty Images, (cl)(cr)The McGraw-Hill Companies, (bl)Glowimages/Getty Images, (br)Sean Justice/Getty Images; **101** (t to b)(2)(3)Hutchings Photography/Digital Light Source, (4)Macmillan/McGraw-Hill; **102** SoloStock Travel/Alamy; **103** (t)McGraw-Hill, (cl)Bob Elsdale, (c)Rod McLean/Alamy, (cr)David Papazian Photography Inc./Jupiterimages, (bl)The McGraw-Hill Companies, (bc)F. Schussler/PhotoLink/Getty Images, (br)Susan E. Degginger/Alamy; **104** (t)Jupiterimages, (c)Digital Vision/Alamy, (b)Andy Aitchison/Corbis; **106** (tl)Matt Carr/Getty Images, (tc)Maurilio Cheli/epa/Corbis, (tr)liquidlibrary/PictureQuest, (b)Jupiterimages/Brand X/Alamy; **108** (t)Mark Douet/Getty Images, (b)Reimar/Alamy; **109** Hutchings Photography/Digital Light Source; **110** (t)imagebroker/Alamy, (b)Brand Z/Alamy; **111** (tl)F. Schussler/PhotoLink/Getty Images, (cl)Digital Vision/Alamy, (r)Corbis, (bl)Mark Douet/Getty Images; **112** (br)(t to b)Hutchings Photography/Digital Light Source, (2)The McGraw-Hill Companies, (3)Macmillan/McGraw-Hill, (4)The McGraw-Hill Companies, (5)Hutchings Photography/Digital Light Source; **113** Hutchings Photography/Digital Light Source; **114** (t)Hutchings Photography/Digital Light Source, (c)Steve Gorton/Dorling Kindersley/Getty Images, (b)Jupiterimages; **117** (l)Hutchings Photography/Digital Light Source, (r)Neil Duncan/photolibrary.com; **156** (t)Jochen Tack/photolibrary.com, (bl)Biophoto Associates/Photo Researchers, Inc., (bc)adrian davies/Alamy, (br)Oleksiy Maksymenko/Alamy; **157** (t)Dennis Kunkel Microscopy, Inc./Visuals Unlimited, Inc., (c)PhotoAlto/PunchStock, (b)Amanda Hall/Robert Harding World Imagery/Getty Images; **158–159** David R. Frazier Photolibrary, Inc./Alamy; **160** Edward Kinsman/Photo Researchers, Inc.; **161** (t)Hutchings Photography/Digital Light Source, (bl)Thinkstock/Corbis, (bc)Sharon Dominick/Getty Images, (br)Dimitri Vervitsiotis/Getty Images; **163** (tl)Hutchings Photography/Digital Light Source, (tr)Hutchings Photography/Digital Light Source, (c)Image Source/Corbis, (b)Thinkstock/Corbis; **164** Hutchings Photography/Digital Light Source; **165** (t to b)David

Credits

Madison/Getty Images, (2)Andrew Lambert Photography/Photo Researchers, Inc., (3)Design Pics/age fotostock, (4)VStock/Alamy, (5)Matthias Kulka/zefa/Corbis, (6)Goodshoot/PunchStock; **166** (t)Thinkstock/Corbis, (c)Hutchings Photography/Digital Light Source, (b)Dimitri Vervitsiotis/Getty Images; **167** (t)Ivy Close Images/Alamy, (b)Lawrence Manning/Corbis, (bkgd)Royalty-Free/Corbis; **168** Alaska Stock LLC/Alamy; **169 171** Hutchings Photography/Digital Light Source; **172** Peter Cade/Getty Images; **173** (t)Hutchings Photography/Digital Light Source, (b)Lorcan/Getty Images; **174** (t)Hutchings Photography/Digital Light Source, (b)Peter Cade/Getty Images; **175** (br)(t to b)Hutchings Photography/Digital Light Source, (2)Macmillan/McGraw-Hill, (3) Hutchings Photography/Digital Light Source, (4)Macmillan/McGraw-Hill, (5)(6)(7)Hutchings Photography/Digital Light Source; **176** Phil Noble/Reuters/Corbis; **177** (t)Hutchings Photography/Digital Light Source, (bl)Jupiterimages, (bc)BrandX/Jupiterimages, (br)Pixtal/age fotostock; **181** Chinch Gryniewicz, Ecoscene/Corbis; **182** Frank Whitney/Getty Images; **184** (t to b)Bill Ross/Corbis, (2)Royalty-Free/Corbis, (3)Courtesy USDA/NRCS, photo by Lynn Betts, (4)Chinch Gryniewicz, Ecoscene/Corbis, (5)Ingram Publishing/SuperStock, (6)Arctic-Images/Corbis, (7)Joel Sartore/National Geographic/Getty Images; **185** (t)Bill Ross/Corbis, (c)Frank Whitney/Getty Images, (b)Getty Images/Steve Allen; **186 187** Hutchings Photography/Digital Light Source; **188** (t)David Madison/Getty Images, (c)Hutchings Photography/Digital Light Source, (b)Royalty-Free/Corbis; **190** Chinch Gryniewicz, Ecoscene/Corbis; **191** David R. Frazier Photolibrary, Inc./Alamy; **194–195** Tyrone Turner/National Geographic Stock; **196** Philip Scalia/Alamy; **197** (t)Hutchings Photography/Digital Light Source, (b)Jamie Sabau/Getty Images; **199** Johner/Getty Images; **201 202** Summer Jones/Alamy; **203** (r)Hutchings Photography/Digital Light Source, (b)Alaska Stock LLC/Alamy; **205** (t)Hutchings Photography/Digital Light Source, (b)The McGraw-HIll Companies; **206** Anthony-Masterson/Getty Images; **208** (t)(c)Matt Meadows, (b)The McGraw-Hill Companies; **209** (t)Corbis (Royalty-Free), (b)Hutchings Photography/Digital Light Source; **210** Matt Meadows; **212** (t)(b)Matt Meadows; **213** (tl)Rachael Bowes/Alamy, (tr)Christopher Weddle/UPI/Landov, (b)Paul Glendell/Alamy, (bkgd)David Papazian/Beateworks/Corbis; **214** Kevin Foy/Alamy; **216** (inset)Thomas Northcut/Getty Images, (bkgd)steven langerman/Alamy; **217** Hutchings Photography/Digital Light Source; **219** (l)steven langerman/Alamy, (r)Thomas Northcut/Getty Images; **220** (r)Hutchings Photography/Digital Light Source, (t to b)(2)Macmillan/McGraw-Hill, (3)Hutchings Photography/Digital Light Source, (4)(5)The Mcgraw-Hill Companies, (6)Macmillan/McGraw-Hill, (7)The Mcgraw-Hill Companies; **221** Hutchings Photography/Digital Light Source; **222** (t)Summer Jones/Alamy, (b)Corbis (Royalty-Free); **224** (l)Anthony-Masterson/Getty Images, (r)Thomas Northcut/Getty Images; **225** (l)Tom Uhlman/Alamy, (r)Tyrone Turner/National Geographic Stock; **270–271** Gregor M. Schmid/Corbis; **272** Atlantide Phototravel/Corbis; **273** Hutchings Photography/Digital Light Source; **275** (t)Royalty Free/Corbis, (c)Steve Hamblin/Alamy, (b)The McGraw-Hill Companies; **276** (t)Vito Palmisano/Getty Images, (b)creativ collection/age fotostock; **277** (l)Mauritius/SuperStock, (r)Hutchings Photography/Digital Light Source; **278** Alberto Coto/Getty Images; **280** (t)Finley—StockFood Munich, (b)The McGraw-Hill Companies, (bkgd)Michael Rosenfeld/Getty Images; **281** Alaska Stock/age fotostock; **282 285** Hutchings Photography/Digital Light Source; **286** (l)Charles D. Winters/Photo Researchers, Inc., (r)Jean du Boisberranger/Getty Images; **288** Hutchings Photography/Digital Light Source; **289** (t)Alaska Stock/age fotostock, (c)Hutchings Photography/Digital Light Source, (bl)Jean du Boisberranger/Getty Images, (br)Charles D. Winters/Photo Researchers, Inc.; **290** Hutchings Photography/Digital Light Source; **291** Check Six/Getty Images; **292 295 298 299** Hutchings Photography/Digital Light Source; **303** Gregor M. Schmid/Corbis; **308** (t)Roger Ressmeyer/Corbis; **308** (c)Getty Images, (b)American Museum of Natural History; **310** (inset)Cern Photo/Frédéric Pitchal/Sygma/Corbis; **310–311** (bkgd)James Brittain/Photolibrary; **312** (inset)Drs. Ali Yazdani & Daniel J. Hornbaker/Photo Researchers, Inc., (bkgd)Alessandro Della Bella/Keystone/Corbis; **313** (t)Hutchings Photography/Digital Light Source, (l to r, t to b)Royalty-Free/Corbis, (2)Creatas/PunchStock, (3)Royalty-Free/Corbis, (4)DAJ/Getty Images; **314** (l)Scala/Art Resource, NY, (r)The Royal Institution, London, UK/The Bridgeman Art Library; **315** (t)Horizons Companies, (b)Drs. Ali Yazdani & Daniel J. Hornbaker/Photo Researchers, Inc.; **320** Hutchings Photography/Digital Light Source; **323** (t)Horizons Companies, (b)Drs. Ali Yazdani & Daniel J. Hornbaker/Photo Researchers, Inc.; **324** Royalty-Free/Corbis; **325** (inset)Derrick Alderman/Alamy, (bkgd)Derrick Alderman/Alamy; **326** Hutchings Photography/Digital Light Source; **329** The McGraw-Hill Companies; **330** (t)SPL/Photo Researchers, Inc., (b)Time Life Pictures/Mansell/Time Life Pictures/Getty Images, (inset)figure 13 John Cancalosi/age fotostock; **334** (t to b)The McGraw-Hill Companies, (2)Hutchings Photography/Digital Light Source, (3)(4)The Mcgraw-Hill Companies; **335** Hutchings Photography/Digital Light Source; **339** James Brittain/Photolibrary; **342–343** Jim Reed/Science Faction/Getty Images; **344** P.J. Stewart/Photo Researchers, Inc.; **345** Hutchings Photography/Digital Light Source; **347** (tl)DEA/A. Rizzi/De Agostini Picture Library/Getty Images, (tr)Astrid & Hanns-Frieder Michler/Photo Researchers, Inc., (cl)Visuals Unlimited/Ken Lucas/Getty Images, (cr)Richard Treptow/Photo Researchers, Inc., (bl)Corbis, (br)ImageState/Alamy; **350** (l)David J. Green/Alamy, (c)Wildlife/Peter Arnold Inc., (r)Mark Schneider/Visuals Unlimited/Getty Images; **351** (l)LBNL/Photo Researchers, Inc., (c)Boyer/Roger Viollet/Getty Images, (r)Ullstein Bild/Peter Arnold, Inc.; **353** Hutchings Photography/Digital Light Source; **354** Paul Katz/photolibrary.com; **355** Hutchings Photography/Digital Light Source; **356** (tl)The McGraw-Hill Companies, (tc)Paul Katz/Getty Images, (tr)Egyptian National Museum, Cairo, Egypt, Photo © Boltin Picture Library/The Bridgeman Art Library International; **356** (bl)NASA, (bc)Hutchings Photography/Digital Light Source, (br)Charles Stirling/Alamy; **357** (l)The McGraw-Hill Companies, Inc./Stephen Frisch, photographer, (c)sciencephotos/Alamy, (r)Martyn Chillmaid/Oxford Scientific (OSF)/photolibrary.com; **358** (l)Royalty-Free/Corbis, (cl)Dr. Parvinder Sethi, (cr)Joel Arem/Photo Researchers, Inc., (r)Ingram Publishing/SuperStock; **359** Hutchings Photography/Digital Light Source; **360** (t)Egyptian National Museum, Cairo, Egypt, Photo © Boltin Picture Library/The Bridgeman Art Library International, (c)The McGraw-Hill Companies, Inc./Stephen Frisch, photographer, (b)Paul Katz/Getty Images; **361** Jeff Hunter/Getty Images; **362** E.O. lawrence Berkely National Laboratory, University of California, U.S. Department of Energy; **363** Hutchings Photography/Digital Light Source; **364** (tl)Ted Foxx/Alamy (tr)Richard Treptow/Photo Researchers, Inc., (c)Hutchings Photography/Digital Light Source, (bl)Photodisc/Getty Images, (br)Charles D. Winters/Photo Researchers, Inc.; **365** sciencephotos/Alamy; **366** NASA-JPL; **367** (l)Ingemar Aourell/Getty Images, (cl)Don Farrall/Getty Images, (cr)Gabe Palmer/Alamy, (r)Henrik Sorensen/Getty Images; **368** (t)PhotoLink/Getty Images, (b)Hutchings Photography/Digital Light Source; **369** (t)Richard Treptow/Photo Researchers, Inc., (c)sciencephotos/Alamy, (b)PhotoLink/Getty Images; **370 371** (b)Hutchings Photography/Digital Light Source; **372** (t)David J. Green/Alamy, (b)Mark Schneider/Visuals Unlimited/Getty Images; **375** Jim Reed/Science Faction/Getty Images; **378–379** altrendo images/Getty Images; **379** Hutchings Photography/Digital Light Source; **380** Douglas Fisher/Alamy; **381 382 386** Hutchings Photography/Digital Light Source; **388** (t)Popperfoto/Getty Images, (c)Underwood & Underwood/Corbis, (bl)John Meyer, (br)Ilene MacDonald/Alamy; **389** Gazimal/Getty Images; **390** 394 Hutchings Photography/Digital Light Source; **396** (l)Macmillan/McGraw-Hill, (r)Hutchings Photography/Digital Light Source; **397** Brent Winebrenner/Photolibrary.com; **398** Hutchings Photography/Digital Light Source; **402** (t)Photodisc/Getty Images, (c)C Squared Studios/Getty Images, (b)Jennifer Martine/Jupiter Images; **404** (tr)(cr) (bl)Macmillan/McGraw-Hill, (br)Hutchings Photography/Digital Light Source, (t to b)(2)Hutchings Photography/Digital Light Source, (3)Macmillan/McGraw-Hill, (4) Hutchings Photography/Digital Light Source,

Credits

(5)(6)Macmillan/McGraw-Hill, (7)(8)Richard Hutchings (see Digital Light Source), (9)Macmillan/McGraw-Hill; **405** Hutchings Photography/Digital Light Source; **409** altrendo images/Getty Images; **414** (t)Paris Pierce/Alamy, (bl)British Library, London, UK/© British Library Board. All Rights Reserved/ The Bridgeman Art Library, (br)University Library, Leipzig, Germany/ Archives Charmet/The Bridgeman Art Library; **415** Ace Stock Limited/Alamy; **416–417** (bkgd)Anton Luhr/Photolibrary; **418** Darwin Dale/Photo Researchers, Inc.C5454; **419** Hutchings Photography/Digital Light Source; **420** (tl)Royalty-Free/Corbis, (tr)Charles D. Winters/Photo Researchers, Inc., (cl)London Scientific Films/photolibrary.com, (cr)sciencephotos/ Alamy, (bl)Brand X Pictures, (br)Dante Fenolio/Photo Researchers, Inc.; **423 424** Hutchings Photography/Digital Light Source; **428** (t to b)(2)The Mcgraw-Hill Companies, (3)(4)Macmillan/McGraw-Hill; **429** sciencephotos/ Alamy; **430** AP Photo/Greg Campbell; **431** (t)Charles D. Winters/Photo Researchers, Inc., (b)The McGraw-Hill Companies, Inc./Jacques Cornell photographer; **432** (t)The McGraw-Hill Companies, Inc./Stephen Frisch, photographer, (c)sciencephotos/Alamy, (b)Park Dale/Alamy; **433** The McGraw-Hill Companies, Inc./Stephen Frisch, photographer; **434** Joel Sartore/National Geographic/Getty Images; **435** Brand X Pictures/ PunchStock; **439** (tl)McGraw-Hill Companies, (tc)Brand X Pictures/ PunchStock, (tr)Tetra Images/Getty Images; **439** (bl)The McGraw-Hill Companies, (bc)Royalty-Free/Corbis, (br)Alexis Grattier/Getty Images; **440** Hutchings Photography/Digital Light Source; **442** (t to b)(2)Hutchings Photography/Digital Light Source, (3)(4)Macmillan/McGraw-Hill, (5)Hutchings Photography/Digital Light Source; **443** Hutchings Photography/Digital Light Source; **444** Park Dale/Alamy; **447** Anton Luhr/Photolibrary; **450-451** Keith Kapple/SuperStock; **452** Visual&Written SL/Alamy; **453** Hutchings Photography/Digital Light Source; **454** (tl)PHOTOTAKE Inc./ Alamy, (tr)Royalty-Free/Corbis, (cl)Reuters/Landov, (cr)Flirt/SuperStock, (bl)Don Farrall/Getty Images, (br)Royalty-Free/Corbis; **455** (tl)liquidlibrary/ PictureQuest, (tcl)Mark Steinmetz, (tcr)Michael Maes/Getty Images, (tr)C Squared Studios/Getty Images, (bl)Dennis Kunkel Microscopy, Inc., (br)Mark Steinmetz; **456** (t)(b)Hutchings Photography/Digital Light Source; **458** (tl)Royalty-Free/Corbis, (tr)C Squared Studios/Getty Images, (b)Hutchings Photography/Digital Light Source; **459** (br)Stockdisc/PunchStock, Getty Images/Photodisc; **460** Chris Cheadle/Getty Images; **461** Hutchings Photography/Digital Light Source; **462** (t)Ingram Publishing/Alamy, (c)Michael Maes/Getty Images, (b)SuperStock/age fotostock; **463** The Mcgraw-Hill Companies; **464** Hutchings Photography/Digital Light Source; **465** Hutchings Photography/Digital Light Source; **467** (t)Stock Connection Distribution/Alamy, (2)(3)(4)Hutchings Photography/ Digital Light Source; **468** Hutchings Photography/Digital Light Source; **469** (br)(t to b)(2)(3)Hutchings Photography/Digital Light Source, (4)Macmillan/McGraw-Hill; **470** Fletcher & Baylis/Photo Researchers, Inc.; **471** (t)Hutchings Photography/Digital Light Source, (b)Horizons Companies; **472** Hutchings Photography/Digital Light Source; **473** (cw from top)Charles D. Winters/SPL/Photo Researchers, (2)Steve Mason/Getty Images, (3)Yasuhide Fumoto/Getty Images, (4)Phil Degginger/Alamy, (5)Westend61/SuperStock, (6)F. Schussler/PhotoLink/Getty Images, (7)Richard Megna, Fundamental Photographs, NYC; **474** (t)Royalty-Free/Corbis, (bl)Busse Yankushev/age fotostock, (br)Spencer Jones/Getty Images; **475** (tl)liquidlibrary/PictureQuest, (tr)Royalty-Free/Corbis, (bl)The McGraw-Hill Companies, Inc./Jacques Cornell photographer, (br)Robert Manella for MMH; **476** L. S. Stepanowicz/ Visuals Unlimited; **478** (t to b)(2)(3)Hutchings Photography/Digital Light Source, (4)(5)Macmillan/McGraw-Hill; **480** The Mcgraw-Hill Companies; **483** Keith Kapple/SuperStock; **486–487** Jean-Paul Ferrero/Minden Pictures; **488** The Natural History Museum/Alamy; **489** (t)Hutchings Photography/ Digital Light Source, (b)Wolfgang Pölzer/Alamy; **491** (tl)(tr)Image Source/ Getty Images, (br)Michael Betts/Getty Images; **494** Hutchings Photography/ Digital Light Source; **496** Michael Betts/Getty Images; **497** (tr)Courtesy of Texas Instruments, (cr)Bettmann/Corbis, (bl)Digital Art/Corbis, (br)Roger Ressmeyer/Corbis, (bkgd)Digital Vision/Getty Images; **498** Kim Taylor/ Minden Pictures; **499** Hutchings Photography/Digital Light Source; **501** (t)AP Photo/Rita Beamish, (c)Franceso Tomasinelli/Photo Researchers, Inc., (b)Jamie Grill/Getty Images; **502** Image Source/Getty Images; **503** Hutchings Photography/Digital Light Source; **504** (t to b)Andy Hibbert, Ecoscene/Corbis, (2)Image Source/Getty Images, (3)Leonard Lessin/Photo Researchers, Inc., (4)Glyn Thomas/Alamy; **505** Glyn Thomas/Alamy; **506** (br)(t to b)(2)(3) Hutchings Photography/Digital Light Source, (4)Macmillan/McGraw-Hill, (5)The McGraw-Hill Companies, (6)(7) Hutchings Photography/Digital Light Source; **507** Seb Rogers/Alamy; **508** Hutchings Photography/Digital Light Source; **510** (l)Imag Source/Getty Images, (t)Richard Megna, Fundamental Photographs, NYC, (r)Brian Hagiwara/Jupiter Images, (bl)(br)Richard Megna, Fundamental Photographs, NYC; **513** Brian Hagiwara/ Jupiter Images; **514** (l to r, t to b) (2)Hutchings Photography/Digital Light Source, (3)Macmillan/McGraw-Hill, (4)(5)(6) Hutchings Photography/Digital Light Source, (7)The McGraw-Hill Companies; **515** Hutchings Photography/ Digital Light Source; **516** Image Source/Getty Images; **519** Jean-Paul Ferrero/Minden Pictures; **525** Corbis; **526-527** Kaz Mori/Getty Images; **528** Deco/Alamy; **529** Hutchings Photography/Digital Light Source; **531** Hutchings Photography/Digital Light Source; **534** (t)Royalty-Free/ Corbis, (c)Jason Hosking/zefa/Corbis, (b)Mark Downey/Lucid Images/ Corbis; **535** (l)Jack Milchanowski/Papilio/Alamy, (r)Edward Kinsman/ Photo Researchers, Inc., Hutchings Photography/Digital Light Source; **537** (inset)NOAA, (bkgd)Comstock/PunchStock; **538** Tony Friedkin/ Sony Pictures Classics/ZUMA/Corbis; **539** Hutchings Photography/ Digital Light Source; **541** Hutchings Photography/Digital Light Source; **545** (br)Hutchings Photography/Digital Light Source, (t to b)Macmillan/ McGraw-Hill, (2)Hutchings Photography, (3)Comstock Images/Alamy, (4) Hutchings Photography/Digital Light Source; **546** Gustoimages/Photo Researchers, Inc.; **547** (t)Hutchings Photography/Digital Light Source, (b)NASA, ESA, and the Hubble Heritage (STScI/AURA)-ESA/Hubble Collaboration; **548** Car Culture/Getty Images; **549** (t)Richard Megna, Fundamental Photographs, NYC, (b)Hutchings Photography/Digital Light Source; **552 553** Richard Megna, Fundamental Photographs, NYC; **554** (t to b)Macmillan/McGraw-Hill, (2) Hutchings Photography/Digital Light Source, (3)Comstock Images/Alamy, (4)(5)Hutchings Photography/ Digital Light Source; **555** Hutchings Photography/Digital Light Source; **556** Richard Megna, Fundamental Photographs, NYC; **559** Kaz Mori/ Getty Images; **562–563** Dan Barba/Photolibrary; **564** Mike Kemp/Getty Images; **565** Hutchings Photography/Digital Light Source; **567** Jens Kuhfs/ SeaPics.com; **568** Michael Mährlein/age fotostock; **569** (t)Hutchings Photography/Digital Light Source, (b)Wave Royalty Free/age fotostock; **570** Michael Mährlein/age fotostock; **571** (t)Michael Newman/PhotoEdit, (b)Steve Hamblin/Alamy; **572** Stefano Cellai/age fotostock; **573** Hutchings Photography/Digital Light Source; **577** 580 581 (t to b)Hutchings Photography/ Digital Light Source, (4)Macmillan/McGraw-Hill; **582** (inset)Emory Kristof/ National Geographic Stock, (bkgd)NOAA; **583 585** Hutchings Photography/ Digital Light Source; **586** (t)John Luke/age fotostock, (bl)View Pictures Ltd/SuperStock, (br)Zave Smith/Getty Images; **587** DAJ/Getty Images; **588** Zephyr/Photo Researchers, Inc.; **589** John Luke/age fotostock; **590** (r)(t to b)Hutchings Photography/Digital Light Source, (2)(3)Macmillan/ McGraw-Hill, (4)(5)(6)Hutchings Photography/Digital Light Source; **591** Hutchings Photography/Digital Light Source; **592** (t)Michael Mährlein/ age fotostock, (b)Zave Smith/Getty Images; **595** Dan Barba/Photolibrary; **598–599** NASA; **600** Menno Boermans/Aurora Photos/Corbis; **601** Hutchings Photography/Digital Light Source; **603** Hutchings Photography/Digital Light Source; **604** X-ray: NASA/CXC/J. Hester (ASU), Optical: NASA/ESA/J. Hester & A. Loll (ASU), Infrared: NASA/JPL-Caltech/ R.Gehrz (Univ. Minn.); **605** (tr)(cr)Bjorn Rorslett/Photo Researchers, Inc., (bl)Hutchings Photography/Digital Light Source, (br)Ralf Nau/Getty Images; **606** X-ray: NASA/CXC/J. Hester (ASU), Optical: NASA/ESA/J. Hester & A. Loll (ASU), Infrared: NASA/JPL-Caltech/R. Gehrz (Univ. Minn.); **607** NASA; **608** Felix Stenson/age fotostock; **609** Hutchings Photography/Digital Light

Source; **610** (t)izmostock/Alamy, (c)PM Images/Getty Images, (b)D. Hurst/Alamy; **611** (l)Digital Vision/PunchStock, (tr)Peter Cade/Getty Images, (br)Richard T. Nowitz/Phototake; **612** (t)Felix Stenson/age fotostock, (c)BananaStock/PunchStock, (b)Richard T. Nowitz/Phototake; **613** (tl)Steve Cole/Getty Images, (cl)The Mcgraw-Hill Companies, (r)(bl)Hutchings Photography/Digital Light Source; **614** Carl Trinkle; **615** Howard Sokol/age fotostock; **617** Joe Drivas/Getty Images; **619** (t)Simon Belcher/Alamy, (c)NASA, (t to b)Stocktrek Images/Getty Images; **620** (t)Daniel L. Geiger/SNAP/Alamy, (b)Corbis/age fotostock, (b)Digital Vision/age fotostock; **621** The McGraw-Hill Companies, Inc./Ken Cavanagh photographer; **623** NASA; **624** (l)Nikreates/Alamy, (tr)NASA/JPL, (br)Getty Images; **625** Hutchings Photography/Digital Light Source; **626** (t)D. Hurst/Alamy, (b)The McGraw-Hill Companies, Inc./Ken Cavanagh photographer; **629** NASA; **632–633** Charles Krebs/Getty Images; **634** Russell Burden/photolibrary.com; **635** (t)Hutchings Photography/Digital Light Source, (b)Andrew Geiger/Getty Images; **636** (t)Hutchings Photography/Digital Light Source, (b)Jasper James/Getty Images; **637** Picturenet/Getty Images; **640** (t)Jasper James/Getty Images, (b)Picturenet/Getty Images; **641** (t)Scott Smith/photolibrary, (c)image100/Corbis, (b)Simon Stuart-Miller/Alamy; **642** Jeremy Horner/Getty Images; **643** Hutchings Photography/Digital Light Source, Hutchings Photography/Digital Light Source; **646** sciencephotos/Alamy; **648** (br)Hutchings Photography/Digital Light Source; **649** Image Source/Getty Images; **650** Hutchings Photography/Digital Light Source; **652** (tl)(tr)Don Farrall/Getty Images, (b)Hutchings Photography/Digital Light Source; **655** (t)Getty Images; **656** Glow Images, Inc/photolibrary.com; **657** MIXA/Getty Images; **658** Getty Images; **659** (r)Hutchings Photography/Digital Light Source, (t to b)Macmillan/McGraw-Hill, (2)Hutchings Photography/Digital Light Source; **659** (3)(4)Macmillan/McGraw-Hill; **660** Michael Kappeler/AFP/Getty Images; **661** (t)Hutchings Photography/Digital Light Source, (b)StockTrek/Getty Images; **663** Hutchings Photography/Digital Light Source; **664** Getty Images; **665** (tl)Phil Degginger/Getty Images, (tr)James Leynse/Corbis, (bl)AP Photo/Thomas Kienzle, (br)Construction Photography/Corbis; **666** Bruce Ando/Getty Images; **667** (t)Phil Degginger/Getty Images, (b)Bruce Ando/Getty Images; **668** (r)Hutchings Photography/Digital Light Source, (t to b)The McGraw-Hill Companies, (2)Hutchings Photography/Digital Light Source, (3)The Mcgraw-Hill Companies, (4)Hutchings Photography/Digital Light Source, (5)Macmillan/McGraw-Hill, (6)Steve Cole/Getty Images; **669** Hutchings Photography/Digital Light Source; **670** (t to b)Hutchings Photography/Digital Light Source, (2)sciencephotos/Alamy, (3)Don Farrall, (4)Bruce Ando/Getty Images; **673** Charles Krebs/Getty Images; **676–677** Bettmann/Corbis; **678** Rick Elkins/Getty Images; **679** NASA; **682-683** Charles Krebs/Getty Images; **683** Hutchings Photography/Digital Light Source; **684** Russell Burden/photolibrary.com; **685** (l)vladimir zakharov/Getty Images, (t)Hutchings Photography/Digital Light Source, (b)Andrew Geiger/Getty Images, (inset)Gary Withey/Bruce Coleman Inc.; **686** (t)Hutchings Photography/Digital Light Source, (b)Alan R Moller/Getty, (b)Jasper James/Getty Images, (bkgd)Ralph H Wetmore II/Getty; **687** Picturenet/Getty Images, Rick Elkins/Getty Images; **689** Construction Photography/Corbis; **690** (t)Jasper James/Getty Images, (b)Picturenet/Getty Images, Hutchings Photography/Digital Light Source; **691** (t)Scott Smith/photolibrary, (c)image100/Corbis, (b)Simon Stuart-Miller/Alamy, (inset) Hutchings Photography/Digital Light Source; **692** Hutchings Photography/Digital Light Source, Jeremy Horner/Getty Images; **693 695** Hutchings Photography/Digital Light Source; **696** Construction Photography/Corbis, sciencephotos/Alamy; **697** (tl)Macmillan/McGraw-Hill, (cl) (r)Hutchings Photography/Digital Light Source, (bl)The Mcgraw-Hill Companies; **698** (other)Hutchings Photography/Digital Light Source, Troy Paiva/LostAmerica.com; **699** Hutchings Photography/Digital Light Source, Image Source/Getty Images; **700** Hutchings Photography/Digital Light Source; **702** (tl)Don Farrall, (tr)Don Farrall/Getty Images, (b)Hutchings Photography/Digital Light Source, Troy Paiva/LostAmerica.com; **703** Tetra Images/Getty Images; **704** Hutchings Photography/Digital Light Source; **705** (t)Getty Images; **705** Hutchings Photography/Digital Light Source; **706** (br) (t to b)(2)Hutchings Photography/Digital Light Source, (3)(4)Macmillan/McGraw-Hill, (5)Hutchings Photography, (6)The Mcgraw-Hill Companies, (7)Macmillan/McGraw-Hill, Glow Images, Inc/photolibrary.com; **707** Hutchings Photography/Digital Light Source, MIXA/Getty Images; **708** Getty Images; **709** (r)Hutchings Photography/Digital Light Source, (t to b)Macmillan/McGraw-Hill, (2)Hutchings Photography/Digital Light Source, Macmillan/McGraw-Hill, (4)Macmillan/McGraw-Hill; **710** Michael Kappeler/AFP/Getty Images; **711** (t)Hutchings Photography/Digital Light Source, (b)StockTrek/Getty Images, Bettmann/Corbis; **713** Hutchings Photography/Digital Light Source; **714** Getty Images; **714-715** age fotostock/SuperStock, (tl)Phil Degginger/Getty Images, (tr)James Leynse/Corbis, (bl)AP Photo/Thomas Kienzle, (br)Construction Photography/Corbis; **716** Bruce Ando/Getty Images, imagebroker/Alamy; **717** (t)Phil Degginger/Getty Images, (tl)Joel Arem/Photo Researchers, Inc., (tr)Clive Streeter/Getty Images, (b)Bruce Ando/Getty Images, (b)Stockbyte/Getty Images; **718** (l)Hutchings Photography/Digital Light Source, (r)Hutchings Photography/Digital Light Source, (t to b)The McGraw-Hill Companies, (2)Hutchings Photography/Digital Light Source, (3)The Mcgraw-Hill Companies, (4)Richard Hutchings (see Digital Light Source), (5)Macmillan/McGraw-Hill, (6)Steve Cole/Getty Images; **719** (tl) (tr) (bl) (br)Cordelia Molloy/Photo Researchers, Inc., Hutchings Photography/Digital Light Source; **720** (t to b)Hutchings Photography/Digital Light Source, (2)sciencephotos/Alamy, (3)Don Farrall, (4)Bruce Ando/Getty Images, Grambo/Getty Images; **721** Hutchings Photography/Digital Light Source; **723** Charles Krebs/Getty Images; **724** (t)Cordelia Molloy/Photo Researchers, Inc., (b)Joel Arem/Photo Researchers, Inc.; **726** Alexis Rosenfeld/Photo Researchers, Inc.; **727** 730 Hutchings Photography/Digital Light Source; **733** (t to b)Macmillan/McGraw-Hill, (2)Hutchings Photography, (3)Macmillan/McGraw-Hill, (4) (5)Hutchings Photography; **734** (inset)bobo/Alamy, BlueMoon Stock/Alamy; **735** Hutchings Photography/Digital Light Source; **738** (t)Lester Lefkowitz/Corbis, (b)RIA Novosti/Science Photo Library; **741** (t)Lester Lefkowitz/Corbis, (b)age fotostock/SuperStock; **742** (r) (t to b)Hutchings Photography/Digital Light Source, (2)Macmillan/McGraw-Hill, (3)Hutchings Photography/Digital Light Source, (4)Macmillan/McGraw-Hill, (5)The Mcgraw-Hill Companies, (6)Hutchings Photography; **743** Hutchings Photography/Digital Light Source; **744** (t)imagebroker/Alamy, (c)Alexis Rosenfeld/Photo Researchers, Inc., (b)bobo/Alamy; **746** Cordelia Molloy/Photo Researchers, Inc.; **747** (l)Cordelia Molloy/Photo Researchers, Inc., age fotostock/SuperStock; **SR-0–SR-1** (bkgd)Gallo Images—Neil Overy/Getty Images; **SR-2** Hutchings Photography/Digital Light Source; **SR-6** Michell D. Bridwell/PhotoEdit; **SR-7** (t)The McGraw-Hill Companies, (b)Dominic Oldershaw; **SR-8** StudiOhio; **SR-9** Timothy Fuller; **SR-10** Aaron Haupt; **SR-12** KS Studios; **SR-13** SR-47 Matt Meadows; **SR-48** (c)NIBSC/Photo Researchers, Inc., (r)Science VU/Drs. D.T. John & T.B. Cole/Visuals Unlimited, Inc., Stephen Durr; **SR-49** (t)Mark Steinmetz, (r)Andrew Syred/Science Photo Library/Photo Researchers, (br)Rich Brommer; **SR-50** (l)Lynn Keddie/Photolibrary, (tr)G.R. Roberts, David Fleetham/Visuals Unlimited/Getty Images; **SR-51** Gallo Images/Corbis.

PERIODIC TABLE OF THE ELEMENTS

Element	Hydrogen
Atomic number	1
Symbol	H
Atomic mass	1.01
State of matter	

Gas
Liquid
Solid
Synthetic

A column in the periodic table is called a **group.**

A row in the periodic table is called a **period.**

Period	1	2	3	4	5	6	7	8	9
1	Hydrogen 1 **H** 1.01								
2	Lithium 3 **Li** 6.94	Beryllium 4 **Be** 9.01							
3	Sodium 11 **Na** 22.99	Magnesium 12 **Mg** 24.31							
4	Potassium 19 **K** 39.10	Calcium 20 **Ca** 40.08	Scandium 21 **Sc** 44.96	Titanium 22 **Ti** 47.87	Vanadium 23 **V** 50.94	Chromium 24 **Cr** 52.00	Manganese 25 **Mn** 54.94	Iron 26 **Fe** 55.85	Cobalt 27 **Co** 58.93
5	Rubidium 37 **Rb** 85.47	Strontium 38 **Sr** 87.62	Yttrium 39 **Y** 88.91	Zirconium 40 **Zr** 91.22	Niobium 41 **Nb** 92.91	Molybdenum 42 **Mo** 95.96	Technetium 43 **Tc** (98)	Ruthenium 44 **Ru** 101.07	Rhodium 45 **Rh** 102.91
6	Cesium 55 **Cs** 132.91	Barium 56 **Ba** 137.33	Lanthanum 57 **La** 138.91	Hafnium 72 **Hf** 178.49	Tantalum 73 **Ta** 180.95	Tungsten 74 **W** 183.84	Rhenium 75 **Re** 186.21	Osmium 76 **Os** 190.23	Iridium 77 **Ir** 192.22
7	Francium 87 **Fr** (223)	Radium 88 **Ra** (226)	Actinium 89 **Ac** (227)	Rutherfordium 104 **Rf** (267)	Dubnium 105 **Db** (268)	Seaborgium 106 **Sg** (271)	Bohrium 107 **Bh** (272)	Hassium 108 **Hs** (270)	Meitnerium 109 **Mt** (276)

The number in parentheses is the mass number of the longest lived isotope for that element.

Lanthanide series	Cerium 58 **Ce** 140.12	Praseodymium 59 **Pr** 140.91	Neodymium 60 **Nd** 144.24	Promethium 61 **Pm** (145)	Samarium 62 **Sm** 150.36	Europium 63 **Eu** 151.96
Actinide series	Thorium 90 **Th** 232.04	Protactinium 91 **Pa** 231.04	Uranium 92 **U** 238.03	Neptunium 93 **Np** (237)	Plutonium 94 **Pu** (244)	Americium 95 **Am** (243)